Fruit Breeding

HANDBOOK OF PLANT BREEDING

Editors-in-Chief:

JAIME PROHENS, *Universidad Politecnica de Valencia, Valencia, Spain*
FERNANDO NUEZ, *Universidad Politecnica de Valencia, Valencia, Spain*
MARCELO J. CARENA, *North Dakota State University, Fargo, ND, USA*

For further volumes:
http://www.springer.com/series/7290

Marisa Luisa Badenes • David H. Byrne
Editors

Fruit Breeding

Springer

Editors
Marisa Luisa Badenes
Instituto Valenciano de Investigaciones
Agrarias (IVIA)
Valencia, Spain
mbadenes@ivia.es

David H. Byrne
Texas A&M University
College Station, TX, USA
d-byrne@tamu.edu

ISBN 978-1-4939-3904-6 ISBN 978-1-4419-0763-9 (eBook)
DOI 10.1007/978-1-4419-0763-9
Springer New York Dordrecht Heidelberg London

Printed on acid-free paper

Springer is part of Springer Science+Business Media (www.springer.com)

Preface

This book begins with a discussion of the overall trends in fruit breeding, intellectual property management, the breeding for cultivars with enhanced health benefits, and an assessment of some of the emerging fruit crops that have great potential for further development. The next three sections: small fruits, tree fruits, and nut crops contain crop-specific chapters describing the economic importance, use, adaptation, origin, domestication, breeding history, accomplishments, goals, breeding techniques, and the advances in the use of biotechnology for each crop. The crops reviewed have domestication history of millennium to decades and breeding activity ranging from thousands of generations to just a few generations. Likewise, their biology and ploidy levels (diploid to octoploid) are diverse which leads to a plethora of approaches to their genetic improvement.

Breeding of perennial fruit species is a long-term activity involving a high investment as compared to annual crops due to two challenges: long juvenile periods and large plant size. In spite of these difficulties, breeding programs have been developed in all important perennial fruit crops, aimed at the improved economic profitability of the crops by increasing yields, altering the harvest window, creating new fruit types, and improving fruit quality while simplifying management. The recent increase in activity has been encouraged by the integration of the intellectual property rights (IP rights) in fruit production which has created substantial research incentive in private and public spheres for innovation in the fruit industry.

Yield is intertwined with the ease of management, as a prerequisite of high yields is excellent adaptation to the environment. This includes the ability to grow and yield under the abiotic conditions of soil, temperature, and humidity and the biotic stresses, such as fungus, bacteria, nematodes, and viruses in the production zone. This later objective has recently increased in importance with the enhanced public awareness of the negative consequences of the use of agrochemicals. This has spurred the dramatic increase of research into the development of sustainable fruit production systems. The globalization of the fruit industry is resulting in increased activity in developing cultivars of temperate fruits adapted to subtropical and tropical environments. Beyond the simplification of management by reducing the use of agrochemicals, work on the modification of tree architecture either through dwarfing rootstock or unique scion growth habits and the conversion of self-incompatible crops to self-compatible or parthenocarpic crops continue to improve the quantity and consistency of yield and the ease of managing the crops.

The value of fruit generally increases when less is available. Thus, much breeding has been done to extend the harvest season both earlier and later when fruit supplies are lower. Consequently, there has been much progress. A good example would be the extension of the peach season from 1–2 months to 6–8 months through the breeding for shorter and longer fruit development periods. In addition to this, the shift of adaptation of cultivars to earlier and later blooming areas has contributed to these extended fruit marketing seasons. Although there has been success, much work needs to be done especially in the improvement of fruit quality at the extremes of the harvest season. Another approach to reduce the availability is to offer something unique. In the US peach industry, this has played out several times starting with the introduction of the nectarine, and then with white fleshed fruit, and now with pantao types. This work continues across all crops and involves traits, including appearance (flesh and skin color, shape, size), quality (flavor, aroma, texture, acidity, sugar, levels of health promoting phytochemicals, storability), and convenience (seedlessness, glabrous skin, ease of peeling, size, shelf life) traits.

The traditional breeding approach is the foundation of our success. Nevertheless, the integration of the new genetic and molecular tools into the breeding programs makes a major impact. These new tools increase the efficiency of the breeding programs by identifying important genes at the molecular level. Molecular markers have been developed for genetic studies and the identification of cultivars in the major fruit species. Genetic linkage maps are available in many perennial species, including stone fruits, pome fruits, strawberry, grapes, chestnut, and walnut. These maps have been key in the identification and selection of the target genes or markers linked to them. The advent of genomics, whole genome sequences (apple, peach, grape, strawberry, and citrus) and the rapidly improving DNA sequencing technologies have opened up new opportunities for developing new markers and for identifying and understanding the gene function which controls the important phenotypes in fruit breeding. In vitro technology has led to improved propagation and virus certification protocols, efficient procedures to grow out unique hybrid seedlings (embryo rescue, in vitro grafting, somatic hybridization), and to create transgenic plants.

This book tries to present a broad vision of fruit breeding to stimulate the thought process and hopefully inspire the next generation of fruit breeders to create the breakthrough cultivars of the future.

Valencia, Spain — Marisa Luisa Badenes
College Station, TX, USA — David H. Byrne

Contents

Contributors

Sandra Agnanostakis The Connecticut Agricultural Experiment Station, New Haven, CT, USA

José Manuel Alonso Unidad de Fruticultura, Centro de Investigación y Tecnología Agroalimentaria de Aragón (CITA), Zaragoza, Spain

Malli Aradhya National Clonal Germplasm Repository, USDA, ARS, University of California, Davis, CA, USA

Amelie Brazelton Aust Fall Creek Farm and Nursery Inc., Lowell, OR, USA

Antonio Ballester Instituto de Investigaciones Agrobiológicas de Galicia, CSIC, Santiago de Compostela, Spain

Daniele Bassi Dipartimento di Produzione Vegetale, Università degli Studi di Milano, Milan, Italy

Ignasi Batlle Departament d'Arboricultura Mediterrània, IRTA-Centre de Mas Bové, Reus-Tarragona, Spain

Gabriele L. Beccaro Department of Colture Arboree, University of Torino, Grugliasco (TO), Italy

Roberto Botta Department of Colture Arboree, University of Torino, Grugliasco (TO), Italy

Giancarlo Bounous Chairman ISHS Group on Chestnut, FAO/CIHEAM Liaison Officer Subnetwork on Chestnut, Department of Colture Arboree, University of Torino, Turin, Italy

Susan Brown Department of Horticulture, Cornell University, New York State Agricultural Experiment Station (NYSAES), Geneva, NY, USA

Lorenzo Burgos CEBAS-CSIC, Murcia, Spain

Patrick Byers Greene County Extension Office, University of Missouri Extension, Springfield, MO, USA

David H. Byrne Department of Horticultural Sciences, Texas A&M University, College Station, TX, USA

Craig K. Chandler Department of Horticultural Sciences, University of Florida, Wimauma, FL, USA

John R. Clark Department of Horticulture, University of Arkansas, Fayetteville, AR, USA

Rafel Socias i Company Unidad de Fruticultura, Centro de Investigación y Tecnología Agroalimentaria de Aragón (CITA), Zaragoza, Spain

Marco Conedera WSL, Swiss Federal Institute for Forest, Snow and Landscape Research, Bellinzona, Switzerland

Patrick J. Conner University of GA, Tifton, GA, USA

Elena Corredoira Instituto de Investigaciones Agrobiológicas de Galicia, CSIC, Santiago de Compostela, Spain

Rita Costa Instituto Nacional de Recursos Biológicos I.P. Quinta do Marquês, Oeiras, Portugal

Peter S. Cousins USDA ARS, Grape Genetics Research Unit, NYS Agricultural Experiment Station, Geneva, NY, USA

Carlos H. Crisosto Department of Plant Sciences, University of California, Davis, CA, USA

Marco A. Dalbó Epagri, Estação Experimental de Videira, Videira, SC, Brazil

Adam Dale Department of Plant Agriculture, University of Guelph, Simcoe Research Station, Simcoe, Canada

Luca Dondini Dipartimento di Colture Arboree, Università degli Studi di Bologna, Bologna, Italy

Louise Ferguson Department of Plant Sciences, University of California, Davis, CA, USA

Chad E. Finn US Department of Agriculture-Agricultural Research Service, Horticultural Crops Research Laboratory, Corvallis, OR, USA

Kevin Folta Department of Horticultural Sciences, University of Florida, Gainesville, FL, USA

Ksenija Gasic Department of Environmental Horticulture, Clemson University, Clemson, SC, USA

Edgardo Giordani Plant, Soil and Environmental Science, University of Florence, Florence, Italy

José Gomes-Laranjo CITAB, University of Trás-os-Montes and Alto Douro, Vila Real, Portugal

Thomas M. Gradziel Department of Plant Sciences, University of California, Davis, CA, USA

Charles J. Graham LSU Agricultural Center, Shreveport, LA, USA

Andrew Granger Plant & Food Research, Mt Albert, New Zealand

Harvey Hall Shekinah Berries Ltd, Motueka, New Zealand

Mark Herrington Queensland Department of Employment, Economic Development and Innovation, Queensland Government, QLD, Australia

Károly Hrotkó Corvinus University of Budapest, Budapest, Hungary

Kim E. Hummer USDA ARS National Clonal Germplasm Repository, Corvallis, OR, USA

USDA ARS Arctic and Subarctic Plant Gene Bank, Palmer, AK, USA

Jennifer Johnson-Cicalese PE Marucci Center, Rutgers University, Chatsworth, NJ, USA

Robert Jondle Jondle and Associates, Castle Rock, CO, USA

Salih Kafkas Department of Horticulture, University of Cukurova, Adana, Turkey

Craig E. Kallsen University of California, Cooperative Extension, Kern County, Bakersfield, CA, USA

Frank Kappel Agriculture and Agri-Food Canada, Summerland, BC, Canada

Chaim Kempler Pacific Agri-Food Research Centre, Agriculture and Agri-Food Canada, Agassiz, BC, Canada

Ossama Kodad Unidad de Fruticultura, Centro de Investigación y Tecnología Agroalimentaria de Aragón (CITA), Zaragoza, Spain

Patrik Krebs WSL, Swiss Federal Institute for Forest, Snow and Landscape Research, Bellinzona, Switzerland

Thomas L. Kubisiak USDA Forest Service, Southern Research Station, Southern Institute of Forest Genetics, Saucier, MS, USA

Craig Ledbetter USDA, ARS, CDP&G, SJVASC, Parlier, CA, USA

Charles Leslie Walnut Improvement Program, Department of Plant Sciences, University of California, Davis, CA, USA

Weisheng Liu Liaoning Institute of Pomology, Xiongyue, Yingkou, Lianoning, People's Republic of China

Gerardo Llácer IVIA, Moncada, Valencia, Spain

Gale McGranahan Walnut Improvement Program, Department of Plant Sciences, University of California, Davis, CA, USA

Eric W. Mercure Paramount Farming Company, Bakersfield, CA, USA

María Angeles Moreno Estación Experimental de Aula Dei (CSIC), Zaragoza, Spain

Jay Morris Department of Cellular and Molecular Pharmacology and Experimental Therapeutics, Hollings Cancer Center, Medical University of South Carolina, Charleston, SC, USA

Vondina Moseley Department of Cellular and Molecular Pharmacology and Experimental Therapeutics, Hollings Cancer Center, Medical University of South Carolina, Charleston, SC, USA

Luis Navarro Centro de Protección Vegetal y Biotecnología, Instituto Valenciano de Investigaciones Agrarias (IVIA), Moncada, Valencia, Spain

Michael Neumüller Technische Universität München, Freising, Germany

Patrick Ollitrault Centre de Coopération Internationale en Recherche Agronomique pour le Développement (CIRAD), Montpellier, Cedex, France

Christopher L. Owens USDA ARS, Grape Genetics Research Unit, NYS Agricultural Experiment Station, Geneva, NY, USA

Dan E. Parfitt Department of Plant Sciences Mail Stop 2, University of California, Davis, CA, USA

Santiago Pereira-Lorenzo Departamento de Producción Vegetal, Universidad de Santiago de Compostela, Lugo, Spain

Salvador Peréz Recursos Genéticos y Mejoramiento de Prunus, Querétaro, Mexico

Maria Claudia Piagnani Dipartimento di Produzione Vegetale, Università degli Studi di Milano, Milan, Italy

Kirk W. Pomper Kentucky State University, Frankfort, KY, USA

Joseph Postman USDA ARS National Clonal Germplasm Repository, Corvallis, OR, USA

Ana M. Ramos-Cabrer Departamento de Producción Vegetal, Universidad de Santiago de Compostela, Campus de Lugo, Lugo, Spain

Maria Bassols Raseira EMBRAPA – Clima Temperado, Pelotas, RS, Brazil

Gregory L. Reighard Department of Environmental Horticulture, Clemson University, Clemson, SC, USA

Bruce I. Reisch Department of Horticulture and Plant Breeding, N.Y.S. Agricultural Experiment Station, Cornell University, Geneva, NY, USA

Dougal M. Russell Horticulture & Forestry Science, Department of Employment and Economic Development, Nambour, QLD, Australia

Silviero Sansavini Dipartimento di Colture Arboree, Università degli Studi di Bologna, Bologna, Italy

Yutaka Sawamura National Institute of Fruit Tree Science, Tsukuba, Ibaraki, Japan

Mirko Schuster Julius Kuehn Institute, Dossenheim, Germany

Ed Stover USDA/ARS Horticulture and Plant Breeding Unit, Horticultural Research Laboratory, Ft. Pierce, FL, USA

Norio Takada National Institute of Fruit Tree Science, Tsukuba, Ibaraki, Japan

Maxine M. Thompson Department of Horticulture, Oregon State University, Corvallis, OR, USA

Tommy E. Thompson USDA-ARS Pecan Genetics and Breeding Program, Somerville, TX, USA

Bruce L. Topp Queensland Alliance for Agriculture and Food Innovation, University of Queensland, Maroochy Research Station, Nambour, QLD, Australia

Francisco J. Vargas Departament d'Arboricultura Mediterrània, IRTA-Centre de Mas Bové, Reus-Tarragona, Spain

Ana M. Vieitez Instituto de Investigaciones Agrobiológicas de Galicia, CSIC, Santiago de Compostela, Spain

Nicholi Vorsa PE Marucci Center, Rutgers University, Chatsworth, NJ, USA

Michael J. Wargovich Department of Cellular and Molecular Pharmacology and Experimental Therapeutics, Hollings Cancer Center, Medical University of South Carolina, Charleston, SC, USA

Rebecca Weber Department of Cellular and Molecular Pharmacology and Experimental Therapeutics, Hollings Cancer Center, Medical University of South Carolina, Charleston, SC, USA

Vance M. Whitaker Department of Horticultural Sciences, University of Florida, Wimauma, FL, USA

Masahiko Yamada National Institute of Fruit Tree Science, Tsukuba, Ibaraki, Japan

Toshiya Yamamoto National Institute of Fruit Tree Science, Tsukuba, Ibaraki, Japan

Keizo Yonemori Graduate School of Agriculture, Kyoto University, Kyoto, Japan

Francis Zee USDA, ARS Pacific Basin Agricultural Research Center (PBARC), Hilo, HI, USA

Tatyana Zhebentyayeva Genetics and Biochemistry, Clemson University, Clemson, SC, USA

Part I
General Chapters

Chapter 1
Trends in Fruit Breeding

David H. Byrne

Abstract Fruit breeding is a long-term process which takes a minimum of about a decade from the original cross to a finished cultivar. Thus, much thought needs to go into which objectives to be emphasized in the breeding. Although certain objectives, such as yield and basic quality, are always important, the overall lifestyle, environmental, marketing, and production trends affect the objectives that breeders emphasize in their programs as they strive to anticipate the future needs of the fruit industry. The importance of each trend varies with the crop and environment. The major trends are to develop cultivars which simplify orchard practices, have increased resistance to biotic and abiotic stress, extend the adaptation zones of the crop, create new fruit types, create fruit cultivars with enhanced health benefits, and provide consistently high quality.

Keywords Food marketing • Carbon foot print • Food for health • Fruit quality • Labor, food safety • Organic, sustainable production • Global warming • Environmental contamination • Host plant resistance

1 Introduction

Fruit breeders need to anticipate cultivar needs at least 10 years into the future, as this is the minimum time that most fruit cultivars take to develop from pollination to release. This chapter explores the larger trends in our lives, such as environmental issues, health consciousness, consumer trends in lifestyle, and the expectations and needs of producers to examine how these affect the objectives of our fruit breeding programs.

D.H. Byrne (✉)
Department of Horticultural Sciences, Texas A&M University,
College Station, TX 77843-2133, USA
e-mail: dbyrne@tamu.edu

M.L. Badenes and D.H. Byrne (eds.), *Fruit Breeding*, Handbook of Plant Breeding 8,
DOI 10.1007/978-1-4419-0763-9_1, © Springer Science+Business Media, LLC 2012

2 Trends in the Business of Plant Breeding

Improved plant protection legislation in the USA, Europe, and throughout the world has stimulated substantial research and the development of new plants for commercial exploitation. This has also tended to shift the breeding into the private sector (Heisey et al. 2001; Frey 1996, 1998; Traxler 1999). This shift was quicker for the annual large acreage crops, such as corn, where public-generated commercial cultivars in the USA disappeared in the 1940s and the use of publically generated inbred lines ceased in the 1970s. Currently, public corn breeders concentrate more on basic research into corn breeding and genetics (Traxler 1999).

In fruit crops, this shift has been slower and dependent on the crop, with those crops with shorter life cycles and larger markets shifting to the private sector more rapidly. Throughout the world, the proportion of peach releases from public programs has decreased from 45% in the 1980s to 34% in the early 1990s (Della Strada et al. 1996; Della Strada and Fideghelli 2003; Fideghelli et al. 1998). During the last decade in the USA, only ~15% of the peach and nectarine cultivars were released by public institutions. Support for the development of apricots, cherries, and apples is still with public institutions, but this is eroding and the private sector is becoming more involved in the release and marketing of new cultivars (Kappel 2008; Fideghelli and Della Strada 2010; Lespinasse 2009). The initial development of many small fruits, such as strawberries, blueberries, blackberries, and raspberries, was done by public breeders, but currently the private breeders are expanding their efforts to develop proprietary cultivars with a marketing advantage (Clark and Finn 2008; Finn et al. 2008; Hancock and Clark 2009).

Another factor is decreased funding for public breeding programs. In the USA, the public funding dedicated to breeding activities has decreased dramatically since the 1970s as the government shifted from a philosophy of completely funding programs to assisting programs with partial funding (Moore 1993; Frey 1996; Heisey et al. 2001). Thus, those programs that were able to develop additional sources of funding were able to survive. Many did not. A similar trend is seen in Europe.

In the early 1980s, most public fruit breeding programs in the USA made public releases without protecting the intellectual property. The idea was to get the cultivar out to the producer without charging twice since tax dollars were used in the development of the new cultivars and to maximize germplasm exchange (Moore 1993). In the present environment, public breeding programs are raising money by patenting their releases and partnering with the private sector to test and market new cultivars. Although these arrangements are working, it has led to less germplasm exchange among the public breeding programs. There is a need to modify the paradigm to encourage germplasm exchange (Hancock and Clark 2009).

The other aspect of this trend is the amount of ongoing research into germplasm development, genetics, and new breeding techniques. In the USA, private fruit breeding programs devote more than 90% of their efforts to the development of new cultivars, whereas public breeding programs only devote 36% of their efforts to developing new cultivars (Table 1.1); the other 64% of their efforts are in germplasm

Table 1.1 Public versus private breeding programs in temperate fruit and nut crops in the USA (Frey 1996, 1998)

Activity	Public	Private
Cultivar development (%)	36	91
Germplasm enhancement (%)	36	6
Genetic research (%)	28	3
Total (scientist-years) effort	73	32

development, genetics, and breeding technology (Frey 1996, 1998). The funding for this type of research which also funds the training of new plant breeders comes mainly from federal grants. This is where private breeding programs need to get more involved because industry support strongly influences the governmental funding decisions (Sansavini 2009; Byrne 2005; Llacer 2009). This research is essential for the long-range success of the breeding programs in the world

3 Broad Trends Affecting Fruit Breeding

Fruit breeders need to be cognizant of the major issues of the day that influence the production, marketing, and consumption of fruit as they are, in part, a predictor of the future. The cultivars that they are developing currently will not be important in the marketplace for about a decade. There are several broad trends that influence the breeding objectives of breeders.

3.1 Environmental Issues

The most important issue is the preservation of our environment. This is a very broad issue that includes a wide range of discussions on environmental contamination, sustainable agricultural development, biodiversity, and global warming.

The environmental contamination discussion considers the use of pesticides, fungicides, fertilizers, and plastics, their role in the contamination of the ground water, soil, and the general environment, their effect on the flora and fauna and on human health, and the ability to recycle. These concerns have launched innumerable studies into integrated pest control, organic farming techniques, recycling, optimization of resource use, biodegradability of agricultural chemicals and other inputs, and the effects of agricultural chemical accumulation on the ecology and biodiversity of the agroecosystem. These studies have led to more restrictions of the use of agricultural chemicals and the development of more environment-friendly and sustainable fruit production and marketing systems.

Global warming relates to agriculture mainly as agriculture replaces the forests and the carbon footprint generated in the production and marketing of fruit. Some have argued that a long-term fruit production system is more sustainable than an

Table 1.2 Relative energy cost of moving freight according to the mode of transportation (Heyes and Smith 2008)

Mode of transportation	Description	Energy (MJ/ton km)
Air	Short haul	23.7
Air	Long haul	8.5
Road	Small van	1.7
Road	Large truck	1.1
Sea	Roll on/roll off	0.55
Sea	Bulk carrier	0.15

annual crop production system which may be true, but in both cases the natural vegetation is replaced by an introduced crop reducing biodiversity tremendously. Although this discussion is important, more pertinent to this article would be the carbon footprint of production and marketing of fruit. In the mid 1990s, the concept of "food miles" was popularized as a tool to measure the environmental consequences of our globalized food system. This approach did not take into account how food was transported or any of the production and postharvest aspects of production and thus was not very accurate in its conclusions (Coley et al. 2009). Since then, there has been a shift toward measuring the "carbon footprint" using a more comprehensive approach, the Life Cycle Assessment, which attempts to calculate the carbon cost of the product from production through harvesting, processing, marketing, consumption, and the disposal of any waste (Brenton et al. 2009; Sim et al. 2007). This type of analysis has indicated that even though a fresh product is produced several thousand miles away it does not mean that its carbon footprint is greater than locally produced product, especially if the production costs are high, the product is not in season, or it needs to be stored for an extended period. Good examples of this would be comparisons of the carbon footprints of apples consumed in Europe and produced in either Europe or the southern hemisphere (Blanke and Burdick 2005; Milà i Canals et al. 2007) and cut flowers for Europe and produced in either the greenhouse in Holland or Kenya (Brenton et al. 2009).

In most cases, it would seem that the carbon footprint of locally produced fruit in season is less than that of imported fruit. Given that the market wants a year-round supply of fresh fruit, the issue becomes how to reduce the carbon footprint of out-of-season fruit. The cost of transportation varies widely depending on the mode of transportation, with air freight being 15 to over 100 times more energy intensive than sea freight (Table 1.2). Among the modes of land transportation, larger trucks are less energy intensive than smaller trucks and freight by train is about 50% more energy efficient than truck transportation (Canning et al. 2010). This cost to transport fresh produce is a critical component of the carbon cost of supplying product in the off season, especially for fruit that is highly perishable.

As global marketers go "green" and reduce their carbon footprint, there is a trend to transport fruit more via boat versus airplane, as this reduces the carbon footprint tremendously. Although this is routinely done with such crops as apples, grapes, nuts, bananas, and citrus, many other crops, such as berries and stone fruit, have short postharvest durability which limits their ability to be shipped consistently via sea freight. This requires improved postharvest characteristics of the fruit cultivars. In addition, there is greater emphasis to produce fruit locally wherever possible which creates a need for more locally adapted cultivars.

The other footprint which needs to be reduced in the future is the water footprint of production. Water quantity and quality are becoming major challenges in many growing regions. Currently, 70% of the world's fresh water supply is used in agriculture (Sansavini 2009). This reality has spurred much research in better delivery (i.e., drip irrigation) and more efficient management techniques (real-time weather monitoring linked to irrigation control). More needs to be done to develop the genetics that perform well under less or with poorer quality water.

3.2 Health Consciousness

As we learn more about the benefits of fruit consumption in human health (Prior and Cao 2000; Wargovich 2000), the demand for healthier foods is increasing. These foods could take the form of fresh fruit with high levels of health-promoting substances or other natural products, such as fruit extracts for natural sources of antioxidants, antimicrobials, or food colorants for the health and food industries (Cevallos-Casals et al. 2002, 2006).

Currently, it seems that no matter where you look there is information on the health benefits (or hazards) of everything. Health concern is one of the major driving forces of the world food market and globally, although it varies by region, is the first or second most important concern of consumers. Consumers see the connection between diet and health and associate their diets with the prevention of cardiovascular disease, vision problems, lack of energy, obesity, arthritis/joint pain, and high cholesterol (Sloan 2006; Dillard and German 2000). Since the early 1990s, the US Government has been promoting the consumption of three to five servings of fruits and vegetables for good health, and recently raised this suggested level to five to nine servings of fruits and vegetables per day which would include three to four fruits or two cups of fruit per day (Wells and Buzby 2008; USDA 2005). Unfortunately, the average per capita consumption of fruits (both fresh and processed) in the USA is only about 1/2 of this with only a 5–6% increase since the mid 1970s (Fig. 1.1). This increase is primarily due to the per capita increase in fresh fruit consumption (~20%) as the consumption of processed (canned, frozen, juice, dried) fruit has decreased about 6% over this same period (Pollack and Perez 2008; Wells and Buzby 2008).

Fruit has been in the forefront of the food for health movement with a proliferation of superfruits which are touted to have exceptional health benefits. Although the best known are blueberries, pomegranate, and several exotics like acai, noni fruit, and mangosteen, many of our temperate fruits have also been claimed to be super fruits as can be easily seen in a quick Internet search for the terms 'superfruit' and your favorite fruit. Such a search quickly determines that someone promotes fruits, such as the apple, plum, prune, blackberry, raspberry, strawberry, grape, black currants,

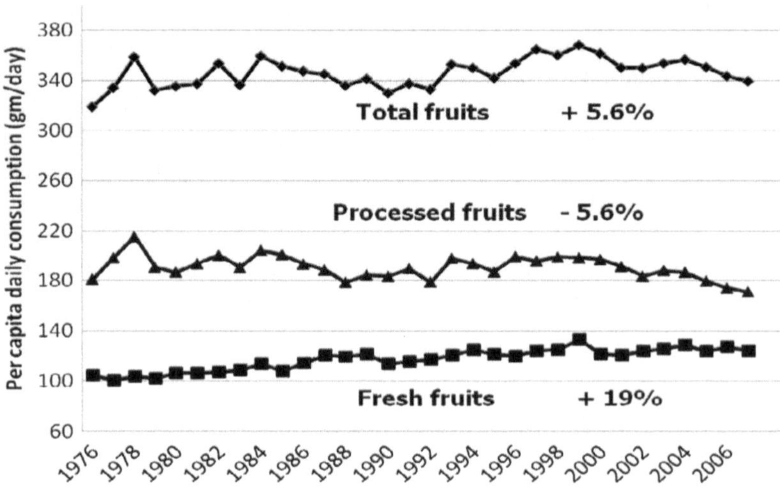

Fig. 1.1 Per capita fruit consumption in the USA (data from Pollack and Perez 2008)

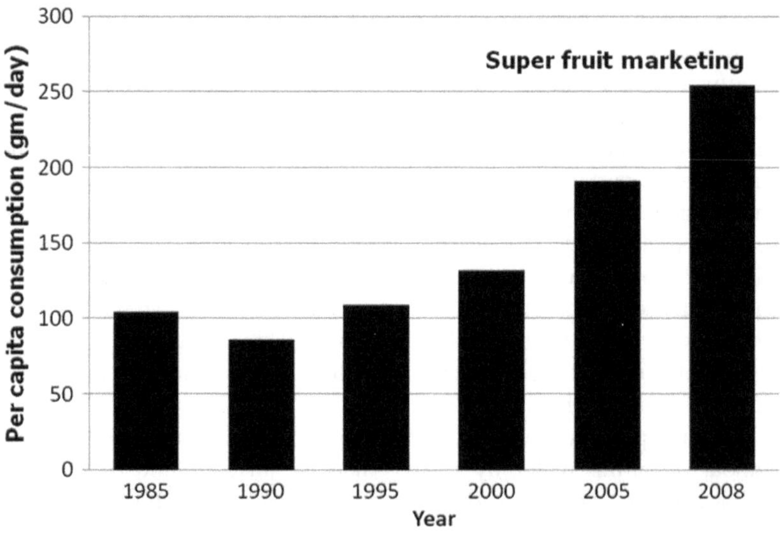

Fig. 1.2 Per capita blueberry consumption in the USA (data from Pollack and Perez 2008)

persimmons, orange, and cherry, and others as superfruits. The term is not well-defined, so it only denotes that a particular fruit is perceived to be particularly beneficial from a health perspective. Thus, it is mainly a marketing term. Nevertheless, the blueberry has seen a distinct increase in per capita consumption in the USA since it was promoted as a superfruit in the late 1990s (Fig. 1.2). This type of marketing has shifted and promoted the consumption of more fruits.

The other side of this health consciousness is the consumer concern over the safety of the food supply and the possible contamination of our fresh fruits with pathogenic agents, pesticides, and fungicides (Johnston and Carter 2000; Batt and Noonan 2009; Sloan 2006; Wei 2001). These concerns have led to stricter regulations and more testing for residues in our produce along with improved systems to trace the source of the produce. This allows excellent enforcement if residues are found, so the potentially tainted produce can be removed from the market and any problems can be corrected (Golan et al. 2004; van Rijswijk et al. 2008).

This food safety concern has led to the greater interest in growing fruit using sustainable or organic production systems which use few or no agrochemicals. This market, although still small, is rapidly growing (20–35% annually) (Delate et al. 2008) with the USA and the EU being the largest consumers of organic produce (Dimitri and Oberholtzer 2005). About 3% of the apples worldwide are being grown organically (Granatstein and Kirby 2007) and 1–5% of the fruit in the EU is certified organic. This is low compared to the 10% market share that organic vegetables have in the EU (Weibel et al. 2007; Sansavini 2009). The rapid growth is also reflected in the mainstreaming of organic produce from a specialty produce category mainly carried by natural food stores to a produce item found in most conventional grocery stores (Dimitri and Greene 2002; Dimitri and Oberholtzer 2005; Granatstein and Kirby 2007; Martinez 2007).

Currently, much of the organic tree fruit production is in semiarid climates with traditional cultivars, where disease control is not the major issue as the disease and pest control procedures are still not reliable. In spite of higher prices (20–40%), the higher risk and lower yields (15–40% less), especially for more humid zones, have discouraged growers from switching from conventional to organic production. In apple production, although the scab-resistant cultivars facilitate organic production, the apple market is cultivar specific and the acceptance of these cultivars in the mainstream market is limited. The potential benefits, both economically and environmentally, have encouraged increased private and public investment to develop better management approaches and disease-resistant cultivars for sustainable and organic agricultural systems throughout the world (Delate et al. 2008; Granatstein and Kirby 2007; Weibel et al. 2007; Sansavini 2009). Whereas public policy in the USA has relied on the free market approach to encourage organic production, in the EU "green payments" are used to subsidize the transition costs from conventional to organic production (Dimitri and Oberholtzer 2005). More common (60–90% fruit sales) in Europe are Integrated Fruit Production systems which are designed to minimize the use of agricultural chemicals.

3.3 Consumer Expectations and Habits

Consumer expectations drive the marketing trends. Thus, beyond the search for products that are "green," healthy and safe as previously discussed, consumers now expect to have produce that is convenient to eat, of consistent quality, good flavor,

Table 1.3 Fresh fruit production of major Southern Hemisphere temperate fruit exporters (http://FAOstat.fao.org, accessed 10 Nov 2010)

Fruit	1970	1975	1980	1985	1990	1995	2000	2005
Strawberry	11	15	19	23	37	46	70	86
Plum	13	13	14	14	19	24	31	45
Cherry	15	18	16	18	22	31	39	48
Pear	447	470	497	547	699	1,019	1,296	1,353
Peach	697	811	798	756	786	881	986	1,139
Apple	1,400	1,560	1,980	2,500	3,190	3,940	4,500	4,990

Figures are 5-year averages in 1,000 mt
Argentina, Australia, Brazil, Chile, New Zealand, Peru, South Africa

and of a wide variety all year round (Byrne 2005; Sloan 2006, 2007, 2008; Lucier et al. 2005; Jaeger 2006; Jaeger et al. 2003; Jaeger and Harker 2005; Blisard et al. 2002).

Globally, there is a shift toward a supermarket distribution system which requires fruit with good storability (Frazão et al. 2008). Furthermore, with the advent of technological advances in transportation, storage, remote monitoring of refrigerated systems, and communications, the global trade of all agricultural products and particularly fresh fruits and vegetables has blossomed. In 1961, the value of the global trade in fruits and vegetables was $360 million, and by 2001 it had grown to a value of $11.8 billion. Since the 1980s, the global trade of fruits and vegetables has increased more rapidly than any other agricultural commodity (Huang 2004; Huang and Huang 2007). This has allowed the long-distance shipment of fruits to the markets, allowing exotic tropical fruits as well as off-season temperate fruits to arrive to a market destination thousands of miles away from the production site in excellent condition. An example of this would be the growth of fruit production in the Southern Hemisphere (Argentina, Australia, Brazil, Chile, New Zealand, Peru, South Africa) to supply the off-season markets in the Northern Hemisphere. The production of these countries increased rapidly beginning in the 1980s (Table 1.3).

Beyond the year-round availability, the diversity of produce items available in supermarkets has increased over the last several decades (Calvin and Cook 2001). This reflects not just an expanded array of cultivars or fruit types available for temperate fruits, but more exotic fruits and a new class of convenience food: the minimally processed products (Handy et al. 2000).

The minimally processed product reflects our ever-increasing tendency to fix meals in less time and to eat out more often (Stewart et al. 2006). The time spent preparing food in the USA has decreased from 65 to 31 min a day from 1965 to 1995 partially due to the use of minimally processed and other prepared foods as well as the increase of food preparation and cleaning appliances in the home. The percent of calories eaten away from home in the USA has increased from 18 to 32% from the mid 1970s until the mid 1990s (Canning et al. 2010). This trend to use minimally processed foods has extended to the food service industry as they strive to cut preparation costs. This is reflected by the decrease of jobs available in the food service industry and the increase of jobs available in the food processing industry in preparing these minimally processed products from 1996 to 2000 (Canning et al. 2010). Unfortunately, this trend to eat out more tends to decrease the consumption of fruits and vegetables (Guthrie et al. 2005), although there are efforts by fast food and other food service venues to develop offerings that are healthier (Martinez 2007; Sloan 2007). Nevertheless, as postharvest and packaging technology improves, more washed, peeled, precut, and packaged produce will be there in our future (Handy et al. 2000; Allende et al. 2006).

Convenience, along with health issues, is a major driving force in the food marketing business, and time constraints are an important barrier to eating healthy. Thus, healthy snacks based on fruits and vegetables that deliver one or several servings are being actively developed (Sloan 2007; Jaeger 2006). A convenient fresh fruit needs to be consistently available, keep well, not be susceptible to bruising or other postharvest damage, not be messy to eat, eaten without a utensil, and be suitable for a range of uses (meals, snacks, desserts). Fruits differ dramatically in their convenience, with apples and bananas being excellent and peaches, melons, and mangoes not very convenient to eat (Jaeger 2006).

Although convenience and health are important desires, fruits also need to have consistent quality and flavor. The difficulty to make good on these requirements varies widely from fruit to fruit. Nuts, citrus, apples, and grapes are easier to deliver with consistently good quality and flavor than stone fruit, strawberries, and blackberries. Surveys have identified the lack of consistent quality as a major reason people do not buy peaches (Byrne 2005). In addition, there is a willingness of consumers to pay more for better quality (Opara et al. 2007), which is the reason for developing branded fruit that consistently delivers quality fruit (Jaeger 2006).

3.4 Producer Expectations: Simplified Management

To stay in business, a producer needs to produce high yields of quality fruit for a minimum of expense both economically and from a management perspective. Thus, any cultivar used needs to be productive and produce quality fruit as has been discussed previously. In fruit and vegetable production, the two largest variable expenses are for labor and for agricultural chemicals to protect the crop from damaging diseases and pests (Lucier et al. 2005).

The high cost and need for trained labor, especially in developed countries, has led to a research emphasis on modifying tree size, growth, and cropping, simplifying training techniques, and mechanization of fruit tree production. Dwarfing rootstocks have been available and commercially used for apple for 60 years to create orchards with smaller, easier-to-handle trees that generally produce more precociously and at a higher yield. Unfortunately, in most crops (i.e., cherries, pears, peaches, plums), dwarfing rootstocks are a relatively new innovation which is currently being researched with renewed excitement (Webster 2006; Reighard 2000; Reighard and Loreti 2008; Lang 2000).

This approach is complemented by developing scion cultivars that do not set excessive fruit, set fruit without cross-pollination or with parthenocarpy (Kappel 2008; Socias i Company 1990, 1998; Sansavini and Lugli 2008; Lespinasse et al. 2008), grow less (spur, compact types), and have unique growth forms that lend themselves to high-density, highly productive plantings (columnar/pillar, weeping) that may simplify or allow the mechanization of pruning, thinning, harvesting, and other processes of orchard management (Webster 2006; Liverani et al. 2004; Scorza et al. 2006).

Beyond the environmental and health costs of using agricultural herbicides, fungicides, and pesticides, their use requires a substantial economic and management cost. Thus, there is an increasing need for scion and rootstock cultivars that are tolerant/resistant to a wide array of nutrient problems, pests, and diseases.

4 Trends in Fruit Breeding Goals

These broad trends influence the objectives of breeding programs in many ways as the breeder is always trying to anticipate the future needs of the fruit industry. The importance of each trend varies with the crop and environment. The major trends are to develop cultivars which simplify orchard practices, have increased resistance to biotic and abiotic stress, extend the adaptation zones of the crop, create new fruit types, create fruit cultivars with enhanced health benefits, and provide consistently high quality.

4.1 Simplifying Orchard Practices

A major driver of this category is the cost of labor and management of fruit crop production. The high cost of labor, especially in developed countries, has led to research emphasis on modifying tree size or growth, simplifying training techniques, and the mechanization of fruit and nut tree production over the last 50 years. The objective of limiting the vegetative growth of tree fruit and nut species is particularly a problem on fertile soils and in lower chill subtropical and tropical zones, where the growing season is greatly extended as compared to temperate production zones. Among tree fruits, the apple has led the way with its use of size-controlling rootstocks, high-density orchards, and specialized pruning techniques to maximize precocity, yields, and quality while minimizing pruning and general management costs. This success has spurred research in other fruit tree crops and substantial progress has been achieved in pears, cherries, peach, and plum (Beckman and Lang 2003; Lang 2000; Fideghelli et al. 2003; Scorza et al. 2006; Reighard 2000; Reighard and Loreti 2008; Webster 2006).

There are two complementary genetic approaches to modify the tree size and architecture. One can work on the rootstock and/or the scion component of the orchard system. In apple, pear, and cherry, all generally large orchard trees, most effort has been invested in developing rootstocks that induce less scion growth and greater precocity. These dwarfing rootstocks were essential in the development of the modern high-density apple orchard by providing an inexpensive approach to control the scion growth as well as improving precocity, light penetration within the canopy, and allowing greater efficiency of pesticide applications. In the last 20 years, especially with stone fruit, there has been a shift from seedling to clonal rootstocks (Beckman and Lang 2003) which has facilitated the use of interspecific hybrids as rootstocks, especially those between distantly related species which are more probable to result in rootstocks that are able to dwarf the scion cultivar.

The approach from a scion perspective has been to modify tree architecture. This ranges from selecting within the standard growth type for better branching habit and increased spur formation to developing cultivars with unique tree architecture. These new growth habits range from dwarf, semi dwarf, compact, pillar, and weeping (Hu and Scorza 2009; Scorza et al. 2006; Liverani et al. 2004; Fideghelli et al. 2003; Webster 2006; Lauri et al. 2008; Segura et al. 2007; Schuster 2009). Between 1990 and 2000, 56 of the 2,700 fruit cultivars released had unique growth types. The most common being dwarves and spur types (apples). Unfortunately, with the exception of the spur-type apples which were mainly bud sports of established cultivars, these releases are mainly for garden use due to their current lack of fruit quality (Fideghelli et al. 2003). More recent work on pillar types in peach has resulted in several new cultivars with improved quality (Scorza et al. 2006; Liverani et al. 2004).

The most promising growth modifications useful for high-density and/or higher yielding capacity appear to be the pillar type and spur growth habit. Both these allow better light penetration, require less pruning, and potentially could deliver greater yield efficiencies (Fideghelli et al. 2003; Kodad and Socias i Company 2006; Scorza et al. 2006; Socias i Company 1998; Kenis and Keulemans 2007). The weeping habit is also being explored by several peach breeding programs as a growth habit that would decrease management costs (Scorza et al. 2006; Bassi and Rizzo 2000). Whatever results from this work, it is clear that the optimal training system needs to be developed for each unique tree architecture (Scorza et al. 2006) and marketing needs to bundle these unique cultivars with the optimal training systems.

Beyond facilitating harvest by modifying tree growth and architecture, there is an increasing interest in mechanical harvesting to reduce labor cost and time required for harvest. There are already mechanical harvesting systems for a range of crops but mainly for processing as the cosmetic appearance requirements are less demanding. Nevertheless, breeding for more uniform ripening, ease of detachment, non-bruising types, and better firmness should lead to cultivars better adapted to mechanical or at least to a once-over harvest approach as compared to the multiple harvests needed with the current cultivars.

4.1.1 Fruiting Stability

All breeding programs select for high fruit set and are always looking for stability of fruit set in spite of the climatic conditions. An important trait to ensure consistent fruit set is self-fertility. Currently, there are various dioecious species (pistachio, kiwi), monecious species (pecan, walnut), and species with perfect flowers that display self-incompatibility (apple, plum, sweet cherry, almond) which require cross-pollination either via wind or insects as pollinators. This need for cross-pollination requires the planting of pollinizers, management of pollinators, and the presence of appropriate weather during the pollination period which complicates management and creates more uncertainty in production. No work is ongoing to transform dioecious or monoecious crops into perfect-flowered, self-compatible, or parthenocarpic crop. This is basically what happened during the development of the modern grape which began as a dioecious species in the Neolithic period and was, over thousands of years, transformed into the current perfect-flower, self-compatible fruit crop (Riaz et al. 2007). Currently, there is active work in the development of sweet cherry, Japanese pear, apricot, and almond cultivars that are self-fertile, and in the development of pear and persimmon cultivars that consistently set fruit parthenocarpically or are self-fertile (Gradziel 2008; Gradziel and Kester 1998; Socias i Company 1990; Apostol 2005; Kappel et al. 2006, 2011; Sansavini and Lugli 2005; Okada et al. 2008; Yamada et al. 1987). These incompatibility systems have been studied genetically, and currently there are markers that can be used for characterizing the incompatibility alleles present in various species (Tao and Iezzoni 2010; Schuster et al. 2007; Kodad and Socias i Company 2009; Guerra et al. 2009; Bokszczanin et al. 2009).

4.2 Resistance to Insect and Disease Problems

Concerns about the safety of agricultural workers, potential of environmental contamination, and safety of the consumer have spurred the development of tighter governmental restrictions on the use of agricultural chemicals and on alternate pest and disease control strategies. This has led to greater governmental and privately funded work in integrated pest and disease management systems to reduce the amount of pesticides and fungicides used in the production of fruit (Dimitri and Greene 2002; Dimitri and Oberholtzer 2005; Weibel et al. 2007). One facet of these management systems is the use of genetic resistance to various diseases and pest problems.

Each crop has multiple important disease/pest problems (Table 1.4), some which are worldwide in distribution while others regional. Throughout the world, there has been an increased emphasis on the development of higher levels of disease and pest resistance in fruit scion and rootstocks. In Europe, there are 64 pome fruit breeding programs of which two-thirds are in apple breeding and one-third in pear breeding. Most of the scion programs are developing new pome cultivars with disease resistance (scab, powdery mildew, fire blight) as important objectives, and from 2000 to 2004 almost half of the apple cultivars released by these programs had resistance to scab and many times to other pathogens as well (Lespinasse 2009). Unfortunately, the vast majority of the apple and pear production does not use disease-resistant cultivars even in IFP because the market demands high quality and consumers

Table 1.4 Disease and pest problems of major tree fruit crops

Crop	Disease	Pathogen/pest	Comments
Pome fruit	Apple scab	*Venturia*	Genes/markers identified, many resistant apple cv.
	Powdery mildew	*Podosphaera*	Genes/markers identified, resistant apple cv.
	Fire blight	*Erwinia*	Active work, resistant apple/pear cv. and rootstock
	Black spot	*Stemphylium*	Little work, widespread on pear
	Psylla	*Cacopsylla*	Transmit pear decline
Stone fruit	Brown rot	*Monolinia* spp.	Little progress, some less susceptible cv.
	Bacterial leaf spot	*Xanthomonas*	Good progress, polygenic, resistant cv.
	Plum pox	*Potyvirus*	Genes/markers identified, active breeding, transgenic resistant plum
	Peach scab	*Cladosporium*	Little work, widespread problem
	Root knot nematodes	*Meloidogyne*	Genes/markers identified, resistant rootstocks
Citrus	Citrus greening	*Candidatus Liberibacter*	No resistance known
	Citrus canker	*Xanthomonas*	Tangerines moderately resistant, polygenic resistance
	Citrus tristeza virus	*Closterovirus*	Genes/markers identified, resistant rootstocks, active breeding
	Phytophthora	*Phytophthora*	Resistant rootstocks
	Nematodes	*Tylenchulus*	Genes/markers identified
Grapes	Powdery mildew	*Erysiphe*	Gene identified, active breeding
	Pierce's disease	*Xylella*	Gene identified, active breeding
	Nematodes	*Meloidogyne*	Dominant gene, resistant rootstocks
	Phylloxera	*Daktulosphaira*	Resistant rootstocks

Source: Brown (2003); Lespinasse (2009); Lespinasse et al. (2008); Fischer et al. (2003); Byrne (2005); Gmitter et al. (2007); Riaz et al. (2007); Ramming et al. (2009)

generally do not sacrifice quality for less pesticide use. In addition, the pome market's cultivar specificity makes it very difficult for a new cultivar to enter the market without a substantial promotion effort (O'Rouke et al. 2003; Weibel et al. 2007; Fischer et al. 2003).

From a breeding perspective, the incorporation of a simply inherited adaptation trait, such as low chilling in peach, Pierce's disease resistance in grape, and scab resistance in apple from a wild germplasm, takes at least three cycles of backcrossing into high-quality genotypes to reach a commercially acceptable fruit quality (Byrne et al. 2000; Ramming et al. 2009; Brown 2003). These disease-resistant cultivars, although acceptable and compete well in the local market, do not necessarily compete well with the quality of the cultivars available in the regional or international markets. Thus, several more generations of breeding are necessary. Unfortunately, multiple resistances are needed in each cultivar, which makes incorporating disease resistance with excellent quality and production a much more challenging goal.

Nevertheless, on the few diseases that have received substantial attention such as apple scab, bacterial leaf spot in peach, plum pox in apricot, fireblight in pear, and Pierce's Disease in grape, rapid progress has been achieved in transferring good resistance into commercially acceptable background. Thus far, the effort expended on developing disease/pest resistance in tree fruit crops has been minimal, and as this effort increases resistant cultivars that have the quality and production characteristics needed for widespread commercial use will emerge as has been seen in the major agronomic and vegetable crops.

Efforts and advances in development of genomic tools facilitate the identification of genes involved in resistance to diseases and the implementation of molecular markers for the selection and introgression of resistance genes into fruit crops. There has been excellent progress in identifying markers for resistance genes to apple scab, various nematode species, plum pox, and powdery mildew (Riaz et al. 2009; Gardiner et al. 2007; Esmenjaud and Dirlewanger 2007), although their incorporation into breeding programs is still in its infancy. The use of transformation to increase the disease resistance of fruit is species specific. Transgenic plants are much easier to generate in species, such as apple, pear, and citrus, than in stone fruits, such as peach, almond, plum, or apricot. Nevertheless, this effort has led to a plum pox-resistant European plum cultivar which is currently being field tested (Scorza 2000; Ravelonandro and Scorza 2009). Once these techniques are better developed, transformed cultivars could lead to reduced pesticide use, but the public acceptance of such cultivars is still not known.

4.3 Expansion of Production Zones

With the strong demand for fruit availability on a year-round basis and with the advances in postharvest, communications, and transportation which have made the sourcing of fruit from any place in the world a possibility, production is shifting into new production regimes and regions. According to FAO figures (FAOSTAT, http://apps.fao.org/), the production of the major fruits in the world has increased two- to threefold over the past 30 years. This production increase has not been even throughout the world, as the fruit industries' importance in developed countries, such as Japan, Canada, the USA, and many European countries, has leveled off or decreased over the last 20–30 years, whereas it has rapidly increased in Asia (mainly China), Africa, and South America.

Temperate-zone breeding programs of most fruits and nut crops have successfully extended the harvest season by developing earlier and later ripening cultivars and by breeding cultivars adapted to the extremes of the temperate zones. Peach breeding efforts in North America and Europe have extended the fruit availability from about 1 month to 6–8 months. The limitation is the climatic cycle. Work to overcome climatic restrictions has led to the production of fruit crops under protection (greenhouses to high tunnels) to extend the harvest season forward or backward. This has been increasingly used in the temperate zone, and with stone fruit can move the

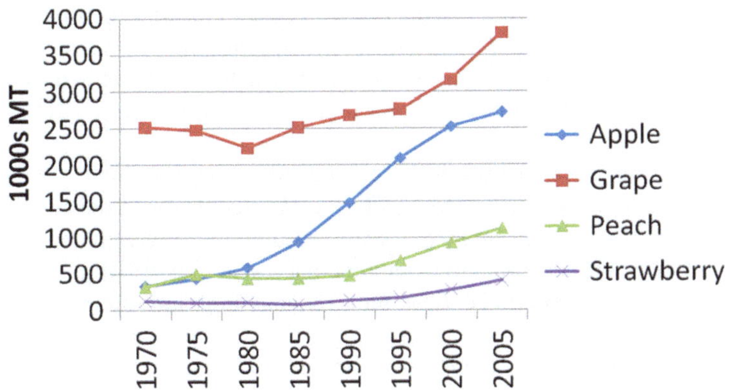

Fig. 1.3 Fruit production in medium- and low-chill zones of the Americas and Northern Africa

harvest forward 30–90 days and with some small fruits could allow year-round production (Jiang et al. 2004; Lang 2009; Gaskell 2004; Demchak 2009). Currently, the cultivars used are those developed for field production. For more efficient production, specialized cultivars adapted to the greenhouse environment would be the best. These would be low- and medium-chill cultivars with a short fruit development period (if the objective was early ripening) and medium vigor, ability to grow well under low light conditions, ability to set fruit under high temperatures, and high quality since the soluble solids of protected culture produce fruit are usually 1–2% Brix lower than field-produced fruit (Jiang et al. 2004; Byrne 2010). These structures are also used to protect the crop from rain to minimize disease issues, avoid fruit cracking, and can generally lead to better and more consistent fruit production and quality.

Within temperate-zone production, off-season fruit can be produced in the opposite hemisphere. Consequently, over the last 30 years, fruit production in the temperate zone of the Southern Hemisphere (Chile, Argentina, South Africa, Australia, and New Zealand (Table 1.3)) has increased to supply the winter fruit demand in the Northern Hemisphere markets (Europe, North America, and Asia). Even so, there are still gaps in the supply of many fruits in March/April and October to December. These gaps are being closed more and more by the production of temperate fruit species in the subtropical and tropical zones.

There is a trend toward increased production in subtropical and tropical regions in the Americas (Brazil, Bolivia, Mexico, Uruguay, and Ecuador), northern Africa (Algeria, Egypt, Morocco, and Tunisia) (Fig. 1.3), and Asia. Although the climates vary tremendously in the subtropical and tropical regions, the major climatic restrictions found in these regions are heat related: chilling requirement and heat tolerance. Initially, production in tropical zones took advantage of the cooler tropical highland conditions, where traditional medium to high chill cultivars could be grown either directly or with some cultural manipulation to compensate for a lack of chilling. These production systems have evolved to include lower chill cultivars, especially as the production moved to warmer zones and the dormancy management systems were improved. These production systems are exemplified by the double cropping systems developed for grapes and peaches in the warm tropics and continuous production possibilities for berry and peach production in the cool tropical highlands (1,500–2,500 m above sea level) (Clark 2005; Lavee 2000; George and Erez 2000).

The pioneering work in low-chill fruit breeding was done with peaches. This began in California and continued in Florida (USA), Texas, Louisiana, Mexico, Brazil, and South Africa. In many low-chill breeding programs, the emphasis is to develop early ripening cultivars to extend the harvest season forward to capture the lucrative early fruit market. In contrast, the Brazilian programs, although early cultivars were developed, many mid-season and late-ripening cultivars were also released to support their local produce/processing industry (Byrne et al. 2000). Currently, most of the peach production in many subtropical and tropical regions is sold in the regional market. This success has encouraged increased activity in stone fruit, pome fruit, and berry medium- and low-chill breeding programs (Byrne et al. 2000; Hauagge and Cummins 2000; Darnell 2000; Hancock 2000; Lyrene 2005) with the resulting commercial development of fruit production enterprises in these zones.

As the production moves out of the tropical highlands to the warmer tropical climates, the tolerance to high heat during bloom and throughout the fruiting cycle becomes critical. Fruit crops vary tremendously in their sensitivity to heat. Among those most sensitive to poor fruit set under high heat conditions (25°C) during flowering are peaches, nectarines, strawberries, and blackberries, whereas apples, plums, and grapes appear to fruit well under warmer conditions (Lavee 2000; Hancock 2000; Clark 2005; Jackson 2000; Byrne 2010). Heat during the growing season can also affect bud initiation and development and fruit quality while the fruit is developing. In many crops, high temperatures (>25°C) can lead to poor fruit bud initiation and development, more rapid fruit development, problems with good fruit sizing, fruit shape, and fruit color (anthocyanin) development (Byrne 2010; Kozai et al. 2004; Hancock 2000; Hauagge and Cummins 2000). Although much more work needs to be done, there has been progress in developing low-chill genotypes that are heat tolerant in peach, apple, strawberries, blackberries, and other crops, and this bodes well for future work.

Beyond the adaptation traits of low chilling requirement and tolerance to heat during bloom and fruit development, there is a need to select genotypes well-adapted to the cultural manipulations used to avoid dormancy and induce the flowering/fruiting cycle in the cropping systems used in the tropics. This would include the ability of the genotype to rapidly develop flower buds to allow a rapid cycling, ease of induction via hormone application or cultural manipulation, and the ability to crop well through multiple cycles of fruiting per year.

Various other abiotic challenges are encountered more as fruit production expands to new regions, where the soil/water combination is nonoptimal for fruit production due to soil pH, salinity, or moisture status. In many fruit crops (peach, pear, citrus, grape), there has been some work to develop rootstocks adapted to calcareous soils which are commonly found in the more arid fruit production zones but much less work on rootstocks adapted to soils that are waterlogged, acid (high aluminum), or heavy textured. These objectives will continue to maintain regional importance, but the major focus will shift to the ability to grow fruit with less quantity or less quality of water in the future. Several of the major arid fruit production areas that depend on irrigation for production, such as the central valley of California, are beginning to experience problems with both the quantity as well as the quality of water. Breeders of agronomic crops (maize, cotton, sorghum) have worked extensively on the development of drought (Cattivelli et al. 2008; Sinclair 2011) and salinity tolerance (Flowers and Flowers 2005; Ashraf and Akram 2009) with moderate success. Although much has been done to increase the efficiency of managing water and salinity among fruit, little has been done to develop rootstocks and/or scion cultivars that use water more efficiently or are tolerant to salinity. At this point, the major emphasis is at the point of identifying differences among germplasm in their response to drought (Grant et al. 2010; Rieger et al. 2003; Kocsis et al. 2009; Cochard et al. 2008) or salinity stress (Musacchi et al. 2006; Syvertsen and Melgar 2010).

4.4 Diversification of Fruit Types

In multiple studies, it has been shown that the consumers throughout the world, especially as their income level rises, are looking for interesting foods that are convenient to consume (Blisard et al. 2002; Frazão et al. 2008). This is reflected in the doubling of items available in the produce section of the grocery store in the USA (Davis and Stewart 2002). This consists of several classes of items: new cultivars of traditional fruit, more exotic fruits, organic versions of traditional fruits, and minimally processed fruits.

Many studies have documented the heterogenous nature of consumers and more recently have been characterizing the various flavor classes within a given fruit (Jaeger et al. 2003; Jaeger and Harker 2005; Tomala et al. 2009; Ross et al. 2010; Crisosto et al. 2006, 2007). With apple and pears, the fruit is sold by the cultivar name, whereas with stone fruit and small fruit this is not generally the case. Thus, as new flavor classes are introduced into the market, the consumer gets confused as it is not obvious from the external appearance what the flavor of the fruit is. Consequently, it has been suggested that stone fruit as well as others are sold in a way that the flavor class is obvious (Byrne 2002; Crisosto et al. 2006, 2007; Ross et al. 2010). Although unique fruit products may not be sold in high volumes, it is clear that if the new offering has sufficient quality there will be consumers willing to pay a premium for it (Jaeger and Harker 2005; Gamble et al. 2006).

In the case of peach, there is a wide diversity of regional peach and nectarine types traditionally grown throughout the world. Most are regional preferences, such as low acid white and pantao peaches in China and Japan, yellow-fleshed acid types in North America, and nonmelting yellow–orange peaches in many regions of Latin America. Now, given the globalization of the produce market and the need for new produce items, more types of these previously regionally grown peaches are being sold in any given market. The nectarine was initially developed in the USA in the

1950s and 1960s, and now nectarine production is approaching the production level of the peach crop in the USA and Europe. Thus, the USA market has evolved over the decades from mainly yellow-fleshed acid peaches to a market that has both peaches and nectarines that are either white or yellow flesh with low or high acidity. Recently, low-acid pantao peaches have been appearing in the market. This offering will expand into a series of pantao cultivars and then will diversify to have the range of flesh colors, acidity, and skin types (peach/nectarine). Other unique types being developed would include cultivars with nonmelting red or orange flesh, skin/flesh without anthocyanins, and enhanced flavor and health properties (Byrne 2005; Pascal et al. 2009; Nicotra and Conte 2003; Monet and Bassi 2008; Vizzotto et al. 2007). Similar emphasis on developing unique shapes, colors, and flavors is seen in other fruits. Examples of this would be a range of colors and flavors among seedless table grapes, development of bright yellow- and red-fleshed plums (Halgryn et al. 2000), work toward developing red-fleshed apples (Volz et al. 2009a, b), and development of a low-acid sweet kiwifruit (Wismer et al. 2005).

Convenience is a major driver of innovation in the food industry and should be considered as new fruit cultivars are developed. There are several factors that influence whether a fruit is a convenient item to consume. These include the following: consistent availability, good postharvest traits, easy to eat and not messy, and suitability for a variety of uses (breakfast, snacks, dessert). Most nut crops qualify as convenient food as do fruits, like apples, grapes, and bananas, while others, such as peaches, mangos, and melons, do not (Jaeger 2006).

Traits that make fruit more difficult to eat would be seeds in the fruit, need or difficulty of peeling the fruit, size of the fruit, need to cut or use utensils to eat the fruit, and juiciness of the flesh. Thus, we want a fruit that is seedless, can be eaten without peeling, is bite size, and does not spurt juice out when eaten. Such innovations are already here for some fruit and being developed for others. In citrus breeding, two essential traits are the ease of peeling and seedlessness (Stover et al. 2005) and table grapes are already bite-size fruits which do not need to be peeled and have no seed. Along these lines, work is active to develop bite-size kiwifruit, stone fruit without a pit, and stone fruit and berries with a longer postharvest life (Clark and Finn 2008; Byrne 2005).

And as people throughout the world eat more of their meals away from home (Normile and Leetmaa2004; Stewart et al. 2006; Gale and Huang 2007; Frazão et al. 2008), the importance of minimally processed foods increases both in the food service business and for personal use (Handy et al. 2000). Products, such as peeled baby carrots, bagged salads, and precut vegetables of many types, are now mainstays in most grocery stores in the USA, but similar products with fruits are still not common. When whole versus fresh cut apples were offered in elementary and middle schools in the USA, more fruit was eaten when offered as a fresh cut product (McCool et al. 2005). This approach would help encourage children and others to eat more fruit. The best example of a fresh cut product being offered in a fast food restaurant would be the fruit salad (sliced apples, grapes, and walnuts) offered by MacDonald's. This demonstrates the effort of fast food and other restaurants to develop healthier menus for a consumer that is increasingly health conscious (Martinez 2007).

The development of healthy snacks, whether they are minimally processed, precut, and peeled fruits, dried, pickled, or juice preparations that supply the equivalent of one serving of fruit, needs to be accelerated to adapt to the new consumption patterns seen in our modern world. There have been impressive advances with postharvest treatment and packaging strategies to prolong the shelf life of these products; nevertheless, the selection of the appropriate cultivars is important as this industry develops and expands into the fruit arena. This requires collaboration among food scientists and plant breeders to best match the genetics and traits of the fruit with the requirements of the processing required. Some work is looking at the suitability of cultivars for this use (DeEll et al. 2009), but no breeding program has yet embraced this objective.

4.5 Health Benefits of Fruit

The health benefits of fruits and other produce always seem to be in the news (Variyam and Golan 2002). The initial work compared different fruit crops for their varying levels of antioxidant activity, carotenoids, phenolics, anthocyanins, and other phytochemicals. At times, this data was contradictory as only one or a few cultivars were generally used to represent the crop (Mattila et al. 2006; Sun et al. 2002; Vinson et al. 2003; Wang et al. 1996). The appearance of this type of information and other studies showing that the consumption of fruits has protective properties against various pathological conditions, such as inflammation, cancer, atherosclerosis, and other circulatory problems (Prior and Cao 2000; Wargovich 2000; Southon 2001) has fed the current interest in the health benefits of consuming fruits. Furthermore, this work also showed that fruits had a higher level of phenol antioxidants than common vegetables (Vinson et al. 2003).

From this work, the concept of a "superfruit" emerged in the marketing world which has encouraged the increased consumption of multiple fruits and fruit products. Thus, as you stroll through the supermarket, it is common to see a range of health claims on fruit products, with the most common being high in antioxidants, high in vitamin C, B6, and B12, heart healthy, low in saturated fats and cholesterol, low sodium, and, for cranberry, promotes urinary tract health.

As the public becomes more aware of the health benefits of fruits and is being told to eat a colorful diet, there is a potential to create a new market for cultivars specifically developed for their health benefits. Such "health-enhanced" cultivars would provide a new product that could be sold fresh or processed (total crop or as an outlet for the cull fruit) into extracts that are natural sources of antioxidants, antimicrobials, and colorants (Byrne 2002). The prerequisite of developing these "health-enhanced" cultivars is that there is genotypic differences in the traits that provide health benefits: i.e., cultivars and selections differ in the bioactivity or phytochemical levels. This has been shown to be the case with peaches, plums (Cevallos-Casals et al. 2002, 2006; Chang et al. 2000; Cantín et al. 2009; Gil et al. 2002; Tomas-Barberan et al.

2001;Vizzotto et al. 2007; Byrne et al. 2009), blueberries (Connor et al. 2002a, b), apples (Yoshizawa et al. 2005; Lata et al. 2005; Lee et al. 2003), blackberries (Wang and Lin 2000; Connor et al. 2005b, d), raspberries (Connor et al. 2005a, c; Weber et al. 2008), grapes (Stringer et al. 2009; Pastrana-Bonilla et al. 2003; Xu et al. 2010; Yang et al. 2009; Vilanova et al. 2009), and many other crops. Although there is still a lack of knowledge of the genetics of these various phytochemicals, the data that exists indicates that this process of developing cultivars with "enhanced" levels of antioxidant activity, polyphenolics, and anthocyanins should be a straightforward process (Connor et al. 2002b, 2005a, c; Cantín et al. 2009).

Anthocyanins are generally reported as high in many berry crops (Mattila et al. 2006), but there is also a great potential to develop tree crops with red flesh. Thus far, it has been shown that some peach and plum cultivars can rival the anthocyanin, total phenolics, and antioxidant activity of blueberries (Cevallos-Casals et al. 2006; Vizzotto et al. 2007; Byrne et al. 2009). In the case of developing red flesh among normally green, yellow, and white tree fruit (apples, pears, peaches, plums) cultivars, there appear to be a few major genes that condition this anthocyanin production in various fruits (Sekido et al. 2010; Werner et al. 1998; Volz et al. 2009a). Currently, several fruit breeding programs are exploring or developing berry and tree fruit crops with greater levels of anthocyanins (Byrne et al. 2009; Connor et al. 2002a, 2005a, c; Cantín et al. 2009; Volz et al. 2009a, b; Sekido et al. 2010).

One very important decision in any plant breeding program is to select the target. In the case of developing a health-enhancing cultivar, one has to decide what chemical(s) and levels to select for. This is not as simple as it may seem. Although there is a substantial body of literature which describes the antioxidant activity, antiproliferative activity to various cancers, ability to inhibit LDL oxidation, anti-inflammatory activity, among many other useful actions of fruits and their extracts, most of this work is done either in cell culture experimental systems or in small animal experimental systems. These approaches are very useful at identifying potential effects but do not necessarily translate well to a human system (Finley 2005). Although there has been a substantial amount of work to establish the antioxidant levels of fruits and it is generally considered that the consumption of more antioxidants is good for one's health, there is not definitive proof to confirm that supplemental antioxidant consumption reduces the incidence of chronic disease (Amiot 2009). Consequently, more research is needed to identify the target phytochemicals and, probably more difficult, the target concentration needed in the fruit to be effective at promoting the long-term health of the consumer as compared to a normal cultivar. Nevertheless, fruit breeding programs are exploring and actively breeding for cultivars with enhanced levels of antioxidants, phenolics, carotenes, and anthocyanins (Vizzotto et al. 2007; Cantín et al. 2009; Connor et al. 2002a, b, 2005a, c; Volz et al. 2009a; Stringer et al. 2009; Weber et al. 2008; Battino and Mezzetti 2006; Khanizadeh et al. 2009) and we are beginning to see the promotion of specific fruit cultivars as health enhanced.

4.6 Consistent High Fruit Quality

For repeat purchasing, a good experience is essential. Surveys with stone fruit in Southeast Asia, the USA, and in Europe have indicated that inconsistent fruit quality is the major impediment to greater sales (Clareton 2000; Crisosto et al. 2003, 2007; Moreau-Rio 2006; Wei 2001). In addition, the earliest fruit to harvest is commonly of lesser quality which has the potential to depress the market as has been seen in citrus (Poole and Baron 1996) and stone fruit. The fruit industry needs to deliver what the consumer wants: an excellent quality piece of fruit every time.

Although the specific quality traits may differ among fruits, for most fruits the most important traits are flavor, most commonly measured as total soluble solids and titratable acidity and texture, measured as firmness, crispiness, and/or juiciness (Poole and Baron 1996; Racskó et al. 2009; Kajikawa 1998; Crisosto et al. 2003, 2006, 2007; Crisosto et al. 2004; Crisosto and Crisosto 2005; Péneau et al. 2006; Harker et al. 2008; Turner et al. 2008). Common complaints for fruit would be the lack of flavor and mealy flesh without juice as seen in stone fruit with internal breakdown (Crisosto et al. 1999; Peace et al. 2005a, b) and in apples (Jaeger et al. 1998; Racskó et al. 2009). Aroma, although important, is difficult to measure and external qualities, such as shape, color, and size, which are easily standardized during packing, are usually of secondary importance to flavor and texture. Other flavor components that consumers complain about would be off flavors and astringency (Crisosto et al. 2007). To maximize the consumption of fruit, high quality needs to be delivered consistently as previous experience influences future purchasing decisions (Racskó et al. 2009; Poole and Baron 1996). There is evidence that consumers are willing to pay more for significantly better tasting fruit (Gamble et al. 2006; Opara et al. 2007).

The ability to produce high-quality fruit depends on many factors, some out of the control of the producer, such as the weather, and others dependent on the production practices, such as irrigation, fertilizer application, pest/disease management, pruning, and fruit thinning (Crisosto et al. 1997; DeJong et al. 2002). Harvest practices are critical in producing high-quality fruit as picking at an immature state results in poor quality (Iglesias and Echeverría 2009). In many crops, the quality increases and firmness decreases during fruit ripening, so a decision to harvest is a compromise between maximizing quality and having sufficient firmness to allow easy fruit handling for cleaning, sorting, packing, and shipping. Once harvested, the postharvest treatment can make or break a shipment of fruit.

Crops differ in their ability to produce a consistent product. Pome fruits, citrus, and table grapes are better at delivering a consistent high-quality product as compared to stone fruit, strawberries, blackberries, and raspberries. Part of this is due to the crop's postharvest behavior. In general, pome, citrus, and grapes can be stored for several months to a year, whereas many stone fruits can be stored for less than 6 weeks and many soft berry crops for less than 3 weeks. There have been great strides made in postharvest handling and transportation technology in the last decades which have made the produce industry a global enterprise with the ability to deliver fresh fruit thousands of miles to the market and still maintain high quality (Huang 2004; Frazão et al. 2008). Although these advances have been critical, the success also depends on the genetics of the fruit cultivar.

In the past, fruit breeders have been criticized for developing productive, large, firm, and very attractive fruit cultivars that were lacking in flavor. These criticisms are being taken seriously by many programs which have increased emphasis on high quality and the postharvest behavior of the cultivars that they are developing. In addition, there are several large international programs that are focusing their efforts on developing better genetic tools to improve the quality of fruits. These include the RosBREED project in the USA (Iezzoni et al. 2009, http://www.rosbreed.org/) and the FruitBreedomics project in Europe (http://fruitbreedomics.com/) among others.

Soluble solids are important in all fruit crops to ensure high quality. In apples, the soluble solids were associated with price in Japan (Kajikawa 1998). Common levels of soluble solids found in fruit range from 8–10 Brix in some blackberry and early-ripening peach and plum cultivars, 15–25 Brix for sweet cherries and table grapes, and over 30 Brix in apricot cultivars from Central Asia. In blackberry breeding, the newer cultivars, such as Navaho and Ouachita, have soluble solid levels of 10–12 Brix which has made this fruit more palatable to a wider audience, and further improvement to 15 Brix appears possible (Clark and Finn 2008).

Among stone fruit, consumer-acceptable levels of soluble solids differ with the fruit and its acidity, with minimum levels of 11 Brix for acid peaches, 12 Brix for low-acid peaches and plums, and 16 Brix for sweet cherries (Crisosto et al. 2003, 2004, 2006, 2007; Ross et al. 2010). Unfortunately, many common peach, plum, and apricot cultivars, especially early-ripening cultivars, have soluble solid levels of 8–10 Brix. Genetic studies in peach have documented a negative genetic correlation between soluble solids and fruit development period (days from full bloom to commercial ripe) and fruit weight (Souza et al. 1998, 2000; Byrne 2005). Thus, it may be difficult to develop high-soluble-solid peaches that are large and early-ripening. Nevertheless, current collaborative work between Texas A&M University and the USDA (Kearney, CA) indicates that it is possible to combine good soluble solids (12–15 Brix) with good fruit size with a fruit development period of less than 100 days.

There has been excellent progress in developing high-soluble-solid peaches/nectarines for the mid- and late-season harvest periods, and levels of 15% or greater should be our goal. There are nectarine cultivars in California (Crisosto et al. 1998; Byrne et al. 2000) and peach cultivars in Italy (Nicotra and Conte 2003) that are reported to be in this range. In the last decade, 172 peaches and 134 nectarines have been patented/released in the USA. Of these, 40% were described as having low- (<11 Brix), 40% medium- (12–15 Brix), and 20% high- (>15 Brix) soluble solids. When the group of releases with high-soluble solids is examined by ripening season, 84% were mid- or late-maturing cultivars (after mid June). Only 16% were those that ripened during the early season and 90% of these early-maturing cultivars were nectarines which tend to have higher soluble solids than peaches (Wen et al. 1995a, b). In addition, the early-season high-soluble solid releases are lower chilling (range 375–700 CU, mean ~500 CU) than those that ripen in mid season (range 500–800 CU, mean ~600 CU) or late season (range 600–850 CU, mean 700 CU). This approach helps because the lower chilling cultivars bloom earlier than the higher chill cultivars. Thus, for a given ripening season, the lower chill cultivars have a longer fruit development period than the higher chilling cultivars which means that the lower chill cultivars have more time to accumulate sugars. Additional challenges are to combine high sugars with high yields (especially early), large size, and for nectarines good skin finish (few speckles and lenticels) and low cracking. As demand for quality increases, there may be some compromises on fruit size and nectarine skin appearance if high quality can be guaranteed.

Beyond high sugars, many other factors are considered in the development of high-quality fruit cultivars, including aromatic components of flavor, relative amounts of specific sugars (sucrose, glucose, fructose, sorbitol), texture, mouthfeel, and acidity. Finally, since the growing practices (pruning, fertility, irrigation, harvesting) have such a great influence on the ultimate quality of the fruit, there is a need to specify the minimal cultural practices to obtain the highest potential quality of the cultivar.

4.6.1 Firmness and Postharvest Competence

Good fruit firmness, beyond being important in consumer-perceived quality while eating, is essential for ease of harvesting, handling, marketing, and for storage of all fruit crops. Firm fruit tends to be more resistant to rain-induced cracking in cherries, allows for more ripening on the tree and consequently better quality, and frequently has a better postharvest life (Kappel 2008; Giovannini et al. 2006a, b; Sherman and Lyrene 2003; Sansavini and Lugli 2008; Oraguzie 2010). Thus, firmness has been an important selection criterion for fruit breeders, and advances in firmness have transformed stone fruit and, more recently, small fruits, such as strawberries, blackberries, and raspberries, from locally marketed crops to fruits with potential to be shipped thousands of miles to the market. Further advances are needed in all crops to facilitate a global sourcing required to supply high-quality fruit throughout the year, as this requires extended storage life to allow the transport in the most carbon-friendly means: by boat.

Fruit ripening has been extensively studied in tomato (Giovannoni 2004) which has aided much of the work in other fruit systems, such as apple and peach. The two major pathways that have been studied extensively would be the ethylene-mediated pathway that induces ripening and the endopolygalactaronase (EndoPG) cell wall softening pathway. Variations of these seem to be well-conserved over a wide range of species, including our common tree and small fruit crops.

Ethylene is known as the ripening hormone and many postharvest procedures focus on reducing the level of ethylene that fruits are exposed to or reducing the response of fruits to ethylene (i.e., 1-methylcyclopropene, 1-MCP) as a protocol to extend the storage life of fruit. In both apples and peaches, the corresponding genes that code for 1-aminocyclopropane-carboxylase (ACC) synthase (ACS), ACC oxidase (ACO), and ethylene receptor (ETR) proteins that are key to the fruit ripening process have been identified (Wang et al. 2009; Marić et al. 2009). Interestingly, apples and peaches do not respond the same when 1-MCP, an ethylene blocker, is applied (Cin et al. 2006) indicating that these systems differ significantly as does their postharvest competence. With apples, various allelic forms of ACO and ACS have been characterized across cultivars, and the allelic states that condition the best firmness have been identified with molecular markers opening up the possibility of using these in the selection for better postharvest quality in apple (Tatsuki et al. 2009; Oraguzie 2010; Zhu and Barritt 2008).

Among stone fruits, peach has been studied the most, but similar systems probably exist across the various species. There are several traits that apparently reduce ethylene production identified in peach: the slow or nonripening genes described in peach (Brecht et al. 1984; Brecht and Kader 1984) and plum (Yamaguchi and Kyotani 1986) and the stony hard (SH) gene described in peach (Haji et al. 2005). Of these, the most studied is the stony hard gene which has the potential to extend the postharvest life of the peach. Various breeding programs are actively working toward and/or have developed cultivars with stony hard flesh (Giovannini et al. 2006a, b; Lu et al. 2008; Byrne 2005).

The low expression of the cell wall degradation enzyme, endopolygalactaronase (PG, EC 3.2.1.15), also seems important in the storage ability of the peach (Wakasu et al. 2006; Peace et al. 2005b). Throughout much of the world, the common flesh type used for fresh market peach is the melting-type flesh. Although much progress has been made at developing firm melting flesh types, it is still difficult to pick them firm enough at a high level of fruit quality. In contrast, the processing industry uses a firmer flesh type: the nonmelting flesh. This is conditioned by an allele at the PG gene which disrupts the activity of EndoPG (Peace et al. 2005b) resulting in a flesh that does not "melt." This firmer flesh type allows the harvesting at a higher quality, tree-ripe stage with enough firmness to the market. These types have been used for centuries for the fresh market in Latin America from Mexico south to Brazil and in Spain. The main objection to these types is that the flesh does not separate from the stone which is preferred for the fresh market. Nevertheless, since early-ripening peaches and nectarines are usually clingstone because they ripen before their pit/flesh separation occurs, many breeders have begun to develop earlier ripening peach and nectarines with nonmelting flesh. In the USA, this approach has been spearheaded by the work in Florida (Byrne 2005) and currently there are multiple fresh market releases with nonmelting flesh in the USA ('UFGold,' 'UFPrince,' 'Springprince,' 'Springbaby,' and 'Crimson Lady'), South America (Raseira and Nakasu 2006), and Europe (Giovannini et al. 2006a, b). Recent work has also reported semifree forms with nonmelting flesh which overcomes a potential problem in the fresh market and would also be a useful trait in the processing market (Beckman and Sherman 1996; Gradziel 2003).

Beyond ensuring that the fruit can maintain its firmness and taste during extended storage, work needs to be done on the genetic basis for the various postharvest disorders that occur in fruits. The most important post harvest physiological disorders seen are internal breakdown problems in stone fruit (Crisosto et al. 1999;

Peace et al. 2006; Ogundiwin et al. 2007, 2009) and bitter pit and superficial scald in pome fruit (Blazek et al. 2007; Pesis et al. 2009). Work has begun to identify the genotypic variation that promotes resistance of cultivars to these disorders (Crisosto et al. 1999; Trivedi et al. 2010; Volz et al. 2006), although, due to the difficulty of these evaluations, work is now focused on parental material and advanced selections. It is not yet sufficiently efficient for primary selection among seedlings. The development of reliable selection criteria for these storage disorders is essential for rapid phenotyping and genetic advance. There are several groups working toward this goal.

Currently, destructive sampling of fruit can detect particularly poor lots of fruit. However, to ensure consistently high-quality fruit, the testing needs to be done on an individual fruit basis. Work on nondestructive systems to measure quality using acoustical and near-infrared systems (Ariana et al. 2006; Nicolai et al. 2006; Valero et al. 2007; Ruiz et al. 2009; Kleynen et al. 2005) has led to commercial use in a packing line situation. This allows the selection of individual fruit for acceptable fruit quality and puts higher quality standards on the cultivars that are developed. Thus, high-quality cultivars are needed; if a cultivar consistently produces poor-quality fruit, it will not be accepted in the marketplace in the future.

References

Allende, A., Tomás-Barberán, A. and Gil, M. J. (2006) Minimal processing for healthy traditional foods. Trends in Food Sci & Technol. 17, 513–519.

Amiot, M. J. (2009) Fruit, vegetables, phytochemicals and human health: Past and future. Acta Hort. 817, 61–69.

Apostol, J. (2005) New sweet cherry varieties and selections in Hungary. Acta Hort. 667, 59–64.

Ariana, D., Guyer, D. E., and Shrestha, B. (2006) Integrating multispectral reflectance and fluorescence imaging for defect detection on apples. Computers and Electronics in Agriculture 50, 148–161.

Ashraf, M. and Akram, N. A. (2009) Improving salinity tolerance of plants through conventional breeding and genetic engineering: An analytical comparison. Biotechnology Advances 27, 744–752.

Bassi, D. and Rizzo, M. (2000) Peach breeding for growth habit. Acta Hort. 538, 411–414.

Batt, P. J. and Noonan, J. (2009) Global trends in food quality: an exploratory study in fresh produce supply chains. Acta Hort. 831, 95–103.

Battino, M. and Mezzetti, B. (2006) Update on fruit antioxidant capacity: a key tool for Mediterranean diet. Public Health Nutrition 9 (8A), 1099–1103.

Beckman, T. G., and Lang, G. A. (2003) Rootstock breeding for stone fruits. Acta Hort. 622, 531–551.

Beckman, T. G., and Sherman, W. B. (1996) The non-melting semi-freestone peach. Fruit Var. J. 50:189–193.

Blanke, M. and Burdick, B. (2005) Food (miles) for thought. Energy balance for locally-grown versus imported apple fruit. Environ. Sci. & Pollut. Res. 12, 125–127.

Blazek, J., Opatova, H., Golias, J. and Homutova, I. (2007) Ideotype of apples with resistance to storage disorders. Hort. Sci. (Prague) 24, 107–113.

Blisard, N., Lin, B-H., Cromartie, J., and Ballenger, N. (2002) America's changing appetite: Food consumption and spending to 2020. Food Review 25(1): 2–9.

Bokszczanin, K., Palucha, A. and Prybyla, A. (2009) Identification of S-alleles in several apple cultivars. Acta Hort. 814, 391–393.

Brecht, J. K., and Kader, A. A. (1984) Ethylene production by fruit of some slow ripening nectarine genotypes. J. Amer. Soc. Hort. Sci. 109:763–767.

Brecht, J. K., Kader, A. A. and Ramming, D. W. (1984) Description and postharvest physiology of some slow-ripening nectarine genotypes. J. Amer. Soc. Hort. Sci. 109:596–600.

Brenton, P., Edwards-Jones, G., and Jensen, M. F. (2009) Carbon labeling and low-income country exports: A review of the development issues. Development Policy Review 27(3), 243–267.

Brown, S. K. (2003) Pome fruit breeding: Progress and prospects. Acta Hort. 622, 19–34.

Byrne, D. H., Sherman, W. B., and Bacon, T. A. (2000) Stone fruit genetic pool and its exploitation for growing under warm climatic conditions, p. 157–230. In: Erez, A. (ed.). Temperate Fruit Crops in Warm Climates. Kluwer Academic Publishers. Dordrecht, The Netherlands.

Byrne, D. H. (2002) Peach Breeding Trends: A world wide perspective. Acta Hort. 592:49–59.

Byrne, D. H. (2005) Trends in stone fruit cultivar development. HortTechnology, 15(3):494–500.

Byrne, D. H. (2010) Environmental challenges of breeding peaches for low chill regions. Acta Hort. 872, 129–138.

Byrne, D. H., G. Noratto, G., Cisneros Zevallos, L., Porter, W. and Vizzotto, M. (2009) Health benefits of peaches and plums. Acta Hort., 841: 267–274.

Calvin, L. and Cook, R. (2001) U. S. fresh fruit and vegetable marketing: Emerging trade practices, trends, and issues. Economic Research Service, USDA. Agric. Econ. Report No. 795.

Canning, P., Charles, A., Huang, S., Polenske, K. R., and Waters, A. (2010) Energy use in the U.S. Food system. USDA. Economic Res. Service, Econ, Res. Rept. 94.

Cantín, C. M., Moreno, M.A. and Gogorcena Y (2009) Evaluation of the antioxidant capacity, phenolic compounds and vitamin C content of different peach and nectarine [*Prunus persica* (L.) Batsch] breeding progenies. J Agric Food Chem 57, 4586–4592.

Cattivelli, L., Rizza, F., Badeck, F., Mazzucotelli, E., Mastrangelo, A., Francia, E., Marè, C. Tondelli, A., and Stanca, A. (2008) Drought tolerance improvement in crop plants: Am integrated view from breeding to genomics. Field Crops Research 105, 1–14.

Cevallos-Casals, B. A., Byrne, D. H., Cisneros-Zevallos, L., and Okie, W. R. (2002) Total phenolic and anthocyanin content in red-fleshed peaches and plums. Acta Hort. 592: 589–592.

Cevallos-Casals, B., Byrne, D., Okie, W. R. and Cisneros-Zevallos, L. (2006) Selecting new peach and plum genotypes rich in phenolic compounds and enhanced functional properties. Food Chem. 96: 273–280.

Chang, S., Tan, C., Frankel, E. N. and Barrett, D. M. (2000) Low-density lipoprotein antioxidant activity of phenolic compounds and polyphenol oxidase activity in selected clingstone peach cultivars. J. Agric. Food Chem. 48:147–151.

Cin, V. D., Rizzini, F. M., Botton, A. and Tonutti, P. (2006) The ethylene biosynthetic and signal transduction pathways are differently affected by 1-MCP in apple and peach fruit. Postharvest Biol Technol. 42, 125–133.

Clareton, M. (2000) Peach and nectarine production in France: Trends, consumption, and perspectives, p. 83–91. Summaries. Prunus Breeders Meeting – 2000. Empresa Brasileira de Pesquisa Agropecuária, Clima Temperado. Pelotas (RS). Brazil. Nov. 29 to Dec. 2, 2000.

Clark, J. R. (2005) Changing times for Eastern United States blackberries. HortTechnology 15 (3), 2–5.

Clark, J. R. and Finn, C. E. (2008) New trends in blackberry breeding. Acta Hort. 777, 41–47.

Cochard, H., Barigah, S., Kleinhentz, M. and Eshel, A. (2008) Is xylem cavitation resistance a relevant criterion for screening drought resistance among *Prunus* species? J. Plant Physiol. 165, 976–982.

Coley, D., Howard, M. and Winter, M. (2009) Local food, food miles and carbon emissions: A comparison of farm shop and mass distribution approaches. Food Policy 34, 150–155.

Connor, A. M., Luby, J. J. and Tong , C. (2002a) Variation and heritability estimates for antioxidant activity, total phenolic content, and anthocyanin content in blueberry progenies. J. Amer. Soc. Hort. Sci. 127:82–88.

Connor, A. M., Luby, J. J., Tong, C. , Finn, C. E. and Hancock , J. F. (2002b) Genotypic and environmental variation in antioxidant activity, total phenolic content, and anthocyanin content among blueberry cultivars. J Amer. Soc. Hort. Sci. 127:89–97.

Connor, A. M., Stephens, M. J., Hall, H. K., and Alspach, P. A. (2005a) Variation and heritabilities of antioxidant activity and total phenolic content estimated from a red raspberry factorial experiment. J. Amer. Soc. Hort. Sci. 130, 403–411.

Connor, A. M., Finn, C. E. and Alspach, P. E. (2005b) Genotypic and environmental variation in antioxidant activity and total phenolic content among blackberry and hybridberry cultivars. J. Amer. Soc. Hort. Sci. 130, 527–533.
Connor, A. M., McGhie, T. K., Stephens, M. J., Hall, H. K. and Alspach, P. A. (2005c) Variation and heritability estimates of anthocyanins and their relationship to antioxidant activity in a red raspberry factorial mating design. J. Amer. Soc. Hort. Sci. 130, 535–542.
Connor, A. M., Finn, C. E., McGhie, T. K. and Alspach, P. A. (2005d) Genetic and environmental variation in anthocyanins and their relationship to antioxidant activity in blackberry and hybridberry cultivars. J. Amer. Soc. Hort. Sci. 130, 680–687.
Crisosto, C. H. and Crisosto, G. M. (2005) Relationship between ripe soluble solids concentrations (RSSC) ans consumer acceptance of high and low acid melting flesh peach and nectarine (Prunus persica (L.) Batsch) cultivars. Postharvest Biol Technol. 38:239–246.
Crisosto, C. H., G. Crisosto, G. M. and Bowerman, E. (2003) Searching for consumer satisfaction: New trends in the California peach industry. Proc. of the First Mediterranean Peach Symposium. Sept. 10, 2003, Arigento, Italy.
Crisosto, C. H., Garner, D., Crisosto, G. M. and Bowerman (2004) Increasing 'Blackamder' plum (Prunus salicina Lindell) consumer acceptance. Postharv. Biol. Technol. 34, 237–244.
Crisosto, C. H., Crisosto, G. M., Echeverria, G. and Puy, J. (2006) Segregation of peach and nectarine (Prunus persica (L.) Batsch) cultivars according to their organoleptic characteristics. Postharvest Biol, Technol. 39:10–18.
Crisosto, C. H., Crisosto, G. M., Echeverria, G. and Puy, J. (2007) Segregation of plum and pluot cultivars according to their organoleptic characteristics. Postharvest Biol, Technol. 44, 271–276.
Crisosto, G. Crisosto C. and M. Watkins. (1998) Chemical and organoleptic description of white flesh nectarines and peaches. Acta Hort. 465:497–505.
Crisosto, C., Johnson, R. S. and DeJong, T. M. (1997) Orchard factors affecting postharvest stone fruit quality. HortScience 32:820–823.
Crisosto, C. H. Mitchell, F. G. and Ju, Z. (1999) Susceptibility to chilling injury of peach, nectarine, and plum cultivars grown in California. HortScience 34:1116–1118.
Darnell, R. L. (2000) Blueberries, p 429–444. In: Erez, A. (ed.). Temperate Fruit Crops in Warm Climates. Kluwer Academic Publishers. Dordrecht, The Netherlands.
Davis, D. E. and Stewart, H. (2002) Changing consumer demands create opportunities for U. S. food system. Food Rev. 25:19–23.
DeEll, J. Toivonen, P., Khanizadeh, S. and Hampson, C. (2009) Browning potential of new apple varieties. Acta Hort. 814, 529–532.
DeJong, T. M., Johnson, R. S., Bryla, D., Doyle, J. F. and Ramming, D. (2002) Evaluation of size controlling rootstocks for California peach production. 2001 Research Report, California Tree Fruit Agreement, p. 113–120.
Delate, K., McKern, A., Turnbull, R., Walker, J., Volz, R., White, A., Bus, V., Rogers, D., Cole, L., How, N., Guernsey, S. and Johnston, J. (2008) Organic apple systems: constraints and opportunities for producers in local and global markets: Introduction to the Colloquim. HortScience 43, 6–11.
Della Strada, G. and Fideghelli, C. (2003) Le cultivar de drupacee introdottee del 1991 al 2001. L'Informatore Agrario 41:65–70.
Della Strada, G. Fideghelli, C. and Grassi, F. (1996) Peach and nectarine cultivars introduced in the world from 1980 to 1992. Acta Hort. 374:43–51.
Demchak, K. (2009) Small fruit production in high tunnels. HortTechnol. 19(1), 44–49.
Dillard, C. J. and German, J. B. (2000) Phytochemicals: nutraceuticals and human health. J. Sci. Food Agric. 80, 1744–1756.
Dimitri, C. and Greene, C. (2002) Recent growth patterns in the U. S. Organic food market. USDA, ERS, Market and Trade Econ. Div. and Resource Econ. Div. Agric. Information Bull No. 777.
Dimitri, C. and Oberholtzer, L. (2005) Market-led versus government-facilitated growth. Development of the U. S. and EU Organic agricultural sectors. USDA.ERS. WRS-05-05 (www.ers.usda.gov).
Esmenjaud, D. and Dirlewanger, E. (2007) Plum, p. 119–135. In: Kole, C. (Ed.) Genome mapping and molecular breeding in plants, Volume 4 Fruit and Nuts. Springer-Verlag, Berlin.

Fideghelli, C. and Della Strada, G. (2010) The breeding activity on apricot in the world from 1980 through today. Acta Hort. 862, 93–98.

Fideghelli, C., Della Strada, G., Grassi, F. and Morico,G. (1998) The peach industry in the world: Present situation and trend. Acta Hort. 465:29–39.

Fideghelli, C., Sartori, A. and Grassi, F. (2003) Fruit tree size and architecture. Acta Hort. 622, 279–293.

Finley, J. W. (2005) Bioactive compounds and designer plant foods: The need for clear guidelines to evaluate potential benefits to human health. Chronica Horticulturae 45(3):6–11.

Finn, C. E., Kempler, C. and Moore, P. P. (2008) Raspberry cultivars: What's new? What's succeeding? Where are breeding programs headed? Acta Hort. 777, 33–40.

Fischer, M., Geibel, M. and Fischer, C. (2003) The future of disease-resistant apples. Acta Hort. 622, 329–334.

Flowers, T. J. and Flowers, S. A. (2005) Why does salinity pose such a difficult problem for plant breeders? Agric. Water Management 78, 15–24.

Food and Agriculture Organization. 2011. FAOSTAT data, accessed in 20 Jan 2011. Http://faostat.fao.org/faostat/form?collection=Production.Crops.Primary&Domain=Production&servlet=1&language=EN&hostname=apps.fao.org&version=default).

Frazão, E., Meade, B. and Regmi, A. (2008) Converging patterns of global food consumption and Food delivery systems. Amber Waves.6(1), 22–29.

Frey, K.J. (1996) National Plant Breeding Study: I. Human and Financial Resources Devoted to Plant Breeding Research and Development in the United States in 1994. Special Report 98. Iowa State University.

Frey, K.J. (1998) National Plant Breeding Study. III. National Plan for Genepool Enrichment of U. S. Crops. Special Report 101. Iowa State University.

Gale, F. and Huang, K. (2007) Demand for food quantity and quality in China. Econ. Res. Report No. 32. U. S. Dept of Agriculture.

Gamble, J., Jaeger, S. and Harker, F. (2006) Preferences in pear appearance and response to novelty among Australian and New Zealand consumers. Postharvest Biol. And Technol 41, 38–47.

Gardiner, S. E., Bus, V. G. M., Rusholme, R. L. Change, D. and Rikkerink, E. H. A. (2007) Apple, p. 1–62. In: Kole, C. (Ed.) Genome mapping and molecular breeding in plants, Volume 4 Fruit and Nuts. Springer-Verlag, Berlin.

Gaskell, M. (2004) Field tunnels permit extended season harvest of small fruits in California. Acta Hort. 659, 425–430.

George, A. P. and Erez, A. (2000) Stone fruit species under warm subtropical and tropical climates, p 231–265. In: Erez, A. (ed.). Temperate Fruit Crops in Warm Climates. Kluwer Academic Publishers. Dordrecht, The Netherlands.

Gil, M. I., Tomas-Barberan, F. A., Hess-Pierce, B. and Kader, A. A. (2002) Antioxidant capacities, phenolic compounds, carotenoids, and vitamin C contents of nectarine, peach, and plum cultivars from California. J. Agric. Food Chem. 50:4976–4982.

Giovannini, D., Liverani, A., Merli, M. and Brandi, F. (2006) Breeding strategies to improve peach fruit quality. Acta Hort. 713, 107–112.

Giovannoni, J. J. (2004) Genetic regulation of fruit development and ripening. Plant Cell 16, S170–S180.

Giovannini, D., Liverani, A., Merli, M. and Brandi, F. (2006) Breeding strategies to improve peach fruit quality. Acta Hort. 713, 107–112.

Gmitter, F. G., Chen, C., Rao, M. N., Soneji, J. R. (2007) Citrus fruits, p. 265–279. In: Kole, C. (Ed.) Genome mapping and molecular breeding in plants, Volume 4 Fruit and Nuts. Springer-Verlag, Berlin.

Golan, E., Krissoff, B.G. and Kuchler, F. (2004) Food traceability, One ingredient in a safe and efficient food supply. Amber Waves 2(2), 14–21.

Gradziel, T. M. (2003) Interspecific hybridizations and subsequent gene introgression within Prunus Subgenus Amygdalus. Acta Hort. 622, 249–255.

Gradziel, T.M. (2008) Almond (*Prunus dulcis*), p. 1–33. In: M. Priyadarshan and S.M. Jain (eds). Breeding of plantation crops. Springer Sci. Publ. Berlin.

Gradziel, T.M. and Kester, D.E. (1998) Breeding for self-fertility in California almond cultivars. Acta Hort. 470:109–117.

Granatstein, D. and Kirby, E. (2007) The changing face of organic tree fruit production. Acya Hort. 737, 155–162.

Grant, O., Johnson, A., Davies, M., James, C. and Simpson, D. (2010) Physiological and morphological diversity of cultivated strawberry (*Fragaria x ananassa*) in response to water deficit. Environ. Exper. Botany 68, 264–272.

Guerra, M. E., Rodrigo, J., López-Corrales, M. and Wünsch, A. (2009) S-allele identification in Japanese plum cultivars by PCR and cross pollination. Acta Hort. 814, 405–409.

Guthrie, J. F., Lin, B-H., Reed, J. and Stewart, H. (2005) Understanding economic and behavioral influences on fruit and vegetable choices. Amber Waves 3:36–41.

Haji, T., Yaegaki, H. and Yamaguchi, M. (2005) Inheritance and expression of fruit texture melting, non-melting and stony hard in peach. Scientia Hort. 105, 241–248.

Halgryn, P. J., Smith, C., von Mollendorff, L. and Labuschangé (2000) Breeding and cultivar development for the South African deciduous fruit industry with special reference to African Carmine™ apple, Rosemarie pear and the yellow plums Sun Kiss™ and Sundew™. Acta Hort. 538, 207–210.

Hancock, J. F. and Clark, J. R. (2009) Intellectual property protection and the funding of blueberry breeding in the future: the new paradigm. Acta Hort. 810, 43–48.

Hancock, J. F. (2000) Strawberries, p 445–455. In: Erez, A. (ed.). Temperate Fruit Crops in Warm Climates. Kluwer Academic Publishers. Dordrecht, The Netherlands.

Handy, C., Kaufmann, P. and Park, K. (2000) Evolving marketing channels reveal dynamic U. S. produce industry. Food Review 23(2), 14–20.

Harker, F. R., Kupferman, E. M., Marin, A. B., Gunson, F. A. and Triggs, F. M. (2008) Eating quality standards for apples based on consumer preferences. Postharv. Biol. Technol. 50, 70–78.

Hauagge, R. and Cummins, J. N. (2000) Pome fruit genetic pool for production in warm climates, p 267–304. In: Erez, A. (ed.). Temperate Fruit Crops in Warm Climates. Kluwer Academic Publishers. Dordrecht, The Netherlands.

Heisey, P. W., Srinivasan, C. S. and Thirtle, C. (2001) Public sector plant breeding in a privatizing world. Resource Economics Division, Economic Research Service, U. S. Department of Agriculture, Agriculture Information Bulletin No. 72.

Heyes, J. A. and Smith, A. (2008) Could "Food Miles" become a non-tariff barrier? Acta Hort. 768, 431–436.

Hu, D. and Scorza, R. (2009) Analysis of the 'A72' peach tree growth habit and its inheritance in progeny obtained from crosses of 'A72' with columnar peach trees. J. Amer. Soc. Hort. Sci. 134, 236–243.

Huang, S. and Huang, K. (2007) Increased U.S. imports of fresh fruit and vegetables. USDA, ERS. FTS-328-01. www.ers.usda.gov.

Huang, S. W. (2004) Global trade patterns in fruits and vegetables. USDA, ERS, Agric. Trade Reprt WRS-04-05.

Iezzoni, A., Peace, C., Bassil, N., Fazio, G., Luby, J., Main, D., Weebadde, C., Yue, C., van de Weg, E., Bink, M., Brown, S., Byrne, D., Clark, J., Crisosto, C., Davis, T., Evans, K., Finn, C., Gallardo, K. Gasic, K., Gradziel, T., Hancock, J., Jussaume, R., McCracken, V., Oraguzie, N., Reighard, G., Stone, A., Taylor, M., Wang, D. and Xu, K. (2009) RosBREED, Enabling marker-assisted breeding in Rosaceae. Abstract. ASHS meeting. Palm Desert, CA. August, 2009

Iglesias, I. and Echeverría, G. (2009) Differential effect of cultivar and harvest date on nectarine colour, quality and consumer acceptance. Scientia Hort. 120, 41–50.

Jackson, J. E. (2000) Apple production at low latitudes, p 305–342. In: Erez, A. (ed.). Temperate Fruit Crops in Warm Climates. Kluwer Academic Publishers. Dordrecht, The Netherlands.

Jaeger, S. R. (2006) Non-sensory factors in sensory science research. Food Qual. Preference 17:132–144.

Jaeger, S., Rossiter, K., Wismer, W. and Harker, F. (2003) Consumer-driven product development in the kiwifruit industry. Food Quality and Preference 14, 187–198.

Jaeger, S. and Harker, F. (2005) Consumer evaluation of novel kiwifruit: willingness-to-pay. J. Sci. Food Agric. 85, 2519–2526.

Jaeger, S., Andani, Z., Wakeling, I. and MacFie, H. (1998) Consumer preferences for fresh and aged apples: A cross cultural comparison. Food Qual. Preference 9(5), 355–366.
Jiang, W., Qu, D., Mu, D. and Wang, L. R. (2004) China's energy saving greenhouses. Chronica Hort. 44:15–17.
Johnston, W. E. and Carter, H. O. (2000) Structural adjustment, resources, global economy to challenge California agriculture. Calif. Agric. 54 (4): 16–22.
Kajikawa, C. (1998) Quality level and price in Japanese apple market. Agribusiness 14 (3), 227–234.
Kappel, F. (2008) Breeding cherries in the "New World". Acta Hort. 795, 59–69.
Kappel, F., Granger, A., Hrotko, K. and Schuster, M. (2011) Cherries. In: Fruit Breeding. Badenes, M, and Byrne, D. H. (eds.) Springer.
Kappel, F., MacDonald, R.A. and Brownlee, R. (2006) 13S2009 (Staccato™) sweet cherry. Can. J. Plant Sci. 86:1239–1241.
Kenis, K. and Keulemans, J. (2007) Study of tree architecture of apple (Malus x domestica Borkh.) by QTL analysis of growth habits. Mol. Breeding 19, 193–208.
Khanizadeh, S., Tsao, R., Rekika, D., Yang, R., Charles, M. T. and Rupasinghe, H. P. V. (2009) Advances in fruit breeding in Eastern Canada – Role of phytochemicals in designing specialty fruits. Acta Hort. 814, 205–207.
Kleynen, O., Leemans, V. and Destain, M. (2005) Development of a multi-spectral vision system for the detection of defects on apples. J. Food Engineering 69, 41–49.
Kodad, O. and Socias i Company, R. (2006) Influence of genotype, year and type of fruiting branches on the productive behaviour of almond. Scientia Hort. 109:297–302.
Kodad, O. and Socias i Company, R. (2009) Review and update of self-incompatibility alleles in almond. Acta Hort. 814, 421–426.
Kocsis, L., Varga, Z. and Pernesz, G. (2009) Introduction od a lime and drought tolerant rootstock variety. Acta Hort. 827, 465–469.
Kozai, N., Beppu, K.., Mochioka, R., Boonprakob, U., Subhadrabandhu, S. and Kataoka, I. (2004) Adverse effects of high temperature on the development of reproductive organs in 'Hakuho' peach trees. J. Hort. Sci. Biotechnol. 79, 533–537.
Lang, G. A. (2000) Precocious, dwarfing, and productive – how will new cherry rootstocks impact the sweet cherry industry? HortTechnology 10:719–725.
Lang, G. A. (2009) High tunnel tree fruit production: The final frontier? HortTechnol. 19(1), 50–55.
Lata, B., Przeradzka, M. and Binkowska, M. (2005) Great differences in antioxidant properties exist between 56 apple cultivars and vegetation seasons. J. Agric. Food Chem. 53,8970-8978.
Lauri, P.E., Bourdel, G., Trottier, C. and Cochard, H. (2008) Apple shoot architecture: evidence for strong variability for bud size and composition and hydraulics within a branching zone. New Phytologist 178, 798–807.
Lavee, S. (2000) Grapevine (Vitis vinifera) growth and performance in warm climates, p 343–366. In: Erez, A. (ed.). Temperate Fruit Crops in Warm Climates. Kluwer Academic Publishers. Dordrecht, The Netherlands.
Lee, K. W., Kim, Y. J., Kim, D., Lee, H. J. and Lee, C. Y. (2003) Major phenolics in apple and their contribution to the total antioxidant capacity. J. Agric. Food Chem. 51, 6516–6520.
Lespinasse, Y. (2009) Review of pome fruit breeding in Europe: Which strategies for the near future? Acta Hort. 814, 865–871.
Lespinasse, Y., Chevalier, M., Durel, C. and Robert, P. (2008) Pear breeding for scab and psylla resistance. Acta Hort. 800, 475–481.
Liverani, A., Giovannini, D. Brandi, F. and Merli, M. (2004) Development of new peach cultivars with columnar and upright growth habit. Acta Hort. 663, 381–386.
Llacer, G. (2009) Fruit breeding in Spain. Acta Hort. 814, 43–56.
Lu, M., Song, C., Huang, C. and Ou, S. (2008) Changes in flesh firmness and ethylene production of different peach types during fruit ripening. Acta Hort. 768, 153–159.
Lucier, G., Pollack, S., Ali, M. and Perez, A. (2005) Fruit and vegetable backgrounder. USDA. ERS. VGS-313-01 (www.ers.usda.gov).
Lyrene, P. M. (2005) Breeding low-chill blueberries and peaches for subtropical areas. HortScience 40, 1947–1949.

Marić, S., Lukić, M., Bošković, R. I. (2009) The polymorphism of the genes involved in ethylene biosynthesis and perception in apple. Acta Hort. 839, 441–448.
Martinez, S. (2007) The U.S. food marketing system: recent developments, 1997–2006. USDA, ERS, Economic Res. Report No. 42.
Mattila, P., Hellström, J. and Törrönen, R (2006) Phenolic acids in berries, fruits, and beverages. J. Agric. Food Chem. 54, 7193–7199.
McCool, A. C., Myung, E. and Chien, T-C. (2005) Modification of the form in which fresh fruit is served as a possible means of increasing the consumption of fruit offered to elementary and middle scholl students. J. Foodservice Bus, Res. 8, 73–85.
Milà i Canals, L., Cowell, S. J., Sim, S. and Basson, L. (2007) Comparing domestic versus imported apples: A focus on energy use. Environ. Sci. Pollut. Res. 14(5), 338–344.
Monet, R. and Bassi, D. (2008) Classical genetics and breeding. In: D. R. Layne and D. Bassi (Eds.), The Peach. Botany, Production and Uses. CAB International, Wallingford, UK, pp. 61–84.
Moore, J. N. (1993) Plant patenting: A public fruit breeder's assessment. HortTechnology 3, 262–266.
Moreau-Rio, M. A. (2006) Perception and consumption of apricots in France. Acta Hort. 701, 31–37.
Musacchi, S., Quartieri, M. and Tagliavini, M. (2006) Pear (Pyrus communis) and quince (Cydonia oblonga) roots exhibit different ability to prevent sodium and chloride uptake when irrigated with saline water. Europ. J. Agronomy 24, 268–275.
Nicolai, B. M., Lötze, E., Peirs, A., Scheerlinck, N. and Theron, K. I. (2006) Non-destructive measurement of bitter pit in apple fruit using NIR hyperspectral imaging. Postharvest Biol. Technol. 40, 1–6.
Nicotra, A. and Conte, L. (2003) Nuove tipologie di frutto per il mercato delle pesche: nascono le serie "UFO" e "Ghiacchio" Frutticoltura 65 (7–8), 20–25.
Normile, M. A. and Leetmaa, S. E. (2004) U.S. – EU Food and agriculture comparisons. USDA. Market and Trade Economics Div., Econ, Res. Serv. Agric. And Trade Report WRS-04-04. October 2008, FTS-2008.
Ogundiwin, E., Peace, C., Gradziel, T., Dandekar, A., Bliss, F. and Crisosto, C. (2007) Molecular genetic dissection of chilling injury in peach fruit. Acta Hort. 738, 633–638.
Ogundiwin, E., Peace, C., Gradziel, T., Parfitt, D., Bliss, F. and Crisosto, C. (2009) A fruit quality gene map of Prunus. BMC Genomics 10:587 doi:10.1186/1471-2164-10-587.
O'Rouke, D. Janick, J. and Sansavini, S. (2003) World apple cultivar dynamics. Chronica Hort. 43, 10–13.
Okada, K., Tonaka, N., Takasaki, T, Sawamura, Y. and Matsumoto, T. (2008) Selection of self-compatible trees by S_4^{sm} haplotype specific marker in Japanese pear. Acta Hort. 800, 401–407.
Opara, L., Al-Said, F. A. and Al-Abri, A. (2007) Assessment of what the consumer values in fresh fruit quality: case study of Oman. New Zealand J Crop Hort. Sci. 35, 235–243.
Oraguzie, N. C. (2010) Fruit softening in pome fruit – the role of ACS genes. Acta Hort. 859, 135–142.
Pascal, T., Iglesias Casellarnau, I. Blanc, P. and C. Pitiot. (2009) Joint experiments in France and Catalonia of new flat peaches-nectarines and canning peaches from INRA. Acta Hort. 814:299–304.
Pastrana-Bonilla, E., h, G. (2003) Phenolic content and antioxidant capacity of muscadine grapes. J. Agric. Food Chem. 51, 5497–5503.
Peace, C. P., Ahmad, R., Gradziel, T. M., Dandekar, A. M. and Crisosto, C. H. (2005a) The use of molecular genetics to improve peach and nectarine post-storage quality. Acta Hort. 682, 403–409.
Peace, C. P., Crisosto, C. H. and Gradziel, T. M. (2005b) Endopolygalacturonase: a candidate gene for Freestone and Melting flesh in peach. Molecular Breeding 16, 21–31.
Peace, C., Crisosto, C. H., Garner, D. T., Dandekar, A. M., Gradziel, T. and Bliss, F. A. (2006) Genetic control of internal breakdown in peach. Acta Hort. 713, 489–496.
Péneau, S., Hoehn, E., Roth, H. Escher, F. and Nuessli, J. (2006) Importance and consumer perception of freshness of apples. Food Qual. Preference 17, 9–19.
Pesis, E., Ibáñez, A. M., Phu, M. L., Mitcham, E. J., Ebeler, S. E. and Dandekar, A. M. (2009) Superficial scald and bitter pit development in cold-stored transgenic apples suppressed for ethylene biosynthesis. J. Agric. Food Chem. 57, 2786–2792.
Pollack, S. and Perez, A. (2008) Fruit and Tree Nuts Situation and Outlook Yearbook. Market and Trade

Poole, N. and Baron, L. (1996) Consumer awareness of citrus fruit attributes. British Food J. 98/1, 23–28.
Prior, R. L. and Cao, G. (2000) Antioxidant phytochemicals in fruits and vegetables: Diet and health implications. Hort Science 35:588–592.
Racskó, J., Miller, D. D., Duarte, E. E., Szukies, J., Szabó, Z., Soltész, M. and Nyéki, J. (2009) Is consumer preference for apple driven only by fruit quality? Acta Hort. 831, 331–337.
Ramming, D. W., Walker, M. A., Tenscher, A. and Krivanek, A. F. (2009) Breeding table and raisin grapes with increased fruit quality while retaining Pierce's Disease resistance Acta Hort 827, 445–450.
Raseira, M.C.B. and Nakasu, B.H. (2006). Peach breeding program in Southern Brazil. Acta Hort. 713, 93–97.
Ravelonandro, M. and Scorza, R. (2009) Silencing in genetically engineered *Prunus domestica* provides durable and safe resistance ot Plum pox virus (Sharka Disease). Acta Hort 814:397–402.
Reighard, G. L. (2000) Peach rootstocks for the United States: Are foreign rootstocks the answer? HortTechnology 10:714–718.
Reighard, G.L. and Loreti, F. (2008) Rootstock development. In: D. Layne, and D. Bassi (Eds.), The Peach, Botany, Production and Uses. CAB International, Wallingford, U.K, pp. 193–220.
Riaz, S., Doligez, A., Henry, R. J. and Walker, M. A. (2007) Grape. p. 63–101. In: Kole, C. (Ed.) Genome mapping and molecular breeding in plants, Volume 4 Fruit and Nuts. Springer-Verlag, Berlin.
Riaz, S., Tenscher, A. C., Graziani, R., Krivanek, A. F., Ramming, D. W. and Walker, W. A. (2009) Using marker-assisted selection to breed Pierce's disease-resistant grapes. Am. J. Enol. Vitic. 60(2), 199–207.
Rieger, M., Lo Bianco, R. and Okie, W. R. (2003) Responses of *Prunus ferganensis, Prunus persica*, and two interspecific hybrids to moderate drought stress. Tree Physiol. 23, 51–58.
Ross , C., Chauvin, M. and Whiting, M. (2010) Assignation of sweet cherry selections to 3 taste groupings based on perceived sweetness and sourness. J. Food Sci. 75, S48–S54.
Ruiz, D., Audergon, J. Bureau, S., Grotte, M., Renard, C., Gouble, B. and Reich, M. (2009) Rapid and non-destructive determination of soluble solids content and titratable acidity in apricot using near-infrared spectroscopy (NIR). Acta Hort. 814: 501–505.
Sansavini, S. and Lugli, S. (2005) New sweet cherry cultivars developed at the University of Bologna. Acta Horticulturae 667: 45–52.
Sansavini, S. and Lugli, S. (2008) Sweet cherry breeding programs in Europe and Asia. Acta Hort. 795, 41–57.
Sansavini, S. (2009) Horticulture in Europe: from history to innovation. Acta Hort. 817, 43–58.
Schuster, M., Flachowsky, H. and Köhler, D. (2007) Determination of self-compatible genotypes in sweet cherry (*Prunus avium* L.) accessions and cultivars of the German Fruit Gene Bank and from private collections. Plant Breeding 126, 533–540.
Schuster, M. (2009) Sour cherries Prunus cerasus L. with columnar tree habit. Acta Hort 814, 325–328.
Scorza, R. (2000) Progress in tree fruit improvement through molecular genetics. HortScience 36:855–858.
Scorza, R., Miller, S., Glenn, D. M., Okie, W. R. and Tworkoski, T. (2006) Developing peach cultivars with novel tree growth habits. Acta Hort. 713, 61–64.
Segura, V., Denance, C., Durel, C.-E. and Costes, E. (2007) Wide range QTL analysis for complex architectural trait in a 1-year-old apple progeny. Genome 50, 159–171.
Sekido, K., Hayashi, Y., Yamada, K., Shiratake, K., Matsumoto, S., Macjima, T. and Komatsu, H. (2010) Efficient breeding system for red-fleshed apple based on linkage with S_3-RNase allele in 'Pink Pearl'. HortScience 45, 534–537.
Sherman, W. B. and Lyrene, P. M. (2003) Low chill breeding of deciduous fruits at the University of Florida. Acta Hort. 622, 599–605.
Sim, S., Barry, M., Clift, R. and Cowell, S. J. (2007) The relative importance of transport in determining an appropriate sustainability strategy for food sourcing. Intern. J. LCA 12 (6), 422–431.
Sinclair, T. R. (2011) Challenges in breeding for yield increase for drought. Trends in Plant Sci. 2011, 1–5. Doi 10.1016/j.tplants.2011.02.008
Sloan, E. (2006) Top 10 functional food trends. Food Technology 04.06, 23–34.

Sloan, E. (2007) Great ideas from around the world. Food Technology 10.07:20–33.
Sloan, E. (2008) The top 10 functional food trends. FoodTechnology 04.08, 25–35.
Socias i Company R. (1990) Breeding self-compatible almonds. Plant Breed. Rev. 8:313–338.
Socias i Company R. (1998) Fruit tree genetics at a turning point: the almond example. Theor. Appl. Genet. 96:588–601.
Souza, V., Byrne, D. H. and Taylor, J. F. (1998) Heritability, genetic and phenotypic correlations, and predicted selection response of quantitative traits in peach: II. An analysis of several fruit traits, J. Amer. Soc. Hort. Sci. 123:604–611.
Souza, V., Byrne, D. H. and Taylor, J. F. (2000) Predicted breeding values for nine plant and fruit characteristics of 28 peach genotypes. J. Amer. Soc., Hort Sci. 125:460–465.
Southon, S. (2001) Increased fruit and vegetable consumption: Potential health benefits. Nutr. Metab. Cardiovasc. Dis. 11(Suppl. To No. 4): 78–81.
Stewart, H., Blisard, N. and Jolliffe, D. (2006) Let's eat out. Americans weigh taste, convenience, and nutrition. USDA. Econ. Inf. Bull. No. 19.
Stover, E., Castle, W. and Chao, C. (2005) Trends in U.S. sweet orange, grapefruit, and mandarin-type cultivars. HortTechnology 15 (3), 12–17.
Stringer, S. J., Marshall, D. A., Cochran, T. and Perkins-Veazie, P. (2009) Nutraceuatical compound concentrations of muscadine (*Vitis rotundifolia* Michx.) grapes cultivars and breeding lines. Acta Hort 841, 553–556.
Sun, J., Chu, Y. F., Wu, X. and Liu, R. H. (2002) Antioxidant and antiproliferative activities of common fruits. J. Agric. Food Chem. 50:7449–7454.
Syvertsen, J. and Melgar, J. (2010) Salinity tolerance and leaf water use efficiency in *Citrus*. J. Amer. Soc. Hort. Sci. 135, 33–39.
Tao, R. and Iezzoni, A. (2010) The S-RNase-based gametophytic self-incompatibility system in Prunus exhibits distanct genetic and molecular features. Scientia Hort. 124, 423–433.
Tatsuki, M., Hayama, H. and Nakamura, Y. (2009) Apple ethylene receptor protein concentrations are affected by ethylene, and differ in cultivars that have different storage life. Planta 230, 407–417.
Tomala, K., Barylko-Pikielna, N., Jankowski, P., Jeziorek, K. and Wasiak-Zys, G. (2009) Acceptability of scab-resistant versus conventional apple cultivars by Polish adult and young consumers. J Sci. Food Agric. 89, 1035–1045.
Tomas-Barberan, F. A., Gil, M. I., Cremin, P., Waterhouse, A. L., Hess-Pierce, B. and Kader, A. A. (2001) HPLC-DAD-ESIMS analysis of phenolic compounds in nectarines, peaches, and plums. J. Agric. Food Chem. 49, 4748–4760.
Traxler, G., (1999) Balancing basic, genetic enhancement and cultivar development research in an evolving US plant germplasm system. AgBioForum 2(1), 43–47.
Trivedi, P, Caridhas, D. and Solomos, T. (2010) Apple scald development and regulation. Acta Hort. 857, 349–358.
Turner, J., Seavert, C., Colonna, A. and Long, L. E. (2008) Consumer sensory evaluation of sweet cherry cultivars in Oregon, USA. Acta Hort. 795, 781–786.
U. S. Department of Agriculture (2005) My Pyramid.gov: Steps to a healthier you, http://www.mypyramid.gov/, Accessed 10 Dec 2011.
Valero, C., Crisosto, C. H. and Slaughter, D. (2007) Relationship between nondestructive firmness measurements and commercially important ropening fruit stages for peaches, nectarines, and plums. Postharvest Biol Technol. 44:248–253.
van Rijswijk, W., Frewer, L. J., Menozzi, D. and Faioli, G. (2008) Consumer perceptions of traceability: A cross-national comparison of associated benefits. Food Qual. And Preference 19, 452–464.
Variyam, J. and Golan, E. (2002) New health information is reshaping food choices. Food Review 25:13–18.
Vilanova, M., Santalla, M. and Masa, A. (2009) Environmental and genetic variation of phenolic compounds in grapes (Vitis vinifera) from northwest Spain. J. Agric. Sci. 147, 683–697.
Vinson, J. A., Su, X. Zubik, L.and Bose, P. (2003) Phenol antioxidant quantity and quality in foods: Fruits. J. Agric. Food Chem. 49, 5315–5321.
Vizzotto, M., Cisneros, L., Okie, W. R., Ramming, D. W. and Byrne, D. H. (2007) Large variation found in the phytochemical content and antioxidant activity of peach and plum germplasm. J. Amer. Soc. Hort. Sci. 132, 334–340.

Volz, R., Alspach, P. A., Fletcher, D. J. and Ferguson, I. B. (2006) Genetic variation in bitter pit and fruit calcium concentrations within a diverse apple germplasm collection. Euphytica 149, 1–10.

Volz, R., Oraguzie, N., Whitworth, C., How, N. Change, D., Carlisle, C., Gardiner, S., Rikkerink, E. and Lawerence, T. (2009a) Breeding for red flesh colour in apple: progress and challenges. Acta Hort. 841:337–342.

Volz, R., Rikkerink, E., Austin, P., Lawrence, T. and Bus, V. (2009b) "Fast-Breeding" in apples: a strategy to accelerate introgression of new traits into elite germplasm. Acta Hort. 814:163–168.

Wakasu, Y., Kudo, H., Ishikawa, R., Akada, S., Senda, M., Niizeki, M. and Harada, T. (2006) Low expression of an endopolygalacturonase gene in apple fruit with long-term storage potential. Postharvest Biol. Technol. 39, 193–198.

Wang, A., Tan, D., Tatsuki, M., Kasai, A., Li, T., Saito, H. and Harada, T. (2009) Molecular mechanism of distinct ripening profiles in 'Fuji' apple fruit and its early maturing sports. Postharvest Biol. Technol. 52, 38–43.

Wang, S. and Lin, H. (2000) Antioxidant activity in fruits and leaves of blackberry, raspberry and strawberry varies with cultivar and developmental stage. J. Agric. Food Chem. 48, 140–146.

Wargovich, M.J. (2000) Anticancer properties of fruits and vegetables. HortScience 35:573–575.

Wang, H., G. Cao, R. L. Prior. 1996. Total antioxidant capacity of fruits. J. Agric. Food Chem. 44:701–705.

Weber, C. A., Perkins-Veazie, P., Moore, P. P. and Howard, L. (2008) Variability of antioxidant content in raspberry germplasm. Acta Hort. 777, 493–497.

Webster, T. (2006) Control of growth and cropping of temperate fruit trees. Chronica Horticulturae 46 (3), 20–26.

Wei, S. (2001) Singapore and Hong Kong market research for early season stone fruit. Austr. Fresh Stone Fruit Qrtly. 3(1):8–12.

Weibel, F. P., Tamm, L., Wyss, E., Daniel, C., Haseli, A. and Suter, F. (2007) Organic fruit production in Europe: Successes in production and marketing in the last decade, perspectives and challenges for the future development. Acta Hort. 737, 163–171.

Wells, H. F. and Buzby, J. C. (2008) Dietary assessment of major trends in U. S. food consumption, 1970–2005. Economic Information Bulletin No. 33 Economic Research Service, U. S. Dept. of Agriculture.

Wen, I-C., Koch, K.E. and Sherman, W.B. (1995a) Comparing fruit and tree characteristics of two peaches and their nectarine mutants. J. Amer. Soc. Hort. Sci. 120:101–106.

Wen, I-C., Sherman, W.B. and Koch, K.E. (1995b) Heritable pleiotropic effects of the nectarine mutant from peach. J. Amer. Soc. Hort. Sci. 120:721–725.

Werner, D. J., Crueller, M. A. and Chaparro, J. X. (1998) Inheritance of the blood-flesh trait in peach. HortScience 33,1243-1246.

Wismer, W. V., Harker, F. R., Gunson, F. A., Rossiter, K. L., Lau, K., Seal, A. G., Lowe, R. G. and Beatson, R. (2005) Identifying flavor targets for fruit breeding: A kiwifruit example. Euphytica 141, 93–101.

Xu, C., Zhang, Y., Cao, L. and Lu, J. (2010) Phenolic compounds and antioxidant properties of different grape cultivars grown in China. Food Chem. 119, 1557–1565.

Yamada, M., Kurihara, A. and Sumi, T. (1987) Varietal differences in fruit bearing in Japanese persimmon (*Diospyros kaki* Thunb.). J. Japan. Soc. Hort. Sci. 56:293–299. (in Japanese with English summary).

Yamaguchi, M. and Kyotani, H. (1986) Differences in fruit ripening patterns of Japanese plum cultivars under high (30°C) and medium (20°C) temperature storage. Bull. Fruit Tree Res. Stn. A 13:1–19.

Yang, J., Martinson, T. E. and Liu, R. H. (2009) Phytochemical profiles and antioxidant activities of wine grapes. Food Chem. 116, 332–339.

Yoshizawa, Y., Sakurai, K., Kawaii, S., Asari, M., Soejima, J. and Murofushi, N. (2005) Comparison of antiproliferative and antioxidant properties among nineteen apple cultivars. HortScience 40, 5, 1204–1207.

Zhu, Y. and Barritt, B. H. (2008) Md-ACS1 and Md-ACO1 genotyping of apple (Malus x domestica Borkh.) breeding parents and suitability for marker-assisted selection. Tree Gen Genomes 4, 555–562.

Chapter 2
Developing Fruit Cultivars with Enhanced Health Properties

Michael J. Wargovich, Jay Morris, Vondina Moseley, Rebecca Weber, and David H. Byrne

Abstract One hypothesis to account for the dramatic increase of inflammatory driven diseases, such as cancer, cardiovascular disease, obesity, diabetes, and others, across the world is the coincidental displacement of fruits and vegetables in the diet with processed foods as populations in the developing world rapidly acculturate to a more affluent lifestyle. Fruits are rich sources of antioxidant and anti-inflammatory natural compounds that offset many of the biological events leading to the development of the above-mentioned chronic diseases. In this review, potentially cancer-protective phytochemicals in fruits are reviewed to describe the research approaches, the range of chemistry and mechanisms seen in the study of the health benefits of fruit phytochemicals. Furthermore, given the rapid increase in research, public's interest in the health benefits of food, and the government's and food industry's efforts to develop and promote healthy foods, fruit breeders have begun to investigate the feasibility of developing health-enhanced fruit cultivars. Thus far, there appears to be ample genetic variability within fruit crops to develop cultivars with higher levels of plant phytochemicals, such as total phenolics, anthocyanins, and antioxidant activity. Nevertheless, selecting breeding targets is elusive as there is little information on which specific phytochemical or combination of phytochemicals and the levels needed to effectively enhance the health of the consuming public.

M.J. Wargovich (✉) • J. Morris • V. Moseley • R. Weber
Department of Cellular and Molecular Pharmacology and Experimental Therapeutics, Hollings Cancer Center, Medical University of South Carolina, 86 Jonathan Lucas Street, Charleston, SC 29424, USA
e-mail: wargovic@musc.edu; morrisjl@musc.edu; browv@musc.edu; webe@musc.edu

D.H. Byrne
Department of Horticultural Sciences, Texas A&M University, College Station, TX 77843-2133, USA
e-mail: dbyrne@tamu.edu

M.L. Badenes and D.H. Byrne (eds.), *Fruit Breeding*, Handbook of Plant Breeding 8, DOI 10.1007/978-1-4419-0763-9_2, © Springer Science+Business Media, LLC 2012

Keywords Phytochemicals • Cancer • Cardio vascular disease • Obesity • Diabetes • Antioxidants • Phenolics • Anthocyanins • Cartenoids • Chronic diseases • Anti-inflammation

1 Introduction

Chronic diseases are on the rise in the developing world. At the core of risk for diseases, such as cancer, heart disease, neurological disorders, obesity, and diabetes, is uncontrolled chronic inflammation deep in the cells of the body. While inflammation is a natural process of healing damage to the body, the genetic and biochemical machinery underpinning inflammation is often corrupted, resulting in the prevalent chronic diseases we recognize today.

One recognized factor in the development of chronic disease is poor nutrition. And in an inverse way, the climb from undeveloped to developed nation status makes us come full circle from inadequate nutrition to super-adequate nutrition, both states that could be characterized as "poor." To explain this conundrum, it is possible that populations may reach a state of affluence, where they displace the fruit and vegetable portion of the diet with super-caloric foods, devoid of natural phytochemicals, that may have helped to offset chronic disease risk. With these natural guardians against oxidative damage and inflammation, affluent societies are now afflicted with epidemics of chronic inflammatory-driven diseases.

Fruits and vegetables have always been considered a foundation of a healthy lifestyle and a healthy diet. Unfortunately, despite the solid research, government and health agency recommendations, and a population that is growing increasingly old, the public health message to eat more fruits and vegetables has fallen on deaf ears. In the USA, most Americans do not come close to the recommended consumption of five to nine servings of fruits and vegetables per day (Pollack and Perez 2008; Wells and Buzby 2008) and this aversion begins in the teen and preteen years, a time when chronic disease risk may be set (Nanney et al. 2007; Cade et al. 2006).

The intent of this chapter is to review the evidence for fruit consumption and health benefits with an emphasis on cancer and evaluate the potential of developing fruit cultivars with enhanced levels of beneficial phytochemicals as an approach to increase the consumption of these useful compounds.

2 Phytochemicals and Cancer

The USA and many developed countries are experiencing an epidemic of diseases which may have chronic, unresolved inflammation as their common etiology (Beaglehole et al. 2007). Clearly, the impact of diet is seminal in establishing protection early in life from chronic disease, and the loss of dietary protectants, by circumstance or will, may now factor into the epidemic facing all societies. In the

Table 2.1 Human evidence for cancer prevention: Fruit consumption

Site of cancer	Types of study	Finding	Reference
Oropharyngeal	2 ECO, 1CO, 35 CC	Probably preventive	AICR (2007)
Esophagus	7 ECO, 4 CO, 36 CC	Probably preventive	AICR (2007)
Lung	7 ECO, 25 CO, 32 CC	Convincingly preventive	AICR (2007)
Stomach	23 ECO, 16 CO, 51 CC	Probably preventive	AICR (2007)
Pancreas	8 ECO, 6 CO, 6 CC	Not plausible	AICR (2007)
Liver	1 CO, 5 CC	Not plausible	AICR (2007)
Prostate	3ECO, 28 CO, 18 CC	Inconsistent	Lewis et al. (2009)
Breast	8 CO, 2CC	Inconsistent	Vainio and Weiderpass (2006)

Abbreviations: *ECO* ecological studies, *CO* cohort studies, *CC* case–control studies

last 40 years, a wealth of epidemiological data, gleaned from over 150 ecological, cohort, and case–control studies, has supported the notion that persistent dietary exposure to fruits and vegetables are salutary for health. While overall evidence is suggestive of protection, evidence for reduction in risk for only a few of the major cancers is considerable enough to be called protective. It should not be concluded that phytochemicals from frequent fruit and vegetable consumption are ineffective for other cancers, rather that there is at present insufficient data to warrant a conclusive protective effect. Table 2.1 lists some of the common sites of cancer and summarizes the available data regarding cancer protection. The conclusions are drawn by an expert panel commissioned by the World Cancer Research Fund/American Institute for Cancer Research in its updated review published in 2007.

2.1 *Phytochemicals in Fruits*

In the last 15 years of research, much of the protective effects for consumption of plant foods have been ascribed to the constituent phytochemicals resident in them (Newman and Cragg 2007). In all fruits and vegetables, the major classes of phytochemicals consist broadly of carotenoids, flavonoids, isoflavonoids, and phenolic acids (Pan et al. 2008). Plant phenolics represent a structurally diverse superclass of compounds possessing one or more aromatic rings, one or more hydroxyl groups, and additional moieties covering over 8,000 unique chemicals (Huang et al. 2010). The flavonoids represent over 4,000 compounds and are an extension of the phenolic group, but have at least two aromatic rings with a variety of additional structural elements. It is this class of natural compounds that have generated so much interest in the cancer prevention research and represents many of the active compounds in fruits. Many of the bioactive agents identified from medicinal herbs and spices are members of this class of phytochemicals. Flavonoids can be further subdivided into flavones, flavonols, flavonones, isoflavones, and anthocyanidins. The latter category is of intense interest. The anthocyanidins broadly account for the red-to-purple pigmenting of many commonly consumed fruits, especially in grapes, plums, cherries, and berries. In many tree fruits, the presence of anthocyanidins in most cultivars is typically concentrated in the skin, although most of these, such as apples, peaches, plums, and kiwis, have genotypes that contain anthocyanins in the flesh as well (Vizzotto et al. 2007; Voltz et al. 2009; Jaeger and Harker 2005).

2.2 *Fruit Phytochemicals: Evidence for Health Benefits*

Taken as whole, the production of fruits and vegetables has been robust with most of the growth in production in vegetables, rather than in fruit. Exports of fruit have grown, especially those from developing countries, and the industry has diversified, ensuring (at least in some developed countries) not only a year-round supply of fresh fruit, but oversupply has led to the marketing of specialized fruits, such as those organically grown. This for a large part has been due to the public perception that organically grown is better for health maintenance (WHO 2005).

Often, the first type of evidence for health benefits of fruit consumption is drawn from *epidemiological* studies. Three types of studies are often conducted: those at the ecological level (comparing types of fruit and quantities across populations), the cohort level (comparing fruit consumption within a population that has been followed for some time), and the case–control level (comparing fruit consumption in those with and without disease). Among tumor types, risk for cancers of the oral cavity, esophagus, and colorectum seems to be less when high amount of fruits and vegetables are in the diet. The evidence for protection is less than certain for cancers of the stomach, lung, breast, and prostate (Key 2011). The overall risk for cancer has been examined in four large and well-conducted prospective studies. In two cohort studies, the Nurse's Health Study and the Health Professionals' Follow-up study, conducted by Harvard, no significant reduction in overall risk was noted, although there was a trend to protection (Hung et al. 2004). These studies were supported by the Japanese Public Health Center prospective study while the European-based EPIC study found a significant reduction in cancer risk for consumption of fruits and vegetables (Takachi et al. 2008; Buchner et al. 2011). The US-based NIH-AARP Diet and Health Study found mixed results with more protection noted for vegetable consumption than fruits (George et al. 2009). These types of studies are notoriously difficult to conduct and to interpret, and it may well be that certain types of cancers are more amenable to prevention by specific fruits or specific vegetables based upon their unique phytochemical signatures. Examples include some of the unique phytochemicals in green tea and the phytoestrogenic compounds in soy.

Basic research into the potential mechanisms by which fruits or vegetables prevent cancer has unveiled an incredible variety of ways in which the cancer process can be interrupted. How does the process of identifying potential benefits of a particular fruit or vegetable begin? The customary protocol for this type of research originates with epidemiology. When a consumption pattern is associated with reduced risk for cancer, the usual first step is to extract the fruit or vegetable in organic solvents or by supercritical CO_2 for testing in in vitro assays to test whether the extracts have cytotoxicity toward human cancer cells. Ideally, these assays detect whether the parent extract kills tumor cells in a dose- and time-related manner. Also, ideally, it should include testing on normal human cells from the same organ, but there are many limitations as these are not available from human cells for many common sites of tumorigenesis. The next step is a process of discovery and employs the concept of structure–activity-guided fractionization. Essentially, the plant extract is further purified leading to identity of specific classes or individual compounds for which the most robust anticancer activity is noted. Thus, a specific fruit can be extracted into specific flavonoid fractions, yielding a specific chemical identified through mass spectrometry. The identified chemical may be the best of the extracted agents that shows robust cytotoxicity as well as other important anticancer features, such as being anti-inflammatory, antiangiogenic, proapoptotic, or activating genes involved in cell regulation (Table 2.2). Often, cell culture studies are used to probe potential mechanisms by which phytochemicals prevent cancer growth or expansion.

After gathering data from in vitro systems, the next step is to evaluate the candidate-preventive phytochemicals in vivo in relevant *animal models* (Table 2.3) that replicate human cancer. Animal models for cancer are usually developed in mice or rats, and can be carcinogen initiated or initiated by altering key genes that have been associated with common human cancers. Typically, animal carcinogenesis assays provide the phytochemical orally either mixed into rodent diet or given in the drinking water. Sometimes, it is necessary to intragastrically intubate the animal with the test agent. In a preventive protocol, the animals are introduced to the test phytochemical prior to or during the time of "initiation" while in a therapeutic protocol, the test agent is administered after the tumorigenic process has advanced. End points typically involve the measure of incidence of cancer in the animals, the tumor burden and severity, as well as the measure of biological markers. One of the newest approaches to the testing of the anticancer capacity of a given phytochemical is to see if it may work additively or synergistically to aid and abet conventional cancer treatment. An added benefit would be to observe an increased therapeutic index while offsetting or reducing the incidence of off-target toxicity, commonly referred to as the side effects of cancer therapy.

3 Phytochemicals and Other Chronic Diseases

The previous section examined the evidence, phytochemicals, mechanisms, and the experimental approaches involved to determine the effect of fruit phytochemicals on the development of cancer. For all diseases, the experimental approaches of epidemiological studies combined with in vitro animal models and human clinical trials are used to identify major risk factors and potential control strategies. Because there is increasing evidence that aberrant inflammation lies at the molecular core of processes involved in more than just cancer, it is possible that fruit consumption will have collateral benefits for prevention of heart disease, obesity, diabetes, Alzheimer's disease, and other neurodegenerative diseases.

Table 2.2 Mechanisms of tumor growth inhibition

Mechanism	Key mediators	Mechanism of action	Outcome	Fruits	References
Cell adhesion molecules	VEGF	Decreased	Decreased angiogenesis	Black raspberries	Liu et al. (2005a, b)
	ICAM-1, VCAM	Reduce upregulation induced by TNFα	Decreased cell migration	Blueberries and cranberries	Youdim et al. (2002)
		Decreased	Could be due to inactivation of NFκβ pathway	Cranberry	Ruel and Couillard (2007)
	MMPs	Expression of 2 and 9 inhibited	Decreased cell migration	Cranberries, raspberries, blackberries, blueberries, muscadine grapes	Neto (2007), Tate et al. (2004), Matchett et al. (2005)
		Decreases 9		Resveratrol	Woo et al. (2004)
	GAGs	Decreases in sulfation	Reduce chronic inflammation	Blueberries	Neto (2007), Tovar et al. (1998)
	MCP-1	Reduce upregulation		Blueberries and cranberries	Youdim et al. (2002), Neto (2007)
Proinflammatory	MAPK/ERK	Increases ERK activation	Increased neurogenesis	Blueberry	Shukitt-Hale et al. (2008)
	IGF-1	Activates		Blueberry	Shukitt-Hale et al. (2008)
	TNFα	Reduced expression	Decreased inflammation	Black raspberries	Montrose et al. (2011), Bodet et al. (2006)
				Cranberries	
	CRP	Decreased expression	Decreased inflammation	Raspberries	Rao and Snyder (2010), Bodet et al. (2006)
				Cranberries	
	IL-1β	Reduced expression	Decreased inflammation	Black raspberries	Montrose et al. (2011)
	NFκβ	Reduced expression via increased Iκβ expression	Decreased inflammation	Black raspberries	Montrose et al. (2011)
		Reduced expression	Decreased inflammation	Berry fruits	Shukitt-Hale et al. (2008)
		Reduced expression	Decreased inflammation	Apple oligogalactan	Liu et al. (2010)
		Activation of pathways	Suppress inflammatory cascade	Resveratrol	Leiro et al. (2005)

	COX2	Reduced expression	Decreased inflammation	Black raspberries	Mallery et al. (2008)
		Reduced expression		Raspberries	Rao and Snyder (2010)
		Inhibit COX1/2		Blueberries and strawberries	Shukitt-Hale et al. (2008)
	JAK/STAT	Activation of pathways	Suppresses inflammatory cascade	Resveratrol	Wung et al. (2005)
	PKC	Increase		Fruit phenolics, berry fruits	Shukitt-Hale et al. (2008)
		Inhibits	Reduced MMP9	Resveratrol	Woo et al. (2004)
	JNK	Inhibits	Reduced MMP9	Resveratrol	Woo et al. (2004)
	IL-8	Reduce upregulation	Decreased inflammation	Blueberries, cranberries	Bodet et al. (2006), Neto (2007)
		Reduced		Reduced	
	IL-6	Reduced	Decreased inflammation	Cranberries	Bodet et al. (2006)
	PGE_2	Reduced expression	Decreased inflammation	Black raspberries	Montrose et al. (2011)
	Arachadonic acid	Suppress pathway		Fruit phenolics	Shukitt-Hale et al. (2008)
				Berry fruits	
Apoptotic		Reduce apoptosis		Blueberries and cranberries	Neto (2007)
Oxidative stress	iNOS	Decreases ROS formation	Protection against oxidative damage	Berry juice blend	Jensen et al. (2008)
		Decreased expression		Raspberries	Rao and Snyder (2010)
	Peroxide	Reduces H202	Reduced necrosis	Cranberry	Neto (2007)
	Glutathione	Upregulation of glutathione synthesis	Decrease oxidative stress and reduced DNA damage	Fruit phenolics	Shukitt-Hale et al. (2008), Weisel et al. (2006)
				Berry fruits, raspberries	

(continued)

Table 2.2 (continued)

Mechanism	Key mediators	Mechanism of action	Outcome	Fruits	References
	Reactive oxygen species	Reduced	Decrease oxidative stress	Blueberries and strawberries, concord grape	Neto (2007), Shukitt-Hale et al. (2008, 2006)
		Reduced	Reactive oxygen species absorbed	Plums	Yang and Gallaher (2005)
		Reduced	Oxidative damage	Apples	Gerhauser (2008)
		Scavenge	Reducing neuronal age-related deficits	Blueberries	Shukitt-Hale et al. (2008, 2006), Joseph et al. (1999)
		Scavenge	Antioxidant radicals	Prunes (plums)	Stacewicz-Sapuntzakis et al. (2001)
		Reduced	Lipid peroxidation	Apple extracts	Fini et al. (2011)
Platelet effects	TXA_2	Inhibits platelet aggregation, calcium mobilization, hydrogen peroxide formation, and TXA_2 production induced from collagen and arachadonic acid	Inhibits platelet activation	Pomegranate	Mattiello et al. (2009)
Lipoprotein effects		Inhibiting oxidation of circulating lipoproteins	Slowed CIMP progression for individuals at risk for CHD	Pomegranate	Davidson et al. (2009)
		Protect against lipid oxidation		Strawberries and blueberries	Shukitt-Hale et al. (2008), Neto (2007)
		Inhibits lipid peroxidase	Protection against oxidative damage	Berry juice blend	Jensen et al. (2008)
		Prevention of obesity-related colon cancer	Protection against oxidative damage	Apple juice	Koch et al. (2009)

Table 2.3 Effect of fruit phytochemicals in animal models of cancer

Organ	Animal model	Phytochemical	Result	Reference
Breast	7,12 DMBA rats	Grape seed extract	Reduction in tumor multiplicity	Kim et al. (2004), Mehta and Lansky (2004)
		Pomegranate seed oil	Tumor reduction	
Skin	DMBA, TPA mouse	Pomegranate seed oil (anthocyanins)	Chemopreventive	Adhami et al. (2009), Hora et al. (2003), Afaq et al. (2005), Jang et al. (1997), Zhao et al. (1999)
		Resveratrol		
		Grape seed powder		
	UV-induced mouse	Resveratrol	Chemopreventive	Aziz et al. (2005)
Esophagus	NMBA F344 rats	Resveratrol	Inhibits tumor multiplicity	Li et al. (2002), Stoner et al. (2010)
		Acai, strawberries, wolfberry, noni		
	NMBA mice	Black raspberries	Limit cancer development	Stoner et al. (2008), Chen et al. (2006)
Colon	AOM rat	Black raspberries	Inhibit carcinogenesis	Harris et al. (2001), Lala et al. (2006), Gosse et al. (2005), Kohno et al. (2004)
		Bilberry	Reduced ACF	
		Chokeberry	Promotes apoptosis	
		Grape	Chemopreventive	
		Apple procyanidins		
		Pomegranates		
	AOM, DMBA rat	Grape seed extract	Decreased ACF	Durak et al. (2005)
	1,2 DMH F344 rats	Resveratrol	Reduced colon tumors	Sengottuvelan et al. (2006), Barth et al. (2005)
		Cloudy apple juice (procyanidins, pectin)	Decreased proliferation, ACF, DNA damage	
	APC Min mice	Black raspberries	Inhibit carcinogenesis	Duncan et al. (2009), Rajakangas et al. (2008)
		White currants (anthocyanins)		
Cheek	7,12 DMBA hamster	Black raspberries		Casto et al. 2002
Prostate	TRAMP model	Grape seed extract	Cell cycle arrest	Raina et al. (2007), Konijeti et al. (2010)
		Tomatoes (lycopene)	Chemopreventive	
Lung	B(a)P and NTCU mice	Pomegranate	Chemopreventive	Khan et al. 2007

Carcinogen key: *AOM* azoxymethane, *DMBA* dimethylbenzanthracene, *TPA* tetradecanoylphorbol acetate, *DMH* dimethylhydrazine, *NMBA* nitrosomethyl benzylamine, *B(a)P* benzo(a)pyrene, *NTCU* *n*-nitroso-tris-chloroethylurea

3.1 *Tunneling Down: An Example of a Phytochemical Class with Promise for Prevention of Disease: Anthocyanins*

With the large array of fruits and the added numbers of beneficial phytochemicals they contain, determining which whole fruit, compound, or extract is the most beneficial for a given modality can be exhausting. We have already discussed epidemiological evidence as well as some basic research involving benefits from fruits. Here, we focus on the class of fruits rich in anthocyanins, a class of phytochemicals which have been heavily studied, and the results in disease prevention have been promising (Hung et al. 2004; Neto 2007; Shukitt-Hale et al. 2008; Rao and Snyder 2010; Kim et al. 2004; Afaq et al. 2005; Jang et al. 1997; Lala et al. 2006; Larsson et al. 2008; Pan et al. 2008; Johnson 2007; Renaud and de Lorgeril 1992; Chou et al. 2001; Freedman et al. 2001; Sautebin et al. 2004; Ilbey et al. 2009; Kim et al. 2008). Anthocyanins are primarily responsible for the red, blue, and purple colors of fruits and over 400 individual compounds have been identified (Mazza and Miniati 1993). The average daily intake of anthocyanins is estimated to be 12.5 mg/day/person in the USA (NHANES 2001–2002). The amount and type of anthocyanin vary for different fruits, but for our purposes we focus on total anthocyanins. For instance, red grapes have 42.7 mg while concord grapes have 192 mg of total aglycone anthocyanins (mg/100 g fresh wt) (Wu et al. 1993). For berries like black, blue, cran, and raspberries, the total aglycone anthocyanin levels are 353, 529, 133, and 116 mg (mg/100 g fresh wt), respectively (Wu et al. 2006). Pomegranate juice has 429.9 mg/l total anthocyanins (Orak 2009), whereas the acai berry has been shown to contain 3.1919 mg/g dry wt total anthocyanins (Schauss et al. 2006). Despite these varying anthocyanin levels beneficial disease-preventative properties have been reported in all of these fruits.

Grapes and other small fruits are the most commonly known anthocyanin-rich fruits. From red and concord grapes to wine to blueberries and raspberries, most people have consumed one or more of these in their diet. For instance, the "French Paradox," first mentioned in 1992, is related to relatively low risk of cardiovascular disease (CVD) in the French despite a diet rich in saturated fats (a risk factor component of CVD (Renaud and de Lorgeril 1992). Years later, evidence still supports that moderate consumption of red wine (one to two drinks per day) contributes beneficial cardiovascular effects in most populations (Lippi et al. 2010). Is this true for wine's predecessors, the grape? Yes. Grapes have beneficial preventative properties too (Table 2.4). Despite the lack of alcohol (an active component in wine), compounds from red and concord grapes displayed numerous preventative effects. Extracts from grapes have been shown to improve cardiovascular health through reduction in cellular oxidation (Bertelli and Das 2009; Rice-Evans et al. 1996) by enhancing nitric oxide release (Freedman et al. 2001) and inhibiting some cholesterol intake (Leifert and Abeywardena 2008). In addition to the heart, grapes have been implicated in improving motor and memory function as well as improving mood (Shukitt-Hale et al. 2006; Krikorian et al. 2010). These are just a few of the human health-related benefits of grape consumption.

Table 2.4 Disease prevention by anthocyanin-rich fruits

Selected fruit	Selected diseases	Preventative properties	Reference
Grape	Cardiovascular disease (CVD)	Antioxidation	Bertelli and Das (2009)
Red	Brain degeneration (CVD)	Inhibits cholesterol uptake and 5-LOX activity	Leifert and Abeywardena (2008)
Concord	Dementia	Protects against decrease in synaptic protein function	Sun et al. (1999)
		Reduces LDL oxidation	Rice-Evans et al. (1996)
		Endothelial function improvement	Chou et al. (2001)
		Enhances nitric oxide release	Freedman et al. (2001)
		Increases dopamine release and motor function	Shukitt-Hale et al. (2006)
		Improves memory function	Krikorian et al. (2010)
Berries	Age-related cognitive decrease	Increase in working and short-term memory	Shukitt-Hale et al. (2009)
Blackberry	Endotoxic shock	Reduced iNOS and COX activity	Sautebin et al. (2004)
Blueberry	Diabetes	Insulin-like active principles and protection against glucose toxicity	Martineau et al. (2006)
Cranberry	Urinary tract infections	Bacterial antiadhesion	Gupta et al. (2007), Howell (2007)
Pomegranate juice	Prostate cancer	Decreases PSA doubling time, decreases cell proliferation, and increases apoptosis	Pantuck et al. (2006)
	Renal tubular cell injury	Reduces oxalate crystal formation	Ilbey et al. (2009)
	Osteoarthritis	Decrease in cell proliferation and inflammatory cells in synovial fluid	Hadipour-Jahromy and Mozaffari-Kermani (2010)
Muscadine grape	Microbial infection	Antimicrobial activity	Kim et al. (2008, 2010)
	Prostate cancer	Chemopreventative agent	God et al. (2007), Hudson et al. (2007)
Acai	Inflammation	Antioxidant	Schauss et al. (2006)

Much like grapes, berries are rich in anthocyanins (Mazza and Miniati 1993; Wu et al. 2006). Almost everyone is familiar with the effects cranberries have on urinary tract infections. This is due to the bacterial antiadhesion properties found within the anthocyanin profile of cranberries (Gupta et al. 2007; Howell 2007). Other berries, like blueberries, have disease prevention properties that differ from cranberries. Blueberries exhibit antidiabetic properties in in vitro assay, such as insulin-like active properties, to protect against toxicity from glucose (Martineau et al. 2006). These two are not the only berries with anthocyanin-mediated health benefits. Blackberries have been reported to increase working and short-term memories which both play roles in age-related cognitive impairment (Shukitt-Hale et al. 2009). These berries can also reduce harmful effects from endotoxic shock (Sautebin et al. 2004). These are numerous other berries with high anthocyanin levels and reported human health benefits.

The historical-, clinical-, and media-driven reports of the health benefits of grapes and berries has led to the emergence of pomegranates, another fruit high in anthocyanins, as another fruit promoted for its health benefits (Wu et al. 2006). This promotion has led to a tripling of pomegranate plantings in California from 2002 to 2007 (USDA 2007). Research has shown that pomegranates are beneficial to prostate cancer prevention (Pantuck et al. 2006), can slow the symptoms of osteoarthritis (Hadipour-Jahromy and Mozaffari-Kermani 2010), and can reduce oxalate crystal formation in renal cells (Ilbey et al. 2009). Much like the grapes and berries, the pomegranate juice offers an easy enjoyable delivery system for the humans to ingest healthy anthocyanin compounds.

Muscadine grapes are common in the southeastern USA due to their ability to handle the humid summers and warmer winters (Olien 1990). They are red to purple in color like other grapes; however, they have higher antioxidant capacity than table grapes. This is due to a different anthocyanin profile, one similar to blackberries and raspberries (Rommel and Wrolstad 1993). Muscadine extracts and powders have an effect against microbial infection (Kim et al. 2008, 2010) and are potential chemopreventative agents in prostate cancer (God et al. 2007; Hudson et al. 2007).

Grapes and berries are not the only anthocyanin-rich fruits around; they are just the most well-known and, for the most part, well-studied. Other temperate fruit crops, such as apples, peach, plum, kiwi, and others, although typically do not have red flesh, have the potential to develop red-fleshed cultivars. In addition, exotic crops, such as acai berry, are starting to gain notoriety as a superfruit. The first research on acai focused on the remarkable antioxidant potential of acai berries and their impact of inflammation reduction (Schauss et al. 2006). Current research is focused on studying the health benefits of acai in animal models (Stoner et al. 2010; de Souza et al. 2010). As the beneficial effects with animal models become well-documented, hopefully the research will expand to human trials. Other fruits of interest are the pitanga (*Eugenia uniflora* L.) which has long been utilized in traditional Brazilian medicine to treat diarrhea (Brandelli et al. 2009). The more understanding of traditional medicine from plants to practice yields even more fruits with health benefits. Researchers making inroads into western Africa, Colombia, and other countries expand the knowledge of anthocyanin-rich fruits.

From this brief highlight of anthocyanin-rich fruits, it can be concluded that their health impact is widespread and varied. Previous sections have focused on specific cellular processes and epidemiological evidence. Here, we have shown how one class of bioactive compounds and fruits rich in anthocyanins are a cornerstone in understanding how specific dietary compounds can impact a myriad of maladies from heart disease and cancer to microbial infections. Research continues to show that fruit and vegetable consumption is beneficial to improved health. This is due to, in part, anthocyanins and the increased protection they provide along with other bioactive compounds in fruits.

4 Genetic Variation Within Fruit Crops

4.1 Trend in Fruit Breeding

Fruit breeders need to anticipate the future as the cultivars they begin to develop now will not enter production for at least 10 years and frequently longer. Their objectives need to reflect the desires of the market (Byrne 2005). The previous section of this chapter has asserted that fruit phytochemicals affect the health of the people that consume them. Most of the studies have dealt with one cultivar and/or focused on a few chemical components of the phytochemicals available in the fruit. Thus, it has been clearly shown that there are differences among crops and that there is strong evidence that phytochemicals from these crops have protective properties against various chronic diseases, such as cancer and cardiovascular disease.

This information has been widely publicized and has created a proliferation of superfruits which are touted for their high level of antioxidants. These would include fruits, such as blueberries, pomegranates, cranberries, plums, acai, and others. This marketing approach has been effective in promoting the increased consumption of blueberries and pomegranates. The consumer makes the connection between food and health as the vast majority of consumers surveyed indicate that they take health into account when choosing food to purchase. This heightened awareness of the health benefits of food has increased the food industry's efforts in the development of foods with health benefits (Sloan 2006, 2008; Dillard and German 2000).

Since the 1990s, the US Government has been working toward convincing people to consume three to four portions or two cups of fruit a day, but still the average fruit consumption is only about half this recommendation (Pollack and Perez 2008; Wells and Buzby 2008). This presents an opportunity to fruit breeders. Since the amount of fruit consumed has not increased, the other approach would be to enhance the health benefits of the fruits that are consumed. As it has been seen with the health-oriented marketing of superfruits (i.e., pomegranate, blueberries), it is possible to increase the consumption of specific fruits by touting their high antioxidant capacity. The next step of this process would be to develop health-enhanced cultivars with a better phytochemical mix for a given crop.

4.2 Phytochemical Profiles Among Crops

The phytochemical profile of various crops and even their parts (peel versus flesh) also differs dramatically (Table 2.5). In apples, peaches, and plums, the peel is 6–9% of the fruit fresh weight, but because it contains from two to about five times the concentration of phenolics than the flesh, the peel is an important source of phenolics.

Table 2.5 Comparative phytochemical profile (% of total for each chemical group) of apple, peach, plum, and blueberry cultivars

Chemical group	Apple	Apple	Apple	Peach	Peach	Plum	Plum	Blueberry
	Fruit	Flesh	Peel	Flesh	Peel	Flesh	Peel	Fruit
Procyanidins	42	53–56	38–60	50–67	40–59	71	56	0
Hydroxycinnamic acids	29	39–40	8–10	30–46	22–36	27	12	30
Flavanols	21	0–2	18–42	2	4–7	0	11	15
Dihydrochalcones	7	4–6	7–12	0	0	0	0	0
Anthocyanins	0	0	0–10	1–2	14–17	7	21	55
Reference	Lata et al. (2009)	Khanizadeh et al. (2008), Tsao et al. (2003)	Lata et al. (2009), Khanizadeh et al. (2008), Tsao et al. (2003)	Tomas-Barberan et al. (2001)	Tomas-Barberan et al. (2001)	Tomas-Barberan et al. (2001)	Tomas-Barberan et al. (2001)	Zheng and Wang (2003)

Procyanidins (catechin, epicatechin, procyanidin B1, procyanidin B2, other procyanidins), hydroxycinnamic acid (chlorogenic acid, neochlorogenic acid, *p*-coumaroyl quinic, caffeic acid, related compounds), flavanols (quercetin-3-rutinoside, quercetin-3-rhamnoside, other quercetin derivatives, myricetin, kaempferol), dihydrochalcones (phloretin-3-xyloglucoside, phloridzin), anthocyanins (cyanidin 3-glucoside, cyandin 3-rutinoside, and glycosides of delphinidin, petunidin, malvidin, and others)

The peel can commonly contain 20–40% of the total phenolics and a major portion of the antioxidant capacity of these large fruited crops (Cevallos-Casals et al. 2006; Drogoudi et al. 2008; Lata et al. 2009; Khanizadeh et al. 2008; Tomas-Barberan et al. 2001). A similar situation exists in small fruits (blueberry, blackberry, raspberry) as seen in the negative correlation between fruit size and total phenolics and antioxidant activity. Although this effect is significant, when the data is adjusted for size, there is still abundant genetic variability for the total phenolic content in the flesh (Connor et al. 2002b, c, 2005a, b).

Among the cultivars of apple, peaches, plums, and blueberries surveyed, the predominance of the various chemical groups varies. All of these fruits have hydroxycinnamic acids as a predominant phenolic among their phytochemical mix. Apple, peach, and plum tend to be high in procyanidins and low in anthocyanins, whereas blueberries are the reverse. Apple is the only fruit of these that contain dihydrochalcones. Thus, the mix of phytochemicals within each crop varies from others which emphasizes the importance of the recommendation of eating a diversity of fruits to maintain good health.

This observation can be taken one step further to look at the composition of the specific compounds within each subclass in each crop. For example, the anthocyanins found in peach are mainly cyanidin 3-glucoside and cyanidin 3-rutinoside (Tomas-Barberan et al. 2001), whereas blueberries contain various forms (mainly 3-galactoside, 3-glucoside, and 3-arabinoside) of delphinidin, petunidin, cyanidin, and malvidin (Zheng and Wang 2003). This is frequently the situation within other classes of phytochemicals between the various crops.

The development of health-enhanced fruit cultivars requires that there is genetic variation for the trait within the crop with which the breeder is working. From a breeding perspective, the next step is to determine if the crop has the genetic variability needed to develop health-enhanced cultivars. Although there are hundreds of phytochemicals found in fruits, most of the literature is focused on the antioxidant bioactivity and the concentration of total phenolics and anthocyanins of fruit crops.

4.3 Antioxidants

The consumption of high levels of antioxidants is promoted as being beneficial to one's long-term health by reducing general oxidative stress within the body. Consequently, there has been interest in exploring the levels of antioxidants in fruits both among crops and more recently among cultivars and breeding materials within a crop (Tables 2.6–2.8). These studies focus on a few classes of compounds with the most frequent being vitamin C, carotenoids, total phenolics, and anthocyanins with a couple of studies looking at the levels of various phenolic compounds among cultivars.

Correlation studies among these various phytochemicals and antioxidant activity have consistently shown that among a range of crops total phenolics and, in berries such as blueberries and blackberries, anthocyanins are well-correlated with antioxidant activity, whereas carotenoids and vitamin C contribute little to the antioxidant

Table 2.6 Antioxidant activity among cultivars within selected fruit crops

Crop	Genotypes	Number	Range of AOA μg Trolox/100 g FW	Reference
Peach/ nectarine	California cultivars	20	46–1,006 (flesh) (DPPH) 230–1,789 (peel) (DPPH)	Gil et al. (2002)
	Red-fleshed peaches	8	440–1,784 (DPPH)	Cevallos-Casals et al. (2006)
	White-fleshed peaches	4	540–1,096 (DPPH)	Vizzotto et al. (2007)
	Yellow-fleshed peaches	6	437–1,128 (DPPH)	Vizzotto et al. (2007)
	Red-fleshed peaches	9	2,787–13,505 (DPPH)	Vizzotto et al. (2007)
	Segregating progeny	218	227–630 (DPPH)	Cantín et al. (2009)
	California cultivars	20	350–2,250 (DPPH)	Byrne et al. (2009)
Japanese plum	California cultivars and breeding selections	45	1,311–6,471 (DPPH)	Vizzotto et al. (2007)
	Red-flesh plums	14	1,254–3,244 (DPPH)	Cevallos-Casals et al. (2006)
	California cultivars	5	205–518 (flesh) (DPPH) 701–1,314 (peel) (DPPH)	Gil et al. (2002)
	California cultivars	6	2,300–8,600 (DPPH)	Byrne et al. (2009)
Blueberries	High-bush cultivars	6	1,700–3,701 (ORAC)	Prior et al. (1998)
	Rabbiteye cultivars	4	1,390–2,550 (ORAC)	Prior et al. (1998)
	V ashei, rabbiteye cultivar and selections	4	11,100–13,000 (ORAC)	Moyer et al. (2002)
	High-bush cultivars and selections	15	1,900–9,600 (ORAC)	Moyer et al. (2002)
	High-bush cultivars	80	332–582 (ORAC)	Kalt et al. (2001)
	Low-bush cultivars	135	515–901 (ORAC)	Kalt et al. (2001)
	High-bush cultivars	4	379–549 (DPPH) 2,130–2,640 (FRAP)	Giovanelli and Buratti (2009)
	High-bush, low-bush cultivars	39		Giongo et al. (2006)
	High-bush and hybrid, rabbiteye cultivars	87	46–311 (ORAC)	Elhenfeldt et al. (2001)
	Breeding materials	52	500–6,300 (MeLO)	Connor et al. (2002b)
	High-bush cultivars	9	2,500–4,300 (MeLO)	Connor et al. (2002a)
	High-bush cultivars	11	2,000–7,900 (FRAP)	Beccaro et al. (2006)

(continued)

Table 2.6 (continued)

Crop	Genotypes	Number	Range of AOA μg Trolox/100 g FW	Reference
	High-bush cultivars	19	2,780–5,060 (FRAP)	Remberg et al. (2007)
Apples	Cider cultivars and selection	8	Peel: 175–452 (FRAP, ASCE) Flesh: 32–125 (FRAP, ASCE)	Khanizadeh et al. (2008)
	Cultivars	6	Fruit: 335–739 (ABTS)	Vieira et al. (2009)
	Cultivars	11	Peel: 1225–4145 (ABTS) Peel: 1004–3878 (DPPH) Peel: 521–1161 (FRAP) Flesh: 380–961 (ABTS) Flesh: 346–891 (DPPH) Flesh: 140–262 (FRAP)	Vieira et al. (2011)

2,2′-azinobis(3-ethylbenzothiazoline-6-sulfonic acid) (ABTS) and 1,1-diphenyl-2-picrylhydrazyl (DPPH) measure the scavenging of free radicals, ORAC measures the oxygen radical absorption capacity using a biologically relevant radical source, FRAP measures the ferric reducing power, MeLO measures the inhibition of peroxyl radical-induced oxidation of linoleic acid. ASCE, measured in ascorbic acid equivalents instead of Trolox equivalents

capacity of the fruit (Vizzotto et al. 2007; Cevallos-Casals et al. 2006; Kalt et al. 2001; Prior et al. 1998; Giovanelli and Buratti 2009; Connor et al. 2002a; Henriquez et al. 2009; Beccaro et al. 2006; Lee et al. 2003).

Antioxidant activity among genotypes has been reported with peach, plum, apple, and blueberry (Table 2.6) using various in vitro methods on phenolic extracts of the fruit. The most commonly used assays are the aqueous-based assays, such as 2,2′-azinobis(3-ethylbenzothiazoline-6-sulfonic acid) (ABTS) and 1,1-diphenyl-2-picrylhydrazyl (DPPH) which measure the scavenging of free radicals, ORAC which measures the oxygen radical absorption capacity using a biologically relevant radical source, and FRAP which measures the ferric reducing power of the extract. Less frequently, MeLO which measures the inhibition of peroxyl radical-induced oxidation of linoleic acid is used. Several studies with fruit crops have shown that these various methods were correlated among themselves (Thaipong et al. 2006; Connor et al. 2002a, b) and correlated similarly with total phenolics and other phytochemical components being studied (Vieira et al. 2011; Wojdylo et al. 2008).

For apple, peach, plum, and blueberry, there is a wide range in measured antioxidant capacity irrespective of the methodology used (Table 2.6), total phenolics (Table 2.7), and anthocyanins (Table 2.8). Although among commercial cultivars the differences were significant, in some studies that examined breeding materials and other noncommercial germplasm, the range of antioxidant capacity, total phenolics, and/or anthocyanins measured were greatly enlarged (Vizzotto et al. 2007; Cevallos-Casals et al. 2006; Moyer et al. 2002; Conner et al. 2002b). Thus, it is clear that there is variation among genotypes within crops. Currently, there are genotypes within the commercial cultivar mix that have higher levels of antioxidants that could

Table 2.7 Total phenolics among cultivars within selected fruit crops

Crop	Genotypes	Number	Range of phenolics mg/100 g FW	Reference
Peach/ nectarine	California cultivars	20	14–111 (CGA)	Gil et al. (2002)
	Processing cultivars	8	48–80 (CGA)	Chang et al. (2000)
	Red-fleshed peaches	8	100–448 (CGA)	Cevallos-Casals et al. (2006)
	White-, yellow-, red-fleshed peaches	19	137–1,260 (CGA)	Vizzotto et al. (2007)
	Yellow peach, nectarines, white nectarine	13	37–73 (GAE)	Vaio et al. (2008)
	Commercial cultivars	11	14–50 (GAE)	Taravini et al. (2008)
	Segregating progeny	218	13–71 (GAE)	Cantín et al. (2009)
Japanese plum	California cultivars and breeding selections	45	182–898 (CGA)	Vizzotto et al. (2007)
	Red-flesh plums	14	298–563 (CGA)	Cevallos-Casals et al. (2006)
	California cultivars	5	42–109 (CGA)	Gil et al. (2002)
Blueberries	High-bush cultivars	6	181–391 (GAE)	Prior et al. (1998)
	Rabbiteye cultivars	4	230–457 (GAE)	Prior et al. (1998)
	V ashei, rabbiteye cultivar and selections	4	717–961 (GAE)	Moyer et al. (2002)
	High-bush cultivars and selections	15	171–868 (GAE)	Moyer et al. (2002)
	High-bush cultivars	80	165–216 (GAE)	Kalt et al. (2001)
	Low-bush cultivars	135	346–412 (GAE)	Kalt et al. (2001)
	Store bought blueberries	5	292–672 (CGA)	Cevallos-Casals et al. (2003)
	High-bush cultivars	4	251–310 (GAE)	Giovanella and Buratti (2009)
	High-bush, low-bush cultivars	39	187–495 (catechin)	Giongo et al. (2006)
	High-bush and hybrid, rabbiteye cultivars	87	25–199 (GAE)	Ehlenfedt et al. (2003)
	Breeding materials	52	150–945 (CGA)	Connor et al. (2002b)
	High-bush cultivars	9	401–604 (CGA)	Connor et al. (2002a)
	High-bush cultivars	11	166–459 (GAE)	Beccaro et al. (2006)
Apples	Cultivars	5	170–212 (GAE)	Henriquez et al. (2009)
	Cider cultivars and selection	8	Peel: 101–214 (GAE) Flesh: 23–52 (GAE)	Khanizadeh et al. (2008)
	Cultivars	10	Flesh: 37–90 (HPLC, epicatechin)	McGhie et al. (2005)
	Cultivars	56	Peel: 48–235 (GAE)	Lata et al. (2005)
	Cultivars	6	105–270 (GAE)	Vieira et al. (2009)

(continued)

Table 2.7 (continued)

Crop	Genotypes	Number	Range of phenolics mg/100 g FW	Reference
	Cultivars	11	Peel: 304–713 (GAE)	Vieira et al. (2011)
			Flesh: 128–212 (GAE)	
	Cultivars	8	102–235 (GAE)	Tsao et al. (2003)
	Cultivars	4	Peel: 309–589 (GAE)	Wolfe et al. (2003)
			Flesh: 75–103 (GAE)	
			Fruit: 119–159 (GAE)	

Total phenolics expressed as equivalents of chlorogenic acid (CGA), gallic acid (GAE), catechin, or epicatechin

Table 2.8 Total anthocyanins among cultivars within selected fruit crops

Crop	Genotypes	Number	Range of phenolics mg C3G/100 g FW	Reference
Peach/ nectarine	California cultivars	20	Flesh: 0–23	Tomas-Barberan et al. (2001)
			Peel: 34–273	
	Red-fleshed peaches	8	1–36	Cevallos-Casals et al. (2006)
	White-, yellow-, red-fleshed peaches	19	1–266	Vizzotto et al. (2007)
	Segregating progeny	218	0.1–31	Cantín et al. (2009)
	California cultivars	20	0.5–7	Byrne et al. (2009)
Japanese plum	California cultivars and breeding selections	45	2–611	Vizzotto et al. (2007)
	Red-flesh plums	14	25–175	Cevallos-Casals et al. (2006)
	California cultivars	5	Flesh: 0–28 (C3R)	Tomas-Berbaran et al. (2001)
			Peel: 129–1,615 (C3R)	
	California cultivars	6	15–105	Byrne et al. (2009)
Blueberries	High-bush cultivars	6	93–235	Prior et al. (1998)
	Rabbiteye cultivars	4	61–187	Prior et al. (1998)
	V ashei, rabbiteye cultivar and selections	4	242–515	Moyer et al. (2002)
	High-bush cultivars and selections	15	73–430	Moyer et al. (2002)
	High-bush cultivars	80	93–148	Kalt et al. (2001)
	Low-bush cultivars	135	127–210	Kalt et al. (2001)
	Store bought blueberries	5	138–385	Cevallos-Casals and Cisneros-Zevallos (2003)
	High-bush cultivars	4	92–129	Giovanella and Buratti (2009)
	High-bush, low-bush cultivars	39	95–445	Giongo et al. (2006)

(continued)

Table 2.8 (continued)

Crop	Genotypes	Number	Range of phenolics mg C3G/100 g FW	Reference
	High-bush and hybrid, rabbiteye cultivars	87	89–331	Ehlenfeldt et al. (2003)
	Breeding materials	52	1–428	Connor et al. (2002b)
	High-bush cultivars	9	105–236	Connor et al. (2002a)
	High-bush cultivars	11	30–231	Beccaro et al. (2006)
Apples	Cider cultivars and selection	8	Peel: 0–29 Flesh: 0	Khanizadeh et al. (2008)
	Cultivars	10	Fruit: 0–3.7	McGhie et al. (2005)
	Cultivars	56	Peel: 1–56	Lata et al. (2005)
	Cultivars	6	Peel: 5–42 (C3Gal)	Vieira et al. (2009)
	Cultivars	11	Peel: 27–117 (C3Gal)	Vieira et al. (2011)
	Cultivars	8	Peel: 4–21	Tsao et al. (2003)
	Cultivars	4	Peel: 2–27	Wolfe et al. (2003)

Anthocyanins measured as equivalents of cyanidin 3-glucoside (C3G), except for plums in Tomas-Barberan et al. 2001, who used equivalents of cyanidin 3-rutinoside (C3R), and on apples in Vieira et al. 2009, 2011, who used cyanidin 3-galactoside (C3Gal)

be promoted as such and this type of marketing has already been initiated. Furthermore, in the case of peaches, plums, and blueberries, there are also genotypes outside the commercial mix of cultivars that have even higher levels of antioxidants than commercial germplasm indicating the possibility of increasing the levels even more.

Beyond examining the variation in general antioxidant activity or levels of the major classes of anitoxidants (total phenolics and anthocyanins), there have been studies examining the ability of genotypes to inhibit proliferation of cancer cells, inhibition of LDL oxidation and other bioactivities in strawberries (Meyers et al. 2003), apples (Yoshizawa et al. 2005; Wolfe et al. 2003; Thompson et al. 2009), blueberries (Yi et al. 2005), peaches, plums (Chang et al. 2000; Byrne et al. 2009), and other fruits. These studies have shown that, as was seen with antioxidant activity and the levels of phytochemicals, genotypes within a crop differed in their bioactivity toward cancer growth or CVD development as measured by various in vitro assays. Another crucial observation is that these various bioactivites are not consistently correlated with antioxidant activity, total phenolics, or total anthocyanin content (Byrne et al. 2009; Sun et al. 2002; Liu 2004; Liu 2003; Meyers et al. 2003). This does not indicate that antioxidant activity is not important in preventing these chronic diseases, but rather that there are other mechanisms by which these diseases are regulated and that the phytochemicals within a fruit work both additively and synergistically to affect disease development (Liu et al. 2005a, b).

5 Breeding for Enhanced Phytochemical Levels

Many of the publications that report variation in antioxidants or bioactivities among genotypes within a crop mention that breeding for enhanced health properties is a goal of the breeding program (Vizzotto et al. 2007; Cantín et al. 2009; Connor et al. 2002a, 2005b, c; Vorsa and Polashock 2005; McDougall et al. 2007; Moyer et al. 2002; Kappel 2008; Khanizadeh et al. 2009). Nevertheless, it is not clear how much work is ongoing in the breeding of health-enhanced fruits as a breeder always has many competing objectives to balance. For a cultivar to be successful, it must be productive for the growers and produce high-quality fruit or it will not sell well. Both these traits are complex and are in turn divided into dozens of well-defined traits that the breeder selects for or against. One thing that is clear from various surveys is that whatever health-enhanced cultivar released also has to taste good (Sloan 2008; Byrne 2005).

5.1 Breeding Studies

As discussed previously, there have been a multiplicity of studies that have examined the genotypic variation of antioxidant activity and the level of phytochemicals in fruits of which some examined differences among years (Lata et al. 2005, 2008, 2009; Wojdylo et al. 2008) and between locations (McGhie et al. 2005; Prior et al. 1998; Connor et al. 2002b, c, 2005b, d). In general, although the cultivar effect was large, the antioxidant activity and phytochemical concentrations seen among cultivars frequently varied from year to year and among locations presumably due to differences in climatic, cultural, edaphic, or some other condition.

Breeding studies with blueberry (Connor et al. 2002a) and red raspberry (Connor et al. 2005a, c) estimated the narrow-sense heritability as moderate for antioxidant activity (0.43 and 0.54 for blackberry and red raspberry, respectively) and total phenolic content (0.46 and 0.48 for blackberry and red raspberry, respectively) and moderate to high for total anthocyanin content (0.56 and 0.74 for blackberry and red raspberry, respectively). In red raspberries, the narrow-sense heritability estimates varied from 0.45 to 0.78 for individual anthocyanins. The anthocyanin with the highest concentration (cyanidin 3-sophoroside) had a heritability of 0.56. These moderate to high heritabilities indicate that good progress can be expected in the breeding of blueberry and red raspberry for higher antioxidants (Connor et al. 2005c). In these crops, the year accounted for little of the variance, whereas the importance of the genotype x year effect differed between the crops with only blueberry having a significant interaction effect.

In peach, a study with 15 progenies done over 3 years indicated that the cross variation explained ~20, ~34, and ~16% of the phenotypic variation seen for antioxidant activity, total phenolics, and total anthocyanins, respectively. In this study, the variation due to the year or the cross x year effects was not significant (Cantín et al. 2009). This study used commercial germplasm which is limited in the amount of antioxidant activity, total phenolics, and total anthocyanins as compared to the breeding germplasm available (Tables 2.6–2.8) and it is likely that the genetic component for these traits would be higher if this high antioxidant/phytochemical material was used in the breeding. Although this would facilitate rapid progress in boosting the antioxidant/phytochemical levels of peaches, further analysis would be needed as these materials are lacking in many important commercial traits.

A novel approach to improve the effective anthocyanin levels in fruit was described in cranberry, where the proportion of specific anthocyanins vary with the species. In the cultivated cranberry (*Vaccinium macrocarpon* Ait.), the major antioxidants are galactosides and arabinosides versus glucosides of cyanidin and peonidin as is found in the related species *V. oxycoccus* L. This is important as the glucoside form is more bioavailable than the galactoside and arabinoside forms. Thus, it was shown that it was possible to dramatically increase the proportion of the more bioavailable glucoside form using interspecific hybridization (Vorsa and Palashock 2005).

5.2 Breeding for Higher Anthocyanins in Tree Fruits

Berries, such as blueberries, blackberries, and red raspberries, have been touted for their high anthocyanin contents and breeding work indicates that in blueberries and red raspberries the total anthocyanin content is moderately to highly heritable (Connor et al. 2002a, 2005c). In contrast, the commercial cultivars of tree fruits, such as apples, peaches, and kiwi among others, generally have little anthocyanin in the flesh of the fruit and what they have is concentrated in the skin (Table 2.8). Nevertheless, there are variants of these fruit that have red flesh (Cevallos-Casals et al. 2006; Vizzotto et al. 2007; Volz et al. 2009; Jaeger and Harker 2005). In fact, there are red-flesh peaches and plums that have anthocyanin levels equal to or even greater than those reported for commercial blueberry cultivars (Cevallos-Casals et al. 2006; Vizzotto et al. 2007; Byrne et al. 2009). In peach and apple and probably in other normally white-, yellow-, or green-fleshed fruit species, there appear to be one or two major genes that allow the development of anthocyanins in the flesh (Sekido et al. 2010; Werner et al. 1997; Volz et al. 2009). As is seen in the work with peaches and plums, the red-fleshed genotypes vary widely in the total anthocyanins in the fruit (Cevallos-Casals et al. 2006; Vizzotto et al. 2007). Thus, once converted into a red-fleshed genotype, further selection would need to be done to optimize the anthocyanin content as well as multiple other traits essential for commercial success. Currently, there are traditional, advanced selections and newly released red-fleshed peach and nectarine cultivars in Asia, North America, and Europe (Byrne et al. 2009; Pascal, personal communication; Ma, personal communicaton), red-fleshed commercial cultivars of Japanese plum (Vizzotto et al. 2007), red-fleshed kiwis developed in New Zealand (Jaeger and Harker 2005), and work toward the development of red-fleshed apples in Japan and New Zealand (Sekido et al. 2010; Volz et al. 2009).

5.3 Breeding Targets: An Assessment

Multiple breeding programs have explored the levels of phytochemicals, antioxidant activity, and other bioactivities among the genotypes that comprise their breeding germplasm (Tables 2.6–2.8). These, combined with a few breeding studies, clearly indicate that there is sufficient genetic variability to develop cultivars with increased levels of antioxidant activity, total phenolics, and anthocyanins.

Epidemiological studies have indicated that low fruit and vegetable consumption is a risk factor for both cancer and CVD (Chong et al. 2010; Danaei et al. 2005). In the case of CVD, evidence supports the assertion that fruits with higher total phenolics reduce the risk of CVD more than low-phenolic fruits (Chong et al. 2010). Unfortunately, in spite of the thousands of studies which identify extracts or specific compounds that affect the development of chronic diseases, it is not clear which chemicals nor what levels of these chemicals should be the target of breeding programs. In part, this is because the bulk of the work has been done in cell culture model systems which serve to identify potentially useful chemicals and study their mechanisms of action but, due to bioavailability and other issues, not to establish the effective levels in animal model systems or for use in humans. Even the work with small animal models, although better than a cell culture protocol, does not necessarily translate well to a human system (Finley 2005). Furthermore, there are potential synergistic interactions among various phytochemicals which make the situation more complex (Liu 2004; Milde et al. 2007) and consequently more difficult to select a breeding target.

It has been frequently asserted that the consumption of higher levels of antioxidants is good for one's health and many products are sold using this claim. Nevertheless, there is not definitive proof to confirm that supplemental antioxidant consumption reduces the development of chronic disease (Amiot 2009). Thus, more research is needed to identify target phytochemicals and the levels needed to have a beneficial effect on long-term health and the development of chronic diseases. These studies need to compare cultivars with varying levels of phytochemicals as well as specific individual or combination of phytochemicals in animal model and human clinical trials to identify the key targets for the development of truly health-enhanced cultivars of fruit.

References

Adhami, V. M., Khan, N., and Mukhtar, H. (2009) Cancer chemoprevention by pomegranate: laboratory and clinical evidence. Nutr Cancer 61, 811–815.

Afaq, F., Saleem, M., Krueger, C. G., Reed, J. D., and Mukhtar, H. (2005) Anthocyanin- and hydrolyzable tannin-rich pomegranate fruit extract modulates MAPK and NF-kappaB pathways and inhibits skin tumorigenesis in CD-1 mice. Int J Cancer 113, 423–433.

American Institute for Cancer Research., and World Cancer Research Fund. (2007) Food, nutrition, physical activity and the prevention of cancer : a global perspective : a project of World Cancer Research Fund International, American Institute for Cancer Research, Washington, D.C.

Amiot, M. J. (2009) Fruit, vegetables, phytochemicals and human health: Past and future. Acta Hort. 817, 61–69.

Aziz, M. H., Afaq, F., and Ahmad, N. (2005) Prevention of ultraviolet-B radiation damage by resveratrol in mouse skin is mediated via modulation in survivin. Photochem Photobiol 81, 25–31.

Barth, S. W., Fahndrich, C., Bub, A., Dietrich, H., Watzl, B., Will, F., Briviba, K., and Rechkemmer, G. (2005) Cloudy apple juice decreases DNA damage, hyperproliferation and aberrant crypt foci development in the distal colon of DMH-initiated rats. Carcinogenesis 26, 1414–1421.

Beaglehole, R., Ebrahim, S., Reddy, S., Voute, J., and Leeder, S. (2007) Prevention of chronic diseases: a call to action. Lancet 370, 2152–2157.

Beccaro, G., Mellano, M., Botta, R., Chiabrando, V. and Bounous, G. (2006) Phenolic and anthocyanin content and antioxidant activity in fruits of bilberry (Vaccinium myrtillus L.) and of highbush blueberry (V. corymbosum L.) cultivars in north western Italy. Acta Hort. 715, 553–557.

Bertelli, A. A., and Das, D. K. (2009) Grapes, wines, resveratrol, and heart health. J Cardiovasc Pharmacol 54, 468–476.

Bodet, C., Chandad, F., and Grenier, D. (2006) Anti-inflammatory activity of a high-molecular-weight cranberry fraction on macrophages stimulated by lipopolysaccharides from periodontopathogens, J Dent Res 85, 235–239.

Brandelli, C., Giordani, R., De Carli, G., and Tasca, T. (2009) Indigenous traditional medicine: in vitro anti-giardial activity of plants used in the treatment of diarrhea. Parasitology Research 104, 1345–1349.

Buchner, F. L., Bueno-de-Mesquita, H. B., Ros, M. M., Kampman, E., Egevad, L., Overvad, K., Tjonneland, A., Roswall, N., Clavel-Chapelon, F., Boutron-Ruault, M. C., Touillaud, M., Kaaks, R., Chang-Claude, J., Boeing, H., Weikert, S., Trichopoulou, A., Naska, A., Benetou, V., Palli, D., Sieri, S., Vineis, P., Tumino, R., Panico, S., van Duijnhoven, F. J., Peeters, P. H., van Gils, C. H., Lund, E., Gram, I. T., Sanchez, M. J., Jakszyn, P., Larranaga, N., Ardanaz, E., Navarro, C., Rodriguez, L., Manjer, J., Ehrnstrom, R., Hallmans, G., Ljungberg, B., Key, T. J., Allen, N. E., Khaw, K. T., Wareham, N., Slimani, N., Jenab, M., Boffetta, P., Kiemeney, L. A., and Riboli, E. (2011) Variety in vegetable and fruit consumption and risk of bladder cancer in the European Prospective Investigation into Cancer and Nutrition. Int. J. Cancer 128(12), 2971–2979.

Byrne, D. H. (2005) Trends in stone fruit cultivar development. HortTechnology, 15(3), 494–500.

Byrne, D. H., Noratto, G., Cisneros Zevallos, L., Porter, W. and Vizzotto, M. (2009) Health benefits of peaches and plums. Acta Hort. 841, 267–274.

Cade, J. E., Frear, L., and Greenwood, D. C. (2006) Assessment of diet in young children with an emphasis on fruit and vegetable intake: using CADET--Child and Diet Evaluation Tool, Public Health Nutr 9, 501–508.

Cantín, C.M., Moreno, M.A. and Gogorcena Y (2009) Evaluation of the antioxidant capacity, phenolic compounds and vitamin C content of different peach and nectarine [Prunus persica (L.) Batsch] breeding progenies. J Agric Food Chem 57, 4586–4592.

Casto, B. C., Kresty, L. A., Kraly, C. L., Pearl, D. K., Knobloch, T. J., Schut, H. A., Stoner, G. D., Mallery, S. R., and Weghorst, C. M. (2002) Chemoprevention of oral cancer by black raspberries, Anticancer Res 22, 4005–4015.

Cevallos-Casals, B., Byrne, D., Okie, W. R. and Cisneros-Zevallos, L. (2006) Selecting new peach and plum genotypes rich in phenolic compounds and enhanced functional properties. Food Chem. 96, 273–280.

Cevallos-Casals, B. and Cisneros-Zevallos, L. (2003) Stochiometric and kinetic studies of phenolic antioxidants from Andean purple corn and red-fleshed sweetpotato. J Agric. Food Chem. 51, 3313–3319.

Chang, S., Tan, C., Frankel, E. N. and Barrett, D. M. (2000) Low-density lipoprotein antioxidant activity of phenolic compounds and polyphenol oxidase activity in selected clingstone peach cultivars. J. Agric. Food Chem. 48, 147–151.

Chen, T., Hwang, H., Rose, M. E., Nines, R. G., and Stoner, G. D. (2006) Chemopreventive properties of black raspberries in N-nitrosomethylbenzylamine-induced rat esophageal tumorigenesis: down-regulation of cyclooxygenase-2, inducible nitric oxide synthase, and c-Jun, Cancer Res 66, 2853–2859.

Chong, M., Macdonald, R. and Lovegrove, J. (2010) Fruit polyphenols and CVD risk: a review of human intervention studies. Brit. J. Nutrition. 104, s28–s39.

Chou, E. J., Keevil, J. G., Aeschlimann, S., Wiebe, D. A., Folts, J. D., and Stein, J. H. (2001) Effect of ingestion of purple grape juice on endothelial function in patients with coronary heart disease. Am J Cardiol 88, 553–555.

Connor, A. M., Luby, J. J. and Tong, C. (2002a) Variation and heritability estimates for antioxidant activity, total phenolic content, and anthocyanin content in blueberry progenies. J. Amer. Soc. Hort. Sci. 127,82–88.

Connor, A. M., Luby, J. J., Tong, C., Finn, C. E. and Hancock, J. F. (2002b) Genotypic and environmental variation in antioxidant activity, total phenolic content, and anthocyanin content among blueberry cultivars. J Amer. Soc. Hort. Sci. 127:89–97.

Connor, A. M., Stephens, M. J., Hall, H. K., and Alspach, P. A. (2005a) Variation and heritabilities of antioxidant activity and total phenolic content estimated from a red raspberry factorial experiment. J. Amer. Soc. Hort. Sci. 130, 403–411.

Connor, A. M., Finn, C. E., and. Alspach, P. E. (2005b) Genotypic and environmental variation in antioxidant activity and total phenolic content among blackberry and hybridberry cultivars. J. Amer. Soc. Hort. Sci. 130, 527–533.

Connor, A. M., McGhie, T. K., Stephens, M. J., Hall, H. K., and Alspach, P. A. (2005c) Variation and heritability estimates of anthocyanins and their relationship to antioxidant activity in a red raspberry factorial mating design. J. Amer. Soc. Hort. Sci. 130, 535–542.

Connor, A. M., Finn, C. E., McGhie, T. K., and Apspach, P. A. (2005d) Genetic and environmental variation in anthocyanins and their relationship to antioxidant activity in blackberry and hybridberry cultivars. J. Amer. Soc. Hort. Sci. 130, 680–687.

Danaei, G., Vander Hoom, S., Lopez. A., Murray, C., Ezzati, M., and the Comparative Risk Assessment collaborating group (Cancers). (2005) Causes of cancer in the world: comparative risk assessment of nine behavioural and environmental risk factors. Lancet 366, 1784–1793.

Davidson, M. H., Maki, K. C., Dicklin, M. R., Feinstein, S. B., Witchger, M., Bell, M., McGuire, D. K., Provost, J. C., Liker, H., and Aviram, M. (2009) Effects of consumption of pomegranate juice on carotid intima-media thickness in men and women at moderate risk for coronary heart disease. Am J Cardiol 104, 936–942.

de Souza, M. O., Silva, M., Silva, M. E., Oliveira Rde, P., and Pedrosa, M. L. (2010) Diet supplementation with acai (Euterpe oleracea Mart.) pulp improves biomarkers of oxidative stress and the serum lipid profile in rats. Nutrition 26, 804–810.

Dillard, C. J. and German, J. B. (2000) Phytochemicals: nutraceuticals and human health. J. Sci. Food Agric. 80, 1744–1756.

Drogoudi, P., Michailidis, Z., and Pantelidis, G. (2008) Peel and flesh antioxidant content and harvest quality characteristics of seven apple cultivars. Scientia Hort. 115, 149–153.

Duncan, F. J., Martin, J. R., Wulff, B. C., Stoner, G. D., Tober, K. L., Oberyszyn, T. M., Kusewitt, D. F., and Van Buskirk, A. M. (2009) Topical treatment with black raspberry extract reduces cutaneous UVB-induced carcinogenesis and inflammation. Cancer Prev Res (Phila) 2, 665–672.

Durak, I., Cetin, R., Devrim, E., and Erguder, I. B. (2005) Effects of black grape extract on activities of DNA turn-over enzymes in cancerous and non cancerous human colon tissues. Life Sci 76, 2995–3000.

Fini, L., Piazzi, G., Daoud, Y., Selgrad, M., Maegawa, S., Garcia, M., Fogliano, V., Romano, M., Graziani, G., Vitaglione, P., Carmack, S.W., Gasbarrini, A., Genta, R.M., Issa, J.P., Boland, C.R., and Ricciardiello, L. (2011) Chemoprevention of intestinal polyps in ApcMin/+ mice fed western or balanced diets by drinking Annurca apple polyphenol extract. Cancer Prev Res (Phila). 2011 Mar 7. [Epub ahead of print] PubMed PMID: 21383028.

Finley, J. W. (2005) Bioactive compounds and designer plant foods: The need for clear guidelines to evaluate potential benefits to human health. Chronica Horticulturae 45(3), 6–11.

Freedman, J. E., Parker, C., 3rd, Li, L., Perlman, J. A., Frei, B., Ivanov, V., Deak, L. R., Iafrati, M. D., and Folts, J. D. (2001) Select flavonoids and whole juice from purple grapes inhibit platelet function and enhance nitric oxide release. Circulation 103, 2792–2798.

George, S. M., Park, Y., Leitzmann, M. F., Freedman, N. D., Dowling, E. C., Reedy, J., Schatzkin, A., Hollenbeck, A., and Subar, A. F. (2009) Fruit and vegetable intake and risk of cancer: a prospective cohort study. Am J Clin Nutr 89, 347–353.
Gerhauser, C. (2008) Cancer chemopreventive potential of apples, apple juice, and apple components. Planta Med 74(13), 1608–24.
Gil, M. I., Tomas-Barberan, F. A., Hess-Pierce, B., and Kader, A. A. (2002) Antioxidant capacities, phenolic compounds, carotenoids, and vitamin C contents of nectarine, peach, and plum cultivars from California. J. Agric. Food Chem 50, 4976–4982.
Giongo, L., Ieri, F., Vrhovsek, U., Grisenti, M., Mattivi, F. and Eccher, M. (2006) Characterization of Vaccinium cultivars: Horticultural and antioxidant profile. Acta Hort. 715, 147–151.
Giovanelli, G. and Buratti, S. (2009) Comparison of polyphenolic composition and antioxidant activity of wild Italian blurberries and some cultivated varieties. Food Chem. 112, 903–908.
God, J. M., Tate, P., and Larcom, L. L. (2007) Anticancer effects of four varieties of muscadine grape. J Med Food 10, 54–59.
Gosse, F., Guyot, S., Roussi, S., Lobstein, A., Fischer, B., Seiler, N., and Raul, F. (2005) Chemopreventive properties of apple procyanidins on human colon cancer-derived metastatic SW620 cells and in a rat model of colon carcinogenesis. Carcinogenesis 26, 1291–1295.
Gupta, K., Chou, M. Y., Howell, A., Wobbe, C., Grady, R., and Stapleton, A. E. (2007) Cranberry products inhibit adherence of p-fimbriated Escherichia coli to primary cultured bladder and vaginal epithelial cells. J Urol 177, 2357–2360.
Hadipour-Jahromy, M., and Mozaffari-Kermani, R. (2010) Chondroprotective effects of pomegranate juice on monoiodoacetate-induced osteoarthritis of the knee joint of mice. Phytother Res 24, 182–185.
Harris, G. K., Gupta, A., Nines, R. G., Kresty, L. A., Habib, S. G., Frankel, W. L., LaPerle, K., Gallaher, D. D., Schwartz, S. J., and Stoner, G. D. (2001) Effects of lyophilized black raspberries on azoxymethane-induced colon cancer and 8-hydroxy-2'-deoxyguanosine levels in the Fischer 344 rat. Nutr Cancer 40, 125–133.
Henriquez, C., Almonacid, S., Escobar, B., Chiffelle, I., Gómez, M. and Speisky, H. (2009) Antioxidant content and activity in different structures of five apple cultivars grown in Chile. Acta Hort 841, 275–280.
Hora, J. J., Maydew, E. R., Lansky, E. P., and Dwivedi, C. (2003) Chemopreventive effects of pomegranate seed oil on skin tumor development in CD1 mice. J Med Food 6, 157–161.
Howell, A. B. (2007) Bioactive compounds in cranberries and their role in prevention of urinary tract infections. Molecular Nutrition & Food Research 51, 732–737.
Huang, W. Y., Cai, Y. Z., and Zhang, Y. (2010) Natural phenolic compounds from medicinal herbs and dietary plants: potential use for cancer prevention. Nutr Cancer 62, 1–20.
Hudson, T. S., Hartle, D. K., Hursting, S. D., Nunez, N. P., Wang, T. T. Y., Young, H. A., Arany, P., and Green, J. E. (2007) Inhibition of prostate cancer growth by muscadine grape skin extract and resveratrol through distinct mechanisms. Cancer Research 67, 8396–8405.
Hung, H. C., Joshipura, K. J., Jiang, R., Hu, F. B., Hunter, D., Smith-Warner, S. A., Colditz, G. A., Rosner, B., Spiegelman, D., and Willett, W. C. (2004) Fruit and vegetable intake and risk of major chronic disease. J Natl Cancer Inst 96, 1577–1584.
Ilbey, Y. O., Ozbek, E., Simsek, A., Cekmen, M., Somay, A., and Tasci, A. I. (2009) Effects of pomegranate juice on hyperoxaluria-induced oxidative stress in the rat kidneys. Ren Fail 31, 522–531.
Jaeger, S. and Harker, F. (2005) Consumer evaluation of novel kiwifruit: willingness-to-pay. J. Sci. Food Agric. 85, 2519–2526.
Jang, M., Cai, L., Udeani, G. O., Slowing, K. V., Thomas, C. F., Beecher, C. W., Fong, H. H., Farnsworth, N. R., Kinghorn, A. D., Mehta, R. G., Moon, R. C., and Pezzuto, J. M. (1997) Cancer chemopreventive activity of resveratrol, a natural product derived from grapes. Science 275, 218–220.
Jensen, G. S., Wu, X., Patterson, K. M., Barnes, J., Carter, S. G., Scherwitz, L., Beaman, R., Endres, J. R., and Schauss, A. G. (2008) In vitro and in vivo antioxidant and anti-inflammatory

capacities of an antioxidant-rich fruit and berry juice blend. Results of a pilot and randomized, double-blinded, placebo-controlled, crossover study. J Agric Food Chem 56, 8326–8333.
Johnson, I. T. (2007) Phytochemicals and cancer, Proceedings of the Nutrition Society. 66, 207–215.
Joseph, J. A., Shukitt-Hale, B., Denisova, N. A., Bielinski, D., Martin, A., McEwen, J. J., and Bickford, P. C. (1999) Reversals of age-related declines in neuronal signal transduction, cognitive, and motor behavioral deficits with blueberry, spinach, or strawberry dietary supplementation. J Neurosci 19, 8114–8121.
Kalt, W., Ryan, D., Duy, J., Prior, R., Ehlenfeldt, M. and Vander Kloet, S. (2001) Interspecific variation in anthocyanins, phenolics, and antioxidant capacity among genotypes of highbush and lowbush blueberries (Vaccinium Section cyanococcus spp.) J. Agric. Food Chem. 49, 4761–4767.
Kappel, F. (2008) Breeding cherries in the 'New World'. Acta Hort. 795, 59–69.
Key, T. J. (2011) Fruit and vegetables and cancer risk. Br J Cancer. 104(1), 6–11.
Khanizadeh, S., Tsao, R., Rekika, D., Yang, R., Charles, M. T., and Rupasinghe, H. P. V. (2008) Polyphenol composition and total antioxidant capacity of selected apple genotypes for processing. J. Food Comp. Anal. 21, 396–401.
Khanizadeh, S., Tsao, R., Rekika, D., Yang, R., Charles, M. T., and Rupasinghe, H. P. V. (2009) Advances in fruit breeding in Eastern Canada – Role of phytochemicals in designing specialty fruits. Acta Hort 814, 205–207.
Khan, N., Afaq, F., Kweon, M. H., Kim, K., and Mukhtar, H. (2007) Oral consumption of pomegranate fruit extract inhibits growth and progression of primary lung tumors in mice. Cancer Res 67, 3475–3482.
Kim, H., Hall, P., Smith, M., Kirk, M., Prasain, J. K., Barnes, S., and Grubbs, C. (2004) Chemoprevention by grape seed extract and genistein in carcinogen-induced mammary cancer in rats is diet dependent. J Nutr 134, 3445S–3452S.
Kim, T. J., Weng, W. L., Silva, J. L., Jung, Y. S., and Marshall, D. (2010) Identification of natural antimicrobial substances in red muscadine juice against Cronobacter sakazakii. J Food Sci 75, M150–154.
Kim, T. J., Weng, W. L., Stojanovic, J., Lu, Y., Jung, Y. S., and Silva, J. L. (2008) Antimicrobial effect of water-soluble muscadine seed extracts on Escherichia coli O157:H. J Food Prot 71, 1465–1468.
Koch, T.C., Briviba, K., Watzl, B., Fähndrich, C., Bub, A., Rechkemmer, G., Barth, S.W. (2009) Prevention of colon carcinogenesis by apple juice in vivo: impact of juice constituents and obesity. Mol Nutr Food Res. 53(10), 1289–302.
Kohno, H., Suzuki, R., Yasui, Y., Hosokawa, M., Miyashita, K., and Tanaka, T. (2004) Pomegranate seed oil rich in conjugated linolenic acid suppresses chemically induced colon carcinogenesis in rats. Cancer Sci 95, 481–486.
Konijeti, R., Henning, S., Moro, A., Sheikh, A., Elashoff, D., Shapiro, A., Ku, M., Said, J. W., Heber, D., Cohen, P., and Aronson, W. J. (2010) Chemoprevention of prostate cancer with lycopene in the TRAMP model. Prostate 70, 1547–1554.
Krikorian, R., Nash, T. A., Shidler, M. D., Shukitt-Hale, B., and Joseph, J. A. (2010) Concord grape juice supplementation improves memory function in older adults with mild cognitive impairment. Br J Nutr 103, 730–734.
Lala, G., Malik, M., Zhao, C., He, J., Kwon, Y., Giusti, M. M., and Magnuson, B. A. (2006) Anthocyanin-rich extracts inhibit multiple biomarkers of colon cancer in rats. Nutr Cancer 54, 84–93.
Larsson, S. C., Andersson, S. O., Johansson, J.E., and Wolk, A. (2008) Fruit and Vegetable Consumption and Risk of Bladder Cancer: A Prospective Cohort Study. Cancer Epidemiol Biomarkers Prev 17, 2519–2522.
Lata, B., Przeradzka, M., and Binkowska, M. (2005) Great differences in antioxidant properties exist between 56 apple cultivars and vegetation seasons. J. Agric. Food Chem 53, 8970–8978.
Lata, B. (2008) Apple peel antioxidant status in relation to genotype, storage type and time. Scientia Hort 117, 45–52.
Lata, B., Trampczynska, A., and Paczesna, J. (2009) Cultivar variation in apple peel and whole fruit phenolic composition. Scientia Hort 121, 176–181.

Lee, K. W., Kim, Y. J., Kim, D., Lee, H. J., and Chang, Y. L. (2003) Major phenolics in apple and their contribution to the total antioxidant capacity. J. Agric. Food Chem. 51, 6516–6520.

Leifert, W. R., and Abeywardena, M. Y. (2008) Grape seed and red wine polyphenol extracts inhibit cellular cholesterol uptake, cell proliferation, and 5-lipoxygenase activity. Nutr Res 28, 842–850.

Leiro, J., Arranz, J. A., Fraiz, N., Sanmartin, M. L., Quezada, E., and Orallo, F. (2005) Effect of cis-resveratrol on genes involved in nuclear factor kappa B signaling. Int Immunopharmacol 5, 393–406.

Lewis, J. E., Soler-Vila, H., Clark, P. E., Kresty, L. A., Allen, G. O., and Hu, J. J. (2009) Intake of plant foods and associated nutrients in prostate cancer risk. Nutr Cancer 61, 216–224.

Li, Z. G., Hong, T., Shimada, Y., Komoto, I., Kawabe, A., Ding, Y., Kaganoi, J., Hashimoto, Y., and Imamura, M. (2002) Suppression of N-nitrosomethylbenzylamine (NMBA)-induced esophageal tumorigenesis in F344 rats by resveratrol. Carcinogenesis 23, 1531–1536.

Lippi, G., Franchini, M., Favaloro, E. J., and Targher, G. (2010) Moderate Red Wine Consumption and Cardiovascular Disease Risk: Beyond the "French Paradox". Semin Thromb Hemost 31, 059,070.

Liu, L., Li, Y.H., Niu, Y.B., Sun, Y., Guo, Z.J., Li, Q., Li, C., Feng, J., Cao, S.S., Mei, Q.B. (2010) An apple oligogalactan prevents against inflammation and carcinogenesis by targeting LPS/TLR4/NF-κB pathway in a mouse model of colitis-associated colon cancer. Carcinogenesis 31(10), 1822–32.

Liu, R. H. (2003) Health benefits of fruits and vegetables are from additive and synergistic combinations of phytochemicals. Am.J. Clin. Nutr. 78 (Suppl.), 517s–520s.

Liu, R. H. (2004) Potential synergy of phytochemicals in cancer prevention: mechanism of action. J. Nutr. 134, 3479S–3485S.

Liu, R. H., Liu, J. and Chen, B. (2005) Apples prevent mammary tumors in rats. J. Agric. Food Chem. 53, 2341-2343.

Liu, Z., Schwimer, J., Liu, D., Greenway, F. L., Anthony, C. T., and Woltering, E. A. (2005) Black raspberry extract and fractions contain angiogenesis inhibitors, J Agric Food Chem 53, 3909–3915.

Mallery, S. R., Zwick, J. C., Pei, P., Tong, M., Larsen, P. E., Shumway, B. S., Lu, B., Fields, H. W., Mumper, R. J., and Stoner, G. D. (2008) Topical application of a bioadhesive black raspberry gel modulates gene expression and reduces cyclooxygenase 2 protein in human premalignant oral lesions. Cancer Res 68, 4945–4957.

Martineau, L. C., Couture, A., Spoor, D., Benhaddou-Andaloussi, A., Harris, C., Meddah, B., Leduc, C., Burt, A., Vuong, T., Mai Le, P., Prentki, M., Bennett, S. A., Arnason, J. T., and Haddad, P. S. (2006) Anti-diabetic properties of the Canadian lowbush blueberry Vaccinium angustifolium Ait. Phytomedicine 13, 612–623.

Matchett, M. D., MacKinnon, S. L., Sweeney, M. I., Gottschall-Pass, K. T., and Hurta, R. A. (2005) Blueberry flavonoids inhibit matrix metalloproteinase activity in DU145 human prostate cancer cells. Biochem Cell Biol 83, 637–643.

Mattiello, T., Trifiro, E., Jotti, G. S., and Pulcinelli, F. M. (2009) Effects of pomegranate juice and extract polyphenols on platelet function. J Med Food 12, 334–339.

Mazza, G., and Miniati, E. (1993) Anthocyanins in fruits, vegetables and grains, CRC Press Inc., Boca Raton.

McDougall, G, Dobson, P., Shpiro, F. Smith, P., Stewart, D. and Fyffe, S. (2007) Assessing bioavailablity of soft frut polyphenols in vitro. Acta Hort. 744, 135–148.

McGhie, T., Hunt, M., and Barnet, L. (2005) Cultivar and growing region determine the antioxidant polyphenolic concentration and composition of apples grown in New Zealand. J. Agric. Food Chem. 53, 3065–3070.

Mehta, R., and Lansky, E. P. (2004) Breast cancer chemopreventive properties of pomegranate (Punica granatum) fruit extracts in a mouse mammary organ culture. Eur J Cancer Prev 13, 345–348.

Meyers, K. J., Watkins, C., Pritts, M. and Lu, R. H. (2003) Antioxidant and antiproliferative activities of strawberries. J Agric. Food Chem. 51, 6887–6892.

Milde, J., Eistner, and Graßmann, J. (2007) Synergistic effects of phenolics and carotenoids on human low-density lipoprotein oxidation, Mol. Nutr. Food Res. 51, 956–961.

Montrose, D. C., Horelik, N. A., Madigan, J. P., Stoner, G. D., Wang, L. S., Bruno, R. S., Park, H. J., Giardina, C., and Rosenberg, D. W. (2011) Anti-inflammatory effects of freeze-dried black raspberry powder in ulcerative colitis. Carcinogenesis 32(3), 343–50.

Moyer, R., Hummer, K., Finn, C., Frei, B. and Wrolstad, R. (2002) Anthocyanins, phenolics, and antioxidant capacity in diverse small fruits: Vaccinium, Rubus, and Ribes. J. Agric. Food Chem. 50:519–525.

Nanney, M. S., Schermbeck, R., and Haire-Joshu, D. (2007) Examination of the adherence to the "5 A Day the Color Way" campaign among parents and their preschool children. J Cancer Educ 22, 177–180.

Neto, C. C. (2007) Cranberry and blueberry: evidence for protective effects against cancer and vascular diseases. Mol Nutr Food Res 51, 652–664.

Newman, D. J., and Cragg, G. M. (2007) Natural products as sources of new drugs over the last 25 years. J Nat Prod 70, 461–477.

Olien, W. C. (1990) Muscadine: A classic southeastern fruit. Hortscience 25, 726–831.

Orak, H. H. (2009) Evaluation of antioxidant activity, colour and some nutritional characteristics of pomegranate (Punica granatum L.) juice and its sour concentrate processed by conventional evaporation. International Journal of Food Sciences and Nutrition 60, 1–11.

Pan, M. H., Ghai, G., and Ho, C. T. (2008) Food bioactives, apoptosis, and cancer. Mol Nutr Food Res 52, 43–52.

Pantuck, A. J., Leppert, J. T., Zomorodian, N., Aronson, W., Hong, J., Barnard, R. J., Seeram, N., Liker, H., Wang, H., Elashoff, R., Heber, D., Aviram, M., Ignarro, L., and Belldegrun, A. (2006) Phase II study of pomegranate juice for men with rising prostate-specific antigen following surgery or radiation for prostate cancer. Clin Cancer Res 12, 4018–4026.

Pollack, S. and A. Perez. 2008. Fruit and Tree Nuts Situation and Outlook Yearbook 2008. Market and Trade Economics Division, Economic Research Service, U. S. Department of Agriculture, October 2008, FTS-2008.

Prior, R., Cao, G., Martin, A., Sofic, E., McEwen, J., O'Brien, C., Lischer, N., Ehlenfeldt, M., Kalt, W., Krewer, G., and Mainland, C. M. (1998) Antioxidant capacity as influenced by total phenolic and anthocyanin content, maturity, and variety of Vaccinium species. J. Agric. Food Chem. 46, 2886–2693.

Raina, K., Singh, R. P., Agarwal, R., and Agarwal, C. (2007) Oral grape seed extract inhibits prostate tumor growth and progression in TRAMP mice. Cancer Res 67, 5976–5982.

Rajakangas, J., Misikangas, M., Paivarinta, E., and Mutanen, M. (2008) Chemoprevention by white currant is mediated by the reduction of nuclear beta-catenin and NF-kappaB levels in Min mice adenomas. Eur J Nutr 47, 115–122.

Rao, A. V., and Snyder, D. M. (2010) Raspberries and human health: a review, J Agric Food Chem 58, 3871–3883.

Remberg, S., Mage, F., Haffner, K. and Blomhoff, R. (2007) Highbush blueberries Vaccinium corymbosum L., raspberries Rubus idaeus L. and black currants Ribes nigrum L. - influence of cultivars on antioxidant activity and other quality parameters. Acta Hort. 744, 259–265.

Renaud, S., and de Lorgeril, M. (1992) Wine, alcohol, platelets, and the French paradox for coronary heart disease. The Lancet 339, 1523–1526.

Rice-Evans, C. A., Miller, N. J., and Paganga, G. (1996) Structure-antioxidant activity relationships of flavonoids and phenolic acids. Free Radic Biol Med 20, 933–956.

Rommel, A., and Wrolstad, R. E. (1993) Ellagic acid content of red raspberry juice as influenced by cultivar, processing, and environmental factors. Journal of Agricultural and Food Chemistry 41, 1951–1960.

Ruel, G., and Couillard, C. (2007) Evidences of the cardioprotective potential of fruits: the case of cranberries. Mol Nutr Food Res 51, 692–701.

Sautebin, L., Rossi, A., Serraino, I., Dugo, P., Di Paola, R., Mondello, L., Genovese, T., Britti, D., Peli, A., Dugo, G., Caputi, A. P., and Cuzzocrea, S. (2004) Effect of anthocyanins contained in

a blackberry extract on the circulatory failure and multiple organ dysfunction caused by endotoxin in the rat. Planta Med 70, 745–752.
Schauss, A. G., Wu, X., Prior, R. L., Ou, B., Patel, D., Huang, D., and Kababick, J. P. (2006) Phytochemical and nutrient composition of the freeze-dried Amazonian palm berry, Euterpe oleraceae Mart. (Acai), Journal of Agricultural and Food Chemistry 54, 8598–8603.
Sekido, K., Hayashi, Y., Yamada, K., Shiratake, K., Matsumoto, S., Macjima, T., and Komatsu, H. (2010) Efficient breeding system for red-fleshed apple based on linkage with S3-RNase allele in 'Pink Pearl'. HortScience 45, 534–537.
Sengottuvelan, M., Viswanathan, P., and Nalini, N. (2006) Chemopreventive effect of trans-resveratrol--a phytoalexin against colonic aberrant crypt foci and cell proliferation in 1,2-dimethylhydrazine induced colon carcinogenesis. Carcinogenesis 27, 1038–1046.
Shukitt-Hale, B., Carey, A., Simon, L., Mark, D. A., and Joseph, J. A. (2006) Effects of Concord grape juice on cognitive and motor deficits in aging. Nutrition 22, 295–302.
Shukitt-Hale, B., Cheng, V., and Joseph, J. A. (2009) Effects of blackberries on motor and cognitive function in aged rats. Nutr Neurosci 12, 135–140.
Shukitt-Hale, B., Lau, F. C., and Joseph, J. A. (2008) Berry fruit supplementation and the aging brain. J Agric Food Chem 56, 636–641.
Sloan, E. (2006) Top 10 functional food trends. FoodTechnology 04.06, 23–34.
Sloan, E. (2008) The top 10 functional food trends. FoodTechnology 04.08, 25–35.
Stacewicz-Sapuntzakis, M., Bowen, P.E., Hussain, E.A., Damayanti-Wood, B.I., and Farnsworth, N.R. (2001)Chemical composition and potential health effects of prunes: a functional food? Crit Rev Food Sci Nutr.41(4), 251–86.
Stoner, G. D., Dombkowski, A. A., Reen, R. K., Cukovic, D., Salagrama, S., Wang, L. S., and Lechner, J. F. (2008) Carcinogen-altered genes in rat esophagus positively modulated to normal levels of expression by both black raspberries and phenylethyl isothiocyanate. Cancer Res 68, 6460–6467.
Stoner, G. D., Wang, L. S., Seguin, C., Rocha, C., Stoner, K., Chiu, S., and Kinghorn, A. D. (2010) Multiple berry types prevent N-nitrosomethylbenzylamine-induced esophageal cancer in rats. Pharm Res 27, 1138–1145.
Sun, J., Chu, Y. -F., Wu, X. and Liu, R. H. (2002) Antioxidant and antiproliferative activities of common fruits. J Agric Food Chem 50, 7449–7454.
Sun, G. Y., Xia, J., Draczynska-Lusiak, B., Simonyi, A., and Sun, A. Y. (1999) Grape polyphenols protect neurodegenerative changes induced by chronic ethanol administration. Neuroreport 10, 93–96.
Takachi, R., Inoue, M., Ishihara, J., Kurahashi, N., Iwasaki, M., Sasazuki, S., Iso, H., Tsubono, Y., and Tsugane, S. (2008) Fruit and vegetable intake and risk of total cancer and cardiovascular disease: Japan Public Health Center-Based Prospective Study. Am J Epidemiol 167, 59–70.
Tate, P., God, J., Bibb, R., Lu, Q., and Larcom, L. L. (2004) Inhibition of metalloproteinase activity by fruit extracts. Cancer Lett 212, 153–158.
Thaipong, K., Boonprakob, U., Crosby, K., Cisneros-Zevallos, L., and Byrne, D. H. (2006) Comparison of ABTS, DPPH, FRAP, and ORAC assays fro estimating antioxidant activity from guava fruit extracts. J. Food Composition and Analysis 19, 669–675.
Thompson, M., Stushnoff, C., McGinley, J., and Thompson, H. (2009) In vitro measures used to predict anticancer activity of apple cultivars and their comparison to outcomes from a rat model of experimentally induced breast cancer. Nutrition and Cancer 61, 510–517.
Tomas-Barberan, F. A., Gil, M. I., Cremin, P., Waterhouse, A. L., Hess-Pierce, B., and Kader, A. A. (2001) HPLC-DAD-ESIMS analysis of phenolic compounds in nectarines, peaches, and plums. J Agric Food Chem 49, 4748–4760.
Tovar, A. M., Cesar, D. C., Leta, G. C., and Mourao, P. A. (1998) Age-related changes in populations of aortic glycosaminoglycans: species with low affinity for plasma low-density lipoproteins, and not species with high affinity, are preferentially affected. Arterioscler Thromb Vasc Biol 18, 604–614.

Tsao, R., Yang, R., Younf, J. C., and Zhu, H. (2003) Polephenolic profiles in eight apple cultivars using high-performance chromatography (HPLC). J Agric. Food Chem. 51, 6347–6353.

Vainio, H., and Weiderpass, E. (2006) Fruit and vegetables in cancer prevention, Nutr Cancer 54, 111–142.

Vieira, F., Borges, G., Copetti, C., Amboni, R. Denardi, F., and Fett, R. (2009) Physico-chemical and antioxidant properties of six apple cultivars (Malus domestica Borkh) grown in southern Brazil. Scientia Hort. 122, 421–425.

Vieira, F., Borges, G., Copetti, C., Di Pietro, P. Nunes, E., and Fett, R. (2011) Phenolic compounds and antioxidant activity of the apple flesh and peel of eleven cultivars grown in Brazil. Scientia Hort. 128, 261–266.

Vizzotto, M., Cisneros, L., Okie, W. R., Ramming, D. W., and Byrne, D. H. (2007) Large variation found in the phytochemical content and antioxidant activity of peach and plum germplasm. J. Amer. Soc. Hort. Sci., 132: 334–340.

Volz, R., Oraguzie, N., Whitworth, C., How, N. Change, D., Carlisle, C., Gardiner, S., Rikkerink, E., and Lawerence, T. (2009) Breeding for red flesh colour in apple: progress and challenges. Acta Hort. 841:337–342.

Vorsa, N. and Polashock, J. (2005) Alteration of anthocyanin glycosylation in cranberry through interpsecific hybridization. J. Amer. Soc. Hort. Sc. 130, 711–715.

Wells, H. F., and Buzby, J. C. (2008) Dietary assessment of major trends in U. S. food consumption, 1970–2005. Economic Information Bulletin No. 33 Economic Research Service, U. S. Dept. of Agriculture.

Weisel, T., Baum, M., Eisenbrand, G., Dietrich, H., Will, F., Stockis, J. P., Kulling, S., Rufer, C., Johannes, C., and Janzowski, C. (2006) An anthocyanin/polyphenolic-rich fruit juice reduces oxidative DNA damage and increases glutathione level in healthy probands. Biotechnol J 1, 388–397.

Werner, D. J., Creller, M. A., and Chaparro, J. X. (1997) Inheritance of blood flesh in peach. HortScience 33, 1243–1246.

Wojdylo, A., Osmianski, J., and Laskowski, P. (2008) Polyphenolic compounds and antioxidant activity of new and old apple varieties. J. Agric. Food Chem. 56, 6520–6530.

Wolfe, K., Wu, X., and Lu, R. H. (2003) Antioxidant activity of apple peels. J. Agric. Food Chem. 51, 609–614.

Woo, J. H., Lim, J. H., Kim, Y. H., Suh, S. I., Min, D. S., Chang, J. S., Lee, Y. H., Park, J. W., and Kwon, T. K. (2004) Resveratrol inhibits phorbol myristate acetate-induced matrix metalloproteinase-9 expression by inhibiting JNK and PKC delta signal transduction. Oncogene 23, 1845–1853.

World Health Organization. (2005) Fruits and Vegetables for Health: Report of a Joint FAO/WHO Workshop.

Wu, X., Beecher, G. R., Holden, J. M., Haytowitz, D. B., Gebhardt, S. E., and Prior, R. L. (2006) Concentrations of anthocyanins in common foods in the United States and estimation of normal consumption. Journal of Agricultural and Food Chemistry 54, 4069–4075.

Wung, B. S., Hsu, M. C., Wu, C. C., and Hsieh, C. W. (2005) Resveratrol suppresses IL-6-induced ICAM-1 gene expression in endothelial cells: effects on the inhibition of STAT3 phosphorylation. Life Sci 78, 389–397.

Yang, Y., and Gallaher, D.D. (2005) Effect of dried plums on colon cancer risk factors in rats. Nutr Cancer 53(1), 117–25.

Yi, W., Fischer, J., Krewer, G., and Akoh, C. (2005) Phenolic compounds from blueberries can inhibit colon cancer cell proliferation and induce apoptosis. J. Agric. Food Chem. 53, 7320–7329.

Yoshizawa, Y., Sakurai, K., Kawaii, S., Asari, M., Soejima, J., ans Murofushi, N. (2005) Comparison of antiproliferative and antioxidant properties among nineteen apple cultivars. HortScience 40, 5, 1204–1207.

Youdim, K. A., McDonald, J., Kalt, W., and Joseph, J. A. (2002) Potential role of dietary flavonoids in reducing microvascular endothelium vulnerability to oxidative and inflammatory insults. J Nutr Biochem 13, 282–288.

Zhao, J., Wang, J., Chen, Y., and Agarwal, R. (1999) Anti-tumor-promoting activity of a polyphenolic fraction isolated from grape seeds in the mouse skin two-stage initiation-promotion protocol and identification of procyanidin B5-3'-gallate as the most effective antioxidant constituent. Carcinogenesis 20, 1737–1745.

Zheng, W. and Wang, S. (2003) Oxygen radical absorbing capacity of phenolics in blueberries, cranberries, chokeberries, and lingonberries. J Agric Food Chem 51, 502–509.

Chapter 3
Intellectual Property Protection and Marketing of New Fruit Cultivars

John R. Clark, Amelie Brazelton Aust, and Robert Jondle

Abstract The most common international protection offered for fruit cultivars is plant breeder's rights (PB rights). The main international intergovernmental regulatory institution which provides for and promotes an international system of plant variety protection is the International Union for the Protection of New Varieties of Plants (UPOV). The UPOV Convention was first written in 1961 and subsequently modified in 1978 and 1991. The intention of the UPOV system is to ensure that germplasm sources such as protected varieties remain accessible to plant breeders. Plant breeder's rights usually include protection of the variety for not less than 20 years from the date of the grant, or 25 years for trees or vines and depend on which act of the UPOV Convention a country follows. In the USA, plant patents are used to protect clonally propagated cultivars of plants. One of the newest movements in intellectual property is the integration of trademarks into the plant protection and commercialization strategy for a new variety. There is also the license agreement which is the vehicle that grants nonowners access to the intellectual property at hand, whether it be PB rights, patent protection, or the use of a trademark. With increased intellectual property issues in fruit breeding, options are being examined concerning the sharing of germplasm for testing and/or breeding. Breeding agreements including public-to-public and public-to-private options are expanding. Marketing and commercialization for

J.R. Clark (✉)
Department of Horticulture, University of Arkansas, 316 Plant Science, Fayetteville, AR 72701-1201, USA
e-mail: jrclark@uark.edu

A.B. Aust
Fall Creek Farm and Nursery Inc., 39318 Jasper-Lowell Road, Lowell, OR 97452, USA
e-mail: ameliea@fallcreeknursery.com

R. Jondle
Jondle and Associates, P.C, Suite 230, 858 Happy Canyon Road, Castle Rock, CO 80108, USA
e-mail: rjondle@jondlelaw.com

M.L. Badenes and D.H. Byrne (eds.), *Fruit Breeding*, Handbook of Plant Breeding 8,
DOI 10.1007/978-1-4419-0763-9_3,

some fruit crops have become much more complex with territorial marketing, club models, and closed commercial systems becoming more common.

Keywords Plant patents • Trademark • Plant breeding rights • UPOV • Plant patent • Plant breeders rights • International Union for the Protection of New Varieties of Plants • Plant Variety Protection • Trademark • Germplasm sharing • Material transfer agreement • Cultivar marketing

1 Introduction

In recent decades, the integration and importance of intellectual property rights (IP rights) have become among the foremost important developments in fruit breeding. This has come about on the back of significant leaps in developments in breeding and genetics, and the increased sophistication of marketing and commercialization schemes based on new plant varieties. IP rights are varied and far-reaching in the world. From utility patents to trademarks, from sui generis plant breeder's rights to plant patents, proprietary protection can be granted for varieties, Cultivar names and trademarks, processes for breeding, genes, and other inventions, depending on the law of the country in question.

There are many reasons why IP rights have become a key factor in the fruit industry's growth. First, IP rights offer the owner of an invention sole proprietorship and the ability to collect royalty payments for the use of the invention. The development of new varieties, done at both a private and public level, requires substantial funding, and IP rights have played key roles in financing these breeding programs. IP rights also offer control to owners and exclusive licensees who benefit from offering something unique to the marketplace, therefore raising demand for the product through limited supply, and in effect, price margins. The exponential increase in the use of IP rights in the fruit industry has created substantial research incentive in private and public spheres to continue advancements in fruit innovation.

The following discussion is for all who are interested in the major types of IP rights regarding fruit crops, along with neighboring topics such as licensing, germplasm sharing, and marketing/commercialization. This information is not intended to be used as legal advice for IP rights protection; legal counsel should be consulted for more detailed information and procedures.

2 Protection Options

2.1 Plant Breeder's Rights

The most common international protection offered for varieties is termed plant breeder's rights (PB rights). The term "plant breeder's rights" is not used in the USA, but PB rights is similar to US Plant Variety Protection. There are international levels of regulation for these types of plant protection, the most prominent being the International Union for the Protection of New Varieties of Plants (UPOV—Union internationale pour la protection des obtentions vègètales). It was established by the UPOV Convention in 1961 with additional acts of the Union in 1978 and 1991. UPOV is an intergovernmental entity that was established to provide for and promote an international system of plant variety protection. UPOV's mission is to "to provide and promote an effective system of plant variety protection, with the aim of encouraging the development of new varieties of plants, for the benefit of society" (UPOV 2008). Both seed and clonally propagated crops are covered by UPOV PB rights.

The UPOV Convention codifies certain standards for IP rights requirements and criteria for plant breeders. Unlike the USA, Australia, and Japan, which allow for the patenting of plants, most countries around the world have a sui generis form of PB rights, which is a term used to describe the country-specific hybrid systems of plant protection and breeders rights. Those countries that are members of UPOV are required to adhere to minimum standards of protection criteria, such as the requirement that a variety be new, distinct, uniform, and stable. These standards have greatly harmonized the plant protection processes around the world.

The original UPOV convention was signed in 1961, but the majority of countries that are UPOV members are parties to the 1978 or 1991 UPOV Conventions. There are several important differences between the 1978 and the 1991 UPOV Conventions, some of which include the following:

1. The protection scope under the 1978 Convention is for the production for purposes of commercial marketing, offering for sale, and marketing of propagated material of a protected variety. The 1991 Convention increased the scope of protection to include production or reproduction, conditioning for the purpose of propagation, exporting, importing, and stocking.
2. Under the 1978 Convention, breeders are free to use a protected variety to develop a new variety, but not if the use requires repeated use of the variety. Conversely, under the 1991 Convention the previous exemption is restricted and, among other provisions, a protected variety is not allowed to be used to produce varieties which are essentially derived from a protected variety or which are not distinguishable from the parent variety.
3. The scope of protection under the 1991 Convention can extend to harvested material—thus, only authorized propagation of the variety allows for fruit or other products to be marketed in a territory where there is protection. Also, protection can be extended to products made directly from the harvested material.

Another breakthrough in the development of international IP rights standards was the Agreement on Trade Related Aspects of IP Rights (TRIPS). Administered by the World Trade Organization, TRIPS was signed in 1994. In essence, TRIPS sets minimum required standards of protections for all forms of intellectual property, in particular for copyrights, patents, and, in Article 27(3)(b), for plant varieties either by way of a sui generis system, a plant patent system, or a combination thereof (World Trade Organization 2008). This requirement led to a substantial expansion in the number of countries that have put variety protection and PB rights into place, as can be seen by the jump in the number of signees to the UPOV Convention, which has more than doubled to 68 members since 1994 (UPOV 2010).

While the requirements for the granting of PB rights in the UPOV system varies from country to country, the minimum requirements have similarities to the US Plant Variety Protection Act in that a variety must be new, distinct, uniform, and stable. The novelty requirement for PB rights requires that a variety, at the minimum, must not have been sold or otherwise disposed of in the territory of the country concerned (i.e., the country where the protection is sought) for more than 1 year prior to application for the right, or more than 4 years (or 6 years for trees and vines) in a country other than that of the member of the Union in which the application was filed.

The 1991 Convention provides that the breeder's rights protection for varieties includes the right to exclude unauthorized entities from the following:

- Propagation
- Conditioning for the purpose of propagation
- Offering for sale
- Selling or other marketing
- Exporting
- Importing
- Stocking for any of the purposes listed above

Plant breeder's rights do not restrict breeding activity with a protected variety, nor do other experimental or private/noncommercial uses. The intention of the UPOV system is to ensure that germplasm sources such as protected varieties remain accessible by plant breeders. Plant breeder's rights usually include protection of the variety for not less than 20 years from the date of the grant, or 25 years for trees or vines, and depend on which act of the UPOV Convention a country has adhered.

A major component of UPOV internationally is to provide for cooperation among UPOV member countries in the use and approval of variety names and examination of new varieties. The concept allows for one member to conduct an examination of a variety for potential granting of PB rights, and another member can choose to accept the evaluations for its grant. The intention is that this system can reduce the cost and complication of attaining protection when applying in multiple countries and territories. Each member defines the criteria for granting, whether this includes a technical description of the new variety, or actual growing of the plants for examination within UPOV standard guidelines.

The European Union (EU) provides a good example of a system administered by a group of UPOV member states whereby PB rights are granted on a territory-wide basis. Based in Angers, France, The "Community Plant Variety Office" has been operating since 1995. This system allows for protection for numerous countries in one application. The application protocol includes an application form, technical questionnaire, proposed variety name, and photographs. Applicants with residence outside the EU are required to appoint a procedural representative residing in the EU for filing. Filing for PB rights in the EU requires that plant material be submitted by the breeder, and the variety is then grown for examination in a selected site with other candidates and/or standard varieties for the EU. Fees include a filing/application fee, in addition to an examination fee for the growing period of the evaluation along with annual fees for the duration of the protection period.

The EU provides guidelines for distinctness, uniformity, and stability tests which include details of about the plant health status of the material submitted along with botanical description guidelines. For the blackberry, as an example, guideline characteristics such as growth habit, number of new canes emerged, dormant cane length, dormant cane diameter, cane branch number, presence and density of spines, leaf characteristics, and several flower and fruit characteristics must be described along with comparisons to standard varieties.

Below are examples of additional PB rights protection distinctions in various countries. Please note that the plant species protected in each country vary, as well as standards regarding distinctiveness, uniformity, and stability, along with the term of the protection and other important aspects. Because of the differences among countries, it is recommended to discuss protection with legal counsel before applying for PB rights in any country.

Australia is a member of the 1991 UPOV Convention and may grant protection for a variety if the variety has a breeder, is distinct, uniform, and stable and has not been exploited or has only been recently exploited (Australian Government 2008b). In regard to the term recently exploited, an application for PB rights must be filed within 1 year of sale of the variety in Australia and within 6 years from sales outside of Australia for trees or vines or 4 years for sales outside of Australia for all other species. The term of protection for trees and vines is 25 and 20 years for all other species.

Canada is a member of the 1978 UPOV Convention and grants a term of 18 years for protected varieties (Canadian Food Inspection Agency 2007). A plant variety may not have been for sale in Canada prior to the filing of a PB rights application and must be filed within 6 years for woody plant varieties and their rootstock for those varieties which have been sold outside of Canada and within 4 years for all other varieties. In order to claim the priority of an application previously filed in another country, the applicant must file an application in Canada within 1 year from the date when the application was originally filed in the UPOV-member country.

Chile is a member of the 1978 UPOV Convention and provides PB rights protection for all botanical genera and species (UPOV 1997a). Protection is applied to the complete plant, including "flowers, fruit, and seed or any part thereof that may be used as propagating material". The term of protection in Chile is 18 years for trees and vines and 15 years for other species. A PB rights application must be applied for in Chile within 1 year of the date of the first sale of the plant variety or propagating material in Chile or within 6 years for any sales outside of Chile. A "variety" does not include the sale of fruit but rather only plants. Where protection of a variety has been applied for in another country, the applicant has 1 year following the filing date in the country of origin for filing an application in Chile. For the application process to begin in Chile, specimens do not have to be in Chile at the time of filing. After the initial application fee has been paid the applicant is also be responsible for the cost of annual maintenance of the variety.

Mexico is also a member of the 1978 UPOV Convention (UPOV 1997b). The term of protection in Mexico is 18 years for perennial species (forest and fruit trees, vines, and ornamentals) and their rootstock, and 15 years for all other species. A PB rights application must be filed in Mexico within 1 year of the plant variety or propagating material being sold in Mexico or within 6 years of the plant variety or propagating material being sold abroad. "Propagating material" does include the sale of fruit. For priority to be claimed to a prior PB rights application that was filed in another country, an application must be filed in Mexico within 1 year of the filing date of the original application. Finally, specimens of the variety do not have to be within the country at the time of filing.

The Republic of South Africa (RSA) is also a member of the 1978 UPOV Convention and provides protection to various plant varieties (UPOV 1997c). The varieties protected under the PB rights of the RSA vary and are limited to certain species. The term of protection is 25 years for vines and trees and 20 years for all other varieties. An application for PB rights should be made within 1 year of any sales of propagating material or harvested material of the variety in the RSA, and within 6 years of sales of vines or tree varieties or 4 years for other species outside of the country.

As more and more fruit genotypes are used on a worldwide basis, familiarity with the UPOV system is important to provide for widespread protection. Costs, timing, choosing of commercial cooperators, and targeted countries for filing must all be considered when planning for broad protection.

2.2 *Plant Patents*

In the USA, one form of protection for a new plant variety is a plant patent. The patenting of plants that can be asexually reproduced has been allowed in the USA since the US Congress passed the Townsend-Purnell Plant Patent Act in 1930 ("Act"). According to the Act, "Whoever invents or discovers and asexually reproduces any distinct and new variety of plant, including cultivated sports, mutants, hybrids, and newly found seedlings, other than a tuber-propagated plant or a plant found in an uncultivated state, may obtain a patent therefore, subject to the conditions and requirements of this title" (US Patent and Trademark Office 2007a). The 1998 amendment to the Act added that "in the case of a plant patent, the grant shall include the right to exclude others from asexually reproducing the plant, and from using, offering for sale, or selling the plant so reproduced, or any parts thereof, into the US" (US Patent and Trademark Office 2007b). This not only emphasized that plant patent protection extended to asexually propagated plants, but also that "parts" of the plants were protected as well, which has important implications for fruit-bearing crops. The 1998 amendment to the Act also put provisions into place restricting the importation of plant parts into the USA. This applies not only to fruits from proprietary plant varieties but also to flowers and leaves. Inherent in this provision is that fruit harvested from plants illegally propagated and/or grown outside of the USA is subject to US plant patent law once it reaches the US border, therefore warranting confiscation. One aspect of "plant parts" is still a hotly debated topic—does this include just fruits, leaves, flowers, and other tissues, or also the gametes and the genome of the protected variety? The answer to this has major implications for possible restrictions on breeding rights concerning proprietary varieties in the USA.

The number of annual grants of plant patents in the USA since 1930 has increased substantially, from 362 in 1996 to 1,067 in 2007. The number of issued plant patents for fruit varieties has also significantly increased, adding up to 50 to 100 fruit patents per year between 1990 and 2007. A range of 7–13% of all plant patents between 2000 and 2007 were for fruit varieties (information gathered from searching http://www.uspto.gov). The passing of the Bayh-Dole Act in 1980, which gave universities in the USA the opportunity to protect inventions that were aided by federal government-funded research, has contributed to this increased filing by these organizations.

There are a handful of essential criteria that a new variety must have to be eligible for a plant patent. This includes: novelty (the variety must be new), utility (the variety must be useful in some way), and nonobviousness (the variety must not be obvious to one skilled in the relevant art). For a complete list of guidelines on filing a plant patent, see the US Patent and Trademark Office (USPTO) Web site at http://www.uspto.gov. Although individual inventors are allowed to complete and file a plant patent application, legal counsel is generally sought for the process.

Once a plant patent is granted, a number of rights are conferred on the owner. First, the term of patent protection is 20 years from the date of the filing of the application. In the application itself, only one claim is allowed, and that must be for the variety. Any inventor may file a plant patent application, regardless of country of citizenship. Such a patent only grants rights within the USA. Lastly, the cost of a plant patent application is generally much less than other forms of protection, and the examination is usually significantly quicker and simpler.

One consideration that practitioners should take into account is that the application requirements, especially the gathering of variety botanical information, take time and planning. Collecting the botanical information that must be submitted with the plant patent application usually requires an entire growing season, thus 1 year or near that. The description might include, for example, descriptive information about canes, branches, and/or buds during dormancy (size, color, surface characteristics, etc.), budbreak and bloom characteristics in the spring, fruit, shoot, and leaf characters early to mid-summer, fruit characteristics at ripening or maturity (color of skin and flesh, size, shape, flavor, and other qualities), and possibly late-season to beginning dormancy observations (fall leaf color or other distinct features during this period). Many issues can arise during this data-collection process, such as lack of labor available to collect data and potential environmental issues that impact the plants (i.e., disease, weather, etc.). Although there is no standard list of botanical characteristics required by the USPTO for each species, helpful resources are recently granted plant patents for the same species and UPOV guidelines for required botanical information.

It is also important to keep in mind that preparing the application should be carefully timed with the first sale or public offer of the variety, because once this occurs, the clock begins to tick for the patent's novelty requirement. In order for a new variety to be considered "novel" by the USPTO, the variety may not have been sold, made publicly available, or offered for sale in the USA more than 1 year before the date of filing the application. This is also the case if the variety is sold, publicly made available, or described in any other country in the world more than 1 year before the US plant patent filing date. These stringent rules on novelty require very careful actions by plant breeders who wish to inform the industry about their efforts or even enter into trialing agreements inside or outside of the USA. Therefore, it is recommended that all plant material made available for testing or propagation before patent filing be accompanied by a testing and/or confidentiality agreement.

2.3 Utility Patent

The term "utility patent" is used in the USA to describe the patent which applies to any useful, new, and nonobvious invention, as opposed to a plant patent, which is a patent particularly for plant varieties. However, utility patents generally confer more rights to an owner than a plant patent, making them more desirable if an owner or inventor desires supplemental or stronger protection. Although a utility patent's term is also 20 years from the date of application, similar to a plant patent, a utility patent generally has many claims, whereas the plant patent application can only have one claim (which is the variety itself). Like a plant patent, a US utility patent also requires that filing occurs within 1 year of first sale or disclosure of the variety or invention. Furthermore, unlike the filing of a PB rights application, a lawyer is required for the filing of a utility patent because of the increased requirements inherent in the application. In addition, a utility patent application can be significantly more expensive to prepare and file than a plant patent or PB rights application.

Utility patents can claim a wide range of inventions such as DNA, pollen, genes, promoters, selectable markers, quantitative trait loci, expressed sequence tags, software, proteins, biological methods, genomes, bioinformatics, and more. Claims also might include a plant variety, an improved method or process for breeding or genetic testing, a new trait (such as resistance to a certain chemical or a disease), or a heightened level of such a trait (such as a higher amount of antioxidants).

Patent protection for seed-propagated plant varieties is also available in Australia and Japan. Both countries have requirements for variety patents that are distinct from those in the USA.

In Australia a utility patent provides protection of the invention for up to 20 years, but unlike the USA, the novelty requirement requires absolute novelty (Australian Government 2008a). This means that the invention may not have been for sale or publicly disclosed anywhere in the world before the time the application was filed. This is significantly different from the novelty requirement of the USA, which allows for an application to be filed within 1 year of sale or public disclosure of the variety/invention.

Japan provides a patent protection term of 20 years for biological patents and as Australia requires absolute novelty (Japan Patent Office 2007). But unlike Australia, there is a 6-month grace period in certain disclosure situations. Patent protection in Japan and Australia is similar to that in the USA and provides stronger protection than that of PB rights. For information regarding filing a patent application in Japan or Australia, please obtain legal counsel.

2.4 *Plant Variety Protection*

The USA is a member of the 1991 UPOV Convention and provides protection for sexually reproduced crops and tuber crops through a system known as plant variety protection (PVP). Plant variety protection provides protection for 20 years for most crops and 25 years for trees, shrubs, and vines. Since fruit crops are normally clonally propagated, this form is usually not an option. If seeds are a common plant part used for propagation then PVP protection might be considered. An example would be a peach or other species rootstock, where seed propagation may be the method of propagation. This type of protection was attained for peach rootstock BY520-9 (PVP 9400013), and also a seed-propagated peach Truegold (PVP 200400055).

For a plant to be granted PVP protection in the USA, the plant must be stable, uniform, and distinct. Furthermore, the plant must not have been for sale in the USA for more than 1 year before the filing date of the application or more than 4 years outside of the USA.

There are two distinct differences between the PVP and utility patent protection. With regard to a PVP, a third party may conduct research on the protected variety (such as breeding) and a farmer can legally save seeds of the protected variety for on-farm use whereas with a utility patent this is not true. For more details on this type of protection, see the Web site for the US Plant Variety Protection office at http://www.ams.usda.gov/science/PVPO/PVPindex.htm.

2.5 *Trademarks*

One of the newest movements in intellectual property concerning plant varieties is the integration of trademarks into the plant protection and commercialization strategy for a new variety. A trademark is a word, symbol, or device, which distinguishes an entity's goods or services in the marketplace, serving as an indicator of source of those goods or services while also distinguishing those from the goods and services of others. Trademarks are often seen as both a protector of traders (from misappropriation of a mark by other traders) and of consumers (to reduce confusion and add transparency as to the origin and quality of goods). Rights conferred to a trademark owner include, in particular, the right to prevent others from using the same or a confusingly similar mark for the same goods or services. Duration of protection in both the USA and in the EU is 10 years from the date of registration of the mark; the trademark term may be indefinitely extended and renewed, as long as certain use and maintenance requirements are met. Initial requirements for protection, though differing slightly from country to country, generally include the following:

1. A mark may not be registered if there is another identical mark, or confusingly similar mark, for the same goods or services, in use by someone else.
2. The trademark being applied for must be distinctive for the goods or services it will be used with, not descriptive or generic. If the mark is descriptive, then it must have achieved "secondary meaning" for those goods or services. For example, attempting to register the trademark "Bright Light" for light bulbs would arguably be refused registration on the grounds that it is descriptive of the goods it will distinguish, and that these words should remain in the public domain for competitors needing to describe their goods (i.e. light bulbs). However, if a company can show over time that consumers have come to associate "Bright Light" as the mark of the company, not as a mere description of the product, then there might be a chance for registration based on the argument that the mark has achieved "secondary meaning".
3. The trademark must be put to use in commerce. Depending on the country, this use must commence either before or after the trademark registration. In the USA, the mark must be used in interstate trade to be federally registered. In order to obtain an application priority date before commencing use, there is the option of placing an "intent to use" application for the mark. This is especially useful when planning the launch of a significant marketing campaign because it is possible to file and get a priority date for the mark before it makes contact with the public. When a mark is filed as "intent to use", the mark must be shown to be used in commerce within 6 months of receipt of the Trademark's Notice of Allowance; if use of the mark cannot be shown within the required 6 months, an extension of time can be filed. More details about the application process in the USA can be found at http://www.uspto.gov. In other regions, such as in the EU, a mark can be registered before use has taken place. The use requirements for a community-wide trademark specify that a trademark must commence "genuine" use somewhere in the EU within 5 years after registration. More details about registering a community trademark in the EU can be found at http://www.oami.europa.eu.

It is important to note that trademark protection does not equate to extended patent or PB rights protection. It does not in any way restrict propagation or use of plant parts, or any other use of plant material. Because of the infinite nature of trademark protection, courts are generally very careful to make sure trademark protection does not go beyond its core function of protecting the mark or name.

One significant difference in trademark law between many countries around the world is whether protection is offered to a mark once it has been used in commerce, or whether there is no protection offered until a mark has officially been filed and registered. One of the most famous use-based systems is in the USA. Though the extent of protection offered depends on the extent of use, rights to a trademark can exist simply by using the mark in commerce. At this point, if use-based or "common law" trademark rights are being claimed, then the TM symbol should be used with the mark. If this protection is given automatically in the USA, why register a mark? Federally registering a trademark at the USPTO offers additional protection benefits, such as the presumption of a mark's validity if the question is brought to court, the ability to register the mark with the US Customs Service for better import/export monitoring, and easier international filing through the Madrid Protocol.

Trademark use in relation to plant varieties varies widely. The most common and traditional trademark strategy is when a company trademarks its name and uses this for a line of products that the company sells. Examples in fruits include Dole®, Driscoll's®, Tropicana®, and Chiquita®.

One of the newest and most innovative trademarking strategies currently taking shape is the pairing of a trademark with a particular new plant variety. This allows for the promotion of a new variety as a specialty food, often bringing about higher prices and margins. Many examples of this approach can be found, especially in the apple industry, such as the Pink Lady® and Jazz® brands. The use of a trademark in conjunction with a plant variety's variety name is allowed under Art. 20(8) of the UPOV 1991 Convention (UPOV 1991), as long as it does not interfere with the public domain's access to the variety name, and as long as the trademark is always used in conjunction with the variety name.

Under Art. 20 (1b) of the 1991 UPOV Convention, "no rights in the designation registered as the denomination of the variety shall hamper the free use of the denomination in connection with the variety, even after the expiration of the breeder's right" (UPOV 1991). Thus, names used as varietal names, including those written in PB rights and plant patent applications, cannot be used as trademarks for that plant or other plants.

Some entities are making key mistakes that may put their trademarks in jeopardy in the coming years. One of the most dangerous current issues is that many entities are not separating the variety or variety name from the trademark name for that variety. As a result, consumers are quickly adopting the trademark name as the name of the variety. This situation describes "genericide," which is an instance where a trademark no longer points to the origin of a product, but rather to the name of the product itself. Once a trademark has become the generic name for a good, it is no longer enforceable as a trademark, and can be canceled by a court, thereby making what was once protected freely available for anyone to use. An example of a court ruling of a trademark cancelation for a fruit is that of Scarlet Spur® Red Delicious apple (Snipes cultivar) in which a judge ruled that the trademark name had become generic for that variety (Warner 2006). Before embarking on a trademark strategy, consult legal advice on how to best maintain the trademark–variety name distinction when presenting the variety to consumers. General recommendations include:

1. Consider using a figurative trademark with special font, or even a logo, to indicate to consumers that the mark is really a trademark, not a variety name.

2. Never use the trademark as a noun or pluralize it (i.e., have you tried a Jazz® yet?); this is key evidence of the trademark being generic.
3. Monitor all trademark use and specify trademark maintenance requirements, including the use of both the variety and trademark names together, in all licensing contracts.

One of the most promising new trademarking strategies which has emerged in recent years has been the use of a trademark as an indicator of special features. This might include taste, size, color, etc. Examples include Flavor Safari® tree fruits from Family Tree Farms in Reedley, California and Super Blues® blueberries from Gourmet Trading Company based in Los Angeles, California. Both trademarks are used for fruit promoted to have exceptional taste and size, respectively. Not only is such trademark use less risky because it automatically links the mark with a special feature rather than with a single product, therefore avoiding the genericide problem, but it also allows companies more flexibility. For example, a company could promote a number of early, mid-, and late-season varieties under one trademark to keep up with supply of a fruit that is recognized by the buyer to be the same or similar in key characteristics. In addition, this also opens up the possibility to rotate in new varieties under the brand without having to design a completely new trademark for each new variety. In terms of risk, having an overarching brand also gives more leeway to shift around products under the brand in case of crop failures, pest susceptibilities, or other issues related to supply.

Important in any trademarking strategy is the consistent use of either the ™ or ® symbol with the trademark. The TM symbol indicates notice of use or ownership of a mark, and ® indicates that the mark has been federally registered with the USPTO or other official governmental organization. One of these symbols should be present on all materials where the trademark appears, including labels, tags, catalogs, Web sites, etc. Details about use requirements and trademark maintenance should be clearly stated in all contracts and licensing agreements. If any of the trademarked products will be sold and shipped internationally, then trademark protection should be sought in those countries, especially if they offer only registration-based protection.

International registration of trademarks can be most easily done through the Madrid Protocol (http://www.wipo.int/madrid), which is a convention regulated by the World Intellectual Property Organization (WIPO) that offers a streamlined application process for international trademark registration. Currently with 84 members (as of 2010), including the USA and EU, one application can be filed through any member's trademark office, or through WIPO in Geneva, Switzerland directly, which will then be forwarded to all indicated countries for processing. Each country's trademark office examines the application as if it had been filed in that country. There are governmental fees for each designated country, but the total cost is significantly lower than filing in each country directly. If a country is not a member of the Madrid Protocol, such as Mexico and Chile (as of 2010), then direct filing in that country is necessary.

The importance of trademarks in the coming years will only escalate for new varieties. As long as proper maintenance is made a priority from the very beginning of a trademark's release, trademarks should offer new possibilities for marketers and commercialization schemes that lead to more specialty products and higher margins for all levels of the supply chain.

2.6 *Trade Secrets*

Trade secrets are important assets to a company, university, or inventor prior to filing for intellectual property protection of a variety or invention. The definition of a trade secret, or "undisclosed information" according to the TRIPS agreement Article 39, is information that is a secret not usually known among or readily accessible to persons within the circles that normally deal with the kind of information in question, information that has commercial value because it is a secret, and information that has been subject to considerable steps to keep it secret (World Trade Organization 2008). An example of a trade secret could be a special method or procedure for propagating plant tissue or increasing the germination rate of a certain type of seed. Trade secrets by definition are not intended to be disclosed to the public. Therefore, agreements are important in maintaining trade secrets and confidentiality, and may include using confidentiality agreements, material use/testing agreements, and production agreements.

2.7 *Contracts and Licensing*

A contract is an agreement between two or more parties that creates an obligation to do or not to do a particular action or activity; the term "contract" also refers to a written document which contains the terms of the agreement, although a written agreement is sometimes not necessary to create the obligation (i.e., a verbal agreement). Licensing is the granting of the rights of the invention through a contract. Contracts are governed by state law (in the USA) with the terms of the license agreement agreed to by both parties.

The license agreement is the vehicle that grants nonowners access to the intellectual property at hand, whether it be patent protection or the use of a trademark. It is very important that the agreement's language addresses an array of issues. Items commonly addressed in fruit crop license agreements include the following:

- The variety being licensed, including information on protection such as patent numbers and/or trademark designations.
- Statements of ownership of the rights to the variety (the licensor) and with whom the agreement is being established (the licensee).
- The definition of the territory where the rights are provided (for sale, propagation, or other use) and any assignment allowance of the rights.
- Definitions of the scope of the rights agreed to, including time period, exclusivity or nonexclusivity, and other items.

- Payments due for the rights initial fee (if any), royalty (per plant, tree, quantity of fruit sold, area of planting, or a combination of items, etc.), minimum royalty due annually (if any), date of payment of royalty, along with provisions providing access to sales records or other information pertaining to proof of payments due the licensor.
- Nonperformance and minimum performance clauses.
- Requirements for protection of the variety by the licensee within the territory.
- Sublicensing requirements (similar to assignability), including whether or not the variety can be sublicensed and what the licensee is required to do prior to sublicensing (such as seek approval from the licensor, provide assurance to licensor of sublicensee terms, providing a copy of the sublicense agreement to the licensor, etc.).
- Language that outlines the progress or goals of the licensee under the agreement.
- Labeling or other required use of variety or institution/originating entity name language required to be used by the licensee.
- Warranty clauses which disclaim any warranty of fitness of the variety for a particular purpose along with the warranty of merchantability of the plant.
- Termination clauses providing for the ending of the agreement by either party.
- Indemnity clauses, which can release either party from liability by any use of the plant.
- Enforcement clauses which spell out each party's role, if any, if a third-party infringement issue should arise which directly affect the licensed rights.
- Venue and choice of law to determine how and where a dispute should be heard.
- Resolution of dispute language defining details of the procedures and rights of each party should differences arise under the agreement.

Numerous issues must be considered in determining how to approach licensing. Exclusive licensing is especially convenient for the licensor, as only one licensee is dealt with in the agreement. However, the risk is also greater because if the exclusive licensee does not perform as hoped, options for looking elsewhere for profit from the invention might be limited. Therefore, minimum and nonperformance clauses are especially important for exclusive licenses. This route is more commonly chosen for international licensees with defined territories, while several licensees may be used in an array of countries or territories. Domestic licensing often involves multiple licensees to ensure that the plant is widely available for growers in the country or region where the variety was developed. Limiting domestic licensees can create concern and political tension between the breeding program and the local industry, especially with a public breeding program where the program is supported by local grower organizations.

For international licenses when choosing the jurisdiction and choice of law that will rule over the contract, consider choosing the law and jurisdiction of the foreign territory. Even though this adds in cost for foreign legal counsel, in addition to the added analysis assessing the strength of the country's intellectual property laws, there can be important benefits if the licensee breaches the contract. If courts in the licensee's country have the competence and jurisdiction to make a ruling within their own country, the intellectual property can be enforced almost immediately. However, if a US court is the jurisdiction for a foreign issue, the power to enforce the judgment over the infringer is limited and often takes a significant amount of time. It is important to consult legal counsel about this matter before infringement arises, and ideally before the contract is even written.

Choosing licensees and developing the strategy for licensing can be major challenges. If a breeding program is known widely due to the value of its prior developments, then potential licensees are usually readily identified and agreements readily executed. If a program is not widely known for success with a crop, or is not located in a region where the crop is particularly important, then licensing opportunities may be limited. In this instance, the variety released may require more promotion by the breeding program (see later discussion on marketing and commercialization).

A licensor may desire to entertain proposals from potential licensees to determine interest and projected use of the variety. Proposals for commercialization and use could be solicited by the licensor by asking potential licensees to address questions such as the following:

- What is the company's current, potential, and projected volume of plant and or fruit sales of the crop?
- What is the company's current, potential, and projected market share in the relevant markets from weeks 1 to 52?
- What is the company's history and current profile in marketing plants or fruit of this species?
- Does the company have its own breeding program for the crop, and if so, have any varieties been released, what variety releases are upcoming, and what are breeding plans for the future?
- If the variety is licensed, what is the potential sales volume projected for the new variety?
- What price is the organization willing to pay for the rights to the variety, in addition to the royalty per plant and/or per unit of fruit?
- What propagation capability does the company have to increase the variety?
- What experience or expertise does the company have in the area of IP rights including applying for, attaining, and managing IP protection in the territory?
- Does the company have references to consult that can comment on their performance in prior licensing agreements?

Another important licensing or contracting aspect to consider is the use of Restrictive Use Language on fruit, seed, or plant material containers. Restrictive Use wording clauses are used extensively with seed-propagated crops but are also used with asexually reproduced crops in notifying the buyer or recipient of plant material of any restrictions associated with using plant material in the container, including limiting the use of the plant material to one harvest, limiting warranties, and any dispute resolution procedures that may be required, including arbitration and mediation requirements.

Confidentiality agreements are also an important form of contracting that can be used to protect vital inventions as well as trade secrets. When using a confidentiality agreement, it is important to identify all relevant parties as well as all relevant material that may be disclosed to the relevant parties. It is also important to determine how, when, and where the disclosed information will be used and by whom. Please consult with an IP rights attorney prior to signing any confidentiality agreement.

3 Material Transfer and Testing Agreements

3.1 Basic Material Transfer Agreements

Testing agreements are very important for the exchange of germplasm among cooperators. These arrangements are usually governed by a Material Transfer Agreement (MTA), which can include terms such as the following:

- Clauses limiting the testing of the genotype including the restriction of sharing the material with other parties, limits on plant testing number allowed, territory restrictions, and use or distribution of any fruit or other products from the material, statement of ownership of the material in that it remains the property of the breeding program and that no grant of ownership or proprietary right is conferred.
- Definition of limitations on where the material can be tested (examples including only experiment station sites, land owned or controlled long term by the cooperator, etc.), requirements of security for the test site (limits of access to the site, etc.), and any restrictions on allowing viewing or examination of the material by third parties.
- Restrictions on any alteration of the material in any way or method, including the use of the material in breeding or other genetic manipulation where the germplasm is moved into other ownership (such as to the cooperator).
- Requirement of reporting any sports or other variants, noteworthy results, or other findings that may have proprietary value in relation to the testing.
- Restrictions on any sharing or publication of results of the testing, reporting results without written permission, describing the material in publication or public presentations, and other items of potential concern in the area of disclosure.
- Provisions for reporting of test results back to the provider of the material including data to be reported, timing, format, or other details of reports.
- Agreement that the material is used in compliance with all applicable statutes and regulations including those related to research involving the use of recombinant DNA.
- Statement indicating that the source (the breeding program providing, university, etc.). Of the material shall in no event be liable for any liability related to the testing including use, loss, claim, damage, or other liability, that may arise from or in connection with the growing or other use, handling, or storage of the material

(this clause would cover concerns of pathogens being brought in with the material or any other potential damaging aspect of the material).

- Termination language and instructions for destruction of the material when the testing is completed.
- Laws of governance in effect for the agreement.

Routine MTA use does not usually provide for fees or other monetary exchange for testing rights; and this is the more common type of agreement used among public breeders. Also, care must be taken in the execution and signing of testing agreements; years ago breeders often signed agreements, while in current times IP technical officers or administrators usually review and approve an MTA.

3.2 *Testing Agreements for Varieties or Selections Involving Fees*

Testing agreements involving fees provide for a different arrangement than the basic MTA and is potentially more complex. In this instance money is exchanged for the rights to test the material. A testing agreement with fee could be done for advanced or otherwise important or unique selections where a commercial tester is interested in identifying the value of new developments prior to release, or for recently released varieties where the testing partner is interested in evaluating the genotype for commercialization in a specified territory where the variety has not been released or commercialized. Another option with this type of agreement is testing of selections that were deemed not worthy of release and thus passed over for release by the breeding program, but may have value in a narrow market such as home gardens or other limited, noncommercial production areas, or possibly commercial use in a different environment. The cooperator usually requires some sort of right to the genotype such as a first right of refusal for licensing and an exclusive agreement for use in commercialization. The value of these agreements to a breeding program include the following: (1) program support, (2) selections might be found to perform better at the cooperator's location (a more desirable genotype x environment interaction might be attained than where the breeding was conducted and otherwise the genotype might be discarded and no value attained for it), (3) commercialization might be done only in a specified territory and not in the area of the breeding program, (4) the cooperator could utilize other growing techniques or cultural management options not available to the breeding program, increasing the chances of maximizing the potential of the material, (5) the cooperator might have commercialization capabilities that exceed those of the institution that developed the material and be able to provide more return for the invention along with broader use, (6) protection costs might be paid for by the cooperator, and (7) if an advanced selection is involved, the determination of commercial value might be determined much earlier than if tested after release, providing time for protection to be filed in the territory (for instance, prior to the expiration of the 4- or 6-year time limitation after first sale of plants outside of the state or territory for UPOV-member countries).

These agreements usually have language similar to the MTA described in the prior section, in addition to items concerning fees, performance requirements of the tester, lack of warranty of the performance of the material, a defined time frame for testing after which the rights to testing expire, rights of access to the test site, payment of travel expenses for the breeder to examine the material, nonassignability of the testing rights, and other items of legal concern. Multiyear testing agreements could be considered where a cooperator attains a specified number of selections annually such as over a 5-year period. This provides for a longer and more sustained relationship and allows testing of new developments during the agreement period.

The choosing of partners for this type of agreement can be more complex than the basic MTA relationship, since a major reason to enter into a testing agreement with a fee is to potentially provide more economic return from the genotype. Therefore, the cooperator should be examined to determine its breadth of commercialization capability, including actual or potential sales of the crop genotype, marketing strategy, and or other aspects that provide for positive use of the genotype if testing indicates commercial value. Also, the evaluation of the capability in facilities and personnel of the cooperator should be examined to ensure reliable and complete testing is possible.

Concerns can exist in testing agreements of this type. One concern is that the breeding program may end up with a genotype in the commercial market that does not meet the standards of the program, and thus reflect negatively on the program. Another could be the choice of the cooperating tester, in that competing entities that did not get access to the genotype may be unhappy and if local and/or politically active in a public arena could contribute to concerns for the public breeding program's administration.

4 Breeding Agreements

In current times, seldom do breeding programs freely share selections, seedlings, or other breeding material. This is due to the economic value of the breeding material and the expense involved in its development in combination with the availability of IP rights to bring about return on the investment. Also, breeding program support is always needed, and it is often considered poor judgment to work on the improvement of a crop or set of traits of a crop, and then simply "give away" the improvements to others in selections, pollen, seeds, or other unreleased genetic material. Further, decisions concerning restrictions of sharing may not be made by the breeder but rather by the administrative or intellectual property officials of the organization. Many feel the restriction on exchange of germplasm has damaged fruit breeding efforts because of the necessity to keep genetic diversity with the germplasms. However, cooperative breeding agreements can provide for sharing of material among breeding programs and should be considered as a way to continue germplasm sharing.

4.1 Public-to-Public Breeding Agreements

Most long-established public fruit breeding programs have a history of freely sharing germplasm in earlier years. If fact, sharing of material was more the rule than the exception until up into the late 1980s to 1990s (and some programs still freely share germplasm such as US Department of Agriculture—Agriculture Research Service programs that work in germplasm development). As sharing has become more limited, there has been a reduction in the breadth of use of genetic advances from other programs (other than named varieties) in breeding. However, genetic diversity is still a top priority for breeding advancement and is critical for breeding progress. Overcoming this limitation can be achieved by formal breeding agreements among public agencies.

Public programs can develop reciprocal agreements which allow for sharing of selections, seedlings, or pollen among programs. Selection of commercial genotypes would occur with the shared material and likely commercial genotypes would result that warrant release. When these genotypes are identified, decisions on release could be made together by the programs involved, and any resulting royalties or other IPR income be shared by the institutions. Sharing could be based on a percentage of royalty income, a per plant basis, or some other formula. One arrangement could be that the program that conducted the initial crossing, seedling evaluation, and/or selection would "own" any resulting variety, arrange for IP rights protection, take steps in commercialization and licensing, collect IP rights income, and handle other IP rights issues (policing, testing agreements, etc.). One area of potential concern is that of how long does the sharing of royalty proceeds continue, as in only first-generation hybrids, or second or later generations? A common approach is to require at least second if not third-generation sharing of royalty proceeds, although on a reduced scale than those from the first-generation use.

There are some potential substantial issues related to public-to-public agreements. Hancock and Clark (2009) suggested that these agreements could result in (1) similar varieties may be released by both programs, and these developments might compete in the marketplace, and (2) if there is a difference in size of the two programs, then the larger program might provide a disproportional gain to the smaller program in royalty income. He further stated that the genetic diversity gained from sharing along with broader testing of germplasm could still be very valuable to both programs and possibly yield more advances in using the breeding germplasm than that of a single program.

4.2 Public-to-Private Breeding Agreements

Expanded activity in private fruit breeding programs has occurred in the past 10–20 years. The concept of sharing germplasm among public and private programs has emerged as an option in germplasm management. Differing from the previously described public-to-public agreements, these arrangements usually move germplasm in one direction, from the public to private program. The main reason for a public program to consider such an arrangement is for program funding support, with the private program paying for access to the public program's germplasm. Also, wider use of the germplasm from the public program could be attained, with the resulting commercialization value of varieties derived from the effort being greater than that of the public program alone. The private program might also have diversity in environments to allow for enhanced chances of uncovering more favorable genotype x environment interactions, and have broader evaluation opportunities (within or outside the region or country where the public program is based). Finally, the public–private program interaction might provide for some unique germplasm blending that would not occur without a breeding agreement.

Disadvantages to the contributing public program include the following: (1) sharing can result in the release of competing varieties from the two programs, (2) there could be increased potential for loss of the germplasm if the commercial partner does not operate fairly and honestly within the terms of the agreement, and (3) the contributing program could have its material genetically "laundered" by the private program using the material, with rapid crossing and subsequent generations of progeny produced which reduce or eliminate any royalty return from the agreement. However, this issue can be addressed to some degree by covering multiple generations of germplasm use in the breeding agreement. Finally, the issue of possible negative implications of the public-to-private relationship with the local industry, program supporters, or other entities that support the public program could develop. For instance, if a breeding agreement was set up with a private entity located in another location that competed with local growers for market share, this could result in difficulties in relations with local growers. This issue is more a concern with a processed crop, however, as the fruit resulting from the private entity development could be stored and shipped long distances and introduced in the market of local growers, negatively impacting the market price.

Conversely, looking from the private program side there are some issues to examine. Initially, due to the history of changes in public programs at times (program cutbacks, shifts in program directives, etc.), it is critical to determine if the public program is stable in funding and committed to the cooperative effort (this is an issue if continued crossing and selection in the public program is essential to produce new genotypes to share). Additionally, if the current public program leader should leave the program, will program activity continue at the same pace and the breeder's position be refilled?

Developing agreements of this type can be complex. Before committing to a breeding agreement, each program should evaluate who would be the best entity to partner with. For the public program, determining which private entity to work with is a top priority. This could be done by surveying potential private partners for interest and their potential use of a variety developed in the agreement. Items to consider in such an inquiry could include the following: (1) sales volume (fruit or plants or other product) of the company, (2) the potential impact a new variety would have on the company's sales, (3) where the company would market the development or product (a defined territory or potentially worldwide?), (4) the research and development capability with personnel and facilities specifically for breeding, (5) past experiences in protection, commercialization, and other aspects of using proprietary developments, (6) propagation resources available to increase the new variety, and (7) monetary outlay the company is willing to pay in access or initial fees and royalties upon commercial use of the developments. Companies with experience in breeding should be familiar with the long-term nature of such an endeavor and understand that the development period and length of the relationship may continue for many years (even after germplasm sharing is complete). By comparison, those that have not conducted breeding activities may be surprised at the cost, personnel, and facilities involved, and other noteworthy components that go into a breeding effort. From the private angle, other than the outlook for the public's program commitment to continued breeding and germplasm development, considerations could include issues such as any ownership or approvals for commercialization of a development that the public program might require (and if so, the timeframe for decisions to be made), freedom of the public program's germplasm from any additional issues of ownership claims, and if the cooperative breeding is exclusive for a territory or other restriction that could limit competitors in attaining the same rights to the public program's germplasm.

Items that could be included in an agreement are the following:

- Length of time and location of the breeding activity.
- Material to be provided from the public program such as pollen, plants, or cuttings of selections, or other parts to be used in breeding, and the time of year the material would be provided.
- Payment amounts for access to the germplasm along with royalty schedule for fruit, plants, or other items of income from the developments (and considerations of subsequent-generation royalties to be paid on second- or later-generation varieties developed).
- Inclusion of selection testing and potential commercialization of any of the public program's material that might be found to perform well (genotypes provided directly, not a result of crossing in the agreement).
- Ownership of developments from the cooperative effort.
- Issues of exclusivity of the agreement and definitions of territory.
- Restrictions on sharing of germplasm with others by the private partner.
- Security at the sites of the breeding and testing activities.
- Definitions of subsequent use by the cooperator in further crossing of selections generated in the program.
- Allowance of germplasm flow back to the public program or not.
- Outlines of what amount, type, and other terms of the germplasm to be shared (selection number volume, seed, or population quantities, etc.).
- IP rights protection requirements for commercialized developments.
- Commercialization or other use limitations of developments outside any defined territory (including issues of introduction of a variety in the territory of the public program, or at any world location).

- Confidentiality language concerning cooperative activities, internal information shared particularly concerning the private partner, or other proprietary information.
- Program access by personnel from the contributing program including visits to breeding and testing sites, funding of travel costs, or other terms of access.
- Assignability of the agreement by the private company.
- Liability and indemnity clauses.
- Agreement termination language.

If an agreement of this type is developed, there will be additional time and resources required by the public program to fulfill the requirements of the agreement. Items to consider include the breeder's time in developing potential germplasm to share, crossing and seed collection costs (if seeds are provided in the agreement), pollen collection expenses, propagation or other activities in producing plants for sharing, and any other aspects to fulfill the terms of the cooperation. Likewise, resources required by the private cooperator must be examined, particularly if the cooperator has not been involved with breeding before. Foremost is the issue of personnel involved, since most commercial-entity staffing is involved with production (nursery, fruit production, or other nonresearch and development activity). Often, when "money is on the table" on the private side, including time in harvesting, marketing, propagation, grower relations, or other routine activities of a company, these items will likely take precedence over research including breeding activities. This issue must be carefully considered, and personnel and other resources committed to the breeding effort should have some separation from routine commercial duties.

5 Marketing and Commercialization of Fruit Varieties

The marketing and promotion of new fruit varieties has drastically changed in recent years. Parallel with IP rights changes, strategies of how to commercialize a variety, such as the simple philosophy of release and "let the variety find its way on its own" to the marketplace is not as commonly practiced as in past times. When fruit breeding programs were first started and for many years following, largely by public institutions, varieties were not often protected, formally marketed, or promoted by the developer. Information about new varieties often came from the following sources:

- Grower meetings where breeders or extension service agents shared early performance of new varieties compared to popular or industry-standard genotypes.
- Extension fact sheets, release notices, research bulletins, or other public-agency sources of new variety information.
- Trade journals or other popular press sources.
- Nursery catalogs and promotional items.
- Word of mouth of performance among growers.

To varying degrees, these sources of information still continue today, in addition to other options.

Another aspect of release was that often the variety was given a minimal to extensive test as an advanced selection before or soon after naming and release. With this system, growers often had documented information about a variety prior to its planting, with testing often done in the area or region where the grower was located. Varieties often took many years to "catch on" and be used extensively, with the longest period for tree fruit or nut crops. A number of factors have contributed to a reduction in testing of advanced selections and new variety introductions. One factor is that public agencies have reduced variety testing programs due to expense, budget limitations, and program priority redirection. Another issue is that in recent years for some crops there have been too many varieties released to conduct thorough testing. Finally, at times proprietary concerns or limitations in attaining test plants (especially advanced selections) have limited testing.

Promotion of public-entity-developed varieties began some years back by "in-house" marketing, such as brochures and other materials, and more recently by Web sites. Most programs have some type of variety development display on-line, with sources of the varieties often provided particularly if the variety is protected. Parallel to this has been the increasing profile of nurseries in promotion. This role was further expanded as protected varieties came on the scene and individual nurseries were the sole or limited source of a new development. Nurseries were often utilized to better manage and monitor the distribution of protected varieties, which was an advantage for the breeding programs, while also benefiting specific nurseries in having access to the new varieties. This "competitive edge" began to play a more substantial role in promotion and marketing, as these efforts had a direct tie back to the nursery's success and the exclusive or limited access to the new variety.

In the last 10–20 years, rapid change has occurred in marketing along with the expansion of IP rights protection, limited and controlled access to new varieties, vertically integrated variety use, "managed" variety program use, trademarks, and other marketing strategies and approaches.

5.1 Territorial Marketing

One of the most important initial decisions that a breeding program faces when considering commercialization strategies for a new variety is how to maximize value, not just in a home country, but potentially all over the world. This is particularly relevant because of the global and year-round nature of food trade and supply today, making it preferable that the variety be made available in many different climates and regions. Because IP rights are territorially regulated, most often by country, protection must be sought in each individual country where the variety will be sold. As a result, having commercialization strategies based on territories, be it a country, a group of countries, or a continent, is a common approach because it parallels the IP rights protection scheme.

Once the countries or territories where the new variety should be commercialized are made, the question how it should be implemented comes to the forefront. The two main approaches include seeking either nonexclusive or exclusive licensing relationships. For example, a single exclusive licensee, perhaps a nursery or a fruit marketing company, might be given the rights to propagate and sell a new variety within a single country or numerous countries within a territory, or, multiple nonexclusive licensees might be given the rights to a variety in a single country. There are many factors that contribute to this decision. One is the question of the strength of IP rights laws and enforcement capabilities in the country in question. In countries with poor enforcement histories, it might make more sense to develop an exclusive "closed" relationship with a party in the territory to lower the risk of the variety being "let loose" and illegally propagated, therefore lowering its value on the market.

Another important consideration when commercializing territorially is that the fruit of many crops is not necessarily sold where the plants are grown, especially when considering the Southern Hemisphere's role in supplying the Northern Hemisphere during off-seasons. Therefore, any exclusive relationship formed within a territory should clarify where the fruit will be sold, as it could affect exclusive relationships in other countries.

5.2 Club Models

A commercialization and marketing approach that has developed in recent years is variety brand management. As opposed to an openly released variety, a managed variety is one whose commercialization is controlled or monitored by a central organization that manages, often on a global level, the main elements of the market strategy, such as plantings, supply, quality criteria, distribution channels, licensing of a trademark associated with a variety, and promotion of the variety. Such an entity might be an independent company formed especially for the management of such a variety, or even an existing fruit marketing company that adds the variety to its portfolio of exclusively managed varieties.

A form of variety brand management that is becoming more and more prevalent is referred to as a "club." Even though the club may have differing models, usually the central management has very tight reigns on the quantity and quality of a variety grown and sold in different regions around the world. In the club model, growers "opt in" to a central marketer's commercialization strategy. The growers usually pay higher royalties and fees to cover marketing costs and IP rights maintenance done by the central authority for the club as a whole, and they are then licensed exclusively to grow a certain number of plants (or allotment of fruit) in a certain way. This theoretically brings extra returns for growers because the price stays high (due to the controlled, limited supply). Such marketing activities quite often include the application of a trademark for the variety at hand, which is then used to distinguish the club's branded product. Some fruit industries are further along than others in terms of implementing such innovative commercialization strategies. Pioneers, such as leaders in the fresh apple industry, are paving the way as a flagship for other industries that are also looking to implement a more controlled, high-margin commercialization system for their new varieties.

The future success of such clubs will rely on (1) maintaining high enough margins to make the high royalties worth it to all players involved, (2) continually keeping track of new sports and family members of the variety and incorporating them into the club in a controlled way to maintain the agricultural sustainability and viability of the club (i.e., disease resistance, expanded growing regions, etc.), and (3) maintaining a healthy and valuable trademark so that it is viable even after PB rights or plant patent expiration. In other words, even if a club is based initially on a single variety, it is important to think long-term about sustainable longevity of the club's branded product to maximize long-term returns.

5.3 Closed Commercialization Systems

A new commercialization strategy quietly developing in the private fruit sector is the "closed" commercialization system. This strategy is called "closed" because the only players that have access to a particular variety, and often its fruit, are individual entities that form a contractual distribution channel which is not available to outside growers, distributors, wholesalers, or retailers. More and more, this closed channel spans from the variety development stage all the way to the retailer shelves. For example, a grocery store retailer might be interested in its own brand of stone fruit or berries, much like many chains have their own brand of consumer goods such as soft drinks or cereals. In order to obtain such an exclusive product, the retailer would have to go to the source and develop a pull-through closed chain involving a new variety, growers, packers, and logistics. Oftentimes, this is managed by a separate company that is either part of the value chain, or one that is a private IP rights management company. The closed nature of this arrangement reduces the visibility of the commercialization along the value chain because the contractual relationships are formed before the fruit ever arrives in the marketplace. This kind of pull-through access to varieties from grocery retailers has been especially prevalent in the UK grocer markets.

Other closed systems include end-product companies that look to a similar pull-through system. An example might be a juice maker that wants exclusive rights to varieties for its end product. Setting up a closed supply chain around an exclusive variety not only allows for the end-user to market the product as its own exclusive good, and therefore implement promotions and theoretically bring in higher margins, but also aligns the breeding program more directly with the end user, which is where higher margins for the breeding program tend to be found. The value of such an exclusive offer for a breeding program can and should be significant to make up for the potential substantial costs and management of licensing a variety to many different entities. Considerable up-front fees, as well as royalties calculated from end-product sales, are common in this type of arrangement. A practical advantage to the closed system for the breeding program is the ease of monitoring the IP rights: infringement is more easily identified because the variety is supposed to only pass through the hands of a few contractually linked players. This is an emerging commercialization system for proprietary fruit varieties, and breeding programs should be aware of this as an option to consider.

5.4 Factors to Consider

Any new variety that is selected for release not only has risks attached to it in terms of its agricultural performance, but also in terms of its relative position in the marketplace regarding other new varieties and breeding programs. Therefore, before setting out and deciding on a commercialization strategy, whether it is one that has always been done in the past, or whether it is a new approach, it is important to consider these factors:

- What is the market? A “market” can be defined in many ways. It could be defined as particular country or climate zone, or even as a pool of similar varieties that meet certain customer needs. By identifying the market, it should become clear where and against what the variety will be competing.
- What is the competition? Within the identified market(s), what other similar varieties answer the same or similar needs that the variety at hand answers? Is this variety significantly better than the others? This will help gauge the variety’s value. If it is the only available variety for a particular growing area, marketplace, or window in time, it is essential to be knowledgeable about these issues and consider negotiating higher fees and royalties. One should also remember to check whether the other “competing” varieties are already “tied up” in an exclusive relationship in the market. This could have a significant impact on the potential value and interest in the new variety.
- Who will make the variety a success? No matter how special a new protected variety is, it is only as successful as the licensee who commercializes it. Due diligence and building incentives into commercialization contracts is essential in lowering the risk of sub-standard performance by the licensee.

Once the market and the competitive situation of the new variety is better understood, the breeding program can gain a more clear understanding of its potential value. Another important factor to consider is how much involvement a breeding program wants to have in the commercialization and IP rights management process. If little involvement is desired, then looking to an exclusive licensee or an IP management company might be key to maximizing return with minimal administrative work. One consideration to make when evaluating commercialization options is the political arena. For public breeders, are there obligations to a university or specific groups of tax payers that should be considered? For private breeders, are there past relationships or conflicts of interest that could compromise the success of the new variety or even future relationships? The more potentially valuable a new variety is, the more important such questions become.

6 Conclusions

Intellectual property protection of fruit varieties has expanded as most public and private breeding programs have increased their filing activity for various kinds of protection in numerous countries around the world. Protection options such as plant breeder's rights, plant patents, utility patents, and trademarks are all being utilized in various locations. Contracts and licensing are playing key roles in the assignment of the protection rights to nurseries, fruit production companies, and others, and a range of options exist for breeding programs to consider as they release new developments. With increased intellectual property issues in fruit breeding, options are being examined concerning the sharing of germplasm for testing and/or breeding. Breeding agreements including public-to-public and public-to-private options are expanding. Marketing and commercialization have become much more complex, compared to earlier times when varieties were released and made their way to commercial use through simpler arrangements. Options such as territorial marketing, club models, and closed commercial systems are becoming more common.

The future of commercialization and marketing options for breeding programs is both exciting and daunting. With the increases in IP rights protection around the world, and with the ever-increasing intensity of competition in the agricultural industry, the potential value of each new variety is considerable. In addition, with the globalization of the perishable food trade, traditional territorial management is being changed substantially because the market for fruit plants is different than the market where the fruit produced is being sold. These prospects bring about an age in new variety marketing where breeding programs must think outside the box, be open to private club and closed-system variety management, and consider global issues to implement effective and profitable commercialization schemes.

References

Australian Government. 2008a. IP Australia, Patents. 26 Feb. 2008. <http://www.ipaustralia.gov.au/patents/index.shtml>.

Australian Government. 2008b. IP Australia, Plant Breeders Rights. 25 Feb. 2008. <http://www.ipaustralia.gov.au/pbr/about.shtml>.

Canadian Food Inspection Agency. 2007. Guide to Plant Breeders Rights. 25 Feb. 2008. <http://www.inspection.gc.ca/english/plaveg/pbrpov/guidee.shtml>.

Japan Patent Office. 2007. Patents. 26 Feb. 2008. <http://www.jpo.go.jp/>.

Hancock, J.F. and J.R. Clark. 2009. Intellectual property protection and the funding of blueberry breeding in the future: the new paradigm. Acta Hort. 810:43–48.

Union for the Protection of New Varieties of Plants. 1991. 1991 Act Chapter 6, Article 20 Variety Denomination. 26 Feb. 2008. <http://www.upov.int/en/publications/conventions/1991/w_up912_.htm#_20>.

Union for the Protection of New Varieties of Plants. 1997a. Chile Law 19.342 On the Rights of Breeders of New Varieties of Plants. 25 Feb. 2008. <http://www.upov.int/export/sites/upov/en/publications/npvlaws/chile/chili.pdf>.

Union for the Protection of New Varieties of Plants. 1997b. Mexico Federal Law on Plant Varieties. 25 Feb. 2008. <http://www.upov.int/export/sites/upov/en/publications/npvlaws/mexico/mexico_law.pdf>.

Union for the Protection of New Varieties of Plants. 1997c. Republic of South Africa Plant Breeders Rights Act. 25 Feb. 2008. <http://www.upov.int/export/sites/upov/en/publications/npvlaws/southafrica/sa-act82.pdf>.

Union for the Protection of New Varieties of Plants. 2010. Members of UPOV. 22 Apr. 2010. <http://www.upov.int/en/about//pdf/pub423.pdf>.

Union for the Protection of New Varieties of Plants. 2008. Mission Statement. 29 Feb. 2008. <http://www.upov.int/en/about/mission.html>.

US Patent and Trademark Office. 2007a. Chapter 15, 35 U.S.C. 161 Patents for Plants. 25 Feb. 2008. <http://www.uspto.gov/web/offices/pac/mpep/consolidated_laws.pdf.>.

US Patent and Trademark Office. 2007b. Chapter 15, 35 U.S.C. 163 Patents for Plants. 25 Feb. 2008. <http://www.uspto.gov/web/offices/pac/mpep/consolidated_laws.pdf>.

Warner, G. 2006. Judge orders cancellation of Scarlet trademark. Good Fruit Grower 57(8). 26 Feb. 2008. http://www.goodfruit.com/issues.php?article=55&issue=3.

World Trade Organization. 2008. Overview: The Trips Agreement. 25 Feb. 2008. <http://www.wto.org/english/tratop_e/trips_e/intel2_e.htm>.

Chapter 4
Emerging Fruit Crops

Kim E. Hummer, Kirk W. Pomper, Joseph Postman, Charles J. Graham, Ed Stover, Eric W. Mercure, Malli Aradhya, Carlos H. Crisosto, Louise Ferguson, Maxine M. Thompson, Patrick Byers, and Francis Zee

Abstract Hundreds of fruit species with commercial potential are currently in a status of low economic importance. Some, such as quince, pomegranate, and figs, have been cultivated for thousands of years. Others have only been locally collected and consumed from wild populations of the fruit. The development of these under-appreciated crops depends on a range of factors including the cultivation limitations, yields, uses of the fruit, and marketing potential. Although initially many crops are developed using selections from the wild, as they are developed, breeding programs work toward improving the crop for both production and quality. This chapter examines nine emerging crops chosen among hundreds of potential crops

K.E. Hummer (✉)
USDA ARS National Clonal Germplasm Repository,
33447 Peoria Road, Corvallis, OR 97333-2521, USA

USDA ARS Arctic and Subarctic Plant Gene Bank, Palmer, AK, USA
e-mail: Kim.Hummer@ars.usda.gov

K.W. Pomper
Kentucky State University, 129 Atwood Research Facility, Frankfort, KY 40601, USA

J. Postman
USDA ARS National Clonal Germplasm Repository,
33447 Peoria Road, Corvallis, OR 97333-2521, USA
e-mail: Joseph.Postman@ars.usda.gov

C.J. Graham
LSU Agricultural Center, Pecan Research/Extension Station,
10300 Harts Island Road, Shreveport, LA, USA
e-mail: CJGraham@agcenter.lsu.edu

E. Stover
USDA/ARS Horticulture and Plant Breeding Unit, Horticultural Research Laboratory,
2001 S. Rock Rd., Ft. Pierce, FL 34945, USA
e-mail: Ed.Stover@ars.usda.gov

M.L. Badenes and D.H. Byrne (eds.), *Fruit Breeding*, Handbook of Plant Breeding 8,
DOI 10.1007/978-1-4419-0763-9_4,

which are currently showing much promise as commercial crops. These include five tree fruits, namely, pawpaw, quince, mayhaw, pomegranate, and fig, and four berry crops, namely, blue honeysuckle, elder, goji, and ‘ōhelo.

Keywords Underutilized genetic resources • Specialty crops • Local crops • Heritage fruit cultivars • Potential new fruit

1 Introduction

As Darrow and Yerks (1937) state, “All of our present cultivated plants, it must be remembered, have been derived from wild plants.” Those that were outstanding or most readily adaptable were taken from forest and field and grown at the dooryard; others were left in the wild so that products could be gathered and used. Yet, the definition of an ‘emerging crop’ is vague from a temporal sense. Some crops require millennia while others centuries or decades to achieve notoriety.

Internationally, economically important fruit crops, such as grapes, apples, cherries, and pears, date to Western antiquity with cultivation over millennia. Quince (*Cydonia oblonga* L.), pomegranates (*Punica granatum* L.), and figs (*Ficus carica* L.) fit that timeframe, but are still ‘emerging’ crops, despite their documentation in ancient references; their development is expanding in today’s markets.

E.W. Mercure
Paramount Farming Company, Bakersfield, CA, USA
e-mail: EricM@paramountfarming.com

M. Aradhya
National Clonal Germplasm Repository, USDA, ARS, University of California, One Shields Ave., Davis, CA 95616, USA
e-mail: Malli.Aradhya@ars.usda.gov

C.H. Crisosto • L. Ferguson
Department of Plant Sciences, University of California, One Shields Ave., Davis, CA 95616, USA
e-mail: lferguson@ucdavis.edu

M.M. Thompson
Department of Horticulture, Oregon State University, Corvallis, OR 97330, USA
e-mail: thompsom@onid.orst.edu

P. Byers
Greene County Extension Office, University of Missouri Extension, 833 North Boonville Street, Springfield, MO 65802, USA
e-mail: byerspl@missouri.edu

F. Zee
USDA, ARS Pacific Basin Agricultural Research Center (PBARC), P.O. Box 4487, Hilo, HI 96720, USA
e-mail: Francis.Zee@ars.usda.gov

The European history of many present-day economically important berry crops, such as raspberries (*Rubus idaeus* L.), blackberries (*Rubus* subgenus *Rubus*), currants and gooseberries (*Ribes* L), and strawberries *Fragaria* × *ananassa* Duchesne ex Rozier, is counted in centuries. Elderberry (*Sambucus* L), among them, now has increasing demand for production in juice, wine, and processed products. American pawpaw (*Asimina triloba* (L.) Dunal) and mayhaw (*Crataegus aestivalis* (Walter) Torr. & A. Gray) were recognized by early European settlers and have continued to emerge as cultivated crops over the past several centuries.

American blueberries (*Vaccinium corymbosum* Ait.) and cranberries (*Vaccinium macrocarpon* Ait.) are recent and have been selected, developed, and bred from the wild over decades. Their relative the Hawaiian ʻōhelo *V. reticulatum* Sm. has now surfaced as another with cultivation potential.

Two Asian berries, the goji (*Lycium barbarum* L.), mentioned in Chinese medicinal texts of antiquity, and the blue honeysuckle (*Lonicera caerulea* L.), also touted in Russian and Chinese folk medicinal traditions, have made a recent splash as new crops for Western production. These two crops, grown near their center of origin by traditional farmers, remain important for the subsistence of local communities. With the advent of the nutraceutical industry, international interest in expanding cultivation for these crops has increased and encouraged breeding and commercial cultivation.

This chapter examines only nine emerging crops: five tree fruits, namely, pawpaw, quince, mayhaw, pomegranate, and fig, and four berry crops, namely, blue honeysuckle, elder, goji, and ʻōhelo. A much more extensive list of neglected berries, potential new berries, and crops with unmet potential has been discussed (Darrow 1975; Finn 1999). Many additional horticultural crops appear on the horizon of development in horticultural compendia such as *Stuartevant's Notes on Edible Plants* (Hedrick 1919), *Hortus Third* (L. H. Bailey Hortorium 1999) or the *Encyclopedia of Fruits and Nuts* (Janick and Paul 2008). These references have a more complete listing and summary of potential crops beyond the scope of this chapter.

Diversification of local production is the key to save small farmers and resolve the food shortage (Lumpkin 2007). Locally produced horticultural crops are the key to success in the United Nations millennium development goals (http://www.un.org/millennium/declaration/ares552e.htm) to eradicate extreme poverty and develop environmental sustainability. Emerging crops will serve to strengthen local economic success through diversification of crop species.

2 The North American Pawpaw

2.1 Botany

The North American pawpaw, *Asimina triloba* (L.) Dunal, grows wild as an understory tree, often in large patches due to root suckering, in hardwood forests in the eastern USA (Kral 1960) (Fig. 4.1). Trees may reach 30 ft in height and assume a

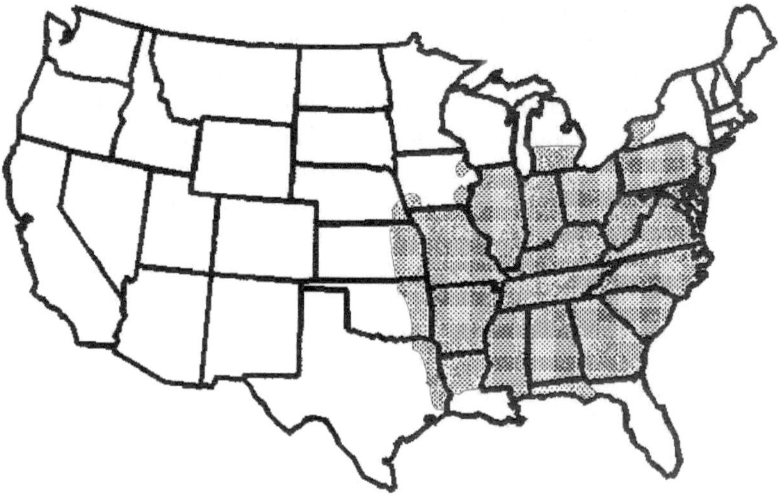

Fig. 4.1 Native distribution of *Asimina triloba* (L.) Dunal, the North American pawpaw, prepared by Kirk Pomper, Kentucky State University

pyramidal habit in sunny locations. This plant can be grown successfully in USDA plant hardiness zones 5 through 8 (Kral 1960). Fruits weigh up to 2 lbs and may be borne in clusters of up to 13 fruit or singly (Fig. 4.2). The fruits are highly nutritious, have a strong aroma, and have a unique flavor that resembles a combination of banana, mango, and pineapple (Pomper and Layne 2005; Duffrin and Pomper 2006). The fruit has both fresh market and processing potential.

2.2 Origin and Domestication

Pawpaw has a well-established place in folklore and American history. The traditional American folk song, "Way down, yonder in the pawpaw patch" is still known to children and fall pawpaw hunting in the woods is still a common tradition for rural families in the eastern USA. In 1541, the Spanish explorer Hernando de Soto reported Native Americans growing and eating pawpaws in the Mississippi valley. Native Americans also used the bark of pawpaw trees to make fishing nets. Daniel Boone and Mark Twain were reported to have been pawpaw fans. In 1806, Lewis and Clark recorded in their journal how pawpaws saved their party from starvation. There was interest in pawpaw as a fruit crop in the early 1900s; however, the rapid perishability of fruit likely decreased interest in this fruit (Peterson 1991). Interest in pawpaw grew between 1950 and 1985, and recently, the appeal of pawpaw as a gourmet food has increased (Pomper and Layne 2005).

Fig. 4.2 Pawpaw fruit. Photocredit: Kirk Pomper, Kentucky State University

2.3 Production and Uses

Pawpaw is in the early stages of commercial production as a new high-value tree fruit crop. The greatest market potential for pawpaw currently is for sales at Farmers' markets and direct sales to restaurants and other gourmet food clientele. Pawpaw fruits are mainly collected from natural stands in the forest or from production from small plantings. Sellers often have difficulty finding sufficient pawpaws to meet the demand. Wild fruits collected from some trees can have a bitter aftertaste, while fruits from grafted trees of named cultivars are of a higher quality, do not have a bitter aftertaste, and have greater market potential.

Pawpaw production challenges have been reviewed by Pomper and Layne (2005) and included the need for high quality cultivars, poor pollination, and fruit perishability issues. A number of high-quality pawpaw cultivars with large fruit (over 5 oz) have been selected since 1950. Some of these cultivars have been evaluated at Kentucky State University and cultivars that can be recommended based on large fruit size and production (about 20 lbs/tree/year) are: 'NC-1,' 'Overleese,' 'Potomac,' 'Shenandoah,' 'Sunflower,' 'Susquehanna,' and 'Wabash.' Grafted trees usually begin reliable fruit production at 5–6 years after planting (Pomper et al. 2003a, b, 2008). Pawpaws need to cross-pollinate and flies and beetles are the main pollinators (Faegri and van der Pijl 1971). Efforts to attract these pollinators to pawpaw plantings can improve fruit set. Perishability is still a problem for pawpaw; ripe fruits soften rapidly and have 5-to-7-day shelf life at room temperature (Archbold and Pomper 2003). However, fruits that are just beginning to soften can be stored for about 3 weeks at 4°C and maintain a good eating quality.

Initial investments include land preparation, purchase of plants, installation of an irrigation system, and tree establishment. The recommended density is 295 pawpaw trees per acre. Grafted trees usually cost between $10 and $25 each. Growers may not recover the cost of establishing a pawpaw planting until about 7 years after planting. According to estimates by the University of Kentucky, production costs for pawpaw are estimated at $884 per half-acre, with harvesting and marketing costs at $720 per half-acre (University of Kentucky 2008). Total expenses per half-acre come to approximately $2,075. Presuming gross returns of $3,500 per half-acre, returns to land, capital and management are approximately $1,490 per half-acre. These returns could be substantially higher than the $1 per pound wholesale price used in estimates; pawpaws sold at the Lexington and Frankfort, Kentucky, farmers' markets for $3 per pound in 2009. Other challenges to expanding a pawpaw industry include: developing a grower base, improving orchard establishment rates, rootstock development, improving clonal propagation methods, new cultivar development, increasing yields, postharvest handling of fruit, and developing an overall marketing strategy.

2.4 *Breeding Potential*

From about 1900 to 1960, at least 56 clones of pawpaw were selected and named. Fewer than 20 of these selections remain, with many being lost from cultivation through neglect, abandonment of collections, and loss of records necessary for identification (Peterson 1991, 2003). Since 1960, additional pawpaw cultivars have been selected from the wild or developed as a result of breeding efforts of hobbyists. More than 40 clones are currently available (Pomper and Layne 2005). The loss of cultivars over the last century may have led to erosion in the genetic base of current pawpaw cultivars (Huang et al. 1997). New breeding efforts by The PawPaw Foundation have led to the release of several new cultivars (Peterson 2003). Additionally, fruit to fruit consistency in ripeness and quality, longer cold storage ability, and higher yields would be desirable in new pawpaw cultivars.

Since 1994, Kentucky State University has served as a satellite site of the USDA National Clonal Germplasm Repository, Corvallis, Oregon, Genebank for *Asimina* species (Pomper et al. 2003a, b). The collection contains over 2,000 accessions from 17 states. Assessing genetic diversity and evaluating pawpaw germplasm for the repository collection is a top priority and will hopefully conserve current pawpaw germplasm and serve as a source of new germplasm for breeding new cultivars in the future.

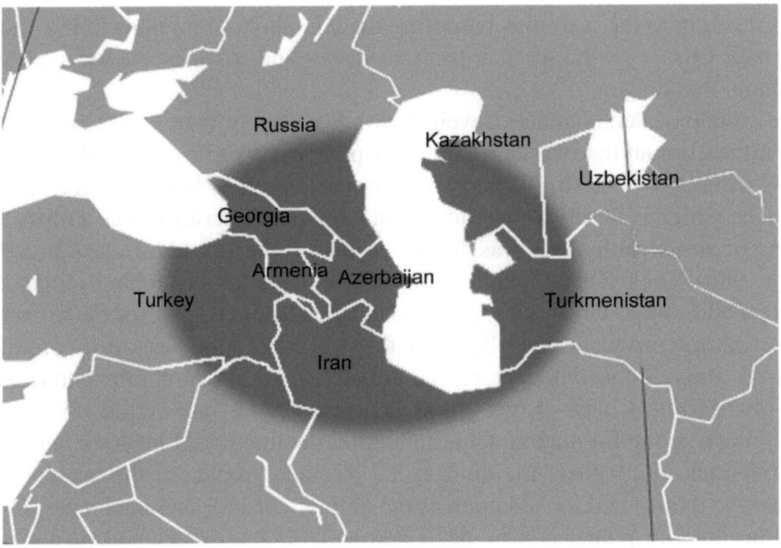

Fig. 4.3 *Cydonia* distribution, prepared by Joseph Postman, USDA ARS

3 Quince

3.1 Botany

Cydonia oblonga Mill. is a monotypic genus belonging to family Rosaceae, subfamily Spiraeoideae, tribe Pyreae and subtribe Pyrinae (USDA 2009a). It grows as a multistem shrub or small tree and has pubescent to tomentose buds, petioles, leaves, and fruit. Leaves are ovate to oblong, about 5 cm across and 10 cm long. The white, solitary flowers are 4–5 cm across, have 5 petals, 20 or more stamens, 5 styles, an inferior ovary with many ovules, and are borne on current season growth. Bloom time overlaps with that of apples, usually beginning in mid April in the middle latitudes of the northern hemisphere. The fruit is a fragrant, many-seeded pome about 8 cm in diameter. Shape ranges from round to pear-like, flesh is yellow, and the Bailys refer to it as 'hard and rather unpalatable' (Bailey and Baily 1976; Rehder 1986). Fruit size and leaf size of cultivated varieties can be many times larger than the wild type described above.

3.2 Origin and Domestication

Cydonia is native to western Asia, and the center of origin is considered to be the Trans-Caucasus region including Armenia, Azerbaijan, Iran, SW Russia, and Turkmenistan (Fig. 4.3; USDA 2009a). During ancient times, it spread from its wild center of origin to the countries bordering the Himalaya Mountains to the east, and throughout Europe to the west. It has many uses and traditions associated with it throughout this range.

The quince of Persia attains a weight of 1.5 kg (more than 3 lbs), ripens on the tree or in the store, and can be eaten like a soft ripe pear; according to the 'Horticulturist' of 1849 (Meech 1908). This is hardly the quince known in America today, or rather the quince which is hardly known today. In Colonial America it was a rarity in the gardens of the wealthy, but was found in nearly every middle class homestead (Roach 1985). The fruit was an important source of pectin for food preservation and a fragrant addition to jams, juice, pies and candies. However, by the early twentieth century quince production declined as the value of apple and pear production increased. Today's consumers prefer the immediate gratification provided by sweet, ready-to-eat fruits. Charles Knox introduced powdered gelatin in the 1890s and the use of quince pectin for making jams and jellies declined. U.P. Hedrick lamented in 1922 (Hedrick 1922) that "the quince, the 'Golden Apple' of the ancients, once dedicated to deities and looked upon as the emblem of love and happiness, for centuries the favorite pome, is now neglected and the least esteemed of commonly cultivated tree fruits."

Luther Burbank took credit for helping to transform this neglected fruit from a commodity that was "altogether inedible before cooking" into a crop he likened to the best apple. He half-jokingly cited a formula to make quince fruits edible: "Take one quince, one barrel of sugar, and sufficient water" (Whitson et al. 1914). Burbank released several improved cultivars in the 1890s that he hoped would raise the status of the fruit. While two Burbank cultivars 'Van Deman' and 'Pineapple' are important commercially in California today, quince fruit production in the USA is so small that it is not even tracked by the USDA National Agricultural Statistics Service (McCabe 1996; USDA 2009b). Both of these Burbank quinces, however, have found their way to other parts of the world where they are among the handful of cultivars considered worthy of production (Campbell 2008).

Meech described 12 varieties important in the USA of 1909, although some 'varieties' such as 'Orange' (syn. = 'Apple') were as often as not grown from seed rather than propagated as clones. Quince is easily grown from either hard-wood or soft-wood cuttings, and is readily grafted onto another quince rootstock. Although quince is an important dwarfing rootstock for pear, the reverse graft is not reliable and therefore pear should not be used as a rootstock for quince.

Quince has a very extensive history in the Middle East, and may have even been the fruit of temptation in the Garden of Eden. The ancient Biblical name for quince translates as 'Golden Apple' and cultivation of *Cydonia* predates cultivation of *Malus* in the region once known as Mesopotamia, now Iraq. Juniper and Mabberly (2006) relate how this region is well adapted to cultivation of quince, pomegranate and other fruits, but is much too hot and dry for the cultivation of all but the most recently developed low-chill apple cultivars. Quince was revered in ancient Greece where a fruit was presented to brides on their wedding day as a symbol of fertility. It was mentioned as an important garden plant in Homer's Odyssey and Pliny the Elder extolled its valuable properties.

3.3 Production and Uses

Worldwide, there are about 43,000 ha of quince in production with a total crop of 335,000 MT. Turkey is the largest producer with about 25% of world production. China, Iran, Argentina, and Morocco each produce less than 10%. The USA is a very minor player in terms of growing quince for fruit with only about 100 ha in production, mainly in California's San Joaquin Valley. Burbank's 'Pineapple' is the most widely grown in that state and is said to be more flavorful than 'Smyrna' (McCabe 1996). Membrillo, or Quince Paste, is popular in several European countries, particularly Spain, and in parts of Latin America. This fragrant, sweet, jelly-like confection is cut into slices and often served with cheese. Quince is also served poached in either water or in wine and develops a rich aroma and deep purple-red color. In Armenia, quince is used in many savory as well as sweet dishes, and is often cooked with lamb (Ghazarian 2009).

While quince is still grown for its fruit in some parts of the world, in other places including England, France, and the USA, it is primarily grown for use as a dwarfing pear rootstock. In the region around Angers, France, quince has been used as a pear rootstock since before 1500. The French were growing quince from cuttings and in stool beds by layering by the early 1600s and France became an important source of rootstocks around the world. Quince rootstocks grown near Angers were known as 'Angers Quince' and those propagated near Fontenay were known as 'Fontenay Quince' (Roach 1985; Tukey 1964). Confusion arose as to the identities of various quince rootstocks, and in the early 1900s researchers at East Malling in England collected rootstocks from various nurseries and designated clones with letters of the alphabet. Quince rootstock clones now available in the USA include Quince A and Quince C, which came from East Malling–Long Ashton (EMLA); and Provence Quince (= Quince BA 29-C) from France. A pear tree grafted onto Quince A will be about half the size of a tree grafted onto pear seedling rootstock. The tree will also be more precocious and fruit size will be larger. Quince C produces a tree slightly smaller and more precocious still. Province Quince rootstock produces a pear tree slightly larger than Quince A or C. Some pear varieties are not graft compatible with quince and require a compatible interstem pear cultivar such as 'Comice,' 'Old Home' or 'Beurre Hardy' as a bridge.

3.4 Breeding Potential

A collection of quince germplasm was established in Izmir, Turkey beginning in 1964 that includes many regionally developed cultivars and landraces (Sykes 1972). In Karaj, Iran a collection of more than 50 *Cydonia* accessions are maintained, including both cultivated and wild types (Amiri 2008). A large fruit tree collection in Kara Kala, Turkmenistan was once a part of the Vavilov Institutes during Soviet times. Many fruit accessions, including quince, were rescued from that station and brought to other genebanks for safe keeping. A dozen quince accessions from that collection are now growing at the USDA genebank in Oregon. The USDA Agricultural Research Service maintains several important fruit germplasm collections at the National Clonal Germplasm Repository (NCGR), in Corvallis Oregon. The NCGR *Cydonia* collection includes more than 100 clones with origins from 15 countries maintained as self-rooted trees in a field collection (Postman 2008). About half of this collection represents cultivars for fruit production, and the other half are pear rootstock selections, wild types and seedlings. Observations made at the genebank have revealed a wide diversity of genotype resistance to Fabraea leaf and fruit spot [*Fabraea maculata* Atk. (anamorph = *Entomosporium mespili* (DC.) Sacc.)], and a range of ripening seasons that may make it possible to produce quince fruit in short season production areas (Postman unpublished). Several recent USDA-funded plant collecting expeditions to Armenia, Georgia, and Azerbaijan returned with quince seeds and cuttings from these countries. The availability of *Cydonia* germplasm available in the USA increased significantly from 2002 to 2006 as a result of these collections (McGinnis 2007). Selections made in Bulgaria after fire blight invaded that country have shown resistance to the disease, and some of this Bulgarian quince germplasm was recently introduced into the USA by NCGR.

Quince is adapted to hot, dry climates and to acid soils. Under favorable conditions ripe fruit can become quite fragrant, juicy and flavorful. When grown in high pH soils, however, trees can become stunted and chlorotic due to iron deficiency, a disorder referred to as "lime induced chlorosis." In northern latitudes or colder climates, the fruit of many cultivars does not fully ripen prior to the onset of winter, and in places where it rains during the time when fruit is ripening, fruit cracking can be a big problem. Quince, whether grown for fruit production or for use as a pear rootstock, is impacted by several disease problems. Fire blight (*Erwinia amylovora* (Burrill) Winslow) limits the cultivation of quince for either fruit or rootstock, especially in regions with warm, humid summers. The genus *Cydonia* is one of the most susceptible to fire blight in the family Rosaceae, which includes many genera that are hosts for this disease (Postman 2008). Leaf and fruit spot caused by can result in tree defoliation and production of disfigured, unmarketable fruits if not controlled. Powdery mildew caused by *Podosphaera leucotricha* (Ell. & Ev.) Salmon and various rust diseases can also impact quince production.

Genetic improvements needed for expanding the use of quince as a dwarfing pear rootstock include increased resistance to fire blight for warm and humid summer climates, and increased winter cold hardiness for northern climates. Adaptation to alkaline soils will allow quince production to expand to more diverse soil conditions both as a rootstock for pear and for production of quince fruit. Very slight progress in soil adaptation was achieved by selecting somoclonal variants of rootstock clone Quince A in high pH tissue culture (Bunnag et al. 1996). Quince for fruit production will benefit from earlier ripening, and elimination of summer 'rat-tail' blooms, which predispose a tree to attack by fire blight. Most quince genotypes are adapted to regions with long, hot growing seasons and will not ripen properly without adequate heat units. Fruits that are picked too green may never ripen in storage

(McCabe 1996). Resistance to the fungal rusts and mildews will allow quince to be produced with fewer pesticide applications.

For nearly a century, the quince has been ignored for fruit production in North America, while many improvements have been made in the Middle East and central Asia. Germplasm is available in the USA for expanding the use of *Cydonia* both as a rootstock for pear and as a fruit producing tree in its own right. As Luther Burbank concluded a hundred years ago, "The quince of today is, indeed, a half wild product that has waited long for its opportunity. It remains for the fruit growers of tomorrow … to see that the possibilities of this unique fruit are realized" (Whitson et al. 1914).

4 Mayhaws: A Multiuse Native Fruit

4.1 Botany

The genus *Crataegus* is a complex group of deciduous shrubs and small trees native to northern temperate zones (Mabberley 1997), mostly between latitudes 30° and 50°N (Phipps 1983). *Crataegus* belongs to the subfamily Maloideae in the Rosaceae, a natural group originally occurring as suppressed, understory trees in the virgin forests with the ability to interbreed (hybridize) freely because they possess the base haploid chromosome number of $x = 17$. Following the clearing of the dominant trees for human colonization, *Crataegus* underwent rapid proliferation and are now abundant in clearings, along streams, sloughs, river bottoms, and abandoned fields (Phipps et al. 1991; Robertson 1974; Robertson et al. 1991).

The genus has vexed so many authors that early experts on the group termed the situation "the *Crataegus* problem" (Eggleston 1910; Palmer 1932). Quantification of hawthorn species is controversial because of hybridization and unusual factors relative to reproduction, including (1) apomixis, (2) polyploidy, and (3) aneuploidy (Duncan and Duncan 2000; Phipps 1988; Talent and Dickinson 2007b). Apomixis, polyploidy, and hybridization blur the boundaries between species. At the end of the nineteenth and the early years of the twentieth century, workers described hundreds of species in ignorance of the occurrence of apomixis and polyploidy (Dickinson et al. 2007). Several recent studies now demonstrate that both apomixis and polyploidy are implicated in the complex variation seen in this genus in North America (Dickinson 1985; Muniyamma and Phipps 1979, 1984, 1985; Phipps 1984).

4.2 Origin and Domestication

Over 100 species of hawthorn have been described from North America (Phipps 1983; Phipps et al. 2003), but only those early ripening, edible southern US

Crataegus species including *Crataegus aestivalis* [Walter] Torrey & Gray, *C. opaca* Hook. & Arn., and *C. rufula* Sarg, are considered mayhaws (Bush et al. 1991; Payne et al. 1990). Mayhaws are atypical among the hawthorns in their early flowering period (from late February through late-March) and their early fruit ripening dates (late April to mid-May, central Louisiana, Zone 9A) (Craft et al. 1996). This arborescent shrub has outstanding ornamental characteristics such as, attractive foliage, showy blossoms, and clusters of brilliantly colored fruits. Mayhaws are native to the alluvial acid soils of rivers, streams and swamps from North Carolina to Florida and west to Arkansas and Texas (Clewell 1985; Craft et al. 1996; Godfrey and Wooten 1981; Phipps 1988; Radford et al. 1974; Sargent 1965; Vines 1977).

4.3 Production and Uses

Mayhaw fruit is a small pome (1.4–3.7 g), yellow to dark burgundy, fragrant, acidic, and juicy, of a high culinary value. Over the last 30 years, more than 70 cultivars have been selected from native stands and seedlings for improved color, fruit size, yield, and ease of harvest. Harvested fruit is processed into marmalades, butters, preserves, jellies, condiments, syrups, wines, and desserts (Gibbons 1974; Morton 1963; Payne et al. 1990; Reynolds and Ybarra 1984; Vines 1977). Superior clones are grafted onto rootstocks, but many orchards still consist of seedling trees. Fruits are hand harvested or mechanically shaken onto tarps and catch frames. The fruits can be processed fresh, refrigerated for a few days, or frozen for several months without loss of quality. Thus, the opportunity exists for a greatly expanded market based upon a consistent supply of fruits (Bush et al. 1991; Payne et al. 1990).

There is limited information on the pest management of mayhaws; however, they are susceptible to many of the insects and diseases that attack other pome fruits (Krewer and Crocker 2000; McCarter and Payne 1993; Moore 2006; Scherm and Savelle 2003). Several insects including plum curculio [*Conotrachelus nenuphar* (Herbst)], flower thrips (*Frankliniella spp.*), roundheaded appletree borer (*Saperda candida* F.), leafminers (many different insect species), terrapin scale [*Mesolucanium nigrofasciatum* (Pergande)], and mealybugs (Pseudococcidae family) feed on the foliage, flower, fruit, and wood of mayhaw. The plum curculio in particular has caused extensive damage to fruit in many locations (Krewer and Crocker 2000; Payne et al. 1990).

There are several major diseases of mayhaw including quince rust (caused by *Gymnosporangium clavipes* Cke. and Pk.), fire blight [caused by *E. amylovora* (Burrill) Winslow], and hawthorn leaf blight [caused by *Monilinia johnsonii* (Ellis & Everh.)]. Quince rust attacks both the leaves and fruit of mayhaw trees. Fire blight can be severe in many parts of North America and many mayhaw growers consider it the most limiting factor in mayhaw production. Blossoms, actively growing shoots, and immature fruits are most readily infected, but the trunks and roots may become infected as well (McCarter and Payne 1993). Lack of available chemicals to control insects and diseases continues to be a major deterrent for growers interested in starting new mayhaw orchards. However, great strides have been made in the labeling of pesticides, especially fungicides, for commercial production in the USA (Graham 2000).

4.4 Breeding Potential

4.4.1 Scion

Many selections of mayhaw have been evaluated for fruit quality, growth habit, and disease resistance (Craft et al. 1996; Krewer and Crocker 2000; Graham et al. 2000). While no single selection is resistant to quince rust, a wide range of tolerance exists among *C. opaca* cultivars. Hybridization of the most tolerant cultivars could lead to selections with a greater disease tolerance. Mayhaw cultivars have exhibited a wide range of relative susceptibility to fire blight. 'Maxine' has shown high resistance to fire blight in Louisiana orchards (Craft Personal communication). It has been used in controlled hybridizations, but the progeny are still in the juvenile stage.

Taxonomists are using flow-cytometric DNA measurements to elucidate relationships within and between *Crataegus* populations (Talent and Dickinson 2007a). Lo et al. (2007) used two nuclear and four intergenic chloroplast DNA regions to clarify the phylogeny of many *Crataegus* species. Species grouped with the cultivated species *C. aestivalis* [Walter] Torrey & Gray and *C. opaca* Hook. & Arn included *C. calpodendron* (Ehrh.) Medik., *C. crus-galli* L., *C. lassa* Beadle, *C. mexicana* DC, *C. mollis* Scheele, *C. punctata* Jacq., *C. triflora* Chapm., *C. uniflora* Munchh., and *C. viridis* L. These related species need to be evaluated for horticultural characteristics and disease resistance. Hybridization of these species with mayhaws may lead to novel fruit types with better disease resistance.

4.4.2 Rootstock

Currently, mayhaws are grown on seedling rootstock of *C. opaca* and *C. aestivalis* in the southern USA. Although mayhaw appears to be initially compatible on most *Crataegus* rootstocks, our knowledge of mayhaw rootstocks is rudimentary at best. 'Royalty' mayhaw was tested on seedlings of 'Annette,' 'Flame,' 'Redskin,' 'Super Spur,' 'Texas Super Berry,' 'Toledo Giant,' 'Turnage #57' and 'Warpaint' in a multiyear, replicated trial in Louisiana. There were no significant differences in trunk caliper, fruit number/tree, fruit weight/tree, fruit size, or yield efficiency detected among rootstocks (Graham et al. 2005). Trials using other hawthorn species for rootstocks in Louisiana has been limited predominantly to observational tests of *C. arnoldiana*, *C. azarolus* L., *C. brachyacantha* Sarg. & Engelm., *C. coccinoides* Ashe, *C. columbiana*, *C. crus-galli* L., *C. cuneata* Siebold., *C. douglasii* Lindl., *C. laevigata* (Poir) DC., *C. marshallii* Eggl., *C. mollis* Scheele, *C. phaenopyrum* (L.f.) Med., *C. punctata* Jacq., *C. uniflora* Munchh. and *C. viridis* L. Species deemed unacceptable and rogued from the test are *C. crus-galli* L., *C. laevigata* (Poir) DC., *C. marshallii* Eggl., *C. monogyna* Jacq., and *C. phaenopyrum* (L.f.) Med. (Craft 2003, Personal communication). In Mississippi, *C. marshallii* Eggl. is considered an excellent rootstock for *C. opaca* (McDaniel 1980). In Georgia, *C. flava* Aiton can be used, but due to its slow growth rate, the mayhaw scions may overgrow the rootstock. Mayhaw seedlings are currently the best choice as a rootstock in damp soils (Payne et al. 1990).

4.5 Breeding History

Improvement of native mayhaws began in the 1970s by selection of superior clones from native seedling stands. Most commercially important cultivars being grown in current orchards have originated in this manner. While selection in native stands for potential cultivars is not as efficient as controlled hybridization, it is still the most common method of introducing new cultivars. However, this practice has led to some confusion in the industry, since more than one person can collect scion wood from the same native tree and introduce it to the industry under different cultivar names.

Following selection from native stands, growers began screening seedlings derived from superior clones in several states. Progress is slowly being made for improved disease resistance and fruit quality. Controlled hybridization for mayhaw improvement was initiated in the late twentieth century (Craft et al. 1996). Initial breeding objectives included (1) late blooming clones, (2) improved fruit size, skin toughness, and flesh firmness, (3) increased fruit quantity per cluster, (4) reduced fruit shattering before maturity, and (5) improved cold hardiness of flowers. Currently, three cultivars developed by controlled hybridization have been released to the public. They are 'Red Majesty' ('Cajun' × 'Texas Star'), 'Abundance' ('Cajun' × 'Texas Star') and 'Double GG' ('Texas Star' × 'Royal Star'). All three of the cultivars have dark red skin, red flesh, bloom late, and are shatter resistant. Unfortunately, all are susceptible to fire blight (Craft Personal communication).

Current mayhaw cultivars are still deficient in many horticultural characteristics and many of the previous goals have not changed. Improvements that would benefit growth of the industry include (1) late blooming selections that reach peak flowering after danger from late frost is past, (2) shatter resistance, (3) high antioxidant levels, (4) uniformity of ripening within a plant for mechanical harvesting, and (5) resistance to fire blight for expansion of the industry.

Little breeding work has been done to improve mayhaw rootstocks. Present trends in the mayhaw industry toward intensive culture with high-density plantings and mechanical harvesting indicate that greater demands will be made for improved mayhaw rootstocks than ever before. To be adapted to the intensive system of culture, mayhaw rootstocks should have certain properties, including broad adaptability to varying climates and soils, resistance to major diseases and pests, good anchorage, compatibility with scion cultivars, dwarfing ability, and the capacity to induce precocious fruiting. No rootstock cultivars are available that possess all of the characters just mentioned, although the genetic resources exist in *Crataegus* to make the required improvements possible. Dwarfing germplasm may be found within *Crataegus*, but it requires a careful search of the available sources and the use of effective test procedures.

4.6 Breeding Methods and Techniques

Mayhaw flowers are produced slightly before or at the same time as the leaves. They are born in 2–5 flowered glabrous corymbs on short pedicels, usually on spurs, but also on terminal or lateral buds of previous season's growth. The flower consists of five white petals, a calyx of five sepals, 20 stamens, and a pistil divided into five styles. Hawthorn fruits are known as pomes, although the seeds and their bony endocarps are termed pyrenes, or nutlets (Vines 1977; Craft et al. 1996).

Flowers of the seed parent are emasculated at the balloon stage. Removal of the petal and stamens prevents self-pollination, exposes the stigmas, and minimizes insect visitation and possible contamination with unknown pollen. Collected pollen of the desired male parent is applied to the stigmatic surface using a brush, pencil eraser, or fingertip. Following pollination, clusters can be bagged to reduce possible pollen contamination and to reduce insect and bird deprivation.

Crossed fruits are harvested at maturity, and the fruits can be macerated to separate the seeds from the fleshy pericarp. The macerated pericarp material can be removed by water flotation; and if seeds are to be stored, they must be dried thoroughly and stored at 5°C. Stored seeds must be stratified at 5°C for 10–16 weeks prior to sowing. Unlike most fall ripening hawthorns, freshly collected mayhaw fruit can be fermented for up to 8 days and planted. Such treatment has resulted in over a 90% seed germination rate (Baker 1991).

Following seed germination, the evaluation procedure is similar to other pome and stone fruits. The seedlings are planted in nursery rows for initial selection, followed by testing as grafted trees at normal orchard spacing in a single location. The final step is to test the grafted selection at normal orchard spacing at multiple locations.

4.7 Integration of New Biotechnologies in Breeding Programs

The vision of orchardists and breeders and their skills of observation have served the mayhaw industry well. Improvements in biotechnology may help overcome some of the limitations of conventional breeding. Development of the technique of marker-assisted selection in other pome fruit, such as apple and pear, may provide an avenue for transferring the technology to mayhaw scion and rootstock breeding. Of course, this technique relies on sufficient markers being identified in mayhaw

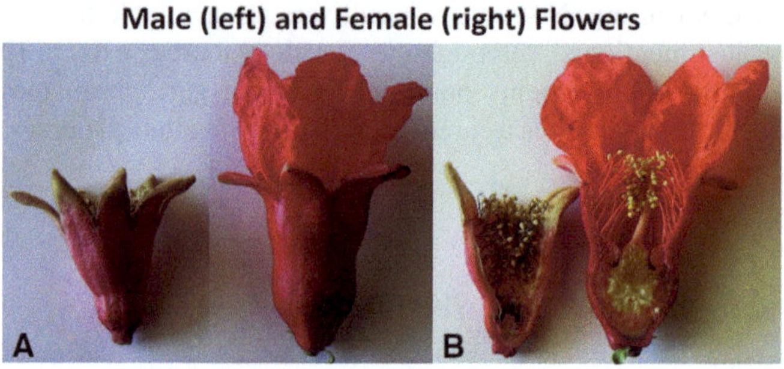

Fig. 4.4 Pomegranate flowers, male (l) and hermaphroditic (r) in (**A**) full and (**B**) cross-section. Photo credit: Jeff Moersfelder, USDA ARS

that are linked to important characteristics. Currently, complex characteristics influenced by many genes such as yield, flavor, enhanced color, and texture cannot be targeted by biotechnology. At present only traditional breeding can effectively manipulate polygenic traits which will ultimately lead to superior cultivar development and release.

5 Pomegranate

5.1 Botany

Pomegranate (*Punica granatum* L.) is subtropical and although naturally grows as a multitrunked small tree or large shrub (3–6 m at maturity), it can be trained to form a single trunk. Plants are typically deciduous, though evergreen types are noted (Singh et al. 2006). Branches are often spiny, with small, narrow, oblong leaves and short stems, and aggressive sprouts often develop from the crown area and the roots (Morton 1987). Flowers occur as single blossoms or in clusters of up to five and are usually borne subterminally on short lateral branches older than 1 year (El-Kassas et al. 1998), but some genotypes flower on spurs. Flowers are heterostylous: larger long-styled perfect flowers set more fruit than short-style types, which are often functionally male. Flowers are typically red to red-orange (Fig. 4.4) and funnel shaped and are self-pollinated or cross-pollinated by insects (Morton 1987). Double and variegated flowers are found in some ornamental selections. Period of bloom may be very prolonged, but most flowering in the Central Valley of California occurs from mid-May to early June.

The fruit is berry-like and has a prominent calyx which is maintained to maturity and contributes to the fruits' distinctive shape as observed in the cultivar 'Wonderful,' (Fig. 4.5) which is an industry standard. The leathery rind includes a pericarp, comprising a cuticle layer and fibrous mat, and the mesocarp which is the inner fruit wall and is further elaborated into membranes dividing a number of locules.

Fig. 4.5 Pomegranate 'Wonderful' fruit. Photo credit: Malli Aradhya, USDA ARS National Clonal Germplasm Repository, Davis, California

The juicy arils are the edible portion of the fruit, are attached to the mesocarp, and are derived from epidermal cells. In different cultivars, arils range from deep red to virtually colorless, seed softness varies greatly based on content of schlerenchyma tissue, and acidity varies from 0.2 to 3% of the expressed juice. At maturity, soluble solids are quite high (15–20%) and differing levels of acid result in fruits which range from sweet to sweet/tart to very tart indeed.

Most pomegranate genotypes root extremely easily and are seldom grafted. Orchards are sometimes established by direct planting of unrooted cuttings (Blumenfeld et al. 2000). Pomegranate is especially well adapted to hot summer/cool winter Mediterranean climates, but can be grown in the humid tropics or subtropics, and is injured by temperatures below −11°C (Morton 1987). Dry summer climates are most conducive to commercial production. While extremely drought-tolerant, pomegranate crops better with regular moisture. Pomegranate has high salinity resistance and is adapted to a wide variety of soils (Melgarejo 2003).

5.2 Origin and Domestication

Pomegranate is one of only two species in its genus, *Punica*, which is the sole genus in the Punicaceae (ITIS 2006). This fruit was likely dispersed by humans in

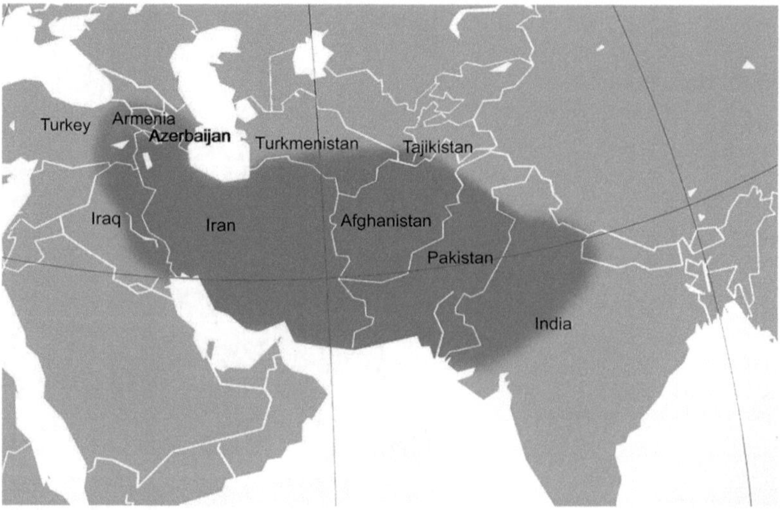

Fig. 4.6 Center of origin for Pomegranate (*Punica granatum* L.). Prepared by Ed Stover, USDA ARS

prehistoric times and interpretation of the original native range varies among authors, but Iran and the surrounding area (Mars 2000) are widely accepted (Fig. 4.6), while others extend the region of origin more broadly (Morton 1987). In general, wild pomegranate fruits have thicker rinds, extremely high acidity, and smaller arils compared to cultivated types (Bist et al. 1994; Kher 1999). Human use of pomegranate has a long history, with cultivation projected as early as 3000 BCE (Stover and Mercure 2007). Pomegranates are important in the symbolism and literature of many middle-eastern cultures. It is one of the symbols of the love goddess Aphrodite (Encyclopedia Britannica 2006), is central to the Greek myth of Persephone, and is mentioned three times in the Qur'an and 23 times in the Hebrew Bible (Janick 2007).

5.3 Production and Use

Pomegranate is widely grown in many countries where it is well-adapted, but no global production estimates are available. India produces pomegranate on more than 100,000 ha and it is considered one of the most important fruits of tropical and subtropical areas of that country (Indian Council of Agricultural Research 2005). In Iran, 600,000 tons of pomegranate are produced annually on 65,000 ha, and 30% is exported (Mehrnews 2006). Turkish production in 1997 was 56,000 tons (Gozlekci and Kaynak 2000). Spain is the largest Western European producer with ~3,000 ha in 1997 and expectation of continued growth (Costa and Melgarejo 2000). Production is also growing in the USA, with 5,600 ha of commercial pomegranate (mostly in the San Joaquin Valley) in 2006, largely dominated by the cultivar 'Wonderful' (Kotkin 2006).

US production of pomegranate has expanded as a result of reported health benefits from consuming the fruit and its juice. Antioxidant content of pomegranate juice is among the highest of any foods (Guo et al. 2003), and it is reported that these polyphenol compounds may lower risk of heart disease (Aviram et al. 2004) and slow cancer progress (Adams et al. 2006). Pomegranate also has a range of wonderful flavors, ranging from mild watermelon or strawberry-like flavors in low-acid types to bright cherry and cranberry-like flavors in sweet/tart cultivars. California commercial orchards are reportedly expected to produce mature yields of up to 33 tons/ha (Karp 2006). With reported health benefits, high juice yields, modest pest pressures, and relatively undemanding handling of fruit for processing, pomegranate juice production should continue to grow in commercial importance.

The pomegranate fruit is not climacteric (Kader et al. 1984), has a storage life equaling the apple, and ships very well (Morton 1987). However, greatly increased consumption of fresh pomegranates may be unlikely, as many consumers are daunted by peeling the fruit and extracting the arils. The greatest fresh fruit potential appears to be in minimally processed arils (Sepulveda et al. 2000), for eating as snacks and use as a garnish.

5.4 Breeding Potential

There are innumerable pomegranate cultivars, and germplasm collections have been established in many countries. More than 1,000 accessions were assembled in the Turkmenistan Experimental Station of Plant Genetic Resources (Levin 1995). Collections of 200–300 accessions are maintained in Azerbaijan, The Ukraine, Uzbekistan, and Tajikistan. Local cultivars have been conserved in many Mediterranean and Middle Eastern countries (Spain, Morocco, Tunisia, Greece, Turkey, Egypt) (Mars 1996). India has three collections containing at least 30 accessions each (Gulick and Van Sloten 1984). There are diverse genotypes (238 reported cultivars) within China (Feng et al. 2006). The US National Clonal Germplasm Repository (NCGR), in Davis, Calif., has almost 200 pomegranate accessions, including many obtained from the Turkmenistan collection. Included in the NCGR are many soft-seeded types, sometimes called 'seedless.' The NCGR policy is to distribute plant material, free of charge, to research interests around the world (see our Web site http://www.ars-grin.gov/dav/).

Most pomegranate cultivars likely arose through selection among chance seedlings through millennia of cultivation. Recently, directed plant improvement efforts have been employed in several countries. India, China, and Israel appear to have the most developed and sustained pomegranate breeding programs (Feng et al. 2006; Jalikop et al. 2006). The characteristics of greatest interest have been similar in most programs: bigger fruit, larger arils, greater juice yield/thinner skins, attractive fruit and aril color, soft seeds, altered time of maturity, good soluble solids and acid levels, less fruit splitting, and sometimes resistance to diseases. In China, 50 cultivars are reportedly being grown from these directed programs and selection of sports (Feng et al. 2006). There is one report of a Chinese cultivar yielding 90% juice (Zhu et al. 2004). 'Mridula' and 'Bhagwa,' two important cultivars exported from India, are products of controlled crossing and selection. Pomegranate cultivars have also been released from breeding programs in the USA. There appears to be a considerable potential for further development of improved pomegranate cultivars, perhaps including selection for higher levels of health-promoting compounds.

6 Fig

6.1 Botany

Cultivated fig (*Ficus carica* L.) trees are deciduous, spreading in habit, and fast-growing. Where freezing or other damage does not disrupt tree structure, figs grow into single-trunked trees with little training. They also root easily and so are seldom grafted.

The fig is a composite 'fruit' called a 'syconium' (reviewed in Condit 1947), comprising a shell of receptacle tissue enclosing hundreds of individual fruits, which are drupelets developing from the female flowers lining the receptacle wall. The syconium has a small opening (called the ostiole or eye) at the distal end. The mature edible fig fruit has a thin skin which may be somewhat tough, a pale interior rind, and a sweet gelatinous pulp comprising the individual ripe drupelets. The seeds (achenes) within the drupelets range from virtually nonexistent to subtly crunchy.

The fig has a distinctive pollination biology which is important in commercial production. It is gynodioecious, with both male and female flowers produced in wild figs and cultivated caprifigs (which are grown to provide pollen), while fruiting cultivars produce only functionally female flowers, though aborted hermaphroditic flowers surround the ostiole (Beck and Lord 1988). The female flowers in edible figs are long-styled and produce a much more succulent drupelet than do the female flowers in the short-styled monoecious wild-type figs. Figs are distinguished by their cropping/ pollination characteristics. The type called 'common figs' are edible figs requiring no pollination to set a commercial crop. The other two types of edible fig require pollination to set the main crop of figs on current season's growth: Smyrna types (e.g. 'Calimyrna') and San Pedro types (e.g. 'King'). The San Pedro types are distinguished by also setting a crop, but without the need for pollination, on previous season growth.

The wasp (*Blastophaga psenes* L.), which has coevolved with the fig (Kjellberg et al. 1987), carries the fig pollen. The protogynous nature of the wild-fig/caprifig is a critical aspect of the wasp–fig coevolution. Female flowers are receptive 6–8 weeks before anthers mature in the same syconium (Condit 1932): this permits wasps to enter, pollinate, and oviposit in syconia which will have mature pollen during the emergence of the next wasp generation. The wasp larvae cannot mature in edible figs, so the life cycle is completed solely within caprifigs.

6.2 Origin and Domestication

The species in the genus *Ficus* ranges in number from 600 to more than 1,900, according to different taxonomists, with most found in the tropics or subtropics and just a few producing palatable fruits (reviewed in Condit 1955). Edible figs reportedly became established across the Mediterranean region around 6,000 years ago (Ferguson et al. 1990). Archeological evidence from the Jordan Valley suggests it is one of the earliest plants domesticated, 11,000 years ago, much earlier than wheat, barley and legumes (Kislev et al. 2006). Storey (1975) proposes that the long-styled pistils and succulent fruitlet of the edible fig resulted from a single mutation in the wild fig, and this trait was key to domestication. The edible fig is well adapted to high temperatures and drought and is commonly planted in home gardens throughout regions with Mediterranean climates. Commercial plantings range in scale from small production for local markets to large mechanized farm operations. Even though most of world's figs are eaten fresh, their short market-life restricts them to largely local consumption. By contrast, high sugar content and stability make dried figs easily transportable, and therefore most fig exports are as dried fruit.

6.3 Production and Uses

Worldwide, over one million MT of figs are harvested annually from 427,000 ha (FAO 2006). The largest fig producer is Turkey with ~26% of the world's figs, and producer price is listed as $892 per ton (for dried product) for 2005 (FAO 2007). Turkey, Egypt, Iran, Greece, Algeria, and Morocco are the top six global producers and account for ~70% of world annual production. Data from 2005 indicate that the USA ranked eighth with 4% of global fig production. Commercial fig production is reported in fourteen US states. However, 98% of the US crop is produced in California, on 5,100 ha, with yields per ha three times the global average. Almost all California production is in the San Joaquin Valley, with ideal conditions for both fig production and the drying of figs under ambient conditions.

Most Western European and US fig production relies on common figs which do not require pollination. However, generally the major world fig producers, and many California orchards (of 'Calimyrna' fig), focus on production of Smyrna-types which do require pollination. Separate orchards of caprifig trees are maintained to control pollen flow. For edible figs requiring pollination, mature caprifig fruits are typically supplied three times in California fig production, at regular intervals in

May–June. Ideally each fig is entered by only one wasp, since fruit splitting can result from excessive pollination and the entry of multiple wasps increases the risk of introducing microorganisms which cause internal defects.

Mature tree height varies by cultivar and typically ranges from 3 to 10 m. Since most fruit is on current season's growth, fig trees can be pruned aggressively and remain productive. Orchards for fresh figs are typically pruned low for ease of harvest. Some growers now produce figs on small plants trellised and pruned like grapevines.

Worldwide, fig production is small compared to major commodities such as apples, bananas, and citrus fruits. However, at a third to half the global production of familiar crops such as apricot and sweet cherry, the fig may have transcended the threshold for being considered a minor crop. Most commercial efforts have focused on dried fig production, and most growth potential for fig appears to revolve around greater consumer access to top-quality fresh figs.

In Mediterranean and Middle-Eastern countries, most consumers already enjoy fresh figs. The interest in commercial fresh fig production in California has increased with consumer demand for diverse premium produce. Total California production of processed fig was fairly constant from 2000 to 2005, but fresh fig production doubled in this period, so that in 2005, fresh figs represented more than 9% of California commercial production (NASS 2006). Consumer prices for fresh figs are quite high and are similar to those for raspberries and blackberries. It seems likely that successfully marketing fresh figs provides greater grower profits than does dried fig production.

The greatest limitation to expansion of fresh fig sales is the very short shelf life of fresh figs. Currently a useable life of 3–10 days from harvest to sale is typical. Advances in postharvest handling and/or development of varieties with better shelf-life are sorely needed. Fresh fig quality is greatest at tree ripeness, when the pedicel begins to sag, but such figs are very soft, sensitive to damage (Chessa 1997) and have very short market life. Therefore, harvest at early ripeness (good color) is essential for current fresh-fig commercial sale.

Leading commercial fig cultivars in the USA have mostly been grown for drying and feature mild honey, melon, or mild berry-like flavors with little balancing acidity. Visitors to the US fig genebank are often delighted by the bright fruity flavors, reminiscent of berries or citrus, of some fig varieties which are not yet grown commercially. The potential for broader fig appreciation may be demonstrated by a quote from the prophet Mohammed indicating, "If I could wish a fruit brought to paradise it would certainly be the fig" (Condit 1947).

6.4 Breeding Potential

The National Clonal Germplasm Repository (NCGR) in Davis, California houses the US collections of most of the Mediterranean-adapted fruit and nut crops. The NCGR fig collection currently includes: 78 named fruiting cultivars, 44 regional

Fig. 4.7 Fruits of Brown Turkey fig. Photo credits: Louise Ferguson, University of California, Davis

selections from diverse areas of the world, 40 advanced selections from plant breeders (mainly from the UC Riverside breeding program), 28 caprifigs, and a small number of species and hybrids. The named cultivars in the NCGR collection represent a fair cross-section of figs from major old-world growing areas and represent the largest collection in North America. It is our policy to distribute plant material, free of charge, to research interests around the world (see our Web site http://www.ars-grin.gov/dav/). Other large international collections include the Conservatoire Botanique National Méditerranéen in Porquerolles, France and the Instituto Valenciano de Investigaciones Agrarias, in Spain. Turkey and many other countries have invaluable collections of local cultivars.

Domestication and early human selection for edible figs from among chance seedlings contributed many of the fig types grown today in gardens and commercial orchards (Fig. 4.7). Few important fig cultivars arose through a planned breeding program. A sustained fig improvement program was maintained by the University of California at Riverside from 1928 to 1980s, by Ira Condit and William Storey. The focus of this effort was development of drying figs with 'Calimyrna'-like quality without the need for pollination and with a small ostiole (reviewed in Storey 1975). Doyle et al. (2003) of the University of California at Davis recently released the 'Sierra' fig for drying and the 'Sequoia' fig, which is expected to find a place in fresh fig production. Louisiana State University has released four new fig cultivars in the last 6 years, continuing their efforts to identify material well adapted to the humid southeatern USA (O'Rourke et al. 2005; Johnson et al. 2010a, b, c). Other efforts are ongoing in the USA and in other countries.

Existing commercial fig cultivars vary markedly in post harvest qualities (Stover et al. 2006), suggesting that breeding efforts to enhance and pyramid desirable traits should provide improved varieties. Dark figs generally show less marking as they pass through steps necessary for marketing, so a focus on highly pigmented varieties may prove desirable.

There are significant opportunities to develop cultivars with enhanced production of fruit for fresh sales both early and late in the production season. Brebas are the first figs of the season, setting on wood from the previous year, and typically maturing in June in the Central Valley of California (vs. August through October for main crop fruit). Brebas tend to be larger than main crop figs, are relatively scarce on the market and tend to get a high price as fresh fruit. The cultivar 'King' is especially noteworthy for producing a high proportion of brebas (with only modest quality) and may prove a useful parent for enhancing breba production. Some varieties tend to be much later than others, with continued production well into November and sometimes December in our collection, and may serve as parents in an effort to enhance late season production.

7 Blue Honeysuckle

7.1 Botany

Blue honeysuckle, honeyberry or haskap (Thompson 2006), *Lonicera caerulea* L., is in the family Caprifoliaceae, section Isika Rehd., subsection Caeruleae Rehd. It is a polymorphic, circumpolar species, with several ecogeographic forms, designated subspecies or sometimes, separate species. Plants are deciduous shrubs, to 2 m or more in height. Leaves are simple, opposite, oval to elongate, 3–5 cm in length. At each node, there are 3 buds, one above the other. Normally the most advanced, lower buds on previous year's growth, develop into one, or rarely 2, shoots the current year, whereas the uppermost buds remain dormant until a few years later, when vigorous shoots may emerge from older wood lower down in the shrub. Pairs of flowers are borne at the lowest one to 4 nodes of current year's shoots. Compared to ornamental *Lonicera* species, flowers are small, about 2 cm long, tubular with flared lobes, pale yellow to cream-colored. Each flower consists of two tubular corollas atop what appears to be a single ovary, but that actually consists of two ovaries surrounded by fleshy bracts. Plants are essentially self-incompatible and require bees for cross-pollination. Because blooming occurs rather early in spring when temperatures are unfavorable for honey bees, bumble bees are the principal pollinators. Also, blue orchard bees (*Osmia* sp.) are used in Japanese plantings. Fruits are dark blue to purple berries, with varying amounts of a white waxy covering, or bloom. Shapes are variable ranging from oval to long and thin. Size ranges from 0.3 g to rarely over 2.0 g. Flavors are unique and vary considerably; from a pleasant mild taste, more sprightly tart-sweet, mildly tart, very tart to slightly or very bitter. There is a maximum of 20, but usually fewer, very small seeds in a fruit.

Climatic adaptation varies with the subspecies. In Russia, cultivars developed from *L. c.* subsp. *kamtschatica*, native to NE Russia, are more successful in NW and

NE regions, whereas cultivars developed from *L. c.* subsp. *edulis* or *L. c.* subsp *boczkarnikovae*, native to SE Russia and Central Siberia perform well in those regions. Russian cultivars are extremely cold-hardy in long severe winters (to −50°C), but are not successful in moderate climates with fluctuating winter temperatures (Plekhanova et al. 1993). Following completion of very short chilling requirements by October–November, plants lose their hardiness when temperatures rise to +5 to 10°C. Then as cold weather resumes, there occurs varying degrees of freeze damage, depending upon the temperature severity. The Japanese *L. c.* subsp. *emphyllocalyx*, native to Hokkaido and northern Honshu, appears to be more adapted to moderate climates. Plants bloom a few weeks later than Russian types. Winter cold hardiness is not known but is under test in Saskatchewan. Over a 6-year period, plants have performed very well in western Oregon and northern Idaho (Thompson and Barney 2007). In spite of very early blooming, spring frosts are not a hazard: Russians claim that flowers are hardy to −7°C. For optimum performance, all forms of blue honeysuckle require good soil moisture conditions and moderately warm summer temperatures. Plants are relatively free of serious pests and diseases and tolerant of a wide range of soil types.

7.2 Origin and Domestication

Historically, blue honeysuckle berries were harvested from wild plants by local people in regions where edible forms exist, primarily in Russia and in Hokkaido, Japan. Folklore in both regions has long attributed high nutritional and medicinal values to these berries, a fact supported by recent phytochemical analyses (Chaovanalakit et al. 2004; Plekhanova et al. 1993; Tanaka and Tanaka 1998). Domestication of this crop occurred only in the twentieth century. Minor efforts to develop this crop in Russia date back to 1913–1915 but not until the 1950–1960s was a serious research program initiated (Plekhanova 2000). Extensive germplasm explorations were conducted by the Vavilov Institute for Plant Industry in Leningrad (now St. Petersburg), and plants were distributed to the Lisavenko Research Institute for Horticulture in Siberia at Barnaul and several other research stations in the USSR for evaluations. From these studies, over 200 cultivars have been named and distributed to farmers and gardeners.

Currently, blue honeysuckle plants are widely grown in Russia, mainly in gardens, but also in some commercial plantings. No production statistics are available. As the first fruit of the season, and because of the widely acknowledged healthful attributes, this berry is very popular in Russia. No doubt, due to the isolation of Russia from the rest of the world during the period of its domestication, this crop was virtually unknown elsewhere. In recent years, with open exchange of information and plant materials, there has been increasing interest in northern European countries. Also, a small industry, based on introduced Russian cultivars and wild-gathered berries, is developing in Jilin province in N.E. China (Huo et al. 2005). In North America, a few Russian cultivars (sold as 'honeyberries') were introduced several years ago and are available in many nurseries but, as yet, these are mainly on trial in home gardens. As expected, when planted in the moderate climatic regions in most of the USA, plants have failed to perform satisfactorily. In the past decade, a research program at the University of Saskatchewan in Canada has stimulated considerable enthusiasm for this new crop. A few selections have been made and a haskap grower's organization established for promotion of this berry that performs well in the severe climate of the northern prairie region (http://www.haskap.ca).

Domestication of the Japanese ssp. *L. c.* subsp. *emphyllocalyx* occurred even more recently. Beginning in the late 1960s and 1970s both the Hokkaido Prefectural Agriculture Experiment Station and a Farmers Coop in Chitose began to make selections from the nearby wild populations in the Yufutsu Plains near Tomakomai city. This region was famous for the abundance of fruiting shrubs from which people had long collected wild berries. Selections were evaluated for several years and the best few distributed to farmers, including one named cultivar, 'Yufutsu' (Tanaka et al. 1994). Small scale commercial production began in the 1970s and increased to 195 ha by 1991. During this initial enthusiasm for haskap (the Ainu word used by the Japanese for this berry) a large array of high quality, high priced processed products were developed and have become popular as gift items. However, due to the high cost of labor, and the lack of mechanical harvesting, by 2005 the area under production had decreased to 85 ha with an estimated 200 tons of berries, an insufficient amount to satisfy the demand that had been created (Lefol 2007).

7.3 Production and Uses

With its unique flavors and high nutritional values this tart/sweet berry (Fig. 4.8) should receive good acceptance by consumers, especially as processed products. Haskap berries are expected to fill a niche market in specialty food stores where they will attract health-conscious customers and those who seek organically grown products. Thus far, because of the lack of significant pests or diseases, haskap appears to be a good candidate for organic growers. As it is the earliest fruit to mature, these plants make a good complement to other berries with similar culture (e.g. blueberries) by spreading the harvest season. When cultivars are developed with milder taste (i.e., higher sugar–acid ratios), and fruits firm enough for prolonged storage, there is also potential for the fresh market. In addition to commercial production, this berry is an excellent plant for the home gardener because of its easy care.

The two major restraints to production in the USA are unfamiliarity with this berry and dearth of well-tested cultivars to recommend to growers. The first step to develop a successful new crop is the selection of superior cultivars. This requires breeding programs. As funding for horticultural research is usually driven by grower and processor demands, without these pressures funds are very difficult to obtain. In order to promote this new berry crop, plants must be available in nurseries and growers must provide berries for processors and consumers. For commercial

Fig. 4.8 Russian blue honesuckle *Lonicera caerulea* 'Morena.' Photo credit: Kim Hummer USDA ARS

production, it is essential that cultivars be suitable for mechanical harvesting. Although it appears feasible, this aspect has not been tested, as yet, but is under consideration for future research.

7.4 Breeding Potential

Within the relatively limited germplasm available in the USA, there is a wide range of variability in all traits so there is good potential to select cultivars that satisfy needs of both commercial production and the home garden or small U-pick farms. The potential of available diversity has not yet been fully exploited. However, additional germplasm from Japan is desirable to increase the range of diversity available for future breeding. In North America, there are two main sources of germplasm. In the USA, the USDA/ARS National Clonal Germplasm Repository in Corvallis, Oregon, has a small collection of Russian cultivars (Hummer 2006). In 2000, at this same location, and with collaboration with the University of Idaho, Sandpoint REC, in Sandpoint, ID, Oregon State University initiated the first, and only, genetic improvement program in the USA. This small breeding program, using primarily the Japanese *L. c.* subsp. *emphyllocalyx*, has several selections under trial. In Canada, the University of Saskatchewan has a much larger collection of Russian cultivars, as well as some Japanese germplasm on trial. In this region, where the climate is similar to that of Siberia, Russian cultivars and hybrids are doing very well. Over the past decade, there has been an active breeding program which has already released a few cultivars, 'Tundra' and 'Borealis,' and there is much enthusiasm among farmers to grow this crop. University scientists have been in discussions with Japanese processors concerning a possible export market (Lefol 2007). At the Vavilov Institute of Plant Industry (VIR) in St. Petersburg Russia, there has been a blue honeysuckle research program for several decades. There, exists the largest collection of blue honeysuckle germplasm in the world, 500 accessions as of 2000 (Plekhanova 2000). There are several other selection programs in Russia; e.g. at the Siberian Horticulture Institute in Barnaul and in VIR, Vladivostok. Recently, in Japan, a breeding program was initiated at Hokkaido University (Takada et al. 2003).

8 Elderberry

8.1 Botany

The genus *Sambucus*, which includes the edible elderberry (Fig. 4.9a, b), is presently classified as a member of the family *Adoxaceae* (Donoghue et al. 2003), though long considered a member of the *Caprifoliaceae*. This small family of five genera and approximately 200 species is distributed in northern and southern hemispheres and is primarily temperate and tropical montane in distribution (Stevens 2007). The genus *Sambucus* includes 9–20 species with nearly worldwide distribution. Three closely related species, *S. canadensis* L., *S. nigra* L., and *S. cerulea* Raf. are of commercial interest for fruit, blossoms, and other plant parts. The relationship of these three species is of some discussion; Bolli (1994) classified all three as subspecies of *S. nigra*. Others classify them as distinct species (Yatskievych 2006), noting differences among the species in leaf form, number of rhizomes formed, berry characteristics, and anthocyanin profiles. The American elderberry, *S. canadensis*, is native to eastern North America from Nova Scotia to Manitoba and south to Florida, Texas, the Caribbean islands, and Mexico. The European or black elderberry, *S. nigra*, is native to Europe, northwest Africa, and western Asia. The blue elderberry, *S. cerulea*, is native to western North America from British Columbia to California and east to Montana and Utah. At present, most commercial interest is centered on the American and European elderberries.

The plant is a medium to large shrub or small tree with spreading roots. American elderberry suckers freely from the root system; European elderberry is less prone to

Fig. 4.9 Elderberry (**a**) flower and (**b**) fruit. Photo credits: Patrick Byers, University of Missouri

suckering. The leaves are opposite, pinnate, and 5–30 cm long, with 5–9 leaflets (usually five leaflets in *S. nigra* and seven leaflets in *S. canadensis*) with serrated margins. The bark is gray to yellowish brown, often appearing roughened or warty. The flowers (Fig. 4.9a) are borne in dense cymes, usually terminal on the branches. Individual blossoms are white to pink, usually 3–5 mm in diameter. The fruits

(Fig. 4.9b) are rounded berry-like drupes, 4–7 mm in diameter, that are orange-red to bluish black at maturity. The plants are hardy and long-lived.

The elderberry is an adaptable plant, as might be inferred from its broad range. The native range of *S. nigra* stretches from Norway (63°N) to the Mediterranean basin (Atkinson and Atkinson 2002). *S. canadensis* is found from eastern Canada (45°N) to subtropical areas of the North American Gulf Coast, Mexico, and the Caribbean Islands. The plant is tolerant to a range of soil types and exposures, but is typically found in moist, well-drained soils in full sun.

8.2 Origin and Domestication

Bolli (1994) proposes a center of diversity for *Sambucus* in central Asia, with the parent type established perhaps as long ago as the Oligocene. Dispersal of the genus possibly took two routes—west to Europe, North America, South America, and northern Asia; and east to southeast Asia and Australia. A second center of diversity is North America (Eriksson and Donoghue 1997). The genus at present is widely distributed; several species have circumboreal ranges. Natural dispersal was likely assisted by birds and other animals. Humans were also important in dispersal as elderberry is naturalized throughout much of the temperate and subtropical regions where humans live (Ritter and McKee 1964).

Although elderberry was widely utilized in traditional medicine and as a food source in the New and the Old World, records of cultivation are scanty and most fruits were probably harvested from the wild. Commercial production of elderberry began in the late nineteenth century.

8.3 Production and Uses

Commercial production of European elderberry is well established. While actual production figures are difficult to obtain, sizeable plantings are found in Austria, Hungary, Denmark, Poland, Switzerland, and Italy (Charlebois 2007; Kaack Personal communication; Lee and Finn 2007). The 2006 Austrian crop was estimated at 7,400 tons (Statistik Austria 2008). A considerable amount of the European elderberry crop is harvested from the wild (Kaack Personal communication). Commercial production of American elderberry is much less, with sizeable plantings reported only from Oregon (Lee and Finn 2007) and Missouri. While commercial scale plantings are increasing, much of the American elderberry crop is also harvested from the wild. Historically, 2,000–2,500 tons of wild fruits were harvested annually in the 1960s in Pennsylvania, Ohio, and New York (Darrow 1975).

The elderberry, though considered a minor fruit crop, is of increasing interest worldwide (Charlebois et al. 2010). The ripe fruit is processed into jelly, juice, and juice blends, wines and other alcoholic beverages, a heat stable colorant, and flavoring for a wide range of products. The blossoms are eaten fresh in various preparations, dried for teas, and used to flavor wines and other products such as enhanced waters and candies. Considerable interest worldwide is focused on elderberry as a nutraceutical (Charlebois 2007). A wide range of health benefits are claimed for elderberry as the ripe fruits are rich in anthocyanins and other substances with antioxidant properties (Lee and Finn 2007). The blossoms and other plant parts also have appreciable amounts of antioxidants (Thomas et al. 2008).

While superior wild plants of both species were likely propagated and cultivated from ancient times, organized efforts to improve the elderberry are recent (Way 1957, 1981). Many of these early wild selections are still currently grown, including the American elderberry cultivars 'Adams 1' and 'Adams 2' and several European elderberry cultivars (Kaack Personal communication). Efforts to select superior wild plants continue at present (Byers and Thomas 2005). Organized breeding efforts included programs at the New York Agricultural Experiment Station; the Kentville, Nova Scotia experiment Station; the Research Center for Horticulture in Arslev, Denmark; and several private breeding efforts.

8.4 Breeding Potential

As might be expected with a genus of worldwide distribution, a considerable amount of variability is present within and among *Sambucus* species. Two basic chromosome karyotypes are recognized in *Sambucus*, $2n=38$ (*Sambucus cerulea* (including synonyms *S. glauca* Nutt. and *S. mexicana* Auct.), *S. racemosa* L., *S. racemosa* f. *stenophylla* (Nakai) H. Hara (syn. *S. sieboldiana* var. *miqueli* [Nakai] H. Hara), *S. racemosa* subsp. *kamtschatica* (E. L. Wolf) Hultén (syn. *S. kamschatca* E.L. Wolf), *S. racemosa* subsp. *sibirica* (syn. *S. siberica* Nakai), *S. racemosa* subsp. *sieboldiana* (Miq.) H. Hara (syn. *S. sieboldiana* [Miq.] Blume ex Graebn), *S. racemosa* var. *arborescens* (Torr. & A. Gray) A. Gray (syn. *S. callicarpa* Greene), and *S. racemosa* var. *melanocarpa* (A. Gray) McMinn) and $2n=36$ (*Sambucus canadensis* var. *laciniata* A. Gray (syn. *Sambucus simpsonii* Rehder, *S. williamsii* Hance), *S. canadensis, S. nigra, and S. ebulis L.*) (Ourecky 1970). Interestingly, *S. racemosa* is reported to have three karyotypes, $2n=36$, 38, and 42 (Chia 1975). The following interspecific hybridizations among *Sambucus* species are reported: *S. canadensis* × *S. cerulea* (Slate 1955), *S. canadensis* × *S. pubens* Michx. (*S. racemosa* subsp. *pubens* (Michx.) House) (Eaton et al. 1959), *S. nigra* × *S. racemosa* (Koncalova et al. 1983), *S. nigra* × *S. ebulis* (Koncalova et al. 1983), *S. canadensis* × *S. nigra* (Chia 1975), and *S. cerulea* × *S. nigra* (Chia 1975).

In a discussion of *S. canadensis* and *S. nigra*, Lee and Finn (2007) note that variability in several traits of interest is present and should allow for selection of superior progeny through traditional breeding. From personal observation, considerable variability is present in wild populations of these species, and selection can often be made for traits of interest among wild plants. *Sambucus canadensis*, for example, is a likely source for large individual fruit clusters and profuse annual suckering from the root system as well as a potential source for acylated anthocyanins (Lee and Finn 2007). Other species may also offer traits of interest to plant breeders, including *S. cerulea* for its large and attractive berries with a heavy layer of surface bloom and *S. pubens* for its early ripening (Eaton et al. 1959).

Over 100 years have passed since the description of one of the first American elderberry cultivars, 'Brainerd' in 1890 (Bailey 1906). Ritter and McKee (1964) describe the development of improved elderberry cultivars. Most early cultivars were selected from the wild, such as the cultivars 'Adams 1' and 'Adams 2' selected by William W. Adams in New York in 1926 and released by the New York Agricultural Experiment Station. 'Ezyoff,' of unknown parentage, was introduced by Samuel H. Graham of Ithaca, New York, in 1938. More recent breeding efforts at the Agriculture and Agri-Foods Canada (Kentville, Nova Scotia) experiment station have resulted in 'Nova,' 'Scotia,' 'Kent,' and 'Victoria,' all released in 1960, as well as the release of an older selection, 'Johns,' in 1954. The more recent Nova Scotia releases are all seedlings of either 'Adams 1' or 'Adams 2.' 'York' (1964), is a cross of 'Ezyoff' and 'Adams 2' and was developed by the New York Agricultural Experiment Station. The University of Missouri/Missouri State University development program has recently released two cultivars, 'Bob Gordon' and 'Wyldewood,' both wild selections (Byers et al. 2010, Byers and Thomas 2011). Although the origins of many European elderberry cultivars are unclear, many undoubtedly are selections from the wild, such as 'Korsor' (Denmark), 'Allesø' (Denmark), and 'Mammoth' (Germany). 'Haschberg' was developed in an Austrian breeding program. Recent breeding efforts at the Research Center for Horticulture in Arslev, Denmark, have produced a series of cultivars particularly suited for juice production, including 'Samyl,' 'Samidan,' 'Sampo,' and 'Samdal' (Kaack 1989). Little improvement is reported for *S. cerulea*; Luther Burbank released the cultivar 'Superb' in 1921.

Breeding objectives for elderberry include large berry size, firmer berry texture, large berry cluster size, small seeds, self fruitfulness, increased productivity (number and size of cymes and berry size), vigorous and strong canes, uniformity of ripening within and among clusters, attractive color (glossy, dark), better fruit and juice quality, increased nutraceutical content, resistance to shattering, resistance to diseases, immunity or tolerance to virus diseases, wider adaptation, and pendulous fruit clusters less prone to bird damage (Darrow 1975; Kaack et al. 2008; Lee and Finn 2007). The Danish breeding program is seeking plants that are low growing with strong upright shoots from the root or lower part of the bush, characteristics that improve harvest efficiency (Kaack 1989). The University of Missouri/Missouri State University development program, in addition to the characteristics mentioned above, is seeking plants with tolerance to leaf diseases and a species of eriophyoid mite that causes a significant economic impact.

9 Gojiberry or Wolfberry

9.1 Botany

The genus *Lycium* L., family *Solanaceae*, was named in 1753, by Carl Linnaeus. He likely chose this name from the ancient southern Anatolian region of Lycia, or from the Latin, lychnus, meaning 'light' or 'lamp,' possibly due to the fruit shape and color. His species *L. barbarum* L., Latin for 'foreign' or 'from the outside,' may refer to the ancient country of Barbary, formerly part of northern Africa (Gross et al. 2006). Stuartevant's list of edible plants of the world included *L. europaeum* L. a native of Asian minor (Hedrick 1919) that escaped through Europe.

The genus includes more than 100 species of deciduous or evergreen woody shrubs, native to tropical or warm temperature parts of mainland East and Southeast Asia, Asia Minor, Europe, South Africa, and North America (Hitchcock 1932; Bailey L. H. Hortorium 1976).

Common names for *Lycium* include box thorn, matrimony vine, bocksdorn, Duke of Argyll's tea tree, gojiberry, and wolfberry. Several species of *Lycium* are now being sold as gojiberry or wolfberry. The names Tibetan goji and Himalayan goji are names applied by the health food promoters for a nomenclatural marketing advantage, though commercial cultivation of the crop does not occur in those regions.

The plant is an erect or clambering, woody perennial shrub. Some species have spines, others do not. The plant, left unattended, can grow to 6 m. Leaves are alternate, often clustered, small, commonly narrow, entire, and are usually grayish-green without stipules. Flowers (Fig. 4.10) are perfect and solitary or clustered in leaf axils. Corolla is funnel form and different species are greenish, whitish, or purplish. Fruits ripen orange to scarlet (Fig. 4.10), sometimes yellow or black, e.g., *L. ruthenicum* Murr.

Some species are considered noxious weeds because of their tendency to sucker (Bailey L. H. Hortorium 1976; GRIN 2009) and because of their potential spread by birds. Like other genera in the Solanaceae, the vegetative plant parts are poisonous (FDA 2009), though the berries are edible.

9.2 Cultivation

Gojiberry plants prefer full sun but can tolerate some shade. Soils in Ningxia are alkaline (pH 7–8), but plants do well in a wide pH range. Soils can be heavy clay loams, but a higher sand ratio in the loam is best. *Lycium* does not grow well in wet soil. Much of the acreage in the Yinchuan is on flat areas in the Yellow River Valley, and plantings are successful on the surrounding hills. Ningxia has a continental climate with severe winters, but damage from winter cold or spring freezes seldom occurs. The plants are hardy to −23°C (−10°F) (Gross et al. 2006). The optimum

Fig. 4.10 *Lycium barbarum* L. flower and fruit. Photo credits: Kim Hummer USDA ARS

fruit quality (chemical content) occurs under hot dry summer conditions, while cooler or cloudy weather diminishes fruit quality. Ripe fruit also tends to crack in rain at maturity.

9.3 Origin and Domestication

Lycium barbarum L. is native to eight autonomous regions and provinces of China (GRIN 2009) (Table 4.1). The largest gojiberry producing area is Ningxia Hui, a small autonomous region on the northwestern loess-soil highlands of China, which used to be part of Gansu Province. The Chinese characters 寧夏枸杞, 'Ningxia wolfberry,' refer to the plant of *L. barbarum*.

A closely related species, Chinese wolfberry, *Lycium chinense* P. Mill., native to Mongolia, China, Japan, Korea, Taiwan, and Thailand is also cultivated (GRIN 2009) (Table 4.1). While *L. barbarum* tends to have more large-sized fruit per plant than does *L. chinese*, both species are labeled and sold as gojiberry or wolfberry. The name 'goji' probably was derived from the Chinese, 枸杞, gǒuqǐ, with the character for 'gǒu' being related to a character for dog or wolf (Dharmananda 2007).

9.4 Production and Uses

An early description of the use of *Lycium* is in, the *Shennong Bencao Jing*, the Divine Farmer's Materia Medica Classic, one of the ten premodern classics of

Table 4.1 Selected *Lycium* species, common name: gojiberry or wolfberry[a]

Species	Native range	Comments
L. barbarum L. (Syn. = *L. halmifolium* P. Mill)	Gansu, Hebei, Nei Monggol, Ningxia, Qinghai, Shanxi, Sichuan, and Xinjiang, China	Erect plant with spreading branches, reaches 6 m without size control; fruit orange to red
L. chinense P. Mill	China—Anhui, Fujian, Gansu, Guangdong, Guangxi, Guizhou, Hainan, Hebei, Heilongjiang, Henan, Hubei, Hunan, Jiangsu, Jiangxi, Jilin, Liaoning, Nei Monggol, Ningxia, Qinghai, Shaanxi, Shanxi, Sichuan, Xinjiang, Yunnan, and Zhejiang Japan—Hokkaido, Honshu, Kyushu, Ryukyu Islands, and Shikoku Mongolia, Korea, Taiwan, Thailand	Prostrate rambler, can grow on itself to 2 m (WPSM p. 694–696); fruit orange to red
L. ruthenicum Murr.	Afghanistan; Iran; Iraq; Turkey; Armenia; Azerbaijan; Kazakhstan; Kyrgyzstan; Tajikistan; Turkmenistan; Uzbekistan; Mongolia; China—Gansu, Nei Monggol, Ningxia, Qinghai, Shaanxi, Xinjiang, Xizang; Pakistan; Russian Federation (European part)	Black-fruited species. The small, sweet, and flavorless berry is eaten in India. Common name: Russian box thorn

[a]Taxonomic and distribution information adapted from (GRIN 2009)

Chinese herbal medicine (Gross et al. 2006). Traditional use of gojiberry in tonics was limited until the end of the Ming Dynasty when production was encouraged (1368–1644) (Dharmananda 2007). Gojiberry species are widely scattered throughout China, and wild plants in fence rows and nonfarmed areas have been picked for family use or sold for about 800 years.

Gojiberry cultivation in Ningxia was promoted beginning in 1987 by government-backed company projects. Since 2005, the production and sales of these products have skyrocketed, because nutritionists have described the berry as an 'exotic superfood' for the polysaccharide, vitamin, and carotenoid content (Dharmananda 2007). Now gojiberries are processed for juice and juice combination drinks, dried in tea, and as nutraceutical supplements. Dried fruits can be eaten directly and used in confectionary goods or in bakery products (Fig. 4.11).

Most of the gojiberries of world commerce are produced in Ningxia Hui Autonomous Region, China. Their products include juice and juice concentrate, dried fruit, goji seed oil, and powdered goji (Dharmananda 2007). Juice types are formulated for marketing in different countries. The Chinese producers expect that the demand for juices will grow most rapidly in the next several years (E. Hanson Personal communication). The berries are sold as dried fruit (Fig. 4.11) to be used in bakery and confectionary products, and the seed oil and powdered gojiberry are prepared for nutritional supplements.

Fig. 4.11 Some gojiberries (*Lycium barbarum* L.) products available for purchase in the USA. *Upper left proceeding clockwise*: natural carbonated juice, packaged tea, dried fruit combination, chocolate covered dried gojiberries, goji cookie bar, dried gojiberries. Photo credits: Kim Hummer, USDA ARS

Commercial plantings have increased recently due to the availability of improved cultivars and the increased demand for health products. The new plantings are composed of clonally propagated improved genotypes, not seedlings.

In 2004, the China Daily (2004) reported that 86 MT (95 tons) of gojiberry were produced worth US$120 million. In 2008, Ningxia Hui grew gojiberries on 72,843 ha (180,000 acres) while about 101,171 ha (250,000 acres) total were grown in China. The maximum yield is about 7,845 kg/ha (7,000 lb/acre) from elite genotypes, while the yield from seedlings is lower (E. Hanson Personal communication). These figures would indicate almost a 10-fold increase for 2008 over the China Daily's report for the 2004 crop.

Individual growers manage between 0.08 and 0.8 ha (0.2–2 acres). They sell their fruit to brokers, who then sell to processors or distributors. Growers can also sell at specialty markets. In 2008, grower prices were about $1.00/kg ($0.45/lb) fresh or $6.61/kg ($3.00/lb) of dried fruit (E. Hanson Personal communication).

One Chinese processor exports to ten countries and their top three customers are in the USA. In 2007, sales for this exporter were $4 million. Between 80 and 90% of their product is from Ninxia Province (E. Hanson Personal communication).

In Ningxia, the plants are grown with 1.5 m between rows and about 1 m between plants. Full production is reached by year 3 or 4. Row middles are cultivated to control suckers. The plants are pruned by removing nonfruitful shoots in

May and June. Dormant pruning is not practiced in China (E. Hanson Personal communication).

Plants can be propagated by softwood cuttings in June or with semi-hardwood cuttings in July to August. These 5–10 cm long cuttings are taken with a heel, i.e., with a piece of the previous year wood, and placed into individual pots in a frame (Sheat 1957). Alternatively, cuttings of mature wood of the current season's growth can be collected in autumn to late winter and placed in a cold frame for rooting. Thirdly suckers can be divided from mother plants in late winter. This technique is very easy because the suckers can be planted out directly into their permanent positions.

In China, plant nutritional requirements are met with manure applied in the spring. Too much fertility results in excess vegetation, shading, and reduced fruit quality. Foliar nutrient sprays are also routinely applied. Plantings are irrigated by surface flooding. Soils are allowed to dry considerably between irrigations. Excess irrigation reduces fruit quality. Growers generally treat plantings with fungicides or insecticides 2–3 times per year.

Plants are pruned in several systems. In the first system, the plants are allowed to grow into a large bush. Pruning is performed annually to encourage more fruit and flowers. If left alone, the bushes will overgrow themselves, causing shading. Pruning is done to prevent overlapping growth. The second method is to shape plants into a small tree. Commercial growers use this technique to allow for easy picking. Finally, the plants can also be trellised to promote a vining growth habit. Growing gojiberry in tropical areas where the plants receive no chilling hours is under research (E. Hanson Personal communication).

9.5 Seeds

Seeds can be extracted from fruits by pressing the pulp through a screen and floating out the fruit flesh (Rudolf and Busing 2002). On a larger scale berries may be fermented, mashed, and run through screens. The seeds can be dried and stored at 5°C. Germination of *L. barbarum* can be hastened and improved by stratification in moist sand for 60–120 days at 5°C. After stratification, the seeds can be germinated at diurnally alternating temperatures of 30 to 20°C. The seeds of *L. barbarum* have about 20 seed per fruit and about 573,000 seed per kg. The seeds of *L. chinense* are larger having about 377,000 seed per kg (Rudolf and Busing 2002).

For nursery practice, the seeds can be sown in the fall as soon as the fruits ripen or can be stratified in the spring and then covered with soil. Two-year-old seedlings are transplanted (Rudolf and Busing 2002).

Most gojiberry genotypes appear to be self-fruitful; cross-pollination is not required for commercial production. Plants are harvested from late June until October, on 5–7 day intervals (E. Hanson Personal communication). Given that the annual yield is 7,845 kg/ha (7,000 lb/acre) and plantings are picked 16 times, less than 560 kg/ha (500 lb/acre) is harvested in each picking.

Mechanical harvesting is not performed, although investigations in China are beginning to address this need since labor is a limitation where large plantings have been established in remote areas. The mechanization will not be simple to develop. A combination of breeding and cultural approaches is needed. One primary issue is the reduction of the fruit ripening period. Shoots of the present genotypes grow continuously and may simultaneously contain flowers, green fruit, and ripe fruit. Machinery that damages shoots will reduce later harvests. The ripe fruit do not readily dehisce when a branch is mechanically shaken unlike blueberries or cherries.

Many traditional medicinal uses of gojiberry have been described in Chinese folk medicine. The berries have been used in tonics to lower cholesterol or blood pressure, to treat kidney disease, to improve vision and eye disease, and to increase longevity. Some Chinese tonic soups combine gojiberries with chicken or pork, vegetables, and other herbs such as wild yam and licorice root. The berries are boiled to make an herbal tea (Facciola 1990), often along with chrysanthemum (*Chrysanthemum* L.) flowers and/or red jujubes (*Zyziphyus jujube* Mill.).

Fresh fruits may be squeezed for juice which is then concentrated for beverages. About 2 kg fruit is needed to produce 1 kg juice. A combination of grape and gojiberry fruit is used to produce wine. At least one Chinese company produces gojiberry beer or ale. Since the early twenty-first century, an instant coffee product containing gojiberry extract has been produced in China.

Alternatively, the fruits are dried to about 15% of the fresh weight. The fruits can be dried with or without sulfur. The fruits are dried in the sun for 7 days, or in driers. Driers are quicker and produce a better quality product. Dried gojiberries are eaten as a snack. Their taste has an accent of tomato and seems similar in flavor to that of dates, dried cranberries, or raisins, though drier, more pungent, less sweet, and with an herbal scent. Some people describe the fruit as having a sweet, licorice-like flavor. The fruits can be added to soups and braised dishes or used to prepare a liqueur (Facciola 1990).

Young goji shoots and leaves are also grown commercially in China as a leaf vegetable; however, FDA, lists the leaves and stems of some *Lycium* species as poisonous to humans and livestock (FDA 2009).

Gojiberry plants are used for land conservation plantings. The plants have an extensive root system and can stabilize sandy river banks. In Europe and Asia, these plants are grown as informal hedges (Hedrick 1919; Rehder 1940) succeeding in desert, subtropical, and maritime exposures.

Lycium fruit is known for its carbohydrate and carotenoid content (Gross et al. 2006). The carotene pigments of *Lycium* fruit include beta-carotene, zeaxanthin, lutein, lycopene, cryptozanthin, and xanthophyll (Gross et al. 2006; Dharmananda 2007). The fruit also contains protein, fiber, minerals (calcium, phosphorus, potassium, iron, zinc, and selenium), and vitamins (C, riboflavin, nicotinic acid, and thiamine) (Gross et al. 2006).

Some fruit marketers promote sugars from goji as having supermedicinal or healthful qualities, but cure-all and extreme longevity claims are undocumented and are under scrutiny from governments in Europe, Canada, and the USA.

9.6 Breeding Potential

The Ningxia Research Center of Wolfberry Engineering Technology in Yinchuan, Ningxia Hui Autonomous Region, China, has a goji breeding program. The Center is the only Chinese national institute devoted to goji (E. Hanson Personal communication).

The objectives of the research institute are

1. Breeding goji to increase yields, fruit size and to improve the quality.
2. Improve the planting technology and culture to increase yield.
3. Improve the postharvest /processing of the fruit.

The Center has 21 full-time staff, several buildings, a goji museum, and several thousand acres of farm land for collections. The land or 'base' is a nationalized farm. It is mostly planted to goji for fruit production, but some wine grape (*Vitis vinifera* L) vineyards are also planted. Proceeds from fruit sales help support the Center functions.

Scientists are performing chemical analyses of the goji fruit in Yinchuan including the measurement of antioxidant activity. The Chinese consider *L. barbarum* to have the highest quality fruit for health, and that Ningxia provides the best climate for optimal health promoting compounds of the fruit.

Lycium species hybridize readily. The Center has developed four cultivars, which contain material from 1 to 3 different species:

1. Ninxia #1. This type comprises 80% of the acreage in Ningxia Province and is grown in other regions as well. It is believed to have the highest antioxidant content. This cultivar is marketed as 'Crimson Star™' in the USA.
2. Ninxia #2. No information given.
3. Ninxia #3. This cultivar is being propagated for distribution now. It is a large-fruited type that is well suited for drying. They hope this fruit will be shipped throughout China.
4. Ninxia #4. This is a unique cultivar developed for production of edible shoots. The tips of the young succulent shoots are cut and eaten steamed or in dishes. The shoots also have high antioxidant content. The taste seems similar to that of steamed spinach.

Gojiberry, like many better known small fruit and berry shrubs and trees, produce nutritious, tasty fruits. The plant has potential for cultivation in environments equivalent to its native environment in China.

Small fruit producers in the northern tier of states and Canada may wish to diversify their present plantings and grow some acres of this crop. Growers should be cautious to guard against escapes of this plant because it has the potential to become a noxious weed. Cultivars should be planted in preference to seedlings and are now available in some American plant nurseries. Additional research needs to be done to improve mechanical harvesting technology and develop cultivars for mechanical harvesting. This would be necessary for any potential North American crop to be competitive with present Chinese production.

10 ‘Ōhelo Berry

10.1 Botany

The ‘ōhelo and closely related species are members of section *Myrtillus* of the genus *Vaccinium* L., family Ericaceae. The genus comprises not only the economically important crops such as the blueberry, cranberry, and lingonberry but also more than 400 berry-producing species distributed the South Pacific, Southeast Asia, and around the world (Vander Kloet 1993).

On Hawai’i, native *Vaccinium* species were called ‘ōhelo’ or ‘ōhelo ‘ai’ by the indigenous people (Table 4.2). The true ‘ōhelo refers to a low growing plant species, *V. reticulatum*, which is distributed in open forests at medium to high elevation on Hawa’i and Maui (Degener 1984). This species is rhizomatous and rarely grows taller than 0.6 m although some plants may reach 1.0 m. A second low-growing shrub, *V. dentatum* Smith, is less common but is also endemic. A high-bush species, ‘ōhelo kau la’au, *V. calycinium* Smith, which can attain a height of 5.0 m (Wagner et al. 2005), and an intermediate form *V. ×pahalae* Skottsberg, are also present (Degener 1984).

Vaccinium reticulatum (Fig. 4.12a, b) thrives on the less weathered lava flows and beds of volcanic ash and cinders (Degener 1984). ‘Ōhelo, a member of the pioneer plant community, is most common on disturbed sites at elevations from 600 to 3,700 m. It is frequently found on Maui and the Island of Hawai’i but only occasionally found on Kaua’i, O’ahu, and Moloka’i (Wagner et al. 1990; Herring 2008).

The ‘ōhelo is common on Kilauea, Hawa’i, on high slopes of Haleakala, Maui, and near the Koolau Gap, Maui. The plant has coriaceous, orbicular, green leaves that overlap when viewed from the stem apex. The leaf attachment and branching structure provide a noteworthy texture to the plant from an ornamental landscape perspective. In optimal Hawai’ian conditions, the plant can have simultaneous flowering and fruiting. Peak flowering season is from April to September and, because the berries take 50–60 days to ripen, mature berries are available from June through November. One plant can produce two crops of fruit in 1 year (Vander Kloet 1993). The flowers, one per pedicel, are epigynous, brilliant red, narrow convolvulate, and cluster near branch apices. Wagner et al. (1990) describes the fruits as being red, reddish purple, bluish purple, dull black, yellow, orange yellow, yellowish green, or pink (Fig. 4.12a, b). The skins of lighter colored berries can have red speckles.

Table 4.2 Distribution of ‘ōhelo species in Hawaii

Vaccinium species	Distribution
V. calycinum Small	Kaua’i, O’ahu, Moloka’i, Lana’i, Maui, Hawai’i
V. dentatum Small	Kaua’i, O’ahu, Moloka’i, Lana’i, Maui, Hawai’i
V. reticulatum Small	Kaua’i, O’ahu, Moloka’i, Maui, Hawai’i
V. ×pahalae Skottsberg	O’ahu, Hawai’i

Fig. 4.12 (**a, b**) 'Ōhelo (*Vaccinium reticulatum* Sm.) (**a**) with red fruit growing out of lava rock on the Big Island, Hawaii (**b**) yellow fruited form. Photo credits: Kim Hummer, USDA ARS

The berries range from 0.6 to almost 1.2 cm (1/4 to almost 1/2 in.) in diameter and contain numerous (70 to over 100) small, brown seeds. The flowers of *V. reticulatum* are self fertile. However, self pollination results in fewer seeds per berry than does cross-pollination (Herring 2008).

Typical fruits are globose, with a flattened top and bottom. The fruit can be covered with a waxy bloom. The 'ōhelo berry is one of the few endemic, edible fruits in Hawai'i, and is an important food for the native and endangered nēnē goose (*Branta hylobadistes* Storrs L. Olson & Helen F. James).

Vaccinium dentatum is a decumbent, sprawling or weakly rhizomatous shrub. Its leaves are elliptic to narrowly elliptic, in contrast with the ovate to obovate leaves of *V. reticulatum*. The leaves are persistent have serrate margins and are usually glabrous at maturity. The berries are bright to scarlet red, 8–10 mm in diameter, round, and slightly smaller in diameter than those of *V. reticulatum*.

The plant form of *V. calycinum* is an understory shrub, somewhat reminiscent in of the flame azalea [*Rhododendron calendulaceum* (Michx.) Torr.] of North Carolina, with a notable difference that the fruits are red berries rather than dry capsules. The high-bush 'ōhelo is distinguished from the low-growing 'ōhelo by its height and leaf morphology. It grows in the open rain forest east of Kilauea and below the Koolau Gap of Haleakala Crater on Maui (Degener 1984). The leaves are deciduous, relatively thin, and lanceolate with serrate margins. The flowers are greenish, and the fruits are red, globose and can be bitter. The fruits are borne singly and occur basal to the flush of newest growth (Wagner et al. 2005).

Vaccinium × *pahalae* is native to the Sulphur Bank near the Tree Fern Forest near Kilauea (Degener 1984). The plants of this species are slender but have hard, leathery, recurved leaves with serrated margins. The berries are elongated. Further study using molecular markers is needed to determine the relationship of the 'ōhelo species and to the others in section *Myrtillus*.

10.2 Origin and Domestication

'Ōhelo (*V. reticulatum* Smith) is a small, native Hawai'ian shrub (Fig. 4.12a, b) commonly found in disturbed, open sites at 640–3,700 m elevation on several islands in the Hawai'ian archipelago. The plant has been significant to native Hawai'ian legends and lore. Local people collect berries of this plant for individual uses. Concerns of the impact of this wild collection on delicate environments might be reduced if 'ōhelo was cultivated and marketed to meet the demand for the fruit.

Some Hawai'ian myths (Beckwith 1940) describe gods who have lived on earth and take the form of a plant at their death. From the body of Kaohelo, sister of Pélé, the Hawai'ian volcano goddess, grew the 'ōhelo bushes which are abundant on Hawai'ian volcanic mountainsides: "the flesh became the creeping vine (*V. reticulatum*) and the bones became the bush plant (*V. calycinum*)."

The 'ōhelo plant was especially sacred to the worshipers of Pélé. Old Hawai'ian law, or kapu, required that upon arriving near the Kilauea crater, a branch bearing 'ōhelo berries, be broken and half of the branch was thrown toward the center of the active volcano while the visitor said, "Pélé here are thy 'ōhelo. I offer some to thee; some I also eat." Only after performing this ritual could the berries be eaten freely without incurring Pélé's wrath. The kapu were officially abolished after 1818, though many people continued the old customs.

In December, 1824, Princess Kapiolani, a devout Christian, set out to break the 'ōhelo kapu. She and her followers walked more than 100 miles over rugged lava flows. Though she was entreated not to, Kapiolani descended to a ledge near the Kilauea volcano and ate 'ōhelo berries without first making the required offering. She defied the old Hawai'ian way and demonstrated to her people the foundation of her new faith. She read passages from the bible and sang a hymn. This was a courageous act considering the reverence and fear with which her contemporaries regarded Pélé. This event was immortalized by the British poet, Alfred, Lord Tennyson (1892) in a poem entitled 'Kapiolani.'

Another association of 'ōhelo is stated in proverb 2044 (Pukui 1983).

"Mai hahaki 'oe I ka 'ohelo o punia I ka ua noe." [Do not pluck the 'ōhelo berries lest we be surrounded by rain and fog.] This is a warning not to do bad things.

10.3 Production and Use

Because of the old Hawaiian traditions and laws, little domestication and even less breeding of 'ōhelo has occurred until the past several years. This plant has the potential for agricultural development and several research and improvement projects have been initiated (Zee et al. 2008). Potential uses include

- Ornamental outdoor landscape plant for cooler climates (best 10–20°C)
- Colorful red and green potted plant for the holiday season
- Berry for fresh eating

- Processed berries for jams, jellies, and the baking industry
- Processed berries for the candy industry
- Dried fruits
- Value added products into chocolates, sauces, or liquors
- Infusion of the leaves for tea
- Extracts or concentrates for the health industry

Chefs and confectionery trades in Hawaii would appreciate a broader availability of this specialty berry for their products, but production has not been sustainable or reliable thus far. The development of this crop could provide an alternative to sugar cane production, which has greatly reduced acreage, for local Hawaiian agriculture.

The plant can be propagated sexually by seed or asexually via cuttings or tissue culture (Zee et al. 2008). The seeds (100 seeds weigh about 0.1 g) are small (Zee et al. 2008). Open pollinated seeds can be harvested from healthy ‘ōhelo plants. Berries are placed in a blender with 3–4 cups of water and blended at medium speed. The viable seed sink to the bottom and the nonviable seed can be decanted off. The cleaned seeds are air-dried on paper towels for 2 days at ambient temperature. ‘Ōhelo seeds are very small, 100 seeds weigh about 0.1 g. Fresh seeds have a high germination rate. Seeds stored at 4°C lose viability after a year (Zee et al. 2008).

Seeds can be germinated in a 1:1:1 mixture of peat, vermiculite and perlite in a greenhouse at 60–80% shade. Seedlings germinate about 40–45 days after sowing. Seedlings younger than 3 months were sensitive to over watering and drying. After 4 months, the seedlings should be transplanted to a 1:1:2 media mixture after 4 months. After establishment, the ‘ōhelo seedlings were very hardy to drought.

Four-month-old ‘ōhelo seedlings can be transplanted to 5-cm pots containing 1:1:2 peat, vermiculite, and perlite, side-dressed with 14–14–14 slow-release fertilizer. Foliar fertilizer every 2 weeks also improves seedling health. Seedlings can be tipped pruned at transplanting to encourage multiple branching with the goal to form a compact crown of reddish new growth for market. At 10 months old, the seedlings can be transplanted into 4 liter containers containing the 1:1:2 medium for foliage plant production. Six to ten month old plants can be field planted for fruit production.

Stem cuttings should be harvested from healthy, upright woody branches (Zee et al. 2008). The cutting should consist of a 2-in.-long internode below a whorl of intact leaves. Rooting is stimulated by dipping the basal end of the cutting into a low concentration of a powdered auxin formulation. The stem is then stuck into a 2 × 2-cm moistened rooting cube. The cuttings should be kept under 60% shade and protected from drying, wind, and heat. Overhead mist or a humidity tent is required. Most of the cuttings should root within 3 months.

Tissue culture procedures have been described (Zee et al. 2008). Explants are placed in sterile solutions and placed on a base medium modified from Lloyd and McCown (1980). Initiation medium containing zeatin (Reed and Abdelnour-Esquivel 1991) and growth/multiplication and rooting media follow. A maintenance medium can be used for medium-term (5-year) storage of the plantlets under refrigerated conditions.

Several disease symptoms have been observed on cultivated highbush blueberries, *V. corymbosum*, growing at the Mealiani Agricultural Research Station in Waimea, Hawai'i (Hummer and Zee 2007; Keith et al. 2008). The main disease pressure that may limit 'ōhelo berry production and ornamental qualities in Hawaii was determined to be powdery mildew (Keith Personal communication). Diseases of *Vaccinium* include *Lasiodiplodia* (an anamorph of *Botryosphaeria*) causing wilting and reddening of the leaves; *Botrytis*, brown lesions and tip dieback; *Phytophthora*, reddening of leaves, discoloration of roots and stems, *Pestalotiopsis*, leafspot; *Fusarium* wilt disease; and foliar rust caused by *Pucciniastrum vaccinii* (Bristow and Stretch 1995).

The Mediterranean fruit fly (*Ceratitis capitata* Wiedemann), oriental fruit fly (*Bactrocera* dorsalis Hendel), melon fly (*B. cucurbitae* Coquillett) and Malaysian fruit fly (*B. latifrons* Hendel) are major pests of fruits and vegetables in Hawai'i. Control measures for these flies should be implemented during the cultivation of *Vaccinium* for fruit (Hummer and Zee 2007). 'Ōhelo berry is a marginal host for *B. dorsalis* and apparently a nonhost for *C. capitata*, *B. cucurbitae*, and *B. latifrons* (Follett and Zee 2011).

While the native 'ōhelo berries are a staple for native and endangered nēnē goose, cultivated berries could be a favorite food of large birds on the islands. The most effective way to exclude birds is to enclose the plants under bird or smaller screened netting on a metal pipe frame.

10.4 Potential Breeding

'Ōhelo species are neither endangered nor threatened (Wagner et al. 2005). The US Department of Agriculture, Agricultural Research Service, National Clonal Germplasm Repository in Corvallis Oregon holds the national *Vaccinium* genebank for the USA. Limited samples of wild and cultivated *V. reticulatum* and wild *V. calycinum* are preserved at this genebank. Additional representatives of *V. dentatum* and *V.* ×*pahalae* are sought. Species are represented by seedlots and selected genotypes are maintained clonally. Selections from wild material have been named. Initial breeding for this crop in Hawaii has shown that from several hundred seedlings from seeds extracted from wild-collected fruits, a few seedlings had an impressive yield per plant while others had high quality in ornamental characteristics (Zee et al. 2008).

References

Adams, L.S., Seeram, N.P., Aggarwal, B.B., Takada, Y., Sand, D., and Heber, D. (2006) Pomegranate juice, total pomegranate ellagitannins, and punicalagin suppress inflammatory cell signaling in colon cancer cells. J. Agr. Food Chem. 54:980–985.

Amiri, M.E. (2008) The status of genetic resources of deciduous, tropical, and subtropical fruit species in Iran. Acta Hort. 769:159–167.

Archbold, D.D. and K.W. Pomper. (2003) Ripening pawpaw fruit exhibit respiratory and ethylene climacterics. Postharvest Biology and Technology. 30:99–103.
Atkinson, M.D and Atkinson, E. (2002) *Sambucus nigra* L. J. Ecol. 90, 895–923.
Aviram, M., Rosenblat, M., Gaitini, D., Nitecki, S., Hoffman, A., Dornfeld, L., Volkova, N., Presser, D., Attias, J., Liker, H., and Hayek, T. (2004) Pomegranate juice consumption for 3 years by patients with carotid artery stenosis reduces common carotid intima-media thickness, blood pressure and LDL oxidation. Clin. Nutr. 23:423–433.
Bailey, L.H. (1906) *Sketch of the Evolution of Our Native Fruits*. The Macmillan Company, New York.
Bailey, L. H., Hortorium. (1976) Hortus Third: A Concise Dictionary of Plants Cultivated in the United States and Canada. John Wiley & Sons, Inc. Hoboken, NJ p. 688.
Bailey, L.H. and E.Z. Baily. (1976) Hortus Third: A Concise Dictionary of Plants Cultivated in the United States and Canada. John Wiley and Sons, Inc.
Baker, M.L. (1991) Increasing seed germination percentage of *Crataegus opaca* (mayhaw) by fermentation. HortScience 26:496 (abstr.).
Beck, N.G. and E.M. Lord. (1988) Breeding system in Ficus carica, the common fig. I. Floral diversity. Am. J. Bot. 75:1904–1912.
Beckwith, M. (1940) Hawaiian Mythology. Yale University Press. p. 99. http://www.sacred-texts.com/pac/hm/hm00.htm. Accessed 29 December 2008.
Bist, H.S., Srivastava, R., and Sharma, G. (1994) Variation in some promising selections of wild pomegranate (*Punica granatum* L.) Hort. J. 7:67–70.
Blumenfeld, A., Shaya, F., and Hillel, R. (2000) Cultivation of pomegranate. Options Méditerranéennes Ser. A 42:143–147.
Bolli, R. (1994) Revision of the Genus *Sambucus*. Diss. Bot. 223.
Bristow, P. and A. W. Stretch. (1995) p. 20 In: Compendium of Blueberry and Cranberry Diseases. F. L. Caruso and D. C. Ramsdell, eds. The American Phytopathological Society, St. Paul, MN.
Bunnag, S., R. Dolcet-Sanjuan, D.W.S. Mok and M.C. Mok. (1996) Responses of two somaclonal variants of quince to iron deficiency in the greenhouse and field. J. Amer. Soc. Hort. Sci. 121:1054–1058.
Bush, E. W, Johnson, C. E. and Payne, J. T. (1991) Commercial nursery production of *Crataegus opaca* in Louisiana. Proc. South. Nurs. Assoc. Res. Conf., 36th Ann. Rep.: 113–115.
Byers, P.L. and Thomas, A.L. (2005) Elderberry research and production in Missouri. Proc. Missouri Small Fruit and Vegetable Conf. 25:91–97.
Byers, P.L., Thomas, A.L., and Millican, M. (2010) 'Wyldewood' Elderberry. Hortscience 45(2): 312–313.
Byers, P.L. and Thomas, A.L. (2011) 'Bob Gordon' Elderberry. J. Am. Pom. Soc. 65(2): 52–55.
Campbell, J. (2001) Quince Growing. New South Wales AgFact H4.1.3.
Chaovanalakit, Arusa, Maxine M.Thompson, and Ronald E.Wrolstad. (2004) Characterization and quantification of anthocyanins and polyphenolics in blue honeysuckle (*Lonicera caerulea* L.). J. Agric. and Food Chem. 52:848–852.
Charlebois, D. (2007) Elderberry as a medicinal plant. In: J. Janick and A. Whipkey (Eds.), *Issues in New Crops and New Uses*. ASHS Press, Alexandria, VA, pp 284–292.
Charlebois, D., Byers, P.L., Finn, C.E. and Thomas, A.L. (2010) Elderberry: horticulture, botany, potential. *Hort. Rev*. (Amer. Soc. Hort. Sci.) 37:213–280.
Chessa, I. (1997) Fig. In: S. Mitra (ed.). Postharvest physiology and storage of tropical and subtropical fruits. CAB International, Wallingford, UK, pp. 245–268.
Chia, C.L. (1975) A chromosome and thin-layer chromatographic study of the genus *Sambucus* L. PhD thesis, Cornell University, USA.
China Daily 2004-07-19. Wolfberry festival to be held in Ningxia. http://www.chinadaily.com.cn/chinagate/doc/2004-07/19/content_349679.htm.
Clewell, A.F. (1985) *Guide to the vascular plants of the Florida panhandle*. Florida State Univ. Press, Tallahassee.
Condit, I.J. (1932) The structure and development of flowers in *Ficus carica* L. Hilgardia 6:443–481.
Condit, I.J. (1947) *The Fig*. Waltham, Mass, USA.

Condit, I.J. (1955) Fig varieties: a monograph. Hilgardia 23:323–538.
Costa, Y. and P. Melgarejo. (2000) A study of the production costs of two pomegranate varieties grown in poor quality soils. Options Méditerranéennes, Ser. A 42:49–53.
Craft, B.R. (2003) Report on dwarfing rootstock evaluations and hybridization of selected mayhaws. LA Mayhaw Assoc.
Craft, B.R., Melcher, G. and Langston, E. (1996) *Mayhaws: A guide to orchard production and propagation*. Morris Publishing, Kearney, NE.
Darrow, G.M. (1975) Minor temperate fruits. In: J. Janick and J.N. Moore (Eds), *Advances in Fruit Breeding*. Purdue University Press, West Lafayette, IN, pp. 271–273.
Darrow, G. M., and Yerks, G. E. (1937) Some unusual opportunities in plant breeding. pp. 545–558 in: *Yearbook of Agriculture*. US Government Printing Office, Washington, D.C.
Degener, O. (1984) Plants of Hawaii National Park Illustrative of Plants and Customs of the South Seas. Braun-Brumfield, Inc. Ann Arbor. pp. 240–245.
Dharmananda, Subhuti. (2007) Lycium Fruit: Food and Medicine, Institute for Traditional Medicine Online Portland, OR. http://www.itmonline.org/arts/lycium.htm (Accessed 16 February 2009).
Dickinson, T. A. (1985) The biology of Canadian weeds. 68. *Crataegus crus-galli* L. *sensu lato. Can. J. Plant Sci.* 65:641–654.
Dickinson, T.A., Lo, E. and Talent, N. (2007) Polyploidy, reproductive biology, and Rosaceae: understanding evolution and making classifications. Pl. Syst. Evol. 266:59–78.
Donoghue, M.J., Bell, C.D. and Winkworth, R.C. (2003) The evolution of reproductive characters in Dipsacales. Int. J. Plant Sci. 164, 453–464.
Doyle, L.F., Ferguson, L., Herman, K., López Corrales, M., and Bernalte García, M.J. 2003. Fig cultivar development and evaluation. Acta Hortic. 605:29–32.
Duffrin M.W. and K.W. Pomper. (2006) Development of Flavor Descriptors for Pawpaw Fruit Puree: A Step Toward the Establishment of a Native Tree Fruit Industry. Family & Consumer Sciences Research Journal. 35:118–130.
Duncan, W.H. and Duncan, M.B. (2000) Trees of the Southeastern United States. Athens: University of Georgia Press.
Eaton, E.L., Aalders, L.E., and Hall, I.V. (1959) Hybrids of an interspecific cross of elder. Proc. Amer. Soc. Hort. Sci. 73, 145–146.
Eggleston, W. W. (1910) *Sketch of the Crataegus problem, with special reference to work in the South*. J. N.Y. Bot. Garden 11:78–83.
El-Kassas, S.E., El-Sese, A.M., El-Salhy, A.M., and Abadía, A.A. (1998) Bearing habits in some pomegranate cultivars. Assiut J. Agr. Sci. 29:147–162.
Encyclopedia Britannica. (2006) Aphrodite. 1 Sept. 2006. http://search.eb.com/eb/article-9008000.
Eriksson, T. and Donoghue, M.J. (1997) Phylogenetic relationships of *Sambucus* and *Adoxa* (Adoxoideae, Adoxaceae) based on nuclear ribosomal ITS sequences and preliminary morphological data. Syst. Bot. 22, 555–573.
Facciola, S. (1990) Cornucopia: a source book of edible plants. Kampong Pub. Vista, CA. p.205.
FDA. (2009) US Food and Drug Association. Poisonous plants database for *Lycium*. http://www.cfsan.fda.gov/~djw/plantox.html (accessed 16 February 2009).
Faegri, K. and L. van der Pijl (1971) The principles of pollination ecology. Pergamon, NY. 2nd ed. p. 112–122.
FAO. (2006) FAOSTAT agricultural data. http://faostat.fao.org/site/408/default.aspx. Dec. 2006.
FAO. (2007) FAOSTAT agricultural data. http://faostat.fao.org/site/570/DesktopDefault.aspx?PageID=570. Jan. 2008.
Feng, Y. Z., Song, M. T. and D. B. Han. (2006) The general status of pomegranate germplasm resources in China. China Fruits 4:57–58.
Ferguson, L., T. J., Michailides and H.H Shorey. (1990) The California fig industry. Hort. Rev. 12:409–490.
Finn, C. (1999) Temperate berry crops. p. 324–334. In: J. Janick (ed.), Perspectives on new crops and new uses. ASHS Press, Alexandria, VA.

Follett, P. A., and Zee, F. T. (2011) Host status of Vaccinium reticulatum to invasive tephritid fruit flies in Hawaii. Journal of Economic Entomology 104:571–573

Ghazarian, B. (2009) *Simply Quince*. Mayreni Publishing, Monterey, CA. 216 pp.

Gibbons, E. (1974) Stalking the healthful herbs. David McKay Co., New York, NY.

Godfrey, R.K. and Wooten, J. W. (1981) Aquatic and wetland plants of southeastern United States: Dicotyledons. Univ. of Georgia Press, Athens.

Gozlekci, S. and L. Kaynak. (2000) Investigations on pollen production and quality in some standard pomegranate (*Punica granatum* L.) cultivars. Options Méditerranéennes, Ser. A 42:71–77.

Graham, C.J. (2000) New millennium brings several new fungicides registered for use on mayhaws. Louisiana Mayhaw Association Newsletter Vol. 3, Number 2, pp. 3–4.

Graham, C. J., R. K. Aulds and B. E. Herrington, Jr. (2000) Tolerance of mayhaw clones to quince rust. LAES Research Summary 118:71–74.

Graham, C. J., S. C. Laws and A. Gibson. (2005) The influence of rootstock on 'Royalty' mayhaw production in (2004) LAES Research Summary 164:17–19.

GRIN. (2009) Taxonomy of *Lycium*. USDA, ARS, National Genetic Resources Program. *Germplasm Resources Information Network - (GRIN)* [Online Database]. National Germplasm Resources Laboratory, Beltsville, Maryland. URL: http://www.ars-grin.gov/cgi-bin/npgs/html/taxon.pl?22939 (accessed 16 February 2009).

Gross, P.M. X. Zhang, MD, R. Zhang. (2006) Wolfberry: Nature's Bounty of Nutrition and Health, Booksurge Pub., Vancouver Island, British Columbia.

Gulick, P. and D.H. Van Sloten. (1984) Directory of Germplasm collections. 6-1- Tropical and subtropical fruits and nut trees. IBPGR, Rome.

Guo, C.J., J.Y. Wei, J.J. Yang, Y.F. Li, J. Xu and Y.G. Jiang. (2003) The antioxidant capacity of 66 vegetables and fruits: a comparative study. *Acta Nutrimenta Sinica* 25:203–207.

Hedrick, U. P. (1919) Sturtevant's notes on edible plants. J. B. Lyon Co. Albany, NY. 686 pp.

Hedrick, U.P. (1922) Cyclopedia of Hardy Fruits.

Herring, E. (2008) Hawaiian Native Plant Propagation Database. *Vaccinium reticulatum*. University of Hawaii at Manoa. http://www2.hawaii.edu/~eherring/hawnprop/vac-reti.htm.

Hitchcock CL. (1932) A monographic study of the genus *Lycium* of the Western Hemisphere. Annals of the Missouri Botanical Garden 19:179–364.

Huang, H., D.R. Layne, and R.N. Peterson. (1997) Using isozyme polymorphisms for identification and assessment of genetic variation in cultivated pawpaws [*Asimina triloba* (L.) Dunal]. J. Amer. Soc. Hort. Sci 122(4):504–511.

Hummer, K.E. (2006) Blue honeysuckle: A new berry crop for North America. J. Amer. Pomol. Soc. 60:3–8.

Hummer, K. and F. Zee. (2007) Evergreen Production of Southern Highbush Blueberries in Hawai'i. J. Amer. Pom. Soc. 61(4): 188–195.

Huo, JunWei, GuoHui Yang, Wei Sui and ZeYuan Yu. (2005) Review of studies on blue honeysuckle (*Lonicera caerulea* L) germplasm resources. Acta Hort. Sinica 32 (1):159–164.

Indian Council of Agricultural Research. (2005) Pawar inaugurates new national research centre on pomegranate. 1 Sept. 2006. http://www.icar.org.in/pr/25092005.htm >.

ITIS. (2006) Integrated Taxonomic Information System. 1 Sept. 2006. http://www.itis.usda.gov/.

Jalikop, S.H., P.S. Kumar, R.D. and Ravindra-Kumar. (2006) Breeding pomegranate for fruit attributes and resistance to bacterial blight. Indian J. Hortic. 63:352–358.

Janick, J. (2007) Fruits of the bibles. HortScience 42:1072–1076.

Janick, J. and Paul, R.E., editors. (2008) *The Encyclopedia of Fruits and Nuts*. Cambridge, MA. CABI.

Johnson, C.E., O'Rourke, E., and Boudreaux, J.E. (2010a) 'Champagne' fig. HortScience 45: 210–311.

Johnson, C.E., O'Rourke, E., and Boudreaux, J.E. (2010b) 'O'Rourke' fig. HortScience 45: 826–827.

Johnson, C.E., O'Rourke, E., and Boudreaux, J.E. (2010c) 'Tiger' fig. HortScience 45: 828–829.

Juniper, B.E. and D.J. Mabberly. (2006) The story of the apple. Timber Press, Portland, OR. 219 pp.

Kaack, K. (1989) New varieties of elderberry (*Sambucus nigra* L.). Tidsskr. Planteavl. 93, 59–65.

Kaack, K., Frette, X. C., Christensen, L.P, Landbo, A.K, and Meyer, A.S. (2008) Selection of elderberry (*Sambucus nigra* L.) genotypes best suited for the preparation of juice. Euro. Food Res. Tech. 226, 843–855.

Kader, A.A., A. Chordas and S.M. Elyatem. (1984) Responses of pomegranates to ethylene treatment and storage temperature. California Agr. 38(7–8):4–15.

Karp, D. (2006) The pomegranate: for one and all. Fruit Gardener 38(5):8–12.

Keith, L., L. Sugiyama, A. Strauss, R. Kai, F. Zee, R. Hamasaki, M. Yamasaki, and S. Nakamoto. (2008) First Report of Leaf Rust of Blueberry Caused by *Pucciniastrum vaccinii* in Hawaii. Plant Disease 92(11):1590.

Kher, R. (1999) A note on the physico-chemical characters of the wild pomegranate (*Punica protopunica* L.). Ann. Biol. Ludhiana 15:231–232.

Kislev, M. E., A. Hartmann and O. Bar-Yosef. (2006) Early domesticated fig in the Jordan Valley. Science 312:1372–1374.

Kjellberg, F., P.H. Gouyon, M. Ibrahim, M. Raymond and G. Valdeyron. (1987) The stability of the symbiosis between dioecious figs and their pollinators: a study of *Ficus carica* L. and *Blastophaga psenes* L. Evolution 41:693–704.

NASS. (2006) Noncitrus fruits and nuts: 2005 summary. http://usda.mannlib.cornell.edu/usda/nass/NoncFruiNu//2000s/2006/NoncFruiNu-07-06-2006_final.pdf. Dec. 2006.

Koncalova, M.N., Hrib, J., and Jicinska, D. (1983) The embryology of the *Sambucus* species and hybrids. In: O. Erdelska (Ed.), *Fertilization and Embryogenesis in Ovulated Plants (Proceedings of the VII International Cytoembryological Symposium)*, Veda, Bratislava, Czechoslovakia, pp. 43–47.

Kotkin, C. (2006) Pomegranates on parade. Wine News. 1 Sept. 2006. http://www.thewinenews.com/decjan0506/cuisine.asp.

Kral. R. (1960) A revision of *Asimina* and *Deeringothamnus* (Annonaceae). Brittonia 12(4):233–278.

Krewer, G. W. and Crocker, T. E. (2000) Experiments and observations on growing mayhaws as a crop in south Georgia and north Florida. Bulletin H-00-053, Georgia Cooperative Extension Service, 12 pp.

Lee, J. and Finn, C.E. (2007) Anthocyanins and other polyphenolics in American elderberry (*Sambucus canadensis*) and European elderberry (*S. nigra*) cultivars. J. Sci. Food and Agri. 87, 2665–2675.

Lefol, Eric. (2007) Haskap market development-the Japanese opportunity. www.parklandagroforestry.com/haskap.htm. 2/28/08.

Levin, G.M. (1995) Genofund of pomegranate in Turkmenistan. Problems Desert Dev. 3:84-89.

L.H. Bailey Hortorium. (1999) Hortus Third: a concise dictionary of plants cultivated in the United States and Canada. Macmillian Pub. Co. NY.

Lloyd, G. and McCown,B. (1980) Commercially feasible micropropagation of mountain laurel, Kalmia latifolia by use of shoot tip culture. Combined Proceedings of the International Plant Propagators Society. 30:421–427.

Lo, E. Y. Y., Stefanovic, S. and Dickinson, T. A. (2007) Molecular reappraisal of relationships between *Crataegus* and *Mespilus* (Rosaceae, Pyreae) – two genera or one? System. Bot. 32(3):596–616.

Lumpkin, T. (2007) Energizing global horticultural research for the developing world. *Acta Hort.* 759:15–28.

Mabberley, D. J. (1997) The plant-book: a portable dictionary of the vascular plants. 2nd ed. Cambridge, UK: Cambridge University Press.

Mars, M. (1996) Pomegranate genetic resources in the Mediterranean region, p. 345–354. In: Proc. First MESFIN Plant Genet. Res. Meeting Tenerife, Spain, 2-4 Oct. 1995.

Mars, M. (2000) Pomegranate plant material: genetic resources and breeding, a review. Options Méditerranéennes, Ser. A 42:55–62.

McCarter, S. M. and Payne, J. A. (1993) Fire blight caused by *Erwinia amylovora* on Mayhaw in Georgia. Plant Dis. 77(12):1262.

McDaniel, J. C. (1980) More on hawthorn as a rootstock. Pamona 13(3):169–170.

McGinnis, L. (2007) Quest for Quince: Expanding the NCGR Collection. Agricultural Research, January 2007:20–21.
McCabe, C. (1996) Enjoying the forbidden fruit. Saveur 14:105–110.
Meech, W.W. (1908) Quince Culture, and illustrated handbook for the propagation and cultivation of the quince, with descriptions of its varieties, insect enemies, diseases and their remedies. Orange Judd Co., New York. 180 pp.
Mehrnews. (2006) Iran, only producer of premium pomegranate. 1 Sept. 2006. http://www.mehrnews.com/en/NewsDetail.aspx?NewsID=216517.
Melgarejo, P. (2003) Tratado de fruticultura para zonas aridas y semoaridas. II. Algarrobo, granado y jinjolero. Mundi-prensa, Madrid.
Moore, L. (2006) Mayhaw: *Crataegus opaca* Hook. & Arn. USDA NRCS National Plant Data Center, Baton Rouge, LA.
Morton, J. F. (1963) Principal wild food plants of the United States excluding Alaska and Hawaii. *J. Econ. Bot.* 17:319–330.
Morton, J. (1987) Fruits of warm climates. Miami, FL.
Muniyamma, M. and Phipps, J. B. (1979) Cytological proof of apomixis in *Crataegus. Amer. J. Bot.* 66:149–155.
Muniyamma, M. and Phipps, J. B. (1984) Studies in *Crataegus.*11. Further cytological evidence for the occurrence of apomixis in North American hawthorns. *Can. J. Bot.* 62:2316–2324.
Muniyamma, M. and Phipps, J. B. (1985) Studies in *Crataegus.* 12. Cytological evidence for sexuality in some diploid and tetraploid species of North American hawthorns. *Can. J. Bot.* 63: 1319–1324.
O'Rourke, E., Johnson, C. E., Boudreaux, J. E, and Bourgeois, W. (2005) 'LSU Gold' fig. HortScience 40: 486–487.
Ourecky, D.K. (1970) Chromosome morphology in the genus *Sambucus.* Am. J. Bot. 57, 239–244.
Palmer, E. J. (1932) The *Crataegus* problem. J. Arnold Arbor. 13: 342–362.
Payne, J.A, Krewer, G.W. and Eitenmiller, R.R. (1990) Mayhaws: trees of pomological and ornamental interest. *HortScience* 25: 246–375.
Peterson, R. N. (1991) Pawpaw (*Asimina*). In: J. N. Moore and J. R. Ballington (eds.). Genetic resources of temperate fruit and nut trees. Acta. Hort. 290:567–600.
Peterson, R.N. (2003) Pawpaw Variety Development: A History and Future Prospects. HortTechnology 13: 449–454.
Phipps, J. B. (1983) Biogeographic, taxonomic, and cladistic relationships between east Asiatic and North American *Crataegus. Ann. Mo. Bot. Gard.* 70:667–700.
Phipps, J. B. (1984) Problems of hybridity in the cladistics of *Crataegus* (Rosaceae). In: Grant WF, ed. *Plant biosystematics.* Toronto: Academic Press: 417–438.
Phipps, J. B. (1988) *Crataegus* Maloideae, Rosaceae) of the southeastern United States: 1. Introduction and series *Aestivales. J. Arnold Arbor.* 69:401–431.
Phipps. J. B., Robertson, K. R., Rohrer, J. R. and Smith, P. G. (1991) Origins and evolution of subfam. Maloideae (Rosaceae). *System. Bot.* 16:303–332.
Phipps, J. B., O'Kennon, R. J. and Lance, R. W. (2003) *Hawthorns and medlars.* Timber Press, Portland, OR.
Plekhanova, M.N. (2000) Blue honeysuckle (*Lonicera caerulea* L.). A new commercial berry crop for temperate climate: Genetic resources and breeding. Acta Hort. 538:159–163.
Plekhanova, M.N., S.A. Streltsyna, and N.S. Rostova. (1993) Phenolic compounds in berries of *Lonicera* subsect. Caeruleae species. Plant Resources 29:16–25. (In Russian).
Pomper, K.W., S.B. Crabtree, S.P. Brown, S.C. Jones, T.M. Bonney, and D.R. Layne. (2003) Assessment of genetic diversity of pawpaw varieties with inter-simple sequence repeat markers. J. Amer. Soc. Hort. Sci. 128:521–525.
Pomper, K.W., S.B. Crabtree, D.R. Layne, R. N. Peterson, J. Masabni, and D. Wolfe. (2008) The Kentucky pawpaw regional variety trial. J. Amer. Pom. Sci. 62:58–69.
Pomper, K.W. and D.R. Layne. (2005) The North American Pawpaw: Botany and Horticulture. Horticultural Reviews. Vol. 31:351–384.

Pomper, K. W., D.R. Layne, R. N. Peterson, and D. Wolfe. (2003) The Pawpaw Regional Variety Trial: background and early data. HortTechnology 13:412–417.

Postman. J. (2008) The USDA Quince and Pear Genebank in Oregon, a World Source of Fire Blight Resistance. Acta Horticulturae 793:357–362.

Pukui. M. K. (1983) 'Ōlelo no 'eau: Hawaiian Proverbs and poetical sayings. Bishop Museum Press. Honolulu.

Radford, A. E., Ahles, H. E. and Bell. C. R. (1974) *Manual of the vascular flora of the Carolinas.* Chapel Hill: University of North Carolina Press.

Reed, B.M., and A. Abdelnour-Esquivel. (1991) The use of zeatin to initiate in vitro cultures of *Vaccinium* species and cultivars. HortScience 26(10):1320–1322.

Rehder, A. (1940) Manual of Cultivated Trees and Shrubs. Macmillian Company, NY.

Rehder, A. (1986) Manual of cultivated trees and shrubs hardy in North America, 2nd edition. Dioscorides Press, Portland.

Reynolds, S. and P. W. Ybarra. (1984) So easy to preserve. Georgia Extension Service, Univ. of Georgia, Athens.

Ritter, C. M. and McKee, G. W. (1964) *Elderberry: History, Classification, and Culture*. Penn. State Univ. Ag. Expt. Sta. Bull. 709.

Roach, F.A. (1985) Quinces. in Cultivated Fruits of Britain: Their Origin and History. Blackwell, London pp. 220–225.

Robertson, K. R. (1974) The genera of Rosaceae in the southeastern United States. *J. Arnold Arbor.* 55: 303-332, 334-401, 611–662.

Robertson, K. R., Phipps, J. B., Rohrer, J. R. and Smith, P. G. (1991) A synopsis of genera in Maloideae (Rosaceae). *System. Bot.* 16:376–394.

Rudolf, P. O., and Busing, R. T. (2002) *Lycium* L. wolfberry. pp 694-696 In: F. T. Bonner and R. T. Nisley (eds.) Woody Plant seed Manual. US Department of Agriculture Misc. Pub. 654. Wash. D. C. and http://www.nsl.fs.fed.us/wpsm/ accessed 20 February 2009.

Sargent, C. S. (1965) Manual of the trees of North America. Vol. 2. Dover Pub., New York, NY.

Scherm, H. and Savelle, A. T. (2003) Epidemic development of hawthorn leaf blight (Monilinia johnsonii) on mayhaw (Crataegus aestivalis and C. opaca) in Georgia. *Plant Dis*. 539–543.

Sepulveda, E., L. Galleti, C, Saenz, and M. Tapia. (2000) Minimal processing of pomegranate var. Wonderful. Options Méditerranéennes, Ser. A 42:237–242.

Sheat. W. G. (1957) *Propagation of Trees, Shrubs and Conifers.* Third printing. MacMillan and Co. Ltd. London.

Singh, D. B., Samadia, D. K. and A. R. P. Kingsly. (2006) Conservation, characterization and evaluation of pomegranate germplasm under arid ecosystem of India. p. 15. In: 1st International Symposium on Pomegranate and Minor Mediterranean Fruits, Abstracts contributed papers, 16-19 Oct, Adana, Turkey.

Slate, G.L. (1955) Minor fruits. National Horticulture Magazine 34, 139–149.

Statistik Austria. (2008) Agriculture and forestry, fruit website. http://www.statistik.at/web_en/statistics/agriculture_and_forestry/farm_structure_cultivated_area_yields/fruit/index.html.

Stevens, P. F. (2007) Angiosperm phylogeny website, version 8. http://www.mobot.org/MOBOT/research/APweb/.

Storey, W.B. (1975) Figs. p. 568–588. In: J. Janick and J.N. Moore (eds.), Advances in fruit breeding. Purdue University Press, West Lafayette, Indiana.

Stover, E., and Mercure, E.W. (2007) The pomegranate: a new look at the fruit of paradise. Hortsci. 42:1088–1092.

Stover, E., M. Aradhya, L. Ferguson, and C.H. Crisosto. (2006) Assessing commercial potential of diverse fig cultivars. Proc. Calif. Fig Res. Inst. Pp. 1–25.

Sykes, J.T. (1972) A description of some quince cultivars from western Turkey. Economic Botany 26:21–31.

Takada, Makiko, Hideki Nakano, Yoichiro Hoshino, and Hiroji Sato. (2003) Evaluation of eating qualities and some horticultural characteristics for selection of elite lines in *Lonicera caerulea* L. Research Bull. Hokkaido University Farm 33:21–38. (In Japanese).

Talent, N. and Dickinson, T. A. (2007a) Apomixis and hybridization in Rosaceae subtribe Pyrineae Dumort.: a new tool promises new insights. Apomixis: Evolution, Mechanisms and Perspectives, Regnum Vegetabile. 301–316.

Talent, N. and Dickinson, T. A. (2007b) The potential for ploidy level increases and decreases in *Crataegus* (Rosaceae, Spiraeoideae, tribe Pyreae). *Can. J. Bot.* 85:570–584.

Tanaka, Shizuyuki, Masashi Kakizaki, Hisaaki Watanabe, Tsuneya Minegishi, Fumio Matsui, Hiroshi Muramatsu, Ryuichi Ogano, Hideo Narita, and Akeo Iwasaki. (1994) New blue honeysuckle (*Lonicera caerulea* L. var. *emphyllocalyx* Nakai) cultivar 'Yufutsu'. Bull. Hokkaido Pref. Agric. Expt. Sta. Bull. 67:30–41. (In Japanese).

Tanaka, Tsuneo and Akira Tanaka. (1998) Chemical composition and characteristics of hasukappu berries in various cultivars and strains. Nippon Shokuhin Kogaku Kaishi 45: 129–133. (In Japanese).

Tennyson, A. (1892) 'Kapiolani' In: *The Death of Oenone,.Akbar's dream, and other poems.* Macmillian and Co. London.

Thomas, A.L., Byers, P.L., Finn, C.E., Chen, Y.-C., Rottinghaus, G.E., Malone, A.M. and Applequist,W.L. (2008) Occurrence of rutin and chlorogenic acid in elderberry leaf, flower, and stem in response to genotype, environment, and season. Acta Hort. 765, 197–206.

Thompson, M.M. (2006) Introducing haskap, Japanese blue honeysuckle. J. Amer. Soc. Pomol. Soc. 60:164–168.

Thompson, M.M. and D.L. Barney. (2007) Evaluation and breeding of haskap in North America. J. Amer. Pom. Soc. 61:25–32.

Tukey, H.B. (1964) Dwarfing rootstocks for the pear. Ch. 11 in: Dwarfed Fruit Trees, The MacMillan Co., New York. pp. 182–199.

University of Kentucky New Crop Opportunities Center. (2008) Pawpaw. http://www.uky.edu/Ag/NewCrops/introsheets/pawpaw.pdf.

USDA. (2009a) Germplasm Resources Information Network - (GRIN) Online Database. National Germplasm Resources Laboratory, Beltsville, Maryland. http://www.ars-grin.gov/cgi-bin/npgs/html/taxon.pl?12779 (05 February 2009).

USDA. (2009b) National Agricultural Statistics Service, U.S. fruit production data. http://www.nass.usda.gov/QuickStats/indexbysubject.jsp (4 February, 2009).

Vines, R. A. (1977) *Trees of East Texas.* University of Texas Press, Austin, TX.

Vander Kloet, S. P. (1993) Biosystematic studies of *Vaccinium* section *Macropelma* (Ericaceae) in Hawaii. *Pacific Science* 47 (1):76–85.

Wagner, W L., D. R. Herbst, and S. H. Sohmer. (1990) *Manual of the flowering plants of Hawai'i.* 2 vols, *Bishop Museum Special Publication 83.* Honolulu: University of Hawaii Press and Bishop Museum Press. p. 593–595.

Wagner, W. L., D. R. Herbst, and D. H. Lorence. (2005) Flora of the Hawaiian Islands website. http://ravenel.si.edu/botany/pacificislandbiodiversity/hawaiianflora/index.htm [accessed 31 December 2008].

Way, R.D. (1957) Cultivated elderberries. New York Farm Research 23, 15.

Way, Roger D. (1981) *Elderberry Culture in New York State.* New York State Agri. Expt. Sta. Food and Life Sciences Bull. No. 91.

Whitson, J., R. John and H.S. Williams (eds.) (1914) The Transformation of the Quince. Chapter 7, Volume 4 in Luther Burbank, His Methods and Discoveries and Their Practical Application. Luther Burbank Press, New York and London pp. 211–240.

Yatskievych, G. (2006) *Steyermark's Flora of Missouri, Vol 2.* Missouri Botanical Garden Press, St. Louis, MO.

Zee, F.T., Strauss, A.J., Arakawa, C.N. (2008) Propagation and Cultivation of 'Ohelo. Cooperative Extension Service, CTAHR, University of Hawaii. Fruits and Nuts F&N-13

Zhu, L-W, Zhang, Y-M, Wang, D-X, and Lou, Z. (2004) 'Baiyushizi', a high quality pomegranate cultivar. South China Fruits 33:69–70.

Part II
Small Fruit

Chapter 5
Blackberry

Chad E. Finn and John R. Clark

Abstract Blackberries are in *Rosaceae* family, the *Rubus* genus and subgenus (formerly *Eubatus*).Commercial cultivars are a multispecies complex and generally do not have a species epitaph. The primary progenitor species for the cultivated blackberries are all perennial plants with biennial canes. In these species, vegetative canes called primocanes are produced the first year and after a dormant period they are called floricanes. The floricanes flower, fruit, and die while new vegetative primocanes are growing. Blackberries can be grown throughout much of the temperate regions in the world. They do best when grown on well-drained, fertile soils with adequate moisture, in regions with moderate or mild winters and moderate summertime conditions. Although blackberries are a minor crop among fruits, there have been hundreds of cultivars named ranging from wild selections to those developed from multiple cycles of selection. Initially, a germplasm pool was assembled that lead to cultivars that were commercially viable and that later had outstanding traits. Then, as sources of thornlessness were identified, breeders incorporated them into this germplasm, and eventually high-quality cultivars were developed. A primary focus of all programs is fruit quality for promoting consumption. Other objectives are disease and pest resistance, primocane-fruiting, productivity, yield, plant architecture, and thornlessness. The use of molecular and other techniques in blackberry has been very limited. The use of simple sequence repeat markers (SSR) was reported for assessing genetic similarity and fingerprinting.

C.E. Finn (✉)
US Department of Agriculture-Agricultural Research Service, Horticultural Crops Research Laboratory, 3420 NW Orchard Avenue, Corvallis, OR 97330, USA
e-mail: finnc@hort.oregonstate.edu

J.R. Clark
Department of Horticulture, University of Arkansas, Fayetteville, AR, USA
e-mail: jrclark@uark.edu

M.L. Badenes and D.H. Byrne (eds.), *Fruit Breeding*, Handbook of Plant Breeding 8,
DOI 10.1007/978-1-4419-0763-9_5,

Keywords *Rubus* • Hybridberry • Raspberry Blackberry Hybrid • Trailing • Erect • Semierect Primocane-fruiting • Specialty Crops • Small Fruits

1 Introduction

Blackberry breeding has taken on greater emphasis as its importance as a crop has dramatically increased in the past 10 years. Previous, detailed reviews of blackberry breeding were done by Hall (1990) and Clark et al. (2007), with other valuable reviews by Darrow (1937; 1967), Waldo (1950a, 1968), Sherman and Sharpe (1971), Ourecky (1975), Moore (1984), Jennings (1988), Daubeny (1996), and Finn (2008).

1.1 Economic Importance and Use

Blackberry (*Rubus* sp.) consumption has increased substantially in the past 20 years (Strik et al. 2007). In 1990, North American production was 4,385 ha, with about 75% of that in the Pacific Northwest (Clark 1992; Strik 1992) and about 90% of the Pacific Northwest production was for processing. In the late 1990s, off-season shipments of fruit into North American markets from Chile, Guatemala, and Mexico began to increase. Since that time, California has become a major fresh market producer, and fresh market production in the South has expanded also with these production regions providing a substantial amount of the domestic crop for shipping. There has also been a rapid expansion of production for processing not only in the Pacific Northwest but also in Serbia and China. In 2005, there was an estimated 20,035 ha of blackberries planted and commercially cultivated worldwide with an additional 8,000 ha of fruit that was harvested from the wild, for a total estimate of 140,292 Mg (Strik et al. 2007).

Blackberries are sold fresh, primarily in clam shell packages, and as a processed product. The primary processed products are individually quick frozen (IQF), bulk frozen (whole fruit, puree, juice), canned, or dried. From these basic wholesale products, a plethora of products are made for the retail market and institutional food service product lines.

1.2 Taxonomy

Blackberries are classified in the *Rubus* subgenus *Rubus* (formerly *Eubatus*). Since most of the cultivated types were derived from two or more species, none of them have a species epitaph. Blackberries, red raspberries (*R. idaeus* L.; *Idaeobatus*), and black raspberries (*R. occidentalis* L.; *Idaeobatus*) are the most widely grown commercial *Rubus* (*Rosaceae*). However, nearly every region of the world where *Rubus* is native has developed thriving local industries based on their local species, a few examples include the following: Mora (*R. glaucus*, Benth.) in Andean South America; wineberry (*R. phoenicolasius* Max.), Korean black raspberry (*R. coreanus* Miq.), and trailing raspberry (*R. parvifolius* L.) in Asia; and cloudberry (*R. chamaemorus* L.) and arctic raspberry (*R. arcticus* L.) native to the far northern regions of Eurasia and North America (Finn 1999, 2008; Finn and Hancock 2008).

The primary progenitor species for the cultivated blackberries are all perennial plants with biennial canes. In these species, vegetative canes called primocanes are produced the first year and after a dormant period they are called floricanes. The floricanes flower, fruit, and die while new vegetative primocanes are growing. Recently, primocane-fruiting cultivars have been developed. Blackberries are generally larger and more vigorous than raspberries, and the cultivated types have prostrate (trailing) to very upright (erect) growth habits with canes up to 5 m tall (Clark et al. 2007).

Blackberry flowers have white or pink petals surrounding a receptacle that has multiple ovaries, styles, and stigmas. The flowers are insect pollinated. If pollination and fertilization are successful, an aggregate fruit is produced that consists of the central torus (receptacle) surrounded by a number of fleshy drupelets that each contains a seed (pyrene). Flowers and fruit are born in a panicle-like or racemose-cymb, with primary fruit ripening prior to secondary, quaternary, or tertiary (Hummer and Janick 2007). At fruit maturity, an abscission zone forms at the base of the blackberry receptacle. If the torus picks with the fruit, it is considered a blackberry, whereas if it remains on the plant it is considered a raspberry.

1.3 Production Zones and Adaptation

Blackberries can be grown throughout much of the temperate regions in the world. They do best when grown on well-drained, fertile soils with adequate moisture, in regions with moderate or mild winters and moderate summertime conditions.

Strik et al. (2007) provides a thorough overview of worldwide production. North America has the greatest production, with 65% of that production in Oregon and 32% in Mexico. Mexican production is rapidly increasing and doubled from 2002 to 2004. Europe was the second most productive region, with Serbia accounting for 69% of European production. Asia was in third place with about half the production as in North America. China accounts for all of the known Asian production. In the 1990s and early 2000s, production rapidly increased worldwide for several reasons including new cultivars making the crop more desirable to customers, the interest by consumers in 'new' crops and in crops with high antioxidant levels, and the recognition that blackberries were more profitable to grow due to longer-lived plantings than some of their *Rubus* relatives such as red raspberry.

Blackberries inevitably are compared to their other commercial *Rubus* brethren and in general are more heat tolerant, less winter cold tolerant, and more tolerant of heavy soils than red raspberries. The primary cultivated types have been limited to temperate regions, although the primocane-fruiting types suggest that chilling is not required for flowering. Imposed drought combined with growth regulators have been used to overcome a lack of chilling and trigger flowering in some production regions. Blackberry fruit are susceptible to sunburn, particularly in regions with intense sunlight and low humidity. While blackberries are generally fairly disease tolerant, there are a few diseases such as double blossom/rosette (*Cercosporella rubi* [Wint.] Plakidas) (see section on "Disease and Pest Resistance") that prevent commercial production in some areas.

2 Origin and Domestication

While *Rubus* is presumed to have been a food source wherever it was found with humans, the Hummer and Janick (2007) review of *Rubus* indicated this genus was used in ancient and historical times by its inclusion in artwork or illustrations of these times. European blackberry and red raspberry plants were mentioned by Ancient Greek and Roman rhyzomotists and were illustrated on lost scrolls of western antiquity. At Newberry Crater near Bend, OR artifacts of food remnants containing *Rubus* date to 8,000 BCE. Aeschylus and Hippocrates from 500 to 400 BCE discussed caneberries with Hippocrates recommending leaves and stems as part of a poultice for wounds. The Hebrew Bible contained many references to thorny plants that some have attributed to *Rubus sanctus* Schreb. or *R. ulmifolius* Schott, which are native to the Holy Land (Hummer and Janick 2007). The term *sĕneh* used to describe these species is also the term used in Exodus 3:1–5 to describe God's appearance to Moses 'in the flame of fire in the bush.' Numerous herbals, particularly Dioscorides' *De Materia Medica* written about 65 CE, included descriptions of how blackberry could be used to benefit health. The first image of *Rubus* that survived antiquity is from the *Juliana Anicia Codex*, an illustrated manuscript based on Dioscorides work from around 512 CE. With the Renaissance's explosion of exploration and flourishing of botanical study, *Rubus* was well represented. Hummer and Janick (2007) cite two paintings by Jan Bourdichon (1503–1508) that illustrate *Horae ad isum Romanum:* a prayer book for Anne of Bretagne including a drawing of a *Rubus* plant by Leonardo da Vinci (1510–1512), and a wood cut by Leonhart Fuchs (1544) from the herbal *De Historia Stiripum* as examples.

By the 1600s, blackberries were being mentioned in gardening books (Jennings 1988). However, since blackberries were so common where people lived, especially *R. argutus* Link, *R. allegheniensis* Porter, and *R. trivialis* Michx. in eastern and *R. ursinus* Cham et Schltdl. in western North America, there seemed to be little interest in domestication and identification of superior genotypes, let alone breeding, until the 1800s. Not surprisingly some of the first recorded selections from the wild were oddities such as albino or pink-fruited selections (Hedrick 1925). 'Dorchester,' a selection from the wild, was the first cultivar named in 1841 and 'New Rochelle' (syn. 'Lawton'), released in 1854, another wild selection, was the first to be widely planted (Hedrick 1925). Several other cultivars that became important commercially were also wild selections named at about the same time and included 'Aughinbaugh,' 'Eldorado,' 'Lucretia,' and 'Snyder' (Hedrick 1925; Ourecky 1975; Jennings 1988; Moore 1984; Clark et al. 2007)

Judge James H. Logan of Santa Cruz, CA is usually credited with having the first documented breeding effort. 'Loganberry,' released in 1890, and 'Black Logan' were the most successful from his program. 'Loganberry' is still grown commercially and was selected from open-pollinated fruit of the pistillate 'Aughinbaugh' presumably crossed with 'Red Antwerp' red raspberry (Logan 1955). The great horticultural personality of the time, Luther Burbank, from the San Jose, CA area was intrigued and developed/found 'Phenomenal'/'Burbank's Logan' that was nearly indistinguishable from 'Loganberry' (Darrow 1925; Clark et al. 2007). Byrnes M. Young in Morgan City, LA could not grow 'Loganberry' or 'Phenomenal,' but was in contact with Burbank. He made a cross between the latter and the better-adapted 'Austin Mayes' to produce 'Youngberry' in 1905 (Christy 2004; Clark et al. 2007). 'Youngberry' is not widely grown but has a lucrative market niche as a juice product and liqueur in South Africa. More importantly it is a parent of the widely grown 'Olallie' and a grandparent of 'Marion.' Another early cultivar of uncertain origin but continuing in use is 'Boysenberry.' The most thorough examination of 'Boysenberry's' history was done by Wood et al. (1999), although others have weighed in on the topic (Darrow 1937; Stellar 1937; Thompson 1961; Jennings 1988). 'Boysenberry' was discovered by Rudolph Boysen on the farm of John Lubben in Napa County, CA. Boysen moved to southern California, and it was there that the genotype grabbed the attention of USDA-ARS plant breeder George Darrow from Beltsville, MD who in turn convinced a local fruit grower and nurseryman, Walter Knott, to put in trials of this selection. Knott and Darrow named the selection after the discoverer. Knott went on to develop a thriving business that started as a farm with a dining room serving 'Boysenberry' pie and became the Knott's Berry Farm empire. While Wood et al.'s (1999) explanation of the historical origins of 'Boysenberry' is well researched there is still no certainty of its genetic origins. 'Boysenberry' is often cited as being from a raspberry x blackberry hybridization; however, its similarity to 'Youngberry' has led some to hypothesize it is a cross of a 'Loganberry-like' genotype with an eastern trailing blackberry such as 'Lucretia' or 'Austin Mayes' (Nybom and Hall 1991; Hall et al. 2002). Similar hybrid berries were also developed in Europe during the same time period including 'Laxtonberry,' 'Veitchberry,' 'Mahdi,' and 'Kings Acre' (Darrow 1937).

'Thornless Evergreen' is another selection from the wild that continues to have a significant commercial presence (Waldo 1977). 'Evergreen' was a selection of *R. laciniatus* that was traced back to the 1800s in Europe and the 1850s in the USA. Since its introduction, it has become widely naturalized along the Pacific Ocean coastal regions. A thornless chimera, 'Thornless Evergreen' was discovered in Stayton, OR in 1926 and quickly became the industry standard. The thornless chimeral form is unstable and commonly reverts to thorny canes with environmental or mechanical injury. The genetically thornless 'Everthornless' was developed from somaclonal plants from 'Thornless Evergreen' (McPheeters and Skirvin 2000).

Amidst this flurry of activity by private breeders/hobbyists, the beginnings of formal public breeding efforts began. Darrow (1937) cites the Texas Agricultural Experiment Station (College Station, TX) as the first blackberry breeding program. Along the lines of Young's efforts, the primary original emphasis was to develop "hybrid berries" that were adapted to hot climates with low chilling requirement. 'Nessberry' developed there using *R. trivialis* germplasm had some popularity, but it was even more valuable as a parent of the low-chill 'Brazos.'

The John Innes Horticultural Institute in England and the New York State Agricultural Experiment Station followed by the USDA-ARS in Georgia were the next to develop programs. The biggest long-term impact of the John Innes program was the development of 'Merton Thornless,' which is the primary source of thornlessness in all tetraploid cultivars. The New York program developed several erect cultivars in the 1950s including 'Bailey,' 'Hedrick,' and 'Darrow,' the last of which is still occasionally grown as it is one of the hardiest developed. The USDA-ARS program in Georgia served as the basis for the Beltsville, MD and Corvallis, OR programs.

While there have been many programs worldwide since these first breeding programs, few are still active (Finn and Knight 2002). The major current breeding efforts worldwide are with the University of Arkansas, the USDA-ARS in Oregon and the private program run by Driscoll's Strawberry Associates (Watsonville, CA).

The USDA-ARS Beltsville program is responsible for incorporating thornlessness from 'Merton Thornless' into the first outstanding thornless cultivars released in the late 1960s and early 1970s including 'Black Satin,' 'Smoothstem,' 'Thornfree,' and 'Dirksen Thornless' (Scott and Ink 1966). The USDA-ARS had a significant effort at their station in Carbondale, IL in the 1960s until it was closed in the early 1970s. 'Hull Thornless' and the very important 'Chester Thornless' came from this effort. The last release from these programs was 'Triple Crown' in the 1990s (Galletta et al. 1998b). This group of breeding material and cultivars is called "semierect' and the plants are characterized as being thornless, with very vigorous, erect canes that grow 4–6 m long from a crown and arch to the ground. Their fruit is similar in quality to the erect blackberries and they are very productive.

The Horticulture and Food Research Institute of New Zealand Ltd. (formerly New Zealand HortResearch Inc.) program was one of the most valuable and aggressive programs in the 1980s and 1990s; however, its ongoing funding is in question. While this program, begun in 1980, had several objectives, the most important was the development of new 'Boysenberry-like' cultivars (Hall et al. 2002). They blended germplasm from the USDA-ARS (Ore.) and the Scottish Crop Research Institute (Dundee) as well as other available cultivars and developed the 'Lincoln Logan' source of spinelessness (S_{fL}) (Hall et al. 1986a; b; c). Their most important releases have been 'Ranui,' 'Waimate,' 'Karaka Black,' and 'Marahau' (Hall and Stephens 1999; Clark and Finn 2002; Hall et al. 2003).

The USDA-ARS program in Oregon was started in 1928 and is the oldest continuously active program. The effort there combined wild selections of the native, trailing, dioecious *R. ursinus* Cham et. Schlt. with a perfect-flowered gene pool including 'Loganberry,' 'Youngberry,' 'Himalaya,' 'Santiam,' and 'Mammoth,' and cultivars from elsewhere, to develop cultivars for a whole new industry based on trailing blackberries. The plants are characterized as crown-forming, have very long canes that trail along the ground if not trained to a trellis, tend to have excellent fruit quality, but have poorer winter hardiness than the other types. The first cultivars from this program included 'Pacific,' 'Cascade,' 'Chehalem,' and 'Olallie' that were released from 1942 to 1950 and these were instrumental in establishing a new industry (Waldo and Wiegand 1942; Waldo 1948, 1950b). These were followed by the release of 'Marion' in 1956 (Waldo 1957), which is still the industry standard in the Pacific Northwest, and 'Kotata' (Lawrence 1984). During the 1970s–1980s, improving the thornless germplasm pool was a program goal that resulted in the release of the first trailing, thornless cultivar Waldo, which carries thornlessness from 'Austin Thornless' (Lawrence 1989). The challenge of thorn contamination in the product pushed the development of thornless cultivars for processing even harder (Strik and Buller 2002). While 'Waldo's' release was important, it was this germplasm pool that led to the thornless 'Black Diamond,' 'Black Pearl,' and 'Nightfall' that have been widely planted (Finn et al. 2005b; d; e). In a moderate climate, trailing blackberries tend to be earlier ripening than the semierect or erect blackberries, and in these climates the earliest ripening trailing genotypes are the earliest ripening of all blackberries. The recently released 'Obsidian' and 'Metolius,' while thorny, are the earliest ripening cultivars available (Finn et al. 2005a; c).

The University of Arkansas is primarily responsible for the development of the erect blackberries from eastern North American blackberry species. They are characterized by plants that produce stiff, upright canes that are 1–4 m tall, and the plants sucker to produce a hedgerow. While there are erect cultivars, such as 'Eldorado,' which can be traced back to the 1800s, the focused effort in Arkansas developed this type as a viable commercial crop. Breeding at the University of Arkansas began in 1964 and continues today. This type is tetraploid and shares a similar genetic background with the semierect cultivars and has comparable fruit characteristics. The 'Merton Thornless' source of thornlessness was incorporated into this gene pool and 'Navaho' was the first thornless, erect cultivar to be released in the late 1980s. Some of the other cultivars that have been released from this program include 'Cheyenne' and 'Cherokee' released in the 1970s, 'Shawnee' in the 1980s, 'Kiowa,' 'Apache,' and 'Chickasaw' in the 1990s, and 'Ouachita' and 'Natchez' in the 2000s. Recently, this program developed a new type known as primocane-fruiting blackberries that flower and fruit very late in the season on current -season canes; 'Prime-Jan'® and 'Prime-Jim'® were the first cultivars of this type followed by 'Prime-Ark®45.' This trait was critical to the worldwide expansion of the red raspberry industry, and it is hoped that it will have a similar impact on blackberry production.

Driscoll Strawberry Associates, Inc. (Watsonville, CA) has been breeding red raspberries in some manner since the 1930s and their blackberry program was started in 1991. The blackberry program is one of the larger efforts in the world. While it may be irrelevant to others what they do as the cultivars they develop are kept within the company, they have played a critical role in the expansion of the fresh raspberry and blackberry industry and the acreage devoted to their cultivars is large.

Smaller sized, productive programs are active elsewhere as demonstrated by the recent development of 'Tupy' from EMBRAPA (Empresa Brasileira de Pesquisa Agropecuária) Brazil (Clark and Finn 2002), 'Loch Maree,' 'Loch Ness,' and 'Loch Tay' from the Scottish Crop Research Institute (Jennings 1989; Clark and Finn 2006; Clark et al. 2008), and 'Čačanska Bestrna' ('Čačak Thornless') from the Serbian Research Institute (Belgrade) (Clark and Finn 1999; Stanisavljevic 1999).

3 Genetic Resources

Rubus is divided into 15 subgenera, and blackberries are classified in the subgenera *Rubus*, which is further divided into 12 sections (USDA-ARS National Genetic Resources Program 2010b). Cultivated types were derived from species in the *Allegheniensis*, *Arguti*, *Rubus,* and *Ursini.* Temperate species from the *Idaeobatus*, which contains raspberry and the Andean blackberry (*R. glaucus*), have also contributed to the cultivated germplasm. Chromosome numbers in *Rubus* range from $2n=2x=14$ to $2n=18x=126$ including odd-ploids and aneuploids (Thompson 1995a, b, 1997; Meng and Finn 1999). The chromosomes are small, 1–3 μm in length, with a nuclear DNA content for the diploid species ranging from 0.56 to 0.59 pg (Lim et al. 1998; Meng and Finn 2002). While manual counting is the most reliable method of determining the ploidy level in *Rubus*, flow cytometry has proven to work well to differentiate ploidy level if not the precise number of chromosomes (Meng and Finn 2002).

All of the cultivated types of blackberries have multiple species in their background, but Clark et al. (2007) laid out the primary groups: (1) European blackberries that were derived from a group of diploid and polyploid species ($2n=28$, 42, and 56). The backgrounds of the European cultivars are so mixed that the designation *R. fruticosus* L. agg. is often used (Daubeny 1996). (2) Erect and semierect blackberries (4x) and trailing dewberries (2x) domesticated from diploid and tetraploid species from eastern America. (3) Trailing blackberries generated from polyploid species from western North America, predominantly *R. ursinus* at $2n=56$, 84, with infusions of 4x blackberry and 2x red raspberry through intersectional hybrids such as 'Logan' and 'Tayberry' ($2n=42$), 'Boysenberry' and 'Youngberry' ($2n=49$). The trailing cultivars can be found at $2n=42$, 49, 56, 63, 72, and 80, along with various aneuploids such as 'Aurora' ($2n=58$) and 'Santiam' ($2n=61$) (Thompson 1997; Meng and Finn 2002) (Fig. 5.1).

Polyploidy has played a significant role in the evolutionary development of *Rubus* (Gustafsson 1942, 1943; Thompson 1997). However, it is uncertain as to whether the polyploid genotypes are allopolyploids or autopolyploids (Einset 1947; Ourecky 1975; Stafne 2005; Clark et al. 2007). Tetrasomic inheritance appears to predominate in the tetraploids, although polysomic and disomic inheritance also

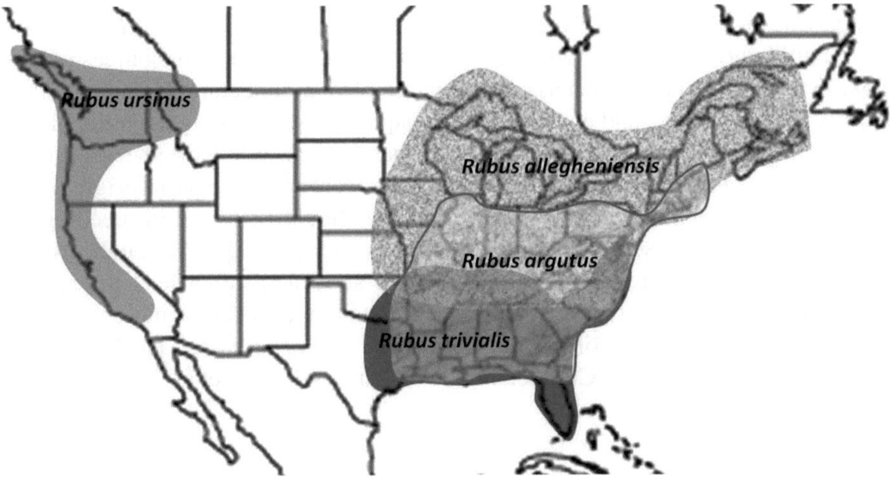

Fig. 5.1 Distribution of blackberry species that have been the major contributors of cultivar development

appears to occur for some traits (Lopez-Medina et al. 2000; Stafne 2005). *Rubus allegheniensis* and *R. argutus* are the predominant species within the eastern North American tetraploid group, and these are from different taxonomic sections; the *Allegheniensis* and *Arguti*, respectively (Stafne and Clark 2004). While these two sections can be separated morphologically, the separation among several sections can be vague (Davis et al. 1969a; b) and affected by environmental conditions (Brainerd and Peitersen 1920). Pamfil et al. (2000) using randomly amplified polymorphic DNA (RAPD) markers, and Stafne et al. (2003) studying genetic distance within the internal transcribed spacer region of the nuclear ribosomal DNA found that these two species have closely related genomes. This information, combined with other studies, suggests that the tetraploid, eastern germplasm pool is made up of autopolyploids or segmental allopolyploids, as opposed to true allopolyploids (Clark et al. 2007).

A great deal of effort has been placed in trying to understand the evolutionary background of the western trailing blackberry, particularly *R. ursinus* (Brown 1943; Jennings 1988; Alice and Campbell 1999; Alice et al. 2001). Alice and Campbell (1999) were able to confidently place *R. ursinus* within the *Rubus* subgenus. Several hypotheses on the origin have been put forth. Brown (1943) proposed that a chromosomal substitution from one parent might be a cause of the variation in the *Ursini* and that an extinct species similar to either *R. allegheniensis* or *R. argutus* led rise to the *Ursini* (Clark et al. 2007). Alice and Campbell (1999) proposed that *R. macraei* from the *Idaeobatus* and an unknown member of the subgenus *Rubus* are the progenitors of *R. ursinus*.

A wide range of species have been identified as being important sources of germplasm in blackberry breeding (Clark et al. 2007; Finn et al. 1999b; Finn et al. 2002a; b; Finn 2008; Jennings et al. 1992). Over 25 species in the *Allegheniensis*,

Arguti, *Caesii*, *Canadenses*, *Flagellares*, *Rubus*, *Ursini*, *Verotriviales*, *Idaeobatus*, and *Lampobatus* were identified by Finn (2008) as being important sources of germplasm in blackberry. Characteristics related to plant architecture, phenology, fruit quality, pest resistance, and environmental adaptation were among the traits identified that might be introgressed into cultivated germplasm. While the species used in developing the cultivated types are largely American or European in origin, Asia, specifically China, has a wealth of diversity that should be useful in breeding for environmental and disease tolerance (Jennings et al. 1992). Although crosses between diploids and tetraploids or with other subgenera often lead to sterility, crosses among other ploidy levels within the *Rubus* subgenera are commonly fertile, and crosses with members of the *Idaeobatus* are often successful and have led to several important cultivars (Clark et al. 2007; Finn 2001; Finn et al. 2002a, b).

4 Major Breeding Achievements

Although blackberries are a minor crop among fruits, there have been hundreds of cultivars named ranging from wild selections to those developed from multiple cycles of selection (Clark et al. 2007). Blackberry cultivation began using wild selections and chance discoveries, and these genotypes provided the basis for genetic improvement in breeding since the early 1900s. One of the early achievements was the mixing of blackberry species in the eastern USA and blackberry and raspberry species in the western USA to develop a tremendously diverse germplasm pool.

The major accomplishments in blackberry followed a similar pattern in each type of blackberry that in turn is linked closely to a specific breeding program. Initially, a germplasm pool was assembled that lead to cultivars that were commercially viable and that later had outstanding traits. Then, as sources of thornlessness were identified, breeders incorporated them into this germplasm, and eventually high-quality cultivars were developed. In the case of the trailing blackberries, the first major accomplishment was the development of trailing blackberries as a crop by the USDA-ARS (Oregon). In turn, breeders developed cultivars with exceptional processing and fruit quality characteristics and that were machine harvestable. Over time, the 'Austin Thornless' source of thornlessness was introgressed into this high ploidy germplasm pool resulting in high-quality thornless cultivars suited for processing. The USDA-ARS in Beltsville developed the semierect cultivars with thornlessness that had been previously isolated in 'Merton Thornless' by the John Innes Institute. Thornless cultivars were later developed that had good winter hardiness, extreme productivity, and good postharvest handling. By merging germplasm from several different sources, the University of Arkansas program developed erect-caned cultivars with improved fruit size and quality with adaptation to the mid-to upper South of the USA. While initially a regional novelty, very quickly a commercially viable industry developed that now has spread worldwide. As the 'Merton Thornless' thornlessness was merged into this germplasm, thornless cultivars with exceptional postharvest handling capacity were developed. The other critical accomplishment that was tied with the incorporation of this thornlessness was that this germplasm had resistance to rosette/double blossom, one of the most limiting factors in blackberry production in the southern USA. Separately, the Horticulture and Food Research Institute of New Zealand Ltd.'s unique development of the 'Lincoln Logan' source of thornlessness being used in the trailing blackberries has been a major accomplishment as it further facilitates breeding thornless types and incorporates more raspberry germplasm into blackberry germplasm.

The original work in Texas and later at EMBRAPA in Brazil to develop low-chill germplasm was a major accomplishment. 'Brazos,' developed in Texas, was the most important cultivar in Mexico for several years and its replacement, 'Tupy,' from Brazil has significantly better quality in a low-chill background. Taking low chilling a step further, the recent development by the University of Arkansas of the primocane-fruiting types that fruit on current season's growth, apparently without the need for chilling, has the potential to expand the industry to new heights as occurred with red raspberry decades ago.

5 Current Goals and Challenges

Blackberry breeding programs are not extensive in the world today compared to many fruit crops, and in 2002 there were 15 programs operating (Finn and Knight 2002). Most of these programs continue in operation and some expansion is likely occurring in breeding activity as blackberries increase in popularity in world markets. For a further review of past and current breeding, see Clark et al. (2007).

Although programmatic activity in blackberry is not extensive, the promise and excitement associated with potential improvement in blackberry is great. In general, goals in various breeding programs have some common and differing objectives depending on the type of blackberry, use and market, and genetic variability available (Clark and Finn 2008). The achievements in breeding along with genetic approaches are discussed for various traits such as fruit quality, architecture, adaptation, and others later in this chapter.

A primary focus of all programs, and the main area that can advance consumption, is fruit quality. Advances in quality from the early wild selections and first improved cultivars have been substantial thus far. The progress made has moved blackberry from being viewed as a fruit harvested from the wild to one that is now routinely found on retail market shelves throughout the world. When quality is discussed, most consumers consider berry sweetness to be foremost in need of enhancement. Progress in this area can be made, and along with manipulation of flavor components, acidity, astringency, and postharvest handling. There is more than adequate genetic variation available to substantially improve quality attributes and displace current cultivars. Breeding blackberries with broader adaptation has an even greater profile today than in prior years. This interest is primarily due to the expanded production from temperate to tropical climates. Until the mid 1990s, little to no interest existed in growing blackberries in areas with low to no chilling. The breakthrough in culture development for Central Mexico, using defoliation, pruning, and growth regulator applications, has been an eye-opening experience for the industry. The cultural method development, plus the use of 'Tupy,' has allowed this area to become the world's largest area for fresh market production. Proximity to USA markets where trucks can be used to transport fruit plus the increased demand for blackberries has advanced the substantial market potential for off-season sales. Breeding in no-chill environments has not been reported, and little is known of the ultimate extremes achievable in reducing chilling requirement to even lower levels than that found with 'Tupy.' Further, the introduction of primocane-fruiting in blackberry offers a method to eliminate chilling concerns completely, since these canes do not go through a dormant period prior to initiating flowers. In an opposite adaptation challenge, primocane fruiting may overcome winter injury to canes as the canes can be fruited with no requirement for overwintering them. It is exciting to envision the cultural advantages that this type of blackberry can provide for industry expansion.

As with any crop, limitations in genetic variability and breeding methodology can provide challenges in improvement. It appears that adequate variability exists for improvement of most major traits such as thornlessness, architecture, disease and insect resistance, adaptation, productivity, fruit quality, and fruit size. Traits that are limiting to some or a great extent include seedlessness, adaptation to heat in primocane-fruiting genotypes, complete resistance of fruit to sunburn, resistance to some viruses and other diseases, hardiness levels adequate for very cold climates, and complete durability of fruit in rainy conditions during ripening.

Another area of limitation in blackberry breeding is the lack of molecular technique development. Since blackberries have been one of the more "minor" crops, the investment in molecular investigations has been minimal. Areas such as mapping, development of molecular markers for seedling and selection screening, genomic investigations, transformation technology development, and other biotechnological procedures lag substantially behind other crops including the major fruits. As advances are made in molecular methods and technology in other *Rubus* species, it is hoped that these can be applied to genetic improvement in blackberry.

6 Breeding Methods and Techniques

6.1 Major Traits and Selection Techniques

6.1.1 Adaptation

Blackberries are generally considered to be broadly adapted to a wide range of climates and soils. A few environments are limiting for blackberries, however, with the two having a major impact being winter low temperatures that contribute to winter injury to the canes and/or flower buds and low-chill environments that do not provide for adequate chilling requirement fulfillment. As production has expanded to lower-chill regions of the world, plus the continued interest in growing blackberries in cold climates, breeding for adaptation has taken on an increased enthusiasm in recent years (Clark et al. 2007).

Winter injury has been a concern for breeding programs in the eastern USA along with those in northern Europe. Moore (1984) shared that a lack of winter hardiness is the major limitation to the expansion of blackberry production in much of North America, particularly in areas where winter temperatures cause damage in the upper South and northward (Warmund et al. 1989; Warmund and George 1990; Warmund and Krumme 2005). Cultivars including 'Illini Hardy' and 'Chester Thornless' (Moore 1997; Galletta et al. 1998a) are more recent releases with improved winter hardiness. Unfortunately, breeding for substantial winter hardiness has largely been discontinued in the USA and limited work is underway elsewhere in the world. Breeding for hardiness has been much like for other quantitative traits, with crossing of the hardiest genotypes in hopes of recovering progeny that are as hardy, or hardier, than the parents. A good location with regular "test" winters for screening parents and seedlings is required for this process. Also, multiple years and locations for evaluation of hardiness are usually needed to have confidence in the ultimate hardiness determination of a new cultivar.

A more recent approach to blackberry breeding for more northern climates involves a cold-injury avoidance mechanism – primocane-fruiting (see section on primocane-fruiting). Since fruiting is on current-season canes, overwintering of canes is not required and therefore injury to the canes is not a factor unless fruiting on the floricanes also is intended. Unfortunately, testing of the first primocane-fruiting cultivars Prime-Jim® and Prime-Jan® in northern USA locations with shorter growing seasons (than Arkansas where these were developed) including St. Paul, MN and Geneva, NY gave less than desirable results. Some crown and root damage was experienced and the full completion of the fruiting cycle was not achieved each year. Since mature fruit was not produced until approximately 1 Sept. at these locations, time was limited to allow a substantial amount of the fruit to ripen prior to frost (J. Luby and C. Weber personal communication). However, breeding to allow for earlier fruiting is being pursued, with the hope that this will allow fruit maturity into August resulting in a longer harvest period.

Although cultivated blackberry production has usually been practiced only in temperate climates, expansion of production to subtropical and tropical climates has greatly increased in the past 15 years. Production in reduced-chilled environments has had the greatest expansion in Central Mexico. In Mexico, the initial success was with the floricane-fruiting 'Brazos.' A system was developed in the 1990s whereby plants were allowed to grow in the traditional rainy season of June through August, and then a series of cultural manipulations was applied including defoliation with chemicals, pruning, and application of growth regulators. These treatments were further refined for commercial production and provide for flowering and fruiting from November until May with fruit shipped primarily to the USA and the EU (Jose Lopez-Medina personal communication). In the early 2000s, the Brazilian cultivar

Tupy was brought to Mexico and tested in the environment using the cultural system developed for 'Brazos.' 'Tupy' has much better fruit quality (firmness, flavor, and productivity) than 'Brazos' and by the mid-2000s it had taken over the majority of the planting area in Central Mexico. The basis for 'Brazos' and 'Tupy' to be adapted there is their development in lower-chill locations (College Station, TX, and Pelotas, Rio Grande do Sul, Brazil, respectively). Further breeding with screening of progeny in low-chill locations will allow for expansion of cultivar choices.

A substantial issue in blackberry production in some regions is sunburn damaged fruit. Although sunburn damage can be seen on fruits in most all environments at high temperatures, the general observation is that sunburn damage is greatest in lower-humidity climates with high light intensities. Examples of these environments include the Willamette Valley of Oregon, Central Valley of California, Australia, and dry climates of Chile (such as north of Santiago near Nogales) (J.R. Clark personal observation). Likely there are many other locations in the world with similar sunlight conditions, and therefore the testing of blackberry genotypes should always include evaluation of sunburn damage susceptibility. Sunburn usually results in drupelets having a white appearance, with either individual or groups of drupelets affected. Occasionally whole plants can be affected by high heat and light, such as in the Central Valley of California where 'Apache' plants were observed to withstand high heat without plant damage while 'Triple Crown' experienced leaf burn and some plant collapse (J.R. Clark personal observation). However, 'Apache' has shown to commonly have problems with white drupelets, severe enough to make fruit unmarketable for the shipping market, and this problem seems to be exacerbated by rainfall during fruit ripening. By comparison, 'Navaho' seldom experiences this problem. Heritability of heat reactions has not been investigated, however, but it is clear that some segregants in breeding populations are markedly more susceptible than others (H. Hall personal communication).

Bloom and harvest season varies substantially in blackberries. As fresh blackberries have become more popular in retail markets, the time of ripening has had increased focus since prices can be substantially different among months of the year. Shippers are interested in having a continuous supply of fruit so that blackberries maintain retail market space year-around. Breeding can play the most important role in achieving this continuous supply of fruit. Within floricane-fruiting genotypes, crossing among the earliest and latest parents can result in progeny that are earlier or later than their parents (transgressive segregants). Also, location of production has substantial impact on cultivars. The best example is in Oregon where the earliest trailing genotypes ripen about 2 weeks earlier than the earliest erect genotypes from Arkansas, while in Arkansas the ripening times are usually much closer. This is apparently due to heat unit response in the spring – more heat units are provided in Arkansas in the spring compared to Oregon (C.E. Finn and J.R. Clark personal observation). The expanded development of primocane-fruiting cultivars can greatly alter harvest times of blackberries. With breeding for earlier- and later-flowering genotypes, plus cultural manipulation including mowing of primocanes and use of high tunnels, the harvest period should be possible over a long period in a single location.

6.1.2 Disease and Pest Resistance

Blackberries are generally subject to far fewer disease and insect problems than red and black raspberries. Historically, the fungal diseases have been more of a problem than bacterial or viral diseases; however, an increased awareness of viruses in blackberries has led to a recognition that they are causing more problems than had been believed earlier (R.R. Martin personal communication). While each production region has unique problems depending on the environmental conditions and the types of blackberry grown, there are a number of common disease problems including anthracnose (*Elsinoe veneta* [Burkholder] Jenk.), cane botrytis and botrytis fruit rot (*Botrytis cinerea* Pers.: Fr.), and cane blight [*Leptosphaeria coniothyrium* (Fuckel) Sacc.] (Ellis et al. 1991).

In the Midwestern and Eastern USA and Eurasia, where continental climates and erect or semierect types of blackberry predominate, in addition to anthracnose and botrytis fruit rot, Botryosphaeria cane canker (*Botryosphaeria dothidea* (Moug.: Fr.) Ces. & De Not) and *Colletotrichum* spp. are common problems (Clark et al. 2007; Finn 2008). Potentially much more devastating is orange rust [*Gymnoconia peckiana* (Howe) Trott.] and, in the southern states, double blossom/rosette that can kill the plant (Marroquin et al. 1990; Ellis et al. 1991; Smith and Diehl 1991; Lyman et al. 2004). Resistance to orange rust is present in most eastern USA developed cultivars; however, 'Navaho,' which is one of the most popular erect cultivars, is susceptible (Clark et al. 2007). Variable resistance to double blossom has been identified. Most of the thornless Arkansas-developed blackberries are resistant to double blossom in Arkansas; however, under the intense disease pressure further south in Mississippi, the resistance does not reliably hold up (Buckley et al. 1995; Gupton and Smith 1997; Gupton 1999). In general, materials derived from 'Merton Thornless' show some resistance to double blossom.

In the maritime and Mediterranean climates of Chile, Mexico, New Zealand, and the western USA, cane botrytis, cane spot (*Septoria rubi* Westend), purple blotch (*Septocyta ruborum* [Lib.] Petr.), and spur blight [*Didymella applanata* (Niessl) Sacc.] are common problems. Fruit rots are not as much of a problem in these climates because much of the ripening season is dry. In these climates, when wet conditions during bloom intersect with downy mildew (*Peronospora sparsa* Berk.) sporulation, this disease can be a serious problem especially on the raspberry-blackberry hybrids such as 'Boysenberry' and 'Loganberry' (Gubler 1991; Breese et al. 1994). As Mexican production has expanded into areas with dry conditions throughout the growing season, powdery mildew [*Sphaerotheca macularis* (Wallr.: Fr) Lind.] has become a significant problem (Clark et al. 2007).

While bacterial diseases, particularly crown gall (*Agrobacterium tumefaciens* [E.V. Smith & Townsend) Conn.], can be problems, they infrequently cause severe crop loss (Ellis et al. 1991). Differences in susceptibility to crown gall, fireblight (*Erwinia amylovora* [Burr.] Winslow et al.), and Pseudomonas blight (*Pseudomonas syringae* van Hall) have been identified and characterized (McKeen 1954; Stewart et al. 2003).

Raspberry bushy dwarf virus (RBDV) along with several other virus diseases have long been known to be a serious problem in raspberry; however, until the past 10–15 years, viruses tended to be considered asymptomatic or not a significant problem in blackberries (Converse 1987; Jennings et al. 1992). New tools used to assess problematic plants identified new viruses and gave some insight into the spread of virus in commercial plantings (Chamberlain et al. 2003; Guzmán-Baeny 2003; Martin et al. 2004; Susaimuthu et al. 2007; Tzanetakis and Martin 2004). RBDV infection has been reported in western and eastern types of blackberry (Wood 1995; Wood and Hall 2001; Strik and Martin 2003). RBDV, while widespread in many native western *Rubus* species (e.g., *R. idaeus*, *R. parviflorus*, and *R. spectabilis* Pursch.) (Martin 2002), was not found in a broad survey of *R. ursinus*, the primary progenitor species of the western trailing blackberry (Finn and Martin 1996). While potentially a serious problem, the erratic nature of transmission and occurrence of RBDV has made it difficult to assess whether breeding for resistance is necessary or possible (Strik and Martin 2003). *Tomato ringspot virus* (ToRSV) and *Tobacco ringspot virus* (TRSV) are commonly identified in most blackberry production regions. More recently *Impatiens necrotic spot virus* (INSV) and *Blackberry yellow-vein associated virus* (BYVaV) have been identified in the southeastern USA (Guzmán-Baeny 2003; Martin et al 2004; Susaimuthu et al. 2007). The primary impact to this point on breeding programs has not been to breed for resistance, as these viral diseases are too poorly understood in blackberry, but rather for breeding programs to clean up their parental material of viruses so that breeding material begins clean and so that the programs are not a vector for the virus.

Each production area has insect problems that may have to be controlled. While there are often no standard insecticide programs (Ellis et al. 1991), some of the common problems can include: raspberry crown borer (*Pennisetia marginata* [Harris]), red-necked caneborer (*Agrilus ruficollis* [Fabricius]), redberry mite (*Acalitus essigi* Hassan), strawberry weevil (*Anthonomus signatus* Say), brown and green stink bugs (*Euschistus* spp. and *Acrosternum hilare* Say, respectively), Japanese beetle (*Popillia japonica,* Newman), thrips (eastern and western flower thrips, *Frankliniella tritici* Fitch and *F. occidentalis* Pergande, respectively), grass grub (*Costelytra zealandia* White), and foliar nematode (*Aphelenchoides ritzemabosi* [Schwartz] Steiner) (Clark et al. 2007). A few production regions have severe pests that require substantial control programs. A good example is in New Zealand where 'Boysenberry,' 'Marion,' and all other *Rubus* are attacked severely by raspberry bud moth (*Heterocrossa rubophaga* Dugdale) and/or blackberry bud moth (*Eutorna phaulacosma* Meyrick). The green vegetable beetle *Nezara viridula* L. and the leaf roller species including *Epiphyas postivittana* Walker, *Planotortrix exessana* Walker, *P. octo* Dugdale, *Ctenopseustis obliquana* Walker, *C. herana* Felder, and Rogenhofer and *Cnephasia jactatana* Walker can also be severe problems in New Zealand. The New Zealand program identified resistance to bud moth and leaf roller species in black raspberry and has attempted to move this in to blackberry (H. Hall personal communication).

Starting with clean plant material along with other cultural and chemical controls are generally effective for economically controlling blackberry pests. We are not aware of any breeding program that actively screens for resistance to insect or disease pests by actually applying the organism to the seedlings. Most breeding programs passively screen for disease resistance by not selecting genotypes that have serious disease symptoms and by discarding selections that develop serious disease symptoms during their evaluation. Some assessment of disease tolerance is sometimes obtained by screening selections in an environment where the disease pressure is intense; this has been used in the evaluation of a selection's response to double blossom (Clark et al. 2007).

6.1.3 Architecture

In nature, blackberries range in plant habit from completely procumbent to very upright. In commercial terms, they are usually classified with three cane types: trailing, semierect, and erect (Strik 1992). Trailing types are crown-forming and grow at or near ground level, and the canes must be bundled and tied to a trellis. Cultivars such as 'Marion,' 'Thornless Evergreen,' and 'Black Diamond' are examples of this type of plant and most commonly have been used for processing. Blackberries with semierect habit are also crown-forming and require a trellis, with the mature canes growing upward about 1 m before arching over to a horizontal orientation. Important semierect cultivars are 'Chester Thornless,' 'Loch Ness,' and 'Triple Crown.' The erect-caned blackberries are the third grouping of commercial types; their canes grow more upright and many of these sucker beneath the soil line and are less crown forming but rather can provide a continuous row of canes. Although erect types can be grown in a free-standing hedgerow, supporting wires are usually used commercially even for erect types. Erect cultivars include 'Navaho,' 'Arapaho,' 'Ouachita,' 'Natchez,' and 'Chickasaw.' Erect and semierect cultivars respond positively to tipping, of the canes while trailing cultivars are usually not tipped in their management. In general, cane growth habit is considered a quantitative trait. Crossing of erect × trailing usually yields semierect progeny while crossing within a cane habit yields plants with similar form as the parents.

Emphasis on erect-caned cultivars has been a major focus of the University of Arkansas breeding program with the original idea to develop cultivars for the fresh and processing markets whose canes required no trellising. Foundation parents used in this program included the erect cultivars Brazos and Darrow and a cross of these resulted in three thorny, erect-caned cultivars, Comanche, Cherokee, and Cheyenne (Clark 1999). The development of erect, thornless plants proved to be much more challenging. The thornless gene chosen for use in breeding was the recessive source derived originally from 'Merton Thornless' (Jennings 1988). In the Arkansas program, 'Thornfree' and 'Smoothstem' and related selections from the USDA-ARS program based at Carbondale, IL were used. Major problems included the quantitative nature of cane inheritance (only an incremental enhancement of erectness with each generation), coupled with associated negative traits that were inherited with this thornlessness source including late-ripening, less cold hardiness, tart flavor, variable drupelet fertility, small fruit size, poor seed germination, and poor adventitious shoot sprouting from roots (Clark 2005b). In 1980, Ark. 1172 was selected, which had erect canes, good fruit quality, and good plant adaptation. It was released in 1989 as 'Navaho,' the first thornless, erect blackberry cultivar (Moore and Clark 1989).

The trailing habit offers some management advantages for machine harvesting and winter hardiness. If the new primocanes are trained along the row and are lying on the ground they are below the catcher plates of a machine harvester. With the primocanes out of the way, the berries fall through the floricane vegetation more easily as the machine passes thereby reducing yield loss and mechanical damage to the fruit. As we move toward mechanization of training and pruning, an additional advantage of trailing types is that since the primocanes grow in a different physical space than floricanes, it will be easier for machines to differentiate the two cane types. The primary cane-related issues in trailing blackberries are thorniness along with cane flexibility. Most trailing blackberries such as 'Marion' have flexible canes that can be untangled, bundled, and trained to the trellis with minimal cane breakage. However, some genotypes, particularly those whose thornlessness is derived from 'Austin Thornless' such as 'Waldo,' are prone to having their canes broken during training. There is a wide range of expression of this brittleness among genotypes in populations and it is easy to select thornless genotypes that have flexible canes.

6.1.4 Primocane-Fruiting

The occurrence of primocane-fruiting, which is the development of flower buds on first-season canes, has been very important in recent years in red raspberry production expansion. This fruiting habit has great potential in blackberry production particularly when winter damage to floricanes limits production, when chilling requirement issues are important or limiting, and where scheduling of production for nontraditional times of ripening (such as the fall of the year) is desired.

The primary source of primocane-fruiting used thus far in breeding has been the wild selection referred to as 'Hillquist.' This source was reported to come from a wild plant found by L.G. Hillquist of Ashland, VA that was provided to the New York State Agricultural Experiment Station, Geneva, NY in 1949 (USDA 2010a). Although not commercialized as a cultivar, it likely had the name assigned to it in New York. The plant was noted to have a "rudimentary" level of primocane-fruiting. 'Hillquist' is a diploid (Thompson 1995b) and the first recorded use of breeding with it was by James Moore of the University of Arkansas (Ballington and Moore 1995). The cross 'Brazos' × 'Hillquist' was made in 1967 and a selection (Ark. 593) was made from this population. It was assumed that 'Hillquist' produced an unreduced male gamete to combine with the female gamete of the tetraploid 'Brazos.' Based on its success in producing consistent fertile offspring in crosses with tetraploids, Ark. 593 was determined to be tetraploid. Ark. 593 did not express the primocane-fruiting trait. James Ballington of North Carolina State University selfed Ark. 593 and recovered primocane-fruiting offspring. Ballington and Moore

(1995) released the germplasm selection NC 194 and hypothesized, later confirmed (Lopez-Medina et al. 2000), that the primocane trait was recessive. 'Prime-Jan'® (cultivar APF-8) and 'Prime-Jim'® (cultivar APF-12) released in 2004 were the first primocane-fruiting blackberry cultivars (Clark et al. 2005). The primary recommended use for these was for home garden planting as they were not deemed suitable for shipping, and had variable productivity depending on location. Subsequent evaluation in climates different from Arkansas provided some evidence of commercial potential for these in California and Oregon (Strik et al. 2008). In 2009, 'Prime-Ark'® 45 was released, providing the first cultivar of this type with shipping-quality fruit (J. R. Clark personal communication).

Early in the evaluations in Arkansas of the first-generation primocane-fruiting selections, it was noted that fruit produced on primocanes was substantially smaller and of lower quality than that borne on floricanes of the same plants. In testing of 'Prime-Jan'® and 'Prime-Jim'® in Aurora, OR they were observed to have large fruit and significant yields on primocanes, with fruit ripening from early September until early November (Clark et al. 2005). Primocane fruit in Oregon were also larger than floricane fruit from Arkansas. This substantial genotype × environment interaction is thought to be due to heat during flowering and fruit development. Temperatures over 30°C commonly occurred during this period in Arkansas while cooler temperatures occurred in Oregon (Clark et al. 2005). This observation was later confirmed in work by Stanton et al. (2007) who showed that "Prime-Jim"® and 'Prime-Jan'® flowering parameters were adversely affected by high temperatures, with the greatest impact at 35°C. Selection at more moderate summer temperature locations could be important to identify the most promising genotypes. Likewise, selection in a hot environment should allow the identification of more heat-tolerant genotypes, and variation for this trait has been observed (J.R. Clark personal observation). Thompson et al. (2008) have also begun to tease apart differences in flowering/fruiting morphology in this type of blackberry.

Another substantial effort was undertaken to move the primocane-fruiting trait into blackberry from red raspberry in the UK (Lim and Knight 2000). They used colchicine to double the chromosome number of red raspberries yielding tetraploid plants. These plants were subsequently crossed to 4x, 6x, and 8x blackberries with the raspberries used as the female Progeny were produced that had large fruit, good flavor, and detached like blackberries. However, most of the progeny did not express the primocane-fruiting trait strongly enough or lacked fruit quality. As of 2011, this material has not been further improved (V. Knight personal communication).

6.1.5 Thornlessness

Blackberry canes range from having no thorns to dense thorns that can have varying forms from small and straight to large and curved. Botanically, blackberry thorns or 'prickles' are spines since they are derived from outside the vascular cortex rather than true thorns that are subtended from vascular tissue. Thornlessness has long been a priority in almost all blackberry breeding programs, and remains a major goal today. Great progress has been made toward this goal with increasing numbers of thornless cultivars available each year. Fortunately there are several sources of thornlessness for use by blackberry breeders. For a more thorough review of thornlessness in blackberries, the reader should consider the compilation by Clark et al. (2007).

One source of thornlessness is the recessive 4*x* source (designated *s*) derived from *R. ulmifolius* in the UK at the John Innes Institute. 'Merton Thornless' was released from this program and later was the source of thornlessness used by the USDA-ARS Maryland breeding program. The first improved cultivars from this effort included 'Thornfree' and 'Smoothstem' (Scott and Ink 1966) and later by the commercially important 'Chester Thornless' (Galletta et al. 1998a). Selections from the USDA-ARS program were used as the thornless gene source in the University of Arkansas program begun in 1964 and from this effort the first erect, thornless cultivar 'Navaho' was released in 1989, followed by the thornless 'Arapaho,' 'Apache,' 'Ouachita,' and 'Natchez' (Clark and Finn 1999; Clark and Finn 2006; Clark and Moore 2008; Moore 1997). Further thornless cultivars using this thornless gene have been released from other programs including the 'Loch-series' from the Scottish Crop Research Institute, 'Cacanska Bestrna' from Serbia, and proprietary cultivars from Driscoll Strawberry Associates. This source of thornlessness is stable with all seedlings carrying the four recessive alleles being consistently and entirely thornless. The major disadvantage in breeding with this source is that due to the recessive nature of the gene, a second generation of crossing is needed to recover thornless progeny if the initial cross is of thorny × thornless parents. However, after substantial thornless genotypes have been generated in a program to serve as a parent base, thornless × thornless crosses allow for rapid numbers of entirely thornless populations. Additionally, in populations segregating for thornlessness, the thornless progeny can be identified at the cotyledon stage by examination of the margins of the cotyledons for the absence of glandular hairs (one or more hairs indicating a thorny plant) thus facilitating the removal of thorny offspring at a very early seedling age.

'Austin Thornless' is an octoploid and provided another source of thornlessness for use at the 6x and higher ploidy levels. This dominant source (designated S_f) has been important in breeding trailing types. With this source of thornlessness, thorns are found at times on the basal 0.3 m of the cane; these same canes are thornless beyond this point and are commercially thornless since fruit is borne only in the thornless area of the cane. The major drawback to breeding with this source is that thornless seedlings cannot be identified until 20–30 cm tall and must be potted prior to thornlessness being verified in segregating populations. Negative associated traits with the dominant thornless trait that have been overcome have included sterility, dwarfed plant habit, brittle canes, and tight fruit clusters contributing to more fruit rot concerns. 'Waldo' was the first cultivar to have this thornless source. Ploidy levels of subsequent releases include 6x, 8x, and 9x and include the cultivars Black Diamond, Black Pearl, and Nightfall (Finn et al. 2005b, d, e).

A newer thornless source was developed by Hall et al. (1986c), and this dominant source is designated as gene S_{ft}. A tissue culture technique in which a

Loganberry-type clone (L654) was used resulted in a spontaneous embryo from callus tissue. The resulting plant was released as 'Lincoln Logan' and was used subsequently in the New Zealand and the USDA-ARS Oregon breeding programs. Early associated limitations with this source of thornlessness included semierect and brittle canes, fruit characters much like red raspberry or 'Loganberry,' disease susceptibility, lack of winter hardiness, and small fruit with tender skins. Many of these limitations have been overcome in subsequent crossing, and the first cultivars with the S_{fl} source are likely to be released in the near future.

The future is bright in thornless breeding since the continued use of thornless genotypes has led to a greatly increased number of parents in all existing programs and thornless progeny are increasing yearly. Also, thornlessness has been incorporated into primocane-fruiting types in Arkansas, and the first cultivars with this unique combination should be released in the near future (J.R. Clark personal communication). On the horizon is a time when only thornless cultivars will comprise new releases.

6.1.6 Productivity, Yield, and Fruit Size

Productivity and yield are complicated traits from genetic, horticultural, and marketing perspectives. Yield components have recently received considerable attention, however, primarily within a few cultivars, particularly 'Marion,' or in context of training and harvesting systems (Bell et al. 1995a; b; Cortell and Strik 1997a, b; Himelrick et al. 2000; Takeda and Peterson 1999; Takeda 2002; Takeda et al. 2002; 2003). While these studies give information from a horticultural and physiological standpoint, they do not give much insight into genetic variability for the traits. While good yields are essential for the economic viability of a cultivar, if fruit quality is sacrificed or fruits cannot be efficiently harvested, the cultivar will not be accepted in the marketplace. In general, increased yields are obtained by crossing complementary parents that are high yielding as would be done for other quantitatively inherited traits (Clark et al. 2007).

Fruit size is an important yield component. For many years, a primary goal of all breeding programs was large fruit size (Darrow 1937; Sistrunk and Moore 1973; Ourecky 1975; Caldwell and Moore 1982; Jennings 1988; Daubeny 1996). Large fruit size was often a primary criterion for selecting genotypes from the wild for inclusion in breeding programs, and the inheritance of the trait has been documented in erect blackberries by Caldwell and Moore (1982). They found that fruit size was quantitatively inherited with partial dominance for small fruit size. In the trailing blackberries, while Strik et al. (1996) did not study inheritance, they did look at the variability present in a range of genotypes for fruit size and drupelet set. However, by the 1990s, cultivars had been developed that regularly weighed 10–15 g and required at least a couple of bites to eat (Hall 1990; Finn et al. 1998). While the novelty aspect of this sort of fruit is appealing, they are too large for most fresh or processed whole-berry applications. Large berries cannot be efficiently packed in the plastic clamshells that are the standard for the wholesale fresh market and currently an 8–10 g berry is ideal. Blackberries that are too large cannot be used in frozen berry mixes as they dwarf the raspberries and blueberries in the mixes.

6.1.7 Fruit Quality

The importance of fruit quality in breeding of blackberries cannot be overemphasized. Quality is the primary limitation that the public views in consideration of purchasing of fresh fruit. Likewise, quality is of ultimate importance in processed blackberries. To increase blackberry in importance in the marketplace, quality must always be a top priority in a cultivar improvement program.

Clark (2005a) shared that enhanced quality, emphasizing sweetness along with an attractive balance of acidity and elimination of astringency or bitterness, is the key to expansion of fresh-market blackberries. Although use of blackberry fruit ranges from processed to fresh, there are a number of traits of primary interest in breeding including fruit flavor (often divided into the components sweetness, acidity, astringency, bitter, aromatic components, etc.), color, firmness, fruit removal ease at harvest, shape, skin strength, texture, nutraceutical and nutritional content, and perception of seediness (size and feel of seeds in the mouth).

There is a wide range of fruit flavors among blackberry genotypes in the world. The distinct flavors of the *Ursini* section are widely desired and bring a premium price. Typically aromatic flavors with a pleasant balance of sweetness and acidity are best evidenced by 'Marion,' which is a standard for quality. High acidity is important for anthocyanin stability in processed products, and when balanced with high soluble solids the berries have a full, intense flavor. Differences in flavor among a number of trailing blackberry genotypes have been evaluated (Kurnianta 2005; Yorgey and Finn 2005). Flavors of blackberries derived from eastern-USA germplasm are distinctly different and are desired by many consumers who are familiar with the flavors of wild eastern species. Kurnianta (2005) found the semierect 'Chester Thornless' was more different from 'Marion' than many of the trailing genotypes for flavor. Aromatic flavor components of a given genotype can vary significantly depending on the environment in which it is grown (Wang et al. 2005).

Along with flavor components, a major point of focus in current breeding is enhancing sweetness—the most common consumer interest with fresh-market blackberries. Soluble solids levels of 10–12% can be found in the erect-caned cultivars such as Navaho and Ouachita. Further enhancements to 15% soluble solids or possibly higher are possible by crossing among high soluble solids parents and selecting desired progeny. Trailing blackberry cultivars including 'Boysen' have high soluble solids levels (11–13%) compared to 'Chester Thornless' (8%) and in some years 'Boysen' can have over 15% soluble solids (Fan-Chiang 1999; Siriwoharn et al. 2004). Berries that have low acidity can be undesirable as they have a 'flat' flavor (Hall 1990). After concerns with sweetness and high acidity, astringency and bitterness may be the most noticeable flavor components by consumers of fresh blackberries and low astringency can be selected for in seedling populations. The future holds the opportunity to combine the *Ursini* and eastern USA-germplasm-derived flavors to expand the flavor of commercial cultivars.

Maturity of blackberry fruit greatly affects fruit quality, particularly the sugar and acid levels. Soluble solids increased while titratable acid decreased as berries matured from underripe to ripe in "Navaho" (Perkins-Veazie et al. 2000), 'Marion,' and 'Thornless Evergreen' (Siriwoharn et al. 2004). Dull-black fruit were found to be the sweetest compared to mottled or shiny black fruit but were also softer. Volatiles were much higher in dull-black compared to shiny-black fruit. The challenge exists in that shiny black fruit are far superior in postharvest handling (Perkins-Veazie et al. 1997), and for a cultivar to have fresh-market potential, it must have high quality including good flavor, high soluble solids content, good acceptable acidity, all in a shippable, shiny-black berry.

Postharvest quality has had tremendous focus in breeding in recent years. The quality of fruits for the fresh market is determined by how a genotype responds to storage and handling practices from the time the fruit is harvested until it is in the consumers' hands. There was a substantial cooperative effort between the University of Arkansas and the USDA-ARS, Lane, OK beginning in 1992 to evaluate postharvest potential of blackberries (Perkins-Veazie and Clark 2005). Prior to the early 1990s, shelf life was usually estimated to be no more than 5 days under the best storage and transport conditions (Perkins-Veazie and Clark 2005) and therefore they were not found in most retail markets. The initial effort focused on evaluation of cultivars in various temperatures and times of storage, and the thornless cultivar Navaho was found superior to the thorny 'Cheyenne,' 'Choctaw,' and 'Shawnee' mainly due to its firm fruit that retained black drupelet color (Perkins-Veazie et al. 1996; Perkins-Veazie et al. 1999). Subsequent thornless cultivars from the Arkansas program were also found to be superior to thorny genotypes (Perkins-Veazie and Clark 2005). 'Navaho' was also found to store well when harvested at the dull-black stage (Perkins-Veazie et al. 1996), could be successfully shipped from the USA to Europe (Perkins-Veazie et al. 1997), and could be stored for up to 21 days (Perkins-Veazie et al. 2000).

Parameters examined in past and ongoing postharvest evaluations of genotypes in the University of Arkansas program include appearance, firmness, and flavor. Limitations such as presence of decay, leakage of juice, obvious mushiness of fruit, or presence of substantial red drupelet color limit consumer appeal, while shiny, fully black berries are desired (Perkins-Veazie and Clark 2005). A complicating factor in evaluations was rainfall during harvest as rain within 4 days of harvest greatly affected subsequent postharvest performance, particularly firmness (Perkins-Veazie and Clark 2005). Multiyear evaluations were essential to fully determine the postharvest potential of new genotypes. One of the most significant findings was that firmness evaluations in the field were not a reliable indicator of potential postharvest handling potential. Firm-rated genotypes in the field were not always found to retain firmness and have adequate postharvest storage potential for commercial shipping. In breeding for fresh-market shipping potential, one must have a uniform system of evaluating genotypes for overall postharvest potential and examining the key components that contribute to postharvest success.

Evaluation for processing quality includes several variables. Ease of separation of the fruit from the plant when shaken by a mechanical harvester is imperative for a processing blackberry, and is a distant goal for the fresh market. Firmness of the fruit is not as critical for fresh market use, but processing berries must have adequate firmness to move through the harvesting and sorting process with minimal visible damage and to maintain good frozen appearance. Some degree of drupelet skin breakage is acceptable in processing berries, but it is not allowed for fresh-market berries. Processing berries must have intense color and flavor, high soluble solids and titratable acidity levels, low pH, and the perception of low "seediness" (Finn et al. 1997; Hall et al. 2002). Maintenance of these qualities when the berries are frozen and subsequently thawed, canned, dried, or juiced is imperative. Genotypes are usually evaluated by machine harvesting (or must be evaluated for this sometime in the testing process), and berries sorted and frozen as individually quick frozen (IQF) fruit. Later, subsamples are made to evaluate "chemistry" and this includes pH, titratable acidity, soluble solids and, when appropriate, total anthocyanins. As selections advance in the breeding program, processed samples are prepared as IQF, pureed, and occasionally juiced for evaluation by panels for appearance, flavor, color, and overall quality (Hall et al. 2002; Finn et al. 2005a, b, c; Yorgey and Finn 2005).

Many genotypes change from black to purple under high temperature stress or when refrigerated or frozen. This character is poorly understood but is becoming more critical in the commercial industry as the public wants a uniform black product. Fruit maturity interacts with this environmental response as mature fruit are less likely to lose their black color than immature fruit. There is genetic variability for this trait, as mature fruit of 'Obsidian,' 'Kotata,' 'Navaho,' and 'Chester Thornless' hold their color during refrigeration and/or freezing (Finn et al. 2005c; C. Finn and J. Clark personal observation). Some of these cultivars may lose some color during high temperature stress but can recover their full black color if the stress is removed in the field.

One variable that many consumers of blackberries notice immediately upon eating fruit is that of seed size or seed "feel" in the mouth. The overall size and presence of seeds in blackberry genotype must be considered by the breeder. Some perceive trailing blackberries as "seedless" or as having low levels of seediness (Finn et al. 1997), a perception apparently due to seed shape and endocarp thickness (Takeda 1993). Erect blackberry seeds were generally ellipsoidal and smaller than those of eastern semierect blackberries that were "clam shaped" (Takeda 1993). Takeda also found that trailing blackberries such as 'Marion' had seeds that were flat with a soft, thin endocarp. Seed size was found to be quantitatively inherited with partial dominance for small size (Moore et al. 1975); therefore, progress in crossing and selecting for small seeds should be successful. Progenies derived from crosses between eastern erect and western trailing blackberries show a range of seediness (C. Finn personal communication). Large fruit size can be attained with moderate to small seed size in breeding and 'Siskiyou' is an excellent example of this (J.R. Clark personal observation; Finn et al. 1999a; Strik et al. 1996).

Tied to these traits is the critical trait of fruit shape. Ideally, a berry has a very uniform, barrel, round or conical shape with uniformly sized and shaped drupelets

(Clark et al. 2007). Many older cultivars were fairly round with variably sized and shaped drupelets. When drupelet sizes are uneven, fruit are less attractive and the skin on the larger drupelets is more likely to be damaged during harvest and handling leading to "leaky" berries that are prone to rot.

Research has investigated nutraceutical/antioxidant levels in blackberries (Bushman et al. 2004; Cho et al. 2004; Cho et al. 2005; Clark et al. 2002; Connor et al. 2005a; b; Moyer et al. 2002; Perkins-Veazie and Kalt 2002; Siriwoharn et al. 2004; Wada and Ou 2002; Wang and Lin 2000). In Arkansas, noteworthy variation was found among cultivars with two-fold and four-fold differences in oxygen radical absorbance capacity (ORAC) depending on the year (Clark et al. 2002). Some genotypes exhibited substantial year-to-year variation. Perkins-Veazie and Kalt (2002) reported no ORAC differences between shiny- and dull-black fruit or for fruit stored for 7 days, although values differed among genotypes. Wang and Lin (2000) found differences in ORAC values between green, red, and ripe fruit of three semierect cultivars as well as significant differences among the cultivars. Cho et al. (2004) found variation among cultivars for ORAC along with differences in anthocyanin and flavanol contents. Moyer et al. (2002) determined anthocyanin, phenolics, and antioxidant capacity across a broad range of *Vaccinium*, *Rubus*, and *Ribes* species. Within *Rubus*, they found substantial variability for all of the traits evaluated. Within blackberries, there was a correlation of antioxidant capacity as measured by ORAC with total anthocyanins ($r=0.70$) and with total phenolics ($r=0.73$). The correlations between antioxidant activity as measured by FRAP was poor for anthocyanin content ($r=0.38$) but fairly good for total phenolics ($r=0.75$). 'Thornless Evergreen,' 'Marion,' and 'Boysen' along with red and black raspberries were analyzed to determine phenolic content, especially ellagic acid, and antioxidant activity as measured by ORAC (Wada and Ou 2002). These genotypes all had high antioxidant activity and were good sources of anthocyanins and phenolics. Connor et al. (2005a, b) examined genotype and environmental variation (years and locations) for anthocyanins, phenolics, and antioxidant activity from cultivars grown in New Zealand and Oregon for two years. Cultivars included two erect ('Navaho,' 'Shawnee'), one semierect ('Hull Thornless'), and 10 trailing ('Chehalem' 'Aurora,' 'Waldo,' 'Black Butte,' 'Ranui,' 'Silvan,' 'Siskiyou,' 'Kotata,' 'Marion,' and ORUS 1826) genotypes, along with three blackberry/raspberry hybrids ('Boysen,' 'Tayberry,' and 'Logan'). Antioxidant activity (AA) as determined by FRAP, total phenolics (TPH), total anthocyanins (ACY) as well as individual anthocyanins were measured. AA and TPH were not significantly different among cultivars and locations but the variation between years within location and the genotype × environment interactions were significant. The genotype × environment interaction was also significant for total and individual ACYs. Correlations between ACY and AA were much lower ($r=0.63$) than they were for TPH and AA ($r=0.97$). Overall, these studies indicate that genetic variation for anthocyanins, total phenolics, and antioxidant levels exist, and that breeding for enhanced levels would likely be possible. This area offers potential for breeders as blackberry health properties and their benefits could be important in the promotion and marketing in the future.

6.2 Breeding Methodology

The first step in any breeding program is to determine the objectives one hopes to accomplish and attach to these some priority. If insufficient variability is available for the traits of interest, then the germplasm base needs to be expanded. Mehlenbacher (1995) describes the most common approach to breeding fruits and nuts as complementary hybridization where the parent clones of each cross are chosen such that the weaknesses of one are matched by the strengths of the other, with the hope that a few of their offspring will have the strengths of both parents and none of the weaknesses. In blackberries, the parents are highly heterozygous and the seedling populations usually have substantial segregation. A stepwise evaluation program is used to winnow thousands of seedlings down to a few selections. Since the selections from one generation serve as the parents for the next generation, the approach is essentially phenotypic recurrent selection (Mehlenbacher 1995). While only additive gene effects respond to selection over cycles, in any one generation, the breeder can take advantage of all types of genetic variance because desirable gene combinations can be fixed by clonal propagation.

While this approach is typical, every breeding program develops a unique set of approaches matched to its location and facilities. Credit for many of the technical details in the following discussions goes to M. Peterson and K. Wennstrom of the USDA-ARS who have fine-tuned these approaches over time; a fuller discussion is available in Clark et al. (2007).

6.2.1 Parental Selection

As described above, based on the set of objectives that a breeder is trying to meet, parents are chosen based on their phenotypic performance and how they may mesh with each other. Once parents are chosen they should be tested for freedom from pollen-borne viruses including RBDV and TSV. While the transmission rate to seedlings of these viruses is low during controlled crossing (0–4% R. Martin personal communication.), it is best to start the process with virus-tested parents.

6.2.2 Emasculation and Pollination

Emasculation and pollination techniques are similar in blackberry to those for other members of the *Rosaceae*. While crossing can be done in the greenhouse or the field, in the field there are usually a large number of flowers at the correct stage to choose from and with copious pollen. Some programs have found that greenhouse-produced seed germinates more readily than field-produced seed (H. Hall personal communication).

As flowers begin to open, buds are collected at the "popcorn" stage with the buds expanding and showing some petal but before they are open to potential contamination from pollinators. Buds are cut in half and put under a low-watt incandescent bulb about 20–24 cm away from them in a protected area to dry overnight. These fully dried buds are then placed in a container (salve tins, film canisters) and into a refrigerated desiccator. Some programs extract the anthers from the flower to dry, and while this yields a tidy product, it is more time consuming to prepare. Pollen handled as described will remain viable for 1–2 weeks or more. If the pollen is held from spring to summer or to the next season, the dried pollen should be frozen in a desiccator.

The flowers to be used as the seed parent must be emasculated. When the primary flowers have bloomed and the secondary buds are reaching the "popcorn" stage is the ideal time to most efficiently emasculate a flower. At this point, the stigmas are not yet mature and the pollen has yet to dehisce. Typically, 3–5 buds on four or more flowering laterals are emasculated for each cross. This should yield a minimum of about 16 fruit with enough seed to produce 100 seedlings or more after taking into account the ways emasculated laterals can be destroyed (curious crows, tractors, wind, etc.).

Buds are emasculated by slicing the underside through the sepal, petal, and stamen whorls simultaneously with a single-edged razor, leaving only the receptacle. Thumbnails, forceps, and scalpels can all be effectively used as well. Once the emasculations are complete, many programs place a waxed paper bag over the laterals. Some breeding programs do not bag emasculated flowers as the emasculated flowers are not attractive to pollinators (Finn 1996). However, in climates where rain is common, bagging keeps the flowers dry, making it easier to return to the field quickly after rain showers. Two to three days after emasculation, the styles mature and spread outward and their color changes from bright-green to pale-yellow indicating receptivity. Pollen that was previously collected is applied with small paint brushes or an index finger. Brushes are sterilized between pollinations or, more ideally, each tin/parent has its own brush. Depending on environmental conditions, the flowers are repollinated 2–3 days later and when possible a third time.

Ripe fruit is harvested and refrigerated until seed extraction. While it is possible to extract seed from moldy fruit, it is much easier if done while the fruit are reasonably sound. Fruit are placed in a small container (small beakers or magenta boxes) and mashed with 2–4 drops of pectinase and enough water to make a slurry. The slurry is left overnight and then poured through a small strainer and rinsed. The pectinase separates the flesh nicely from the seed and is greatly preferred to blenders with padded blades as the potential for damage to the seeds is eliminated. Seed is spread on paper towels and dried overnight and then placed in labeled envelopes for storage. Seed can be held at room temperature for several weeks without loss in viability. However, for long-term storage, seeds should be kept in a refrigerated desiccator where they can be kept for 10+ years (Clark et al. 2007).

6.2.3 Germination

Scarification followed by stratification is generally required for germination of seed lots. With wide genetic crosses or for small seed lots, an in vitro procedure can be used to maximize seedling production. However, for most crosses the following standard procedure works well to produce field-ready seedlings in as little as 18–22 and as long as 28 weeks. The standard germination-to-field protocol consists of acid scarification, a water and sodium bicarbonate rinse, a calcium hypochlorite soak, another rinse, overnight warm stratification, several weeks in cold stratification, germination and transplanting, growing larger plants in the greenhouse, acclimation to outdoor conditions, and finally, field planting.

In preparation for scarification with concentrated sulfuric acid, seeds are placed in 100-mL test tubes. To ensure even distribution of the acid and to prevent clumping, the number of seeds per tube should be less than about 300. Because most breeding programs have such a wide variety of *Rubus* germplasm, seed lots vary tremendously in seed size and thickness of the pericarp. In general, trailing blackberries require 1–4 h scarification and semierect/erect blackberries might require 3–4 h. Seed must be dry prior to scarification. Approximately 10 mL acid is poured into each tube, and then stirred using a vortex mixer to coat the seeds. The tube is placed in a rack immersed in an ice bath. The seeds should be stirred periodically and monitored to see if the white embryos become visible at which point the seeds should be removed from the acid. When the time in scarification is completed, ice water is poured quickly into the tubes and stirred rapidly to dilute the acid and slow the reaction. The seeds are then poured through a strainer and rubbed to remove some of the charred surface as they are rinsed under tap water for a few minutes. Seeds are placed in a saturated solution of sodium bicarbonate for 5 min and then rinsed again. Finally, they are placed put in a 1% calcium hypochlorite solution (3 gL^{-1}; based on formulation with 70% active chlorine) with excess calcium hydroxide for 5–6 days at 4°C to complete acid neutralization and to remove the carbon layer.

A wide variety of germination flats can be used and are commonly filled with vermiculite, watered, and topped with 0.7 cm sphagnum peat and then misted. Seeds are spread on the surface and pressed in but not covered, and then are left under mist overnight before placing in clear plastic bags and stored at 4°C with 16 h of light for 6–10 weeks. Stratification time varies from cross to cross depending on the genetic background, so flats should be checked regularly to see if the flats are still moist and whether any seedlings have started to emerge. Typically, stratification requirements are satisfied in 4–6 weeks for trailing types and 12–15 weeks in erect and semierect types.

After stratification, the flats are moved to the mist bench under intermittent mist and bottom heat (24°C). Seedlings generally begin to emerge in less than a week, and germination is mostly complete within 4 weeks although germination can extend over 12 or more weeks. When seedlings have developed two true leaves they are pricked out and transplanted into 50- to 72-cell plug trays filled with a bedding plant mix. Deeper cells are preferred for better root development. Plugs are watered in and grown in the greenhouse at 22–24°C under 16-h daylength. They are initially fertilized with a balanced fertilizer at 1–2 times per week with 100 ppm N for 2–3 weeks, then 200 ppm N for 3–4 weeks. When roots fill the plugs and outdoor temperatures allow, the flats are moved outdoors under shade cloth for 1 week, then moved to full sun to await field planting after the last frost date.

While most seed lots are germinated using the basic procedure just described, in vitro procedures are used for small seed lots that are typically from wide crosses (Galletta and Puryear 1983; Galletta et al. 1986; Hall 1990; Clark et al. 2007; Finn 2008). An in vitro germination protocol (Clark et al. 2007) involves surface sterilization with ethanol and bleach, cold stratification, repeat surface sterilization, dissection, germination on media, and the typical procedures are then followed until field planted.

Seed is surface-sterilized prior to stratification using 1 min in 70% ethanol while swirling by hand, then into a 20–25 mL solution of 10% bleach + 1–2 drops surfactant with agitation on a shaker table at 300+ rpm for 60 min. Seed and bleach are poured through a strainer, and seeds are then placed into the sterilized Petri dishes, sealed with Parafilm® (Pechiney Plastic Packaging Co., Chicago) and stratified at 4°C for 6–10 weeks. They should be checked every few weeks and the filter paper remoistened with sterilized water if necessary.

When the stratification is complete the seed is removed from the Petri dish and surface sterilized using 70% ethanol for 1 min, followed by bleach + surfactant for 1 h and placed into a tube of sterile water to await dissection. Using a dissection microscope with backlighting, it is easy to identify the radicle end and to visually inspect the seeds for viability. Viable seed will be uniformly yellowish or tan, with no blotchiness or variability among seeds, while underdeveloped seed might be dark, grayish, reddish, or black. Using forceps, grasp the radicle end of each seed (identified by its more pointed shape in contrast to the more rounded edge of the cotyledon end) and with a scalpel, sever and remove the half of the seed containing the tips of the cotyledons. Make sure to remove at least half of the seed. Embryos will begin to germinate as quickly as 2–4 h after initial cutting. The prepared seeds are left in the sterile water for 4 h, or overnight, most of the embryos will expand enough to expel themselves from the seed coat thereby separating the embryo from a major source of contamination. The seeds are drained and transferred to germination medium in a 48-well (0.4 mL) sterile culture plate. Wells are 2/3 filled with autoclaved, 1/2 strength MS media with 100 mg L^{-1} myoinositol, 10 mg L^{-1} sucrose, and 7 mg L^{-1} agar.

Germination follows quickly at room temperature although best results have been realized with 16-h daylength with a temperature around 25°C. Within 10 days the embryos develop green color and root growth will begin. Once germination has started the embryos can be transferred to a test tube for further growth or, if they are allowed to grow in the culture plates until the first true leaves appear, they can be transferred directly to small plug trays with germination in soilless media. The flats are started in a mist bench and slowly acclimated first to the greenhouse and then outdoors.

6.2.4 Seedling Care, Planting, and Field Plot Design

The number of seedlings that are planted in the field is determined by an understanding of the inheritance of the traits of interest, the objectives of the cross, and land and labor resources. Typically, 100–200 seedlings per cross are established in the field at a plant spacing from 0.25 to 1.0 m apart within the row and 2–4 m between rows. Populations are often planted in a serpentine pattern. While it is possible to grow large enough primocanes the first year to attain a good crop the second year and make selections 14 months after planting, it is hard to get the entire field at that level. Therefore, breeding programs generally manage the plants intensively and evaluate the seedlings two years after planting. While it is not uncommon for programs to evaluate seedlings a second time in the fourth year, the tremendous expense of doing so leads many programs to limit selection to only the third year in the field. Primocane-fruiting, erect blackberries are usually evaluated in the second year as only limited Primocane-flowering is seen in the planting year.

6.2.5 Evaluation of Seedling Populations

Evaluation typically is stepwise, with the number of selected genotypes decreasing in each step while the number of vegetatively propagated plants of each increases. Selections are made in the seedlings based initially on the subjective evaluation of yield, plant health, and fruit quality, traits that can be evaluated quickly and inexpensively, with few notes or detailed evaluations. Most breeding programs save 0.5–1.0% of their seedlings as selections. At the time of selection, the breeder must assess the use or value of the selection. Selections that are deemed most outstanding can be designated for immediate replicated trial (i.e., three to four replications of three to five plants/plot). More commonly selections are deemed to have great promise but the breeder has some uncertainty and these are typically marked for observation plots; single, three- to five-plant plots for trailing or semierect types and 6 m plots for erect genotypes with plantings at one or two locations. The final group is those selections that were made as part of a germplasm development program, and these selections are marked also for an observation plot but they may not need to be put in a situation where yield and intensive evaluations will be made. Other than in the case of unique characteristics (extreme size, very early/late ripening, etc.), it almost never works out to make a selection only with the intention of using it as a parent; these are invariably passed over in preference of elite clones, with outstanding characteristics and that have been more intensively evaluated.

6.2.6 Evaluation of Test Plots

Observation and replicated plots are established and managed as closely to commercial standards as possible and they are evaluated intensively. As the season begins, the breeder must evaluate the plots quickly to determine whether any genotypes can be discarded before the expensive harvest begins. Depending on environmental conditions, the plots are evaluated about once per week during the fruiting season. Each breeding program has a suite of traits that are important, with more common ones beyond yield and fruit size including fruit firmness/skin toughness, color, shape, and flavor, ease of fruit separation, and plant vigor. The most promising selections are harvested for yield and often for postharvest fresh-market storage or processing evaluation. The fruit is frozen, pureed, and/or juiced for an evaluation of processing quality. Storage trials to assess fresh fruit quality involve evaluation of refrigerated fruit stored in clamshells under refrigeration and room temperature regimes that parallel handling in the commercial chain; leaky, soft, discolored, or moldy fruit are scored. Commonly about 10% of the advanced selections are identified that combine good yield and horticultural traits combined with excellent fruit quality and these are propagated for further trial with cooperators. During this time, programs that protect their cultivars begin assembling botanical data needed for filing for plant patent or plant breeder's rights applications.

6.2.7 Breeding Cycle: Yearly Activities from Pollination to Cultivar Release

Beginning the first of the year in the northern hemisphere, seedling germination begins and crosses are planned. Plants in all plantings are evaluated for winter damage as bud break commences and soon thereafter, crossing begins with flowering. New plantings are established as soon as the ground is ready and the danger of frost has past. As fruit begins to ripen, evaluations of seedlings and genotypes in trial intensify and continue until harvest is complete. Fruit from successful crosses are harvested as they ripen. Genotypes identified as selections begin to be propagated in late summer. Seed is extracted, scarified, and placed into stratification in later summer/fall in preparation for the cycle to begin again.

The length of time from pollination to a naming a cultivar can be as little as 9 years if a cultivar has outstanding and unique characteristics, but can commonly take 15–17 years. One of the major changes over the past 20 years has been the shifting of risk from breeding programs to the industry. Historically, breeders tested at a number of sites, with a number of planting years, resulting in a substantial amount of data to use in judging release. While unbiased evaluation is still critical, the industry usually prefers to get selections earlier in the process as they can provide a better 'acid test' of commercial viability and better identify unique market niches.

7 Integration of New Biotechnologies in Blackberry Breeding

The use of molecular and other techniques in blackberry has been very limited. More work has been done with red raspberry than any other *Rubus* species, and more thorough discussion of this can be found in the chapter on raspberries, and also in Clark et al. (2007). Reasons for this minimal work include the lesser economic importance of blackberries compared to other *Rosaceous* crops, limited number of programs to consider including molecular techniques, and the polyploid nature of most genotypes.

The use of minisatellite DNA probes in *Rubus* was found to be useful in identifying raspberry and blackberry cultivars by Nybom et al. (1989). Nybom et al. (1990) detected genetic variation among blackberries and raspberries and found inter- and intraspecific variation along with some identical fingerprints among genotypes. Nybom and Hall (1991) later confirmed that minisatellite DNA fingerprints were useful for evaluating genetic relatedness and for distinguishing genotypes. Further, Kraft et al. (1996) investigated facultatively apomictic blackberries from three countries and found differing fingerprints among countries within a species. They recommended that DNA fingerprinting should be used with morphological characterization due to environmental impacts along with somatic mutations that could occur in the phenotypes but not be evident in the DNA fingerprint. The use of restriction fragment length polymorphisms, RAPD, and amplified random length polymorphisms in *Rubus* has been restricted mainly to raspberry.

The use of simple sequence repeat markers (SSR) was reported by Stafne (2005) in which he used this type of marker to differentiate progeny and assess genetic similarity within a segregating blackberry population. His results indicated a similarity coefficient averaged over all individuals of 73% for SSR markers. The average similarity coefficients ranged from a high of 80% to 57% for SSR markers. Comparison of the parents ('Prime Jim'® and 'Arapaho') indicated a similarity of 62% for SSR markers. Recently, SSRs have been used to effectively genetically fingerprint blackberry accessions in the USDA-ARS, National Clonal Germplasm Repository and these approaches have worked to differentiate genotypes based on leaf tissue as well as on the torus tissue of IQF berries that had been frozen and then thawed but not on frozen and thawed pureed fruit (Bassil et al. 2010).

A substantial advance was made by Lewers et al. (2008) where they reported the first work in developing an expressed sequence tag library for blackberry. A cDNA library of 18,342 clones was generated from young leaf tissue of the common thornless source 'Merton Thornless.' A total of 667 primer pairs were designed from individual sequences containing SSRs. In additional work in this report, 33 randomly chosen primer pairs were tested with two blackberry cultivars ('Prime Jim'® and 'Arapaho') and 10 of the primer pairs detected an average of 1.9 polymorphic PCR products. Their research could lead to the implementation of marker development for use in breeding programs.

No genetic mapping has been done for blackberries, although limited research using SSR markers in *Rubus* found primers that could be useful for mapping (Stafne et al. 2005). Graham et al. (2002) developed the first SSR markers for *Rubus* from red raspberry and tested the markers on blackberries and blackberry × raspberry hybrids. They found that all 10 fluorescently labeled primer pairs amplified polymorphisms suggesting their usefulness in further molecular analysis and genetic mapping. Stafne et al. (2005) evaluated SSR primers from Graham et al. (2002), Amsellem et al. (2001) (derived from *R. alceifolius* Poir.), Lewers et al. (2005) (derived from *Fragaria* × *ananassa* Duch.), and *Rosa*. Their results indicated that 29–30% of 'Glen Moy'-derived SSRs amplified a product in 'Arapaho' and 'Prime-Jim'®, while 25% of the *R. alceifolius* and 19% of the *Fragaria* SSRs amplified a product in the blackberries. No *Rosa*-derived SSRs amplified a product in the blackberries. These preliminary results indicate that blackberry-specific SSR primers are needed to make substantial progress in mapping research. Lopes et al. (2006) identified microsatellite loci in *R. hochstetterorum* Seub., a species native to the Azorean Islands, and 41 SSR markers were identified in a genomic library of this species. These markers achieved cross-species amplification in at least one of the other three tested species of *Rosaceae* including blackberry (*Rubus fruticosus* aggr.).

No studies of blackberry marker assisted selection have been published. Stafne (2005) investigated RAPD and SSR markers for linkage to floricane/primocane-fruiting and thorny traits, but none were found that were adequately linked for use as markers in breeding.

While regeneration systems have been developed for blackberries (Swartz and Stover 1996; Meng et al. 2004), no transgenics have been produced to date and no active breeding programs are working in this area. The highest regeneration efficiency (70% of explants) was accomplished when leaves were incubated in TDZ pretreatment medium for 3 weeks before culturing them on regeneration medium (Woody Plant Medium with 5uM BA and 0.5 uM IBA) in darkness for a week, and then transferring them to a 16-h light photoperiod at 23°C for 4 weeks (Meng et al. 2004). Limited work has been done in red raspberry (see Chap. 8).

References

Alice, L.A. and Campbell C.S. (1999) Phylogeny of *Rubus* (Rosaceae) based on nuclear ribosomal DNA internal transcribed spacer region sequences. Am. J. Bot. 86:81–97.

Alice, L.A., Eriksson, T., Eriksen, B., and Campbell, C.S. (2001) Hybridization and gene flow between distantly related species of *Rubus* (Rosaceae): Evidence from nuclear ribosomal DNA internal transcribed spacer region sequences. Syst. Bot. 26:769–778.

Amsellem L., Dutech, C., and Billotte, N. (2001) Isolation and characterization of polymorphic microsatellite loci in *Rubus alceifolius* Poir (Rosaceae), an invasive weed in La Reunion Island. Mol. Ecol. Notes 1:33–35.

Ballington, J.R. and Moore, J.N. (1995) NC 194 primocane-fruiting, thorny, erect tetraploid blackberry germplasm. Fruit Var. J. 49:101–102.

Bassil, N.V., Muminova. M., and Njuguna W. (2010) Microsatellite-based fingerprinting of western blackberries from plants, IQF berries and puree. *Acta Horticulturae,* In Press.

Bell, N.C., Strik, B.C., and Martin L.W. (1995a) Effect of primocane suppression date on 'Marion' trailing blackberry. II. Cold hardiness. J. Am. Soc. Hort. Sci. 120:25–27.

Bell, N.C., Strik, B.C., and Martin, L.W. (1995b) Effect of primocane suppression date on 'Marion' trailing blackberry. I. Yield components. J. Am. Soc. Hort. Sci. 120:21–24.

Brainerd, E. and Peitersen, A.K. (1920) Blackberries of New England - their classification. Vermont Agr. Expt. Sta. Bul. 217.

Breese, W.A., Shattock, R.C., Williamson, B., and Hackett C. (1994) In vitro spore germination and infection of cultivars of *Rubus* and *Rosa* by downy mildews from both hosts. Ann. Appl. Biol. 125:73–85.

Brown, S.W. (1943) The origin and nature of variability in the Pacific Coast blackberries. Am. J. Bot. 30:686–697.

Buckley, B., Moore, J.N. and Clark, J.R. (1995) Blackberry cultivars differ in susceptibility to rosette disease. Fruit Var. J. 49:235–238.

Bushman, B.S., Phillips, B., Isbell, T., Ou, B., Crane, J.M., and Knapp, S. (2004) Chemical composition of caneberry (*Rubus* spp.) seeds and oils and their antioxidant potential. J. Agric. Food Chem. 52:7982–7987.

Caldwell, J.D. and Moore J.N. (1982) Inheritance of fruit size in the cultivated tetraploid blackberry. J. Am. Soc. Hort. Sci. 107:628–631.

Chamberlain, C.J., Kraus, J., Kohnen, P.D., Finn, C.E., and Martin, R.R. (2003) First report of raspberry bushy dwarf virus in *Rubus multibracteatus* from China. Plant Dis. 87:63.

Cho, M.J, Howard, L.R., Prior, R.L., and Clark J.R. (2004) Flavonoid glycosides and antioxidant capacity of various blackberry, blueberry, and red grape genotypes determined by high-performance liquid chromatography/mass spectrometry. J. Sci. Food Agric. 84:1771–1782.

Cho, M.J, Howard, L.R., Prior, R.L., and Clark J.R. (2005) Flavonol glycosides and antioxidant capacity of various blackberry and blueberry genotypes determined by high-performance liquid chromatography/mass spectrometry. J. Sci. Food Agric. 85:2149–2158.

Christy, J.C. (2004) The Young brothers of Morgan City, Morgan City Archives Publications, Morgan City, LA.

Clark, J.R. (1992) Blackberry production and cultivars in North America east of the Rocky Mountains. Fruit Var. J. 46:217–222.

Clark, J.R. (1999) The blackberry breeding program at the University of Arkansas: thirty-plus years of progress and developments for the future. Acta Hort. 505:73–77.

Clark, J.R. (2005a) Changing times for eastern United States blackberries. HortTechnology 15:491–494.

Clark, J.R. (2005b) Thoughts on breeding intractable traits in eastern U.S. blackberries. HortScience 40:1954–1955.

Clark, J.R and Finn, C.E. (1999) Blackberry and hybridberries. In: W.R. Okie (Ed.). Register of new fruit and nut varieties Brooks and Olmo list 39. HortScience 34:183–184.

Clark, J.R. and Finn, C.E. (2002) Blackberry. In W.R. Okie (Ed.). Register of new fruit and nut varieties, list 41. HortScience 37:251.

Clark, J.R and Finn, C.E. (2006) Blackberry and hybrid berry. In: J.R. Clark and C.E. Finn (Eds.). Register of new fruit and nut cultivars. HortScience 41:1104–1106.

Clark, J.R and Finn, C.E. (2008) Trends in blackberry breeding Acta Hort. 777:41–48.

Clark, J.R. and Moore. J.N. (2008) 'Natchez' thornless blackberry. HortScience 43:1897–1899.

Clark, J.R., Howard, L., and Talcott, S. (2002) Antioxidant activity of blackberry genotypes. Acta Hort. 585:475–479.

Clark, J.R., C. McCall and C.E. Finn (2008) Blackberry, p. 1323–1324. In: C.E. Finn and J.R. Clark (eds.). Register of new fruit and nut cultivars, list 44. HortScience 43.

Clark, J.R., Moore, J.N., Lopez-Medina, J., Perkins-Veazie, P., and Finn, C.E. (2005) 'Prime Jan' (APF-8) and 'Prime-Jim' (APF-12) primocane-fruiting blackberries. HortScience 40:852–855.

Clark, J.R., Stafne, E.T., Hall, H., and Finn, C.E. (2007) Blackberry Breeding and Genetics. Plant Breeding Reviews, Timber Press, Portland, OR. 29:19–144.

Connor, A.M, Finn, C.E., and. Alspach, P.A. (2005b) Genotypic and environmental variation in antioxidant activity and total phenolic content among blackberry and hybridberry cultivars. J. Am. Soc. Hort. Sci. 130:527–533.

Connor, A.M, Finn, C.E., McGhie, T.K, and. Alspach, P.A. (2005a) Genetic and environmental variation in anthocyanins and their relationship to antioxidant activity in blackberry and hybridberry cultivars. J. Am. Soc. Hort. Sci. 130:680–687.

Converse, R.H. (1987) Virus diseases of small fruits. U.S. Dept. of Agri. Agric. Hdbk. no. 631.

Cortell, J.M. and Strik B.C. (1997a) Effect of floricane number in 'Marion' trailing blackberry. I. Primocane growth and cold hardiness. J. Am. Soc. Hort. Sci. 122:604–610.

Cortell, J.M. and Strik B.C. (1997b) Effect of floricane number in 'Marion' trailing blackberry. II. Yield components and dry mass partitioning. J. Am. Soc. Hort. Sci. 122:611–615.

Darrow, G.M. (1925) The Young dewberry, a new hybrid variety. American Fruit Grower 45 (9):33.

Darrow, G.M. (1937) Blackberry and raspberry improvement. p. 496–533, *USDA Yearbook of Agriculture, Yearbook 1937*. United States Department of Agriculture. Washington, D.C.

Darrow, G.M. (1967) The cultivated raspberry and blackberry in North America - breeding and improvement. Am. Hort. Mag. 46:203–218.
Daubeny, H.A. (1996) Brambles. In: J. Janick and J.N. Moore (Eds.). *Fruit breeding. Volume II. Vine and Small Fruits.* John Wiley & Sons, Inc., New York. pp. 109–190.
Davis, H.A., Fuller, A.M. and Davis, T. (1969a) Contributions toward the revision of *Eubati* of eastern North America. IV. Castanea 34:157–179.
Davis, H.A., Fuller, A.M. and Davis, T. (1969b) Contributions toward the revision of *Eubati* of eastern North America. V. Arguti. Castanea 34:235–266.
Einset, J. (1947) Chromosome studies in *Rubus*. Gentes Herbarum 7:181–192.
Ellis, M.A., Converse, R.H., Williams, R.N., and Williamson B. (1991) *Compendium of Raspberry and Blackberry Diseases and Insects*. APS Press, St. Paul, MN.
Fan-Chiang, H.J. (1999) Anthocyanin pigment, nonvolatile acid and sugar composition of blackberries. M.S. Thesis, Oregon State University, Corvallis.
Finn. C.E. (1996) Emasculated trailing blackberry (*Rubus* sp.) flowers set drupelets when not protected from cross pollination by bagging. HortScience 31:1035.
Finn. C.E. (1999) Temperate berry crops. In: J. Janick (Ed.) *Perspectives on New Crops and New Uses.* ASHS Press, Alexandria, VA. pp 324–333.
Finn, C.E. (2001) Trailing blackberries: From clear-cuts to your table. HortScience 36:236–238.
Finn, C.E. (2008) Blackberries. In: J. F. Hancock (Ed.), *Temperate Fruit Crop Breeding: Germplasm to Genomics.* Springer Science +Business Media, pp. 83–114.
Finn, C.E. and Hancock J.F. (2008) Raspberries. In: J. F. Hancock (Ed.), *Temperate Fruit Crop Breeding: Germplasm to Genomics.* Springer Science +Business Media, pp. 359–392.
Finn, C.E. and Knight V.H. (2002) What's going on in the world of *Rubus* breeding? Acta Hort. 585:31–38.
Finn, C.E. and Martin, R.R. (1996) Distribution of tobacco streak, tomato ringspot, and raspberry bushy dwarf viruses in *Rubus ursinus* and *R. leucodermis* collected from the Pacific Northwest. Plant Dis. 80:769–772.
Finn, C.E., Lawrence, F.J., and Strik, B.C. (1998) 'Black Butte' trailing blackberry. HortScience 33:355–357.
Finn, C.E., Lawrence, F.J., Strik, B.C., Yorgey, B.M., and DeFrancesco, J. (1999) 'Siskiyou' trailing blackberry. HortScience 34:1288–1290.
Finn, C., Strik, B.C., and Lawrence, F.J. (1997) Marion trailing blackberry. Fruit Var. J. 51: 130–132.
Finn, C., Swartz, H., Moore, P.P., Ballington, J.R., and Kempler, C. (2002a) Use of 58 *Rubus* species in Five North American Breeding Programs- Breeders Notes. Acta Hort. 585: 113–120.
Finn, C., Swartz, H., Moore, P.P., Ballington, J.R., and Kempler, C. (2002b) Use of 58 *Rubus* species in Five North American Breeding Programs- Breeders Notes. http://www.ars-grin.gov/cor/rubus/rubus.uses.html (3 March 2008).
Finn, C.E., Wennstrom, K., and Hummer K. (1999). Crossability of Eurasian *Rubus* species with red raspberry and blackberry. Acta Hort. 505:363–367.
Finn, C.E., Yorgey, B.M., Strik, B.C., and Martin, R.R. (2005a) 'Metolius' trailing blackberry. HortScience 40:2189–2191.
Finn, C.E., Yorgey, B.M., Strik, B.C., Hall, H.K., Martin, R.R., and Qian, M. (2005b) 'Black Diamond' trailing thornless blackberry. HortScience 40:2175–2178.
Finn, C.E., Yorgey, B.M., Strik, B.C., Martin, R.R., and C. Kempler. (2005c) 'Obsidian' trailing blackberry. HortScience 40:2185–2188.
Finn, C.E., Yorgey, B.M., Strik, B.C., Martin, R.R., and Qian, M. (2005d) 'Black Pearl' trailing thornless blackberry. HortScience 40:2179–2181.
Finn, C.E., Yorgey, B.M., Strik, B.C., Martin, R.R., and Qian, M. (2005e) 'Nightfall' trailing thornless blackberry. HortScience 40:2182–2184.
Galletta, G.J., A.D. Draper, and R.L. Puryear. (1986) Characterization of Rubus progenies from embryo culture and from seed germination. Acta Hort. 183:83–89.
Galletta, G.J. and R.L. Puryear. 1983. A method for *Rubus* embryo culture. HortScience 18:588.

Galletta, G.J., Draper, A.D., Maas, J.L., Skirvin, R.M., Otterbacher, A.G., Swartz, H.J., and Chandler C.K. (1998a) 'Chester Thornless' blackberry. Fruit Var. J. 52:118–122.

Galletta, G.J., Maas, J.L., Clark, J.R., and Finn, C.E. (1998b) 'Triple Crown' thornless blackberry. Fruit Var. J. 52:124–127.

Graham, J., Smith, K., Woodhead, M., and Russell, J. (2002) Development and use of simple sequence repeat SSR markers in *Rubus* species. Mol. Ecol. Notes 2:250–252.

Gubler, W.D. (1991) Downy mildew. In: M.A. Ellis, R.H. Converse, R.N. Williams, and B. Williamson (Eds.). *Compendium of Raspberry and Blackberry Diseases and Insects*. APS Press, St. Paul, MN pp. 15–16.

Gupton, C.L. (1999) Breeding for rosette resistance in blackberry. Acta Hort. 505:313–322.

Gupton, C.L. and Smith, B.J. (1997) Heritability of rosette resistance in blackberry. HortScience 32:940.

Gustafsson, A. (1942) The origin and properties of the European blackberry flora. Hereditas 28:249–277.

Gustafsson, A. (1943) The genesis of the European blackberry flora. Acta Univ. Lund. 39:1–199.

Guzmán-Baeny, T.L (2003) Incidence, distribution, and symptom description of viruses in cultivated blackberry (*Rubus* subgenus Eubatus) in the Southeastern United States. M.S. Thesis, North Carolina State University, Raleigh.

Hall, H.K. (1990) Blackberry breeding, In: J. Janick (Ed.). *Plant Breeding Reviews*. Timber Press, Inc, Portland, OR, pp. 249–312.

Hall, H.K. and Stephens, J. (1999) Hybridberries and blackberries in New Zealand - breeding for spinelessness. Acta Hort. 505:65–71.

Hall, H.K., Cohen, D., and Skirvin, R.M. (1986a) The inheritance of thornlessness from tissue culture-derived Thornless Evergreen blackberry. Euphytica 35:891–898.

Hall, H.K., Brewer, L.R., Langford, G., Stanley, C.J., and Stephens, M.J. (2003) 'Karaka Black': Another 'Mammoth' blackberry from crossing eastern and western USA blackberries. Acta Hort. 626:105–110.

Hall, H.K., Quazi, M.H., and Skirvin, R.M. (1986b) Isolation of a pure thornless Loganberry by meristem tip culture. Euphytica 35:1039–1044.

Hall, H.K., Skirvin, R.M., and Braam, W.F. (1986c) Germplasm release of 'Lincoln Logan', a tissue culture-derived genetic thornless 'Loganberry'. Fruit Var. J. 40:134–135.

Hall, H.K., Stephens, M.J., Stanley, C.J., Finn, C.E., and Yorgey, B. (2002) Breeding new 'Boysen' and 'Marion' cultivars. Acta Hort. 585:91–96.

Hedrick, U.P. (1925) *The Small Fruits of New York*. J.B. Lyon. Albany, NY.

Himelrick, D.G., Ebel, R.C., Woods, F.M., Wilkins, B.S., and Pitts, J.A. (2000) Effect of primocane topping height and lateral length on yield of 'Navaho' blackberry. Small Fruits Rev. 1:95–101.

Hummer, K.E., and Janick, J. (2007) R*ubus* iconography: Antiquity to the Renaissance. Acta Hort. 759:89–106.

Jennings, D.L. (1988) Raspberries and blackberries: Their breeding, diseases and growth. Academic Press, London.

Jennings, D.L. (1989) United States Plant Patent: Blackberry plant-Loch Ness cultivar, Plant Patent 6,782. Washington D.C.

Jennings, D.L., Daubeny, H.A., and Moore, J.N. (1992) In: J.N. Moore and J.R. Ballington (Eds.). Blackberries and raspberries (*Rubus*). *Genetic Resources of Temperate Fruit and Nut Crops*. Acta Hort. 290:331–389.

Kraft, T., Nybom, H., and Werlemark, G. (1996) DNA fingerprint variation in some blackberry species (*Rubus* subg. *Rubus*, *Rosaceae*). Pl. Syst. Evol. 199:93–108.

Kurnianta, A.J. (2005) Descriptive sensory analysis of thornless blackberry selections to determine sensory similarity to 'Marion' blackberry flavor. M.S. Thesis Oregon State University, Corvallis.

Lawrence, F.J. (1984) In: T.B. Kinney and J.R. Davis (eds.). Naming and release of blackberry cultivar Kotata. USDA-ARS Release Notice.

Lawrence, F.J. (1989) Naming and release of blackberry cultivar 'Waldo'. U.S. Dept. of Agr., Oregon Agr. Expt. Sta. Release notice.

Lewers, K.S., Saski. C.A., Cuthbertson, B.J., Henry, D.C., Staton, M.E., Main, D.S., Dhanaraj, A.L., Rowland, L.J. and Tomkins, J.P. (2008) A blackberry (*Rubus* L.) expressed sequence tag library for the development of simple sequence repeat markers. BMC Plant Biology 83:543–548.

Lewers, K.S., Styan, S.M.N., Hokanson, S.C., and Bassil, N.V. (2005) Strawberry GenBank-derived and genomic simple sequence repeat (SSR) markers and their utility with strawberry, blackberry, and red and black raspberry. J. Am. Soc. Hort. Sci. 130:102–115.

Lim, K.Y., Leitch, I.J., and Leitch, A.R. (1998) Genomic characterization and the detection of raspberry chromatin in polyploid *Rubus*. Theo. Appl. Genet. 97:1027–1033.

Lim, Y.K. and Knight V.H. (2000) The successful transfer of primocane fruiting expression from raspberry to *Rubus* hybrid berry. Euphytica 116:257–263.

Logan, M.E. (1955) *The Loganberry.* Mary E. Logan (Mrs. J.H. Logan) Publisher, Oakland, Calif.

Lopes, M.S., Belo Maciel, G., Mendonca, D., Sabino Gil, F., and Da Camara Machado A. (2006) Isolation and characterization of simple sequence repeat loci in *Rubus hochstetterorum* and their use in other species from the Rosaceae family. Molecular Ecol. Notes. 6:750–752.

Lopez-Medina, J., Moore, J.N., and McNew, R.W. (2000) A proposed model for inheritance of primocane fruiting in tetraploid erect blackberry. J. Am. Soc. Hort. Sci. 125:217–221.

Lyman, M.R., Curry, K.J., Smith, B.J., and Diehl, S.V. (2004) Effect of *Cercosporella rubi* on blackberry floral bud development. Plant Dis. 88:195–204.

Marroquin, E., Matta, F.B., Graves, C.H., and Smith, B. (1990) Relationship between flower/fungal development in blackberry infected with *Cercosporella rubi*. HortScience 25:1448.

Martin, R.R. (2002) Virus diseases of *Rubus* and strategies for their control. Acta Hort. 585:265–270.

Martin, R.R., Tzanetakis, I.E., Gergerich, R., Fernandez, G.E., and Pesic, Z. (2004) Blackberry yellow vein associated virus: A new crinivirus found in blackberry. Acta Hort. 656:137–142.

McKeen, W.E. (1954) A study of cane and crown galls on Vancouver Island and a comparison of the causal organisms. Phytopathology 44:651–655.

McPheeters, K.D. and Skirvin R.M. (2000) 'Everthornless' blackberry. HortScience 35:778.

Mehlenbacher, S.A. (1995) Classical and molecular approaches to breeding fruit and nut crops for disease resistance. HortScience 30:466–477.

Meng, R. and Finn, C.E. (1999) Using flow cytometry to determine ploidy level in *Rubus*. Acta Hort 505:223–227.

Meng, R., and Finn, C.E. (2002) Determining ploidy level and nuclear DNA content in *Rubus* by flow cytometry. J. Am. Soc. Hort. Sci. 127:767–775.

Meng, R., Chen, T.H.H., Finn, C.E., and Li, Y. (2004) Improving in vitro plant regeneration from leaf and petiole explants of 'Marion' blackberry. HortScience 39:316–320.

Moore, J.N. (1984) Blackberry breeding. HortScience 19:183–185.

Moore, J.N. (1997) Blackberries, In: *The Brooks and Olmo Register of Fruit and Nut Varieties*. 3rd ed. ASHS Press, Alexandria, VA, pp. 161–173.

Moore, J.N. and J.R. Clark, J.R. (1989) Navaho thornless blackberry. HortScience 24:863–865.

Moore, J.N., Lundergan, C., and Brown, E.D. (1975) Inheritance of seed size in blackberry. J. Am. Soc. Hort. Sci. 100:377–379.

Moyer, R., Hummer, K., Finn, C., Frei, B., and Wrolstad R. (2002) Anthocyanins, phenolics and antioxidant capacity in diverse small fruits: *Vaccinium*, *Rubus* and *Ribes*. J. Agric. Food Chem. 50:519–525.

Nybom, H., and Hall, H.K. (1991) Minisatellite DNA 'fingerprints' can distinguish *Rubus* cultivars and estimate their degree of relatedness. Euphytica 53:107–114.

Nybom, H., Rogstad, S.H., and Schaal, B.A. (1990) Genetic variation detected by use of the M13 'DNA fingerprint' probe in *Malus*, *Prunus*, and *Rubus* (Rosaceae). Theor. Appl. Genet. 79:153–156.

Nybom, H., Schaal, B.A., and Rogstad, S.H. (1989) DNA 'fingerprints' can distinguish cultivars of blackberries and raspberries. Acta Hort. 262:305–310.

Ourecky, D.K. (1975) Brambles. In: J. Janick and J.N. Moore (Eds.). *Advances in Fruit Breeding*. Purdue Univ. Press, West Lafayette, IN, pp. 98–129.

Pamfil, D., Zimmerman, R.H., Naess, K., and Swartz, H.J. (2000) Investigation of *Rubus* breeding anomalies and taxonomy using RAPD analysis. Small Fruits Rev. 1:43–56

Perkins-Veazie, P. and Clark, J.R. (2005) Blackberry research in Arkansas and Oklahoma. Proc. N. Amer. Bramble Growers Assn. Ann. Mtg. p. 39–42.

Perkins-Veazie, P., and Kalt, W. (2002) Postharvest storage of blackberry fruit does not increase antioxidant levels. Acta Hort. 585:521–524.

Perkins-Veazie, P., Collins, J.K., and Clark, J.R. (1996) Cultivar and maturity affect postharvest quality of fruit from erect blackberries. HortScience 31:258–261.

Perkins-Veazie, P., Collins, J.K., and Clark, J.R. (1999) Cultivar and storage temperature effects on the shelflife of blackberry fruit. Fruit Var. J. 53:201–208.

Perkins-Veazie, P., Collins, J.K., and Clark, J.R. (2000) Shelflife and quality of 'Navaho' and 'Shawnee' blackberry fruit stored under retail storage conditions. J. Food Qual. 22:535–544.

Perkins-Veazie, P., Collins, J.K., Clark, J.R., and Risse, L. (1997) Air shipment of 'Navaho' blackberry fruit to Europe is feasible. HortScience 32:132.

Scott, D.H. and Ink, D.P. (1966) Origination of 'Smoothstem' and 'Thornfree' blackberry varieties. Fruit Var. Hort. Dig. 20:31–33.

Sherman, W.B. and R.H. Sharpe (1971) Breeding *Rubus* for warm climates. HortScience 6:147–149.

Siriwoharn, T., Wrolstad, R.E., Finn, C.E., and Pereira C.B. (2004) Influence of cultivar, maturity and sampling on blackberry (*Rubus* L. hybrids) anthocyanins, polyphenolics, and antioxidant properties. J. Agric. Food Chem. 52:8021–8030.

Sistrunk, W.A. and Moore J.N. (1973) Progress in breeding blackberries. Ark. Farm Res. 22(3):5.

Smith, B.J. and Diehl, S.V. (1991) A scanning electron microscope study of blackberry flowers infected with *Cercosporella rubi*. Phytopathol. 81:1232.

Stafne, E.T. (2005) Characterization, differentiation, and molecular marker analysis of blackberry germplasm. Ph.D. dissertation. University of Arkansas, Fayetteville.

Stafne, E.T. and Clark, J.R. (2004) Genetic relatedness among eastern North American blackberry cultivars based on pedigree analysis. Euphytica 139:95–104.

Stafne, E.T., Clark, J.R., Pelto, M.C., and Lindstrom, J.T. (2003) Discrimination of *Rubus* cultivars using RAPD markers pedigree analysis. Acta Hort. 626:119–124.

Stafne, E.T., Clark, J.R., Weber, C.A., Graham, J., and Lewers K.S. (2005) Simple sequence repeat (SSR) markers for genetic mapping of raspberry and blackberry. J. Am. Hort. Soc. 103:722–728.

Stanisavljevic, M. (1999) New small fruit cultivars from Cacak: 1. The new blackberry *[Rubus* sp.] cultivar 'Cacanska Bestrna'. Acta Hort. 505:291–295.

Stanton, M.A., Scheerens, J.C., Funt, R.C., and Clark, J.R. (2007) Floral competence of primocane-fruiting blackberries Prime-Jan® and Prime-Jim® blackberries grown at three temperature regimes. HortScience 42:508–513.

Stellar, O.A (1937) The giant Boysenberry goes national-the brambleberry page. Better Fruit 32 (February):20.

Stewart, P.J., Clark, J.R., and Fenn, P. (2003) Evaluation of resistance to *Erwinia amylovora* and *Botryosphaeria dothidea* in eastern U.S. Blackberry cultivars. In: J.A. Robbins, B. Murphy, and M. Richardson (Eds.). Hort. Studies 2003. Ark. Agr. Exp. Sta. Res. Ser. 520: 32–34.

Strik, B. (1992) Blackberry cultivars and production trends in the Pacific Northwest. Fruit Var. J. 46:202–205.

Strik, B. and Buller G. (2002) Reducing thorn contamination in machine-harvested 'Marion' blackberry. Acta Hort. 585:677–681.

Strik, B.C. and Martin, R.R. (2003) Impact of *Raspberry bushy dwarf virus* on 'Marion' blackberry. Plant Dis. 87:294–296.

Strik, B.C., Clark, J.R. Finn, C.E., and Bañados, M.P. (2007) Worldwide blackberry production. HortTechnology 17:205–213.

Strik, B.C., Finn, C.E., Clark, J.R., and Buller, G. (2008) Management of primocane-fruiting blackberry to maximize yield and extend the fruiting season. Acta Hort. 777:423–428.

Strik, B.C., Mann, J., and Finn, C. (1996) Percent drupelet set varies among blackberry genotypes. J. Am. Soc. Hort. Sci. 12:371–373.

Susaimuthu, J., Gergerich, R.C., Bray, M.M., Dennis, K.A., Clark, J.R., Tzanetakis, I.E., and Martin, R.R. (2007) Incidence and ecology of blackberry yellow vein associated virus. Plant Dis. 91:809–813.

Swartz, H.J. and Stover, E.W. (1996) Genetic transformation in raspberries and blackberries (*Rubus* species). In: Bajaj YPS (Ed.), *Biotechnology in Agriculture and Forestry,* vol 38. Springer-Verlag, Berlin. pp 297–307.

Takeda, F. (1993) Characterization of blackberry pyrenes. HortScience 28:488 (Abstract).

Takeda, F. (2002) Winter pruning affects yield components of 'Black Satin' eastern thornless blackberry. HortScience 37:101–103.

Takeda, F. and Peterson, D.L. (1999) Considerations for machine harvesting fresh-market eastern thornless blackberries: Trellis, cane training systems, mechanical harvester developments, HortTechnology 9:16–21.

Takeda, F., Strik, B.C., Peacock, D., and Clark J.R. (2002) Cultivar differences and the effect of winter temperature on flower bud development in blackberry. J. Am. Soc. Hort. Sci. 127:495–501.

Takeda, F., Strik, B.C., Peacock, D., and Clark, J.R. (2003) Patterns of floral bud development in canes of erect and trailing blackberries. J. Am. Soc. Hort. Sci. 128:3–7.

Thompson, E., Clark, J.R., Strik, B.C., and Finn, C.E. (2008) Flowering and fruiting morphology of primocane-fruiting blackberries. Acta Hort. 777:281–288.

Thompson, M.M. (1961) Cytogenetics of *Rubus* II. Cytological studies of the varieties 'Young', 'Boysen', and related forms. Am. J. Bot. 48:667–673.

Thompson, M.M. (1995a) Chromosome numbers of *Rubus* cultivars at the National Clonal Germplasm Repository. HortScience 30:1453–1456.

Thompson, M.M. (1995b) Chromosome numbers of *Rubus* species at the National Clonal Germplasm Repository. HortScience 30:1447–1452.

Thompson, M.M. (1997) Survey of chromosome numbers in *Rubus* Rosaceae: Rosoideae. Ann. Rpt. Mo. Botanical Garden 84:128–163.

Tzanetakis, I.E. and Martin, R.R. (2004) First report of beet pseudo yellows virus in blackberry in the United States. Plant Dis. 88:223.

USDA, ARS, National Genetic Resources Program. (2010a) *Germplasm Resources Information Network - (GRIN).* [Online Database] National Germplasm Resources Laboratory, Beltsville, MD. URL: http://www.ars-grin.gov/cgi-bin/npgs/acc/search.pl?accid=Hillquist (8 March 2010).

USDA, ARS, National Genetic Resources Program. (2010b) *Germplasm Resources Information Network - (GRIN)* [Online Database]. National Germplasm Resources Laboratory, Beltsville, MD. URL: http://www.ars-grin.gov/cgi-bin/npgs/html/genus.pl?10574 (3 March 2008).

Wada, L. and Ou, B. (2002) Antioxidant activity and phenolic content of Oregon caneberries. J. Agric. Food Chem. 50:3495–3500.

Waldo, G.F. (1948) The Chehalem blackberry. Oregon Agr. Expt. Sta. Circ. 421.

Waldo, G.F. (1950a) Breeding blackberries. Oregon Agr. Expt. Sta. Bul. 475:3–38.

Waldo, G.F. (1950b) Notice of naming and release of a new blackberry adapted to the Pacific Coast region. U.S.D.A. Release Notice.

Waldo, G.F. (1957) The Marion blackberry. Oregon Agr. Expt. Sta. Circ. 571.

Waldo, G.F. (1968) Blackberry breeding involving native Pacific Coast parentage. Fruit Var. J. 22:3–7.

Waldo, G.F. (1977) Thornless Evergreen - Oregon's leading blackberry. Fruit Var. J. 31:26–30.

Waldo, G.F. and Wiegand, E.H. (1942) Two new varieties of blackberry the Pacific and the Cascade. Oregon Agr. Expt. Sta. Circ. 269.

Wang, S.Y., and Lin, H.S. (2000) Antioxidant activity in fruits and leaves of blackberry raspberry, and strawberry varies with cultivar and developmental stage. J. Agric. Food Chem. 48:140–146.

Wang, Y., Finn, C. and M.C. Qian. (2005) Impact of growing environments on 'Chickasaw' blackberry (*Rubus* L.) aroma evaluated by gas chromatography olfactory dilution analysis. J. Agric. Food Chem. 53:3563–3571.

Warmund, M.R. and George, M.F. (1990) Freezing survival and supercooling in primary and secondary buds of *Rubus* spp. Can. J. Plant Sci. 70:893–904.

Warmund, M.R., George, M.F., Ellersieck, M.R., and Slater, J.V. (1989) Susceptibility of blackberry tissues to freezing injury after exposure to 16C. J. Am. Soc. Hort. Sci. 114:795–800.

Warmund, M.R., Krumme, J. (2005) A chilling model to estimate rest completion in erect blackberries. HortScience 1259–1262.

Wood, G.A. (1995) Further investigations of raspberry bushy dwarf virus in New Zealand. N.Z. J. Crop Hort. Sci 23:273–281.

Wood, G.A. and Hall, H.K. (2001) Source of raspberry bushy dwarf virus in *Rubus* in New Zealand, and the infectibility of some newer cultivars to this virus. N.Z. J. Crop Hort. Sci 29:177–186.

Wood, G.A., Andersen, M.T., Forster, R.L.S., Braithwaite, M., and Hall H.K. (1999) History of Boysenberry and Youngberry in New Zealand in relation to their problems with Boysenberry decline, the association of a fungal pathogen, and possibly a phytoplasma, with this disease. N.Z. J. Crop Hort. Sci 27:281–295.

Yorgey, B. and Finn, C.E. (2005) Comparison of 'Marion' to thornless blackberry genotypes as individually quick frozen and puree products. HortScience 40:513–515.

Chapter 6
American Cranberry

Nicholi Vorsa and Jennifer Johnson-Cicalese

Abstract Cranberry breeding has undergone relatively few breeding and selection cycles since domestication in the nineteenth century. The first cranberry breeding program's objective was to develop varieties with a reduced feeding preference to the blunt-nosed leafhopper, the vector of the phytoplasma 'false-blossom' disease. From this program, six varieties were released, of which 'Stevens,' released in 1950, became the most widely planted cultivar. Improved consistent yields, fruit color, and season of ripening continue to be objectives of breeding efforts. However, disease resistance, especially against the fruit rot disease complex, and insect resistance are increasingly necessary objectives. Much of the cranberry germplasm has not been fully explored for disease and insect resistance, and other traits of interest. Recent development of genomic resources in cranberry will provide for innovative plant breeding systems that will reduce the time and field space required and facilitate the breeding of unique superior cranberry cultivars to meet the current and future challenges of this important American crop. The cranberry industry continues to be a strong supporter of genetic enhancement efforts, providing land space and funding.

Keywords *Vaccinium macrocarpon* • American Cranberry • Yield • Fruit set • Flavonoids • Anthocyanin • Proanthocyanidin • Flavonol • Fruit rot resistance • Disease resistance • *Vaccinium oxycoccus* • Heritability

N. Vorsa (✉) • J. Johnson-Cicalese
PE Marucci Center, Rutgers University, 125A Lake Oswego Rd, Chatsworth, NJ 08019, USA
e-mail: vorsa@AESOP.Rutgers.edu

M.L. Badenes and D.H. Byrne (eds.), *Fruit Breeding*, Handbook of Plant Breeding 8,
DOI 10.1007/978-1-4419-0763-9_6, © Springer Science+Business Media, LLC 2012

1 Introduction

1.1 Economic Importance and Uses

In the early 1800s, the American cranberry (*Vaccinium macrocarpon* Ait) became the first species of *Vaccinium* to come under cultivation. Boston was the major cranberry market for fruit harvested from the first cultivated plantings, as well as fruit gathered from native populations (Eck 1990). During this time, the fruit was shipped to Europe and to domestic markets along the east coast to New Orleans. Up until the 1950s, the fruit was sold fresh and as sauces. The popularity of cranberry suffered a setback during the "great cranberry scare" of 1959, when the Secretary of the US Department of Health, Education and Welfare, Arthur Fleming, announced nationally that some cultivated cranberries were contaminated with the herbicide aminotriazole, a suspected carcinogen (Eck 1990). The industry recovered slowly after the "scare," and the breakthrough for the commercial success of cranberry occurred in the 1960s with the development and marketing of cranberry juice cocktail by Ocean Spray Cranberries, Inc., a processed product that made cranberry a year-round commodity. Up until 2000, the juice cocktail continued to be the major cranberry product. During the first decade of the twenty-first century, sweetened-dried cranberry became a major product of cranberry, being utilized in cereals, mixes, etc. Today, cranberries are mostly consumed through processed products including juice, juice cocktail, sauces, and sweetened-dried cranberries. Some portion of the crop is also made into nutraceutical products, e.g., cranberry extract tablets.

The USA is the largest commercial producer of cranberry, with a value over 300 million dollars annually. In 2010, the USA produced 680 million pounds on 38,500 acres. Canada is the second biggest producer with approximately 8,000 acres. Chile is currently producing on approximately 1,000 acres. Some minor production occurs in Eastern Europe. In the USA, Wisconsin is the largest producer of cranberries with 47% of harvested acreage. In Canada, British Columbia and Quebec provinces are the largest producers.

1.2 Adaptation and Morphology

The American cranberry is a diploid fruit species ($2n=2x=24$) endemic to North America, and a member of the *Vaccinium* section *Oxycoccus*, meaning "sour berry." *V. macrocarpon* is an evergreen, woody perennial, with a trailing vine growth habit. A member of the *Ericaceae* (Heath family), cranberry is adapted to moist acidic soils containing high levels of organic matter, which are typically found in bogs, marshes and swamps with a temperate climate. The requirement of an acidic media or soil (maximum pH 5.5) limits the American cranberry's adaptation. Other factors that constrain its adaptation include a fine root system lacking root hairs, making the best suited soils to be sands, loamy sands, and organic soils consisting of course peat or muck. Being a temperate woody perennial, cranberry requires a minimum of

Fig. 6.1 Inflorescence of cranberry developing acropetally, lowest attached pedicel flower (1) has style exerted, second lowest pedicel (2) flower is at anthesis, and upper most attached pedicel flower (3) is unopened

800–1,000 h of chilling to break winter dormancy, preventing its culture in warm climates. In areas which have severe winter freezes, inflorescence buds and leaf tissues are typically protected with a winter flood.

In nature, cranberry reproduces both sexually and asexually through stolons. Stolon sections in contact with soil root readily. Ascending shoots, colloquially referred to as "uprights," are produced along the length of the stolon and are terminated by an inflorescence bud. Typically, inflorescence buds are initiated in late summer and early fall, remaining dormant through winter. After receiving chilling (>1,000 h), the bud breaks dormancy in mid to late spring, forming an inflorescence of 3–7 flowers with acropetal development. Flowering occurs in early summer with flowers borne along the rachis of the upright, which terminates in a leafy shoot. Flowers are 4-merous, perfect, having eight anthers with an inferior four-locule ovary (>20 ovules). Flowers are protandrous, with the style 6–7 mm in length at anthesis inside the anther whorl, then elongating to 8–10 mm, extending 2–3 mm beyond the anther whorl 2–3 days post-anthesis (Fig. 6.1). The stigma appears most receptive 3–5 days after anthesis, producing an exudate. Characteristic of Ericaceae species, pollen is shed as a tetrad with the four pollen grains of a meiotic event held in a tetrahedral formation. All four pollen grains of the tetrad are potentially viable. Anthers of one cranberry flower shed over 7,000 pollen tetrads (Cane et al. 1996). Typically, 1–3 fruit are set per upright, but varietal variation likely plays a role.

Cranberry is an asexually propagated crop, with varieties typically being propagated from material collected from producing commercial beds. Until recently, the method of bed establishment was pressing or "discing-in" dormant vines into the bed media in early spring. Vine material could either be prunings or mowings from existing beds. Pruning the bed yields largely stolons, while mowing the bed yields both runners and uprights. These traditional propagation methods have resulted in compromised varietal identity and problems with genetic heterogeneity commonplace. A more recent approach has been the utilization of rooted cuttings in a nursery system with DNA-fingerprinted and virus indexed material. Rooted cuttings are usually planted at a density of one per square foot or greater.

2 Origin and Domestication

2.1 Native Distribution and Domestication

The natural distribution of *V. macrocarpon* (Fig. 6.2) ranges from Newfoundland west throughout the Great Lakes region to Minnesota, and south to the coast of Delaware; and at higher elevations in the Appalachian Mountains the distribution ranges to North Carolina and Tennessee (Vander Kloet 1988). The main distribution lies between 40 and 50°N latitude (Vander Kloet 1988). Cranberry also colonizes floating sphagnum tufts in Long Island lakes, river banks in northeastern Pennsylvania and New Jersey, and rocky outcrops along the Maine coast. The only other species in the same section as American cranberry, sect. *Oxycoccus*, is *Vaccinium oxycoccus*. Its distribution is circumboreal in the northern hemisphere; in North America it is absent from the Arctic Archipelago and extends southwards to Oregon and into the Appalachians (Fig. 6.2).

The first attempt to cultivate the American cranberry was made in 1810 by Henry Hall, a Revolutionary War veteran of Dennis, Massachusetts (Eck 1990). Domestication proceeded in the early 1800s, by selection of cranberry varieties from native stands in Massachusetts, Michigan, New Jersey, and Wisconsin. Selection criteria were likely large fruit size, good fruit color, early fruit ripening, and fruiting productiveness. The variety 'Early Black' was selected in 1835 by a Cape Cod cranberry grower from a native cranberry stand, and is still cultivated in the twenty-first century. Over 100 varieties have been selected from the wild and named, and are listed by Dana (1983), but most are no longer available in existing collections or farms. Over the past few decades, there has been a transition to varieties developed from breeding programs.

2.2 Breeding History

The USDA, in cooperation with the New Jersey and Massachusetts State Agricultural Experiment Stations, initiated the first cranberry breeding program in 1929 (Chandler et al. 1947). The major objectives of the program were to develop varieties resistant

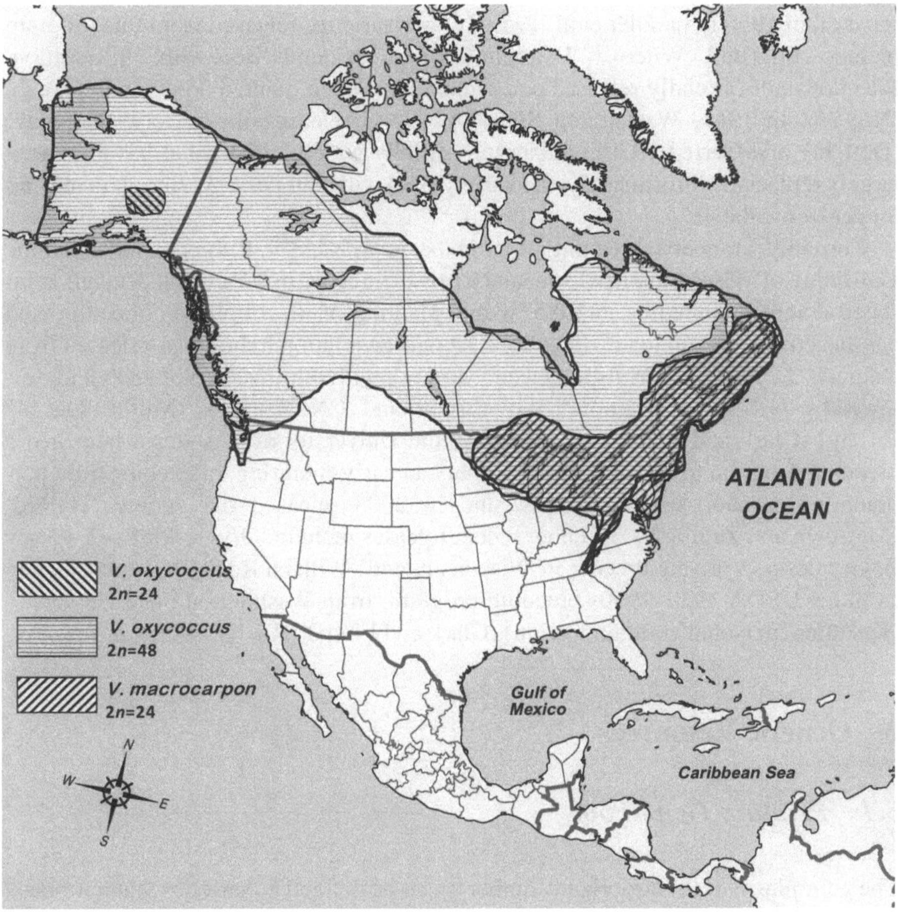

Fig. 6.2 Native distribution of *Vaccinium macrocarpon*, *V. oxycoccus* ($2n=24$), *V. oxycoccus* ($2n=4x$) in North America

to "false-blossom" disease (a phytoplasma vectored by the blunt-nosed leafhopper), along with higher productivity, good fruit color and superior fruit. Over 10,685 seedlings were produced from over 30 crosses, and led to "the 40 selections" for further testing. The 40 selections were further evaluated for sauce and cocktail quality, specific gravity and overall appearance (Chandler et al. 1947). Traits evaluated included susceptibility to leafhopper feeding, date of harvest, size of fruit, decay, yield, and shape (Chandler et al. 1947). In 1939, the Wisconsin Agricultural Experiment Station, in cooperation with the USDA, WI Dept. of Agri., and WI Cranberry Sales Co., established a nursery to test some of these selections at another location, and became active participants in the breeding program. This program resulted in the release of six varieties, including the most widely grown cultivar 'Stevens.' Stevens was derived from a 'McFarlin' × 'Potter' cross and was field selected in 1940 at J.J. White Co., Whitesbog, Burlington Co., New Jersey, and released in 1950 (Chandler et al. 1950). Other varieties released from this program include 'Pilgrim,' 'Wilcox,' 'Franklin,' 'Bergman,' and 'Beckwith.' In addition, selections not officially released and named have been planted commercially, e.g., 'No. 35.' In 1961, Washington State University released the variety, 'Crowley' (Doughty and Garren 1970), which was initially widely planted but has now been largely replaced. Unfortunately, the variety Beckwith and possibly Bergman may no longer be available.

Currently, cranberry breeding programs exist at NJAES, Rutgers University, the University of Wisconsin–Madison, and a private breeder in Wisconsin. Rutgers reinstated cranberry breeding in 1985, with a planting of a germplasm collection, and has an active program today. In 2006, a second-generation hybrid was released from Rutgers, 'NJS98-23' (Crimson Queen® variety), with improved color and yield, followed by 'NJS98-35' (Demoranville® variety) and 'CNJ97-105-4' (Mullica Queen® variety) (Clark and Finn 2010). In 1990, the University of Wisconsin launched a breeding program to develop varieties that had early-maturing, high color fruit, particularly for short-season regions, and in 2003 released the variety 'HyRed' (McCown and Zeldin 2003). Other recent releases include 'Grygleski#1, #2, #3' by E. Grygleski, a private breeder in Wisconsin, and 'Willapa Red' (BE4), a selection from the USDA 1930–1950s breeding program, from Washington State University (K. Patten, personal communication) (Clark and Finn 2010).

3 Genetic Resources

3.1 *Primary Gene Pool*

The germplasm of the American cranberry can be defined as varieties that have been domesticated from native populations over the last 200 years, and that that exists currently in native populations. During the domestication of cranberry in the 1800s on the east coast of North America, the initial cultivated varieties were selected from native populations largely from Massachusetts, Wisconsin and New Jersey, yielding 127 named varieties (Chandler and Demoranville 1958). About 50 of these have been described in morphology and growth habit, but only four, Early Black, 'Howes,' McFarlin, and 'Searles,' became widely grown in the mid 1900s. A few additional native clones have been propagated and widely planted, including 'Ben Lear' and to a lesser degree 'Lemunyon.' Other varieties were identified as natives from a particular location, e.g., 'Jerseys.' Cranberry germplasm in field plots has been maintained largely in State Agricultural Experiment Station programs of Massachusetts, Wisconsin, Washington and New Jersey. The University of Massachusetts Cranberry Research Center, East Wareham, MA; Washington State University, Long Beach, WA; University of Wisconsin located at DuBay Cranberry Co., Junction City, WI; and the PE Marucci Center, New Jersey Agricultural Experimental Station, Rutgers University, Chatsworth, NJ, currently maintain variety plots. Clones of most major cultivars, assorted varieties and seed collections from open-pollination of *V. macrocarpon* are maintained by the USDA National Clonal Germplasm Repository, Corvallis, Oregon. The repository also maintains clonal material and seed from related species collected from North America, Europe, and Asia.

In 1985, the NJAES/Rutgers program assembled collections from the other programs, as well as collected wild germplasm from extant native populations across the geographic distribution of the American cranberry (Bruederle et al. 1996). A survey of the pollen of the germplasm collection with acetocarmine staining of pollen[1] indicated that the majority of variety plots were genetically heterogeneous (N. Vorsa unpublished data). Subsequent SCAR fingerprinting confirmed the genetically heterogeneous state of the collection (Polashock and Vorsa 2002b). Since 1988, all new variety plots at the Rutgers program are established from a single vine, or from multiple vines that have matching SCAR fingerprints. After 20 years, the germplasm collection plots had become increasingly heterogeneous, even though they were established from a single propagule. Therefore, vine was reselected from each plot in the collection based on fruit characteristics, and then SCAR fingerprinted to reestablish the germplasm collection in 2010. The University of Massachusetts program maintains 50 variety plots that were established with multiple vines having fruit matching the varietal phenotype.

Cranberry varieties have been and continue to be identified by phenotype, relying largely on fruit characteristics. Fruit traits or characteristics that are measured and or described include the following: color intensity, berry size (cup count), calyx and calyx lobe features, stem end morphology, predominant fruit shape, bloom (whitish waxy coating), season of ripening, seed number, coloring in storage, and keeping quality. Vegetative traits include vine texture, e.g., fine, medium, or coarse, upright length, leaf shape, and leaf size. However, the quantitative nature of the traits used as variety descriptors, and the significant environmental variation component has led to multiple genotypes being represented by a variety name. DNA fingerprinting with RAPDs and SCAR data has provided some clarification. A common fingerprint has been identified for varieties such as McFarlin (Novy et al. 1996), Early Black, Howes (Novy and Vorsa 1995), and Ben Lear. However, consensus SCAR fingerprints for varieties such as Searles have been problematic (Novy and Vorsa 1995). Thus, the identity of varieties, e.g., Searles, Potter, 'Prolific,' which were utilized by the 1940s USDA breeding program and gave rise to the popular cultivars Stevens and Pilgrim, are ambiguous. Current programs in cranberry genetics and breeding may be able to develop markers for their eventual parental identification (J. Zalapa, personal communication).

The lack of a certified nursery system, and the intrinsic propensity for asexual (stolon) reproduction of cranberry provided the opportunity for off-type varieties to

[1] Aceto-carmine staining of pollen provides a measure of pollen viability. The cranberry pollen stain survey found that percent stainable pollen from different flowers within an upright was similar, whereas between uprights in a germplasm plot, it was variable. Pollen stainability within *Vaccinium* has a low environmental effect, thus the most likely cause of variable pollen stainability was due to genotypic variation, i.e., multiple varieties within a plot.

be introduced and increase over time. This has resulted in cranberry growers identifying "strains" of a variety, or "good" versus "bad" strains. One study of the variety McFarlin in Washington State identified at least 15 RAPD fingerprints, with one fingerprint associated with good production and fruit matching the original McFarlin variety description (Novy et al. 1996). Furthermore, DNA fingerprints (and phenotype) suggest that the majority, if not all, off-types were not genetically related to McFarlin. Being that *V. macrocarpon* is not native to Washington State, it is likely these off-types were introduced early in the cultivation of cranberry in Washington, and were extremely vegetatively competitive. The ambiguous identity of cranberry varieties is not only a problem with named wild selections. Varieties released from the first breeding programs were likely released as mixtures or were contaminated very early. Accessions from various commercial beds of the cultivar Crowley yielded many SCAR fingerprints and a consensus fingerprint was not obtained. Most beds across many growing regions of the cultivar Pilgrim were found to contain an off-type, suggesting the original release may have been contaminated (N. Vorsa and J. Polashock unpublished data).

Native undomesticated germplasm is still available for genetic enhancement. Endemic to North America, the American cranberry has a natural distribution from Newfoundland, west through the Great Lakes to eastern Minnesota, and south through the Appalachian Mountains to North Carolina and Tennessee (Vander Kloet 1983). Genetic diversity of native populations is extremely low relative to other *Vaccinium* sect. *Cyanococcus* species, as determined by allozyme analysis (Bruederle et al. 1996). Expected heterozygosity based on 23 loci was low, but most loci in populations did not significantly deviate from Hardy–Weinberg expectations indicating fairly high panmixis. Unexpected was that a few loci, e.g., glucose-6-hosphate isomerase, were fixed for rare alleles, where one would expect heterozygosity. About 79% of allelic diversity exists within populations, with only 21% between populations. However, populations from various sites exhibit distinct growth and morphology indicating population differentiation. One can observe variation between populations for plant structure, stolon production, fruit traits, etc. (N. Vorsa, unpublished data). Environmental parameters, including light and temperature, have manifested in phenotypic variation (Vander Kloet 1983), suggesting variation exists for local and climatic adaptation within the temperate climatic range that cranberry currently inhabits.

3.2 *Secondary Gene Pools: Related Cranberry Species and Interspecific Hybridization*

American cranberry is a member of the *Vaccinium* section *Oxycoccus*. Galletta (1975) provides a summary of the previous taxonomic and biosystematic literature of cranberry (Camp 1944, 1945). The most recent taxonomic treatment of the section *Oxycoccus* recognizes only two species (Vander Kloet 1983): these are the large-fruited, exclusively diploid American cranberry, *V. macrocarpon* Ait., which is endemic to North America; and *V. oxycoccus* L., a northern hemisphere, circumboreal, polyploid complex existing as diploids ($2n=24$), tetraploids ($2n=48$) and hexaploids ($2n=72$). Others consider diploids, tetraploids, and hexaploids of *V. oxycoccus* as morphologically distinct species, and identify diploids as *V. microcarpon* (Turcz. ex Rupr.) Schmalh., the tetraploid as *V. oxycoccus* L. (Jacquemart 1997), and the hexaploid as *V. hagerupii* (L. & L.) Ahokas (Camp 1944; Ravanko 1990; Jacquemart 1997). Diploid *V. oxycoccus* and *V. macrocarpon* are readily discriminated from another based on allozyme variation (Mahy et al. 2000). Allozyme analysis suggests an autoploid origin of tetraploid *V. oxycoccus* arising from diploid *V. oxycoccus*. But allelic composition of North American tetraploid *V. oxycoccus* suggests introgression of *V. macrocarpon* alleles has occurred (Mahy et al. 2000).

Outside the *Oxycoccus* section, species that may offer desirable traits include species within the sect. *Cyanococcus*, true-cluster fruited blueberry, sect. *Vitis-idaea*, lingonberry, and sect. *Batodendron*, creeping blueberry.

4 Major Breeding Achievements

The American cranberry has undergone relatively few breeding and selection cycles since domestication during the nineteenth century. The major achievements of the first breeding and selection cycle have been increased yield and more reliable production potential in cultivars such as Stevens and Pilgrim. These first-generation hybrids also have more stable production under higher nitrogen environments (Davenport and Vorsa 1999). These varieties were selected in New Jersey on organic, likely muck soils, which are higher in nitrogen. Recently released second generation hybrids have even higher yield potential, earlier season, and especially, higher anthocyanin content (Clark and Finn 2010; McCown and Zeldin 2003).

5 Current Goals and Challenges

Current breeding goals continue to include: (1) higher, consistent production, (2) vegetative vigor for bed establishment, and (3) high anthocyanin content. However, although currently grown cultivars have manageable disease and insect susceptibility, greater disease and insect resistance is emerging to the forefront of principal objectives. The restriction and loss of the broad spectrum organophosphate insecticides, along with the transition to insecticides targeting insect development, have altered the ecology of insect pests in cranberry. False-blossom disease, largely controlled by organo-phosphate insecticides, has recently emerged once again in cranberry culture. Resistance to tipworm, cranberry girdler, sparganothis fruitworm would be desirable.

During the last two decades, the major cranberry growing areas have experienced a warmer climate. Fruit rot diseases, such as early rot, *Phyllosticta vaccinii*, typically relegated to warmer growing areas, e.g., New Jersey and Massachusetts, have been experienced in Wisconsin since 2005, particularly in young plantings of newer varieties. Fruit rot resistance is a major objective of the Rutgers/NJAES program. Since cranberry is typically grown in wetland areas and subjected to considerable exposure to water, resistance to *Phytophthora* root rot species is also desirable.

Typically higher fruit anthocyanin (TAcy) has been and continues to be an objective of most cranberry breeding programs. However, with the health attributes of cranberry products being featured prominently, breeding for specific flavonoid profiles, and/or levels of various fruit constituents may be desirable. These could include: proanthocyanidins, flavonols, hydroxycinnamates, etc. Also, many cranberry products are formulated according to certain Brix specifications. Brix of crop loads is routinely measured, and may be used in determination of grower compensation. Thus, increasing Brix may become another fruit quality objective. Physiological and morphological attributes would include greater heat tolerance, and frost tolerance of flowers, as well as fruit, since fruit is harvested late in the fall season.

Although the majority of cranberry fruit is directed at processed products, there is a fresh fruit market. Varieties differ in their fresh fruit quality. Breeding objectives towards fresh fruit include: (1) uniform, fully colored berry, (2) longer storage-life, (3) resistance to storage fruit rots, e.g., black rot, and (4) round berry facilitating easier sorting.

5.1 Yield

Because cranberry is largely a processed crop, yield is a trait of major commercial consideration. Fruit yield is a complex trait which reflects the outcome of numerous genetic (e.g., varietal) and environmental factors (Roper and Vorsa 1997). Plant parameters include vegetative vigor and biomass, upright density, inflorescence bud set, flower number/upright, gametic fertility, fruit set/upright, berry weight, and seed number. Roberts and Struckmeyer (1942) found that upright density and upright length were also correlated to crop yield in cranberries.

The majority of fruit and vegetative traits of economic importance appear to follow quantitative inheritance, including yield. Accurate estimates of yield and yield components, for parental selection, would facilitate cranberry breeding efforts. Typically, yield is estimated by harvesting all fruit in a representative unit area, usually a square foot. The fruit weight, in grams per square foot (multiplied by 0.958) translates to a barrel/acre estimate. An 8-year yield trial with ten cultivars illustrates the difficulty in assessing yield differences (Fig. 6.3). Cultivar by year interaction effects were significant for yield, fruit set, and berry weight, indicating yield potential should be assessed over multiple years. Fruit set (berry/unit area) accounted for more of the variation in yield than did berry weight (g/berry) in this trial. Berry weight variation is evident

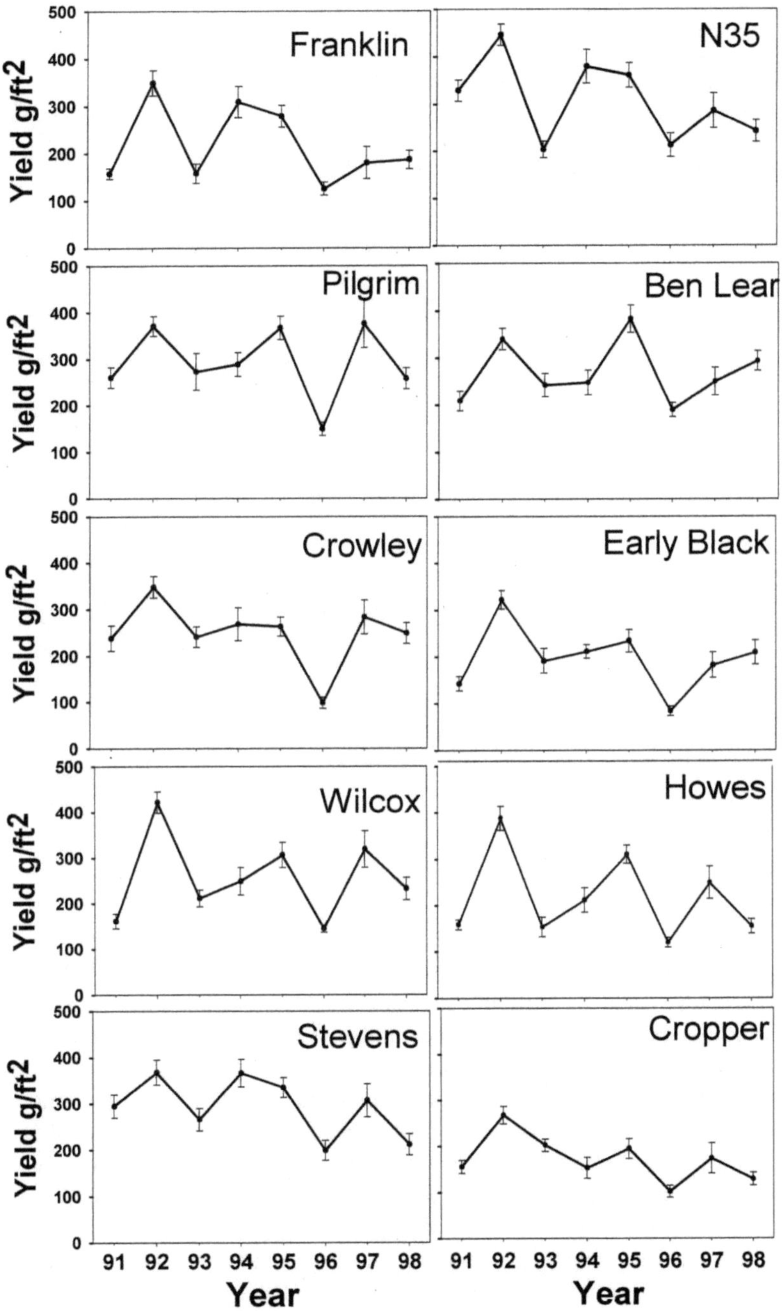

Fig. 6.3 Mean yield performance of ten cranberry cultivars over an 8-year period; 1993 and 1996 were low yield years, with recovery differing among cultivars. *Bars* represent standard error of the mean

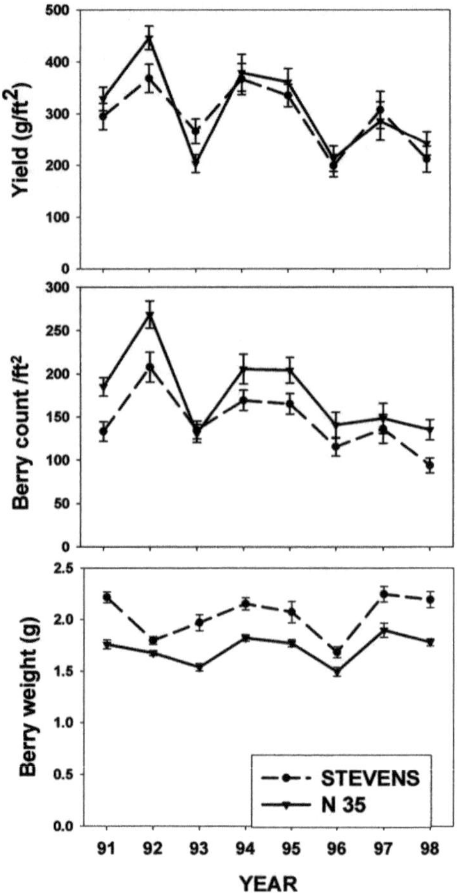

Fig. 6.4 Yield, berry count per sq. ft., and mean berry weight of Stevens and No. 35 over an eight year period. *Bars* represent standard error of the mean

between cultivars. Two of the highest yielding cultivars in this trial, No. 35 and Stevens, achieve yield by different means (Fig. 6.4), Stevens having greater berry weight and No. 35 having higher berry number per unit area. The cultivars released from the first breeding and selection cycle, Stevens and Pilgrim, have mean berry weights greater than the parents. Variation for yield components existed among varieties tested, indicating genetic gain is possible for yield with additional breeding efforts. In particular, greater fruit set should be emphasized as a breeding objective.

Consistency in yield from year to year is another important consideration. Since fruit load consumes plant resources during the period when inflorescence buds are initiated for the following year's crop, biennial bearing is not an uncommon feature in cranberry production. For example in Fig. 6.3, 1992 was a high crop year, followed by a decline in all cultivars, some more severely than others. The first-generation hybrids developed by the USDA breeding program appeared to improve the year-to-year productivity of cranberry, by being more tolerant of environmental stresses, e.g., high nitrogen environment (Davenport and Vorsa 1999).

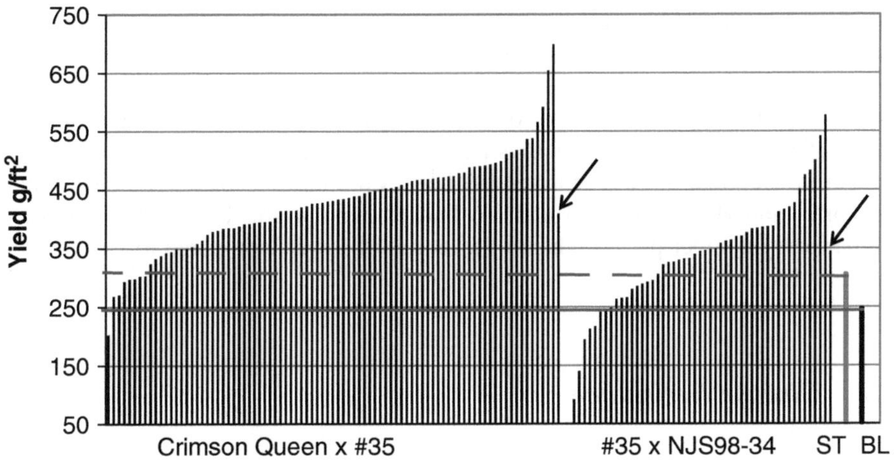

Fig. 6.5 Distribution of mean (2 year) yield of progeny from two populations, Crimson Queen × No. 35 and No. 35 × NJS98-34, as compared to two standards, Stevens (ST) and Ben Lear (BL). *Arrows* represent population means

Since the release of the first hybrids, several additional breeding and selection cycles have been achieved and further genetic gains have been realized. Figure 6.5 presents two populations segregating for yield, relative to the yield of the two current standard cultivars, Stevens and Ben Lear. Progeny yields represent about a four- and tenfold variation for the two populations. Population means are also greater than the standards, indicating population improvement. The variety No. 35, in this example and many others, appears to be a particularly good parent for yield.

Environmental factors impacting yield include plant nutrition, pollination, water relations, and climatic stresses. Environmental effects which impact upright health and physiology (leaf area), carbohydrate movement, and photosynthesis also impact yield (Roper et al. 1992, 1995; Roper and Klueh 1994; Roper 2006). Cranberry is relatively self-fertile and does not require cross-pollination (Sarracino and Vorsa 1991). However, pollination does require bee visitation, for nectar and sometimes pollen, and varietal variation was found in secretion of nectar sugar (Cane and Schiffhauer 1997). In addition, pollinator species may differ in their cue selectivity (Cane and Schiffhauer 2001). The relative attractiveness of varieties to bees is currently unknown.

From an evolutionary perspective, the main purpose of the fruit is for seed dispersal. In cranberry if the ovules are fertilized, the developing embryos and seed stimulate the flower ovary to increase in size and eventually form a mature fruit. Significant varietal variation was found for ovule number in seven varieties evaluated (Sarracino and Vorsa 1991). In controlled crosses, Franklin had the highest mean ovule number per ovary ($n=35$), while Pilgrim the lowest ($n=29$). Developed seed number also varied significantly between the cultivars. Howes and Wilcox had the lowest seed number owing to translocation heterozygosity (Ortiz and Vorsa 1998). Franklin had the highest seed number. Fruit weight was significantly correlated with seed set in six of the eight varieties. In open-pollinated field conditions, similar differences were observed (Elle 1996). In addition, 'Centerville' and 'Stanley' were reported to produce more weight per seed than Early Black and 'Bugle' (Chandler and Demoranville 1958).

Varietal differences have been observed in the number of fruit set per upright. Bain (1933) reported 0.8–0.9 berries per flowering upright for Searles, Howes and McFarlin. Bergman (1950) reported 0.9 berries per upright for Early Black, while McFarlin and Howes had 1.3 and 1.6 berries per upright. Elle (1996) observed higher berry number per upright for Stevens and Howes, compared to Early Black and Franklin.

5.2 Disease Resistance

False-Blossom. In 1929, the USDA embarked on a cranberry breeding program with a major objective to develop cranberry varieties resistant to the phytoplasma disease "false-blossom" (Chandler et al. 1947). This disease devastated the cranberry industry in New Jersey in the early 1900s. Developing resistance to "false-blossom" was directed toward resistance to its vector, the blunt-nosed leafhopper. Early Black and McFarlin were used in many crosses as a source of resistance. Progeny were subjected to "cafeteria" feeding trials; seedlings with the fewest leafhoppers feeding on them, compared to standard varieties, were considered the most resistant. From this breeding program, six varieties were released, with three identified as having greater resistance to false-blossom: Pilgrim, Beckwith and Franklin (see Sect. 5.3, Insect Resistance).

Phytophthora Root Rot. Accessions from native populations and ten cultivars were screened for *Phytophthora* root rot (*Phytophthora* spp.) in greenhouse and field trials. Differences in susceptibility were found, with No. 35 consistently showing better resistance (P. Oudemans, personal communication).

Fruit Rot. Currently, disease resistance work is focused primarily on fungal fruit rots. Over 15 fungal species are known to infect cranberry fruit and incite fruit rot crop loss. Fungal fruit rot species include *Glomerella cingulata* (bitter rot), *Colletotrichum acutatum*, *Phyllosticta vaccinii* (early rot), *Fusicoccum putrefaciens* (end rot), *Phomopsis vaccinii* (viscid rot), *Physalospora vaccinii* (blotch rot), *Allantophomopsis lycopodina* (black rot), and *Coleophoma empetri* (ripe rot) (Oudemans et al. 1998). Most fruit rotting organisms also infect vegetative tissues, providing a source of inoculum. Postharvest fungal rots occur during storage of fresh market cranberries, with black rot causing significant damage. New Jersey growing conditions offer the greatest fruit rot pressure of all North American growing areas. Omission of fungicide application will usually result in total crop loss. To identify potential sources of field fruit rot resistance (FFRR) in cranberry germplasm, fungicide treatments were withheld in 2003 and 2004 on germplasm plots located at the PE Marucci Center, Rutgers University, Chatsworth, NJ (Johnson-Cicalese et al. 2009). The plots were given a visual rating for fruit rot infection, using a 1–5 scale (1 = no rot and 5 = 100% rot). The distribution of FFRR ratings

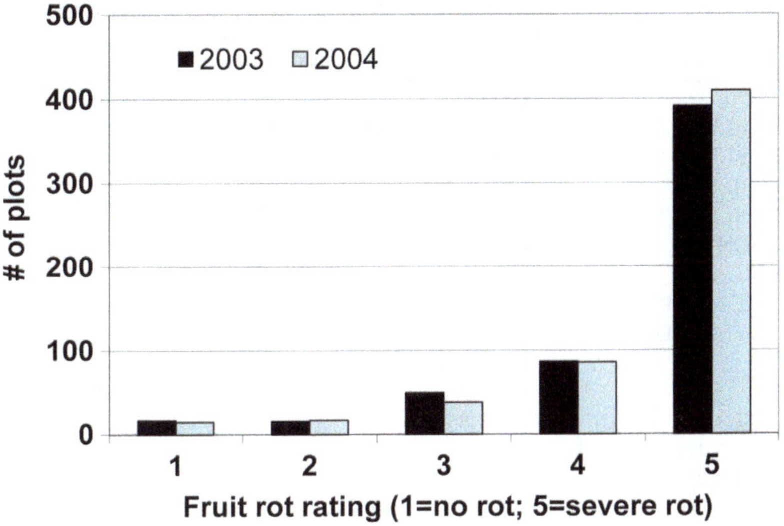

Fig. 6.6 Distribution of fruit rot ratings in a germplasm collection, 22 September 2003 and 7 September 2004; most accessions are highly susceptible. Rating 5 is essentially 100% rot

indicates severe disease pressure, with some selections consistently showing resistance (Fig. 6.6, Table 6.1).

DNA fingerprinting of fruit rot-resistant accessions identified several genetically distinct-types, including 'Holliston-types' (US88-1 and US88-68), 'Budd's Blues' (US88-30), a number of accessions with a Budd's Blues phenotype (US94-176 and US94-161), 'Cumberland' (US88-79), and US89-3 (Fig. 6.7). Budd's Blues had previously been recognized as having fruit rot resistance and is unique because of the heavy waxy bloom on the fruit (A.W. Stretch, personal communication). Unfortunately, it has very poor yield so is not commercially viable. In addition, Budd's Blues progeny are generally not productive, although some can have moderate yields. 'Cumberland' (US88-79), on the other hand, typically has better yields. US89-3 is also of interest to us because of high total phenolics in mature fruit. The genetic diversity found among the 'resistant' varieties suggests potentially different mechanisms of resistance, and might afford the opportunity to make crosses among them to 'pyramid genes' for resistance. In addition, new sources of FFRR are being evaluated. The variety Bugle is considered to have some level of fruit rot resistance (F. Caruso, personal communication).

5.3 *Insect Resistance*

The very first cranberry breeding program was largely directed towards developing varieties with blunt-nosed leafhopper resistance in an effort to reduce false-blossom

Table 6.1 Fruit rot ratings of cultivars and selections in a germplasm evaluation trial planted in 1995 at PE Marucci Center, Rutgers University, Chatsworth, New Jersey

Cultivar or selection	Code[a]	22 September 2003	7 September 2004	Mean
DREVER	US88-1	1.0	1.0	1.0
HAINES BLUES-1	US94-176	1.0	1.0	1.0
HAINES BLUES-2	US94-181	1.0	1.0	1.0
BUDD'S BLUES	US88-30	1.0	1.0	1.0
BUDD'S BLUES-TYPE	US93-34	1.0	1.0	1.0
CHAMPION	US88-116	2.0	1.0	1.5
CUMBERLAND	US88-79	2.0	1.0	1.5
HOLLISTON-TYPE	US88-68	2.0	1.0	1.5
PARADISE MEADOW-1	US88-97	1.0	3.0	2.0
US88-121	US88-121	2.0	2.0	2.0
US89-3	US89-3	2.0	2.0	2.0
PARADISE MEADOW-2	US88-85	3.0	2.0	2.5
GRYGLESKI HYPBRID #3	US94-6	3.0	2.0	2.5
WALES HENRY	US88-67	2.0	3.0	2.5
AR2	US88-43	2.0	3.0	2.5
CUTTS BOG TETPLD B	US94-57	3.0	2.0	2.5
US94-93	US94-93	3.0	2.0	2.5
GEBHARDT'S BEAUTY	US88-115	2.0	3.0	2.5
US94-12	US94-12	2.0	3.0	2.5
HOLLISTON-TYPE	US88-59	3.0	3.0	3.0
GRYGLESKI HYBRID #2	US94-5	3.0	3.0	3.0
PILGRIM LAKE, MASS	NJ91-13-7	3.0	3.0	3.0
WI TETRAPLOID B	US94-67	3.0	3.0	3.0
HOLLISTER RED	US88-70	3.0	4.0	3.5
LEMUNYON		3.8	3.9	3.9
FRANKLIN		4.0	4.0	4.0
WILCOX		4.0	4.5	4.3
#35		4.5	4.5	4.5
PILGRIM		4.0	5.0	4.5
EARLY BLACK		4.5	4.6	4.6
POTTER		4.6	4.6	4.6
STEVENS		4.8	4.5	4.6
BERGMAN		4.3	5.0	4.7
SEARLES		4.8	4.6	4.7
SHAW'S SUCCESS		5.0	4.5	4.8
CROPPER		4.6	5.0	4.8
HOWES		4.8	4.9	4.8
MCFARLIN		5.0	4.8	4.9
AVIATOR		5.0	5.0	5.0
BEN LEAR		–	5.0	5.0
BLACK VEIL		5.0	5.0	5.0
EARLY RICHARD		5.0	5.0	5.0
Mean of 562 accessories		4.5	4.5	4.5

[a]Code is the designation given to each accession when collected in 1988–1994. Cultivars without codes are the means of multiple plots of that cultivar (mean taken only when plots were identical by DNA fingerprinting)

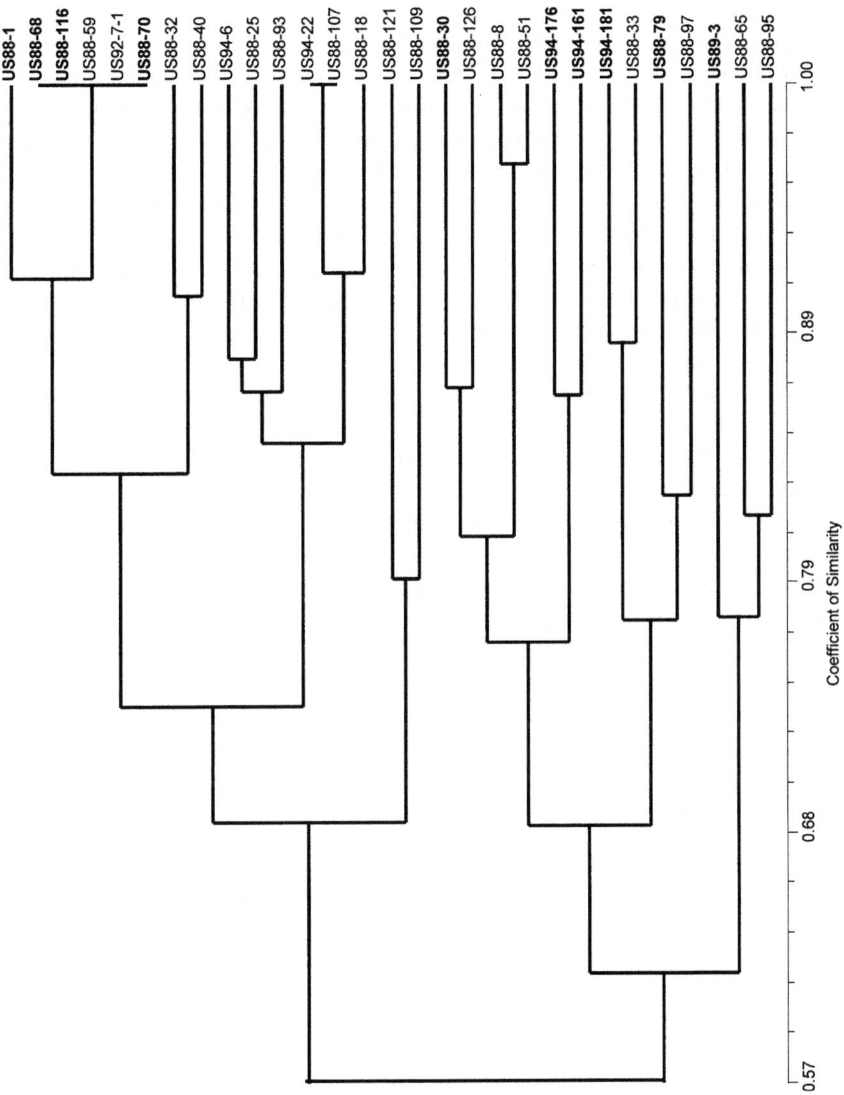

Fig. 6.7 Phenogram based on SCAR markers illustrating the genetic diversity of fruit rot-resistant accessions (resistant accessions in **bold**, see Table 6.1)

disease (Chandler et al. 1947). Based on field observations and "feeding preference" tests, Wilcox and Beckwith (1933) reported Early Black and McFarlin to be less preferable to blunt-nosed leafhopper than Howes. Wilcox (1951) identified the cultivars Early Black, 'Plum,' McFarlin, and 'Shaw's Success' as being most resistant to feeding by the blunt-nosed leafhopper. Other varieties identified as having blunt-nosed leafhopper resistance are Bergman, Franklin, Pilgrim and Wilcox (Dana 1983).

Early Black was also reported to be less susceptible to black-headed fireworm than Howes and 'Smalley Howes,' as well as tipworm (Franklin 1948, 1950). More recently, Neto et al. (2010) reported that gypsy moth larvae (*Lymantriadispar*) exhibited a significant feeding preference for Howes over Early Black. For red-headed flea beetle adults (*Systena frontalis*), Early Black had significantly less feeding damage than Howes ($P<0.053$), whereas cranberry weevil (*Anthonomus musculus*) feeding damage was similar between these cultivars. Resistance may be associated with phenolic content. Phenolic concentration was significantly greater in Early Black than Howes on one of three sampling dates during the growing season. While Early Black appears to be relatively resistant to foliage feeding insects such as blunt-nosed leafhopper, tipworm and black-headed fireworm, Early Black was reported to be susceptible to cranberry fruitworm, along with varieties 'Black Veil' and 'Pride' (M. Dana, unpublished manuscript).

The chemical defenses of five cranberry varieties were examined by Rodriguez-Saona et al. (2011). Significant differences in gypsy moth (*Lymantria dispar*) performance were found among the five varieties, as well as differences in levels of leaf phenolic compounds, although resistance did not correlate with the phenolics measured. Indirect defenses were measured by assaying induced leaf volatile emissions; gypsy moth feeding increased sesquiterpenes in three of the five varieties. Selection for desirable horticultural attributes such as yield and early season may be associated with predisposition to insect susceptibility (Rodriguez-Saona et al. 2011).

5.4 Fruit Quality Traits

Fruit characteristics currently measured by the cranberry processing industry include: percent fruit rot (discussed under Sect. 5.2, Disease Resistance) and unusable fruit, season of harvest, total anthocyanin content (TAcy), soluble solids (Brix), and titratable acidity (TA). In recent years, the flavonoid content of cranberry has received considerable attention in relation to human health benefits. The majority of the focus has been on the three most abundant flavonoid classes, anthocyanins, flavonols, and particularly proanthocyanidins. For the fresh fruit industry, storage life is a major consideration. Fruit appearance is also important, with moderately dark, even coloration being desirable.

Brix and TA. Brix is measured as percent soluble solids using a refractometer (Sapers et al. 1983). The two principal sugars in cranberry are glucose and fructose. TA, expressed as milliequivalents of citric acid, is determined by titrating to a pH 8.1 endpoint with 0.1 N NaOH. The organic acids contributing to TA in cranberry are quinic, citric, and malic acids; each acid occurs at about 1% levels in fruit

(Coppola et al. 1978). The ratio of acid-to-sugar is an important consideration in commercial juice production. Significant genetic and environmental variability exists in cranberry for sugar and acid levels. Schmid (1977) investigated total acids, sugars, vitamin C, and benzoic acid in 12 cultivars of American cranberry in Germany and reported varietal variation for total sugars and benzoic acid. However, year-to-year variation was apparent for sugars.

Anthocyanins. TAcy, milligram anthocyanin per 100 g fresh fruit, is generally measured spectrophotometrically (Sapers and Hargrave 1987; Sapers et al. 1983). The traditional cultivars for early season and good color are Early Black, Ben Lear, Franklin and Bergman. Recently released high color varieties include HyRed, NJS98-23 (Crimson Queen® variety), and NJS98-35 (Demoranville® variety). Anthocyanins in cranberry occur largely in the fruit epidermis, so there is a negative relationship between fruit size and anthocyanin content (Vorsa and Welker 1985). Color development begins when the seed reach maturity, when apparently hormone production subsides. Tacy typically increases over the harvest season in all varieties. However, cultivar by year interaction is also apparent for color development, indicating genetic variation in response to various environmental effects.

The anthocyanin profile of American cranberry fruit has six anthocyanins, composed largely of 3-*O*-galactosides and 3-*O*-arabinosides, and to a lesser amount, 3-*O*-glucosides of the aglycones cyanidin and peonidin. Negative relationships exist between the proportions of cyanidin versus peonidin, and arabinosides versus glucosides, and galactosides versus arabinosides/glucosides (Vorsa et al. 2003). The majority of the varietal variation in profiles arises from cyanidin versus peonidin proportions, with cyanidin to peonidin ratios ranging from 3.6:1 to 0.5:1. Variation for glycosylation profiles is also present, with galactoside proportions ranging from 64 to 75%, arabinoside proportions ranging from 20 to 33%, and glucoside proportions ranging from 3 to 9%. Evidence for both significant qualitative and quantitative genetic variation exists for the methoxylation of cyanidin to peonidin. Significant quantitative genetic variation is also apparent for glycosylation within *V. macrocarpon*.

Qualitative alteration of anthocyanin glycosylation is also possible. The diploid *V. oxycoccus* produces largely glucosides of cyanidin and peonidin. Segregation of anthocyanin glycosylation in *V. macrocarpon* × *V. oxycoccus* hybrids, i.e., *V. macrocarpon* phenotype (galactosides and arabinosides) versus *V. oxycoccus* phenotype (>95% glucosides), is consistent with single locus codominant inheritance (Vorsa and Polashock 2005).

Proanthocyanidins and flavonols. Cranberry proanthocyanidins occur primarily as polymers of epicatechin and are classified as A-type, where two epicatechin units are linked by a double linkage (Foo et al. 2000). Quantification of proanthocyanidins in cranberry is of interest due to their potential health benefits, particularly urinary tract health (Vorsa et al. 2002; Foo et al. 2000). Total proanthocyanidins can be quantified using two different spectrophotometric assays. Initially we used a vanillin-sulphuric acid assay which reads at 490–520 nm wavelengths (wavelengths in the red spectrum). One problem with this assay is that cranberry anthocyanins (with an absorbance of 560 nm), interfere with readings and need to be removed through column chromatography, adding time and expense to the assay. Another spectrophotometric assay now available utilizes 4-dimethylaminocinnamaldehyde (DAC) as a reagent (McMurrough and McDowell 1978), with absorbance determined at the 640 nm wavelength. We also have established that cranberry anthocyanins and flavonols react minimally with the DAC reagent suggesting that removal of these constituents is not necessary (Vorsa and Johnson-Cicalese 2005). HPLC methods have also been developed for evaluating individual proanthocyanidins (Singh et al. 2009).

A survey of cranberry germplasm and breeding populations found a sixfold variation in proanthocyanidin content (Vorsa and Johnson-Cicalese 2005). In a comparison of two widely grown cranberry cultivars, Stevens had higher proanthocyanidin concentrations than Ben Lear, over both the fruit growth phase and during fruit ripening (Vvedenskaya and Vorsa 2004).

Flavonols in cranberry occur primarily as quercetin glycosides, and quantification methods have been developed using HPLC-PDA analysis (Vvedenskaya and Vorsa 2004; Singh et al. 2009). Limited screening of cranberry accessions has found significant genetic variation in flavonol content, but it appears to be less variable than for proanthocyanidins or anthocyanins.

6 Breeding Methods and Techniques

Cranberry is highly self-fertile (Sarracino and Vorsa 1991), necessitating emasculation 3–5 days prior to anthesis. Pollen sheds with minimal agitation through terminal poricidal openings. Reports on stigma receptivity differ. Stigma of the variety 'Stevens' appears to be receptive from anthesis through petal drop (Rigby and Dana 1972), whereas Bain (1933) reported receptivity occurs 2–3 days postanthesis. Roberts and Struckmeyer (1942) reported receptivity occurs when style reaches the length of the anther whorl. Pollen is easily collected from flowers by holding flower between thumb and forefinger, rolling and gently squeezing the flower. Deep well microscope slides, with the depression covered with a tape secured cover-slip, provide a convenient vessel to collect and store pollen. Pollen can be stored at 2–4°C for up to year or more. Pollen can be applied to stigma by dipping the eraser end of a pencil into the pollen and gently touching the stigma 2–5 days post-emasculation. Or pollen can be collected on any smooth surface, such as a metal spatula, and transferred directly. Seeds can be harvested once fruit is ripe, generally 1–2 weeks following color development. Seeds should be maintained in moist conditions at 1–4°C for 2–3 months, and then sown on a moist acidic surface. Seeds can also be dried and stored refrigerated until sowing, although viability may be reduced. Milled sphagnum peat moss is optimal media for seed germination, providing fungistatic properties. Cranberry seedlings can be transplanted to potted culture in peat/sand 1/1 (v/v) media. Irrigation should be done with neutral to low pH water. Seedlings can be field planted directly, or propagated through cuttings to establish 1.5 m × 1.5 m plots of 24 plants. Plots will require 2–3 years to be fully colonized. Seedling beds are maintained under similar regimen as commercial beds with winter flood, fertilization, irrigation and pesticide schedule.

Cranberry requires hymenopteran pollinators, and most commercial beds utilize honeybees at about 1–2 hives/acre. Some growers will utilize up to five hives/acre. In the northern hemisphere, cranberry pollination usually begins in early to mid June and completes by early to mid July. Majority of fruit growth is completed by mid September. Fruit evaluation is initiated in late summer, identifying early ripening progeny. Plots are typically rated for yield, fruit rot, vegetative cover, "runnering" (stolon production), upright density, vegetative diseases, fruit traits such as size, color, etc. Yield is usually estimated by harvesting fruit from square foot samples, where grams per square fruit translates approximately to barrels/acre, the standard commercial parameter for cranberry production. Fruit traits of economic importance include total anthocyanins (TAcy), percent soluble solids (Brix), titratable acidity (citric acid equivalents). TAcy is measured in mg/100 g fruit fresh weight by water extraction, filtration and absorbance at 520 nm (Vorsa et al. 2003).

Cranberry is subjected to both disease and insect pressure, and seedling and variety plots are evaluated for disease, insect and abiotic stresses. Cranberry diseases of economic impact (Caruso et al. 2000) include vegetative diseases, *Phytophthora* root rot (Oudemans 1999; Caruso and Wilcox 1990; Jeffers 1988), false-blossom (a phytoplasma), upright dieback, and field and storage fruit rots (Oudemans et al. 1998). Insect stresses include foliage feeders such as cutworms (Lepidoptera: Noctuidae), spanworms (Lepidoptera: Geometridae), fireworms (Lepidoptera: Tortricidae), and a fleabeetle (Coleoptera: Chrysomelidae) (Averill and Sylvia 1998). Fruit feeding insects include Sparganothis fruitworm, cranberry fruitworm, and cranberry blossomworm. In certain growing areas, cranberry tipworm (*Dasineura oxycoccana*), cranberry weevil (*Anthonomus musculus*), and cranberry girdler (*Chrysoteuchia topiaria*) also cause damage.

6.1 Interspecific Crosses

Generally, homoploid interspecific crosses within a *Vaccinium* section result in fertile or partially fertile offspring. Previously attempts were made to cross a native North American tetraploid *V. oxycoccus* L. cranberry with *V. macrocarpon* (Kust 1965). Since polyploids generally have larger organ structure, the objective was to develop larger fruited varieties, along with increased color. Because of the heteroploid nature of the cross, American cranberry tetraploid clones were developed by treating with colchicine, developing periclinal chimeral tissues and eventually recovering fully tetraploid clones (Bain and Dermen 1944; Derman 1947). However, the first-generation interspecific hybrids, although relatively vegetatively vigorous, were less hardy and grew more slowly than diploids. Additionally, the fertility (seed set) was reduced, possibly by numerically unbalanced chromosome segregation during meiotic anaphase.

Diploid *V. oxycoccus* crosses readily with *V. macrocarpon*, bilaterally. Hybrids between *V. macrocarpon* and diploid *V. oxycoccus*, in either species' cytoplasm, are vigorous and produce abundant flowers. Fertility of hybrids is somewhat variable, but some are quite fertile. Backcrosses to either parental species as well as F_2 populations are readily produced. Traits that diploid *V. oxycoccus* offers *V. macrocarpon* include early flowering and phenology, unique anthocyanin glycosylation profile (Vorsa and Polashock 2005) (see Sect. 5.4 Fruit Quality Traits *Anthocyanins*), frost tolerance, and adaptation to more polar latitudes. *V. oxycoccus* will produce a "second bloom" in July in New Jersey, which is undesirable.

6.2 Intersectional Hybridization

Intersectional hybrids have been produced with *V. macrocarpon*. These include the following: *V. macrocarpon* × *V. vitis-idaea*, sect. *Vitis idaea* (Zeldin and McCown 1997); *V. macrocarpon* × *V. crassifolium*, sect. *Batodendron*; *V. macrocarpon* × *V. reticulatum*, sect. *Macropelma* (Zeldin and McCown 1997); and *V. macrocarpon/oxycoccus* × *V. darrowii*, sect. *Cyanococcus* (Vorsa et al. 2009). Most intersectional hybrids are highly sterile and have not allowed for advanced generations. The *V. macrocarpon/oxycoccus* × *V. darrowii*, sect. *Cyanococcus* hybrid, although sterile in backcrosses to cranberry, has yielded a few seedlings in crosses to tetraploid *V. corymbosum*.

6.3 Inheritance

Heritability. In cranberry, varietal or broad sense heritability for yield, season of harvest, fruit color, Tacy, Brix, and berry size (weight) is apparent when contrasting the broadest range of varietal variation. Most traits of horticultural and economic importance are quantitatively inherited and have a significant environmental variance component. The weather during a given growing season, year, growing region, soils, horticultural management, etc., all have an effect on these traits. Genetic gain for various traits has been realized after one breeding and selection cycle, such as larger fruit size, e.g., Stevens and Pilgrim, relative to the native selections. Season of ripening and fruit color differences are also obvious. However, differences between varieties quantitatively closer in phenotype will not necessarily be consistent across years. For example, the differences in fruit weight between Stevens and Pilgrim are observed some years, and not others, and may differ between growing regions.

Due to the need for considerable field space to assess traits of economic significance, e.g., yield, experiments specifically for obtaining heritability estimates are lacking. In one Rutgers breeding project, 16 crosses were replicated in the field, along with parental plots to provide a mid-parent to progeny mean regression value. The progeny from a five-parent diallel crossing scheme (ten crosses with six reciprocal crosses, using the cultivars Ben Lear, Franklin, Pilgrim, Stevens and Wilcox), were represented in two replicate groups and planted in 1.5 m × 1.5 m

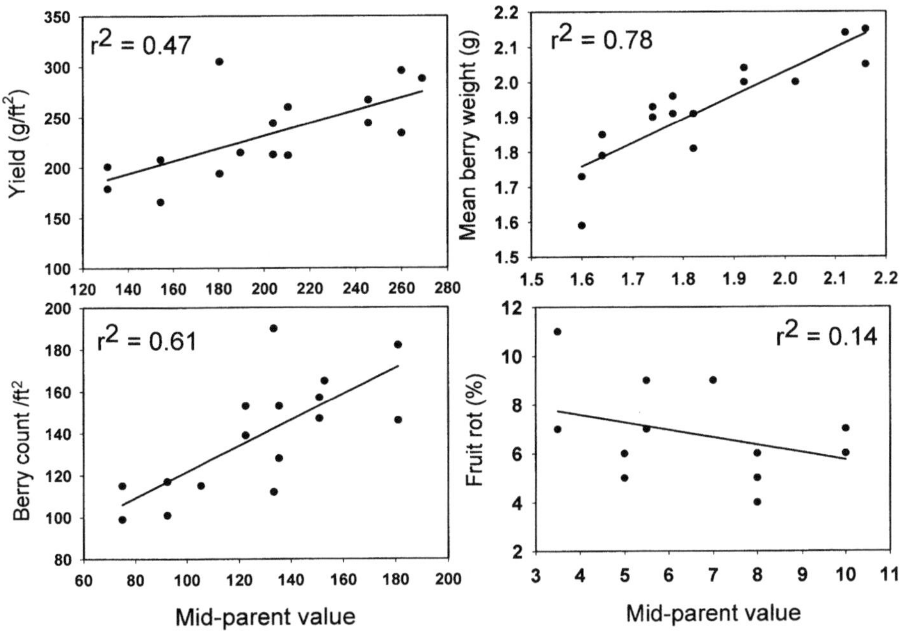

Fig. 6.8 Heritability estimates as determined by mid-parent-mean progeny regression for yield, berry count, berry weight, and fruit rot for year 2000 data

squares. Mid-parent-progeny mean regression of 16 crosses allowed us to estimate heritability for yield, berry count, berry weight, fruit rot, TAcy, Brix, and proanthocyanidins over 3 years, TA over 2 years. In 2000, heritability estimates for yield, berry count, berry weight and fruit rot were 0.47, 0.61, 0.78, and 0.14, respectively (Fig. 6.8). Over the 3 years, heritability for yield ranged from 0.29 to 0.47, berry count 0.53–0.61, and berry weight 0.73–0.92. Additive genetic variance appears to be significant, and genetic gain for these traits would be predicted in future breeding and selection cycles.

Heritability for Brix and TA was variable across years. Brix heritability ranged from 0.05 to 0.51, and TA heritability ranged from 0 to 0.34.

Heritability of anthocyanin content was fairly high and consistent, ranging from $r^2 = 0.61–0.80$, and was relatively consistent from year to year, whereas for proanthocyanidin content, heritability was variable across years ranging from 0 (1998) to 0.42 (1999) (Fig. 6.9) (Vorsa and Johnson-Cicalese 2005). Parental proanthocyanidin values were lowest for Ben Lear, Pilgrim and Stevens, and Franklin and Wilcox represented cultivars with higher proanthocyandin content (Fig. 6.10). TAcy and PAC levels are negatively correlated to fruit size, and factors affecting fruit size would contribute to reduced heritability. Furthermore environmental stresses such as drought, heat stress, insect and disease incidence also contribute to large environmental variance and low heritability. Transgressive segregation for certain traits,

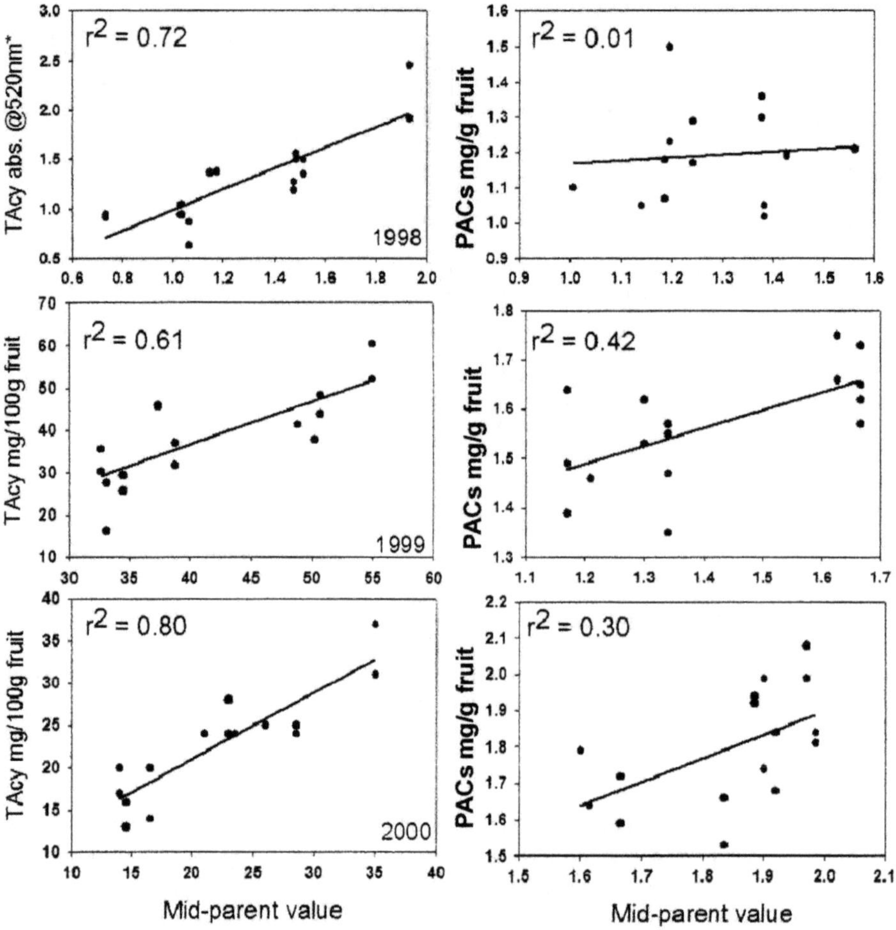

Fig. 6.9 Heritability estimates as determined by mid-parent-mean progeny regression over 3 years (1998, 1999, and 2000) for TAcy (*left column*) and proanthocyanidins (PAC; *right column*)

such as TAcy, is also apparent. Progeny exhibiting both lower and higher Tacy than the parental range have been observed.

Little is known regarding the inheritance of insect resistance. Wilcox (1951) stated "…that in hybridization, varieties of known susceptibility to vector attack tend to contribute their respective susceptibility to the progeny." Resistance to leafhopper apparently was neither recessive nor dominant, suggesting largely additive variance. When five related varieties were evaluated for gypsy moth performance and chemical defenses, differences in resistance were found. The relatively

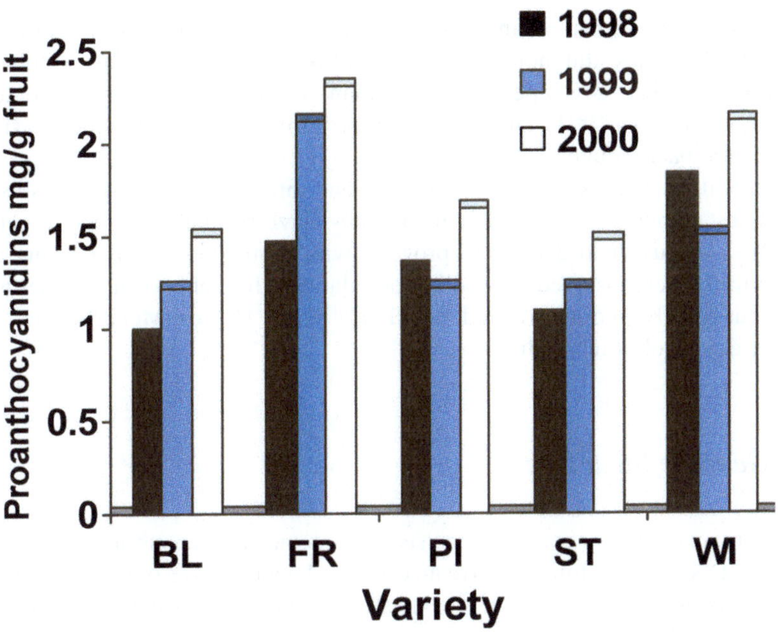

Fig. 6.10 Proanthocyanidin (PAC) content for five cultivars Ben Lear (BL), Franklin (FR), Pilgrim (PI), Stevens (ST) and Wilcox (WI) over 3 years. Values provided for the mid-parent values for heritability estimates in Fig. 6.9

susceptible variety, NJS98-23, was derived from a cross between a susceptible parent, Ben Lear, and a resistant parent, Stevens (Rodriguez-Saona et al. 2011)

Qualitative Inheritance. A few traits have been discovered that segregate consistent with Mendelian inheritance. 'Yellow Bell,' a wild clone discovered in Maine, has fruit lacking anthocyanins. Open pollinated seed from Yellow Bell segregated for red (21 progeny) and yellow (three progeny) fruited progeny, and crosses of Yellow Bell with red fruited cranberry gave all red fruited progeny, indicating the trait is recessive and under the control of one or few loci (N. Vorsa and J. Johnson-Cicalese unpublished data). Another variant found in a commercial bed in Massachusetts, referred to as 'Murphy's Green,' lacks anthocyanin development in foliage, stems, flower pedicels, and stamen, and segregates as a single locus recessive trait (N. Vorsa and J. Johnson-Cicalese unpublished data).

6.4 Breeding System

Although cranberry flower development is protandrous, cranberry is highly self-fertile. We have developed a seventh generation selfing line of the cultivar Ben Lear, sixth generation selfing lines of Stevens and Pilgrim, and fifth generation selfing line of Wilcox. However, selfing lines (after the second generation selfing) of the cultivar Early Black did not thrive. In addition, cross-pollinations exhibited a significantly higher developed seed set than self-pollinations in ten cultivars (Sarracino and Vorsa 1991), suggesting at least a low level of genetic load exists in some genetic backgrounds.

In general, backcrosses to a noninbred parent, intercrosses among half-sibs, and intercrosses among full-sibs result in progeny having relatively good vegetative vigor. No obvious yield decline in progeny with inbreeding levels ranging up to $F=0.25$ have been observed in seedling populations. It is unknown whether native clones that have been domesticated, e.g., Early Black, Howes, McFarlin, Ben Lear, etc. have any level of inbreeding.

6.5 Genome Structure

The American cranberry genome consists of 12 metacentric or submetacentric chromosomes. DNA flow cytometry gave an estimated genome size of 608 Mbp/haploid genome (Costich et al. 1993), but recent estimates place the size closer to 568 Mpb (see Sect. 7.3 Molecular Tools).

There appears to be two configurations, i.e., two genetic maps, for the cranberry genome. The cultivar Howes, and a progeny of Howes, Wilcox, were identified to be translocation heterozygotes (Ortiz and Vorsa 1998).[2] Pollen tetrad analysis of these translocation heterozygotes was used to study the coorientation of centromeres during meiosis 1, and centromere orientation in relation to frequency of interstitial chiasma (Ortiz and Vorsa 1998). Translocations may offer an advantage by maintaining heterozygosity, or a block of genes as a linkat, in a self-pollinated crop. Pollen from Wilcox and Howes were crossed to normal cultivars, and a Wilcox × Howes cross was analyzed to study transmission of the translocated configuration (Ortiz and Vorsa 2004). When crossed to normal varieties, Wilcox and Howes gave ratios of 71 translocated: 31 normal, and 79 translocated: 37 normal, respectively. Segregation deviated from the expected one translocated: one normal progeny ratio, but fit either a 3:1 or 2:1 ratio. The altered segregations may indicate the presence of a balanced lethal system located in the translocated segments of both Howes and Wilcox. Sterile individuals were found in the progeny of Wilcox × Howes, which could indicate that the two parents have nonidentical translocations. The translocated progeny of both cultivars had a normal distribution for pollen stainability, which indicated that both the occurrence of crossing over in the interstitial region and the segregation of chromosomes are under polygenic control.

[2] Individuals heterozygous for a translocation exhibit reduced gametic viability due to recombination within the interstitial region followed by chromosome segregation which results in generation of genetically unbalanced gametic constitutions (Burnham 1984).

7 Integration of Biotechnology

7.1 Tissue Culture

Tissue culture methods for micropropagation and genetic engineering of cranberry have been developed, and are reviewed in McCown and Zeldin (2005). Tissue culture provides for a method which enables embryo rescue, and could be useful in cases where endosperm breakdown occurs, e.g., heteroploid crosses with unbalanced endosperm balance number. Developing seed can be removed 8 weeks after pollination, and embryos are excised by slicing off the radical end and squeezing out the embryo (McCown and Zeldin 2005). The embryos are placed on a hormone-free medium since even low levels of cytokinin will stimulate callus formation.

7.2 Genetic Transformation

Genetic transformation of cranberry is possible with particle bombardment (Serres et al. 1997). Genes that have been used in other plants, such as herbicide tolerance and *Bt* endotoxin, could be of potential commercial value to cranberry (McCown and Zeldin 2005). Using particle bombardment of stem sections with a construct that contained genes for GUS, NPTII and *Bacillus thuringiensis Bt* endotoxin, Serres et al. (1992) successfully developed the first transgenic cranberry. Although successful, this system generated a low frequency (0.15%) of recovered transclones. However, *Bt* transclones did not consistently deter feeding by the black-headed fireworm (McCown and Zeldin 2005). Transformants with the *Bar* gene, which provides resistance to the herbicide L-Phosphinothricin were also developed by McCown and Zeldin (2005). The *Bar* gene, derived from *Streptomyces hygroscopius*, encodes an enzyme which inactivates the herbicide. Resistance was sexually transmissible, and some progeny exhibited greater resistance than the original transformed plants. *Agrobacterium*-mediated transformation has been problematic (McCown and Zeldin 2005; Polashock and Vorsa 2002a).

The utilization of plants derived from genetic engineering methods for cranberry crop improvement, however, faces acceptance and environmental obstacles. The commercialization of an herbicide-tolerant cranberry has been hampered by issues of acceptance, and thus was not pursued (McCown and Zeldin 2005). While the integration of herbicide resistance into cranberry does not appear to have any obvious environmental risks, other genes, e.g., *Bt*, may pose environmental risks. Cranberry has a host of insect pests within the Order *Lepidoptera*. Thus, engineering cranberries to express the *Bt* gene obviously would be most useful for cranberry insect management. However, the 'Bog Copper' or 'Cranberry-Bog Copper' (*Lycaena epixanthe* (=*Epidemia epixanthe*)) is a North American butterfly in the family *Lycaenidae* whose adults feed almost exclusively on cranberry nectar, and thus, generally spend their entire lives within the area of a single acid bog

(Cech 2005). The distribution range of *L. epixanthe* is largely that of cranberry, from Maine south to New Jersey and West Virginia, and in the west through northern Great Lake states and Ontario. There are many commercial cranberry beds adjacent to native cranberry populations. It is conceivable, and probable, that gene flow from commercial to native populations occurs at some level. In contrast to crops such as maize, where sexually compatible wild populations are usually lacking, gene flow between domesticated and wild cranberry will be unimpeded.

7.3 Molecular Tools

In contrast to maize and other major crops, genetic tools are limited in cranberry. Randomly amplified polymorphic DNA (RAPDs) were the first markers developed and have been useful in initial genotyping and cultivar identification, as well as determining off-types in cranberry beds (Novy et al. 1994, 1996; Novy and Vorsa 1996; Debnath 2006). However, due to difficulties inherent with RAPDs, Sequence Characterized Amplified Region (SCAR) markers were subsequently developed from RAPDs. Nine SCAR primer sets, for use in two multiplex PCR reactions, currently provide for the genotyping of cranberry (Polashock and Vorsa 2002b). Heteroduplex formation with RAPDs for some markers generates additional characteristic bands from heterozygous individuals, facilitating their detection (Novy and Vorsa 1996). Although the SCAR markers currently in use for genotyping have been useful for practical considerations, their quantity is insufficient for high density genetic mapping.

Currently, there are two laboratories developing molecular markers for cranberry genetic studies and enhancement; the NJAES, Rutgers University, Chatsworth, NJ program (N. Vorsa); and the USDA-ARS Cranberry Breeding, Genetics, and Genomics program in Madison, Wisconsin (J. Zalapa, initiated in 2010). Industry stakeholders have realized the value of genomic data, and support activities towards the development of fundamental genomic resources for cranberry, as well as for the *Vaccinium* and *Ericaceae* research communities. Cranberry genetics and breeding are hampered by a long generation interval, an especially long interval in assessing yield (7–8 years), a long chilling requirement, and large field space requirement to assess agronomic traits. In addition, few populations are available that are optimal for genetic studies, which relegates genetic analysis largely to populations developed for breeding objectives. However, working with cranberry does offer some advantages, it is a long-lived perennial, easy to propagate, self-compatible, and diploid.

Microsatellite markers (also called simple-sequence repeats or SSRs) offer a number of desirable characteristics, including reproducibility, abundance in the genome, high levels of polymorphism, codominance, and transferability among crosses and also between related species [see e.g., Morgante and Olivieri (1993) and Varshney et al. (2005)]. Cranberry microsatellite markers are therefore being developed in both programs. The USDA (J. Zalapa) program is also developing Single

Nucleotide Polymorphic (SNP) markers which are present in even higher frequency in the genome and can be identified in nearly any gene of interest, although the number of possible alleles is obviously more limited.

The NJAES-Rutgers University program has exploited the interspecies transferability of microsatellites using 29 blueberry microsatellite primer pairs (Boches et al. 2005; Bassil et al. 2009; Bassil personal communication; Rowland et al. 2003) to begin genotyping a cranberry mapping population with the goal of identifying fruit rot resistance markers. In addition, next-generation sequencing technology has provided an abundance of cranberry sequence data from a fifth-generation selfed inbred cranberry derived from the cultivar Ben Lear. 'Mining' of this assembled sequence for microsatellite markers is now underway. The first draft assembly was based on mate-paired SOLiD 3 plus (Applied Biosystems) sequence. It presently comprises 68,498 scaffolds with a total length of 568 Mbp, which is close to the genome size based on flow cytometry; however, the assembled length includes gaps totaling 258 Mbp (Georgi et al. 2011). Since the genome was sequenced at approximately 66× coverage, these gaps are most likely due to difficulties assigning short sequence reads to unique locations in the genome (Miller et al. 2010). Since microsatellites are simple sequence repeats that are highly abundant throughout the genome, it is not unexpected for the assembly to frequently contain gaps next to microsatellites. Thus, mate pair information is vital for identifying unique flanking sequence. Numerous cranberry microsatellite fragments have been successfully amplified using primers designed from the assembled sequence, providing evidence that the assembly is basically accurate. Once the microsatellite markers have been placed on a genetic map, it will provide information about the relative positions of the corresponding sequence scaffolds in the genome. Initially, the largest scaffolds were mined for microsatellites, but more recently, the focus has shifted to scaffolds that contain particular sequences of interest, such as genes involved in responses to necrotrophic pathogens (Laluk and Megiste 2010) and flavonoid biosynthesis (Winkel-Shirley 2001; Koes et al. 2005; Jaakola et al. 2010). To the extent that the assembly is accurate, mapping microsatellites from scaffolds containing these genes will also provide the presumptive genetic map locations of the genes.

The primary goal of the USDA-ARS program in Madison, Wisconsin is the development of genetic tools to aid cranberry genetic improvement, utilizing the latest high-throughput DNA sequencing technologies (J. Zalapa, personal communication). This program is using Roche 454 pyrosequencing to generate ~620 Mb of sequence per run in read lengths reaching 400–500 bp, with the expectation of isolating sequences containing molecular markers such as usable microsatellite repeats. Unique cranberry genotypes will be sequenced to identify SNPs and to develop transcriptome profiles of cultivars with superior trait characteristics for functional and comparative genomic studies. Genomic information generated will be immediately useful for the genetic characterization of cultivated and undomesticated germplasm, polymorphism screening for linkage map development, and the discovery of complementary gene pools in cranberry for controlled crosses.

The ultimate goal of both programs is the development of genetic and physical maps of cranberry along with the QTL and association mapping information that will be essential for marker-assisted selection (MAS), comparative genomics, and positional gene cloning and identification of genes for superior productivity, improved environmental adaptation, enhanced fruit quality traits, and increased disease and insect resistance. Ideally, the USDA-ARS program will facilitate and improve connections between existing germplasm resources (e.g., National Clonal Germplasm Repository), breeding programs (e.g., Wisconsin and New Jersey), and genomic projects, to create an integrated effort towards the development of enhanced cranberry germplasm and cultivars. Collaboration with other genetics and breeding programs nationally and internationally should enable a comprehensive sequencing of the cranberry genome. The development of genomic resources in cranberry will provide for innovative plant breeding systems that will reduce the time and field space required and facilitate the breeding of unique superior cranberry cultivars to meet the current and future challenges of this important American crop.

Acknowledgments The authors thank Laura Georgi and Juan Zalapa for their contribution to the Biotechnology and Molecular Tools section. Funding sources: Ocean Spray Cranberries, Inc.; USDA-NIFA Research Initiative Grant No. 2009-34155-19957; USDA-CSREES SCRI Grant No. 2008-51180-04878.

References

Averill, A.L. and Sylvia, M.M. (1998) Cranberry insects of the northeast: A guide to identification, biology, and management. Gazette Printing, Easthamptom, MA.

Bain, H.F. (1933) Cross pollinating the cranberry. Proc. Wisc. State Cranberry Growers' Assoc. 47,7–11.

Bain H.F. and Dermen, H. (1944) Sectorial polyploidy and phyllotaxy in the cranberry (*Vaccinium macrocarpon* Ait.). Amer. J. Bot. 31,581–587.

Bassil, N., Oda, A., and Hummer, K.E. (2009) Blueberry microsatellite markers identify cranberry cultivars. Acta Horticulturae 810,181–186.

Bergman, H.F. (1950) Cranberry flower and fruit production in Massachusetts. Cranberries 15(4),6–10.

Boches, P.S., Bassil, N.V., and Rowland, L.J. (2005) Microsatellite markers for *Vaccinium* from EST and genomic libraries. Molecular Ecology Notes 5,657–660.

Bruederle, L.P., Hugan, M.S., Dignan, J.M. and Vorsa, N. (1996) Genetic variation in natural populations of the large cranberry, *Vaccinium macrocarpon* Ait. (Ericaceae). Bull. Torrey Bot. Club 123,41–47.

Burnham, CR. (1984) Discussion in Cytogenetics. Burgess Publishing Company, St. Paul, Minnesota.

Camp, W.H. (1944) A preliminary treatment of the biosystematy of Oxycoccus. Bul. Torrey Bot. Club 71,426–437.

Camp, W.H. (1945) The North American blueberries with notes on other groups of Vacciniaceae. Brittonia 5,203–275.

Cane, J.H., Schiffhauer, D., and Kervin, L.J. (1996) Pollination, foraging, and nesting ecology of the leaf-cutting bee Megachile (Delomegachile) addenda (Hymenoptera: Megachilidae) on cranberry beds. Ann. Entomol. Soc. Am. 89(3),361–367.

Cane, J.H. and Schiffhauer, D. (1997) Nectar production of cranberries: genotypic differences and insensitivity to soil fertility. J. Amer. Soc. Hort. Sci. 122,665–667.

Cane, J.H. and Schiffhauer, D. (2001) Pollinator genetics and pollination: do honey bee colonies selected for pollen-hoarding field better pollinators of cranberry Vaccinium macrocarpon? Ecol. Ent. 26,117–123.

Caruso, F.L., Bristow, P.R. and Oudemans, P.V. (2000) Cranberries: The Most Intriguing Native North American Fruit. APSnet Features. Online. doi:10.1094/APSnetFeature-2000-1100.
Caruso, F. L., and Wilcox, W. F. (1990) Phytophthora cinnamomi as a cause of root rot and dieback of cranberry in Massachusetts. Plant Disease. 74,664–667.
Cech, R., and Tudor, G. (2005) Butterflies of the East Coast. Princeton Univ. Press, Princeton, NJ.
Chandler, F.B., Wilcox, R.B., Bain, H.F., Bergman, H.F. and Dermen H. (1947) Cranberry breeding investigation of the U.S. Dept. of Agriculture. Cranberries 12, 6–9 (May); 12,6–10 (June).
Chandler, F.B., Bain, H.F., and Bergman, H.F. (1950). The Beckwith, the Stevens and the Wilcox cranberry varieties. Cranberries 14(11), 6–7 (March).
Chandler, F.B. and Demoranville, I. (1958) Cranberry varieties of North America. Exp. Sta. College of Agric. Univ. Mass. Bull. 513.
Clark, J.R. and Finn, C.E. (2010) Register of New Fruit and Nut Cultivars List 45. HortSci. 45,716–756.
Coppola, E.D., Conrad, E.C., and Cotter, R. (1978) High pressure liquid chromatographic determination of major organic acids in cranberry juice. JAOAC 61,1490–2.
Costich, D.E., Ortiz, R., Meagher, T.R., Bruederle, L.P. and Vorsa, N. (1993) Determination of ploidy level and nuclear DNA content in blueberry by flow cytometry. TAG 86,1001–1006.
Dana, M. N. (1983) Cranberry cultivar list. Fruit Var. J. 37:88–95.
Davenport, R.J. and Vorsa, N. (1999) Cultivar fruiting and vegetative response to nitrogen fertilizer in cranberry. J. Amer. Soc. Hort. Sci. 124,90–93.
Debnath, S.C. (2006) An assessment of the genetic diversity collection within a collection of wild cranberry (*Vaccinium macrocarpon* Ait.) clones with RAPD-PCR. Genetic Resources and Crop Evolution 54,509–517.
Derman, H. (1947) Periclinal cytochimeras and histogenesis. Amer. J. Bot. 34,32–43.
Doughty, C.C. and Garren, R. (1970) 'Crowley' a new early maturing cranberry variety for Washington and Oregon. Fruit Varieties and Horticulture Digest 15,65.
Eck, P. (1990) The American Cranberry. Rutgers University Press. New Brunswick and London.
Elle, E. (1996) Reproductive trade-offs in genetically distinct clones of *Vaccinium macrocarpon*, the American cranberry. Oecologia 107,61–70.
Foo, L.Y., Lu, Y., Howell, A.B. and Vorsa, N. (2000) Structural characterization of A-Type proanthocyanidin trimers from cranberry that inhibit adherence of uropathogenic P-fimbriated Escherichia coli, Phytochemistry 54(2),173–181.
Franklin, H.J. (1948) Cranberry insects in Massachusetts. Mass. Bul. 445, Part I, pp.64.
Franklin, H.J. (1950) Cranberry insects in Massachusetts. Mass. Bul. 445, Parts II–VII, pp.88.
Galletta, G.J. (1975) Blueberries and cranberries. In Advances in Fruit Breeding. Eds. J. Janick and J.N. Moore, Purdue Univ. Press, West Lafayette, pp.154–196.
Georgi, L., Herai, R.H., Vidal, R., Falsarella Carazzolle, M., Guimarães Pereira, G., Polashock, J., and Vorsa, N. (2011) Cranberry microsatellite marker development from assembled next-generation genomic sequence. Molecular Breeding: DOI 10.1007/S11032-011-9613-7.
Jaakola, L., Poole, M., Jones, M.O., Kämäräinen-Karppinen, T., Koskimäki, J.J., Hohtola, A., Häggman, H., Fraser, P.D., Manning, K., King, G.J., Thomson, H., and Seymour, G.B. (2010) A SQUAMOSA MADS box gene involved in the regulation of anthocyanin accumulation in bilberry fruits. Plant Physiology 153,1619–1629.
Jacquemart, A.-L. (1997) *Vaccinium oxycoccos* L. (*Oxycoccus palustris* Pers.*) and Vaccinium microcarpum* (Turcz. ex Rupr.) *Schmalh. (Oxycoccus microcarpus* Turcz. ex Rupr.). J. Ecol. 85,381–396.
Jeffers, S. N. (1988) Phytophthora Species Associated with a Cranberry Decline Syndrome in Wisconsin. Phytopathology. 78,1572.
Johnson-Cicalese, J., Vorsa, N., and Polashock, J. (2009) Breeding for fruit rot resistance in *Vaccinium macrocarpon*. Acta Hort. 810,191–198.
Koes R., Verweij, W., and Quattrocchio, F. (2005) Flavonoids: a colorful model for the regulation and evolution of biochemical pathways. Trends in Plant Science 10,236–242.
Kust, T. (1965) The need for a cranberry breeding program. Univ. Wis. Dept. Hort. Mimeo 627.
Laluk, K., and Megiste, T. (2010) Necrotroph attacks on plants: Wanton destruction or covert extortion? The Arabidopsis Book 8:e0136. doi:10.1199/tab.0136

Mahy, G.; Bruederle, L. P.; Connors, B.; Hofwegen, M. V. and Vorsa, N. (2000) Allozyme evidence for genetic autopolyploidy and high genetic diversity in tetraploid cranberry, *Vaccinium oxycoccus* (Ericaceae). Amer J Bot. 87, 1882–1889.

McCown, B.H., and Zeldin, E.L. (2003) 'HyRed', an early, high fruit color cranberry hybrid. HortScience 38,304–305.

McCown, B.H. and Zeldin, E.L. (2005) *Vaccinium* spp. Cranberry. In: Biotechnology of Fruit and Nut Crops. Ed. R.E. Litz. Biotechnology Series No. 29. CABeBooks. pp.247–261.

McMurrough, I. and McDowell, J. 1978. Chromatographic separation and automated analysis of flavanols. Anal. Biochem. 91,92–100.

Miller, J., Koren, S, and Sutton, G. (2010) Assembly algorithms for next-generation sequencing data. Genomics 95,315–327.

Morgante, M., and Olivieri, A.M. (1993) PCR-amplified microsatellites as markers in plant genetics. Plant Journal 3,175–182.

Neto, C.C., Dao, C.A., Salvas, M.R., Autio, W.R., and Vanden Heuvel, J.E. (2010) Variation in concentration of phenolic acid derivatives and quercetin glycosides in foliage of cranberry that may play a role in pest deterrence. J. Amer. Soc. Hort. Sci. 135,494–500.

Novy, R.G., Kobak, C., Goffreda, J., and Vorsa, N. (1994) RAPDs identify varietal misclassification and regional divergence in cranberry (*Vaccinium macrocarpon* Ait.). Theor. and Appl. Genet. 88(8),1004–1010.

Novy, R.G. and Vorsa, N. (1995) Identification of intracultivar genetic heterogeneity in cranberry using silver-stained RAPDs. HortSci. 30,600–604.

Novy, R.G., Patten, K. and Vorsa, N. (1996) Identifying genotypic heterogeneity in the 'McFarlin' cranberry: A randomly-amplified polymorphic DNA (RAPD) and phenotypic analysis. J. Amer. Soc. Hort. Sci. 121,210–215.

Novy, R.G. and Vorsa, N. (1996) Evidence for RAPD heteroduplex formation in cranberry: Implications for pedigree and genetic relatedness studies and a source of codominant RAPD markers. Theor. and Appl. Genet. 92,840–849.

Ortiz, R. and Vorsa, N. (1998) Tetrad analysis with translocation heterozygotes in cranberry (*Vaccinium macrocarpon* Ait.): Interstitial chiasma and directed segregation of centromeres. Hereditas 129,75–84.

Ortiz, R. and Vorsa, N. (2004) Transmission of a cyclical translocation in two cranberry cultivars. Hereditas 140,81–86.

Oudemans, P. V. (1999) *Phytophthora* species associated with cranberry root rot and surface irrigation water in New Jersey. Plant Disease. 83,251–258.

Oudemans, P. V., Caruso, F. L., and Stretch, A.W. (1998) Cranberry fruit rot in the northeast: A complex disease. Plant Dis. 82,1176–1184.

Polashock, J., and Vorsa, N. (2002a) American Cranberry (*Vaccinium macrocarpon*) Transformation and Regeneration. In: Transgenic Fruit Crops. Eds. Khachatourians, G.G., McHughen, A., Scorza, R., Nip, W.K., and Hui, Y.H. Marcel Dekker Inc., New York, NY.

Polashock, J. and Vorsa, N. (2002b) Development of SCARs for DNA fingerprinting and germplasm analysis of cranberry. J. Amer. Soc. Hort. Sci. 127(4), 677–684.

Ravanko, O. (1990) The taxonomic value of morphological and cytological characteristics in *Oxycoccus* (subgenus of *Vaccinium*, Ericaceae) species in Finland. Annales Botanici Fennici, 27,235–239.

Rigby, B. and Dana, M.N. (1972) Flower opening, pollen shedding, stigma receptivity and pollen tube growth in the cranberry. HortSci. 7,84–85.

Roberts, R.H. and Struckmeyer, B.E. (1942) Growth and fruiting of the cranberry. Proc. Amer. Soc. Hort. Sci. 40,373–379.

Rodriguez-Saona, C., Vorsa, N., Singh, A.P., Johnson-Cicalese, J., Szendrei, Z., Mescher, M.C. and Frost, C.J. (2011) Tracing the history of plant traits under domestication in cranberries: potential consequences on anti-herbivore defenses. J. Exper. Bot. 62(8), 2633–2644.

Roper, T.R. (2006) The physiology of cranberry yield. Wisconsin Cranberry Crop Management Newsletter, Vol. XIX. www.hort.wisc.edu/cran.

Roper, T.R., Stang, E.J., and Hawker, G.M. (1992) Early season leaf removal reduces fruit set and size in cranberry. HortScience 27,75.

Roper, T.R. and Klueh, J.S. (1994) Removing new growth reduced fruiting in cranberry. HortScience 29,199–201.

Roper, T.R., Klueh, J., and Hagidimitriou, M. (1995) Shading timing and intensity influences fruit set and yield in cranberry. HortScience 30,525–527.

Roper, T.R. and Vorsa, N. (1997) Cranberry:Botany and Horticulture. Hort. Rev. 21,215–249.

Rowland, L.J., Dhanaraj, A.L., Polashock, J.J., and Arora, R. (2003) Utility of blueberry-derived EST-PCR primers in related Ericaceae species. HortSci. 38:1428–1432.

Sapers, G.M. and Hargrave, D.L. (1987) Proportions of individual anthocyanins in fruits of cranberry cultivars. J. Amer. Soc. Hort. Sci. 112,100–104.

Sapers, G.M., Phillips, J.G., Rudolf, H.M., and DiVito, A.M. (1983) Cranberry quality: selection procedures for breeding programs. J. Amer. Soc. Hort. Sci. 108,241–246.

Sarracino, J.M. and Vorsa, N. (1991) Self and cross fertility in cranberry. Euphytica 58, 129–136.

Schmid, P. (1977) Long term investigation with regard to the constituents of various cranberry varieties (*Vaccinium macrocarpon* Ait.). Acta Hort. 61,241–254.

Serres, R., Stang, E., McCabe, D., Russell, D., Mahr, D. and McCown, B. (1992) Gene transfer using electric discharge particle bombardment and recovery of transformed cranberry plants. J. Amer. Soc. Hort. Sci. 117,174–180.

Serres, R.A., Zeldin, E.I., and McCown, B.H. (1997) Applying biotechnological approaches to Vaccinium improvement: A review. Acta Hort. 446, 221–226.

Singh, A. P., Wilson, T., Kalk, A.J., Cheong, J., and Vorsa, N. (2009) Isolation of specific cranberry flavonoids for biological activity assessment. Food Chem. 116, 963–968.

Vander Kloet, S.P. (1983) The taxonomy of Vaccinium & Oxycoccus. Rhodora 85,1–43.

Vander Kloet, S.P. (1988) The genus *Vaccinium* in North America. Res. Branch Agric. Can. Publ. 1828.

Varshney, R.K., Graner, A., and Sorrells, M.E. (2005) Genic microsatellite markers in plants: features and applications. Trends in Biotechnology 23,48–55.

Vorsa, N. and Welker, W.V. (1985) Relationship between fruit size and extractable anthocyanin content in cranberry. HortScience 20,402–403.

Vorsa, N., Polashock, J., Howell, A., Cunningham, D., and Roderick, R. (2002). Evaluation of fruit chemistry in cranberry germplasm: potential for breeding varieties with enhanced health constituents. Acta Hort. 574,215–219.

Vorsa, N., Polashock, J., Cunningham, D., and Roderick, R. (2003) Genetic inferences and breeding implications from analysis of cranberry germplasm anthocyanin profiles. J Amer Soc Hort Sci 128, 691–697.

Vorsa, N. and Polashock, J. (2005) Alteration of anthocyanin glycosylation in cranberry through interspecific hybridization. J. Amer. Soc. Hort. Sci. 130,711–715.

Vorsa, N. and Johnson-Cicalese, J. (2005) Breeding the American cranberry for health constituents: genetic variation for proanthocyanidin content. Acta Hort. 715,243–251.

Vorsa, N., Johnson-Cicalese, J., and Polashock, J.J. (2009) A blueberry by cranberry hybrid derived from a *Vaccinium darrowii* x (*V. macrocarpon x V. oxycoccus*) intersectional cross. Acta Hort. 810,187–189.

Vvedenskaya, I.O. and Vorsa, N. (2004) Flavonoid composition over fruit development and maturation in American cranberry, *Vaccinium macrocarpon* Ait. Plant Sci. 167,1043–1054.

Wilcox, R.B. (1951) Tests of cranberry varieties and seedlings for resistance to the leafhopper vector of false blossom disease. Phytopath 41,722–735.

Wilcox, R.B. and Beckwith, C.S. (1933) A factor in the varietal resistance of cranberries to the false blossom disease. J. Agric. Res. 47,583–590.

Winkel-Shirley, B. (2001) Flavonoid Biosynthesis. A colorful model for genetics, biochemistry cell biology, and biotechnology. Plant Physiology 126,485–493.

Zeldin, E.L. and McCown, B.H. (1997) Intersectional hybrids of lingonberry *Vaccinium vitis-idaea*, sect. Vitis-idaea) and cranberry (*V. macroacrpon*, sect. Oxycoccus to *V. reticulatum*, sect. Macropelma). Acta Hort. 446,235–238.

Chapter 7
Grape

Bruce I. Reisch, Christopher L. Owens, and Peter S. Cousins

Abstract Grapes are grown worldwide, on about 7.9 million ha, and are used to produce wine, raisins, juice, jam, concentrate, and seed oils, as well as fresh fruit. Grapes (*Vitis* sp.) are members of the Vitaceae. *Vitis* includes two subgenera, *Euvitis* (38 chromosomes) and *Muscadinia* (40 chromosomes), with about 60 species in total. The primary centers of species diversity are North America and East Asia. Scion cultivars are derived chiefly from the European grape, *Vitis vinifera*, which was domesticated ca. 6,000–10,000 years ago in the region between the Black and Caspian Seas. Grapes spread east into Asia and west into the Mediterranean region. Rootstocks were developed from North American species, including *V. riparia*, *V. rupestris*, and *V. berlandieri*. Scion breeding programs focus on the development of cultivars adapted to biotic and abiotic stress, with high fruit quality, and time of ripening during desirable periods of market demand. Fungal disease resistance is a primary goal of many programs, while cold hardy cultivars help extend the limits of grape cultivation. Rootstock breeding focuses on providing protection against phylloxera and nematodes as well as adaptation to high pH, low pH, and/or water-stressed conditions. Rootstocks should propagate easily by grafting and cuttings. New cultivars are more rapidly adapted in the raisin and table grape sectors than in the wine industry, although there are several notable examples of successful wine grape cultivars developed by breeding. The availability of two published genomic DNA sequences has stimulated numerous projects to further understand the function of the ca. 30,000 grapevine genes. Marker-assisted selection, primarily for disease resistance and seedlessness, is being applied in many breeding programs. Projects that focus on breeding seedless

B.I. Reisch (✉)
Departments of Horticulture and Plant Breeding, N.Y.S. Agricultural Experiment Station, Cornell University, 630 W. North St, Geneva, NY 14456-1371, USA
e-mail: bruce.reisch@cornell.edu

C.L. Owens • P.S. Cousins
USDA ARS, Grape Genetics Research Unit, NYS Agricultural Experiment Station, Geneva, NY, USA
e-mail: chris.owens@ars.usda.gov; peter.cousins@ars.usda.gov

M.L. Badenes and D.H. Byrne (eds.), *Fruit Breeding*, Handbook of Plant Breeding 8, DOI 10.1007/978-1-4419-0763-9_7,

cultivars commonly use embryo rescue techniques, enabling the crossing of two seedless parents, to increase the percentage of seedlings that are seedless. Genetic transformation is a routine procedure and is being used for both functional analysis of gene action as well as directly for cultivar improvement (both scions and rootstocks), although transgenic grape cultivars currently are not in commercial production.

Keywords Grape Breeding • Downy Mildew • Nematode • Phylloxera • Powdery Mildew • Fruit Quality • Raisin • Seedless • Wine • *Vitis* • *Vitis vinifera*

1 Introduction

1.1 Economic Importance

Grapes are among the most extensively cultivated fruits, grown on about 7.9 million ha. Annual global grape production is about 674 million quintals (Office International de la Vigne et du Vin 2006).

1.2 Uses

Grapes are processed to make wine and other fermented beverages, eaten fresh and dried, and used as unfermented juice and concentrate. Wine is the most important use of grapes by both tonnage and production area. Wine grapes cultivars usually have relatively small seeded berries. Important wine grape cultivars include 'Cabernet Sauvignon' and 'Pinot noir,' used for red wine production, and 'Chardonnay' and 'Sauvignon blanc,' used for white wine production. Table grapes are consumed fresh. Table grape cultivars have relatively large berries and seedlessness is valued by many consumers. Most dried grapes, often called raisins, are made from seedless grapes. Unfermented juice is manufactured from cultivars with distinctive flavors and aromas. Varieties with relatively heat stable flavors and aromas, such as 'Concord' and 'Niagara,' are used in the production of pasteurized juices. Cultivars such as 'Chasselas' with flavors and aromas that are noticeably altered by pasteurization are processed for unfermented juice production using ultrafiltration for juice sterilization. Jams, jellies, and other spreads are made from juice grape cultivars. Grape concentrate is juice with some water removed; it is used as a natural sweetener and coloring agent for beverages and foods. The concentrate market is an outlet for excess grapes in all market classes and is a target market for certain cultivars; 'Rubired,' a highly pigmented cultivar, is used in red concentrate.

Grape cultivars may be used in one or several market classes. For example, 'Sultanina' (known as 'Thompson Seedless' in the United States) is the dominant raisin cultivar worldwide and is also an important table grape, wine grape, and concentrate cultivar. In contrast, 'Cabernet Sauvignon' is used for wine but is not desirable as a table or raisin grape. Premium wine and table grape cultivars are more specialized in their utilization than are raisin, juice, and concentrate varieties.

1.3 Taxonomy

The grape is a member of the Vitaceae, commonly called the grape family. The genus *Vitis* consists of about 60 species, plus some natural interspecific hybrids (Wen 2007). Nearly all grapes cultivated for fruit production are of the species *V. vinifera* or are hybrids that include *V. vinifera* in their parentage.

Vitis species are found across the temperate zones of the Northern Hemisphere. The genus has the highest species diversity in east Asia and in eastern and southern North America, with about 30 species in each region. *Vitis* is separated into two subgenera, *Euvitis* and *Muscadinia*; some authorities treat the sections as the genera *Vitis* and *Muscadinia*. The subgenera are separated by morphological, anatomical, and cytological characteristics. Subgenus *Euvitis* species have $2n=2x=38$ chromosomes, forked tendrils, striate bark, pyriform seeds, and nodal diaphragms. These species and their hybrids are called bunch grapes. Subgenus *Muscadinia* species have $2n=2x=40$ chromosomes, unforked tendrils, stellate bark, naviform seeds, and lack diaphragms at the nodes; they are known as muscadine grapes. Within a subgenus, species are maintained in nature by range and flowering time and can be considered ecospecies. Hybrids between species within a subgenus are typically fully fertile and many interspecific hybrids between *Euvitis* species have been developed as scion and rootstock cultivars. Hybrids between the subgenera are usually sterile due to the difference in chromosome number; two have been commercialized as rootstocks (Walker et al. 1991; Lider et al. 1988) and backcrossing with partially fertile intersubgeneric hybrids has led to the introduction of disease resistance from *V. rotundifolia* into bunch grape gene pools (Pauquet et al. 2001).

Subgenus *Euvitis* species (about 57 species) are the most important in viticulture. Most grape cultivars belong to the species *V. vinifera,* which is a native of the Mediterranean basin, southern and central Europe, northern Africa, and southwest and central Asia. *V. vinifera* cultivars are grown worldwide and account for the overwhelming majority of cultivated area and grapes produced. Interspecific hybrid cultivars, which are selected from crosses of *V. vinifera* with other species, including *V. labrusca*, *V. amurensis*, *V. riparia*, *V. rupestris*, and *V. aestivalis*, are important locally, but are mostly minor components of world viticulture and enology. Rootstocks, which are used exclusively for bunch grape varieties, are mostly interspecific hybrids or selections of North American *Euvitis* species.

The subgenus *Muscadinia* includes only three species. The range of the subgenus is limited to the southeastern United States and eastern Mexico. Muscadine grape cultivars, primarily *V. rotundifolia* and a few interspecific hybrids, are grown commercially only in the native region of *V. rotundifolia* in the southeastern United States.

Grapevines are indeterminate woody perennial tendril-bearing tree-climbing vines. Ordinarily deciduous, in tropical regions grapevines may be evergreen. Grapevine bark is usually shed in long strips. The leaves are alternate, with each leaf consisting of a blade, a petiole, and a pair of stipules. Each leaf axil bears a complex lateral bud; primary latent buds bear inflorescences. The inflorescences are initiated in latent buds between bud break and bloom.

The inflorescences are found near the base of the shoot, usually in a zone between nodes three and nine. The grape inflorescence is a panicle found opposite the leaf, at the node. Usually one to three inflorescences are borne per shoot, but there may be six or more, with variation due to cultivar and environmental conditions. Flowers per inflorescence vary from fewer than 60 to more than 1,000. Grape flowers range from about 2 to 7 mm long. Each flower bears a minute calyx of five rudimentary sepals, five petals fused at their tips to form a calyptra, five stamens, and a pistil. The pistil has a superior ovary, usually with two carpels. The calyptra falls off completely at bloom by abscising at the base of each petal. The fruit is a fleshy berry, usually with not more than five seeds.

Grapes require pollination for fruit set and most cultivars require fertilization. Nearly all important cultivars are perfect flowered and are both self-pollinating and self-fruitful. Pollination occurs before or during calyptra abscission and no pollen vector is required. A few *V. vinifera* and interspecific hybrid bunch grape cultivars are pistillate flowered, but these are mostly archaic cultivars—pistillate flowered cultivars have been largely replaced by perfect flowered cultivars, which tend to be higher yielding and do not require interplanting with a pollinizer cultivar or hand pollination. Only in *V. rotundifolia* table grape production are pistillate cultivars still commercially dominant (Basiouny and Himelrick 2001). Parthenocarpic cultivars require pollination but not fertilization for fruit set and development. While nearly all scion cultivars are perfect flowered, wild grape species are dioecious and bear functionally imperfect flowers, with individual vines bearing either staminate or pistillate flowers. Many rootstock cultivars are imperfect flowered. The pollen vectors for wild vines are not known; insects and wind both are considered to have roles in pollination.

1.4 Where Grown

Major wine producing countries grow the most grapes. Countries bordering the Mediterranean Sea, where grapes have been grown for thousands of years, are leading grape growers and wine producers. Italy, France, and Spain are major grape and wine producing countries, each producing over 65 Mqx grapes annually (OIV 2005); Turkey is a leading grape grower. Other regions with a Mediterranean climate are also leading production zones, including the western United States, Australia, and southern South America and Africa; major producing countries are the United States, China, South Africa, Chile, Argentina, Australia, Iran, Germany, Romania, Portugal, and India. Commercial production of muscadine grapes is limited to the southeastern United States and the total area cultivated is about 1600 ha.

1.5 Limits on Adaptation

Grapes are grown in regions where there is adequate growing season, heat accumulation and sufficiently moderate winter low temperatures. Grape growing is most successful in areas that receive at least 1,700 Winkler-Amerine growing degree days

(Mullins et al. 1992). Most grapes are produced in areas where the mean temperature of the warmest month exceeds 18°C and the mean temperature of the coldest month exceeds −1°C (Prescott 1965). The primary production regions for grapes are between about 30°N and 50°N and between about 30°S and 40°S.

Regional environmental effects and specialized viticultural practices allow cultivation beyond the zone. Grapevine cultivation in southern Germany, at about 51°N, is conditioned by the warming influence of the Gulf Stream. Tropical and subtropical cultivation is carried out at high altitudes to achieve temperate zone conditions, or management practices are modified to encourage vine productivity despite lack of adequate chilling (such as use of plant growth regulators and modified pruning that promote uniform budbreak). Tropical and subtropical countries cultivate mostly table and juice grapes, since wine quality is typically lower in very hot regions. In regions with very cold winters, such as Scandinavia and the northern interior of North America, very hardy varieties are grown, which may be cold hardy to winter minimums of −35°C or colder. Despite the adaptation of these varieties, grape cultivation in these regions is minimal.

2 Origin and Domestication

Grapes were first domesticated approximately 6,000–10,000 years ago (Levadoux 1956; McGovern 2003; Zohary and Hopf 2000). There are several morphological and biochemical traits associated with the domestication of *V. vinifera* that were derived from the progenitor species *V. vinifera* subsp. *sylvestris*. Significant differences are the emergence of perfect flowers, greater uniformity of berry maturity within clusters, higher sugar content, and the selection for a wide range of fruit colors (Levadoux 1956; Olmo 1995; Zohary and Spiegel-Roy 1975). Extant, isolated patches of *V. vinifera* ssp. *sylvestris* can be found from Western Europe to central Asia and North Africa. Archaeological evidence suggests that the early domestication of grapes spread first from the mountainous regions between the Caspian and Black Seas to regions southwards in the Jordan Valley, Egypt, and the western side of the Fertile Crescent by 5,000 B.P. (McGovern 2003; McGovern and Michel 1995; Zohary and Hopf 2000). Continued western expansion of viticulture occurred in Crete and both coasts of the Iberian and Italian peninsulas by approximately 2,800 B.P. (McGovern 2003) (Fig. 7.1).

Historically, geographical origins and morphological characteristics have been used to sub-divide *V. vinifera* into three morphotypes: *occidentalis*, *pontica*, and *orientalis* (Negrul 1938). The *occidentalis* group is characterized by small berries, small clusters, highly fruitful shoots, and is associated with cultivars of Western European origin. The *orientalis* group consists of large berried, loose clustered cultivars from Central Asia. The *pontica* group comprises an intermediate grouping of cultivars from Eastern Europe and the Black Sea Basin. Debate exists concerning the number of domestication events and the location of their occurrence, as *V. vinifera* ssp. *sylvestris* had a wide geographic range, and wild populations were likely used as a food source across much of that range. Evidence from the use of chloroplast

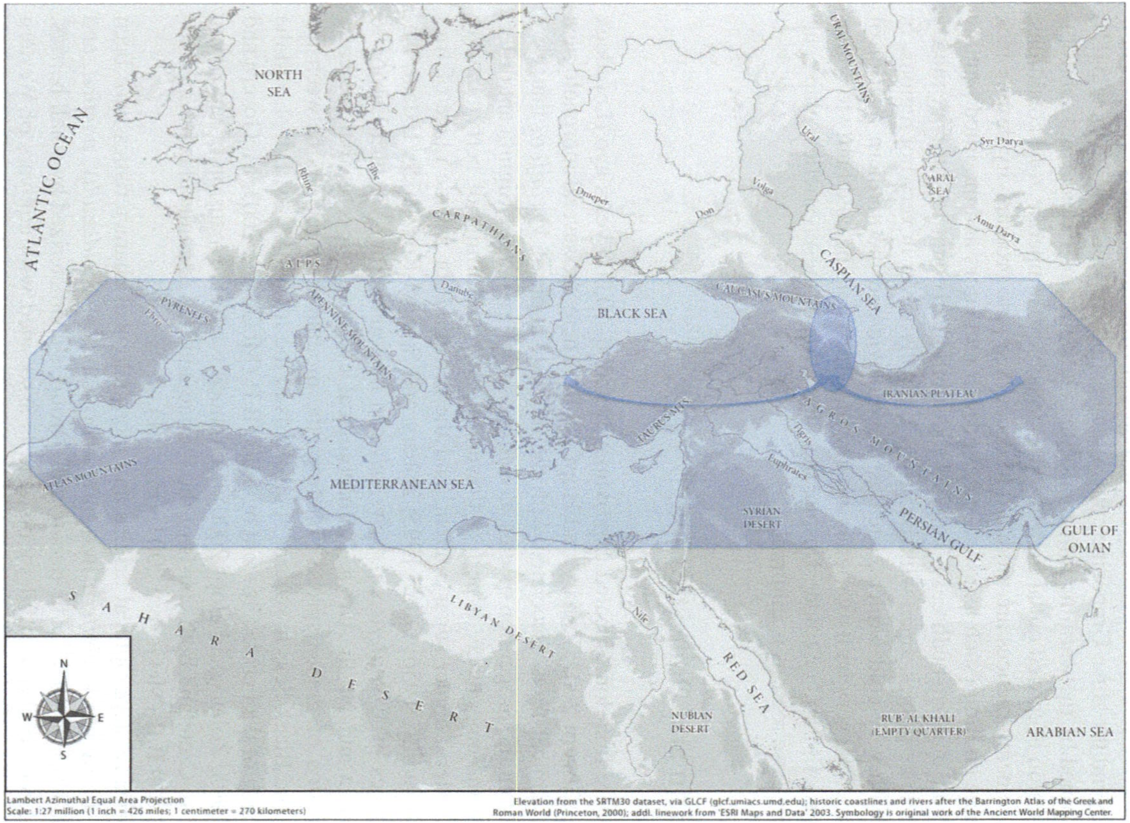

Fig. 7.1 Natural range of the European grape, *Vitis vinifera* subsp. *sylvestris*. Domestication most likely occurred first in the region indicated by the oval area between the Black and Caspian Seas. Cultivated grapes spread to the east and west (arrows). Secondary domestication centers occurred at additional locations within the natural range of *V. vinifera* subsp. *sylvestris*. (Map Base © 2011, Ancient World Mapping Center (www.unc.edu/awmc))

molecular markers supports the presence of at least two major domestication centers, approximately corresponding with Negrul's *occidentalis* and *orientalis* group (Arroyo-Garcia et al. 2006). Additional attempts at finding genetic relationships between cultivars have provided only weak discrimination among geographic groupings and the presence of secondary domestication centers have been proposed based on evidence from nuclear markers (Aradhya et al. 2003; Grassi et al. 2003). Recent results utilizing over 6,000 SNP markers distributed across the grape genome provide strong support for a Near Eastern origin of cultivated grapes, but also finds support for a limited amount of gene flow between western *V. vinifera* ssp. *sylvestris* and *V. vinifera* (Myles et al. 2011).

Controlled grape breeding is thought to have occurred for close to 200 years. Henri and Louis Bouschet de Bernard are believed to have begun generating hybrids between 'Teinturier du cher' and 'Aramon' in 1824 in southern France (Paul 1996). These crosses led to the intensely pigmented varieties possessing color within the berry flesh as well as the skin. The birth of modern grape breeding is strongly connected with the arrival of North American diseases and insects to Europe. In successive waves in the mid nineteenth-century the root louse, phylloxera (*Daktulosphaira vitifoliae* Fitch), powdery mildew (*Uncinula necator* Burr), downy mildew (*Plasmopara viticola* Berl.), and black rot (*Guignardia bidwellii* Ellis) were exported to European vineyards where they caused substantial losses on the highly susceptible *V. vinifera* vines planted there.

Several major advances in viticulture and grape breeding occurred as a result of the epidemics spreading through Europe in the late nineteenth and early twentieth centuries. The first was the advent of rootstock breeding as an effective and immediate means to control phylloxera. Successive waves of wild vines from North America were first imported to be used as rootstocks, principally cuttings of *V. riparia* and *V. rupestris*, that provided phylloxera resistance. Subsequent importations of *V. cinerea* var. *helleri (V. berlandieri)* vines for their combined resistance to phylloxera and adaptation to calcareous soils provided much of the initial genetic material, along with selections of *V. aestivalis* var. *lincecumii*, in the earliest wave of grape rootstock breeding (Campbell 2005).

Breeding programs to develop cultivars that possessed resistance to phylloxera as well as the fungal pathogens in one vine were begun as early as 1874. Collectively, these hybrid vines became known as 'Les hybrides producteurs directes' (the hybrid direct producers, HPDs) (Cahoon 1998). Many of the initial HPDs were imported from the United States and would later be outlawed in France: 'Clinton,' 'Noah,' 'Herbemont,' 'Othello,' and others. These cultivars were primarily hybrids of *V. labrusca*, *V. aestivalis*, *V. riparia*, and *V. vinifera*. Due to the unpopularity of flavors associated with *V. labrusca*, breeders in France attempted to produce HPDs without utilizing this species.

The early French breeders primarily relied on *V. rupestris*, *V. riparia*, and *V. aestivalis* var. *lincecumii*. These breeders included Eugene Contassot, Albert Seibel, Georges Couderc, Fernand Gaillard, Francois and Maurice Baco, Bertille Seyve, Eugene Kuhlmann, Pierre Castel, and Christian Oberlin. Additional French grape breeders during the twentieth century continued efforts with the previous generation's parental material: Bertille Seyve-Villard, Joannes Seyve, J-F Ravat, Joanny Burdin, Jean-Louis Vidal, Alfred Galibert, Pierre Landot, and Eugen Rudelin.

Grape breeding in North America is thought to have begun in the early nineteenth century. William Valk named the first reported cultivar as the result of a cross between a native American cultivar and *V. vinifera*, 'Ada,' in 1852 (Cattell and Miller 1980). Other notable grape breeders of the mid-nineteenth century in the United States included E.S. Rogers of Roxbury, Massachusetts, J.H. Ricketts of Newburgh, NY, and Jacob Moore of Brighton, NY, who developed the important early varieties 'Brighton', 'Diana', 'Hamburg', and 'Diamond.' Ephraim Bull of Concord, Massachusetts, developed the highly successful juice, jelly, and wine grape, 'Concord.' Hermann Jaeger and Jacob Rommel in Missouri (Rommel produced 'Elvira') also developed many cultivars and had a direct influence on Thomas Volney Munson.

T.V. Munson of Denison, TX, became one of the most significant early grape hybridizers and botanists in the United States (McLeRoy and Renfro Jr. 2004) and a leading figure for future viticulturists through the publication of his influential book *Foundations of American Grape Culture* (Munson 1909). Munson also had a significant role in providing rootstock material to French breeders and viticulturists looking for parental material for phylloxera-resistant rootstocks, particularly those that would be adapted to highly calcareous soils. Some of Munson's more notable cultivars are 'America,' 'Bailey,' 'Brilliant,' 'Headlight,' and 'President.'

3 Genetic Resources

3.1 Scions

Grape species of the same chromosome number are highly interfertile. Geographical isolation and differences in flowering time appear to be the primary forces in maintaining species identity in natural environments, although interspecific hybrids can be observed when species boundaries do overlap. Selfing of hermaphroditic cultivars is possible, although inbreeding depression is typically observed and can be severe. Crosses between sections have had limited success due to the differences in chromosome number. However, a small number of viable offspring can be recovered and utilized in breeding programs (Bloodworth et al. 1980; Bouquet 1986; Olmo 1971; Ramming et al. 2000). *V. rotundifolia*, a 40-chromosome member of the section *Muscadinia*, has been identified as a source of dominant resistance to the primary fungal disease of grape worldwide, powdery mildew. Crosses between *V. rotundifolia* and *V. vinifera* have yielded breeding lines and genetic resources that have been useful in determining the nature of this resistance (Bouquet 1986; Doligez et al. 2002; Donald et al. 2002).

The number of existing cultivars of *V. vinifera* has been estimated to be approximately 5,000 (Alleweldt and Dettweiler 1994; This et al. 2006). Due to the ease of asexual propagation, the age of some cultivars, the ease by which desirable cultivars can be transported, and the importance of viticulture in many regions, a situation has arisen in which there are a large number of synonyms and homonyms of cultivar names. Most of the grape-growing countries of the world maintain grape germplasm collections, and microsatellite markers have been extensively used to better characterize and inventory those collections (Aradhya et al. 2003; Lopes et al. 1999; Martin et al. 2003; Sefc et al. 2000). A reference set of cultivars and markers has been put forth to ease comparisons among locations (This et al. 2004). Larger data sets comprised of SNP markers are now being assayed on germplasm collections, included the majority of the US national grape collection (Myles et al. 2011).

Considering the thousands of cultivars of *V. vinifera*, there has been substantial interest in utilizing molecular markers for germplasm management, assessment of genetic diversity, and determination of degrees of relatedness among cultivars and wild accessions (Dangl et al. 2001; Lopes et al. 1999; Thomas et al. 1994). Molecular markers, primarily microsatellites, have been used to identify the parents of many major cultivars of *V. vinifera*, including 'Syrah,' 'Cabernet Sauvignon,' 'Müller-Thurgau,' 'Muscat Hamburg,' and 'Petite Sirah' (Bowers and Meredith 1997; Cervera et al. 1998; Crespan 2003; Dettweiler et al. 2000; Lopes et al. 2006; Meredith et al. 1999; Vouillamoz and Grando 2006). Notably, the cultivars 'Pinot' and 'Gouais' have been shown to be the parents of a large number of important European cultivars, including 'Chardonnay,' 'Auxerrois,' 'Gamay noir,' and 'Melon' (Bowers et al. 1999a). Similarly, microsatellites have been employed to trace the geographic origin of cultivars that have been introduced to areas outside the region of initial cultivation (Maletic et al. 2004). Recent evidence suggests that a complex network of close pedigree relationships exist with *V. vinifera* and that although substantial genetic diversity is present that diversity has not been well explored (Myles et al. 2011).

Molecular markers have been used to better understand the relationships among autochthonous cultivars and to identify synonyms and homonyms within numerous collections of cultivars around the world, including Italy (Labra et al. 2003; Labra et al. 2001; Rossoni et al. 2003), Iran (Fatahi et al. 2003), Spain (Martin et al. 2003), Portugal (Lopes et al. 2006) Albania (Ladoukakis et al. 2005), Turkey (Ergul et al. 2006), Japan (Goto-Yamamoto et al. 2006), and Bulgaria (Hvarleva et al. 2004) as well as groups of ambiguous cultivar names, such as the Pinots (Regner et al. 2000) and 'Trebbiano' (Labra et al. 2001).

An important source of genetic variation in *V. vinifera* is the presence of numerous bud sports, or somatic mutations. Due to the ease of clonal propagation in grapevine, it is conceivable that a substantial proportion of phenotypically recognized mutants are chimeric in nature (Einset and Lamb 1951; Thompson and Olmo 1963). Molecular markers have been utilized to confirm the presence of chimerism in several cultivar groupings (Franks et al. 2002; Hocquigny et al. 2004; Pelsy 2010; Riaz et al. 2002). Polyploid sports and periclinal chimeras containing tissue layers of differing ploidy level have been reported for grapevine (Einset and Lamb 1951; Einset and Pratt 1954; Sauer and Antcliff 1969). The utilization of naturally occurring mutants of grapevine for the dissection of individual genes controlling important phenotypic traits has recently begun primarily through candidate gene analysis (Boss and Thomas 2002; Fernandez et al. 2006; Fernandez et al. 2010; Kobayashi et al. 2004).

3.2 Rootstocks

Phylloxera *(D. vitifoliae)* protection is the most important reason that rootstocks are used in viticulture. Phylloxera is an aphid-like insect that can feed on and damage grapevine roots. The roots of *V. vinifera* cultivars are highly susceptible to damage resulting from phylloxera feeding (Viala and Ravaz 1903). In regions where phylloxera is not present or is not important, grapevines are routinely grown on their own roots (ungrafted), even if other damaging soil pests are present that could be managed with rootstocks (Walker and Stirling 2008). Cultivars with some degree of phylloxera resistance or tolerance, such as Concord, Niagara, and many interspecific hybrids, are similarly frequently grown on their own roots, as own rooted vines are less expensive to propagate and simpler and cheaper to manage in colder viticultural regions (where graft unions should be protected). Because phylloxera is the most important grape root pest, American species of *Vitis*, which evolved with phylloxera pressure and which demonstrate resistance or tolerance to phylloxera are the most important genetic resource for rootstocks. Old World species of *Vitis* (which evolved without phylloxera pressure) are susceptible to this insect pest. The same grape species used in rootstock breeding are used in breeding disease resistant and cold hardy scion varieties.

Most of the 25–30 American grape species are resistant or tolerant to phylloxera, with some species of western North America, such as *V. californica,* as possible exceptions (Viala and Ravaz 1903). Despite the widespread resistance and tolerance to phylloxera, only a few species are suited for direct use as rootstocks, because many grape species do not root easily from dormant cuttings, which are the basis for commercial vine propagation through grafting. The wild grapes, *V. riparia* and *V. rupestris*, were an important source of early rootstock selections. These two species root easily from dormant cuttings and provide protection against phylloxera. Selections of *V. riparia* and *V. rupestris* are used directly as rootstocks and these species have been hybridized with other North American species to introduce facile root strike with adaptation to calcareous soils and resistance to other pests and diseases. With the exception of *V. vinifera* hybrid rootstocks, all commercially grown rootstocks are either selections of *V. rupestris* and *V. riparia* or hybrids, which include one or both of these species in their background.

Adaptation to calcareous soils, characterized by high pH, is an important attribute of many rootstocks because of the prevalence of these soils in many European viticultural regions. Most selections of *V. riparia* and *V. rupestris* are not well adapted to calcareous soils and at high soil pH show iron-deficiency chlorosis. *Vitis berlandieri* (synonym *V. cinerea* var. *helleri*) has been used extensively in rootstock breeding as a source of adaptation to calcareous soils. Numerous rootstocks used on calcareous soils are hybrids of *V. berlandieri*. *V. vinifera* is very well adapted to calcareous soils and has also been used in rootstock breeding to introduce the adaptation to such sites, but must be used cautiously as *V. vinifera* rootstock hybrids are more susceptible to phylloxera than rootstocks bred or selected exclusively from North American species.

Vitis riparia, V. rupestris, and *V. berlandieri* have been the most important sources of phylloxera protection used in rootstock breeding. Other North American species have had minor roles. *Vitis cinerea* and *V. cordifolia* (syn. *V. vulpina*) are difficult to propagate species with poor adaptation to calcareous soils and have been used as parents exclusively in hybridization with easy to propagate species.

Resistance to nematodes has been identified in multiple grape species and incorporated into rootstock breeding programs. 'Freedom' and 'Harmony' derive resistance to root-knot nematodes from *V. x champinii* and *V. solonis* (syn. *V. acerifolia*); *V. x champinii* is a natural *V. rupestris/V. mustangensis* hybrid complex, with two selections of *V. x champinii*, 'Dog Ridge' and 'Ramsey' used directly as rootstocks. *Vitis cordifolia, V. aestivalis, V. nesbittiana, V. monticola, V. mustangensis, V. rotundifolia, V. rupestris,* and *V. acerifolia* (Boyden and Cousins 2003; Cousins and Lauver 2003; Walker et al. 1994a; Firoozabady and Olmo 1982; Bloodworth et al. 1980; Lider 1954) are all reported to be sources of resistance to root-knot nematodes. Some may be suited to direct use as rootstocks. Sources of resistance to the dagger nematode, *Xiphinema index,* were found in *V. arizonica, V. rufotomentosa, V. rotundifolia, V. solonis, V. x slavinii,* and *V. mustangensis* (Walker et al. 1998; Meredith et al. 1982).

V. rotundifolia has resistance to many important grape root pests, including phylloxera (Viala and Ravaz 1903), root-knot nematodes (Walker et al. 1994a), and dagger nematodes (Walker et al. 1998). Although *V. rotundifolia* can be grafted, it is commercially cultivated ungrafted on its own roots. The utilization of *V. rotundifolia* in breeding scion varieties in bunch grapes has been limited by the intersubgeneric fertility barrier imposed by different chromosome numbers. However, sterile hybrids between *V. rotundifolia* and *V. vinifera* have been developed as rootstocks (Walker et al. 1994b; Lider et al. 1988). Although the sterility of these intersubgeneric hybrids makes difficult their use as parents in further breeding, the sterility does not pose a barrier to their utilization as rootstocks.

4 Major Breeding Achievements

4.1 Scions

Scion grapes are used to make wine, juice, jelly, jam, pie, raisins, and other processed products. Fresh grapes are also sold for direct consumption as table grapes and are often seedless. Grape seed oil products have also become more common in recent years. The wine industry tends to be highly conservative and the most widely grown grape varieties originated long ago, sometimes many centuries ago, rather than as a product of defined breeding efforts. On the other hand, the table and raisin grape markets are very receptive to new cultivars, and many have rapidly gained market share.

Grape breeding predates the rediscovery of Mendel's laws (ca. 1900) with the development of 'Alicante Bouschet' and 'Petit Bouschet.' Both resulted from controlled crosses of *V. vinifera* cultivars beginning in 1824 by Louis Bouschet and his son in southern France (Paul 1996). Not long thereafter (1830–1860) significant efforts to breed grapes adapted to North America took place in New York, Massachusetts, and Missouri (Owens 2008), and were followed (late 1800s) by the notable and monumental efforts of T.V. Munson (1909) to develop over 300 new cultivars for the southwestern United States.

In the mid-1800s, viticulture in Europe was afflicted with numerous grapevine pests that originated in North America. These included phylloxera (*D. vitifoliae*), as well as powdery mildew (*E. necator*), downy mildew (*P. viticola*), and black rot (*G. bidwellii*). There were multiple responses to combat these problems, as described in Sect. 2 above. Vineyardists sought new wine varieties that were phylloxera-resistant, and resistant to the newly introduced fungal diseases, as well. Nurserymen and researchers responded with the development of the so-called hybrides producteurs directes (hybrid direct-producers), also known as French–American hybrids, as they resulted from crosses between American species and the European *V. vinifera* cultivars (Cahoon 1998).

These new cultivars became exceedingly popular in France but only for table wine, not quality wine production. In 1958, there were over 400,000 ha of French American hybrids grown in France. However, they were rapidly removed as the French passed laws in 1953 restricting their planting and sale (Cahoon 1998). Very few plantings of French–American hybrids remain in France today; one notable exception is Baco 22A, which is permitted for use in Armagnac production. Selected French–American hybrid grapes are still in commercial use in the eastern United States and Canada.

Today, there is an increasing trend in Europe to allow the cultivation of disease-resistant interspecific hybrid grape cultivars. 'Regent,' released in 1996 by the Julius Kühn-Institut, Bundesforschungsinstitut für Kulturpflanzen, Institut für Rebenzüchtung Geilweilerhof, Germany, is now grown on over 2000 ha. It has been followed (2009–2010) by a number of new disease-resistant hybrid releases, such as 'Felicia,' 'Villaris,' 'Calandro,' and 'Orion.' In Hungary, commercialized interspecific hybrids include the white wine grape, 'Bianca.'

In North America, breeding programs in New York, Florida, Minnesota, and Ontario (Canada) have been responsible for a number of commercially successful interspecific hybrid introductions. Notable among these are 'Conquistador,' 'Stover, and 'Orlando Seedless' (Florida); 'Traminette,' 'Cayuga White,' and 'Chardonel' (New York); 'La Crescent,' 'Frontenac,' and 'Marquette' (Minnesota); and 'L'Acadie' and 'Ventura' (Ontario, Canada).

While there has been much activity worldwide in breeding interspecific hybrid scion cultivars, there has also been notable success in the development of new cultivars of *V. vinifera*. Among wine grapes, 'Müller-Thurgau' was developed and released in 1882 (Reisch and Pratt 1996) and is now one of the most widely grown cultivars in Germany. 'Dornfelder,' a red wine cultivar, was developed at the research institute in Weinsberg, Germany, and released for cultivation in 1979. It is now widely grown in northern Europe as well as in colder regions of the United States.

The 'Pixie' grape (Boss and Thomas 2002; Cousins 2007), originating from a layer of the chimeric cultivar, 'Pinot Meunier', represents an achievement in the development of a small, short-lifecycle grapevine suitable for genetic studies. Inflorescences are produced in place of tendrils.

The seedless *V. vinifera* table grape market has grown rapidly over the past 50 years, and the general public (as well as market buyers) are much more accepting of new varieties since they do not rely as much on name recognition as they do on visual and sensory appeal to be successfully sold in the market. Luigi and Alberto Pirovano bred and released a series of table grapes in the early twentieth century, including 'Italia' and 'Sultana Moscato.' Harold Olmo of the University of California, Davis, released the highly successful 'Perlette' and 'Redglobe' table grapes. Another major development came from the USDA in Fresno, California, with their release of the crisp-textured red seedless grape, 'Flame Seedless' (1973). It is widely planted in California and around the world. The first major cultivar to overcome the apparent association between small berry size and seedlessness was 'Fantasy Seedless' (1994), a naturally large-berried (7–8 g) black seedless from the USDA Fresno program. Others have followed since. A number of private companies around the world have large and very successful table grape breeding programs. Notable among them are Sun World International, LLC (Bakersfield, California) and Sunview Vineyards (Delano, California).

During the past 50+ years, Japan and Korea have led the way in the development of large-berried seeded and seedless table grapes. Most are large berried due to tetraploidy, leading to an increase in cell size and a resulting increase in berry size. Cultivars, such as 'Kyoho,' 'Pione,' 'Olympia,' and 'Heukgoosul' are examples of seeded tetraploid cultivars grown in Asia. All originated from 4x × 4x crosses. There are also recent examples of seedless cultivars due to triploidy, such as 'Honey Seedless,' 'King Dela,' and 'Mirei' (Morinaga 2001). Most, if not all, of the triploid and tetraploid grapes of Asia are derived from both *V. vinifera* and *V. labrusca*. One Japanese cultivar, 'Takao,' is seedless due to aneuploidy at the tetraploid level ($2n = 4x - 1 = 75$) (Ashikawa 1972).

Among technical advances, the use of embryo culture is one of the most important contributions to grape breeding in the twentieth century. This subject was recently reviewed by Burger et al. (2009). Embryos are rescued in tissue culture prior to abortion, and grown to a seedling stage before transplantation to soil. This technique is mostly used to enable crossing between seedless grapes, thereby leading to very high percentages of seedless progeny. It is widely used among table grape breeding programs in the United States, Israel, South Africa, Chile, and Australia. Embryo culture is also used to rescue triploid seedlings from crosses between tetraploids and diploids, and to enable improved germination rates among seedlings derived from early ripening parents.

Major achievements have also impacted the development of new raisin grape cultivars. Grape breeders working on raisin grape improvement usually seek types that are seedless, have little tendency to become sticky, have a pleasing flavor, and large berry size leading to large raisin size. Some new cultivars are suitable for natural 'dried-on-the-vine' (DOV) raisin production. These have the advantage of being suitable for drying without the need to cut canes, thereby reducing production costs, alleviating the dependency on labor, and averting the risk from late season rains. DOV raisins can also be harvested mechanically. Some examples of these new raisin grapes include 'DOVine' and 'Selma Pete' from the USDA breeding program at Parlier, CA.

4.2 Rootstocks

Rootstocks primarily are used in viticulture to provide protection against soil-borne pests and diseases, especially phylloxera, an aphid-like insect that can damage grapevine roots. The central accomplishments in rootstock breeding have been the identification and deployment of rootstocks that a) provide protection against phylloxera, b) are easily propagated by dormant cuttings, c) are graft compatible with important scion varieties, and d) are adapted to a range of viticultural soils, particularly the calcareous soils prevalent in many European viticultural regions.

Rootstocks provide durable protection against phylloxera. Species selections, which were the very first rootstocks and were not the result of breeding programs, are still used as rootstocks. *Vitis rupestris* 'du Lot' has been used as a rootstock since 1879; *V. riparia* 'Gloire de Montpellier' has been in constant use as a rootstock since the late nineteenth century as well (Viala and Ravaz 1903). Both of these rootstocks are in widespread commercial use internationally. First generation interspecific hybrids, such as 3309 Couderc, 101–14 Mgt, 420A, and 1103 Paulsen all were developed before 1900 and remain important in viticulture. Their continuous use since introduction reflects the durability of the phylloxera protection they provide. Rootstocks that are selections of North American species or interspecific hybrids of North American species provide durable, long-term protection against phylloxera, which completely enable viticulture in phylloxera infested regions. Without phylloxera protective rootstocks, it would be essentially impossible to grow *V. vinifera* cultivars in infested areas.

Subsequent breeding achievements in rootstocks center on the role of rootstocks in providing protection against other soil-borne pests and diseases. 'Freedom' and 'Harmony' were the first rootstocks bred and introduced specifically to provide protection against nematodes (Clark 1997; Weinberger and Harmon 1966), here the root-knot nematode (*Meloidogyne*), although root-knot nematodes were recognized as damaging to grapevines and rootstocks were suggested as early as 1889 (Neal 1889). Selection for dual resistance to the dagger nematode *Xiphinema index* and the virus disease fanleaf degeneration produced the rootstocks VR O39-16 and VR O43-43 (Walker et al. 1994b, 1991; Lider et al. 1988). These rootstocks are significant because they show the first integration of virus resistance into a new grapevine cultivar of any type.

5 Current Goals and Challenges

5.1 Goals Common to the Improvement of Wine, Table, and Raisin Grapes

Many breeding programs work toward the development of disease-resistant cultivars in response to the diseases prevalent in a given region. While vinifera grapes constitute more than 95% of the world market, they are, in general, highly susceptible to many diseases. It is expected that as the interest in organic and sustainable viticulture increases, demand for resistant cultivars will increase. Evidence for this trend can be seen in Germany ('Regent'), eastern North America (hybrids from Cornell, Minnesota, Florida, Arkansas, and Ontario as well as French–American hybrids), in Hungary ('Bianca'), and elsewhere. Hybrids with *V. labrusca* and Asian species are used to develop resistant cultivars in Japan, Korea, China, and Thailand.

In areas where winter minimum temperatures dip to −20 to −35°C, significant efforts to develop winter hardy cultivars are taking place, notably at the University of Minnesota and at Cornell University. Private breeders in many states in the United States and in the provinces of Canada are also active, as are public and private breeders in Russia and the Baltic countries. Components of cold hardiness include the degree of survival of primary buds follow low temperature episodes; the ability of primary buds to remain hardy during fluctuating winter temperatures; the ability of phloem tissue to survive low temperatures; and the ability of emerging buds and shoots to avoid frost damage in the spring.

In recent years, considerable attention has been focused upon the health benefits of grapes and grape products, primarily due to antioxidant activity of a variety of phenolic compounds (Pezzuto 2008; Waffo-Téguo et al. 2001), especially flavonoids and stilbenoids. With this in mind, and with public interest in health-promoting foods, some breeders have begun programs to elevate levels of health-promoting substances in grape products.

The time of ripening is also an important consideration in breeding programs. For wine grapes, it is important that cultivars in a given region not ripen at the same time; use of winery equipment and labor resources is improved by the harvest of wine grapes over an extended time period. The same is true for table grapes, as breeders develop grapes for different market niches at different times during the ripening season, and with a range of colors, flavors, and shapes. In fact, the earliest ripening grapes have a considerable price advantage when grapes on the market are in short supply.

5.2 Wine Grape Breeding

Essential to any wine grape breeding program is the incorporation of marketable to superior wine quality in new cultivars. There would be no interest in cultivars that produce unmarketable wine. New seedlings in a breeding program are typically grown as single vines and later propagated to second test blocks in four to six vine plots. Breeders at many locations in North America and Europe typically test wine potential using microvinification techniques that start with the grapes available from the original seedling vine and later from multiple vine plots. Fermentations may be carried out with 1–25 L of must, or more (Reisch and Mansfield, pers. comm.; Ewart 1988). In later stages, the most promising selections may be grown in research and semicommercial trials and the fermentation parameters (yeast strains, fermentation temperature profiles, malo-lactic fermentation, etc.) can be optimized prior to potential market release.

Breeders using species other than *V. vinifera* as sources of abiotic and biotic stress tolerance traits face obstacles in the development of quality wine cultivars due to the unfavorable genes affecting wine aromas and flavors common in these species. For instance, breeders using *V. labrusca* seek to avoid selecting vines that produce β-damascenone, *o*-aminoacetophenone, and methyl anthranilate (Shure and Acree 1994). Other species may harbor compounds such as *cis*-3-hexenol, responsible for green and grassy aromas (Polášková et al. 2008; Chisholm et al. 1994) and excessive amounts should be avoided. Muscat aromas and other positive aroma attributes are also goals in wine grape development programs. Numerous studies have identified compounds responsible for muscat aroma, including terpene alcohols such as linalool and geraniol (Hardy 1970). Inheritance via five complementary genes was proposed (Wagner 1967). A single gene is now thought to be responsible for the accumulation of a variety of monoterpenes in Muscat grapes (Battilana et al. 2009; Emanuelli et al. 2010). Breeders seeking to enhance quality may also focus on *cis*-rose oxide, a compound related to the aroma typical of 'Gewürztraminer' grapes (Ong and Acree 1999).

5.3 Table Grape Breeding

Seedlessness is the essential focus of all table grape breeding efforts, though there have been some seeded cultivars released, as well, over the past 30 years. Efforts are underway to develop new cultivars in a range of colors with attractive berry shapes, firm to crisp texture, attractive clusters of reasonable size for packing, and ripening at a range of time points throughout the growing season. Breeders in important table grape growing areas focus efforts on grapes suitable for distinct market niches, based on availability at a unique time of year, with quality traits that are superior to cultivars that might already be on the market. Suitability for storage as well as avoidance of postharvest problems (shatter, rot, brown rachises) are important as well, especially for later ripening cultivars.

Flavor is of some importance to table grape breeders. In eastern North America as well as Asia, there has been considerable use made of the fruity and aromatic flavors of *V. labrusca*. Some California grape breeders are using *V. labrusca* as well. Other species are also being used now in table grape breeding in an attempt to backcross powdery mildew (Coleman et al. 2009; Ramming et al. 2011; Riaz et al. 2011) and Pierce's disease-resistant alleles (Riaz et al. 2009) into table grape selections.

5.4 Raisin Grape Breeding

Raisins should possess certain characteristics, namely, a soft texture, little tendency to become sticky, seedlessness, a pleasing flavor, and either large or very small size (Winkler 1949). Since much of the harvest in California is concentrated on

'Sultanina' ('Thompson Seedless'), early maturing types are needed to better utilize the labor required. In addition, the development of raisins which dry on the vine (DOV) with or without cuttings canes has the potential to reduce vineyard production costs, allow for mechanical harvesting, alleviate the dependency on labor, and avert the risk from late season rains in California. Early ripening raisin cultivars also help avoid risks associated with late season rain. Seedless muscat raisins are also under development, and cultivars such as 'Summer Muscat' (1999) and 'Diamond Muscat' (2000) have been released from USDA-ARS-Parlier.

5.5 *Scion Grape Cultivars*

Alleles conferring resistance to diseases have been identified in numerous species. Breeders in many locations are seeking to combine resistance alleles for each disease of importance into elite selections and cultivars and to utilize the tools of molecular breeding to track resistance alleles (Di Gaspero and Cattonaro 2010; Eibach and Töpfer 2010). The challenge is not just to combine multiple resistance sources for long lasting, stable resistance, but to backcross those alleles into a *V. vinifera* background while also separating the resistance alleles from low fruit quality genes also typical of nonvinifera species.

Many traits are considered to have quantitative inheritance and/or to be controlled by the coordinated expression of gene networks. Breeders are interested in not only alleles for major gene traits but also in the manipulation of quantitative traits. Environmental interactions are also important, as many genes express different phenotypes in different environments. Cold hardiness, yield, berry, and cluster shape are examples of traits likely to be quantitatively controlled and subject to genotype × environment interactions.

Public breeders also face the challenge of long-term funding for their scion cultivar development efforts in the face of diminishing government support. Though germplasm preservation efforts are generally well supported, the breeding programs that utilize germplasm resources face long-term operating budget difficulties. As a result, the number and quality of public grape breeding efforts in North America and elsewhere continue to decline.

5.6 *Rootstocks*

Grape rootstock improvement is focused on enhancing resistance to soil-borne pests and diseases and broadening environmental adaptation while retaining protection against phylloxera and ease of propagation (both rooting and grafting ability). Phylloxera protection is essential, but many rootstocks are available that provide protection against phylloxera yet are not commercially propagated because they are not substantially different enough from other rootstock varieties. New rootstock varieties in the future will be more narrowly adapted, addressing problems that are important regionally rather than globally. Pest management through rootstocks is expected to become more desirable as pesticides become more expensive and are increasingly regulated due to the risk they pose to human and animal health and to the environment broadly. Use of methyl bromide, a broad spectrum pesticide, has been substantially reduced following international agreement, and the use of other soil pesticides has been curtailed.

Following phylloxera, nematodes are the primary soil-borne pest of grapevines (Nicol et al. 1999). Although many species of nematodes feed on grapevine roots, rootstock breeding has focused on incorporating resistance to root-knot (*Meloidogyne*) species and the dagger nematode *Xiphinema index*. Cain et al. (1984) reported that virulent populations of root-knot nematodes emerged in vineyards where nematode-resistant rootstocks were used. Selection for virulent nematode populations should be anticipated and new sources of resistance (Boyden and Cousins 2003; Cousins and Lauver 2003; Walker et al. 1994a) identified and deployed to enhance the resistance breadth and durability. Recessive resistance to root-knot nematodes has been reported in grape (Cousins et al. 2007); recessive resistance genes provide a durable approach to disease control against biotrophic plant pathogens including nematodes (Wang and Goldman 1996; Qiu et al. 1997; Walters et al. 1997) and may be preferable to race-specific R-gene resistance. Other nematode species, including ring (*Mesocriconema xenoplax*), citrus (*Tylenchulus semipenetrans*), and root lesion (*Pratylenchus* species), can cause substantial damage by feeding on grape roots and locally may be more important than root-knot nematodes or *X. index* (Walker and Stirling 2008, Pinkerton et al. 2005). Identifying and utilizing sources of resistance to these nematodes will become more important as root-knot nematode and *X. index* management through rootstocks becomes widespread.

Dagger nematode *X. index* resistance should be viewed in the context of protection against fanleaf degeneration, caused by grapevine fanleaf virus. Resistance to the dagger nematode vector is insufficient to provide protection against the disease (Walker et al. 1994b). At present, only two rootstocks that provide protection against fanleaf degeneration, O39-16 and O43-43, are available. These rootstocks are half *V. vinifera* and O43-43 is considered prone to phylloxera damage (Walker et al. 1994b), although O39-16 has not shown phylloxera susceptibility even though it is half *V. vinifera*. Introducing rootstocks that provided resistance or tolerance to fanleaf degeneration and have no *V. vinifera* parentage will help ensure durable phylloxera protection. Wider environmental adaptation is needed as well, since O43-43 and O39-16 are poorly adapted to calcareous soils (Bavaresco et al. 2005). Other nematode transmitted virus diseases are targets of rootstock breeding. Ringspot declines, caused by tomato ringspot virus and tobacco ringspot virus, are vectored by nematodes of the *X. americanum* species complex. Field resistance to the virus has been identified in rootstocks (Stobbs et al. 1988), although the interaction of rootstocks with different virus and nematode populations has not been determined.

The rootstock interaction with the scion is influenced by scion virus status. Some rootstock varieties are used for woody virus indexing (Rowhani et al. 2005) because of their dramatic intolerant response to particular virus disease isolates.

Golino (1993) described the relationship between rootstock cultivar and virus disease isolate in the context of rootstock choice for vineyards. Rootstocks that are intolerant of a scion infected with a particular virus disease can contribute to severe and rapid vine decline and even death (Golino 1993). Spread of corky bark and leafroll diseases through the vectoring of their causal viral agents by mealybugs and other insects increases the value of tolerant rootstocks, since scions may become infected with a virus disease either after planting or through infected scion propagation material. AXR#1, a phylloxera susceptible *V. vinifera* × *V. rupestris* hybrid, is very tolerant of virus infected scions (Golino 1993), but provides insufficient protection to phylloxera. Since rootstock responses to virus diseases vary from highly tolerant to highly intolerant, it should be possible to select for rootstocks that are tolerant of virus infected scions and retain the phylloxera protection required.

Enhancing rootstock resistance to other pests and diseases should be expected to emerge as improved evaluation methods are developed, and the economic value of rootstocks as a management tool is demonstrated. Rootstock resistance to crown gall, caused by the bacterium *Agrobacterium vitis*, has been identified, although the role of rootstocks in management of the disease is not fully determined and rootstocks are not yet practically used to provide protection against crown gall (Burr et al. 1998). Interactions with scion cultivar susceptibility and pathogen strain apparently contribute to the challenge of managing crown gall with rootstocks. Armillaria root disease, caused by the fungus *Armillaria mellea*, is important where vineyards are planted on land converted from forests or orchards. Baumgartner et al. (2008) and Baumgartner and Rizzo (2006) evaluated rootstocks and identified varieties with resistance and tolerance. New techniques in screening rootstocks for resistance and tolerance to Armillaria root disease indicate that varieties can be screened to identify candidate rootstocks for use in Armillaria root disease prone vineyards and to identify resistant germplasm for use in breeding. Resistance and tolerance to other fungal pathogens of roots can be expected to follow a similar pattern; cotton root rot (*Phymatotrichum omnivorum*) is identified as a serious threat to viticulture in the southern United States, especially Texas, and a potential threat in central and southern California as well as other arid and semiarid regions (Walker 1992). Margarodes (*Eurhizococcus brasiliensis*), an insect pest important in Brazil, feeds on grapevine roots and causes damage similar to phylloxera, although common phylloxera-resistant rootstocks do not provide protection against margarodes (Camargo and Ritschel 2008); a breeding program for resistant rootstocks is underway in Brazil and demonstrates the possibilities for selecting for resistance against regionally important insect pests.

Rootstocks differ widely in the level of vigor contributed to their scions. Because pest and disease resistance are the primary reason that rootstocks are used in viticulture, vigor induction has not been the chief selection driver. Smaller vine size is related to earlier fruit and dormant bud maturation and improved winter hardiness. Rootstocks in grapevines do not demonstrate dramatic dwarfing effects as are seen in some other fruit crops, notably apple. Some rootstocks are devigorating, reducing trunk diameter and shoot growth. Growers presently choose the rootstocks first based on pest and disease resistance and then select among rootstock varieties to complement the scion, site, and management practices. Developing rootstock selections with known pest and disease resistance but reduced vigor induction would benefit grape growers by providing a broader spectrum of vigor induction from rootstocks.

Comparison of several autotetraploid grape rootstocks demonstrated that they induce less vigor in scions than the diploid varieties from which they are derived (Motosugi et al. 1999). This is also true of autotetraploid apple (Beakbane 1967) and citrus (Lee 1988) rootstocks. Autotetraploid grape rootstocks produced smaller vines when grown ungrafted (Motosugi et al. 2002a), but demonstrated the same high level of resistance to phylloxera as their diploid progenitors (Motosugi et al. 2002b). Autotetraploid rootstocks may provide an opportunity to select for a particular level of vigor reduction while maintaining pest and disease resistance.

Adaptation to abiotic stress and soil conditions will continue to be an important factor in rootstock selection, although as now, pest resistance will be foremost. Rootstocks that improve vine productivity and fruit quality with reduced water quality and quantity would be especially useful in irrigated regions, where the cost of water is increasing and availability is decreasing (Carbonneau 1985). While extensive characterization and selection of rootstocks for adaptation to high pH calcareous soils have been accomplished due to the prevalence of these soils in important European grape-producing regions, acidic vineyard soils (pH 5.5 and below) have not received as much attention in rootstock breeding. Adaptation to acidic vineyard soils should be an area for rootstock evaluation and improvement, as many tropical agricultural soils are acidic as are many vineyard soils in the northeastern United States. Repeated use of nitrogen fertilizers increases soil acidity. The rootstock cultivar 'Gravesac' was selected for acidic soils (Delas 1992; Pouget and Ottenwalter 1984), but rootstock trials for acidic soil adaptation often vary from region to region (Fráguas 1999; Conradie 1983), limiting the transferability of rootstock recommendations. Techniques in evaluation of other crop plants for adaptation to abiotic stresses and soil conditions (Raman et al. 2005; Raman et al. 2002) may be useful for developing new approaches in grapevine rootstock evaluation and improvement for such challenging conditions.

6 Breeding Methods and Techniques

6.1 Parental Selection

Parents are usually chosen based on the traits determined to be most important to the goals of a breeding program. Two parents are paired in crossing when each one harbors complementary desirable phenotypes that one seeks to combine in a new cultivar. These traits may be assessed through field observations or laboratory tests. Increasingly, parental selection is also based on knowledge of molecular markers linked to major genes as well as QTL affecting traits of interest (Eibach and Töpfer 2010). It is also possible to select two parents harboring different desirable genes affecting the same trait, where the breeder seeks to combine those genes/alleles into an elite selection.

Once the objectives of a grape breeding program are determined, and the parent vines are chosen, controlled pollinations are made and the resulting seeds harvested and grown. The traditional techniques of breeding depend upon basic knowledge of flower development and seed germination (Reisch and Pratt 1996).

6.2 *Pollination and Seedling Production*

Controlled pollination may be done by cutting a previously bagged, freshly blooming cluster and tapping it lightly against an emasculated cluster, which is immediately bagged again (Burger et al. 2009). Pollen that has been stored can be applied with a camel's hair brush.

Pollen is collected from newly opened flowers. Barrett and Arisumi (1952) stripped flowers from the cluster, dried them on a glass plate, then sifted out the pollen. The dry pollen can be scraped up with a razor blade and put into small vials or gelatin capsules. Pollen (often mixed with anthers and other flower parts) can also be stored in screw cap vials with a layer of cotton on top of desiccant. Equipment is cleaned with alcohol to kill unwanted pollen.

Grape pollen may be stored to use on a later blooming female parent or for other purposes. It has been kept for 4 years at low temperature (−12°C optimum) and low relative humidity (28% optimum) maintained by the appropriate mixture of sulfuric acid and water in a desiccator; pollen, which showed a germination percentage of 6% or better gave as good a set in the field as fresh pollen (Olmo 1942). Some researchers routinely store pollen at −20°C for 12 months (Boyden 2005) or longer.

Since hermaphroditic grapes are self-fertile, the buds must be emasculated prior to anther dehiscence for use in controlled crosses. The cap and the stamens are removed by forceps. A pair of eyebrow tweezers with broad ends can be notched with a file and the ends bent slightly inward. The calyptra is grasped between the notches and removed with one motion. Another emasculation tool was devised by Barrett and Arisumi (1952); small, sharp-pointed scissors were notched on the inside of the blades and the degree of closure regulated by a thumb screw. Breeders use a variety of types of forceps from straight fine-tipped to curved forceps, to blunt-end forceps. The tools chosen are based on personal experience.

Grape seeds often germinate poorly. Research on germination and seedling growth has been largely the by-product of breeding programs, rather than systematic physiological studies. A more comprehensive review was presented by Reisch and Pratt (1996).

Vitis seeds are usually extracted from the berries at or soon after fruit maturity by manual pressing, by slicing open individual berries, or in a laboratory blender operated at low speed to avoid chipping the seeds. The seeds have a hard seed coat of variable thickness (Pratt 1971).

Stratification at 0–10°C under moist conditions for about 3 months is essential for quick, uniform, and high germination (Flemion 1937; Scott and Ink 1950;

Harmon and Weinberger 1959; Rives 1965). Fungicides can be used to reduce fungal growth while in storage. To distribute moisture among stored seed, filter paper, sphagnum, or peat moss can be used. There is also evidence that 24 h treatment in 1.5% H_2O_2 followed by 24 h in 1,000 ppm GA_3 can reduce the chilling requirement to 21 days (Ellis et al. 1983). Hydrogen cyanamide was effective in promoting seed germination of four cultivars of *V. vinifera* (Spiegel-Roy et al. 1987), a 5-min soak substituted for chilling, allowing immediate germination of harvested seed.

Stratified seeds are planted in a variety of media, but results are best when the soil is light, well drained, and well aerated (e.g., addition of extra perlite). It is best to use supplementary light (e.g., 16:8 L:D photoperiod) when days are cloudy. Daytime temperatures of 28–32°C followed by 22°C at night is a good guideline to encourage rapid seedling growth with warm temperatures.

Once germinated, further care of seedlings depends on the climate as well as plans for early screening and selection. In some regions, seedlings are moved outdoors and planted to permanent vineyard locations (own-rooted) soon thereafter. In tropical regions, it may not be possible to grow seedlings own-rooted due to soil pathogens. In areas with short growing seasons, or where irrigation is not available at permanent vineyard sites, seedlings are grown in a field nursery for 1–2 years to attain sufficient size prior to planting to a permanent vineyard site.

6.3 Breeding Strategies

Grapevines are vegetatively propagated, and therefore the approach used in breeding aims to select single elite genotypes combining sets of desirable traits from both parents. Once a seedling with potential is selected, it is then propagated to other vineyards and other locations for replicated testing. Breeders usually employ a modified pedigree breeding scheme, where in every generation elite parents with complementary traits are crossed to produce the next generation of seedlings. Recurrent selection (Bouquet et al. 1981) as well as modified backcross breeding (Bouquet 1986) are also used. Because grapes are highly heterozygous, and it is desirable to maintain heterozygosity among new cultivars, when the latter technique is used, the recurrent parent varies in every generation. For instance, to introgress the *Run1* allele for powdery mildew resistance from *V. rotundifolia*, a cross was first made between *V. rotundifolia* and a *V. vinifera* wine grape parent. A resistant seedling from that cross was then backcrossed to a different *V. vinifera* wine grape, and the same was done with a third *V. vinifera* wine grape in the next generation; each time a resistant seedling was chosen for backcrossing (Bouquet 1986).

Inbreeding has been used little due to severe inbreeding depression (Burger et al. 2009), however, lines of use for genetic studies have been developed by inbreeding. 'Pinot noir' was self-pollinated for six generations to produce a highly homozygous line (Bronner and Oliveira 1990) later used for genomic sequencing (Jaillon et al. 2007). Some breeders carry out limited inbreeding for the development of parental lines better able to transmit desired traits to progeny populations.

7 Integration of New Biotechnologies in Breeding Programs

7.1 Molecular Marker Maps

Many molecular markers have been developed over the last couple of decades for use in grapevine. These molecular markers have multiple uses in grape breeding and genetics, including cultivar identification and germplasm management; mapping of traits of interest; and estimation of genetic diversity (Bowers et al. 1993; Bowers and Meredith 1996; Dalbó et al. 2000; Ye et al. 1998). Since the early 1990s, microsatellites have been one of the most widely used molecular marker systems in grapevine (Thomas et al. 1993; Thomas and Scott 1993) and to a lesser extent AFLPs (Cervera et al. 1998). The number of publicly available microsatellite markers has greatly expanded (Bowers et al. 1999b; Bowers et al. 1996), including descriptions of multiplexes (Merdinoglu et al. 2005); reference sets of alleles and accessions (This et al. 2004; Laucou et al. 2011); the development of microsatellites directly from transcribed sequence of ESTs (Scott et al. 2000); and linkage maps based on microsatellite markers (Adam-Blondon et al. 2004; Riaz et al. 2004). With the rapid decrease in the cost of DNA sequencing, molecular markers for grape have begun to shift to large datasets of SNP markers (Myles et al. 2011; Myles et al. 2010) and we are now entering the era of genotyping by whole-genome sequencing (Elshire et al. 2011).

Many genetic linkage maps have been constructed for grapevine since the first reported genetic map utilizing DNA-based markers (Lodhi et al. 1995). These maps represent *V. vinifera* intraspecific crosses (Adam-Blondon et al. 2004; Doligez et al. 2006; Fanizza et al. 2005; Riaz et al. 2004) as well as interspecific crosses utilizing *V. vinifera* (Grando et al. 2003), and more complex interspecific crosses (Doucleff et al. 2004; Fischer et al. 2004; Lodhi et al. 1995; Lowe and Walker 2006; Mandl et al. 2006).

In 2007, the whole genome sequence of two grapevine genomes was reported (Jaillon et al. 2007; Velasco et al. 2007). The release of these genome sequences marked a significant watershed moment in the history of grape genetics and breeding. The availability of genomic sequence data is already having a significant impact upon grapevine improvement efforts (Di Gaspero and Cattonaro 2010).

7.2 Disease and Abiotic Stress Resistance

Powdery mildew is the most significant fungal pathogen of grape as it affects many production regions worldwide. Several sources of powdery mildew resistance have been identified among North American *Vitis* species. Among the 38 chromosome species of the *Euvitis* section, resistance is quantitatively inherited, in which narrow-sense heritability estimates have been made ranging from 0.31 to 0.51 (Eibach et al. 1989). A single, dominant locus for resistance to powdery mildew, *Run1*, has been identified from the 40 chromosome *V. rotundifolia*. The locus has been introgressed into a *V. vinifera* background in which multiple generations of back-crossing have now occurred (Bouquet 1986). The *Run1* locus has now been mapped, first by identifying candidate genes in the region showing similarity to conserved plant-resistance genes (Donald et al. 2002; Pauquet et al. 2001) and subsequently, through the fine genetic and physical mapping of this locus (Anderson et al. 2011; Barker et al. 2005). Additional major loci for powdery mildew resistance have been identified and mapped, including a locus from within *V. vinifera* and from an Asian species, *V. romanetii* (Coleman et al. 2009; Ramming et al. 2011; Riaz et al. 2011).

QTL for powdery mildew and downy mildew resistance have been identified in multiple interspecific crosses, which have utilized *Euvitis* sources of resistance (Dalbó et al. 2001; Fischer et al. 2004; Moreira et al. 2011; Welter et al. 2007). Additional efforts to identify candidate genes that have a high probability of being linked to disease resistance loci has been conducted by identifying resistance gene analogs and resistance gene-like genes from numerous grape species (Di Gaspero and Cipriani 2002; Di Gaspero and Cipriani 2003; Welter et al. 2007).

Downy mildew resistance has been identified in several North American species and is quantitatively inherited. Loci for downy mildew resistance (Bellin et al. 2009), phylloxera resistance (Zhang et al. 2009), and dagger nematode resistance (Hwang et al. 2010; Xu et al. 2008) have also been mapped.

Pierce's disease has traditionally limited the cultivation of *V. vinifera* in the south-eastern United States and in recent years has become a more serious concern in California due to the spread of insect vectors capable of spreading the causal bacterium to wider production regions. Mortensen (1968) estimated the resistance to Pierce's disease to be a dominant trait, and qualitatively controlled by three independent loci upon observing resistance in several segregating populations derived from *V. aestivalis* var. *aestivalis*, *V. cinerea* var. *floridana*, and *V. shuttleworthii* in Florida under field conditions for 5 years. More recently, the narrow-sense heritability of Pierce's disease resistance was estimated to range from 0.37 to 0.63 for different populations of the pathogen, *Xylella fastidiosa*, in a hybrid population derived from *V. rupestris* × *V. arizonica* (Krivanek et al. 2005). These results indicated the existence of a major gene for Pierce's disease resistance, *PdR1*, which has now been placed on a genetic linkage map of this cross (Krivanek et al. 2006; Riaz et al. 2006, 2008).

Several genetic sources of resistance for abiotic stress are known, yet, little work has been reported on the genetic mapping of abiotic stress resistance genes. QTL for magnesium deficiency were identified and placed on a map of 'Welschriesling' × 'Sirius' (Mandl et al. 2006). Also, QTL for a photoperiod-induced growth cessation derived from *V. riparia* have been identified (Garris et al. 2009).

7.3 Fruit Quality

The genetic control and inheritance of fruit color or anthocyanin production in grapevine is not fully understood despite evidence that the primary determination of anthocyanin production in berries appears to be controlled by a single dominant locus in *V. vinifera* (Doligez et al. 2002; Riaz et al. 2004) with white fruit being a recessive character. This observation is supported by numerous reports showing that controlled crosses between white-fruited vines universally result in white-fruited progeny (Barritt and Einset 1969; Hedrick and Anthony 1915; Madero et al. 1986; Snyder and Harmon 1939; Snyder and Harmon 1952; Wellington 1939).

The presence of *Gret1*, a Ty3-*gypsy*-type retrotransposon in the promoter region of *Vvmyba1*, a *myb*-like regulatory gene showing sequence similarity to previously described anthocyanin regulators from maize and other plants, is present in white-fruited cultivars of *V. vinifera* (Kobayashi et al. 2004). White-fruited grapes are linked to the homozygous presence of *Gret1* in the promoter region of *Vvmyba1* as well as mutations in the tightly linked gene *Vvmyba2* (This et al. 2007; Walker et al. 2007). Pigmented cultivars possess at least one allele at the *VvmybA1* locus not containing this large insertion (Kobayashi et al. 2004). Evidence shows that *VvmybA1* co-segregates with the morphological marker for berry color (Lijavetzky et al. 2006) and that mutations in *VvmybA1* are associated with the vast majority of white-fruited *V. vinifera* accessions and many pink and red accessions as well (Lijavetzky et al. 2006; This et al. 2007). Additional polymorphisms within this cluster of closely related *myb*-genes are also significantly associated with quantitative variation in anthocyanin content of berries (Fournier-Level et al. 2009).

Genetic mapping for muscat flavor and specific monoterpenes has been completed (Doligez et al. 2006) identifying a significant QTL. The gene 1-deoxy-D-xylulose 5-phosphate synthase co-segregates with QTL controlling monoterpene production and has been further validated by association mapping (Battilana et al. 2009; Emanuelli et al. 2010).

Many differing hypotheses have been put forth over the last 70 years to explain the inheritance of stenospermocarpic seedlessness in grapes. Many of these hypotheses are based on small population sizes and limited numbers of populations. The most recent hypothesis to explain the inheritance of seedlessness attempts to take into consideration all prior reports on the segregation patterns of this trait and concludes seedlessness is controlled by three independent recessive genes plus one dominant acting regulatory gene (Bouquet and Danglot 1996).

Several markers linked to seedlessness have been identified (Adam-Blondon et al. 2001; Lahogue et al. 1998) and the gene *VvAGL11* co-localizes with a major QTL for stenospermocarpy (Mejia et al. 2011). Additional fruit quality traits are just beginning to be genetically analyzed at the molecular level. Other recent progress is reported for mapping of QTL associated with yield components (Cabezas et al. 2006; Doligez et al. 2006; Fanizza et al. 2005) and vine phenology (Cabezas et al. 2006; Costantini et al. 2008).

7.4 Association Mapping

Linkage disequilibrium (LD)-based association mapping is of interest to grape geneticists considering the potentially high resolution and the time savings associated with the utilization of existing germplasm collections (Owens 2011;

This et al. 2006). Significant haplotypic LD was observed over 30 cm in a *V. vinifera* core collection when estimating LD with 38 microsatellite markers scattered among the 19 *Vitis* linkage groups (Barnaud et al. 2005). Utilizing a subset of the pigmented accessions from the same core collection, a much more rapid decay in LD at the single locus level was observed (This et al. 2007). High-density SNP analysis has shown the LD decays rapidly in *V. vinifera* and is at background levels within only approximately 2 kb (Myles et al. 2011). A candidate gene-based association mapping strategy was also employed to test candidates within a major QTL for monoterpene production (Emanuelli et al. 2010).

7.5 Tissue Culture

Somatic embryogenesis has been documented in grapevine for over 30 years (Hirabayashi et al. 1976; Mullins and Srinivasan 1976). Success has been reported primarily for the use of sporophytic anther (Rajasekaran and Mullins 1979) and ovary tissues (Kikkert et al. 2005), although leaves, petioles, and stem segments have also been used to establish embryogenic calli (Krul and Worley 1977). Success in regeneration of grapevine has been limited to a relatively small number of cultivars, but the list of successful source material has been steadily increasing (Perrin et al. 2004; Torregrosa 1998). Many modifications and improvements have been reported (Iocco et al. 2001; Perl and Eshdat 1998; Perl et al. 1995; Perrin et al. 2001; Wang et al. 2004).

7.6 Genetic Transformation

Early attempts to transform grape using *Agrobacterium tumefaciens* met with difficulty despite the bacterium being a naturally occurring pathogen of the species. The use of high-quality embryogenic suspension cell cultures has allowed the transformation of grape using *Agrobacterium* to become routine in many laboratories around the world (Perl and Eshdat 1998). Biolistic transformation using DNA-coated microprojectiles has been reported in grape since the early 1990s (Hébert et al. 1993) for the interspecific hybrid 'Chancellor' (Kikkert et al. 1996) and has expanded over time to successfully include cultivars of *V. vinifera* (Vidal et al. 2003). Presently, *Agrobacterium*-mediated methods are the prominently employed protocols for grape transformation worldwide and continued improvements in protocols have been made (Dhekney et al. 2009; Dutt et al. 2008; Li et al. 2008).

The most notable success in grapevine scion transformation is the insertion of antimicrobial genes (Rosenfield et al. 2010; Vidal et al. 2003) and antifungal genes (Yamamoto et al. 2000) to potentially confer greater bacterial and fungal disease resistance. Transgenic vines of 'Chardonnay' with the ability to produce magainins, short peptides with broad-spectrum antimicrobial activity, were tested for resistance to crown gall and powdery mildew (Vidal et al. 2006). Lines expressing magainins had significantly reduced crown gall symptoms under controlled conditions, and had only a limited ability to reduce powdery mildew disease symptoms. To date, testing of disease resistance of transgenic grapevines under field conditions has not been reported. The advantages of reduced pesticide use and stable, long-term disease resistance are counter-balanced by public concern over the release of transgenic grapevines, and it is difficult to foresee when, or if, this promising technology will be successfully commercialized.

Transformation of grapevine has not just been restricted to scion varieties but has also been used to transfer useful traits to rootstocks. *Vr-ERE*, a gene encoding a NADPH-dependent aldehyde reductase, which coverts eutypine to the alcohol eutypinol, has been shown in laboratory tests with both cultured *V. vinifera* cells and whole vines to have some efficacy in detoxifying the toxin produced by the fungus *Eutypa lata* (Guillen et al. 1998; Legrand et al. 2003). Rootstocks transformed with a chimeric gene containing the alfalfa PR 10 promoter and a *Vitis* stilbene synthase gene (*Vst1*) showed enhanced foliar resistance to *Botrytis cinerea*, primarily a pathogen of fruit (Coutos-Thevenot et al. 2001). Attempts to transform rootstocks with genes potentially capable of conferring resistance to crown gall and multiple viruses have been reported (Fuchs et al. 2007; Valat et al. 2006; Xue et al. 1999).

One concern with the release of transgenic grapevines is the ease with which pollen flow may occur and the existence of wild grape species in many production regions. Assessment of the field safety of transgenic vines containing the coat protein gene from grapevine fanleaf virus has recently been conducted (Fuchs et al. 2007; Valat et al. 2006; Vigne et al. 2004).

References

Adam-Blondon, A.-F., Lahogue-Esnault, F., Bouquet, A., Boursiquot, J.-M. and This, P. (2001) Usefulness of two SCAR markers for marker-assisted selection of seedless grapevine cultivars. Vitis 40, 147–155.

Adam-Blondon, A.-F., Roux, C., Claux, D., Butterlin, G., Merdinoglu, D. *et al.*, (2004) Mapping 245 SSR markers on the *Vitis vinifera* genome: a tool for grape genetics. Theor. Appl. Genet. 109, 1017–1027.

Alleweldt, G. and Dettweiler, E. (1994) *The genetic resources of Vitis - world list of grapevine collections*. Geilweilerhof, Germany.

Anderson, C., Choisne, N., Adam-Blondon, A.-F. and Dry, I.B. (2011) Positional cloning of disease resistance genes in grapevine. In: A.-F. Adam-Blondon and J. M. Martinez Zapater (Eds.), *Genetics, Genomics and Breeding of Grapes*. Science Publishers, St. Helier, Jersey, British Isles, pp. 186–210.

Aradhya, M. K., Dangl, G.S., Prins, B.H., Boursiquot, J.-M., Walker, M.A. *et al.*, (2003) Genetic structure and differentiation in cultivated grape, *Vitis vinifera* L. Genet. Res. 81, 179–182.

Arroyo-Garcia, R., Ruiz-Garcia, L., Bolling, L., Ocete, R., Lopez, M.A. *et al.*, (2006) Multiple origins of cultivated grapevine (*Vitis vinifera* L. ssp. *sativa*) based on chloroplast DNA polymorphisms. Molec. Ecol. 15, 3707–3714.

Ashikawa, K. (1972) New grape variety 'Takao'. Bul. Tokyo-to Agr. Expt. Sta. 7:1–9.

Barker, C. L., Donald, T., Pauquet, J., Ratnaparkhe, M.B., Bouquet, A. *et al.*, (2005) Genetic and physical mapping of the grapevine powdery mildew resistance gene, *Run1*, using a bacterial artificial chromosome library. Theor. Appl. Genet. 111, 370–377.

Barnaud, A., Lacombe, T. and Doligez, A. (2005) Linkage disequilibrium in cultivated grapevine, *Vitis vinifera* L. Theor. Appl. Genet. 112, 708–716.

Barrett, H.C. and Arisumi, T. (1952) Methods of pollen collection, emasculation and pollination in fruit breeding. Proc. Amer. Soc. Hort. Sci. 59, 259–262.

Barritt, B. H. and Einset, J. (1969) Inheritance of 3 major fruit colors in grapes. J. Amer. Soc. Hort. Sci. 94, 87–89.

Basiouny, F. M. and Himelrick, D. G. (2001) *Muscadine Grapes*. ASHS Press, Alexandria, Virginia.

Battilana, J., Costantini L., Emanuelli, F., Sevini, F., Segala, C., Moser, S., Velasco, R., Versini, G. and Grando, M.S. (2009) The 1-deoxy-d-xylulose 5-phosphate synthase gene co-localizes with a major QTL affecting monoterpene content in grapevine. Theor. Appl. Genet. 118, 653–669.

Baumgartner, K. and Rizzo D. M. (2006) Relative resistance of grapevine rootstocks to *Armillaria* root disease. Amer. J. Enol. Vitic. 57, 408–414.

Baumgartner, K., Bhat, R. and Fujiyoshi P. (2008) Characterizing resistance to infection by the root pathogen *Armillaria mellea* in tolerant and susceptible grapevine rootstocks. Phytopathology 98, S22.

Bavaresco, L., Presutto, P., and Civardi, S. (2005) VR 043–43: a lime susceptible rootstock. Amer. J. Enol. Vitic. 56, 192–195.

Beakbane, A. B. (1967) The dwarfing effect of a tetraploid sport of M.XIII apple rootstock. Rep. East Malling Res. Sta. for 1966, 96–97.

Bellin, D., Peressotti, E., Merdinoglu, D., Wiedemann-Merdinoglu, S., Adam-Blondon, A.-F. *et al.*, (2009) Resistance to *Plasmopara viticola* in grapevine 'Bianca' is controlled by a major dominant gene causing localised necrosis at the infection site. Theor. Appl. Genet. 120, 163–176.

Bloodworth, P.J., Nesbitt, W.B. and Barker, K.R. (1980) Resistance to root knot nematodes in *Euvitis* x *Muscadinia* hybrids, In: *Proceedings of the 3rd International Symposium on Grape Breeding*, Davis, CA. pp. 275–292.

Boss, P.K., and Thomas, M.R. (2002) Association of dwarfism and floral induction with a grape 'green revolution' mutation. Nature 416, 847–850.

Bouquet, A., (1986) Introduction dans l'espéce *Vitis vinifera* L. d'un caractére de résistance à l'oidium (*Uncinula necator* Schw. Burr.) issu de l'espéce *Muscadinia rotundifolia* (Michx) Small. Vignevini 13, Suppl. 12, 141–146.

Bouquet, A., and Danglot, Y. (1996) Inheritance of seedlessness in grapevine (*Vitis vinifera* L). Vitis 35, 35–42.

Bouquet, A., Truel, P. and Wagner, R. (1981) Recurrent selection in grapevine breeding (in French, English summary). Agronomie 1, 65–73.

Bowers, J., Boursiquot, J.-M., This, P., Chu, K., Johansson, H. *et al.*, (1999a) Historical Genetics: the parentage of Chardonnay, Gamay, and other wine grapes of Northeastern France. Science 285, 1562–1565.

Bowers, J.E., Dangl, G.S. and Meredith, C. P. (1999b) Development and characterization of additional microsatellite DNA markers for grape. Amer. J. Enol. Vitic. 50, 243–246.

Bowers, J.E., and Meredith, C. P. (1997) The parentage of a classic wine grape, Cabernet Sauvignon. Nature Genetics 16, 84–87.

Bowers, J.E., and Meredith, C.P. (1996) Genetic similarities among wine grape cultivars revealed by restriction fragment-length polymorphism (RFLP) analysis. J. Amer. Soc. Hort. Sci. 121, 620–624.

Bowers, J.E., Bandman, E. B. and Meredith, C. P. (1993) DNA fingerprint characterization of some wine grape cultivars. Amer. J. Enol. Vitic. 44, 266–273.

Bowers, J.E., Dangl, G.S., Vignani, R. and Meredith, C. P. (1996) Isolation and characterization of new polymorphic simple sequence repeat loci in grape. Genome 39, 628–633.

Boyden, L.E. (2005) Allelism of root-knot nematode resistance and genetics of leaf traits in grape rootstocks. Ph.D. Thesis. Cornell University, Ithaca.

Boyden, L.E. and Cousins, P. (2003) Evaluation of *Vitis aestivalis* and related taxa as sources of resistance to root-knot nematodes. Acta Horticulturae 623, 283–290.

Bronner, A. and Oliveira, J. (1990) Creation and study of the Pinot noir variety lineage. *Vitis* (special issue) Proc. 5th Intern. Symp. Grape Breeding, St. Martin/Pfalz, Germany, 12–16 September 1989, pp. 69–80.

Burger, P., Bouquet, A. and Striem, M.J. (2009) Grape breeding. In: S.M. Jain and P.M. Priyadarshan (Eds.). *Breeding Plantation Tree Crops: Tropical Species*. Springer, pp. 161–189.

Burr, T. J., Bazzi, C., Süle, S., and Otten, L. (1998) Crown gall of grape: biology of *Agrobacterium vitis* and the development of disease control strategies. Plant Dis. 82, 1288–1297.

Cabezas, J.A., Cervera, M. T., Ruiz-Garcia, L., Carreno, J. and Martinez-Zapater, J. M. (2006) A genetic analysis of seed and berry weight in grapevine. Genome 49, 1572–1585.

Cahoon, G.A. (1998) French hybrid grapes in North America, In: D.C. Ferree (Ed.), *A history of fruit varieties*. Good Fruit Grower Magazine, Yakima, Washington. pp. 152–168.

Cain, D. W., McKenry, M. V., and Tarailo, R. E. (1984) A new pathotype of root-knot nematode on grape rootstocks. J. Nematol. 16, 207–208.

Camargo, U. A. and Ritschel, P. S. (2008) New table and wine grape cultivars: world scenario with emphasis on Brazil. Acta Horticulturae 785, 89–95.

Campbell, C. (2005) *The Botanist and the Vintner: How Wine Was Saved for the World*. Algonquin Books of Chapel Hill, Chapel Hill.

Carbonneau, A. (1985) The early selection of grapevine rootstocks for resistance to drought conditions. Amer. J. Enol. Vitic. 36, 195–198.

Cattell, H., and Miller, L. S. (1980) *The Wines of the East. Vol. III. Native American Grapes*. L& H Photojournalism, Lancaster, PA.

Cervera, M.-T., Cabezas, J. A., Sancha, J. C., Martinez de Toda, F. and Martinez-Zapater, J.M. (1998) Application of AFLPs to the characterization of grapevine *Vitis vinifera* L. genetic resources. A case study with accessions from Rioja (Spain). Theor. Appl. Genet. 97, 51–59.

Chisholm, M.G., Guiher, L.A., Vonah, T.M. and Beaumont, J.L. (1994) Comparison of some French-American hybrid wines with White Riesling using Gas Chromatography-Olfactometry. Amer. J. Enol. Vitic. 45, 201–212.

Clark, J. R. (1997) Grape. In: The American Society for Horticultural Sciences, (Ed.). The Brooks and Olmo Register of Fruit and Nut Varieties. ASHS Press, Alexandria, Virginia. pp 248–299.

Coleman, C., Copetti, D., Cipriani, G., Hoffmann, S., Kozma, P., Kovács, L., Morgante, M., Testolin, R. and Di Gaspero, G. (2009) The powdery mildew resistance gene *REN1* co-segregates with an NBS-LRR gene cluster in two Central Asian grapevines. BMC Genet. 10, 89.

Conradie, W. J. (1983) Liming and choice of rootstocks as cultural techniques for vines in acid soils. S. Afr. J. Enol. Vitic. 4, 39–44.

Costantini, L., Battilana, J., Lamaj, F., Fanizza, G. and Grando, M. (2008) Berry and phenology-related traits in grapevine (*Vitis vinifera* L.): From Quantitative Trait Loci to underlying genes. BMC Plant Biol. 8, 38.

Cousins, P. (2007) Tiny grape could do big things. Agric. Res. 55, 23.

Cousins, P., Johnston, D., Switras-Meyer, S. and Meyer, C. (2007) Recessive resistance to the root-knot nematode *Meloidogyne incognita* derived from the grapevine rootstock 3309 C. J. Nematology 39, 70–71.

Cousins, P. and Lauver, M. (2003) Segregation of resistance to root-knot nematodes in a *Vitis vulpina* hybrid population. Acta Horticulturae 623, 313–318.

Coutos-Thevenot, P., Poinssot, B., Bonomelli, A., Yean, H., Breda, C. *et al.*, (2001) In vitro tolerance to *Botrytis cinerea* of grapevine 41B rootstock in transgenic plants expressing the stilbene synthase *Vst1* gene under the control of a pathogen-inducible PR 10 promoter. J. Exp. Bot. 52, 901–910.

Crespan, M. (2003) The parentage of Muscat of Hamburg. Vitis 42, 193–197.

Dalbó, M. A., Ye, G.N., Weeden, N.F., Steinkellner, H., Sefc, K.M. and Reisch, B.I. (2000) A gene controlling sex in grapevines placed on a molecular marker-based genetic map. Genome 43, 333–340.

Dalbó, M. A., Ye, G.N., Weeden, N.F., Wilcox, W.F. and Reisch, B.I. (2001) Marker-assisted selection for powdery mildew resistance in grape. J. Amer. Soc. Hort. Sci. 126, 83–89.

Dangl, G. S., Mendum, M.L., Prins, B.H., Walker, M. A., Meredith, C. P. *et al.*, (2001) Simple sequence repeat analysis of a clonally propagated species: A tool for managing a grape germplasm collection. Genome 44, 432–438.

Delas, J. J. (1992) Criteria used for rootstock selection in France. In: J.A. Wolpert, M.A. Walker and E. Weber. (Eds.). *Proceedings Rootstock Seminar: A Worldwide Perspective, Reno, Nevada, June 24, 1992.* The American Society for Enology and Viticulture, Davis, California. pp. 1–14.

Dettweiler, E., Jung, A., Zyprian, E. and Töpfer, R. (2000) Grapevine cultivar Müller-Thurgau and its true to type descent. Vitis 2, 63–65.

Dhekney, S. A., Li, Z.T., Zimmerman, T.W. and Gray, D.J. (2009) Factors influencing genetic transformation and plant regeneration of *Vitis*. Amer. J. Enol. Vitic. 60, 285–292.

Di Gaspero, G. and Cattonaro, F. (2010) Application of genomics to grapevine improvement. Aust. J. Grape Wine Res. 16 (supplement S1), 122–130.

Di Gaspero, G. and Cipriani, G. (2002) Resistance gene analogs are candidate markers for disease-resistance genes in grape (*Vitis* spp.). Theor. Appl. Genet. 106, 163–172.

Di Gaspero, G. and Cipriani, G. (2003) Nucleotide binding site/leucine-rich repeats, *Pto*-like and receptor-like kinases related to disease resistance in grapevine. Molec. Genet. Genomics 269, 612–623.

Doligez, A., Audiot, E., Baumes, R. and This, P. (2006) QTLs for muscat flavor and monoterpenic odorant content in grapevine (*Vitis vinifera* L.). Molec. Breeding 18, 109–125.

Doligez, A., Bouquet, A., Danglot, Y., Lahogue, F., Riaz, S. *et al.*, (2002) Genetic mapping of grapevine (*Vitis vinifera* L.) applied to the detection of QTLs for seedlessness and berry weight. Theor. Appl. Genet. 105, 780–795.

Donald, T. M., Pellerone, F., Adam-Blondon, A.-F., Bouquet, A., Thomas, M.R. *et al.*, (2002) Identification of resistance gene analogs linked to a powdery mildew resistance locus in grapevine. Theor. Appl. Genet. 104, 610–618.

Doucleff, M., Jin, Y., Gao, F., Riaz, S., Krivanek, A.F. *et al.*, (2004) A genetic linkage map of grape, utilizing *Vitis rupestris* and *Vitis arizonica*. Theor. Appl. Genet. 109, 1178–1187.

Dutt, M., Li, Z.T., Dhekney, S.A. and Gray, D.J. (2008) A co-transformation system to produce transgenic grapevines free of marker genes. Plant Sci. 175, 423–430.

Eibach, R. and Töpfer, R. (2010) Progress in grapevine breeding. In: 10th International Conference on Grapevine Breeding and Genetics, Geneva, New York. New York State Agricultural Experiment Station. (abstract).

Eibach, R., Diehl, H. and Alleweldt, G. (1989) Untersuchungen zur Vererbung von Resistenzeigenschaften bei Reben gegen *Oidium tuckeri*, *Plasmopara viticola* und *Botrytis cinerea*. Vitis 28, 209–228.

Einset, J. and Lamb, B. (1951) Chimeral sports of grapes. J. Hered. 42, 158–162.

Einset, J. and Pratt, C. (1954) Giant sports of grapes. Proc. Amer. Soc. Hort. Sci. 63, 251–256.

Ellis, R.H., Hong, T.D. and Roberts, E.H. (1983) A note on the development of a practical procedure for promoting the germination of dormant seed of grape (*Vitis* spp.). Vitis 22, 211–219.

Elshire, R., Glaubitz, J., Sun, Q., Poland, J., Kawamoto, K., Buckler, E. and Mitchell S. (2011) A robust, simple genotyping-by-sequencing (GBS) approach for high diversity species. PLoS ONE 6(5), e19379. doi:10.1371/journal.pone.0019379.

Emanuelli, F., Battilana, J., Costantini, L., Le Cunff, L., Boursiquot, J.M., This, P. and Grando, M.S. (2010) A candidate gene association study on muscat flavor in grapevine (*Vitis vinifera* L.). BMC Plant Biol. 10, 241.

Ergul, A., Kazan, K., Aras, S., Cevik, V., Celik, H. *et al.*, (2006) AFLP analysis of genetic variation within the two economically important Anatolian grapevine (*Vitis vinifera* L.) varietal groups. Genome 49, 467–475.

Ewart, A.J.W. (1988) Sources of variation: Vineyard to wine judging. In: R. Smart, R. Thornton, S. Rodriguez and J. Young (Eds). Proc. 2nd Int. Cool Climate Viticulture and Oenology

Symposium, 11–15 January 1988; New Zealand Society for Viticulture and Oenology, Auckland, New Zealand, pp. 209–210.

Fanizza, G., Lamaj, F., Costantini, L., Chaabane, R. and Grando, M.S. (2005) QTL analysis for fruit yield components in table grapes (*Vitis vinifera*). Theor. Appl. Genet. 111, 658–664.

Fatahi, R., Ebadi, A., Bassil, N., Mehlenbacher, S. A. and Zamani, Z. (2003) Characterization of Iranian grapevine cultivars using microsatellite markers. Vitis 42, 185–192.

Fernandez, L., Romieu, C., Moing, A., Bouquet, A., Maucourt, M. *et al.*, (2006) The grapevine fleshless berry mutation. A unique genotype to investigate differences between fleshy and non-fleshy fruit. Plant Phys. 140, 537–547.

Fernandez, L., Torregrosa, L., Segura, V., Bouquet, A., and Martinez-Zapater, J.M. (2010) Transposon-induced gene activation as a mechanism generating cluster shape somatic variation in grapevine. Plant J. 61, 545–557.

Firoozabady, E. and Olmo, H. P. (1982) The heritability of resistance to root-knot nematode (*Meloidogyne incognita acrita* CHIT.) in *Vitis vinifera* x *V. rotundifolia* hybrid derivatives. Vitis 21, 136–144.

Fischer, B. M., Salakhutdinov, I., Akkurt, M., Eibach, R., Edwards, K.J. *et al.*, (2004) Quantitative trait locus analysis of fungal disease resistance factors on a molecular map of grapevine. Theor. Appl. Genet. 108, 501–515.

Flemion, F. (1937) After-ripening at 5°C favors germination of grape seeds. Contrib. Boyce Thompson Inst. 9, 7–15.

Fournier-Level, A., Le Cunff, L., Gomez, C., Doligez, A., Ageorges, A. *et al.*, (2009) Quantitative genetic bases of anthocyanin variation in grape (*Vitis vinifera* L. ssp *sativa*) berry: A quantitative trait locus to quantitative trait nucleotide integrated study. Genetics 183, 1127–1139.

Fráguas, J. C. (1999) Tolerância de porta-enxertos de vidiera ao alumínio do solo. Pesq. Agropec. Bras. 34, 1193–1200.

Franks, T., Botta, R. and Thomas, M.R. (2002) Chimerism in grapevines: implications for cultivar identity, ancestry and genetic improvement. Theor. Appl. Genet. 104, 192–199.

Fuchs, M., Cambra, M., Capote, N., Jelkmann, W., Kundu, J. *et al.*, (2007) Safety assessment of transgenic plums and grapevines expressing viral coat protein genes: New insights into real environmental impact of perennial plants engineered for virus resistance. J. Plant Path. 89, 5–12.

Garris, A., Clark, L., Owens, C., McKay, S., Luby, J. *et al.*, (2009) Mapping of photoperiod-induced growth cessation in the wild grape *Vitis riparia*. J. Amer. Soc. Hort. Sci. 134, 261–272.

Golino, D. A. (1993) Potential interactions between rootstocks and grapevine latent viruses. Amer. J. Enol. Vitic. 44, 148–152.

Goto-Yamamoto, N., Mouri, H., Azumi, M., and Edwards, K.J. (2006) Development of grape microsatellite markers and microsatellite analysis including oriental cultivars. Amer. J. Enol. Vitic. 57, 105–108.

Grando, M. S., Bellin, D., Edwards, K. J., Pozzi, C., Stefanini, M. *et al.*, (2003) Molecular linkage maps of *Vitis vinifera* L. and *Vitis riparia* Mchx. Theor. Appl. Genet. 106, 1213–1224.

Grassi, F., Labra, M., Imazio, S., Spada, A., Sgorbati, S. *et al.*, (2003) Evidence of a secondary grapevine domestication centre detected by SSR analysis. Theor. Appl. Genet. 107, 1315–1320.

Guillen, P., Guis, M., Martinez-Reina, G., Colrat, S., Dalmayrac, S. *et al.*, (1998) A novel NADPH-dependent aldehyde reductase gene from *Vigna radiata* confers resistance to the grapevine fungal toxin eutypine. The Plant J. 16, 335–343.

Hardy, P.J. (1970) Changes in volatiles of muscat grapes during ripening. Phytochem. 9, 709–715.

Harmon, F.N. and Weinberger, J.H. (1959) Effects of storage and stratification on germination of vinifera grape seeds. Proc. Amer. Soc. Hort. Sci. 73, 147–150.

Hébert, D., Kikkert, J. R., Smith, F. D. and Reisch, B. I. (1993) Optimization of biolistic transformation of embryogenic grape cell suspensions. Plant Cell Rep. 13, 405–409.

Hedrick, U. P. and Anthony, R. D. (1915) Inheritance of certain characters of grapes. New York State Agricultural College Technical Bulletin No. 45, 3–19.

Hirabayashi, T., Kozaki, I. and Akihama, T. (1976) *In vitro* differentiation of shoots from anther callus in *Vitis*. HortScience 11, 511–512.

Hocquigny, S., Pelsey, F., Dumas, V., Kindt, S., Heloir, M.-C. *et al.*, (2004) Diversification within grapevine cultivars goes through chimeric states. Genome 47, 579–589.

Hvarleva, T., Rusanov, K., Lefort, F., Tsvetkov, I., Atanassov A., *et al.*, (2004) Genotyping of Bulgarian *Vitis vinifera* L. cultivars by microsatellite analysis. Vitis 43, 27–34.

Hwang, C. F., Xu, K. N., Hu, R., Zhou, R., Riaz, S. *et al.*, (2010) Cloning and characterization of *XiR1*, a locus responsible for dagger nematode resistance in grape. Theor. Appl. Genet. 121, 789–799.

Iocco, P., Franks, T. and Thomas, M. R. (2001) Genetic transformation of major wine grape cultivars of *Vitis vinifera* L. Transgenic Res. 10, 105–112.

Jaillon, O., Aury, J. M., Noel, B., Policriti, A., Clepet, C. *et al.*, (2007) The grapevine genome sequence suggests ancestral hexaploidization in major angiosperm phyla. Nature 449, 463–467.

Kikkert, J. R., Hébert-Soule, D., Wallace, P. G., Striem, M. J. and Reisch, B. I. (1996) Transgenic plantlets of 'Chancellor' grapevine (*Vitis* sp.) from biolistic transformation of embryogenic cell suspensions. Plant Cell Rep. 15, 311–316.

Kikkert, J.R., Striem, M.J., Vidal, J.R. Wallace, P.G., Barnard, J. and Reisch, B.I. (2005) Long-term study of somatic embryogenesis from anthers and ovaries of 12 grapevine (*Vitis* sp.) genotypes. In Vitro Cell. Dev. Biol. - Plant 41, 232–239.

Kobayashi, S., Goto-Yamamoto, N. and Hirochika, H. (2004) Retrotransposon-induced mutations in grape skin color. Science 304, 982.

Krivanek, A. F., Famula, T.R., Tenscher A., and Walker, M.A. (2005) Inheritance of resistance to *Xylella fastidiosa* within a *Vitis rupestris* x *Vitis arizonica* population. Theor. Appl. Genet. 111, 110–119.

Krivanek, A. F., Riaz, S. and Walker, M. A. (2006) Identification and molecular mapping of *PdR1*, a primary resistance gene to Pierce's disease in *Vitis*. Theor. Appl. Genet. 112, 1125–1131.

Krul, W. R., and Worley, J. F. (1977) Formation of adventitious embryos in callus cultures of 'Seyval', a French hybrid grape. J. Amer. Soc. Hort. Sci. 102, 360–363.

Labra, M., Imazio, S., Grassi, F., Rossoni, M., Citterio, S. *et al.*, (2003) Molecular approach to assess the origin of cv. Marzemino. Vitis 42, 137–140.

Labra, M., Winfield, M., Ghiani, A., Grassi, F., Sala, F. *et al.*, (2001) Genetic studies on Trebbiano and morphologically related varieties by SSR and AFLP markers. Vitis 40, 187–190.

Ladoukakis, E. D., Lefort, F., Sotiri, P., Bacu, A., Kongjika E., *et al.*, (2005) Genetic characterization of Albanian grapevine cultivars by microsatellite markers. Journal International Des Sciences De La Vigne Et Du Vin 39, 109–119.

Lahogue, F., This, P. and Bouquet, A. (1998) Identification of a codominant scar marker linked to the seedlessness character in grapevine. Theor. Appl. Genet. 97, 950–959.

Laucou, V., Lacombe, T., Dechesne, F., Siret, R., Bruno, J.-P., Dessup, M., Dessup, T., Ortigosa, P., Parra, P., Roux, C., Santoni, S., Vares, D., Peros, J.-P., Boursiquot, J.-M. and This, P. (2011) High throughput analysis of grape genetic diversity as a tool for germplasm collection management. Theor. Appl. Genet. 122, 1233–1245.

Lee, L. S. (1988) Citrus polyploidy—origins and potential for cultivar improvement. Aust. J. Agric. Res. 39, 735–747.

Legrand, V., Dalmayrac, S., Latche, A., Pech, J.-C., Bouzayen, M. *et al.*, (2003) Constitutive expression of *Vr-ERE* gene in transformed grapevines confers enhanced resistance to eutypine, a toxin from *Eutypa lata*. Plant Sci. 164, 809–814.

Levadoux, L. (1956) Les populations sauvages et cultivees de *Vitis vinifera* L. Ann. Amelior. Plantes 6, 59–118.

Li, Z. J. T., Dhekney, S. A., Dutt, M. and Gray, D. J. (2008) Improved protocol for *Agrobacterium*-mediated transformation of grapevine (*Vitis vinifera* L.). Plant Cell Tissue and Organ Culture 93, 311–321.

Lider, L. A. (1954) Inheritance of resistance to a root-knot nematode (*Meloidogyne incognita* var. *acrita* Chitwood) in *Vitis* spp. Proc. Helminthol. Soc. Wash. 21, 53–60.

Lider, L. A., Olmo, H. P. and Goheen, A. C. (1988) Hybrid grapevine rootstock named 'VR O43-43'. United States Plant Patent 6, 319.

Lijavetzky, D., Ruiz-Garcia, L., Cabezas, J. A., De Andres, M. T., Bravo, G. *et al.*, (2006) Molecular genetics of berry colour variation in table grape. Molec. Genet. Genomics 276, 427–435.

Lodhi, M. A., Daly, M. J., Ye, G. N., Weeden, N. F. and Reisch, B. I. (1995) A molecular marker based linkage map of *Vitis*. Genome 38, 786–794.

Lopes, M. S., dos Santos, M. R., Dias, J. E. E., Mendonca, D. and da Camara Machado, A. (2006) Discrimination of Portuguese grapevines based on microsatellite markers. J. Biotech. 127, 34–44.

Lopes, M. S., Sefc, K. M., Eiras Dias, E., Steinkellner, H., Laimer da Camara Machado, M. *et al.*, (1999) The use of microsatellites for germplasm management in a Portuguese germplasm grapevine collection. Theor. Appl. Genet. 99, 733–739.

Lowe, K. M., and Walker, M. A. (2006) Genetic linkage map of the interspecific grape rootstock cross Ramsey (*Vitis champinii*) x Riparia Gloire (*Vitis riparia*). Theor. Appl. Genet. 112, 1582–1592.

Madero, E., Boubals, D. and Truel, P. (1986) Transmission hereditaire des principaux caracteres des cepages Cabernet Franc, Cabernet Sauvignon et Merlot (*V. vinifera* L.). Vignevini 13, Suppl. 12, 209–219.

Maletic, E., Pejic, I., Kontic, J. K., Piljac, J., Dangl, G. S. *et al.*, (2004) Zinfandel, Dobricic, and Plavac mali: The genetic relationship among three cultivars of the Dalmatian Coast of Croatia. Amer. J. Enol. Vitic. 55, 174–180.

Mandl, K., Santiago, J. L., Hack, R., Fardossi, A. and Regner, F. (2006) A genetic map of Welschriesling x Sirius for the identification of magnesium-deficiency by QTL analysis. Euphytica 149, 133–144.

Martin, J. P., Borrego, J., Cabello, F. and Ortiz, J. M. (2003) Characterization of Spanish grapevine cultivar diversity using sequence-tagged microsatellite markers. Genome 46, 10–18.

McGovern, P. E. (2003) *Ancient wine: the search for the origins of viticulture*. Princeton University Press, Princeton.

McGovern, P. E. and Michel, R. H. (1995) The analytical and archaeological challenge of detecting ancient wine: two case studies from the ancient Near East, In: P. E. McGovern, S. J. Fleming and S. H. Katz (Eds.) *The Origins and Ancient History of Wine*. Gordon and Breach, Amsterdam. pp. 57–67.

McLeRoy, S. S. and Renfro, R. E. Jr., (2004) *Grape Man of Texas*. Eaking Press, Austin, TX.

Mejia, N., Soto, B., Guerrero, M., Casanueva, X., Houel, C. *et al.*, (2011) Molecular, genetic and transcriptional evidence for a role of *VvAGL11* in stenospermocarpic seedlessness in grapevine. BMC Plant Biol. 11, 57.

Merdinoglu, D., Butterlin, G., Bevilacqua, L., Chiquet, V., Adam-Blondon, A.-F. *et al.*, (2005) Development and characterization of a large set of microsatellite markers in grapevine (*Vitis vinifera* L.) suitable for multiplex PCR. Molec. Breeding 15, 349–366.

Meredith, C. P, Lider, L. A., Raski, D. J. and Ferrari, N. L. (1982) Inheritance of tolerance to *Xiphinema index* in *Vitis* species. Amer. J. Enol. Vitic. 33, 154–158.

Meredith, C. P., Bowers, J. E., Riaz, S., Handley, V., Bandman, E. B. *et al.*, (1999) The identity and parentage of the variety known in California as Petite Syrah. Amer. J. Enol. Vitic. 50, 236–242.

Moreira, F. M., Madini, A., Marino, R., Zulini, L., Stefanini, M. *et al.*, (2011) Genetic linkage maps of two interspecific grape crosses (*Vitis* spp.) used to localize quantitative trait loci for downy mildew resistance. Tree Genet. Genomes 7, 153–167.

Morinaga, K. (2001) Grape Production in Japan. In: M.K. Papademetriou and F.J. Dent (Eds.), *Grape Production in the Asia-Pacific Region*. Food and Agriculture Office of the United Nations, Regional Office for Asia and the Pacific, Bangkok, Thailand. pp. 38–69.

Mortensen, J. A. (1968) The inheritance of resistance to Pierce's disease in *Vitis*. J. Amer. Soc. Hort. Sci. 92, 331–337.

Motosugi, H., Naruo, T. and Kataoka, D. (1999) The growth of diploid and tetraploid grape rootstocks and 'Kyoho' grape grafted on them. J. Japan. Hort. Sci. 68 (Suppl. 2), 112.

Motosugi, H., Okudo, K., Kataoka, D. and Naruo, T. (2002a) Comparison of growth characteristics between diploid and colchicines-induced tetraploid grape rootstocks. J. Japan. Hort. Sci. 71, 335–341.
Motosugi, H., Naruo, T., Komazaki, S., and Yamada, M. (2002b) Resistance of autotetraploids of grape rootstock cultivars to phylloxera (*Daktulosphaira vitifoliae* Fitch). Vitis 41, 103–106.
Mullins, M. G., and Srinivasan, C. (1976) Somatic embryos and plantlets from an ancient clone of the grapevine (cultivar Cabernet Sauvignon) by apomixis *in vitro*. J. Exp. Bot.27, 1022–1030.
Mullins, M. G., Bouquet, A. and Williams, L. E. (1992) *Biology of the Grapevine*. Cambridge University Press, Cambridge.
Munson, T. (1909) *Foundations of American Grape Culture*. T.V. Munson & Son, Denison, Texas.
Myles, S., Boyko, A.R., Owens, C.L., Brown, P.J., Grassi, F., Aradhya, M.K., Prins, B., Reynolds, A., Chia, J.-M., Ware, D., Bustamante, C.D. and Buckler, E.S. (2011) Genetic structure and domestication history of the grape. Proc. Natl. Acad. Sci. (USA) 108, 3530–3535.
Myles, S., Chia, J.-M., Hurwitz, B., Simon, C., Zhong, G. Y., Buckler, E.S. and Ware, D. (2010) Rapid Genomic Characterization of the Genus *Vitis*. PLoS ONE 5, e8219.
Neal, J. C. (1889) The root-knot disease of peach, orange, and other plants in Florida. U.S. Department of Agriculture, Division of Entomology, Bulletin 20, 1–31.
Negrul, A. M. (1938) Evolucija kuljturnyx from vinograda. Doklady Akademii nauk SSSR 8, 585–588.
Nicol, J.M., Stirling, G.R, Rose, B.J., May, P. and Heeswijck, R.V. (1999) Impact of nematodes on grapevine growth and productivity: current knowledge and future directions, with special reference to Australian viticulture. Austr. J. Grape Wine Res. 5, 109–127.
Office International de la Vigne et du Vin. (2006) *Situation Report for the World Vitivinicultural Sector in 2005*. Paris.
Olmo, H.P. (1942) Storage of grape pollen. Proc. Amer. Soc. Hort. Sci. 41, 219–224.
Olmo, H.P. (1971) *Vinifera rotundifolia* hybrids as wine grapes. Amer. J. Enol. Vitic. 22, 87–91.
Olmo, H.P. (1995) The origin and domestication of the *Vinifera* grape, In: P. E. McGovern (Ed.). *The origins and ancient history of wine*. Gordon and Breach, Amsterdam. pp. 31–43.
Ong, P.K.C. and Acree, T.E. (1999) Similarities in the aroma chemistry of Gewürztraminer variety wines and lychee (*Litchi chinesis* Sonn.) fruit. J. Agric. Food Chem. 47, 665–670.
Owens, C. L. (2011) Linkage disequilibrium and prospects for association mapping in *Vitis*. In: A.-F. Adam-Blondon and J.M. Martinez Zapater (Eds.). *Genetics, Genomics and Breeding of Grapes*. Scientific Publishers, St. Helier, Jersey, British Isles. pp. 93–110.
Owens, C.L. (2008) Grapes. In: J.F. Hancock (Ed.), *Temperate Fruit Crop Breeding*. Springer, pp. 197–233.
Paul, H. W. (1996) *Science, vine, and wine in modern France*. Cambridge University Press, Cambridge.
Pauquet, J., Bouquet, A., This, P. and Adam-Blondon, A.-F. (2001) Establishment of a local map of AFLP markers around the powdery mildew resistance gene *Run1* in grapevine and assessment of their usefulness for marker assisted selection. Theor. Appl. Genet. 103, 1201–1210.
Pelsy, F. (2010) Molecular and cellular mechanisms of diversity within grapevine varieties. Heredity 104, 331–340.
Perl, A. and Eshdat, Y. (1998) DNA transfer and gene expression in transgenic grapes. Biotech. Genet. Engineering Rev. 15, 365–386.
Perl, A., Saad, S., Sahar, N. and Holland, D. (1995) Establishment of long-term embryogenic cultures of seedless *Vitis vinifera* cultivars -- a synergistic effect of auxins and the role of abscisic acid. Plant Sci. 104, 193–200.
Perrin, M., Gertz, C. and Masson, J. E. (2004) High efficiency initiation of regenerable embryogenic callus from anther filaments of 19-grapevine genotypes grown worldwide. Plant Sci. 167, 1343–1349.
Perrin, M., Martin, D., Joly, D., Demangeat, G., This, P. *et al.*, (2001) Medium-dependent response of grapevine somatic embryogenic cells. Plant Sci. 161, 107–116.
Pezzuto, J.M. (2008) Grapes and human health: A perspective. J. Agric. Food Chem. 56, 6777–6784.

Pinkerton, J. N, Vasconcelos, M. C., Sampaio, T. L. and Shaffer, R. G. (2005) Reaction of grape rootstocks to ring nematode *Mesocriconema xenoplax*. Amer. J. Enol. Vitic. 56, 377–385.

Polášková, P., Herszage, J. and Ebeler, S.E. (2008) Wine flavor: chemistry in a glass. Chem. Soc. Rev. 37, 2478–2489.

Pouget, R. and Ottenwalter, M. (1984) Recherche de nouveaux porte-greffes adaptés aux sols acides. Prog. Agric. Vitic. 101, 73–75.

Pratt, C. (1971) Reproductive anatomy in cultivated grapes--a review. Amer. J. Enol. Vitic. 22, 92–109.

Prescott, J. A. (1965) The climatology of the vine: The cool limits of cultivation. Trans. Roy. Soc. South Aust. 89, 5–23.

Qiu, B.X., Sleper, D.A. and Arelli, A.P.R. (1997) Genetic and molecular characterization of resistance to *Heterodera glycines* race isolates 1, 3, and 5 in Peking. Euphytica 96, 225–231.

Rajasekaran, K. and Mullins, M.G. (1979) Embryos and plantlets from cultured anthers of hybrid grapevines. J. Exp. Bot.30, 399–407.

Raman, H., Moroni, J. S., Sato, K., Read, B. J. and Scott, B. J. (2002) Identification of AFLP and microsatellite markers linked with an aluminum tolerance gene in barley (*Hordeum vulgare* L.). Theor. Appl. Genet. 105, 458–464.

Raman, H., Zhang, K., Cakir, M., Appels, R., Garvin, D.F., Maron, L.G., Kochian, L.V., Moroni, J.S., Raman, R., Imtiaz, M., Drake-Brockman, F., Waters, I., Martin, P., Sasaki, T., Yamamoto, Y., Matsumoto, H., Hebb, D.M., Delhaize, E. and Ryan, P.R. (2005) Molecular characterization and mapping of *ALMT1*, the aluminum-tolerance gene of bread wheat (*Triticum aestivum* L.). Genome 48, 781–791.

Ramming, D. W., Emershad, R. L. and Tarailo, R. (2000) A stenospermocarpic, seedless *Vitis vinifera* x *Vitis rotundifolia* hybrid developed by embryo rescue. HortScience 35, 732–734.

Ramming, D.W., Gabler, F., Smilanick, J., Cadle-Davidson, M., Barba, P., Mahanil, S. and Cadle-Davidson, L. (2011) A single dominant locus, *Ren4*, confers rapid non-race-specific resistance to grapevine powdery mildew. Phytopathol. 101, 502–508.

Regner, F., Stadlbauer, A., Eisenheld, C. and Kaserer, H. (2000) Genetic relationships among Pinots and related cultivars. Amer. J. Enol. Vitic. 51, 7–14.

Reisch, B.I. and Pratt, C. (1996) Grapes. In: J. Janick and J.N. Moore (Eds.), *Fruit Breeding. Volume II. Vine and Small Fruits.* John Wiley and Sons, New York. pp. 297–359.

Riaz, S., Dangl, G. S., Edwards, K. J. and Meredith, C. P. (2004) A microsatellite marker based framework linkage map of *Vitis vinifera* L. Theor. Appl. Genet. 108, 864–872.

Riaz, S., Garrison, K. E., Dangl, G. S., Boursiquot, J.-M. and Meredith, C. P. (2002) Genetic divergence and chimerism within ancient asexually propagated winegrape cultivars. J. Amer. Soc. Hort. Sci. 127, 508–514.

Riaz, S., Krivanek, A. F., Xu, K. and Walker, M. A. (2006) Refined mapping of the Pierce's disease resistance locus, *PdR1*, and *Sex* on an extended genetic map of *Vitis rupestris* x *V. arizonica.* Theor. Appl. Genet. 113, 1317–1329.

Riaz, S., Tenscher, A. C., Ramming, D. W. and Walker, M. A. (2011) Using a limited mapping strategy to identify major QTLs for resistance to grapevine powdery mildew (*Erysiphe necator*) and their use in marker-assisted breeding. Theor. Appl. Genet. 122, 1059–1073.

Riaz, S., Tenscher, A. C., Rubin, J., Graziani, R., Pao, S. S. *et al.*, (2008) Fine-scale genetic mapping of two Pierce's disease resistance loci and a major segregation distortion region on chromosome 14 of grape. Theor. Appl. Genet. 117, 671–681.

Riaz, S., Tenscher, A.C., Graziani, R., Krivanek A.F. and Walker, M.A. (2009) Using marker assisted selection to breed for Pierce's disease resistance in grapevine. Amer. J. Enol. Vitic. 60, 199–206.

Rives, M. (1965) The germination of grape seeds. l. Preliminary experiments (in French, English summary). Ann. Amélior. Plantes 15, 79–91.

Rosenfield, C. L., Samuelian, S., Vidal, J. R. and Reisch, B. I. (2010) Transgenic disease resistance in *Vitis vinifera*: Potential use and screening of antimicrobial peptides. Amer. J. Enol. Vitic. 61, 348–357.

Rossoni, M., Labra, M., Imazio, S., Grassi, F., Scienza, A. *et al.*, (2003) Genetic relationships among grapevine cultivars grown in Oltrepo Pavese (Italy). Vitis 42, 31–34.

Rowhani,, A., Uyemoto, J. K., Golino, D. A. and Martelli, G. P. (2005) Pathogen testing and certification of *Vitis* and *Prunus* species. Annu. Rev. Phytopathol. 2005. 43, 6.1–6.18.

Sauer, W. and Antcliff, A. J. (1969) Polyploid mutants of grapes. HortScience 4, 226–227.

Scott, D.H. and Ink, D.P. (1950) Grape seed germination experiments. Proc. Amer. Soc. Hort. Sci. 56, 134–139.

Scott, K. D., Eggler, P., Seaton, G., Rossetto, M., Ablett, E. M. *et al.*, (2000) Analysis of SSRs derived from grape ESTs. Theor. Appl. Genet. 100, 723–726.

Sefc, K. M., Lopes, M. S., Lefort, F., Botta, R., Roubelakis-Angelakis, K. A. *et al.*, (2000) Microsatellite variability in grapevine cultivars from different European regions and evaluation of assignment testing to assess the geographic origin of cultivars. Theor. Appl. Genet. 100, 498–505.

Shure, K.B. and Acree, T.E. (1994) Changes in odor-active compounds in *Vitis labruscana* Cv. Concord during growth and development. J. Agric. Food Chem. 42, 350–353.

Snyder, E. and Harmon, F. N. (1939) Grape progenies of self-pollinated vinifera varieties. Proc. Amer. Soc. Hort. Sci. 37, 625–626.

Snyder, E. and Harmon, F. N. (1952) Grape breeding summary 1923–1951. Proc. Amer. Soc. Hort. Sci. 60, 243–246.

Spiegel-Roy, P., Shulman, Y., Baron, I. and Ashbel, E. (1987) Effect of cyanamide in overcoming grape seed dormancy. HortScience 22, 208–210.

Stobbs. L. W., Potter, J. W., Killins, R. and Van Schagen, J. G. (1988) Influence of grapevine understock in infection of De Chaunac scion by tomato ringspot virus. Can. J. Plant Pathol. 10, 228–231.

This, P., Jung, A., Boccacci, P., Borrego, J., Botta, R. *et al.*, (2004) Development of a standard set of microsatellite reference alleles for identification of grape cultivars. Theor. Appl. Genet. 109, 1448–1458.

This, P., Lacombe, T. and Thomas, M. R. (2006) Historical origins and genetic diversity of wine grapes. Trends Genet. 22, 511–519.

This, P., Lacombe, T., Cadle-Davidson, M. and Owens, C. L. (2007) Wine grape (*Vitis vinifera* L.) color associates with allelic variation in the domestication gene *VvmybA1*. Theor. Appl. Genet. 114, 723–730.

Thomas, M. R. and Scott, N. S. (1993) Microsatellite repeats in grapevine reveal DNA polymorphisms when analyzed as Sequence-Tagged Sites (STSs). Theor. Appl. Genet. 86, 985–990.

Thomas, M. R., Cain, P. and Scott, N. S. (1994) DNA typing of grapevines - a universal methodology and database for describing cultivars and evaluating genetic relatedness. Plant Molec. Biol. 25, 939–949.

Thomas, M. R., Matsumoto, S., Cain, P. and Scott, N. S. (1993) Repetitive DNA of grapevine: classes present and sequences suitable for cultivar identification. Theor. Appl. Genet. 86, 173–180.

Thompson, M. M. and Olmo, H. P. (1963) Cytohistological studies of cytochimeric and tetraploid grapes. Amer. J. Bot. 50, 901–906.

Torregrosa, L. (1998) A simple and efficient method to obtain stable embryogenic cultures from anthers of *Vitis vinifera* L. Vitis 37, 91–92.

Valat, L., Fuchs, M. and Burrus, M. (2006) Transgenic grapevine rootstock clones expressing the coat protein or movement protein genes of grapevine fanleaf virus: Characterization and reaction to virus infection upon protoplast electroporation. Plant Sci. 170, 739–747.

Velasco, R., Zharkikh, A., Troggio, M., Cartwright, D. A., Cestaro, A. *et al.*, (2007) A high quality draft consensus sequence of the genome of a heterozygous grapevine variety. PLoS ONE 2, e1326.

Viala, P. and Ravaz, L. (1903) *American Vines*. 2nd ed. (Translated from French by R. Dubois and E.H. Twight). Freygang-Leary, San Francisco.

Vidal, J. R., Kikkert, J. R., Malnoy, M. A., Wallace, P. G., Barnard, J. and Reisch, B.I. (2006) Evaluation of transgenic 'Chardonnay' (*Vitis vinifera*) containing magainin genes for resistance to crown gall and powdery mildew. Transgenic Res. 15, 69–82.

Vidal, J. R., Kikkert, J. R., Wallace, P. G. and Reisch, B.I. (2003) High-efficiency biolistic co-transformation and regeneration of 'Chardonnay' (*Vitis vinifera* L.) containing npt-II and anti-microbial peptide genes. Plant Cell Rep. 22, 252–260.
Vigne, E., Komar, V. and Fuchs, M. (2004) Field safety assessment of recombination in transgenic grapevines expressing the coat protein gene of *Grapevine fanleaf virus*. Transgenic Res. 13, 165–179.
Vouillamoz, J. F. and Grando, M. S. (2006) Genealogy of wine grape cultivars: 'Pinot' is related to 'Syrah'. Heredity 97, 102–110.
Waffo-Téguo, P., Hawthorne, M.E., Cuendet, M., Mérillon, J.-M., Kinghorn, A.D., Pezzuto, J.M. and Mehta, R.G. (2001) Potential cancer-chemopreventive activities of wine stilbenoids and flavans extracted from grape (*Vitis vinifera*) cell cultures. Nutrition and Cancer 40, 173–179.
Wagner, R. (1967) Study of some segregation in progenies of Chasselas, Muscat Ottonel and small-berried Muscat (in French). Vitis 6, 353–363.
Walker, A.R., Lee, E., Bogs, J., McDavid, D. A. J., Thomas, M. R. *et al.*, (2007) White grapes arose through the mutation of two similar and adjacent regulatory genes. Plant J. 49, 772–785.
Walker, G. E. and Stirling, G. R. (2008) Plant-parasitic nematodes in Australian viticulture: key pests, current management practices and opportunities for future improvements. Australasian Plant Pathol. 37, 268–278.
Walker, M. A. (1992) Future directions for rootstock breeding. In: J.A. Wolpert, M.A. Walker and E. Weber (Eds.). *Proceedings Rootstock Seminar: A Worldwide Perspective, Reno, Nevada, June 24, 1992.* The American Society for Enology and Viticulture, Davis, California. pp 60–66.
Walker, M. A., Ferris, H. and Eyre, M. (1994a) Resistance in *Vitis* and *Muscadinia* to *Meloidogyne incognita*. Plant Dis. 78, 1055–1038.
Walker, M. A., Wolpert, J . A. and Weber, E. (1994b) Viticultural characteristics of VR hybrid rootstocks in a vineyard site infected with grapevine fanleaf virus. Vitis 33, 19–23.
Walker, M. A., Lider, L. A., Goheen, A. C., and Olmo, H. P. (1991) VR O39-16 grape rootstock. HortScience 26, 1224–1225.
Walker, M.A, Jin, Y., Min, B.E. and Hajdu, E. (1998) Development of resistant rootstocks to control *Xiphinema index* and fanleaf degeneration. Acta Horticulturae 473, 113–120.
Walters, S.A., Wehner, T.C., and Barker, K.R. (1997) A single recessive gene for resistance to the root-knot nematode (*Meloidogyne javanica*) in *Cucumis sativus* var *hardwickii*. J. Hered. 88, 66–69.
Wang, M. and Goldman, I.L. (1996) Resistance to root knot nematode (*Meloidogyne hapla* Chitwood) in carrot is controlled by two recessive genes. J. Hered. 87, 119–123.
Wang, Q., Mawassi, M., Sahar, N., Li, P., Violeta, C.-T., Gafny, R., Sela, I., Tanne, E. and Perl, A. (2004) Cryopreservation of grapevine (*Vitis* spp.) embryogenic cell suspensions by encapsulation-vitrification. Plant Cell, Tissue and Organ Culture 77, 267–275.
Weinberger, J. H. and Harmon, F. N. (1966) Harmony, a new nematode and phylloxera resistant rootstock for vinifera grape. Fruit Var. Hort. Dig. 20, 63–65.
Wellington, R. (1939) The Ontario grape and its seedlings as parents. Proc. Amer. Soc. Hort. Sci. 37, 630–634.
Welter, L. J., Gokturk-Baydar, N., Akkurt, M., Maul, E., Eibach, R. *et al.*, (2007) Genetic mapping and localization of quantitative trait loci affecting fungal disease resistance and leaf morphology in grapevine (*Vitis vinifera* L). Molec. Breeding 20, 359–374.
Wen, J. (2007) Vitaceae. In: K. Kubitzki (Ed.). *The Families and Genera of Vascular Plants. Volume IX: Flowering Plants-Eudicots*. Springer-Verlag, Berlin. pp. 467–479.
Winkler, A.J. (1949) Grapes and wine. Econ. Bot. 3, 46–70.
Xu, K., Riaz, S., Roncoroni, N. C., Jin, Y., Hu, R. *et al.*, (2008) Genetic and QTL analysis of resistance to *Xiphinema index* in a grapevine cross. Theor. Appl. Genet. 116, 305–311.
Xue, B., Ling, K.-S., Reid, C. L., Krastanova, S., Sekiya, M. *et al.*, (1999) Transformation of five grape rootstocks with plant virus genes and a *virE2* gene from *Agrobacterium tumefaciens*. In Vitro Cell. Dev. Biol.--Plant 35, 226–231.
Yamamoto, T., Iketani, H., Ieki, H., Nishizawa, Y., Notsuka, K. *et al.*, (2000) Transgenic grapevine plants expressing a rice chitinase with enhanced resistance to fungal pathogens. Plant Cell Rep. 19, 639–646.

Ye, G. N., Soylemezoglu, G., Weeden, N. F., Lamboy, W. F., Pool, R. M. *et al.*, (1998) Analysis of the relationship between grapevine cultivars, sports and clones via DNA fingerprinting. Vitis 37, 33–38.

Zhang, J. K., Hausmann, L., Eibach, R., Welter, L. J., Töpfer, R. *et al.*, (2009) A framework map from grapevine V3125 (*Vitis vinifera* 'Schiava grossa' x 'Riesling') x rootstock cultivar 'Börner' (*Vitis riparia* x *Vitis cinerea*) to localize genetic determinants of phylloxera root resistance. Theor. Appl. Genet. 119, 1039–1051.

Zohary, D. and Hopf, M. (2000) *Domestication of Plants in the Old World.* Oxford University Press, London.

Zohary, D. and Spiegel-Roy, P. (1975) Beginnings of fruit growing in the old world. Science 187, 319–327.

Chapter 8
Raspberry

Chaim Kempler, Harvey Hall, and Chad E. Finn

Abstract The red raspberry, *Rubus idaeus* L., is a valuable crop that has recently increased in production, generating a large interest in commercial ventures and in research. Traditionally, most of the crop has been sold to processors for freezing, jam production, canning, juice and flavorings for ice cream, yogurt, and other products, but in recent years fresh market production has increased and become a very important sector of this industry. There has been an increased interest in black, purple, and Arctic raspberries because of their high nutraceutical value. *R. idaeus,* a diploid ($2n=14$), is included in the Idaeobatus and is the most important commercial species in this subgenus. The flowers are hermaphroditic; however, in some cases, they are unisexual, especially among wild species. Domestication of raspberries is comparably recent as it occurred less than 500 years ago. Red raspberries are widely distributed in all temperate regions of Europe, Asia, and North America with the greatest diversity in China. Enriching the cultivated gene pool by incorporating the unique genetics from wild germplasm to meet the challenges that lie ahead is desired. Breeding goals are the improvement of fruit quality which includes selection for better preharvest hanging ability and postharvest shelf life and processed quality. Resistance to heat and cold and resistance to pests and disease are also important, as well as large fruit size, good presentation, and ease of harvest. Fruit color of the newer cultivars varies from very dark red to a light orange–red and there has become a tradition of cultivar selection

C. Kempler (✉)
Pacific Agri-Food Research Centre, Agriculture and Agri-Food Canada,
P.O. Box 1000, 6947 #7 Highway, Agassiz, BC, Canada V0M 1A0
e-mail: Chaim.kempler@agr.gc.ca

H. Hall
Shekinah Berries Ltd, Motueka, New Zealand
e-mail: hkhall@clear.net.nz

C.E. Finn
US Department of Agriculture-Agricultural Research Service, Horticultural Crops Research Laboratory, Corvallis, OR, USA
e-mail: Chad.finn@ars.usda.gov

M.L. Badenes and D.H. Byrne (eds.), *Fruit Breeding*, Handbook of Plant Breeding 8,
DOI 10.1007/978-1-4419-0763-9_8, © Springer Science+Business Media, LLC 2012

specifically for processing or for fresh market. There are now approximately 50 active raspberry breeding programs in 26 countries, mostly in Europe and North America. Use of molecular markers for genetic studies and mapping is referenced; however, in this crop, it is at an early stage with only a few genes mapped.

Keywords *Rubus idaeus* • Breeding • Resistance • Breeding history • World production • Cold hardiness • *Rubus* • Idaeobatus • *Phytophthora rubi* (root rot) • *Amphorophora* • *Raspberry bushy dwarf virus* (RBDV) • Primocane-fruiting • floricane-fruiting

> "In spite of the fact that a good deal has already been accomplished, the possibilities of improving the red raspberry by utilizing the available cultivated varieties in further breeding work are still enormous. Some of the qualities, now found separately, that may be combined in raspberries of the future are the very large fruit size of European varieties and newer American production, immense fruit clusters, great productiveness, firmness, vigor, and resistance to diseases. But there is also a large reservoir of germplasm, hardly yet touched by raspberry breeders, in the wild species of Asia and elsewhere, some of which resemble the grape, hawthorn, bamboo, maple, and apple in their leaf forms, and vary from low and soft-stemmed plants to plants with stems 3 inches thick and 14 feet high". George M. Darrow 1937

1 Introduction

The red raspberry, *Rubus idaeus* L., is a valuable crop that has recently increased in production, generating a large interest in commercial ventures and in research. There has also been an increase in the number of symposia and small fruit meetings, and the last 2–3 years have yielded a number of published reviews on raspberry breeding and culture and on nutraceutical benefits of raspberry consumption (Bañados and Dale 2008; Finn and Hancock 2008; Hall et al. 2009). World production in 2005 was estimated to be more than 616,000 metric tons, which is about a 2.3-fold increase over the last 25 years (Fig. 8.1).

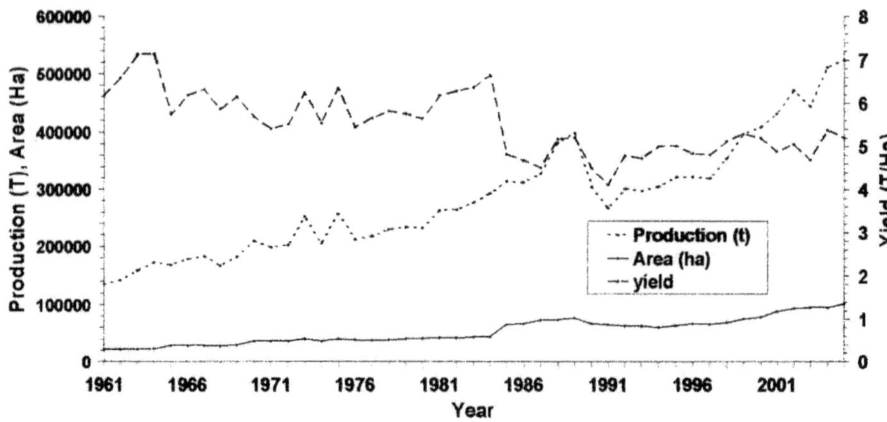

Fig. 8.1 World raspberry production, growing area, and yield (FAO 2009)

Traditionally, most of the crop has been sold to processors for freezing, jam production, canning, juice and flavorings for ice cream, yogurt, and other products but in recent years fresh market production has increased and become the dominant sector of this industry. Handpicking is the rule for fresh market production as it is for the processing market with the exceptions being in the Pacific Northwest of the USA and Canada (Oregon, Washington, and British Columbia) where about 95% of the crop is machine harvested. Raspberries ideally are grown in regions where winters are mild and the summers are moderate. However, with increased demand for a year-round supply of fresh fruit, production has shifted to areas where conditions are marginal. Raspberries are now among the most important of temperate berry fruit crops and are also grown in areas with no chilling, where summers are very hot and soils are alkaline (Oliveira et al. 2002). The major production areas for the processing market are Serbia, the Russian Federation, China, Poland, Chile, and the Pacific Northwest (Table 8.1). Fresh market production has expanded significantly in traditional production areas in the last decade and has also expanded into nontraditional areas where the climate is warmer in the winter. These include coastal California, high elevation locations in central Mexico, and Chile in the New World, as well as Southern Europe and North Africa. Out-of-season produce is shipped to the market from warmer climates in the Southern Hemisphere in Australia, New Zealand, and South Africa. Raspberries are still restricted to climates or environments that are moderate during fruit ripening.

1.1 World Production

Reliable estimates of world production are difficult to obtain as the Food and Agriculture Organization (FAO) of the United Nations just recently separated raspberry from blackberry production in its statistical reporting (FAO 2009). Raspberries are produced in at least 39 countries worldwide on about 114,600 ha (Table 8.1). Since 1992, production has increased by about 38%, with about 20% of the increase attributed to increased hectarage and the rest to increases in yield per hectare. Average yields are about 5.6 t/ha ranging from less than 1.68 to about 10.0 t/ha (Table 8.1; Fig. 8.1).

In recent years, there has been an increased interest in black, purple, and Arctic raspberries because of their high nutraceutical value (Stoner et al. 2002). Black raspberry (*R. occidentalis* L.) production has been largely concentrated in western Oregon, with recent production increases in South Korea. Production of purple raspberries, which are hybrids of red and black raspberries, is scattered in small plantings throughout North America and in northeastern China. Arctic raspberries (*R. arcticus* L.) and cloudberries (*R. chamaemorus* L.) are largely wild harvested in Scandinavia.

More than half of the red raspberry crop comes from Europe that includes the Russian Federation, Ukraine, Germany, France, Hungary, the UK, and Spain. Very little information in English is available on production and research in the Russian Federation. The second largest raspberry producer in the world is the Republic of

Table 8.1 World production of raspberries 2005 (source FAO 2009)

Country	Area harvested (ha)	Production (mt)	Yield (mt/ha)
Russian Federation[a]	34,000	175,000	5.1
Former Serbia and Montenegro	16,500	84,331	5.1
The USA	6,840	82,826	12.1
Poland[a]	17,200	65,000	3.8
Chile	10,500	64,000	6.1
Ukraine[a]	5,000	27,000	5.4
Germany[b]	5,900	20,000	3.4
China	2,200	15,300	7.0
Canada[b]	2,958	15,000	5.1
The UK	1,430	12,200	8.5
Spain[b]	1,400	7,000	5.0
Hungary	1,200	6,724	5.6
Azerbaijan[a]	1,400	6,300	4.5
France	1,303	5,742	4.4
Korea	1,000	4,700	4.7
Mexico	380	4,253	11.2
Romania[b]	200	4,200	2.1
Bulgaria[b]	1,200	3,000	2.5
Norway	282	1,719	6.1
Kyrgyzstan[a]	600	1,700	2.8
Bosnia and Herzegovina[b]	427	1,700	4.0
Moldova[a]	300	1,500	5.0
Italy	178	1,421	8.0
Switzerland	158	1,285	8.1
Croatia[b]	278	800	2.9
Finland	418	608	1.5
Australia[b]	230	600	2.6
The Netherlands[b]	50	500	10.0
New Zealand[b]	300	390	1.3
Estonia[b]	400	300	0.8
Belgium	30	275	9.2
Slovakia[b]	80	200	2.5
Sweden[b]	130	190	1.5
Ireland[b]	44	100	2.3
Zimbabwe[b]	50	80	1.6
Denmark[b]	30	65	2.2
Morocco[b]	16	50	3.1
The Czech Republic[b]	25	28	1.1
Slovenia[b]	2	4	2.0
Total	114,639	616,091	

[a]Unofficial figure
[b]FAO estimate

Serbia, which produces about 65,000 metric tons mainly for the frozen export market. The farms are usually small family-owned operations and range in size between 0.05 and 1 ha. More than 90% of the plantings are the 'Willamette' cultivar, the crop is harvested by hand, and the average yield is low. Raspberry is the most profitable exported agricultural commodity for Serbia. In North America, production in California, the largest growing area, is directed to the fresh market while production in the Pacific Northwest, where the most important growing area is Washington State, followed by British Columbia, and Oregon, is directed to the processing market. The main cultivar in this region is 'Meeker', which was released from the Washington State University (WSU) breeding program more than 40 years ago. Almost 40% of the new plantings are still of this cultivar. 'Meeker' is not resistant to root rot and the disease cannot be controlled effectively by chemicals and soil fumigation. A replacement is desperately needed as 'Meeker' also has low tolerance to winter injury, is frequently damaged by frost, and lacks resistance to *Raspberry bushy dwarf virus* (RBDV). Recently released cultivars, Chemainus, Saanich, and Cascade Bounty, from the Agriculture and Agri-Food Canada–Pacific Agri-Food Research Centre (AAFC-PARC) and WSU programs are increasing in popularity and more than 20% of the new plantings are of these three new cultivars (P. Moore, pers. comm.). Production for the fresh market has expanded rapidly in California, and in recent years it also has extended to Southern California, Mexico, and Central America, where low-chill cultivars are used. Several primocane-fruiting cultivars (e.g., 'Caroline,' 'Summit,' Himbo Top™) have proved to work well in this system. Recently, cultivars increasingly more successful have been selected for this environment by private breeding programs. These proprietary cultivars have been selected from large seedling populations produced from intercrossing existing private and public cultivars and through the introduction of superior traits into this germplasm from leading cultivars from other parts of the world (Fear 1992).

1.2 Uses

Red raspberries are widely used fresh and as a processed product in Europe and North America. As more information on health benefits has been published, there has been an increase in consumption and commercial use of the fruit. Fresh fruit are consumed as snacks, desserts, in fruit salads, and with ice cream, yogurt, or breakfast cereals. In recent years, there has been a large increase in fresh fruit consumption and now most supermarkets carry fresh raspberries year round. Fruit is shipped by ground, sea, and air transport all over the world. Raspberries are also grown out of season in greenhouses and tunnels and marketed locally or internationally.

Most of the raspberry crop is block frozen as puree, juice concentrate, or individually quick frozen (IQF) fruit. Other forms of processing include the production of fruit leathers and dried fruit by heating, freeze drying, and/or using microwaves. Fruits are block frozen in packs, pails, or drums, but IQF fruits are considered to be of the highest quality and value. Cultivars that can be mechanically harvested and produce fruits that are suitable for production of IQF fruit are essential. High-quality IQF raspberries must be clean, have good appearance, and contain less than 5% broken berries. Much of the world production of IQF raspberries is with the cultivar

'Meeker' and sometimes 'Willamette' (often, only early in the harvest season). From the basic frozen industrial formulation of IQF fruit, purees, and juices, a myriad of retail products can be found in all sections of a grocery store, including medicines and nonedible consumer products (e.g., shampoos, lip gloss, etc.).

1.3 The Genus Rubus of the Rosaceae Family

Rubus idaeus is included in the diploid ($2n = 14$) subgenus Idaeobatus whose species are distinguished by the ability of drupelets to abscise at the base, enabling the mature fruit to separate from the receptacle when picked. The picked fruit is a conical-thimble shape and has a cavity, where it was attached to the receptacle. Drupelets adhere to one another by small hairs. In contrast, the abscission zone in blackberries is at the base of the fleshier receptacle enabling it to be harvested and consumed with the fruit. This botanical difference has been the basis for taxonomically separating raspberry types among *Rubus* fruits, although it may be somewhat arbitrary as it splits a complex genus on the basis of only one morphological difference. Nevertheless, this feature is a reasonably effective division of the genus and only a few species are notable as blackberries with raspberry-type abscission or raspberries with blackberry-like abscission (Jennings 1988). The Idaeobatus, which contains the red raspberry, has a northerly distribution mainly in Asia but it also is found in Australia, Africa, Europe, and North America. Plants in the Idaeobatus are deciduous perennial shrubs with trailing to erect canes, where the canes are typically biennial and the roots are perennial. Canes are glabrous, hairy, glandular, bristly or prickly. In the second year, in most genotypes, short lateral branches grow from the nodes of canes and these bear flowers and fruit. After fruiting, the canes die to be replaced by the new canes that have grown during the same period. Some genotypes fruit on the first-year primocanes as well as second-year floricanes and are often called primocane- or fall-fruiting, remontant or everbearing. Leaves are alternate pinnate with usually five leaflets on the primocanes and three leaflets on the laterals (second-year growth). Petioles and petiolules usually resemble the canes, and stipules are always present at the base of the petioles. The flowers are hermaphroditic; however, in some cases, they are unisexual, especially in the wild. The most important species in the Idaeobatus that have been domesticated are *R. idaeus* var. *vulgatus* Arrhen and *R. idaeus* var. *strigosus* Michx (red raspberries). Other domesticated species are *R. occidentalis* L. (black raspberries), principally grown in North America; *R. glaucus* Benth (Andean blackberry) widely cultivated in Central and South America; *R. coreanus* Miq., *R. crataegifolius* Bunge, *R. niveus* Thunb., and *R. parvifolius* L. grown in China; *R. phoenicolasius* Miq. grown in Japan; and *R. arcticus* L. in the Cylactis and *R. chamaemorus* L. in the Chamaemorus grown in Scandinavia (Fig. 8.2; Finn 1999).

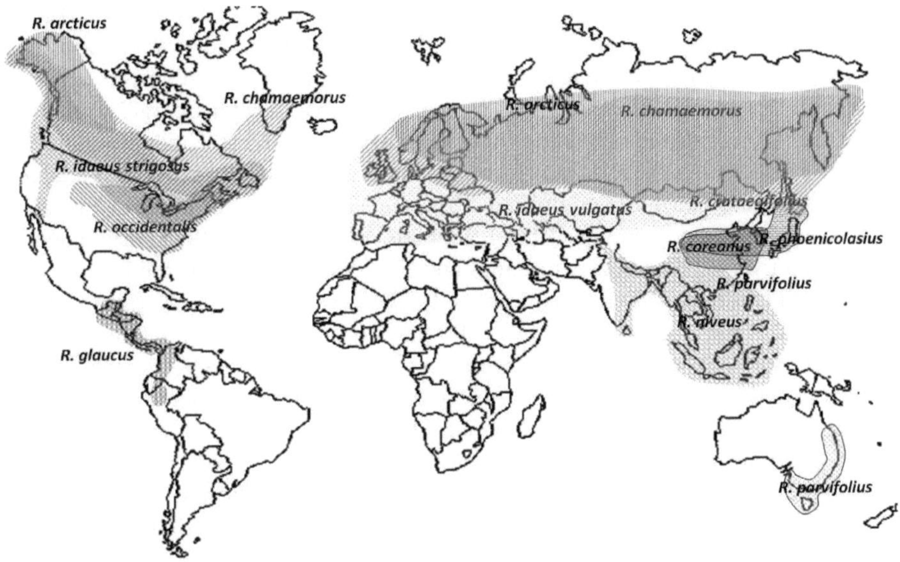

Fig. 8.2 The distribution of *Rubus* in the world

2 Origin and Domestication

Raspberries have been recognized as a crop of value for human consumption with archeological evidence going back to 45 AD. The first historical record of the European red raspberry, *R. idaeus*, was by Pliny the Elder who wrote how the people of Troy at the base of Mount Ida gather ‘Ida’ fruits. At the time, the plant was more important as a medicine than as a food, as the blossom was used to make an eye ointment or stomach draught (Jennings 1988). However, it is likely that they originally came from the Ide Mountains of Turkey and not from Greece. By the fourth century, they were already mentioned as a cultivated fruit. The name ‘ida’ was later used by Linnaeus for the species name *idaeus* and for the genus he used the name *Rubus* derived from the Latin word *Ruber* meaning red. In his book, Johnson (1829) listed 23 cultivated varieties growing in English gardens and 20 varieties were listed by Prince (1832) as growing in North America. Most of the cultivars dating from this period are hybrids of the European and North American species *R. idaeus* and *R strigosus* (Daubeny 1983). It is very clear that the domestication of raspberries is comparably recent to most fruits and did not start more than 500 years ago (Hedrick 1925).

In the early 1900s, in Britain, George Pyne was probably the most successful nurseryman to obtain new cultivars by transplanting self-sown seedlings. His successes included ‘Park Lane’ and its derivative ‘Mayfair’ that had outstanding flavor, ‘Devon’ for its firm fruit, and ‘Pyne’s Royal’ for its high yield, large fruit, and outstanding flavor. His most successful cultivar was 'Pynes Royal.' 'Lloyd George,' a floricane bearing cultivar with some primocane fruit, was found in the woods of Kent by J.J. Kettle and introduced in 1919, after he had moved to Corfe Mullen in Dorsetshire (Roach 1985). 'Pyne's Royal' and 'Lloyd George' were used extensively for controlled breeding in the East Malling (UK) research program.

In North America, the *R. idaeus* form was preferable, but they proved less adapted than *R strigosus*. The use of controlled crosses in North America started at an earlier date than in Europe (Jennings 1988). 'Latham,' introduced in 1912, originated in Minnesota from a cross between 'King' and 'Loudon' and became the leading cultivar east of the Rocky Mountains (Darrow 1937). It was the leading North American cultivar until the 1929 introduction of 'Newburgh' (Brooks and Olmo 1946). However, the greatest advances improving raspberry cultivars occurred when European and North American species *R. idaeus* and *R strigosus* were crossed. The red raspberry did not become commercially important in North America until after 1865, when an industry was founded on the famous 'Cuthbert' cultivar that was discovered as a chance seedling in what is now a part of New York City. It was probably a cross of the European cultivar 'Hudson River Antwerp' with the wild native North American raspberry *R. strigosus* (Darrow 1937). Some of the best cultivars at that time were 'Latham,' 'Chief,' 'Ohta,' 'King,' and 'Viking.' 'Ohta' was developed by the renowned plant explorer N. Hansen of the South Dakota Experiment Station and was a cross of 'Minnetonka' and a wild selection of *R. strigosus* from North Dakota (Hedrick 1925). Breeding work in the 1930s in North America was carried out in Colorado, Georgia, Hawaii, Maine, Massachusetts, Montana, Maryland, Oregon, Wyoming, New York, South Dakota, Illinois, Washington, Minnesota, Tennessee, North Carolina, and Ontario (Darrow 1937). Several of these programs also worked with black and purple raspberries and made crosses between cultivars and species from elsewhere in the world, particularly Asia. These breeding efforts resulted in the development of primocane-fruiting cultivars and improvements to fruit size that would have been thought impossible earlier, and considerable success was also achieved with development of disease resistance (Darrow 1937).

3 Genetic Resources

Red raspberries are widely distributed in all temperate regions of Europe, Asia, and North America with the greatest diversity in China, the likely center of origin of the subgroup (Fig. 8.2; Jennings 1988). There are 15 recognized subgenera within *Rubus*; the domesticated raspberries are part of the *Idaeobatus* subgroup that contains more than 200 wild species (USDA-ARS GRIN 2009). Cultivated red raspberries are derived mainly from two subspecies of *R. idaeus* var. *vulgatus* from Europe and *R. idaeus* var. *strigosus* from North America. In this review, they are referred to as *R. idaeus* and *R. strigosus*. Dale et al. (1993) examined the diversity of a large number of raspberry cultivars released between 1960 and 1993 and concluded that the genetic base from which the improvements were made was very narrow, and that only five ancestral parent cultivars dominated the ancestry of red raspberry. These were 'Lloyd George' and 'Pyne's Royal,' derived from *R. idaeus,* and 'Preussen,' 'Cuthbert,' and 'Newburgh' that are derived from *R. idaeus* and *R. strigosus.* Extensive accounts of early domestication of raspberry are given by Daubeny (1996), Jennings (1988), Roach (1985), Finn and Hancock (2008), and Hall et al. (2009). Cultivated forms of raspberries are very different from their wild relatives. Wild forms produce large numbers of canes that are shorter and thinner than the cultivated forms. The cultivated forms produce large fruit while the wild forms produce small, soft, crumbly fruit with fewer but larger drupelets (Jennings 1988).

4 Major Breeding Achievements

Darrow (1937) wrote that there remains a large reservoir of germplasm that has been hardly touched in terms of its utilization in plant improvement. Now, more than 70 years later, we can still say that very little of this germplasm diversity has been used for breeding. Despite this, big improvements have been made to yield components and fruit quality and major achievements have been made in pest and disease resistance and adaptation, as well as in the development of new cultivars with primocane-fruiting production.

Of particular importance has been the use of 'Lloyd George' with ancestry that is presumably from North American and European genetics. This cultivar contributed very important traits, including primocane-fruiting, large conical fruit size, and resistance to aphids (Finn and Hancock 2008). 'Willamette,' which is a hybrid of a cross made by George Waldo in 1933 between 'Newburgh' × 'Lloyd George,' dominated the industry for more than half a century and is still an important cultivar that is grown in many raspberry-growing regions around the world, including PNW, Chile, and Serbia (Daubeny et al. 1989). There are interrelated reasons for 'Willamette's success: among them are the fruit traits, dark purple–red color, ease of harvest, which made it suited for hand and machine harvesting, and resistance to cane Botrytis, cane spot, powdery mildew, crown gall, and RBDV (Daubeny et al 1989).

Notable for cold hardiness are the Scandinavian cultivars 'Veten', 'Norna', and 'Asker', 'Preussen' from Germany, the Minnesota cultivars 'Latham' and 'Viking', 'Boyne' from Manitoba, and the cultivar 'Malling Exploit' that also has tolerance to hot dry summers (Jennings et al. 1992). In addition, 'Ottawa' and 'Honeyking' are particularly cold hardy. While 'Boyne' is hardy enough for most demanding parts of Canada and northeastern states of the USA and is the leading cultivar in many areas, it does not tolerate fluctuating temperatures as well as 'Latham' and 'Nova' (Jennings et al. 1992).

Important cultivars have been released from the Horticultural Research International (HRI)-East Malling program, starting in the early 1950s with the releases of the 'Malling series' and the primocane cultivars, among them 'Autumn Bliss.' The Scottish program at Scottish Crop Research Institute (SCRI) released the

'Glen series' with 'Glen Moy' and 'Glen Prosen' being spineless and offering nice fruit appearance and excellent fruit quality. 'Glen Ample', released in 1994, has become the standard throughout Europe for fresh market production alongside 'Tulameen' (Finn and Hancock 2008).

The breeding programs in the Pacific Northwest of North America at WSU (Puyallup, Wash.), AAFC (Agassiz, BC), and the US Department of Agriculture-Agricultural Research Service in Oregon (USDA-ARS; Corvallis) have benefited from many years of collaboration with one another and with the UK programs. The USDA-ARS floricane releases, 'Willamette' and 'Canby,' and the primocane-fruiting cultivars, 'Summit' and 'Amity,' are still commercially important. 'Meeker,' developed by WSU and released in the 1960s, is still the processing industry standard in the PNW and many other growing regions (Moore and Daubeny 1993; Malowicki et al. 2008a). The WSU program has recently released two cultivars, 'Cascade Delight', with excellent fruit quality suited for the fresh market, and 'Cascade Bounty', that is suited for mechanical harvesting and the processing market, both of which have excellent root rot tolerance.

The AAFC program has been one of the most prolific and important programs in the world over the past 30 years. The breeders there took full advantage of germplasm exchanges with the UK and were very successful at identifying outstanding selections from crosses between British Columbia selections and some of the 'Glen series,' particularly 'Glen Prosen' (Finn 2006). The 1977 releases, 'Chilcotin,' 'Skeena,' and 'Nootka,' had excellent fruit quality and high yields for a fresh market berry. The program followed these releases with 'Chilliwack' in the mid 1980s and the very significant cultivar 'Tulameen' in 1989. 'Tulameen' set new standards for fresh market quality, especially with outstanding flavor. This program remains active with the recent cultivar releases of 'Cowichan,' 'Chemainus,' and 'Saanich,' being widely planted (Kempler et al. 2005a, b, 2006, 2007).

'Heritage,' the most important primocane-fruiting cultivar, was released in 1969 from New York Agricultural Experiment Station (Geneva), and it became the standard in growing regions, where cold winter temperatures caused damage to canes of floricane-fruiting raspberries (Daubeny 1996). 'Latham' and 'Chief' released from the University of Minnesota program are valued by breeding programs for their root rot resistance (Finn and Hancock 2008). The primocane-fruiting cultivars, Caroline, Anne, and Josephine, released from the cooperative program centered at the University of Maryland in cooperation with Virginia Tech University, Rutgers University, and the University of Wisconsin River – Falls offers high production and improved fruit quality from fall-producing cultivars. This type of production became the standard in California, where companies, such as Driscoll's Strawberry Associates, developed cultivars and a production system where the plants were in the ground for only 18 months (Finn and Knight 2002).

In Russia, the use of Eastern Bloc floricane-fruiting genetics and primocane-fruiting genetics from East Malling and the Eastern Bloc has resulted in outstanding new primocane-fruiting cultivars with extremely large fruit, very high yields per cane, and strong, upright growth, including 'Bryanskaya Divo', 'Penguin', 'Atlant', and 'Gerakl'. These cultivars are very early, allowing the majority of fruit to be produced before the onset of winter frosts and enabling a long period between harvests for fresh market (Hall, personal observation).

5 Current Goals and Challenges of Breeding

Enriching the cultivated gene pool by incorporation of new genetics from the wild in order to meet the challenges that lie ahead is desired. However, to fully capitalize on the extensive reservoir of alleles within wild germplasm, some advances are still needed, including increasing our understanding of the molecular basis for key traits, expanding the phenotyping and genotyping of germplasm collections, improving our molecular understanding of recombination in order to enhance rates of introgression of alien chromosome regions, and developing new breeding strategies that permit introgression of multiple traits.

Raspberry cultivars grown in a high latitude climate need cold tolerance that is associated with deep and prolonged dormancy, but resistance to fluctuating temperature is also important. Heat and drought tolerance is a requirement for cultivars grown in the Mediterranean, Australia, Africa, South America, and other warm climate regions. These regions require cultivars with low chill adaptation to prevent 'blind bud' syndrome, where, because of inadequate chilling, numbers of buds fail to grow.

There are several reviews that discuss in detail plant characteristics of value in breeding for adaptability to mechanical harvest (Dale et al. 1994; Hall et al. 2002, 2009). Nevertheless, there remains much to be learned about the means of detachment of berries from the receptacle, different mechanisms controlling detachment and mechanisms controlling release of fruit from the plant by different forms of shaking. In addition, there appears to be variation in the timing of release versus the physiological development and ripening of the fruit. A fuller understanding of this variability and timing of the release process should enable the development of high-quality cultivars suitable for the machine harvest of fresh fruit for the future.

Primary improvement criteria for fruit quality include selection for better preharvest hanging ability and postharvest shelf life and resilience under a range of influences from environmental pressures, including physical damage from wind and hail, as well as damage from harvesting machines and sprayers. Resistance to heat and cold and resistance to pests and disease are also important, as well as large fruit size, good presentation, and ease of harvest. Fruit color of newer cultivars varies from very dark red to a light orange–red and there has become a tradition of cultivar selection specifically for processing or for fresh market, particularly in North America, and there is little place for dual-purpose cultivars. Dark fruit color is required by the processing market, but bright, light-red color without blueness is required for the fresh market. Fruit flavor is an important selection criterion, especially for particular flavor volatiles. Seed size contributes to the quality of the product and while seedlessness or parthenocarpy has been identified in some genotypes, it has not been successfully incorporated into any commercial cultivars.

The yield components of fruit size, number of fruit per lateral, number of fruiting laterals per unit of cane length or unit of row area, and number of strong, healthy new canes are important in determining total yield (Dale and Daubeny 1985). Much improvement has been achieved in numbers of fruit per lateral and numbers of fruit from a single fruiting node, but there remains considerable scope for further increasing fruit numbers per lateral and per cane. This has been demonstrated particularly well in Russia with the selection of new primocane-fruiting cultivars with fruit numbers of over 600 on a single cane (Hall, personal observation).

Damage through pests and diseases are significant in reducing productivity in raspberry plantings. Effective chemical controls are being withdrawn, leaving development of resistance to be the primary avenue in dealing with these issues. Sources of resistance to the serious *Rubus* pests and diseases are listed by Jennings et al. (1992). Best resistance is achieved when we place an increased emphasis on diversification of the genetic base of resistance by utilizing indigenous populations of *R. idaeus, R. strigosus*, and other related species. Daubeny (1996) and Finn et al. (2002) list species with the most useful donor traits that successfully cross with *R. idaeus*. Future exploitation of species germplasm has the potential of introducing valuable resistant traits.

Some diseases are widespread, including Botrytis fruit rot and cane disease that is ubiquitous, Phytophthora root rot, which is found in most production regions around the world, and diseases, such as cane blight, spur blight, and *RBDV*, that are also very widespread. Other pests and diseases are localized in their area of economic significance and breeding for resistance to them is only needed in a few programs around the world.

6 Breeding Methods and Techniques

There are now approximately 50 active raspberry breeding programs in 26 countries, mostly in Europe and North America. In the last 30 years, they have released about 160 cultivars (Finn and Hancock 2008). The objectives of all the breeding programs are the development of higher yielding cultivars with improved fruit quality that are suitable for hand harvest, fresh market, or for machine harvest processing. Cultivars need to have some improved pest and disease resistance, mainly to *Phytophthora rubi* (Wilcox and Duncan) Man in 't Veld, cane diseases (*Botrytis cinerea* Pers.: Fr., *Didymella applanata* [Niessl] Sacc. and *Elsinoë veneta* [Burkolder] Jenk), and RBDV. Historically, producers relied heavily on the use of chemicals to manage and control pests. Consumer demands are now for lower residues and banning older formulations that are harmful to the environment and to human health. Because raspberry is a minor crop, the agrochemical industry is reluctant to register new formulations because of the cost. With the lack of chemicals available for control of pests and diseases, developing cultivars with resistance using standard plant breeding methods is the most sustainable way to combat them. The option of using resistance produced in transgenic cultivars is not favored because of consumer resistance, although a transgenic RBDV-resistant 'Meeker' has been developed (Martin et al. 2004; Malowicki et al. 2008a, b). Molecular tools are being developed by some programs mainly for marker-assisted selection, genetic fingerprinting, mapping, and disease diagnostics.

In recent years, the raspberry industry has diversified, with the advent and increase in tunnel production, year-round production for the fresh market, machine harvesting for processing, IQF storage, demand for nutraceuticals, and significant increase in production and competition from China, Chile, Poland, and Serbia. This has placed a demand on breeding programs to foresee changes in the market and develop cultivars that can address industry needs. A significant number of private programs have in recent years entered into *Rubus* breeding and development of proprietary cultivars. The protection of the intellectual property rights of new cultivars, whether through plant breeder's rights in Europe and Canada or patenting in the USA, is rapidly becoming the standard for all breeding programs. The ability to charge royalties on protected plant material has secured income and sustained some programs, but it has reduced the level of germplasm exchange between breeders and limited the cultivars available to growers (Hall et al. 2009).

6.1 Adaptation

Breeding programs select for adaptation to a wide range of environmental conditions including some that are less than optimal for production of older raspberry cultivars. Typical optimum conditions for raspberries are deep, well-drained, mildly acid soils in mild maritime climates with cool to moderate summer daytime temperatures (17–23°C). A Mediterranean-type climate is ideal, with ample rain during the winter months and springtime, when plant growth and development are very rapid, followed by a dry period, supplemented by drip irrigation during the harvest season. Winter months need to have sufficiently low temperatures to meet chilling requirements but not too low to result in winter damage to canes and flower buds (Daubeny 1996). The expansion of raspberry growing to suboptimal environments and less favorable conditions has meant more emphasis on breeding for adaptation to adverse soil conditions, a greater temperature range, adverse winter conditions, and low chill conditions.

The ability of plants to withstand cold winter temperatures is complex and has been studied on numerous plant species, including *Rubus* (Palonen and Buszard 1997). Adaptation to low winter temperature in raspberry involves several factors: the ability to harden in the fall, ability to withstand cold temperatures throughout the winter, a dormancy that cannot be easily broken by fluctuating temperatures, and bud break late enough to avoid frosts (Jennings 1988; Daubeny 1996). During acclimation, hardy cultivars have a high concentration of soluble carbohydrates, high carbohydrate reserves especially sucrose, and a high ratio of sucrose to glucose and fructose (Lindén et al. 1999; Palonen 1999a, b). Cold injury can occur at different physiological states of the plants that makes breeding and genotype evaluation more difficult. Genotypes that acclimatize early in the fall are less likely to be injured early in the fall. Relative hardiness is most easily assessed after naturally occurring low temperatures using a rating system or objective methods, such as tetrazolium staining or conductivity (Quamme and Stushnoff 1983). Very few winter hardy cultivars have been released in the last 20 years (Jennings 1988; Daubeny 1996; Hall et al. 2009). 'Boyne,' 'Killarney,' and 'Nova' are the leading cultivars grown in colder raspberry growing regions of North America and 'Ottawa,' 'Muskoka,' and the more recently introduced 'Jenkka' are grown in Finland. While winter hardy cultivars from the Russian Federation and other northern Asian countries may have been released, that information is not available in Western literature. The challenge for breeders is to introduce more winter hardy cultivars with improved fruit size, firmness, and flavor. Many *Rubus* species are sources of hardiness, including *R. arcticus, R. crataegifolius, R. deliciosus* Torr., *R. hirsutus* Thunb, *R. idaeus, R innominatus, R. odoratus, R. occidentalis, R. pungens* Cambess (syn. *R. oldhamii), R. sachalinensis* H. Lév, and *R. strigosus* (Daubeny 1996; Finn et al. 2001; Hall et al. 2009). Because winters are variable in their effect, it is very difficult to develop reliable testing procedures that can identify genotypes that perform well under variable field conditions. An untested screening protocol has been suggested by Hall et al. (2009), where the plants go through a cycle of endodormancy initiation and cold temperature acclimatization that is followed by two cycles of cold temperatures, tested with dehardening and ending with growing the plants until bud break. After screening, the plants are then evaluated for horticultural traits. However, the test may not represent typical winter conditions, and it still may not cover factors that were listed by Daubeny (1996) and the difference in response between first-year seedlings and adult plants. Adaptation to high summer temperatures and low chilling requirement is important in newer, more marginal raspberry-growing areas.

6.2 Productivity

Higher yield involves the interaction of many factors and has been extensively reviewed (Dale 1989; Daubeny 1996; Jennings 1988; Finn and Hancock 2008; Hall et al. 2009). Yield is influenced by cane number, diameter, vigor, height, internode length, lateral number per cane, percentage bud break, number of fruit per lateral, and fruit size (drupelet size, weight, and number) (Jennings and Dale 1982; Dale 1989). In floricane-fruiting types, a key to reliable yield is the ability of the genotype to balance yield and vegetative production of primocanes. Cultivars with compact growth habit and shorter canes produce more nodes per cropping area and cultivars with high cane numbers are more productive (Daubeny 1996; Kowalenko et al. 2008). In primocane-fruiting types, the yield is dependent on the number of canes and degree of branching which affect the number of fruiting nodes (Hoover et al. 1988). Large fruit size is easily identified by breeders and is correlated closely with high yield, allowing breeders to make steady improvements in yield by selecting for larger fruit size (Cormack and Woodward 1977).

6.3 Resistance to Diseases

Root rot diseases are among the most important limiting production factors in most raspberry regions. Root rot, usually caused by *P. rubi*, results in significant losses for growers and without proper control makes the production of raspberries impossible. It usually occurs in heavy, moisture-saturated soils when excessively irrigated or too much rain has fallen. Often, the disease starts in low-lying parts of the field and spreads to the rest of the field with cultivation or water movement. Typical disease symptoms include reduced vigor, wilting, and a collapse of the canes and water-soaked lesions on the roots (Wilcox 1989). Different species of *Phytophthora* have been isolated from infected plants, but *P. rubi* was found to be the most virulent (Wilcox 1989). There are several control measures that growers can take to limit damage and the spread of root rot, including improving drainage, planting on raised beds, application of gypsum, soil solarization, use of high-health-certified planting stock, and fungicide application (Hall et al. 2009). However, the most effective control is the use of root rot-resistant cultivars. Sources of resistance have been identified in 'Latham,' 'Asker,' 'Boyne,' 'Newburgh,' 'Durham,' 'Chief,' 'Chilliwack,' 'Cherokee,' 'Pathfinder,' 'Sumner,' 'Sunrise,' and 'Cascade Bounty' (Hall et al. 2009). Strong resistance has been identified in *Rubus* species material, including *R. crataegifolius, R. coreanus, R. glaucus, R. lasiostylus* Focke, *R. odoratus* L., *R. phoenicolasius*, *R. pileatus* Focke, *R. spectabilis* Pursh., *R. strigosus*, and *R. sumatranus* Miq as well as the blackberry *R. ursinus* Cham. et Schlecht, which can be successfully hybridized to red raspberry (Barritt et al. 1981; Seemüller et al. 1986; Bristow et al. 1988; Kennedy and Duncan 1993; Finn 2008; Finn and Hancock 2008; Knight and Fernández Fernández 2008b).

Breeding for root rot resistance is well established in several programs. The WSU program has a long history of screening for *P. rubi* in greenhouses and under high disease pressure in the field. These efforts have resulted in the release of 'Cascade Delight,' 'Cascade Dawn,' and 'Cascade Bounty' (Moore 2004, 2005, 2007). The sources of resistance used in this breeding are from 'Cherokee' and 'Latham' and, by selecting in fields heavily infested with *P. rubi*, they have identified individuals with high field resistance. The New York program is screening for *P. rubi* in the field and in hydroponic systems in the greenhouse (Pattison et al. 2004). 'Prelude,' 'Heritage,' and 'Taylor' show good field resistance to *P. rubi*. Their sources of resistance come from 'Latham' and accessions of *R. occidentalis* and *R. strigosus*. Further sources of resistance from *R. strigosus* have also been incorporated in the AAFC–PARC program and selections derived from these appear highly resistant in trials at Puyallup, WA.

Kennedy and Duncan (1993) reported the existence of several *P. rubi* races in North America and Europe. This is not a surprise since there are a number of races in the closely related fungus, *P. fragariae*, that affect strawberry (Kennedy and Duncan 1988). They also reported that 'Latham' is very resistant to all of the raspberry *P. rubi* races. It is possible that because of the narrow source of resistance new races that infect resistant cultivars may appear and overcome this resistance.

Therefore, breeders need to identify and develop broader resistance by incorporating new sources of resistance into the germplasm. In addition, nurseries and producers need to include integrated control systems and not rely solely on resistant cultivars.

Verticillium wilt (*Verticillium albo-atrum* Reinke and Berthier and *V. dahliae* Kleb.) is a minor disease on red raspberry that can cause severe injuries in black raspberry, blackberry, and some blackberry cultivars (Finn 2008; Finn and Hancock 2008). With the increase of production in southern climates and tunnel production, the disease may become more common.

Crown gall [*Agrobacterium tumefaciens* (E.F. Smith and Townsend) Conn] poses a serious threat to the production of susceptible cultivars. The causal bacteria infect plants through cuts and injuries to the roots, often at planting or during cultivation or through wounds caused by nematode feeding. Symptoms are swelling or galls on the crowns and the roots that range from the size of a pea to the size of a tennis ball. They weaken the plant and cause wilting, especially in warm weather. Planting clean, certified, disease-free nursery stock is the most effective control measure as there is no chemical control for infected plantings. Cultivars vary in susceptibility with 'Qualicum,' 'Skeena,' and 'Chilliwack' being susceptible while 'Willamette' has a useful degree of resistance and 'Meeker' and 'Nootka' do not develop galls (Daubeny 1996). Products that are applied at planting that contain the naturally occurring avirulent strain of *Rhizobium radiobacter* (Beijerinck and van Delden) Young comb. nov [syn. *A. radiobacter* (Beijerinck and van Delden)] that is antagonistic to the crown gall bacterium (Deacon et al. 2009) have been developed but with mixed results and consequently have not been widely adopted by the industry.

Gray mold (*B. cinerea*) is the most serious fruit rot disease of raspberry. It causes significant losses to production, reduces shelf life of harvested fruit, and is particularly a problem in wet or humid environments. This disease is the main reason for the increase in the use of tunnels for fresh market production (Hall et al. 2009). Studies show that infection starts at flowering when conidia of the *B. cinerea* grow through the styles and form a mycelium in the carpel while also colonizing the senescing styles and stamens (Daubeny and Pepin 1981; McNicol et al. 1985; Williamson et al. 1987). This led to the introduction of fungicide spray programs that focus on protective sprays during flowering. *Botrytis cinerea* also infects canes causing cane Botrytis. Preharvest control of fruit rot is especially important for machine harvesting as fruit must reach an advanced ripening stage before developing the abscission zone essential for the fruit to be shaken free of the receptacle by the harvester (Hall et al. 2009). Cultivars that are leafy, that have drooping laterals, or whose fruit are tightly clustered show a higher incidence of fruit rot than those with open plant habit, upright laterals, and widely spaced fruit (Daubeny 1996; Hall et al. 2009).

Early work identified sources of resistance to Botrytis fruit rot in 'Cuthbert,' *R. pileatus*, *R. occidentalis, R. crataegifolius*, and *R. coreanus* (Finn and Hancock 2008; Hall et al. 2009). Stephens et al. (2002) suggested that germplasm with *R. pileatus* and *R. occidentalis* in its ancestry has improved shelf life. Factors, like fruit firmness, small drupelet size, and stronger skin, that improve fruit quality also tend to improve genotype resistance to fruit rot. Methods to screen germplasm for resistance to postharvest rot and for improved shelf life are described by Daubeny and Pepin (1974), Barritt et al. (1980), and Stephens et al. (2002).

Spur blight (*Didymella applanata*) is a serious disease that infects leaves on the primocanes and spreads down the leaf infecting the nodes, reducing lateral vigor, and causing large yield losses (Ellis et al. 1991). Resistance to spur blight and cane Botrytis can be improved by selecting for pubescent canes (major gene H) along with the spine-free and dense, waxy bloom traits (Jennings 1983, 1988; Jennings and Ingram 1983). Although it was generally accepted that the H gene also conferred susceptibility to anthracnose, yellow rust [*Phragmidium rubi-idaei* (D.C.) Karst.] and powdery mildew [*Podosphaera macularis* (Wallr.) U. Braun and S. Takam. syn. *Sphaerotheca macularis* (Fr.) Jaczewski] (Jennings and Brydon 1989), recent work in molecular mapping did not support this even though it confirmed the association of gene *H* with resistance to spur blight and cane Botrytis (Graham et al. 2006).

Cane Botrytis (*Botrytis cinerea*) is the same fungus that causes gray mold on fruit and it can be especially destructive in wet seasons when the growth is lush and dense. Cultural practices and spray programs reduce fruit rot. Resistance to cane Botrytis is correlated with spur blight resistance, cane pubescence, and gene *H*. 'Chief,' 'Chilcotin,' 'Meeker,' 'Nootka,' and 'Willamette' are sources of resistance.

Cane blight (*Leptosphaeria coniothyrium* [Fuckel] Sacc.) enters the primocanes through wounds and potentially can cause significant damage in fields that are mechanically harvested, where the spring-loaded catcher plates rub against the new primocanes. Sources of resistance are found in *R. coreanus, R. mesogaeus* Focke, *R. pileatus*, and *R. odoratus* (Finn and Hancock 2008). Resistance is also associated with the spinelessness gene *s* from the old cultivar 'Burnetholm' and it is found in 'Helkal,' 'Julia,' 'Pocahontas,' and 'Tomo' (Hall et al. 2009).

Anthracnose [*Elsinoe veneta* (Burkh.) Jenkins] is a serious disease, also known as cane spot, which in years when weather remains wet can cause considerable cane damage. The first symptoms are small, purplish, circular spots on the cane that become sunken. Infected canes are more prone to winter damage and have reduced and uneven bud break. The disease is easy to control with adequate cultural practices and the normal spray program to control fruit rot. Large differences in resistance are found in red raspberry cultivars with 'Willamette,' 'Nootka,' 'Meeker,' 'Lauren,' 'Vene,' and 'Heritage' being resistant while 'Glen Clova,' 'Glen Moy,' 'Leo,' 'Skeena,' and 'Qualicum' are susceptible (Jennings 1988).

Midge blight is a disease complex involving several fungal pathogens and the larvae of the raspberry cane midge (*Resseliella theobaldi* [Barnes]) that is restricted to Europe (Gordon and Williamson 1991a). Damage caused by the cane midge larvae feeding site becomes infected by *Didymella applanata* or *Fusarium avenaceum* (Fr.:Fr.) (Weber and Entrop 2007).

Fusarium wilt (*Fusarium avenaceum*) is a disease that can cause extensive damage to the floricanes, and increases susceptibility to winter damage. Also the shift of producing raspberries in warmer climates and tunnels increases pressure from this disease. Damage caused by the cane midge (R*esseliella theobaldi*) larvae feeding site becomes infected by the fungus. There are no reports of cultivars that are resistant to Fusarium wilt. 'Tulameen' and 'Glen Ample' are reported as very susceptible to Fusarium wilt in Germany (Weber and Entrop 2007).

Yellow rust (*Phragmidium rubi-idaei* (D.C.) Karst. Syn. *P. imitans* Arth.) is a relatively minor problem that occurs in wet growing seasons when all succulent plant parts are infected and vigor is reduced. 'Glen Clova,' 'Malling Delight,' 'Malling Joy,' 'Cuthbert,' and 'Marlboro' are susceptible (Zeller and Lund 1933; Anthony and Shattock 1983; Anthony et al. 1985a, b). Resistance in 'Latham,' 'Chief,' and 'Boyne' is conferred by gene *Yr* that prevents sporulation. A second source of resistance is found in 'Meeker,' where a polygenic incomplete resistance causes a delay in the appearance of pustules and reduction in their size and number (Jennings 1988).

Late leaf rust [*Pucciniastrum americanum* (Farl. Arth.)], also called autumn rust or late yellow rust, occurs in California, British Columbia, and northern part of central and eastern North America. Infection causes defoliation which reduces vigor and increases susceptibility to winter injury. 'Festival and Heritage' cultivars are particularly susceptible and 'Nova,' 'Chilliwack,' 'Comox,' 'Esta,' 'Hollins,' 'K81-6,' 'Lawrence,' 'Malling Joy,' 'Malling Orion,' 'Ruby,' Tola,' and black raspberries are resistant (Nickerson 1991; Hall et al. 2009).

Raspberry leaf spot (*Sphaerulina rubi* Demi. & Wilc.) is a damaging disease at the southern limits of the raspberry-growing regions in the USA and Europe, where under warm humid conditions plants can be killed. Most cultivated raspberries are susceptible to the disease, with the exception of the red raspberries 'Ranere,' 'Dixie,' 'Pyne's Royal,' 'Bath Perfection,' 'Citria,' 'Fertodi Rubina,' and 'Iskra' and the purple/black raspberries 'Potomac' and 'Evens' (Darrow 1937; Hall et al. 2009). Further resistance has been identified in the Asiatic species *R. biflorus* Buch.-Ham. ex Sm., *R. microphyllus* L. f., *R. inopertus* (Focke) Focke, *R. innominatus* S. Moore, *R. mesogaeus, R. crataegifolius* (syn. *R. morifolius* Siebold ex Franch & Sav, *R.* wrightii Gray), *R. niveus, R. parvifolius, R. phoenicolasius, R. rosifolius* Sm., and *R. thibetanus* Franch. (syn. *R. veitchii* Rolfe) (Keep 1989). However, few modern cultivars can withstand pressure from this disease under warm humid conditions and little effort has been put into breeding for resistance.

Powdery mildew [*Podosphaera macularis* (Wallr.) U. Braun & S. Takam. syn. *Sphaerotheca macularis* (Fr.) Jaczewski and *S. humili* (DC.) Burr.] is a widespread disease that reduces fruit quality of infected fruit. Screening for the disease can be achieved very easily during the early stages in the greenhouse propagation process when susceptible individuals segregate. Breeding for resistance to this disease becomes important when breeding for fresh market cultivars grown in tunnels, where conditions are favorable for the disease infection of plants and fruit. Sources of resistance include most black and purple raspberries (with the exception of 'Black Hawk,' 'Dundee,' and 'Munger' black raspberries and 'Cardinal' purple raspberry that are susceptible), as well as several *Rubus* sp. (Keep 1989; Finn and Hancock 2008; Hall et al. 2009).

Root-lesion nematode [*Pratylenchus penetrans* (Cobb) Schuurmans-Stekhoven] is a pest that feeds on raspberry roots resulting in root lesions and cell death causing poor plant establishment, replant problems, and root rots (McElroy 1991). 'Nootka' appears to be resistant while 'Glen Clova' and 'Chilcotin' are not susceptible.

Inheritance studies showed that 'Chilliwack' gave progenies with the highest resistance (Vrain et al. 1994).

Dagger nematodes (*Xiphinema* species). *Xiphinema bakeri* (Williams) is limited to the Pacific Northwest feeds on root meristems and can cause significant stunting of root systems. *Xiphinema americanum* (Cobb) and *X. diversicaudatum* (Micoletzky) Thorne are vectors for *tomato ringspot virus* (TomRSV) and *strawberry latent ringspot virus* (SLRV). Sources of resistance have been little investigated, but some host plant resistance was reported by Jones et al. (1989).

Virus diseases cause some of the most damaging diseases of crop plants. Perennial and vegetatively propagated crops like raspberry are particularly vulnerable to virus diseases. To maintain economically acceptable levels of yield, it may be necessary to replant at frequent intervals with virus-free planting stock. The causes of losses due to virus diseases are twofold. First, there are the losses that result directly from the effect of the disease on growth and yield of the host plant. Second, there are the costs of attempting to control the diseases, like applying pesticides to control the vector or replanting with virus-free stock. The use of resistant cultivars is the most effective and cheapest way of reducing damage by viruses. In recent years, a marked improvement has been made in virus detection with the introduction and widespread use of enzyme-linked immunosorbent assay (ELISA) and polymerase chain reaction (PCR) techniques and these procedures have become important means of virus detection.

Raspberries, like other vegetatively propagated crops, are subject to attack by a very large number of viruses. These viruses can conveniently be considered in several main groups: aphid-borne, nematode-borne, leafhopper-borne, and pollen-borne viruses, as well as virus diseases with unknown vectors, and virus-like conditions and disorders (Converse 1987).

Aphid-borne raspberry viruses that are damaging include the *raspberry mosaic virus* complex (RMD), *raspberry leafspot virus* (RLSV), *raspberry leaf mottle virus* (RLMV), *Rubus yellow net virus* (RYNV), *black raspberry necrosis virus* (BRNV), *raspberry vein chlorosis virus* (RVCV), and *raspberry leaf curl virus* (RLCV) (Keep 1989). Sources of aphid resistance that in turn impart virus resistance have been well-documented (Jennings et al. 1991). The large raspberry aphid, *Amphorophora idaei*, transmits several viruses that are referred to as mosaic viruses. The vector resistance approach to control mosaic viruses has been used by the East Malling breeders for more than 50 years (Knight and Keep 1958). Mosaic viruses were once widespread along the Pacific Northwest of North America, but are now rarely a serious problem as older cultivars have been replaced by newer aphid-resistant cultivars and virus-free planting stock is used (Stace-Smith 1987).

Pollen-transmitted RBDV is spread rapidly in susceptible cultivars and can also be seed borne. RBDV has been reported in red raspberry, black raspberry, and blackberry (Converse 1988). Symptoms of the disease include drupelet abortion leading to crumbly fruit and sometimes leaf yellows on lower primocane leaves in the early spring, but often plants have no vegetative symptoms. The virus is found in almost all raspberry-growing regions and a more virulent resistance-breaking strain named RB-RBDV is found in Russia, Serbia, England, and Wales (Barbara et al. 2001). Table 8.2 lists 87 cultivars that are resistant to RBDV, conferred by

Table 8.2 List of RBDV-resistant cultivars and their parents

Cultivar	Female parent	Male parent	Year of release	Origin	Breeder
Amethyst	Robertson	Cuthbert	1968	Ames, Iowa	E.L. Denisen
Avon	Malling Promise	Cuthbert	1967	AAFC, Kentville, Nova Scotia	D.L. Craig and L.E. Alders
Black Hawk	Quillan	Black Pearl	1955	Ames, Iowa	T. Maney
Black Pearl	*R. occidentalis*	*R. occidentalis*	1905	St. Joseph, Missouri	H. Krumrei
Boyne	Chief	Indian Summer	1960	Morden, Manitoba	C.R. Ure
Bristol	Watson Prolific	Honeysweet	1934	Geneva, New York	R. Wellington
Burnetholm	*R. idaeus*	*R. idaeus*	1927	Geneva, New York	R. Wellington
Carnival	Ottawa	Rideau	1955	AAFC, Ottawa, Ontario	AAFC
Cayuga	June	Cuthbert	1922	Geneva, New York	R. Wellington
Chief	Latham	Latham	1930	Excelsior, Minnesota	M.J. Dorsey and A.N. Wilcox
Chilcotin	Sumner	Newburgh	1977	AAFC, PARC, B.C.	H.A. Daubeny
Citadel	Mandarin	Md S420-5	1966	College Park Maryland.	I.C. Haut
Citria	Cayuga	Orr's Seedling	1996	Pitesti-Maracineni, Ages, Romania	P. Mladin
Clyde	Bristol	NY 17861	1961	Geneva, New York	G.L. Slate and J. Watson
Columbian	Cuthbert	Gregg	1891	Geneva, New York	
Cowichan	Newburgh	Qualicum	2001	AAFC, PARC, B.C.	C. Kempler and H.A. Daubeny
Cumberland	Gregg	Gregg	1896	Camp Hill, Pennsylvania	D. Miller
Cuthbert	*R. strigosus* 15	Hudson River Antwerp	1865	New York State	T. Cuthbert
Dormanred	*R. parvifolius*	Dorsett	1972	State College, Mississippi	J.P. Overcash
Fairview	ORUS 782	Washington	1961	ORSU, Corvallis, Oregon	G.F. Waldo
Fillbasket = Superlative	*R. idaeus*	*R. idaeus*	1855	England	
Gertrudis	*R. idaeus*	*R. idaeus*	1940	Breda, the Netherlands	I. Rietsema
Glen Clova	SCRI 11/510	SCRI S29/97	1969	Scottish Crop Research Institute	D.L. Jennings
Glen Magna	Meeker	SCRI 7719B11	1994	Scottish Crop Research Institute	R.J. McNicol and D.L. Jennings
Golden Queen	Cuthbert	Cuthbert	1882	Berlin, New Jersey	E. Stokes

Gregg	*R. occidentalis*	*R. occidentalis*	1866	Ohio County, Indiana	
Haida	Malling Promise	Creston	1973	AAFC, PARC, B.C.	H.A. Daubeny
Hailsham	*R. idaeus*	*R. idaeus*	1916	Hailsham, Sussex	Mr. Dann.
Heritage	Milton	Cuthbert	1969	Geneva, New York	L Slate
Heija	Malling Promise	Merva	1975	Piikkiö, Finland	H. Hiirsalmi and J Säkö
Herbert	*R. strigosus*	*R. strigosus*	1887	Ottawa, Ontario	R.B. Whyte
Hilton	Newburgh	St Walfried	1965	Geneva, New York	G.L. Slate and J. Watson
Honeysweet	*R. idaeus*	*R. idaeus*	1925	France	
Jatsi	Ottawa	Malling Promise × Merva	1997	Agric. Res. Centre for Finland	P. Dalman, H. Hiirsalmi, T. Hietaranta and M. Linna.
Jenkka	Malling Promise × Merva	Ottawa	1997	Agric. Res. Centre for Finland	P. Dalman, H. Hiirsalmi, T. Hietaranta and M. Linna.
June	Louden	Marlboro	1909	Geneva, New York	
Killarney	Indian Summer	Chief	1961	Morden, Manitoba	C.R. Ure
King	Thompson	Thompson	Early 1900s		
Kiwigold	Sport of Heritage		1988	Morrinsville, New Zealand	C.E.H. and M.H. Thomas
Krupna Dvorodna	Malling Exploit	Rubin	1973	Fruit Research Institute, Cacak, Yugoslavia	P.D. Misic, V.Z. Bugarcic, and M.B. Tesic.
La France	*R. idaeus*	*R. idaeus*	1912	Stamford, Connecticut	A. Alivs
Latham	King	Louden	1914	Excelsior, Minnesota	M.J. Dorsey
Lord Lamborne	*R. idaeus*	*R. idaeus*	Pre 1900		J. Kettle
Malling Admiral	EM 91/161	EM 121/8	1978	East Malling, England	E. Keep
Malling Augusta	EM 2819/10	Malling Joy	1989	East Malling, England	E. Keep and V.H. Knight
Malling Delight	EM 759/16	EM 542/16	1978	East Malling, England	E. Keep
Malling Enterprise	Preussen	EM 23/50	1943	East Malling, England	N.H. Grubb
Malling Exploit	Newburgh	EM 30/8	1937	East Malling, England	N.H. Grubb
Malling Gaia	EM 2497/47	EM 2180/57	1992	East Malling, England	V.H. Knight and E. Keep
Malling Jewel	Preussen	EM 23/50	1932	East Malling, England	N.H. Grubb
Malling Joy	EM 1317/42	EM 1302/48	1980	East Malling, England	E. Keep, V.H. Knight and J.H. Parker

(continued)

Table 8.2 (continued)

Cultivar	Female parent	Male parent	Year of release	Origin	Breeder
Malling M	Preussen	Lloyd George	1946	East Malling, England	N.H. Grubb
Malling Minerva	EM 5030/3	SCRI 7269/67	2005	East Malling, England	V.H. Knight
Malling Orion	EM 55/6	EM 277/4	1978	East Malling, England	E. Keep
Malling Promise	Newburgh	EM 30/8	1937	East Malling, England	N.H. Grubb
Mandarin	*R. parvifolius* × Taylor	Newburgh	1955	Raleigh, North Carolina	C.F. Williams
Motueka	B257	F29	2002	Nelson Research Centre, NZ	H.K. Hall
Moutere	Haida	Qualicum	2008	Nelson Research Centre, NZ	H.K. Hall and M.J. Stephens
Newburgh	Newman	Herbert	1929	Geneva, New York	R. Wellington
Nootka	Carnival	Willamette	1977	AAFC, PARC, B.C.	H.A. Daubeny
Novost Kuzmina					
Octavia	Glen Ample	EM 5928/114	2002	East Malling, England	V.H. Knight
Okawa	86105N68	Selwyn	1990	Nelson Research Centre, NZ	H.K. Hall
Phoenix	*R. idaeus*	*R. strigosus*	1896	Little Silver, New Jersey	J.T. Lovett
Preussen	Superlative	Marlboro	1919	Eisleben, Germany	F. Fromme
Pynes Royal	*R. idaeus*	*R. idaeus*	1913	Topshan, Devon, England	G. Pyne
Quillan	*R. occidentalis*	*R. occidentalis*	Pre 1900		
Ranere	*R. strigosus*	*R. strigosus*	1912	Little Silver, New Jersey	J.T. Lovett
Rannaya Sladkaya	Usanka	Usanka			
John Robertson	*R. occidentalis*	*R. occidentalis*	1935	Hot Springs, South Dakota	J. Robertson
Rubin Bulgarski	Preussen	Lloyd George			
Ruvi	Rubin Bulgarski	Viking	1996	Fruit Research Institute Pitesti-Maracineni, Ages, Romania	P. Mladin.

Schonemann	Lloyd George	Preussen	1950	Germany	W. Schönemann
Selwyn	Marcy	Malling Delight	1992	Nelson Research Centre, NZ	H.K. Hall
Sentinel	Sunrise	Milton	1966	College Park, Maryland	I.C. Haut
September	Marcy	Ranere	1947	Geneva, New York	G.L. Slate
Spirina Belaya					
Star	Cayuga	Orr's Seedling	2000	Fruit Research Institute Pitesti-Maracineni, Ages, Romania	P. Mladin.
Sumner	Washington	Tahoma	1956	WSU, Puyallup, Washington	C.D. Schwartze and A.S. Myhre
Sunrise	Latham	Ranere	1939	Glen Dale, Maryland	G.M. Darrow
Tadmor	Lewis	Waimea	2002	Nelson Research Centre, NZ	H.K. Hall
Valentina	EM 6225/11	EM 5588/81	2005	East Malling, England	V.H. Knight
Waiau	Fairview	Marcy	1990	Nelson Research Centre, NZ	H.K. Hall
Waimea	SCRI 795B10	SCRI 82224D4	2002	Nelson Research Centre, NZ	H.K. Hall
Washington	Cuthbert	Lloyd George	1938	WSU, Puyallup, Washington	C.D Schwartze
Watson Prolific	*R. idaeus*	*R. idaeus*	pre 1915		
Willamette	Newburgh	Lloyd George	1943	ORSU, Corvallis, Oregon	G.F. Waldo

the *Bu* gene, but only 'Haida' and 'Schönemann' are known to show some field resistance to RB-RBDV. Breeding for resistance to RBDV by the Pacific Northwest breeding programs is of major importance. The main cultivar grown, 'Meeker,' is susceptible and becomes 100% infected within 5–6 years after planting, producing low-grade, crumbly fruit and forcing the grower to replant with virus-free plants (Martin 2003). The AAFC–PARC breeding fields in Abbotsford, BC, are within the commercial production area with high RBDV disease pressure, resulting in about 10–20% of the plants becoming infected in the selection block each year. However, experience shows that it may take an additional 10 years to confirm actual resistance. For example, 'Saanich,' from a 1989 cross, tested positive for RBDV for the first time in 2004 more than 10 years after being selected (Kempler et al. 2007). Developing molecular markers to identify RBDV-resistant genotypes could be used for screening seedlings prior to planting in the field or for screening selections with other desired characteristics (Martin and Mathews 2001; Malowicki et al. 2008a, b). Martin's USDA-ARS lab in Oregon has developed resistant plants via transformation, but because of public and industry concerns with genetically modified organisms (GMOs) these have never been released to producers (Martin and Mathews 2001; Martin et al. 2004).

6.4 Resistance to Pests

Aphids Amphorophora idaei Börner and *A. agathonica* Hottes are found in Europe and North America, respectively, and are primarily a concern as vectors of several viruses. Breeding for resistance to the vectors became a major objective of several breeding programs, including East Malling, SCRI, and AAFC–PARC, and was recently adopted by the USDA-ARS program in Corvallis in its black raspberry breeding efforts (Dossett and Finn 2008). Since the 1960s, this approach has been effective in preventing virus spread by controlling the vector in the UK (Birch et al. 2002). Currently, five biotypes of *A. idaei* and several genes that differ in their effectiveness against them have been identified. Over the years, a number of cultivars containing genes for resistance have been commercialized. The A_1 gene derived from *R. idaeus* is inherited as a single dominant allele. The gene confers resistance to biotypes 1 and 3 and was bred into several commercial cultivars released from SCRI. By the 1990s, the A_1 base gene was overcome by aphid populations and the A_{10} gene derived from *R. occidentalis*, which confers resistance to *A. idaei* biotypes 1–4, was bred into 'Malling Leo,' 'Malling Joy,' 'Autumn Bliss,' 'Gaia,' 'Glen Rosa,' 'Glen Doll,' and 'Glen Fyne' developed at EMR and SCRI (Birch et al. 1994; Jennings et al. 2008). Studies have shown that the A_{10} gene affects the chemical composition of the leaf surface wax components that interfere with the initial settling behavior of *A. idaei* (Birch and Jones 1988; Robertson et al. 1991). In North America, breeding for resistance to *A. agathonica* has been based almost exclusively on a single gene, although A_{g2} and A_{g3} have been described but not used in breeding (Daubeny and Stary 1982). It appears that selection pressure has resulted in the appearance of diverse *A. agathonica* biotypes, but they do not colonize resistant plants. One possible reason for this may be that the majority of the fields in the Pacific Northwest are planted to nonresistant cultivars, like 'Meeker' and 'Willamette' (Kempler, personal observation). Several aphid-immune cultivars have been released from the AAFC–PARC program, including: 'Haida,' 'Nootka,' 'Skeena,' 'Qualicum,' 'Malahat,' 'Cowichan,' 'Esquimalt,' 'Chemainus,' and 'Saanich' (Daubeny 1973, 1978a, b; Daubeny and Kempler 1995; Kempler and Daubeny 2000; Kempler et al. 2005a, b, 2006, 2007). 'Algonquin,' also from the AAFC–PARC program, has been identified as homozygous for gene A_{g1} with apparent lack of segregation when 'Algonquin' is used as parent. Its resistance was inherited from 'Haida' and 'Canby,' which are heterozygous for gene A_{g1} (Daubeny and Sjulin 1984).

It is essential that further genetic sources of aphid resistance are identified to ensure continued success of this control stratagem. The combination of several resistance genes provides more robust long-lasting resistance. Toward these goals in red raspberry and even more so in black raspberry, work in Oregon has uncovered strong new sources of aphid resistance in two populations collected in Ontario and Maine. Two selections from each of these populations were crossed with the susceptible 'Black Hawk' and 'Munger' and all of the resulting progeny showed strong aphid resistance. Subsequent tests with the closely related nonvectoring aphid *A. rubitoxica* Knowlton showed that the population from Maine was resistant to this species while the Ontario population was not. The data suggest that aphid resistance in these two populations is controlled by different genes and each is inherited as a dominant trait (Dossett and Finn 2010).

Cane midge (*Resseliella theobaldi* Barnes) is a small fly that is rarely noticed as an adult and the larvae are creamy white and up to 0.5 cm in length. They bore inside the shoot and girdle it causing it to wilt and die. The damage sites often become infected by a range of fungi and these become more of a problem than the damage from the midge larvae. Cultivars that show good cane vigor and the ability to produce a new flush of cane growth appear to withstand the damage more effectively. Cultivars that show no cane splitting also show less infestation (Gordon and Williamson 1991a, b).

Mite species, especially the two-spotted spider mite (*Tetranychus urticae* Koch), affect raspberries mainly during dry weather and can cause severe defoliation. The shift to tunnel production has increased the incidence of two-spotted spider mite as a problem. Breeding for resistance has not been reported, but there is large variability among cultivars in susceptibility and the ability to sustain large populations without damage. While not commercialized, efforts to use molecular genetics to engineer resistance in cultivars have been attempted (Vrain 1997).

6.5 Nutraceutical Properties

Wide diversity exists within the *Rubus* species for micronutrients, vitamins, and health-beneficial compounds (Stewart et al. 2007; Seeram 2008). Conventional breeding methodology assisted by molecular marker tools can effectively be employed to develop raspberry genotypes containing significantly higher amounts of critical and beneficial nutraceuticals. Developing GMO raspberries with improved nutraceutical compounds is a feasible option as technology has been already successfully used in developing GMO rice with incorporated genes for lysine, iron, zinc, and β-carotene (Krishnan et al. 2003). The public is not ready to accept transgenic fruit crops, but it is possible that developing transgenic crops with health benefits could change public opinion and acceptance of transgenic fruits. Developing cultivars with higher nutritional benefits requires a one-time investment with results that would be self-sustaining. It takes time before it will make an impact, but on the long-term basis can be a very effective health strategy. Once introduced into the working germplasm, these traits will remain in all future cultivars.

Recent publications dealing with cancer prevention frequently point to the importance of fruits and vegetables for diverse health benefits. Anthocyanins and polyphenols, such as ellagic acid, are shown in vitro and in vivo to be beneficial in protecting cells from various health injuries, such as ageing and different forms of cancer. The cancer prevention and suppression action by ellagic acid have been reported in many papers as has the very high ellagic acid content of raspberries (Mullen et al. 2002).

While a tremendous effort has focused on the value of fruits and vegetables as sources of antioxidants, recent research has found that while many fruits and vegetables are rich sources of antioxidants the human body is not capable of absorbing them in sufficient quantities to have a direct antioxidant effect in human cells (B. Frei pers. comm.). Instead, compounds, such as the polyphenolics, associated with berries seem to have a cell modulation effect. While there is little doubt that over time a raspberry genotype could be developed with higher nutritional or antioxidant values using traditional or molecular approaches, it would seem to make more sense to focus on making the fruits taste better so that they are more desirable and on making the genotypes more efficient to produce, thereby making them more affordable. These last two approaches are tried and true, are already emphasized in many programs, are not subject to the ebb and flow of the hype surrounding the latest and greatest compound that solves all of your health problems, do not risk alienating consumers, and do not require any new techniques to be developed.

6.6 Generation of Genetic Variation

Interspecific hybridization between cultivated raspberries and wild *Rubus* germplasm frequently exposes a large number of gene and chromosome organization differences. This leads to a bewildering complexity of variation in the segregating generations. Moreover, many of the recombinations are disharmonious ones, neither having the ability to survive in the wild nor to be selected by the plant breeder.

Substantial reviews of the use and value of *Rubus* species other than the primary progenitor species of red and black raspberries have been made by Jennings (1988), Jennings et al. (1991), Finn et al. (2002), and Finn and Hancock (2008). The greatest effort and success have been achieved using related species in the Idaeobatus and with the highly polyploid *R. ursinus* blackberry. This diverse germplasm has been a source of altered plant architecture and phenology, biotic and abiotic stress resistance, and improved fruit quality. While it generally takes many generations after the initial hybridization to achieve commercial genotypes, 'Loganberry,' 'Chehalem,' and 'Boysen' are examples of first-generation hybrids with a different species, in this case *R. ursinus* or *R. armeniacus*, resulting in commercial genotypes (Clark et al. 2007). While there are probably numerous examples, two recent releases have novel species in their background with 'Malahat' tracing back to *R. phoenicolasius* and *R. occidentalis* and 'Malling Juno' to *R. crataegifolius* (Kempler and Daubeny 2000; Knight and Fernández Fernández 2008a).

The AAFC–PARC program successfully utilized several *R. strigosus* lines as sources of resistance to root rot caused by *P. rubi (*syn. *fragariae* var. *rubi)* and as alternative sources of resistance to the big aphid (*A. agathonica*) and possibly resistance to the resistant-breaking aphid biotype. Hybrids between *R. idaeus* and *R. parviflorus* Nutt. show that the morphological differences between the species were not associated with chromosome differences, although low fertility is observed in the F_1 generation (Jennings and Ingram 1983; Daubeny 1996). Full fertility may be restored within one or two backcross generations. Fertility levels of crosses between non-Idaeobatus species and a cultivated red raspberry are generally lower than crosses within the Idaeobatus species (Ourecky 1975).

Polyploidy has generally not been important in raspberry breeding. Raspberry species are most commonly diploid ($2n=14$), the basic chromosome number $x=7$, with some tetraploid ($2n=28$) cultivars and species (Thompson 1997; Hall et al. 2009). Chromosomes are small (1–2 μm in length) and nuclear DNA content ranges from 0.56 to 0.59 pg in diploid species (Lim et al. 1998; Meng and Finn 2002). Tetraploidy in red raspberries does not give any significant adaptive value as fruits are irregular with large drupelets and pyrenes and with reduced drupelets set possibly due to abortion during meiosis (Jennings 1988; Hall et al. 2009). In raspberries, reduplication of the chromosome set has not been important in the development of cultivars, and when it does occur it arises from a single species and is referred to as simple polyploidy or autopolyploidy. Naturally occurring triploid (e.g., 'Erskine Park') and tetraploid cultivars (e.g., 'LaFrance,' 'Hailsham,' 'Colossus,' and 4× forms of 'Heritage' and 'Autumn Bliss') have been reported, but they have reduced fertility (Ourecky 1975; Jennings 1988).

Induced mutation through irradiation methods has been considered, but no beneficial mutations were ever reported. Spontaneous mutations for yellow fruit and large fruit size have been reported. Two yellow fruiting mutations are 'Kiwigold' from 'Heritage' and 'Allgold' from 'Autumn Bliss' (Daubeny 1996). 'Glen Garry's' large fruit size traces to a spontaneous mutation of 'Malling Jewel' (Knight et al. 1989). This mutation is attributed to a single dominant gene (L_1), which influences the development of the fruiting laterals and increases the numbers and the size of individual drupelets (Jennings 1988). The L_1 gene was later found to be unstable and has had to be rooted out of breeding programs.

6.7 Breeding Methodology

Raspberries have biennial canes that require a dormant period prior to flowering. Flower bud initiation starts in late summer to early fall when day length becomes shorter and temperatures are lower than 13°C (Dale and Daubeny 1987). Exceptions to the biennial trait are the primocane-fruiting types that initiate flowers under long day conditions in the spring and flower and fruit late in the summer and the fall. While raspberry is a protandrous species, a significant level of self-pollination occurs (Daubeny 1971).

Crosses can be made in the field or in a protected environment like a greenhouse or growth chamber. Individual plants to be used as parents are tested to be free of *tobacco streak virus* (TSV) and all strains of RBDV. Flower buds just beginning to show petals are emasculated using a pointed forceps to cut a complete circle into the base of the sepals that, when pulled away, removes the sepals, petals, and anthers, leaving the gynoecium with the styles and stigmas unharmed. Paper, glassine, or other semitransparent bags that are weatherproof are placed on the laterals covering all the flowers. Plastic bags are not used as they can cause excessive heat buildup. Additional mature flower buds can be emasculated in 2–3 days, but any small buds are removed. Flowering laterals on the male parent are also bagged to provide flowers as a pollen source that is not contaminated with unknown pollen. Two to three days after emasculation, open flowers from bagged laterals on the pollen parent are harvested and placed in Petri plates to dry. Later, these can be used directly as a 'brush' to transfer pollen to the stigmatic surfaces or the dishes with the dried flowers can be shaken and the pollen that collects on the plate surface transferred to the female flowers with a camel hair brush or a glass rod. The process is repeated at 2–3-day intervals until it appears that the flowers are not receptive anymore, when the stigma and style start to brown. While weather dependent, the flowers can be receptive for 7–10 days. Laterals are rebagged after each pollination and 70% alcohol is used to clean hands and tools and prevent pollen contamination. Variations of this process are reported by Ourecky (1975), Jennings (1988), Daubeny (1996), and Finn and Hancock (2008). Pollen can be extracted by removing anthers and drying them under incandescent light. The dry anthers are then crushed with a glass rod to release the pollen. The dry pollen can be stored for four or more weeks in a desiccator with calcium chloride at 5°C. For out-of-season crosses, dormant plants may be brought into a warm environment (15°C with 16-h day length) and the same process as for field pollination is followed.

Bags are kept on the laterals covering the fruit until it ripens. Harvested ripe fruit can be placed in the refrigerator until the whole cross has been harvested or sufficient fruits have been harvested for seed extraction. Fruit is covered with water and about ten drops of pectinase are added to the slurry. The fruit may be simply mashed with a fork or, if very carefully done, the fruit can be pureed with a few quick pulses with a low-speed blender with reversed or protected blades to prevent damaging the seeds. The slurry is kept at room temperature for 12–24 h, more water is added, and the viable seeds settle to the bottom of the container and the pulp and hollow seeds can be decanted off. Seeds may be placed on paper or in cups to dry before being stored in seed envelopes. Seed may be stored for a few months at room temperature before sowing as this seems to keep the seed from going into a deeper dormancy. Seed that needs to be stored longer may be refrigerated at a temperature between 1 and 5°C or at −18°C until sowing. Refrigerated seed stored in a desiccator remains viable for many years (Ourecky 1975; Daubeny 1996).

Rubus seeds require scarification treatment that involves cutting the seed coat using abrasion, thermal stress, or chemicals to physically remove much of the pericarp, making it permeable and encouraging germination. Seeds also require stratification to simulate winter conditions so that germination may occur. The procedure described by Daubeny (1996) and Ourecky (1975) has been used by the AAFC–PARC breeding program and other programs with satisfactory results. Dry seeds are placed in glass test tubes kept in a crushed ice bath and treated for 15–20 min with enough concentrated H_2SO_4 to cover the seeds. The tube is filled with water to dilute the acid and the contents are poured into filter mesh and washed for 5 min under running tap water. The seeds are then immersed for 1 week in a 1% solution of calcium hypochlorite followed by a wash in running water for 5 min. Seed may be stored in moist sand or directly on moistened peat in the germination flat at 5°C for 6 weeks. Seed is sown on light soilless potting medium and covered with a small amount of sand and placed in a 25°C, 16-h day length, and high humidity environment. Seedlings are more often killed by excessive than too little watering. Intermittent mist that keeps the seeds damp but not soaked is ideal. Six weeks of stratification are not needed if seeds are treated immediately after harvest with sulfuric acid and calcium hypochlorite (Dale and Jarvis 1983). When only small amounts of seeds are available, it is possible to nick through the seed coats and expose the embryo or to use in vitro germination procedures (Ke et al. 1985; Nesme 1985; Finn and Hancock 2008).

Germination begins within 3–4 weeks (Dale and Jarvis 1983). When the first true leaves appear, seedlings are ready to be transplanted into larger pots. At this time, selection for spineless canes expressed by gene *s* can be made, where spineless segregates are devoid of stalked glands at the edge of the cotyledon leaves while on the spiny plants glandular hairs are present (Hall et al. 2009). Seedlings may be screened at this early stage for resistance to the large raspberry aphid, the vector of the RMD. The aphid vectors are *Amphorophora idaei* in Europe and *A. agathonica* in North America. *Amphorophora idaei* has several biotypes that are differentiated by their abilities to overcome plant-resistance genes. Bioassays of aphid field populations showed a strong shift toward A_1 resistance-breaking biotypes since the 1960s (Jones et al. 2001; Birch et al. 2002). *Amphorophora agathonica* that attacks plants previously identified as being resistant have been found; however, these 'biotypes' have not reproduced well on resistant plants and, so far, have not been shown to be a threat in the field (Daubeny et al. 1992; Kempler, personal observation). Screening for reaction to *A. agathonica* in the AAFC–PARC program is done prior to field planting on seedlings with at least three leaves. Aphids are reared according to Forbes et al. (1985) and three aphids are placed on each plant every 3–4 days for 3 weeks. Plants that are not colonized by aphids are classed as resistant.

Additional observations are made in the field to identify escapes from the common biotype or susceptibility to a resistance-breaking biotype. Young seedlings have also been screened for reactions to root rot caused by *Phytophthora rubi* in the AAFC–PARC program. The seedlings are grown in individual pots of substrate, and at the five true leaf stage the roots are inoculated with a mycelial suspension of the pathogen. Above-ground symptoms of root rot usually appear within 10 weeks and susceptible genotypes are dead within 15 weeks while resistant seedlings grow vigorously (Daubeny 1996). Pattison et al. (2004) developed an effective hydroponic procedure to conduct screening for resistance. Young seedlings can be pre-field screened for resistance to other diseases and pests according to the relative importance of the problem and the practicality of the screening.

Seedlings are then planted in the field typically at 75–150 cm within the row. The AAFC–PARC program plants at 90 cm within the row as this allows us to reasonably distinguish between individual plants. However, at this close spacing propagation stock must be harvested carefully to ensure genotype integrity. The between-row spacing depends on local farming practices. The 240 cm between row spacing used by the AAFC–PARC program allows for a tractor to pass between rows. For most crosses, a progeny size of 100 seedlings gives a good representation of the potential of the specific combination. Larger progeny size may be valuable when the parents involved are especially genetically diverse or when primocane-fruiting segregates are sought from crosses between floricane-fruiting and primocane types, as in the seedling populations that produced 'Erika' and 'Sugana,' each from around 6,000 plants of the cross 'Tulameen' × 'Autumn Bliss.' Each year, the AAFC–PARC program plants 2,000–5,000 seedlings after prescreening for susceptibility to the aphid vector *A. agathonica*, which eliminates about 30% of the seedlings. The AAFC program makes 25–60 crosses annually and field plants an average of 60–80 seedlings per cross. Seedlings are planted early in the spring and immediately irrigated. Some programs plant in the fall to reduce problems with weed control. Dormant primocanes of young plants that are evaluated for their floricane crop may be cut back in order to save on labor in pruning and training. If grown well and under ideal conditions, some programs may successfully make selections of superior primocane-fruiting genotypes toward the end of the growing season. However, since developing cultivars that are resistant to RBDV is a main objective of the AAFC–PARC program, the canes are left to flower and to add another year of exposure to RBDV infection. Usually, selection takes place in the second or occasionally the third year after planting. During the fruiting season, fields are walked every 2–5 days and plants are selected according to the objectives of the program. In the AAFC–PARC program, notes are collected only on the selected plants and selection is done according to the desired plant habit, fruit characteristics (especially flavor), and suitability for the fresh or processing markets. DNA screening of the plant population can be used to identify individuals with desirable traits. A selection rate of 0.5–1.5% in the seedling field is common in most breeding programs, but occasionally it may be up to 10% (Hall et al. 2009). Leaf tissue from each selection is tested for the presence of RBDV. Selections that test positive are discarded mainly because selections that became infected in the field after short exposure (two to three seasons) are very susceptible but also because it is time consuming to use heat therapy to produce a virus-free clone. This would not be appropriate if the parents had not been tested prior to using them to produce the cross. Soon after a selection is made, stem nodes are collected from the primocanes to establish the genotype *in vitro*. Nodes that are collected late (September–October in British Columbia) have already initiated flowers and produce no vegetative buds (Sønsteby and Heide 2008; Kempler, personal observation). In the AAFC–PARC program, selections are also transplanted into a 'repository field,' where they can be used as parents for crossing and where they are also tested for RBDV every year. Enough plants are propagated in tissue culture over the winter for early-spring planting in first-year trials. If the selection has the potential to be suited for mechanical harvesting, ten plants are planted in an unreplicated plot at 75–90 cm between plant spacing, where they are harvested in the second and third year after planting with a commercial harvester. A gap between plots allows excellent separation of the harvested fruit between the selections and collection of the fruit into separate trays. Machine harvest evaluation early in the evaluation of several genotypes was critical in allowing for the relatively rapid release of the AAFC–PARC cultivars 'Chemainus,' 'Saanich,' 'Nanoose,' and 'Ukee.' The selections are assessed weekly and rated numerically for yield, overripe fruit, unripe and green fruit, fruit color and firmness, fruit integrity, and suitability for mechanical harvest (plant growth habit). Their possible suitability for IQF processing is also inferred from fruit qualities. Clones that show promise are propagated and planted in large-scale growers' trials.

Three-plant plots that are replicated three times are planted with promising selections along with standard commercial cultivars. Two years after planting, when the plants are well-established, they are evaluated for horticultural parameters, like total yield, fruiting season, fruit size, firmness, soluble solids concentration (Brix), flavor, and pre- and postharvest fruit rot that is mostly caused by *Botrytis cinerea*. If sufficient labor is not available for harvest, yield estimates may be made using yield component estimates (Daubeny et al. 1986). Fruit samples are collected and frozen immediately after harvest. They are used to determine titratable acidity, pH, soluble solids, and anthocyanin concentration during the winter months. In recent years, there has been an increased interest in the health benefits of the fruit and so the anthocyanins, ellagic acid content, and level of antioxidant activity (e.g., ORAC, TEAC, or FRAP) and other traits may be measured.

Plant growth habit is also evaluated throughout the growing season to assess whether the clones have a desirable growth habit. Ideally, the plants have upright spine-free or nearly spine-free canes that carry strongly attached, short-to-medium length, upright laterals with fruit that is well-spaced and not bunched. Plants should have a sufficient number of new replacement primocanes that are strong, straight, and long enough to reach the trellis wires. During the late winter months and before bud break, selections are examined for their reaction to various cane diseases, including spur blight, cane Botrytis, and anthracnose, and during the summer months the foliage is inspected for cane blight, powdery mildew, yellow rust, and other diseases that are present in the area. The plots are rated on a numerical scale and compared against standard cultivars. This information is used to help choose the parents for crosses during the process of introducing improved resistance into the germplasm. To assist with the process of identifying resistance and susceptibility, the AAFC–PARC program does not apply any field spray program to control diseases or pests. Although 'Qualicum' was identified to be winter hardy in BC, it showed significant winter injury when tested in other production areas because it is very susceptible to anthracnose. This was not considered important as commercial growers in BC routinely use a spray program to control fruit rot. This fungicide spray program is also very effective in controlling cane diseases, including anthracnose, and the release of 'Qualicum,' an anthracnose susceptible cultivar, was therefore not a concern in BC (Daubeny and Kempler 1995). Most breeding programs follow a minimal spray program in their selection trials. A typical breeding program might have the basic dormant sprays for cane diseases and a reduced Botrytis fruit rot program. This is important for two primary reasons: (1) sometimes, genotypes with some tolerance to biotic stress are overwhelmed with inocula from nearby plots of very susceptible genotypes and (2) for some diseases, despite tremendous efforts on the part of breeders and pathologists, no good resistance has been uncovered. *Botrytis* fruit rot is a very good example of this.

7 Integration of New Biotechnologies in Breeding Programs

7.1 Potential

Biotechnology applications to raspberry breeding have resulted in a significant change in the methods of determining genetic variation in raspberry breeding and allied genetic studies (Hall et al. 2009). Red raspberry has had a significant amount of basic work in molecular genetics, including genomics. While this work is promising, it has yet to deliver any improvements in cultivars or commercial production. Recently, strong efforts within the Rosaceae have been implemented to try to tie this great laboratory information with practical tools that assist plant breeding. Marker-assisted breeding potentially opens the way for quickly and precisely incorporating genes targeted for specific resistances and for quality- or production-influencing traits, as well as expanding the germplasm base from new genetics resources that had not previously made a contribution to the development of modern raspberry cultivars (Graham et al. 2007). Careful observation and recording of trait segregations in seedling populations are being correlated with genetic variability found at the molecular level, in proteins, DNA, and RNA. Detection of genome-wide variability has led to the characterization of genetic variation throughout the entire raspberry genome, for assessment of germplasm and development of genetic linkage maps. Genetic linkage maps have been constructed containing numerous markers for polygenic traits that can be used to identify genomic regions or genes controlling complex phenotypes. Understanding the genetic control of commercially and nutritionally important traits and the linkage of these characteristics to molecular markers on chromosomes will hopefully play a role in future plant breeding (Graham et al. 2007).

In addition, biotechnology has developed the ability to incorporate genes from other species of plants, animals, or even from bacteria, fungi, viruses, and other sources of genetic variability. This latter technology has possibilities for incorporation of many new traits, but it is bounded by ethical and moral concerns and in some locations disdain and distrust by the public at large.

7.2 Molecular Markers

Molecular markers (random amplified polymorphic DNA(RAPD) and SSR) have been shown to distinguish between cultivars and to group cultivars of similar origin, closely following pedigree relationships, similar origins, and cultivars from the same breeding program (Badjakov et al. 2006; Fernández et al. 2008). In addition, it has been possible to determine the genetic diversity among cultivars. The use of markers and the development of DNA fingerprints for each cultivar have particular value as they are independent of environmental factors, the vegetative stage of the plant, and the plant tissue source. Use of these markers also has been adopted for identification of cultivars in tissue culture.

Once markers had been identified, it became possible for traits to be selected in the juvenile stage of growth or before plants needed planting in the field. Markers for root rot resistance, aphid resistance, growth habit, fruit size, and other fruiting characters can now be routinely screened among seedling populations, as long as they are related to the population in which the markers were initially developed.

The use of marker-assisted selection of seedlings bearing a desired trait will hopefully soon be routinely possible before the plant has been established in the field or a lot of resources have been used to grow large populations for evaluation under field conditions. While possible, to this point, MAS has not moved from theoretical into the breeders toolbox. For MAS to be effective, it is necessary for the markers to be closely associated with the desired trait, with little crossing over between the gene of interest and the site of the marker on the chromosome so that the number of false ‘identifications’ can be minimized and that the use of markers can reliably be used in a breeding program. The usefulness of MAS is theoretically limited to the population in which the markers were developed. When populations with different genetic backgrounds are examined, the same markers may not have any correlation with the trait desired. To identify markers for a trait, reasonably large populations have to be grown and these need to be closely scrutinized to identify the individuals with the desired trait, and plants where the identification is unclear need to be eliminated from the study. This has been done with raspberries very effectively with the ‘Glen Moy’ × ‘Latham’ population grown at Dundee by SCRI, when the entire segregating population of 300 individuals was replicated and planted at two locations, in randomized complete block trials, with three replicates and two plant plots at each of the two locations (Graham et al. 2004a, b). While the future cannot be predicted, at least in the short run, it is likely that marker-assisted breeding will be valuable in discarding a portion of the seedlings that are inferior rather than identifying those that are superior; it gets rid of the 'junk,' so the breeder can focus on the germplasm that is more likely to contain improved genotypes.

7.3 State of the Map

Mapping in raspberry is at an early stage with only a few genes mapped. The first genetic linkage map was constructed using a cross between the North American cultivar Latham and the European raspberry cultivar Glen Moy (Graham et al. 2004a, b). SSR markers were developed from genomic and cDNA libraries from 'Glen Moy.' The SSR markers, along with amplified fragment length polymorphism (AFLP) markers, were used to generate a linkage map and the map was later enhanced with further SSR and expressed sequence tag (EST)-SSR markers (Graham et al. 2006, 2007). Gene *H*, controlling pubescence, has been mapped to group 2 of the raspberry map from the 'Latham' × 'Glen Moy' population and further mapping of resistance genes for root rot and other diseases is underway or near publication (Graham et al. 2007; Graham pers. commun.). Genes *A1* and *Dw* controlling aphid resistance and dwarfing have been mapped from a population of 'Malling Jewel' × 'Malling Orion' (Sargent et al. 2007). Further genes for root rot resistance have been identified in populations involving 'Latham,' 'Titan,' and NY00-34, a 'Titan' × 'Latham' hybrid (Pattison et al. 2007).

7.4 Traits Marked with Molecular Markers

Quantitative trait loci (QTL) data have been collected on cane spininess and root sucker density and diameter in the 'Latham' × 'Glen Moy' population grown in two different environments (Graham et al. 2004a, b). Eight linkage groups were identified from this cross, six common to both cultivars, and a different one from each parent. There are several genes conferring spinelessness, but an examination of data on spines from this cross, which did not segregate for spinelessness, showed that 98% of the variation in spines was associated with three or more genes on two linked regions on linkage group 2. QTLs for density and spread of suckers were overlapped and located on linkage group 8. QTLs for fruit quality parameters have also been identified on the raspberry maps and some genes associated with these traits have been identified, including a QTL for fruit size (Graham et al. 2007).

With the 'Malling Jewel' × 'Malling Orion' population, a smaller number of seedlings were screened for 24 AFLP primer combinations, giving a total of 114 segregating products that were scored in the parents of the cross. Forty-five dominant markers segregated in 'Malling Jewel' and 47 in 'Malling Orion' while 22 were in both. Of the 52 SSR markers tested, a total of 22 were in the progeny, 3 in 'Malling Jewel,' 7 in 'Malling Orion,' and 12 in both parents. The *A1* gene mapped to linkage group 3 and the *dw* gene mapped to linkage group 6 (Sargent et al. 2007).

The 'Malling Jewel' × 'Malling Orion' map covers a total distance of 505 cM, significantly shorter than the 636 cM of the 'Latham' × 'Glen Moy' map.

With the Phytophthora root rot resistance populations in New York, the resistant 'Latham' and susceptible 'Titan' were used to create F_1, F_2, B_1, B_2, and S_1 populations for analysis. Inheritance of root rot resistance was investigated using classical and molecular methodologies. The latter approach constructed linkage maps of 'Latham' and 'Titan' from AFLP, RAPD, and uncharacterized resistant gene analog polymorphism (RGAP) markers. Seven linkage groups were found with a total length of 440 cM for 'Latham' and 370 cM for 'Titan' (Pattison et al. 2007). In the B_2 population, several RAPD markers were identified in two linkage groups associated with root rot resistance. QTL analysis identified two similar genomic regions on each map that explained much of the variation observed in disease symptoms. This observation supports the dominant two-gene model developed from the analysis of segregation ratios. The results indicate that durable resistance to Phytophthora root rot is available, and show the value of recurrent selection for the development of resistant cultivars.

7.5 Genomics

Considerable progress has been achieved in comparative and functional genomic studies for other members of the Rosaceae, including the development of ESTs, bacterial artificial chromosome libraries, physical and genetic maps, and molecular markers, combined with genetic transformation protocols and bioinformatic tools. In 2010, genome sequencing was completed in apples (*Malus* × *domestica*), a draft of the peach genome (*Prunus persica*) was released (see: http://www.rosaceae.org/peach/genome), and work had begun with strawberry (*Fragaria* × *ananassa*) and raspberry (Shulaev et al. 2008; Velasco et al. 2010). Breeding raspberries is a time-consuming process, but genomic technologies have the potential to speed up the process and allow for the improvement of targeted traits. Technology for sequencing has given considerable genomic and EST information and this is being applied alongside endeavors to locate, explain, and assign biological function (Graham et al. 2004a, b). Traits targeted in raspberries include fruit quality, architecture, firmness, shelf life, aroma, flavor, suitability for processing, and freedom from processing defects, pest, and disease resistances (Graham et al. 2004b) as well as antioxidant components of fruit (D'Amico and Perrotta 2005). Some exploratory studies have also been done in metabolomics and proteomics with raspberries, but detailed analyses have not yet been published.

7.6 Transgenics

The science of biotechnology has further possibilities in the incorporation of novel genes into the raspberry genome to produce genetically transformed or transgenic new plants. This technology has been used to generate clones of 'Meeker' raspberry with resistance to RBDV through the incorporation of genetic constructs for the coat protein or movement protein of the virus (Martin and Mathews 2001). Many of the transgenic plants had poor fruit set or other issues in the plants, but some selections were obtained with good resistance to the virus and potential for commercial development. However, public distrust of this technology has prevented this transgenic 'Meeker' from being released and commercialized.

Raspberry breeders have made enormous progress since George Darrow wrote the quote in the foreword of this chapter. Nevertheless, it has given us just the skeleton of what we will see in the future as we are able to utilize the reserves of germplasm that have been and will be collected and stored in the germplasm repositories around the world. Raspberry breeders of the twenty-first century have greater germplasm resources, new tools, better training, advanced cultivar development, and the benefits of the insights of over a century of breeding by more than 100 breeders around the world.

Considerable developments in the West have filtered very slowly to the Eastern bloc countries and information and materials from there even more slowly to the West. Introductions of genetics from East Malling and the SCRI to the former USSR in the 1980s have enabled great advances in production, fruit size, and fruit quality. They also have been able to incorporate cold hardiness from older cultivars that used to be widely grown in the former USSR as well as shelf life and fruit firmness from *R. crataegifolius*. Advances around the world will be accelerated when there is improved freedom of movement of plant materials between the East and the West. Unfortunately, in the West, the current trend is to privatize breeding programs and for private companies to initiate their own plant improvement programs with the intention to tie up all the new cultivars with Plant Patents, Plant Variety Rights, and Plant Breeders Rights around the world. This is likely to impede genetic progress through secrecy and the unavailability of new cultivars to other breeders for incorporation in their improvement programs.

References

Anthony, V.M. and R.C. Shattock (1983) Resistance of raspberry cultivars to yellow rust. Ann. Appl. Biol. 102:136–137

Anthony, V.M., R.C. Shattock and B. Williamson (1985a) Interaction of red raspberry cultivars with isolates of *Phragmidium rubi-idaei*. Pl. Path. 34:521–527

Anthony, V.M., R.C. Shattock, B. and Williamson (1985b) Life-history of *Phragmidium rubi-idaei* on red raspberry in the United Kingdom. Pl. Path. 34:510–520

Badjakov, I., E. Todorovska, V. Kondakova, R. Boicheva, and A. Atanassov (2006) Assessment the genetic diversity of Bulgarian raspberry germplasm collected by microsatellite and RAPD markers. Journal of Fruit and Ornamental Plant Research 14 (Suppl. 1):61–76

Bañados, M.P. and A. Dale (2008) Proceedings of the Ninth International Rubus and Ribes Symposium, Acta Horticulturae. Vol. 777. 540pp Drukkerij Geers, Gent (Ostakker), Belgium

Barbara, D.J., A. Morton, S. Ramcharan, I.W. Cole, A. Phillips and V.H. Knight (2001) Occurrence and distribution of raspberry bushy dwarf virus in commercial *Rubus* plantations in England and Wales. Pl. Path. 50:747–754

Barritt, B.H., P.C. Crandall, and P.R. Bristow (1981) Red raspberry clones resistant to root rot. Fruit Var. J. 35:60–62
Barritt, B.H., L.C. Torre, H.S. Pepin, and H.A. Daubeny (1980) Fruit firmness measurements in red raspberry. HortScience 15:38–39
Birch, A.N.E. and A.T. Jones (1988) Levels and components of resistance to *Amphorophora idaei* in raspberry cultivars containing different resistance genes. Ann. Appl. Biol. 113:567–578
Birch, A.N.E., B. Fenton, G. Malloch, A.T. Jones, M.S. Phillips, B.E. Harrower, J.A.T. Woodford, and M.A. Cately (1994) Ribosomal spacer length variability in the large raspberry aphid, *Amphorophora idaei* (Aphidinae: Macrosiphini). Insect Molecular Biology 3:239–245
Birch, A.N.E., A.T. Jones, B. Fenton, G. Malloch, I. Geoghegan, S.C. Gordon, J. Hillier, and G. Begg (2002) Resistance-breaking raspberry aphid biotypes: Constraints to sustainable control through plant breeding. Acta Hort. 585:315–317
Bristow, P.R., H.A. Daubeny, T.M. Sjulin, H.S. Pepin, R. Nestby, and G.E. Windom (1988) Evaluation of *Rubus* germplasm for reaction to root rot caused by *Phytophthora erythroseptica*. J. Am. Soc. Hort. Sci. 113:588–591
Brooks, R.M. and H.P. Olmo (1946) Register of new fruit and nut varieties list No. 2. Proc. Am. Soc. Hort. Sci. 47:544–569
Clark, J.R., E.T Stafne, H.K Hall and C.E Finn (2007) Blackberry breeding and genetics. Plant Breeding Reviews 29:19–144
Converse, R.H. (1987) Virus diseases of small fruit. USDA, Agriculture Handbook No. 631
Converse, R.H. (1988) Tobacco streak and Raspberry bushy dwarf virus in California Boysen fields. Plant Disease Reporter 72:175
Cormack, M.R. and P.J. Woodward (1977) Raspberry cultivar assessments at the National Fruit trials and the Scottish Horticultural Research Institute. Expt. Hort. 29:1–14
D'Amico, E. and G. Perrotta (2005) Genomics of berry fruits antioxidant components. BioFactors 23:179–187
Dale, A. (1989) Productivity in red raspberries. Horticultural Reviews 11:185–228
Dale, A. and B.C. Jarvis (1983) Studies on germination in raspberry (*Rubus idaeus* L). Crop. Res. (Hort. Res.) 23:73–81
Dale, A. and H.A. Daubeny (1985) Genotype-environment interactions involving British and Pacific Northwest red raspberry cultivars. HortScience 20:68–69
Dale, A. and H.A. Daubeny (1987) Flower-bud initiation in Red Raspberry (*Rubus idaeus* L.) in two environments. Crop Res. (Hort Res.) 27:61–66
Dale, A., P.P. Moore, R.J. McNicol, T.M. Sjulin, and L.A. Burmistrov (1993) Genetic diversity of red raspberry varieties throughout the world. J. Am. Soc. Hort. Sci. 118:119–129
Dale, A., E.J. Hanson, D.E. Yarborough, R.J. McNicol, E.J. Stang, R.M. Brennan and J.R. Morris (1994) Mechanical harvesting of berry crops. Hort. Rev. 16:255–382
Darrow, G.M. (1937) Blackberry and raspberry improvement. p. 496–533. In: USDA Yearbook of Agr. Government Printing Office, Washington D.C.
Daubeny, H.A. (1971) Self-fertility in red raspberry cultivars and selections. J. Am. Soc. Hort. Sci. 96:588–591
Daubeny, H.A. (1973) Haida red raspberry. Can. J. Plant. Sci. 53:345–346
Daubeny, H.A. (1978a) Skeena red raspberry. Can. J. Plant. Sci. 58:565–568
Daubeny, H.A. (1978b) Nootka Red Raspberry. Can. J. Plant. Sci. 58:899–901
Daubeny, H.A. (1983) Expansion of genetic resources available to red raspberry breeding programs. Proceedings of the 21st. International Horticultural Congress 1:150–155
Daubeny, H.A. (1996) Brambles. In: Janick J., J.N. Moore (eds.) Fruit Breeding. vol. II. Vine and small fruit. John Wiley & Sons, New York
Daubeny, H.A. and H.S. Pepin (1974) Variations among red raspberry cultivars and selections in susceptibility to the fruit rot causal organisms *Botrytis cinerea* and *Rhizopus* spp. Can. J. Plant. Sci. 54:511–516
Daubeny, H.A. and H.S. Pepin (1981) Resistance of red raspberry fruit and canes to *Botrytis*. J. Am. Soc. Hort. Sci. 106:423–426

Daubeny, H.A. and D. Stary (1982) Identification of resistance to *Amphorophora agathonica* in the native North American red raspberry. J. Am. Soc. Hort. Sci. 107:593–597

Daubeny, H.A. and T.M. Sjulin (1984) BC 72-1-7 Red Raspberry. HortScience 19:733–734

Daubeny, H.A. and C. Kempler (1995) 'Qualicum' red raspberry. HortScience 30:1470–1472

Daubeny, H.A., A. Dale and G.R. McGregor (1986) Estimating yields of red raspberries in small research plots. HortScience 21:1216–1217

Daubeny, H.A., F.J. Lawrence, and G.R McGregor. (1989) 'Willamette red raspberry' Fruit Var. J. 43:46–48

Daubeny, H.A., H.S. Pepin and C.A. Levesque. (1992) Breeding for resistance to aphids and root rot in red raspberry. Acta Hort. 317:187–190

Deacon, J., A. Robertson, and A. Isbister (2009) The microbial world: Biology and control of Crown gall (*Agrobacterium tumefaciens*). url: http://www.biology.ed.ac.uk/research/groups/jdeacon/microbes/crown.htm (Accessed: 24-1-09.) The University of Edinburgh, Institute of Cell and Molecular Biology, Edinburgh, UK

Dossett, M. and C.E. Finn (2008) Variation and inheritance of vegetative characteristics and reproductive traits in black raspberry (*Rubus occidentalis* L.). Acta Hort. 777:147–152

Dossett, M. and C.E. Finn (2010) Identification of resistance to the large raspberry aphid in black raspberry. J. Amer. Soc. Hort. Sci. 135:438–444

Ellis, M.A., R.H. Converse, R.N. Williams, and B. Williamson (1991) Compendium of raspberry and blackberry diseases and insects. APS Press, St. Paul, Minn.

FAO. 2009. World production quantity fruits (1000 tonnes) for 2005. URL: http://faostat.fao.org/site/336/DesktopDefault.aspx?PageID=336 (Accessed: 13/03/2009.) Food and Agr. Organization of the United Nations

Fear, C.D. (1992) Trends in the California Raspberry industry. Proceedings of the Western Washington Horticulture Association 118–123

Fernández, M.P., S. Hernáiz, and J. Ibáñez (2008) Genetic characterization of raspberry cultivars using molecular markers. Acta Hort. 777:125–132

Finn, C.E. (1999) Temperate berry crops. In: J. Janick (ed.), Perspectives on new crops and new uses. ASHS Press, Alexandria, Virg. p. 324–333.

Finn, C.E., (2006) Caneberry breeders in North America. HortScience 41:22–24

Finn, C.E. (2008) Blackberries. In: J.F. Hancock (ed.), Temperate fruit crop breeding: Germplasm to genomics . Kluwer Academic Publishers, Dordrecht, The Netherlands. p. 83–114

Finn, C.E. and J.E. Hancock (2008) Raspberries. In: J.F. Hancock (ed.), Temperate fruit crop breeding: Germplasm to genomics. Kluwer Academic Publishers, Dordrecht, The Netherlands. p. 359–392

Finn, C.E., and V.H. Knight. 2002. What's going on in the world of *Rubus* breeding? Acta Hort. 585:31–38

Finn, C.E., H.J. Swartz, P.P. Moore, J.R. Ballington, and C. Kempler (2001) Breeders experiences with Rubus species. url: http://www.ars-grin.gov/cor/rubus/rubus.uses.html (21 Mar. 2009)

Finn, C.E., H.J. Swartz, P.P. Moore, J.R. Ballington, and C. Kempler (2002) Use of 58 *Rubus* species in five North American breeding programs- breeders notes. Acta Hort. 585:113–119

Forbes, A.R., B.D. Frazer, and C.K. Chan (1985) Aphids vol. 1. Elsevier Science Publishers, Amsterdam

Gordon, S.C. and B. Williamson (1991a) Midge blight. p. 7. In: M.A. Ellis, R.H. Converse, R.N. Williams, and B. Williamson (eds.), Compendium of raspberry and blackberry diseases and insects. APS Press, St Paul, Minnesota

Gordon, S.C. and B. Williamson (1991b) Raspberry cane midge. p. 75–76. In: M.A. Ellis, R.H. Converse, R.N. Williams, and B. Williamson (eds.), Compendium of raspberry and blackberry diseases and insects. APS Press, St Paul, Minnesota

Graham, J., I. Hein, and W. Powell (2007) Chapter 9: Raspberry. p. 207–216. In: C. Kole (ed.), Genome mapping and molecular breeding in plants. Vol 4: Fruits and nuts. Springer-Verlag, Berlin

Graham, J., K. Smith, I. Tierney, K. MacKenzie, and C.A. Hackett (2006) Mapping gene H controlling cane pubescence in raspberry and its association with resistance to cane botrytis and spur blight, rust and cane spot. Theor. Appl. Genet. 112:818–831

Graham, J., K. Smith, K. MacKenzie, L. Jorgensen, C.A. Hackett, and W. Powell (2004a) The construction of a genetic linkage map of red raspberry [*Rubus idaeus* subsp. *idaeus*] based on AFLPs, genomic-SSR and EST-SSR markers. Theor. Appl. Genet. 109:740–749

Graham, J., I. Hein, J. Russell, M. Woodhead, S.C. Gordon, K. Smith, L. Jorgensen, R. Brennan, and W. Powell (2004b) The use of genomics technologies in contemporary *Rubus* and *Ribes* breeding programmes. Acta Hort. 649:319–322

Hall, H.K., M.J. Stephens, P.A. Alspach, and C.J. Stanley (2002) Traits of importance for machine harvest of raspberries. Acta Hort. 585/2:607–610

Hall, H.K., K.E. Hummer, A.R. Jamieson, S.N. Jennings, and C.A. Weber (2009) Raspberry Breeding. p. 39–353. In: J. Janick (ed.), Plant Breeding Reviews vol. 32. Timber Press, Inc, Portland, Ore.

Hedrick, U.P. (1925) The small fruits of New York. J.B. Lyon Co., Albany, N.Y.

Hoover E., J. Luby, D. Bedford and M. Pritts (1988) Vegetative and reproductive yield components of primocane-fruiting red raspberries. J. Am. Soc. Hort. Sci. 113:824–826

Jennings, D.L. (1983) Inheritance of resistance to *Botrytis cinerea* and *Didymella applanata* in canes of *Rubus idaeus*, and relationships between these resistances. Euphytica 32:895–901

Jennings, D.L. (1988) Raspberries and blackberries: Their breeding, diseases and growth. Academic Press, London

Jennings, D.L. and A. Dale (1982) Variation in the growth habit of red raspberries with particular reference to cane height and node production. J. Hort. Sci. 57:197–204

Jennings, D.L. and R. Ingram (1983) Hybrids of *Rubus parviflorus* (Nutt.) with raspberry and blackberry, and the inheritance of spinelessness derived from this species. Crop Res. 23:95–101

Jennings, D.L. and E. Brydon (1989) Further studies on breeding for resistance to *Botrytis cinerea* in red raspberry canes. Ann. Appl. Biol. 115:507–513

Jennings, D.L., H.A. Daubeny, and J.N. Moore (1991) Blackberries and raspberries (*Rubus*). In Genetic resources of temperate fruit and nut crops. Acta Hort. 290:331–389

Jennings, D.L., Daubeny, H.A. and J.N. Moore (1992) In: J.N. Moore and J.R. Ballington (Eds.). Blackberries and raspberries (*Rubus*). Genetic resources of temperate fruit and nut crops. Acta Hort. 290:331–389

Jennings, S.N., L. Ferguson, and R.M. Brennan (2008) New prospects from the Scottish raspberry breeding programme. Acta Hort. 777:203–206

Johnson, G.W. (1829) A history of English gardening, chronological, biographical, literary, and critical, tracing the progress of the art in this country from the invasion of the Romans to the present time. Baldwin & Cradock, London

Jones, A.T., W.J. McGavin and A.N.E. Birch (2001) Some factors influencing the effectiveness of resistance genes to the aphids virus-vector, *Amphorophora idaei* Borner in raspberry. Acta Hort. 551:39–44

Jones, A.T., M.J. Mitchell, and D.J.F. Brown (1989) Infectibility of some new raspberry cultivars with arabis mosaic and raspberry ringspot virus and further evidence for variation in British isolates of these two nepoviruses. Ann. Appl. Biol. 115:57–69

Ke, S., R.M. Skirvin, K.D. McPheeters, A.G. Otterbacher, and G.J. Galletta (1985) In vitro germination and growth of *Rubus* seeds and embryos. HortScience 20:1047–1049

Keep, E. (1989) Breeding red raspberry for resistance to diseases and pests. Plant Breeding Reviews 6:245–321

Kempler, C. and H.A. Daubeny (2000) 'Malahat' red raspberry. HortScience 35:783–785

Kempler, C., H.A. Daubeny, B. Harding, and C.E. Finn (2005a) 'Esquimalt' red raspberry. HortScience 40:2192–2194

Kempler, C., H.A. Daubeny, B. Harding, and C.G. Kowalenko (2005b) 'Cowichan' red raspberry. HortScience 40:1916–1918

Kempler, C., H.A. Daubeny, L. Frey, and T. Walters (2006) 'Chemainus' red raspberry. HortScience 41:1364–1366

Kempler, C., H.A. Daubeny, B. Harding, T.E. Baumann, C.E. Finn, P.P. Moore, M. Sweeney, and T. Walters (2007) 'Saanich' red raspberry. HortScience 42:176–178

Kennedy, D.M. and J.M. Duncan (1988) Frequency of virulence phenotypes of Phytophthora fragariae in the field. Pl. Path. 37:397–406

Kennedy, D.M. and J.M. Duncan (1993) Occurrence of races of *Phytophthora fragariae* var *rubi* on raspberry. Acta Hort. 352:555–562

Knight, R.L. and E. Keep (1958) Developments in soft fruit breeding at East Malling. Report of East Malling Research Station for 1957:62–67

Knight, V.H. and F. Fernández Fernández (2008a) New summer fruiting red raspberry cultivars from East Malling. Acta Hort. 777:173–181

Knight, V.H. and F. Fernández Fernández (2008b) Screening for resistance to *Phytophthora fragariae* var. *rubi* in *Rubus* germplasm at East Malling. Acta Hort. 777:353–359

Knight, V.H., D.L. Jennings, and R.J. McNicol (1989) Progress in the UK raspberry breeding programme. Acta Hort. 262:93–103

Kowalenko, C.G., C. Kempler and S. Bittman (2008) Do differences in floricane and primocane growth characteristics of raspberry cultivars influence the recycling of nitrogen in the soil-plant system? Acta Hort. 777:453–458

Krishnan, S., K. Datta, N. Baisakh, M. Vasconcelos, and S.K. Datta (2003) Tissue-specific localization of β-carotene and iron in transgenic Indica rice (*Oryza sativa* L.). Current Science 84:1232–1234

Lim, K.Y., I.J. Leitch, and A.R. Leitch (1998) Genomic characterisation and the detection of raspberry chromatin in polyploid *Rubus*. Theor. Appl. Genet. 97:1027–1033

Lindén, L., P. Palonen, M. Seppänen, and A. Väinölö (1999) Cold hardiness research on agricultural and horticultural crops in Finland. Agr. Food Sci. Fin. 8:459–477

Malowicki, S.M.M., R.R. Martin, and M.C. Qian (2008a) Comparison of sugar, acids, and volatile composition in Raspberry bushy dwarf virus-resistant transgenic Raspberries and the wild type 'Meeker' (Rubus idaeus L.). J. Agricult. Food Chem. 56:6648–6655

Malowicki, S.M.M., M.C. Qian, and R.R. Martin (2008b) Fruit quality of transgenic 'Meeker' Red Raspberry with resistance to Raspberry bushy dwarf virus. Acta Hort. 780:41–48

Martin, R.R. (2003) Economic significance of RBDV http://www.geocities.com/martinrr_97330/RBDVweb/significance.htm (Accessed: 30/8/06)

Martin, R.R. and H. Mathews (2001) Engineering resistance to raspberry bushy dwarf virus. Acta Hort. 551:33–37

Martin, R.R., K.E. Keller and H. Mathews (2004) Development of resistance to *raspberry bushy dwarf virus* in 'Meeker' red raspberry. Acta Hort. 656:165–169

McElroy, F.D. (1991). Nematode parasites, p. 59–62. In: M.A. Ellis, R.H. Converse, R.N. Williams, and B. Williamson (eds.). Compendium of raspberry and blackberry diseases and pests. APS Press, St. Paul, MN.

McNicol, R.J., B. Williamson and A. Dolan (1985) Infection of red raspberry styles and carpels by *Botrytis cinerea* and its possible role in post-harvest grey mould. Ann. Appl. Biol. 106:49–53

Meng, R. and C.E. Finn. (2002) Determining ploidy level and nuclear DNA content in *Rubus* by flow cytometry. J. Am. Soc. Hort. Sci. 127:767–775

Moore, P.P. (2004) 'Cascade Delight' Red Raspberry. HortScience 39:185–187

Moore, P.P. (2005) 'Cascade Nectar' red raspberry. HortScience 40:256–257

Moore, P.P. (2007) 'Cascade Bounty' red raspberry. HortScience 42:393–396

Moore, P.P., and H.A. Daubeny (1993) 'Meeker' red raspberry. Fruit Var. J. 47:2–4

Mullen, W., J. McGinn, M.E.J. Lean, M.R. MacLean, P.T. Gardner, G.G. Duthie, T. Yokota, and A. Crozier (2002) Ellagitannins, flavonoids, and other phenolics in red raspberries and their contribution to Antioxidant capacity and vasorelaxation properties. J Agric Food Chem. 50:5191–5196

Nesme, X. (1985) Respective effects of endocarp, testa and endosperm, and embryo on the germination of raspberry seeds. Can. J. Plant Sci. 65:125–130

Nickerson N.L. (1991) Late leaf rust p. 30–32. In: M.A. Ellis, R.H. Converse, R.N. Williams, and B. Williamson (eds.). Compendium of raspberry and blackberry diseases and pests. APS Press, St. Paul, MN.

Oliveira, P.B., L.L. da-Fonseca, and A.A. Monteiro (2002) Combining different growing techniques for all year round red raspberry production in Portugal. Acta Hort. 585/2:545–553

Ourecky, D.K. (1975) Brambles. p. 98–129. In: J. Janick, and J.N. Moore (eds.), Advances in Fruit Breeding. Purdue Univ. Press, West Lafayette, Ind.

Palonen, P. (1999a) Relationship of seasonal changes in carbohydrates and cold hardiness in canes and buds of three red raspberry cultivars. J. Am. Soc. Hort. Sci. 124:507–513

Palonen, P. (1999b) Carbohydrate concentrations and dormancy as related to winter hardiness in red raspberry (*Rubus idaeus* L.). Publ. No. 36. (Academic Dissertation), Helsinki, University of Helsinki, Helsinki

Palonen, P. and D.J.I. Buszard (1997) Current state of cold hardiness research on fruit crops. Can. J. Plant. Sci. 77:399–420

Pattison, J.A., S.K. Samuelian, and C.A. Weber (2007) Inheritance of *Phytophthora* root rot resistance in red raspberry determined by generation means and molecular linkage analysis. Theor. Appl. Genet. 115:225–236

Pattison J.A., W.F. Wilcox and C.A. Weber (2004) Assessing the resistance of red raspberry (*Rubus idaeus* L.) genotypes to *Phytophthora fragariae* var, *rubi* in hydroponic culture. HortScience 39:77

Prince, W.R. (1832) The pomological manual: or, a treatise on fruits: containing descriptions of a great number of the most valuable varieties for the orchard and garden. T. & G. Swords. New York

Quamme, H.A. and C. Stushnoff (1983) Resistance to environmental stress. p. 242–266. In: J.N. Moore and J. Janick (eds.), Methods in fruit breeding. Purdue Univ. Press, W. Lafayette, IN

Roach, F.A. (1985) Raspberries. p. 263–273, Cultivated fruits of Britain. Their origin and history. Basil Blackwell, Oxford and New York

Robertson, G.W., D.W. Griffiths, A.N.E. Birch, A.T. Jones, R.J. McNicol and J.E. Hall (1991) Further evidence that resistance in raspberry to the virus vector aphid, Amphorophora idaei, is related to the chemical composition of the leaf surface. Ann. Appl. Biol. 119:443–449

Sargent, D.J., F. Fernández-Fernández, A. Rys, V.H. Knight, D.W. Simpson and K.R. Tobutt (2007) Mapping of A1 conferring resistance to the aphid *Amphorophora idaei* and dw (dwarfing habit) in red raspberry (*Rubus idaeus* L.) using AFLP and microsatellite markers. BMC Plant Biology 7:15

Seemüller, E., J.M. Duncan, D.M. Kennedy and M. Riedel (1986) Phytophthora sp. Als ursache einer wurzelfäule an himbeere. (German) Nachrichenblatt des Deutschen Pflanzenschutzdienstes 38:17–21

Seeram, N. (2008) Berry fruits: Compositional elements, biochmical activities, and the impact of their intake on human health, performance, and disease. J.Agric. Food Chem. 56:627–629

Shulaev, V., S.S. Korban, B. Sosinski, A.G. Abbott, H.S. Aldwinckle, K.M. Folta, A.F. Iezzoni, D. Main, P. Arús, A.M. Dandekar, K.S. Lewers, S.K. Brown, T.M. Davis, S.E. Gardiner, D. Potter and R.E. Veilleux (2008) Multiple models for Rosaceae genomics. Plant Phys. 147:985–1003

Sønsteby, A. and O.M. Heide (2008) Environmental control of growth and flowering of Rubus idaeus L. Cv. Glen Ample. Scientia Horticulturae 117:249–256

Stace-Smith, R. (1987) Virus and viruslike diseases of *Rubus* in: Virus diseases of small fruits. R.H. Converse editor. USDA Handbook No. 631:167–254

Stephens, M.J., P.A. Alspach and H.K. Hall (2002) Red raspberry- grey mould resistance from *Rubus* species. Acta Hort. 585:349–353

Stewart, D., G.J. McDougall, J. Sungurtas, S. Verrall, J. Graham and I. Martinussen (2007) Metabolomic approach to identifying bioactive compounds in berries: advances toward fruit nutritional enhancement. Mol. Nutr. Food Res. 51:645–651

Stoner, G.D., L.A. Kresty and P. Carlton (2002) The nutraceutical value of strawberries and black raspberries to inhibit cancer. Ohio State University Seeds. 1–2

Thompson, M.M. (1997) Survey of chromosome numbers in *Rubus* Rosaceae: Rosoideae. Ann. Missouri Bot. Gard. 84:128–163

USDA-ARS-GRIN (2009) Catalogue of raspberry cultivars and selections url: http://www.ars-grin.gov/cor/catalogs/rubredrasp.html (Accessed: 13/3/09.) USDA-ARS, Corvallis, OR

Velasco R, A. Zharkikh, J. Affourtit, A. Dhingra, A. Cestaro, A. Kalyanaraman, P. Fontana, S.K. Bhatnagar, M. Troggio, D. Pruss, S. Salvi, M. Pindo, P. Baldi, S. Castelletti, M. Cavaiuolo, G.

Coppola, F. Costa, V. Cova, A. D. Ri, V. Goremykin, M. Komjanc, S. Longhi, P. Magnago, G. Malacarne, M. Malnoy, D. Micheletti, M. Moretto, M. Perazzolli, A. Si-Ammour, S. Vezzulli, E. Zini, G. Eldredge, L. M. Fitzgerald, N. Gutin, J. Lanchbury, T. Macalma, J.T. Mitchell, J. Reid, B. Wardell, C. Kodira, Z. Chen, B. Desany, F. Niazi, M. Palmer, T. Koepke, D. Jiwan, S. Schaeffer, V. Krishnan, C. Wu, V.T. Chu, S.T. King, J. Vick, Q. Tao, A. Mraz, A. Stormo, K. Stormo, R. Bogden, D. Ederle, A. Stella, A. Vecchietti, M.M. Kater, S. Masiero, P. Lasserre, Y. Lespinasse, A.C. Allan, V. Bus, D. Chagné, R.N. Crowhurst, A.P. Gleave, E. Lavezzo, J.A. Fawcett, S. Proost, P. Rouzé, L. Sterck, S. Toppo, B. Lazzari, R.P. Hellens, C.-E. Durel, A. Gutin, R.E. Bumgarner, S.E. Gardiner, M. Skolnick, M. Egholm, Y. Van de Peer, F. Salamini, and R. Viola (2010) The Genome of the Domesticated Apple (*Malus* × *domestica* Borkh.) Nature Genetics 42: 833–839

Vrain, T.C. (1997) Engineering genetic resistance to root weevils, two-spotted spider mites, and root-lesion nematodes in red raspberry. Ann. Rept. to Wash. Red Raspberry Commission, pp 52–56

Vrain, T.C., H.A. Daubeny, J.W. Hall, R.M. DeYoung, and A.K. Anderson (1994) Inheritance of resistance to root lesion nematode in red raspberry. HortScience 29:1340–1341

Weber, R.W.S. and A.P. Entrop (2007) *Fusarium avenaceum*, a new raspberry cane disease in northern Germany which is possibly synergistic with the raspberry cane disease *Leptosphaeria coniothyrium*. (*Fusarium avenaceum*, ein neuartiger erreger der himbeerrutenkrankheit in Norddeutschland, und sein möglicher synergismus mit *coniothyrium fuckelii* die himbeerruten-krankheit). (German) Mitteilungen des Obstbauversuchsringes 62:53–58

Wilcox, W.F. (1989) Identity virulence and isolation frequency of seven *Phytophthora* spp. Causing root rot of raspberry in New York State. Phytopathology 79:93–101

Williamson, B., R.J. McNicol and A Dolan (1987) The effect of inoculating flowers and developing fruits with Botrytis cinerea on post-harvest grey mould of red raspberry. Ann. appl. Biol. 111:285–294

Zeller, S.M. and W.T. Lund (1933) Yellow rust of Rubus. Phytopathology 24:257–265

Chapter 9
Strawberry

Craig K. Chandler, Kevin Folta, Adam Dale, Vance M. Whitaker, and Mark Herrington

Abstract The cultivated strawberry, *Fragaria × ananassa* Duch., is a versatile crop in terms of its adaptability to various locations and cultural systems. Breeding efforts started in the early 1800s and continue today in numerous public and private programs. Among these programs and in germplasm repositories, there is still considerable variation available in traits of economic interest. Currently, the biggest opportunity in strawberry breeding is the development of day-neutral cultivars for cool summer climates outside of California while the biggest challenge facing strawberry breeders may be the development of cultivars that can produce fruit with consistent size, appearance, and flavor over an extended period of time. To accomplish this challenge, breeders need to stay focused on these traits as their primary screens. Despite the complexity of the octoploid strawberry genome, new genomics knowledge and biotechnologies make increasing contributions to strawberry breeding.

C.K. Chandler (✉) • V.M. Whitaker
Department of Horticultural Sciences, University of Florida, GCREC, 14625 CR 672, Wimauma, FL 33598-6101, USA
e-mail: ckc@ufl.edu; vwhitaker@ufl.edu

K. Folta
Department of Horticultural Sciences, University of Florida, Gainesville, FL, USA
e-mail: kfolta@ufl.edu

A. Dale
Department of Plant Agriculture, University of Guelph, Simcoe Research Station, 1283 Blueline Road, Simcoe, ON N3Y 4N5, Canada
e-mail: adale@uguelph.ca

M. Herrington
Queensland Department of Employment, Economic Development and Innovation, Queensland Government, Brisbane City East, QLD, Australia
e-mail: mark.herrington@deedi.qld.gov.au

M.L. Badenes and D.H. Byrne (eds.), *Fruit Breeding*, Handbook of Plant Breeding 8, DOI 10.1007/978-1-4419-0763-9_9,

Keywords Fragaria ×ananassa • Rosaceae • polyploidy • Day neutrality • Trait variability • Fruit quality • Vegetative generations • Cropping systems • Vernalization

1 Introduction

The cultivated strawberry *Fragaria* × *ananassa* Duch. is the most widely distributed fruit crop in the world. It is grown in every country with a temperate or subtropical climate and even in many tropical countries in highland areas, where the climate is mild. Strawberry fruits are highly prized for their universal appeal to the human senses of sight, smell, and taste. Favorite uses of fresh strawberries include sliced on cakes, breakfast cereal, and in salads (both mixed fruit and green) and dipped whole in melted chocolate. Processed strawberries are used in ice cream, jam, fruit leather, and mixed drinks. Fresh strawberries can be a valuable component of a healthy diet. Strawberries are low in calories, but high in fiber, folic acid, vitamin C, and several other antioxidants. The USA, the world's largest strawberry-producing nation, produces over one million tonnes of fruit per year, of which over 80% is consumed fresh (Sjulin 2007).

Strawberries are in the rose family (Rosaceae). The strawberry genus, *Fragaria*, is distributed predominantly in the Northern Hemisphere and contains 20 named species: 12 diploids ($2n=14$), 2 tetraploids, 1 pentaploid, 1 hexaploid, and 4 octoploids (Hancock et al. 2008). The 12 diploid species are present throughout Europe

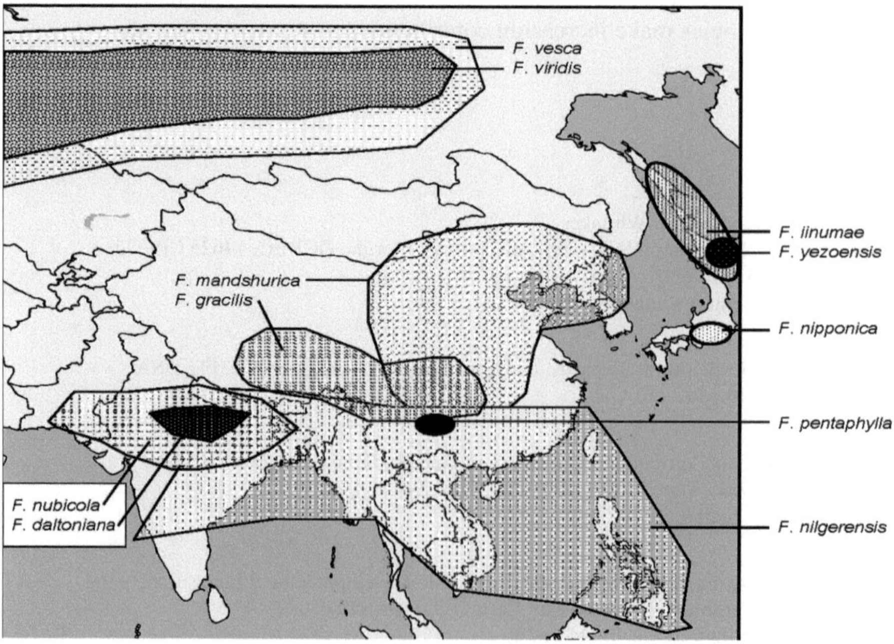

Fig. 9.1 The approximate geographical distribution of wild diploid strawberry species

and Asia, with the most speciation present in the Western Himalayas, through China and into Japan (Fig. 9.1). These areas represent the natural distribution of species, like *F. bucharica* Losink, *F. mandshurica* Staudt, and *F. iinumae* Makino, respectively. The diploid *F. vesca* Coville is found throughout Europe, in Asia west of the Ural Mountains, and throughout North America. Octoploid wild species are almost exclusively limited to the New World. *Fragaria virginiana* Mill has a range stretching across almost all of North America. *Fragaria chiloensis* (L.) Mill, by contrast, is adapted to a coastal habitat, radiating along the western seaboard of North America, in South America, and in Hawaii, with substantial populations occurring in Chile. Another high-ploid strawberry occurs exclusively on the side of an active volcano, Atsunupuri, on the Kurile Island of Iturup. This plant, known as *F. iturupensis* Staudt, has been reported correctly as both octoploid and decaploid and represents the only known Asian octoploid. *Fragaria* × *ananassa* ($2n=8\times=56$) is an interspecific hybrid between two of the octoploids, *Fragaria virginiana* Mill. and *F. chiloensis* (L.) Mill., both of which are native to the Americas. *Fragaria* × *ananassa* is generally considered to be fully diploidized (i.e., its chromosomes pair bivalently and its molecular markers exhibit disomic inheritance) (Hancock et al. 2008).

The strawberry inflorescence is a cyme, with one primary flower subtended by up to 14 low-order flowers. The flowers are complete, usually self-fertile, and composed of numerous pistils. Each pistil, following pollination and fertilization, develops into a single-seeded fruit (achene). These true fruits are distributed in a Fibonacci spiral pattern on the outside of the receptacle (edible fleshy tissue) (Darnell 2003). There are two main types of strawberries in terms of flowering response to photoperiod: the short day type, which typically only initiate flowers when photoperiod is less than 14 h, and the day neutral type, which initiate flowers under any photoperiod as long as the temperature does not reach a critical maximum for a particular genotype (≤25°C).

Many *Fragaria* species are dioecious. However, hermaphrodism is also present and ranges in expression from plants with entire cymes that are completely self-fertile to plants where individual flowers are occasionally self-fertile (Hancock 1999). Almost all cultivars are now fully hermaphroditic, although occasionally pistillate cultivars are released (e.g., Pegasus, Orleans).

In addition to crowns (compressed stems), strawberry plants can produce runners (stolons). Daughter plants arise from these runners and are the means by which strawberry cultivars are clonally propagated. Strawberry cultivars vary widely in their ability to produce runners, with day-neutral types typically producing fewer runners than short-day types.

Strawberry is a short-statured (<30 cm tall) perennial plant. Nevertheless, it is increasingly being cropped for less than a year and, at most, no more than 3–4 years. This is because young plants tend to have larger, higher quality fruit than older, highly branched (multiple crowned) plants. Also, growing strawberries as an annual, with some break between crops, allows for better disease and pest control.

Many cultural practices are used to grow strawberries. These include spaced plants with runners removed or matted rows in which runners are not removed, bare soil or polyethylene-covered beds (plasticulture), ground culture or raised hydroponic channels, and open-field culture or polyethylene-covered protective tunnels or greenhouses.

2 Origin and Domestication

F. × *ananassa* originated in European gardens during the eighteenth century when female clones of *F. chiloensis* from Chile were interplanted with clones of *F. virginiana* that had previously been imported from eastern North America. By chance, hybrid seedlings appeared that combined the hardiness and productivity of *F. virginiana* with the relatively large fruit size of *F. chiloensis*. Breeding activities accelerated in the early 1800s, particularly in Great Britain. 'Keen's Seedling,' developed by Michael Keen, a market gardener near London, dominated English strawberry acreage in the late nineteenth century and is in the pedigree of many modern cultivars. By the middle of the twentieth century, breeding programs were also active in other European countries, such as Scotland, Germany, and the Netherlands (Hancock 1999).

As European *F.* × *ananassa* clones were brought to North America in the middle of the nineteenth century, breeding activities on the continent (reviewed in detail by Hancock et al. 2008) increased dramatically in both public and private spheres. Until about 1930, most work was done by private breeders and involved both intraspecific crosses and hybridization of European material with wild species. Charles Hovey of Massachusetts produced the first cultivar from a controlled hybridization in North America which he named 'Hovey' (1836), which was a cross between a European cultivar and a native *F. virginiana* selection. Albert Etter carried on an active breeding program in California using native *F. chiloensis* clones in the early twentieth century, and his cultivars (especially 'Ettersburg 80') are in the background of many modern cultivars.

In the middle of the century, most cultivars were released from public breeding programs. The famous George Darrow began his career in the 1920s at the USDA in Beltsville, MD, where he released strawberry cultivars that formed the foundational germplasm of the USDA and other public breeding programs. Royce Bringhurst and Victor Voth from UC Davis produced a bevy of important cultivars from the 1950s onward. Their cultivars, particularly Chandler and Camarosa, quickly spread throughout the world.

In recent years, there has been a resurgence of private breeding, as large fruit companies have sought proprietary cultivars that give them a marketing advantage. Driscoll Strawberry Associates conducts its major efforts in California but has recently expanded into Florida, Great Britain, and other areas of the European continent. At the present time, proprietary cultivars produced by Plant Sciences, Inc. hold a dominant market share in southern California.

3 Genetic Resources

There are currently over 40 strawberry breeding programs worldwide, with most of them located in North America and Europe. However, there are also active programs in Asia, Australia, and South America. Each of these programs contains unique genotypes of *F.* × *ananassa,* and some of the programs also contain unique genotypes of *F. virginiana*, *F. chiloensis*, or other *Fragaria* spp. Several germplasm repositories hold many older cultivars and wild germplasm, including the USDA National Clonal Germplasm Repository (NCGR) at Corvallis, Oregon, the Canadian National Clonal Genebank at Harrow, Ontario, and the Fruit Genebank at the Institute of Fruit Breeding in Dresden, Germany. These public repositories are able to maintain many more genotypes than working breeding programs. For example, the NCGR currently holds more than 1,700 accessions, representing 17 species and genotypes from 37 countries.

Among the public and private collections, there is still significant variation in traits of economic interest, including plant habit, production pattern, fruit size, flavor, texture, and resistance to important biotic and abiotic disorders (Hancock et al. 2008). However, there is increasing use of wild octoploid germplasm in breeding, as it has been established that the genetic base of cultivated strawberry is fairly narrow. Indeed, most strawberry cultivars arise from only seven nuclear sources (Sjulin and Dale 1987) and ten cytoplasmic sources (Dale and Sjulin 1990). A collection of 38 elite species clones representing major subspecies and native regions has been assembled at the NCGR as a 'supercore' collection. These accessions are being crossed to *F.* × *ananassa* and also among themselves to 'reconstitute' the cultivated strawberry, as the octoploids are fully interfertile with one another. These accessions contain a wealth of disease-resistance traits, multiple sources of day neutrality (*F. virginiana* only), large fruit size (*F. chiloensis*), winter hardiness, drought tolerance, heat tolerance, and many other traits of interest to breeders. Furthermore, there is still ample opportunity for the collection of octoploid progenitor species throughout their native ranges. *F. chiloensis* is found in Hawaii and the west coast of the Americas from Chile to British Columbia (Hancock 1999). *F. virginiana* is widespread throughout central and eastern North America from Mississippi to Ontario and west to the Rocky Mountains (Hancock 1999).

4 Major Breeding Achievements

Since the inception of formal strawberry hybridization, breeding progress has been dramatic. The most obvious gains have been in fruit size and firmness. Average fruit size for *F. virginiana* is typically 1–3 g; for current cultivated cultivars, average fruit weights are commonly in excess of 20 g. Examples of large-fruited cultivars include 'Jewel,' developed at Cornell University, NY, for the northeastern USA and southern Canada, and 'Camino Real,' developed by University of California-Davis for southern California. 'Sabrosa' (Planasa Nursery, Spain) is a dominant cultivar in southwest Spain due to its large fruit size in combination with excellent fruit shape and uniformity.

As the expectation of a year-round fruit supply has increased and strawberries are being shipped thousands of miles to grocery store chains, concomitant gains in firmness have been realized through breeding. 'Strawberry Festival,' developed at the University of Florida in 2000, has excellent firmness and tough skin that allow shipping of winter fruit as far as southern Canada. Its postharvest qualities combined with consistently uniform shape and color have rendered it the dominant cultivar in central Florida and the Queensland growing area of Australia as well as an important cultivar in southern Spain. It also possesses long fruit stems (pedicels) which dramatically increase harvest efficiency. The dominant cultivar in Western Europe, 'Elsanta,' also has excellent harvest efficiency mainly due to a low proportion of cull fruit because of its firm flesh and strong skin compared to other cultivars in the region.

Large gains in amount and timing of yield have been realized over the past few decades. Released in 1993 from UC-Davis, 'Camarosa' quickly became the dominant cultivar in Mediterranean regions around the world and is still grown in significant acreage. This is undoubtedly due in large part to its firmness and postharvest qualities, but its high total yield potential is perhaps its most impressive attribute when compared with other short-day germplasm. Earlier yielding cultivars have replaced 'Camarosa' over time in subtropical regions. The cultivars, 'Sweet Charlie' and 'Winter Dawn', are very early producers for subtropical and Mediterranean regions. In the mid-Atlantic region, 'Earliglow' (1975, USDA-ARS, Beltsville, MD) was a transformative cultivar due to its early yield potential. In northern regions, winter hardiness is vital to protect and preserve yields of short-day cultivars grown in the perennial matted row system. Released in 1981, 'Kent' (Kentville, Nova Scotia) became a standard for cold hardiness. Breeding work at the University of Minnesota has also resulted in cultivars, such as 'Mesabi,' that survive minimum winter temperatures below −34°C.

The development of day-neutral cultivars was pivotal, allowing fruit production throughout the summer and early fall in regions with moderate climates. This has allowed extended season harvest as day lengths increase, resulting in high total yields in coastal California. Cultivars, such as 'Seascape' and 'Albion,' have been important in this region and have also been adapted to annual plasticulture systems in northern temperate regions, though further breeding is needed to adapt day-neutral germplasm to other regions.

Flavor continues to be an important breeding objective, and it is expected that greater attention will be paid to this complex trait in the coming years. 'Chandler,' released in 1983 from UC-Davis, is still used for annual production systems in the mid-Atlantic region due to its exceptional eating quality. Researchers at the Queensland Horticultural Institute (now Department of Employment, Economic Development and Innovation) have given special attention to flavor traits, resulting in the release of 'Rubygem,' a cultivar with consistently excellent flavor, even as temperatures fluctuate throughout the course of the growing season.

Disease resistance remains important, though the resistances incorporated into cultivars in each region differ widely depending on the pathogens that are present and prevalent in those locales. Nevertheless, important resistances can be found throughout the cultivated gene pool. Examples include 'Allstar' (red stele root rot resistance), 'Governor Simcoe' (Verticillium root rot resistance), 'Sweet Charlie' (Anthracnose fruit rot resistance), and 'Totem' (virus tolerance).

5 Current Goals and Challenges of Breeding

High total yield continues to be a goal among strawberry breeders; however, in some subtropical and short day environments, current cultivars may be approaching a physiological limit for this trait. For example, the average seasonal yield for Florida strawberries is currently about 27,500 kg/ha, which is substantially higher than that for other fresh fruit crops grown in the southeastern USA. The average yield for high-density blueberries in Florida is only about 6,600 kg/ha, and the average yield for peaches in Georgia is about 8,800 kg/ha [http://www.usda.gov/nass/. Also, the wisdom of developing cultivars with greater total yield per plant may be questionable, since high yield can result in low-soluble solids (an important component of flavor).

The fruit yield of a strawberry plant can be considered to be the product of its dry matter production and harvest index (that proportion of the dry matter contained in the fruit) (Moore and Janick 1983). Over the years, increased yields have resulted primarily from a change in the harvest index brought about by more inflorescences and larger fruit with a more or less constant dry matter content. In the future, yield in many regions will be increased as day-neutral cultivars, which tend to produce more inflorescences per plant, are developed for those environments (Dale 2005). A greater challenge is to increase dry matter production. This is possible, as both *F. chiloensis* and *F. virginiana* vary in their photosynthetic rates (Hancock 1999).

Even if higher total yield is not possible or advisable, it is possible (and highly desirable) to increase *marketable* yield. Marketable yield can be increased by eliminating or reducing losses due to diseases, arthropod damage, fruit malformation, and cracking. Certainly, great strides have been made in the areas of disease and pest control, and, as for fruit malformation and cracking, breeders are working to develop cultivars that produce symmetrically shaped and crack-resistant fruit over a range of environmental conditions.

When selection is applied to improve the economic value of the strawberry plant, it is generally applied to several traits simultaneously (Hancock et al. 1996; Herrington et al. 2002). When several traits are selected simultaneously, the objective is to achieve maximum genetic progress toward a stated (economic) goal. When a strawberry genotype is selected for commercial production or as a parent, the value of each trait is at least informally evaluated by breeders, producers, retailers, and others in the strawberry production and marketing chain. Their relative value may vary with the position of the evaluator in the chain. This evaluation requires the study of the economic system that uses the cultivars because the economic values and nature of the costs and incomes are specific to the industry structure. With the current emphasis on the developed world on sustainable agriculture, strawberry breeders need to incorporate disease and pest resistance into their cultivars. However, they need to decide which organisms out of a myriad of pathogens deserve their interest. This decision is governed by several factors: the pests and diseases most prevalent in a given region; pathogenic race variability for those pathogens; efficiency of cultural methods to minimize the disease or pest; screening methods available to detect resistant genotypes; and genetic variability available for resistance traits (Dale 2005). Inevitably, because of the complexity of the decision, only those pests and diseases that cause the largest economic damage are actively screened in a breeding program.

Currently, the biggest opportunity in strawberry breeding is to develop day-neutral cultivars adapted to cool summer climates outside of California (i.e., southern Canada, northern Europe, and certain highland regions in South America, Africa, Asia, and southern Europe). Such cultivars would allow these areas to greatly extend their fruiting season, thus increasing production efficiency and marketing options.

The biggest challenge facing strawberry breeders may be to develop cultivars that can produce fruit with consistent size, appearance, and flavor over an extended period of time (i.e., 3–4 months or longer). Breeders that develop cultivars for mild winter climates (e.g., Florida, Queensland Australia, southern Spain) already face this problem (Chandler et al. 2003), and as breeders in mild summer areas come closer to releasing finished day-neutral cultivars they face this problem as well.

To meet this challenge, it is important to remember the words of Royce Bringhurst (1983): "If specific traits of secondary importance (e.g., resistance to root rotting pathogens) are used as the primary screen in fruit breeding …, the valuable combinations of essential fruit traits may never be observed, much less selected. Undesirable linkages and simple chance discards of the individuals with rare combinations of desirable genes for superior fruit quality, production, etc. may limit the possibilities to the extent that nothing of real value will ever turn up." Therefore, breeders need to evaluate the current and likely future production and marketing systems in light of plant characteristics and genetic mechanisms to identify traits of high priority.

6 Breeding Methods and Techniques

The breeding method predominately used for strawberry is the one described by N.W. Simmonds for clonal crops in his book Principles of Crop Improvement (Simmonds 1979). Breeding of strawberry involves the crossing of heterozygous clonal parents and making selections in the F_1 seedlings and in subsequent vegetative generations, with the objective of determining the most desirable genotypes. As material proceeds from one vegetative generation to the next (e.g., in some breeding programs, these generations are designated stage 1, 2, and 3), the number of genotypes are reduced while the number of plants per genotype is increased. Typically, strawberry breeders plant thousands of genotypes in their stage 1 trial, with each genotype represented by 1 or 2 plants (i.e., the seedling or one or two clonal daughter plants from the seedling); several hundred genotypes in their stage 2 trial, with each genotype represented by 10 or more plants planted in 1 or 2 plots; and a dozen or more genotypes in their stage 3 trial, with each genotype represented by 40 or more plants planted among two or more plots. The breeder often works closely with a food scientist, plant pathologist, and cultural management specialist to assess the desirability of potential cultivars. Finally, the decision to release a genotype as a new cultivar is based on trials replicated over sites and seasons.

A typical scheme to create an initial 10,000 genotypes is to grow out an average of 100 seedlings from each of 100 different controlled crosses. These crosses often result from using 30 or more parents in various combinations. In strawberries, 20 parents is considered to be the minimum number needed to prevent inbreeding depression (Sjulin and Dale 1987). Generally, parents are chosen on a phenotypic basis from among the most desirable genotypes in the stage 2 and 3 trials.

Crossing techniques and procedures for handling seedlings vary somewhat between breeding programs and depend on the time of year the crosses are made. However, what is done in the University of Florida program is typical of the steps that must be taken. In the morning, flowers whose petals are partially open (but whose anthers have not dehisced) are collected from the field and placed in open, brown paper bags. The bags are then placed on a greenhouse bench, and if conditions are cold or cloudy the bags are placed under an incandescent light bulb. The light bulb provides supplemental heat, which accelerates drying of the anthers. Pollination is generally done late in the afternoon. Pollen is transferred to an emasculated flower by carefully brushing the anthers of the detached flower onto the receptacle of the intact flower. Usually, one flower is collected from the field for each flower to be pollinated.

Potted plants in a greenhouse are used as females. Flowers are emasculated 1 day before anthesis. Tweezers are used to remove stamens, petals, and sepals in one operation. This operation should be done carefully to avoid injuring the receptacle.

When the pollinated fruit is ripe, it is harvested and placed in the pitcher of a food blender with other ripe fruits that have resulted from the same cross. The fruits are covered with water (to a level of 1/4th the volume of the pitcher), and then the blender is turned on for 10–15 s. The pitcher is then filled with water and time allowed for viable seeds to settle to the bottom of the pitcher. The pulp and water are poured off, leaving the viable seeds on the bottom of the pitcher. The seeds are rinsed clean and then placed on a paper towel to dry. Once the seeds are completely dry, they are placed in labeled glass vials and stored in a refrigerator.

Before sowing, seeds are scarified in concentrated sulfuric acid for 15 min, and then rinsed thoroughly with distilled water. Seeds are then redried before scattering on the surface of a moist, peat-based germination medium at a density of about 1 per square cm. The germination media can be in pots or trays. Periodic misting is used to keep seeds and germination media moist. Scarified strawberry seeds typically germinate in waves over a 4–6-week period. The range of germination is 25 to over 50%.

After most of the seedlings have produced at least one or two true leaves, they are transplanted into peat pellets that have been saturated with water. Then, after establishment in the peat pellets, seedlings are transplanted into a field nursery. Fruits from these seedlings are evaluated in the first year; therefore, a breeding cycle of one generation per year is possible. In practice, the average cycle is longer, since parents are usually chosen on the basis of their performance in replicated trials over multiple years.

7 Integration of New Biotechnologies in Breeding Programs

The application of biotechnology to cultivated strawberry has been hindered by issues related to its higher ploidy. As mentioned previously, the cultivated strawberry is octoploid ($2n=8\times=56$), adding considerable difficulty to deciphering most of the common outputs of biotechnological inquiry. *F.* × *ananassa*'s reticulate ancestry precludes the agile application of nondominant molecular markers, as information stemming from a chromosome complement from four individual genomes clouds simple interpretations from even the most fundamental inheritance patterns. Directed tests of gene function that could aid in the development of molecular markers are confounded again by the complicated genome. Furthermore, the potential gains of transgenic enhancement have practical barriers. Strawberry transformation and regeneration efficiency are highly genotype specific, so protocols must be constantly tailored and adjusted to fit a specific cultivar. Once a transgenic genotype is developed, it must be maintained into perpetuity, as seed propagation does not produce progeny identical to the parental genotype. All of these challenges have slowed the application of biotechnology to strawberry.

Despite these barriers, progress in strawberry biotechnology has accelerated greatly in the past decade, facilitated by the use of diploid strawberry as a genetically tractable system to begin to understand the molecular basis for important traits in the cultivated genotypes. The hypothesis is that findings in diploid strawberry species will translate to cultivated types, as gene structure, organization, and sequence are most likely highly conserved between cultivated strawberry and its antecedents. The basic findings in the diploid strawberry should prove useful in addressing questions in the octoploids, as the fundamental system informs developments in the genetically complex. Ironically, many of the first biotechnology breakthroughs to assist breeding efforts originate in the diploid strawberry, a plant with little direct economic value.

7.1 Genome Structure

Implementation of biotechnology in breeding requires a fundamental handle on the nature of the strawberry genome, one of the most complex genomes of cultivated crops. Strangely, the complex octoploid genome comprises a relatively simple fundamental unit, as the haploid genome ($n=7$) of strawberry is relatively small. At ~200 Mb, it is among the smallest of flowering plants (Akiyama et al. 2001; Antonius and Ahokus 1996; Bennett et al. 2003). Recent studies have shown that this space is confined to seven linkage groups with a total genetic distance of 606.8 cM.

Experiments dating back to the early twentieth century have examined the basic structure and composition of the cultivated strawberry genome. Most studies examined mitotic configurations of strawberry chromosomes when octoploid chromosomes were complexed with those from the diploid *Fragaria vesca* Coville.

F. vesca was a logical choice as a basis of the octoploid genome, as it shares a common geographical distribution with *F. virginiana* and is the dominant New World diploid. Ichijima (1926) and Fedorova (1946) observed bivalent formation between chromosomes from *F. vesca* and *F. virginiana*, strongly indicating that this subgenome was at least a partial contributor to the octoploid. The corresponding genome composition models reflected an autopolyploid component (AAAABBCC and AAA′A′BBBB, respectively). More recent constructions note the homology between A and C subgenomes and have described the cultivated genome as allopolyploid with a composition of AAA′A′BBB′B′, indicating the presence of four distinct subgenome donors (Bringhurst 1990). New evidence now presents genetic and molecular overlays to support and/or challenge these constructions.

The cultivated strawberry genome is undoubtedly complex, but the behavior of the subgenomes during meiosis would thicken the complexity of inheritance if subgenomes are intermingling with each other. Therefore, a comprehensive understanding of inheritance is prerequisite to deploying genome-enabled breeding strategies. Several studies have addressed inheritance in the octoploid strawberry. Observations of meiotic behavior indicate that the subgenomes (for the most part) segregate within themselves. Evidence of disomic segregation dates back almost 30 years when Arulsekar and colleagues (1981) identified patterns of disomic inheritance in isozyme variants. A study of Amplified Fragment Length Polymorphisms (AFLP) products in octoploid crosses again suggested that inheritance was mostly disomic, with some evidence of polysomic behavior (Lerceteau-Kohler et al. 2003). Segregation of Simple Sequence Repeat (SSR) and Cleaved Amplified Polymorphic Sequence (CAPS) markers was traced in relevant populations and demonstrated again that individual plants had no more than two alleles for any one locus and that alleles were transmitted in a manner consistent with disomic inheritance (Ashley et al. 2003; Kunihisa et al. 2005). Further refinements to the octoploid linkage map and comparative mapping efforts have since confirmed that allele segregation in octoploid strawberry is, indeed, mainly disomic (Rousseau-Gueutin et al. 2008).

Several contemporary findings have hinted at the origin of the subgenome constituents. Analysis of alcohol dehydrogenase intron size indicates that the octoploid maintains haplotypes reminiscent of *F. iinumae* Makino, a Japanese diploid. Analysis of intergenic regions from diploid and octoploid strawberry further supports this interpretation, as octoploid sequence samples typically contain members that best match *F. iinumae* (Tombolato, Davis and Folta, unpublished obs.). Studies of octoploid-based SSR marker transferability to various other species demonstrated that octoploid-based SSRs could be amplified in *F. vesca* most readily, followed by amplification in *F. iinumae*, followed by lesser frequencies in other diploids, again bolstering the vesca-iinumae associations with the octoploid genome. Ongoing studies of highly variable intergenic regions promise to further illuminate the identity of the octoploid subgenome constituents. Once the nature of the subgenomes is understood, it may facilitate the development of subgenome-specific markers. Examination of wild octoploids and diploids reveals basic phenotypes that are consistent with subgenome dominance, as various octoploid genotypes retain morphological characters that are shared with various diploids. Strawberry phenotypes are highly plastic, so these are by no means hard, scientific observations. However, it is likely that further examination of the diploid strawberry will provide discrete markers that translate to the octoploid allowing breeders to select traits associated with discrete subgenomes.

Studies in other polyploids, such as cotton, indicate that epigenetic mechanisms govern the expression of specific subgenomes, oftentimes relegating subgenome traits to specific tissues or developmental contexts. These trends are likely to be apparent in octoploid strawberry, possibly making simple translation of traits associated with diploid markers less clear. However, this apparent hurdle is also an opportunity, as it offers another control point, another opportunity to regulate the end products most prized by the breeder, the industry, and the consumer.

7.2 Genetic Linkage Mapping

To the breeder, it would be useful to know which gene (or genes) underlies a specific trait of interest, and then where that gene is located in the genome. This information would place a physical address on a given quality and permit another means of selection using genomics-enabled tools. In cases where single Mendelian traits vary between diploid parents of proven interspecific crosses (such as those used by Sargent and colleagues at East Malling Research, UK), it is possible to assign them to locations on the *Fragaria* linkage map. However, contemporary studies in cultivated strawberries are constrained by the limited genetic variation in breeding populations and the need to assay relatively large numbers of markers to identify subgenome-specific polymorphisms and assign them to linkage groups. Some clear identification of inheritance patterns in octoploid strawberry has been observed, but for the most part linking trait to genetic locus has been accomplished in the diploid species.

The variability within *F. vesca* and that of other diploid species has been used to devise an increasingly dense genetic linkage map. Genetic studies relating back to the early twentieth century described evidence of genetic linkage between various morphological traits in the diploid strawberry (Richardson 1914). The genetic loci regulating runnering or nonrunnering habits (*r*), precocious flowering (*semperflorens*; *s*), and yellow fruit color (*c*) were defined by Brown and Wareing (Brown and Wareing 1965a, b). Additional morphological characters have been slow to assign, with the 'pale-green' leafy phenotype being the most recent addition to linkage group IV (Sargent et al. 2004).

The sparse linkage assignments have been made possible by intraspecific crosses of the diploid strawberry *F. vesca* (Davis and Yu 1997) as well as interspecific crosses with other interfertile diploids, like *F. bucharica* Losinsk (formerly *F. nubicola* Lindh ex. Lacaita) (Sargent et al. 2004, 2006) or *F. viridis* Duchesne (Nier et al. 2006). The first measurable genetic linkage (1.1 cM) was observed between an SKDH isozyme variant and the yellow fruit (c) locus (Williamson et al. 1995). This report was soon followed by the observation of cosegregation of the *Pgi-2* and nonrunnering (*r*) locus (Yu and Davis 1995). Since then, and especially in the last 5 years, the map has grown increasingly dense and complex. The majority of markers are based on polymorphism in SSRs. These islands of di-, tri-, or polynucleotide repeats expand and contract with evolutionary time, and serve as a valuable tool for linkage mapping. The other advantages of microsatellite markers are that they tend to be highly reproducible between laboratories and are often transferable between species, e.g., an SSR that amplifies in strawberry also amplifies in closely related species, such as brambles (Davis et al. 2006; Lewers et al. 2005).

The *Fragaria vesca* 815 × *Fragaria bucharica* 601 (FV×FB) cross at East Malling Research (UK) segregates for 3 morphological traits and approximately 340 other proven markers (D. Sargent, personal comm.). A set of eight representative progeny allows efficient assignment to linkage maps using a bin mapping strategy (Sargent et al. 2008). Using this approach, an additional 103 markers were added to the linkage map.

An additional study examined SSRs derived from expressed gene sequences. These markers not only provide further resolution associated with a highly variable sequence, but also are directly related to discrete genes, some with potential roles in processes of horticultural interest (Gil-Ariza et al. 2006; Sargent et al. 2006). Strawberry EST resources are increasingly prevalent (Folta et al. 2005; Slovin and Rabinowicz 2007), and as these resources continue to emerge it enables the evaluation of specific genes for allelic characters and then correlation of these variants with phenotypes of interest. In this way, it may be possible to mobilize molecular tools to directly assess the likelihood of a given trait to present favorably in a specific genotype.

The foundation of genetic linkage mapping in diploid species has been extended to the cultivated strawberry. Analysis of segregation of SSR markers in both species permits comparative characterizations to be made (Davis et al. 2006). Recent results indicate strong colinearity between the diploid linkage assignments and homeologous linkage groups in the octoploid strawberry (Rousseau-Gueutin et al. 2008). The findings suggest minimal chromosomal rearrangement and strongly related subgenomes.

7.3 Molecular Markers

The promise of molecular markers that segregate faithfully with traits of interest stands to propel breeding efforts. However, these tools have been slow to develop in strawberry again due to the complicated genome. The most useful markers to date are dominant PCR products that are amplifiable in genotypes containing a trait of interest. The ideal marker-assisted scenario would permit a breeder to perform the first round of evaluations in a Petri dish or small flat, selecting for (or in some cases eliminating) plant genotypes presenting a given molecular indication. Such markers would be inexpensive, simple, and rapid to generate. The goal would be to pipeline a set of plants with a higher likelihood of exhibiting favorable disease resistance, yield, flavor, etc., long before they were ever evaluated in the field. Such efforts would focus breeder space and time on those candidates likely containing favorable traits. Certain private breeding programs have initiated marker development programs, and today use molecular markers as a cornerstone of their selection strategy (Tom Sjulin, personal comm.). Some well-characterized marker types have the potential to amplify a rich array of products, many that are genotype specific (as in Govan et al. 2008). Here, the complicated subgenome structure provides the researcher with greater marker resolution, as more specific character states are available to relate to traits of interest. These molecular markers have been useful to construct phylogenetic trees that display the relative relatedness of various cultivars. These simple analyses may be useful when constructing crosses, developing pedigrees, or establishing hard mechanisms to protect intellectual property.

Description of valued genotypes with comprehensive sets of molecular markers also assists in the protection of breeders' rights and patent enforcement. Yet unrealized benefits undoubtedly arise as marker saturation in breeding materials matures with the power of computational analysis.

Several classes of molecular markers relevant to traits of interest have been employed in strawberry and their specifics have been discussed in two reviews (Folta and Davis 2006; Hokanson and Maas 2001). Their principal use has been in genotyping cultivars (Arnau et al. 2003; Congiu et al. 2000; Degani et al. 1998; Kunihisa et al. 2003, 2005) or estimating genetic diversity (Debnath et al. 2008; Degani et al. 2001; Graham et al. 1996) rather than defining associations with favorable traits. Part of the problem comes from the fact that most markers used in these studies are based on Randomly Amplified Polymorphic DNA (RAPD), AFLP, or Intersimple Sequence Repeat (ISSR)-based haplotypes, and these are traditionally difficult to apply across genotypes or between laboratories.

These arbitrary markers can be made more reproducible by converting them to Sequence-Characterized Amplified Region (SCAR) markers. For instance, an AFLP marker appeared in bulked cultivated material resistant to *Colletotrichum acutatum*, the causative agent of the symptom spectrum known as strawberry anthracnose. Early analysis of the genetics of anthracnose sensitivity and resistance indicated that the disease presentation was affected by several loci quantitatively (Gimenez and Ballington 2002). In the AFLP study, a cosegregating amplicon was converted to a SCAR marker representing the *Rca2* locus (Lerceteau-Kohler et al. 2005). The product segregated with resistance in 81.4% of the accessions tested, indicating its utility as a predictor of disease sensitivity. Because it is a DNA-based marker, it is not influenced by environment, development, or gene expression.

Other markers have been identified, but perhaps did not prove applicable across many populations or with specific variants of particular pathogens. They are worthy of mention because they may be useful in some applications. For instance, other disease resistance markers were characterized, including a set associated with resistance to *Phytophthora fragariae* C. J. Hickman var. *fragariae*, the causative agent of red stele root rot disease. There are believed to be at least five avirulence genes in European accessions (Haymes et al. 1997, 2000; van de Weg 1997a, b) and approximately ten in New World races (Maas 1987). Again, analysis of RAPD markers in bulked segregant populations identified a DNA region that generally associated with the *Rfp1* locus. Although RAPD markers are difficult to reproduce between laboratories or genotypes, they are frequently a best first step to identification of more durable markers, and it is possible that those associated with the *Rfp1* locus may find favor in future breeding strategies.

Marker-assisted selection would greatly benefit the identification of germplasm likely to respond to specific photoperiodic signals, and such markers are currently implemented in commercial breeding programs (Folta and Davis 2006). Photoperiodic flowering is controlled by a well-defined pathway in model systems. At least a subset of the same components exists in cultivated strawberry, and these candidates may contribute to the flowering response. Various photoperiodic sensitivities are important to breeding programs in specific locales, as plants grown in some regions benefit from a short-day phenotype while others require a day-neutral phenotype. Genetic studies have shown that the control of day neutrality is based on a major controlling locus with several modifiers (Shaw 2003; Shaw and Famula 2005). A pair of SCAR markers has been mapped in close proximity to the *Seasonal Flowering Locus* (*SFL*) in a diploid population (Albani et al. 2004). This locus maintains genetic elements relevant to the ever-bearing phenotype segregating in the progeny from this cross. Other efforts to develop markers associated with ever-bearing flowering habits by converting RAPD markers have been attempted, but the markers identified show only loose associations with the locus of interest (11.8–15.8 cM; Sugimoto et al. 2005), so they are not likely practically deployable. Other studies have defined multiple QTL intervals that together underlie the phenotypic variation associated with the trait (Weebadde et al. 2008).

Ultimately, the deployment of molecular markers for marker-assisted strawberry breeding is hastened from advances in sequencing techniques. It may soon be possible to obtain complete genomic sequence information from a single strawberry genotype for a cost of under $1,000. Here, the newest technologies meet the available body of carefully scored phenotypic characters, identifying correlations between the presence/absence of candidate genes, meaningful alleles, and large-scale variations in subgenome structure. The relationship between bioinformatics and characters directly related to traits of interest speeds breeding efforts, and is a useful (if not indispensable) tool for the breeder. A decade from now, the first selections may be pulled from Petri dishes, as germplasm containing suites of markers segregating faithfully with a likelihood of producing favorable traits is quickly moving from science fiction to science fact. Such progress would save massive resources or, at best, allow the breeder's art to focus the same resources on larger quantities of materials with a high likelihood of commercial success. While true for just about any crop species, the complexity of the strawberry genome benefits much more substantially from implementation of these technologies.

7.4 Transgenic Technologies

How can transgenic plants help breeders? Transgenics can have important impacts in several ways, as the genes installed in the laboratory may satisfy various practical applications. When considering transgenic technologies aimed at horticultural applications, we typically think of plants that have been genetically enhanced via transformation to meet a given cultivation challenge. While there is no evidence of harm, there remains a strong negative public sentiment toward the consumption of transgenic crops, particularly in the European Union. Therefore, the deployment of transgenic plants has not been approached in *Fragaria*. However, there is an important reason to support the development of transgenic strawberry. Both diploid and octoploid strawberries are readily transformable, progressing from explant, through regeneration to rooted plant on the scale of weeks or months. Therefore, plant genotypes never destined for the field may still inform and guide traditional breeding efforts and marker-assisted selection from the standpoint of gene validation. If the impact of a given gene or genes can be identified in plants in an actual production setting, the marriage of genomicists and breeders can then identify favorable alleles in wild or breeding populations, and then introgress them into elite genotypes. In this way, transgenics does not represent a hard-field-present end point for some time in strawberry. Instead, such tools enable the deployment of sophisticated marker strategies with direct impacts.

Both the octoploid and diploid strawberries are readily transformed (Folta and Dhingra 2006; Mezzetti and Costantini 2006). Transformation and regeneration remain highly genotype specific, as the culture conditions that work well for one genotype do not translate well to others. Over 20 octoploid accessions have been successfully transformed and regenerated (Folta and Dhingra 2006), along with a substantial set of diploid accessions (Oosumi et al. 2006). Various explant sources have been tested, and again vary in usefulness from genotype to genotype. All *Fragaria* accessions are remarkably sensitive to antibiotics and herbicides, so the use of such markers requires careful attention to dosage and the construction of phytotoxicity curves prior to intensive attempts to regenerate organs. Many researchers have initiated selection on relatively low doses of antibiotics, gradually increasing the amount until a high selective pressure is obtained (as in Alsheikh et al. 2002; Mathews et al. 1998).

While the value of transgenics has been realized in many plant systems, like Roundup Ready Soybean (Funke et al. 2006) and virus-resistant papaya (Chiang et al. 2001; Gonsalves 2002), the technologies have not yet made their way to the strawberry field. However, necessity is the mother of invention and it is wholly possible that a gene fostering incremental increases in yield, flavor, disease resistance, pest tolerance, or lower agricultural inputs could gain favor among growers and even consumers. There are several examples with outstanding potential. Mezzetti et al. (2004) showed that a transgenic construct of the ovule/placenta-specific snapdragon *DefH9* promoter driving *iaaM*, an auxin biosynthetic gene from the olive pathogen *Pseudomonas syringae*, results in a substantial increase in the number of strawberry flowers. There also are significant changes in fruit size. The transgenic plants produced show higher yields of well-proportioned fruits. Jiménez-Bermúdez et al. (2002) used antisense technology to suppress the expression of pectate lyase, a protein central to the softening process. The study showed that all characteristics of yield and fruit quality were unaffected, yet the fruits were firmer. This technology could become commonplace in the near future as the industry seeks solutions to the high cost of shipping delicate produce.

The enzymes that control the production of volatile compounds in strawberry have been exquisitely studied (Aharoni et al. 2000, 2004). Variants in some of the critical enzymes that delineate the favorable and unfavorable compounds produced have been well-defined, especially with regard to differences between wild and cultivated plants (Aharoni et al. 2004). The identification of these genes as critical differences inspired by cultivation allows the development of markers that could allow favorable traits to be introgressed from wild plants, without significant linkage drag affecting flavor. Future efforts may target flavor, disease resistance, nematode resistance, yield, nutrient efficiency, and a host of other horticulturally relevant traits.

Currently, the consumer acceptance of genetically enhanced foods is relatively low, especially in the European Union. New technologies will circumvent these barriers. Some use marker-free processes, where chemically inducible recombinase can excise marker genes, producing plants that cannot be easily distinguished as transgenic. Such strawberry genotypes have been developed (Schaart et al. 2004). These plants can be engineered to contain an advantageous transgenic construct targeting a trait of interest, yet do so without any excess molecular baggage that some may find objectionable. The future application of 'cis-genic' approaches, i.e., using only strawberry genes and regulatory sequences to transform strawberry, is also a promising way to bring potential benefits to the industry (Jacobsen and Schouten 2007; Schouten et al. 2006).

References

Aharoni A, Giri AP, Verstappen FW, Bertea CM, Sevenier R, Sun Z, Jongsma MA, Schwab W, Bouwmeester HJ (2004) Gain and loss of fruit flavor compounds produced by wild and cultivated strawberry species. Plant Cell 16:3110–3131

Aharoni A, Keizer LC, Bouwmeester HJ, Sun Z, Alvarez-Huerta M, Verhoeven HA, Blaas J, van Houwelingen AM, De Vos RC, van der Voet H, Jansen RC, Guis M, Mol J, Davis RW, Schena M, van Tunen AJ, O'Connell AP (2000) Identification of the SAAT gene involved in strawberry flavor biogenesis by use of DNA microarrays. Plant Cell 12:647–662

Akiyama Y, Yamamoto Y, Ohmido N, Oshima M, Fukui K (2001) Estimation of the nuclear DNA content of strawberries (*Fragaria spp.*) compared with Arabidopsis thaliana by using dual-stem flow cytometry. Cytologia 66:431–436

Albani MC, Battey NH, Wilkinson MJ (2004) The development of ISSR-derived SCAR markers around the SEASONAL FLOWERING LOCUS (SFL) in Fragaria vesca. Theoretical and Applied Genetics 109:571–579

Alsheikh MK, Suso HP, Robson M, Battey NH, Wetten A (2002) Appropriate choice of antibiotic and Agrobacterium strain improves transformation of anti biotic-sensitive *Fragaria vesca* and *F.v. semperflorens*. Plant Cell Rep 20:1173–1180

Antonius K, Ahokus H (1996) Flow-cytometric determination of polyploid level in spontaneous clones of strawberries. Hereditas 124:285

Arnau G, Lallemand J, Bourgoin M (2003) Fast and reliable strawberry cultivar identification using inter simple sequence repeat (ISSR) amplification. Euphytica 129:69–79

Arulsekar S, Bringhurst R, Voth V (1981) Inheritance of PGI and LAP isozymes in octoploid cultivated strawberries. . J Am Soc Hort Sci 106:679–683

Ashley MV, Wilk JA, Styan SM, Craft KJ, Jones KL, Feldheim KA, Lewers KS, Ashman TL (2003) High variability and disomic segregation of microsatellites in the octoploid *Fragaria virginiana* Mill. (Rosaceae). Theor Appl Genet 107:1201–1207

Bennett MD, Leitch IJ, Price HJ, Johnston JS (2003) Comparisons with *Caenorhabditis* (approximately 100 Mb) and *Drosophila* (approximately 175 Mb) using flow cytometry show genome size in Arabidopsis to be approximately 157 Mb and thus approximately 25% larger than the Arabidopsis genome initiative estimate of approximately 125 Mb. Ann Bot (Lond) 91:547–557

Bringhurst, R.S. (1983) Breeding strategy. In J.N. Moore and J. Janick (Eds.), *Methods in Fruit Breeding*. Purdue Univ Press, West Lafayette, Ind.

Bringhurst, R.S. (1990) Cytogenetics and evolution in American Fragaria. Hortscience 25:679–683

Brown T, Wareing PF (1965a) Genetical control of everbearing habit and 3 other characters in varieties of *Fragaria vesca*. Euphytica 14:97–112

Brown T, Wareing PF (1965b) Genetical control of flowering and runnering in varieties of *Fragaria vesca*. Heredity 20:651–653

Chandler, C.K., M. Herrington, and A. Slade (2003) Effect of harvest date on soluble solids and titratable acidity in fruit of strawberry grown in a winter, annual hill production system. Proc 26th Intl Hort Congr pp. 353–354

Chiang CH, Wang JJ, Jan FJ, Yeh SD, Gonsalves D (2001) Comparative reactions of recombinant papaya ringspot viruses with chimeric coat protein (CP) genes and wild-type viruses on CP-transgenic papaya. J Gen Virol 82:2827–2836

Congiu L, Chicca M, Cella R, Rossi R, Bernacchia G (2000) The use of random amplified polymorphic DNA (RAPD) markers to identify strawberry varieties: a forensic application. Mol Ecol 9:229–232

Dale, A. (2005) Future trends in strawberry breeding in North America. in: S.Khanizadeh and J. DeEll (Eds.) Our strawberries – Les Fraziers de cheznous. Agriculture and Agri-Food Canada, St.Jean-sur-Richelieu, Quebec, Canada pp 70–83

Dale, A. and T.M. Sjulin (1990) Few cytoplasms contribut to North American strawberry cultivars. HortScience 25:1341–1342

Darnell, R.L. (2003) Strawberry growth and development. In: N.F. Childers (Ed.), *The Strawberry*. Dr. Norman F. Childers Publications, Gainesville, Fla., pp. 3–10

Davis T, DiMeglio L, Yang R, Styan S, KS L (2006) Assessment of SSR marker transfer from the cultivated strawberry to diploid strawberry species: functionality, linkage group assignment, and use in diversity analysis. Journal of the American Society of Horticultural Science:506–512

Davis TM, Yu H (1997) A linkage map of the diploid strawberry, *Fragaria vesca*. J Hered 88:215–221

Debnath SC, Khanizadeh S, Jamieson AR, Kempler C (2008) Inter Simple Sequence Repeat (ISSR) markers to assess genetic diversity and relatedness within strawberry genotypes. Can J Plant Sci 88:313–322

Degani C, Rowland LJ, Levi A, Hortynski JA, Galletta GJ (1998) DNA fingerprinting of strawberry (Fragaria x ananassa) cultivars using randomly amplified polymorphic DNA (RAPD) markers. Euphytica 102:247–253

Degani C, Rowland LJ, Saunders JA, Hokanson SC, Ogden EL, Golan-Goldhirsh A, Galletta GJ (2001) A comparison of genetic relationship measures in strawberry (Fragaria x ananassa Duch.) based on AFLPs, RAPDs, and pedigree data. Euphytica 117:1–12

Fedorova N (1946) Crossability and phylogenetic relations in the main European species of *Fragaria*. Comp Rend Acad Sci USSR 53:545–547

Folta K, Staton M, Stewart P, Jung S, Bies D, Jesudurai C, Main D (2005) Expressed sequence tags (ESTs) and simple sequence repeat (SSR) markers from octoploid strawberry (*Fragaria* x *ananassa*). BMC Plant Biology 5:12

Folta KM, Davis TM (2006) Strawberry genes and genomics. Critical Reviews in Plant Sciences 25:399–415

Folta KM, Dhingra A (2006) Transformation of strawberry: The basis for translational genomics in Rosaceae. InVitro Cellular and Developmental Biology-- Plant 42:482–490

Funke T, Han H, Healy-Fried ML, Fischer M, Schonbrunn E (2006) Molecular basis for the herbicide resistance of Roundup Ready crops. Proc Natl Acad Sci U S A 103:13010–13015

Gil-Ariza DJ, Amaya I, Botella MA, Blanco JM, Caballero JL, Lopez-Aranda JM, Valpuesta V, Sanchez-Sevilla JF (2006) EST-derived polymorphic microsatellites from cultivated strawberry (Fragaria x ananassa) are useful for diversity studies and varietal identification among Fragaria species. Mol Ecol Notes 6:1195–1197

Gimenez G, Ballington JR (2002) Inheritance of resistance to Colletotrichum acutatum Simmonds on runners of garden strawberry and its backcrosses. Hortscience 37:686–690

Gonsalves D (2002) Coat protein transgenic papaya: "acquired" immunity for controlling papaya ringspot virus. Curr Top Microbiol Immunol 266:73–83

Govan C, Simpson D, Johnson A, Tobutt KR, Sargent DJ (2008) A reliable multiplexed microsatellite set for genotyping Fragaria and its use in a survey of 60 F.xananassa cultivars. Mol Breeding 22:649–661

Graham J, McNicol RJ, McNicol JW (1996) A comparison of methods for the estimation of genetic diversity in strawberry cultivars. Theoretical and Applied Genetics 93:402–406

Hancock, J.F. (1999) *Strawberries.* CAB Publishing, New York

Hancock, J.F., T.M. Sjulin, and G.A. Lobos (2008) Strawberries. In J.F. Hancock (Ed.), *Temperate Fruit Crop Breeding*. Springer, New York

Hancock, J. F., Scott, D. H., and Lawrence F. J. (1996). Strawberries. In Fruit Breeding, Volume II: vine and small fruits crops. Ed Jules Janick and James Moore John Wiley and Sons, NY p 419–470

Haymes KM, Henken B, Davis TM, van deWeg WE (1997) Identification of RAPD markers linked to a *Phytophthora fragariae* resistance gene (*Rpf1*) in the cultivated strawberry. Theoretical and Applied Genetics 94:1097–1101

Haymes KM, Van de Weg WE, Arens P, Maas JL, Vosman B, Den Nijs APM (2000) Development of SCAR markers linked to a Phytophthora fragariae resistance gene and their assessment in European and North American strawberry genotypes. J Am Soc Hortic Sci 125:330–339

Herrington, M., Moisander, J. and Chandler, C. 2002. Choosing and using selection criteria in a horticultural crop - an example from the Better Berries Strawberry breeding program for the LAWS market. Proc. 12th Australasian Plant Breeding Conference, Perth, W. Australia, 15–20 September 2002. p. 601–604

Hokanson SC, Maas JL (2001) Strawberry Biotechnology. Plant Breeding Reviews. John Wiley and Sons, Inc., pp 139–180

Ichijima K (1926) Cytological and genetic studies on *Fragaria.* Genetics 11:590–603

Jacobsen E, Schouten HJ (2007) Cisgenesis strongly improves introgression breeding and induced translocation breeding of plants. Trends Biotechnol 25:219–223

Jiménez-Bermúdez S, Redondo-Nevado J, Muñoz-Blanco J, Caballero JL, López-Aranda JM, Valpuesta V, Pliego-Alfaro F, Quesada MA, Mercado JA (2002) Manipulation of strawberry fruit softening by antisense expression of a pectate lyase gene. Plant Physiol 128:751–759

Kunihisa M, Fukino N, Matsumoto S (2003) Development of cleavage amplified polymorphic sequence (CAPS) markers for identification of strawberry cultivars. Euphytica 134:209–215

Kunihisa M, Fukino N, Matsumoto S (2005) CAPS markers improved by cluster-specific amplification for identification of octoploid strawberry (Fragaria x ananassa Duch.) cultivars, and their disomic inheritance. Theoretical and Applied Genetics 110:1410–1418

Lerceteau-Kohler E, Guerin G, Denoyes-Rothan B (2005) Identification of SCAR markers linked to *Rca 2* anthracnose resistance gene and their assessment in strawberry germ plasm. Theor Appl Genet 111:862–870

Lerceteau-Kohler E, Guerin G, Laigret F, Denoyes-Rothan B (2003) Characterization of mixed disomic and polysomic inheritance in the octoploid strawberry (Fragaria x ananassa) using AFLP mapping. Theor Appl Genet 107:619–628

Lewers KS, Styan SMN, Hokanson SC, Bassil NV (2005) Strawberry GenBank-derived and genomic simple sequence repeat (SSR) markers and their utility with strawberry, blackberry, and red and black raspberry. J Am Soc Hortic Sci 130:102–115

Maas JL (1987) Availability of Strawberry Germplasm Resistant to 10 United-States Races of Phytopthora Fragariae and Several Isolates of Verticillium. Phytopathology 77:1774–1774

Mathews H, Dewey V, Wagoner W, Bestwick RK (1998) Molecular and cellular evidence of chimaeric tissues in primary transgenics and elimination of chimaerism through improved selection protocols. Transgenic Res 7:123–129

Mezzetti B, Costantini E (2006) Strawberry (Fragaria x ananassa). Methods Mol Biol 344:287–295

Mezzetti B, Landi L, Pandolfini T, Spena A (2004) The defH9-iaaM auxin-synthesizing gene increases plant fecundity and fruit production in strawberry and raspberry. BMC Biotechnol 4:4

Moore, J. N. and J. Janick. 1983. Methods in fruit breeding. Purdue Univ. Press West Lafayette, Indiana. 464 pp.

Nier S, Simpson DW, Tobutt KR, Sargent DJ (2006) A genetic linkage map of an inter-specific diploid Fragaria BC1 mapping population and its comparison with the Fragaria reference map (FV X FN). J Hortic Sci Biotech 81:645–650

Oosumi T, Gruszewski HA, Blischak LA, Baxter AJ, Wadl PA, Shuman JL, Veilleux RE, Shulaev V (2006) High-efficiency transformation of the diploid strawberry (Fragaria vesca) for functional genomics. Planta 223:1219–1230

Richardson C (1914) A preliminary note on the genetics of Fragaria. Journal of Genetics 3:171–177

Rousseau-Gueutin M, Lerceteau-Kohler E, Barrot L, Sargent DJ, Monfort A, Simpson D, Arus P, Guerin G, Denoyes-Rothan B (2008) Comparative genetic mapping between octoploid and diploid Fragaria species reveals a high level of colinearity between their genomes and the essentially disomic behavior of the cultivated octoploid strawberry. Genetics 179:2045–2060

Sargent DJ, Cipriani G, Vilanova S, Gil-Ariza D, Arus P, Simpson DW, Tobutt KR, Monfort A (2008) The development of a bin mapping population and the selective mapping of 103 markers in the diploid Fragaria reference map. Genome 51:120–127

Sargent DJ, Clarke J, Simpson DW, Tobutt KR, Arus P, Monfort A, Vilanova S, Denoyes-Rothan B, Rousseau M, Folta KM, Bassil NV, Battey NH (2006) An enhanced microsatellite map of diploid *Fragaria*. Theor Appl Genet

Sargent DJ, Davis TM, Tobutt KR, Wilkinson MJ, Battey NH, Simpson DW (2004) A genetic linkage map of microsatellite, gene-specific and morphological markers in diploid Fragaria. Theoretical and Applied Genetics 109:1385–1391

Schaart JG, Krens FA, Pelgrom KTB, Mendes O, Rouwendal GJA (2004) Effective production of marker-free transgenic strawberry plants using inducible site-specific recombination and a bifunctional selectable marker gene. Plant Biotechnol J 2:233–240

Schouten HJ, Krens FA, Jacobsen E (2006) Do cisgenic plants warrant less stringent oversight? Nat Biotechnol 24:753

Shaw DV (2003) Heterogeneity of segregation ratios from selfed progenies demonstrate polygenic inheritance for day neutrality in strawberry (Fragaria xananassa Duch.). J Am Soc Hortic Sci 128:504–507

Shaw DV, Famula TR (2005) Complex segregation analysis of day-neutrality in domestic strawberry (*Fragaria xananassa* Duch.). Euphytica 145:331–338

Simmonds, N.W. (1979) *Principles of Crop Improvement*. Longman, London

Sjulin, T.M. (2007) United States strawberry marketing and production trends. Proc 2007 North Amer. Strawberry Symp pp. 2–3

Sjulin,T.M.,and A. Dale (1987) Genetic diversity of North American strawberry cultivars J. Amer. Soc. Hort. Sci. 112: 375–385

Slovin J, Rabinowicz P (2007) Fragaria vesca, a useful tool for Rosaceae genomics. American Society for Horticultural Science, Ventura, CA

Sugimoto T, Tamaki K, Matsumoto J, Yamamoto Y, Shiwaku K, Watanabe K (2005) Detection of RAPD markers linked to the everbearing gene in Japanese cultivated strawberry. Plant Breeding 124:498–501 van de Weg WE (1997a) A gene-for-gene model to explain interactions between cultivars of strawberry and races of Phytophthora fragariae var fragariae. Theoretical and Applied Genetics 94:445–451

van de Weg WE (1997b) Resistance to *Phytophthora fragariae* var. fragariae in strawberry: The *Rpf2* gene. Theoretical and Applied Genetics 94:1092–1096

Weebadde CK, Wang D, Finn CE, Lewers KS, Luby JJ, Bushakra J, Sjulin TM, Hancock JF (2008) Using a linkage mapping approach to identify QTL for day-neutrality in the octoploid strawberry. Plant Breeding 127:94–101

Williamson SC, Yu H, Davis TM (1995) Shikimate dehydrogenase allozymes -inheritance and close linkage to fruit color in the diploid strawberry. J Hered 86:74–76

Yu HR, Davis TM (1995) Genetic-linkage between runnering and phosphoglucoisomerase allozymes, and systematic distortion of monogenic segregation ratios in diploid strawberry. J Am Soc Hortic Sci 120:687–690

Part III
Tree Fruits

Chapter 10
Apple

Susan Brown

Abstract The cultivated apple, *Malus xdomestica* Borkh., is a interspecific hybrid complex of allopolyploid origin. The progenitor species is thought to be *M. sieversii* (Lodeb.) Roem., which hybridized with both European and Asian species throughout its domestication. Modern breeding continues to employ relatively few of the 25–30 species of *Malus* from throughout the northern hemisphere for both scion and rootstock development. The apple is the most produced temperate tree crop and is widely grown throughout the temperate zone and recently it has been expanding into subtropical and tropical zones. Major goals of scion breeding programs include fruit quality, disease resistance (scab, fire blight, powdery mildew), nutritional components and excellent postharvest traits to allow long storage and use as a fresh-cut product. Rootstock breeding efforts emphasize resistance to abiotic and biotic stress as well as plant vigor control. Much progress has been seen in the integration of biotechnology with the development of transformation systems, multiple maps, a large number of markers, extensive EST libraries and, most recently, with the whole genome sequencing of apple. Research has identified marker–traits associations for various disease resistance, plant architecture, postharvest, and flavor traits. International collaborative efforts are actively working to exploit the biotechnological approaches to understand the genetic basis of a range of commercially important traits to improve the efficiency of breeding programs.

Keywords *Malus x domestica* • Pome fruit • Pip fruit • Allopolyploid • Origin • Apple scab • *Venturia* • Powdery mildew • *Podosphaera* • Fire blight • Rootstock • Dwarfing • Allergenicity • Fresh-cut • Domestication • Post harvest • Antioxidants • Marker traits association • Incompatibility

S. Brown (✉)
Department of Horticulture, Cornell University,
New York State Agricultural Experiment Station (NYSAES), Geneva, NY 14456, USA
e-mail: skb3@cornell.edu

M.L. Badenes and D.H. Byrne (eds.), *Fruit Breeding*, Handbook of Plant Breeding 8,
DOI 10.1007/978-1-4419-0763-9_10,

1 Introduction

1.1 Economic Importance

Apple production has increased by more than 50% over the last 20 years with the bulk of this growth in China (Table 10.1). Currently, China is the largest producer of apples, the USA second, followed by apple producers in the European Union (with Poland, Italy, and France being the largest producers). In the USA, apple production is valued at more than $2.5 billion dollars annually. The apple is the third most valuable fruit crop in the USA, following grapes and oranges. More than 60% of apple production is marketed as fresh fruit.

Apples are used for fresh consumption or processed into a number of different products such as apple sauce, apple slices, baby food, juice, cider, brandy, and distilled spirits. The markets for fresh-cut and organic apple are also increasing and expanding the availability of fresh apples in the fast food industry and for school lunch programs.

1.2 Taxonomy, Basic Botany and Description of the Crop

Luby (2003) provided an excellent review of the taxonomy of apple. Apple is an interspecific hybrid complex that usually is designated as *Malus* x*domestica* Borkh. or *Malus domestica* Borkh. Apple is a member of the subfamily Maloideae of the Rosaceae family. The haploid chromosome number is x = 17. Apple is an allopolyploid, but behaves like a diploid. Gametophytic self-incompatibility and inbreeding depression encourages outcrossing in nature and in breeding programs.

While diploids are frequent, triploids can occur spontaneously in crosses between diploids. Such triploids have larger leaves and fruit than their diploid relatives but are pollen sterile and cannot supply pollen for fertilization. Many popular cultivars ('Jonagold,' 'Mutsu') are triploids and prized for their quality and fruit size. Some breeders have tried to use triploids in breeding with mixed results (Sato et al. 2007).

There are 25–30 species of apple reported (Table 10.2). The four species native to North America formed distinct groups on the basis of simple sequence repeats (SSRs).

Table 10.1 Apple production (1,000 MT) in the world (data from http://faostat.fao.org)

	1986–1990	1996–2000	2005–2008
Americas	7,434	8,988	9,257
Asia	11,451	28,912	38,259
Europe	21,088	17,097	15,456
World	41,451	57,408	65,631

Table 10.2 Major species of *Malus* found in the northern hemisphere (Way et al. 1990)

Region	Common species found	
North America	*M. angustifolia*	*M. fusca*
	M. coronaria	*M. ioensis*
Europe	*M. florentina*	*M. sylvestris*
	M. pumila	
Asia Minor	*M. pumila*	*M. trilobata*
Himalaya	*M. sikkimensis*	
SW China	*M. prattii*	*M. yunnanesis*
SE China	*M. micromalus*	
Central China	*M. honanensis*	*M. hupehensis*
NW China	*M. kansuensis*	*M. sieversii*
N & NE China	*M. asiatica*	*M. prunifolia*
	M. baccata	
Taiwan	*M. doumen*	
Japan	*M. baccata*	*M. sieboldii*
	M. halliana	*M. tschonoskii*
	M. sargentii	
Korea	*M. asiatica*	*M. prunifolia*
	M. baccata	*M. sieboldii*
	M. micromalus	

Many *Malus* species have been used and continue to be used in breeding, with the increased recognition of the value of diversity and a means to study genes present in these relatives of cultivated apple.

1.3 Adaptation

Apples are grown in most temperate climates and they require a period of cold (temperatures below 45°F/7°C) to bloom and grow normally. For standard cultivars chill units of 500–1,000 are needed, while low chill cultivars require 400–600 h. Heat units are also needed.

Since several apple-producing areas require cultivars with low chilling hour requirements, research on low chilling has expanded (Labuschagne et al. 2001). Broad sense heritability values of 30% were calculated for total variation in number of buds sprouting and 62% for time of bud sprouting (Labuschagne et al. 2002, 2003). In areas of adequate winter chilling, cold hardiness is often a concern as is later blooming to avoid spring frosts. The issue of climate change, although controversial, is offering new challenges to apple producers worldwide, with more erratic climatic conditions, new pathogens and in some areas an increased frequency of hail. Heat tolerance and sunburn susceptibility are being investigated as major issues as is drought susceptibility.

2 Origin and Domestication of Scion Cultivars

2.1 Center of Origin

Although Malus species are found throughout the northern hemisphere, the center of origin of apple includes Asia Minor, the Caucasus, central Asia, Himalayan India and Pakistan and western China, areas where at least 25 native species of *Malus* occur. The Old Silk Road crossing from the Black Sea region to western China was important in the evolution of cultivated apple (Juniper et al. 1998; Zhou 1999; Luby et al. 2001) (Table 10.2).

Malus sieversii (Lodeb.) Roem. is thought by many to be the progenitor species of apple hybridizing with *M. prunifolia* Borkh., *M. baccata* Borkh., and *M. sieboldii* (eastern species) and with *M. turkmenorum* and *M. sylvestris* Mill. (western species). Selected cultivars from such random hybridizations were established and disseminated through grafting. There are reports of apples from 4,000 BC and later Roman authors documented apple culture. Apple cultivars being grown in Western Europe were cut off from their parental origins and evolved in relative isolation (Luby 2003).

However, research on chloroplast diversity raised interesting questions about the relationship of European wild apple *Malus sylvestris* and domesticated apple (Coart et al. 2006). A close relationship between the two was established by the existence of natural hybrids between the wild and cultivated forms and at the cytoplasmic level with the detection of eight shared chloroplast haplotypes.

North America became a "melting pot" not only for settlers but also for apple genetic diversity. Settlers brought apple seeds or grafts when they arrived, but many of these apples were not well adapted to the "new world." Settlers quickly established apple orchards for a source of apple cider, as the safety of drinking water was a concern. Settlers soon learned to propagate the seedlings best suited to the new climate. Meanwhile, individuals such as John Chapman 'Johnny Apple seed' disseminated apple seeds as new territories expanded to the west. Thousands of new cultivars were established and named (Luby 2003).

2.2 Domestication of Crop

Andrew Knight was the first documented apple breeder. The establishment of apple breeding programs worldwide often coincided with the establishment of research stations in growing regions that often partnered with universities. The history of apple is very rich and is detailed in both popular press and in the scientific literature.

Overviews of breeding and cultivar releases over time are found in an overview of the Brown and Maloney (2003), Laurens' (1999) review of breeding programs and objectives and in Knight et al. (2005) survey of breeding methodology and accomplishments.

3 Genetic Resources

The importance of conservation and characterization of germplasm is recognized worldwide (Büttner et al. 2004). Information from the large center of origin of apple increasingly is being published and studied. Zhou (1999) detailed apple genetic resources in China as Forsline et al. (2003) reviewed the collection, maintenance, characterization, and utilization of wild apples of central Asia.

3.1 Scion

The European Cooperative Program for Plant Genetic Resources has a Malus/Pyrus working group (http://www.wcpgr.cgiar.org/workgroups/malus_pyrus/malus-pyrus.html), and many of its members are in charge of their country's germplasm collection. The group represents a total of over 20,000 accessions in 13 countries. Büttner et al. (2004) evaluated the use of *Malus* germplasm in Germany, while Fischer and Dunemann (2000) used the collection to search for scab and mildew resistance. In Spain, Pereira-Lorenzo et al. (2008) evaluated local Spanish cultivars and used simple sequence repeat (SSR) markers for discrimination and to eliminate duplicates from the collection.

The US Department of Agriculture's Agricultural Research Service is responsible for the Clonal repository of Apples, Grapes and Tart Cherry in Geneva, NY. Researchers at this unit have been very active in the acquisition of materials from the center of origin (Central Asia) and in making material, especially *Malus sieversii*, available for study by researchers worldwide (Forsline et al. 2003). Volk and Richards (2008) detailed the availability of information on apple germplasm, including genotypic information, via the use of GRIN (Genetic Resources Information Network) database. In addition, for ex situ conservation of apple, a seed-based core collection for *Malus sieversii* has been established (Volk et al. 2005).

3.2 Rootstock

Rootstock germplasm includes many of the same *Malus* species used as sources of resistance in scion breeding. Some novel objectives include the use of apomictic species for seed production of clonal stocks, the selection for resistance/tolerance to specific environmental rigors of the root such as drought, water logging, salinity, nutrient deficiency, and other challenges.

3.3 Germplasm Diversity

Genetic diversity and population structure has been examined in *Malus sieversii*, a wild progenitor species of domesticated apple (Richards et al. 2008). Examination of almost 950 individuals from 88 half-sib families from eight *M. sieversii* populations from Kazakhstan revealed that differentiation was mostly congruent with geographical location. Among the eight collection sites there were two narrow and two broadly distributed clusters, with the southwestern collection sites more admixed and more diverse than the northern sites.

Korban's (1986) review of interspecific hybridizations documented early studies in this area and researchers continue to try to exploit genes from apple's wild relatives for everything from genes for resistance to studying drought, winter hardiness (Luby et al. 1999) and nutrient uptake. Many researchers are concentrating on genetic studies of *Malus sieversii* as the probable progenitor species of apple (Coart et al. 2003, 2006). *Malus orientalis* Uglitzk. Ex Juz. from Turkey and southern Russia is also under investigation (Volk et al. 2008).

The inbreeding concerns expressed by Noiton and Alspach (1996) remained the same or may be worse. The common progenitors of the past ('Golden Delicious,' 'Delicious' 'McIntosh,' 'Cox's Orange Pippin,' 'Jonathan') have declined in use as parents, only to be replaced by cultivars only one generation removed. Most related species of apples have been used in resistance breeding, with *Malus floribunda* prevalent in scab resistant material.

3.4 *Major Traits and Sources for Traits*

Sources of some traits are listed briefly, with more detailed information in the section on breeding and markers.

3.4.1 Quality

Improving quality is an objective for breeders worldwide, yet defining quality, quantifying quality and its components and minimizing environmental influence is a huge challenge. Apple juiciness, firmness, crispness, and aroma are crucial to quality but are complex characteristics that are affected by many environmental factors, making their study and improvement difficult. Contrasting instrumental tests with sensory perception is important, but equally complex and expensive. Trained taste panels are best, however most breeding programs find this extra cost prohibitive and rely on staff within the research program. Yet as advances are made in quality, new techniques are being developed and used to dissect components of these traits and identify significant marker–trait associations.

3.4.2 Apple Scab (*Venturia inaequalis*)

In breeding for resistance to apple scab x, breeders mainly focused on the V_f gene from *Malus floribunda* 821. Recently other sources of resistance are being targeted such as Russian seedling R1270-4A and *Malus sieversii*, especially for their prospects for pyramiding resistance and for understandings of race specificity especially susceptibility to races 2 and 4 of scab. The V_m from *Malus micromalus* Makino and *M. atrosanguinea* 804 (Spaeth) C. Schneider confers resistance to all but race 5 of scab. V_b from Hansen's baccata # 2 and V_{bj} from *Malus baccata jackii* Rehder have not been used extensively in cultivar development, but their resistance to other diseases is generating interest in their use. Numerous reports of additional sources of scab resistance provide many leads for future study (Gessler et al. 2006).

3.4.3 Powdery Mildew (*Podosphaera leucotricha*)

Sources of mildew resistance include Pl_1 from *Malus × robusta* Carr Rehd. and Pl_2 from *M. zumi* (Matsum.) Rehder (Knight and Alston 1968), *Plw* from White Angel' (Gallot et al. 1985; Batlle and Alston 1996), *Pld* from D12 a selection from open pollinated crabapple seed (Visser and Verhaegh 1976), and Pl_m, which has been eroded (Dayton 1977). Other sources of resistance have been identified (Fischer and Dunemann 2000; Schuster 2000), including quantitative resistance from U211 (Stankiewicz-Kosyl et al. 2005).

3.4.4 Fire Blight

Malus robusta 5 is the major source of resistance used in rootstock and scion breeding. Commercial cultivars with fairly good resistance include 'Delicious' and the scab resistant variety 'Liberty.'

3.4.5 Vitamin C and Antioxidants

Improving and documenting the nutritional components of apple cultivars is important yet also very complex. There are many publications on individual cultivars and several germplasm screens done in apple including Stushnoff et al. (2003) and Nybom et al. (2008a, b, c) that provide evidence of the wide range of variation for these compounds.

3.4.6 Red Pigmentation for Ornamentals and for Enhancing Antioxidants

Red pigmentation in apple flesh and foliage is derived primarily from *Malus pumila* var. *niedzwetzkyana* and its derivatives. Although a dominant gene for anthocyanin production was proposed, a deficiency of red plants is often noted. Highly pigmented cultivars, especially with red flesh, are of interest due to their ornamental and nutraceutical properties, with researchers in New Zealand emphasizing this as one of their breeding goals. The genetics, genomics and complexity of apple skin and flesh color is detailed more extensively in the section on genomics.

4 Major Breeding Achievements

4.1 Scion

Reviews of breeding and program objectives and achievements include Laurens (1999), Knight et al. (2005), Brown and Maloney (2003, 2004) and Gardiner et al. (2007), but a literature review of apple breeding or a scan of recent cultivar releases or commercialization is evidence of the ultimate accomplishment: new and improved varieties and knowledge to be used in the improvement process.

In reviewing the literature, advances in our knowledge of quality and how to measure it, the factors affecting of flesh browning and the complexity and intrigue of enhancing total or specific antioxidants is evident.

Despite the challenges, disease resistance breeding has widened in scope and added to our knowledge of sources of resistance, their location in the genome and their interactions. Interest in insect resistance has experienced a revival and this research area will advance our ability to produce fruits suitable to the organic market.

Food safety has become important and apple allergens are one aspect of this topic. Apple allergens have been identified and their mode of action and type have been confirmed, the effect of processing on the different allergens has been studied, and the location of these allergens have been documented. This knowledge has largely been collected over the last 10 years and it promises to aid breeders in the choice of parents and breeding strategies in developing low allergen apples.

4.2 Rootstock

Reviews of apple rootstocks and their breeding include those by Webster and Wertheim (2003), Cummins and Aldwinckle (1983), and Ferree and Carlson (1987). Dwarfing, induction of precocity, disease resistance (*Phytophthora,* fire blight, scab), and climatic adaptation remain important goals. Breeding programs and releases include Ag-Canada in Quebec (Khanizadeh et al. 2000), Japan (Soejima et al. 2000), the 'Supporter' series of rootstocks from Pillnitz, Germany (Fischer 2001), and rootstocks developed in Poland (Jakubowski and Zagaja 2000).

Among the best-known rootstock breeding programs, the East Malling program in England has continued the tradition of releasing rootstocks with some recent releases that have been patented in the USA. Attributes of the joint Cornell/USDA apple breeding program are detailed in Robinson et al. (2003) and Fazio et al. (2006b).

5 Current Goals and Challenges of Breeding

One of the biggest challenges for breeding programs is to find funding to maintain active programs of sufficient capacity to develop and maintain large populations for varietal development. Funding to conduct sufficient phenotyping and genotyping to develop robust markers for marker-assisted breeding is also needed.

5.1 *Disease Resistance Breeding*

European researchers have taken a lead in the testing of organic apple production and all the complexities involved. Learning more about larger scale organic production will help in devising strategies to produce apple varieties with suitable resistance and fruit quality so that reduced sprays are a reality (Weibel and Haseli 2003).

5.2 *Low Allergenicity Apples*

Tremendous progress has occurred in advancing our knowledge of allergens in apple. Four major allergens in apple have been identified, researched and mapped: Mal d 1, Mal d 2, Mal d 3, and Mal d 4 (Gao et al. 2005a; 2005b; 2005c). Lipid transfer proteins (LTP) have been implicated in fruit allergies. When over 80 cultivars from two countries were evaluated for LTP there was about a 100-fold difference in LTP among cultivars (Sancho et al. 2008). Cultivars with low levels of Mal d 1, previously designated as low allergenic, did not always have low levels of LTP. LTPs need to be tested and confirmed using oral challenges before discussions of allergenicity can be made.

5.3 *Processing and Fresh-Cut Markets*

Apple cultivars suited to the fresh-cut market are needed; not just nonbrowning apples, but fruit that maintain firmness, are not prone to microbial growth and has no flavor change with time. The fresh-cut industry offers convenience to consumers, but the producers have food safety concerns. Toivonen and Brummell (2008) reviewed the biochemical bases of appearance and texture changes in fresh-cut fruits and vegetables.

5.4 Rootstocks

Resistances to biotic and abiotic stresses continue to be a primary goal of apple rootstock breeding. Rootstock induced reduction of plant vigor not only is important culturally but also holds promise of elucidating scion-stock interactions. Replant disease remains a complex but real problem, with several new stocks performing well in old orchard sites with this disorder. Many traits related to plant propagation or orchard performance are objectives, but some can be negatively correlated. One example is that while breeders are selecting against tendency for burr knot production, reduction or elimination of this trait can make propagation and rooting of liners more difficult. The development of rootstock maps should allow marker-assisted breeding to become a reality in rootstocks.

6 Breeding Methods and Techniques

6.1 Major Traits and Selection Techniques

Future challenges associated with adaptation involves climate change, with the resulting introduction of diseases and insects associated with warmer climates, changing bloom times and the chance for frost with earlier blooming. The incidence of hail is also increasing; making the use of hail nets a prospect for regions that have never needed protected cultivation.

Research on quantifying bud break, prolonged dormancy and chilling requirement is critical to understanding and producing apple trees for low-chill regions. Progress in this area is evident (Labuschagne et al. 2001, 2002, 2003).

6.1.1 Disease Resistance

Disease resistance ideally encompasses more than one resistance, as a scab resistant apple will still be prone to powdery mildew and fire blight. However, as the number of resistances increases, so does the challenge of obtaining commercially acceptable fruit quality. With climate change, comes the introduction of new pathogens in regions formerly inhospitable to their spread. Partial resistances or field resistance to diseases are also being targeted to provide producers with a less intense spray strategy. However, several groups are still investigating how to screen for partial resistance and what this will mean under commercial production. Fischer and Fischer (2008) provided an excellent overview of the challenges of trying to incorporate multiple resistance yet obtain quality. Despite these challenges there are many recent scab resistant apples in the industry that are well received by consumers.

6.1.2 Apple Scab (*Venturia inaequalis*)

An excellent review of apple scab and genes for resistance was published by Gessler et al. (2006). The most widely used gene V_f, from *Malus floribunda* 821 is the most likely to erode. Thus it is important to use other genes such as genes in the Russian Seedling R12740-7A (Hemmat et al. 2002; Bus et al. 2005a, b) and the V_m from *Malus micromalus* and *M. atrosanguinea* 804 which confers resistance to all but race 5 of scab. V_b from Hansen's baccata # 2 and V_{bj} from *Malus baccata jackii* have not been used extensively in cultivar development, but their resistance to other diseases is generating interest in their use. Recently Soufflet-Freslon et al. (2008) reported on a new gene and QTLs from 'Dulmener Rosenapfel' that confer resistance to scab.

6.1.3 Fire Blight

Sources of resistance include Robusta 5 and wild *Malus* species including *Malus sieversii* (Fazio et al. 2006a). Unfortunately many of the new popular cultivars are very susceptible. Breeding for resistance to fire blight is more challenging than for other pathogens as there is differential resistance which is hard to measure as the environment and the growth status of the plant can impact the screening procedures. Furthermore since different strains are being used by different researchers, the results are difficult to compare. There are both shoot and blossom infection in scions and scion infection can travel to the rootstock.

6.1.4 Powdery Mildew (*Podosphaera leucotricha*)

Pl_1 from *Malus × robusta* Carr Rehd. and Pl_2 from *M. zumi* (Knight and Alston 1968), *Plw* from White Angel' (Gallot et al. 1985; Batlle and Alston 1996), *Pld* from D12 a selection from open pollinated crabapple seed (Visser and Verhaegh 1976), and Pl_m (Dayton 1977) have been used in breeding. Other sources of resistance have been identified (Fischer and Dunemann 2000; Schuster 2000), including quantitative resistance from U211 (Stankiewicz-Kosyl et al. 2005). Bus (2006) used a partial diallel design used to study resistance in six apple progeny and found that parental performance was not a good indication of the performance of its progeny.

There is a concern about the potential erosion of major genes for resistance due to the breakdown of Pl_2 observed in France (Caffier and Parisi 2007; Caffier and Laurens 2005). Sources of resistance other than Pl-2 [*M. hupehensis* (Pampan.) Rehder,

M. mandshurica (Maxim) V. Komarov, *M. robusta* Rehder, *M. sargentii* Rehder, *M. sieboldii* (Regel) Rehder, D12, Mildew Immune Selection, and 'White Angel' remained resistant to the virulent population. The combination of Pl-2 with quantitative resistance genes resulted in a high level of resistance (Caffier and Parisi 2007).

6.1.5 Valsa Canker (*Valsa ceratosperma*)

An inoculation protocol developed by Abe et al. (2007) was effective in screening for resistance to Valsa canker and this research identified *M. sieboldii* as having a high level of resistance that was effective against several different isolates.

6.1.6 Alternaria

Resistant cultivars are homozygous for the recessive gene *alt alt*. Apple cultivars reported to have resistance include: 'Indo,' 'Red Gold,' 'Raritan,' 'Delicious,' 'Fuji,' 'Golden Delicious,' 'Ralls,' 'Toko,' 'Tsugaru,' 'Mutsu,' 'Jonagold,' and 'Jonathan' (Sawamura 1990). Another source of resistance is a Korean cultivar *M. bacatta* cv. Jeongsean (Heo et al. 2006).

6.1.7 National Variety and Rootstock Testing

In the USA, regional projects such as the NE183 "Multidisciplinary evaluation of apple varieties," similar to the NC 140 apple rootstock trials, have added to our knowledge of rootstocks, rootstock/scion interactions and scion varieties (Miller et al. 2005). The NE-183 trials also had separate plantings for insect and diseases studies. There have been studies on cultivar susceptibilities to some of the less studied diseases, such as *Colletotrichum acutatum* (Biggs and Miller 2001). The EUFRIN (European Fruit Research Institutes Network) has an extensive network of multisite tests to trial new scion cultivars and clones (Stehr 2009).

6.1.8 Insect Resistance

Aphids: Stoeckli et al. (2008) identified molecular markers associated with QTLs for resistance to the rosy apple aphid (*Dysaphis plantaginea* Passerini) and the leaf curling aphid (*Dysaphis* cf. devecta) and confirmed the presence of resistance alleles in cultivars like 'Wagener' and 'Cox's Orange Pippin' that have been reported to confer resistance to their progeny.

Miñaro and Depena (2008) evaluated tolerance of some scab-resistant apple cultivars to the rosy apple aphid (RAA), *Dysaphis plantaginea*, a major apple pest.

The use of tolerant cultivars would contribute to nonchemical crop protection. The susceptibility of nine scab-resistant apple cultivars to RAA was evaluated in greenhouse trials and field observations conducted over 2 years. Significant differences were observed among cultivars in aphid abundance and damage level 21 days after an infestation in the greenhouse. 'GoldRush' and 'Galarina' were considered tolerant, and 'Jonafree' and 'Redfree' were highly susceptible.

Woolly apple aphid (*Eriosoma lanigerum* Hausm.) resistance is an important breeding objective, especially in rootstocks. The *Er1* and *Er2* genes derived from 'Northern Spy' and 'Robusta 5,' respectively, are the two major sources used. The gene *Er3*, from 'Aotea 1' (an accession classified as *Malus sieboldii*), is a new major gene for WAA resistance. Genetic markers linked to the *Er1* and *Er3* genes were identified by screening RAPD markers across resistant and susceptible DNA bulks. The closest RAPD markers were converted into sequence-characterized amplified region (SCAR) markers and Er1 and Er3 were assigned to LG 08 of 'Discovery,' while the *Er2* gene was mapped on LG 17 of 'Robusta 5.' Markers for each gene were validated for their utility for marker-assisted selection in separate populations (Bus et al. 2007).

Germplasm screens for resistance to plum curculio, were not promising for genetic solutions to this problem (Meyers et al. 2007), yet tests for resistance to apple maggot (Meyers et al. 2008) have yielded some interesting sources of resistance to investigate further.

6.1.9 Plant Architecture

Groups of researchers are collaborating to advance research on plant architecture and modeling. This has resulted in more in-depth studies of branching and genotypic/phenotypic difference (Costes et al. 2006; Lauri et al. 2008; Kenis and Keulemans 2007; Segura et al. 2007). An emphasis has been placed on the geometry of plant architecture but also the topography. QTL analysis for complex architectural traits was conducted using progeny of 'Starkrimson' × 'Granny Smith' (Segura et al. 2007). This research across disciplines and germplasm promises to advance our understanding of plant architecture and its manipulation.

6.1.10 Columnar

Combining columnar habit with resistance to apple scab has been a goal of several programs and has resulted in the release of several scab resistant columnar apples in Romania (Braniste et al. 2008) and in Latvia (Ikase and Dumbras 2004). The release of nonresistant columnar releases has also accelerated.

6.1.11 Genetic Parameters

Genetic studies have also increased in apple. Tancred et al. (1995) studied the inheritance patterns and heritability of ripening date and suggested that determining the mean harvest of the two parents was a good way to predict offspring harvest date, while fruit shape was found to have a narrow sense heritability of 0.79 (Currie et al. 2000). The effects of a recurrent selection program was examined by Oraguzie et al. (2001), who also looked at the heritability of fruit quality in open-pollinated families and found that heritability estimates of harvest date and fruit weight were high (>0.70) but sensory traits were moderate; with russet 0.34–0.54 and firmness 0.26–0.59 (Alspach and Oraguzie 2002). Softening has been the subject of recent studies by Iwanami et al. (2005; 2008), who suggest that at least two harvest dates are needed in studying apple softness.

The presence of an open calyx in disease resistant apples is a large concern due to secondary pathogens entering the core early in fruit formation. The incidence of core rot may be low (1–3%), but is enough for fruit to be rejected due to contamination concerns.

Sensory and consumer testing have helped provide a better understanding of what consumers want (Harker et al. 2008). New instruments will aid our ability to quantify important components of apple quality. Collaborative research efforts on an international scale will also further progress.

6.1.12 Harvest Determination

A generic starch iodine chart is a simple way to assess the stage of maturity (Blanpied and Silsby 1992). Although not all cultivars have harvest stages that correspond to the iodine staining, it provides breeders with a quick and low-cost measure of relative stage of maturity or staining. Several harvest dates should be assessed to best judge the recommended harvest maturity and quality at those dates. As selections advance through trials more detailed measurements of maturity, involving ethylene production would be helpful.

Fruit softening and its challenges: Fruit softening is important to breeders, producers and consumers, but it seems the more we learn, the more questions we have and the more clues we obtain on this complex phenomenon. Knowledge of ACS (1-aminocyclopropane 1-carboxylate synthase gene) and ACO (1-aminocyclopropane 1-carboxylic acid oxidase gene) have added to our understanding of the importance of ethylene and the many steps involved in its production and perception, but more genes are being implicated as important components of softening. Transgenic studies, functional analyses and the testing of markers associated with ACS and ACO across a wider range of populations will enhance our understanding of genotypic differences.

1-methylcyclopropene (1-MCP), marketed as Smartfresh, blocks the effect of ethylene on apples. Its use is adding to our knowledge of cultivar variation in response to this treatment as well as the effect of ethylene on volatiles that contribute to perception of flavor.

6.1.13 Storage Disorders

Our ability to understand genetic susceptibility to storage disorders will be aided by advances in genomics and by the use of well-characterized populations. The use of parents susceptible to specific disorders, such as 'Honeycrisp' and its susceptibility to bitter pit, soft or ribbon scald and rots will add to our knowledge of the inheritance of such disorders. Resistance to superficial scald continues to be the focus of several groups, especially in relation to alpha-farnescene.

A better understanding of various rots and chilling disorders is also needed.

Fruit mineral nutrition: In the mineral nutrition of apples, the site is important for evaluation of some nutrients (Volz et al. 2006). It is best to select across sites and seasons to assess susceptibility to bitter pit. While fruit calcium might be a useful means of indirect selection for bitter pit susceptibility, this worked within, but not among, families (Volz et al. 2006). Korban and Swiader (1994) suggested that two dominant genes were responsible for resistance to bitter pit, but this finding needs additional confirmation.

Sensory testing and understanding consumer preferences and satisfaction: The importance of quality, sensory testing, and obtaining a better understanding of consumer perceptions and preferences has expanded greatly over the last few decades. Means of quantifying components of fruit quality, such as firmness (Harker et al. 1996), texture (Harker et al. 2002a), and sweetness and acidity (Harker et al. 2002b) have been researched extensively. Such studies have contrasted trained sensory panelists and objective measurements of Brix and acid. Titratable acidity was the best predictor of acidity and values need to differ by 0.08% titratable acidity to be perceived. For Brix, sensory panels were only able to detect a difference with a change of more than 1°Brix. Sensory panels were recommended for differentiation of sweetness and flavor. Harker et al. (2008) determined that increasing firmness usually was associated with increased preference although some people prefer soft apples. Higher Brix and acidity can improve preference for apples that are firm, but not if they are soft. Improved protocols for quality evaluations and sensory testing are needed, although great progress has been made in this area.

It is best to evaluate fruit softening after at least two harvest dates for obtaining a genotypic mean for softening (Iwanami et al. 2005). Iwanami et al. (2008) obtained narrow sense heritability for postharvest fruit softening. Softening rate as measured by parent–offspring regression was high ($h^2=0.93$), but as estimated by sib analysis it was only moderately high ($h^2=0.55$).

6.1.14 Physiological Studies Coupled with Genetic Investigations

Physiology is a focus as researchers examine some of the challenges in apple production. Fruit thinning, fruit set, and abscission are being evaluated. Apple germplasm with different rates of abscission varied for internal ethylene concentration by three orders of magnitude (Sun et al. 2009). A genomic study of shade-induced apple abscission revealed 66 unique genes involved and suggested that better methods of thinning might be a future outcome (Zhou et al. 2008).

6.1.15 Storage and Storage Disorders

Blazek et al. (2007) studied cultivars and selections over a 3-year period as to characteristics related to good storage and freedom from storage diseases. An ideotype was proposed and important characteristics identified. Thresholds were established for some of the parameters; higher skin toughness and thickness, low ethylene production, naturally high calcium content, high total phenolics and antioxidants, high flesh firmness, and high fruit acidity as expressed by pH. Inoculation with one of the bitter rot pathogens, *Pezicula alba* Guthrie was suggested as a final screening on selections with the ideal ideotype.

6.1.16 Health and Antioxidant Research

Reviews by Boyer and Liu (2004) and Biedrzycka and Amarowicz (2008) illustrated the importance and complexity of antioxidants in apple. Assessing apple germplasm collections for antioxidants were the focus of studies by Stushnoff et al. (2003) and Nybom et al. (2008a, b, c). The complexity of antioxidants in apple has generated controversy over which phenolics are most bioavailable or the best to target for improvement (Lee et al. 2003). Eberhardt et al. (2000) suggested that certain apple phenolics had antioxidants equivalent to 1,500 mg of vitamin C and were more important to target for improvement. Lata (2008) stressed the importance of site and season on efforts to quantifying antioxidants in apple. Khanizadeh et al. (2007) reported polyphenol composition and total antioxidant capacity of selected apple genotypes for processing. Davey and Keulemans (2004) and Davey et al. (2006) has concentrated on the importance of vitamin C, including QTL studies. Planchon et al. (2004) explained that some of the variability in vitamin C content was due to sampling.

6.2 Breeding methodology

Methodology in apple breeding has been reviewed in different chapters of Moore and Janick's "Methods in Fruit Breeding" (1983), in Janick et al. (1996) and breeding programs worldwide were reviewed in Brown and Maloney (2003). Unfortunately, very few of the studies each breeder routinely makes within their program are published and those that are published may be hard to find as they published in many different journals.

6.2.1 Rootstock Propagation

While propagation of seedlings onto rootstocks provides an estimation of performance on clonal stocks, the added cost and record-keeping has some of the largest apple breeding programs preferring to plant seedlings on their own roots and fast-tracking promising selections by doing rapid propagation of promising selections after one or a few years of fruiting. This reduces the costs substantially. However programs that have partnered with nurseries that provide propagation have an advantage and lower costs. The Cornell program has found that optimizing seedling growth in the first several years enables us to have fruit in many progenies 4 years after planting without the added expense of rootstocks.

6.2.2 Parental Selection

In planning crosses, attention must be paid to the parents carrying recessive genes for pale green lethal, genetic dwarfs and also sublethals (Alston et al. 2000; Gao and van de Weg 2006), as these will greatly reduce the number of usable progeny obtained. Crosses of heterozygotes for these traits will reduce populations by 25% for each trait.

Although there are still gaps in our knowledge of S-alleles in apple, fully compatible or semi-compatible matches should be targeted when possible (Broothaerts 2003; Matsumoto et al. 2007; Nybom et al. 2008a). The use of markers, especially S-alleles, to verify proposed parentage, has resulted in the discovery of quite a few faulty pedigrees. Surprisingly, even the seed parent of crosses has been in error.

6.2.3 Pollen Collection, Emasculation, Pollination and Fruit Set

There are cultivars or selections that are very sensitive to emasculation, perhaps causing abscission from the wounding that occurs. Other selections/cultivars may have pistils with curled styles that are injured during the emasculation process. No more than two flowers per cluster are recommended for pollination as greater numbers usually result in some of them abscising. To avoid contamination between crosses, 70% alcohol should be used to kill pollen on any surfaces used in pollination such as fingers or brushes.

Screening methods have been established for many of the more problematic pathogens (scab, mildew, rusts, fire blight and for the rootstock pathogens such as Phytophthora species). Breeders must ensure that they know the races they are using in inoculations.

The ability to apply preselection to breeding populations has long been a goal, but screening for disease resistance has been the primary application. Slowly molecular markers are starting to be used, but while markers are being developed they still need to be tested for their validity and their robustness. Markers must be tested in different genetic backgrounds and the populations where they are used should be maintained to test for any juvenile/adult interactions.

6.2.4 Testing and Replication

Many programs have their own systems of replication number and testing strategies. A study on replication in the initial selection trials of clonally propagated crops suggests that any increase in trial area for initial selection is best used for increasing the number of genotypes tested and growing just one plant per genotype (Aikman and Langton 1983). How many at each stage often is a function of funding. Some programs have research stations willing to act as test sites, providing valuable performance information to the breeder without substantial costs. Other programs rely on testing with cooperative growers, but this system is not without its risks of lost data. There are programs that offer advanced testing for a fee.

6.2.5 Record Keeping

There are almost as many record keeping systems, as there are breeding programs. Although a universal system would be highly desirable, each breeder has different priorities and systems of evaluation. While effective phenotyping is important in breeding and in genomic research, the reality of many breeding programs is that efficiency in breeding often requires few detailed records on individual seedlings, but more detailed records and phenotyping on promising individuals in that cross. Genetic studies can be detailed in records, but if each population was thoroughly characterized, breeders could not go through the large number of seedlings that need to be evaluated to discern the few desirable segregants.

6.2.6 Statistics

The use of unbalanced designs common to most fruit breeding programs was addressed by (Durel et al. 1998). Genetic parameters (narrow-sense heritabilities and genetic correlations) were estimated for major traits in apple using large unbalanced data sets, aided by the use of wide-pedigree information. The software REML VCE took into account the complex pedigrees of the apple-breeding populations, by combining the restricted maximum likelihood procedure with the construction of the entire relationship matrix between hybrids planted in the field and their ancestors. Narrow-sense heritability estimates ranged from 0.34 to 0.68 for traits exhibiting a normal distribution. Heritability values (~0.35–0.40) were obtained for fruit size, texture, flavor, juice content, attractiveness, and russeting. Higher values of heritability were obtained for vigor, as assessed by trunk circumference (0.51) and powdery mildew resistance (0.68). Additive genetic correlations between traits were estimated and showed a very high relationship between fruit-quality traits.

6.2.7 Ploidy Manipulation

A comparison among reciprocal diploid × triploid crosses suggested that 2x × 3x (but not 3x × 2x) can be used in apple breeding, with trunk circumference index (circumference relative to average circumference of diploid progeny of the same age) used as an early indicator of whether the seedlings will flower and bear fruits (Sato et al. 2007).

In any breeding program, the establishment of clear objectives per cross and and culling thresholds and/or agreement on limiting factors for discarding selections need to be made.

6.2.8 Propagation and Release of Varieties

With the advent of the 'Pink Lady' model of exclusive licensing and trademarking to ensure quality and control demand, the apple industry entered into the era of controlled management of new varieties. The following are some of the cultivars marketed under some type of exclusive or controlled management system, often with production royalties: 'Pink Lady,' 'Pacific Rose.' 'Jazz,' 'Delblush,' 'Ambrosia,' 'Sonya,' 'Cameo,' and 'SweeTango' (MN 1914). The scab resistant cultivars 'Ariane' and 'Juliet' are also controlled and trademarked.

Innovative partnerships among breeding program and nurseries have resulted in multisite testing, providing important information on genotype by environment interactions. One example is the company Novadi in France that partners with breeding programs and nurseries in the pursuit and testing of new cultivars (Laurens and Pitiot 2003).

7 Integration of New Biotechnologies

7.1 State of the Map(s)

Tremendous progress has been made in map construction since the first maps of apple ('White Angel' × 'Rome Beauty') were published (Hemmat et al. 1994). Currently maps of differing marker density are available for at least 50 scion cultivars: including 'Prima' × 'Fiesta' (Maliepaard et al. 1998; Liebhard et al. 2003b).

A linkage map of the columnar, reduced branching mutation, 'Wijcik McIntosh' was constructed along with maps of two scab resistant selections by Conner et al. (1998), followed by maps of 'Braeburn' × 'Telamon,' a columnar genotype (Kenis and Keulemans 2005), and 'Fiesta' × 'Totem' (a columnar genotype) (Fernandez-Fernandez et al. 2008).

Maps of 'Delicious' and 'Ralls Janet' were constructed from progeny of each parent crossed with 'Mitsubakaido' (*Malus sieboldii*) as the pollen parent for each. (Igarashi et al. 2008). N'Diaye et al. (2008) developed a consensus map using four different populations ('Discovery' × TN 10-8, 'Fiesta' × 'Discovery,' 'Discovery' × 'Prima,' and 'Durello di Forli' × 'Prima'). Additional mapping populations are being developed and used for fine scale mapping, synteny studies, and in attempts to locate and clone genes of interest. Populations include 'Royal Gala' × A689-24 in New Zealand, several for physical map construction and in Italy, eight populations are being used.

Maps have also been constructed for three rootstock cultivars/selections (Malling 9, Robusta 5, Ottawa 3) (Celton et al. 2009; Fazio, personal communication).

7.1.1 Marker Development

Evolution of marker use in apple mirrors that in many plants, starting with isozymes, then RAPDS (random amplified polymorphic markers), RFLPs (restriction fragment length polymorphic markers), AFLPs, SCARs (sequence characterized amplified region) and progressing to SSRs (simple sequence repeats) and SNPs (single nucleotide polymorphisms).

7.1.2 Simple Sequence Repeats

Over 300 simple sequence repeats (SSRs) have been developed and tested in apple (Liebhard et al. 2002; Silfverberg-Dilworth et al. 2006) and a set of recommended SSRS to be used in PCR multiplex was recently released (Patocchi et al. 2008).

7.1.3 Universal Primers in the Rosaceae

In an effort to develop more "universal" markers across the Rosaceae, Sargent et al. (2008) contrasted *Malus* cDNA sequences with homologous *Arabidopsis* sequences to identify putative intron–exon junctions and conserved flanking exon sequences. Primer pairs were designed from the conserved exon sequences flanking predicted intron–exon junctions. Eleven loci polymorphisms in 'Fiesta' × 'Totem' mapped to seven LGs. 38% of these genes were successfully mapped in *Fragaria* and *Prunus* revealing some patterns of synteny across genera. Similarly, Gasic et al. (2009) developed markers from apple ESTs that were then tested on 50 individual members of the Rosaceae, representing 3 genera and 14 species). They found that transferability ranged from 25% in apricot to 59% in the more closely related pear.

After analyzing over 350,000 EST sequences in apple, a set of 93 new markers was mapped in apple that coded for 210 single nucleotide polymorphisms (SNP) (Chagne et al. 2008). This demonstrates the potential for SNP discovery and utilization in apple. Several research groups in the USA and Italy are developing additional SNPs and pursuing pedigree based association studies (Oraguzie et al. 2007a).

7.2 *Traits Marked with Molecular Markers*

7.2.1 Scab Resistance Genes

Many groups have developed markers for V_f, (reviewed in Gardiner et al. 2007). Markers for V_r, V_x, V_{h2}, V_{h4}, (Bus et al. 2005b; Hemmat et al. 2002), Vh_8 (Bus et al. 2005a), V_m from *Malus micromalus* and *M. atrosanguinea* 804 (Cheng et al. 1998; Patocchi et al. 2005), V_b (Erdin et al. 2006), V_{bj} from *Malus baccata* jackii (Gygax et al. 2004), V_a from 'Antonovka' (Hemmat et al. 2003) have also been developed.

Some clones of 'Antonovka' possess the V_a gene, but some clones only transmit polygenic inheritance (Quamme et al. 2003). QTLs for scab resistance were reported by Liebhard et al. (2003c), Calenge et al. (2004) and Schouten and Jacobsen (2008).

7.2.2 Powdery Mildew (*Podosphaera leucotricha*)

Markers exist for *Plw* from 'White Angel' (Evans and James 2003), *Pld* from an open pollinated crabapple selection (James et al. 2004), Pl_1 from *Malus robusta* (Markussen et al. 1995; Dunemann et al. 2007), Pl_2 from *Malus zumi* (Dunemann et al. 1999), and quantitative resistance from clone U211 (Stankiewicz-Kosyl et al. 2005). Field studies conducted over 4 years have identified some stable and unstable QTLs for mildew resistance (Calenge and Durel 2006).

7.2.3 Fire Blight

The marker CHO3E03 was useful for resistance from *Malus robusta* 5 (Peil et al. 2007a) but it needs to be tested on populations with other resistance donors to assess its utility. Peil et al. (2007b) established strong evidence for this fire blight resistance gene from *Malus robusta* location on linkage group 3. A major QTL for resistance was identified on LG 7 of 'Fiesta' in two progenies and four minor QTLs were also found on LG2 3, 12 and 13 (Calenge et al. 2005). Several significant digenic interactions were also identified, suggesting putative epistatic QTLs. Two distinct major QTL for fire blight were found to colocalize on linkage group 12 in apple genotypes 'Evereste' and *Malus floribunda* clone 821, carrying distinct QTL alleles at that genomic position (Durel et al. 2009).

7.2.4 Alternaria

Markers linked to Alternaria blotch resistance (Soejima et al. 2000) and susceptibility (Heo et al. 2006) have been reported. More testing of these markers is required.

7.2.5 Columnar

Columnar or reduced branching apples provided breeders with a means to study the genetics of plant form. The columnar trait is dominant, but there is usually a deficiency of columnar types in the progenies studied. Hemmat et al. (1997) found a DNA marker for columnar growth habit that contained a simple sequence repeat. Conner et al. (1997) developed a genetic linkage map for the source of columnar, 'Wijcik McIntosh,' and conducted a QTL study on its effect (Conner et al. 1998).

A population of standard 'Fuji' by the columnar genotype 'Tuscan' was used to identify RAPD markers linked to the columnar gene. From the closest RAPD marker, a SCAR (sequence characterized amplified region) marker was developed. This marker produces a 670 bp product in columnar material that is absent in non-columnar plants (Kim et al. 2003). Next, a population of 'Spur Fuji' × 'Telamon' allowed Tian et al. (2005) to map the *Co* gene between the SSR markers CH03d 11 and COL on linkage group 10. The region around the *Co* gene was constructed using nine new markers and three markers developed earlier. Inter-simple sequence repeat (ISSR) markers were used by Zhu et al. (2007) in an effort to find markers closer to the Co gene, but although additional markers were mapped the closest was 10 cM away.

QTL studies have also targeted some populations with one columnar parent in an effort to learn more about the effect of the Co gene on branching and other components of plant architecture (Kenis and Keulemans 2007; 2008).

Increasingly, derivatives from 'Wijcik McIntosh,' such as 'Telamon' and others are being used in genetic studies of architecture and branching. There is a great degree of variation in columnar form in different clones heterozygous for the columnar gene.

7.2.6 Dwarfing Genes

Pilcher et al. (2008) used Celton et al.'s (2009) mapping population of 'Malling 9' × 'Robusta 5' to map markers associated with dwarfing genes. Markers need to be tested on other rootstock populations and also on scion material to see how the markers perform in populations with different genetic backgrounds.

7.2.7 S-Incompatibility

This area of marker research is readily applicable and has aided breeders in their design of crosses and has also indicated where parentage is not as documented. While more *S* alleles need to be resolved due to high homology with existing *S*-alleles, the progress in this area has been excellent (Broothaerts 2003; Matsumoto et al. 2007). When Nybom et al. (2008a) evaluated a collection of cultivars in Sweden they found that five alleles, *S1–S3*, *S5*, and *S7*, had frequencies ranging from 11 to 18%, whereas the remaining 9 alleles were below 6%. Additional studies have revealed the need for studies on S alleles outside of *Malus domestica*.

7.2.8 Softening (ACS, ACO, and Ethylene)

Research on ACS (1-aminocyclopropane-1carboxylate synthase gene) has evolved rapidly. In 2000, Harada et al. identified an allele associated with low ethylene production in apple cultivars. Later, an allelotype of a ripening-specific

1-aminocyclopropane-1-carboxylate synthase gene was found to define the rate of fruit drop in apple (Sato et al. 2004). Oraguzie et al. (2007b) studied the influence of Md-ACS1 allelotype and harvest season within an apple germplasm collection on fruit softening during cold air storage, finding that Md-ACS1-2/2 allelotypes had a slower rate of softening than the other genotypes. In another study, the amount of MdACO transcripts in seeds was found to be a good indicator of abscission following benzylaminopurine application (Dal Cin et al. 2007).

Genotyping of Md-ACS1 and Md-ACO1 for parents and their suitability for marker-assisted selection was assessed by Zhu and Barritt (2008) who found only 8 of 95 cultivars homozygous for ACS-2 or AC0-1. Such homozygotes had firmer flesh at harvest and after 1 month storage at 0°C. The eight homozygotes included four breeding selections and 'Delblush,' 'Fuji,' 'Pacific Beauty,' and 'Sabina.' Later, characterization of cultivar differences in alcohol acyltransferase and 1-aminocyclopropane-1carboxylate synthase gene expression and volatile compound emission during apple fruit maturation and ripening was evaluated (Zhu et al. 2008a).

Nybom et al. (2008b) determined that modern apple breeding is associated with a significant change in the allelic ratio of ethylene production gene Md-ACS1 with a shift towards the allele associated with less ethylene production.

7.2.9 Flavor (Volatiles)

Improvement of apple flavor by breeding or biotechnology is a complex problem and has many challenges (Brown 2008). Research on QTL mapping of aroma compounds (Dunemann et al. 2009) is an important first step, as is the discovery via genomics that showed that aroma production in apple is controlled by ethylene primarily at the final step in each biosynthetic pathway (Schaffer et al. 2007). Rowan et al. (2009) also examined volatiles in a 'Royal Gala' × 'Granny Smith' cross and used principal component analysis to discriminate progeny as to level and type of esters. Dunemann et al. (2011) found that functional diversity of the alcohol acyl-transferase gene (MdAAT1) was associated with fruit ester volatile content in apple cultivars.

7.3 MAS (Seedling and Parental Selection)

A challenge to developing effective molecular markers for marker-assisted breeding is accurate phenotyping and the funding to conduct such phenotyping on sufficient individuals and populations. While many molecular markers have been developed and identified, few programs use them extensively due to cost, lack of funds, or inadequate knowledge about the robustness of these markers. Apple breeders need to have markers that are easy to use and inexpensive. There is a clear need for better, high throughput DNA extractions and multiplexing would be an advantage (Patocchi et al. 2008).

Breeders must assess if the use of markers is both effective in breeding and cost effective (Luby and Shaw 2001). Pyramiding several genes for resistance is one example where markers would be very useful and the objective would very difficult to achieve readily without markers. The ability to differentiate sports by use of markers and to identify the mechanisms responsible for their activation would also be invaluable to nurseries interested in intellectual property right protection and to breeders in assessing how to manipulate key traits.

QTL studies in disease resistance have been highlighted in the section on disease resistance, but QTL studies in other areas have also progressed. Conner et al. (1998) study of QTL of tree growth and development has been followed by QTL studies of fruit texture and firmness (King et al. 2000; 2001), physiological attributes (Liebhard et al. 2003a), QTLS for plant form and fruit quality (Kenis and Keulemans 2007; 2008), aphid resistance (Stoeckli et al. 2008), vitamin C (Davey et al. 2006), and plant architecture (Segura et al. 2007).

Association mapping and linkage disequilibrium are both challenges and advances in our approach to understanding the apple genome (reviewed in Oraguzie et al. 2007a). Pedigree assisted breeding is starting to be utilized across several programs which will help understand its use in breeding programs and genetic studies (van de Weg et al. 2004).

7.4 Genomics

Gardiner et al. (2007) reviewed genomics research in apple. The database availability of expressed sequences has accelerated genomic (Newcomb et al. 2006; Park et al. 2006; Wisniewski et al. 2008), microarray (Lee et al. 2007; Pichler et al. 2007), and functional genomics studies (Janssen et al. 2008) in apple.

Over 150,000 expressed sequence tags in apple collected from 43 different cDNA libraries, representing 34 different tissues and treatments, were analyzed by Newcomb et al. (2006). Clustering of these sequences resulted in a set of 42,938 nonredundant sequences (17,460 tentative contigs and 25,478 singletons), representing about one-half the expressed genes from apple. Park et al. (2006) used a more targeted approach to large-scale statistical analysis of expressed sequence tags by targeting biochemical pathways for precursors to volatile ester production to identify genes with potential roles in apple fruit development and biochemistry.

Lee et al. (2007) used a microarray from young and mature fruits of ‘Fuji’ and determined that many of the genes involved in early fruit development were also active in other organs. When global gene expression analysis of apple fruit development from the floral bud to ripe fruit (eight time points) was examined using a 13,000 gene microarray and compared with a microarray on tomato, 16 genes were identified in both apple and tomato that may have important roles in ripening (Janssen et al. 2008).

Gleave et al. (2008) examined over 120,000 ESTs to find 10 sequences that could be classified into seven plant miRNAs (microribonucleic acids). These small, non-coding RNAs play important regulatory roles.

Wisniewski et al. (2008) studied the response of 'Royal Gala' apple to low temperature and water deficit using expressed sequence tag analysis. This study provided more detailed information based on different source tissue (bark versus xylem, leaf versus root) and two different stresses, one short term (24 h cold) and one chronic (2 weeks of drought).

The use of cDNA suppression subtractive hybridization analysis revealed rapid transcriptional response of apple to fire blight disease (Norelli et al. 2009).

The proteomic analysis of the major soluble components in 'Annurca' apple flesh by Guarino et al. (2007) is the first of many such analyses. Increasingly, metabolomic research is being conducted, including examination of metabolomic changes that precede the development of apple superficial scald (Rudell et al. 2009).

7.4.1 Color as an Example of Progress in Genetics and Genomics

While fruit color was always discussed as a qualitative trait, breeders realized that red versus yellow was only one attribute of color, given that color intensity, pattern and the percent surface covered were also variables. Cheng (1996) identified a RAPD marker linked to red color in 1996. Advances in genomics in apples resulted in numerous groups reporting progress in this area at nearly the same time. Takos et al. (2006) reported that light induced expression of a MYB gene was what regulated anthocyanin biosynthesis in red apples. Chagne et al. (2007) mapped a candidate gene (MDMYB10) for red flesh and foliage color in apple and Espley et al. (2007) reported that red coloration in apple fruit was due to the activity of the MYB transcription factor, MdMYB10. Then, Ban et al. (2007) isolated and conducted functional analysis of this MYB transcription factor gene. In silencing anthocyanidin synthase in apple, Szankowski et al. (2009a) found a shift in polyphenol profile and a sublethal phenotype, emphasizing the importance of anthocyanin in apple.

7.4.2 Development of Research Communities and Databases

The AppleBreed Database was envisioned as an easily accessible way to link molecular and phenotypic data from multiple, pedigree-verified populations including crosses, breeding selections and cultivars (Antofie et al. 2007). This database was developed as part of the European HIDRAS project, but has applications beyond that project. The HIDRAS: High Quality Disease Resistant Apples for a Sustainable Agriculture (users.unimi.it/hidras/) was reviewed by Gianfranceschi and Soglio (2004). HIDRAS was preceded by the collaborative project DARE (Durable resistance to scab and mildew in apple) (Evans et al. 2000) and followed by ISAFRUIT, a European project involving over 200 researchers and 60 Research units, that is aimed at quality fruit from "the seed to consumption" (http://www.isafruit.org). Recently a large project called 'Fruitbreedomics' has been funded that combines breeding in collaboration with genomics across many institutions and countries.

In the USA, the GDR (Genome Database for Rosaceae) is an integrated Web database for Rosaceae genomics and genetics data (Jung et al. 2007) at http://www.bioinfo.wsu.edu/gdr/.

Shulaev et al. (2008) reviewed multiple models of genomics in the Rosaceae and shows the community building evident among researchers in this family. There is a Rosaceae white paper at http://www.Rosaceaewhitepaper.com, which represents efforts of the community to document issues across the Rosaceae. These activities resulted in the RosBREED project (http://www.rosbreed.org) that has as a goal enabling marker-assisted selection in the Rosaceae (Iezzoni et al. 2010).

7.5 Transgenics

Reviews of transgenics in apple include Brown and Maloney (2004), Sansavini et al. 2005; Gardiner et al. (2007) and Gessler and Patocchi (2007). Transgenics with fire blight resistance were discussed by Malnoy and Aldwinckle (2007) and transgenic rootstocks were reviewed by Dolgov and Hanke (2006).

Numerous scion cultivars have been transformed, starting with the transformation of 'Greensleeves' apple in 1989. Many of the top commercial varieties and some of their sports ('Cox's Orange Pippin,' 'Elstar,' 'Fuji,' 'Gala,' 'Greensleeves,' 'Jonagold,' 'McIntosh,' 'Orin') and a scab resistant variety ('Florina') have been transformed. Many transgenes have been targeted, with a priority on imparting resistance to disease and insects.

7.5.1 Resistance to Apple Scab

The synergistic activity of endochitinase and exochitinase from *Trichoderma harzianum* against the pathogenic fungus (*Venturia inaequalis*) in transgenic apple plants revealed that expression of some genes had a fitness cost, plants could be resistance but of extremely low vigor (Bolar et al. 2001). However overexpression of the apple MpNPR1 gene conferred increased disease resistance without a loss of vigor (Malnoy et al. 2007).

The transformation of apple with the cloned scab resistance gene (HcrVf2) from *Malus floribunda* provided the first functional confirmation of a cloned apple gene (Belfanti et al. 2004). This research has progressed to expression profiling in Hcr-Vf-2-transformed apple plants in response to *Venturia inaequalis,* with 523 unigenes identified (Paris et al. 2009). Recently, Malnoy et al. (2008) demonstrated that two receptor-like genes, *Vfa*1 and *Vfa*2, conferred resistance to apple scab. Szankowski et al. (2009b) found that varying the length of the native promoter of HcrVF2 influenced the degree of resistance expressed.

7.5.2 Resistance to Fire Blight

Rapid transcriptional response of apple to fire blight disease was revealed by cDNA suppression subtractive hybridization analysis (Norelli et al. (2009). Malnoy and Aldwinckle (2007) provide an overview of the many transgenic approaches to conferring resistance to this pathogen.

7.5.3 Fungal Resistance

Szankowski et al. (2003) transformed 'Holsteiner Cox' and 'Elstar' with the stilbene synthase gene from grapevine (*Vitis vinifera* L.) and a PGIP gene from kiwi (*Actinidia deliciosa*) to try to impart resistance to fungal diseases.

7.5.4 Modification of Plant Growth or Architecture (Rootstock and Scion)

Zhu et al. (2000) found that integration of the rolA gene into the genome of the vigorous apple rootstock A2 reduced plant height and shortened internodes. In 2005, Bulley et al. reported that the modification of gibberellin biosynthesis in the grafted apple scion allowed the control of tree height independent of the rootstock. Overexpression of the *Arabidopsis gai* (gibberellin-insensitive gene) in apple also significantly reduced plant size (Zhu et al. 2008b).

7.5.5 Flowering Genes

Kotoda et al. (2006) reported that antisense expression of the terminal flowering gene (MdTFL1) reduced the juvenile phase in apple. Overexpression of an FT-homologous gene of apple induced early flowering in *Arabidopsis*, poplar and apple (Tränkner et al. 2010).

7.5.6 Rootstock Transformation

Rootstocks have been transformed with the *rol* genes from *Agrobacterium* (Zhu et al. 2000) and with genes to impart resistance to fire blight (reviewed in Malnoy and Aldwinckle 2007). The rootstocks transformed include A2, M.7, M.26, M.9, 'Marubukaido,' and *Malus micromalus* Makino.

7.5.7 Anti-sense or Silencing

Transgenic approaches to silencing genes often reveal important information about the interaction of genes. Dandekar et al. (2004) found that down regulation of ethylene had a strong effect on the apple fruit flavor complex. Silencing leaf sorbitol synthesis altered long-distance partitioning and apple fruit quality (Teo et al. 2006).

7.5.8 Cisgenesis

Currently cisgenesis, defined as genetic modification of plants inserting genes from the plant itself or from crossable relatives, is a focus of several apple projects (Schouten and Jacobsen 2008). Such programs often also target "markerless technology."

The whole genome sequencing of a clone of 'Golden Delicious' has been completed at the Istituto Agrario San Michele all'Adige (IASMA) in Italy (http://www.ismaa.it) (Velasco et al. 2010) and researchers at Washington State University and South Africa are now partnering with these researchers and others in France in the sequencing of a doubled haploid of 'Golden Delicious.' Enhanced collaboration among breeding programs and genomic groups is providing the integration crucial to future success and application. Genomics has opened up new opportunities for improving apple and for learning some of the issues involved in the many complex traits being targeted. Enhanced collaboration on an international scale is an excellent way for us to further our breeding goals and to add to the database of phenotypic and genotypic information.

References

Abe, K., Kotoda, N., Kato, H. and Soejima, J. (2007) Resistance sources to Valsa canker (*Valsa ceratosperma*) in a germplasm collection of diverse *Malus* species. Plant Breeding 126, 449–453.

Aikman, D.P. and Langton, F.A. (1983) Replication in initial selection trials of clonally propagated crops. Euphytica 32, 821–829.

Alspach, P.A. and Oraguzie, N.C. (2002)Estimation of genetic parameters of apple (*Malus domestica*) fruit quality from open pollinated families. New Zealand J. Crop Hort. Sci. 30, 219–228.

Alston, F.H., Philipps, K.L. and Evans, K.M. (2000) A *Malus* gene list. Acta Hort. 538, 561–570.

Antofie, A., Lateur, M., Oger, R., Patocchi, A., Durel, C.E. and van de Weg, W.E. (2007) A new versatile database created for geneticist and breeders to link molecular and phenotypic data in perennial crops: the AppleBreed DataBase. Bioinformatics 23, 882–891.

Ban, Y., Honda, C., Hatsuyama, Y., Igarashi, M., Bessho, H. and Moriguchi, T. (2007) Isolation and functional analysis of a myb transcription factor gene that is a key regulator for the development of red coloration in apple skin. Plant Cell Physiol. 48, 958–970.

Batlle, I. and Alston, F. H. (1996) Genes determining leucine aminopeptidase and mildew resistance from ornamental apple, 'White Angel'. Theor. Applied Genetics 93, 179–182.

Belfanti, E., Silfverberg-Dilworth, E., Tartarini, S., Patocchi, A., Barbieri, M., Zhu, J., Vinatzer, B.A., Gianfranceschi, L., Gessler, C. and Sansavini, S. (2004) The HcrVf2 gene from a wild apple confers scab resistance to a transgenic cultivated variety. Proc. Natl. Acad. Sci. USA 101, 886–890.

Biedrzycka, E. and Amarowicz, R. (2008) Diet and Health: Apple polyphenols as antioxidants. Food Reviews Internat. 24, 235–251.

Biggs, A.R. and Miller, S.S. (2001) Relative susceptibility of selected apple cultivars to *Colletotrichum acutatum.* Plant Disease 85, 657–660.

Blanpied, D. and Silsby, K. (1992) Predicting harvest date windows for apples. Cornell Cooperative Extension Information Bulletin 221.

Blazek, J., Opatova, H., Golias, J. and Homutova, I. (2007) Ideotype of apples with resistance to storage disorders. Hort. Sci. (Prague) 24, 107–113.

Bolar, J.P., Norelli, J., Harman, G.E., Brown, S.K., and Aldwinckle, H.S. (2001) Synergistic activity of endochitinase and exochitinase from *Trichoderma harzianum* against the pathogenic fungus (*Venturia inaequalis*) in transgenic plants. Transgenic Research 10, 533–543.

Boyer, J. and Liu, R.H. (2004) Apple phytochemicals and their health benefits. Nutrition J. 3:5 (www.nutritionj.com/content/3/1/5).

Braniste, N., Militaru, M. and Budan, S. (2008). Two scab resistant columnar apple cultivars. Acta Hort. 767, 351–354.

Broothaerts, W. (2003) New findings in apple S-genotype analysis resolve previous confusion and request the re-numbering of some S-alleles. Theor. Appl. Genet. 106, 703–714.

Brown, S.K. (2008) Breeding and biotechnology for flavor development in apple (*Malus* x *domestica* Borkh.) pp. 147–156. In: Havkin-Frenkel, D. and Belanger, F.C. (eds.): *Biotechnology in flavor production.* Blackwell Publishing.

Brown, S.K. and Maloney, K. E. (2004) *Malus* x *domestica* Apple. pp. 475–511. In: R. Litz (ed.) *Biotechnology of Fruit and Nut Crops.* CAB International, Oxon, United Kingdom.

Brown, S.K. and Maloney, K.E. (2003) Genetic improvement of apple: Breeding, markers, mapping and biotechnology. pp. 31–59. In: Ferree, D. and Warrington, I. (eds.) *Apples: Botany, Production and Uses.* CAB International, Cambridge, MA, USA.

Bulley, S.M., F.M. Wilson, P. Hidden, A.L. Phillips, S.J. Croker and D.J. James. (2005) Modification of gibberellin biosynthesis in the grafted apple scion allows control of tree height independent of the rootstock. Plant Biotech. J. 3, 215–223.

Bus, V. G. M., Chagne, D., Bassett, C.M., Bowatte, D., Calenge, F., Celton, J.M., Durel, C.E., Malone, M.T., Patocchi, A., Ranatunga, A.C., Rikkerink, E.H.A., Tustin, D.S., Zhou, J. and Gardiner, S.E. (2007) Genome mapping of three major resistance genes to woolly apple aphid (*Eriosoma lanigerum* Hausm.). Tree Genetics Genomics 4, 223–236.

Bus V.G.M., Laurens F.N.D., van de Weg W.E., Rusholme R.L., Rikkerink E.H.A., Gardiner S.E., Bassett H.C.M., Kodde L.P. and Plummer K.M. (2005a) The *Vh8* locus of a new gene-for-gene interaction between *Venturia inaequalis* and the wild apple *Malus sieversii* is closely linked to the *Vh2* locus in *M. pumila* R12740-7A. New Phytologist 166,1035–1049.

Bus V.G.M., Rikkerink E.H.A., van de Weg E.W., Gardiner S.E., Bassett H.C.M., Kodde L.P., Parisi L., Laurens F.N.D., Rusholme R., Meulenbroek B. and Plummer, K.M. (2005b) The Vr and Vx scab resistance genes in two differential hosts derived from Russian apple R12740-7A map to the same linkage group of apple. Mol. Breeding 15, 103–116.

Bus, V. G. M. (2006) A partial diallel study of powdery mildew resistance in six apple cultivars under three growing conditions with different disease pressures. Euphytica 148, 235–242.

Büttner, R., Fischer, M., Forsline, P.L., Geibel, M. and Ponomarenko, V.V. (2004) Gene banks for the preservation of wild apple genetic resources. J. Fruit Ornamental Plant Research 12, 99–104.

Caffier, V. and Laurens, F. (2005) Breakdown of *Pl2*, a major gene of resistance to apple powdery mildew, in a French experimental orchard. Plant Path. 54, 116–124.

Caffier, V. and Parisi, L. (2007) Development of apple powdery mildew on sources of resistance to *Podosphaera leucotricha*, exposed to an inoculum virulent against the major resistance gene Pl-2. Plant Breeding 126, 319–322.

Calenge, F., Drouet, D., Denance, C., van de Weg, W.E., Brisset, M.-N., Paulin, J.P. and Durel, C.-E. (2005) Identification of a major QTL together with several minor additive or epistatic QTLs for resistance to fire blight in apple in two related progenies. Theor. Appl. Genet. 111,128–135.

Calenge, F. and Durel, C.-E. (2006) Both stable and unstable QTLs for resistance to powdery mildew are detected in apple after four years of field assessments. Mol. Breeding 17, 329–339.

Calenge F., Faure A., Goerre M., Gebhardt C., Van De Weg W.E., Parisi L., Durel C-E. (2004) Quantitative Trait Loci (QTL) analysis reveals both broad-spectrum and isolate-specific QTL for scab resistance in an apple progeny challenged with eight isolates of *Venturia inaequalis*. Phytopath. 94, 370–379.

Celton, J.M., Tustin, S., Chagne, D. and Gardiner, S. (2009) Construction of a dense genetic linkage map for apple rootstocks using SSRs developed from *Malus* ESTs and *Pyrus* genomic sequencing. Tree Genetics and Genomes 5, 93–107.

Chagne, D., Carlisle, C.M., Blond, C., Volz, R.K., Whitworth, C.J., Oraguzie, N.C., Crowhurst, R.N., Allan, A.C., Espley, R.V., Hellens, R.P. and Gardiner, S.E. (2007) Mapping a candidate gene (MDMYB10) for red flesh and foliage color in apple. BMC Genomics 8, 212.

Chagne, D., Gasic, K., Crowhurst, R.N., Han, Y., Bassett, H.C., Bowatte, D.R., Lawrence, T.J., Rikkerink, E.H.A., Gardiner, S.E. and Korban, S.S. (2008) Development of a set of SNP markers present in expressed genes of the apple. Genomics 92, 353–358.

Cheng, F.S., Weeden, N.F., Brown, S.K., Aldwinckle, H.S., Gardiner, S.E. and Bus, V.G. (1998) Development of a DNA marker for *Vm*, a gene conferring resistance to apple scab. Genome 41, 208–214.

Cheng, F.S., Weeden, N.F. and Brown, S.K. (1996) Identification of co-dominant RAPD markers tightly linked to fruit skin color in apple. Theor. Appl. Genet. 93, 222–227.

Coart, E., Vekemans, X., Smulders, M.J.M., Wagner, I., Van Huylenbroek, J., Van Bockstaele, E. and Roldan-Ruiz, I. (2003) Genetic variation in the endangered wild apple (*Malus sylvestris* (L.) Mill.) in Belgium as revealed by amplified fragment length polymorphism and microsatellite markers. Mol. Ecology 12, 845–857.

Coart, E.L.S., Van Glabeke, S., DeLoose, M., Larsen, A.S. and Roldan-Ruiz, I. (2006) Chloroplast diversity in the genus *Malus*: new insights into the relationship between the European wild apple (*Malus sylvestris* (L.) Mill.) and the domesticated apple (*Malus domestica* Borkh.). Mol. Ecology 15, 2171–2182.

Conner, P.J., Brown, S.K. and Weeden, N.F. (1997) Randomly amplified polymorphic DNA-based genetic linkage maps of three apple cultivars. J. Amer. Soc. Hort. Sci. 122, 350–359.

Conner, P.J., Brown, S.K. and Weeden, N.F. (1998) Molecular-marker analysis of quantitative traits for growth and development in juvenile apple trees. Theor. Appl. Genet. 96, 1027–1035.

Costes, E., Lauri, P.E. and Regnard, J.L. (2006) Analyzing fruit tree architecture, implication for tree management and fruit production. Hort. Rev. 32, 1–61.

Cummins, J.N. and Aldwinckle, H.S. (1983) Breeding apple rootstocks. Plant Breeding Rev. **x**, 294–394.

Currie, A. J., Ganeshanandam, S., Noiton, D. A., Garrick, D., Shelbourne, C. J. A. and Orgaguzie, N. (2000) Quantitative evaluation of apple fruit shape (*Malus* × *domestica* Borkh.) by principal component analysis of Fourier descriptors. Euphytica 11, 221–227.

Dal Cin, V., Boschetti, A., Dorigoni, A. and Ramina, A. (2007) Benzylaminopurine application on two different apple cultivars (*Malus domestica*) displays new and unexpected fruit abscission features. Ann. Bot. 99, 1195–1202.

Dandekar, A.M., Teo, G., Defillippi, B.G., Uratsu, S.L., Passey, A.J., Kader, A.A., Stow, J.R., Colgan, R.J. and James, D.J. (2004) Effect of down-regulation of ethylene biosynthesis on fruit flavor complex in apple fruit. Transgenic Research 13, 373–384.

Davey, M.W., Kenis, K. and Keulemans, J. (2006) Genetic control of fruit vitamin C contents. Plant Physiol. 142, 343–351.

Davey, M.W. and Keulemans, J. (2004) Determining the potential to breed for enhanced antioxidant status in *Malus*: Mean inter- and intra-varietal fruit vitamin C and glutathione contents at harvest and their evolution during storage. J. Agric. Food Chem. 52, 8031–8038.

Dayton, D .F. (1977) Genetic immunity to apple mildew incited by *Podosphaera leucotricha*. Hortscience 12, 225–226.

Dolgov, S.V. and Hanke, V. (2006) Transgenic temperate fruit tree rootstocks. p. 335–350. In; Fladung, M. and Ewald, E. (eds.). Tree Transgenesis. Recent Developments. Springer-Verlag.

Dunemann, F., Bracker, G., Markussen, T. and Roche, P. (1999) Identification of molecular markers for the major mildew resistance gene Pl2 in apple. Acta Hort. 484, 411–416.

Dunemann, F., Peil, A., Urbanietz, A., and Garcia-Libreros, T. (2007) Mapping of the powdery mildew resistance gene *Pl*1 and its genetic association with an NBS-LRR candidate resistance gene. Plant Breeding 126, 476–481.

Dunemann, F., Ulrich, D., Boudichevskaia,A.. Grafe, C. and Weber, W.E. (2009) QTL mapping of aroma compounds analysed by headspace solid-phase microextraction gas chromatography in the apple progeny 'Discovery' x 'Prima'. Mol. Breeding 23, 501–521.

Dunemann, F., Ulrich, D., Malysheva-Otto, L., Weber, W.E., Longhi, S. Velasco, R. and Costa, F. (2011) Functional allelic diversity of the apple alcohol acyl-transferase gene mdAAT1 associated with fruit ester volatile contents in apple cultivars. Mol. Breeding (on-line early).

Durel, C.-E., Denance, C. and Brisset, M.-N. (2009) Two distinct major QTL for fire blight co-localize on linkage group 12 in apple genotypes 'Evereste' and *Malus floribunda* clone 821. Genome 52, 139–147.

Durel, C.E., Laurens, F., Fouillet, A. and Lespinasse, Y. (1998) Utilization of pedigree information to estimate genetic parameters from large unbalanced data sets in apple. Theor. Appl. Genet. 91, 1077–1085.

Eberhardt, M.V., Lee, C.Y. and Liu, R.H. (2000) Antioxidant activity of fresh apples. Nature 405, 903–904.

Erdin, N., Tartarini, S., Broggini, G.A.L., Gennari, F., Sansavini, S., Gessler, C., and Patocchi, A. (2006) Mapping of the apple scab-resistance gene *Vb*. Genome 49, 1238–1245.

Espley, R.V., Hellens, R.P., Putterill, J., Stevenson, D.E., Kutty-Amma, S., and Allan, A.C. (2007) Red coloration in apple fruit is due to the activity of the MYB transcription factor. MdMYB10. The Plant Journal 49, 414–427.

Evans, K.M. and James, C.M. (2003) Identification of SCAR markers linked to *Pl-w* mildew resistance in apple. Theor. Appl. Genet. 106, 1178–1183.

Evans, K., Lespinasse, Y. and Durel, C. (2000) Durable resistance to scab and mildew in apple- A European project. Pesticide Outlook, 84–87.

Fazio, G., Aldwinckle, H., Mcquinn, R., Robinson, T. (2006a) Differential susceptibility to fire blight in commercial and experimental apple rootstock cultivars. Acta Hort. 704, 527–530.

Fazio, G., Robinson, T., Aldwinckle, H., Mazzola, M., Leinfelder, M., Parra, R. (2006b) Traits of the next wave of Geneva apple rootstocks. Compact Fruit Tree 38, 7–11.

Fernandez-Fernandez, F., Evans, K.M., Clarke, J.B., Govan, C.L., James, C.M., Maric, S. and Tobutt, K.R. (2008) Development of an STS map of an interspecific progeny of *Malus*. Tree Genetics & Genomes 4, 469–479.

Ferree, D. C. and Carlson, R.F. (1987) Apple Rootstocks. pp. 107–143. In: Rom, R.C. and Carlson, R.F. (eds.). *Rootstocks for Fruit Crops*. John Wiley & Sons. US.

Fischer, M. (2001) New dwarfing and semi-dwarfing apple and pear rootstocks. Acta Hort. 557, 55–62.

Fischer, M. and Dunemann, F. (2000) Search for polygenic scab and mildew resistance in apple varieties cultivated at the Fruit Genebank Dresden-Pillnitz. Acta Hort. 538, 71–77.

Fischer, M. And Fischer, C. (2008) The Pillnitz Re-series of apple cultivars-Do they hold promise? -80 years of professional German fruit breeding. Erwerbs-Obstbau 50, 63–67.

Forsline, P.L., Aldwinckle, H.S., Dickson, E.E., Luby, J.J. and Hokanson, S.C. (2003) Collection, maintenance, characterization and utilization of wild apples of central Asia. Hort. Rev. 29, 1–62.

Gallot J.C., Lamb, R.C. and Aldwinckle, H.S. (1985). Resistance to powdery mildew from some small-fruited *Malus* cultivars. Hortscience 20, 1085–1087.

Gao, Z.S. and van de Weg, W. E. (2006) The Vf gene for scab resistance is linked to sub-lethal genes. Euphytica 151, 123–132.

Gao, Z.S., van de Weg, W.E., Schaart, J.G., Schouten, H.J., Tran, D.H., Kodde, L.P., van der Meer, I.M., van der Geest, A.H.M., Kodde, J., Breiteneder, H., Hoffmann-Sommergruber, K., Bosch, D. and Gilissen, L.J.W.J. (2005a) Genomic cloning and linkage mapping of the *Mal dl (PR-10)* gene family in apple (*Malus domestica*). Theor. Appl. Genet. 111, 171–183.

Gao, Z.S., van de Weg, W.E., Schaart, J.G., van der Meer, I.M., Kodde, L.P., Laimier, M., Breiteneder, H., Hoffmann-Sommergruber, K., and Gilissen, L.J.W.J. (2005b) Linkage map positions and allelic diversity of two Mal d 3 (non-specific lipid transfer protein) genes in the cultivated apple (*Malus domestica*) Theor. Appl. Genet. 110, 479–491.

Gao, Z.S., van de Weg, W.E., Schaart, van Arkel, G., Breiteneder, H., Hoffmann-Sommergruber, K., and Gilissen, L.J.W.J. (2005c) Genomic characterization and linkage mapping of the apple allergen genes *Mal d 2* (thaumatin-like protein) and *Mal d 4* (profilin). Theor. Appl. Genet. 111, 1087–1097.

Gardiner, S.E., Bus, V.G.N., Rusholme, R.L., Chagne, D., and Rikkerink, E. (2007) Apples: pp.1-62. In: Kole, C. (ed.) *Genome Mapping and Molecular Breeding in Plants: Fruits and Nuts.* Springer, NY.

Gasic, K., Han, Y., Kertbundit, S., Shulaev, V., Iezzoni, A.F., Stover, E.W., Bell, R.L., Wisniewski, M.E., and Korban, S.S. (2009) Characterization and transferability of new apple EST-derived SSRs to other Rosaceae species. Mol. Breeding (on-line early).

Gessler, C., Patocchi, A., Sansavini, S., Tartarini, S. and Gianfranceschi. L. (2006) *Venturia inaequalis* resistance in apple. Critical Reviews Plant Sci. 25, 473–503.

Gessler, C. and Patocchi, A. (2007) Recombinant DNA technology in apple. Adv. Biochem. Engin./Biotechnology 107, 113–132.

Gianfranceschi, L. and Soglio, V. (2004). The European project HIDRAS: innovative multidisciplinary approaches to breeding high quality disease resistant apples. Acta Hort. 663, 327–330.

Gleave, A.P., Ampomah-Dwamena, C., Berthold, S., Dejnoprat, S., Karunairetnam, S., Nain, B., Wang, Y.Y., Crowhurst, R.N. and MacDiarmid, R.M. (2008) Identification and characterisation of primary microRNAs from apple (*Malus domestica* cv. Royal Gala) expressed sequence tags. Tree Genetics & Genomes 4, 343–358.

Guarino, C., Arena, S., De Simopne, L., D'Ambrosia, C., Sanotora, Simona, Rocco, M., Scaloni, A. and Marra, M. (2007) Proteomic analysis of the major soluble components in Annurca apple flesh. Mol. Nutr. Food Res. 51, 255–262.

Gygax, M., Gianfranceshi, L., Liebhard, R., Kellerhals, M., Gessler, C. and Patocchi, A. (2004) Molecular markers linked to the apple scab resistance gene V_{bj} derived from *Malus baccata jackii*. Theor. Appl. Genet. 109, 1702–1709.

Harada, T., Sunako, T., Wakasa, Y., Soejima, J., Satoh, T. and Niizeki, I. (2000) An allele of the 1-aminocyclopropane-1-carboxylate synthase gene (Md-ACS1) accounts for the low level of ethylene production in climacteric fruits of some apple cultivars. Theor. Appl. Genet. 101, 742–746.

Harker F.R., Kupferman, E.M., Marin, A.B., Gunsun, F.A., and Triggs, C.M. (2008) Eating quality standards for apples based on consumer preferences. Postharvest Biol. Technol. 50, 70–78.

Harker, F.R., Maindonald, J.H. and Jackson, P.J. (1996) Penetrometer measurement of apple and kiwi firmness: operator and instrument differences. J. Amer. Soc. Hort. Sci. 121, 927–936.

Harker, F.R., Maindonald, J.H., Murray, S.H., Gunson, F.A., Hallet, I.C. and Walker, S.B. (2002a) Sensory interpretation of instrumental measurements 1: texture of apple fruit. Postharvest Biol. Technol. 24, 225–239.

Harker, F.R., Marsh, K.B., Young, H., Murray, S.H., Gunson, F.A. and Walker, S.B. (2002b) Sensory interpretation of instrumental measurements 2: sweet and acid taste of apple fruit. Postharvest Biol. Technol. 24, 241–250.

Hemmat, M., Brown, S.K., Aldwinckle, H.S., Mehlenbacher, S.A. and Weeden, N.F. (2003) Identification and mapping of markers for resistance to apple scab from 'Antonovka' and 'Hansen's baccata #2'. Acta Hort. 622, 153–162.

Hemmat, M., Brown, S.K. and Weeden, N.F. (2002) Tagging and mapping scab resistance genes from R12740-7A apple. J. Amer. Soc. Hort. Sci. 127, 365–370.

Hemmat, M., Weeden, N.F., Conner, P.J. and Brown, S.K. (1997) A DNA marker for columnar growth habit in apple contains a simple sequence repeat J. Amer. Soc. Hort. Sci. 122, 347–349.

Hemmat, M., Weeden, N.F., Manganaris, A.G. and Lawson, D.M. (1994) Molecular marker linkage map for apple. J. Heredity 85, 4–11.

Heo, S., Kim, D., Yun, H.R., Hwang, J.H., Lee, H.J. and Shin, Y.U. (2006) Development of AFLP markers linked to resistance against Alternaria blotch in apple (*Malus domestica*). Hort. Environ. Biotech. 47, 324–328.

Iezzoni, A., Weebadde, C., Luby, J., Chengyan, Y., Van de Weg, E., Fazio, G., Mann, D., Peace, C.P., Bassil, N.V. and McFerson, J. (2010) RosBREED: Enabling marker assisted breeding in Rosaceae. Acta Hort. 859, 389–394.

Igarashi, M., Abe, Y., Hatsuyama, Y., Ueda, T., Fukasawa-Akada, T., Kon, T., Kudo, T., Sato, T. and Suziki, M. (2008) Linkage maps of the apple *(Malus* x *domestica* Borkh.) cvs. 'Ralls Janet' and 'Delicious' include newly developed EST markers. Mol. Breeding 22, 95–118.

Ikase, L and Dumbras, R. (2004) Breeding of columnar apple trees in Latvia. Biologia 2, 8–10.

Iwanami, H., Ishiguro, M., Kotoda, N., Takahashi, S. and Soejima, J. (2005) Optimal sampling strategies for evaluating fruit softening after harvest in apple breeding. *Euphytica* 144, 169–175.

Iwanami, H., Moriya, S., Kotoda, N., Takahashi, S. and Abe, K. (2008) Estimations of heritability and breeding value for postharvest fruit softening in apple. J. Amer. Soc. Hort. Sci. 133, 92–99.

Jakubowski, T. and Zagaja, S.W. (2000) 45 years of apple rootstock breeding in Poland. Acta Hort. 538, 723–727.

James, C.M., Clarke, J.B. and Evans, K.M. (2004) Identification of molecular markers linked to the mildew resistance gene *Pl-d* in apple. Theor. Appl. Genet. 110, 175–181.

Janick, J., Cummins, J.N., Brown, S.K., Hemmat, M. (1996) Apples. pp 1–77. In: Janick J., Moore J.N. (eds). *Fruit Breeding Volume 1. Tree and Tropical Fruits.* John Wiley, New York, NY.

Janssen, B.J., Thoday, K., Schaffer, T.J., Alba, R., Balakrishan, L., Bishop, R., Bowen, J.H., Crowhurst, R.N., Gleave, A.P., Ledger, S., McArtney, S., Pichler, F.B., Snowden, K.C. and Ward, S. (2008) Global gene expression analysis of apple fruit development from the floral bud to ripe fruit. BMC Plant Biology 8, 16.

Jung, S., Staton, M., Le, T., Blenda, A., Svancara, R., Abbott, A. and Main, D. (2007) GDR (Genome Database for Rosaceae): integrated web database for Rosaceae genomics and genetics data. Nucleic Acids Research 1–7.

Juniper, B.E., Watkins, R. and Harris, S.A. (1998) The origins of apple. Acta Hort. 484, 27–33.

Kenis, K. and Keulemans, J. (2005) Genetic linkage maps of two apple cultivars (*Malus domestica* Borkh.) based on AFLP and microsatellite markers. Mol. Breeding 15, 205–219.

Kenis, K. and Keulemans, J. (2007) Study of tree architecture of apple (*Malus* x *domestica* Borkh.) by QTL analysis of growth traits. Mol. Breeding 19, 193–208.

Kenis, K. and Keulemans, J. (2008) Identification and stability of QTLs for fruit quality traits in apple. Tree Genetics & Genomes 4, 647–661.

Khanizadeh, S., Groleau, Y., Granger, R., Cousineau, J. and Rousselle, G.L. (2000) New hardy rootstocks from the Quebec apple breeding program. Acta Hort. 538, 719–721.

Khanizadeh, S., Tsao, R., Rekika, D., Yang, R., Charles, M.T. and Vasantha Rupasinghe, H.P. (2007) Polyphenol composition and total antioxidant capacity of selected apple genotypes for processing. J. Food Comp. Analysis 21, 396–401.

Kim, M.Y., Song, K.J., Hwang, J.-H., Shin, Y.U. and Lee, H.J. (2003) Development of RAPD and SCAR markers linked to the Co gene conferring columnar growth habit in apple (*Malus pumilla* Mill.). J. Hort.Sci. Biotech 78, 512–517.

King, G.J., Lynn, J.R., Dover, C.J., Evans, K.M. and Seymore, G.B. (2001) Resolution of quantitative trait loci for mechanical measures accounting for genetic variation in fruit texture of apple (*Malus pumila* Mill.). Theor. Appl Genet. 102, 1227–1235.

King, G.J., Maliepaard, C., Lynn, J.R., Alston, F.H., Durel, C.E., Evans, K.M., Griffon, B., Laurens, F., Manganaris, A.G., Schrevens, E., Tartarini, S. and Verhaegh, J. (2000) Quantitative genetic analysis and comparison of physical and sensory descriptors relating to fruit flesh firmness in apple (*Malus pumila* Mill.). Theor. Appl. Genet. 100, 1074–1084.

Knight, R. L. and Alston, F.H. (1968) Sources of field immunity to mildew (*Podosphaera leucotricha*) in apple. Can. J. Genet. Cytol. 10, 294–298.

Knight, V.H., Evans, K.M., Simpson, D.W. & Tobutt, K.R. (2005) Report on a desktop study to investigate the current world resources in Rosaceous fruit breeding programmes. Submitted to Defra, August 2005.

Korban, S.S. (1986) Interspecific hybridization in *Malus*. HortScience 21, 41–48.

Korban, S.S. and Swiader, J.M. (1994) Genetic and nutritional status in bitter pit resistant and – susceptible apple seedlings. J. Amer. Soc. Hort. Sci. 109, 428–432.

Kotoda, N., Iwanami, H., Takahashi, S., and Abe, K. (2006) Antisense expression of MdTFL1-like gene, reduces the juvenile phase in apple. J. Amer. Soc. Hort. Sci. 131, 74–81.

Labuschagne, I. F., Louw, J. H., Schmidt, K., Sadie, A. (2003) Budbreak number in apple seedlings as selection criterion for improved adaptability to mild winter climates. HortScience 386, 1186–1190.

Labuschagne, I.F., Louw, J.H., Schmidt, K. and Sadie, A. (2001) Genotypic variation in prolonged dormancy symptoms in apple families. HortScience 37, 157–163.

Labuschagne, I.F., Louw, J.H., Schmidt, K. and Sadie, A. (2002) Genetic variation in chilling requirement in apple progenies. J. Amer. Soc. Hort. Sci. 127, 663–672.

Lauri, P.E., Bourdel, G., Trottier, C. and Cochard, H. (2008) Apple shoot architecture: evidence for strong variability for bud size and composition and hydraulics within a branching zone. New Phytologist 178, 798–807.

Laurens, F. (1999) Review of the current apple breeding programs in the world: Objectives for scion cultivar improvement. Acta Hort. 484, 163–170.

Laurens, F. and Pitiot, C. (2003) French apple breeding program: A new partnership between INRA and the nurserymen of Novadi. Acta Hort. 622, 575–582.

Lata, B. (2008) Apple peel antioxidant status in relation to genotype, storage type and time. Scientia Hort. 117, 45–52.

Lee, K.W., Kim, Y.J., Kim, D-O, Lee, H.J. and Lee, C.Y. (2003) Major phenolics and their contribution to the total antioxidant capacity. J. Agric. Food Chem. 51, 6516–6520.

Lee, Y.-P., Yu, G.-H., Seo, Y.S., Han, S.E., Choi, Y-O, Kim, D., Mok, I-G., Kim, W.T. and Sung, S.-K. (2007) Microarray analysis of apple gene expression engaged in early fruit development. Cell Biology Morphogenesis 26, 917–926.

Liebhard, R., Gianfranceschi, L., Koller, B., Ryder, C.D., Tarchini, R., Van de Weg, E. and Gessler, C. (2002) Development and characterization of 140 new microsatellites in apple (*Malus* x *domestica* Borkh.). Mol. Breeding 10, 217–241.

Liebhard, R., Kellerhals, M., Pfammatter, W., Jertmini, M. and Gessler, C. (2003a) Mapping quantitative physiological traits in apple (*Malus* x *domestica* Borkh.). Plant Mol. Biology 52, 511–526.

Liebhard, R., Koller, B., Gianfranceschi, L. and Gessler, C. (2003b) Creating a saturated reference map for the apple (*Malus* x *domestica* Borkh.) genome. Theor. Appl. Genet. 106, 1497–1508.

Liebhard, R., Koller, B., Patocchi, A., Kellerhals, M., Pfammatter, W., Jermini, M. and Gessler, C. (2003c) Mapping quantitative field resistance against apple scab in a 'Fiesta' x 'Discovery' progeny. Phytopath. 93, 493–501.

Luby, J. (2003) Taxonomic classification and brief history. pp. 1–14. *Apples. Botany, Production and Use.* In: Ferree, D.C. and Warrington, I.J. (eds.) CABI Publishing, Cambridge, MA.

Luby, J., Forsline, P., Aldwinckle, H., Bus, V. and Geibel, M. (2001) Silk-road apples – Collection, evaluation and utilization of *Malus sieversii* from central Asia. HortScience 36, 225–231.

Luby, J., Hoover, E., Paterson, M., Larson, D., and Bedford, D. (1999) Cold hardiness in the USDA *Malus* core germplasm collection. Acta Hort. 484, 109–114.

Luby, J.J. and Shaw, D.V. (2001) Does marker-assisted selection make dollars and sense in a fruit breeding program? HortScience 36, 872–879.

Maliepaard, C., Alston, F., van Arkel, G., Brown, L.M., Chevreau, E., Dunemann, F., Evans, K.M., Gardiner, S., Guilford, P., van Heusden, A.W., Janse, J., Laurens, F., Lynn, J.R., Manganaris, A.G., den Nijs, A.P.M., Periam, N., Rikkerink, E., Roche, P., Ryder, C., Sansavini, S., Schmidt, H., Tartarini, S., Verhaegh, J.J., Vrielink-van Ginkel, M. and King, G.J. (1998) Aligning male and female linkage maps of apple (*Malus pumila* Mill.) using multi-allelic markers. Theor. Appl. Genet. 97, 60–73.

Malnoy, M. and Aldwinckle, H.S. (2007) Development of fire blight resistance by recombinant DNA technology. Plant Breeding Rev. 26, 315–358.

Malnoy, M., Jin, Q., Borejsza-Wysocka, E. E. and Aldwinckle, H. S. (2007) Overexpression of the apple MpNPR1 gene confers increased disease resistance in *Malus* x *domestica*. Molecular Plant Microbe Interactions 20, 1568–1580.

Malnoy, M., Xu, M., Borejsza-Wysocka, E. E., Korban, S. and Aldwinckle, H. S. (2008) Two receptor-like genes, *Vfa1* and *Vfa2*, confer resistance to the fungal pathogen *Venturia inaequalis* inciting apple scab disease. Mol. Plant-Microbe Interactions 21, 448–458.

Markussen T., Krüger J., Schmidt H. and Dunemann F. (1995) Identification of PCR-based markers linked to the powdery-mildew-resistance gene *Pl1* from *Malus robusta* in cultivated apple. Plant Breeding 114, 530–534.

Matsumoto, S., Eguchi, T., Bessho, H. and Abe, K. (2007) Determination and confirmation of S-RNase genotypes of apple pollinators and cultivars. J. Hort. Sci. Biotech. 82, 323–329.

Meyers, C.T., Leskey, T.C. and Forsline, P.T. (2007) Susceptibility of fruit from diverse apple and crabapple germplasm to attack by plum curculio (Coleoptera: Curculionidae). J. Econ. Entomol. 100, 1663–1671.

Meyers, C.T., Reissig, W., Forsline, P.L. (2008) Susceptibility of fruit from diverse apple and crabapple germplasm to attack from apple maggot, *Rhagoletis pomonella* (Walsh) (Diptera: Tephritidae). J. Econ. Entomol. 101, 206–215.

Miller, S., C. Hampson, R. McNew, L. Berkett, S. Brown, J. Clements, R. Crassweller, E. Garcia, D. Greene and G. Greene. (2005) Performance of apple cultivars in the 1995 NE-183 Regional project planting: 111. Fruit sensory characteristics. J. Amer. Pomol. Soc. 59, 28–43.

Miñaro, and Depena, E. (2008) Tolerance of some scab-resistant apple cultivars to the rosy apple aphid, *Dysaphis plantaginea* Crop Protection 27, 391–395.

Moore, J.N. and Janick, J. (1983) Methods *in Fruit Breeding*. Purdue University Press.

N'Diaye, A., Van de Weg, E., Kodde, L.P., Koller, B., Dunemann, F., Thiermann, M., Tartarini, S, Gennari, F. and Durel, C.E. (2008) Construction of an integrated consensus map of the apple genome based on four mapping populations. Tree Genetics & Genomes 4, 727–743.

Newcomb, R.D., Crowhurst, R.N., Gleave, A.P., Rikkerink, E.H.A., Allan, A.C., Beuning, L.L., Bowen, J.H., Gera, E., Jamieson, K.R., Janssen, B.J., Laing, W.A., McArtney, S., Nain, B., Ross, G.S., Snowden, K.C., Souleyre, E.J.F., Walton, E.F., and Yauk, Y.-K. (2006) Analyses of expressed sequence tags from apple. Plant Physiol. 141,147–166.

Noiton, D.A.M. and Alspach, P.A. (1996) Founding clones, inbreeding, coancestry, and status number of modern apple cultivars. J. Amer. Soc. Hort. Sci. 121, 773–782.

Norelli, J.L., Farrell, R.E., Bassett, C.L., Baldo, A.M., Lalli, D.A., Aldwinckle, H.A., and Wisniewski, M. E. (2009) Rapid transcriptional response of apple to fire blight disease revealed by cDNA suppression subtractive hybridization analysis. Tree Genetics & Genomes 5, 27–41.

Nybom, H., Rumpunen, K., Persson Hovmalm, H., Marttila, S., Rur, M., Garkava-Gustavsson, L. and Ollsson, M. (2008) Towards a healthier apple – Chemical characterization of an apple gene bank. Acta Hort. 765, 157–164.

Nybom, H., Sehic, J. and Garkava-Gustavsson, L. (2008a) Self-incompatibility alleles of 104 apple cultivars grown in northern Europe. J. Hort. Sci. Biotech. 83, 339–344.

Nybom, H., Sehic, J. and Garkava-Gustavsson, L. (2008b) Modern apple breeding is associated with a significant change in the allelic ratio of ethylene production gene Md-ACS1. J. Hort. Sci. Biotech. 83, 673–677.

Oraguzie, N.C., Hofstee, M.E., Brewer, L.R. and Howard, C. (2001) Estimation of genetic parameters in a recurrent selection program in apple. Euphytica 118, 29–37.

Oraguzie, N.C., Rikkerink, E.H.A., Gardner, S.E. and De Slva, H.N. (2007a) Association Mapping in Plants. Springer, New York, NY.

Oraguzie, N.C., Volz, R.K., Whitworth, C.J., Bassett, H.C.M., Hall, A.J. and Gardiner, S.E. (2007b) Influence of Md-ACS1 allelotype and harvest season within an apple germplasm collection on fruit softening during cold air storage. Postharvest Biol. Technol. 44, 212–219.

Paris, R., Cova, V., Pagliani, G, Tartarini, S., Komjanc, M. and Sansavini, S. (2009) Expression profiling in Hcr-Vf-2-transformed apple plants with in response to *Venturia inaequalis*. Tree Genetics and Genome 5, 81–91.

Park, S., Sugimoto, N., Larson, M.D., Beaudry, R. and van Nocker, S. (2006) Identification of genes with potential roles in apple fruit development and biochemistry through large-scale statistical analysis of expressed sequence tags. Plant Physiol. 141, 811–824.

Patocchi, A. , Walser, M., Tartarini, S., Broggini, G.A.L., Gennari, F., Sansavini, S. and Gessler, C. (2005) Identification by genome scanning approach (GSA) of a microsatellite tightly associated with the apple scab resistance gene *Vm*. Genome 48, 630–636.

Patocchi, A., Fernandez-Fernandez, F., Evans, K., Gobbin, D., Rezzonico, F., Boudichevskaia, A., Dunemann, F., Stankiewicz-Kosyl, M., Mathis-Jeanneteau. F., Durel, C.E., Gianfranceschi, L., Costa, F., Toller, C., Cova, V., Mott, D., Komjanc, M., Barbaro, E., Kodde, L., Rikkerink, E., Gessler, C. and van de Weg, W.E. (2008) Development and testing of 21 multiplexed PCRs composed of SSRs spanning most of the apple genome. Tree Genetics & Genomics 5, 211–223.

Peil, A., Garcia-Libreros, T., Richter, K., Trognitz, B., Hanke, M.V. and Flachowsky, H. (2007a) Strong evidence for a fire blight resistance gene of *Malus robusta* located on linkage group 3. Plant Breeding 126, 470–475.

Peil, A., Hanke, M.V., Flachowsky, H., Richter, K., Garcia, T. and Trognitz, B. (2007b) Developing molecular markers for marker assisted selection of fire blight resistant apple seedlings. Acta Hort. 763, 117–122.

Pereira-Lorenzo, S., Ramos-Cabrer, A.M., Gonzalez-Diaz, A.J. and Diaz-Hernandez, M.B. (2008) Genetic assessment of local apple cultivars from La Palma, Spain, using simple sequence repeats (SSRs). Scientia Hort. 117, 160–166.

Pichler, F.B., Walton, E.F., Davy, M., Triggs, C., Janssen, B., Wunsche, J.N., Putterill, J. and Schaffer, R. J. (2007) Relative developmental, environmental, and tree-to-tree variability in buds from field-grown apple trees. Tree Genetics & Genomes 3, 329–339.

Pilcher, R.L., Celton, J.-M., Gardiner, S.E. and Tustin, D.S. (2008) Genetic markers linked to the dwarfing trait of apple rootstock 'Malling 9'. J. Amer. Soc. Hort. Sci. 133, 100–106.

Planchon, V., Lateur, M., Dupont, P. and Lognay, G. (2004) Ascorbic acid level of Belgian apple genetic resources. Scientia Hort. 100, 51–61.

Quamme, H.A. Hampson, C.R., Hall, J.W., Sholberg, P.L, Bedford, K.E. and Randall P. (2003) Inheritance of apple scab resistance from polygenic sources based on greenhouse and field evaluation. Acta Hort. 622, 317–321.

Richards CM, Volk GM, Reilley AA, Henk AD, Lockwood D, Reeves PA, Forsline PL. (2008) Genetic diversity and population structure in *Malus sieversii*, a wild progenitor species of domesticated apple. Tree Genetics & Genomics (on-line early).

Robinson, T.L, Aldwinckle, H.S., Fazio, G. and Holleran, T. (2003) The Geneva series of apple rootstocks from Cornell: Performance, disease resistance, and commercialization. Acta Hort. 622, 513–520.

Rowan, D.D., Hunt, M.B., Dimourot, A., Alspach, P.A., Weskett, R., Volz, R.K., Gardiner, S.E. and Chagne, D. (2009) Profiling fruit volatiles in the progeny of a 'Royal Gala' x 'Granny Smith' apple (*Malus* x *domestica*) cross. J. Agric. Food Chem. 57, 7953–7961.

Rudell, D., Mattheis, J.P., Maarten, L.A. and Hertog, T.M. (2009) Metabolomic change precedes apple superficial scald symptoms. J. Agric. Food Chem. 57, 8459–8466.

Sancho, A.I., van Ree, R., van Leeuwen, A., Meulenbroek, B.J., van de Weg, E., Gilissen, L.J.W.J., Puehringer, H., Laimer, M., Martinelli, A., Zaccharini, M., Vazquez-Cortes, S., Fernandez-Rivas, M., Hoffmann-Sommergruber, K., Mills, E.N.C., and Zuidmeer, L. (2008) Measurement of lipid transfer protein in 88 apple cultivars. Allergy Immunology 146, 19–26.

Sansavini, S., Belfanti, E., Costa, F. and Donati, F. (2005) European apple breeding programs turn to biotechnology. Chronica Hort. 45, 16–19.

Sargent, D.J., Marchese, A., Simpson, D.W., Howad, W., Fernandez-Fernandez, F., Monfort, A., Arus, P., Evans, K.M. and Tobutt, K.R. (2008) Development of "universal" gene-specific markers from *Malus* spp. cDNA sequences, their mapping and use in synteny within Rosaceae. Tree Genetics & Genomes (on-line early).

Sato, T., Harada, T., Niizeki, M., Kudo, T., Akada, T., and Wakasa, Y. (2004) Allelotype of a ripening-specific 1-aminocyclopropane-1-carboxylate synthase gene defines the rate of fruit drop in apple. J. Amer. Soc. Hort. Sci. 129, 32–36.

Sato, M., Nyui, T., Takahashi, H. and Kanda, H. (2007) Comparison of flowering and fruiting of seedlings from reciprocal crosses between diploid and triploid apple cultivars. J. Japan Soc. Hort Sci. 76, 97–102.

Sawamura K, 1990. Alternaria blotch. In: Jones AL, Aldwinckle HS, eds. *Compendium of Apple and Pear Diseases*. St. Paul, Minnesota, USA: APS Press, 24–25.

Schaffer, R.J., Friel, E.N., Souleyre, E.J.F., Bolitho, K., Thodey, K., Ledger, S., Bowen, J. -H., Ma, J.H., Nain, B., Cohen D., Gleave, A.P., Crowhurst, R.N., Janssen, B.J., Yao, J.L. and Newcomb, R.D. (2007) A genomics approach reveals that aroma production in apple is controlled by ethylene predominantly at the final step in each biosynthetic pathway. Plant Physiology 144, 1899–1912.

Schouten, H.J. and Jacobsen, E. (2008) Cisgenesis and intragenesis, sisters in innovative plant breeding. Letters. Trends Plant Science 13, 260–261.

Schuster, M. (2000) Genetics of powdery mildew resistance in *Malus* species. Acta Hort. 583, 593–595.

Sedov, E.N., Salina, E.S., Levgerova, N.S., Serova, C.M. (2007) Breeding apple varieties for orchards producing raw materials. Russian Agric. Sci. 33, 89–91.

Segura, V., Denance, C., Durel, C.-E. and Costes, E. (2007) Wide range QTL analysis for complex architectural trait in a 1-year-old apple progeny. Genome 50, 159–171.

Shulaev, V., Korban, S.S., Sosinski, B., Abbott, A.G., Aldwinckle, H.S., Folta, K.M., Iezzoni, A., Main, D., Arús, P., Dandekar, A.M., Lewers, K., Brown, S.K., Davis, T.M., Gardiner, S.E., Potter, D. and Veilleux, R.E. (2008) Multiple models for Rosaceae genomics. Plant Physiology 147, 985–1003.

Silfverberg-Dilworth, E., Matasci, C.L., Van de Weg, W.E., Van Kaauwen, M.P.W., Walser, M., Kodde, L.P., Soglio, V., Gianfranceschi, L., Durel, C.E., Costa, F., Yamamoto, T., Koller, B., Gessler, C. and Patocchi, A. (2006) Microsatellite markers spanning the apple (*Malus* x *domestica* Borkh.) genome. Tree Genetics & Genome 2, 202–224.

Soejima, J., Abe, K., Kotoda, N. and Kato, H. (2000). Recent progress of apple breeding at the apple research center in Morioka. Acta Hort. 538, 211–214.

Soufflet-Freslon, V., Gianfranceschi, L., Patocchi, A., and Durel, C. -E. (2008) Inheritance studies of apple scab resistance and identification of Rvi14, a new major gene that acts together with other broad-spectrum QTL. Genome 51, 657–667.

Stankiewicz-Kosyl, M., Pitera, E. and Gawronski, S.W. (2005) Mapping QTL involved in powdery mildew resistance of the apple clone U 211. Plant Breeding 124, 63–66.

Stehr, R. (2009) Standard testing agreement for plant material developed by EUFRIN working group. Acta Hort. 814, 333–336.

Stoeckli, S., Mody, K., Gessler, C., Patocchi, A., Jermini, M. and Dorn, S. (2008) QTL analysis of aphid resistance and growth traits in apple. Tree Genetics & Genome 4, 833–847.

Stushnoff, C., McSay, A.E., Forsline, P.L., and Luby, J. (2003) Diversity of phenolic antioxidant content and radical scavenging capacity in the USDA apple germplasm core collection. Acta Hort. 623, 305–311.

Sun, L., Bukovac, M.J., Forsline, P.L., van Nocker, S. (2009) Natural variation in fruit abscission related traits in apple (*Malus*). Euphytica 165, 55–67.

Szankowski, I., Briviba, K., Fleschhut, J., Schonherr, J., Jacobsen, H.J. and Kiesecher, H. (2003) Transformation of apple (*Malus domestica* Borkh.) with the stilbene synthase gene from grapevine (*Vitis vinifera* L.) and a PGIP gene from kiwi (*Actinidia deliciosa*). Plant Cell Report 22, 141–149.

Szankowski, I., Flachowsky, H., Li, H., Halwirth, H., Treutter, D., Regos, I., Hanke, M.-V., Stich, K. and Fischer, T.C. (2009a) Shift in polyphenol profile and sublethal phenotype caused by silencing of anthocyanidin synthase in apple (*Malus* sp.). Planta 229, 681–692.

Szankowski, I., Waldmann, S., Degenhardt, J., Patocchi, A., Paris, R., Silfverberg-Dillworth, E., Broggini, G. and Gessler, C. (2009b) Highly scab-resistant transgenic apple lines achieved by introgression of HcrVF2 controlled by different native promoter lengths. Tree Genetics & Genomes 5, 349–358.

Tancred, S.J., Zeppa, A.G., Cooper, M. and Stringer, J.K. (1995). Heritability and patterns of inheritance of the ripening date of apples. HortScience 30, 325–328.

Takos, A.M., Jaffee, F.W., Jacob, S.R., Bogs, J., Robinson, S.P. and Walker, A.R. (2006) Light induced expression of a MYB gene regulates anthocyanin biosynthesis in red apples. Plant Physiol. 142, 1216–1232.

Teo G., Suzuki Y., Uratsu S.L, Lampinen B., Ormonde N., Hu W.K., DeJong T.M. and Dandekar A.M. (2006) Silencing leaf sorbitol synthesis alters long-distance partitioning and apple fruit quality. Proc. Natl. Acad. Sci. USA 103, 18842–18847.

Tian, Y.-K., Wang, C.-H., Zhang, J.-S., James, C. and Dai, H.-Y. (2005) Mapping Co, a gene controlling the columnar phenotype of apple, with molecular markers. Euphytica 145, 181–188.

Toivonen, P.AQ.A. and Brummell, P.A. (2008) Biochemical bases of appearance and texture changes in fresh-cut fruit and vegetables. Postharvest Biology Technol. 48, 1–14.

Tränkner, C., Lehmann, S., Hoenicka,H., Hanke, M.V., Fladung, M., Lenhardt, D., Dunemann, F., Gau, A., Schlangen, K., Malnoy, M. and Flachowsky, H. (2010) Over-expression of an FT-homologous gene of apple induces early flowering in annual and perennial plants. *Planta* 232, 1309–1324.

van de Weg, W.E., Voorrips, R.E., Finkers, R., Kodde, L.P., Jansen, J. and Bink, M.C.A.M. (2004) Pedigree genotyping: A new pedigree based approach of QTL identification and allele mining. Acta Hort. 663, 45–50.

Velasco, R., Zharkikh, A., Affourtit, J., Dhingra, A., Cestaro, A., Kalyanaraman, A., Fontana, P., Bhatnagar, S. K., Troggio, M., Pruss, D., Salvi, S., Pindo, M., Baldi, P., Castelletti, S., Cavaiuolo, M., Coppola, G., Costa, F., Cova, V., Dal Ri, A., Goremykin, V., Komjanc, M., Longhi, S., Magnago, P., Malacarne, G., Malnoy, M., Micheletti, D., Moretto, M., Perazzolli, M., Si-Ammour, A., Vezzulli, S., Zini, E., Eldredge, G.,Fitzgerald, L. M., Gutin, N., Lanchbury, J., Macalma, T., Mitchell, J. T., Reid, J., Wardell, B., Kodira, C., Chen, Z., Desany, B., Niazi, F., Palmer, M., Koepke, T., Jiwan, D., Schaeffer, S., Krishnan, V., Wu, C., Chu, V. T., King, S. T., Vick, J., Tao, Q., Mraz, A., Stormo, A., Stormo, K., Bogden, R., Ederle, D., Stella, A., Vecchietti, A., Kater, M. M., Masiero, S., Lasserre, P., Lespinasse, Y., Allan, A. C., Bus, V., Chagné, D., Crowhurst, R. N., Gleave, A. P., Lavezzo, E., Fawcett, J. A., Proost, S., Rouzé, P., Sterck, L., Toppo, S., Lazzari, B., Hellens, R. P., Durel, C. E, Gutin, A., Bumgarner, R. E., Gardiner, S. E., Skolnick, M., Egholm, M., Van de Peer, Y., Salamini, F., and Viola, R. (2010) The genome of the domesticated apple (*Malus xdomestica* Borkh.). Nature Genetics 42, 833–839.

Visser, T.and Verhaegh, J.J. (1976) Review of tree fruit breeding carried out at the Institute for Horticultural Plant Breeding at Wageningen from 1951 to 1976. Proc. Eucarpia tree fruit breeding, Wageningen. pp. 113–132.

Volk, G.M. and Richards, C.M. (2008) Availability of genotypic data for USDA-ARS National Plant Germplasm System accessions using the Genetic Resources Information Network (GRIN) database. HortScience 43, 1365–1366.

Volk, G. M., Richards, C. M., Reilley, A. A., Henk, A. D., Forsline, P. L., Aldwinckle, H. S. (2005) Ex situ conservation of vegetatively propagated species: development of a seed-based core collection for *Malus sieversii*. J. Amer. Soc. Hort. Sci. 130, 203–210.

Volk, G.M., Richards, C.M., Reilley, A., Henk, A.D., Reeves, P.A., Forsline, P.L., Aldwinckle, H. (2008) Genetic diversity and disease resistance of wild *Malus orientalis* from Turkey and southern Russia. J. Amer. Soc. Hort. Sci. 133, 383–389.

Volz, R.K., Alspach, P.A., Fletcher, D.J. and Ferguson, I.B. (2006) Genetic variation in bitter pit and fruit calcium concentrations within a diverse germplasm collection. Euphytica 149, 1–10.

Way, R. D., Aldwinckle, H.S., Lamb, R. C., Rejman, A., Sansavini, S., Shen, S., Watkins, R., Westwood, M.N. and Yoshida, Y. (1990) Apples. (*Malus*). In: J. N. Moore, and J. R. Ballington Jr (eds), Genetic Resources of Temperate Fruit and Nut Crops, 3—62. ISHS, Leuven, Belgium. Acta Hort 290.

Webster, A.D. and Wertheim, S.J. (2003) Apple rootstocks. pp. 91–124. In: Ferree, D. and Warrington, I (eds.): *Apples: Botany, Production and Uses*. CABI Publishing, Cambridge, MA.

Weibel, F. and Haseli, A. (2003) Organic apple production – with emphasis on European experiences. pp, 551–583. *Apples. Botany, Production and Use.* In: Ferree, D.C. and Warrington, I.J. (eds.) CABI Publishing, Cambridge, MA.

Wisniewski, M., Bassett, K., Norelli, J., Macarisin, D., Artlip, T., Gasic, K., and Korban, S. (2008) Expressed sequence tag analysis of the response of apple (*Malus* x *domestica* 'Royal Gala') to low temperature and water deficit. Physiologia Plantarum 133, 298–317.

Zhou, C., Lakso, A., Robinson, T. and Gan, S. (2008) Isolation and characterization of genes associated with shade-induced apple abscission. Mol. Genetics Genomics 280, 83–92.

Zhou, Z.-Q. (1999) The apple genetic resources in China: The wild species and their distributions, informative characteristics and utilization. Genetic Resources Crop Evolution 46, 599–609.

Zhu, L.H., Ahlman, A., Li, X.Y. and Welander, M. (2000) Integration of the rolA gene into the genome of the vigorous apple rootstock A2 reduced plant height and shortened internodes. *J. Hort. Sci. Biotech.* 76, 758–763.

Zhu, Y. and Barritt, B.H. (2008) Md-ACS1 and Md-ACO1 genotyping of apple (*Malus* x *domestica* Borkh.) breeding parents and suitability for marker assisted selection. Tree Genetics & Genomes 4, 555–562.

Zhu, Y., Rudell, D and Mattheis, J. (2008a) Characterization of cultivar differences in alcohol acyltransferase and 1-aminocyclopropane-1carboxylate synthase gene expression and volatile compound emission during apple fruit maturation and ripening. Postharvest Biol. Technol. 49, 330–339.

Zhu, L.H., Li, X.Y. and Welander, M. (2008b) Overexpression of the *Arabidopsis gai* gene in apple significantly reduces plant size. Plant Cell Rep. 27, 289–296.

Zhu, Y.D., Zhang, W., Li, G.C. and Wang, T. (2007) Evaluation of inter-simple sequence repeat analysis for mapping the Co gene in apple (*Malus pumila* Mill.). J. Hort. Sci. & Biotech. 82, 371–376.

Chapter 11
European Pear

Luca Dondini and Silviero Sansavini

Abstract Among the fruit tree species, the European pear (*Pyrus communis* L.) has the most stable cultivar structure. Although the selection activity in the last several centuries has produced several hundred cultivars, only a few pear cultivars are currently grown. Within the Pomoideae (Pyrinae) there are 22 *Pyrus* species, along with another ten or so that have been variously described and assignable to synonyms of the more important species. Perhaps the most widely known species, if not the most widely cultivated, is *P. communis* L. The European pear is essentially the only *Pyrus* species currently grown in Europe while in North America both the European and the Oriental pear are grown. The European pear and its ancestral species, *P. pyraster* Burgsd., grow wild throughout Europe, and it was here where it was domesticated as early as 300 BC. The pear and apple appear to be amphidiploid or allotetraploid species, i.e., those formed by the gametic union of two Rosaceae species of 8 and 9 chromosomes. The high level of genetic recombination combined with selection for fruit size, appearance, flavor, postharvest storability, and resistance to pathogens and diseases has resulted in a diverse array of cultivars. There has been major advances in fruit appearance (shape, color, attractiveness), size, ripening season (summer and fall are predominant) and postharvest traits. Much effort is being invested by researchers to find resistance genes to the main biotic adversities of pear: the fire blight bacterium (*Erwinia amylovora*), the European pear psylla (*Cacopsilla pyri*), which is the vector of the phytoplasma causing pear decline, the scab causing fungi *Venturia pyrina*, and the black spot fungus *Stemphylium vesicarium*.

Keywords *Pyrus communis* • Origin • Varieties • Breeding Goals • Breeding, Biotechnology • Self incompatibility • Genetic transformation • Pome fruit • Pip fruit

L. Dondini (✉) • S. Sansavini
Dipartimento di Colture Arboree, Università degli Studi di Bologna,
via Fanin 46, Bologna 40127, Italy
e-mail: luca.dondini@uinibo.it; fruitseg@agrsci.unibo.it

M.L. Badenes and D.H. Byrne (eds.), *Fruit Breeding*, Handbook of Plant Breeding 8,
DOI 10.1007/978-1-4419-0763-9_11, © Springer Science+Business Media, LLC 2012

1 Introduction

Among the fruit tree species, the European pear (*Pyrus communis* L.) has the most stable cultivar structure. Although the selection activity in the last several centuries has produced several hundred cultivars, only a few pear cultivars are currently grown. In Europe just eight cultivars ('Conference,' 'William' = 'Bartlett' = 'Bon Chrétien' and its red sports, 'Abbé Fétel,' 'Blanquilla'/'Spadona,' 'Doyenne du Comice,' 'Kaiser,' 'Dr. Jules Guyot,' and 'Coscia') represent 80% of the production.

Of the more than 20 million tons of pears (2007) produced in the world, the vast majority, about 19 million tons, are produced in the northern hemisphere. The most important producer countries are China (about 60%, mainly *P. pyrifolia* Nakai and *P. ussuriensis* Maxim pears), Italy (4%, main European pear producer) and United States (3.8%). In the southern hemisphere, Argentina, South Africa, and Chile are the major producers (WAPA 2009). Even though the fruit quality of the most important pear and apple cultivars is comparable, apple production is about 3 times that of pear production. This, in part, is due to the better postharvest storage capabilities of apple cultivars as compared to pear cultivars. Pear production is mainly used for fresh consumption, but a significant part is processed as nectar, juice, canned (almost only the cv. 'William') or dried pears as well as other typical products as beverages or distilled spirits. Cooked pears are also appreciated in very popular recipes.

Within the Pomoideae (Maloideae) there are 22 *Pyrus* species (Bell 1996), along with another ten or so that have been variously described and assignable to synonyms of the more important species. Perhaps the most widely known species, if not the most widely cultivated, is *P. communis* L., whose roots are to be found in the Caucasus and Asia Minor and whose ancestral species, *P. pyraster* Burgsd., can still be found growing wild in these areas. This latter species gave rise to all the European pear cultivars that are grown today in the world (Fig. 11.1). Worth mentioning too in the European context is *P. nivalis* Jacq., or perry pear, which by tradition is found in England and northern France and used for making a moderately alcoholic cider.

There are two other important areas for the origin of pear. One is China, which is the home of the species *P. pyrifolia* Nakai (also known as *P. serotina*) and *P. ussuriensis* Maxim. The former, which is called "Chinese sand pear" or "White pear" (Bao et al. 2007), is best known internationally by its Japanese name "nashi." In contrast, *P. ussuriensis* is virtually unknown outside the Orient. The second area is in western Asia in the area comprising Afghanistan, India, and the Asian republics of the former Soviet Union (Fig. 11.1). The cultivars grown here appear to be intermediate types falling somewhere between *P. communis* and *P.* × *bretschneideri* Rehd., the latter in turn being a hybrid of *P. ussuriensis* × *P. pyrifolia* (Vavilov 1951; Bell 1996), while *P. calleryana* Decne. is used only as a pear rootstock.

The European pear is essentially the only *Pyrus* species currently grown in Europe while in North America both the European and the Oriental pear are grown. This reflects the more globalized consumer market and greater multiethnic (especially Asian) population of North America as compared to Europe. Asia's consumers, on the other hand, even in those countries that are more receptive to international

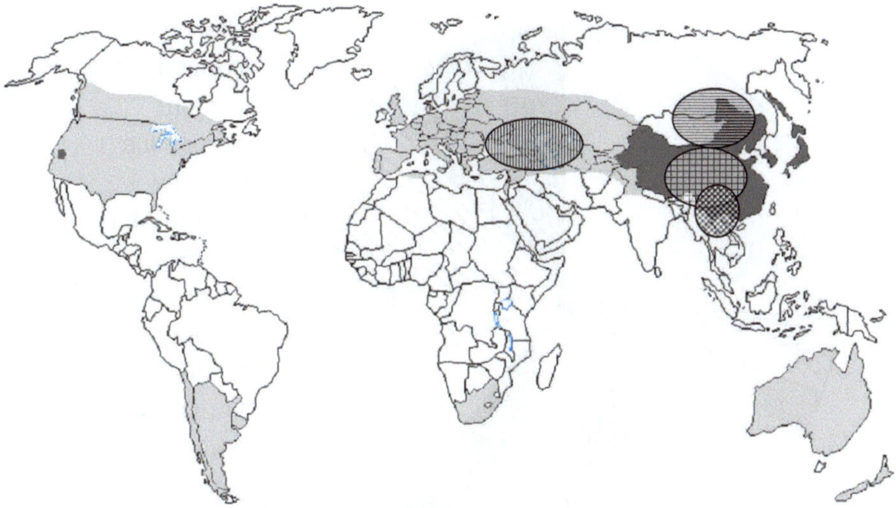

Fig. 11.1 Cultivated and wild species distribution: European (*light gray*) and Asian (*dark gray*) cultivated pears. Origins of *Pyrus* ssp are indicated by circles: *Pyrus communis* (*vertical lines*), *P. pyrifolia* (*squared*), *P. ussuriensis* (*horizontal lines*), and *P. calleryana* (*dotted*)

tastes, continue to prefer nashis and other indigenous cultivars. While hybrid cultivars of *P. communis* × *P. pyrifolia* like the recent cv. 'Maxie' from New Zealand have yet to gain widespread acceptance, many breeding programs throughout the world are working on novel hybrid cultivars. That said, it seems best to confine our discussion and remarks in the rest of the chapter to *P. communis*, with an occasional glance at *P. pyrifolia* for comparative purposes.

The distribution of European pear production (Fig. 11.2) reflects the environmental limitations of this species. It is usually less cold hardy than apple and can have its fruit quality dramatically affected by excessively high temperatures during summer. Moreover, the same environmental limits are typical of the quince; the most used rootstock for pear production. Breeding is the major approach to overcome these climatic bottlenecks and expand pear production.

2 Origin and Domestication

The European pear (*P. communis*) and its ancestral species, *P. pyraster* Burgsd., grow wild throughout Europe and it was here where it was domesticated as early as 300 BC (Hedrick 1914). The genus *Pyrus* has clearly developed through natural hybridization and subsequent environmental and human selection to give us the cultivars and morphologically distinct fruits we see today. These events in turn underlie the diversity of the worlds' pear growing districts as well as that of its markets and preferences of its consumers.

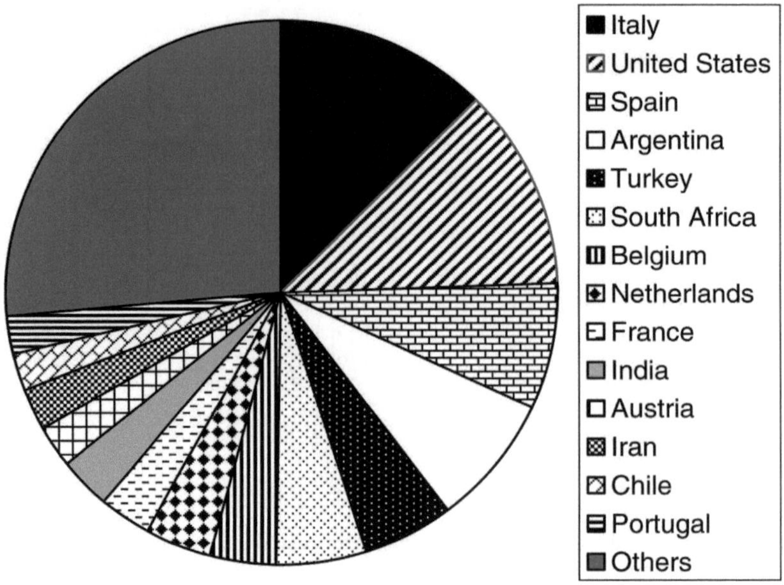

Fig. 11.2 Production of European pears (6,741,000 tons on a total of 20,579,492 tons) in the world (year 2007). (Source WAPA 2009)

The golden age of pear breeding was the century from 1750 to 1850 when amateur naturalists in the service of noble and rich households or even in religious communities began selecting among large numbers of seedlings derived from open-pollinated flowers. One of the first breeders to perform controlled crosses of pear was the Czech monk Gregor Mendel in Brno, whose efforts received public recognition in 1883 (Vavra and Orel 1971).

The cultivar assortment of European pear is a fairly stable mix, with a dozen time-honored cultivars in Europe and four cultivars in the USA (Table 11.1) accounting for over 80% of the market in each region (Seavert 2005; Mielke 2008). What is perhaps most striking, is that some of these cultivars date back more than 150–200 years. One might rightly be tempted to wonder whether the world's pear breeders failed. Yet such a surmise would in truth be wide of the mark. While the number of new pear releases have been far fewer than those of apple, there are a good number of breeding stations throughout the world that work on pear, including 15 or so in Europe, ten in North America and several others in South Africa, Australia and New Zealand (Table 11.2). In the last 15 years, nearly 300 novel cultivars, including about 200 of European pear and a hundred Asian pear cultivars have been released (Fig. 11.3). In the last decades the European countries (France, Germany, Italy, Romania, Spain, Switzerland, and United Kingdom) developed breeding programs mainly focused on fruit quality or on fruit type diversification (shape, maturity date, flesh type, or other traits) while in North America breeding was focused mainly on resistance to pathogen and pests (USDA and Kerneysville in USA; Harrow Station

Table 11.1 Pear production from 1999 to 2006 in Europe (15 countries) and separately, for Spain, Italy (Source: Eurofel), and USA (Mielke 2008)

	Europe %			Spain %			Italy %			USA %		
	99-00	03-04	06-07	99-00	03-04	06-07	99-00	03-04	06-07	99-00	03-04	06
Conference	24.5	30.5	31.6	24.1	32.6	33.5	16.2	14.8	13.6	–	–	–
William (Bartlett)	12.3	13.1	13.1	3	7.4	8.3	20.5	20.8	21	–	–	–
Total summer pears[a]	–	–	–	–	–	–	–	–	–	63.8	59.2	52.1
Max Red Bartlett	1.3	1.2	1.2	–	–	–	3.5	3.2	3.1	–	–	–
Abbé Fétel	9.7	10.9	12.9	–	–	–	27.3	30.8	34.6	–	–	–
Blanquilla—Spadona E.	9.6	7.8	5.4	36.1	30.7	26.9	–	–	–	–	–	–
Doyenné du Comice	5.1	4.9	5.1	–	–	–	7	6.1	5.6	0.5	0.6	0.8
Coscia—Ercolini	4.8	4.2	4.2	9.1	7.2	8.2	6.4	6.3	6.7	–	–	–
Dr. J. Guyot (Limonera)	5.1	4.5	4.2	7.7	7.6	8.7	0.7	0.5	0.3	–	–	–
Kaiser (B. Bosc)	2.2	2.5	2.4	–	–	–	6.1	6.8	6.5	7.8	8.4	10.6
Passe Crassane	1.6	1.5	1.2	0.6	0.5	0.5	2.3	1.8	0.9	–	–	–
Beurré d'Anjou	–	–	–	–	–	–	–	–	–	26.3	29.2	33.4
Other	23.9	19.1	18.7	19.4	13.9	13.9	9.9	8.9	7.8	1.5	2.5	3.1
*Total (000*t)	*2,351*	*2,301*	*2,472*	*627.5*	*580.5*	*494*	*831*	*829.5*	*922*	*809.5*	*742.5*	*627*

[a]USA data included "Bartlett" and "Max Red Bartlett" in the group of "Total summer pears" in which they are the main cultivars

Table 11.2 Main cultivars released in Europe in the last 20 years

Country	Institute	City	Cultivars released
Belgium*	Gebroeders Saels,	Herk de Stad	Corina (Vroege Conference Saels—mutation of Conference)
France	INRA	Angers	Angelys®, Bautomne
	G. Delbard	Commentry	Delsavor, Delbuena
Germany	Inst. Fruit Res.	Pillnitz	Isolda®[a]; Hermann®[a]; Hortensia®[a]; Uta®[a]; David®[a]; Gerburg®[a]; Elektra®[a]
England	East Malling Research Station	Maidstone	Concorde
	John Innes Institute	Norwich	Jowil®Dolacomi*
Italy	CRA—Ist. Sperimentale Frutticoltura.	Forlì	Carmen*; Aida*[b]; Boheme*[b]; Turandot*
	Istea—CNR	Bologna	Abate Light; Conference Light, William Ramada
	DOFFI	Florence	Etrusca; Sabina
Moldova	Instit. de Cercetari pentru Pomicultura	Chisinau	Xenia (Noiabriskaia)
Netherland	CPRO-DLO	Wageningen	Verdi* (Sweet Blush®)
	S.K. Broertjes	Wijdenes	Sweet sensation®
Poland	Skierniewice Res. Inst.	Skierniewice	Hnidzik
Czech Republic	Breeding Station of Fruit Tree	Techobuzice	Dicolor; Bohemica; Delta[a]; Dita; Erika; Omega
Romania*	Fruit Research Station	Voinesti	Corina[b,c]; Tudor[c]; Euras[b]
	Fruit Tree Institute	Pitesti-Maracineni	Daciana; Carpica; Getica[b,c]; Monica[a–c]
	Fruit Tree Res. Station	Cluj	Ina-Estival[c]; Haydeea[b,c]
Spain	IRTA	Lleida	IGE 2002 (mutation of Dr. J. Guyot)
Switzerland	Swiss Federal Research Station	Changins	Champirac; Valerac

N.B. Two cultivars (Belgium and Romania) have been released with the same name: Corina
[a]Resistance/tolerance to scab
[b]Resistance/tolerance to fire blight
[c]Resistance/tolerance to psylla
®Registered trademarks
*Patented varieties

in Canada) as well as on selection for red-skinned (both by breeding as ‘Cascade’ or by natural mutation selection as the case of ‘Doyenne du Comice’ and ‘Beurré d’Anjou’) or bronze-skinned fruits (sports of Beurré Bosc).

The chief aim of breeders and growers is to develop cultivars that bear well and are tolerant to the swings of the seasonal weather that can favor fruit drop and physiological disorders in trees or fruit. Cultivars like the English-bred ‘Concorde’ and the Belgian ‘Beurré d’Anjou’ although well adapted to their places of origin cannot be grown in southern countries, like Italy, because of the high incidence of fruit disorders like corky spot.

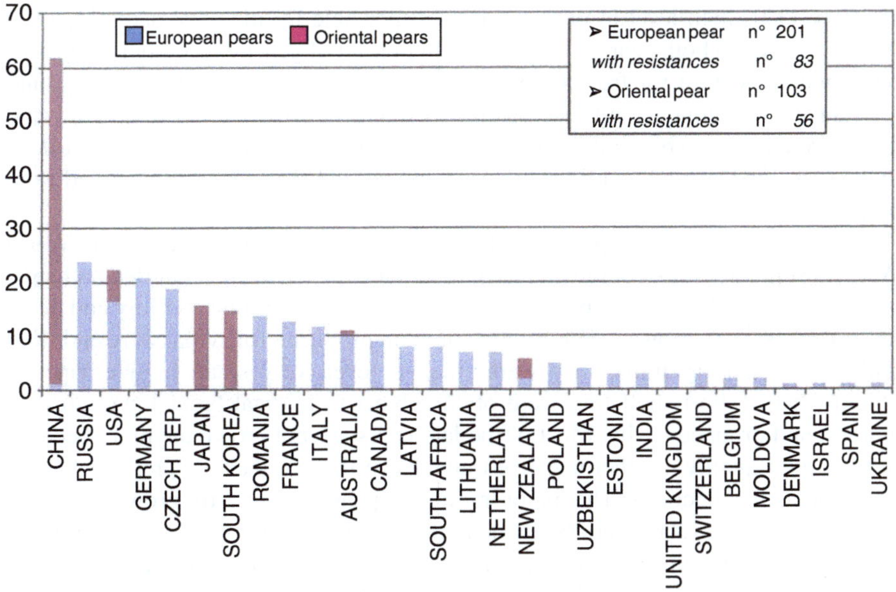

Fig. 11.3 Number of new cultivars released in the world (from 1991 to 2007; source CRA–ISF Rome). Of the cultivars released, 83 of the 201 European pears and 56 of the 103 Oriental pears released were reported to have resistance against a pathogen or pest

3 Genetic Resources

The chromosomal formula for pear, as for the other Pomoideae, is complex: $2n=34$, where $n=17$ chromosomes. While several explanations have been advanced to account for this natural polyploidism, the one most widely accepted is that pear and apple are amphidiploid or allotetraploid species, i.e., those formed by the gametic union of two Rosaceae species of 8 and 9 chromosomes that subsequently led to diploidism. In effect, the behavior of these species and trait segregation in their progeny is typically diploid (Crane and Lewis 1942). The commercial genotypes most commonly found today are diploids although there are triploids ($2n=51$ chromosome) and even tetraploids. Commercial triploid cultivars produce little good pollen, so that cropping in an orchard requires at least two intercompatible cultivars. The tetraploids have aroused interest because of their extra large fruit and leaves although generally, they yield poorly (Zielinski and Thompson 1967).

While hybridization is easy within *Pyrus* species, intergeneric hybridization within Pomoideae is not. The few *Pyrus* × *Malus* hybrids that do exist are not cultivated. Both *Pyrus* and *Malus* have a system of gametophytic self-incompatibility determined by the S-locus, which enforces natural outcrossing with another interfertile genotype. This high level of outcrossing is responsible for the wide natural genetic variability seen in both pear and apple. The high level of genetic recombination combined with selection for fruit size, appearance, flavor, postharvest storability, and resistance to pathogens and diseases has resulted in a diverse array of cultivars. Thus, even today the best cultivars, such as the 'William' pear in England ('Bartlett' in the U.S. and Canada) and 'Doyenne du Comice' in France are those derived from natural crosses.

When today's intensive pear industry began to take root after the Second World War, it swept away most of the old native pear cultivars growing in garden orchards, especially the early ripening pears that were picked daily in June and July and taken to local markets by growers. While some of them produced fruits that were endowed with prized sensory traits, they were small or even very small in size, ripened quickly, had a short shelf life, were often subject to internal breakdown, grew in clusters and dropped easily. Most of the late pears are not grown anymore because of their poor quality or other negative traits. Nevertheless, they are preserved today in germplasm collections. The major traditional pear cultivars still grown today are (Table 11.3; Fig. 11.4) generally divided into four major groups according to ripening season: early summer, mid-season summer, autumn, and winter pear cultivars (Baldini and Scaramuzzi 1957; Nicotra et al. 1979).

3.1 Early Summer Pears

Among the early-season cultivars are 'Coscia' ('Ercolini'), 'Spadona' ('Blanquilla'), 'Dr. J. Guyot' ('Limonera'), and 'Butirra Precoce Morettini,' which are still grown in the southern European pear districts (Table 11.3). 'Coscia,' which is grown in Italy, Spain, and Greece, is an excellent quality early cultivar although its cropping is variable and it is slightly susceptible to internal breakdown. 'Butirra Precoce Morettini' is one of the country's few cultivars from the Italian breeding work by Morettini in the 1950s that proved successful in Italy as well as in Greece, Spain, Eastern Europe, and the USA (California). Although it has been phased out in many orchards, there are still old plantings that remain commercially viable because the cultivar has higher yields and larger size than 'Coscia.' 'Spadona' is a very old cultivar that is largely grown in southern districts inhospitable to 'William.' Currently it is losing importance in most orchards because of crop management problems. It used to be treated with the now banned bioregulator Cycocel in Catalonia districts to control shoot and tree vigor and increase fertility and fruit size. 'Dr. J. Guyot' was once a very popular early-season cultivar but is currently no longer of commercial interest and it is being phased out in Italy and in France. Recently, the precocious Italian cultivar 'Carmen,' which is qualitatively better than 'Coscia,' is being planted more in Europe.

Table 11.3 Traditional and commercial pear cultivars grown in Europe

Cultivar	Origin	Year	Tree vigor[a]	Production[b]	Quince compatibility	Scab resistance[c]	Fire blight resistance[c]	Psylla resistance[c]	Fruit size[d]	Harvest[e]	Fruit shape	Fruit skin[f]	Texture	Quality[g]	Storage[h]
Coscia (Ercolini)	Italy	1800s	M	H (s)	Fair	S	–	S	S–M	E	Short pyriform	DY	Fine	Juicy arom.	–
Butirra Precoce Morettini	Italy	1956	M	G	–	–	–	–	S–M	E	Pyriform	GY	Melting	–	IB
Spadona (Blanquilla)	–	–	VH	H (s)	Exc	–	–	–	–	E	Pyriform	GY, blush	Med firm	M	–
Dr. J. Guyot (Limonera)	France	1870	L–M	H (ch)	–	MR	MR	–	L	E	Short pyriform	GY	–	F	–
Clapp's Favorite	USA	–	M	M	–	–	–	S	L	M	Roundish	GY 30–50% blush	–	M	IB
Santa Maria	Italy	1951	H	H (s/n)	Fair	S	–	S	M–L	M	Roundish pyriform	GY, Pink blush	Fine melting	M	–
William (Bartlett)	England	1756	M	M	Poor	MS	–	S	M–L	M	Pyriform to roundish	Y, sm lenticels sl pink blush	Fine firm	Exc arom.	M
Rocha	Portugal	1840	M	H (p)	Exc	VS	–	MS	VL	A	Turbinate	GY, some russet		Exc juicy	Exc
Conference	England	Late 1800s	L–M	VH (p,pr)	Exc	MS	S	S	M–L	A	Pyriform	GY, blush, high russet	Fine melting	Exc juicy	Exc
Abbé Fétel	France	1866	M	VH (p,pr)	Poor	–	–	–	L	A	Pyriform	Y, lenticels some russet	Firm crispy	Neutral flavor	M
Doyenne du Comice	France	1849	VH	M (i, np)	–	S	–	–	VL	A	Turbinate	GY some russet	–	Exc juicy	Exc

(continued)

Table 11.3 (continued)

Cultivar	Origin	Year	Tree vigor[a]	Production[b]	Quince compatibility	Scab resistance[c]	Fire blight resistance[c]	Psylla resistance[c]	Fruit size[d]	Harvest[e]	Fruit shape	Fruit skin[f]	Texture	Quality[g]	Storage[h]
Packham's Triumph	Australia	Late 1800s	M	F–M (pr)	None	S	–	S	L	A	Pyriform	G to Y, bumpy, russet/calyx	Firm	Exc	M
Beurré Bosc	Belgium	1830	H	H (np)	None	S	S	S	L	A	Bulging pyriform	Bronze, russet	Firm sl grainy	M aromatic	M
Beurré d' Anjou	Belgium	–	M	L–M (np)	–	–	–	S	M–L	A	Doliform	Bronze, some russet	Fine melting	M	M
Bonne Louise d' Avranches	France	Late 1700s	H	M (pr)	–	–	–	VS	S–M	A	Pyriform	G, RO flecks/stripes	Grainy	F–M	M
Passe Crassane	France	1855	H	H (pr)	–	S	S	MS	L	W	Oblate	G-YG, Y lenticels, some russet	Firm melting sl grainy	M aromatic	M
Florelle	Germany	Late 1600s	H	H	–	–	–	–	S–M	W	Ovoid	–	Crispsl grainy	M aromatic	–
Doyenne d' Hiver	Belgium	1800	H	F–M (pr)	–	S	S	S	VL	W	Globular to oblate	G	Grainy	M	–

[a]Vigor: *L* low, *M* medium, *H* high, *VH* very high
[b]Production: *F* fair, *M* medium, *H* high, *pr* precocious, *np* not precocious, *p* parthenocarpic, *i* inconsistent, *ch* cold hardy, *s* southern adaptation, *n* northern adaptation
[c]Resistance: *VS* very susceptible, *S* susceptible, *MS* medium susceptible, *MR* medium resistance, *R* resistant
[d]Fruit size: *S* small, *M* medium, *L* large, *VL* very large
[e]Harvest season: *E* early, *M* mid season, *A* autumn, *W* winter
[f]Skin color: *G* green, *GY* greenish yellow, *Y* yellow
[g]Quality: *P* poor, *F* fair, *M* medium, *Exc* excellent, *arom* aromatic
[h]Storage: *IB* internal breakdown, *M* medium, *Exc* excellent

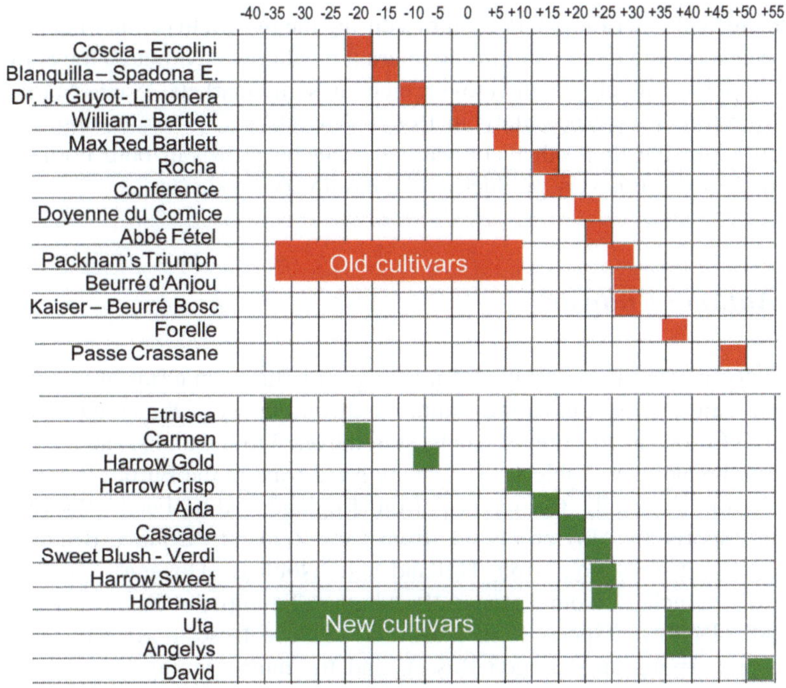

Fig. 11.4 Harvest time of pear cultivars (expressed as days before and after 'William,' syn 'Bartlett'). In the Po valley (Italy) 'William' is picked from the 10th to the 20th of August

3.2 *Mid-Season Summer Cultivar*

Unlike their early-season counterparts, these pears have a longer shelf life and some can be stored for several months (Table 11.3), which allow their use in international trade. Some are subjected to ripening treatments at about 20°C in an ethylene-enriched atmosphere for early market delivery, although this practice can adversely affect their quality, if applied too early.

The main cultivar (~13% European production) 'William,' an old English cultivar (1765) that was renamed 'Bartlett' when it was introduced into North America in the early 1800s, is still unsurpassed as the best summer pear cultivar in both Europe and the Americas. It is also the only one used by the canning industry, juice making and fresh-cut slices alone or in fruit salads. 'Max Red Bartlett' is the most widely grown red sport of 'William.' Unfortunately, the mutation is chimeric and develops numerous chromatic variations from the original red to a greenish yellow with discolored stripes. Other red sports such as 'Sensation,' 'Rosired,' and 'Homored' are used to a lesser extent or not at all.

The other mid season pear cultivars are 'Clapp's Favourite' and 'Santa Maria.' 'Clapp's Favourite' and its dark red fruited sport 'Starkrimson' were widely grown in Germany and other European countries during the twentieth century but have lost favor because of their susceptibility to internal breakdown. Although the cultivar is not being planted much, it still enjoys a niche in some markets. 'Santa Maria' bred by A. Morettini at Florence, Italy, in 1951 ('William' × 'Coscia') is more environmentally adaptable than other cultivars and can be grown in both northern and southern districts. It is very productive and has a very good size.

3.3 Autumn Cultivars

These cultivars are the mainstay of Europe's pear industry in terms of volume, fruit size, and fruit quality. Unlike the summer cultivars, most of the autumn pears are familiar to European consumers, can be stored from 3 to 6 months, and are resistant to internal breakdown (Table 11.3). Common cultivars in the European market are 'Rocha' from Portugal; 'Conference' from England; 'Abbé Fétel,' 'Doyenne du Comice,' (sports are 'Taylor's Gold' from New Zealand, 'Comice Bronzé' from France, 'Sweet Sensation' from Holland and 'Noblesse Doyenne') and 'Bonne Louise d'Avranches' from France; 'Packham's Triumph,' ('Bingo' (1993) and 'Serenad' (1999) are sports from Australia) 'Beurré Bosc,' ('Kaiser' or 'Kaiser Alexander') (sports are 'Golden Russet' and the later-ripening 'Bronzé Beauty' from USA) and 'Beurré d'Anjou' (sport is 'Columbia Red Anjou' from the USA) from Belgium. Concerning their characteristics, 'Conference' is the European pear par excellence (~32% of European production), suitable for production in northern and southern districts, very sweet, with melting flesh, optimum flavor, and long shelf life; 'Abbè Fètel' is the pear rediscovered for the original elongated shape, taste, and its recent market claim for which it actually excels in the southern orchards. 'Doyenne du Comice' is a classical French pear, very juicy with melting flesh, excellent in terms of flavor. 'Beurré Bosc' is a pear of tart-to-sweet taste with a characteristic totally bronze fruit skin. Other pear cultivars are 'Rocha' (90% of Portugal production) and 'Bonne Louise d'Avranches' (grown in central and northern Europe due to cold hardiness but with poor quality). 'Conference,' 'Abbè Fétel,' and 'Rocha' have been described also to produce via parthenocarpy. Cultivars such as 'Doyenne du Comice,' 'Packham's Triumph,' and 'Beurré d' Anjou' are losing favor in European orchards because of management difficulties, size, or productivity. Nevertheless, a few of these ('Packham's Triumph' and 'Beurré d' Anjou') are grown extensively in North America (Oregon), South America, or South Africa.

3.4 Winter Cultivars

Many of the once widely grown winter cultivars like 'Doyenne d'Hiver,' 'Alexandre Lucas,' 'Bergamotte Esperen,' have lost ground in the market place either to autumn cultivars because of their longer storability or to imports of freshly picked pears from South America and South Africa. One exception to this trend is the recently released French 'Angélys,' which joins the winter stalwart 'Passe Crassane' a cultivar released in Franc in 1855. 'Passe Crassane' peaked in popularity in the 1960s—but has declined dramatically in production because it has a lower overall quality than the best autumn cultivars and runs into storage problems if kept longer than 4 months. 'Forelle,' a crisp flesh cultivar that has always enjoyed a niche market in Germany and northern Europe has recently found renewed favor because its fruit is pleasingly different from the melting flesh of traditional pear cultivars (Table 11.3).

4 Major Breeding Achievements

4.1 Recent European Pear Releases

Of the hundred or so cultivars introduced over the past 30 years, few have had a lasting impact on today's modern pear industry (Table 11.3). These cultivars, from both public- and private-sector breeders, reflect the trends the nursery sector exhibits for new orchard plantings, including those covered by exclusive propagation and marketing rights. The new cultivars listed represent one or two of the most notable cultivars from each public-sector breeding program that has been tested in field trials by the pear working group set up by Italy's Agriculture Ministry (Table 11.4). Unfortunately, the range of adaptability of these are not known as most of these have not been widely tested across the range of environment in Europe. For example, there are cultivars like the U.K. 'Concorde,' which crop poorly in Southern Europe and other cultivars like 'Abbé Fétel,' which produce well in southern districts but not in northern ones. Except for 'William,' we have decided not to place too much emphasis on mutations, or sports, of the cultivars reviewed in that almost all of them are spontaneous and either unstable chimeras or crop less than the original cultivar. Not to mention the fact that several red mutant clones of 'Doyenne du Comice' are not being grown by growers because they are less vigorous and productive.

The major advances in the new cultivars depend mostly on the market traits that are still good appearance (shape, color, attractiveness), size (bigger than in the past but, differently from the Asiatic markets, very big fruits are not requested by the consumers), ripening date (summer and fall are predominant), and storability (longer shelf-life). While in the past most of the cultivars were melting and juicy, the new cultivars have firmer flesh, crispness associated with juiciness, few or no sclereids, and lack the astringent aftertaste of the older cultivars. For this reason winter cultivars are very limited and pear industry would prefer fall cultivars with long storability and shelf life. Another advance is the environmental adaptability of new cultivars, which makes them suitable for the sustainable agriculture. Unfortunately, commerce still prefers the older traditional cultivars.

Table 11.4 Main traits of several new pear cultivars, from the main breeding programs

				Tree traits		Resistance		Fruit		
Cultivar	Country	Year	Origin	Vigor	Production	Fire blight	Scab	Harvest	Size	Color
Aida	I	2003	Coscia × Dr. J. Guyot	M	H	T		+15	4	Yellow/green/red
Angelys	F	1999	Doyenne d'Hiver × Doyenne du Comice	M	H	–	–	+40	5	Bronze
Bautomne (Serenade)	F	1988	Conference × D. d'Hiver	M	H	–	–	+30	4	Green/red
Black Pride	USA	1998	US446 × US 505	M	M–H	T/R	–	+21	3	Yellow
Calired (Zaired)	USA	1997	Red Bartlett × Starkrimson	V	M	–	–	+5	3	Red
Carmen	I	1999	Dr. J. Guyot × Bella di Giugno	M	M	–	–	–20	3/4	Green/red
Cascade (Lombacad)	USA	1986	Max Red Bartlett × D. Comice	M	M	–	–	+20	5	Red
Crispie	NZ	1999	Niisseiki (*P. pyrifolia*) × Max Red Bartlett	V	–	–	–	–8	3/4	Yellow
Concorde	GB	1988	Conference × D. Comice	M–V	H			+20	4	Green/russet
David	D	1998	Dr. J. Guyot × D. Comice	W	H	R	–	+55	3	Green
Delsavor (Gourmande)	F	1997	Delbias × Conference	M	M	–	–	+25	3	–
Delta	Cz	1995	Kaiser × Comtesse de Paris	–	H	–	T	+56	–	–
Electra	D	1998	Forelle × Clapp's Favorite	–	–	–	T/R	+39	–	Red
Elliot (Selena)	USA	1988	Elliot n. 4 × Vermont Beauty	M	M	R	–	+20	3	Bronze
Erica	Cz	1995	Kaiser × President Drouard	M	H	–	–	+58	–	–
Etrusca	I	1991	Coscia × Gentile	–	H	–	–	–35	2–3	Green/red
Euras	R	1994	(*P. pyrifolia* × O. de Serres) × D. d'Hiver	–	–	–	T	+35	3	–
Harrow Crisp	CDN	1999	William × US 56112-146	M	H	R	–	+10	3	Yellow
Harrow Gold	CDN	1999	Harvest Queen × Harrow Delight	V	M	R	–	–12	3	Yellow

Harrow Sweet	CDN	1991	William × (Old Home × Early Sweet)	V	M	R	–	+26	3	Yellow
Hortensia	D	1998	Forelle × Clapp's Favorite	–	H	T	R/T	+25	3	Red
Manon	D	1999	Kaiser, open impollinated	–	–	T/R	R/T	+36	4	Bronze
Maxie	NZ	1999	Nijisseiki (*P. pyrifolia*) × Max Red Bartlett	V	M	–	–	+5	3	Yellow
Monica	R	1994	Santa Maria × Principessa Gonzaga	M	M	R	R/T	+25	4	Yellow
Red Silk	USA	1997	Red Bartlett × B. Anjou	W	H	–	–	+25	4	Red
Rosemarie	SA	1990	Bon Rouge × Forelle	–	–	–	–	–10	3	Yellow/red
Shenandoah	USA	2002	Red Bartlett × US56112-146	M	H	T/R	–	+28	4	–
Uta	D	1998	M.me Verté × Kaiser	M/W	H	T	T/R	+40	5	Bronze
Valerac	CH	1997	Conference × President Héron	–	–	–	–	+26	–	–
Verdi (Sweet Blush)	NL	1992	Bonne Louise d'Avranches × D. Comice	–	–	–	–	+20	4/5	Green
Xenia (Noiabriskaja)	M	2005	Triomphe de Vienne × Krier	–	–	–	–	+35	4	–

Legend: *V* vigorous, *M* medium, *W* weak, *R* resistant, *T* tolerant, harvest (number of days before or after William); Fruit size (from 1 to 5), Color (main skin color); – no information

4.2 Rootstocks

Almost all of the rootstocks employed in today's intensive pear orchards are clonal, with the few seedling stocks left like Kirchensaller being phased out. These clonal stocks are widely used because they control tree growth habit, induce early bearing, and promote consistent and high yields of larger and better quality fruit, all factors that encourage a more uniform and easily managed orchard. Pear can also be self-rooted via micropropagation, but, as Wertheim (1998) has observed, cultivars on "their own roots do not perform as well as on a suitable rootstock."

There are only two species of stocks in Europe and the Americas where *P. communis*, the common pear, is grown: selected clones of common pear and quince (*Cydonia oblonga* Mill.). Since the growers in Asia, the home of the nashi *P. pyrifolia*, cannot use quince because it is incompatible with nashi, they use *P. calleryana*, *P. pyrifolia*, *P. betulaefolia*, *P. ussuriensis*, and other oriental species of pear. In actual fact, quince is not fully compatible with common pear either and this partial incompatibility accelerates cropping while restraining shoot and root growth. When quince-pear incompatibility is most severe, as with cultivars like 'Packham's Triumph,' 'William,' and 'Buerré Bosc,' a common pear stock or an interstem with quince, such as 'Buerré Hardy,' is used. An interstem is also used with partially compatible cultivars like 'Conference' and 'Abbé Fétel' to boost tree efficiency, fruit size, and orchard life (Sansavini 2007).

Thanks largely to the selection of quince clones at the research stations of East Malling in the U.K. and of INRA at Angers in France, there are a number of clonal lines that can be employed in plantings with densities as high as 2,000–5,000 trees/ha in a range of environmental conditions, orchard designs and end-market uses. Nevertheless, quince is not well suited for heavy or calcareous soils which are iron deficient (inducing leaf chlorosis), in districts without irrigation, or in climates that are too hot in the summer creating graft compatibility problems or too cold in the winter In addition some clones have proved to be quite susceptible to viruses and phytoplasms (i.e., BA29 is more susceptible than Sydo to pear decline) and others not easily rooted from hardwood cuttings. All of these drawbacks are the reasons why nurseries often employ interstocks even with quince-compatible cultivars and usually propagate via stool beds, which is the most economic method and provides healthy plants with proper virus-indexing.

Micropropagation, albeit easy to do, is only used for those hybrid pear seedling stocks of poor rhizogenesis like the OHxF series and for pear cultivars without stocks for adverse soil conditions such as with cv. 'William' processing orchards in some lowland areas with calcareous soils in Italy.

Thus today's intensive pear industry is still in search of the ideal rootstock. Currently, nearly 70% of the clones used by European growers are one of four quince rootstocks. In conventional orchards, Sydo and Provence BA29 are used. Adams and MC rootstocks are used for high-density orchards (up to 3,000–4,000 trees/ha) in newly planted fertile soils. The most frequent scion-stock combinations employed by Italian pear growers are (a) Sydo followed by MC, the latter with or without an interstock, for 'Abbé Fétel'; (b) BA29, Sydo and MC, equally for

'Conference' (Massai et al. 2008); (c) BA29 and MA for 'Doyenne du Comice'; (d) OHxF clonal seedlings Farold 40 and 69 followed by BA29 with a 'Beurré Hardy' interstem (and/or 'Curato') for 'William,' and (e) Sydo and BA29, both with a 'Beurré Hardy' interstem, for 'Beurré Bosc.' Adams is especially popular in Belgium, Holland, and in Central Europe, though absent in Italy. Alternative stocks for the latter two cultivars, largely quince-incompatible cultivars are OHxF40 and 69, Kirchensaller seedling and, less commonly, Fox 11 and new French and German pear clonal rootstock selections.

Quince. The most popular quince clones in Europe are MA and MC bred at Horticulture Research International East Malling station and Sydo and Provence clone BA29 selected at INRA's Angers station for their dwarfing and graft compatibility. The most recent addition to this list is East Malling's MH (QR-193.16), whose field performance is similar to that of Sydo and MC.

MA is the oldest, having been bred before WW II to improve upon the quince rootstock of the Angers type. Despite its propagation via stoolling, it has been surpassed in popularity by Sydo for pear orchards in Spain, France, and Italy because, although it has the same vigor control as Sydo, it is more susceptible to winter chill and pear decline.

Sydo is the most widely grown quince clone in Europe today because it induces higher yields of rootstock liners than MA in stool-bed propagation and appears to be less susceptible than the latter to pear decline. While it is susceptible like all quince stocks to fire blight and may not outperform MA when grafted to certain cultivars, Sydo has proven its worth in a number of field trials, especially ones conducted in Belgium and Italy. It is also chosen over BA 29 in districts plagued by pear decline and in high-density orchards with intrarow spacing of less than 1–1.5 m.

BA29 is a Provence quince selection of INRA's Angers station whose name reflects the Bois Abbé trial orchard of origin at Beaucouzé, near Angers. Released in the late 1960s, it is best suited to plantations in southern Europe and became very popular during the 1980s and 1990s because it is easy to propagate, and the quince stock most tolerant to high lime soil and only moderately susceptible to chlorosis. Nevertheless, it is not the ideal stock for use in soils of heavy clay or poor fertility. Most cultivars grafted to BA29 in France, but not always in Italy, have proved to be 10–15% more vigorous and higher yielding than when grafted to Sydo or MA. The major drawback of BA29, aside from its higher vigor in fertile soils, is its low tolerance to infectious viruses and pear decline. It is only moderately susceptible to fire blight.

MC is the most dwarfing of the major commercial quince stocks and is well suited to orchards with densities as high as or greater than 3,000–4,000 trees/ha. It is easy to propagate and induces a 20–40% lower vigor than MA giving trees no taller than 2–2.2 m but requires careful soil management because its root system grows close to the surface. It tolerates graft incompatibility as well as MA but the symptoms can appear more readily when the trees have been infected by a disease or are grown in chlorotic soils. Although it is less cold hardy than other quinces, it is usually employed in central and northern European orchards of high density. It is not used in southern Europe because the summer temperatures can heat the soil too much, which can lead to both root and shoot growth being halted.

MC endows trees with high cropping efficiency, a factor that gives high yields but that can limit fruit size if trees are not in perfect condition. While lower fruit size is indeed a risk with cv. 'Conference' in older trees, it can also be an advantage with very large-sized fruit with cultivars like 'Doyenne du Comice.' Orchards with MC rootstock require less investments than with other stocks because its lower tree height translates into easier canopy management. While an interstem is advisable with cultivars having notable graft-incompatability like 'William' and 'Beurré Bosc,' cultivars like 'Abbé Fétel,' 'Conference' and 'Doyenne du Comice' can be grafted without one so long as they grow in the best well-structured soils. Despite these advantages, the major disadvantage is that trees on MC generally have a shorter (15 years versus 20–25 years) economic life than do trees grafted onto Sydo or BA29.

Adams is a quince stock named for the Belgian nurseryman who bred it in the 1970s. While it is not used in Italy despite the positive performance results in trials at Bologna University in the 1980s, it is extensively employed in Belgium and Holland. Its vigor is intermediate to MC, Sydo, and MA and has a tree efficiency as high or superior to MA. In the Netherlands, Adams is the quince stock that has the best post-transplant root growth and promotes the largest fruit size, although like MC it is susceptible to low winter temperatures. It propagates well via stool bed but is not sufficiently compatible with 'William.' There is also a virus-indexed French clone of Adams called C332.

MH (selection QR 193–16) is the newest clone bred at the UK's HRI station at East Malling. In England, this quince rootstock induced a vigor between MC and MA, was slower to initiate bearing than MC, had good yield efficiency and improved fruit size over MC. Although there are no trial data available for Italy, it has been included in the country's nursery certification process.

Clonal seedlings. The most successful clonal seedlings of *P. communis* in the global pear industry have been the US-bred OHxF (Old Home × Farmingdale) series, the most popular being OHxF 40, 69 and 87. There are also the South African-bred BP1, 2, and 3 released in the 1970s, although they never were spread in Europe, and another Swedish BP series that is used for cold tolerance. More recent releases include Bologna University's series Fox 9, Fox 11, and Fox 16, and the Geisenheim Station in Germany's Pyrodwarf. Perhaps the most important rootstock breeding program in Europe has been at INRA's Angers station in France: in the 1960s and 1970s the RV and the Réturière clonal seedling series, all selected for their pronounced dwarfing capacity, were produced. Unfortunately, their sanitary status was poor and most had to be abandoned even after indexing and reselection. The only clone to be released was Pyriam, which today is marketed only in France.

OHxF series. Oregon State University in the US embarked on its breeding program under M. N. Westwood and released in the 1960s and 1970s a series of clonal seedlings from crosses of 'Old Home' × 'Farmingdale,' cultivars that had been grown in orchards prior to WW II and were notably resistant to fire blight. In Europe, these clones have had their ups and downs. Breeders initially dropped the dwarfing

OHxF51 and 333 because of poor field performance and concentrated on several semidwarfing stocks in the series, especially Farold® 40. Although more vigorous than BA29 and less productive and yield efficient than MC, the latter became widely employed for low- or medium-density plantings of 'William' because of good graft compatibility, good cropping, and fruit size, and its resistance to fire blight.

Another OHxF clone that has had some popularity is Farold 69. Although its yield efficiency is lower than that of quince and pear seedling stocks, it is as vigorous as a seedling and has notable resistance to winter cold, fire blight, and, apparently, pear decline. Farold 87, a third clone, is less vigorous, induces precocity, is resistant to fire blight, induces better yield efficiency with 'William' than a seedling and is graft-compatible with all tested cultivars. Farold 87 is the preferred stock with 'William' and the other cultivars grown in the US Northwest. However, it is not as easy to propagate either by cuttings or by micropropagation as the other two OHxF clones.

Fox is the series recently bred at Bologna University (Bassi et al. 1994). Fox 11 (Sel. A28) and 16 (Sel. B21) were selected in the 1980s from an open pollinated progeny of the cooking cultivar Volpina, and are multiplied only by micropropagation. Both induce slightly less vigor than a seedling and are alternatives to quince where the latter exhibits mediocre performance on lime or poor fertility soils. Trials with 'William' in plantings of 800–1,000 trees/ha indicate that Fox 11 outperforms Fox 16, inducing fairly early bearing, good cropping and good fruit shape, although it is slower to bear with other cultivars like 'Beurré Bosc.' Both Fox 11 and 16 perform on par with the Farold stocks. The Fox 9 (Sel. E110) clone, released in 2008 induces medium vigor (slightly more than BA29) and the highest yield of the Fox series according to preliminary field trials (Quartieri et al. 2008).

Pyrodwarf. The Geisenheim station developed this dwarfing stock (size between that of MC and of MA) in the 1980s from a cross of cvs. 'Old Home' and 'Bonne Louise d'Avranches.' It is readily multiplied by cuttings and micropropagation and is suited to high-density plantings. Thus far, its field performance has been inconsistent, with poor to excellent performance depending on the trial (Colombo and Bolognesi 2008).

5 Current Goals and Challenges of Breeding

All novel fruit cultivars must respond to the demands of both growers and consumers. The pear breeding programs at Europe's 15 or so major research stations have similar objectives dictated largely by two driving forces.

1. Improved cultivar traits, which give growers a competitive market advantage. Given the globalization of markets, new cultivars need to have high or excellent quality that, although corresponding to the locally grown type, have marketing possibilities as a recognizable fruit that is marketed with a quality-guaranteed seal.
2. Sustainable eco-production systems. The pear in Europe comes from traditional, fairly limited areas that are often subject to integrated production technology (IFP)

and, even more restrictive, organic systems. Thus, breeders in selecting new genotypes must consider environmental adaptability and tolerance to biotic and abiotic stresses such as fire blight (*Erwinia amylovora* Burrill), black spot (*Stemphylium vesicarium* Wallr. or *Alternaria alternata* Keissler.), and psylla (*Cacopsylla pyri* L.) to reduce the need for chemical treatments. Trees must be more efficient in their use of inputs, including such renewable resources as radiant energy and water, be more readily trained and managed to reduce costs and deliver uniform ripening to reduce picking runs.

These factors have spurred a search for new breeding strategies as well as the exploration of old germplasm and of the ex situ genetic heritage of pear specimens collected throughout the world. Two of the most important pear breeding and repository stations today are at Corvallis, Oregon, in the U.S. headed by K. Hummer and at Gembloux, Belgium, under the direction of M. Lateur. The latter exceeds 5,000 accessions and is currently being screened via molecular markers for genes encoding resistance or tolerance to diseases and adaptability to environmental adversities.

Nevertheless, despite the efforts of pear breeders, these new cultivars have not yet replaced the most popular old cultivars. This, in part, is due to the pear's self-incompatibility, and consequent high degree of heterozygosity, which was exploited by breeders and amateur gardeners throughout the centuries. Indeed, the historical record shows that the seeding and selecting of pear was intense in Belgium, France, the Netherlands, Germany, and the U.K. in the 1700s and 1800s.

It is hard for breeders to upgrade the quality (size, flavor, fruit shape) and cropping standards achieved by the most popular cultivars developed over several centuries of selection. These traits are of the utmost importance in the marketplace as no other fruit is perhaps so readily recognizable by cultivar than is the pear. In fact, when a new cultivar is introduced into the market, it is always greeted with a certain diffidence, making it difficult to become a standard offering. Nevertheless, much work still needs to be done as not all objectives have been achieved by breeders and new ones keep cropping up on the industry agenda (Tables 11.5 and 11.6).

5.1 Fruit

Common objectives are the extension of the harvest season (earlier for southern and later for northern districts), red skin color to enhance consumer appeal, improved flesh structural and sensory traits, and enhanced postharvest qualities. Breeding for greater red skin color is not as easy as it may seem at first glance. The trick here is to breed the red color into fruit via conventional reproduction and not through chance mutations as these often are chimeric and unstable.

Work on flesh structural and sensory traits has expanded beyond melting, butter-like flesh texture as in the past to include fine, compact, juicy and aromatically sweet flesh like 'William' and the crisp flesh of Asian nashi pears as seen in 'Abbé Fétel.' Favored flesh flavors include sweet like 'Conference,' sweet-and-tart like

Table 11.5 Main goals of the European pear breeding programs (from Sansavini and Ancarani 2008)

- Extension of ripening, mainly early and late periods
- Compact and spur tree habits of trees
- Gametophytic alleles of S-incompatibility and self-fertility
- Resistance to pathogen and pests
  - *Erwinia amylovora*
  - *Venturia pyrina*
  - *Cacopsylla pyri*
- Environmental adaptability
  - Winter cold temperatures
  - Late and bloom frost
  - Summer hot temperatures
  - Lime and saline soils
- Fruit Quality
  - Appearance, skin color, flesh texture
  - Organoleptic traits
  - Storability and shelf life
- New fruit typologies
  - European × nashi hybrids (shape, texture, crispness)
  - Red-skinned pears
- Germplasm maintenance
  - Old traits

Table 11.6 Main goals and parental lines employed in breeding programs at the CMVF of Bologna (Italy) (from Sansavini and Ancarani 2008)

Goals	Parental lines
Harvest data and fruit quality	Bartlett, Abbé Fétel, Conference, Passe Crassane
Red fruit and quality	William, Max Red Bartlett, Rosired, California, Canal Red, Cascade
Psylla tolerance	Sel. Geneva 10353, Sel. Geneva 10355
Fire blight resistance	Harvest Queen, Harrow Delight, HW 605, Harrow Sweet, US309
Hybrids with *P. pyrifolia*	Hosui, Nijisseiki, Shinseiki

'Buerré Bosc' and aromatic like 'William.' Although, it should be noted that flesh with excessive sclereids is generally not acceptable with consumers.

Currently, there are many pear cultivars that ripen quickly at climacteric and, hence, have a short shelf life (like 'Clapp's Favourite'). Thus pear breeders are developing cultivars that are resistant to the internal or core breakdown that afflicts many early-ripening cultivars, to corky spot of the flesh like 'Concorde' and 'Beurré Anjou' and to flesh browning that can affect pears like 'Passe Crassane' during cold storage. Breeding novel pears that have a long shelf life is essential to expand the marketability of pears in our global fruit market. A joint IRTA-HortResearch breeding program started in 2002 has the aim to combine the high fruit quality with good size, good handling, and storage performance as well as a long shelf life and early harvest (Batlle et al. 2008).

There are pear breeding programs in Italy, Germany, New Zealand, Japan, and China that focus on interbreeding of *P. communis* and *P. pyrifolia* to combine the flavor and taste traits of European pears with the crisp texture of the Asian nashis as well as to transfer traits like tolerance to certain pests and other disorders from the Asian species. While it is easy to cross pears of the two species, especially given the high fertility of *P. pyrifolia*, achieving results that are in line with expectations is quite another matter, at least for the moment.

5.2 Tree

The most important trait in tree development is good *environmental adaptability*, which along with phenotypic plasticity would extend the pear's growing range further north and south. At higher latitudes tolerance to cold damage is essential. In the lower latitude warmer regions, pears with limited chilling requirement and greater heat tolerance are needed. Such genotypes would not suffer from insufficient chilling, which induces bud drop, staggered bloom, delayed shoot growth, and lower cropping nor from high summer temperatures that slow root growth. Furthermore, these genotypes would also avoid the scion-stock graft incompatibility problems accentuated by insufficient chilling and excessive heat.

Pear, despite its broad-based genetic variability, is less ecologically malleable and environmentally adaptable than apple. Apple production in Europe, for example, extends to higher latitudes than does pear production, which are predominantly found in the northern areas of mid latitude countries, like the Po River lowlands in Italy, Catalonia in Spain, the Loire Valley in France, Bavaria in Germany, Rio Negro valley in Argentina, and California in the U.S., and in the milder area or southerly areas of high-latitude countries, like Kent in England, Oregon and British Columbia in the Pacific Northwest, and Ontario in North America. Pear is more susceptible than apple to low winter temperatures (15–20°C below zero). Given that it blooms before apple, pear is even susceptible to late spring frosts after bloom and requires protection in some districts. Thus, while the country's pear industry produces many good cultivars, there are many cultivars like 'William' that can be grown only in northern districts and those like 'Spadona' and 'Coscia' that can only be grown in southern orchards.

Fortunately, pear cultivars like 'Conference,' 'Abbé Fétel,' and 'William' are parthenocarpic and will develop a fruit without fertilization or when the embryo or seed is damaged by cold weather or other environmental adversities. Whence the common management practice of treating pear before and right after bloom with gibberellins A_3 or A_{4+7} to induce parthenocarpy even when no damage has occurred.

Yet, fruit morphogenesis in the most widely grown pear cultivars also harbors risks that can limit bearing. Fruit thinning is far less frequent, or even unnecessary, in European pear as compared to nashi pear and apple. When in bloom, pear flowers, which are cone-shaped inflorescences with 6–8 flowers, are much less attractive to bees, and fruit set is frequently low because of poor pollination. In addition, unlike apple, pear is often subject to postbloom and even subsequent June fruit drop, which can reduce bearing even after an apparently high initial fruit set. This is the reason why growers often add auxins and other organic, nutritional, and hormone compounds called "retainers" to boost fruit growth and limit drop to the parthenocarpy-inducing treatments with gibberellins.

A number of studies have shown that source–sink competition between shoots and fruitlets during the first 2–3 months of fruit development can engender nutritional or hormonal deficiencies or morphogenetic stress in trees that causes the fruitlets to drop. There are even cultivars like 'Doyenne du Comice' in which fruit drop poses a real threat right up to harvest, even to the point that growers will treat the fruits with preharvest, antidrop auxins in areas where it is permitted.

Yield depends not only on the fertility of individual cultivars and on the intercompatibility of the associated cultivars in orchards but also on the genotype × environment interaction as shown by management practices. The pruning regime, for example, can increase fruit set of spurs on branches most subject to apical dominance (i.e., 'Passe Crassane'). Heading back twigs and fruiting branches (2–3-year-old wood) thus reduces the number of flower buds and competing sinks and, hence, differently from apple, increases fruit set or prevents fruitlet drop (Sansavini 1969).

Breeders are also interested in *cropping habits* that could make a tree, as well as the orchard more efficient and easier to manage. While one seldom sees the spur habit in pear, canopies differ in their formation of spur-like limbs or the ratio of the different types of fruiting branches (Sansavini 2002). For example, a cultivar like 'William' crops mainly on brindles (1-year shoots) and others like 'Beurré Bosc' ('Kaiser') that crop almost entirely on spur-like limbs or spur clusters. In the case of *P. pyrifolia* it is worthwhile noting that trees that are productive and precocious often crop on spurs and 1-year-old shoots.

Another trait that varies notably among cultivars is feathering (lateral summer shoots), a tendency that facilitates tree formation during training (Sansavini and Zocca 1965). This trait ranges from abundant feathering as in 'Conference' and 'Abbé Fétel' to few or no lateral branching as in 'Passe Crassane.' The red-skin mutants of 'William' like 'Max Red Bartlett' have a very compact, upright canopy, with longer but fewer erect limbs and branches than in 'William.' This semi-spur habit enables higher-density plantings. Although the dwarfing trait of 'Nain Vert' has been used in crosses in Italy and France, the resulting cultivars like 'Grand Pearl' have productivity and fruit quality incompatible with today's pear industry and are of interest only to amateur horticulturists.

5.3 Resistance to Biotic and Abiotic Stress

Much effort is being invested by researchers to find resistance genes to the main biotic adversities of pear: the fire blight bacterium (*Erwinia amylovora*), the European pear psylla (*Cacopsilla pyri*), which is the vector of the phytoplasma causing pear decline, the scab-causing fungi *Venturia pyrina*, and the black spot fungus *Stemphylium vesicarium*. While the resistance genes to fire blight have yet to be identified, researchers have long known the sources of resistance and began the breeding for fire blight resistance at Geneva in New York State, Harrow Station in Canada, and Angers in France in the 1960s and later at Pillnitz (Dresden) Station, the ISF Station at Forlì in the 1980s and, more recently, at the DCA University of Bologna in Italy. The results of these and other efforts to date have been the release of such partially resistant or tolerant cultivars such as 'Harrow Sweet,' 'Harrow Crisp,' 'Harrow Gold,' and 'Harrow Delicious' (all from Canada), 'Blake's Pride' and 'Shenandoah' (USDA Kearneysville, WV, USA), 'Aida' and 'Boheme' (ISF Forlì, Italy), and the identification and maintenance mainly by Angers of old cultivars like 'Pierre Corneille' that tolerate the pathogen. There are also cultivars that are tolerant like 'Coscia' and 'Dr Jules Guyot' grown in southern Europe.

Research on the European pear psylla has been less focused with little, if any, progress. While there are cultivars like 'Spina Carpi' and the hybrids such as Sel. 10305 from the Geneva Station in New York that are reportedly resistant and have been used for resistance breeding in Europe, fruit quality of these genotypes is poor so several more generations of crosses are needed to combine high resistance and high fruit quality.

Efforts to combat the two fungi have been more successful. There are several cultivars in Europe that are resistant to *V. pyrina*, 'Dr Jules Guyot' being perhaps the most well known. Researchers at the Pillnitz Station in Germany have bred the resistant cultivars 'Herman' and 'Uta,' the former ripening earlier in the season and the latter at the end. The Pitesti-Maracineni and Voinesti Stations in Romania have also developed several resistant cultivars that have had some success in local markets.

Breeding resistance to black spot is somewhat more complicated since there are pathogenic races, which are currently being cataloged. One of the better known attempts to breed hybrids with Asian nashis, which are usually not susceptible to this fungus, is the New Zealand cultivar 'Crispie,' although it seems not to have proven completely successful.

Self-fertility. There are few self-compatible cultivars within the European pear group. Achieving this goal requires silencing the S-locus, which researchers in Japan have accomplished in nashi with the self-fertile mutant cv. 'Osa-Nijisseiki.'

Cold resistance. The degree of susceptibility to winter cold depends on both scion and stock. Most European pear cultivars can withstand temperatures as low as 10–15°C below zero once the tree is dormant. While breeders can select genotypes that are less susceptible than others to winter cold damage, there is little they can do to protect trees against the spring low temperatures (3–5°C below zero) that cause necrosis of gametic cells or seed embryos and, hence, fruitlet drop. Nevertheless, several parthenocarpic cultivars like 'Conference' and 'Abbé Fétel' can escape this damage with applications of gibberellins A3 or, better, A4+7, because of their ability to develop fruit without seed set (Sansavini et al. 1986a).

Winter hardiness is a complex trait with some genes active during bud dormancy while others act directly on cell cytoplasm. There does not appear to be any correlation between xylem, or young wood, and flower bud resistance. Unfortunately the pear cultivars that are held to be fairly resistant to winter are, however, of little or no commercial interest. The most resistant of the Asian pears to winter injury are *P. ussuriensis* and *P. pashia* P.Don., which can withstand temperatures of 30°C below zero and as low as 16°C below zero, respectively. Quince, the most common stock used in Europe, is susceptible to winter cold and is thus rarely used in the more northerly pear districts like those in Poland and Russia.

6 Breeding Methods and Techniques

6.1 Major Traits and Selection Techniques

Growth habit. The modern trend in fruit production is to breed trees that bear early and are easy to prune, spray, and harvest. These characteristics can be achieved by reducing the usually vigorous size towards dwarf trees while maintaining high crop production and excellent use of light (Tukey 1964). Quince rootstocks are usually used to obtain trees with reduced size but they have limitations in adaptability and graft incompatibility. True dwarf-type growth habit similar to the apple spur-type is rare in pears (Bell 1996). Smaller tree size in pear is seen cultivar 'Nain Vert' (Decourtye 1967), which has a short internode trait conditioned by a monogenic dominant character associated with the polygenic trait of plant vigor (Bagnara and Rivalta 1989) and in two compact Italian clones ('Abate Light' and 'Conference Light'), which were produced by mutagenesis mainly by γ-ray irradiation (as reviewed by Predieri 2001). The Italian cultivars have been shown in field trials to combine compact habit and high productivity (Predieri 2001; Bellini and Nin 2002).

S-locus and gametophytic self incompatibility. Gametophytic self-incompatibility (GSI) is a mechanism triggered by proteins coded by the S locus that determine the inhibition of self-incompatible pollen tube growth without damaging the self-compatible ones. The *Pyrus* genus carries the S-RNase-based self-incompatibility typical of the Rosaceae. In this system pollen tube recognition is triggered by the interaction between stylar determinants, the S-RNases, and pollen determinants, the F-box proteins SLF (S-Locus F-box) or SFB (S-locus F-Box) (Sijacic et al. 2004). Because of self-incompatibility (SI), pear orchards must contain at least two cultivars with the S-genotype compatible for pollination and an overlapping bloom date.

From the molecular point of view, S-genotyping in pear is determined by the identification of the S-RNase alleles. Nineteen European pear S-RNases alleles (S101 to S119 as renumbered by Goldway et al. 2009) have been cloned and sequenced, and used to characterize more than 130 cultivars (Goldway et al. 2009). The most frequent alleles are S101, S102, S104, and S105. Of the 133 cultivars analyzed, 75 carry the S101 allele, 41 the S102, 27 the S104, 20 the S105 allele, and 12 the S103 allele. This reflects the intensive use of 'William' (S101/S102) and 'Coscia' (S103/S104) as parental genotypes in the development of European cultivars (Sanzol and Herrero 2002). S-genotyping is the most powerful support for breeding programs seeking to identify the interfertility groups among European pear cultivars (Table 11.7; from Goldway et al. 2009) and, in this perspective, all novel cultivars should be S-genotyped for efficient fruit production and breeding.

Table 11.7 Distribution of European pear cultivars according to their S-alleles (Goldway et al. 2009)

Variety	Alleles	Variety	Alleles
Ayers	S101/S102	Besi de Saint-Waast	S101/S118
Bartlett/William's/William's Bon-Chrétien		Bon-Chretien d'Hiver	
Bon Rouge		Covert	
Délices d'Hardenpont		Pierre Comelle	
Harvest Queen		Ballad	S101/S119
Louise Bonne d'Avranches		Doyenné d'hiver	
Max Red Barlett		Idaho	
Napoleon		La France	
Orient		Verdi/Sweet Blush	
Pera d'Agua		Santa Maria	S102/S103
Precoce du Trevoux		Spadoncina	
Red Jewell		Beurré Jean Van Geert	S102/S104
Rosired		Canal Red	
Seckel		Honey Sweet	
Seigneur d'Espéren		Joséphine de Malines	
Béurré Precoce Morettini	S101/S103	Tosca	
Fondante Thirriot		Harrow Sweet	S102/S105
Packham's Triumph		Koonce	
Precoce di Fiorano		Marguerite Marillat	
Spadona/Spadona estiva/ Blanquilla		Pierre Tourasse	
Washington		Beurré de l'Assomption	S102/S106
Beurré Lubrum	S101/S104	Michaelmas Nelis	S102/S107
California		Doyenné Gris	S102/S108
Cascade		Akça	S102/S109
Grand Champion		Blickling	S102/S110
Hartman		Comte de Lambertye	
Highland		Comte de Flandre	S102/S111
Howell		Ewart	S102/S114
Jeanne d'Arque		Chapin	S102/S115
Norma		General Leclerc	S102/S118
Onwards		Ovid	
Dagan		Bristol Cross	S102/S119
Aurora	S101/S105	Emile d'Heyst	
Docteur Jules Guyot/Limonera		Kieffer	
Duchesse d'Angouleme		Koshisayaka	
Harrow Crisp		Alexandrine Douillard	S103/S104
Harrow Delight		Coscia/Ercolini	
Magness		Winter Nelis	S103/S107
Rocha		Ankara	S103/S119
Tyson		Abbé Fétel	S104/S105
Beurré Giffard	S101/S106	Doyenné du Comice	
Gentile		Concorde	S104/S108
Summer Doyenne		Glou Morceau	S104/S110

(continued)

Table 11.7 (continued)

Variety	Alleles	Variety	Alleles
El Dorado	S101/S107	Turnbull Giant	S104/S113
Sirrine		Reimer Red	S104/S114
Winter Cole		Le Lectier	S104/S118
Bautomne/Serenade	S101/S108	Condo	S104/S119
Clapp's Favorite		Urbaniste	
Clapp's Rouge/Kalle/Red Clapp's/ Starkrimnson		Charles Ernest	S105/S110
Flemish Beauty		Triomphe de Vienne	
Sierra		Eletta Morettini	S105/S114
Star		Rogue Red	
Delbard première/Delfrap	S101/S109	Beurré Clairgeau	S105/S118
Beurré Superfin	S101/S110	Angelys	S105/S119
Espadona		Kaiser/Beurré Bosc	S107/S114
Oliver de Serres		Nouveau Poiteau	
Dana's Hovay	S101/S111	Garbar	S107/S115
Wilder		Fertility	S107/S118
Old Home	S101/S113	Beurré Hardy	S108/S114
Starking Delicious/Maxine		Royal Red/Red Hardy	
Beurré d'Anjou	S101/S114	Devoe	S108/S118
Moonglow		Conference	S108/S119
Red Anjou		President Héron	S110/S118
Colorée de Juliet	S101/S115	Passe Crassane	S110/S119
Forelle	S101/S116	Silver Bell	
Rosemarie		Saint Mathieu	S114/S116
		Lawson	S115/S117

Cultivars that share the same S-alleles are incompatible in crosses

Overcoming self-incompatibility is one of the most important aims of pear breeding. All the efforts to introduce the Japanese pear S4-RNase deletion, which confers self-compatibility to cv. 'Osa-Nijisseiki' (Sassa et al. 1997), have been unsuccessful. Several studies have reported occasional self-fertility and/or self-fruitfulness to some degree in certain cultivars (Griggs and Iwakiri 1954; Callan and Lombard 1978; Vasilakakis and Porlingis 1985; Sanzol et al. 2006) and a first mutated S-allele conferring self-compatibility to the European pear varieties 'Abugo' and 'Ceremeño' (a retrotransposon insertion within the intron of S121 allele and indels at the 3'UTR) was identified (Sanzol 2009). In spite of this perhaps the best chance to develop new self-compatible pear cultivars is offered by genetic engineering, an approach that was used in apple to silence a gene coding for an S-RNase (Broothearts et al. 2004).

Fruit quality. The concept and, hence, the perception of quality is not the same in every country or every market. Historically speaking, Europe's pear cultivars have gained widespread consumer acceptance because of their typical pyriform shape, weight exceeding 180–200 g, juicy and fine flesh of high-quality flavor, and good shelf life. The most prized pear has melting, butter-like textured flesh without stone cells, whence the term Beurré prefixed to the name of many cultivars with tender, juicy and sweet-tart aromatic taste. 'Conference,' 'Doyenne du Comice' and 'Beurré d'Anjou' are good examples. While the ideotype of Asian pear like nashis has a large, globose-oblate-shaped fruit weighing 250–350 g with crispy, juicy, sweet, and slightly aromatic taste, which can be eaten right off the tree. This nashi gustatory profile has recently had an influence on European expectations to the extent that cultivars like 'Abbé Fétel' have proven successful because of their compact, almost crispy, not-quite-ripe flesh. Many of these traits are polygenic and the possibilities to select high quality genotypes by crossing two cultivars with a high level of heterozygosity are low. At present, sensory evaluation plays a key role in characterizing cultivars for fruit quality but little is known of the genetic basis of the quality traits. Flavor in pear fruit is the sensory perception of sweetness, acidity-tartness, aroma, astringency, and bitterness that is composed by the set of sugars, organic acids, phenolics, and volatile compounds. Genetic studies thus far have indicated that the soluble solid content, juice pH, and sugar–acid balance are controlled by multiple genes (Visser et al. 1968; Zielinski et al. 1965). The inheritance of the phenolics responsible for astringency and bitterness are still unknown.

Among the promising cultivars released in recent years all with improved eating quality, Bellini and Nin (2002) reported: the French 'Angélys' ('Doyenne d'Hiver' × 'Doyenne du Comice'; Le Lézec et al. 2002), the Swiss 'Valérac' ('Conférence' × 'Président Héron') and 'Champirac' ('Grand Champion' × 'Président Héron'; ACW, activity report 2000–2006), the New Zealand 'Crispie' ('Nijisseiki' × 'Max Red Bartlett') and the Swedish 'Ingeborg' and 'Fritjof.' The Naumburg/Pillnitz pear breeding programs used the cultivars 'Doyenne du Comice,' 'Kaiser Alexander,' 'Dr J. Guyot,' 'President Drouard' and 'William' for sources of quality. The released cultivars (i.e., 'Isolda®,' 'Tristan®,' 'Armida®,' 'Elektra®,' 'Hortensia®,' 'Manon®,' 'Agata®,' 'David®,' 'Eckehard®,' and 'Uta®') have good to excellent quality and high to very high yield capacity (Fischer and Mildenberger 2002).

Skin color. Skin color in European pear cultivars ranges from the golden-yellow of 'William,' the greenish-yellow of 'Packham's Triumph' and 'Santa Maria,' the greenish of 'Doyenne d'Hiver' and the russety-bronze of 'Beurré Bosc' ('Kaiser') and 'Angelys' to the striated red of 'Max Red Bartlett,' the multicolored reddish-brown-yellow of 'Cascade,' the yellow with ample, smoky orange-red blush of 'Hortensia' and 'Santa Lucia,' and the all-over red of 'Calired' ('Zaired') and 'Homored.' The Asian nashis on the other hand range in skin color from russetless light green-yellow to smooth light bronze-brown, occasionally reddish and usually with lenticels.

Most recent work has focused on red skin color which can be bred either via crosses, as with 'Red Silk,' 'Canal Red' and 'Calired' ('Zaired'), all deriving from Red Bartlett ('Max Red Bartlett,' 'Sensation,' 'Rosired,' 'Red Princess'), or like 'Starkrimson' from 'Clapp's Favourite,' or via mutagenesis from 'Bartlett,' whether artificial like 'Homored' or natural as in all the rest.

The red skin of most of these cultivars is a chimeric mutation and frequently unstable as in 'Max Red Bartlett' and 'Sensation,' although it can occasionally be stable as with 'Homored' and 'Rosired.' Despite the good market response, nearly all of these red mutants have seen limited success because it is often associated with poor tree vigor and cropping as seen in the cultivars 'Crimson Gem' from 'Doyenne du Comice' and 'Red d'Anjou.'

The pear's red skin-color is under single-gene control (Brown 1966). Brown crossed red and non-red pears and concluded that anthocyanin pigmentation is dominant over nonpigmentation and that 'Max Red Bartlett' was a red–green chimera heterozygous for red. This genetic control has been confirmed recently with this and other sources of red skin color by Booi et al. (2005) and Dondini et al. (2008). In contrast, the red blush from 'Huobali' (*P. pyrifolia*) is controlled independently from the red skin coloration from 'Max Red Bartlett.' Families created by crossing descendents of 'Max Red Bartlett' and 'Huobali' together produced 30–37% of seedlings with significant red skin color (Volz et al. 2008).

Harvesting time. Another important breeding target is extending the harvest time. Although extensive in traditional European pear germplasm, the extremes of the harvest season lacked in important qualities. The very early cultivars have small fruit with poor postharvest qualities, which tended not to ripen uniformly whereas the traditional late ripening cultivars had longer seasonal management, marginal eating quality (high sclereids and a bit tart or astringent) but very firm flesh that had a long storage life. The long storage life allowed a few of these cultivars to be marketed until spring without cold storage which made these quite popular with past generations. By contrast, almost all nashis are confined to the summer or beginning autumn and, hence, appear to offer little potential for calendar expansion. Seasonality thus needs to be thoroughly revisited today by combining lateness and the best flavor traits of the summer–autumn cultivars into novel cultivars that ripen appropriately and, hence, extend the market calendar into spring of the next year with the help of controlled atmosphere (C.A.) and new storage technology. Unlike that of apple, which can cover the year from harvest to harvest, the marketing of pear in Europe continues until the end of winter and imports from the southern hemisphere largely cover the following months.

Several early ripening pears have been released in Italy (DOFFI and ISF-FO). These include the cultivars 'Etrusca' ('Coscia' × 'Gentile'), 'Sabina' ('Santa Maria Morettini' × 'Doyenne du Comice') (Bellini and Nin 2002), 'Tosca' ('Coscia' × 'William'), 'Turandot,' 'Norma' and 'Carmen' (all the three genotypes derived from 'Dr. J. Guyot' × 'Bella di Giugno') (Rivalta and Dradi 1998; Rivalta et al. 2002). These new cultivars have generally improved flavor and other traits as compared to the existing cultivars ripening in the same season.

The Naumburg/Pillnitz pear breeding programs used the cultivars 'Forelle' ('Nordhauser Winterforelle'), 'Madame Verté' and 'Paris' to develop the autumn ripening cultivars 'Armida®,' 'Elektra®,' 'Hortensia®' ('Nordhauser Winterforelle' × 'Clapps Liebling'), 'Manon®,'® and the winter cultivars 'Agata®,' 'David®' ('Dr J. Guyot' × 'Doyenne du Comice'), 'Reglindis,' 'Eckehard®' ('Nordhauser Winterforelle' × 'Clapp's Favourite'), 'Uta®' ('Madame Verté' × 'Kaiser Alexander') (Fischer and Mildenberger 1999).

In France, breeding to replace 'Passe Crassane' has resulted in the late ripening cultivars 'Angélys ('Doyenne d'Hiver' × 'Doyenne du Comice'; Le Lézec et al. 2002),' 'Delmoip' and 'Bauroutard' (Durel et al. 2004).

Disease and pest resistance. Breeders the world over have very ambitious plans for the development of disease- and pest-resistant pear cultivars. However, efforts in this field are constrained by limits of our knowledge the sources of resistance needed for the various important diseases and pests. Indeed, it is probably better to use the term tolerance rather than speak of resistance.

Fire blight. Few pathogens are as devastating as *Erwinia amylovora* for Maloideae. Despite quarantine measures in several countries, the disease continues to spread throughout western, central and southern Europe (Jock et al. 2002). The bacteria can enter a host plant by natural openings (flowers) or wounds caused by hail, pruning activities and insects. It spreads quickly along the stems to the main branches, producing the characteristic symptoms of necrotic shoot blight called 'shepherd crook' (Thomson 2000).

The lack of completely effective control measures has accentuated the importance of resistant cultivars as a promising tool of an integrated disease-management program. While most European cultivated pears (*P. communis* L.) are susceptible to *E. amylovora*, there are several known sources of resistance in pear germplasm such as 'Old Home,' 'Seckel,' 'US309,' and 'Michigan 437' (van der Zwet and Bell 1984; Thibault and Paulin 1984; Thibault et al. 1989; Lespinasse and Aldwinckle 2000; Bell et al. 2002; Rivalta et al. 2002; Durel et al. 2004; Hunter and Layne 2004).

Breeding programs to develop resistant cultivars were initiated in the 1920s and 1930s, developing by the 1960s into two impressive programs: one at Harrow, Canada, and one at Kearneysville, West Virginia, in the United States. Both efforts were based on hybridization of cultivars and selections from *P. ussuriensis* and *P. pyrifolia* and recovering fruit characteristics of *P. communis* by backcrossing to selected *P. communis* cultivars (Bellini and Nin 1997). The results of this work were then employed in other programs to breed new fire blight-resistant pear cultivars in the USA (USDA, Cornell University, Geneva, N.Y.), Canada (AAFC Research Centre, Harrow, Ontario), Italy (DCA of Bologna, DOFFI Florence and CRA of Forlì Italy), England (HRI East Malling, UK), Switzerland (Faw, Wadenswil), Germany (GODP Dresden), Romania, Poland, Russia and France (INRA, Angers France). Commercially available fire blight-resistant pear cultivars include the Canadian cultivars 'Harrow Sweet' ('Bartlett' × 'Purdue 80-51') and 'Harrow Delight' ('Old Home' × 'Early Sweet'), the Kearneysville USDA cultivars 'Moonglow' ('Michigan 436' × 'Roi Charles de Wurttemberg'), 'Potomac' ('Moonglow' × 'Beurré d'Anjou'), 'Magness' ('Seckel' × 'Doyenne du Comice') and 'Blake's Pride' ('US 446' × 'US 505') and the Italian tolerant cultivars 'Aida' ('Coscia' × 'Dr. J. Guyot') and 'Boheme' ('Conference' × 'Dr. J. Guyot').

Fire blight resistance in pear is a quantitative trait (Le Lézec et al. 1985). Thus far, four QTLs linked to it in the tolerant cultivar 'Harrow Sweet' on linkage groups (LG) 2, 4 and 9 (Dondini et al. 2004) and one QTL in the progeny 80.115.69 × 80.91.01 [('Dr. J. Guyot' × 'Bella di Giugno') × open pollination of 'US 309'] located in LG2 have been described (ISF, Forlì). Since a QTL from each study was found on LG2, it is possible that 'Harrow Sweet' ['Bartlett' × 'Purdue 80-51' ('Old Home' × 'Early Sweet')] and 'US309' [('Roi Charles de Wurttemberg' × 'Michigan 437')] have a QTL for fire blight resistance located in the same position.

Black spot. Much attention is also being paid in certain regions to developing cultivars with tolerance and resistance to the fungal pathogen *Stemphilium vesicarum*, the agent of black (or brown) spot. This fungus is taxonomically similar to *Alternaria* sp., which attacks nashi orchards in Asia. *S. vesicarum* is endemic in the Po valley (Italy) and is particularly damaging to 'Abbé Fétel' and 'Conference.' In years when pathogen epidemiology is favored by the weather, standard approaches to control are all but useless. Thus genetics could play a key role in its control.

Venturia pyrina. Scab caused by *V. pyrina* Adher., is one of the most serious fungal diseases affecting the European pear. Despite its importance, the available literature is scanty. The most commonly grown pear cultivars are susceptible and no commercial cultivar is completely resistant. Chevalier et al. (2004) report that the notable variability in cultivar response depends on both environmental conditions and the wide variability of biotype distribution of *V. pyrina* in the world's growing areas. More information is available about *Venturia nashicola* Tanaka, the pathogen of nashi pear (*Pyrus pyrifolia*). Abe et al. (2000) described the inheritance of resistance to *V. nashicola* in European pear cultivars by examining intra- and interspecific hybrids, concluding that European pear ('La France') possesses a single dominant gene that confers resistance to pear scab incited by *V. nashicola*.

Resistance to *V. pyrina* has been reported in the cultivars 'Navara,' 'Delice d'Avril,' 'Winter Nelis,' 'Muscat,' 'Wilder,' 'Madame Favre,' 'D'Aout Amer' and 'Abbé Fétel' (Brown 1960; Chevalier et al. 2004; Villalta et al. 2004; Postman et al. 2005; Lespinasse et al. 2008). While a dominant gene for *V. pyrina* resistance has never been found in European pear cultivars, there is evidence of polygenic resistance (Chevalier et al. 2004) and recently two QTLs linked to scab resistance have been found on linkage groups 3 and 7 of the cultivar 'Abbé Fétel' (Pierantoni et al. 2007).

Psylla. Psylla (*Cacopsylla pyri*) is a serious problem in pear orchards that is difficult to control because of the insect's prolific nature, overlapping generations, and its ability to develop resistance to insecticides. Most commercial pear cultivars are highly susceptible to *C. pyri* which transmits the phytoplasma that incites pear decline and results in crop loss. Psylla resistance is found in *P. calleryana*, *P. fauriei* (Westigard et al. 1970; Quamme 1984) and *P. ussuriensis* (Harris and Lamb 1973). Since *P. ussuriensis* has better fruit quality than either of the other species, it has been used in breeding (Harris and Lamb 1973). The inheritance of psylla resistance seems to be a polygenic trait (Lespinasse et al. 2008). In North America pear breeding to introduce resistance to psylla from *P. ussuriensis* was started in 1920 and resulted in various resistant selections: NY 10352, NY 10353, and NY 10355 (Cornell University, Geneva, NY, USA) which were used in the breeding programs in Italy and France (Rivalta and Dradi 1998; Pasqualini et al. 2006; Lespinasse et al. 2008). In Romania, *P. pyrifolia* and *P. ussuriensis* are being used as sources of resistance to psylla in their pear (*P. communis*) breeding program at the Pitesti-Maracineni Fruit Research Institute (Braniste 2000).

Within *P. communis* there are a few old cultivars in France ('Doyenne de Poitiers,' 'D'Août Lamer') (Robert and Raimbault 2005), eastern Europe ('Karamanka,' 'Jerisbasma,' and 'Vodenjac'; Bell 2003) and Italy ('Spina Carpi') (Rivalta and Dradi 1998) that are resistant to psylla, although 'Spina Carpi' does not transfer it to its progeny (Rivalta and Dradi 1998). It is, however, probably safe to say that genomics is the best approach to furthering these efforts if the QTLs for this pest resistance can be identified and associated markers used for Marker Assisted Selection (MAS).

6.2 Breeding Methods and Techniques

Pears are characterized by a high level of genetic variability, high allelic heterozygosity and a gametophytic system of self-incompatibility controlled by a series of S alleles. The general approach to breeding is phenotypic mass selection consisting of cycles of hybridization and selection among the seedlings to identify both better parents and new cultivars. Hybridizations are planned not only for the specific traits of the parents but also their incompatibility phenotypes and bloom sequence. Given the range of bloom times among cultivars, pollen can be made available for specific crosses by either forcing the flowers in the greenhouse 2 weeks before the expected flowering date in field or with stored pollen as it can stored for up to 2 years under dry conditions at 2–4°C (Bell 1996).

Techniques for emasculation and pollination vary among programs. Emasculation is not always considered necessary due to the self-incompatibility of the species. When done, flowers are emasculated by removing the anthers with special scissors or with the fingers on flowers at the balloon stage. This operation prevents flower visitation by bees and contamination with foreign pollen. Pollen can be applied on the stigmas with small paint brushes or with one's finger tip cleaned each time by ethyl alcohol to avoid pollen contamination (Bell 1996).

Pear seeds are generally stratified for 2 or 3 months at 0–7°C in a moist, well-aerated medium (Hartmann 1990). After dormancy, seeds can be planted in individual peat pots containing an equal mixture of sand and peat moss (Matkin and Chandler 1957) and will begin to germinate in about 10 days at 20°C. These juvenile seedlings usually take 4 or more years to begin fruiting (Bell 1996). Grafting pear seedlings at the time of transition from the juvenile to the adult phase on quince rootstock can shorten the time to fruiting.

Some traits such as resistance to pathogens and pests can be evaluated as young seedlings in the greenhouse under controlled conditions. Young seedlings in the field can be evaluated for vigor, early flowering and precocity (Visser and De Vries 1970). Once fruiting, selections are assessed for fruit size and quality, productivity, ripening uniformity, storability, shelf life and, if needed, processing qualities. The three latter traits due to the expense of the evaluation are only assessed in the most promising seedlings. Because of the long-term nature of the pear breeding it is important to improve the efficiency of the selection strategies by reducing the juvenile stage duration and accelerating the fruiting phase of the tree.

Most breeders evaluate in two phases: first screening in the greenhouse or nursery and a second test in which seedlings are grafted on a dwarfing rootstock like quince. Such field tests require 8 or more years. Selections that remain after the second test field trial are further trialed in multiple environmental conditions to assess the environmental suitability of the selected genotypes.

Beyond using hybridization, cultivars are developed by the identification of spontaneous or induced mutations. Radiation treatments are the most common mutagenic approach for inducing pomologically useful variants in plant size, ripening time, fruit color, and self-fertility (Predieri 2001; Spiegel-Roy 1990; Predieri and Zimmerman 2001). In European pear, mutations affecting bloom time, blossom color, ripening time, fruit color (Decourtye 1970; Roby 1972a, b), and growth habit (compact) (Lacey 1975; Visser et al. 1971) and in Japanese pear (*P. pyrifolia*) mutations affecting disease resistance (Masuda and Yoshioka 1997) and self-compatibility (Hirata 1989) have been reported. The frequent occurrence of chimeras, which are often unstable as well as undesirable traits such as reduced fertility, irregular cropping, and poor fruit attractiveness limits the usefulness of mutagenesis for pear breeding. The risk of chimeras can be reduced by irradiating in vitro-developed buds (Decourtye 1982; Broertjes 1982; Lacey and Campbell 1982).

Somaclonal variation, the recovery of variants produced during in vitro culture has been studied as a potential tool for the selection of fire blight resistance (Duron et al. 1987; Brisset et al. 1988, 1990) and adaptation to abiotic stress such as to a calcareous soil (Fe uptake efficiency, tolerance to high pH soils) (Marino et al. 2000; Palombi et al. 2007) and salinity (NaCl) (Marino and Molendini 2005). Thus far, in vitro procedures have been developed for several European pears ('Durondeau,' 'Conference' and 'Abbé Fetel'), the rootstock 'BA 29' and *Pyrus pyraster* (Viseur 1990; Marino et al. 2000; Palombi et al. 2007; Marino and Molendini 2005), and for the development of transgenic plants (Chevreau et al. 2007). Thus far, somaclonal selection has not produced any clones interesting for field applications.

Among the breeding techniques it has to be mentioned that genetic transformation in pear has accomplished some excellent results (see Sect. 7), which has shown the potential of this methodology of achieving breeding aims that would be very difficult the traditional way.

6.3 Propagation

Pear cultivars are routinely asexually propagated by budding or grafting on a selected rootstock of the same species (*P. communis*) or other compatible species such as *C. oblonga* (quince), *P. calleryana*, *P. betulaefolia* (Stern 2008), or others. When quince is used as rootstock, the possibility of graft incompatibility has to be taken into account and, in this case, a compatible interstem can be used to overcome this problem. The cultivars 'Beurré Hardy' and 'Beurré d'Anjou' are very compatible with quince. Seedlings rootstocks (i.e., Kirchensaller in Europe and Bartlett in the USA) have been used for propagating pear but the trend toward dwarf trees and high-density plantings has resulted in increased use of clonally propagated dwarfing rootstocks (Bell 1996).

The development of in vitro micropropagation techniques (Howard 1987) has facilitated the production of clonal rootstocks that are difficult to propagate by conventional means (Bell 1996) as well as to produce self-rooted pathogen-free scions and as stock material for the cryopreservation of pear germplasm. Micropropagation protocols have been published, beginning in the late 1970s, for over 20 cultivars of pear, including the major *P. communis* cultivars, several Japanese cultivars of *P. pyrifolia* and genotypes of the other *Pyrus* species (Bell and Reed 2002; Sansavini 1994). Among the most common and commercially used strategy is a double-phase technique that combined a liquid and an agar-solidified phase able to enhance the shoot proliferation (Viseur 1987). This technique has been used with the cultivars 'Durondeau,' 'Conference,' 'Doyenne du Comice,' 'Professeur Molon,' 'Abbé Fetel,' 'Dr J. Guyot,' and 'Butirra Precoce Morettini' (Viseur 1987; Rodriguez et al. 1991).

7 Integrated Breeding: Conventional and Molecular Driven Tools

7.1 Molecular Markers

Despite its importance as crop, little molecular work has been done with *P. communis*. Monte-Corvo et al. (2000) analyzed 25 *P. communis* cultivars (among the most cultivated ones) and four commercial *P. pyrifolia* cultivars by RAPD (randomly amplified polymorphic DNA) and AFLP (amplified fragment length polymorphism) techniques focusing on their molecular discrimination and the assessment of their genetic relatedness. The first approaches by SSR (simple sequence repeats) for pear (both Japanese and European) cultivar genotyping were performed by Yamamoto et al. (2001, 2002a) who identified a number of markers suitable for the analysis of the genetic diversity among *Pyrus* spp. and able to confirm the synteny among *Malus* and *Pyrus* genomes. The most extensive study to investigate genetic diversity by SSRs was performed by Wunsch and Hormaza (2007), who described the genetic relationships among 63 European pear cultivars (Fig. 11.5). All the investigated cultivars were unequivocally identified while only two sports could not be distinguished from the original cultivar. Cluster analysis of the estimated genetic similarity grouped the cultivars into three clusters according to their pedigree and geographic origin. The largest cluster (Group A) contained the cultivars 'Dr. Jules Guyot,' 'Comice,' 'Passa Crassana,' 'Conference' and 'Williams' and most of the rest of the cultivars included in this group are derived from crosses involving those genotypes. The two other clusters (Groups B and C) included a more heterogeneous group of ancient cultivars that are currently cultivated to a lesser extent and originated in Southern Europe. Cluster B includes 'Coscia Precoce,' 'Roma' and 'Spina Carpi' from Italy, 'Blanquilla' and 'Abugo' from Spain, and 'Bonne Louise d'Avranches,' 'Cure,' and 'Beurré Giffard' from France, while the French cultivars 'Beurré Hardy' and 'Noveau Poiteau' as well as 'Castell' and 'Magallon' from Spain are clustered in Group C.

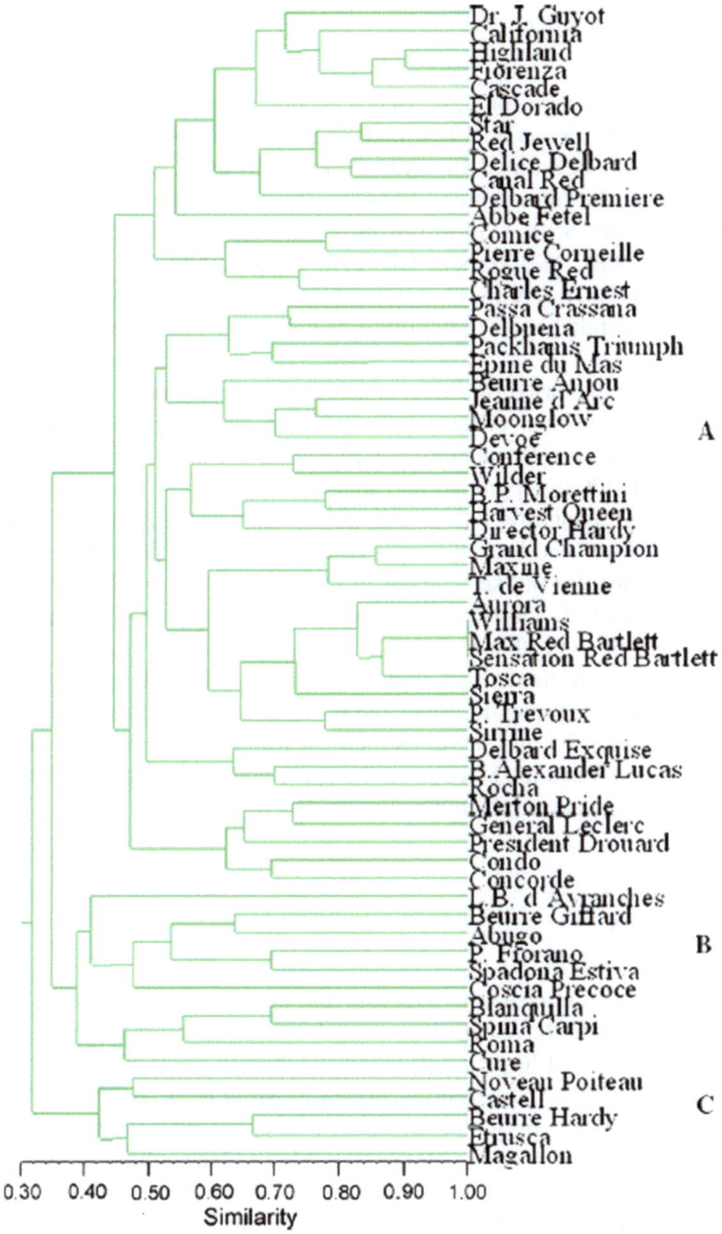

Fig. 11.5 UPGMA analysis of 63 European pear cultivars based on data from seven SSR primers (Wunsch and Hormaza 2007)

7.2 State of the Map

The first molecular maps for pear used a F_1 mapping population and dominant RAPD markers (Weeden et al. 1994; Iketani et al. 2001). Recently about 100 SSR markers have been developed for pear (Yamamoto et al. 2002a, b; Inoue et al. 2007) and due to the high synteny between apple and pear species, most of the 300 available apple SSRs (Guildford et al. 1997; Gianfranceschi et al. 1998; Liebhard et al. 2002; Silfverberg-Dilworth et al. 2006) can be used in pear (Yamamoto et al. 2001, 2002b; Dondini et al. 2004; Pierantoni et al. 2004, 2007). Apple SSRs are fundamental to denoting pear linkage groups and aligning apple and pear maps.

Several pear maps based on SSRs and MFLPs (microsatellite-anchored length polymorphism) have been constructed analyzing several F_1 populations. These include the interspecific (*P.communis* × *P pyrifolia)* population 'Bartlett' × 'Hosui' (Yamamoto et al. 2002b, upgraded in Yamamoto et al. 2004), the progeny derived from the 'Passe Crassane' × 'Harrow Sweet' cross and the progeny derived from 'Abbé Fétel' × 'Max Red Bartlett' cross (Pierantoni et al. 2007). These maps lead to a panel of molecular markers linked to the S-locus, fire blight, and scab resistance. Yamamoto et al. (2007) integrated the information from the progenies of 'Bartlett' × 'Hosui' with those of 'Hosui' × 'La France' to construct a map that could be aligned with the densest apple map of 'Fiesta' × 'Discovery' (Liebhard et al. 2002, 2003). This map describes the position of more than 130 SSRs in pear including 66 apple SSRs and serves as the Pear Reference Map. The colinearity of these 66 apple SSRs and the S-locus on the apple and pear maps confirms the high level of synteny between apple and pear.

7.3 Genomics

Functional genomics in pear also suggests that the S-locus is similar to the one in apple. While S-RNases have been known and studied for more than a decade, with 25 alleles identified in 130 pear cultivars (Zuccherelli et al. 2002; Zisovich et al. 2004; Sanzol et al. 2006; Takasaki et al. 2006; Mota et al. 2007; Goldway et al. 2009; Sanzol 2009), the pollen determinant F-Box has only recently been identified. The sequencing of genomic clones in the Maloideae subfamily has led to the identification of two F-box genes inside the S locus in apple (*Malus* × *domestica*) and three in Japanese pear (*P. pyrifolia*) called SFBB, or S-locus F-Box Brothers (Sassa et al. 2007). These sequences display a pattern of conserved and variable domains most likely involved in biochemical recognition. The first SFBB sequences of European pear confirmed the former data reported in *P. pyrifolia* (Di Sandro et al. 2008). In more recent papers it is reported that S-locus region of *P. communis* contains no less than six SFBB members surrounding S-RNases and that its structure seems to be rather conserved between apple and pear species (De Franceschi et al. 2011a; De Franceschi et al. 20011b).

Fischer et al. (2007) characterized the flavonoid biosynthesis pathway by cloning the main pear flavonoid cDNAs[1] and elucidated gene functions, gene copy numbers, and gene relationships within the Maloideae using their high homology with apple sequences. This work developed a panel of functional markers specific to the biosynthetic pathway of phenols, which are fundamental for the accumulation of anthocyanins in the skin of red cultivars.

A first indication of the "red" gene position in LG4 of the mutated sport 'Max Red Bartlett' was found by Dondini et al. (2008), although no data are reported about anthocyanin accumulation in pear fruit skin. Several studies of apple have indicated that a transcription factor of the Myb family acting as single gene controls the red skin trait (Takos et al. 2006; Espley et al. 2007) and the color of flesh and foliage (Espley et al. 2007; Chagné et al. 2007). Analogously to apple, a pear Myb factor is expressed 25-fold more in the fruit skin of 'Max Red Bartlett' than in 'William' (Pierantoni et al. 2009 and 2010). Chagné et al. (2007) mapped the MdMyb10 factor on LG9 in the apple progeny 'Discovery' × 91.136 B6-77. This apparent discrepancy may derive from the mutational origin of 'Max Red Bartlett.'

7.4 Transgenics

Genetic engineering represents an alternative strategy to introduce new traits (Table 11.8). In the past decade, several approaches have been pursued to introduce genes conditioning resistance to *E. amylovora*, other pathogens and psylla. Various genes such as attacin E from *Hyalophora cecropia* L, D5C1, whose action is similar to attacin E (Puterka et al. 2002), plant defensins (Lebedev et al. 2002a, b), hairpins (HrpN), a family of bacterial genes known as inducers of systemic resistance (Malnoy et al. 2005), and a depolymerase from the phage ΦEa1h, which causes the degradation of *Erwinia*'s capsular exopolysaccharide (EPS) have been inserted into pear and caused a significant reduction in the cultivar's susceptibility to fire blight as compared to the non-transformed plants (Table 11.8). One gene, the D5C1 also enhanced resistance to psylla in the transformed plants (Puterka et al. 2002). All these approaches use foreign genes to upgrade resistance. No resistance sources have been found within pear germplasm for use in genetic engineering. The same holds true for GM approaches to produce dwarfing rootstocks. Some success was achieved by integrating *rolC* from *A. rhizogenes* (Bell et al. 1999) and the *rolB* gene in the dwarfing rootstock BP10030 (Zhu et al. 2003).

The only pear gene to be employed in pear transformation thus far has been an ACC oxidase (ACO), which has been used in sense and antisense constructs.

[1] Phenylalanine ammonia lyase (PAL), chalcone synthase (CHS), chalcone isomerase (CHI), flavonol synthase (FLS), leucoanthocyanidin reductase (LAR1, LAR2), anthocyanidin synthase (ANS), anthocyanidin reductase (ANR), and UDP-glucose: flavonoid 7-O-glucosyltransferase (F7GT).

Table 11.8 Genetically modified pear cultivars and rootstocks

Genotype	Gene	Effect	Reference
William (Bartlett)	D5C1	Fire blight and psylla resistance	Puterka et al. (2002)
Passe Crassane	AttE	Fire blight resistance	Reynoird et al. (1999)
Passe Crassane	HrpN	Fire blight resistance	Malnoy et al. (2005)
Passe Crassane	Eps depolimerase	Fire blight resistance	Malnoy et al. (2005)
Conference	GUS (under inducible promoters)	Color expression induced by *E. amylovora*	Malnoy et al. (2003)
BP10030 (rootstock)	rol B	Rooting	Zhu et al. (2003)
Beurré Bosc (Kaiser)	rol C	Rooting	Bell et al. (1999)
Spadona E. (Blanquilla)	Gfp	GM plant selection	Yancheva et al. (2006)
Burakovka	Thaumatin II	Taste	Lebedev et al. (2002a)
Burakovka	Plant defensin	Fungal and microbial resistance	Lebedev et al. (2002b)
La France	ACO (sense and antisense)	Ethylene metabolism	Gao et al. (2003, 2007)

The ethylene production in transgenic shoots was consistent with the expression of sense-strand ACO transcription when the samples were incubated in 1 mM ACC, which is a unique substrate of ACO. Ethylene production in in vitro shoots was reduced by 85% in an antisense line where in vitro flowering and abnormal rooting were observed (Gao et al. 2007).

The aversion of governments and consumers in Europe to accept the cultivation of GM plants and the use of GM foods suggests that in future we should focus our research to understand the pear's functional genome. Given the synteny between apple and pear, the availability of genes and markers in the future will be assured because the sequencing of the apple genome is nearing completion.

References

Abe, K., Kotobuki, K., Saito, T. and Terai, O. 2000. Inheritance of resistance to pear scab from European pears to Asian pears. *J Jpn Soc Hort Sci*, 1, 1–8.

Bagnara, G.L. and Rivalta, L. 1989. Dwarfing in pear. *Acta Hort. (ISHS)*, 256, 103–110.

Baldini, E. and Scaramuzzi, F., 1957. Contributo allo studio delle cultivar di pero. *In* Baldini and Scaramuzzi eds. *La coltura del pero in Italia, Rivista Ortoflorofrutticoltura Italiana*, Firenze 1957, pp 255–342.

Bao, L., Chen, K., Zhang, D., Cao, Y., Yamamoto, T. and Teng, Y. 2007. Genetic diversity and similarity of pear cultivars native to East Asia revealed by SSR (simple sequence repeat) markers. *Genet Resour Crop Evol*, 54, 959–971.

Batlle, I., Lozano, L., Iglesias, I., Carbó, J., Bonany, J., White, A.G., Volz, R.K. and Brewer, L.R. 2008. The IRTA-HR pear scion breeding programme: aiming for high fruit quality under warm growing conditions. *Acta Hort. (ISHS)*, 800, 455–460.

Bassi, D., Tagliavini, M. and Marangoni, B. 1994. Selection of clonal rootstocks of *Pyrus communis* (L.). *Acta Hort.(ISHS)*, 367, 364–371.
Bell, R.L. 1996. Pears. *In* Janick, J. and Moore, J.N. eds. *Fruit Breeding Volume 1: Tree and Tropical Fruit.* John Wiley and Sons, New York, 441–514.
Bell, R.L., Scorza, R., Srinivasan, C. and Webb, K. 1999. Transformation of 'Beurré Bosc' pear with the *rolC* gene. *J Am Soc Hortic Sci*, 124, 570–574.
Bell, R.L. 2003. Resistance to pear psylla nymphal feeding of germplasm from central Europe. *Acta Hort. (ISHS)*, 622, 343–345.
Bell, R.L. and Reed, B.M. 2002. In vitro tissue culture of pear: advances in techniques for micropropagation and germplasm preservation. *Acta Hort.* (ISHS) 596, 412–418.
Bell R.L. ; van der Zwet T. and Blake, R.C. 2002; 'Blake's pride'. *Pear 2002*, 37, 711–713.
Bellini, E. and Nin, S. 1997. The breeding of pear tree worldwide (*Pyrus communis*). *Rivista di Frutticoltura,* 59, 19–30.
Bellini, E., and Nin, S. 2002. Breeding for new traits in pear. *Acta Hort*, 596, 217–224.
Booi, S., van Dyk, M.M., du Preez, M.G., Rees, D.J.G. and Labuschagné I. 2005. Molecular typing of red and green phenotypes of bon rouge pear trees, with the use of microsatellites. *Acta Hort. (ISHS).*, 671, 293–297.
Braniste, N. 2000. Collection, preservation and estimation of germplasm fund for Malus spp. and Pyrus spp. in Romania. *Acta Hort. (ISHS)*, 538, 91–94.
Brisset, M.N., Paulin, J.P. and Duron M. 1988. Feasibility of rating fire blight susceptibility of pear cultivars (*Pyrus communis*) on *in vitro* microcuttings. *Agronomie*, 8, 707–710.
Brisset, M. N., Ochatt, S. J. and Paulin, J. P. 1990. Evidence for quantitative responses during co-culture of *Pyrus communis* protoplasts and *Erwinia amylovora.* Plant Cell Reports, 9, 272–275.
Broertjes, C. 1982. Significance of in vitro adventitious bud techniques for mutation breeding of vegetatively propagated crops. *In:* Induced Mutations in Vegetatively Propagated Plants, II, IAEA, Vienna, 1–9.
Broothearts, W., Keulemans, J., Van Nerum,·I. 2004. Self-fertile apple resulting from S-RNase gene silencing. *Plant Cell Rep* 22, 497–501.
Brown, A.G. 1960. Scab resistance in progenies of varieties of cultivated pear. *Euphytica:* 247–253.
Brown, A. G., 1966. Genetical studies in pears V. Red mutants. *Euphytica* 15, 425–429.
Callan, N.W. and Lombard, P.B. 1978. Pollination effects on fruit set and seed development in 'Comice' pear. *J. American Society for Horticultural Science,* 103, 496–500.
Chagné, D., Carlisle, C.M., Blond, C., Volz, R.K., Whitworth, C.J., Oraguzie, N.C., Crowhurst, R.N., Allan, A.C., Espley, R.V., Hellens, R.P., and Gardiner, S.E. 2007. Mapping a candidate gene (MdMYB10) for red flesh and foliage colour in apple. *BMC Genomics*. 8: 212.
Chevalier, M., Bernard, C., Tellier, M., Lespinasse, Y., Filmond, R. and Le Lezec, M. 2004. Variability in the reaction of several pear (*Pyrus communis*) cultivars to different inocula of *Venturia pirina. Acta Hort. (ISHS),* 663, 177–182.
Chevreau, E., Taglioni, J.P., Cesbron, C., Dupuis, F., Sourice, S., Berry, I., Bersegeay, A., Descombin, J. and Loridon, K. 2007. Feasibility of alternative selection methods for transgenic apple and pear using the detoxification gene *Vr-ERE. Acta Hort. (ISHS),* 738, 277–281.
Colombo, R. and Bolognesi S, 2008. Risultati della sperimentazione sui principaliportinnesti per il pero. *Italus Hortus*, 15, 22–26
Crane, M. B., and D. Lewis, 1942: Genetical studies in pears III. Incompatibility and sterility. *J. Genet.*, 43, 31–44.
Decourtye, L. 1967. Etude de quelques caractères à contrôle génétique simple chez le pommier (*Malus* sp.) et le poirier (*Pyrus communis*). *Ann. Amélior. Pl.*, 17, 243–265.
Decourtye, L., 1970. Methodology in induced mutagenesis and results. *In Proc. Eucarpia Fruit Breed Symp Angers*, Eucarpia Fruit Sect S.E.I., C.N.R.A., Versailles pp. 161–174.
Decourtye, L., 1982. Bilancio di 20 anni di miglioramento delle specie legnose da frutto mediante la mutagenesi e prospettive attuali. Atti della giornata di studio sull'uso di tecniche nucleari per il miglioramento genetico dei fruttiferi, Roma, pp. 21–40.

De Franceschi, P., Pierantoni, L., Dondini, L., Grandi, M., Sanzol, J., Sansavini, S. 2011a. Cloning and mapping multiple S-locus F-box genes in European pear (*Pyrus communis* L.). *Tree Genetics & Genomes*, 2, 231–240.

De Franceschi, P., Pierantoni, L., Dondini, L., Grandi, M., Sansavini, S., Sanzol, J. 2011b. Evaluation of candidate F-box genes for the pollen S of gametophytic self-incompatibility in the Pyrinae (Rosaceae), on the basis of their phylogenomic context. *Tree Genetics & Genomes*, 4, 663–683.

Di Sandro, A., Serafini-Fracassin,i D., Del Duca, S., Faleri, C., Cai, G., De Franceschi, P., Dondini, L., Sansavini, S. 2008. Pollen transglutaminase in pear self incompatibility and relationships with S-RNases and S-allele variability. *Acta Hort.,* 800, 423–430.

Dondini, L., L. Pierantoni, F. Gaiotti, R. Chiodini, S. Tartarini, C. Bazzi, and S. Sansavini, 2004. Identifying QTLs for fire-blight resistance via a European pear (*Pyrus communis* L.) genetic linkage map. *Mol. Breed.*, 14, 407–418.

Dondini, L., Pierantoni, L., Ancarani, V., D'Angelo, M., Cho, K.-H., Shin, I.-S. , Musacchi, S., Kang, S.-J. and Sansavini, S. 2008. The inheritance of the red colour character in European pear (*Pyrus communis* L.) and its map position in the mutated cultivar 'Max Red Bartlett'. *Plant Breeding*., 127, 524–526.

Durel, C.E., Guérif, P., Belouin A., and Le Lezec M. 2004. Estimation of fire blight resistance heritability in the french pear breeding program using a pedigree-based approach *Acta Hort*., 663, 251–256.

Duron, M., Paulin, J.P. and Brisset, M.N. 1987. Use of in vitro propagated plant material for rating fire blight susceptibility. *Acta Hort. (ISHS)*, 217, 317–324.

Espley, R.V., Hellens, R.P., Putterill, J., Stevenson, D.E., Kutty-Amma, S. and Allan, A.C. 2007. Red colouration in apple fruit is due to the activity of the MYB transcription factor, MdMYB10. *The Plant Journal* 49, 414–427.

Fischer, TC, Gosch, C., Pfeiffer, J., Halbwirth, H., Halle, C., Stich, K. and Forkmann, G. 2007. Flavonoid genes of pear (*Pyrus communis*). *Trees*, 21, 521–529.

Fischer, M. and Mildenberger, G. 1999. The Naumburg/Pillnitz pear breeding programme results. *Acta Hort. (ISHS)*, 484, 135–138.

Fischer, M. and Mildenberger, G. 2002. New Naumburg/Pillnitz pear breeding results. *Acta Hort. (ISHS)*, 596, 225–231.

Gao, M., Matsuta, N., Nishimura, K., Tao, R., Murayama, H., Toyomasu, T., Mitsuhashi, W. and Dandekar, A. M. 2003. Transformation of pear (*Pyrus communis* cv. 'La France') with genes involved in ethylene biosynthesis. *Acta Hort.*, 625, 387–393.

Gao, M., Matsuta, N., Murayama, H., Toyomasu, T., Mitsuhashi, W., Dandekar, A.M., Tao, R. and Nishimura, K. 2007. Gene expression and ethylene production in transgenic pear (*Pyrus communis* cv. 'La France') with sense or antisense cDNA encoding ACC oxidase. *Plant Science*, 173, 32–42.

Gianfranceschi, L., Seglias, N., Tarchini, R., Komjanc, M. and Gessler, C. 1998. Simple sequence repeats for the genetic analysis of apple. *Theor Appl Genet*, 96,1069-1076.

Goldway, M., Takashi, T.Y., Sanzol, J., Mota, M., Zisovich, A., Stern, R.A. and Sansavini, S. 2009. Renumbering the S-RNase alleles of European pears (*Pyrus communis* L.) and cloning the S109 RNase allele. *Scientia Horticulturae,* 4, 417–422.

Griggs, W.H. and Iwakiri, B.T. 1954. Pollination and parthenocarpy in production of 'Barlett' pears in California. *Hilgardia***,** 22, 643–678.

Guildford, P., Prakash, S., Zhu, J-M., Rikkerink, E., Gardiner, S., Basset, H. and Forster, R. 1997. Microsatellites in Malus x domestica (apple) abundance, polymorphism and cultivar identification. *Theor Appl Genet*., 94, 249–254.

Harris, M.K. and Lamb, R.C. 1973. Resistance to the pear psylla in pears with *Pyrus ussuriensis* lineage.- *J. of the American Society for Horticultural Science*, 98, 378–381.

Hartmann, H.T. 1990. Plant Propagation: Principles and Practices. *In* H.T. Hartmann, D.E. Kester and F.T. Davies, Jr., eds., Prentice-Hall, Englewood Cliffs. NJ.

Hedrick, U.P. 1914. The pears of New York. In, New York Agricultural Experiment Station, Geneva, New York, NY.

Hirata, N., 1989. Self-compatible mutant in Japanese pear. Gamma Field Symposia. Production of mutants in tree crops. July 20–21, Institute of Radiation Breeding, NIAR MAFF, Ibaraki, Japan 28: 71–78.

Howard, B. H. 1987. Propagation. *In* Rom, R. C. és Carlson, R. F. (Eds.). *Rootstocks for fruit crops*. John Wiley and Sons, New York, pp. 29–77.

Hunter, D.M. and. Layne, R.E.C. 2004. Recent pear and apricot introductions from the AAFC-Harrow tree fruit breeding programs. *Acta Hort.* 663, 907–910.

Iketani, H., Abe, K., Yamamoto, T., Kotobuki, K., Sato, Y., Saito, T., Terai, O., Matsuta, N. and Hayashi, T. 2001. Mapping of disease-related genes in Japanese pear using a molecular linkage map with RAPD markers. *Breed Sci*, 51, 179–184.

Inoue, E., Matsuki, Y., Anzai, H. and Evans, K. 2007. Isolation and characterization of micro-satellite markers in Japanese pear (*Pyrus pyrifolia* Nakai). *Molecular Ecology Notes*, 7, 445–447.

Jock, S., Donat, V., Lòpez, M.M., Bazzi, C. and Geider, K. 2002. Following spread of fire blight in Western, Central and Southern Europe by molecular differentiation of *Erwinia amylovora* strains with PFGE analysis. *Environ. Microbiol*, 4, 106–114.

Lacey, C.N.D. 1975. Induction and selection of mutant form of fruit plants. *Long Ashton Ann Rep*, 22–24.

Lacey, C.N.D. and Campbell, I.A. 1982. Progress in mutation breeding of apples (*Malus pumila* Mill.) at Long Ashton Research Station, Bristol, United Kingdom. *In*: Induced Mutations in Vegetatively Propagated Plants, II. International Atomic Energy Agency, Wien, pp. 11–28.

Lebedev, V.G., Taran, S.A., Shmatchenko, V.V. and Dolgov, S.V. 2002. Pear transformation with the gene for supersweet protein thaumatin ii. *Acta Hort. (ISHS)*, 596, 199–202.

Lebedev, V.G., Dolgov, S.V., Lavrova, N., Lunin, V.G. and Naroditski, B.S. 2002. Plant-defensin genes introduction for improvement of pear phytopathogen resistance. *Acta Hort. (ISHS)*, 596, 167–172.

Le Lézec, M., Belouin, A., Guérif, P. and Lespinasse, Y. 2002. "ANGELYS", a new winter pear to replace "Passe Crassane". *Acta Hort. (ISHS)*, 596, 265–269.

Le Lézec, M., Thibault, B., Balavoine, P. and Paulin, J.P. 1985. Sensibilité varietale du pommier et du poirier au feu bacterien. *Phytoma,* 365, 37–44.

Lespinasse, Y. and Aldwinckle, H.S. 2000. Breeding for resistance to fire blight. *In:* J.L. Vanneste ed. *Fire Blight: The Disease and its Causative Agent, Erwinia amylovora*, CABI Publishing, Wallingford, UK, 253–273.

Lespinasse, Y., Chevalier, M., Durel, C.H.-.E., Guérif, P.H., Tellier, M., Denancé, C., Belouin, A. and Robert, P.H. 2008. Pear breeding for scab and psylla resistance. *Acta Hort. (ISHS)*, 800, 475–482.

Liebhard, R., Gianfranceschi, L., Koller, B., Ryder, R., Tarchini, E., van de Weg, E. and Gessler, C. 2002. Development and characterisation of 140 new microsatellites in apple (*Malus x domestica* Borkh.). *Mol. Breed*, 10, 217–241.

Liebhard, R., Kellerhals, M., Pfammatter, W., Jertmini, M. and Gessler, C. 2003. Mapping quantitative physiological traits in apple (Malus × domestica Borkh.). *Plant Molecular Biology*, 52, 511–526.

Malnoy, M., Venisse, J.S., Reynoird,·J.P. and Chevreau, E. 2003. Activation of three pathogen-inducible promoters of tobacco in transgenic pear (*Pyrus communis* L.) after abiotic and biotic elicitation. *Planta,* 216, 802–814.

Malnoy, M. Faize, M. Venisse, J.S. Geider, K. and Chevreau E. 2005. Expression of viral EPS-depolymerase reduces fire blight susceptibility in transgenic pear. *Plant Cell Rep*, 23, 632–638.

Marino, G., Beghelli, F., Rombolà, A.D. and Cabrini, L. 2000. *In vitro* performance at high pH and *in vivo* responses to Fe-deficiency of leaf-derived quince BA29 (*Cydonia oblonga*) somaclones regenerated at variable medium pH. *J Hort Sci Biotechnol*, 75, 433–440.

Marino, G. and Molendini, L. 2005. *In vitro* leaf-shoot regeneration and somaclone selection for sodium chloride tolerance in quince and pear. *J Hort Sci Biotechnol,* 80, 561–570.

Matkin, O.A. and Chandler, P.A. 1957. The U. C. -type soil mixes. *In* The U. C. System for Producing Healthy Container-grown Plants through the Use of Clean Soil, Clean Stock and Sanitation. California Agricultural Experiment Station Manual 23, Berkeley.

Massai, R., Loreti, F. and Fei, C. 2008. Growth and yield of 'Conference' pears grafted on quince and pear rootstocks. *Acta Hort. (ISHS)* 800, 617–624.

Masuda, T. and Yoshioka, T. 1997. *In vitro* selection of a mutant resistant to *Alternaria* blotch disease in 'Indo' apple. *Tech. News Inst. Rad. Breed.*, 56, 1–2.

Mielke, E.A. 2008. Status of the North American pear industry. *Acta Hort. (ISHS),* 800, 83–88.

Monte-Corvo, L., Cabrita, L., Oliveira, C., and Leitão, J. 2000. Assessment of genetic relationships among *Pyrus* species and cultivars using AFLP and RAPD markers. *Genetic Resources and Crop Evolution,* 47, 257–265.

Mota, M., Tavares, L. and Oliveira, C.M. 2007. Identification of S-alleles in pear (*Pyrus communis* L.) cv. 'Rocha' and other European cultivars. *Scientia Horticultura,* 113, 13–19.

Nicotra, A., Cobianchi, D., Faedi, W. and Manzo, P., 1979. Monografia di cultivar di pero, M.A.E, Istituto Sperimentale per la Frutticoltura, Roma 1979.

Palombi, M.A., Lombardo, B. and Caboni, E. 2007. *In vitro* regeneration of wild pear (*Pyrus pyraster* Burgsd) clones tolerant to Fe-chlorosis and somaclonal variation analysis by RAPD markers. *Plant Cell Reports,* 26, 489–496.

Pasqualini, E., Civolani, S., Musacchi, S., Ancarani, V., Dondini, L, Robert, P. and Baronio, P. 2006. *Cacopsylla pyri* behaviour on new pear selections for host resistance programs. *Bulletin of Insectology*, 59, 27–37.

Pierantoni, L., Cho, K.-H., Shin, I.-S., Chiodini, R., Tartarini, S., Dondini, L., Kang, S.-J., and Sansavini, S. 2004. Characterisation and transferability of apple SSRs to two European pear F1 population. *Theor. Appl. Genet.*, 109, 1519–1524.

Pierantoni, L., Dondini, L., Cho, K.-H.,. Shin, I.-S, Gennari, F., Chiodini, R., Tartarini, S., Kang, S.-J., and Sansavini, S. 2007. Pear scab resistance QTLs via a European pear (Pyrus communis) linkage map. *Tree Genetics & Genomes,* 3, 311–317.

Pierantoni, L., Dondini, L., Musacchi, S., Wilkel, B.S. and Sansavini, S. 2009. Gene expression patterns of red color skin in Bartlett pear (*Pyrus communis*) and its bud mutation Max Red Bartlett. *Acta Hort. (ISHS)*, 814, 567–570.

Pierantoni, L., Dondini, L., De Franceschi, P., Musacchi, S., Winkel-Shirley, B.J., Sansavini, S. 2010. Mapping of an anthocyanin-regulating MYB transcription factor and its expression in red and green pear, *Pyrus communis. Plant Physiology and Biochemistry*, 48, 1020–1026.

Postman, J.D., Spotts, R.A. and Calabro, J. 2005. Scab resistance in *pyrus* germplasm. *Acta Hort. (ISHS)*, 671, 601–608.

Predieri, S. (2001) Mutation induction and tissue culture in improving fruits. *Plant Cell Tissue Org Cult*, 64, 185–210.

Predieri, S. and Zimmerman, R.H. 2001. Pear mutagenesis: *in vitro* treatment with gamma-rays and field selection for productivity and fruit traits. *Euphytica*, 117, 217–227.

Puterka, G.J., Bocchetti, C., Dang, P., Bell, R.L. and Scorza, R. 2002. Pear transformed with a lytic peptide gene for disease control affects nontarget organism, pear psylla (*Homoptera: Psyllidae*). *Journal of Economic Entomology*, 95, 797–802.

Quamme, H.A. 1984. Observations of Psylla resistance among several pear cultivars and species. *Fruit Varieties Journal*, 38, 34–36.

Quartieri M., Tagliavini, M., Schiavon, L., Bassi, D. and Marangoni B. 2008. Nuove selezioni di portinnesti franchi (*Pyrus communis*). *Italus Hortus*, 6, 27–35.

Reynoird, J.P., Mourgues, F., Norelli, J., Aldwinckle, H.S., Brisset, M.N. and Chevreau, E. 1999. First evidence for improved resistance to fire blight in transgenic pear expressing the attacin E gene from *Hyalophora cecropia*, *Plant Sci.*, 149, 23–31.

Rivalta, L. and Dradi, M. 1998. Miglioramento genetico del pero presso l'Istituto Sperimentale per la Frutticoltura di Forlì.- *Rivista di Frutticoltura*, 9, 51–57.

Rivalta, L., Dradi, M. and Rosati, C. 2002. Thirty years of pear breeding activity at ISF Forlì, Italy. *Acta Hort. (ISHS)*, 596, 233–238.

Robert P. and Raimbault T. 2005. Resistance of some *Pyrus communis* cultivars and *Pyrus* hybrids to the pear psylla *Cacopsylla pyri (Homoptera, Psyllidae). Acta Hort.*, 671, 571–575.

Roby, F. 1972a. Doce mutaciones en el peral William obtenidas por injertos de ramitasirradiadas. *Rev Invest Agropec Ser 2*, 9, 55–64.

Roby, F. 1972b. Mutaciones inducida por irradiacion en el peral Packham's Triumph. In: Induced Mutation and Plant Improvement, Buenos Aires, 1970, International Atomic Energy Agency, Vienna, pp. 475–483.

Rodriguez, R., Diaz-Sala, C., Cuozzo, L. and Ancora, G. 1991. Pear *in vitro* propagation using a double-phase culture system. *Hort Science*, 26, 62–64.

Sansavini, S. 1969. Ricerche sulla potatura di produzione del pero Passa Crassana. Atti "Giornate di studio sulla potatura degli alberi da frutto". CNR- Firenze, 267–302.

Sansavini, S. 1994. Performance of micropropagated pear trees. *Acta Hort (ISHS)*, 367, 260–266.

Sansavini, S; 2002. Pear fruting-branch models related to yield control and pruning. *Acta Hort (ISHS)*, 596, 627–633.

Sansavini, S. 2007. I portinnesti. In Angelici C. ed. *Il pero*, Bayer Cropscience. Collana Coltura& cultura, Milano, pp. 270–282.

Sansavini, S and Ancarani V., 2008. Miglioramento genetico del pero e nuove varietà in Europa. *Riv. Frutticoltura*, 10, 28–36.

Sansavini, S, Corelli, L. and Ragazzini, D. 1986a. Influenza delle gibberelline A3 e A 4+7 e di altri alleganti sulla partenocarpia e sulla ritenzione dei frutto di pero 'Conference? Con impollinazione controllata. *Riv. Frutticoltura*, 6, 65–81.

Sansavini, S., and Zocca, A., 1965. La diversa attitudine ad emettere. rami anticipati nelle cultivar di pero. *Riv. Frutticoltura*, 3, 233–241.

Sanzol, J. and Herrero, M. 2002: Identifi cation of self-incompatibility alleles in pear cultivars (*Pyrus communis* L.). *Euphytica* 128, 325–331.

Sanzol, J. 2009. Pistil-function breakdown in a new S-allele of European pear, S21°, confers self-compatibility. *Plant Cell Rep*, 28, 457–467.

Sanzol, J., Sutherland, B.G. and Robbins, T.P. 2006. Identification of genomic DNA sequences of the S-ribonuclease gene associated with self incompatibility alleles S1 to S5 in European pear. *Plant Breeding*, 125, 513–518.

Sassa, H., Kakui, H., Miyamoto, M., Suzuki, Y., Hanada, T., Ushijima, K., Kusaba, M., Hirano, H. and Koba, T. 2007. S locus F-box brothers: multiple and pollen-specific F-box genes with S haplotype-specific polymorphisms in apple and Japanese pear. *Genetics*, 175, 1869–1881.

Sassa, H., Hirano, H., Nishio, T. and Koba, T. 1997. Style-specific self-compatible mutation caused by deletion of the S-RNase gene in Japanese pear (*Pyrus serotina*). *Plant Journal*, 12, 223–227.

Seavert, C.F. 2005. Pear production in the North America. *Acta Hort. (ISHS)*, 671, 45–46.

Sijacic, P., Wang, X., Skirpan, A.L., Wang, Y., Dowd, P.E., McCubbin. A.G., Huang, S. and Kao, T.H. 2004. Identification of the pollen determinant of S-RNase-mediated self-incompatibility. *Nature* 429, 302–305.

Silfverberg-Dilworth, E., Matasci, C., van de Weg, W.E., Van Kaauwen, M.P.W., Walser, M., Kodde, L.P., Soglio, V., Gianfranceschi, L., Durel, C.E., Costa, F., Yamamoto, T., Koller, B., Gessler, C. and Patocchi, A. 2006. Microsatellite markers spanning the apple (*Malus* x *domestica* Borkh.). *Tree Genet Gen*, 2, 202–224.

Stern, R.A. 2008. *Pyrus betulifolia* is the best rootstock for the 'Coscia' pear in the warm climate of Israel. *Acta Hort. (ISHS)* 800, 631–638.

Spiegel-Roy, P. 1990. Economic and agricultural impact of mutation breeding in fruit trees. *Mutation Breeding Rev*, 5, 1–26.

Takasaki, T., Moriya, Y., Okada, K., Yamamoto, K., Iwanami, H., Bessho, H. and Nakanishi, T. 2006. cDNA cloning of nine S-alleles and the establishment of a PCR-RFLP system for genotyping European pear cultivars. *Theoretical and Applied Genetics*, 112, 1543–1552.

Takos, A.M., Ubi, B.E., Robinson, S.P. and A.R. Walker, 2006. Condensed tannin biosynthesis genes are regulated separately from other flavonoid biosynthesis genes in apple fruit skin. *Plant Sci*, 170, 487–499.

Thibault, B. and Paulin, J.P. 1984. Pear breeding and selection for fire blight resistance. *Acta Hort. (ISHS)*, 161, 141–146.

Thibault, B., Belouinm A. and Lecomtem P. 1989. Sensibilité variétale du poirier au feu bactérien. *L'Alboriculture fruitière*, 421, 139–148.

Thomson, S.V. 2000. Epidemiology of fire blight. *In*: Vanneste J.L. ed. *Fire blight The Disease and its Causative Agent, Erwinia amylovora.* CABI Publishing, Wallingford, United Kingdom. pp. 9–36.

Tukey, H.B. 1964. Dwarfed fruit trees. Macmillan. New York.

van der Zwet, T. and Bell, R.L. 1984. Comparative evaluation of the degree of fire blight resistance in various pears cultivars and selections. *Acta Hort.* 151, 267–275.

Vasilakakis, M.D., Porlingis, I.C. 1985. Effect of temperature on pollen germination, pollen tube growth, effective pollination period and fruit set of pear. *HortScience*, 20, 733–735.

Vavilov, N. 1951. The Origin, Variation, Immunity and Breeding of Cultivated Plants. *Chronica Botanica*, 13, 1–366.

Vavra, M. and Orel, V. 1971. Hybridization of pear varieties by Gregor Mendel. *Euphytica*, 20, 60–67.

Villalta, O.N., Washington W.S. and McGregor G. 2004. Susceptibility of European and Asian pears to pear scab. *Plant Prot. Q.*, 19, 2–4.

Viseur, J. 1987. Micropropagation of pear, *Pyrus communis* L., in a double-phase culture medium. *Acta Hort. (ISHS)*, 212, 117–124.

Viseur, J. 1990. Evaluation of fire blight resistance of somaclonal variants obtained from the pear cultivar 'Durondeau'. *Acta Hort. (ISHS)*, 273, 275–284.

Visser, T., Schaap, A.A. and De Vries, D.P. 1968. Acidity and sweetness in apple and pear. *Euphytica*, 17. 153–167.

Visser, T. and De Vries, D.P. 1970. Precocity and productivity of propagated apple and pear seedlings as dependent on the juvenile period. *Euphytica*, 19, 141–144.

Visser, T., Verhaegh, J.J and De Vries D. 1971. Pre-selection of compact mutants induced by X-ray treatment in apple and pear. *Euphytica*, 20, 195–207.

Volz, R.K., White, A.G. and Brewer, L.R. 2008. Breeding for red skin colour in interspecific pears. *Acta Hort. (ISHS)*, 800, 469–474.

WAPA. 2009. Southern hemisphere fresh apple and pear crop forecast, February 2009. Eds. World Apple and Pear Association, Brussels (Belgium), 1–21.

Weeden, N.F., Hemmat, M., Lawson, D.M., Lodhi, M., Bell, R.L., Manganaris, A.G., Reisch, B.I. and Brown, S.K. 1994. Development and application of molecular marker linkage maps in woody fruit crops. *Euphytica,* 77, 71–75.

Wertheim, S.J. 1998. Pear rootstock. In Wertheim S.J. ed. *Rootstock Guide*. Wilhelminadorp, Netherlands, pp. 61–82.

Westigard, P.H., Westwood, M.N. and Lombard, P.B. 1970. Host preference and resistance of *Pyrus* species to the pear psilla *Psylla pyricola* Foerster. *Journal of the American Society for Horticultural Science*, 95, 34–36.

Wunsch, A. and Hormaza, J.I., 2007. Characterization, variability and genetic similarity of European pear with SSRs. *Sci. Horticult.*, 113, 37–43.

Yamamoto, T., Kimura, T., Sawamura, Y., Kotobuki, K., Ban, Y., Hayashi, T., Matsuta, N. 2001 SSRs isolated from apple can identify polymorphism and genetic diversity in pear. *Theor Appl Genet.*, 102, 865–870.

Yamamoto, T., Kimura, T. Sawamura, Y., Manabe, T., Kotobuki. K,, Hayashi. T,, Ban. Y. and Matsuta N. 2002a. Simple sequence repeats for genetic analysis in pear. *Euphytica*, 124, 129–137.

Yamamoto, T., Kimura, T., Shoda, M., Imai, T., Saito, T., Sawamura, Y., Kotobuki, K., Hayashi, T. and Matsuta, N. 2002b. Genetic linkage maps constructed by using an interspecific cross between Japanese and European pears. *Theor Appl Genet.*, 106, 1–18.

Yamamoto, T., Saito, T., Kotobuki, K., Matsuta, N., Liebhard, R., Gessler, C., Van de Weg, W.E. and Hayashi, T. 2004. Genetic linkage maps of Japanese and European pears aligned to the apple consensus map. *Acta Hort.,* 663, 51–56.

Yamamoto, T., Kimura, T., Terakami, S., Nishitani, C., Sawamura, Y., Saito, T., Kotobuki, K. and Hayashi, T. 2007. Integrated Reference Genetic Linkage Maps of Pear Based on SSR and AFLP Markers. *Breeding Science*, 57, 321–329.

Yancheva, S.D., Shlizerman, L.A., Golubowicz, S., Yabloviz, Z., Perl, A., Hanania, U. and Flaishman, M.A. 2006. The use of green fluorescent protein (GFP) improves *Agrobacterium*-mediated transformation of 'Spadona' pear (*Pyrus communis* L.). *Plant Cell Report*, 25, 183–189.

Zielinski, Q.B., Reimer, F.C. and Quackenbush, V.L. 1965. Breeding behaviour of fruit characteristics in pears, *Pyrus communis* L. *Proceedings of the American Society for Horticultural Science,* 86, 81–87.

Zielinski, Q.B. and Thompson, M.M. 1967. Speciation in *Pyrus*: Chromosome number and mitotic behavior. *Bot. Gaz.,* 128,109–112.

Zisovich, A.H., Stern, R.A., Shafir, S. and Goldway, M. 2004. Identification of seven *S*-alleles from the European pear *(Pyrus communis)* and the determination of compatibility among cultivars. *Journal of Horticultural Science and Biotechnology*, 79, 101–106.

Zhu, L.H., Li, X.Y., Ahlman, A. and Welander, M. 2003. The rooting ability of the dwarfing pear rootstock BP10030 (*Pyrus communis*) was significantly increased by introduction of the *rolB* gene. *Plant Science,* 165, 829–835.

Zuccherelli, S., Tassinari, P., Broothaerts, W., Tartarini, S., Dondini, L. and Sansavini, S. 2002. *S*-allele characterization in self-incompatible pear (*Pyrus communis L.*). *Sexual Plant Reproduction*, 15, 153–158.

Chapter 12
Apricot

Tatyana Zhebentyayeva, Craig Ledbetter, Lorenzo Burgos, and Gerardo Llácer

Abstract Apricot is in the Rosaceae family within the genus *Prunus* L., subgenus Prunophora Focke, and the section Armeniaca (Lam.) Koch. Depending on the classification system, the number of apricot species ranges from 3 to 12. Six distinct species are usually recognized: *P. brigantina* Vill., *P. holosericeae* Batal, *P. armeniaca* L., *P. mandshurica* (Maxim), *P. sibirica* L., Japanese apricot *P. mume* (Sieb.) Sieb. & Succ. Vavilov placed apricot in three centers of origin: the Chinese center (Central and Western China), the Central Asiatic center (Afghanistan, northwest India and Pakistan, Kashmir, Tajikistan, Uzbekistan, Xinjing province in China and western Tien-Shan), and the Near-Eastern center (interior of Asia Minor). Kostina further divided the cultivated apricot according to their adaptability into four major ecogeographical groups: (1) the Central Asian group, (2) the Iran-Caucasian group, (3) the European group, and (4) the Dzhungar-Zailij group. Many local cultivars are grown in the different areas and producing countries; however, these cultivars lack important traits that needed by modern production and marketing systems. Breeding programs have and continue to develop cultivars with improved adaptability to the environment (temperature requirements, water deficit), extension of the harvest season, fruit quality for fresh consumption and processing, productivity, adequate tree size, and resistance to biotic stresses. The major objectives in apricot breeding

T. Zhebentyayeva (✉)
Genetics and Biochemistry, Clemson University, 116 Jordan Hall, Clemson, SC 29634, USA
e-mail: tzhebe@clemson.edu

C. Ledbetter
USDA, ARS, CDP&G, SJVASC, Parlier, CA, USA
e-mail: craig.ledbetter@ars.usda.gov

L. Burgos
CEBAS-CSIC, Murcia, Spain
e-mail: burgos@cebas.csic.es

G. Llácer
IVIA, Moncada, Valencia, Spain
e-mail: gllacer@ivia.es

M.L. Badenes and D.H. Byrne (eds.), *Fruit Breeding*, Handbook of Plant Breeding 8,
DOI 10.1007/978-1-4419-0763-9_12,

programs are resistance to sharka caused by *Plum Pox Virus*, brown rot caused by *Monilinia* spp., bacterial diseases caused by *Pseudomonas* spp. and *Xanthomonas arboricola* pv. *pruni* (Smith), Chlorotic Leaf Roll Phytoplasma, and Apricot Decline Syndrome. Among these, PPV is the most limiting factor in Europe and much work has to be invested in developing PPV-resistant apricot cultivars. Molecular markers have been developed in apricot and used mainly for construction of linkage maps and genetic diversity studies.

Keywords *Prunus armeniaca* • Centers of origin • Domestication • Eco-geographical groups • Breeding goals • Breeding methods • Marker Assisted Selection • PPV resistance • Fruit quality • Inheritance • Genetic maps • Molecular markers • Genomic resources • Structural and functional genomics • Transgenics

1 Introduction

Apricot is a Rosaceae family member and belongs to section Armeniaca (Lam.) Koch in subgenus Prunophora Focke, genus *Prunus* L. (Rehder 1940). All apricot species thus studied are regular diploids, with eight pairs of chromosomes ($2n=16$), and all can be intercrossed in either direction, making their classification confusing. Depending on the classification system, the number of apricot species ranges from 3 to 12. Six distinct species are usually recognized: Briancon apricot or Alpine plum *P. brigantina* Vill., Tibetan apricot *P. holosericeae* Batal., common apricot *P. armeniaca* L., Manchurian apricot *P. mandshurica* (Maxim), Siberian apricot *P. sibirica* L., Japanese apricot *P. mume* (Sieb.) Sieb. & Succ. (Kryukova 1989; Faust et al. 1998; Bortiri et al. 2001). Three other often recognized species *Prunus × dasycarpa* Ehrh., *P. armeniaca* var *ansu* (Maxim.) Kost., and *P. sibirica* var *davidiana* (Carrière) are apparently of hybrid origin. Most apricot cultivars grown for fruits belong to the species *P. armeniaca* though introgression of *P. mume* and, to the less extent *P. mandshurica* and *P. sibirica*, into cultivated germplasm is a commonly acknowledged fact among apricot breeders. Cultivation of Japanese apricot, *P. mume*, for fruit production has a much shorter history compared with its ornamental flower use (Mega et al. 1988). This review is written with emphasis on apricot species significant for fruit production.

Despite their many positive fruit attributes, namely, attractiveness, tasty flavor and ease of eating, as well as their multiple-use functionality and a nonsurplus production, apricots suffer from several weak points. As compared with the other summer fruits, apricots have a higher sensitivity to diseases. Fluctuating crop levels lead to an irregular market supply, and the narrow range of cultivars allow for only a brief market presentation. Furthermore, all too often consumers are displeased by an insufficient fruit quality and ripeness, leading to a rather low consumption rate compared to the other summer fruits (Moreau-Rio 2006; Audergon et al. 2006a).

In the last 20 years, world production has increased 85%, mainly due to the large plantings made in Asia (Turkey, Iran, Pakistan, Uzbekistan) and Africa (Algeria, Morocco, Egypt). In Europe, production increased at a lower rate, while

Table 12.1 Apricot production (MT × 1,000) from main producer countries (FAO 1989, 2008)

Country	Average production 1985–1987	Average production 2004–2006
Turkey	271	547
Iran, Islamic Republic	56	239
Italy	191	223
Pakistan	61	201
Uzbekistan	–	193
France	104	174
Algeria	40	134
Spain	148	133
Japan	–	119
Morocco	69	106
Syrian Arab Republic	57	101
China	58	86
Ukraine	–	85
Greece	112	79
South Africa	50	75
Egypt	–	73
The USA	91	65
Russian Federation	–	63
Continent		
Africa	213	437
Asia	624	1,731
Europe	745	926
Northern America	99	67
Oceania	37	21
Southern America	30	53
World	1,748	3,235

in North America and Oceania production has decreased. Near 50% of world production is concentrated in Mediterranean countries (Table 12.1; FAO 1989, 2008). The germplasm diversity that will be reported later indicates that apricots can be grown much more widely; hence the species can become a greater part of the world's fruit production. However, the limited ecological adaptation at the genotype level is the main challenge to apricot breeders. The introduction of new cultivars from foreign sources may often give disappointing results, with unpredictable variability depending on the environment (Pennone et al. 2006). Consequently, cultivars must be bred for each producing area and for each marketing opportunity (Layne et al. 1996).

The uses of apricot are multiple and diverse: fresh fruit, processed fruit for drying, canning, jam, juice, sauce, puree for baby food, wine, liquor, and vinegar (Maikeru Shoji 1994; Han 2001; Bala et al. 2005; Bassi and Audergon 2006). Traditional Chinese medicine uses bitter apricot kernels in different preparations for treating asthma and coughs, infant virus pneumonia, and disease of the large intestine (Li 1997; Chen et al. 1997). Dried fruit or fruit juice concentrate of Japanese apricot (*Prunus mume*) are used to prepare a beverage capable of preventing and curing cancer (Fang 1995; Otsuka et al. 2005). Apricot kernel oil is used in a liquid soap composition for dermatitis treatment (Harbeck 2001). In some Asian regions, apricots used for their edible seed and seed oil are more important than apricots grown for fruit (Layne et al. 1996). The use of crushed shells of apricot stones instead of anthracite coal in filters for water treatment is investigated (Aksogan et al. 2003). The ornamental use of apricot trees is discussed later.

2 Origin and Domestication of Scion Cultivars

Some of the most significant evolutionary trends in apricot domestication have been related to fruit quality enhancement, selection of cultivars with nonbitter seeds, adaptation to a greater range of environments (i.e., development of cultivars with lower or higher chilling requirements), and a gradual change in biology of sexual propagation from self-incompatible to self-fertile.

2.1 Centers of Origin

N. Vavilov (1951) placed apricot in three centers of origin for cultivated plants (1) the Chinese center that comprises mountainous regions of Central and Western China together with the adjacent lowlands, (2) the Central Asiatic center that includes Afghanistan, northwest India and Pakistan, Kashmir, Tajikistan, Uzbekistan, Xinjing province in China and western Tien-Shan, and (3) the Near-Eastern center including the interior of Asia Minor, Transcaucasia, Iran, and Turkmenistan. After Vavilov, many discussions ensued regarding the sizes and boundaries of the proposed centers of origin (reviewed by Zeven and de Wet 1982), and in reference to apricot, it is important to mention the revision by Zhukovsky (1971), who included Turkmenistan in the Central Asiatic center and set the boundaries between the Central Asian and Near Eastern centers (Fig. 12.1).

Most contemporary authors support the antiquity of apricot in Central Asia and China and recognize them as independent centers of domestication (Bailey and Hough 1975; Kryukova 1989; Layne et al. 1996; Faust et al. 1998; Hormaza et al. 2007). However, the early history of apricot is still not completely clear. The major question whether or not its cultivation in Central Asia preceded or came after Chinese culture remains to be elucidated (Zohary and Hopf 2001). Apparently, apricot was first brought under cultivation in China. There is Chinese written evidence of apricot cultivation cited by De Candolle (1886) dating from the end of III millennium BC. In Central Asia, apricot cultivation was introduced more recently, around I–II millennia BC (Sinskaya 1969). In accordance with this dating, modern excavations in southern Turkmenistan and Uzbekistan lack evidence for use of fruit and nuts in western Central Asia before 1500 BC (Miller 1999).

In spite of a longer history of cultivation, the typical eastern cultivars did not seem to move far away from the Chinese center and remained preserved in their

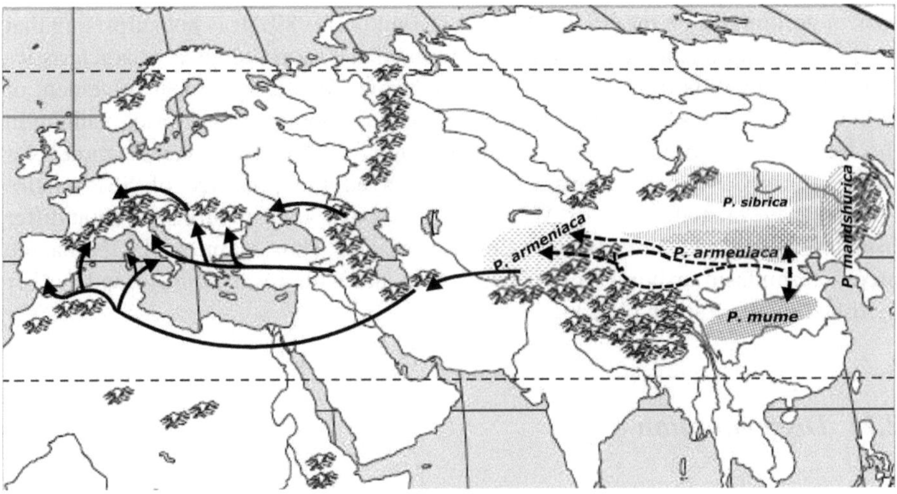

Fig. 12.1 Apricot dissemination from the primary centers of domestication (adapted from Faust et al. 1998). Outline world physical map is courtesy of Houghton Mifflin Educational Place®

native environment of Eastern Asia. It is likely that a germplasm exchange between the Chinese and Central Asian primary centers of cultivation was restricted to the first global trade route, the Silk Road, established in II–III millennia BC. Owing to the practice of seed propagation in Central Asia, the Chinese germplasm delivered through the Silk Road was assimilated and absorbed by local apricots. As a result, some aboriginal varieties grown in the Zeravshan valley and Khorezm oasis have some fruit characteristics resembling typical eastern Chinese apricots (Kovalev 1963). Molecular marker analysis supported an introgression of Chinese germplasm into the Central Asian assortment in zones of admixture linked to the Silk Road (Zhebentyayeva et al. 2003).

In studies on the origin of apricot, Kostina (1946) emphasized the importance of the Central Asian center for its spread worldwide. She definitively distinguished the apricots from Central Asia and Xinjing province in China, genetically linked to wild Tien-Shan *P. armeniaca*, from the Eastern Asian apricots related to East Asian wild species. As a result, she probably missed the Chinese group in apricot classification (Kostina 1964). A survey of indigenous Chinese germplasm (Zhao et al. 2005), as well as the noted population structure of wild apricots in the Ily valley of West China (He et al. 2007) and molecular data on crop-wide germplasm diversity (Zhebentyayeva et al. 2003), all support the theory of western Tien-Shan wild populations as being a major ancestral gene pool for apricot domestication in Central Asia and responsible for its spread from this area to more westerly regions.

In agreement with De Candolle (1886), Vavilov (1951, p. 34) and Kostina (1946) considered the Near Eastern center as a secondary center for cultivated apricots. Historically connected with China, Samarkand (Sogdiana) was the farthest reach of the Persian Empire, the Empire of Alexander the Great and the Chinese Empire. This fact was probably of critical importance for a secondary diversification of apricot germplasm on the Iranian Plateau (Kryukova 1989). It is not surprising that a principal component analysis of 47 anatomical and morphological characteristics of the local apricots from Iran and Armenia provided evidence for involvement of both Central Asian and Chinese apricots in the development of an apricot culture in the Near Eastern center (Rostova and Sokolova 1992). Moreover, in molecular studies, Iran-Caucasian cultivars never displayed the presence of SSR alleles (Zhebentyayeva et al. 2003) or AFLP loci (Zhebentyayeva unpublished) that differ from those among Central Asian or Chinese apricots. Thus, it appears likely that the mixed germplasm arriving from Central Asia and China was adopted and further modified on the Iranian plateau.

2.2 *Dissemination*

Spread of apricot from its centers of first cultivation was discussed in great detail by Faust et al. (1998). In the Mediterranean basin, characteristic large apricot stones were found in several archeological sites from classical times onward (Zohary and Hopf 2001). A few hundred years later, apricot was a well-established fruit tree species in Syria, Turkey, Greece, and Italy.

Several routes have been assigned relative to the dissemination of apricot from the Near East to other regions:

1. Apricots were dispersed to the Middle East, Egypt, and North Africa, and later to Spain. This African route produced cultivars known for their low chilling requirements. Genetic structure of Tunisian apricots and their similarity to Central Asian and Iran-Caucasian apricots confirmed this dissemination route (Khadari et al. 2006).
2. A second dissemination route went north from the Black Sea, extending from Turkey or directly from Iran.
3. There was a central dissemination route to the Danube River valley and Germany. Roman soldiers and Turkish landowners were greatly involved in dissemination via this route. Probably in the Danube valley, European cultivars were originally selected for their size and adaptation to the new environment. The first forms of European apricots with mutated haplotypes conferring self-compatibility might also have originated here (Halász et al. 2007).
4. A more southerly dissemination route was assigned to Greece, and both Middle and Southern Europe, emanating north from the Mediterranean Sea. Most likely, Southern European cultivars were developed due to movement in this direction. One could consider the Italian germplasm as a secondary center of apricot diversification. In a molecular study by Geuna et al. (2003), the high level of diversification in Italian germplasm might reflect an iterative direct introgression of plant material from primary centers of origin.

The apricot spread from China and Central Asia to Europe during last 3,000–4,000 years and was subsequently taken to North America and other parts of the world. Actually, apricot arrived to North America from two opposite directions, from Europe across the Atlantic Ocean and from China across Pacific Ocean (Faust et al. 1998). In North America, the apricot's dissemination route ended with distinct cultivars characteristic of the region: highly desirable fruit appearance (big size, orange flesh color, and firm texture), but with poor flavor and low sugar content. Perhaps due to their Chinese ancestral background, some North American cultivars developed natural resistance to a major pathogen of the *Prunus* species: plum pox virus (Zhebentyayeva et al. 2008). At the end of twentieth century, we observed the movement of North American apricot germplasm back to its Eurasian homeland for the purpose of stopping the spread of the virus in the major apricot production regions.

3 Genetic Resources

3.1 Scion

Based on a genetic approach to descriptions of morphological traits and adaptation to specific ecogeographical environments, Kostina (1936, 1964) developed a successful dichotomous classification of apricot germplasm (other classifications are reviewed in Faust et al. 1998). This classification left room for further amendments and has survived to date without major changes. Her description of diversified apricot germplasm relied on discrete qualitative traits with discrete inheritance such as seed taste (sweet/bitter), fruit skin (glabrous/pubescent), fruit adherence to stone (freestone/clingstone), fruit flesh color (orange/yellow/white), and tree architecture (upright/spreading). These oligogenic traits provided a solid framework for germplasm analysis. Quantitative traits such as chilling requirements (early/late blooming), fruit size (small/medium/large) and resistance to major diseases in specific environments along with emphasis on genetic contributions of nondomesticated species, were important for exploitation of apricot germplasm in breeding programs.

Kostina recognized four major ecogeographical groups of apricots (1) the Central Asian group with five regional subgroups: Fergana, Zeravshan, Shakhrisyabz, Khorezm, and Kopet-Dag, (2) the Iran-Caucasian group, (3) the European group subdivided for eastern, western and northern subgroups, and (4) the Dzhungar-Zailij group closely linked to the wild Tien-Shan apricot. Kovalev (1963) added the Chinese group to this classification and singled out the Southern European and North American apricots into two subgroups of European apricots. Bailey and Hough (1975) separated North African apricots from the Iran-Caucasian group, while Nyujtó and Suránui (1981) recognized only two subgroups within the European group: the continental and Mediterranean. Kryukova (1989) made the most careful revision of Kostina's classification by adding the Chinese group and incorporating the Dzhungar-Zailij apricots into the Central Asian group.

The *Chinese group* of cultivars is the oldest, the most diversified, and currently underexplored. Perhaps this group is the last world resource for apricot improvement using traditional breeding techniques. In China, six commonly accepted apricot species are endemic: *P. armeniaca*, *P. sibirica*, *P. mandshurica*, *P. holosericeae*, *P. mume*, and *P. dasycarpa*. There are also 13 subspecies of Siberian, Manchurian and common apricots resulting from sporadic cross-hybridizations in overlapping areas (Zhao et al. 2005). More than 2,000 cultivars and life-forms have been described in China, and about one third of them are maintained at Liaoning Research Institute of Pomology, Xiongyue. The wealth of this germplasm represents the Chinese (Eastern Asian) center of apricot diversity.

In Eastern Asia, the apricot was brought under cultivation in two geographical regions (Kostina 1964; Kovalev 1963; Layne et al. 1996). *P. armeniaca* cultivars from Eastern and Central China grow in the same areas as wild *P. mume*. They adapted to a warm humid climate and developed resistance to fungal diseases. In Northern and Northeastern China, the distribution of wild *P. armeniaca* overlaps with that of *P. sibirica* and *P. mandshurica*. Northern Chinese cultivars are adapted to severe winter conditions. In molecular studies Chinese cultivars revealed their relatedness to the northeastern species *P. mandshurica* and *P. sibirica* or to *P. mume* and its interspecific hybrid with common apricot, *P. armeniaca* var *ansu* (Zhebentyayeva et al. 2003, 2008).

In China apricot production is focused on the development cultivars for fresh market, kernel production and ornamental use. The local cultivars recommended for fresh market have some individual outstanding traits, but the overall quality of these cultivars is not very good, as most of them are self-incompatible, have a short shelf life, and have limited environmental adaptation. In spite of using *P. sibirica* for apricot propagation, the fruit set and tree productivity are often low due to late season frosts. Cultivar recommendations for fresh market are as follows: early maturation season—'Luotuohuang' (earliest apricot, FDP 55 days, mean weight 51 g), 'Maihuang,' 'Hebao,' 'Shisheng'; for mid-season apricots—'Huaxiangdajiexing,' 'Shajinhong,' 'Yinxiangbai,' 'Jidanxing'; for late-season—'Chuanzhihong' (FDP 95 days, very productive, mean weight 80 g), 'Wanxing,' and 'Badou.' The best cultivar for kernel use, 'Longwangmo,' has high productivity (about 1,500 kg/ha) and seeds (~2 g) with thin shells. Apricots for ornamental use derived from the interspecific hybridization *P. armeniaca* × *P. mume* have 30–70 petals and bloom as early as *P. sibirica* 'Liaomei' and as late as *P. armeniaca* 'Shanmei' (Byrne et al. 2000).

The *Central Asian* ecogeographical group is one of oldest and richest in diversity. This group includes apricots endemic to Afghanistan, Baluchistan, Kashmir, Xinjing, Uzbekistan, Tadjikistan, Kyrgystan, and Turkmeniastan. They grow in regions that overlap with the wild Tien-Shan apricots. Owing to seed propagation and a wealth of wild germplasm, the Central Asian apricots are highly diversified. The trees are vigorous and long-lived, with an extended juvenile growth period. Most cultivars are self-incompatible. They are well adapted to a dry atmosphere and susceptible to fungal diseases.

Central Asian apricots produce fruits from small to medium in size, and without specific aroma or mealiness. The maturation season is long (from May to the end of September), perhaps the longest of the various ecogeographical groups. Skin color varies from white to intensive orange and almost red. Often fruits do not have skin pubescence. In general, the fruits have a high soluble solids content (20–30%). Acidity is usually low, in the range of 0.6–0.8% on a fresh weight basis (Kovalev

1963). Fruits are well attached and often dry (raisin) on the tree. Apricots are eaten fresh or dried. Apricot kernel production is limited to local markets.

Fergana subgroup. Apricots from the Fergana valley are of the most authentic Central Asian type (Kostina 1936; Kovalev 1963; Kryukova 1989; Rostova and Sokolova 1992). In molecular studies, this subgroup is the closest to nondomesticated *P. armeniaca* (Zhebentyayeva et al. 2003, 2008). Apricot production is predominantly for use as dry fruit. Fruits generally have a weak pubescence. There are not many glabrous cultivars (about 5%). Major cultivars of the Fergana subgroup: 'Mirsandzheli,' 'Kandak,' 'Khurmai,' 'Babai,' 'Supkhoni,' 'Isfarak,' and 'Tadzhabai.'

Zeravshan subgroup. Apricots of this subgroup grow in the Zeravshan basin (from highland to Samarkand). This group is more diversified as compared to apricots of the Fergana subgroup. Some popular landraces such as 'Arzami' and 'Akhrori' are somewhat reminiscent of Eastern Asian apricots (Kovalev 1963). Apricot production is aimed at both dried fruit and fresh market consumption. Cultivars for fresh market have an excellent fruit quality, and often open the harvest season. Occurrence of glabrous forms (lyuchaks) is high (up to 40%), and frost resistance is slightly lower than that of Fergana's apricots. Typical cultivars are as follows: glabrous forms—'Maftobi lyuchak,' 'Gulyungi lyuchak,' 'Badami'; pubescent forms—'Maftobi,' 'Gulyungi,' 'Kursadyk,' 'Khodzhendi,' 'Iskaderi,' 'Koshfi,' 'Shirpaivan.'

Shakhrisyabz subgroup. These apricots are native to Southern Uzbekistan and the Kashka-Darya basin. This group is extremely diversified and represented by cultivars for drying. As a whole, apricots of the Shakhrisyabz subgroup are small-fruited and of poor fruit quality for the fresh market.

Khorezm subgroup. The lowlands of the Amu Darya basin are the home of this Central Asian apricot subgroup. The majority of the Khorezm apricots are propagated by seed. The fruit quality is generally poor in comparison with the apricots from the Fergana and Zeravshan subgroups. However, Khorezm's apricots are more resistant to spring frosts, and can withstand both high temperatures and unfavorable soil salinity. About 10% of the cultivars in this subgroup are glabrous-skinned. Major cultivars: 'Kzyl nukul,' 'Ak nukul,' 'Kuzgi khorezmli,' 'Kzyl Khorezmskii,' 'Paivandy Bucharskii.'

Kopet-Dag subgroup. Apricots of this subgroup grow in Central and Southwestern Kopet-Dag and are characterized as being of a primitive Iran-Caucasian type. Some experts consider this semiwild population as a primary relic microcenter of the Near Eastern apricots (Avdeev 1992). However, the isozyme and DNA marker analyses support the scenario of apricot dissemination from a Central Asian center, rather than confirm the originality of this group (Zhebentyayeva et al. 2003; Zhebentyayeva and Ageeva 2004). In this subgroup, fruits are small (10–35 g) and sweet-kerneled, and have skin pubescence with a light yellow color. Taste and fruit texture are good. Kopet-Dag apricots are mainly of the fresh market type (Avdeev 1992; Kryukova 1989).

Dzhungar-Zailii subgroup. This is the youngest of the Central Asian subgroups, endemic to the most northern distribution (up to 44° north) of apricot in Dzhungar and Zailij Alatau, as well as in the Ily valley of western China. The group is comprised of the seed propagated forms selected from wild *P. armeniaca*. Cold hardiness and resistance to fluctuating winter temperatures are the most valuable characteristics of this subgroup. Generally, fruits have a light yellow color, small size and are acidic with bitter kernels. However, some forms have large fruits and are self-fertile.

The *Iran-Caucasian group* is represented by local cultivars from Armenia, Azerbaijan, Georgia, Dagestan, Iran, Iraq, Syria, and Turkey. Some Mediterranean-type cultivars in Europe have similar characteristics. Every country possesses its own germplasm resources, often the same genotypes under different names. For example, one of the best fresh market cultivars from this group is propagated under different names in Turkey ('Aprikoz,' 'Şalak') and Armenia ('Shalakh,' 'Erevani'). Generally, the apricots from this group have lower chilling requirements and bloom early in the spring. Most cultivars are self-incompatible, but self-compatible forms are not uncommon. Apricot maturation season is not as lengthy in comparison with those from the Central Asian group. The predominant fruit color is light yellow, white or creamy with sweet kernels. Glabrous-skinned fruits are rare (up to 4% cultivars).

Apricot germplasm in Turkey deserves special comment as more than 80% of the world's dried apricot originates from this region. Morphological and pomological characteristics of 128 local Anatolian cultivars provide insight on apricot germplasm of the Iran-Caucasian type (Asma and Ozturk 2005; Asma et al. 2007; Kayisi çeşit Kataloğu 1996). About one third of the Turkish cultivars bear small fruit (≤30 g). The same proportions of them have bitter kernels. Cultivars with large fruit (>50 g) are rare. Cultivars for fresh market have a high flesh to pit ratio. Prevailing fruit colors are yellow and orange, 62% and 37%, respectively. White-fleshed cultivars are rare (1%). Mid-season cultivars have high brix (>20%) that naturally contributes to high quality of the dried product. However, the fruit quality of early- and late-season varieties is poor. Major cultivars are as follows: 'Aprikoz,' 'Çataloğlu,' 'Çöloğlu,' 'Hacihaliloglu,' 'Hasanbey,' and 'Kabaaşı.'

Iran-Caucasian subgroup. Tree size and longevity of these cultivars are less as compared with those of Central Asia. However, vigorous trees with a spreading growth habit of a 'Shalakh' type (divergence angle close to 180°) occur as well. Branches are thicker with large and shiny leaves. The leaf anatomy of some typical Iran-Caucasian cultivars shares common characteristics with Chinese apricots (Rostova and Sokolova 1992).

North African subgroup Layne et al. (1996) proposed this subgroup to distinguish apricots from North Africa (Tunisia, Morocco, Libya, Algeria and Egypt). The apricots in this region grow in a climate with very mild winters and very warm summers with low rainfall. Local cultivars from this region have low chilling requirements and some have developed resistance to *Monilia* spp. (Bassi and Pirazzoli 1998). A highly likely scenario for diversification of apricots in North Africa, particularly in Tunisia, implies a bottleneck effect at the initial step of apricot cultivation followed by seed propagation (Khadari et al. 2006).

The *European group* is the best characterized of the ecogeographical groups, and is considered the youngest in origin (Kostina 1964; Layne et al. 1996; Faust et al. 1998). Under controlled propagation by grafting, practiced in Europe since its introduction, the apricot lost its variability in bloom time and maturation season, as well as other characteristics such as tree architecture. The trees are less vigorous, with open-erect growth habits, and have higher chilling requirements as compared with the Central Asia apricots. Naturalized forms of a "zherdeli" type from northern Europe can withstand very low winter temperatures while they are dormant. Most cultivars are self-compatible, but self-unfruitful varieties exist as well. European apricots, especially the newly bred varieties, have larger fruit (up to 70 g and higher) with yellow/orange color, and a characteristic apricot aroma. Glabrous forms are rare. The soluble solids content (SSC) is lower (around 12–17%) while acidity is higher (above 1.3–1.5%) compared with Central Asian varieties (Kovalev 1963; Badenes et al. 1998; Ruiz and Egea 2008). Under a Mediterranean climate, some cultivars accumulate more than 17% dry matter and are acceptable for drying. Apricot for kernel production has never been important in Europe and most cultivars have bitter kernels. It is commonly accepted that European apricots are more tolerant to fungi than Central Asian and Iran-Caucasian cultivars.

Molecular analyses of European germplasm have provided some support for diversification of the apricot in east to west direction (Hagen et al. 2002; De Vicente et al. 1998; Hormaza 2002; Geuna et al. 2003; Romero et al. 2003; Maghuly et al. 2005). Adaptation to continental or Mediterranean climatic zones was a major factor for crop evolution in European countries. The use of a few basic cultivars from clonal selection and their propagation by seedlings from open pollination led to the development of landraces of related cultivars that have a narrow genetic background and are highly specific to their ecological requirements. Apricot germplasm collections in Hungary and Italy are historically the richest in diversity and number of accessions (Zanetto et al. 2002).

By origin, commercial cultivars of North America also belong to the European ecogeographical group. Owing to the "natural" resistance to plum pox virus (PPV) uncovered in this group (for review, Martínez-Gómez et al. 2000), North American resistant cultivars were incorporated into almost all diversity studies with the use of isozymes and molecular markers (Badenes et al. 1996; De Vicente et al. 1998; Hagen et al. 2002; Hormaza 2002; Geuna et al. 2003; Romero et al. 2003). Several sources of introduced germplasm were hypothesized to explain the presence of "non-European" alleles in genotypes of PPV-resistant cultivars. More recent molecular data (Zhebentyayeva et al. 2008) provided evidence that germplasm of Chinese origin was most likely involved in diversification of North American apricots.

3.2 Rootstocks

Given the relative importance of apricot throughout the world, there is a surprisingly small amount of research and development to date for apricot specific rootstocks. Stocks have been used by growers since the discovery of grafting as a means of saving or multiplying valuable clones. The top-working of unselected forest trees to elite selections has been practiced for centuries, and is still practiced in regions where native apricot resources exist in the wild.

Commercial canning operations and drying yards have been traditional sources of large quantities of apricot pits that could then be utilized in the production of nursery trees. While many diverse rootstock choices exist today, seedling apricot is still utilized and recommended as a first choice for new apricot orchards in various growing regions (Slingerland et al. 2002; Khadari et al. 2006). Local cultivars 'Alfred,' 'Goldcot,' 'Manchurian' and 'Veecot' were deemed the most reliable as seedling rootstocks for apricot in the growing regions surrounding Ontario, Canada. Precocity of bearing, tree longevity, and universal graft compatibility with known apricot cultivars were the reasons for the recommendation. The abundance of 'Blenheim,' 'Early Golden,' and 'Royal' pits from drying yards led Wickson (1891) to a similar recommendation for apricot seedling rootstocks in California. As the California apricot industry expanded in the first half of the twentieth century, nurseries discovered that while varieties such as 'Alexander,' 'Catherine,' and 'Tilton' produced seedlings more vigorous than those from 'Blenheim,' the 'Blenheim' variety imparted far greater vigor and longevity in the scion apricot to which it was grafted as compared with many other trialed varieties (Day 1953).

P. armeniaca is considered to be immune to root-knot (*Meloidogyne* spp.) nematode, and several studies have demonstrated its resistance to the root lesion (*Pratylenchus vulnus* Allen and Jensen) nematode as well (Day and Serr 1951; Culver et al. 1989). With resistance to these major orchard pests, one could imagine apricot rootstocks playing a major role in the production of new apricot nursery stock. However, rooting ability of both hardwood and semisoftwood cuttings is below the level of economic feasibility for commercial nurseries (Reighard et al. 1990), limiting apricot rootstocks to only those produced by seed propagation. Furthermore, graft compatibility is limited for both peach/nectarine and almond on apricot root, with specific combinations being deemed safe to producers only after empirical testing.

Besides general recommendations for use of certain varieties of apricot pits from agricultural operations, only limited research has focused on identifying superior germplasm for use as apricot rootstock. In Hungary and Bulgaria, apricot mother trees have been selected that produce seedlings with good nursery performance and broad adaptation to both pathogens and environmental problems (Mády et al. 2007; Dimitrova 2006). Specific apricot seedling rootstocks were observed to be superior by Son and Küden (2003) at imparting larger fruit size and total fruit yield in eight apricot cultivars as compared with GF 31 rootstock in Turkish orchards. In France, INRA selected and introduced 'Manicot GF 1236' as a seed propagated apricot rootstock for well-drained soils (Lichou and Audubert 1989). It is said to have very good seedling vigor and nursery homogeneity, although it is sensitive to both crown gall and bacterial canker. Seedlings from this rootstock cultivar are also susceptible to PPV (Guillet-Bellanger et al. 2006), a fact that may limit its desirability to those regions not yet affected by this important disease. Other apricot seedling populations from various 'Canino' clones have been examined as possible rootstocks for apricot. Clone Canino 9–7 yielded seedlings with higher germination and better vegetative growth than other apricot seedling populations (Orero et al. 2004).

4 Major Breeding Achievements

4.1 European Programs

Although the number of fruit cultivars available in the world is very high, there is a continuing need to develop new cultivars as industry requirements change. In the last 20 years, cultivar development by private breeding programs has increased with a corresponding decrease by publicly funded programs (Byrne 2005). In Europe, the number of breeding programs specific to apricot and new varietal releases is much lower than those focused on other fruit species. The Community Plant Variety Office (CPVO) received in 1994–2001 period over 730 new fruit cultivar applications for Community rights. Only 5% of these applications were new apricot cultivars, while new peach, apple and strawberry cultivars accounted for 20% each (Semon 2006).

The major objectives in European publicly funded apricot breeding programs are resistance to biotic stresses (sharka caused by Plum Pox Virus, brown rot caused by *Monilinia* spp., bacterial diseases caused by *Pseudomonas* spp. and *Xanthomonas arboricola* pv. *pruni* (Smith), Chlorotic Leaf Roll Phytoplasma, and Apricot Decline Syndrome), adaptability to the environment (temperature requirements, water deficit), extension of the harvest season, fruit quality for fresh consumption and for processing, productivity, and adequate tree size and structure (Bassi and Audergon 2006).

Sharka disease caused by Plum Pox Virus (PPV) is the most limiting factor for apricot production in European countries (Cambra et al. 2006a). Many of the apricot breeding programs in these countries encounter two major limitations relative to the development of new PPV-resistant varieties: PPV-resistant genitors have high chilling requirements and a medium to late harvest period (characteristics that are far from the program objectives in the southern countries), and the procedure for screening PPV resistance is a lengthy and laborious biological test involving many plants and lasting a minimum of 2 years (Badenes and Llácer 2006; Llácer et al. 2008). Taking these problems into account, it is difficult for the breeding programs in Southern European countries (Italy, France, Spain, and Greece) to reach the goal of developing new well-adapted high-quality PPV-resistant varieties.

In Italy, there are three publicly funded apricot breeding programs. The "Dipartimento di Produzione Vegetale" at Milano and Bologna Universities has recently introduced three new cultivars ('Boreale,' 'Ardore,' and 'Pieve') with better fruit quality (flavor and aroma) and an extended ripening season (Pellegrino 2006). The 'Dipartimento di Coltivazione e Difesa delle Specie Legnose' at Pisa University also recently offered three new cultivars. The first one, 'Angela,' is an early-maturing cultivar which ripens around 3 weeks before 'Canino' and a few days before 'Priana.' The second one, 'Gheriana' ripens at the same time as 'OrangeRed.' It is a cross between 'Portici' and 'Harcot,' with the best traits of both parents. The third one, 'Silvana,' is a late-maturing cultivar that ripens 25 days after 'Canino' and 10 days later than 'Fantasme.' It is a cross between 'Bergeron' and 'Canino Tardivo,' and is heavy-cropping (Guerriero et al. 2006a). Finally, the "Istituto Sperimentale per la Frutticoltura" at Caserta has produced selections that extend the ripening season and possess interesting characteristics for specific processing products (dry fruit, canning, juice). Tests are in progress to determine the agronomic behavior of these selections in different soil and climatic conditions of southern Italy (Pennone and Abbate 2006).

In France, after a first set of 11 apricot cultivars released since 1982 for each of the three main areas of production, a new set of three cultivars has been released by CEP Innovation under the frame of a national agreement with the "Institut National de la Recherche Agronomique" (INRA) and Agri-Obtentions. 'Solédane' is adapted to Mediterranean coastal areas, 'Florilège' is suitable for the lower part of the Rhone valley, and 'Bergarouge®' Avirine is well adapted to all the French area of cultivation. The three apricot cultivars are registered in the French national catalog and protected under the UPOV rights (Audergon et al. 2006b). Some new recent selections are described by Audergon et al. (2009).

Among the main European producer countries only Spain and Greece have decreased their production in the last 20 years (Table 12.1). These countries are the most affected by PPV in the European Community (Cambra et al. 2006b; Varveri 2006). In Spain, there are two institutions that carry out apricot breeding programs aimed at producing PPV-resistant cultivars: the "Centro de Edafología y Biología Aplicada del Segura" (CEBAS-CSIC), in Murcia, and the "Instituto Valenciano de Investigaciones Agrarias" (IVIA) in Valencia. The first crosses were made in 1991 at CEBAS-CSIC and in 1993 at IVIA. The main donors of PPV resistance in these two programs were 'Stark Early Orange,' 'Goldrich,' 'Orange Red,' 'Harcot,' and 'Lito' (Badenes and Llácer 2006). The program from Murcia has already released six cultivars: 'Rojo Pasión,' 'Selene,' 'Murciana,' 'Dorada,' 'Estrella,' and 'Sublime.' The first three cultivars are PPV resistant with good fruit quality. 'Murciana' is also characterized by its good aptitude for canning. 'Dorada' is a late-ripening cultivar well adapted to the climatic conditions in the mountains of Spain (Egea et al. 2004a, b, 2005a, b, 2009). In Valencia, four varieties that fulfill the objectives of the program (PPV resistance, precocity, fruit quality, and good adaptability) are registered and 11 advanced selections are under study in several apricot regions (Martínez-Calvo et al. 2009).

In Greece, a large apricot breeding program for the control of sharka disease has been administered since 1989 at The National Agricultural Research Foundation, Pomology Institute, at Naoussa-Makedonia. Ten apricot cultivars of North American origin: 'Stark Early Orange,' 'Stella,' 'NJA2,' 'Sunglo,' 'Veecot,' 'Harlayne,' 'Henderson,' 'Goldrich,' 'OrangeRed,' and 'Early Blush,' selected for their resistance to the highly virulent local strain of PPV-M (Marcus), have been used as parents in crosses with very-high-quality cultivars, but mainly with the local cv. 'Bebecou.' Nine new apricot varieties have been introduced based on their resistance to PPV, fruit quality for fresh consumption or for canning, and other desirable characteristics (Karayiannis 2006; Karayiannis et al. 2006).

In Romania, Dr. Cociu started apricot breeding activities in 1951 within the Agronomic Research Institute in Bucharest. The main objective was to modernize the whole apricot assortment in his country. Among 29 cultivars now officially recommended for propagation and planting in Romanian orchards, 21 are new

Romanian cultivars, from which seven are early or very early and nine are later than the latest cultivar from the old assortment. They are more resistant to low winter and early spring temperatures and they have better fruit quality as expressed by size, appearance, and taste. Only six recommended cultivars are of foreign origin, and two are from the old Romanian germplasm (Cociu 2006).

In Bulgaria, a breeding program was developed at the Apricot Research Station in Silistra, with the main objective of enriching the genetic diversity in this species. Over 3,600 seedlings were obtained from 72 intraspecific crosses and 58 open pollinated cultivars. Approximately 1,300 hybrids that reached the adult phase were studied for ten important biological and pomological characteristics during 1989–1999. After a complete evaluation, nine elite genotypes, which combined the greatest number of valuable traits expressed at the highest level, were selected and recommended for cooperative trial plantings and further commercial development (Coneva 2003).

Besides PPV, apricot production in Central Europe has many risks, mainly during the postdormancy period. Particularly in these countries, the apricot decline syndrome is manifested by the emission of gum from woody organs that ends in the sudden death of branches or the apoplexy of the entire tree. Both abiotic (poor adaptation to environmental conditions) and biotic (sensitivity to different pathogens, especially bacteria and fungi) conditions seem to be the most likely causes (Bassi and Audergon 2006). These difficulties have been overcome in the breeding program initiated in 1981 at the Horticultural Faculty in Lednice, the Czech Republic. In a first generation, this program has registered seven cultivars with extended ripening times and increased frost hardiness, and another four promising cultivars have been submitted for registration. Several more hybrids with a high level of PPV tolerance have been selected for further investigation (Krska et al. 2006). On the other hand, the Research Institute of Plant Production at Piestany, Slovak Republic, has registered ten cultivars with late bloom periods, better fruit quality and extended maturation seasons (Benedikova 2006).

4.2 Non-European Programs

Outside of the European Community, numerous apricot breeding efforts have stood in regions where apricots are important, or where new apricot culture would be desirable. The oldest ongoing program for apricot improvement started in 1925 at the Nikita Botanical Gardens in Yalta, Crimea, Ukraine. Publicly funded breeding efforts exist in both New Zealand (HortResearch, Hawke's Bay, NZ) and Australia (South Australian Research and Development Industries, Loxton, South Australia) of Oceania, as well as in South Africa (Agricultural Research Council of South Africa) and Tunisia (Institut National de Recherche Agronomiques de Tunisia) of Africa, China (Liaoning Institute of Pomology, Xiongyue, Peoples Republic of China), and Japan (National Institute of Fruit Tree Science, Tsukuba, Ibaraki, Japan) of Asia, and in the USA (Rutgers University, New Brunswick, NJ and the USDA/Agricultural

Research Service, Parlier, CA). Among these breeding institutions, nearly 50 new apricot cultivars have been introduced since 1990. Numerous private apricot breeding efforts have also provided new cultivars to interested producers. With the inclusion of a recent breeding effort initiated by the University of Santiago in Chile, apricot breeding is occurring on all continents having temperate growing regions.

The development of PPV-resistant cultivars is of lesser importance to many of these programs where the virus does not pose a current threat to apricot growers. Hence, in PPV-free production regions, breeding efforts have focused on other objectives such as higher fruit quality, extended maturity season, or better environmental adaptation. The program at Yalta has had a long history of hybridization between apricots from the different ecogeographical groups with the objective being the selection of those types having high fruit quality as well as broader environmental adaptation. This hybridization scheme has been long suggested as a means of improving a cultivar's adaptation to different growing regions (Kostina 1936).

The Australian program introduced three new apricots in 2005 ('River Ruby,' 'Riverbrite' and 'Rivergold') to complement 'Rivergem,' introduced in 1995. With these new introductions, the program at Loxton hopes to revitalize the Australian drying industry. The new introductions represent marked increases in fruit quality and cropping over the industry mainstay 'Moor Park.' Furthermore, a mechanized drying industry is envisioned, with increased fruit firmness of the newer varieties now allowing experimental mechanical harvesting and cutting. Through several rounds of selection, this program has improved its overall precocity as compared with the Syrian and Turkish progenitor germplasm on which the program was originally based. Selected Chinese germplasm having novel flavors has also been incorporated into breeding lines that are adapted to the Australian growing regions. New Zealand's HortResearch program at Hawke's Bay has also been very active in variety introductions with nearly a dozen releases since 1990. 'Cluthagold' is the current top selling apricot variety, but newer releases may surpass its production as growers begin to develop new acreage. In contrast with the Australian program, HortResearch apricot development is primarily targeting the fresh market. The newer New Zealand-bred apricots have recently been dispersed to selected North American nurseries where they will be trialed.

The Agricultural Research Council of South Africa has introduced six new cultivars from their breeding effort in the last 6 years, and well over 200 advanced selections are being evaluated currently. The program has been actively importing and evaluating newly introduced apricot cultivars for their potential use as parental stock. With concern for the future, imported PPV-resistant cultivars are being bred with local adapted varieties to incorporate resistance with adaptation to the country's growing regions. In the very different environment of North Africa, Tunisian breeders have recently introduced six cultivars ('Asli,' 'Atef,' 'Fakher,' 'Meziane,' 'Ouafer,' and 'Raki') adapted to lower-chill conditions. The new cultivars show marked improvements in fruit quality (higher color, flesh firmness) over locally selected 'mesh-mesh' apricot germplasm.

Chinese breeders at the Liaoning Institute of Pomology have had several recent noteworthy achievements in expanding the apricot ripening season. Newly introduced

'Luotuohuang' is approximately 50% larger in fruit size and ripens 10 days earlier than 'Meihuang,' the previous early season industry standard. At the tail end of the Chinese ripening season, the standard cultivar 'Jinxidahongxing' has been replaced by 'Chuanzhihong,' introduced in 1997. 'Chuanzhihong' ripens 5 days later and is equal to 'Jinxidahongxing' in fruit size; however, 'Chuanzhihong' can be used as a fresh market apricot or for processing. The Liaoning Institute has also been developing apricots specifically for kernel use, with cultivars 'Fenren' and 'Guoren' both representing increased kernel size and production over the industry standard 'Longwangmao.' Liaoning Institute's newest introduction was selected from a local Chinese landrace. 'Shajinhong,' introduced in 2007, ripens mid-season and is large fruited (80–90 g), with firm flesh and very good traditional flavor/aroma characteristics. Japan's breeding effort at Tsukuba has yielded two new *P. mume* cultivars: 'Hachirou' and 'Kagajizou,' both introduced in 1997 (Yamaguchi et al. 2002a, b). The self-compatible 'Hachirou' has demonstrated a high yield of medium-sized clingstone fruit, suitable for processing into pickles. By contrast, 'Kagajizou' is pollen sterile, large fruited, and with good texture. 'Kagajizou' has been recommended for both pickling and fresh marketing.

Publicly funded apricot breeding in North America involves Rutgers University on the eastern seaboard (Cream Ridge, New Jersey) and the USDA/Agricultural Research Service in Parlier, California. The Rutgers breeding effort has achieved success in dispersing their new cultivars to both Europe and North Africa. Five new cultivars have been introduced by the Rutgers program since the mid 1990s. While not a particularly new cultivar, 'OrangeRed' (*syn.* Bhart, NJA32) has had considerable success as a fresh market apricot in medium to high chill European growing regions, and has also been used extensively as a source of resistance in developing new PPV-resistant cultivars (Karayiannis et al. 2008). 'OrangeRed' has been used as a parent in the USDA/ARS breeding effort, and is the seed parent of 'Robada' apricot. Just as with apricots from the Rutgers program, 'Robada' is being grown successfully in both France and Spain, as well as in Australia and New Zealand. Five other apricots have been introduced from the USDA/ARS program since 1994. Among them, 'Helena' (1994) has achieved considerable success in Chile as a high value fresh market export apricot. 'Apache,' released by USDA/ARS in 2001, is currently the earliest-ripening commercial apricot grown in North America.

5 Current Goals of Breeding

5.1 European Programs

This topic has been recently reviewed by Bassi and Audergon (2006). The major objectives in European programs are:

PPV resistance. PPV is the strongest obstacle for the cultivation of apricots in Europe. In the near future it will probably be impossible to grow cultivars sensitive to PPV due to the extensive diffusion of the virus. All apricot cultivars of European origin are susceptible to PPV. Resistance has been found only in some North American cultivars. Badenes et al. (1996) were the first to suggest the role of Eastern Asiatic species, particularly *P. mandshurica*, as a potential source of PPV resistance into North American germplasm. The results from Karayiannis (2006) and Karayiannis et al. (2008) give more support to this idea, even if not all the accessions of *P. mandshurica* are PPV resistant (Rubio et al. 2003). Curiously, North American selections derived from *P. mandshurica* were introduced for their cold hardiness in midwinter and spring, late blooming and the ability to set fruit under adverse conditions for pollination (Bailey and Hough 1975). Besides *P. mandshurica*, other Eastern Asian species such as *P. sibirica* var *davidiana* and *P. mume* could also have been involved in the pedigree of PPV-resistant North American apricots. A likely scenario for introgression of resistance into North American germplasm might include hybridization of European apricots with Northern Chinese varieties cultivated in overlapping areas of *P. armeniaca* and Eastern Asian apricot species (Zhebentyayeva et al. 2008). Currently most apricot breeding programs in Europe use the PPV-resistant North American cultivars to introduce this trait into European germplasm.

Resistance to Monilinia spp. Brown rot caused by *Monilinia laxa* (Aderhold & Ruhland) Honey, *M. fructigena* Honey in Whetzel and *M. fructicola* (G. Wint.) Honey can produce notable economic damage to the apricot as well as to the other stone fruits, attacking flowers, young shoots, branches and fruits. The disease virulence and the severity of the damage are strictly related to the climatic conditions, and several fungicide treatments are often necessary to limit the damage. Therefore, the creation of new resistant varieties is one of the most important objectives of the apricot breeding programs in several European countries (Italy, France, the Czech Republic, Slovak Republic, Romania, and Bulgaria) to avoid damage to trees and yields and to reduce chemical spraying. This goal will allow reductions in both production costs and fungicide residues and will demonstrate a better respect for the environment. In Italy, a breeding program for *M. laxa* resistance reported 11 advanced selections that were evaluated as resistant to the damaging fungus (Nicotra et al. 2006).

Resistance to bacterial diseases. The bacterial diseases in apricot are mainly caused by *Pseudomonas* spp. (wood cankers) and *Xanthomonas arboricola* pv. *pruni* (Smith) (leaf necrosis). The spreading of the first pathogen is facilitated by cold winters and humid climates. The lesions produced by exposure to cold temperatures can be easily infected and develop cankers that may lead to loss of branches, scaffolds or even of the whole tree. *Pseudomonas syringae* pv. *syringae* van Hall along with the fungus *Leucostoma cincta* (Fr.:Fr.) Höhn and winter injury are the major contributors to apricot decline or apoplexy syndrome in central Europe (Layne et al. 1996). *Prunus dasycarpa* sel. P 2315 and the Japanese apricot (*P. mume*) have been described as immune or highly resistant to *Pseudomonas* spp. On the other hand, *X. arboricola* pv. *pruni* spreads in warm and humid climates and affects shoots, leaves and fruits. The cultivars 'Adedi,' 'Alfred,' 'Polonais,' and 'Tirynthos' are recorded as tolerant or not very sensitive to leaf necrosis. Selection can be made for tolerance by artificial inoculation methods on progenies from crosses with highly tolerant parents.

Resistance to Apricot Chlorotic Leaf Roll (ACLR). ACLR is caused by European Stone Fruit Yellows Phytoplasma. It produces a progressive decline of the tree due to obstruction of the vessels. It is transmitted by grafting and some insects, and its diffusion is very serious in Southern France. Some tolerant sources are known that develop symptoms only after a prolonged time period after infection. No sources of immunity are known.

Adaptability to the environment. This trait is still one of the main factors limiting a cultivar's introduction outside the environment where it was selected. The temperature regime during summertime can affect bud flower differentiation and during the winter can alter the morphological completion of the ovary. Another aspect is the sensitivity to spring frosts. This sensitivity cannot only be attributed to early blooming. In experiments carried out in northern Italy (Bassi et al. 1995), it was shown that the early blooming genotypes are not always less productive than those with middle or late blooming times. These genotypes are characterized by very abundant bloom, a phenomenon that often compensates for the effects of late frosts. Parents with spring frost tolerance in apricot and the use of artificial cold-stress methods, eliminating the multiple variables of the field, have been described by Guerriero et al. (2006b).

On the other hand, the tendency to search for genotypes with late blooming time may lead to the introduction of cultivars characterized by high chilling requirements, difficult to satisfy in regions with mild winters. Better results can be obtained by breeding genotypes with high heat requirements, which causes late blooming without negative effects. The cultivars adapted to northern environments show very poor growth with scarce floral induction when grown in mild conditions. This is the problem for southern European countries when using North American genotypes to introduce PPV resistance.

Another phenomenon in terms of adaptation is fruit cracking. While there is certainly a genetic predisposition in specific apricot cultivars, fruit cracking is also strongly dependent on the tree's water balance when the fruit is close to maturity. Rain during this stage, especially when following a dry period, causes an abundant absorption of water by the fruit which can then crack the skin or even split the mesocarp. Among the tolerant cultivars in Italy or France are 'Boreale,' 'Fournes,' 'Goldrich,' 'Moniquí,' and 'San Castrese.'

Extension of the ripening time. In all apricot-producing areas, a frequent goal is the extension of the ripening season to allow better packing house use efficiency and a larger presence in the marketplace. Moreover, earlier and later harvested fruits are often marketed with higher prices (Llácer 2009). There are many potentially useful genotypes for extending the ripening season. Concerning a very early ripening time, there are germplasm resources from Mexico (Pérez-González personal communication), from northern Africa and selections from the program at Rutgers University (the USA), ripening 11–18 days before 'Tyrinthos.' These genotypes were obtained from progenies using American cold resistant cultivars crossed with high fruit quality Central Asian germplasm. Considering the great genetic diversity between the parents, this germplasm can be outstanding breeding stock for future crosses. In relation to late ripening, many potentially interesting parents are also available: 'Reale di Imola,' 'Boccuccia Spinosa,' 'Baracca,' 'San Francesco,' 'Fracasso,' and 'Pisana' from Italy, 'Bergeron' and 'Tardif de Bordaneil' from France, and many of the Eastern European and middle-Asian genotypes. All this germplasm, however, shows lack of adaptation outside its place of origin and some poor traits related to late ripening.

Fruit quality. All cultivar programs point out fruit quality as a priority, but this is a complex trait that needs to be defined for every situation and use. Sensory fruit quality concerns consumer perception of color, shape, size, aroma, flavor, texture, and freshness. There is an external fruit quality, which is perceived by sight, and an internal fruit quality where perception occurs during fruit consumption (Llácer 2009). External fruit quality has been a priority in the past. Owing to consumer preferences, this trend is changing and the internal fruit quality is becoming a priority goal. The lack of internal fruit quality is the main reason claimed by consumers for buying less fresh fruits (Byrne 2002). Additionally, nutritional quality, the content of polyphenols and carotenoids, and food safety are becoming important factors in determining the level of fruit consumption (Ruiz et al. 2005; Badenes et al. 2006). Other traits are also important both for field and postharvest operations: the uniformity and speed of ripening, resistance to handling and transportation, sensitivity to internal browning and adhesion to the pit. For canning apricots, good orange skin and flesh are desired, as well as a uniform medium size, regular shape, resistance to pit burn during high temperatures just before harvest, good texture (freedom from fibers and vascular bundles), small pit, high sugar content, and a good balance of acid and sugar (Layne et al. 1996). For drying, subacid fruits (acidity lower than 0.5%) with high soluble solids (20–25% of sugar) are needed. In general, it can be said that the objectives of the different processing destinations, being rather specific, can be easily achieved by traditional breeding programs, although most of the important traits are quantitatively inherited.

Productivity. This is a basic goal in any breeding program. From a purely economic viewpoint, a consistently productive cultivar of medium fruit quality is generally more profitable in comparison to a high-quality cultivar prone to alternate yields. Productivity depends on several factors: the adaptability to the environment (discussed above), the proportion of normally differentiated flowers and the self-compatibility status of the tree. A rather high percentage of self-incompatible genotypes exist in cultivated apricots. The need for reliable pollinizers to avoid erratic fruit set in these types of apricot cultivars was emphasized by Rodrigo and Herrero (1996). It is very important to carefully evaluate the floral compatibility before introducing a new cultivar, keeping in mind that it would be very difficult to evaluate this trait in a cultivar collection where many pollinizers are usually available. The use of molecular markers has greatly facilitated the identification of self-(in)compatible genotypes, as will be discussed later.

Tree size and structure. Presently, the integration of morphological and architectural traits in fruit tree breeding programs is an important goal in France and Italy. The use of small stature trees as parents could result in progenies characterized by short internodes, fruiting branches and/or spurs (Moser et al. 1999). In many genotypes the fruits obtained from spurs are of better quality than those obtained from standard branches. On the other hand, apricot trees are not very adaptable to formal or rigid training systems and they do not tolerate drastic pruning, particularly in the dormant season. Consequently, it is important to develop tree forms that require infrequent management of the vegetation while producing consistent and adequate yields. A review of these traits has been presented by Costes et al. (2004).

Adaptability to various soil conditions. A large genetic diversity of rootstocks used for apricot is employed in Europe, depending on the various soil conditions of growing areas. Apricot, peach and plum seedlings, clones of different plum species or interspecific hybrids are currently used in apricot orchards. Nevertheless, graft incompatibility, exhibited by many *Prunus* rootstocks with most apricot cultivars, is one of the major problems for rootstock usage and improvement. Interspecific hybridizations between myrobalan plum (*P. cerasifera*) and apricot (*P. armeniaca*) have been undertaken in France and Spain to create hybrids that combine graft compatibility with apricot, favorable rootstock traits from myrobalan plum (adaptation to heavy soils, rooting ability) and resistance to pests and diseases from both species. The first results obtained show that the creation and selection of these interspecific hybrids seems to be a very promising way to improve apricot rootstocks (Poëssel et al. 2006; Arbeloa et al. 2003, 2006).

5.2 Non-European Programs

The lack of PPV in California orchards has allowed the USDA/ARS breeding program to focus effort on specific fruit quality characteristics. Repeat consumer sales throughout the apricot marketing season are hurt by the abundance of low quality (immature, high acidity, low Brix) fruit during the early season. In order to increase the overall fruit quality, numerous California adapted apricots have been hybridized with apricots from Central Asia (Ledbetter and Peterson 2004). The use of Central Asian parents added a great deal of genetic diversity to the program. Novel and useful characteristics obtained from Central Asian parents include late bloom period, high Brix, long fruit development period, glabrous skin, modified sugar profile and diverse skin and flesh colors. The USDA/ARS breeding goals involve new cultivar development for the fresh and processing markets. The expansion of the fruit maturity season is an overall goal, with the current season being only 5 weeks. Numerous crosses have been made to incorporate glabrous skin into California adapted apricots. White flesh apricots are being selected for flesh firmness and high Brix. For processing apricots, the major emphasis is in identifying high Brix freestone drying types whose flesh color will not darken in storage after sulfur/sun drying. At the Rutgers University breeding program, improved cold hardiness is a major goal as inclement springtime weather conditions can limit apricot production. Like the

USDA/ARS program, Rutgers' breeders are actively selecting for high fruit quality and attractiveness, and to lengthen the ripening season. Cream colored flesh and glabrous skin are two novel characters currently under selection at Rutgers.

The breeding work at Tsukuba, Japan has goals for both *P. armeniaca* and *P. mume*. Objectives for Japanese apricots are focused on the fruit's processing ability into "pickles," with particular importance being placed on low gumming of the fruit. Selections are made for a later flowering season and early fruit maturity is desirable as well. Plum × *P. mume* hybrids are also being evaluated for juice and liquor production. Pigments in the hybrid flesh impart a bright red color to products produced from them, providing novel and potential value-added benefits. The Tsukuba team's goals for *P. armeniaca* selections are very low acidity and high Brix in apricots for the fresh market. Tree longevity is desirable, as is a late bloom period, given the propensity of late frosts throughout the Japanese growing regions. Self-fruitfulness and disease resistance are breeding goals in both *P. armeniaca* and *P. mume*. Having a series of sequentially ripening apricots with abundant flavor and firm flesh is the overall goal of the Chinese breeders at Liaoning Institute. To achieve this goal, numerous firm-fleshed North American apricots have been imported for evaluation and for hybridizations with local Chinese landraces having strong aroma/flavor. Future selections will be made where these important traits are combined throughout the fruit maturity season (Weisheng Liu personal communication).

Tunisia's geographic location provides the potential for having available apricots in the earliest possible season, given the availability of adapted germplasm. Breeders at the Institut National de Recherche Agronomiques de Tunis have endeavored to combine several fruit quality traits (orange color, firm flesh, large fruit size, enhanced sugar and aroma) with early-ripening, hoping to produce export-quality cultivars that are ready for harvest and marketing prior to when the first European Community apricots are ready. The Agricultural Research Council of South Africa is evaluating apricot selections for both fresh marketing and processing potentials. With exportable fruit being an important percentage of South Africa's apricot tonnage, postharvest cold storage ability is one of the major breeding objectives. The program continues to import, under quarantine, high fruit quality and PPV-resistant cultivars from other breeding programs for use in hybridizations with locally adapted selections. Thus, this program demonstrates forethought in their breeding goals relative to the nearly inevitable future introduction of PPV into South African growing regions.

New Zealand's HortResearch breeding program desires to develop well-adapted and precocious cultivars that are productive, large-fruited and have both good eating quality and high flavor. Breeders there are also attempting to develop early maturing cultivars for the Hawke's Bay growing region (lower chill area) and late maturing cultivars for the growing regions of Central Otago (higher chill area). A more immediate goal for HortResearch breeders is the replacement of the 'Sundrop' cultivar, an industry standard for both growing regions, due to both cropping concerns and insufficient fruit size (Mike Malone personal communication). Similar breeding objectives exist for the program at Loxton, South Australia; however, Australian breeders are selecting apricots for the drying and processing markets as well as for fresh fruit. With similarities to the program in South Africa, postharvest researchers are assisting in the evaluation of elite fresh market selections to identify those most suitable for export marketing. In addition to high Brix and good product color in dried apricots, Australian breeders aim to automate their drying industry by supplying new cultivars capable of mechanical harvest and fruit cutting, and with low drying ratios.

6 Breeding Methods and Techniques

6.1 Genetics

Breeders generally agree that most apricot traits are quantitative, suggesting a polygenic inheritance (Table 12.2). Although only a few inheritance studies have been done on apricot traits, the high or very high heritability values of most of the traits studied indicate the suitability of choosing parents based on their phenotype and also the high potential for genetic improvement in this species (Couranjou 1995). Crosses made between Asian and European genotypes suggested that traits from the Asian group such as small fruit, large pit, high soluble solids content, long dormancy period and late flowering season have dominant inheritance, while the complementary traits from the European groups are recessive. Similarly, results from crosses between Iran-Caucasian and European apricots suggested that flesh and skin color, and extent of red blush on the fruit are independently inherited (Badenes et al. 2006).

Table 12.2 Quantitative traits suggesting a polygenic inheritance

Trait	Reference
Flowering date	Couranjou (1995)
Maturity date	Couranjou (1995)
Yield	Couranjou (1995)
Fruit size	Couranjou (1995)
Fruit weight	Signoret et al. (2004), Chen et al. (2006)
Fruit skin background color	Couranjou (1995)
Flesh color	Couranjou (1995)
Skin overcolor	Couranjou (1995)
Fruit firmness	Couranjou (1995), Signoret et al. (2004), Peace et al. (2007)
Fruit flavor	Couranjou (1995)
Fruit aroma	Couranjou (1995)
Fruit juiciness	Couranjou (1995)
Self-pollinated fruiting rate	Chen et al. (2006)
Fertile flower rate	Chen et al. (2006)
Fruit sugar content	Signoret et al. (2004)
Fruit acid content	Signoret et al. (2004)
Resistance to *Monilinia laxa*	Conte et al. (2004), Nicotra et al. (2006)

Ripening of climacteric fruits is a complex process that includes many changes in gene expression, especially for enzymes involved in cell wall modifications. Two expansion cDNAs from apricot expressed during fruit ripening are each regulated differently by ethylene (Mbeguie et al. 2002; Mita et al. 2006). Ethylene also regulates the carotenoid accumulation and the carotenogenic gene expression in apricot varieties (Marty et al. 2005; Kita et al. 2007). In peach, Peace et al. (2005) identified endopolygalacturonase (endoPG) as the gene controlling the major fruit firmness and texture traits. Given the close synteny within *Prunus*, endoPG may play a similar role in apricot (Peace et al. 2007).

Regarding resistance to *Monilinia laxa*, the results from Nicotra et al. (2006) indicate that the characteristics "branch resistance" and "fruit resistance" are controlled by different genes, without correlation between them.

The inheritance of chilling requirement for dormancy completion in apricot was studied by Tzonev and Erez (2003). They concluded that this characteristic represents two distinct genetically controlled traits, the first one is a "switch" for bud break and the second is the vigor of the ensuing bud growth. In terms of the inheritance patterns for these traits, low chilling seems to be dominant over high chilling, whereas the second trait exhibits a nondominant intermediate response between the parents.

Some traits inherited in a discrete manner, suggesting an oligogenic inheritance pattern, are seed bitterness (Gómez et al. 1998), male sterility (Burgos and Ledbetter 1994), self-incompatibility (Burgos et al. 1997, 1998), and PPV resistance (Karayiannis et al. 2008).

6.2 Breeding Strategies

Intraspecific hybridization is the most widespread method for apricot scion breeding, while interspecific crossing between *Prunus* species is common for rootstocks breeding or for novel trait improvement in scion cultivars (Bassi and Audergon 2006). For very old cultivars (especially those locally propagated by seeds), screening their natural variability could lead to the selection of improved phenotypes (clonal selection). Physical mutagenesis with gamma-rays or ^{60}Co has been used to increase variability in apricot (Legave and Garcia 1988; Balan et al. 2006), while in vitro cultured anthers were utilized to produce haploid plants (Peixe et al. 2004).

Given a long juvenile period and large plant size, the cost of growing each seedling is high and consequently, when planning a breeding program it is very important to clearly define the objectives, carefully select the parents and specifically define the selection criteria accordingly. Seedling evaluation is based on a two-stage procedure (1) observation of the hybrid on its own roots during 3 consecutive years of production and (2) evaluation of the best hybrids after grafting in several representative areas of production during 3 consecutive years. Considering the juvenile periods in the two stages, the length of the breeding cycle is at least of 12 years. A third stage for assessment of the agronomic and commercial interest of the "elite" hybrids in precommercial orchards is often carried out, particularly in public programs (Audergon et al. 2009; Llácer 2007).

The hybridization techniques (pollination, seed handling, and seedling evaluation) have been extensively described by Layne et al. (1996). However, some improvements can be reported. Mistakes in assigning seedling paternity are more frequent than it seems. When there is a period of cool weather during the blooming season the anthers of some cultivars may dehisce before the petals open. In controlled crosses, when using a self-compatible female parent, all or part of the seedlings may not come from the cross but they may come from selfing (Llácer et al. 2008). Likewise, the incidence of accidental pollination with undesired pollen on interspecific hybridizations was studied by Arbeloa et al. (2006). The percentage of desired hybrids was lower than expected. In these situations, molecular characterization of the progeny should be carried out for paternity assessment.

Regarding seed handling, apricot embryos (seed-coat removed) stratified for 15 days at 4°C have higher germination percentages and seedling growth than those stratified by the standard procedure (pits stratified at 4°C for 2–3 months). This procedure allows plants to get ready for testing (PPV or other pathogen resistance) as soon as possible (Badenes et al. 2000). Embryo culture in vitro can be successfully used as a tool in an apricot breeding program to obtain higher percentages of seedlings or to overcome a lack of seed germination, as occurs with very early-ripening female parents that may not have fully mature embryos (Burgos and Ledbetter 1993; Arbeloa et al. 2003).

Relative to seedling evaluation, methods of screening for resistance to *Monilinia* spp. (Walter et al. 2004) and to PPV (Karayiannis et al. 2008; Llácer et al. 2008) have been recently reported.

The most important progress has been achieved in determination of fruit chemical profiles. High-performance liquid chromatography (HPLC) combined with other identification methods have been applied to the evaluation of vitamins, selenium, carotenoids, polyphenols, and total antioxidant capacity (Munzuroglu et al. 2003; Radi et al. 2004; Veberic and Stampar 2005; Scalzo et al. 2005; Dragovic-Uzelac et al. 2007). Cyanogenic glycosides have been analyzed by HPLC in sweet and bitter kernelled apricot varieties in relation to the resistance to *Capnodis tenebrionis* L. (Sefer et al. 2006).

Near (NIR) and middle (MIR) infrared reflectance spectroscopy have been used for the rapid determination of fruit quality traits such as soluble solids content and titratable acidity (Bureau et al. 2006), while headspace-solid phase microextraction combined with gas chromatography–olfactometry has been applied for aroma characterization (Guillot et al. 2006).

7 New Biotechnology Techniques Available for Fruit Breeding

Traditional fruit tree breeding is a time consuming process in which progress is dependent on a favorable environment during the annual bloom period. Implementation of molecular markers linked to traits of interest, is a direct way to accelerate new cultivar development in deciduous plants with a long juvenile period. Discovery of the nearly complete synteny of genetic linkage maps between *Prunus* species was the major achievement in the area of fruit tree genetics that led to recognizing genomes of all diploid species, including apricot, as a single genetic entity (Arús et al. 2006). This new vision of the *Prunus* genome organization will have an impact on all areas of fruit tree research from classical botany (how many distinct species?) to a modern transgenic study (why are all diploid *Prunus* species so tough to transform?).

A saturated reference map for *Prunus* (Dirlewanger et al. 2004), further enriched with bin mapped markers (Howad et al. 2005), allows an easy transmission of the genetic and genomic information across the genera, i.e., in-between apricot and peach, almond, diploid plums, or cherry. The recent development of centralized bioinformatics resources, Genome Database for Rosaceae (GDR), facilitates this process (Jung et al. 2008). GDR is a major repository of curated and integrated genetics and genomics data of Rosaceae that contains annotated databases of all publicly available *Prunus* ESTs (Expressed Sequence Tags), including those derived from the apricot fruit and leaf cDNA libraries. A genetically anchored peach physical map, apricot genetic maps and comprehensively annotated markers and traits will enable the acceleration of a comparative map of complex traits, and support map-based cloning genes of horticultural importance in apricot (Fig. 12.2).

7.1 Molecular Markers Available for Breeding in Apricot

In apricot, molecular markers were employed for cultivar fingerprinting and to evaluate variability across the crop, for construction of molecular genetic maps, and to develop markers for parental analysis and marker-assisted breeding (complementary review by Hormaza et al. 2007). The list of markers includes isozymes, Randomly Amplified Polymorphic DNAs (RAPDs), Restriction Fragment Length Polymorphisms (RFLPs), Amplified Fragment Length Polymorphisms (AFLP) and Simple Sequence Repeats (SSRs). More sophisticated marker systems include AFLP markers targeting the Resistance Gene Analogs (AFLP-RGAs) or differently expressed cDNAs (AFLP-cDNA), candidate genes for particular traits such as self-incompatibility and resistance to PPV, and EST-SSRs, the SSR markers from the annotated EST (Expressed Sequence Tag) database.

Isozymes, RAPDs, RFLPs, SSRs. The first publications on isozyme analyses in apricot are attributed to Byrne and Littleton (1989). Based on mean heterozygosity at 10 isozyme loci and mixed mating system, apricot was considered as a suitable crop for diversity studies (Byrne 1990). In spite of the limited number of loci, isozymes proved to be reliable markers for genetic variability assessment (Badenes et al. 1996) and cultivar identification (Zhebentyayeva et al. 2001). In a short time, isozymes were replaced with more efficient DNA based markers such as RAPDs (Gogorcena and Parfitt 1994; Takeda et al. 1998; Mariniello et al. 2002), RFLPs (de Vicente et al. 1998) and SSRs (Hormaza 2002; Romero et al. 2003; Zhebentyayeva et al. 2003; Krichen et al. 2006; He et al. 2007). Owing to dominant inheritance and low reproducibility, the application of RAPD markers was limited

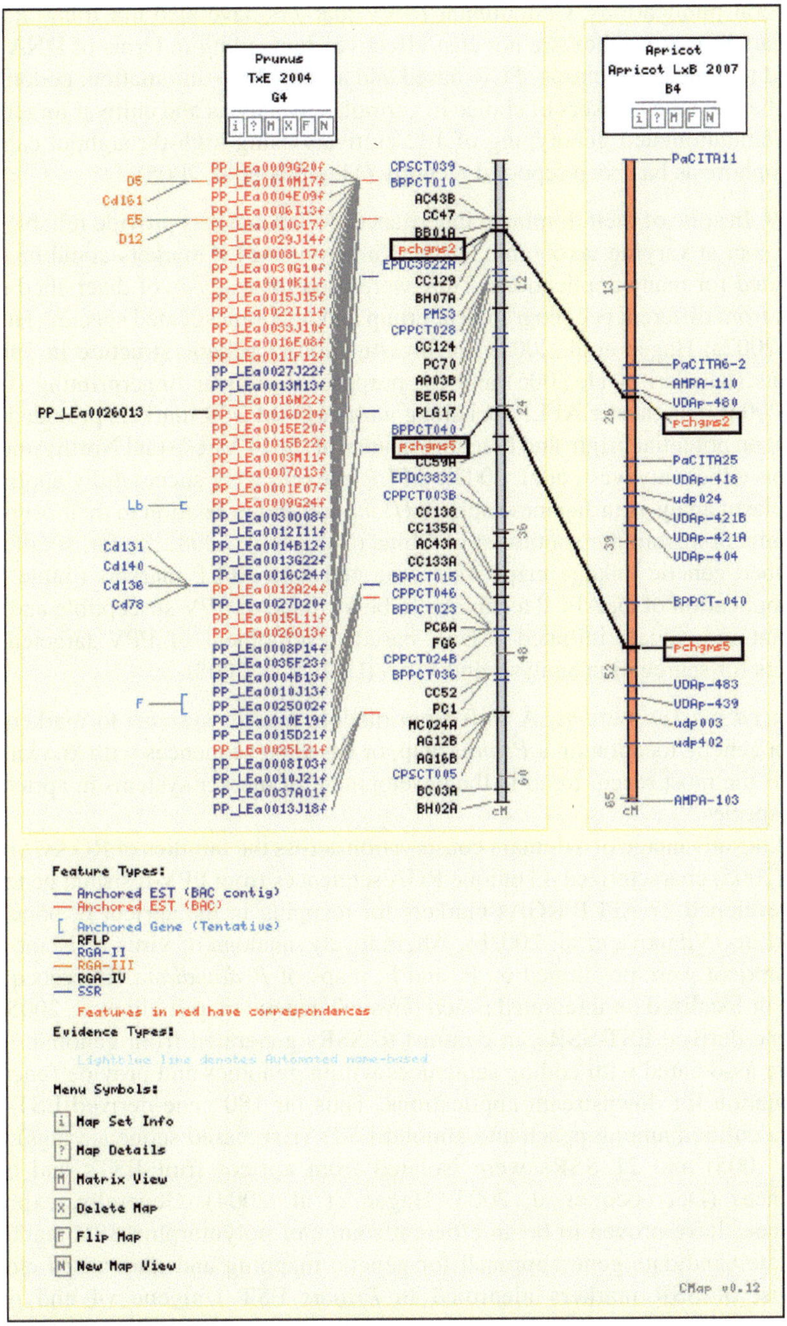

Fig. 12.2 CMap alignment using GDR tools. A screenshot of a CMap page that shows the comparison between G4 of the *Prunus* map and male parent BO81604311 of the apricot L×B map by Dondini et al. (2007). The peach ESTs (PP_LEa), candidate genes representing of RGAs (Cd) and the anchored trait positions are shown on the left. Two SSR loci pchgms2 and pchgms5 (in *boxes*) were used for alignment between two maps

to several publications. Codominant RFLP markers have also not found a broad application because they are not cost-effective, demanding in terms of DNA quality and tedious in execution. PCR-based and amenable to automation, codominant SSRs became the markers of choice in germplasm analysis and cultivar fingerprinting. Semiautomated genotyping of 132 cultivars using high-throughput capillary electrophoresis has been reported recently (Maghuly et al. 2005).

AFLPs. In spite of their dominant inheritance, AFLP markers provide reliable diagnostic loci at varying taxonomic levels. Numerous AFLP markers could be easily generated for many applications. They were used for analysis of diversified germplasm from different ecogeographical group and nondomesticated species (Hurtado et al. 2002a; Hagen et al. 2002), for investigation of genetic structure in Tunisian apricots (Khadari et al. 2006) and for purpose of cultivar fingerprinting (Geuna et al. 2003). Diagnostic AFLP loci along with targeted SSR markers provided more insight on potential origin and breeding history of the PPV-resistant North American apricots (Zhebentyayeva et al. 2008). AFLP markers were successfully applied for germplasm analysis in Japanese apricot (*P. mume*) and in relation to their origin and dissemination from the southwest of China (Yang et al. 2008). So far, six of seven published genetic linkage maps were saturated with AFLP markers (Table 12.3). The application of an AFLP technique to bulks made of PPV susceptible and PPV-resistant individuals initiated a BAC-based development of PPV targeted SSR markers for segregation analysis and MAS (Lalli et al. 2008).

Advanced marker systems. A shift from random marker systems to markers of a known genetic location on a *Prunus* map, or based on sequences with known functions is the most recent trend in the development of marker systems in apricot and other species.

Taking advantage of a domain conservation across the families of RGAs, Soriano et al. (2005) characterized 43 unique RGA sequences from PPV-resistant genotypes and developed 27 AFLP-RGAs markers for mapping in the apricot F_2 population Lito×Lito (Vilanova et al. 2003b). Alternatively, analogs of virus resistant genes from apricot were positioned on F_1 and F_2 maps of *P. davidiana* (Decroocq et al. 2005) or localized on integrated peach physical/genetic map (Lalli et al. 2005).

Gene-derived EST-SSRs, in contrast to SSRs generated from genomic libraries, are associated with coding sequences within genomes and provide functional information for downstream applications. Thus far 180 gene-derived EST-SSRs were identified among peach and almond ESTs (Expressed sequence tags) (Jung et al. 2005) and 21 SSRs were isolated from apricot fruit ESTs and cDNA sequences (Decroocq et al. 2003; Hagen et al. 2004). Generally, expressed sequences have proven to be an efficient source of polymorphic SSR markers to facilitate candidate gene approach for genetic mapping and map-based cloning. The list of SSR markers identified in *Prunus* EST Unigene_v4 and primer sequences automatically designed using GDR tools are available at: http://www.bioinfo.wsu.edu/gdr/projects/prunus/unigeneV4/downloads/PrunusContigsV4_SSR_ORF_PRIMER.xls.

Table 12.3 List of the published apricot maps

Cross	Progeny	Markers (total); trait	Parent and average distance (cM)	Reference
Goldrich × Valenciano (F1 G × C map)	81	AFLP, RAPD,RFLP, SSR (132); PPV	Goldrich-511 Valenciano-467	Hurtado et al. (2002b) (First generation map)
Lito × Lito (F2 L × L map)	76	AFLP, SSR (211); PPV, self-incompatibility	602	Vilanova et al. (2003b) (First generation map)
Polonais × SEO (F1 P × S map)	142	AFLP, RFLP,SSR, cDNA-SSR (141); PPV	Polonaise-538 SEO-699	Lambert et al. (2004, 2007)
Lito × Lito (F2 L × L map)	76	AFLP, SSR, AFLP-RGA (231); PPV, self-incompatibility	615	Soriano et al. 2008 (Second generation map)
Goldrich × Currot (F1 G × C map)	81	AFLP, RAPD,RFLP, SSR (139); PPV	Goldrich-468 Valenciano-451	Soriano et al. (2008) (Second generation map)
LE3246 [SEO × Vestar] × Vestar (BC1 LE × V map)	67	AFLP, SSR (357); PPV	523	Lalli et al. (2008)
Lito × BO81604311 (F1 L × BO map)	125	SSR (185)	Lito-504 BO81604311-620	Dondini et al. (2007)

7.2 State of the Maps

Resistance to PPV was a major focus in all mapping projects published to date. Complete maps were generated for five segregating crosses and, currently, the PPV resistance trait is mapped in four of them: 'Goldrich' × 'Valenciano' (syn. 'Currot'), 'Lito' × 'Lito,' 'Polonais' × 'Stark Early Orange,' (SEO) and LE3246 × 'Vestar' (Table 12.3). Two of the listed maps, 'Lito' × BO81604311 (Dondini et al. 2007) and 'Polonais' × 'SEO' by Lambert et al. (2004) were established using the codominant markers only. Apricot maps are organized in eight linkage groups. The reported total lengths of approximately 500–600 cM are close to that of the *Prunus* map. The mean densities of markers are about 2–4 cM. The highest marker density of 0.92 cM was obtained in G1 on LE3246 × 'Vestar' map (Lalli et al. 2008).

Self-incompatibility and PPV resistance are two traits positioned on the apricot maps. The self-incompatibility locus was located at the end of G6 in agreement with the *Prunus* map. Mapping of PPV resistance is still underway (Hurtado et al. 2002b; Vilanova et al. 2003b) and its control is not completely understood, mainly due to trait complexity and differences in phenotype scoring. The most comprehensive discussions on testing different hypotheses for control of PPV resistance were reported recently (Rubio et al. 2007; Karayiannis et al. 2008; Sicard et al. 2008; Soriano et al. 2008; Lambert et al. 2007; Lalli et al. 2008). At least one genetic location in the upper part of G1 found consensus across the mapping community and was accepted as the major locus conferring the dominant resistance to PPV. On the G × V, L × L, and P × S maps, resolution of the G1 region was increased by mapping PCR-based markers derived from apricot candidate genes potentially involved in resistance to virus (Sicard et al. 2008). Two additional putative QTL loci, including the one detected during the early stages of infection, were localized in the P × S population on G3 of 'Polonais' and G5 of both 'Polonais' and 'SEO' (Lambert et al. 2007).

7.3 Marker-Assisted Selection

Marker-assisted selection (MAS) is the most efficient application of molecular tools and markers to improve apricot cultivars using traditional hybridization techniques. This is especially true in the case of interspecific crosses, when desirable fruit quality often appears as early as a second backcross generation.

Owing to the high level of synteny, markers linked to simple horticultural traits such as fruit color, nonacidic fruit taste, glabrous skin, sweet kernel, and stone adherence can be easily verified and adopted across all *Prunus* species. The same is true for qualitative traits such as bloom time, ripening period, and fruit quality characteristics (Dirlewanger et al. 2004).

Breeding for PPV-resistant cultivars. Evaluation of PPV resistance is the major limitation for apricot breeding programs in many countries. Generational genetic linkage maps for crosses segregating for PPV resistance have located several markers on G1 that were potentially useful in breeding programs. Associations with PPV resistance were reported for markers ssrPaCITA5 and ssrPaCITA17 in Soriano et al. (2008), aprigms18 and EPDCU5100 in Lalli et al. (2008), and pchcms4, RFLP marker AG51, AFLP E37-M13-208 in Lambert et al. (2007). The four markers cd83SSR, cd93SSR, cd195SSR, and cd211SSR were developed with genes potentially involved in plant–virus interactions in Sicard et al. (2008). Altogether, 11 markers are potential candidates for the use of MAS in breeding. Three of them (ssrPaCITA5, ssrPaCITA 17, and aprigms18) were tested for MAS in several crosses from the breeding program at IVIA, Valencia, Spain (Soriano et al. 2008). Depending on the particular population, the proportion of misclassified susceptible seedlings varied from 40 to 69%, while more than 90% of the most resistant plants were preserved in F_1 and F_2 progenies. Further saturation of the PPV resistance region is needed to improve the efficiency of MAS for this trait.

Breeding for self-compatibility. Self-incompatibility (SI) in apricot is another important target for the application molecular technologies. Theoretical background and proposed mechanisms for gametophytic SI (GSI) that apricot shares with other Rosaceae species is thoroughly reviewed by De Nettancourt (2001). In common apricot and Japanese apricot, SI is determined by a single, multiallelic, S-locus, which contains two genes, the stylar S-RNAse gene and the pollen-expressed SFB/SLF (S-haplotype-specific F-box/S-locus F-box) gene (Entani et al. 2003; Romero et al. 2004; Ushijima et al. 2004; Zhang et al. 2008). Both genes exhibit the high polymorphism typical of plant SI loci, but the function of F-box is still unclear. Additional factors not linked to the *S*-locus could also be involved in the breakdown of SI in pollen-part mutants of apricots (Vilanova et al. 2006a). Inheritance of stylar S-RNAse was analyzed in common apricot by Burgos et al. (1998) and in Japanese apricot by Tao et al. (2002). Initially, cultivar genotyping was accomplished using a stylar ribonucleases analysis (Alburquerque et al. 2002). The development of PCR-based markers derived from both genes, S-RNAse and SFB, allowed the discrimination of three cultivar groups: SI group, one universal donor group and SC (self compatible group). This information was incorporated into breeding schemes for producing only self-compatible seedlings (Vilanova et al. 2005). Novel methods of S-allele screening (Vaughan et al. 2006) and dot-blot-*S*-genotyping (Kitashiba et al. 2008) were developed recently for large-scale S-haplotype detection and analysis.

7.4 Genomics

Structural genomics. Large-insert libraries and the physical genetic map developed for model peach genome are indispensable tools for map-based cloning of Mendelian loci in *Prunus*. However, some apricot specific genes such as those involved with self-incompatibility and PPV resistance could not be isolated from peach genomic libraries. To support apricot oriented projects, a BAC library derived from cultivar "Goldrich" was cloned into HindIII site of pBeloBAC11. The library containing

101,376 clones with an average insert size of 64 kb provides 22-fold apricot genome coverage (Vilanova et al. 2003a). The apricot genomic library facilitated a BAC-based cloning of the S-locus genes (Vilanova et al. 2005) and the saturation with SSRs markers in the upper portion G1 associated with PPV resistance (Vilanova et al. 2006b). Currently, this library is being used to sequence the PPV resistance region in 'Goldrich' genotype.

Functional genomics. Three sequenced cDNA libraries from different stages of fruit development (green, half-ripe and ripe mesocarp tissue) were sequenced, annotated and submitted to GeneBank (Grimplet et al. 2005). The total of 13,006 apricot ESTs represents transcriptional profiles of the apricot mesocarp tissue. They supported identification of gene transcripts differently expressed during fruit development in apricot (Geuna et al. 2005). In the *Prunus* database, EST collections from green and half-ripe apricots are the only source of genes expressed at early stages of mesocarp development in stone fruits. About 20% of the ESTs assembled into *Prunus* Unigene set_v 4 are of an apricot origin (Jung et al. 2008). So, apricot functional genomic resources were essential in the development of EST-SSRs and SNPs subsets available from GDR.

A proteomic study (a large-scale protein analysis) was applied for transcript profiling of F_1 individuals derived from crosses between SC and SI apricots. Qualitative and quantitative analyses of parental cultivars revealed 35 proteins with different expression patterns in SC and SI pistils and detected a posttranscriptional regulation of S-RNAse in SI apricots (Feng et al. 2006, 2007).

7.5 Transgenics

Genetic transformation allows discrete alteration of one or more traits in existing crop cultivars if an efficient tissue culture system is available. Transgenic apricot plants may be used as a tool to analyze individual traits through the identification of the corresponding genes and to study their regulation and expression. Understanding gene regulation at the cellular and whole plant level, and identifying and evaluating agriculturally useful genes, should also be possible.

Agrobacterium-mediated transformation of apricot. The virulence of the *Agrobacterium* strain varies with plant species (Cervera et al. 1998), and virulence can be stimulated by the presence of additional copies of the *virG* gene (Ghorbel et al. 2001). In apricots, variation in bacterial virulence between three wild-type *Agrobacterium* strains was not observed in greenhouse evaluation. However, differences in the number of Green Fluorescent Protein (GFP) spots per transformed explant were found when two disarmed strains were compared (Petri et al. 2004). Several environmental factors, including pH, temperature and osmotic stress, have been shown to affect *vir* gene expression (Alt-Mörbe et al. 1989). Stachel et al. (1985) reported that the addition of the phenolic compound acetosyringone (3′, 5′-dimethoxy-hydroxyacetophenone) to the culture medium also stimulated transcription of virulence genes in *Agrobacterium.*

Similar stimulatory effect of acetosyringone on bacterial virulence has been observed in apricot (Laimer da Câmara Machado et al. 1992; Petri et al. 2004). Apricot transformation can also be affected by the duration of cocultivation of inoculated explants with *Agrobacterium*. In general, the transformation frequency is increased with prolonged cocultivation, but a period longer than 3–4 days may cause problems of *Agrobacterium* overgrowth (Petri et al. 2004).

Selectable markers used for transformation. In apricot, GFP has been very useful to optimize early transformation steps. However, its expression is lost with increased plant development due to autofluorescence from chlorophyll, and can only be seen again in roots of developed transgenic shoots (Petri et al. 2008a, b).

Over-expression of regeneration-promoting genes may be a useful selection system as only transformed, but not nontransformed cells, can be regenerated into plants in the absence of growth regulators. The *ipt* gene from *Agrobacterium* (encoding isopentenyl transferase), a key enzyme of cytokinin biosynthesis, is a classical example of a regeneration-promoting gene. Constitutive expression of *ipt* can adversely affect plant growth and development. This can be prevented by placing the gene under the control of an inducible promoter (Kunkel et al. 1999) or in a MAT (multiautonomous transformation) vector, leading to its elimination from the transgenic plants (Ebinuma et al. 1997). Transformation of apricot with a MAT vector containing an *ipt* gene could notably improve the transformation efficiency (López-Noguera et al. 2009) compared to a standard transformation procedure (Petri et al. 2008a, b). Apart from *ipt*, information regarding other regeneration-promoting genes has been virtually lacking. Major efforts are being devoted to identify these genes, whose translation products may be associated with cytokinin synthesis and its recognition, or involved in promoting the vegetative-to-embryogenic or organogenic transition (Zuo et al. 2002).

Selection of transformed plants. Selection of transformed regenerants is a critical step in plant transformation. Antibiotics have been used most commonly as selection agents after integration of genes that confer antibiotic resistance. The concentration of the selective agent and timing of application must be optimized for each plant species. In apricot regeneration-inhibitory concentrations of the antibiotics kanamycin and paromomycin prevented regeneration of transformed plants and a progressive selection pressure with paromomycin, which has been shown to allow a better growth of transformed apricot tissues (Petri et al. 2005a), had to be used to recover transformed plants (Petri et al. 2006, 2008a).

Improvements in the genetic engineering of apricot. The regeneration of adventitious plants from seed-derived apricot tissues was first reported 20 years ago (Lane and Cossio 1986; Pieterse 1989; Goffreda et al. 1995). Using this approach the first apricot plants transformed with the gene encoding the coat protein (CP) of the plum pox virus (PPV) were obtained (Laimer da Câmara Machado et al. 1992). A similar approach was shown to be useful in plum (Ravelonandro et al. 1997), where a post-transcriptional gene silencing phenomenon was responsible for the acquired resistance in the transformed plums (Scorza et al. 2001) and it was shown to remain stable under field conditions (Hily et al. 2004). Unfortunately, there is no further information available on the evaluation of the apricot plants transformed with the CP gene.

Transformation of seed-derived tissues for plants that are vegetatively propagated and with long generation cycles has a limited interest since agronomic characteristics of these plants are unknown and further breeding to introduce the transgene in commercially accepted cultivars needs many years of intensive work. Hence, much effort has been devoted to develop regeneration procedures from clonal tissues of commercial cultivars or new improved selections from breeding programs. The first report on adventitious regeneration from apricot leaves (Escalettes and Dosba 1993) found little reproducibility between experiments. A more effective and reproducible regeneration method from apricot leaves was established (Pérez-Tornero et al. 2000) and optimized latter, increasing regeneration percentages 200% by using ethylene inhibitors and specific gelling agents (Burgos and Alburquerque 2003).

Using the regeneration procedure developed for apricot leaves, an *Agrobacterium*-based transformation procedure was established for apricot leaves that yielded transgenic calluses, expressing *gfp* and *nptII* genes (Petri et al. 2004). The effect of aminoglycoside antibiotics for selection of apricot *nptII*-transformed leaf tissues was studied (Burgos and Alburquerque 2003; Petri et al. 2005a) and the transformation procedure optimized by adding 2,4-D during the cocultivation period (Petri et al. 2005b). However, transformed plants were not obtained.

Coupling transformation with different strategies to select transgenic cells and regenerate plants was necessary to obtain transformed plants. Regeneration inhibitory antibiotic concentrations applied after the coculture period did not allow regeneration of transformed plants and it was necessary to delay the selection pressure or reduce the antibiotic concentration during the first days after coculture before applying regeneration-inhibitory concentrations (Petri et al. 2008a). The first 14 days, including the coculture period, are a regeneration-induction period (in dark) and it is critical to obtain any regenerations from apricot leaves during this time (Pérez-Tornero et al. 2000). This key period probably allows dedifferentiation of leaf cells and differentiation again of those cells into meristems, which may explain the importance of the timing in the application of the selective agent.

Unfortunately, transformation procedures developed for apricot to date are very genotype-dependent, which does not allow using them as an efficient breeding tool.

Shortcomings in the transformation of apricot. Conventional breeding of apricot has been constrained by the long reproductive cycle of the species, with an extended juvenile growth phase, complex reproductive biology and high degree of heterozygosity. New technologies have the potential to reduce the time for cultivar development and offer alternative breeding strategies that are not available to breeders. Progress has been made for apricot in the areas of regeneration, *Agrobacterium*-mediated transformation, gene isolation and mapping, but several obstacles remain to be overcome. This is especially true for the development of a genotype-independent system for tissue culture and genetic transformation, which may be achieved by the transformation of meristematic cells with a high regeneration potential and/or the use of regeneration-promoting genes (Petri and Burgos 2005). Also, the constraint should be addressed that European laws allow neither the deliberate release of plants carrying antibiotic resistance genes used in medicine or veterinary after 2004, nor their commercialization after 2008 (Directive 2001/18/EEC of the European Parliament and the Council of the European Union). The development of a selectable marker-free transformation system for apricot is therefore a priority in future studies.

References

Aksogan S, Basturk A, Yuksel E, Akgiray O (2003) On the use of crushed shells of apricot as the upper layer in dual media filters. Water Science and Technology 48: 497–503

Alburquerque N, Egea J, Pérez-Tornero O, Burgos L (2002) Genotyping apricot cultivars for self-(in) compatibility by means of RNAses associated with S alleles. Plant Breed 121: 343–347

Alt-Mörbe J, Kühlmann H, Schröder J (1989) Differences in induction of Ti plasmid virulence genes virG and virD and continued control of virD expression by four external factors. Mol Plant-Microbe Interact 2: 301–308

Arbeloa A, Daorden ME, García E, Marín JA (2003) Successful establishment of in vitro cultures of Prunus cerasifera hybrids by embryo culture of immature fruits. Acta Hort 616: 375–378

Arbeloa A, Daorden ME, Garcia E, Wunsch A, Hormaza JI and others (2006) Significant effect of accidental pollinations on the progeny of low setting Prunus interspecific crosses. Euphytica 147: 389–394

Arús P, Yamamoto T, Dirlewanger E, Abbott AG (2006) Synteny in the Rosaceae. In: J. Janick (ed) Plant Breeding Reviews, v 27, John Wiley & Sons, Inc, pp 175–211

Asma BM and Ozturk K (2005) Analysis of morphological, pomological and yield characteristics of some apricot germplasm in Turkey. Genetic Resources and Crop Evolution 52: 305–313

Asma BM, Kan T, Birhanlı O (2007) Characterization of Promising Apricot (Prunus armeniaca L.) Genetic Resources in Malatya, Turkey. Genetic Resources and Crop Evolution 54: 205–212

Audergon JM, Blanc A, Gilles F, Gouble G, Grotte M, Reich M, Bureau S, Clauzel G, Pitiot C, Lafond S, Broquaire JM (2009) New recent selections from INRA's apricot breeeding program. Acta Hort 814: 221–226

Audergon JM, Duffillol JM, Gilles F, Giard A, Blanc A, Clauzel G, Chauffour D, Broquaire JM, Moulon B (2006b) 'Soledane', 'Florilege' and 'Bergarouge (R)' Avirine: Three new apricot cultivars for French country. Acta Hort 701: 395–398

Audergon JM, Giard A, Lambert P, Blanc A, Gilles F, Signoret V, Richard JC, Albagnac G, Bureau S, Gouble B, Grotte M, Reich M, Legave JM, Clauzel G, Dicenta F, Scortichini M, Simeone AM, Guerriero R, Viti R, Monteleone P, Bartolini S, Martins JMS, Tsiantos J, Psallidas P (2006a) Optimisation of apricot breeding by a joint conventional and molecular approach applied to the main agronomic traits - ABRIGEN project. Acta Hort 701: 317–320

Avdeev VI (1992) On the centers of provenance of cultivated apricot. Bulletin of applied botany, genetics and plant breeding 146: 33–35 (in Russian)

Badenes MA, Asins MJ, Carbonell EA, Llácer G (1996) Genetic diversity in apricot, Prunus armeniaca, aimed at improving resistance to plum pox viruw. Plant Breeding 115: 133–139

Badenes ML, Llácer G (2006) Breeding for resistance: breeding for Plum pox virus resistant apricots in Spain. Bulletin OEPP/EPPO Bulletin 36: 323–326

Badenes ML, Llácer G, Crisosto C (2006) Mejora de la calidad de frutales de hueso. p 551–578. In: G Llácer, MJ Díez, JM Carrillo and ML Badenes (eds), Mejora Genética de la Calidad en Plantas. Sociedad Española de Ciencias Hortícolas y Sociedad Española de Genética. Ed Universidad Politécnica de Valencia

Badenes ML, Martínez-Calvo J, Llácer G (1998) Analysis of apricot germplasm from the European ecogeographical group. Euphytica 102: 93–99

Badenes ML, Pastor I, Martínez-Calvo J, Llácer G (2000) Improved efficiency in apricot breeding: earlier assessment of seedling progeny for resistance to Plum pox virus. J Hort Sci & Biotechnol 75 (4): 459–464

Bailey CH, Hough LF (1975) Apricots. In: J. Janick and JN Moore (eds). Advances in fruit breeding. Purdue University Press, West Lafayette, IN. pp. 367–383

Bala A, Kaushal BBL, Joshi VK (2005) Utilization of plum and apricot fruits in tomato based sauces. Acta Hort 696: 541–545

Balan V, Tudor V, Petrisor C (2006) Providing the quality features variability of apricot descendants: F-1, F-2, back-cross and V-2. Acta Hort 717: 175–178

Bassi D, Andalò G, Bartolozzi F (1995) Tolerance of apricot to winter temperature fluctuation and spring frost in northern Italy. Acta Hort 384: 315–321

Bassi D, Audergon JM (2006) Apricot breeding: update and perspectives. Acta Hort 701: 279–294

Bassi D, Pirazzoli C (1998) The stone fruit industry in the Mediterranean region: agronomical and commercial overview. Options Méditerranéennes, Série B/n°19, Stone fruit viruses and certification in the Mediterranean: problems and prospects. P. 3–38

Benedikova D (2006) Gene pool utilisation in apricot breeding in Slovak Republic. Acta Hort 717: 173–174

Bortiri E, Oh S-H, Jiang J, Baggett S, Granger A, Weeks C, Buckingham M, Potter D, Parfitt DE (2001) Phylogeny and Systematics of Prunus (Rosaceae) as Determined by Sequence Analysis of ITS and the Chloroplast trnL-trnF Spacer DNA. Systematic Botany 26: 797–807

Bureau S, Reich M, Marfisi C, Audergon JM, Albagnac G (2006) Application of Fourier-transform infrared (FT-IR) spectroscopy for the evaluation of quality traits in apricot fruits. Acta Hort 717: 347–349

Burgos L, Alburquerque N (2003) Low kanamycin concentration and ethylene inhibitors improve adventitious regeneration from apricot leaves. Plant Cell Rep 21: 1167–1174

Burgos L, Ledbetter CA (1993) Improved Efficiency in Apricot Breeding - Effects of Embryo Development and Nutrient Media on In-Vitro Germination and Seedling Establishment. Plant Cell Tissue and Organ Culture 35: 217–222

Burgos L, Ledbetter CA (1994) Observations on inheritance of male sterility in apricot. Hortscience 29: 127

Burgos L, Ledbetter CA, Pérez-Tornero O, Ortín-Párraga F, Egea J (1997) Inheritance of sexual incompatibility in apricot. Plant Breeding 116: 383–386

Burgos L, Pérez-Tornero O, Ballester J, Olmos E (1998) Detection and inheritance of stylar ribonucleases associated with incompatibility alleles in apricot. Sex Plant Reproduction 11: 153–158

Byrne DH (1990) Isozyme variability in four diploid stone fruits compared with other woody plants. J Heredity 81: 68–71

Byrne DH (2002) Peach breeding trends: a worlwide perspective. Acta Hort 592: 49–59

Byrne DH (2005) Trends in stone fruit cultivar development. Horttechnology 15: 494–500

Byrne DH, Littleton TG (1989) Interspecific hybrid verification of Plum x Apricot hybrids via isozyme analyses. HortScience 24: 132–134

Byrne DH, Ramming DW, Topp B (2000) China germplasm collection trip report. August7-August 25, 2000

Cambra M, Capote N, Myrta A, Llácer G (2006a) Plum pox virus and the estimated costs associated with sharka disease. Bulletin OEPP/EPPO Bulletin 36: 202–204

Cambra MA, Serra J, Cano A, Cambra M (2006b) Plum pox virus in Spain. Bulletin OEPP/EPPO Bulletin 36: 215

Cervera M, López MM, Navarro L, Peña L (1998) Virulence and supervirulence of Agrobacterium tumefaciens in woody fruit plants. Physiol Mol Plant P 52: 67–78

Chen XS, Wu Y, Chen MX, He TM, Feng JR and others (2006) Inheritance and correlation of self-compatibility and other yield components in the apricot hybrid F1 populations. Euphytica 150: 69–74

Chen Y, Gong Z, Ye M (1997) Sugar-free instant powder for asthma and cough due to lung-heat in children. Patent Num CN1097320-A

Cociu V (2006) 50 years of apricot breeding in Romania. Acta Hort 701: 355–358

Coneva E (2003) New apricot germplasm selected by ten characteristics. Acta Hort 622: 465–472

Conte L, Nicotra A, Corazza L (2004) New apricot selections resistant to Monilinia laxa (Aderh. et Ruhl.). Acta Hort 663: 245–249

Costes E, Lauri PE, Laurens F, Moutier N, Belouin A, Delort F, Legave JM, Regnard JL (2004) Morphological and architectural traits on fruit trees which could be relevant for genetic studies: a review. Acta Hort 663: 349–356

Couranjou J (1995) Genetic-Studies of 11 Quantitative Characters in Apricot. Scientia Horticulturae 61: 61–75

Culver DJ, Ramming DW, McKenry MV(1989) Procedures for field and greenhouse screening of Prunus genotypes for resistance and tolerance to root-lesion nematode. J Amer Soc Hort Sci 114: 30–35

Day LH (1953) Rootstocks for stone fruits. Observations and experiments with plum, peach, apricot and almond roots for stone fruits. California Agricultural Experiment Station Extension Service. Bulletin 736

Day LH, Serr EF (1951) Comparative resistance of rootstocks of fruit and nut trees to attach by a root-lesion or meadow nematode. Proc Amer Soc Hort Sci 57: 150–154

De Candolle A (1886) Origin of cultivated plants. D. Appleton and Company, 1, 3 and 5 bond street, New York, USA, 468 pp

De Nettancourt D (2001) Incompatibility and Incongruity in Wild and Cultivated Plants. 2nd totally rev, Springer, New York, pp 322

De Vicente MC, Truco MJ, Egea J, Burgos L, Arús P (1998) RFLP variability in apricot (*Prunus armeniaca* L) Plant Breeding 117: 153–158

Decroocq V, Favé MG, Hagen L, Bordenave L, Decroocq S (2003) Development and transferability of apricot and grape EST microsatellite markers across taxa. Theor Appl Genet 106: 912–922

Decroocq V, Foulongne M, Lambert P, Le Gall O, Martin C, Pascal T, Schurdi-Levraud V, Kervella J (2005) Analogues of virus resistance genes map to QTLs for resistance to sharka disease in Prunus davidiana. Mol Genet Genomics 272: 680–689

Dimitrova M (2006) 45 years of apricot rootstock breeding in Bulgaria. Acta Hort 701: 321–323

Dirlewanger E, Graziano E, Joobeur T, Garriga-Caldere F, Cosson P, Howad W, Arús P (2004) Comparative mapping and marker-assisted selection in Rosaceae fruit crops. Proc Natl Acad Sci USA. 101: 9891–9896

Dondini L, Lain O, Geuna F, Banfi R, Gaiotti F, Tartarini S, Bassi D, Testolin R (2007) Development of a new SSR-based linkage map in apricot and analysis of synteny with existing Prunus maps. Tree Genetics & Genomes 3: 239–249

Dragovic-Uzelac V, Levaj B, Mrkic V, Bursac D, Boras M (2007) The content of polyphenols and carotenoids in three apricot cultivars depending on stage of maturity and geographical region. Food Chemistry 102: 966–975

Ebinuma H, Sugita K, Matsunaga E, Yamakado M (1997) Selection of marker-free transgenic plants using the isopentenyl transferase gene. Proc Natl Acad Sci USA 94: 2117–2121

Egea J, Dicenta F, Burgos L (2004a) 'Rojo Pasión' apricot. HortScience 39: 1490–1491

Egea J, Martínez-Gómez P, Dicenta F, Burgos L (2004b) 'Selene' apricot. HortScience 39: 1492–1493

Egea J, Ruiz D, Burgos L (2005b) 'Dorada' apricot. HortScience 40: 1919–1920

Egea J, Ruiz D, Dicenta F, Burgos L (2005a). 'Murciana' apricot. HortScience 40: 254–255

Egea J, Campoy JA, Dicenta F, Burgos L, Patiño JL, Ruiz D (2009) 'Estrella' and 'Sublime' apricot cultivars. HortScience 44: 469–470

Entani T, Iwano M, Shiba H, Che F-S, Isogai A, Takayama S (2003) Comparative analysis of the self-incompatibility (S-) locus region of Prunus mume: identification of a pollen-expressed F-box gene with allelic diversity. Genes Cells 8: 203–213

Escalettes V, Dosba F (1993) In vitro adventitious shoot regeneration from leaves of *Prunus* spp. Plant Sci 90: 201–209

Fang Y (1995) Preparation of anticancer drink. Patent Num CN1094923-A

FAO (1989) FAO Production Yearbook

FAO (2008) http://faostatclassic.fao.org

Faust M, Surányi D, Nyujtó F (1998) Origin and dissemination of apricot, p. 225–266. In: J. Janick (ed.), Horticultural Reviews, vol. 22. John Wiley & Sons, Inc., New York, Chichester, Weinheim, Brisbane, Singapore, Toronto

Feng J, Chen X, Yuan Z, He T, Zhang L, Wu Y, Liu W, Liang Q (2006) Proteome comparison following self- and across-pollination in self-incompatible apricot (*Prunus armeniaca* L.). Protein J 25: 1572–3887

Feng JR, Chen XS, Yuan ZH, Zhang LJ, Ci ZJ, Liu XL, Zhang CY (2007) Primary molecular features of self-incompatible and self-compatible F_1 seedling from apricot (*Prunus armeniaca* L.) Katy × Xinshiji. Mol Biol Rep 36: 263–272

Geuna F, Banfi R, Bassi D (2005) Identification and characterization of transcripts differently expressed during development of apricot (*Prunus armeniaca* L.) fruit. Tree Genetics & Genomes 1: 69–78

Geuna F, Toschi M, Bassi D (2003) The use of AFLP markers for cultivars identification in apricot. Plant Breeding 122: 526–531

Ghorbel R, La-Malfa S, López MM, Petit A, Navarro L, Peña L (2001) Additional copies of virG from pTiBo542 provide a super-transformation ability to Agrobacterium tumefaciens in citrus. Physiol Mol Plant P 58: 103–110

Goffreda JC, Scopel AL, Fiola JA (1995) Indole butyric acid induces regeneration of phenotypically normal apricot (*Prunus armeniaca* L.) plants from immature embryos. Plant Growth Regul 17: 41–46

Gogorcena Y, Parfitt DE (1994) Evaluation of RAPD marker consistency for detection of polymorphism in apricot. Sci Hortic: 163–167

Gómez E, Burgos L, Soriano C, Marín J (1998) Amygdalin content in the seeds of several apricot cultivars. J Sci Food Agric 77: 184–186

Grimplet J, Romieu C, Audergon J-M, Marty I, Albagnac G, Lambert P, Bouchet J-P, Terrier N (2005) Transcriptomic study of apricot fruit (*Prunus armeniaca*) ripening among 13,006 expressed sequence tags. Physiol Plant 125: 281–292

Guerriero R, Viti R, Bartolini S, Iacona C (2006b) Parents for spring frost tolerance in apricot. Acta Hort 717: 153–156

Guerriero R, Viti R, Monteleone P, Iacona C, Gentili M (2006a) Pisa University's contribution to the national apricot breeding programme: Three new apricot cultivars for Tuscan fruit growers. Acta Hort 717: 137–140

Guillet-Bellanger I, Saussac P. Audergon JM (2006) Characterization and inheritance of apricot leaf necrosis observed on 'Manicot' cultivar after sharka inoculations. Acta Hort 701: 493–496

Guillot S, Peytavi L, Bureau S, Boulanger R, Lepoutre JP, Crouzet J, Schorr-Galindo S (2006) Aroma characterization of various apricot varieties using headspace-solid phase microextraction combined with gas chromatography–mass spectrometry and gas chromatography-olfactometry. Food Chemistry 96: 147–155

Hagen LS, Khadari B, Lambert P, Audergon J-M (2002) Genetic diversity in apricot revealed by AFLP markers: species and cultivar comparison. Theor Appl Genet 105: 298–305

Hagen S, Chaib J, Fady B, Decroocq V, Bouchet P, Lambert P, Audergon JM (2004) Genomic and cDNA microsatellites from apricot (Prunus armeniaca L.). Mol Ecol Notes 4: 742–745

Halász J, Pedric A, Hegedüs A (2007) Origin and dissemination of the pollen-part mutated Sc haplotype which confers self-compatibility in apricot (*Prunus armeniaca*). New Phytologist 176: 792–803

Han Z (2001) Fruit wine continuous production. Patent Num CN1172851-A

Harbeck M (2001) Liquid cleansing composition useful for the treatment of dermatitis. Patent Num US2001014316-A1

He T, Chen X, Xu Z, Gao J, Lin P, Liu W, Liang Q, Wu Y (2007) Using SSR markers to determine the population genetic structure of wild apricot (*Prunus armeniaca* L.) in the Ily Valley of West China. Genetic Resources and Crop Evolution 54: 563–572

Hily JM, Scorza R, Malinowski T, Zawadzka B, Ravelonandro M (2004) Stability of gene silencing-based resistance to Plum pox virus in transgenic plum (*Prunus domestica* L.) under field conditions. Transgenic Res 13: 427–436

Hormaza JI (2002) Molecular characterization and similarity relationships among apricot (*Prunus armeniaca* L.) genotypes using simple sequence repeats. Theor Appl Genet 104: 321–328

Hormaza JI, Yamane H, Rodrigo J (2007) Apricot, p 171–185. In: C. Kole (ed) Genome mapping and molecular breeding in plants. V 4, Fruits and nuts. Springer, Berlin, Heidelberg, New York

Howad W, Yamamoto T, Dirlewanger E, Testolin R, Cosson P, Cipriani G, Monforte AJ, Georgi L, Abbott AG, Arús P (2005) Mapping with a few plants: using selective mapping for microsatellite saturation of the *Prunus* reference map. Genetics 171: 1305–1309

Hurtado MA, Romero C, Vilanova S, Abbott AG, Llácer G, Badenes ML (2002b) Genetic linkage map of two apricot cultivars (*Prunus armeniaca* L.) and mapping of PPV (sharka) resistance. Theor Appl Genet 105: 182–192

Hurtado MA, Westman A, Beck E, Abbott A, Llácer G, Badenes ML (2002a) Genetic diversity of apricot based on AFLP markers. Euphytica 127: 297–301

Jung S, Abbott A, Jesudurai C, Tomkins J, Main D (2005) Frequency, type, distribution and annotation of simple sequence repeats in Rosaceae ESTs. Funct Int Genome 5:136–143

Jung S, Staton M, Lee T, Blenda A, Svancara R, Abbott A, Main D (2008) GDR (Genome Database for Rosaceae): integrated web-database for Rosaceae genomics and genetics data. Nucl Acid Res 36: D1034–D1040

Karayiannis I (2006) Breeding for resistance: conventional breeding for *Plum pox virus* resistant apricots in Greece. Bulletin OEPP/EPPO Bulletin 36: 319–322

Karayiannis I, Mainou A, Stylianidis D, Thomidis T, Karayiannis NI, Tsaftaris A(2006) Resistant to sharka disease apricot hybrids of high quality selected in Greece. Acta Hort 701: 337–340

Karayiannis I, Thomidis T, Tsaftaris A (2008) Inheritance of resistance to *Plum pox virus* in apricot (*Prunus armeniaca* L.). Tree Genetic & Genomes 4: 143–148

Kayisi çeşit Kataloğu (1996) Catalog of Turkish apricot cultivars (ed) M. Akçay. Meyvecilik Araştirma Enstitüsu Müdürlüğu, Ankara, Turkey, 93 pp (in Turkish)

Khadari B, Krichen L, Lambert P, Marrakchi M, Audergon JM (2006) Genetic structure in Tunisian apricot, *Prunus armeniaca* L., populations propagated by grafting: a signature of bottleneck effects and ancient propagation by seedlings. Genetic Resources and Crop Evolution 53: 811–819

Kita M, Kato M, Ban Y, Honda C, Yaegaki H, Ikoma Y, Moriguchi T (2007) Carotenoid accumulation in Japanese apricot (*Prunus mume* Siebold & Zucc.): Molecular analysis of carotenogenic gene expression and ethylene regulation. J Agric Food Chem 55: 3414–3420

Kitashiba H, Zhang SL, Wu J, Shirasawa K, Nishio T (2008) S-genotyping and S-screening utilizing SFB gene polymorphism in Japanese plum and sweet cherry by dot-blot analysis. Molecular Breeding 21: 339–349

Kostina KF (1936) The Apricot. Bull Appl Bot Genet and Plant Breeding, Suppl. 83. Institute of Plant Industry, Leningrad (in Russian)

Kostina KF (1946) The origin and evolution of cultivated apricot. Proceedings (Trudi) of the Nikita Botanical Garden 24: 25–31 (in Russian)

Kostina KF (1964) Application the phytogeographical method to apricot classification (in Russian). Proceedings (Trudi) of the Nikita Botanical Garden. Kolos, Moscow, v 24

Kovalev N.V. (1963) Apricot.Selkhozizdat, Moskow. 288p (in Russian)

Krichen L, Mnejja M, Marrakchi M, Trifi-Farah N (2006) Use mocrosatellite polymorphisms to develop an identification key for Tunisian apricots (2006) Genetic Resources and Crop Evolution 53: 1699–1706

Krska B, Vachun Z, Necas T (2006) The apricot breeding programme at the Horticulture Faculty in Lednice. Acta Hort 717: 145–148

Kryukova IV (1989) Botanical classification and geographical distribution, p 9–23. In: VK Smykov (ed), Apricot, Agropromizdat, Moscow, USSR (in Russian)

Kunkel T, Niu QW, Chan YS, Chua NH (1999) Inducible isopentenyl transferase as a high-efficiency marker for plant transformation. Nat Biotechnol 17: 916–919

Laimer da Câmara Machado M, da Câmara Machado A, Hanzer V, Weiss H, Regner F, Steinkeliner H, Mattanovich D, Plail R, Knapp E, Kalthoff B, Katinger HWD (1992) Regeneration of transgenic plants of *Prunus armeniaca* containing the coat protein gene of plum pox virus. Plant Cell Rep 11: 25–29

Lalli DA, Abbott AG, Zhebentyayeva TN, Badenes ML, Damsteegt V, Polák J, Krška B, Salava J (2008) A genetic linkage map for an apricot (*Prunus armeniaca* L.) BC1 population mapping plum pox virus resistance. Tree Genetics & Genomes 4: 481–493

Lalli DA, Decroocq V, Blenda AV, Schurdi-Levraud V, Garay L, Le Gall O, Damsteegt V, Reighard GL, Abbott AG (2005) Identification and mapping of resistance gene analogs (RGAs) in *Prunus*: a resistance map for *Prunus*. Theor Appl Genet 111: 1504–1513

Lambert P, Dicenta F, Rubio M, Audergon JM (2007) QTL analysis of resistance to sharka disese in the apricot (*Prunus armeniaca* L.) 'Polonais' x 'Stark Early Orange' F1 progeny. Tree Genetics & Genomes 3: 299–309

Lambert P, Hagen LS, Arús P, Audergon JM (2004) Genetic linkage maps of two apricot cultivars (*Prunus armeniaca* L.) compared with the almond Texas x peach Earlygold reference map for *Prunus*. Theor Appl Genet 108: 1120–1130

Lane WD, Cossio F (1986) Adventitious shoots from cotyledons of immature cherry and apricot embryos. Can J Plant Sci 66: 953–959

Layne R.E.C., Bailey C.H., Hough L.F (1996) Apricots, p. 79–111. In: J. Janick and J.N. Moore (eds.), Fruit breeding, vol. 1: Tree and Tropical Fruits, John Wiley & Sons, Inc., New York

Ledbetter, C.A. and S.J. Peterson. 2004. Utilization of Pakistani apricot (*Prunus armeniaca* L.) germplasm for improving Brix levels in California adapted apricots. Plant Genetic Resources Newsletter 140: 14–22

Legave JM, Garcia G (1988) Radiosensitivity of Apricot Budsticks Exposed to Acute Gamma-Rays and Nursery Observations on the 2nd Vegetative Generation from Irradiated Buds. Agronomie 8: 55–59

Li S (1997) Feiyangling medicine for curing infantile virus pneumonia. Patent Num CN1105570-A and CN1048882-C

Lichou J. and Audubert A (1989) L'abricotier. Centre Technique Interprofessionnel des Fruits et Légumes. (CTIFL). ISBN : 2-901002-69-2

Llácer G (2007) Apricot breeding program from the IVIA: first results (in Spanish). Proc. II Int. Fruit Congress 'Ciutat de Carlet', Valencia, Spain, p 13–36

Llácer G (2009) Fruit breeding in Spain. Acta Hort 814: 43–56

Llácer G, Badenes ML, Romero C (2008) Problens in the determination of inheritance of *Plum pox virus* resistance in apricot. Acta Hort 781: 263–268

López-Noguera S, Petri C, Burgos L (2009) Combining a regeneration-promoting gene and site-specific recombination allows a more efficient apricot transformation and the elimina-tion of marker genes. Plant Cell Rep 28: 1781–1790

Mády, R., Klincsek, P., Szani, Z.S., Szabó, T., Erdós, Z. and I. Skola. 2007. Hungarian seedling rootstocks for apricot. Acta Horticulturae (ISHS) 732: 297–302

Maghuly F, Fernandez EB, Ruther S, Pedryc A, Laimer M (2005) Microsatellite variability in apricot (*Prunus armeniaca* L.) reflects their geographic origin and breeding history. Tree Genetics & Genomes 1: 155–163

Maikeru Shoji K (1994) Japanese apricot seasoning. Patent Num JP6062790-A

Mariniello L, Sommella MG, Sorrentino A, Forlani M (2002) Identification of *Prunus armeniaca* cultivars by RAPD and SCAR markers. Biotech Letter 24: 749–755

Martínez-Calvo J, Font A, Llácer G, Badenes ML (2009) Apricot and peach breeding programs from the IVIA. Acta Hort (ISHS) 814: 185–188

Martínez-Gómez P, Dicenta F, Audergon J-M (2000) Behavior of apricot (*Prunus armeniaca* L.) cultivars in the presence of sharka (*plum pox potyvirus*): A review. Agronomie-Paris 20: 407–422

Marty I, Bureau S, Sarkissian G, Gouble B, Audergon JM, Albagnac G (2005) Ethylene regulation of carotenoid accumulation and carotenogenic gene expression in colour-contrasted apricot varieties (*Prunus armeniaca*). J Exp Bot 56: 1877–1886

Mbeguie AM, Gouble B, Gomez RM, Audergon JM, Albagnac G, Fils-Lycaon B (2002) Two expansin cDNAs from Prunus armeniaca expressed during fruit ripening are differently regulated by ethylene. Plant Physiology and Biochemistry 40: 445–452

Mega K, Tomita E, Kitamura S, Saito S, Mizukami S (1988) Ume, p 289–300. In: Aoba T (ed.) The Grand Dictionary of Horticulture, Shogakukan, Tokyo

Miller NF (1999) Agricultural development in western Central Asia in the Chalcolithic and Bronze Ages. Vegetation History and Archaeobotany 8: 13–19

Mita S, Nagai Y, Asai T (2006) Isolation of cDNA clones corresponding to genes differentially expressed in pericarp of mume (*Prunus mume*) in response to ripening, ethylene and wounding signals. Physiologia Plantarum 128: 531–545

Moreau-Rio MA (2006) Perception and consumption of apricots in France Acta Hort 701: 31–37

Moser L, Conte L, Nicotra A (1999) A description of some dwarf or compact genotypes of apricot (in Italian). Italus Hortus 6 (3): 33–34

Munzuroglu O, Karatas F, Geckil H (2003) The vitamin and selenium contents of apricot fruit of different varieties cultivated in different geographical regions. Food Chemistry 83: 205–212

Nicotra A, Conte L, Moser L, Fantechi P, Barbagiovanni I, Corazza ML, Vitale S, Magnotta A (2006) Breeding programme for *Monilinia laxa* (Aderh. et Ruhl.) resistance on apricot. Acta Hort 701: 307–311

Nyujtó F, Suránui D (1981) Kajszibarack.Mezögazd. Kiadó, Budapest

Orero G, Cuenca J, Romero C, Martínez-Calvo J, Badenes ML, Llácer G (2004) Selection of seedling rootstocks for apricot and almond. Acta Hort 658 (2): 529–533

Otsuka T, Tsukamoto T, Tanaka H, Inada K, Utsunomiya H and others (2005) Suppressive effects of fruit-juice concentrate of *Prunus mume* Sieb. et Zucc. (Japanese apricot, Ume) on Helicobacter pylori-induced glandular stomach lesions in Mongolian gerbils. Asian Pac J Cancer Prev 6: 337–341

Peace CP, Callahan A, Ogundiwin EA, Potter D, Gradziel TM, Bliss FA, Crisosto CH (2007) Endopolygalacturonase genotypic variation in *Prunus*. Acta Hort 738: 639–646

Peace CP, Crisosto CH, Gradziel TM (2005) Endopolygalacturonase: a candidate gen for freestone and melting flesh in peach. Molecular Breeding 16: 21–31

Peixe A, Barroso J, Potes A, Pais MS (2004) Induction of haploid morphogenic calluses from in vitro cultured anthers of Prunus armeniaca cv. 'Harcot'. Plant Cell Tissue and Organ Culture 77: 35–41

Pellegrino S (2006) Apricot industry in Italy (in Spanish) 'Updating the apricot production technology Course' Escuela Agraria de Cogullada, Zaragoza, Spain

Pennone F, Abbate V (2006) Apricot breeding in Caserta: New perspectives of apricot growing in Southern Italy. Acta Hort 717: 157–161

Pennone F, Guerriero R, Bassi D, Borraccini G, Conte L, De Michele A, Mattatelli B,Ondradu G, Pellegrino S, Pirazzini P.(2006) Evolution of the apricot industry in Italy and the national program (MIPAF-regions) "List of recommended fruits varieties". Acta Hort 701: 351–354

Pérez-Tornero O, Egea J, Vanoostende A, Burgos L (2000) Assessment of factors affecting adventitious shoot regeneration from in vitro cultured leaves of apricot. Plant Sci 158: 61–70

Petri C, Alburquerque N, Burgos L (2005a) The effect of aminoglycoside antibiotics on the adventitious regeneration from apricot leaves and selection of nptII-transformed leaf tissues. Plant Cell, Tiss Org Cult 80: 271–276

Petri C, Alburquerque N, García-Castillo S, Egea J, Burgos L (2004) Factors affecting gene transfer efficiency to apricot leaves during early Agrobacterium-mediated transformation steps. J Hortic Sci Biotech 79: 704–712

Petri C, Alburquerque N, Pérez-Tornero O, Burgos L (2005b) Auxin pulses and a synergistic interaction between polyamines and ethylene inhibitors improve adventitious regeneration from apricot leaves and Agrobacterium-mediated transformation of leaf tissues. Plant Cell Tissue Organ Cult 82: 105–111

Petri C, Burgos L (2005) Transformation of fruit trees. Useful breeding tool or continued future prospect? Transgenic Res 14: 15–26

Petri C, López-Noguera S, Alburquerque N, Burgos L (2006) Regeneration of transformed apricot plants from leaves of a commercial cultivar. Acta Hort 717: 233–235

Petri C, López-Noguera S, Alburquerque N, Egea J, Burgos L (2008a) An antibiotic-based selection strategy to regenerate transformed plants from apricot leaves with high efficiency. Plant Sci 175: 777–783

Petri C, Wang H, Alburquerque N, Faize M, Burgos L (2008b) Agrobacterium-mediated transformation of apricot (*Prunus armeniaca* L) leaf explants. Plant Cell Rep . 27: 1317–1324

Pieterse RE (1989) Regeneration of plants from callus and embryos of 'Royal' apricot. Plant Cell Tissue Organ Cult 19: 175–179

Poëssel JL, Faurobert M, Esmenjaud, D, Dirlewanger E, Lemoine MC, Gurrieri F, Michelot P, Lafond S (2006) Breeding for compatible apricot rootstocks cumulating resistance to *Plum Pox Virus* and root-knot nematodes: the *P x dasycarpa* way. Acta Hort 701: 333–336

Radi M, Mahrouz M, Jaouad A, Amiot MJ (2004) Characterization and identification of some phenolic compounds in Apricot fruit (*Prunus armeniaca* L.). Sciences des Aliments 24: 173–183

Ravelonandro M, Scorza R, Bachelier JC, Labonne G, Levy L, Damsteegt VD, Callahan AM, Dunez J (1997) Resistance of transgenic Prunus domestica to plum pox virus infection. Plant Dis 81: 1231–1235

Rehder A (1940) Manual of cultivated trees and shrubs, 2nd edn. Macmillan, New York

Reighard GL, Cain DW, Newall WC (1990) Rooting and survival potential of hardwood cuttings of 406 species, cultivars and hybrids of *Prunus*. HortScience 25(5): 517–518

Rodrigo J, Herrero M (1996) Evaluation of pollination as the cause of erratic fruit set in apricot Moniquí. J Hort. Sci 71 (5): 801–805

Romero C, Perdic A, Muñoz V, Llácer G, Badenes ML (2003) Genetic diversity of different apricot geographical groups determined by SSR markers. Genome 46: 244–252

Romero C, Vilanova S, Burgos L, Martínez-Calvo J, Vicente M, Llácer G, Badenes ML (2004) Analysis of the S-locus structure in *Prunus armeniaca* L. Identification of S-haplotype specific S-RNase and F-box genes. Plant Mol Biol 56: 145–157

Rostova IS, Sokolova EA (1992) Variability of anatomical and morphological leaf characters in apricot (*Armeniaca* Scop.) species and varieties. Bulletin of applied botany, genetics and plant breeding 146: 74–86

Rubio M, Audergon JM, Martínez-Gómez P, Dicenta F (2007) Testing genetic control hypothesis for *Plum pox virus* (sharka) resistance in apricot. Scientia Horticulturae112: 361–365

Rubio M, Dicenta F, Martínez-Gómez P (2003) Susceptibility to sharka (*Plum pox virus*) in *Prunus mandshurica* x *P. armeniaca* seedlings. Plant Breeding 122: 465–466

Ruiz D and Egea H (2008) Phenotypic diversity and relationships of fruit quality traits in apricot (*Prunus armeniaca* L.) germplasm. Euphytica 163: 143–158

Ruiz D, Egea J, Tomás-Barberán FA, Gil MI (2005) Carotenoids from new apricot (*Prunus armeniaca* L.) varieties and their relationship with flesh and skin color. J Agric Food Chem 53: 6368–6374

Scalzo J, Politi A, Pellegrini N, Mezzetti B, Battino M (2005) Plant genotype affects total antioxidant capacity and phenolic contents in fruit. Nutrition 21: 207–213

Scorza R, Callahan A, Levy L, Damsteegt VD, Webb K, Ravelonandro M (2001) Post-transcriptional gene silencing in plum pox virus resistant transgenic European plum containing the plum pox potyvirus coat protein gene. Transgenic Res 10: 201–209

Sefer F, Misirli A, Gülcan R (2006) A research on phenolic and cyanogenic compounds in sweet and bitter kernelled apricot varieties. Acta Hort 701: 167–169

Semon SFA (2006) Community plant variety rights and new apricot cultivars. Acta Hort 701: 39–42

Sicard O, Marandel G, Soriano JM, Lalli DA, Lambert P, Salava J, Badenes ML, Abbott AG, Decroocq V (2008) Flanking the major Plum pox virus resistance locus in apricot with co-dominant markers (SSRs) derived from candidate resistance genes. Tree Genetics & Genomes 4: 359–365

Signoret V, Bureau S, Reich M, Gouble B, Clauzel G, Albagnac G, Audergon JM (2004) Inheritance of organoleptic quality traits of apricot. Acta Hort 663: 275–282

Sinskaya E.N. (1969) Historical geography of cultivated floras (at the dawn of agriculture). Kolos, Leningrad, USSR (in Russian)

Slingerland, K., Fisher, H. and D. Hunter. 2002. Apricot cultivars. Factsheet No. 214. Ontario Ministry of Agriculture, Food and Rural Affairs. http://www.omafra.gov.on.ca/english/crops/facts/02-035.htm#f

Son L and Küden A (2003) Effects of seedling and GF-31 rootstocks on yield and fruit quality of some table apricot cultivars grown in Mersin. Turkish J Agric Forestry 27 (5): 261–267

Soriano J M, Vilanova S, Romero C, Llácer G, Badenes M L (2005) Characterization and mapping of NBS-LRR resistance gene analogs in apricot (*Prunus armeniaca* L.) Theor Appl Genet 110: 980–989

Soriano JM, Vera-Ruiz EM, Vilanova S, Martínez-Calvo J, Llácer G, Badenes ML, Romero C (2008) Identification and mapping of a locus conferring plum pox virus resistance in two apricot-improved linkage maps. Tree Genetics & Genomes 4: 391–402

Stachel SE, Messens E, Van Montagu M, Zambryski P (1985) Identification of the signal molecules produced by wounded plant cells that activate T-DNA transfer in *Agrobacterium tumefaciens*. Nature 318: 624–629

Takeda T, Shimada T, Nomura K, Ozaki T, Haji T, Yamaguchi M, Yoshida M (1998) Classification of apricot varieties by RAPD analysis. J Jpn Soc Hort Sci 67: 21–27

Tao R, Habu T, Namba A, Yamane H, Fuyuhiro F, Iwamoto K, Sugiura A (2002) Inheritance of Sf-RNase in Japanese apricot (*Prunus mume*) and its relation to self-incompatibility. Theor Appl Genet 105: 222–228

Tzonev R, Erez A (2003) Inheritance of chilling requirement for dormancy completion in apricot vegetative buds. Acta Hort 622: 429–436

Ushijima K, Yamane H, Watari A, Kakehi E, Ikeda K, Hauck NR, Iezzoni AF, Tao R (2004) The S haplotype-specific F-box protein gene, SFB, is defective in self-compatible haplotypes of *Prunus avium* and *P. mume*. Plant J 39: 573–586

Varveri C (2006) *Plum pox virus* in Greece. Bulletin OEPP/EPPO Bulletin 36: 209

Vaughan SP, Russell K, Sargent DJ, Tobutt KR (2006) Characterization of pollen S alleles in *Prunus avium* and their application in a novel method suitable for large-scale population studies of self-incompatibility in *Prunus* species. Theor Appl Genet (2006) 112: 856–866

Vavilov NI (1951) The phytogeographic basis of plant breeding. In: Chronica Botanica, an international collection of studies in the method and history of biology and agriculture (Ed) F. Verdoorn, translation from Russian by K.S. Chester, v 13, N1/6, pp 13–54

Veberic R, Stampar F (2005) Selected polyphenols in fruits of different cultivars of genus *Prunus*. Phyton-Annales Rei Botanicae 45: 375–383

Vilanova S, Badenes ML, Burgos L, Martínez-Calvo J, Llácer G, Romero C (2006a) Self-compatibility of two *Prunus armeniaca* selections is associated with two pollen-part mutations of different nature. Plant Physiology 142: 629–641

Vilanova S, Romero C, Abernathy D, Abbott AG, Burgos L, Llácer G, Badenes ML (2003a) Construction and application of a bacterial artificial chromosome (BAC) library of *Prunus armeniaca* L. for the identification of clones linked to the self-incompatibility locus. Mol Genet Genomics 269: 685–691

Vilanova S, Romero C, Burgos L, Llácer G, Badenes ML (2005) Identification of self-(in)compatibility alleles in apricot (*Prunus armeniaca* L.) by PCR and sequence analysis. J Am Soc Hortic Sci 130: 893–898

Vilanova S, Soriano M, Lalli DA, Romero C, Abbott AG, Llácer G, Badenes M L(2006b) Development of SSR markers located in the G1 linkage of apricot (*Prunus armeniaca* L.) using a bacterial artificial chromosome library. Mol Ecol Notes 6: 789–791

Vilanova S, Romero C, Abbott AG, Llácer G, Badenes M L (2003b): An apricot (*Prunus armeniaca* L.) F2 progeny linkage map based on SSR and AFLP markers mapping Plum box virus resistance and self-incompatibility traits. Theor Appl Genet 107: 239–247

Walter M, McLlaren GF, Fraser JA, Frampton CM, Boyd-Wilson KSH, Perry JH (2004) Methods of screening apricot fruit for resistance to brown rot caused by *Monilinia* spp. Australasian Plant Pathology 33: 541–547

Wickson EJ (1891) The Apricot. In: The California Fruits and how to grow them. 2nd Edition, Dewey & Co., San Francisco, CA. Chapter XVII pp. 254–271

Yamaguchi M, Kyotani H, Yoshida M, Haji T, Nishimura K, Nakamura Y, Miyake M, Yaegaki H, Asakura T (2002a) New Japanese apricot cultivar 'Kagajizou.' (in Japanese) Bulletin of the National Institute of Fruit Tree Science 1: 23–33. English abstract: http://sciencelinks.jp/j-east/article/200219/000020021902A0645500.php

Yamaguchi M, Kyotani H, Yoshida M, Haji T, Nishimura K, Nakamura Y, Miyake M, Yaegaki H, Asakura T (2002b) New Japanese apricot cultivar 'Hachirou.' (in Japanese) Bulletin of the National Institute of Fruit Tree Science 1: 35–46. English abstract: http://sciencelinks.jp/j-east/article/200219/000020021902A0645501.php

Yang CD, Zhang YW, Yan XL, Bao MZ (2008) Genetic relatedness an genetic diversity of ornamental mei (*Prunus mume* Sieb.&Zucc.) as analyzed by AFLP markers. Tree Genetics & Genomes 4: 255–262

Zanetto A, Maggioni L, Tobutt R, Dosba F (2002) *Prunus* genetic resources in Europe: Achievements and perspectives of a networking activity. Genetic Resources and Crop Evolution 49: 331–337

Zeven AC, de Wet JMJ (1982) Dictionary of cultivated plants and their regions of diversity. Excluding most ornamentals, forest trees and lower plants. Center for Agricultural Publishing and Documentation, Wageningen, Netherlands. 263 pp

Zohary D. and Hopf M. (2001) Domestication of plants in the Old World. 3 rd ed. Oxford University Press, Oxford, UK. 334 pp

Zhang L, Chen X, Chen X-L, Zhang C, Liu X, Ci X, Zhang H, Wu C, Liu C (2008) Identification of self-incompatibility (S-) genotypes of Chinese apricot cultivars. Euphytica 160: 241–248

Zhao F, Liu W, Liu N, Yu X, Sun M, Zhang Y, Zhou Y (2005) Reviews of the apricot germplasm resources and genetic breeding in China. J Fruit Science, 22: 687–690 (in Chinese)

Zhebentyayeva TN, Ageeva NG (2004) Intraspecific component composition of peroxidase in apricots of a different eco-geographical origin. Proceedings of Nikita Botanical Garden 122: 64–70 (in Russian)

Zhebentyayeva TN, Ageeva NG, Gorina V (2001) Identification of apricot cultivars by isozyme composition. Cytology and Genetics (Kiev) 35: 46–51

Zhebentyayeva TN, Reighard GL, Gorina VM, Abbott AG (2003) Simple sequence repeat (SSR) analysis for assessment of genetic similarity in apricot germplasm. Theor Appl Genet 106: 435–444

Zhebentyayeva TN, Reighard GL, Lalli D, Gorina VM, Krška B, Abbott AG (2008) Origin of plum pox virus resistance in apricot: what new AFLP and targeted SSR data analyses tell. Tree Genetics & Genomes 4: 403–417

Zhukovsky PM (1971) Cultivated plants and their wild relatives. Systematics, geography, cytogenetics, resistance, ecology, origin and use. Kolos, Leningrad, 751 pp (in Russian)

Zuo J, Niu QW, Ikeda Y, Chua NH (2002) Marker-free transformation: increasing transformation frequency by the use of regeneration-promoting genes. Curr Opin Biotechnol 13: 173–180

Chapter 13
Cherry

Frank Kappel, Andrew Granger, Károly Hrotkó, and Mirko Schuster

Abstract The two major species of cherries in world trade are the diploid *Prunus avium* L. (sweet cherries) and the tetraploid *Prunus cerasus* L. (sour cherries). The sour cherry is an allopolyploid species, probably as a result from a natural hybridization between ground cherry, *P. fruticosa*, and unreduced pollen of the sweet cherry, *P. avium*. In rootstock breeding, major species include *P. avium*, *P. cerasus*, *P. canescens* Bois, *P. fruticosa* Pall., and *P. mahaleb* L. Sweet cherries are divided into four groups based on fruit color, shape, and texture: black geans, amber geans, hearts, and bigarreaux, whereas sour cherries are divided into two groups: Morellos (Griottes, Weichsel) with red to dark red colored juice and Amarelles (Kentish) with colorless juice. It has been suggested that sweet cherry originated in an area south of the Caucasian mountains with a secondary dissemination into Europe. The sour cherry, *Prunus cerasus* L., is native from middle and south Europe to north India, Iran, and Kurdistan, and its center of origin extends from the south border of the Black Sea along Anatolia and the south Caucasus to Iran. Major breeding objectives are fruit size, firmness, fruit quality, self-fertility, extended harvest season, and adaptability to mechanical harvest. Recently, precocity, and productivity, resistance to rain-induced cracking, resistance to diseases and insects are additional goals.

F. Kappel (✉)
Agriculture and Agri-Food Canada, 11305 Dale Meadows Rd., Summerland, BC, Canada V0H 1Z8
e-mail: Frank.Kappel@shaw.ca

A. Granger
Plant & Food Research, Mt Albert, New Zealand
e-mail: Andrew.Granger@plantandfood.co.nz

K. Hrotkó
Corvinus University of Budapest, Budapest, Hungary
e-mail: karoly.hrotko@uni-corvinus.hu

M. Schuster
Julius Kuehn Institute, Dossenheim, Germany
e-mail: mirko.schuster@jki.bund.de

M.L. Badenes and D.H. Byrne (eds.), *Fruit Breeding*, Handbook of Plant Breeding 8, DOI 10.1007/978-1-4419-0763-9_13,

Breeding for rootstocks are focused in the effect of the rootstock on the scion in traits as vigor, growth habit, precocity, and fruit quality. Graft compatibility and good propagation for nurseries are important questions along with pest resistance and adaptability to soil and environmental conditions. Genetic linkage maps are being developed for sour cherry. There is a growing body of work in other *Prunus* species, particularly peach and almond, that have great potential for application to cherry. Transformation protocols have been applied to sour cherry, but sweet cherry has been proven very difficult to transform.

Keywords *Prunus cerasus* and *P. avium* • Stone fruit • Drupe • Sweet cherry • Sour cherry • Origin • Fruit breeding • Breeding goals • Rootstocks • Cultivars • Genetic resources • Quality traits • Resistance breeding • Molecular markers • Incompatibility

1 Introduction

World production of cherries has been increasing steadily in many new and traditional regions. Sweet cherries are one of the few remaining seasonal fruit crops, and in many markets no other item creates as much seasonal in-store activity as fresh cherries (Perishable Group 2007). Potential health benefits of sour cherries have enhanced the economic outlook for sour cherries. Much of the increase in production is taken up by new cultivars, many developed by fruit breeding programs from around the world.

World production of cherries is around three million metric tons (5-year average 2001–2005) and in 2005 the world production of all deciduous fruits was greater than 478 million metric tons. The top five cherry-producing countries are Turkey, the USA, Russia, Iran, and Ukraine (Table 13.1). World cherry production has steadily increased from 1990 to 2005. Turkey has moved from being the third largest producer

Table 13.1 Top ten cherry[a] producers in the world (2001–2005) FAO Statistics at http://faostat.fao.org/site/291/default.aspx September 2007

	Average production within designated time period (1,000 tons)		
Country	2001–2005	1996–2000	1990–1995
Turkey	378.6	329.2	250.5
USA	299.6	312.6	293.1
Russian Federation	294.6	241.2	199.6
Iran	259.3	258.4	144.4
Ukraine	230.2	196.4	213.6
Poland	220.2	182.7	139.1
Germany	121.5	220.3	272.9
Italy	114.5	134.7	126.5
Spain	98.4	85.6	71.4
Romania	85.0	77.3	72.9
World	2913.9	2841.1	2687.7

[a]Includes both sweet and sour cherries

Table 13.2 Top ten cherry[a] producers based on crop value (2001–2005). FAO Statistics at http://faostat.fao.org/site/291/default.aspx September 2007

	Average value[b] within designated time period (million US $)		
Country	2001–2005	1996–2000	1991–1995
USA	399.6	273.9	227.6
Turkey	343.9	210.8	177.2
Iran	249.7	238.6	300.5
Japan	246.9	223.4	221.5
Italy	231.6	294.7	283.9
Germany	207.8	349.5	479.9
Spain	181.6	125.2	102.9
Syria	158.0	122.8	63.5
France	128.3	104.6	118.8
Poland	99.9	102.9	51.9

[a]Includes sweet and sour cherries
[b]Average value calculated from production data and producer price in US $

to the largest over that period, whereas Germany went from second most important to seventh. The value of the US cherry crop was highest in 2001–2005 (Table 13.2). World trade in cherries was about $500 million (US currency), with Japan and Germany importing the greatest value of cherries (Table 13.3). The USA was the largest exporter of cherries with three times the value of second place Turkey.

Sweet cherries are chiefly produced for the fresh market, whereas sour cherries are largely processed (Kaack et al. 1996). Cherries are frozen in bulk or individually quick-frozen (IQF) and can be further processed. Canned sweet cherries are primarily consumed as a substitute for fresh fruit, whereas sour cherries are used for pie-fillings. Dried sweet or sour cherries are used as it is, included in dried fruit-and-nut mixes, or enrobed in chocolate. Jams and jellies are other processed cherry products manufactured from whole or crushed fruit (sweet or sour). Maraschino, glace and candied cherries are sweet cherry fruit that have been bleached, recolored, and sweetened in sucrose solutions and used as garnishes for drinks and desserts. Cherry fruit is also used for juice, nectar, liqueurs, and wines.

A number of reports have suggested various taxonomic groupings of cherries (Hedrick 1915; Zielinski 1977; Iezzoni et al. 1990; Brown et al. 1996; Watkins 1976; Webster 1996). It is now generally accepted that the two species of cherries in world trade are *Prunus avium* L. (sweet cherries) and *Prunus cerasus* L. (sour cherries). Other cherry species that have occasionally been grown for their fruit include the following: *P. fruticosa* Pall., *P. tomentosa* Thunb., and *P. pseudocerasus* Lindl. (Webster 1996). The species *P. besseyi* Bailey, *P. pumila* L., and *P. humilis* Bge. have also occasionally been raised for their fruit; however, they are more closely related to plums than cherries (Webster 1996). In rootstock breeding, the parental material has come predominantly from the subgenus *Cerasus* (Rehder 1974). Major species in the parentage of rootstocks include *P. avium*, *P. cerasus*, *P. canescens* Bois, *P. fruticosa*, and *P. mahaleb* L. Other species that have either been used as rootstocks or used in rootstock breeding programs include: *P.* x *dawyckensis* Sealy, *P. incisa* Thunb., *P. concinna* Koehne, *P. serrulata* Lindl., *P. subhirtella* Miq., *P. pseudocerasus*, *P. tomentosa*, and *P. serrula* (Webster and Schmidt 1996). *P. canescens*, which

Table 13.3 World trade in cherries[a]; top ten importing and exporting countries (2001–2005). FAO Statistics at http://faostat.fao.org/site/291/default.aspx September 2007

	Average value (million US $)		
Country	2001–2005	1996–2000	1990–1995
Imports			
Japan	104.1	94.0	86.7
Germany	99.5	94.2	67.4
China	53.0	32.7	12.9
UK	50.6	34.1	19.2
Canada	42.3	18.1	14.2
Italy	23.1	8.9	7.6
The Netherlands	22.4	17.1	16.7
USA	22.3	7.1	3.7
RussianFederation	17.1	3.0	1.3
Austria	16.6	10.9	4.0
World	554.7	392.2	292.5
Exports			
USA	179.7	131.6	95.6
Turkey	52.0	24.0	8.7
Chile	36.5	15.4	8.1
Spain	31.6	23.8	11.6
Austria	27.1	4.3	0.4
France	21.9	16.2	21.4
Italy	19.6	27.1	16.2
Hungary	13.1	20.8	16.3
The Netherlands	11.1	6.5	3.2
Belgium	9.5	1.6	0.5
World	476.5	323.1	231.4

[a]Includes sweet and sour cherries.

is native to Central and Western China, has scarcely been used as cherry rootstock. *P. fruticosa* is a small shrub from Eastern Europe and its native area overlaps with *P. avium* and *P. mahaleb*, which in certain years allows for natural hybridization (Kárpáti 1944; Wojcicki 1991; Hrotkó and Facsar 1996). *P. mahaleb* occurs in great diversity and has been classified by Terpó (1968) into several subspecies. These include the subspecies mahaleb, which is known as small-leaved Mahaleb, and the subspecies simonkaii (Pénzes) Terpó, known as broad-leaved Mahaleb, which is more adapted to a continental climate. One rootstock for cherries, Adara (*P. cerasifera* Ehrh.), is from the subgenus Prunophora (Moreno et al. 1996).

Sweet cherries have been further divided into subgroups based on fruit color, shape, and texture (Webster 1996). The groups are black geans, amber geans, hearts, and bigarreaux. Geans have heart-shaped fruit with tender flesh, with black geans having dark colored flesh and amber geans light yellow and translucent flesh and skin. Bigareaux have light-colored skin with hard, cracking flesh. Hearts are dark in color with flesh texture in between Geans and Bigarreaux. Sour cherries can be divided based on skin and juice color and fruit shape, into either Morellos or Amarelles. Fruit with red to dark red-colored juice are described as Morellos

(Griottes, Weichsel). The Morello fruit is also very dark red with spherical or cordate shape. The cherries with colorless juice are the Amarelles (Kentish) and they have pale red fruits with more or less flattened shape. Duke cherries are thought to be hybrids between sweet and sour cherries with dark red skins and semiacid juice. They have been currently named *P.* x *gondouinii* Rehd. (Faust and Suranyi 1997; Saunier and Claverie 2001; Tavaud et al. 2004). An additional division described by Hedrick (1915) are the Marasca cherries. This cherry is native to Dalmatia in Croatia, where the tree grows wild and now is sparingly cultivated. The tree is vigorous and the fruit are small, deep red or almost black in color and have deep red flesh and juice. The tree and fruit characteristics of the Danish local cultivar Stevnsbear are very similar, and it is possible that Stevnsbear originated from the Marasca cherry (Stainer 1975).

In general, cherry root systems are not well adapted to poorly drained or wet soils. Some soil-related issues can be managed by choice of rootstock. For example Mahaleb is used where drought tolerance is required, and Mazzard is often used where poorer drainage is known to occur. Cherries require a warm growing period with minimal rain during the fruit ripening period, especially for sweet cherries to reduce the amount of rain-induced fruit cracking. Cherries also need a cool period to allow trees to meet their chilling requirement. Chilling requirements for cherry cultivars generally range between 750 and 1,400 h (Seif and Gruppe 1985). In general, sour cherries tend to have somewhat higher chilling requirements than sweet cherries (Thompson 1996). No sources of low-chilling sweet cherries for subtropical production have been found in *P. avium* (Sherman and Lyrene 2003). In areas of inadequate chilling, temperate fruit culture has depended on chemical sprays to stimulate bud burst and thus compensate for incomplete chilling. Extremely cold winter temperatures can limit production of cherries. Fully dormant sweet cherries can withstand temperatures as low as −29°C (Proebsting 1970). Sweet cherry cultivars vary in their susceptibility to low winter temperature damage (Kadir and Proebsting 1994). Spring frosts also limit areas that are suitable for growing cherries. Killing temperatures vary depending on bud development and cultivar (Ballard et al. 1997). High summer temperatures at the time of transition from sepal to petal differentiation will lead to double pistils (Beppu et al. 2001) and will result in fruit doubles or spurs. Southwick et al. (1994) suggest a temperature above 22°C during this sensitive stage of floral differentiation is associated with abnormal fruit the following season for 'Bing' sweet cherries. Doubling potential varies among sweet cherry cultivars (Micke et al. 1983).

2 Origin and Domestication of Scion Cultivars

Numerous reviews and reports outline the origins and domestication of sweet and sour cherries (Hedrick 1915; Faust and Suranyi 1997; Webster 1996; Watkins 1976; Iezzoni et al. 1990; Brown et al. 1996). It has been suggested that sweet cherry originated in an area south of the Caucasian mountains with a secondary dissemination into Europe [De Candolle (1886) in Faust and Suranyi 1997]. Watkins (1976)

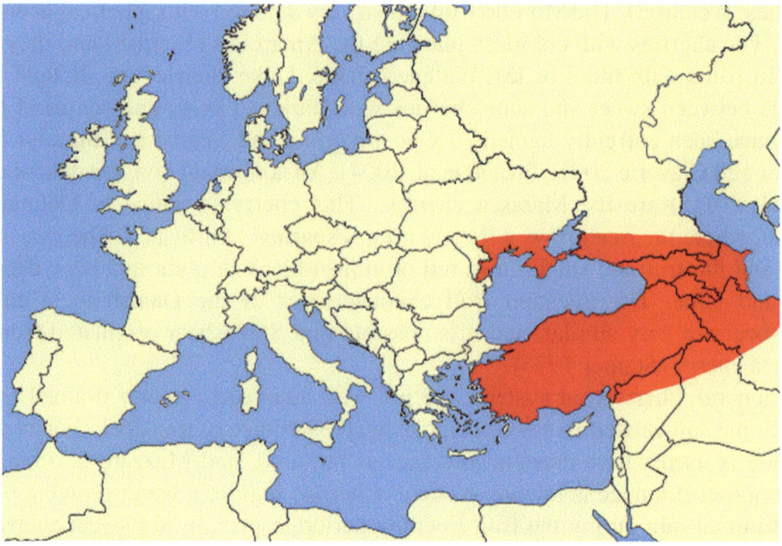

Fig. 13.1 Center of diversity of sweet (*Prunus avium*) and sour (*P. cerasus*) cherry. Centered around Asia minor, northern Iran, Iraq, Syria, the Ukraine, and other countries south of the Caucasus mountains

suggests that the first diploid *Prunus* species occurred in central Asia. Webster (1996) has reported that sweet cherries are indigenous to northern Iran, the Ukraine and other countries to the south of the Caucasus Mountains (Fig. 13.1). Also, it is native to Europe and originated in an area close to the Caspian and Black Seas. Webster (1996) further reports that there are conflicting opinions on the origin of sour cherries. He cites De Candolle (1886) suggesting that sour cherry originated from the same area as sweet cherries. Other authorities (Hedrick 1915) suggested that the area should include the area from Switzerland to the Adriatic Sea, and from the Caspian Sea to the north of Europe. Olden and Nybom (1968) suggest that the-sour cherry originated as a hybrid between ground cherry (*P. fruticosa*) and sweet cherry. Isozyme analysis, genomic in situ hybridization and karyotype analysis support the hybrid origin of *P. cerasus* (Hancock and Iezzoni 1987; Santi and Lemoine 1990; Schuster and Schreiber 2000). Beaver and Iezzoni (1993) investigated the inheritance of seven enzyme loci in sour cherry and confirmed the disomic inheritance as a feature of the allotetraploid hypothesis for sour cherry. Brettin et al. (2000) reported that for the most part the chloroplast genome of sour cherry, which is maternally inherited, is derived from ground cherry. Brown et al. (1996) reported that sweet, sour, and ground cherry originated in the area that includes Asia Minor, Iran, Iraq, and Syria. The spread of both sweet and sour cherry from the centers of origin was accomplished by animals, birds, and humans. Hedrick (1915) writes that Theophrastus was the first of the Greek writers to mention cherry about 300 years before the Christian era. Pliny suggests that Lucullus brought cherries back to Italy when he returned from the Pontus region in Turkey.

Historically most of the sweet cherry cultivars were developed by astute growers and nurseries in the various sweet cherry growing regions of the world (Bargioni 1996). More recently, the use of cultivars from controlled crosses of known parents has gained increasing importance (Table 13.4). Early sour cherry cultivars were developed from superior selections that were propagated by suckers and the primary emphasis was on collecting better strains of local cultivars (Iezzoni 1984). This led to the development of land races in Eastern Europe that include 'Cigany,' 'Pandy,' 'Oblačinska,' 'Mocanesti,' 'Strauchweichsel,' 'Weinweichsel,' 'Stevnsbaer,' and 'Vladimirskaya' (Faust and Suranyi 1997). These local landraces indicate the rich genetic diversity in Europe (Iezzoni 1996).

Table 13.4 Cherry cultivars released since the since the mid-1990s

Cultivar	Country	Parentage/origin
Sweet cherry		
Aida	Hungary	Moldvai fekete × seedling of Germersdorfi open pollinated
Alex	Hungary	Van × John Innes 2420
Alma	Germany	Rube × Allers Späte
Andersen™ (NY9295)	The USA	Wederscher Markt open pollinated
Andy G's Son	The USA	Sport of Early Burlat
Anita	Hungary	Trusenszkaja × seedling of Germersdorfi open pollinated
Anu	Estonia	Leningradskaya Chernaya open pollinated open pollinated
Aranka	The Czech Republic	Early Rivers × Moreau
Arthur	Estonia	Krasavitsa open pollinated
Bellise	France	Starking Hardy Giant × Burlat
Bianca	Germany	Rube × Allers Späte Knorpel
BlackGold™ (Ridgewood)	The USA	Starks Gold × Stella
Black York™ (cv. Haas)	The USA	Giant × Emperor Francis
Black Star	Italy	Lapins × Burlat
Blaze Star	Italy	Lapins × Durone compatto di Vignola
BlushingGold™ (cv. Pendleton)	The USA	Yellow Glass × Emperor Francis
Carmen	Hungary	Sárga Dragán × seedling of Germersdorfi open pollinated
C t lina	Romania	Parents unknown
Cashmere	USDA	Stella × Early Burlat
Celeste (*Sumpaca* Celeste)	Canada	Van × Newstar
Cet uia	Romania	Parents unknown
Chelan	The USA	Stella × Beaulieu
Christiana	The Czech Republic	Van × Kordia
Columbia™	The USA	Stella × Beaulieu
Cristalina (*Sumnue* Cristalina)	Canada	Star × Van
Dame Nancy	Australia	Stella open pollinated
Dame Roma	Australia	Black Douglas × Stella
Early Bigi® Bigi Sol	Italy	Parents unknown

(continued)

Table 13.4 (continued)

Cultivar	Country	Parentage/origin
Earlise® Rivedel	France	Starking Hardy Giant × Burlat
Earlisweet	The USA	Stella open pollinated
Early Garnet™ (Early Red)	The USA	Garnet × Ruby
Early King	The USA	Sport of King
Early Korvik	The Czech Republic	Mutant of Korvik (Kordia × Vic)
Early Robin™ (cv. Doty)	The USA	Whole tree mutation of Rainier
Early Star® Panaro 2	Italy	Burlat × Stella
Elle	Estonia	Juku open pollinated
Erika	Germany	Rube × Steckmanns Bunte
Fabiola	The Czech Republic	Van × Kordia
Ferbolus	France	Hedelfinger × Reverchon
Fercer	France	Starking Hardy Giant × o.p.
Ferdelice	France	Parents unknown
Ferdiva	France	Fercer × o.p.
Ferdouce	France	Rainier × Fercer
Ferlizac	France	Parents unknown
Fermina	France	Vittoria × clone INRA
Ferobi	France	Burlat × Fercer
Ferpin	France	Parents unknown
Fertard	France	Sunburst open pollinated
Folfer	France	Fercer × o.p.
Giant Ruby ™ (Giant Red)	The USA	Large Red × Ruby
Glacier™	The USA	Stella × Burlat
Glenare	The USA	Tulare open pollinated
Glenred	The USA	Tulare × Brooks
Glenrock	The USA	Tulare open pollinated
Golia	Romania	Parents unknown
Grace Star	Italy	Burlat open pollinated
Gronkavaya	Estonia	Severnaya × pollen mixture
Halka	The Czech Republic	Van × Stella
Horka	The Czech Republic	Van open pollinated
Index™	The USA	Stella × unknown
Irma	Estonia	Leningradskaya Chernaya open pollinated open pollinated
Jaama Maguskirss	Estonia	Dönissens Gelbe Knorpelkirsche × Kozlovskaya
Johanna	Germany	Schneiders Späte Knorpel × Rube
Justyna	The Czech Republic	Kordia × Starking Hardy Giant
Jacinta	The Czech Republic	Veag × o.p.
Karmel	Estonia	Norri open pollinated
Kasandra	The Czech Republic	Burlat × Sunburst
Kavics	Hungary	Germersdorfi óriás × Budakalászi
Kiona	The USA	Glacier × Cashmere
Kordia	The Czech Republic	Parents unknown
Kristiina	Estonia	Krasavita × unknown
Krupnoplodnaja	Ukraine	Big. Napoleon blanc × mix (Valerij Tschkalov + Elton + Jaboulay)

(continued)

Table 13.4 (continued)

Cultivar	Country	Parentage/origin
Lala Star	Italy	Compact Lambert × Lapins
Late Garnet™ (Firm Red)	The USA	Large Red × Garnet
Liberty Bell™	The USA	(Rainier × Bing) × Stella
Lodi™ (Large Red)	The USA	Hardy Giant × Berryessa
Lovranska	Croatia	Local selection
Maria	Romania	Parents unknown
Marina	Romania	Parents unknown
Marta	The Czech Republic	Kordia × Early Rivers
Meelika	Estonia	Leningradskaya tschernaya × unknown
Minnie Royal	The USA	Seedling 6HB480 open pollinated
Nadino	Germany	Spansche Knorpel open pollinated
Namare	Germany	Große Schwarze Knorpel open pollinated
Namati	Germany	Bopparder Kracher open pollinated
Namosa	Germany	Farnstädter Schwarze oen pollinated
Naprumi	Germany	Hedelfinger × St. Charmes
Nies Red	The USA	Parents unknown
Nord	Estonia	Leningradskaya open pollinated
Norri	Estonia	Leningradskaya Chernaya open pollinated open pollinated
Nugent™ (NY518)	The USA	Germorsdorfer open pollinated
Oktavia	Germany	Schneiders Späte Knorpel × Rube
Olympus™	The USA	Lambert × Van
Paulus	Hungary	Burlat × Stella
Penny	UK	Colney × Inga
Petrus	Hungary	Burlat × Stella
Piret	Estonia	Norri open pollinated
Red Crystal	The USA	Chance seedling
Redlac	The USA	Budsport of Rainier
Regina	Germany	Schneiders Späte Knorpel × Rube
Rita	Hungary	Trusenszkaja 2 × seedling of Germersdorfi open pollinated
Royal Dawn	The USA	Seedling 32 G500 open pollinated
Royal Kay	The USA	Seedling 13HA431 open pollinated
Royal Rainier	The USA	Stella open pollinated
Samba (*Sumste* Samba)	Canada	2S-84-10 × Stella
Sándor	Hungary	Burlat × Stella
Sandra Rose	Canada	2C-61-18 × Sunburst
Santina	Canada	Stella × Summit
Satin	Canada	Lapins × 2N-39-05
Scarlet	The USA	Chance seedling
Sentennial	Canada	Sweetheart open pollinated
Sequoia™ (cv. Glenoia)	The USA	Unnamed seedling open pollinated
Simcoe	The USA	Stella × (Hollander or Starking Hardy Giant)
Sir Don	Australia	Black Douglas × Stella

(continued)

Table 13.4 (continued)

Cultivar	Country	Parentage/origin
Sir Douglas	Australia	Stella × Vega
Sir Hans	Australia	Stella × Vega
Sir Tom	Australia	Black Douglas × Stella
Skeena	Canada	2N-60-07 × 2N-38-32
Sonata (*Sumleta* Sonata)	Canada	Lapins × 2N-39-05
Sovereign	Canada	Sweetheart open pollinated
Staccato™ (cv. 13S2009)	Canada	Sweetheart open pollinated
Stardust™	Canada	2N-63-20 × Stella
Sweet Early® Panaro 1	Italy	Burlat × Sunburst
Sweetheart	Canada	Van × Newstar
Summer Jewel	Canada	2C-61-18 × 2D-28-30
Sunset Bing™ (cv. Brown)	The USA	Branch mutation of Bing
Sylvana	The Czech Republic	Parents unknown
Symphony	Canada	Lapins × Bing
Tamara	The Czech Republic	Krupnoplodnaja × Van
Těchlovan	The Czech Republic	Van × Kordia
Tehranivee	Canada	Van × Stella
Tieton	The USA	Stella × Early Burlat
Tontu	Estonia	Norri open pollinated
Vanda	The Czech Republic	Van × Kordia
Valerij Tschkalov	Ukaine	Rozovaya open pollinated
Valeska	Germany	Rube × Stechmanns Bunte
Vandalay	Canada	Van × Stella
Vera	Hungary	Ljana [Trusenszkaja 6] × Van
Vilma	The Czech Republic	Kordia × Vic
Viola	Germany	Schneiders Späte Knorpel × Rube
WhiteGold™ (Newfane)	The USA	Emperor Francis × Stella
0900 Ziraat	Turkey	Local selection
Sour Cherry		
Achat	Germany	Köröser × (Fanal × Kelleriis 16)
Balaton™ (Bunched of Újfehértói or Ujfehertoi fürtös)	The USA/Hungary	Local selection
Ciganymeggy clones 7, 59, 404(syn. Gypsy)		Selection of Cigany
Coralin	Germany	Kelleriis 16 × (Köröser × Schattenmore lle)
Csengödi		Landrace selection
Danube™ (Erdi bötermö)	The USA/Hungary	Pandy × Nagy Angol
De Botoşani	Romania	Parents unknown
Debreceni bõtermõ	Hungary	Landrace selection
Eva	Hungary	Local selection
Fanal (syn. Heimanns Konservenkirsche)	Germany	Local selection
Gerema	Germany	Kelleriis 14 × open pollinated
Habunt	Germany	Valeska × Sunburst
Hamid	Germany	Kordia × Regina

(continued)

Table 13.4 (continued)

Cultivar	Country	Parentage/origin
Ideal	Russia	*P. chamaecerasus* × *P. pennsylvanica*
Jachim	Germany	Köröser × Safir
Jade	Germany	Köröser × Röhrigs Weichsel
Jagoli	Estonia	Kose Kirss open pollinated (treated with colchicine)
Jubiljenaja	Russia	Ostheimer Weichsel open pollinated
Kantorjanosi	Hungary	Local selection
Karneol	Germany	Köröser × Schattenmorelle
Komsomolskaja	Russia	Ideal × Tschernij Orel (sweet cherry)
Korund	Germany	Köröser × Schattenmorelle
Krassa sewera	Russia	Vladimirskaja ranaja × Winklers Weiße (*P. avium*)
Kütahya	Turkey	Local selection
Lucyna	Poland	English Morello × Shirpotreb
Mailot	Germany	Große Lange Lotkirsche × Rote Mai
Maliga emleke	Hungary	Pandy × Eugenia
Mari Timpuri	Romania	Local selection
Morina	Germany	Köröser × Reinhardts Ostheimer
Narana	Germany	Knauffs × Souvenir de Charmes
Nesjabkaja	Russia	Ideal × Krassa sewera
Nordia	Sweden	Tschernokorka × BPr24179 (Vladimir O-241 × Brysselska Bruna)
Pamjat Vavilova	Russia	Seedling of unknown cultivar
Petri	Hungary	Local clone selection Ujfehertoi fütös
Piramis	Hungary	[Pandy × a Hungarian local sweet cherry] × Meteo korai
Pitic de Iasi	Romania	Parents unknown
Plodorodnaja Mitschurina	Russia	Selection of Mitschurinskaja karlikowaja
Poljevka	Russia	Ideal open pollinated
Polshir	Russia	Ideal × Plodorodnaja
Rubellit	Germany	Köröser × Schattenmorelle
Sabina	Poland	English Morello × Shirpotreb
Safir	Germany	Schattenmorelle × Fanal
Schukovskaja	Russia	seedling of unknown cultivar
SK Carmine Jewel	Canada	Kerr's Easypick × Northstar
Spinell	Germany	Köröser × (Fanal × Kelleriis 16)
Standart Ural	Russia	Parents unknown
Studentskaja	Russia	Schukovskaja × Schirpotrep tschernaj
Suda	The USA	Schattenmorelle open pollinated
Surefire	The USA	Borchert Black Sour × (Pichmorency × Schattenmorelle)
Tamaris	Russia	Schirpotrep tschernaj open pollinated
Timpuruiu de Osoi	Romania	Parents unknown
Topas	Germany	Fanal × Kelleriis 16
Turgenjevka	Russia	Schukovskaja open pollinated

(continued)

Table 13.4 (continued)

Cultivar	Country	Parentage/origin
Tschernokorka	Russia	local selection
Uralnaja Rubinovaja	Russia	seedling of unknown cultivar
Wanda	Poland	Nefris × Wolynska
Rootstock		
GiSelA® 1	Germany	*P. fruticosa* Klon 64 × *P. avium* (tested as Gi 172–9)
GiSelA® 4	Germany	*P. avium* × *P. fruticosa* (tested as Gi 473–10)
GiSelA® 5	Germany	*P. cerasus* Schattenmorelle × *P. canescens* (tested as Gi 148–2)
GiSelA® 6	Germany	*P. cerasus* Schattenmorelle × *P. canescens* (tested as Gi 148–1)
GiSelA® 7	Germany	*P. cerasus* Schattenmorelle × *P. canescens* (tested as Gi 148–8)
GiSelA® 8	Germany	*P. cerasus* Schattenmorelle × *P. canescens* (tested as Gi 148–9)
GiSelA® 10	Germany	*P. fruticosa* Klon 64 × *P. cerasus* (tested as Gi 173-9_
GiSelA® 11	Germany	*P. canescens* × *P. cerasus* Leitzkauer (tested as Gi 195–1)
GiSelA® 12	Germany	*P. canescens* × *P. cerasus* Leitzkauer (tested as Gi 195–2)
GiSelA® 3 (GI®2091)	Germany	*P. cerasus* Schattenmorelle × *P. canescens* (tested as Gi 209–1)
Krymsk® 5 (cv. VSL-2)	Russia	*P. fruticosa* × *P. serrulata* var. lannesiana
Krymsk® 6 (cv. LC-52)	Russia	*P. cerasus* × (*P. cerasus* × *P. maackii*)
Myrobalan RI-I	The USA	*P. cerasifera* open pollinated
P-HL-A	The Czech Republic	*P. avium* (Mazzard open pollinated)
P-HL-B	The Czech Republic	*P. avium* (Mazzard open pollinated)
P-HL-C	The Czech Republic	*P. avium* (Mazzard open pollinated)
Piku® 1	Germany	*P. avium* × (*P. canescens* × *P. tomentosa*)
Piku® 3	Germany	*P. pseudocerasus* × (*P. canescens* × *P. incisa*)
Piku® 4	Germany	*P. Schattenmorelle* × *P. Kursar*
UCMH 55	The USA	*P. mahaleb* open pollinated
UCMH 56	The USA	*P. mahaleb* open pollinated
UCMH 59	The USA	*P. mahaleb* open pollinated
Victor®	Italy	*P. cerasus*

3 Genetic Resources

The most productive cherry trees with the highest qualities were selected through the ages by peasants and gardeners (Iezzoni et al. 1990). These were propagated by root suckers and eventually grafting. These trees represent a great deal of genetic diversity, especially for adaptation and have been used in European and other breeding programs. Cai et al. (2007) report a rich source of cherry germplasm in the mountainous areas of China; however, there is little to no information available regarding this resource. Evaluation of *P. avium* and *P. cerasus* germplasm in their center of origin also needs to be completed. Kolesnikova (1975) reports two ecological groups of sour cherries, the Western Europe group, characterized by lower winter hardiness, and the Russian group that is better adapted to colder winters. Hillig and Iezzoni (1988) maintained that there were not two distinct groups, but rather a continuous range of variation.

Wünsch and Hormaza (2002) grouped 23 ancient sweet cherry cultivars using SSR sequences into two main clusters. One group contained the genotypes from southern Europe and the other from northern Europe. A low level of polymorphism in sweet cherry has been detected using RAPD markers (Stockinger et al. 1996; Gerlach and Stösser 1998), isozyme markers (Beaver et al. 1995; Boškoviú and Tobutt 1998), AFLP analysis (Zhou et al. 2002), and SSR sequences (Wünsch and Hormaza 2002), which probably reflects a narrow genetic base in sweet cherry germplasm. Choi and Kappel (2004) have shown that the four North American breeding programs are based on only five founding cultivars. These results suggest that the genetic base of sweet cherry breeding in North America has been narrowed to an alarming level. Beaver et al. (1995) suggest that sour cherry and other tetraploid cherry species are more polymorphic than sweet cherry. Further they suggest that sweet, sour, and ground cherry share a common gene pool and share alleles through introgression. Arús (2007) suggests that there is a single *Prunus* genome shared by all the species studied to date.

Currently, sour cherry growing is dominated by a small collection of cultivars. In most cases these cultivars are landraces or clonal selections of regional cultivars. In Middle Europe the main sour cherry cultivar is 'Schattenmorelle' with various local synonyms, ('Łutovka' in Poland, 'Griotte du Nord' or 'Griotte Noir Tardive' in France, and Benelux and occasionally 'English Morello' in Great Britain).

This cultivar is self-compatible and highly productive with dark red fruits and juice. The origin of this cultivar is likely the Chateau de Moreille in France. The cultivar 'Montmorency' dominates sour cherry production in the USA. The origin of this 400-year-old cultivar is France. 'Montmorency' is self-compatible and highly productive with bright red fruit with clear juice. The landrace cultivar 'Pandy' (syn. 'Crisana,' 'Köröser') and related cultivars are most popular in Hungary and Romania. 'Pandy' is self-sterile and has excellent fruit quality with light-red skin and juice.

New sour cherry cultivars in Germany were selected in two different breeding programs. The program at the Max Planck Institute in Köln-Vogelsang, Zwintzscher (1973) selected cultivars from self-pollinated 'Schattenmorelle' seedling populations. Wolfram (2000) and Schuster and Wolfram (2004) in Dresden successfully selected cultivars of which 'Köröser' was one parent. In Hungary, Romania, and Serbia, many new cultivars resulted from regional clonal selections of the landraces 'Pandy,' 'Mocanesti' and 'Oblačinska' respectively, or are hybrids between landraces. Cultivars released in Russia and Canada may be interspecific hybrids with

P. fruticosa because of the need to incorporate cold hardiness (Zhukov and Charitonova 1988; Bors 2005). In the USA new cultivars were realized from crosses of European sour cherry cultivars.

P. avium rootstock breeding has produced selected seed trees and vegetatively propagated clones such as F12/1. The original selections were made from progenies of native forest trees (Mazzard) (Webster and Schmidt 1996). Selected clones of landraces and cultivars of *P. cerasus* are in use as rootstocks. Many dwarfing rootstocks belong to this species or have been derived from it as hybrids. *P. canescens* has proven to be a promising parent for rootstock breeding (Trefois 1980; Gruppe 1985; Wolfram 1996). *P. fruticosa* has been utilized in several rootstock breeding projects (Cummins 1972; Plock 1973; Hein 1979; Gruppe 1985; Hrotkó 2004; Rozpara and Grzyb 2005). The subgenus *Mahaleb* contributes one species to rootstock development, *P. mahaleb*. It is a major rootstock in Central and Southern European countries as well as in Asia Minor, Central Asia, and China.

Various *Prunus* species have potential usefulness in breeding programs (Table 13.5), but there has been very little interspecific hybridization for scion cultivars. Any interspecific hybrids that have been made have been limited to the following crosses between *Prunus* species to develop rootstocks (Iezzoni et al. 1990; Webster and Schmidt 1996): *P. avium* × *P. pseudocerasus*; *P. incisa* × *P. serrula*; *P. cerasus* × *P. maackii*; *P. cerasus* × *P. avium*; *P. cerasus* × *P. canescens*; *P. cerasus* × *P. fruticosa*.

4 Major Breeding Achievements

4.1 Scion

Self-fertility: The development of self-fertility has had significant impact in the development of production around the world and must be considered one of the major achievements in the breeding of sweet cherries. Sweet cherries are normally self-incompatible, having a gametophytic self-incompatibility system that requires cross-pollination with a cultivar from a different incompatibility group. Crane and Brown (1937) first identified 11 incompatibility groups and growers needed to plant suitable numbers of an appropriate pollinizing (cross compatible) cultivar to ensure adequate cropping. With the development of self-fertile cherries (Lewis and Crowe 1954) the possibility of larger blocks of single cultivars became a reality. However, the greatest benefit of self-fertile cultivars is the potential for consistent cropping, even in years when pollination conditions may not be favorable. The first commercial self-fertile cultivar was ‘Stella’ and was released in 1968 by the Canada Department of Agriculture Research Station at Summerland, British Columbia (Lapins 1971). Currently, all commercially released self-fertile cultivars, except the Hungarian cultivar ‘Alex’ (S_3S_3'), can trace their ancestry back to Stella (Sansavini and Lugli 2005; Lang et al. 1998b; Granger 1998; Kappel 2002; Kappel et al. 2000a, b, 2006; Apostol 2005).

Table 13.5 Systematic classification of cherry species that may have potential for genetic improvement of cherry (scions and rootstocks) according to Rehder (1974) and Iezzoni et al. (1990)

Subgenus	Section	Species	Distribution	Most frequent chromosome number
CERASUS Pers.	*Microcerasus* Webb	*P. besseyi* Bailey	Canada, the USA	16
		P. japonica Thunb.	C. China, E. Asia	16?
		P. pumila L.	The USA	16
		P. tomentosa Thund.	N. & W. China, Japan, and the Himalayas	16
	Pseudocerasus Koehne	*P. incisa* Thunb.	Japan	16
		P. kurilensis (Miyabe) Wils.	Japan	16
		P. nipponica Matsum	Japan	16
		P. serrulata Lindl.	Japan, China, Korea	
		P. subhirtella Miq.	Japan	
	Lobopatalum Koehne	*P. pseudocerasus* Lindl.	N. China	32
	Cerasus Koehne	*P. avium* L.	Europe, W. Asia, Caucasus	16
		P. cerasus L.	W. Asia, S.E. Europe	32
		P. fruticosa Pall.	C. & E. Europe, Siberia	32
		P. canescens Bois	C. & W. China	16
	Mahaleb Focke	*P. mahaleb* L.	Europe, W. Asia	16
		P. pensylvanica L.	Canada, the USA	16
PADUS (Moench) Koehne		*P. maackii* Rupr.	Manchuria, Korea	32
		P. padus L.	Europe, N. Asia, Korea, Japan	32
		P. serotina Ehrh.	Canada, the USA	32
		P. virginiana L.	Canada, the USA	32

P. fruticosa × *P. avium*; *P. subhirtella* × *P. yedoensis*; *P. mahaleb* × *P. avium; P. avium* × *P. kurilensis; P. avium* × *P. incisa; P. canescens* × *P. incisa; P. canescens* × *P. tomentosa; and P. cerasus* × *P. pensylvanica*

Fruit size: Fruit size has become a determinant of price in today's global sweet cherry market and growers now consider fruit size of new sweet cherry cultivars as a priority attribute (Omeg and Omeg 2005). Recent introductions by fruit breeders have achieved significant fruit size gains (Lapins 1974; Lane and Schmid 1984; Lang et al. 1998b; Lang 1999; Sansavini and Lugli 2005; Kappel et al. 2000a, b, 2006; Lang 2002). These include 'Glacier,' 'Grace Star,' 'Regina,' 'Summit,' 'Sunburst,' 'Skeena,' 'Samba,' and 'Tieton.'

Firmness: The level of firmness that a new cultivar requires now is significantly higher than standard cultivars (Looney et al. 1996). Many of the new cultivars being released are firmer than traditional standard cultivars (Sansavini and Lugli 2005; Kappel and Lane 1998; Kappel 2005). 'Bing,' a standard cultivar, has a fruit firmness of 170 g/mm, whereas 'Sweetheart' has a firmness of 299 g/mm (Kappel 2005).

Extending maturity date: Growers wish to produce cherries outside the peak production periods to take advantage of higher market prices. This has been a high priority for many breeding programs and a wider maturity range has supported increased planting of cherries (Sansavini and Lugli 2005; Lang et al. 1998b; Kappel et al. 1998, 2006).

4.2 Rootstocks

Cherry rootstocks are predominantly seedling rootstocks but there is growing interest in new clonal rootstocks with the potential of greater vigor control of the scion. The eventual goal is to develop productive orchards that are "pedestrian orchards", that is, orchards where the bulk of the work can be completed without the use of ladders.

Seed tree selections: Seed sourced/mother trees selected for superior phenotypic traits have been released from several countries (Table 13.6). Advantages of seed orchards include the potential for a virus-free seed source, higher germination capacity, hybrid seed of known parents, and improved uniformity of orchard trees compared to open-pollinated seeds.

Clonal Rootstocks: Most vegetatively propagated rootstocks have been derived from three cherry species, *P. avium*, *P. mahaleb*, and *P. cerasus* (Tables 13.7–13.9). The main advantage of clonally selected rootstocks of *P. avium* such as F12/1 and Charger was uniform plant material for fruit growers and ease of propagation in stoolbeds for nurseries. These rootstocks do not improve vigor control or precocity of scions (Webster and Schmidt 1996). Interspecific hybrids with *P. avium* (Table 13.7) provide a wide range of vigor, adaptability, and tolerance to diseases (James et al. 1987; Grzyb et al. 2005; Wolfram 1996; Hrotkó et al. 2009). Vegetatively propagated, interspecific *P. cerasus* clones have proven to be the most promising. Major disadvantages of this group of rootstocks are graft incompatibility and root suckering (Perry 1987; Granger 2005). Recent breeding projects in several countries selected clones from landraces of sour cherry and resulted in nonsuckering

Table 13.6 Selected seed tree clones of Mazzard and Mahaleb cherry

Country	Selected seed tree clones	References
Mazzard		
Bulgaria	IK (Plovdiv), N 123 (Dryanovo)	Webster and Schmidt (1996)
The Czech Rep.	P-TU 1, 2, 3	Anonymous (2003)
France	Pontavium (Fercahun), Pontaris (Fercadeu)	Charlot, et al. (1998)
Germany	Hz 170, Hz 53 (clonal derivatives of Limburger); Gi 81, 84, 90, 94; Alkavo (K 2/4, 4/2, 4/23, 5/28, 5/38)	Funk (1969)
		Küppers (1978)
		Webster and Schmidt (1996)
Hungary	C 2493	Nyújtó (1987)
Ukraine	Mazzard Nr. 3,4,5; Susleny and Napoleon	Yoltuchovski (1977)
The USA	Mazzard Nr 570, Sayler, OCR 1	Perry (1987)
Romania	F 12/1, Dönissens Gelb (cross-pollinated)	Webster and Schmidt (1996)
Mahaleb		
France	SL 405 (self-fertile)	Claverie (1996)
Germany	Heimann X. (self-fertile); Alpruma (AF 5/19, AF 3/9, AF 6/16 and PB 9)	Heimann (1932), Funk (1969), Küppers (1978)
Hungary	C 500 (Cema), C 2753 (Cemany), Érdi V. (cross-pollinated); Korponay (self-fertile)	Nyújtó (1987), Hrotkó (1990, 1993, 1996)
Ukraine	Mahaleb N 24	Tatarinov and Zuev (1984)
The USA	Nos 902, 904, 908, 916, (as Mahaleb 900)	Perry (1987)
Moldavia	Rozovaya prodolgovataya, Chernaya Kruglaya iz Bykovtsa, Nr 1 iz Solonchen	Yoltuchovski (1977), Tatarinov and Zuev (1984)

Table 13.7 Vegetatively propagated *P. avium* rootstocks and derivatives (*P. avium* as female parent)

Cultivar	Brief description	References
Alkavo	Vigorous *P. avium* selection	Webster and Schmidt (1996)
F 12/1	Vigorous, resistant to bacterial canker	
Charger	Vigorous resistant to bacterial canker	Webster and Schmidt (1996)
Cristimar	Land race selection, reduced vigorous	Cireasa et al. (1993)
Interspecific Hybrids		
Colt	*P. avium* × *P. pseudocerasus*, $2n=24$; easy to propagate, 80% vigor, flat branching, limited adaptability to drought and lime soils	Webster (1980)
Hexaploid Colt	*P. avium* × *P. pseudocerasus*, $6n=48$; easy to propagate, 75% vigor.	James, et al. (1987), Webster et al. (1997)
P-HL-A	Supposedly *P. avium* × *P. cerasus*; promising dwarf rootstock in the Czech Republic and Poland, limited soil adaptability	Blažková and Hlušičkova (2004), Grzyb et al. (2005)
Piku 1	*P. avium* × (*P. canescens* × *P. tomentosa*), moderate vigor, high productivity, adaptability. Tolerant to PDV and PNRSV.	Wolfram (1996), Hilsendegen (2005), Lankes (2007), Hrotkó et al. (2009)
GiSelA 4	Gi 473/10, *P. avium* × *P. fruticosa*, dwarf, suckers badly	Gruppe (1985), Stehr (2005), Kappel et al. (2005), Hrotkó et al. (2006)

Table 13.8 Vegetative propagated *P. cerasus* rootstocks and derivatives (*P. cerasus* as female parent)

Cultivar	Brief description	References
CAB—6 P	Land race selection from Italy, moderate dwarfing, few suckers, shallow roots	Faccioli et al. (1981)
CAB 11—E		Sansavini and Lugli (1996)
Masto de Montagna	Land race selection from Spain, moderate dwarfing, few suckers	Jiménez et al. (2004)
Weiroot 10 Weiroot 13	Land race selection from Germany, vigorous, few suckers, adaptability to clay soils, good compatibility and fruit size	Schimmelpfeng (1996), Treutter et al. (1993), Hrotkó et al. (2006)
Weiroot 154 Weiroot 158	Hybrids of land races selected in Germany, semidwarf, few suckers, adaptability to clay soils, good compatibility and fruit size	Treutter et al. (1993), Schimmelpfeng (1996), Stehr (2005), Bujdosó et al. (2004)
Weiroot 72 Weiroot 53	Hybrids of land races selected in Germany, dwarf, few suckers, variable compatibility, low soil adaptability, poor anchorage	Schimmelpfeng (1996), Treutter et al. (1993), Bujdosó et al. (2004)
Edabriz	Selected from Iranian wild genotypes, dwarfing, suited on fertile loam and clay	Edin et al. (1996), Hrotkó et al. (2007), Hilsendegen (2005)
Victor	Selected in Italy, semidwarf	Battistini and Berini (2004)
Interspecific hybrids of *P. cerasus*		
IP-C1	*P. cerasus* × *P. avium*, sel. Romania, moderate vigorous, less suckers, tolerates wet soil	Parnia et al. (1997)
Piku 4	*P. cerasus Schattenmorelle* × *P. Kursar* moderate vigor, high productivity, adaptability, especially with respect to yield and fruit size on dry and sandy sites without additional irrigation	B. Wolfram (pers. comm.)
GiSelA 3 (Gi 209/1)	*P. cerasus* × *P. canescens*, very dwarf, partially sensitive to PDV and PNRSV	Gruppe (1985); Franken-Bembenek (2004), Lankes (2007)
GiSelA 5 (Gi 148/2)	*P. cerasus* × *P. canescens*, dwarf, tolerates PDV and PNRSV, suited on fertile loam, good compatibility and productivity, precocious, early senescence	Gruppe (1985), Walther and Franken-Bembenek (1998), Franken-Bembenek (2005), Lankes (2007), Hrotkó et al. (2007)
GiSelA 6 (Gi 148/1)	*P. cerasus* × *P. canescens*, semidwarf, partially sensitive to PDV and PNRSV, suited on fertile loam, needs irrigation, precocious	Gruppe (1985), Kappel et al. (2005), Stehr (2005), Lankes (2007)
Gi 195/20	*P. cerasus* × *P. canescens*, semidwarf, good precocity and productivity	Hilsendgen (2005)
GiSelA 7 (Gi 148/8)	*P. cerasus* × *P. canescens*, moderate vigorous, higher soil adaptability, good precocity and productivity	Gruppe (1985), Walther and Franken-Bembenek (1998), Kappel et al. (2005) Hrotkó et al. (2006)

Table 13.9 Vegetatively propagated *P. mahaleb* rootstocks and derivatives (*P. mahaleb* as female parent)

Cultivar	Brief description	References
SL 64	Selected in France from wild genotypes, vigorous, easy to propagate, good compatibility and productivity with sweet and sour cherries	Thomas and Sarger (1965), Claverie (1996), Edin et al. (1996), Hrotkó et al. (1999)
Bogdany	Selected as root sucker of an old and productive sweet cherry tree, vigorous, wide crotch angles, good compatibility and productivity	Hrotkó (1993), Hrotkó and Magyar (2004), Hrotkó et al. 2007)
Egervár Magyar SM 11/4	Moderate vigorous clones selected in Hungary, good compatibility and productivity, wide crotch angles	Hrotkó (1993), Hrotkó and Magyar (2004)
Bonn 60	Vigorous clones selected in Germany, did not get into commercial propagation	Baumann (1977)
Bonn 62		
Interspecific hybrids. *P. mahaleb* × *P. avium*		
MxM 2 and MxM 60	The USA, Oregon, very vigorous, adaptability like on Mahaleb, good compatibility, resistant to *Phytophtora*, narrow crotch angle, more precocious than seedling, good productivity	Westwood (1978), Perry (1987), Hrotkó et al. (2006)
MxM 14 and MxM 97	The USA, Oregon, moderate vigorous, adaptability like on Mahaleb, good compatibility, resistant to *Phytophtora*, narrow crotch angle, more precocious than seedling, good productivity	Westwood (1978), Perry (1987), Edin et al. (1996), Hrotkó et al. (1999, 2006, 2007)

P. cerasus rootstocks with a wide range of vigor and scion compatibility (Table 13.8). In addition, I.P. C1, VG1, and V.V.1 from Romania (Parnia et al. 1997) and VP 1 (a hybrid of *P. cerasus* × *P. maacki*) from the former USSR are being developed. The most productive and extensive interspecific hybridisation project was carried out in Giessen, Germany (Gruppe 1985). Recent results from various national rootstock trials suggest that the most promising hybrids are from the *P. cerasus* × *P. canescens* and reciprocal crosses. Agronomic shortcomings are limiting the commercial potential of a number of candidates, but they are an important genetic resource.

The first *P. mahaleb* clonal rootstock, Sainte Lucie 64 (SL 64), (Table 13.9), was selected in France for its ease of propagation, compatibility with sweet cherries, and productivity in orchard conditions. Further, *P. mahaleb* selections that are low in vigor and are precocious have been developed in Hungary (Hrotkó 1982; Hrotkó and Magyar 2004); Italy including REAL 19, 24, 27B, 48, and 52 (Giorgio and Standardi 1996). Successful hybridization of *P. mahaleb* with *P. avium* was carried out in Oregon (Westwood 1978; Perry 1987) resulting in the MxM series and OCR 2, 3. Crosses between *P. mahaleb* and *P. fruticosa* have been reported by De Palma et al. (1996) and Hrotkó (2004); testing is in early stages.

5 Current Goals/Challenges of Breeding

5.1 Scion

Precocity and productivity: Modern orchard production requires a quicker return on investment, and precocity can help achieve this need. Some of the new dwarfing rootstocks are much more precocious than the traditional standard rootstocks, Mazzard and Mahaleb. Cultivars such as 'Sweetheart' are also very precocious even on the standard rootstocks.

Improved productivity is also extremely important. Many cultivars produce low yields and this makes the economics of cherry orchards difficult. Increased yield must be counterbalanced by fruit quality (fruit size) and pack-outs (Omeg and Omeg 2005); that is, very high yields of small poor quality fruit are uneconomical.

Sour cherries are predominantly used for processing, so high productivity is extremely important. Consequently high fertility is important for a good fruit set. Sour cherries are frequently considered to be self-compatible, although self-incompatible and partially self-compatible cultivars do exist; sour cherry cultivars may have a reduced incompatibility reaction. Redalen (1984) regarded cultivars with a final fruit set of more than 15% as self-compatible because self-incompatible cultivars may set a few fruit. Cultivars with an intermediate final fruit set have been characterized as partly self-compatible. Certain pairs of cultivars are cross-incompatible, reciprocally or unilaterally (Boškovi?et al. 2006). The field of sour cherry fertility is not completely understood. Recent investigations demonstrated that a gametophytic self-incompatibility (GSI) exists in sour cherry (Yamane et al. 2001; Tobutt et al. 2004). This GSI illustrated the occurrence of self-compatible and self-incompatible cultivars in sour cherry, as in sweet cherry. Self-compatibility in sour cherry requires the loss of function for a minimum of two *S*-haplotype specificity components (Hauck et al. 2006).

Resistance to rain-induced cracking: Rain-induced cracking is a major problem in many cherry growing regions of the world. Even in drier growing areas, it can be a major problem in occasional years. Development of cultivars truly resistant to rain-induced cracking has been hampered in the past by a lack of understanding of the mechanisms of cracking. This is compounded by the lack of a good selection tool to evaluate seedlings. The cracking index developed by Verner and Blodgett (1931) or the modified cracking index (Christensen 1972) have not had a consistent relationship with experiences in the field and have been set aside as a selection tool. Brown et al. (1996) suggest that there appears to be a positive relationship between fruit firmness and susceptibility to cracking; however, at Summerland, we have found no such relationship in the field (Kappel et al. 2000c). Over the years, certain cultivars have demonstrated a level of resistance to rain-induced cracking (e.g., 'Summit,' 'Regina,' and 'Lapins'). Recent investigations of Knoche et al. (2001) and Beyer et al. (2005) demonstrated the importance of fruit development and the permeability and water transport through the fruit surface to fruit cracking in sweet cherry.

Unfortunately, a tool to aid in selecting genotypes with reduced susceptibility to cracking has not yet been developed.

Resistance to diseases and insects: Integrated pest management programs and organic production of cherries would benefit from genetic pest and disease resistance. Major diseases include powdery mildew (*Podosphaera oxyacanthae* (DC) d By.), brown rot (*Monilinia* spp.), leaf spot (*Blumeriella jaapii* (Rehm)), bacterial canker (*Pseudomonas* spp.), Cytospora canker (*Leucostoma* spp.), and various viruses. Key insect pests include cherry fruit fly (*Rhagoletis* spp.), black cherry aphid (*Myzus cerasi* Fab.), and cherry slug (*Caliroa cerasi* L.).

Cultivars or wild cherry species resistant to *Monilinia* spp. are unknown. The symptoms caused by *Monilinia* spp. and the degree of susceptibility depend on climatic conditions and the virulence of specific local races (Biggs and Northover 1988; Budan et al. 2005a). Kappel and Sholberg (2008) evaluated a number of cultivars from the breeding program at Summerland, British Columbia and found some slight differences in susceptibility; however, the level of resistance is not high enough for any to be used as parents. Leaf spot is common in most cherry growing areas in North America and Europe. Only a few sour cherry cultivars are tolerant to leaf spot (e.g., 'Morina,' 'Köröser Gierstädt,' 'Hartai,' and 'Karneol'). Wild cherry species (*P. sargentii, P. serrulata* var. *spontanea, P. subhirtellapendula rosea, P. insisa, P. canescens, P. kurilensis, P. nipponica,* and *P. maackii*) and interspecific hybrids showed a high level of resistance to *B. jaapii* (Wharton et al. 2003; Schuster 2004; Budan et al. 2005b). Bacterial canker caused by *Pseudomonas* spp. is an important problem in a number of growing regions. Variability in the pathogen is slowing any progress in breeding for resistance to bacterial canker (Iezzoni et al. 1990). The John Innes Institute released a number of cultivars resistant to bacterial canker caused by *Pseudomonas syringae* pv. *morspunorum* (Wormald) Young et al., including 'Merla,' 'Mermat,' and 'Merpet' (Matthews and Dow 1978, 1979) and 'Inge' (Matthews and Dow 1983). Theiler-Hedtrich (1985) found that 'Vittoria' was the most resistant cultivar to *P. morspunorum*. He also found that 'Rigikirsche,' 'Heidegger,' and 'Schauenburger' can be regarded as moderate to highly resistant.

Improved fruit quality: Large size, firmness, and sweetness are all considered important fruit quality traits for sweet cherries (Proebsting 1992; Christensen 1995; Ystaas and Frøynes 1990). Kappel et al. (1996) demonstrated that the optimum fruit size is $\approx$ 12 g. Large fruit size is a major contributor to consumer perception of a high-quality sweet cherry (Facteau 1988; Proebsting 1992; Christensen 1995). Cultivar plays a major role as do crop load, leaf area, and production practices. Work with sensory panels showed that there was a linear relationship between fruit firmness and panelist's perception of ideal fruit firmness, that is, the panelists continued to favor cherries with higher firmness (Kappel et al. 1996). It is unknown from this work if cherries can be too firm or too hard. Ross et al. (2009) have reported a negative correlation between sensory firmness and sensory juiciness and analytical firmness and perceived juiciness in cherries. They further reported that a minimum analytical value differing 40 g/mm was required before a trained sensory panel could determine a difference in firmness. The sensory evaluation work performed by Kappel et al. (1996) suggests that a minimum soluble solids content of 17–19% was required, and that there is also a close relationship between the sweet–sour balance and the sensory rating for fruit sweetness. This indicates that titratable acidity is important for the perception of fruit sweetness and flavor impact.

The major quality parameters in sour cherry are soluble solids, titratable acidity, fruit and juice color, firmness and good taste. The relative importance of these characteristics depends on whether the fruit are processed or for the fresh market. The ideal fruit characteristics for most processing uses are fruit diameter from 21 to 24 mm, dark red colored juice, high sugar and acidity content (Brix 16–20%, titratable acidity >25 g/L malic acid) combined with good aroma. For juice production and fresh consumption, larger fruit size is preferred. Recent studies investigated anthocyanins and aroma components in sour cherry during the ripening season (Schmid and Grosch 1986; Poll et al. 2003; Šimunic et al. 2005). Anthocyanins from sour cherry have been shown to possess strong antioxidant and anti-inflammatory activities (Wang et al. 1999).

Extension of the ripening season: Cherries ripening at either end of the season have the greatest chance of obtaining high prices if other quality attributes are in place. O'Rourke (2005) suggests that opportunities remain to increase sales by lengthening the Northern Hemisphere season. Similar objectives for sour cherries can be made, thereby utilizing mechanical harvesters more efficiently and potentially lowering labor costs.

Mechanical harvesting: Concern over labor costs and availability for harvesting sweet cherries has created an interest in mechanical harvesting of cherries destined for the fresh market. Mechanical harvesting has been used for processing cherries, but cherries intended for the fresh market are required to meet higher quality standards. A market for stemless sweet cherries in the fresh produce trade would need to be developed. Issues that come into play are tree architecture, stem retention force, and resistance to bruising, all preferably without the use of growth regulators such as ethephon. Cultivars such as 'Vittoria' (Bargioni 1970) and 'Cristalina' (Kappel et al. 1998) that develop a dry abscission zone between the fruit and stem may be best suited to mechanical harvesting. Other cultivars have reduced stem retention (e.g., 'Symphony').

Resistance to environmental stress: With the increased interest in cherry plantings, more cherries are being planted at the margins of traditional production areas. Driving forces include cost and availability of land and extending harvest season either earlier or later. This places cherries at risk to winter injury, spring frosts, and heat stress. Low temperatures during late autumn and early winter adversely affect production of sweet cherries (Caprio and Quamme 2006). Spring frosts also limit production, and temperatures that can kill flower buds vary depending on bud development and cultivar (Ballard et al. 1997; Kappel 2010). Not only can heat stress affect the formation of fruit doubles and spurs the following year, it can also affect fruit quality of the current crop. Micke et al. (1983), reported that the cultivars 'Vernon,' 'Sam,' 'Sue,' 'Black Republican,' 'Black Tartarian,' 'Rainier,' and

'Jubilee' appeared to have low doubling potential; therefore, sources of resistance appear to be available.

Early blossoming cherries can have buds, flowers, and young fruit killed by spring frosts. Selection of late blooming genotypes with longer chilling requirements and with tolerance to spring frost can reduce the risk. In Russia and Canada, interspecific hybrids with *P. fruticosa* were used to increase the blossom frost tolerance in sour cherry (Zhukov and Charitonova 1988; Bors 2005).

5.2 Rootstocks

Cherry rootstock traits can be divided into two broad groups. The first group includes those traits that are regulated by the rootstock genome and are expressed in the rootstock. The second group of traits is determined by the genome of the scion or interstock and is expressed in the interaction between the scion or interstock and rootstock. Perry (1987) listed the following breeding objectives for cherry rootstocks: tree size reduction; increased scion precocity and cropping; wide range of scion compatibility; uniformity in performance (asexual propagation); cold hardiness; adaptation to a wide range of soils; and disease and pest tolerance. These breeding objectives are still of importance, although the order of priorities may vary.

Cherry rootstock literature illustrates the determined effort to find a dwarfing rootstock with the vision that it would help solve the problems of intensive cherry production in high density orchards similar to apples. Currently there is a range of vigor among cherry rootstocks, but they are not all truly satisfactory. With variable soil fertility in the various cherry growing regions and the uncertainties surrounding climate change, greater emphasis needs to be placed on soil and climate adaptability without losing sight of the need for vigor control.

Effect of rootstock on scion vigor and growth habit: The overriding factors responsible for rootstock control of scion vigor have not been clarified. Interactions of the various growth regulators are similar to those of other composite fruit trees (rootstock-scion) involving localized production of auxin and cytokinins. Even though many of the elements of growth control can be explained by hormonal control, full understanding of rootstock effects on assimilate partitioning, water and nutrient uptake and translocation, and the mechanism of rootstock effects on precocity and cropping efficiency require continued study.

Gruppe (1985) attempted to apply the phloem–xylem ratio model (Beakbane 1941) as a preselection method for vigor control but the results proved to be unsuitable. The use of hormone levels or interactions was also unsuccessful. Misirli et al. (1996) related the vigor of the tree to sieve tube size (Tanrisever and Feucht 1978) when selecting for low-vigor Mahaleb rootstocks. They found a direct relationship between vigor and the size of sieve tube elements in wood of old trees, but no correlation in young trees; therefore, they concluded that it cannot be used as a criterion for predicting vigor.

Based on the results of various breeding projects (Trefois 1980; Gruppe 1985; Wolfram 1971, 1996), there is no doubt that within the section Eucerasus, *P. fruticosa*, *P. cerasus*, and *P. canescens* are major sources for vigor control. Further sources may be found in the species of section Pseudocerasus (*P. pseudocerasus* and *P. serrulata*), but the hardiness and drought tolerance of these hybrids may not be acceptable (Cummins 1979a, b). No dwarfing effect has been found in *P. avium* genotypes that have been used as rootstocks, although the possible utilization of genetic dwarfs in further breeding (Webster and Schmidt 1996) or the effect of inbreeding has not been fully investigated. A range of vigor from standard to moderate is found in *P. mahaleb* (Hrotkó 2004; Hrotkó and Magyar 2004) and genetic dwarf genotypes of Mahaleb are signs that these may be sources of scion vigor control.

Branch angle can also be affected by rootstock. Webster and Schmidt (1996) reported that some *P. avium* and *P. pseudocerasus* clones cause the scion to develop wide branch angles. Hrotkó et al. (1999) also observed that scions on *P. mahaleb* 'Magyar' had wide crotch angle, whereas scions on MxM 14 and MxM 97 had narrow crotch angles.

Effect of rootstocks on precocity, cropping, and fruit quality of scion cultivar: Precocity, abundance, and consistency of yield as well as fruit quality are affected by rootstocks, but there is considerable interaction between rootstock, training and pruning, tree spacing and nutrition. Perry (1987) reported that scions on Mahaleb seedling rootstock produced fruit 1–2 years earlier than on Mazzard rootstocks. Intensive orchards with close spacing of trees and fruiting wood management can also contribute to precocity (Hrotkó et al. 2009). Rootstocks in each vigor class can improve precocity but it is not necessarily linked to dwarfing. In the European Cherry Rootstock Trial the semidwarf rootstock, Damil, produced only 50% of the yield of the rootstock Weiroot 158 in early years (Hrotkó et al. 2006). This confirms the report by Webster and Schmidt (1996) that yield efficiency may not be linked with vigor control in cherry. Webster (1980) and Gruppe (1985) reported that many dwarfing rootstocks had poor yield efficiencies. Further, many of the scions on dwarfing rootstocks had abundant flowering that did not translate into adequate cropping. Triploid and tetraploid crosses within Eucerasus were more productive than diploids (Webster and Schmidt 1996). Screening for scion productivity can only be determined with field trials. A relationship between yield efficiency, crop load and leaf area can affect fruit size (Edin et al. 1996; Simon et al. 2004; Cittadini et al. 2007). Highly efficient dwarfing rootstocks can increase the fruit to leaf area ratio and thereby reduce fruit size and quality.

Graft compatibility of rootstocks: Intraspecific grafts of *P. avium* (i.e., sweet cherry cultivars on Mazzard) are usually compatible. Graft incompatibility occurs only when the composite tree is produced from two or more species (e.g., sweet cherry/*P. cerasus*, *P. mahaleb* or interspecific hybrids). Sour cherry cultivars usually show good compatibility on *P. mahaleb* and *P. avium*. Incompatibility symptoms may not manifest themselves under optimal growth conditions; however, when the tree is overcome by environmental stress the underlying incompatibility will be revealed. Incompatibility symptoms can include: poor bud take; the scion snapping off at the bud union; small yellow leaves; stunted growth; early reddening and fall of leaves in the autumn; scion or rootstock overgrowth; excessive rootstock suckering; and excessive early fruiting (Perry 1987; Webster and Schmidt 1996). Rootstocks may not be compatible with all cultivars in a species (Webster and Schmidt 1996). Wolfram (1971) has found that when *P. canescens* and *P. avium* are used as parents with Asian *Prunus* species the graft compatibility with sweet cherry scions is improved. This does not appear to be the case with *P. tomentosa* though. To date, there is no generally applicable preselection method to test for incompatibility of the graft union.

Propagation opportunities and nursery value of rootstock plants: Selected seed orchards produce more uniform populations when compared to seedlings of unknown origin. Seed propagation is relatively straightforward when using appropriately selected seed sources of *P. avium* or *P. mahaleb*. The germination capacity of *P. avium* scion cultivars is very low, and it is variable in *P. cerasus*. Vegetative propagation provides uniform rootstock material, and therefore, the adventitous root production capacity becomes an essential trait for rootstock candidates. The rooting capacity of *P. avium* is low, only the clones F12/1 and Charger have been successfully propagated by layering (Webster 1996). Colt, a hybrid of *P. avium*, propagates readily as hardwood cuttings or by layering. Hybrids of *P. cerasus* and *P. canescens* with *P. wadai* (*P. pseudocerasus* x. *P. subhirtella*) (Wolfram 1971; Gruppe 1985) are also readily propagated as cuttings or by layering. Softwood cuttings of *P. mahaleb* clonal rootstocks form adventitious roots easily (Sarger 1972; Hrotkó 1982) but hardwood cuttings and layering have failed. Many of the commercially available rootstocks from interspecific hybrids are micropropagated; however, growth rate in the nursery has been an issue (i.e., length of time before shoots can be budded).

Tolerance to environmental conditions (climate, soil, water supply): Cold hardiness is an important attribute of rootstocks and rootstocks can also affect the response of the scion to cold temperatures (Howell and Perry 1990). *P. cerasus* and *P. fruticosa* are considered the hardiest rootstocks and Mahaleb is hardier than Mazzard. Within the Mahaleb species, the broad-leaved subspecies is hardier than the small leaved subspecies (Hrotkó 2004). *P. avium* is the least hardy species (Perry 1987) within Eucerasus, although Küppers (1978) reports differences in hardiness of Mazzard selections. In the nursery, Colt can show sensitivity to early frost, but no injury has been observed on sweet cherry trees budded on Colt.

Drought and heat tolerance of rootstocks is essential in many cherry growing regions and this attribute may be linked to root depth. Shallow-rooted dwarfing rootstocks (some dwarfing interspecific hybrids, *P. cerasus* and *P. fruticosa*) are more susceptible to drought and heat injury. The most tolerant rootstocks appear to be the *P. mahaleb* selections and hybrids (MxM series).

Adaptability to different soil conditions is considered an important rootstock trait. *P. mahaleb* and derivatives tolerate light sandy and gravel soils with high lime content and pH levels of 7.8–8.2. Mahaleb seedlings (Cema) proved to be tolerant to the calcareous and high pH soils in Shaanxi province of China, where in the summer, during the rainy season, anaerobic conditions may cause iron chlorosis when using *P. pseudocerasus* as rootstock (Faust et al. 1998; Cai et al. 2007).

Tolerance or resistance of rootstocks to pests and diseases: Several nematode species attack the roots of cherry trees. *P. avium* and *P. cerasus* are sensitive to root lesion nematodes (*Xiphinema* and *Ptratylenchus* species), and Zepp and Szczygiel (1985) found that *Pratylenchus penetrans* Cobb attacks *P. mahaleb* roots more readily than Mazzard and *P. cerasus*. Mazzard and *P. cerasus* are more tolerant than *P. mahaleb* to *Meloidogyne incognita* (Kofoid and White) Chitwood (Webster and Schmidt 1996).

Phytophthora species may cause serious tree decay on heavy soils with low drainage capacity, and *P. cerasus* and Mazzard are more tolerant than the susceptible *P. mahaleb* (Wicks et al. 1984; Cummins et al. 1986). *P. canescens* (Camil) and its hybrids are also sensitive to *Phytophthora* (Webster and Schmidt 1996). All rootstocks are sensitive to *Verticillium* and there is no known source of resistance. In the USA, where *Armillaria mellea* (Vahl ex Fr.) Kummer can cause root damage, *P. mahaleb*, *P. cerasus*, Colt and Inmil were found to be sensitive, Mazzard less sensitive, and MxM 60 showed the least sensitivity (Proffer et al. 1988). Leaf spot caused by *Blumeriella jaapii* can cause severe leaf fall in nursery liners. Only *P. mahaleb* is tolerant, whereas *P. avium*, *P. cerasus*, and their derivatives are more or less sensitive. VP1 (*P. cerasus* × *P. maacki*) is reported to be tolerant (Yoltuchovski 1977; Micheyev et al. 1983). As reported previously in this chapter, there are a number of wild cherry species with a high level of resistance to leaf spot that could be used in a breeding program (Wharton et al. 2003; Schuster 2004; Budan et al. 2005b). In Northwest Europe *Thielaviopsis basicola* (Berk. and Br.) Ferraris, a fungal disease, can cause severe replant problems and the hybrids of *P. avium* × *P. pseudocerasus* can be used as a source of resistance for this threat.

Bacterial diseases that create problems include crown gall (*Agrobacterium tumefaciens* Smith and Townsend (Conn) which infects trees in the nursery as well as in orchards where it can reduce growth and productivity. Colt and the Mazzard clone F12/1 are both sensitive whereas the Mahaleb rootstocks and *P. fruticosa* hybrids are less sensitive. *Pseudomonas s.* pv *mors-prunorum* and *Pseudomonas syringae* pv *syringae* van Hall or bacterial canker is a particularly damaging disease in the humid zone of temperate areas. The Mahalebs are known to be tolerant, whereas Mazzard genotypes are considered susceptible. The clonal rootstock Charger and some *P. avium* × *P. pseudocerasus*, or *P. avium* × *P. incisa* hybrids can be used as a source of resistance (Webster and Schmidt 1996).

There is no known resistance to viruses or phytoplasmas, although there are considerable differences in sensitivity. *P. fruticosa* and its derivatives are particularly hypersensitive to viruses. Some clones of *P. cerasus* have shown higher sensitivity to Prune Dwarf Virus (PDV) whereas *P. canescens* were more sensitive to *Prunus Necrotic* Ringspot Virus (PNRS) (Lang et al. 1998a). Lankes (2007) found that Colt, GiSelA 5, and Piku 1, 3, and 4 are tolerant, while GiSelA 3 and GiSelA 6 are partially sensitive to PDV and PNRS viruses. In France *P. mahaleb* rootstocks are more susceptible to leafhopper transmitted phytoplasma (Moliére's decline) compared to trees on Mazzard. Western X Disease in the USA, which is also transmitted by leafhoppers, can infect trees on Mazzard and Colt, whereas *P. mahaleb* rootstocks are hypersensitive.

6 Breeding Methods and Techniques

The following quote by Janick et al. (1996) best describes the process for sweet cherry breeding at the moment: "the traditional strategy for fruit breeding has been to identify superior phenotypes, propagate the best selections, develop cultural practices that enhance the performance of the selected cultivars, hybridize among the best selections, and then continue the cycle. This breeding method may be considered a form of recurrent mass selection in which the key concept is selection of the best individuals and continual recombination over many cycles".

6.1 Improved Fruit Quality

At Summerland fruit samples for fruit quality determinations are harvested at a similar maturity for all selections. For dark cherries the Centre Technique Interprofessionel des Fruits et Légume (Ctifl) color chart is used to determine maturity and the #6 color chip is used as a standard. For blushed or bicolored cherries, skin color, firmness and taste are used to determine an appropriate sampling time. A 2–2.5 kg sample of fruit is harvested from all parts of the tree and these fruits are used for all quality measurements.

Fruit size is highly dependent on leaf area per fruit or crop load (Roper and Loescher 1987; Proebsting 1990) which is important to remember when evaluating fruit size. However, Olmstead et al. (2007) have shown that mesocarp cell number is under genetic control and that cell number was the major contributor to fruit diameter. Fruit size can be determined a number of ways including fruit weight, fruit diameter or fruit size distribution. With the need to handle a large number of fruit samples during a short ripening season, fruit weight for each selection or cultivar is determined by weighing two samples (100 fruit each) per selection to arrive at fruit weight in grams per fruit. This value is then used to calculate an average fruit weight using previous years' data for fruit weight. Calculating these values over many years provides a fairly good idea of sizing potential for the selections in the program at Summerland.

Fruit firmness is a combined measure of both skin and flesh firmness (Brown and Bourne 1988). Initially, a Shore Durometer was used to measure fruit firmness. A durometer is an instrument that is used to indicate the hardness of material such as rubber and plastic. Results from the durometer have a good positive relationship with human sensory perception of fruit firmness (Kappel et al. 1996). However, the durometer is subject to variability. Currently, an instrument (FirmTech®) that measures fruit firmness by measuring the compression force required to depress fruit 1 mm has been adopted as a standard by many in the industry. A 25 fruit subsample is used to gain an average firmness reading.

Fruit taste is determined by a combination of sensory panels and objective measures, namely total soluble solids, titratable acidity and pH. Work by Kappel et al. (1996) suggests that as soluble solids content increases so does the overall acceptability, and that a soluble solids content of 17–19% should be considered as the minimum for sweet cherries. The perception of sweetness is closely related to the sweet–sour balance, and cherries with a lower soluble solids reading can be considered acceptable if titratable acidity levels are also lower. Sensory evaluations have been used to describe an "ideal" sweet cherry (Kappel et al. 1996) and to profile selections and cultivars and compare them to industry standards (Dever et al. 1996). This work suggests that an "ideal" red sweet cherry would have an optimum color represented by the #5 color chip from the Centre Technique Interprofessionnel des Fruit et Légumes (Ctifl), a fruit firmness between 70 and 75 using a Shore Instrument durometer, a minimum soluble solids concentration between 17 and 19%, optimum pH of the juice of 3.8, and an optimum sweet–sour balance between 1.5 and 2 (SSC/ml NaOH).

6.2 Precocity and Productivity

Precocity is determined by the number of years from planting to first flowering and fruiting. 'Sweetheart' can be used as a standard for a sweet cherry cultivar that is considered quite precocious. Productivity initially is rated by cropping level on individual seedlings and then by eventually advanced selections planted in replicated trials to gain more confidence in yield measurements.

6.3 Resistance to Rain-Induced Cracking

The cracking index developed by Verner and Blodgett (1931) and modified by Christensen (1972) has been used to compare susceptibility to cracking of cultivars and selections. However, the results obtained seldom appear to match what happens to cherries in the field. In general, breeding programs do not use this index as a selection tool for evaluating sweet cherries' susceptibility to rain-induced cracking. Another more reliable albeit slower method is to evaluate field cracking each year by determining the proportion of cracked fruit and calculating an average cracking percentage over a number of years. Once several years of data have been collected a sense of the level of resistance or susceptibility of the selections can be developed. The number of years evaluation spans is dependent on conditions encountered each year. For example when a dry, low rainfall year is encountered, another year of evaluation with good rainfall and cracking conditions is necessary.

6.4 Resistance to Diseases and Insects

Brown Rot: Two organisms cause brown rot, *Monilinia laxa* (Aderh. & Ruhl) Honey and *M. fructicola* (Wint.) Honey. *M. laxa* is more of a problem in Europe. Sour and sweet cherry cultivars resistant to blossom blight and brown rot are unknown. The control of blossom blight with fungicides is difficult. The selection of cultivars tolerant to the infection could be the most durable method of control. Schmidt (1937) described artificial inoculation tests of twigs in the laboratory and in the field. The artificial inoculation of flowers with a conidial suspension of *Monilinia* spp. is more successful and less labor intensive than the in vitro test (Schmidt 1937). To evaluate resistance to fruit infections, Brown and Wilcox (1989) tested a range of sweet and sour cherry cultivars and compared susceptibility at the green and ripe fruit stages. They concluded that differences in disease susceptibility were more pronounced at the ripe fruit stage and this stage should be used to evaluate cultivars. However, there appeared to be little resistance in the cultivars evaluated.

Bacterial canker: Two related bacteria, *Pseudomonas syringae* pv *syringae* and *P. s. morsprunorum* causes bacterial canker which causes cankers on branches and twigs. When cankers girdle the limbs dieback can occur. Leaves, blossoms, and fruit can also be infected especially during cold wet springs. The blossom infections show similar symptoms to brown rot and are also referred to as blossom blight. Dormant buds can be killed by the bacteria and fail to open in the spring leading to the term "dead bud". Two methods of inoculation can be used to test for resistance. These include the leaf node method where young seedlings are inoculated at a number of leaf nodes with a mixture of the bacteria and then rated some time later. A bark inoculation can be used on older trees. A piece of bark can be cut out using a cork borer and replaced with an agar plug containing the inoculum. Growth of the canker can be measured a number of months later. Bargioni (1996) suggests that even if it is not possible to have fully resistant cultivars a worthy goal would be to have cultivars with field tolerance.

Leaf spot: Cherry leaf spot, caused by the fungus *Blumeriella jaapii* (Rehm) v. Arx. (syn. *Coccomyces hiemalis* Higgins), is one of the most serious fungal diseases of sour cherry. Leaf spot is common in cherry growing areas in North America and Europe. Most sour cherry cultivars are susceptible to leaf spot. Only a few cultivars show tolerance (Budan et al. 2005b). The tetraploid interspecific hybrids 'Almaz' and 'Paljus' have resistance to leaf spot but show low fertility. The tetraploid wild species *P. maackii* and the diploid species *P. canescens* could be used as resistance donors for sour and sweet cherry breeding (Schuster 2004). Wharton et al. (2003) and Schuster (2004) established artificial inoculation test methods with leaf disks in the laboratory. These test methods could be incorporated into cherry breeding programs to evaluate sources of resistance in sweet and sour cherry breeding populations.

Powdery Mildew: Resistance to powdery mildew caused by *Podosphaera clandestina* (Waller.: FR) Lev., exists in the sweet cherry (Toyama et al. 1993; Olmstead et al. 2000). It appears that the resistance in the selection PMR-1 is due to a single gene (Olmstead et al. 2001b). Olmstead et al. (2000; 2001a) have developed a leaf disk procedure to assess resistance which will be useful to test parental material, but may not be useful for mass screening of seedlings. A molecular screening procedure appears to be possible as a resistance map for *Prunus* that a number of putative resistance regions has been generated (Lalli et al. 2005). These may assist in the development of molecular markers to use in a seedling screening procedure.

6.5 Resistance to Environmental Stress

Assessing winter hardiness is possible using differential thermal analysis (Kadir and Proebsting 1994). However, the technique is not suitable for screening large numbers of genotypes. It is more suitable for assessing the hardiness of the most advanced selections. The appropriate period in the fall and winter needs to be determined. Since low temperatures during late autumn and early winter adversely affect production of sweet cherries the following season (Caprio and Quamme 2006), both mid-winter hardiness and late fall-early winter hardiness need to be determined.

6.6 Interspecific Hybridization

Interspecific hybridization could be used to introduce characteristics from different *Prunus* species that are not currently in the commercial species *P. cerasus* or *P. avium*. Crosses between *P. cerasus* and *P. avium* have been made primarily to imporve the fruit quality in sour cherry. Although most of the seedlings resulting from these crosses are triploid and usually sterile, some fertile tetraploid genotypes were selected (Enikeev et al. 1979). According to Turovtseva et al. (1996), 64% of seedlings from crosses between sour cherry and sweet cherry are dominated by sour cherry characteristics. In reciprocal crosses, 50% of the seedlings showed characteristics of the Duke-cherry, 24% of sweet cherry, 11% of sour cherry and 14% showed new attributes. In Russia and Canada, hybrids between *P. cerasus* and *P. fruticosa* were created to increase the winter hardiness (Zhukov and Charitonova 1988; Bors 2005). To improve resistance to diseases, Mitschurin (1951) developed Cerapadus hybrids from crosses between *P. fruticosa* × *P. maackii*. Zhukov (1979) produced Podocerus hybrids which are crosses between *P. maackii* × *P. cerasus* cv. Plodorodnaja Mitschurina. Schuster (unpublished) developed new interspecific hybrids of the tetraploid combination between *P. cerasus* and *P. maackii, P. padus, P. serotina, P. spinosa* and additionally between *P. avium* and the diploid species *P. canescens, P. cerasifera*, and *P. tomentosa*.

6.7 Breeding Methodology

The breeding methods and techniques in sweet and sour cherry are very similar. When considering parents, not only the traits need to be considered but also the

S-allele complement of the parents, and their maturity date and virus status. The *S*-alleles of the parents will determine the compatibility of the cross. Seeds of early maturing cultivars may have lower germination rates. Therefore, it may be preferable to use the early maturing cultivar as the pollen parent. Otherwise, embryo rescue techniques may be necessary. Prune Dwarf Virus and Prunus Necrotic Ringspot Virus and others are transmitted by pollen; therefore, consideration to the use of virus-free parents is important.

Thompson (1996) has provided an in-depth review of the floral biology of cherry blossoms. Normally pollen matures shortly before anthesis; therefore, pollen can be collected just before flowers begin to open. Flowers are emasculated by pinching off the stamens and petals and dried pollen can be applied using a glass rod or a camel hair brush. However, Hedhly et al. (2009) have reported that flower emasculation reduced fruit set by more than half in two consecutive years. Generally, it is not necessary to enclose the flowers in a cage to keep out bees for bees will not visit emasculated flowers (Fogle 1975). If the purpose of the cross is the development of new cultivars, a small level of contamination from outcrossing may be acceptable. If on the other hand, any pollen contamination is unacceptable (such as for genetic studies), then the branches or trees will need to be covered with material that will exclude pollinating insects. Another successful approach if the female parent is not self-fertile is to enclose the tree in a cage of material that will exclude bees. Then introduce a small bee hive and bouquets of the pollen parent. The bees will then pollinate the flowers and this technique can result in a large number of seeds. We have used a variation of this procedure whereby non-self-fertile trees are enclosed in material that excludes bees. Then, as the flowers begin to open, compatible pollen is gently placed on stigmas using a glass rod. Seed set has increased using this technique.

Fruit are harvested when they are mature and the seeds are extracted. If seeds are removed from the stony endocarp, increased germination is observed during the stratification period providing the seed is protected from fungal infection. The length of the stratification period is dependent on the cultivar and can last from 90 to 120 days. A second stratification period for seeds that do not germinate will lead to a second flush of seedlings.

Once seeds show signs of germination they are transferred to the greenhouse and grown on until they reach suitable planting size, they are then hardened off and planted into a nursery or field plots. This should be done after all threats of spring frost have passed and before the heat of summer arrives. If planted into a nursery bed first, seedlings are planted into seedling orchards the following year. Depending on location, autumn planting of seedling trees has worked well as long as early winter freezes are not likely to occur. Optimum spacing in the seedling orchard is dependent on quality of soil and availability of land. A spacing of 4 m between rows and 1 m within rows is functional as long as unsuitable seedlings are rogued ruthlessly.

Evaluation of fruit in the seedling orchard usually can begin in the fourth to sixth year after planting. Very little information is usually gained from the very first crop other than maturity date; quality of fruit from the first crop does not reflect fruit quality of more mature trees. The seedling orchard should be "walked" at least once a week to ensure fruit is not missed. Potentially interesting selections can be brought into the lab for further evaluation of fruit size, level of natural cracking, fruit firmness, total soluble solids, pH, and titratable acidity. Remaining fruit can be placed in a plastic bag and placed into storage at 1°C for 14 days to observe preliminary storage potential.

Seedling selections that display good fruit quality traits for a number of years can be propagated onto rootstocks and placed in second test trials. The number of trees propagated depends on the breeder's speculation of the potential impact of the selection. The increased number of trees provides larger amounts of fruit that can then be used for more comprehensive storage and shelf life trials. Selections that continue to exhibit superior characteristics can be repropagated to include replicated trials and grower evaluations. Grower evaluations are extremely important to provide information related to commercial handling of the fruit and uncover any potential defects. If defects are uncovered, it can be determined whether they are fatal or manageable.

Once trees are propagated for advanced testing (third tier) the virus status of the tree needs to be determined. It would be ideal to test selections as they are placed into second test. However, the cost of indexing and heat treatment only allows a limited number of selections to be evaluated for virus status and cleaned of known viruses.

At some stage during the evaluation process a decision needs to be made regarding the protection of intellectual property and the commercialization of the new cultivar. Options can include following the traditional path of releasing a cultivar through a nursery and receiving a royalty for each tree sold. A more recent development has been to release the cultivar to a packing/sales entity to limit the production of the cultivar and thereby increase the value to the growers that are licensed to grow the cultivar. It then is possible for the breeding program to benefit from the increased value by receiving a "fruit royalty". Each process has pros and cons and needs to be carefully evaluated.

6.8 Rootstock

Seed tree selection and establishing of seed orchards: Seed orchards with selected seed trees are planted to provide the nursery industry with a regular supply of high quality seeds that are either hybrid seeds or inbred lines. The early seed orchards which were initially selected from wild populations based on the phenotype characteristics (e.g., healthy, vigorous tree, regular seed crop) now have given way to evaluation of the progeny of the seed trees from known pollinations (Hrotkó and Erdős 2006). The flower fertility of these genotypes determines the mating options within the orchard and accordingly the genetic composition of the seedling progeny. The flowers in the majority of seed producing clones for successful seed production need cross-pollination; therefore, the seedling progeny is a hybrid with all attributes of hybrid vigor and greater homogeneity in the phenotype of the F1 population than former seed sources. Such F1 populations are produced for rootstock use e.g., *P. avium* (Küppers

1978; Claverie 1996). Several seed orchards with cross-pollination can consist of three to five clones, each pollinating the other (Funk 1969; Nyújtó 1987; Perry 1987). In this type of hybrid mating system, the progeny represents a hybrid family of different mating combinations. Progenies of these seed orchards are usually evaluated for their nursery and orchard value.

Inbreeding: Heimann (1932), Claverie (1996), Hrotkó (1996, 2004), and Hrotkó and Magyar (1998) reported on self-fertile types of *P. mahaleb*. Self-fertile seed trees may produce a diversity in seedling characters and segregation in the population (Hrotkó and Magyar 1998; Hrotkó 2004), which is more or less tolerable among the seedlings used for rootstocks. Seedlings of the self-fertile genotype of *P. mahaleb* 'Heimann X' was known for having very uniform progeny (Heimann 1932; Küppers 1978). Fischer (1985) achieved no progress in tree size reduction and yield efficiency of 'Schattenmorelle' trees budded on the inbred seedling lines. In France, Claverie (1996) and in Hungary Hrotkó and Magyar (1998) reported on utilization of inbreeding of *P. mahaleb* seed tree selection with the aim of producing less vigorous and more uniform seedling populations. The inbreeding of self-fertile seed trees could provide a useful tool for rootstock breeding. Despite these opportunities no clonal selection among inbred populations has been reported.

Clonal rootstock selection from different cherry species: For the selection of clonal cherry rootstocks, wild populations as well as land-races provide appropriate genetic diversity. Major selection criteria are: ease of vegetative propagation, cold hardiness, adaptability to different soil and climatic conditions, tolerance or resistance to pests and diseases, freedom from suckering, graft compatibility, tree longevity, tree size control, effect on scion precocity, increased productivity, and fruit quality.

Creation of interspecific hybrids: Most breeding projects have produced interspecific hybrids to overcome the graft incompatibility of many species in Subgenus Cerasus and improve vegetative propagation. Several crossing partners were selected simply by chance or because of their availability in collections or botanical gardens (Wolfram 1971; Cummins 1979a, b). For the creation of interspecific hybrids, huge efforts were made to synchronize blossoming in greenhouses (Gruppe 1985) and to overcome low flower fertility (De Palma et al. 1996).

Dwarf mutants from irradiation breeding and polyploid clones: Dwarf *P. avium* mutants have been produced by Walther and Sauer (1985), using cobalt-60 but most of the mutants were unstable chimeras. From Mazzard F12/1, dwarf mutants were produced by Theiler-Hedtrich (1990) that were successfully propagated in vitro and planted in the nursery. Unfortunately there is no information about their outcome. Similarly, dwarf mutants were produced from Colt rootstock (James et al. 1987) using colchicine. The hexaploid ($6n=48$) Colt is now involved in several tests plantings in Europe. In a Hungarian test plot, 'Lapins' on hexaploid Colt was about 10% less vigorous and showed no difference in yield efficiency (Hrotkó et al. 2006).

Rootstock breeding is a particularly long term endeavor and progress tends to be slow regardless of which method used to develop the candidate rootstocks. At the moment, preselection techniques are not available. Therefore, to determine if the selection has any potential, the candidate rootstock will need to be propagated to obtain sufficient plant material to provide for an adequate test. Then a range of cultivars will need to be propagated on these candidates for field evaluations. A number of years are required to determine if there is any effect on scion precocity, vigor, yield, and fruit quality. However, at each stage (multiplication, scion propagation, etc.), unsuitable candidates can be discarded. For example, if propagation is extremely difficult or costly for a particular candidate, it can be discarded.

7 Integration of New Biotechnologies in Breeding Programs

The potential of molecular markers to facilitate selection on the basis of genotype rather than phenotype is particularly appealing to cherry breeders because the generation time of a cherry is at the very least 3 years but longer in practice (seed to first fruit on a seedling). Cherry trees are large perennials requiring large areas of land for seedling populations and germplasm repositories. Thus selections made on small seedlings grown in a greenhouse would provide savings in land area and maintenance of plantings. Savings in time (that is from crossing to eventual release and naming) are debatable because the need to evaluate preferred genotypes in a range of environments still prevails. For which traits should the scientific community be developing markers? The immediate response is usually "the traits of most importance to a breeding program." These are many and varied and change over time. At present fruit quality traits are most important and these include fruit size and firmness in North American sweet cherry programs, while in France, taste is considered most important and in Australia resistance of fruit to rain-induced fruit cracking is the priority. Many of the fruit quality traits are quantitative trait loci (QTLs) and heritabilities are unknown, and to date markers for the above-mentioned traits have not been published.

The most extensively studied trait in cherry at the genetic and molecular level is self-incompatibility. In the wild, sweet cherry is an obligate outbreeder by way of gametophytic self-incompatibility and sour cherry exhibits both self-incompatible and self-compatible types. Determination of incompatibility groups through test crosses has therefore been an important part of cultivar characterization. The large amount of work required and the frequently inconsistent results of these test crosses make molecular markers attractive. The methods of linked isozymes and stylar protein *S*-RNase isozymes have been largely replaced by PCR based methods of *S*-allele detection using consensus and specific primers (Sonneveld et al. 2001, 2003, 2005; Tao et al. 1999, Tsukamoto et al. 2008a, b, 2010). Characterization of commercial cultivars and final selections have been genotyped by PCR from a number of collections of sweet cherry from around the world (Canadian by Wiersma et al. 2001; US by Choi et al. 2002; Hungarian by Bekefi et al. 2003; English, Belgian by De Cuyper et al. 2005; German by Schuster et al. 2007; and Latvian; and Swedish by Lacis, et al. 2008). Self-compatible cherry cultivars have become increasingly important in recent years. Many new sweet cherry releases have a mutant *S*-allele, originally induced by radiation at the John Innes Institute (Lewis and Crowe 1954). There are also reports of spontaneous self-compatible mutants (Wünsch and Hormaza 2004; Marchese et al. 2007). In 2004, two groups reported the development of a molecular marker for self-compatibility (Zhu et al. 2004; Ikeda et al. 2004). Sonneveld et al. (2005) examined two pollen-part mutant haplotypes of self-fertile sweet cherry. Both were found to retain the *S*-RNase, which determines stylar specificity, but one (S_3' in JI2434) has a deletion including the haplotype-specific SFB gene and the other (S_4' in JI2420) has a frame-shift mutation of the haplotype-specific SFB gene, causing amino acid substitution and premature termination of the protein. Markers or primers will be most cost-effective when applied to large populations and this will require high throughput methods for DNA extraction and analysis. Efficiency gains will be made by increasing the number of markers or primer sets used (multiplex reactions) (Vaughan and Russell 2004; Hayden et al. 2008). While markers for *S*-genotype will continue to be important for final characterization of new cultivars before release it is unlikely that they will be used within large populations as a screening tool. Markers for self-compatibilty, on the other hand, could be effectively applied as a breeding selection tool.

Markers for self-incompatibility were developed by the candidate gene approach (using knowledge of genes and their functions). Practically every available marker system has been applied to sweet and sour cherries. Markers such as RAPD, isozymes, RFLP, AFLP, SSR, cpDNA, and cDNA have been applied to gene flow studies (Granger 2004), paternity analysis (Schueler et al. 2003), cultivar fingerprinting (Zhou et al. 2002; Boritzki et al. 2000; Cantini et al. 2001), Plant Breeders Rights issues (Congiu et al. 2000; De Rick 2001), maternal inheritance (Mohanty et al. 2001; Brandt et al. 1999), phylogeny and kinship studies (Wünsch and Hormaza 2002; Lee and Wen 2001), characterization of sour cherry genome (Schuster and Schreiber 2000), diversity (Struss et al. 2001) and identification of accessions (Granger 1993). Also, species and rootstock identification (Bortiri et al. 2001; Struss et al. 2002), determination of genomic contribution in hybrids (Brettin et al. 2000), evolutionary biology (Mariette et al. 1997) and pedigree analysis have used molecular biology techniques.

Although linkage maps for cherry are incomplete at this time there is a growing body of work in other *Prunus* species, particularly peach and almond, that have great potential for application to cherry. Genetic linkage maps are being developed for sour cherry (Wang et al. 1998; Canli 2004) and sweet cherry (K. Tobutt pers. comm.; Stockinger et al. 1996). Dirlewanger et al. (2004) used data from different *Prunus* linkage maps, including cherry, anchored by the reference *Prunus* map to establish a general map. Key to this was *Prunus* marker colinearity and a high level of synteny among the *Prunus* maps allowing transfer of markers between the genera (Arús et al. 2006). This work is an indication of the collaboration of the international community in the area of Rosaceae genomics and population mapping. This is a rapidly changing field and up-to-date news, and data, including linkage maps, DNA sequence,s and bioinformatics tools for Rosaceae species are available at the Genomics Database for Rosaceae (GDR; Jung et al. 2008).

Newcomb et al. (2006) released the New Zealand apple EST database and the genomic sequencing of apple has been completed (S. Gardiner pers. comm.). However, of greater importance for *Prunus* breeders is the completion of the genome sequence of peach (Genome Database for Rosaceae 2010). Transformation protocols have been applied to sour cherry (Song and Sink 2006) and *Prunus* in general and sweet cherry specifically have proven very difficult to transform. Government regulations surrounding the development of genetically modified plants and consumer uncertainty toward genetically modified food and patents on transformation techniques have meant that no transformed commercial cherry cultivars have been released. Nonetheless, transformation is a useful research tool; it has been used to verify the function of genes. A transient assay involving transformation of tobacco has been used to assess the function of over 160 apple genes (Hellens et al. 2005). Or more directly related to Rosaceous fruit crops is the use of a strawberry-based system developed for rapid transformation and regeneration to analyze gene function (Folta et al. 2006). This allows for a forward look at traits, and in the future, transformation of cherry with novel genes will allow for evaluation of a gene and its interaction with other genes.

References

Anonymous. 2003. National list of varieties inscribed in the State Cultivar Book of the Czech Republic by August 1, 2003. Central Institute for Supervising and Testing in Agriculture, Brno, 136 p.

Apostol, J. 2005. New sweet cherry varieties and selections in Hungary. Acta Hort. 667:59–64.

Arús, P. 2007. Integrating genomics into rosaceae fruit breeding. Acta Hort. 738:29–35.

Arús, P., Yamamoto, T., Dirlewanger, E., Abbott, A.G. 2006. Synteny in the Rosaceae. Plant Breeding Reviews.25:175–211.

Ballard, J.K., Proebsting, E.L., and Tukey, R.B. 1997. Critical temperatures for blossom buds: cherries. Extension Bulletin EB1128. Washington State University, Pullman. Washington.

Bargioni, G. 1970. 'Vittoria', a new sweet cherry cultivar (in Italian). Revistadella Otro. Italiana 63:3–12.

Bargioni, G. 1996. Sweet cherry scions: characteristics of the principal commercial cultivars, breeding objectives and methods. In: Webster, W.D. and Looney, N.E. 1996. Cherries: crop physiology, production and uses. CAB International, Wallingford, UK.

Battistini, A. and Berini, E.S. 2004. Agronomic Results of Victor, a Semi-Dwarf Cherry Rootstock. Acta Hort. 658:111–113.

Baumann, G. 1977. Clonal selection in *Prunus mahaleb* rootstocks. Acta Hort. 75:139–148.

Beakbane, A.B. 1941. Anatomical studies of stems and roots of hardy fruit trees. III. The anatomical structure of some clonal and seedling apple rootstocks stem- and root-grafted with a scion variety. J. Pomol. 18:344–367.

Beaver, J.A. and Iezzoni, A.F. 1993. Allozyme inheritance in tetraploid sour cherry (*Prunus cerasus* L.) J. Amer. Soc. Hort. Sci. 118:873–877.

Beaver, J.A., Iezzoni, A.F., and Ramm, C.W. 1995. Isozyme diversity in sour, sweet and ground cherry. Theor. Appl. Genet. 90:847–852.

Bekefi, Z., Tobutt, K.R., and Sonneveld, T. 2003. Determination of (in)compatibility genotypes of Hungarian swet cherry (*Prunus avium* L.) accessions by PCR based methods. Inter. J. Hort. 9:37–42.

Beppu, K., Ikeda, T., and Kataoka, I. 2001. Effect of high temperature exposure time during flower bud formation on the occurrence of double pistils in 'Satonishiki' sweet cherry. Scientia Hort. 87:77–84.

Beyer, M., Lau, S., and Knoche, M. 2005. Studies on water transport through the sweet cherry fruit surface: IX. Comparing permeability in water uptake and transpiration. Planta 220:474–485.

Biggs, A.R. and Northover, J. 1988. Influence of temperature and wetness duration on infection of peach and sweet cherry fruits by Monilinia fructicola. Phytopathology 78:1352–1356.

Blažková, J., and Hlušičkova, I. 2004. First results of an orchard trial with new clonal sweet cherry rootstocks at Holovousy. Hort. Sci. (Prague) 31:47–57.

Boritzki, M., Plieske, J., and Struss, D. 2000. Cultivar identification in sweet cherry (*Prunus avium* L.) using AFLP and microsatellite markers. Acta Hort. 538:505–510.

Bortiri, E., Ott, S.H., Jiang, J., Baggett, S., Granger, A., Weeks, C., Buckingham, M., Potter, D., and Parfitt, D.E. 2001. Phylogeny and systematics of *Prunus* (Rosaceae) as determined by sequence analysis of ITS and the chloroplast *trnL-trnF* spacer DNA. Systematic Bot. 26:797–807.

Bors R.H. (2005) Dwarf sour cherry breeding at the University of Saskatchewan. Acta Hort. 667: 135–140.

Bošković, R., and Tobutt, K.R. 1998. Inheritance and linkage relationships of isoenzymes in two interspecific chery progenies. Euphytica 103:273–286.

Bošković, R., Wolfram, B., Tobutt, K.R., Čerovic, R., and Sonneveld, T. (2006) Inheritance and interactions of incompatibility alleles in tetraploid sour cherry. Theor. Appl. Genet. 112:315–326.

Brandt, B., Witherspoon, J., Granger, A.R., and Collins, G.G. 1999. Identification of pollen donors for sweet cherry cultivars 'Stella' and 'Summit' by isozyme analysis. Aust. J. Exp. Agric. 39:473–477.

Brettin, T.S., Karle, R., Crowe, E.L., and Iezzoni, A.F. 2000. Chloroplast inheritance and DNA variation in sweet, sour, and ground cherry. J. Heredity 91:75–79.

Brown, S.K. and Bourne, M.C. 1988. Assessment of components of fruit firmness in selected sweet cherry genotypes. HortScience 23:902–904.

Brown, S.K. and Wilcox, W.F. 1989. Evaluation of cherry genotypes for resistance to fruit infection by *Monilinia fructicola* (Wint.) Honey. HortScience 24:1013–1015.

Brown, S.K., Iezzoni, A.F., and Fogle, H.W. 1996. Cherries. In: Janick, J. and Moore, J.N. Fruit breeding. Volume 1: Tree and tropical fruits. John Wiley and Sons. New York.

Budan, S., Mutafa, I., Stoian, I., and Popescu, I. 2005a. Screening of 100 sour cherry genotypes for *Monilia laxa* field resistance. Acta Hort. 667:145–151.

Budan, S., Mutafa, I., Stoian, I., and Popescu, I. 2005b. Field evaluation of cultivar susceptibility to leaf spot at Romania's sour cherry genebank. Acta Hort. 667:153–157.

Bujdosó G., Hrotkó K. and Stehr R. 2004. Evaluation of sweet and sour cherry cultivars on German dwarfing rootstocks in Hungary. J. Fruit Ornamental Plant Res. 12:233–244.

Cai, Y.L., Cao, D.W., and Zhao, G.F. 2007. Studies on genetic variation in cherry germplasm using RAPD analysis. Scientia Hort. 111:248–254.

Canli, F.A. 2004. Development of a second generation genetic linkage map for sour cherry using SSR markers. Pakistan J. Biol. Sci. 7:1676–1683.

Cantini, C., Iezzoni, A.F., Lamboy, W.F., Boritzki, M., and Struss, D. 2001. DNA fingerprinting of tetraploid cherry germplasm using simple sequence repeats. J. Amer. Soc. Hort. Sci. 126:205–209.

Caprio, J.M. and Quamme, H.A. 2006. Influence of weather on apricot, peach and sweet cherry production in the Okanagan Valley of British Columbia. Can. J. Plant Sci. 86:259–267.

Charlot, G., Edin, M., Charmont, S., and Blanc, P. 1998. Infos Cerise. Mai 98.

Choi, C. and Kappel, F. 2004. Inbreeding, coancestry, and founding clones of sweet cherries from North America. J. Amer. Soc. Hort. Sci. 129:535–543.

Choi, C., Tao, R., and Anderson, R.L. 2002. Identification of self-incompatibility alleles and pollen incompatibility groups in sweet cherry by PCR based S-allele typing and controlled pollination. Euphytica 123:9–20.

Christensen, J.V. 1972. Cracking in cherries III. Determination of cracking susceptibility. Acta Agriculturae Scandinavica 22:129–136.
Christensen, J.V. 1995. Evaluation of fruit characteristics of 20 sweet cherry cultivars. Fruit Var. J. 49:113–117.
Cireasa, V., Cireasa, E., and Gavrilescu, C. 1993. 'Cristimar' the latest cultivar of dwarfish cherry tree. Acta Hort. 349:283–284.
Cittadini, E.D., van Keulen, H., Peri, P.L. and Ridder, N. 2007. Designing a "Target-Tree" for Maximizing Gross Value of Product in Patagonian Sweet Cherry Orchards . Inter. J. Fruit Sci. 6:3–22.
Claverie, J. 1996. New selections and approaches for the development of cherry rootstocks in France. Acta Hort. 373–375.
Congiu, L., Chicca, M., Cella, R., Rossi, R., Bernacchia, G. 2000. The use of random amplified polymorphic DNA (RAPD) markers to identify strawberry varieties: a forensic application. Mol. Ecol. 9:229–232.
Crane, M.B. and Brown, A.G. 1937. Incompatibility and sterility in the sweet cherry, *Prunus avium* L. J. Pom. Hort. Sci. 15:86–116.
Cummins, J.N. 1972. Vegetatively propagated selections of *Prunus fruticosa* as dwarfing stocks for cherry. Fruit Var. Hort. Dig. 26:76–79.
Cummins, J.N. 1979a. Exotic rootstocks for cherries. Fruit var. J. 33.3.74–84.
Cummins, J.N. 1979b. Interspecific hybrids as rootstocks for cherries. Fruit Var. J. 33.3.85–89.
Cummins, J.N., Wilcox, W.F. and Forsline, P.L. 1986. Tolerance of some new cherry rooootstocks to December freezing and to Phytophtora root rots. Compact Fruit Tree 19:90–93.
De Candolle, A. 1886. Origin of cultivated plants. Hafner, New York.
De Cuyper, B., Sonneveld, T., and Tobutt, K.R. 2005. Determining self-incompatibility genotypes in Belgian wild cherries. Molecular Ecology 14:945–955.
De Palma, L., Palasciano, M. and Godini, A. 1996. Interspecific hybridization program aimed at obtaining dwarfing and non-suckering rootstocks for sweet cherry. Acta Hort. 410:177–181.
De Rick, J. 2001. Are molecular markers strengthening plant variety registration and protection? Acta Hort. 552:215–223.
Dever, M.C., MacDonald, R.A., Cliff, M.A., and Lane, W.D. 1996. Sensory evaluation of sweet cherry cultivars. HortScience 31:150–153.
Dirlewanger, E., Graziano, E., Joobeur, T., Garriga-Calderé, F., Cosson, P., Howard, W., and Arús, P. 2004. Comparative mapping and marker-assisted selection in Rosaceae fruit crops. Proc. Nat. Academy Sci. 101:9891–9896.
Edin, M., Garcin, A., Lichou, J., and Jourdain, J.M. 1996. Influence of dwarfing cherry rootstocks on fruit production. Acta Hort. 410:239–243.
Enikeev, C.K., Mladentseva, M.S., and Jurtsev, V.N. (1979) Some peculiarities of sweet cherry x sour cherry hybrids (*C. vulgaris* x *C. avium*) in the nonchernozem zone. Tag,-Ber., Akad. Landwirtsch.-Wiss. DDR, Berlin 174, 123–130.
Faccioli, F., Interieri, G., and Marangoni, B. 1981. Portinnesti nanizzanti del ciliego: le selezione CAB. Atti. Giorn. Sulle scelte varietali in frutticoltura. 19.12.125–128.
Facteau, T. 1988. Improving cherry quality. Proc. Pacific Northwest Cherry Production Shortcourse, Washington State Univ.Pullman. p. 216–230.
Faust, M., Deng, X., and Hrotkó, K. 1998. Development project for cherry growing in Shaanxi province of China P.R. Acta Hort. 468:763-769.
Faust, M. and Suranyi, D. 1997. Origin and dissemination of cherry. Hort. Rev. 19:263–317.
Fischer, M. 1985. Selektionsarbeiten an *Prunus mahaleb* L. als Unterlagen für Süß- und Sauerkirschen. Arch. Gartenbau. 33(2):75.
Folta, K., Dhingra, A., Howard, L., Stewart, P., Chandler, C. 2006. Characterization of LF9, an octoploid strawberry genotype selected for rapid regeneration and transformation. Planta 224:1058–1067.
Fogle, F.W. 1975. Cherries. In: Janick, J. and Moore, J.N. Advances in fruit breeding. Purdue University Press. West Lafayette, Indiana.
Franken-Bembenek, S. 2004. GiSelA 3 (209/1) – A New Cherry Rootstock Clone of the Giessen Series. Acta Hort. 658:141–143.

Franken-Bembenek, S. 2005. Gisela® 5 Rootstock in Germany. Acta Hort. 667:167–172.

Funk, T. 1969. Virusgetestete'Kaukasische Vogelkirsche' als neue Unterlage. Obstbau. 9. 140–142.

Genome Database for Rosaceae. 2010. http://www.rosaceae.org/peach/genome.

Gerlach, H.K. and Stösser, R. 1998. Sweet cherry cultivar identification using RAPD-derived DNA fingerprints. Acta Hort. 468:63–69.

Giorgio, V. and Standardi, A. 1996. Growth and production of two sweet cherry cultivars grafted on 60 ecotypes of *Prunus mahaleb*. Acta Hort. 410:471-475.

Granger, A.R. 1993. Sweet cherry cultivar identification by leaf isozyme polymorphism. Theor. Appl. Genet. 86: 458–464.

Granger, A.R. 1998. Sweet cherry breeding in Australia. Acta Hort. 468:111–114.

Granger, A.R. 2004. Gene flow in cherry orchards. Theor. Appl. Genet. 108: 497–500.

Granger, A.R. 2005. The effect of three rootstocks on yield and fruiting of sweet cherry. Acta Hort. 667:233–238.

Gruppe, W. 1985. An overview of the cherry rootstock breeding program at Giessen. Acta Hort. 169:189–198.

Grzyb, Z.S., Sitarek, M. and Guzowska-Batko, B. 2005. Results of a sweet cherry rootstock trial in Northern-Poland. Acta Hort. 667:207–210.

Hancock, A.M. and Iezzoni, A.F. 1987. Malate dehydrogenase isozyme patterns in seven *Prunus* species. HortScience 23:381–383.

Hauck, N.R., Yamane, H., Tao, R. and Iezzoni A.F. 2006. Accumulation of non-functional *S*-haplotypes results in the breakdown of gametophytic self-incompatibility in tetraploid *Prunus*. Genetics 172:1191–1198.

Hayden, M.J., Nguyen, T.M., Waterman, A., and Chalmers, K.J. 2008. Multiplex-Ready PCR: a new method for multiplexed SSR and SNP gene typing. BMC Genomics 9:80. http://www.biomedcentral.com/1471-2164/9/80.

Hedhly, A., Hormaza, J.I., and Herrero, M. 2009. Flower emasculation accelerates ovule degeneration and reduces fruit set in sweet cherry. Scientia Hort. 119:455–457.

Hedrick, U.P. 1915. The cherries of New York. J.B. Lyon. Albany, NY.

Heimann, O.R. 1932. Zur Frage der Selektion der Steinweichsel *Prunus mahaleb* als Veredlungsunterlage für Kirschen. Der Obst und Gemüsebau Bln. 78:138.

Hein, K. 1979. Zwischenbericht über eine Prüfung der Steppenkirsche (*P. fruticosa*) und anderen Süsskirchenunterlagen und Unterlagenkombinationen. Erwerbsobstbau 21:219–219.

Hellens, R.A. Allan, A.C. Friel, E.N. Bolitho, K., Grafton, K., Templeton, M.D., Karunairetnam, S., Gleave, A.P., and Laing, W.A. 2005. Transient expression vectors for functional genomics, quantification of promoter activity and RNA silencing in plants. Plant Methods 1:13.

Hillig, K.W. and Iezzoni, A.F. 1988. Multivariate analysis of a sour cherry germplasm collection. J. Amer. Soc. Hort. Sci. 113:928–934.

Hilsendgen, P. 2005. Preliminary results of a National German sweet cherry rootstock trial. Acta Hort. 667:179–187.

Howell, G.S. and Perry, R.L. 1990. Influence of cherry rootstocks on the cold hardiness of twigs of the cherry scion cultivar. Scientia Hort. 43:103–108.

Hrotkó, K. 1982. Sajmeggy alanyklónok szaporítása zölddugványozással (Propagation of Mahaleb cherry by leafy cuttings), Kertgazdaság, 4:45–50.

Hrotkó, K. 1990. The effect of rootstocks on the growth and yield of 'Meteor korai' sour cherry variety. XXIII. Int. Hort. Congress. Abstr. 2. No. 4165.

Hrotkó, K. 1993. *Prunus mahaleb* Unterlagenselektion an der Universitätür gartenbau und Lebensmittelindustrie in Budapest. Erwerbsobstbau. 35:39–42.

Hrotkó K. 1996. Variability in *Prunus mahaleb* L. for rootstock breeding. Acta Hort. 410: 183–188.

Hrotkó, K. 2004. Cherry rootstock breeding at the department of Fruit Science, Budapest. Acta Hort. 658. 491–495.

Hrotkó, K. and Erdős, Z. 2006. Floral biology of tree fruit rootstocks. Inter. J. Hort. Sci. 12: 153–161.

Hrotkó, K. and Facsar, G. 1996 . Taxonomic classification of Hungarian populations of *Prunus fructicosa* (Pall.) Woronow hybrids. Acta Hort. 410:495–498.
Hrotkó, K., Gyeviki, M., and Magyar L. 2006. A'Lapins' cseresznyefajta növekedése és term?re fordulása 22 alanyon. (Growth and yielding of Lapins on 22 rootstocks.) Kertgazdaság, 38(2) 14–21.
Hrotkó, K. and Magyar, L. 1998. Inbreeding of *Prunus mahaleb*. Acta Hort. 468: 393–400.
Hrotkó, K. and Magyar, L. 2004. Rootstocks for cherries from Department of Fruit Science, Budapest. Inter. J. Hort. Sci. 10:63–66.
Hrotkó, K., Magyar L. and Gyeviki, M. 2009. Effect of rootstocks on vigor and productivity in high density cherry orchards. Acta Hort. 825:245–250.
Hrotkó, K., Magyar, L., and Simon, G. 1999. Growth and yield of sweet cherry trees on different rootstocks. Inter. J. Hort. Sci. 5:98–101.
Hrotkó, K., Magyar, L., and Simon, G., and Gyeviki, M. 2007. Development inintensive orchard systems of cherries in Hungary. Inter. J. Hort. Sci. 13:79–86.
Iezzoni, A.F. 1984. Sour cherry breeding in eastern Europe. Fruit Var. J. 38:121–125.
Iezzoni, A.F. 1996. Sour cherry cultivars: objectives and methods of fruit breeding and characteristics of principal commercial cultivars. In: Eds. Webster, A.D. and Looney, N.E. Cherries: crop physiology, production and uses. CAB International, Wallingford, UK.
Iezzoni, A.F., Schmidt, and H., and Albertini, A. 1990. Cherries (*Prunus*). In: Moore, J. N., Ballington, J.R. genetic resources of temperate fruit and nut crops. ISHS. Wageningen, the Netherlands.
Ikeda, K., Watari, A., Ushijima, K., Yamane, H., Hauck, N.R., Iezzoni, A.F., and Tao, R. 2004. Molecular markers for the self-compatible *S*4'-haplotype, a pollen-part mutant in sweet cherry (*Prunus avium* L.). J. Amer. Soc. Hort. Sci. 129:724–728.
James, D.J., MacKenzie, K.A.D. and Malhotra, S.B. 1987. The induction of hexaploidy in cherry rootstocks using *in-vitro* regeneration techniques. Theor. Appl. Genet. 73:589–594.
Janick, J., Cummins, J.N., Brown, S.K., and Hemmat, M. 1996. Apples. P 1–77. In: Janick, J. and Moore, J.N. (Eds.) Fruit Breeding. Vol. I. Tree and Tropical Fruits. John Wiley and Sons, New York.
Jiménez, S., Garin, A., Albás, E.S., Bterán, J.A., Gogorcena, Y. and Moreno, M. 2004. Effect of Several Rootstocks on Fruit Quality of Sunburst Sweet Cherry. Acta Hort. 658:353–358.
Jung, S., Staton, M., Lee, T., Blenda, A., Svancara, R., Abbott, A., and Main, D. 2008. GDR (Genome Database for Rosaceae): integrated web-database for Rosaceae genomics and genetics data. Nucleic Acids Res. 36:D1034-D1040.
Kaack, K., Spayd, S.E., and Drake, S.R. 1996. Cherry processing. In: Eds. Webster, A.D. and Looney, N.E. Cherries: crop physiology, production and uses. CAB International, Wallingford, UK.
Kadir, S.A. and Proebsting, E.L. 1994. Screening sweet cherry selections for dormant floral bud hardiness. HortScience 29:104–106.
Kappel, F. 2002. 'Stella' sweet cherry. J. Amer. Pom. Soc. 56:130–131.
Kappel, F. 2005. New sweet cherry cultivars from the Pacific Agr-Food Research Centre (Summerland). Acta Hort. 667:53–57.
Kappel, F. 2010. Sweet cherry cultivars vary in their susceptibility to spring frost. HortScience 45:176–177.
Kappel, F., and Lane, W.D. 1998. Recent sweet cherry introductions from the breeding program at Summerland, British Columbia, Canada. Acta Hort. 468:105–109.
Kappel, F. and Sholberg, P.L. 2008. Screening sweet cherry cultivars from the Pacific Agri-Food Research Centre Summerland breeding program for resistance to brown rot (*Monilinia fructicola*). Can. J. Plant Sci. 88:747–752.
Kappel, F., Fisher-Fleming, B., and Hogue, E. 1996. Fruit characteristics and sensory attributes of an ideal sweet cherry. HortScience 31:443–446.
Kappel, F., Lane, W.D., MacDonald, R., Lapins, K., and Schmid, H. 1998. 'Santina', 'Sumpaca Celeste', and 'Sumnue Cristalina' sweet cherries. HortScience 33:1087–1089.
Kappel, F., Lane, W.D., MacDonald, R., Lapins, K., and Schmid, H. 2000a. 'Sumste Samba', 'Sandra Rose', and 'Sumleta Sonata' sweet cherries. HortScience 35:152–154.

Kappel, F., Lane, W.D., MacDonald, R.A., and Schmid.H. 2000b. 'Skeena' sweet cherry. HortScience 35:306–307.
Kappel, F., Lang, G., Anderson, L., Azarenko, A., Facteau, T., Gaus, A., and Southwick, S. 2005. NC-140 Regional cherry rootstock trial (1998): results from western North America. Acta Hort. 667:223–232.
Kappel, F., MacDonald, R.A., and Brownlee, R., 2006. 13S2009 (Staccato™) sweet cherry. Can. J. Plant Sci. 86:1239–1241.
Kappel, F., MacDonald, R., and McKenzie, D.-L. 2000c. Selecting for firm sweet cherries. Acta Hort. 538:355–358.
Kárpáti Z., 1944. Vizsgálatok a Cerasus alnemzetségbe tartozó hazai Prunusokon, Bulletin of the Hungarian College for Horticulture and Vineculture, Budapest, X., 66–80.
Knoche, M., Peschel, S., Hinz, M., and Bukovac, M.J. 2001. Studies on water transport through the sweet cherry fruit surface. II. Conductance of the cuticle in relation to fruit development. Planta 213:927–936.
Kolesnikova, A.F. 1975. Breeding and some biological characteristics of sour cherry in central Russia. Orel, U.S.S.R.: Priokstoc izdatel'stvo. In: Iezzoni, A.F., Schmidt, H., Albertini, A. 1990. Cherries (*Prunus*). In: Moore, J. N., Ballington, J.R. genetic resources of temperate fruit and nut crops. ISHS. Wageningen, the Netherlands.
Küppers, H. 1978. Problematik der Veredlungsunterlagen für Sauer- und Süßkirschen im Spiegel von 250 Jahren. Deutsche Baumschule. 11. 350–359.
Lacis, G., Kaufmane, E., Rashal, I., Trajkovski, V., and Iezzoni, A.F. 2008. Identification of self-incompatibility (S) alleles in Latvian and Swedish sweet cherry genetic resources collections by PCR based typing. Euphytica 160:155–163.
Lalli, D.A., Decroocq, V., Blenda, A.V., Schurdi-Levraud, V., Garay, L., Le Gall, O., Damsteegt, V., Reighard, G.L., and Abbott, A.G. (2005) Identification and mapping of resistance gene analogs (RGAs) in *Prunus*: a resistance map for *Prunus*. Theor. Appl. Genet. 111:1504–1513.
Lane, W.D. and Schmid, H. 1984. Lapins and Sunburst sweet cherry. Can. J. Plant Sci 64:211–214.
Lang, G.A. 1999. Cherry – Sweet. HortScience 34:186–187.
Lang, G.A. 2002. Cherry – Sweet. HortScience 37:253–255.
Lang, G., Howell, W. and Ophardt, D. 1998a. Sweet cherry rootstock/virus interactions. Acta Hort. 307–314.
Lang, G., Ophardt, D., and Olmstead, J. 1998b. Sweet cherry breeding at Washington State University. Acta Hort. 468:97–104.
Lankes, C. 2007. Testing of *Prunus* rootstock clones for virus tolerance. Acta Hort. 732: 351–354.
Lapins, K.O. 1971. Stella, a self-fruitful sweet cherry. Can. J. Plant Sci. 51:252–252.
Lapins, K.O. 1974. Summit sweet cherry. Can. J. Plant Sci. 54:851.
Lee, S. and Wen, J. 2001. A phylogenetic analysis of *Prunus* and the Amygdaloideae (Rosaceae) using ITS sequences of nuclear ribosomal DNA. Am. J. Bot. 88:150–160.
Lewis, D. and Crowe, L.K. 1954. The induction of self-fertility in tree fruits. J. Hort. Sci. 29:220–225.
Looney, N.E., Webster, A.D. and Kupferman, E. 1996. Harvest and handling sweet cherries for the fresh market. In: Webster, W.D. and Looney, N.E. 1996. Cherries: crop physiology, production and uses. CAB International, Wallingford, UK.
Marchese, A., Bošković, R.I., Caruso, T., Raimondo, A., Cutuli, M., and Tobutt, K.R. 2007. A new self-compatibility haplotype in the sweet cherry 'Kronjo', S_5', attributable to a pollen-part mutation in the SFB gene. J. Exp. Bot. 58:4347–4356.
Mariette, S., Lefranc, M., Legrand, P., Taneyhill, D., Frasczria-Lacoste, N., and Machaon, N. 1997. Genetic variability in wild cherry populations in France. Effects of colonizing process. Theor. Appl. Genet. 94:904–908.
Matthews, P. and Dow, P. 1978. Cherry breeding. John Innes Institute 68th Annual Report for 1977. p 34.
Matthews, P. and Dow, P. 1979. Cherry breeding. John Innes Institute 69th Annual Report for 1978. p 38.
Matthews, P. and Dow, P. 1983. Cherries. John Innes Institute 72nd Report for the two years – 1981–1982. p 151.

Micheyev, A.M., Revyakina, N.T., and Drozdova, L.A. 1983. Klonoviye podvoyi vishnyi I osobennosty ich razmnosheniya. Sadovodstvo, Moscow, 7. 28–29.
Micke, W.C., Doyle, J.F., and Yeager, J.T. 1983. Doubling potential of sweet cherry cultivars. Calif. Agriculture, March-April, 24–25.
Misirli, A., Gülcan, R. and Tanrisever, A. 1996. The relation between tree vigour of *Prunus mahaleb* L. Types and Sieve Tube Size in Phloem Tissue. Acta Hort. 410:227–232.
Mitschurin I.W. (1951) Ausgewählte Schriften. Verlag Kultur und Fortschritt, Berlin.
Mohanty, A., Martin, J.P., and Aguinagalde, I. 2001. Chloroplast DNA study in wild populations and some cultivars of *Prunus avium* L. Theor. Appl. Genet. 103:112–117.
Moreno, M.A., Montanes, L., Tabuenca, M.C., and Cambra, R. 1996. The performance of Adara as a cherry rootstock. Scientia Hort. 65 (1):85–91.
Newcomb, R.D., Crowhurst, R.N., Gleave, A.P., Rikkeriuk, E.H.A., Allan, A.C., Beuning, L.L., Bowen, J.H., Gera, E., Jamieson, K.R., Janssen, B.J., Laing, W.A., McArtney, S., Nain, B., Ross, G., Snowden, K.C., Souleyre, E.J.F., Walton, E.F., and Yauk, Y.-K. 2006. Analysis of expressed sequence tags from apple. Plant Physiol. 141:147–166.
Nyújtó F. 1987. Az alanykutatás hazai eredményei. Kertgazdaság 19(5):9–34.
Olden, E.J. and Nybom, N. 1968. On the origin of *Prunus cerasus* L. Hereditas 59:327–345.
Olmstead, J. W., Iezzoni, A.F., and Whiting, M.D. 2007. Genotypic differences in sweet cherry fruit size are primarily a function of cell number. J. Amer. Soc. Hort. Sci. 132:697–703.
Olmstead, J.W., Lang, G.A., and Grove, G.G. 2000. A leaf disk assay for screening sweet cherry genotypes for susceptibility to powdery mildew. HortScience 35:274–277.
Olmstead, J.W., Lang, G.A., and Grove, G.G. 2001a. Assessment of severity of powdery mildew infection of sweet cherry leaves by digital image analysis. HortScience 36:107–111.
Olmstead, J.W., Lang, G.A., and Grove, G.G. 2001b. Inheritance of powdery mildew resistance in sweet cherry. HortScience 36:337–340.
Omeg, M. and Omeg, L. 2005. Economics of fruit quality. In: Whiting, M.D. ed. Producing premium cherries: Pacific Northwest fruit school cherry shortcourse proceedings. Good Fruit Grower. Yakima, Washington.
O'Rourke, D. 2005. World sweet cherry review: 2005 Edition. Belrose Inc., Pullman, Washington.
Parnia, P., Mladin, G., Dutu, I., Movileanu, M., and Slamnoiu, T. 1997. Ameliorarea portaltoilor pentru cires si visin. in Braniste N. és Dutu, I. (Eds.) Contributii romanesti la ameliorarea genetica a soiurilor si portaltoilor de pomi, arbusti fructiferi si capsuni . Institutul de cercetare – dezvoltare pentru pomicultura, Pitesti-Maracineni. 211 pp. 141–146.
Perishable Group. 2007 http://www.perishablegroup.com/dnn/Portals/17/articles/Cherry-June.pdf
Perry, R. L. 1987. Cherry rootstocks. In: Rom, R. C., Carlson, R. F.: Rootstocks for Fruit crops. John Wiley & Sons, New York. 217–264.
Plock, H. 1973. Die Bedeutung der *Prunus fruticosa* Pall. als Zwergunterlage für Süss- und Sauerkirschen. Mitt. Klosterneuburg. 23. 137–140.
Poll, L., Peterson, M.B., and Nielson, G.S. 2003. Influence of harvest year and harvest time on soluble solids, titrateable acid, anthocyanins content and aroma components in sour cherry (*Prunus cerasus* L. cv. "Stevnsbaer"). Eur. Food Res. Technol. 216:212–216.
Proebsting, E.L., Jr. 1970. Relation of fall and winter temperatures to flower bud hardiness of deciduous fruit trees. HortScience. 5:422–424.
Proebsting, E.L. 1990. The interaction between fruit size and yield in sweet cherry. Fruit Var. J. 44:169–172.
Proebsting, E. 1992. Pruning for higher quality cherries, crop regulation. Proc. Wash. State Hort. Assn. p. 314–315.
Proffer, T.J., Jones, A.L., and Perry, R.L. 1988. Testing of cherry rootstocks for resistance to infection by species of *Armillaria*. Plant Disease 72:488–490.
Redalen G. 1984. Fertility in Sour Cherries. Gartenbauwissenschaft 49:212–217.
Rehder, A. 1974. Manual of cultivated trees and shrubs – hardy in North America. Macmillan, New York.
Roper, T.R. and Loescher, W.H. 1987. Relationships between leaf area per fruit and fruit quality in Bing sweet cherry. HortScience 22:1273–1276.

Ross, C.F., Chauvin, M.A., and Whiting, M. 2009. Firmness evaluation of sweet cherries by a trained and consumer sensory panel. J. Text. Studies 40:554–570.

Rozpara, E. and Grzyb, Z.S. 2005. Frutana – A New Interstock for Sweet Cherry Trees. Acta Hort. 658:247–250.

Sansavini, S. and Lugli, S. 1996. Performance of the sweet cherry cultivar 'Van' on new clonal rootstocks. Acta Hort. 410:363–371.

Sansavini, S., and Lugli, S. 2005. New sweet cherry cultivars developed at the University of Bologna. Acta Hort. 667: 45–52.

Santi, F. and Lemoine, M. 1990. Genetic markers for *Prunus avium* L. 2. Clonal identification and discrimination from *P. cerasus* and *P. cerasus* x *P. avium*. Ann. Sci. For. 47:219–227.

Sarger, J. 1972. Le bouturage ligneux de l'espece *Prunus mahaleb*. Convegno del ciliego, Verona, 1972 June.

Saunier, R., and Claverie, J. 2001. Le cerisier: évolution de la culture en France et dans le monde. Point sur les variétés, les portegreffe. La Fruit Belge 490:50–62.

Schimmelpfeng, H. 1996. Unterlagenzüchtung für Süßkirschen in Deutschland – die Weihenstephaner Arbeiten. Schweiz. Z. Obst- u. Weinbau. 132:331–334.

Schmidt, M. 1937. Infektionversuche mit *Sclerotina cinerea* an süß- und sauerkirschen. Gartenbauwisswissenschaft 11:167–182.

Schmid, W. and Grosch, W. 1986 Identifizierung flüchtiger Aromastoffe mit hohen Aromawerten in Sauerkirschen (*Prunus cerasus* L.). Z Lebensm Unters Forsch 182:407–412.

Schueler, S., Tusch, A., Schuster, M., and Ziegenhagen, M. 2003. Characterization of microsatellites in wild and sweet cherry (*Prunus avium* L.) – markers for individual identification and reproductive processes. Genome 46:95–102.

Schuster, M. 2004. Investigation on resistance to leaf spot disease, *Blumeriella jaapii* in cherries. J. Fruit Ornam. Plant Res. 12, 275–279.

Schuster, M. and Schreiber, H. 2000. Genome investigation in sour cherry, *P. cerasus* L. Acta Hort. 538:375–379.

Schuster, M. and Wolfram, B. 2004. Results of sour cherry breeding in Dresden-Pillnitz. Acta Hort. 663:911–914.

Schuster, M., Flachowsky, H., and Köhler, D. 2007. Determination of self-incompatible genotypes in sweet cherry (*Prunus avium* L.) accessions and cultivars of the German Fruit Gene Bank and from private collections. Plant Breeding 126:533–540.

Seif, S. and Gruppe, W. 1985. Chilling requirements of sweet cherries (*Prunus avium*) and interspecific hybrids (*Prunus* X ssp.). Acta Hort. 149:71–75.

Sherman, W.B. and Lyrene, P.M. 2003. Low chill breeding of deciduous fruits at the University of Florida. Acta Hort. 622:599–605.

Simon, G., Hrotkó, K., and Magyar, L. 2004. Fruit quality of sweet cherry cultivars grafted on four different rootstocks. Inter. J. Hort. Sci. 10:59–62.

Šimunic, V., Kovač, S., Gašo-Sokač, D., Pfannhauser, W., and Murkovic, M. (2005) Determination of anthocyanins in four Croatian cultivars of sour cherries (*Prunus cerasus*). Eur. Food Res. Technol. 220:575–578.

Song, G.Q. and Sink, K.C. 2006. Transformation of Montmorency sour cherry (*Prunus cerasus* L.) and Gisela 6 (*P. cerasus* x *P. canescens*) cherry rootstock mediated by *Agrobacterium tumefaciens*. Plant Cell. Rep. 25:117–123.

Sonneveld, T., Robbins, T.P., Bošković, R., and Tobutt, K.R. 2001. Cloning of six cherry self-incompatibility alleles and development of allele-specific PCR detection. Theor. Appl. Genet. 102:1046–1055.

Sonneveld, T., Tobutt, K.R., and Robbins, T.P. 2003. Allele-specific PCR detection of sweet cherry self-incompatibility (*S*) alleles S1 to S6 using consensus and allele-specific primers. Theor. Appl. Genet. 107:1059–1070.

Sonneveld, T., Tobutt, K.R., Vaughan, S.P., and Robbins, T.P. 2005. Loss of pollen-S function in two self-compatible selections of *Prunus avium* is associated with deletion/mutation of an S haplotype-specific F-box gene. The Plant Cell 17:37–51.

Southwick, S.M., Shackel, K., and Yeager, J.T. 1994. Relationship between summer temperature and deep suture formation in 'Bing' sweet cherry. HortScience 29:440 (abstr.)

Stainer, R. 1975 "Stevnsbaer" – eine interessante Sauerkirsche für die Safterzeugung. Obstbau/ Weinbau 5:142–145.

Stehr, R. 2005. Experiences with dwarfing sweet cherry rootstocks in Northern Germany. Acta Hort. 667:173–177.

Stockinger, E.J., Mulinix C.A. Long, C.M., Brettin T.S., and Iezonni, A.F. 1996. A linkage map of sweet cherry based on RAPD analysis of a microspore-derived callus culture population. J. Hered. 87:214–218.

Struss, D., Boritzki, M., Glozer, K., and Southwick, S.M. 2001. Detection of genetic diversity among populations of sweet cherry (*Prunus avium* L.) by AFLPs. J. Hort. Sci. Biotech. 76:362–367.

Struss, D., Boritzki, M., Karle, R., and Iezzoni, A.F. 2002. Microsatellite markers differentiate eight Giessen cherry rootstocks. HortScience 37:191–193.

Tanrisever, A. and Feucht, W. 1978. Gehalt an Flavanen im Phloem von Kirschartigen Prunusgehölzen. Mitt. Klosterneuburg 28. 72–74.

Tao, R, Yamane, H., Sugiura, A., Murayama, H., Sassa, H., and Mori, H. 1999. Molecular typing of S-alleles through identification, characterization and cDNA cloning for S-RNases in sweet cherry. J. Am. Soc. Hort.l Sci. 124:224–233.

Tatarinov, A.N. and Zuev, V.F. 1984. Pitomnyik plodovüh I jagodnüh kultur. Rosszelhozizdat. Moszkva.

Tavaud, M., Zanetto, A., David, J.L., Laigret, F., and Dirlewanger, E. 2004. Genetic relationships between diploid and allotetraploid cherry species (*Prunus avium*, *Prunus* x *gondouinii* and *Prunus cerasus*. Heredity 93:631–638.

Terpó, A. 1968. A sajmeggy (*Cerasus mahaleb* (L.) Mill.) taxonomiai problémái és a gyakorlat. (Problems of taxonomy of *Cerasus mahaleb* (L.) Mill.). Szőlő- és gyümölcstermesztés, Budapest 4:103–131.

Theiler-Hedtrich, R. 1985. Sweet cherry breeding programme at the swiss federal research station II. Results of bacterial canker resistance and seedling vigour. Acta Hort. 169:63–72.

Theiler-Hedtrich, R.1990. Induction of dwarf F12/1 cherry rootstocks by in vitro mutagenesis. Acta Hort. 280:367–371.

Thomas, M. and Sarger, J. 1965. Selection de *Prunus mahaleb* porte greffe du cerisier. Rapport General du Congress Pomologique de Bordeaux, 175–201.

Thompson, M. 1996. Flowering, pollination and fruit set. 1996. In: Webster, W.D. and Looney, N.E. 1996. Cherries: crop physiology, production and uses. CAB International, Wallingford, UK.

Tobutt, K.R., Bošković, R., Čerovic, R., Sonneveld, T., and Ružić, D. 2004. Identification of incompatibility alleles in tetraploid species sour cherries. Theor. Appl. Genet. 108:775–785.

Toyama, T.K., Ophardt, D.R., Howell, W.E., and Grove, G.G. 1993. New powdery mildew resistant sweet cherry. Fruit Var. J. 47:234–235.

Trefois, R. 1980. New Dwarfing Rootstocks for Cherry Trees. Acta Hort. 114:208–217.

Treutter, D., Feucht, W., and Liebster, G. 1993. 40 Jahre Wissenschaft für den Obstbau in Weihenstephan. Obst- und Gartenbauverlag, Müchen, 170 pp.

Tsukamoto, T., Tao, R., and Iezzoni, A.F. 2008a. PCR markers for mutated *S*-haplotypes enable discrimination between self-incompatible and self-compatible sour cherry selections. Mol. Breeding 21:67–80.

Tsukamoto, T., Potter, D., Tao, R., Vieira, C.P., Vieira, J., Iezzoni, A.F. 2008b. Genetic and molecular characterization of three novel *S*-haplotypes in sour cherry (*Prunus cerasus* L.). Journal of Experimental Botany 59:3169–3185.

Tsukamoto, T., Hauck, N.R., Tao, R., Jiang, N., Iezzoni, A.F. 2010. Molecular and Genetic Analyses of Four Nonfunctional *S* Haplotype Variants Derived from a Common Ancestral *S* Haplotype Identified in Sour Cherry (*Prunus cerasus* L.). Genetics 184:411–427.

Turovtseva, V.A., Turovtseva, N.N., and Turovtsev, N.I. 1996 Distant hybridization in breeding sour cherry. (Russ.) SAdovodstvo I Vinogradarstvo 5–6, 16–17.

Vaughan, S.P. and Russell, K. 2004. Characterisation of novel microsatellites and development of multiplex PCR for large-scale population studies in wild cherry, *Prunus avium* L. Mol. Ecol. Notes 4; 429–431.

Verner, L. and Blodgett, E.C. 1931. Physiological studies of the cracking of sweet cherries. University of Idaho Agriculture Experiment Station Bulletin No. 184.

Walther, E. and Franken-Bembenek, S. 1998. Evaluation of interspecific cherry hybrids as rootstocks for sweet cherries. Acta Hort. 468:285–290.

Walther, E. and Sauer, A. 1985. Analysis of radiosensitivity – a basic requirement for in vitro somatic mutagenesis. I. *Prunus avium* L. Acta Hort. 169:97–104.

Wang, D., Karle, R., Brettin, T.S., and Iezzoni, A.F. 1998. Genetic linkage map in sour cherry using RFLP markers. Theor. Appl. Genet. 97:1217–1224.

Wang, H., Nair, M.G., Strasburg, G.M., Chang,, Y., Booren, A.M., Gray, J.I., and DeWitt D.L. (1999) Antioxidant and antiinflammatory activities of anthocyanins and aglycon, cyanidin, from tart cherries. J. Nat Prod. 62, 294–296.

Watkins, R. 1976. Cherry, plum, peach, apricot and almond. In: Simmonds, N.W. Evolution of crop plants. Longman. New York. 242–247.

Webster, A.D. 1980. Dwarfing rootstocks for plums and cherries. Acta Hort. 114:201–207.

Webster, A.D. 1996. The taxonomic classification of sweet and sour cherries and a brief history of their cultivation. In: Webster, W.D. and Looney, N.E. 1996. Cherries: crop physiology, production and uses. CAB International, Wallingford, UK.

Webster, A.D. and Schmidt, H. 1996. Rootstocks for sweet and sour cherries. In: Webster, W.D. and Looney, N.E. 1996. Cherries: crop physiology, production and uses. CAB International, Wallingford, UK.

Webster, A.D., Tobutt, K.R., and James, D.J. 1997. Rootstock breeding and orchard testing at Horticulture Research International – East Malling. Acta Hort. 451:83–88.

Westwood, M. N. 1978. Mahaleb x Mazzard hybrid cherry stocks. Fruit Var.J. 32–39.

Wharton, P.S., Iezzoni, A., and Jones A.L. 2003. Screening cherry germ plasm for resistance to leaf spot. Plant Dis. 87, 471–477.

Wicks, T.J., Bumbieris, M. Warcup, J.H. and Wallace, H.R. 1984. *Phytophtora* in fruit orchards in South Australia. Biennial Rep. of the Waite Agricultural Research institute. 147.

Wiersma, P.A., Wu, Z., Zhou, L., Hampson, C., and Kappel, F. 2001. Identification of new self-incompatibility alleles in sweet cherry (*Prunus avium* L.) and classification of incompatibility groups by PCR and sequencing analysis. Theor. Appl. Genet. 102:700–708.

Wojcicki, J.J. 1991. *Prunus* x *stacei* (Rosaceae), a new spontaneous hybrid of *P. fruticosa*, *P. cerasus* and *P. avium*, Fragm.Flor. Geobot. 35 (1–2):139–14.

Wolfram, B. 1971. Unterlagenzüchtung für Süsskirschen. Der Neue Deutsche Obstbau. 17.1. 3–4.

Wolfram, B. 1996. Advantages and Problems of Some Selected Cherry Rootstocks in Dresden-Pillnitz. Acta Hort. 410:233–237.

Wolfram B. 2000 Sour cherry breeding at Dresden-Pillnitz. Acta Hort. 538:359–362.

Wünsch, A. and Hormaza, J.I. 2002. Molecular characterization of sweet cherry (*Prunus avium* L.) genotypes using peach [*Prunus persica* (L.) batsch] SSR sequences. Heredity 89:56–63.

Wünsch, A. and Hormaza, J.I. 2004. Genetic and molecular analysis in Cristobalina sweet cherry, a spontaneous self-compatible mutant. Sex. Plant Reprod. 17:203–210.

Yamane H., Tao R., Sugiura A., Hauck N.R., and Iezzoni A.F. 2001. Identification and characterization of S-RNases in tetraploid sour cherry (*Prunus cerasus*). J. Amer. Soc. Hort. Sci. 126: 661–667.

Yoltuchovski, M.K. 1977. Proisvodstvenno-biologicheskaya characteristka raslychnych form podvoyev vishnyi. In: Plodovoje pitomnikovodstvo Moldavii. Redact. Andryushtchenko. Isdatelstvo Krtya Moldavenyanske, Kishinau. 39–69.

Ystaas, J. and Frøynes, O. 1990. An evaluation based on the field performance of 38 cultivars of sweet cherries. Norsk Landbruksforsking 4:115–126.

Zepp, L. and Szczygiel, A. 1985. Pathogenicity of *Pratylenchus crenatus* and *Pratylenchus neglectus* to three fruit tree seedling rootstocks. Fruit Science Reports. 12.3. 109–117.

Zhou, L., Kappel, F., Hampson, C., Wiersma, P.A., and Bakkeren, G. 2002. Genetic analysis and discrimination of sweet cherry cultivars and selections using amplified fragment length polymorphism fingerprints. J. Amer. Soc. Hort. Sci. 127:786–792.

Zhu, M., Zhang, X., Zhang, K., Jiang, L., and Zhang, L. 2004. Development for a single molecular marker specific for detecting the self-compatible *S4*' haplotype in sweet cherry (*Prunus avium* L.) Plant Molecular Biology Reporter 22:387–398.

Zhukov, O.S. 1979. Anwendung neuer genetischer Methoden in der Sauerkirschenzüchtung. Tag.-Ber. AdL DDR. 174:203–206.

Zhukov, O.S. and Charitonova, E.N. 1988. Selektzia vishi. Agroproizdat, Moscow.

Zielinski, Q.B. 1977. Modern systematic pomology. Pomona Books. Rockton, Ont.

Zwintzscher, M. 1973. Die Variabilität in Nachkommenschaften einer Sauerkirschsorte nach Selbstung und Kreuzung und deren Bedeutung für die Sortenzüchtung. Estratto dagli Atti del 2. Convegno del Ciliegio Verona 14/16 giugno 1973, 3–19.

Chapter 14
Peach

David H. Byrne, Maria Bassols Raseira, Daniele Bassi, Maria Claudia Piagnani, Ksenija Gasic, Gregory L. Reighard, María Angeles Moreno, and Salvador Pérez

Abstract The peach is the third most produced temperate tree fruit species behind apple and pear. This diploid species, *Prunus persica*, is naturally self-pollinating unlike most of the other cultivated *Prunus* species. Its center of diversity is in China, where it was domesticated. Starting about 3,000 years ago, the peach was moved from China to all temperate and subtropical climates within the Asian continent and then, more than 2,000 years ago, spread to Persia (present day Iran) via the Silk Road and from there throughout Europe. From Europe it was taken by the Spanish and Portuguese explorers to the Americas. It has an extensive history of breeding

D.H. Byrne (✉)
Department of Horticultural Sciences, Texas A&M University,
College Station, TX, USA
e-mail: dbyrne@tamu.edu

M.B. Raseira
EMBRAPA – Clima Temperado, BR 392 km 78 – Cx.,
Postal 403, Pelotas 96001-970, RS, Brazil
e-mail: bassols@cpact.embrapa.br

D. Bassi • M.C. Piagnani
Dipartimento di Produzione Vegetale, Università
degli Studi di Milano, Milan, Italy
e-mail: daniele.bassi@unimi.it; claudia.piagnani@unimit.it

K. Gasic • G.L. Reighard
Department of Environmental Horticulture,
Clemson University, Clemson, SC, USA
e-mail: kgasic@clemson.edu; grghrd@clemson.edu

M.A. Moreno
Estación Experimental de Aula Dei (CSIC), Zaragoza, Spain
e-mail: mmoreno@eead.csic.es

S. Pérez
Recursos Genéticos y Mejoramiento de *Prunus,*
Guillermo Prieto 14, Centro, Querétaro Qro. 76000, Mexico
e-mail: spg1948@hotmail.com

M.L. Badenes and D.H. Byrne (eds.), *Fruit Breeding*, Handbook of Plant Breeding 8,
DOI 10.1007/978-1-4419-0763-9_14,

that has resulted in scion cultivars with adaptability from cold temperate to tropical zones, a ripening season extending for 6–8 months, and a wide range of fruit and tree characteristics. Peach has also been crossed with species in the *Amygdalus* and *Prunophora* subgenera to produce interspecific rootstocks tolerant to soil and disease problems to which *P. persica* has limited or no resistance. It is the best known temperate fruit species from a genetics perspective and as a model plant has a large array of genomics tools that are beginning to have an impact on the development of new cultivars.

Keywords *Prunus persica* • History • Genetic resources • Breeding • Biotechnology • Interspecific • Hybrids • Model plant • Stone fruit • Drupe

1 Introduction

1.1 Economic Importance

The peach is the third most important temperate tree fruit species behind apples and pears. This total production is estimated at over 17.8 million tons. The production has more than doubled since 1980, from 7.7 to 17.8 million tons, mainly due to the rapid production increases seen in China. Production in the Americas and Europe has remained fairly steady with only small increases since 1980. Other countries that have more than doubled their production over the last 30 years are Korea, Chile, Spain, Egypt, Tunisia, and Algeria. The five largest producer countries are China, which accounts for ~46% of the world production, followed by Italy (~9%), Spain (~7%), the USA (~7%), and Greece (~4%) (USDA/ARS 2008; FAOSTAT 2010) (Table 14.1). Over 90% of this production is for the fresh market. Only nine countries (the USA, South Africa, Australia, Argentina, Chile, China, Spain, Greece, and Italy) are significant producers of processed peaches with the two largest producers, Greece and China, with an estimated production of 338,000 and 206,500 mt, respectively, in 2005 (FAS, USDA World and Export Opportunities 2006).

Table 14.1 World peach production (1,000 MT) from 1980 to 2008 (FAOSTAT, http://www.fao.org accessed 2 March 2010)

Region	1980–1984	1985–1989	1990–1994	1995–1999	2000–2004	2005–2008
World	7,679	8,335	10,434	11,758	14,746	17,840
Asia	1,433	1,832	3,062	4,657	7,179	10,106
Americas	2,060	2,033	2,248	2,244	2,509	2,407
Europe	3,827	4,115	4,637	4,048	4,208	4,319
Africa	261	282	408	710	725	867
Oceania	121	120	88	110	137	149

1.2 Uses

All the economically important cultivars belong to *Prunus persica* (L.) Batsch. The fruit may have melting, nonmelting, or stony hard flesh and varies in color from green to white to yellow and orange to red and purple, with various gradations and combinations of these tonalities. Peaches are mainly used as fresh fruit and processed to produce canned fruit, jellies, jams, juice, pulp for yogurts, and liquors. In some production regions, the seeds are utilized as rootstocks and the hard endocarp is used for charcoal production. The ornamental use of peach flowers is also significant, especially in China and Japan (Yulin 2002; Hu et al. 2005, 2006).

1.3 Taxonomy, Botany, and Basic Description of the Species

The peach belongs to the Rosaceae family, subfamily Prunoideae, genus *Prunus* (L.), subgenus *Amygdalus*, section Euamygdalus. Other subgenera besides *Amygdalus* within the genus *Prunus* are *Prunophora* (plums), *Cerasus* (cherries), *Padus*, and *Laurocerasus*. Commercial peach cultivars belong to the species *Prunus persica* (L.) Batsch. Related interfertile species include *P. dulcis* (Mill.) D. A. Webb, *P. davidiana* (Carr.) Franch, *P. ferganensis* (Kost and Rjab) Kov. & Kost, *P. kansuensis* Rehd, and *P. mira* Koehene. These species have primarily been used directly as or in the development of rootstocks and ornamentals but not in the development of scion cultivars. All originate from China with some range extension into Nepal and India (*P. mira*) and in the countries which previously formed the Soviet Union (*P. ferganensis*) (Scorza and Sherman 1996). *Prunus persica* can be hybridized with *P. dulcis*, *P. davidiana*, *P. ferganensis*, *P. kansuensis*, and *P. mira*, producing, in most cases, fertile hybrids (Watkins et al. 1995; Scorza and Okie 1990). Crosses between almond (*P. dulcis*) and peaches have been produced with several objectives, but mainly for rootstock development (Moreno 2004; Zarrouk et al. 2005; Felipe 2009; Pinochet 2009; Gradziel 2003; Martínez-Gómez and Gradziel 2002; Martínez-Gómez et al. 2004).

1.4 Distribution and Limits on Adaptation

Although the main production areas for the peach are located in both hemispheres between 30 and 45° latitude (Scorza and Sherman 1996), production is also found throughout the subtropics and tropical regions (Byrne et al. 2000). Disease and insect incidence is a limiting factor favored by conditions of high humidity. Windy, spring weather particularly favors the spread and infection by bacteria such as *Xanthomonas arboricola* (syn. campestris) pv. *pruni* ((Smith) Vauterin et al.), which is one of the most important bacterial disease of peach in the world. High humidity and warm temperature can also favor the incidence of fungal diseases, such as brown rot (*Monilinia* spp.) and anthracnose (*Colletotrichum acutatum* Simmonds), whereas cooler conditions favor powdery mildew (*P. pannosa* (Wallr.: Fr.)) and peach leaf curl (*Taphrina deformans* (Berk.) Tul.).

Beyond the humidity related problems encountered throughout the latitudinal range of the peach, temperature related challenges are seen at the extreme latitudes at which peaches are grown. At high latitudes (45°N and S or above), minimum winter temperatures and spring frosts are the limiting factors. In those areas, flower bud death and consequently crop losses are not uncommon due to cold temperatures. The peach flower is bud hardy, depending on the cultivar, to about −25 to −30°C (Layne 1984). The northern range is extended where large bodies of water, such as the Great Lakes, and the Caspian and Black Seas, ameliorate the minimum temperatures. In latitudes lower than 20° such as Australia, Brazil, Thailand, and Taiwan, the lack of consistent chilling and high temperatures during bloom are important limitations. High temperatures during bloom increases the rate of pollen tube growth, stigma maturation and degeneration leading to poor fruit set (Burgos et al. 1991; Egea et al. 1991; Kozai et al. 2002). Highland tropical zones, which have cool and nonfreezing temperatures year round such as the cool highland mountains of Mexico, allow the possibility of manipulating flower induction, to have off-season harvest (Byrne 2010).

Thus, there is a great opportunity for breeders to improve cultivars especially for these marginal areas. However, even in the temperate zone, where adaptation may not be a problem, there is still much to improve, since market, climate, and consumer preferences change over the time.

2 Origin and Domestication

2.1 Origin and Evolution

The origin of peach in Asia and its domestication in China from where it was dispersed to Europe, Africa and America has been widely reported (Hedrick 1917; Hesse 1975; Westwood 1978; Scorza and Sherman 1996). However, little is known about the evolutionary history of the genus although homogamy studies suggest that the speciation of *P. persica* occurred from an allogamous (outcrossing) species such as *P. scoparia* (Spach) C.K. Schneider and *P. dulcis* (Weinbaum et al. 1986). It appears probable that *P. persica* and other species such as *P. dulcis*, *P. kansuensis*, *P. ferganensis*, *P. scoparia*, *P. mira*, and *P. davidiana* evolved from a common ancestor and are all closely related, as interspecific hybridization among them is common (Meader and Blake 1940; Knight 1969).

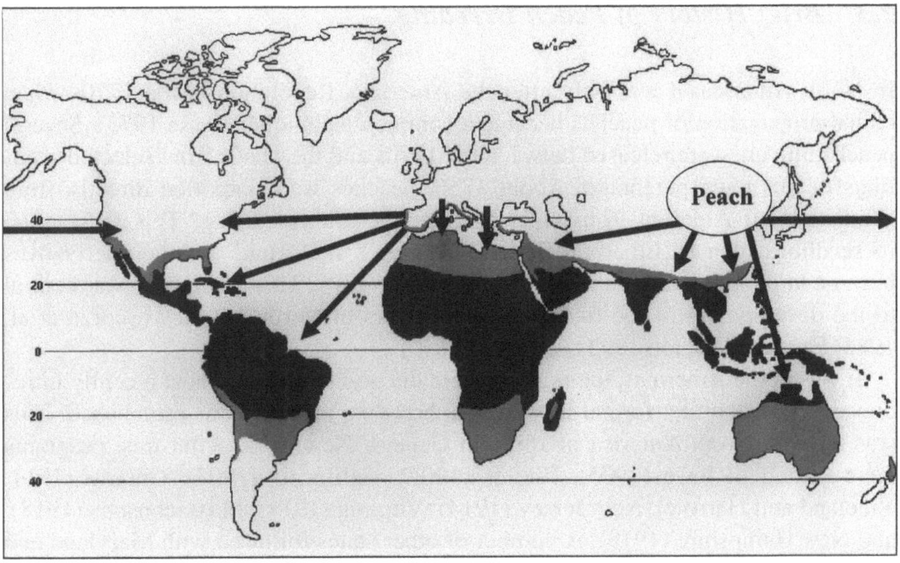

Fig. 14.1 Early dispersal of the peach. The peach dispersed throughout mainland Asia starting about 3,000 years ago and then to Japan and to Persia via the Silk road about 2,000 years ago. From there it was spread throughout Europe and northern Africa and eventually to the Americas by the Spanish and the Portuguese explorers in the sixteenth and seventeenth centuries. *White*, *gray*, and *black* areas are high chill, medium to low chill, and tropical zones, respectively. Modified from Byrne et al. 2000

2.2 Dispersal and Domestication

Starting about 3,000 years ago, the peach was moved from China to all temperate and subtropical climates within the Asian continent and about 1,500–2,000 years ago to Japan (Yamamoto et al. 2003). From Asia, the peach spread to Persia (present day Iran) via the Silk Road and from there throughout Europe more than 2,000 years ago. It was introduced to the Americas by the Spanish and Portuguese during the sixteenth century, where it was rapidly adopted by the Indians and spread to a wide range of environments (Hedrick 1917; Hesse 1975; Scorza and Okie 1990; Faust and Timon 1995), from the tropical highlands of South and Central America, to humid subtropics of Florida and southern Brazil and to the coldest regions in northern USA and southern Canada. There were probably several introductions from different parts of Spain, the Canary Islands, Portugal, and even from the South Pacific, since there were genotypes that adapted well to the humid subtropics (Fig. 14.1).

Seed propagation was the main source of plants up until the first half of the nineteenth century in the USA and Europe and to the middle of the last century in Central and South America. Thus, there are numerous landraces of peaches that have undergone several centuries of selection for adaptation and other characteristics throughout Europe, the Americas, Asia (Byrne et al. 2000; Bouhadida et al. 2007b, 2011; Pérez 1989; Pérez et al. 1993) and Japan (Yamamoto et al. 2003). Some of these traditional cultivars, propagated either by seed or budding, are still used today.

2.3 Brief History of Peach Breeding

In North America, it was only after the American Revolution in the 1770s when clonal propagation of peaches became a common technique (Hesse 1975). Several peach cultivars were released between the 1770s and the 1860s from selected seedlings of unknown parentage. About 1850, peaches were imported directly from China to North America, from which emerged the 'Chinese Cling.' This cultivar and its seedlings such as 'Elberta,' 'Belle of Georgia,' 'J. H. Hale' and their derivatives became important peach cultivars throughout the USA. This germplasm was central to the development of the fresh market cultivars in North America (Scorza et al. 1985; Faust and Timon 1995).

It was in the Americas, the region where the peach has been most recently introduced, where the first formal institutional breeding program was established. This was done in North America in 1895, in Geneva, New York. After this, programs were started in Iowa (1905), Illinois (1907), California (1907), Ontario (1911, Vineland and Harrow), New Jersey (1914), Virginia (1914), Massachusetts (1918), and New Hampshire (1918). A number of other states followed with Maryland and Michigan in the 1920s, Georgia and Texas in the 1930s, Louisiana, Florida, and North Carolina in the 1950s and Arkansas in the mid-1960s (Okie et al. 2008). Private breeding programs were established in California beginning in the 1930s (Okie et al. 2008; Faust and Timon 1995). Most of these programs emphasized the development of locally adapted peaches and nectarines with melting flesh for the fresh market. In Latin America, breeding programs were initiated in southern Brazil at two locations (Pelotas and Sao Paulo) to develop both nonmelting and melting flesh cultivars for both the fresh and processing outlets in the 1950s and in Mexico to develop their nonmelting peaches for the fresh market in the 1980s (Byrne et al. 2000; Byrne and Raseira 2006). Other smaller efforts in developing well adapted peach cultivars are ongoing in Chile, Uruguay, and Argentina.

In Europe, even though peach culture was widespread back in the Middle Ages in France, the first peach breeding program was begun in Italy in the 1920s and later in the 1960s in France. Subsequently, additional programs were established in Spain, Romania, Serbia, Greece, Bulgaria, Ukraine, and Poland (Okie et al. 2008; Llácer 2009). Much of the initial work was based on the cultivars developed in the USA so many of the European cultivars are closely related to North American cultivars (Faust and Timon 1995). These programs include both privately and publically funded programs.

In Asia, where peaches have been cultivated for several thousands of years, the earliest formal breeding program was started 50–60 years ago in Japan followed by multiple breeding efforts in China (1970s) and most recently in Korea, India, and Thailand (Byrne et al. 2000; Okie et al. 2008; Raseira et al. 2008). It is interesting to note that 'Shanghai Suimitsuto' (='Chinese Cling') has also played a key role in the breeding of Japanese cultivars as was seen with North American cultivars (Ma et al. 2006; Xu et al. 2006; Yamamoto et al. 2003).

In South Africa and Australia, the emphasis has historically been on nonmelting flesh peaches. Subsequently, these efforts have expanded to fresh market peaches in both melting and nonmelting flesh (Byrne et al. 2000; Topp et al. 2008).

As the programs evolved, the basic objectives such as productivity, size, excellent appearance, season extension, and firmness are uniform throughout the programs. The major change has been in an increased emphasis in fruit quality, postharvest life, disease and pest resistance, a greater diversity of fruit types, and adaptation to low-chill zones (Byrne 2005). The most dramatic change in peach breeding programs, however, has been the reduction of public breeding and an increase in the private sector breeding programs, which now release the majority of the peaches and nectarines in the USA, France, and Spain. In the USA, about 50% of the public stone fruit breeding programs have closed since 1970. Most of the remaining public breeding programs release new cultivars with patent protection to generate funding for their programs. Even if this is currently a viable approach, in the long term it can create problems by limiting fundamental research in genetics and germplasm resources as well as germplasm exchange among programs (Byrne 2005; Okie et al. 2008). This lack of germplasm exchange is partially counterweighed by the fact that the UE legislation allows the free use of pollen from patented cultivars.

3 Genetic Resources

3.1 Geographic Germplasm Groups

Early domesticates used for fruit were most likely seedlings and coexisted with wild peach seedlings in several geographic regions in China (Wang 1985; Scorza and Okie 1990; Faust and Timon 1995). In China, there are several regional groups of fruiting cultivars: the northern and northwest group, the southern group, and the low-chill group (Yoon et al. 2006; Anderson 2009). The northern and northwestern group includes genotypes adapted to cold winters and hot dry summers and includes the Miantao and Mintao white peach groups, which are drought and cold tolerant, as well as yellow flesh peaches and a few nectarines. The southern group is adapted to a humid subtropical to temperate climate and relatively mild winters. These are generally white, subacid and include many pantao cultivars. It is represented by the 'Shanghai Shuimi' and 'Chinese Cling' peaches which were central in the development of the cultivated peaches developed in Japan (Yamamoto et al. 2003), the USA, and Europe (Scorza et al. 1985; Byrne et al. 2000; Aranzana et al. 2003a). The low-chill group, represented by landraces from Taiwan, Thailand, and southern China, is generally small fruited peaches of low quality. Several of these ('Okinawa' and 'Hawaiian') served as a source of the low-chill trait in the development of the low-chill germplasm in the Florida and Texas breeding programs (Byrne and Bacon 1999; Byrne et al. 2000).

As peaches were moved throughout the world and seed propagated, a series of landraces were developed outside of China that were adapted to a diverse range of climates and selected for regional quality preferences. It appears that many of these landraces and consequently commercial cultivars outside China are derived from the Southern China geographic group as indicated by inbreeding analyses (Scorza et al.

1985, 1988; Byrne and Bacon 1999; Byrne 2003; Byrne and Raseira 2006) and studies with molecular markers (Warburton and Bliss 1996; Yoon et al. 2006; Anderson 2009). Nevertheless, within these groups there are clustering of the genotypes by regional groups (Anderson 2009; Marchese et al. 2005; Badenes et al. 1998; Bouhadida et al. 2007a, b, 2011) and breeding history (Anderson 2009; Yoon et al. 2006; Warburton and Bliss 1996; Aranzana et al. 2003a; Bouhadida et al. 2011). In general, the highest genetic diversity seen was among the northern and northwestern and the low-chill groups and the least among the highly bred cultivars from the USA and Europe (Anderson 2009; Yoon et al. 2006; Warburton and Bliss 1996; Chen et al. 2007).

3.2 Related Species in Breeding

Prunus persica is interfertile with its related species *P. dulcis*, *P. kansuensis*, *P. ferganensis*, *P. scoparia*, *P. mira*, and *P. davidiana*, and interspecific hybridization among them is common (Meader and Blake 1940; Knight 1969). Nevertheless, scion cultivars are almost exclusively developed from *Prunus persica* although there is some work with some interspecifics especially within the Amygdalus section as a source of PPV, powdery mildew, and aphid resistance and for several growth and adaptation traits in scion breeding (Gradziel 2003; Martínez-Gómez et al. 2004; Byrne et al. 2000; Foulongne et al. 2003a, b). The major reason for this is that once the interspecific cross is made it takes from three to five generations to recover the necessary commercial fruiting traits. This is not necessary for the development of ornamental cultivars (Hu et al. 2006) and rootstocks, and thus, a wider range of species have been used.

The most common rootstocks are those derived from species within the section *Euamygdalus* Schneid including peach seedlings (*P. persica*), closely related species to peach (*P. dulcis*, *P. davidiana*, *P. ferganensis*, *P. kansuensis*, and *P. mira*) and interspecific hybrids of peach × almond and peach × *P. davidiana*. Peach is generally graft-compatible with itself and most species within its taxonomic Section *Euamygdalus* (Zarrouk et al. 2006). Peach seedlings have been the main rootstock source for peach on a worldwide basis. Seeds from wild types, commercial cultivars (from canning industry) and special rootstock selections are easily obtained and multiplied in the nursery. In China, seeds of *P. davidiana*, *P. ferganensis*, *P. kansuensis*, and *P. mira* have been also used as rootstocks (Wang et al. 2002; Yulin 2002). The peach × almond hybrids are primarily used in calcareous soils, since they tolerate iron chlorosis well and are graft compatible with peach. They are also vigorous and therefore, appropriate for use in poor, dry soils and in fruit tree replanting situations (Bernhard and Grasselly 1981; Kester and Assay 1986; Egilla and Byrne 1989; Moreno et al. 1994; Felipe 2009). The peach × *P. davidiana* hybrids induce in general good productivity to peach scions, and their selections are resistant to root-knot nematodes (Edin and Garcin 1994).

Although graft compatibility can be an issue, rootstocks from various *Euprunus* species have also been employed as peach rootstocks (Layne 1987; Reighard and

Loreti 2008). This group includes the hexaploid plums (European plums—*P. domestica* L., or St. Julien and 'Pollizo de Murcia' plums—*P. insititia* L.) because the graft compatibility with peaches is generally good. It also includes the diploid (Myrobalan or cherry plum—*P. cerasifera* Ehrh. and Japanese plums—*P. salicina* Lindl.) and tetraploid plums (Sloe—*P. spinosa* L.). In addition, there are numerous interspecific hybrids with different ploidy levels such as the Marianna plums (*P. cerasifera*×*P. munsoniana* W. Wight & U.P. Hedrick). Peach compatibility on fast-growing plums (*P. cerasifera* and interspecific hybrids with this species) differs substantially depending on the evaluated genotype (Zarrouk et al. 2006) and typical "translocated" incompatibility symptoms are frequently seen (Moreno et al. 1993). In the case of 'Damas GF 1869' (a pentaploid rootstock, probably *P. domestica*×*P. spinosa*), at least two dominant alleles are responsible for the incompatibility (Salesses and Alkai 1985), but another type of genetic control might be involved in the case of Myrobalan (Salesses and Bonnet 1992; Pina and Errea 2005). Excessive suckering may occur with several plum rootstocks, mainly if they are micropropagated.

Plum rootstocks are more tolerant to compact soils and waterlogging than other species of *Prunus* L., a fundamental reason for their use (Rowe and Catlin 1971; Salesses and Juste 1970; Xiloyannis et al. 2007). In addition, some of them provide greater tolerance to fungal diseases (*Phytophthora* crown rot, *Armillaria* root rot) favored by waterlogged and/or replant problems in the soil. A more stable resistance to root-knot nematodes (*Meloidogyne* species) can also be found in plum, when compared with resistant peach and almond sources that express a near-complete or incomplete spectrum of resistance (Pinochet et al. 1999; Dirlewanger et al. 2004a, c). Moreover, some Myrobalans are highly resistant or immune to all root-knot nematode species, even under high and continuous inoculum pressure and high temperatures (Esmenjaud et al. 1996). This resistance is attributed to three major genes, *Ma1*, *Ma2*, and *Ma3* (Lecouls et al. 1997; Rubio-Cabetas et al. 1998).

More recently, especially because of improved propagation techniques, there has been the development of interspecific hybrids between species from Sections *Euamygdalus* and *Euprunus*, and rootstocks or hybrids from Sections *Prunocerasus* Koehne and *Microcerasus* Webb. In spite of the sterility these interspecific hybrids have been made with the purpose of bringing together the desirable traits of plum, almond, and peach species (Hesse 1975; Scorza and Okie 1990; Pérez and Moore 1985; Moreno 2004). Once the hybrids are created, they are selected for their ease of propagation as well as the adaptation traits of interest.

3.3 Germplasm Collections

The most extensive collections of peach germplasm have been assembled in China. Since the 1960s most of China has been explored and collections made of the peach germplasm including several of the related species. Thus, these collections include many of the local cultivars and landraces from China where this crop was domesticated as well as introduced cultivars from throughout the world. In the 1980s, China established three national peach repositories in Nanjing, Zhengzhou, and Beijing. The collection in Nanjing has 560 accessions and is focused on the southern germplasm and resistance to various diseases and waterlogging. The Beijing collection (280 accessions) houses the northern peach germplasm and the Zhenghou repository (650 accessions) focuses on the germplasm collected from the northwest of China including accessions from the five related species (Wang and Zhang 2001; Wang et al. 2002). Other significant national collections would be those in Japan (600 accessions), Korea (300 accessions), the USA (280 accessions including four related species and an almond germplasm collection of about 100 accessions), Brazil (732), Ukraine (~1,500 accessions), and over 2,000 accessions in Europe with the largest collections in France, Spain (Bouhadida et al. 2011) and Italy. Beyond the collections in China, these collections tend to consist primarily of commercial cultivars with some accessions to represent wild seedlings, rootstocks, traditional cultivars, and landraces.

4 Major Breeding Achievements

4.1 Scion Cultivars

There are hundreds of peach and nectarine cultivars used commercially throughout the world (Ctifl 1994; Brooks and Olmo 1997; Okie 1998; Yulin 2002). In fact, the international peach breeding community has been very active and over the past several decades have released about 100 new cultivars per year (Della Strada and Fideghelli 2003; Fideghelli et al. 1998). The three most important achievements in peach breeding have been the expansion of its adaptation, the extension of its harvest period, and the diversification of its market.

The first step in this expansion of its adaptation was the dispersal of the peach via seed from its origin in north and northwest China to southern China and then throughout the world. During this early dispersal, the peach was selected for local adaptation from tropical to high latitude temperate zones over a period of centuries. Once breeding programs were initiated this raw germplasm was used to develop better commercial cultivars. Currently, the most active of this breeding is the development of early ripening medium and low-chill peach and nectarine cultivars mainly driven by the desire to have fruit available year round. Beyond, adaptation to temperature variations, work has resulted in peach cultivars resistant to bacterial leaf and fruit spot (*Xanthomonas arboricola* pv. *pruni* (Smith) Vauterin et al.). Unfortunately little work has been done on other major diseases such as brown rot, powdery mildew, peach scab, rust, anthracnose among others because they were either only regionally important, caused occasional damage, or could be easily controlled by chemical applications. Currently, given more restrictions on chemical use, approaches to minimize the use of chemicals via cultural control and the development of disease resistance are being emphasized (Byrne et al. 2000; Byrne 2005).

The extension of the harvest season has been the objective of countless breeding programs and has resulted in expanding a 1- to 2-month harvest season to one that can be as long as 8 months. Much of this was done by manipulating the fruit development period but this was also supplemented by selecting for earlier blooming genotypes. Thus, in regions where spring frosts are not a production limitation, the earliest ripening genotypes are also the earliest blooming. Beyond this, cultivars were also selected for adaptation to lower chill zones where the bloom occurred earlier and thus had the potential of earlier ripening as well.

Finally, the market for peaches has been expanded to two ways. First, the locally marketed peach of the 1900s was transformed into a peach suitable for national and international markets by significantly improving fruit size, appearance and firmness. Unfortunately, the progress in raising the internal qualities such as sugar and antioxidants content, tolerance to internal breakdown (IB) and other postharvest traits has lagged behind, but recently there has been an increased emphasis on these factors in several breeding efforts (Byrne 2005; Peace et al. 2006; Cantín et al. 2009a, b, 2010b). The other strategy for increasing its market share has been the development of new products. The best example of this would be the development of the nectarine as another fruit. This process began in the 1950s in the USA and now nectarine production is about 40% of the fresh peach production. This diversification of the fresh peach products available continues today (Byrne 2005).

4.2 Rootstocks

The range of rootstocks now available for peach worldwide has increased dramatically in the last few decades (Table 14.2). With the improvement of vegetative propagation technology for *Prunus*, including tissue culture, many of the breeding programs have focussed on the generation of complex *Prunus* species hybrids to overcome soil and disease problems to which *P. persica* has limited or no resistance (Reighard 2002; Moreno 2004; Reighard and Loreti 2008).

Considerable progress has been made in developing iron chlorosis tolerant rootstocks, using peach × almond hybrids, first from open pollinated or wild germplasm sources, and in the last two decades with controlled interspecific hybridization. Research on peach–almond hybrid rootstocks tolerant to iron-chlorosis, ease of vegetative propagation and graft compatibility with peach led to the selection of highly vigorous rootstocks such as 'GF 677' (Bernhard and Grasselly 1981), which have been widely adopted in the Mediterranean basin countries. Other regional selections are 'Adafuel' (Cambra 1990; Moreno et al. 1994), 'Mayor' (Cos et al. 2004) and 'Sirio' (Loreti and Massai 1994). Unfortunately all of these are susceptible to root-knot nematodes. Recently, three high-vigor peach–almond hybrids (e.g., 'Monegro,' 'Garnem,' and 'Felinem') have been derived from a cross between the almond

Table 14.2 List and description of commercial or released peach rootstocks

Rootstock[a]	Species	Origin	Calcareous soil tolerance	Waterlogging tolerance	Root-knot nematodes resistance	Other characteristics	References
Peach-based rootstocks (Section Euamygdalus)							
GF 305	*P. persica*	INRA, France	Susceptible	Susceptible	Susceptible (Ma, Mi) Resistant (Mj, Mh)	Easy propagation	Grasselly (1983); Salesses et al. (1970); Esmenjaud et al. (1994)
Guardian™	*P. persica*	USDA-ARS and Clemson U., The USA	Susceptible	Susceptible	Resistant (Mi, Mj)	Tolerant to PTSL, bacterial canker complex	Beckman et al. (1997); Nyczepir et al. (2006); Reighard and Loreti (2008)
Lovell, Halford	*P. persica*	UC, The USA	Susceptible	Susceptible	Susceptible	Easy propagation	Egilla and Byrne (1989); Lu et al. (2000); Reighard and Loreti (2008)
Missour	*P. persica*	Unknown, Morocco	Susceptible	Susceptible	Unknown	Easy propagation	Tagliavini and Rombolà (2001)
Montclair	*P. persica*	INRA, France	Moderately tolerant?	Moderately tolerant	Susceptible	Easy propagation	Grasselly (1988); Fernández et al. (1994); Shi and Byrne (1995)
P.S.B2	*P. persica*	U Pisa, Italy	Susceptible	Susceptible	Resistant	Yield efficiency	Loreti and Massai (2006); Reighard and Loreti (2008)
Rubira	*P. persica*	INRA, France	Susceptible	Moderately tolerant	Susceptible	Red leaf	Grasselly (1988); De Salvador et al. (2002)
Siberian C	*P. persica*	ACRS, Canada	Susceptible	Susceptible	Susceptible	Cold hardiness	Layne (1987); Reighard and Loreti (2008)
Barrier 1	*P. persica* × *P. davidiana*	CNR, Italy	Moderately tolerant	Moderately tolerant	Resistant	High vigor	De Salvador et al. (1991, 2002)
Cadaman™	*P. persica* × *P. davidiana*	France-Hungary	Moderately tolerant	Moderately tolerant	Resistant (Mj)	Yield efficiency, high vigor	Edin and Garcin (1994); Pinochet et al. (1999); Zarrouk et al. (2005)

Flordaguard	*P. persica* × *P. davidiana*	U. Florida, The USA	Highly susceptible	Susceptible	Resistant (Mj, Mi)	Red leaf, high vigor	Pinochet et al. (2002)
Nemaguard	*P. persica* × *P. davidiana*	USDA-ARS, The USA	Highly susceptible	Susceptible	Resistant (Ma, Mi, Mj)	Easy propagation, high vigor	Layne (1987); Shi and Byrne (1995)
Nemared	(*P. persica* × *P. davidiana*) × *P. persica*	USDA-ARS, The USA	Highly susceptible	Susceptible	Resistant (Ma, Mi, Mj)	Red leaf, easy propagation	Ramming and Tanner (1983); Marull et al. (1994); Lu et al. (2000)
Adafuel[PVP]	*P. dulcis* × *P. persica*	CSIC, Spain	Highly tolerant	Susceptible	Susceptible (Mj)	High vigor, easy propagation	Cambra (1990); Moreno et al. (1994); Albás et al. (2004); Zarrouk et al. (2005)
Adarcias[PVP]	*P. dulcis* × *P. persica*	CSIC, Spain	Tolerant	Moderately tolerant	Susceptible	Control of tree vigor, higher fruit quality	Moreno et al. (1994); Zarrouk et al. (2005); Albás et al. (2004)
Felinem[PVP]	*P. dulcis* × *P. persica*	CITA, Spain	Highly tolerant	Susceptible	Resistant (Ma, Mi, Mj, Mhi)	Red leaf, high vigor	Felipe (2009); Dichio et al. (2004); Zarrouk et al. (2005)
Garnem[PVP]	*P. dulcis* × *P. persica*	CITA, Spain	Tolerant	Susceptible	Resistant (Ma, Mi, Mj, Mhi)	Red leaf, high vigor	Felipe (2009); Pinochet et al. (1999); Marull et al. (1994); Dichio et al. (2004); Zarrouk et al. (2005)
Monegro[PVP]	*P. dulcis* × *P. persica*	CITA, Spain	Highly tolerant	Susceptible	Resistant (Ma, Mi, Mj, Mhi)	Red leaf, high vigor	Felipe (2009)
GF 677	*P. dulcis* × *P. persica*	INRA, France	Highly tolerant	Susceptible	Susceptible (Ma, Mi, Mj, Mh)	High vigor, easy propagation	Bernhard and Grasselly (1981); Salesses et al. (1970); Dichio et al. (2004); Zarrouk et al. (2005); Jiménez et al. (2008)

(continued)

Table 14.2 (continued)

Rootstock[a]	Species	Origin	Calcareous soil tolerance	Waterlogging tolerance	Root-knot nematodes resistance	Other characteristics	References
GF 557	*P. dulcis* × *P. persica*	INRA, France	Highly tolerant	Susceptible	Resistant (Mi) Susceptible (Mj)	High vigor	Bernhard and Grasselly (1981); Salesses et al. (1970); Esmenjaud et al. (1994)
Castore, Polluce	*P. dulcis* × *P. persica*	U Pisa, Italy	Tolerant	Susceptible	Unknown	Control of vigor, high fruit quality	Loreti and Massai (2006); Reighard and Loreti (2008)
Sirio	*P. dulcis* × *P. persica*	U Pisa, Italy	Tolerant	Susceptible	Susceptible (Mj)	Control of vigor	Loreti and Massai (1994); Pinochet et al. (1999)
Hansen 536	*P. dulcis* × *P. persica*	UC, The USA	Tolerant	Susceptible	Resistant (Ma, Mi, Mj)	High vigor	Kester and Assay (1986); Felipe et al. (1997b)
Hansen 2168	*P. dulcis* × *P. persica*	UC, The USA	Tolerant	Susceptible	Resistant (Ma, Mi)	High vigor	Kester and Assay (1986); Felipe et al. (1997b)
Nickels	*P. dulcis* × *P. persica*	UC, The USA	Tolerant	Susceptible	Resistant	High vigor	Reighard and Loreti (2008)
Plum-based rootstocks (Section Euprunus)							
Adesoto 101[PVP]	*P. insititia*	CSIC, Spain	Highly tolerant	Highly tolerant	Immune (Ma, Mi, Mj)	Yield efficiency	Moreno et al. (1995); Pinochet et al. (1999); Jiménez et al. (2008)
Montizo™	*P. insititia*	CITA, Spain	Tolerant	Tolerant	Resistant (Ma, Mi, Mj)	Yield efficiency	Felipe et al. (1997a)
Monpol™	*P. insititia*	CITA, Spain	Tolerant	Tolerant	Resistant	Medium vigor	Felipe et al. (1997a)
St. Julien A	*P. insititia*	East Malling, UK	Moderately tolerant	Tolerant	Unknown	Medium to low vigor	Okie (1987)
GF 655/2	*P. insititia*	INRA, France	Moderately tolerant	Susceptible	Resistant (Ma, Mi, Mj)	Excessive suckering	Grasselly (1988); Salesses et al. (1970); Layne (1987)

Brompton	*P. domestica*	East Malling, UK	Moderately tolerant	Moderately tolerant	Resistant (Ma, Mi, Mj)	Medium vigor	Salesses et al. (1970); Layne (1987); Okie (1987)
Penta	*P. domestica*	CRA-FRU, Italy	Tolerant	Tolerant	Resistant (Mj)	Medium vigor	Nicotra and Moser (1997); Pinochet et al. (2002)
Tetra	*P. domestica*	CRA-FRU, Italy	Tolerant	Tolerant	Resistant (Mj)	Medium vigor	Nicotra and Moser (1997); Pinochet et al. (1999)
Damas GF 1869	*P. domestica* × *P. spinosa*	INRA, France	Highly tolerant	Highly tolerant	Resistant (Mi)	Excessive suckering and graft incompatibility	Grasselly (1988); Zarrouk et al. (2006)
Mr.S.2/5	*P. cerasifera*	U Pisa, Italy	Tolerant	Tolerant	Resistant (Mi, Mj)	Medium vigor	Reighard and Loreti (2008)
Jaspi™	*P. salicina* × *P. spinosa*	INRA, France	Tolerant	Tolerant	Resistant	Peach compatibility?	Bernhard and Renaud (1990); De Salvador et al. (2002)
Rootstocks based on hybrids among sections Euamygdalus, Euprunus, Prunocerasus, and Microcerasus							
Controller 5™	*P. salicina* × *P. persica*	UC, The USA	Susceptible	Susceptible	Susceptible	Control of peach vigor	DeJong et al. (2004); Reighard and Loreti (2008)
Ishtara™	(*P. cerasifera* × *P. salicina*) × (*P. cerasifera* × *P. persica*)	INRA, France	Susceptible	Moderately tolerant	Immune (Mj)	Tolerance to *Armillaria*? Highly susceptible to bacterial canker	Grasselly (1988); Renaud et al. (1988); Guillaumin et al. (1991); Pinochet et al. (1999)
Hiawatha	*P. besseyi* × *P. salicina*	Unknown, Canada	Susceptible?	Unknown	Resistant (Mj)	Peach graft compatibility?	Pinochet et al. (2002)
Evrica	(*P. besseyi* × *P. salicina*) × *P. cerasifera*	KEBS, Russia	Susceptible	Moderately tolerant	Resistant (Mj)	Peach graft incompatibility	Zarrouk et al. (2006)

(continued)

Table 14.2 (continued)

Rootstock[a]	Species	Origin	Calcareous soil tolerance	Waterlogging tolerance	Root-knot nematodes resistance	Other characteristics	References
Bruce	*P. salicina* × *P. angustifolia*	Texas A&M, The USA	Unknown	Unknown	Immune (Mj)	Peach graft incompatibility	Pinochet et al. (1999, 2002); Zarrouk et al. (2006)
Pumiselect	*P. pumila*	Geissenheim, Germany	Susceptible	Susceptible	Resistant (Mj)	Cold hardiness, Scion dwarfing	Jacob (1992); Reighard and Loreti (2008); Pinochet et al. (2002)
Krymsk-86™	*P. cerasifera* × *P. persica*	KEBS, Russia	Tolerant	Moderately tolerant	Susceptible (Mj)	Cold hardiness, tolerance to Pv	Jiménez et al. (2008); Reighard and Loreti (2008)

Root-knot nematodes (*Meloidogyne* spp.): Ma: *M. arenaria*; Mi: *M. incognita*; Mj: *M. javanica*; Mh: *M. hapla*; Mhi: *M. hispanica*. Lesion nematodes (*Pratylenchus vulnus*): Pv

ACRS Agriculture Canada Research Station at Harrow (Canada), *CITA* Centro de Investigación y Tecnología Agroalimentaria de Aragón (Spain), *CNR* Centro Nacionale della Recerca Consiglio Nazionale delle Ricerche (Italy), *CSIC* Consejo Superior de Investigaciones Científicas (Spain), *GB* Gregory Brother's, California (The USA), *Geissenheim* Geissenheim Research Station (Germany), *INRA* Institut National de la Recherche Agronomique (France), *CRA-FRU* Centro di Recerca per la Frutticoltura (Italy), *UC* University of California (USA), *USDA-ARS* US Department of Agriculture–Agricultural Research Service (The USA), *U Pisa* University of Pisa (Italy), *Texas A&M* University of Texas, College Station (The USA), *KEBS* Krymsk Experimental Breeding Station (Russia), *PVP* Plant Variety Protection by Community Plant Variety Office in the European Union

[a]Next the Rootstock

'Garfi' × 'Nemared' peach that are resistant to root-knot nematodes and tolerant to calcareous soils have been released (Felipe et al. 1997b; Felipe 2009). These rootstocks have red leaves, a desirable nursery character in rootstocks because of the ease with which failed grafts can be discarded. Other peach–almond and peach × *P. davidiana* hybrids resistant to root-knot nematodes are 'Barrier 1,' 'Cadaman' (Edin and Garcin 1994), 'Hansen 536,' and 'Hansen 2168' (Kester and Assay 1986), but these are less tolerant to iron-chlorosis than 'GF 677' (Jiménez et al. 2008).

Other advances have been made in developing waterlogging and compact soil tolerant plum based rootstocks that are graft compatible with peach. Furthermore, some are tolerant to iron-induced chlorosis, are more precocious, and produce fruits of higher quality (Moreno et al. 1995; Felipe et al. 1997a; Nicotra and Moser 1997). Rootstocks tolerant to waterlogged soils include 'Adesoto 101,' 'Jaspi,' 'Julior,' 'Montizo,' 'Mr.S. 2/5,' 'Penta,' 'Tetra,' and 'Krymsk 86' (Table 14.2). The Tsukuba series of rootstocks from Japan and several peach × *P. davidiana* hybrids have been reported to show some tolerance to waterlogging (Reighard 2002; Zarrouk et al. 2005; Xiloyannis et al. 2007).

There are extensive efforts in Europe and in the USA to obtain resistance to root-knot nematodes (*Meloidogyne* spp.), which cause serious growth reduction in peach trees grown in warmer regions. There are at least five species of root-knot nematodes (*M. arenaria*, *M. incognita*, *M. javanica*, *M. hapla*, and *M. floridensis*) as well as a number of races within each species that feed on peach. Acceptable resistance for the predominant species has been incorporated into rootstock cultivars in different programs from several countries (Fernández et al. 1994; Pinochet 2009; Pinochet et al. 1999; Moreno 2004; Reighard and Loreti 2008: the USA ('Nemaguard,' 'Nemared,' 'Flordaguard,' 'Guardian®,' 'Hansen 536,' and 'Hansen 2168'), Spain ('Adesoto 101,' 'Adara,' 'Monegro,' 'Garnem,' 'Felinem,' and 'Greenpac'), France ('Myran,' 'Ishtara,' 'Cadaman,' and 'Julior'), Germany ('PumiSelect'), Italy ('Barrier 1,' 'Penta,' and 'Tetra'), Japan ('Juseitou' and 'Okinawa'), and China ('Gansutao 1' and 'Shouxingtao 1').

Considerable efforts have been undertaken to find a resistant or tolerant rootstock for peach in areas where peach tree short life (PTSL) syndrome is limiting tree longevity in the southeastern USA. In South Carolina and Georgia, a rootstock with acceptable survival in field tests has been developed and released under the name Guardian™® (Okie et al. 1994; Reighard et al. 1997).

Recently, an increased emphasis has been placed on developing dwarfing or semidwarfing rootstocks adapted to different soil fertilities and allowing higher density in the orchard. Several promising size-controlling clonal rootstocks have been released. These include the peach–almond hybrids 'Adarcias' (Moreno et al. 1994), 'Castore,' 'Polluce,' and 'Sirio' (Loreti and Massai 1994; 2006); the *P. salicina* × peach hybrid 'Controller 5' (DeJong et al. 2004); the complex plum–peach hybrid 'Ishtara' (Renaud et al. 1988), and the plum rootstocks 'Adesoto 101,' 'Montizo,' 'Penta,' and 'Tetra' (Moreno et al. 1995; Felipe et al. 1997a; Nicotra and Moser 1997).

5 Current Goals and Challenges of Breeding

5.1 Scion Cultivars

The ultimate goal of the breeder is to develop cultivars that have superior and consistent fruit production, quality and market appeal. This involves combining a range of adaptation, tree growth/fruiting, and fruit traits into one cultivar that will satisfy the producer, the packer, the merchandiser, and ultimately the consumer. Production consistency relies on excellent adaptation to the regions especially with respect to the yearly variations in temperature and humidity. Major objectives for adaptative traits include cold hardiness, chilling requirement and bloom time, and the tolerance to high heat during bloom in the lower chill zones. In the more humid regions, there is an increasing pressure to reduce the use of crop protectants, and consequently many of these programs breed for resistance to the common diseases such as brown rot, bacterial leaf and fruit rot, powdery mildew, peach leaf curl, and the plum pox virus (PPV).

Given that high tree productivity has been obtained in new cultivars, the next goal would be a tree architecture that is easy to manage but remains very productive. Labor is a major limiting input for fruit production in many production areas and consequently there has been substantial work in developing specific growth types such as pillar and weeping forms as well as in developing growth controlling rootstocks that will contribute to better designed and/or smaller trees that require less pruning, less time to manage and are more efficient producers of quality fruits (Byrne 2005; Sansavini et al. 2006).

The peach fruit can have a range of colors, textures and rate of softening, shapes, sizes, and flavors. Furthermore, what is preferred by the consumer changes with region although there is a trend to make a greater range of fruit types available in any given market. This diversification of the fruit types available will continue as many breeding programs are working toward this objective (Byrne 2005; Sansavini et al. 2006). Specific objectives include orange and red flesh colors, the lack of anthocyanins, higher sugar content, and better health promoting properties such as high levels of antioxidant phytochemicals (Vizzotto et al. 2007; Cantín et al. 2009a, b).

Another increasingly important objective is the improved postharvest behavior of the fruit. This has been a focus of breeding in regions such as Chile and South Africa where the fruit is routinely exported and is becoming increasingly important in other major production areas especially in breeding programs which are global in scope (Infante et al. 2008; Byrne 2005; Okie et al. 2008; Cantín et al. 2010b). The major impediment is the cost of evaluating selections for major postharvest traits such as the resistance to IB and specific flesh types, though good progress is being made to find molecular markers for these traits (Iezzoni et al. 2009; Ogundiwin et al. 2009; Cantín et al. 2010b; Peace et al. 2005, 2007).

5.2 *Rootstocks*

In the Mediterranean countries, where the European peach industry is primarily located, a new generation of peach rootstocks is being developed with the collaboration of different groups from France, Italy, and Spain. The objectives are to obtain genotypes with greater resistance to abiotic (iron chlorosis, waterlogging, and drought) and biotic stresses (*Meloidogyne* spp. nematodes, *Phytophtora* and *Armillaria* fungal diseases, replant disorders), and to improve peach graft compatibility and control of scion vigor (Salesses et al. 1998; Dirlewanger et al. 2004c; Moreno 2004; Pinochet et al. 2005). Controlled interspecific crosses have been undertaken with the purpose of bringing together the desirable traits of different *Prunus* species. Thus, some Myrobalan genotypes were chosen as parents for their high level and wide spectrum of root-knot nematode resistance, and tolerance to waterlogging. Additionally, peach, almond, peach–almond, and peach × *P. davidiana* hybrids have been used as a different source of nematode resistance, tolerance to iron-chlorosis, drought, replant problems, and compatibility with peach.

Within the USA, considerable efforts are devoted to develop a resistant or tolerant rootstock to the peach tree short life (PTSL) syndrome in the southeastern USA and the bacterial canker complex (*Pseudomonas syringae* pv. *syringae* van. Hall) in California, both of them linked with the ring nematode (*Mesocriconema xenoplax* (Raski) Loof & deGrosse). Research to find resistance to other harmful nematodes of the peach industry, such as the root lesion (*Pratylenchus vulnus* Allen and Jensen and *Pratylenchus penetrans* Cobb) and dagger (*Xiphinema americanum* Cobb) nematodes, is in progress because finding a broadly adapted and nematode-resistant rootstock that is also compatible with peach has been unsuccessful until now (Reighard and Loreti 2008). Rootstocks are also being developed for replant sites to reduce incidence of perennial canker (*Leucostoma* spp.) and the bacterial canker (*Pseudomonas syringae*) complexes found in peach production regions having light textured soils.

6 Breeding Methods and Techniques

6.1 *Major Traits in Peach Scion Breeding*

Adaptation is key in the development of consistently high-yielding cultivars. All breeding programs select for various adaptation traits as they select among their progenies for high bud density and fruit set.

Final productivity is dependent on several major adaptation traits: chilling and heat requirements, heat and cold tolerance, and resistance/tolerance to various biotic (disease and pest) and abiotic stresses.

Bloom time for peaches is determined by both the chilling and heat requirements of the flower buds. Given that the bloom order of peaches is consistent from year to year and over environments (Scorza and Sherman 1996), the most important determinant of bloom time is the chilling requirement, although there are some peaches that require more heat to bloom than the majority (Byrne et al. 2000; Citadin et al. 2001; 2003).

Lower chilling requirement is a priority trait for a significant number of breeders. This trend toward lower chilling cultivars is evident in the fact that 50 years ago 90% of the peach cultivars required more than 800 chilling hours to break dormancy, whereas now only 20% of new cultivars require this much chilling (Sansavini et al. 2006). This has occurred inadvertently as breeders selected early ripening cultivars with the largest fruit size which tended to be the lowest chill and earliest blooming as well as purposely selected cultivars adapted to warmer regions or protected culture to expand the production zone of peach. This selection is best done in a low-chill zone as opposed to selecting early blooming (and presumably lower chill) selections in a high-chill zone as in many low-chill zones the warmer temperatures during the dormant and bloom periods dramatically change the fruit quality especially with respect to fruit size and shape (Topp and Sherman 1989; Byrne et al. 2000; Byrne 2010; López et al. 2007). Research into low-chill cultivars has been accelerated recently by the increasing emphasis put on a year-round supply of produce. This is possible with lower-chill cultivars with short development periods and complementary production in both the northern and southern hemispheres (Byrne 2005). Very late ripening cultivars also play a role in this goal.

Chilling requirement as estimated by bloom dates is a moderately to highly heritable (Souza et al. 1998a, 2000; Mowrey and Sherman 1986; Hansche et al. 1972; Hansche 1990). Thus breeders can achieve rapid genetic gain through selection of parents based on phenotype and recurrent mass selection (Topp and Sherman 2000). Low-chill cultivars have prompted most of the interest of peach breeders working in warm environments, starting from southern China germplasm in the late 1940s (Byrne et al. 2000; Byrne and Bacon 1999; Byrne 2003; Topp et al. 2008). Breeding in low-chill regions implies selecting against some common problems such as excessive blind nodes (Boonprakob et al. 1994, 1996; Richards et al. 1994) and bud drop and poor fruit shape which are traits whose expression is amplified by the inconsistent winter chilling and warm spring conditions frequently experienced in the low-chill zones (Byrne 2010). Breeding for low chilling in the last few decades has allowed the peach to be cultivated in many subtropical regions, from the southern states in the USA to Brazil, southeast Asia, Australia, South Africa, and most of the countries facing the warmest shores of the Mediterranean basin (Topp et al. 2008; Byrne et al. 2000; Sherman and Lyrene 2003; Raseira and Nakasu 2006).

High temperatures during bloom can have a negative effect on fruit set and consequently yield. Reports indicate that night temperatures above 15–18°C and day temperatures above 22–25°C are detrimental to fruit set in low-chill peach cultivars (Edwards 1987; Rouse and Sherman 2002b; Couto 2006; Couto et al. 2007). Recent work in Japan with the high-chill cultivar "Hakuho," indicated that as the temperature was raised during flowering from 15 to 30°C, there was a decrease in percent pollen germination, flower and ovule size, and fruit set. The most abrupt changes occurred between 20 and 25°C (Kozai et al. 2002, 2004). In addition, cultivar differences are evident in the tree's ability to set fruit under warm bloom time conditions (Rouse and Sherman 2002b; Couto et al. 2007). As low-chill cultivars are developed, it is important to select them for their tolerance to high temperatures during bloom, as good tolerance to this stress will allow for more consistent production. This is especially important in the warmest production areas but also in areas where peaches can be produced in protected culture, double cropping or forced cropping systems (George et al. 1988; Sherman and Lyrene 1984; Jiang et al. 2004; Byrne 2010). An ability to set under a wider range of temperature conditions would give the producer more flexibility in the timing of the harvest seasons.

Tolerance to freezing temperatures during bloom can also be an important objective in some breeding programs in regions that are subject to crop losses from spring frost and/or freezes during bloom. Several approaches are possible to obtain cultivars tolerant to bloom freezes: late blooming, high bud density, and inherent bud resistance to colder temperatures. The first two approaches are avoidance approaches and represent traits that are moderately to highly heritable (Souza et al. 1998a; Citadin et al. 2003). Thus late blooming cultivars with high bud set have been developed. Unfortunately, little is known about the genetic variation of inherent resistance of deacclimating flower buds transitioning out of dormancy to freezing temperatures.

Extreme low temperatures represent a limiting factor in plant survival (Quamme and Sushnoff 1983). Consequently, breeding programs in cold regions, especially in the northern hemisphere, are focused on developing peaches with greater winter cold hardiness, which extends peach cultivation to higher latitude zones (Callahan et al. 1991). Peach flower and vegetative buds of some cultivars can withstand −30 and −35°C, respectively (Layne 1984). Hardy parents should be chosen among those accessions whose resistance to winter cold is consistent over rootstock, soil and temperature fluctuations, as reported in some Chinese germplasm. However, attention should be paid to bloom time of these accessions to eliminate early blooming progeny that would be susceptible to spring frost damage (Layne 1982, 1984).

Since hardiness is a quantitative trait (Mowry 1964), resistance to low temperatures would be improved by crossing very hardy parents with commercial peaches, and then selecting within the F_1 progeny followed by back crossing to improve fruit quality of the most hardy selections. Selection strategies for developing hardy peaches, other than relying on test winters and assessing the degree of twig xylem and dieback (Myeki and Sazabó 1989; Layne 1982; Szabó 1992), are based on artificially induced low temperatures in portable field chambers or in a cold chamber on winter dormant potted trees (Stushnoff 1972; Quamme and Sushnoff 1983). The threshold of resistance (lowest temperature killing the flower bud) is checked directly or by methods such as exothermal analysis in which death is determined by the sudden temperature rise at the bud base, corresponding to ice formation in bud tissue. Alternatively, the cold treatment could be applied on 1-year old shoots harvested in mid winter. This is more efficient when assessing large progenies. Interestingly, hardy peaches usually possess high flower bud density (Werner et al. 1988), a possible mechanism for spring freeze avoidance (Byrne 1986) even in low-chilling peaches (Sherman and Lyrene 2003).

Disease and pest resistance. The consumers' concern about chemical residues on fruits and vegetables has increased considerably. Numerous disease organisms and pests attack peach and nectarine cultivars. Some, such as the brown rot, are of worldwide distribution, whereas others have regional importance (Scorza and Sherman 1996; Byrne et al. 2000). Breeding programs all over the world, especially the ones located in humid areas have disease resistance as one of their top priorities. The lack of good known resistance sources and the fact that little is known about the inheritance of the disease and pest resistance of peaches is limiting the advances toward this objective.

One of the most serious diseases of peach worldwide is brown rot (*Monilinia fructicola* (Wint.) Honey and *M. laxa* (Aderh & Rull) Honey). Despite its importance, there has been relatively little work done on the development of brown rot resistant stone fruit cultivars because a small infection to the fruit results in complete loss of that fruit and so far the disease has been reliably controlled by fungicides. Nevertheless, several breeders (Brazil, California, Italy, and USA) either individually or associated with pathologists have concentrated efforts on obtaining new cultivars resistant to this pathogen. There are numerous reports of resistance (feral Mexican and Brazilian peaches) or tolerance (peaches from Florida, New Jersey, and Harrow programs) to fruit brown rot (*M. fructicola*) within peach (Feliciano et al. 1987; Scorza and Okie 1990; Scorza and Sherman 1996; Byrne et al. 2000). In general, the level of resistance reported is low to moderate and the screening techniques are not highly reliable. The Brazilian cv. Bolinha is considered to have a certain level of horizontal resistance to *M. fructicola* (Feliciano et al. 1987) as do a few newer Brazilian selections (Wagner et al. 2005a). However, the resistance is only in the epidermis (Gradziel et al. 1997; Lee and Bostock 2007), thus any disruption (such as insect damage) of the skin, will allow the fungus penetration and disease development.

In tests done in Italy, the level of resistance to fruit rot caused by *M. laxa* was assessed in 27 peach and nectarine cultivars. Of these, only four ('Contender,' 'Glohaven,' 'Maria Aurelia,' and 'Maria Bianca') had less than 60% diseased fruits. 'Contender' also had very high level of field resistance and when crossed to very sensitive cultivars (e.g., 'Elegant Lady') yielded seedlings more resistant than itself (Bassi et al. 1998). Artificial inoculation on unwounded fruits was found to be a reliable method in evaluating for brown rot (field) resistance, although the procedure is lengthy and affected by season and year variability.

Beyond attacking the developing fruit, this pathogen also attacks young shoots and flowers. The breeding work in southern Brazil (Pelotas, Rio Grande do Sul) selects for resistance to flower blight in their field plots. Although differences in the level of resistance to flower blight is seen, there seems to be no correlation between flower and fruit resistance and selection needs to be done for flower blight as well as for fruit reaction (Wagner et al. 2005b).

Bacterial leaf spot (*Xanthomonas arboricola* pv. *pruni*) is a disease particularly important in areas of high humidity accompanied by wind and sandy soils. Since chemical control efficacy is not always high, several breeding programs in Brazil, South Africa, and the USA have routinely selected for resistance to bacterial spot in peaches. Little is known about the genetics of resistance to this disease; however, Sherman and Lyrene (1981) suggested that resistance was controlled by a few major genes.

Cultivars of peach vary widely in their resistance to bacterial leaf spot with the more resistant cultivars being developed in humid areas (south and eastern North America, Brazil, and South Africa) where screening is done in the field with the existing pathogen pressure. Unfortunately their resistance may differ dramatically in different geographic regions (Byrne et al. 2000) due to unique pathogenic races of the bacteria (du Plessis 1988; Martins 1996) in different geographic regions. This makes the development of stable resistance to bacterial spot more difficult.

Other wide spread fungal diseases subject to some breeding or selection efforts are peach leaf curl (*Taphrina deformans* (Berk.) Tul.), rust (*Transchelia discolor (Fuckel) Transchel & Litv.*) (Pérez et al. 1993; Rouse and Sherman 2002a; Topp et al. 2008), and powdery mildew (*Sphaeroteca pannosa* (Wall. FR. Lev.); *Podosphaera pannosa* (Wallr.:Fr.) Braun & Takamatsu) (Rodríguez et al. 1992; Pérez 1997; Pascal et al. 2010). The most studied of these diseases are peach leaf curl and powdery mildew which are both cool season pathogens. These are generally adequately, but not always, controlled by a few sprays per growing season. Given the biology of the two fungi, in vitro or artificial inoculation is not easy to do and selection must rely on natural infection, either on the young seedlings in the green house or in the field.

Resistance to peach leaf curl is determined by a polygenic system (Ritchie and Werner 1981; Monet 1985; Viruel et al. 1998). Various sources of resistance have been reported, e.g., the peach seedlings 'GF 305,' 'Redhaven,' and 'Cresthaven' (bearing up to 50% of resistant seedlings in their progeny (Todorovic and Misic 1982), the Italian white fleshed "Cesarini" (Bellini et al. 1993) and *Prunus davidiana* (Pisani and Roselli 1983).

The inheritance of powdery mildew resistance varies with its source. It has been described as a single dominant gene from the peach 'Pamirskij 5' (Pascal et al. 2010), to two loci, one controlling high resistance, the other medium and low resistance from *P. ferganensis* (D'Bov 1975) and polygenically from other peach cultivars (Pérez 1997) and *P. davidiana* P1908 (Dirlewanger et al. 1996). For the latter parent (Pascal et al. 1997), resistance has been introgressed to peach and molecular markers for various QTLs for resistance useful in selection have been identified (Foulongne et al. 2002, 2003a). Although the eglandular leaf phenotype is associated with a strong susceptibility to powdery mildew (Rivers 1906; Saunier 1973), both globose and reniform accessions can also show high susceptibility to this pathogen (Rodríguez et al. 1992). The results from greenhouse screening and field screening for powdery mildew resistance are both equally reliable (Rodríguez et al. 1992; Pérez 1997).

The major virus issue for the European peach and other stone fruit industry is the Sharka disease caused by the Plum Pox Virus (PPV) and transmitted by grafting and several species of very mobile aphids with the green peach aphid (*Myzus persicae* (Sulz.)) among the most important. It was originally described on peach in Greece and now it is reaching a pandemic diffusion in several peach growing countries in Europe and elsewhere (e.g., the USA and Canada). Breeding has been challenging because the assessment for resistance to PPV is a very lengthy procedure and requires artificial infection in insect-proof environments (either screen houses or isolated places with no *Prunus* trees or possible source of PPV infection). Progeny to be tested have to be budded on test rootstocks, e.g., 'GF 305' peach seedlings, to check for possible tolerance mechanism (plant infected but without symptoms). If no symptoms appear on either the rootstock or scion over at least three vegetative cycles, ELISA followed by a PCR test are run to check for possible low concentrations of the virus (Rubio et al. 2009).

Although field resistance and tolerance to PPV has been reported in peach, the best source of resistance found is from a related species, *Prunus davidiana* which is being incorporated into peach by several Italian and French institutions. Resistance to PPV from *P. davidiana* is conditioned oligogenically and is syntenic to PPV resistance in apricot (*P. armeniaca* L.). Recently, QTLs associated with PPV resistance have been mapped, which should facilitate the development of a marker-assisted selection (MAS) approach (Foulongne et al. 2003b; Quilot et al. 2004; Decroocq et al. 2005; Bassi 2006) although this may be complicated by the report that not all the QTLs are stable over all the genetic backgrounds tested (Rubio et al. 2010).

Peaches are attacked by a range of nematodes including root knot (*Meloidogyne* spp.), ring (*Mesocriconema xenoplax* (Raski) Loof & de Grasse), root lesion (*Pratylenchus* spp.), and dagger (*Xiphinema americanum* Cobb) nematodes. Of these, the most important are the root knot nematodes and the ring nematode (Reighard and Loreti 2008). The most extensive work has been done with the *Meloidogyne* species of root knot and several dominant resistance genes have been identified for resistance to *M. incognita* (Kofoid and White) Chitwood and *M. javanica* (Traub.) Chitwood, the two most important species (Sharpe et al. 1970; Yamamoto and Hayashi 2002; Gillen and Bliss 2005; Claverie et al. 2004a, b; Esmenjaud 2009). In addition, a gene conditioning a broad spectrum resistance has been identified in plum and is being used in rootstock breeding (Esmenjaud 2009). Furthermore, markers associated with these various genes for root knot nematode resistance have been identified and are being used for selection of resistant rootstocks (Lu et al. 1998; Wang et al. 2002a; Lecouls et al. 2004; Gillen and Bliss 2005; Esmenjaud 2009).

No clear resistance has been found to *Mesocriconema xenoplax*, a nematode associated with peach tree short life (PTSL). However, Guardian® rootstock is considered to be tolerant to the nematode, since it is less susceptible to peach tree short life and causes the scion to be less susceptible to cold injury and bacterial canker, the main causes of PTSL, than any other rootstock tested thus far (Okie et al. 1994). Screening for resistance to lesion nematodes (*Pratylenchus penetrans* (Cobb) Filipjev and Schuurmans Stekhoven and *P. vulnus* Allen and Jensen) among *Prunus* has shown a range of susceptibility in peach and a source of broad based resistance in plum (McFadden-Smith et al. 1998; Pinochet et al. 2000). Unfortunately, there was a wide range of pathogenicity among *P. vulnus* races which creates difficulties in breeding for resistance (Pinochet et al. 2000).

Thus far, no high level of resistance has been found to the oak root rot fungus (*Armillaria mellea* (Vahl: Fr.) P. Kumm. and *Armillaria tabescens* (Scop.) Dennis et al.) although there has been resistance reported in some plum rootstocks to *A. mellea* in Europe (Guillaumin et al. 1991; Jiménez et al. 2011) and plum germplasm to *A. tabescens* in the USA (Beckman et al. 1998; Beckman 1998; Beckman and Pusey 2001). Unfortunately, some plum rootstocks reported as resistant to *A. mellea* were found to be susceptible to *A. tabescens*. The progress in the development of *Armillaria* resistant rootstocks is expected to be slow due to a lack of an excellent source of resistance and the long and tedious procedure needed to quantify their resistance (Beckman and Pusey 2001).

Even though peaches are attacked by several insect pests, few breeding programs work with insect resistance. The most active programs for insect resistance are those run by INRA in France and by the Centro di Recerca per la Frutticoltura (CRA-FRU) in Italy. These programs focus on green peach aphid (*Myzus persicae*) resistance (Liverani and Giovannini 2000; Sauge 1998; Monet et al. 1998) because of its importance in Europe due to both the direct damage (leaf curl and stunting) it causes but also because it is the vector for Plum pox virus. Green peach aphid resistance has been described in three sources: a weeping peach tree (Weeping Flower Peach), *P. davidiana* and 'Rubira' rootstock (Massonie et al. 1982). This resistance is a hyper-sensitivity reaction to the aphid testing probe on young shoots or leaves which causes a necrotic zone to develop around the puncture hole, thereby isolating the neighboring leaf cells (Sauge 1998). A dominant mode of action for aphid resistance has been identified in the resistance from 'Weeping Flower Tree Peach' (Monet and Massonié 1994; Monet et al. 1998; Monet 1985) and 'Rubira' (Pascal et al. 2002), although it is not known if these are allelic or not.

Resistance to abiotic stresses. Resistance to calcareous high pH soils is an important trait for peach production regions with calcareous soils found most commonly in semi arid and arid zones. High pH causes iron deficiency, which lowers leaf chlorophyll, fruit yield, fruit size and soluble solids content according to the degree of chlorosis (Razeto and Valdés 2006). Tolerance has been identified among peach, plum and particularly almond (Shi and Byrne 1995; Jiménez et al. 2008). Presently peach–almond hybrid rootstocks are commonly used in calcareous soils to ensure sufficient iron uptake by the plant (Reighard and Loreti 2008). Selection procedures include field evaluation in calcareous soils, greenhouse evaluation at various levels of bicarbonate (Shi and Byrne 1995) and most recently via laboratory measurements of root iron reductase activity on hydroponically grown plants (Jiménez et al. 2008).

A soil pH below 5.5 is deleterious to peach tree growth, fruit yield and size, and tree longevity. There is an improved performance of trees when soil pH is maintained above pH 6.0. Deleterious effects of soil pH below 5.5 may be related to the toxicity of Al or low Ca availability (Cummings 1989). Unfortunately, no source of tolerance to aluminum toxicity has been identified (Chibiliti and Byrne 1989). Consequently, this issue is managed by lime application to raise the soil pH.

Peach seedling rootstocks are not tolerant to waterlogging and thus grow poorly or die when planted in even seasonally waterlogged soils. The intensity of the waterlogging effect is more pronounced if the plant is actively growing as compared to dormant trees. The difference in flooding tolerance found among *Prunus* species other than peach is based on complex anatomical processes such as aerenchyma formation and biochemical adaptation involving the fermentative pathways to obtain energy. Several candidate genes have been identified to be involved in the tolerance in two *Prunus* genotypes (Amador et al. 2009; Amador 2010). Various plum and interspecific hybrids have been reported to be tolerant of waterlogged soils (Table 14.2; Moreno 2004; Reighard and Loreti 2008).

Tree architecture. Peach productivity is relatively low and pruning costs are relatively high as compared to other tree fruit such as apples. Higher production efficiency could be obtained with higher cultivation density using modified growth types and dwarfing rootstock. Several breeding programs have worked toward the development of growth habit modification to increase yields with decreased management costs. There are a number of mutations differing from standard growth that could be exploited, ranging from brachytic dwarf to weeping and columnar (pillar) (Bassi 2003; Fideghelli et al. 1979; Mehlenbacher and Scorza 1986; Scorza et al. 1989). Interestingly, some interaction occurs between phenotypes, thus several intermediate growth architectures can be obtained (Bassi and Rizzo 2000; Scorza et al. 2002; Werner and Chaparro 2005; Hu and Scorza 2009). Given the simple inheritance of these traits, selection for a given tree structure is easily performed in one or two generations, depending on the dominance of the trait sought (Monet and Bassi 2008). Since segregation will occur for all of the other traits, several cycles of recurrent selection has to be applied to recover the commercially useful fruiting phenotype. Some recent commercially available introductions are already featuring growth habits different from the standard growth such as the upright 'Sweet-N-UP' and the columnar types 'Crimson Rocket' and 'Alice-col' (Liverani et al. 2004; Scorza et al. 2006).

The modifications for controlling size of trees necessary to satisfy the criteria for modern fruit-culture are aimed at smaller plants more suitable for high density plantings and reduction of the pruning needed to promote new fruiting wood in peach (Scorza and Sherman 1996). However, while these strategies have been largely successful in the apple industry, the peach tree seems more recalcitrant, probably due to the positive relationship between branch or tree vigor and fruit size (Manaresi and Draghetti 1915; Marini and Sowers 1994; Moreno et al. 1994). Although most dwarfing rootstocks for peach runted the trees and negatively affected fruit size, they did generally induce better peach fruit organoleptic quality (Albás et al. 2004; Mathais et al. 2008). Work continues to develop rootstocks that induce precocity, larger fruit size and quality as well as yield.

Fruit traits. The harvest season in the major production zones of the northern hemisphere can range from mid-April to mid-November (Llácer et al. 2009). However, extension of the harvest season remains an important trait in many programs in different growing regions due to market opportunities (Raseira et al. 1992; Byrne et al. 2000) and because of the quality deficiencies of existing cultivars at the extremes of the harvest season (Scorza and Sherman 1996). Various studies on the inheritance of the ripening time and fruit development period (FDP) have shown that these traits are highly heritable and mainly additive though there is evidence of a few genes with relatively large effects (French 1951; Bailey and Hough 1959; Souza et al. 1998a; Yu et al. 1997). Consequently, rapid genetic gains for short FDP are possible in breeding programs (Hansche et al. 1972), though this is limited by a negative genetic correlation with fruit size and fruit quality (Souza et al. 1998b).

Large fruit size is also an important goal in most peach breeding programs. Furthermore, the achievement of large fruit size is more difficult in germplasm with short FDP (Souza et al. 1998a, b, 2000) and in regions with warm temperatures during fruit development (Topp and Sherman 1989; López, et al. 2007). Thus it is an especially challenging objective in warm subtropical and tropical production zones where early ripening is also a major objective (Byrne et al. 2000). Fruit size is a polygenic trait with a low to moderate heritability (Souza et al. 1998b; Hansche et al. 1972) due to the large influence that environment conditions, plant nutrition, and cultural practices (pruning and thinning) have on its expression.

Fruit firmness is essential for efficient handling and marketing. Whereas most fresh market peach breeding programs have traditionally emphasized the development of melting flesh type fruits, some such as the Brazilian (Pelotas), Mexican, and Spanish programs and more recently, Florida and some California programs, in the USA, have worked with nonmelting types. These genotypes are firm enough to harvest at a more mature stage, which allows for better quality (Brovelli et al. 1995, 1998; Beckman and Sherman 1996; Robertson et al. 1992) and larger size. Examples of this are 'Eldorado,' 'Maciel,' and 'Granada' in Brazil (Raseira and Nakasu 2003), 'UFPrince,' 'Gulfking,' 'Springprince,' 'Springbaby,' and 'Crimson Lady' in the USA (Byrne 2005), and 'Calante,' 'Evaisa,' 'Jesca,' and 'Miraflores' in Spain (Bouhadida et al. 2007a; Espada et al. 2009). The melting (M) and nonmelting (NM) flesh types are controlled by four alleles at the *F* locus. The nonmelting clingstone trait is recessive to the various melting flesh types (Peace et al. 2005, 2007; Monet 1989).

Another type of flesh with potential in the development of firmer freestone peaches with tree ripe flavor and longer storage life is the stony hard (SH) flesh found in cultivars such as 'Jingsu' from China (Byrne 2005), 'Yinggetao' from Taiwan (Lu et al. 2008), 'Hakuto' from Japan and 'Yumyeong' from Korea (Liverani et al. 2002; Haji et al. 2005). It is a monogenic recessive trait (Yoshida 1976; Haji et al. 2005; Liverani et al. 2002) that gives the fruit a very firm crunchy flesh which ripens more slowly due to suppressed ethylene production (Hayama et al. 2006). The stony hard trait is inherited independently of the melting flesh/nonmelting flesh trait and is epistatic to this trait (Haji et al. 2005). Unfortunately, it is difficult to identify in the field thereby making reliable selection difficult. Examples of cultivars with stony hard flesh are three of the 'Ghiacco' series of peaches developed in Italy, which were selected from a open pollinated population of 'Yumyeong.' They all have sparse pubescence, white flesh (with a red vein in 'Ghiaccio 22'), juicy but very firm flesh, with a texture similar but not equal to a clingstone peach, and good flavor with high sugar content (Nicotra et al. 2002).

Within the melting texture there is a very interesting phenotype, resembling the SH flesh in firmness and crispiness, but becoming melting when fully ripe and showing a prominent delay in softening, and ethylene production. This flesh texture is found in recently developed cultivars, both nectarines (e.g., 'Big Top') and standard peaches (e.g., 'Rich Lady' and 'Diamond Princess'). Its remarkable keeping quality, particularly on tree, is of primary importance for both growers and consumers. However, it is very difficult to assess on the tree when scoring segregating progenies, as is the SH flesh phenotype. The physiological basis and inheritance of this trait are being actively investigated (Tatsuki et al. 2006; Begheldo et al. 2008).

Flesh color varies in peach, from white to yellow to dark red, with variations in tonalities, greenish-white, light yellow, orange yellowish, and orange (Cevallos-Casals et al. 2005; Vizzotto et al. 2007). Traditionally white flesh peaches were preferred in Asia and in some European countries (e.g., France, Italy) until the 1960, and yellow-fleshed peaches preferred in the Americas and Europe, but recently, there has been an expansion of the use of white-fleshed peaches and nectarines in non-Asian markets. Thus, several programs outside Asia have worked intensively to develop white flesh peaches and nectarines for the American and European markets (Argentina, Brazil, Italy, Taiwan and in the USA the programs of Arkansas, California, Georgia, North Carolina, Texas, among others).White flesh is dominant over the yellow (Connors 1920), but there are variations in tonalities of white as well as yellow.

Blood flesh peaches and nectarines are sought in breeding programs in France (T. Pascal, personal communication), the USA (Okie 1988; Vizzotto et al. 2007; Cevallos-Casals et al. 2005), China (R. Ma, Nanjing, personal communication), Italy, and Spain (Cantín et al. 2009b) for their novelty and potential health benefits of the enhanced levels of anthocyanins (Vizzotto et al. 2007; Cantín et al. 2009b). Both of the sources of this blood flesh trait appear to be inherited independently of yellow/white flesh color locus. Most of this breeding has thus far worked with the recessive blood red gene which was characterized from 'Harrow Blood' and many landraces in France and Italy. This gene induces the early development of anthocyanin in the fruit pulp beginning at the pit hardening stage and is associated with red leaf veins (Werner et al. 1998; Gillen and Bliss 2005). Another source of red flesh in peach has been found in China (T. Pascal, personal communication) and among some local peach selections in Georgia (W. R. Okie, personal communication). This red flesh trait, which appears to be inherited as a dominant trait, is characterized by a late anthocyanin development in the mesocarp and is associated with green veins. On the other extreme, Italian breeders have released two cultivars, 'Ghiacco 1' and 'Ghiacco 3,' without any anthocyanins (Nicotra et al. 2002).

Skin color is not important for cultivars used in the processing industry; nevertheless it is a very important component of appearance when the fruits are produced for fresh market. Most European and American markets prefer a red over color superior to 80% of the skin surface, whereas other markets such as in Asia, Brazil, Mexico, and Spain accept fruit with less than this and even 20% red blush over a bright yellow or white background are well accepted by consumers. In a few specific markets with nonmelting flesh peaches in southern Brazil and parts of Mexico, southern Italy, and Spain, a completely yellow skin associated with nonmelting flesh is preferred. A skin and flesh cream-yellow uniform color is preferred in the very late ripening cultivars grown in the Ebro Valley in Spain (Espada et al. 2009). This peach industry is based on high quality nonmelting fruits individually bagged during their development on the tree.

The expression of a red skin color is difficult to categorize and has a high degree of environment interaction especially with respect to light exposure (altered by climate, growth, position of the fruit in the canopy and pruning practices) and nutrition (Luchsinger et al. 2002: Trevisan et al. 2008). Red skin color is generally controlled by multiple gene action (Hansche 1986; Scorza and Sherman 1996; Souza et al. 1998b) although there also appears to be several qualitative recessive genes controlling skin color: one controlling full red skin color, even on shaded portions of the fruit surface in some germplasm (Beckman and Sherman 2003) and another that suppresses red skin color (Beckman et al. 2005).

Fruit shape is an important fruit quality attribute, since it influences consumer's acceptance and postharvest handling. In addition, protruding tips and sutures can be bruised during handling and shipping of fruit and are, therefore, undesirable traits for commercial peaches (Kader 2002). Fruit shape is moderately heritable (Souza et al. 1998b), but is also influenced by the temperatures during winter and/or early fruit development with warmer temperatures conditioning the development of larger tips and more irregular shapes (Topp and Sherman 1989; Byrne et al. 2000). This represents a production problem especially under tropical and subtropical conditions. Breeding programs have been selecting for rounder shapes and some new cultivars, even in the subtropics, no longer have the problem, such as the cv. 'Rubimel,' released by Embrapa in 2007, that has a very small or no tip, even when cultivated at 23–24° latitude in São Paulo State, Brazil.

Some of the most common complaints by consumers are the presence of off flavors, flesh mealiness, flesh browning and black pit cavity due to IB (Crisosto 2002) and inconsistent quality in stone fruit (Byrne 2005). This is, in part, related to the production techniques which emphasize yield and inadequate postharvest handling protocols but also to the cultivars produced by breeders who focused on external quality at the expense of internal quality. Recently, many breeding programs have shifted their focus on increasing the internal quality of the cultivars that they develop. Although peach flavor is quite complex and preferred profile varies with regional and personal customs (Crisosto et al. 2006), the major easily measured traits are the sugar (total soluble solids, total sugars, sucrose, fructose, glucose, and sorbitol) and acid content (titratable acidity, malic, citric, quinic, and shikimic acids) as well as the ratio between these (Colaric et al. 2005; Crisosto et al. 2006; Cantín et al. 2009a).

Peaches are expected to be sweet and to be readily accepted by consumers, acid and low-acid fruits need to have more than 10 and 11°Brix of soluble solids content (SSC), respectively (Crisosto and Crisosto 2005). Currently, there are selections and cultivars with fruits close to or even higher than 20°Brix such as some nectarines from the private and USDA programs in California and the 'Ghiaccio' series in Italy. Total SSC has a low to moderate heritability, which should allow steady improvement of fruit sugar levels in spite of the variations caused by environmental, maturity, and production differences between regions and years (Cantín et al. 2009a). Although many mid- and late-ripening cultivars already have these minimum levels of SSC, they can be improved. Unfortunately, this process will be more difficult with early ripening genotypes with a very short fruit development period (FDP) due to an association between low FDP and low SSC (Souza et al. 2000).

The acidity levels in peach are controlled by both qualitative and quantitative genes (Connors 1920; Souza et al. 1998b). The dominant allele of gene D conditions low acidity (Connors 1920) and colocalizes with QTLs which affect pH, titratable acidity, and organic acid contents (Boudehri et al. 2009). These low-acid peaches have a higher pH (more than 3.9) and a total acidity 2–4 times lower than standard cultivars due to lower concentrations of citric, malic (about 50%), and quinic (about 20%) acids (Byrne et al. 1991; Brooks et al. 1993; Crisosto et al. 2006). The dominant nature of the low acid and the white flesh traits has made the conversion of superior acid yellow flesh materials traditionally preferred by many American and European markets into low-acid white genotypes preferred by many Asian markets and now with increasing popularity in American and European markets, a relatively easy process. In addition, the low-acid trait allows the earlier harvest of melting flesh fruit without affecting the taste, but if total sugars are below 11–12°Brix, then a very bland flavor is experienced (Crisosto et al. 2001, 2006).

High dietary consumption of fruits and vegetables particularly those with antioxidant activity has been linked to reduced risks of many chronic diseases including cancer and cardiovascular diseases (Wargovich 2000). The phytochemicals in stone fruit have been linked to inhibiting the development of cardiovascular disease and the growth of various cancers (Byrne 2007; Lea et al. 2008; Noratto et al. 2010) and may also extend the shelf life and reduce the incidence of diseases of fruits (Khanizadeh et al. 2007). There is a broad genotype variation in the content of these phytochemicals with some peach selections and many plums having a similar antioxidant activity as blueberry (Byrne et al. 2009; Vizzotto et al. 2007). The antioxidant levels were well correlated with total phenols although not necessarily with anthocyanin content of the fruits (Cevallos-Casals et al. 2005; Vizzotto et al. 2007; Cantín et al. 2009b). Thus far, no stone fruit cultivars have been developed specifically for higher levels of these phytochemicals; however, such cultivars would provide a new product that could be sold fresh or processed into extracts (Byrne 2005). This possibility has guided peach breeders to consider antioxidant compounds and other nutritional properties as interesting targets in breeding programs (Cevallos-Casals et al. 2005; Vizzotto et al. 2007; Cantín et al. 2009b). More research in the health effects of various stone fruit phytochemicals is needed to better define the specific phytochemicals and the quantities desired.

Poor postharvest quality due to the harvesting of hard unripe fruit and IB, a fruit disorder that develops in cold storage, is the main limitation to the marketing of some peach cultivars. Although the symptoms of IB (e.g., mealiness, flesh browning, loss of flavor, and bleeding) can be minimized by storing below 5°C, ethylene application or intermittently raising the temperature during cold storage or by preconditioning fruit prior to storage or shipping, the best approach is to breed cultivars resistant to it (Crisosto et al. 1999; Crisosto 2006; Peace et al. 2006; Cantín et al. 2010b). We know little about the inheritance of IB, but it appears that only a few genes control each of the symptoms (Peace et al. 2006). Given the fact that it is expensive to measure a genotype's susceptibility for IB (Crisosto et al. 1999), there is considerable work trying to identify molecular markers associated with these traits (Peace et al. 2006; Ogundiwin et al. 2009; Cantín et al. 2010b). Although the evaluation techniques for postharvest traits are cumbersome, much emphasis has gone to these objectives. In the development of fresh market cultivars, there is also considerable effort to incorporate nonmelting flesh to increase fruit firmness, which may have the additional effect of improving resistance to IB as peaches with nonmelting flesh tend to be more tolerant to IB than those with melting flesh (Brovelli et al. 1998; Crisosto et al. 1999; Peace et al. 2006).

6.2 Breeding Methods and Techniques

Although the difficulties related to fruit tree genetics (long generation time and large plant size) have slowed genetic investigations on fruit crops, much information on character inheritance has been collected for peach. This is because this species has a shorter generation time and smaller plant size than other major fruit crops, as well as has a small chromosome number, is self-fertile, is tolerant of inbreeding depression, and many important qualitative traits are transmitted according to simple Mendelian inheritance. Mendelian traits in peach, association to specific genomic linkage groups and the estimates of heritability of major quantitative traits have recently been reviewed (Monet and Bassi 2008). Quantitative genetics considers continuously variable traits such as fruit size, fruit skin color, firmness, and taste that are both polygenic and influenced by environment factors (multifactorial traits) and consequently they are more difficult to improve because their level of heritability is relatively low.

In the last century thousands of novel cultivars have been released especially in the USA and Europe. Most of them come from cross breeding, either via controlled crosses (~50%) or via open pollination (~20%) and only around 4% from bud sports (Della Strada et al. 1996). Other possible breeding techniques are somaclonal variation, mutation breeding, and transformation.

Intraspecific crossing is the most common method for peach breeding and still continues to supply the vast majority of the new cultivars worldwide. Variable strategies may be followed according to the available germplasm and goals.

Highly valuable cultivars derive either from self-pollination or from crossing between related parents. This strategy allows the combination of several quantitative traits of horticultural and market importance. It is well known that despite of the very few genotypes used at the origin of peach breeding in the USA and the high degree of inbreeding, most of the cultivar improvement comes from this apparently small gene pool (Scorza et al. 1985) and a continued improvement of quality traits have been made in spite of this high degree of inbreeding. In part, this continuous improvement is due to outcrossing breeding populations with unrelated genotypes to incorporate desirable characters, such as fruit quality, diverse chilling requirements, and pest or disease resistance (Cantín et al. 2010a).

Since peach is tolerant to inbreeding depression (Lesley 1957; Monet and Bassi 2008), it is possible to develop seed propagated genotypes that would breed true-to-type, which is essentially what has been done in the development of seed propagated rootstocks as well as fruiting cultivars in Central America (Pérez 1989). Beyond the ease of handling seed versus budded trees, another advantage of seed propagated cultivars would be the freedom from diseases as most are not transmitted via pollen or seed. It has also been suggested that inbred lines could be developed via several generations of selfing or by doubling haploid lines (Hesse 1971; Toyama 1974; Scorza and Pooler 1993) to create seed propagated hybrids as is done with maize. Unfortunately, a lack of a heterotic effect (Monet and Bassi 2008) would make this approach less useful.

When the desired characters are not to be found within the breeding populations of *P. persica*, related species are employed, usually for incorporating oligo- or monogenic traits. For scion cultivar breeding, the two species worked with most are *P. davidiana* and *P. dulcis*. *P. davidiana* has been used as a donor for resistance to green peach aphid, powdery mildew, peach leaf curl, and PPV (Viruel et al. 1998; Sauge 1998; Foulongne et al. 2003a, b; Decroocq et al. 2005; Rubio et al. 2010), whereas in almond the focus is on the introgression of genes for kernel quality, drought resistance, growth habit (e.g., spur bearing), low bruising, flowering habits of cleistogamy, and resistance to some diseases into peach germplasm (Martínez-Gómez et al. 2004; Gradziel 2003). Although there are few fertility barriers in developing these hybrids and creating subsequent breeding populations, several generations of backcrossing are needed to restore fruit quality (Foulongne et al. 2003b; Pascal et al. 1997).

In breeding for rootstocks, the selection for the desired trait(s) could be pursued within the F_1 progeny and the high level of heterozygosity, sometimes involving floral sterility, does not hamper clonal propagation. Consequently, interspecific hybridization with related species for useful traits such as tolerance to calcareous or droughty soils (almond), nematode resistance (*P. davidiana*, various plum species), waterlogging tolerance (various plum species) and dwarfing (various plum species) is quite common (Table 14.3; Reighard and Loreti 2008; Bouhadida et al. 2007b).

6.3 Breeding Methodology

Criteria for choosing the best parent are particularly critical. While traits under simple Mendelian inheritance can be easily traced within a given progeny and through generations, quantitative traits, controlled by polygenic systems, require a different approach.

Parents may be superior to commercial cultivars characterized by high productivity and fruit quality. This method is simple, fast and offers a good chance to get desired combinations, but the repeated use of the best cultivars as parents leads to high phenotypic homogeneity. In other cases an advanced selection based on one or more useful traits, such as those related to specific resistance or fruit quality, is chosen

Table 14.3 Single gene traits described in peach and their position on the *Prunus* reference map[a]

Character	Gene[b]	References	LG[c]
Tree			
Anthocyanins/anthocyaninless	An/an	Monet (1967)	
Normal/albino (no chlorophyll)	C/c	Bailey and French (1949)	
Tall, normal/pillar (broom)	**Br**/br or Pi/pi	Lammerts (1945)	G2
Tall, normal/bushy	Bu1/bu1	Lammerts (1945)	
	Bu2/bu2		
Normal shape/compact shape	Ct/ct	Mehlenbacher and Scorza (1986)	
Tall, normal/brachytic dwarf	**Dw**/dw	Lammerts (1945)	G6
	Dw2/dw2	Hansche (1988)	
	Dw3/dw3	Chaparro et al. (1994)	
Normal shape/weeping shape	Pl/pl	Monet et al. (1996)	
	We/we	Chaparro et al. (1994)	
Leaves			
Leaf color (red/green)	**Gr**/gr	Blake (1937)	G6–G8
Glandular/eglandular	**E**/e	Connors (1922)	G7
Deciduous/evergreen	**Evg**/evg	Rodríguez et al. (1994)	G1
Leaf shape (narrow/wide)	**Nl**/nl	Yamamoto et al. (2001)	G6
Leaf margin (smooth/wavy)	Wa/wa	Scott and Cullinan (1942)	
	Wa2/wa2	Chaparro et al. (1994)	
Flowers			
Single/double flower	**Dl**/dl	Lammerts (1945)	G2
Pollen (fertile/sterile)	**Ps**/ps	Scott and Weinberger (1944)	G6
	Ps2/ps2	Chaparro et al. (1994)	
Petal color (colored/white)	W/w	Lammerts (1945)	
Petal color (pink/red)	R/r	Lammerts (1945)	
Petal color (dark pink/light pink)	P/p	Lammerts (1945)	
Petal color (pink/pale pink)	**Fc**/fc	Yamamoto et al. (2001)	G3
Showy flowers size (large/small)	L/l	Lammerts (1945)	
Type (nonshowy/showy)	Sh/sh	Bailey and French (1949)	
Fruit			
Monocarpel/polycarpel	**Pcp**/pcp	Bliss et al. (2002)	G3
Anthocyanin (normal/blood flesh)	**Bf**/bf	Werner et al. (1998)	G4
Sweet fruit/normal fruit	**D**/d	Monet (1979)	G5
Freestone/clingstone	**F**/f	Bailey and French (1949)	G4
Pubescent skin/glabrous	**G**/g	Blake (1932)	G5
Saucer shape/nonsaucer	**S**/s	Lesley (1939)	G6
Nonaborting/aborting fruit	**Af**/af	Dirlewanger et al. (2006)	G6
Kernel (bitter/sweet)	**Sk**/sk	Werner and Creller (1997)	G5
Flesh color (white/yellow)	**Y**/y	Connors (1920)	G1
Skin color (red/green)	**Sc**/sc	Yamamoto et al. (2001)	G6–G8
Flesh color around stone (red/white)	**Cs**/cs	Yamamoto et al. (2001)	G3

(continued)

Table 14.3 (continued)

Character	Gene[b]	References	LG[c]
Flesh texture and pit adherence (F)[d]	M/m or F	Bailey and French (1933; 1949); Monet (1989); Peace et al. (2005)	G4
Melting freestone	F/-	Peace et al. (2005)	
Melting clingstone	f/f		
	f/f1		
	f/n	Peace et al. (2005)	
Nonmelting clingstone	f1/f1		
	f1/n		
	n/n	Peace et al. (2005)	
Stony hard flesh (Hd)	hd/hd	Yoshida (1976)	
Stony hard, melting[e]	hd hd/F-	Bailey and French (1949); Haji et al. (2005)	
Stony hard, melting	hd hd/f_1 f_1	Haji et al. (2005)	
Disease or pest resistance			
Myzus persicae resistant/susceptible	Rm1/rm1	Massonie et al. (1982); Monet (1985)	
Powdery mildew resistant/susceptible	Sf/sf	Dabov (1983)	
M. incognita resistant/susceptible	**Mi**/mi	Weinberger et al. (1943)	G2
M. javanica resistant/susceptible	Mj/mj	Sharpe et al. (1970)	

[a]Updated from Dirlewanger and Arús (2005)
[b]Mapped genes in bold
[c]Located on T × E map; G6–G8 genes located close to the translocation breakpoint between these two linkage groups
[d]Four alleles at the same locus controlling both flesh texture (endopolygalacturonase enzyme expression) and pit adherence; the fourth, null allele (n), has the same effect as the f_1 allele (nonmelting clingstone) (Peace et al. 2005)
[e]Independent inheritance of this trait was demonstrated, also suggesting an epistatic influence on the *F* locus, since when exogenous ethylene is applied, the stony hard-melting (*hdhd/f-*) phenotype is induced to soften (Haji et al. 2005)

to introduce the desired trait into a commercially important cultivar. The choice of two parental individuals with complementary phenotypic characters has led to the improvement of most of the commercially important fruit characters (Monet and Bassi 2008). Data on the heritability of quantitative traits confirm that parents could be chosen on the basis of their phenotype to yield rapid gains (Hansche et al. 1972; Souza et al. 2000). However, if the expression of a given trait is influenced by dominance or epistasis, the choice of a parent on a phenotypic basis could be misleading and lead to a worthless progeny.

For the above reason, the genetic value of a given parent should be assessed through a progeny study (Monet 1995). The simplest way is to perform a self-pollination: the more heterozygous the progeny, the more heterozygous the parent. This method gives valuable information particularly on simple traits, unveiling recombination and recessive characters. However, for traits under polygenic control, the evaluation of the prepotency, or combining ability, is better suited to rate the potential of a given genotype in yielding superior progenies (Fogle 1974). The simplest progeny test would be to compare several populations sharing a common parent (Cantín et al. 2009a, 2010a). The evaluation could involve one or more traits and has the advantage that could be done within a given breeding program design, thus not requiring additional studies or plantings.

The number of seedlings required for a given progeny may vary considerably. If segregation is sought for a genetic study on simple traits, just one or very few F_1 individuals are required to obtain an informative F_2 generation. If quantitative traits are to be studied, at least 100 seedlings per progeny are needed to assess variability and linkage relationships (e.g., when searching molecular markers for MAS), but larger numbers, around 1,000 seedlings, will assure sounder results. For heritability estimates, more than 100, even if small-sized, diverse progenies are needed to mimic the panmictic distribution of genes. For breeding purposes the progeny size for selecting new cultivars depends on the commercial cultivars already available, goals sought, and prepotency of the parents (Fogle 1974). Thus, an acceptable size of a progeny with a good probability to yield a new cultivar may vary from a few hundred to a thousand seedlings.

Given the size of the trees, it is common to do pollinations on trees in the field although some programs grow trees for breeding in large pots and move them in and out of a greenhouse for pollination. A major problem in the production of hybrid seed are cold temperatures during bloom which can be protected against by overhead sprinklers, orchard heating, or individual tree protection by covering with plastic films or fabrics and providing an heat source inside.

The hermaphroditic flowers of peach are easy to emasculate by cutting the calyx below the anther attachment with various notched sharpened devices, tweezers, or one's fingernails. This is done at the flower balloon stage, a few days before full bloom. In the case of exposed anthers of a nonshowy flower, it is important to check that the anthers are reddish and not dehicsing when the flower is emasculated.

For pollen, flowers at the balloon stage, before the anthers dehiscence, are collected and taken into the laboratory where the anthers are removed by cutting by hand or via rubbing the flowers either whole or cut in half transversely on a sieve. The detached anthers are allowed to dry at room temperature on an aluminum or paper tray or Petri dish for 24–48 h. Pollen needs to be maintained on a desiccant in cool conditions for current year use. Extra pollen batches or pollen collected for a next season pollinations can be stored desiccated at −18°C for 2–3 years or at −80°C for a longer time. Liquid nitrogen will ensure an almost indefinite storage.

Pollen is taken to the field in a vial, test tube or small jar. It is applied to the pistil with a pencil eraser, small camel hair brush, or one's finger tip and should be done either immediately or within 24–48 h after emasculation. Up to about 2 weeks after pollination the tree has to be checked for any unemasculated flowers that need to be removed to prevent the development of these unpollinated and probably self-pollinated fruit.

For normal breeding operations the flowers are generally not protected because emasculated flowers do not attract pollinating insects and peach pollen is heavy. If the progeny is to be investigated for genetic studies, a fine grid cage can be used to protect the tree from pollen moved by insects or wind from the neighboring trees. An insect-proof cage, usually made from an 80–90% shading net, should be provided to protect the mother tree where self-pollination has to be made. Fruit set is improved when self-pollination is done by hand at full bloom. The cage can be removed after petal fall.

Fruits from pollinated flowers should be harvested when ripe and the seed extracted from the pit to facilitate seed germination. Peach seeds need stratification to overcome dormancy and thus to germinate fully developed plantlets. The chilling requirement is positively related to that of the mother tree (Pérez 1990). Seed coat removal can speed up germination, unless it should be kept to avoid cotyledons splitting before germination occurs. If chilling is not satisfied, germination would be delayed and rosetting will occur. Seeds should be stored at −1 to 1°C in sterilized moist sand, in perlite moistened with a fungicide solution, or in sealed Petri dishes with a filter paper disk wetted by a fungicide solution. After 1–5 months, or as soon as the radicle tip emerges from the seed when still in storage, they can be planted in the greenhouse. Higher stratification temperatures (up to 4°C), although equally effective in overcoming dormancy, may not be low enough to stop seed rot caused by bacteria or fungi that can develop in the cold room.

In temperate climates the seedlings are grown in the greenhouse during the winter, then either transplanted in a nursery plot or directly in the field the following spring although in some programs the seedlings are grown outside just after germination to reduce greenhouse-related disease problems. Seeds collected from low-chilling genotypes in warm winter regions that can be successfully stratified in 3–4 weeks and then germinated, can be grown large enough in the same season to transplant in the field the same year of the cross.

Viability is poor in early ripening genotypes (less than 100–120 FDP) and aseptic culture is needed to ensure germination (Tukey 1934). Generally, embryos with a seed dry weight of less than 30% need to be put through in ovulo and/or embryo rescue procedures for consistent seed germination success (Bacon and Byrne 2005). The fruits of these genotypes should be harvested well before full ripening, not later that the veraison stage, to avoid contamination from juice exposure or fruit rot. The smallest embryos (<5 mm in length and as young as 50 days of development) require 4–8 weeks of in ovulo culture to enlarge the embryos sufficiently before the embryos can be successfully rescued (Ramming 1985; Pinto et al. 1994). The larger embryos (>10 mm in length) are explanted after seed coat removal and placed in a sterile culture tube containing a suitable nutritive medium (i.e., sugars, minerals, and vitamins; growth regulators are usually not needed) (Ramming 1990; Sinclair and Byrne 2003), incubated in a cold room at 0–4°C for 1–2 months to overcome dormancy and germinated in a growth room at 18–24°C (Ramming 1990; Anderson et al. 2002). Germination at the cooler range will give more consistent germination over a range of genotypes (Anderson et al. 2002). Once the seeds have germinated, the plantlets are transplanted into a sterile soil mix and are slowly acclimated to the low humidity and higher temperature regime of the greenhouse. These are grown in the greenhouse until large enough to transplant to the field.

At the end of the dormant season, seedlings can be transplanted in the orchard at densities ranging from 33,000 (0.3 m × 1 m) to 1,000 (2 m × 5 m) plants per hectare. The higher density approach is possible in low-chilling environments, where long growing season conditions favor rapid tree growth and early fruiting, i.e., from the second season after planting. Owing to this very early selection, seedlings can be pulled out before competition between neighboring trees occurs. The highest density tested so far is the "fruiting nursery" (Sherman et al. 1973) where seedlings are planted 13 cm apart and 1 m between rows. This method proved very effective for breeding goals but not for assessing the genetic nature of many quantitative traits. Lower densities, used in environments featuring short growing seasons or when prolonged life of the trees is envisaged to reduce tree competition, allow normal fruiting and make the choice of the best seedling easier. Also, it is best suited for genetic studies since trees can be grown to their full size.

Selection is usually made in the first good cropping year which varies from the second to the third or fourth year from planting, and from warm to temperate and cold environments, respectively. Usually one year of observation is enough to evaluate most of the progeny, given the phenotypes are a good estimate of the genotype (Hansche et al. 1972) as discussed above.

The evaluation method depends on the goals. When the main goal is market-driven, i.e., the release of a new cultivar, the choice of the best recombinants (seedlings) to be propagated as advanced selections should be mainly based on breeder experience and a sound knowledge of the available commercial cultivars. A common mistake would be to keep (and propagate) too many individuals that do not represent a real improvement toward the present cultivar array. However, some seedlings could be selected if they represent valuable genetic material for further crosses; even if per se they do not bring full commercial value, they will be kept to improve the breeding stock. When selecting for new cultivars, data are taken on only the main traits (bloom and ripening date, flower and leaf traits, fruit type and estimate of the yield potential) of selected seedlings. The others are simply discarded without taking data. In the past, field data were taken manually but considering the large number of seedlings often involved in today's breeding activity, data are frequently collected directly into a digital format to save time and avoid transcription errors. Nevertheless, paper and pencil can still prove as effective and are more user-friendly in the field under some situations.

When the evaluation of the progeny is focused on genetic investigations more detailed and accurate data should be taken in accordance with the aim of the studied trait(s). For those under simple Mendelian inheritance, data collection is rather trivial, and the data can be evaluated with the chi-square test. For quantitative traits the record keeping is more laborious and the measuring criteria need to be well defined in advance to maximize the usefulness of the data collected. Furthermore, since multigenic traits are influenced by environmental variability, it is advisable to randomize the seedlings (Okie 1984; Quilot et al. 2004), extend the observations for at least two or even more years and/or plant the population at multiple sites, particularly when linkage studies between QTLs and molecular markers are an objective. When studying the genetic determinants of fruit quality (size, appearance and composition), only a limited number of fruit per tree should be left to allow for the maximum fruit growth and avoid source competition among fruits (Quilot et al. 2004; Cantín et al. 2009a).

After a seedling has been chosen for further evaluation in a test plot, its sanitary status should be checked to exclude viruses, particularly the *Plum pox virus* (PPV), *Prunus necrotic ring spot virus* (PNRSV), *Prune dwarf virus* (PDV), and other intracellular pathogens (e.g., mycoplasms) that may hamper yield and/or fruit quality and exclude its introduction into the nursery system. Several diagnostic tests are available, such as ELISA, indicator host plants, and finally, the most sensitive, PCR-based techniques. If the selected seedling is virus-free, some mother trees should then be established in an insect-proof screen house to be kept as the source of clean propagation material for subsequent propagation for testing and possible release.

The advanced selections should be submitted to a testing procedure in comparison with other concurrent selections (e.g., from other breeding programs) and commercially established cultivars according to the ripening season and fruit type. In many breeding programs this is done in collaboration with commercial growers. To this end, trees are grafted on a given rootstock or, better, two or three common rootstocks and in several locations to collect more data prior to the possible release of a new cultivar. While a perfectly sound statistical design with replications is economically impractical in most situations, an experimental design should be planned to collect objective data not biased by the subjective evaluation of the breeder. From a number of studies, plots with a tree number variable from 6 to 8 are enough for yield records and from 15 to 30 fruits per tree are sufficient for quality assessment (Scorza and Sherman 1996). These test plots require at least 2–3 fruiting years of data before a good decision can be made on its commercial potential.

The superior selections from the second testing stage are then entered into the final stage of evaluation, i.e., the growers' acceptance trial. The market success of a putative new cultivar depends mainly on the acceptance of the growers and the retail distribution chain. Frequently, growers in the main fruit growing districts, even from distinct environments, are eager to test promising selections even at no cost for the breeder. At this point the tests are run under a nonpropagation agreement to avoid unintended or illegal propagation of the advanced selections. These final trials, even though performed informally without a statistically sound design, produce much information a breeder has no means to obtain from his formal tests, i.e., the selection's performance under diverse management (tree training and pruning, thinning) and different soils as well as its fruiting and postharvest behavior under a large field harvest operation. An additional 3–5 fruiting years are needed in this final test to raise enough confidence for the introduction of a selection as a new cultivar.

6.4 Release of Cultivars

The creation of a new cultivar is very expensive and a return on the investment is needed, thus legal protection is required. In the past, cultivar protection was sought only by private breeders but today even cultivars from public programs are being protected. The requirements for protection are different from one country to another. In the USA, patenting a cultivar is equivalent to patenting an industrial process. In the European Union, the Community Plant Variety Office (CPVO) manages a system of plant cultivar rights covering the 27 member states (http://www.cpvo.europa.eu). The applicant files an application for protection either directly through the CPVO or through one of the national Plant Breeder's Rights offices that subsequently transfers it to the CPVO. If no obstacle prevents a grant of Community protection, the CPVO takes the necessary measures for organizing the conducting a technical examination of the candidate cultivar. The aim of this is to verify that the cultivar is distinct from others, uniform in its characteristics and stable in the long run (DUS). Once the CPVO considers that the examination results are satisfactory and that all the other requirements have been fulfilled, it grants a Community Plant Variety Right for a period of 30 years for vines, fruit trees, grape, and potatoes. In Europe, a new cultivar receives a certificate that has approximately the same value as a patent in the USA. For a patent to be issued in the USA, a cultivar must be original and healthy (virus free). The legal protection lasts 20 years and covers its phenotype only, fruit included (see Chap. 3 on Intellectual Property).

New peach cultivars have a relatively short market life: 10–20 years at most with a life of a few years not being uncommon. If we compare this duration to what is needed to create a truly innovative cultivar (15–20 years on average), it can be said that this job is not really rewarding. However, some cultivars retain their commercial value for many years, e.g., 'Redhaven' peach worldwide, and now 'Big Top' nectarine in Italy, but they tend to be the exception rather than the rule. The problem lies in the fact that the breeder is often aiming at a moving target. While the new cultivar may have successfully combined the desired characters that were sought when the program was initiated, the cultivar requirements of the market may have changed during the 15–20 year period in which the cultivar was being developed. Thus, the new cultivar may not meet the existing market requirements when released. This is an inherent risk in fruit breeding, but given the genetic advances seen over the last 50 years, it appears to be a risk well worth taking (Monet and Bassi 2008).

It is becoming increasingly common to see new cultivars released after less than 10 years from the original pollination as the nursery industry push for quicker returns from their investment, and growers and their organizations compete for exclusive cultivation rights on new cultivars. This creates a situation in which these are released with minimal testing. This is why tens of newly introduced cultivars are entering the European and USA market every year. The best of these still remain to be identified and proven, often at grower expense.

6.5 Rootstocks

In the last half of the twentieth century, the selection of peach rootstocks was often begun with the identification and collection of spontaneous peach seedlings, wild plums and/or natural peach–almond hybrids, which were incorporated into *Prunus* collections (Bernhard and Grasselly 1981; Indreias et al. 2004; Moreno 2004). In the first phase, the work basically focused on establishing mother plants and studying their aptitude for sexual or vegetative propagation. For the outstanding clones, their sanitary status was determined and propagation conditions were optimized. In many cases, micropropagation procedures were established, which also accelerated the breeding process by allowing the rapid clonal propagation of *Prunus* hybrids from controlled interspecific crosses to produce plants for evaluation.

To assess scion-rootstock compatibility, experimental nurseries are established to ascertain good graft compatibility of the new rootstocks, mainly when species from botanical sections different from *Euamygdalus* are used. Cases of "translocated" incompatibility in peach are usually expressed during the first year of scion growth, but the occurrence of the "localized" cases may be delayed, and subsequently, more years are necessary to evaluate this feature (Zarrouk et al. 2006). To determine the influence of the outstanding clones on the productive characteristics of peach cultivars (e.g., vigor, yield, and fruit quality), orchard trials are established to assess their performance in the most important areas of production, including a range of soils and pathological challenges. During the last half of the twentieth century, this selection process usually took 20–40 years before a new peach rootstock could be released and widely used into the peach industry.

Traditional selection procedures used to detect tolerance to abiotic stresses (iron chlorosis and waterlogging) are based on field evaluation and usually requires several years. Therefore, new evaluation methods using hydroponic culture have been also developed to select new genotypes tolerant to iron chlorosis based on the root capacity to reduce Fe-chelates (Cinelli and Loreti 2004; Jiménez et al. 2008). Similarly, evaluation for tolerance to waterlogging have been also conducted in specially designed tanks where the soil is flooded and selection is based on the rate at which plants develop symptoms of waterlogging and root asphyxia (Salesses et al. 1970; Amador et al. 2010). In the case of nematodes, tests are usually carried out with plants growing in infected pots established in greenhouses. With these procedures, rootstock evaluation to these stresses can be carried out in several months (Pinochet et al. 1999; 2002).

6.6 *Propagation*

Peach seedling rootstocks have been primarily used in the world because of the availability of inexpensive seeds, the ease of sexual propagation and the good compatibility with budded peach cultivars. However, the horticultural advantages of peach–almond hybrids and plum rootstocks for peaches led to the development of new methods of vegetative propagation. Hardwood and softwood cutting propagation were first established by defining the most appropriate auxins (type and concentration) and timing of propagation during the year (Howard 1987; Webster 1995). At present, all these methods are being replaced by tissue culture of clonally micropropagated selections to produce thousands or millions of plants annually

(Battistini and De Paoli 2002), although micropropagated rootstocks frequently sucker more profusely than those from conventional cutting techniques (Webster 1995).This technique also has value in facilitating the movement of healthy materials over national borders while satisfying plant importation and health regulations. These successful propagation techniques developed for *Prunus* clonal rootstocks and interspecific hybrids has further accelerated interest and research into molecular genetics and MAS in peach rootstocks (Lu et al. 2000; Dirlewanger et al. 2004a).

7 Integration of New Biotechnology in Breeding Programs

7.1 Molecular Markers

Molecular markers have been used in peach for genotyping and genetic diversity analysis (Dirlewanger et al. 2002; Aranzana et al. 2003a; Riaz et al. 2004; Yoon et al. 2006; Bouhadida et al. 2007a, b, 2009, 2011), development of linkage maps (Chaparro et al. 1994; Rajapakse et al. 1995; Dirlewanger et al. 1998; Yamamoto et al. 2001; 2005), trait tagging and MAS (Foulongne et al. 2002; Lecouls et al. 2004; Blenda et al. 2007), and for quantitative trait loci (QTL) positioning (Dirlewanger et al. 1999, 2006; Quilot et al. 2004; Cantín et al. 2010b). A number of molecular marker systems, such as isoenzymes, restriction fragment length polymorphism (RFLP), random amplified polymorphic DNA (RAPD), fragment length polymorphism (AFLP), and simple sequence repeats (SSRs) have been used in peach for the identification of markers tightly linked to traits of interest (Chaparro et al. 1994; Sosinski et al. 1998; Quarta et al. 1998; Joobeur et al. 1998; Dirlewanger et al. 1998; Dettori et al. 2001; Verde et al. 2005). Owing to their abundance, high polymorphism, codominance, reproducibility, and transferability to related species, SSRs are emerging as a marker of choice for linkage and comparative mapping, genotype identification, QTL tagging, and MAS (Cipriani et al. 1999; Aranzana et al. 2002; 2003a, b; Dirlewanger et al. 2004b; Liu et al. 2007). Moreover, the large expansion of DNA databases, particularly those containing EST sequences, has now opened the opportunity for the identification of single nucleotide polymorphisms or (SNPs) in peach (Lazzari et al. 2008). Typically, however, RFLP, RAPD, AFLP, and SSR markers are only genetically linked to the trait of interest, and no functional relationship can be inferred. Therefore, a candidate gene/QTL approach is necessary to associate major genes and QTLs involved in expression of traits of interest to structural genes in peach.

7.2 State of the Map

Chaparro et al. (1994) developed the first genetic map for peach using molecular markers. Since then, nine linkage maps have been constructed for peach (Dirlewanger and Bodo 1994; Dirlewanger et al. 1998; Rajapakse et al. 1995; Abbott et al. 1998), and six interspecific maps between peach and other members of the genus *Prunus*, namely, peach × almond (Joobeur et al. 1998; Foolad et al. 1995; Jáuregui et al. 2001), peach × *P. davidiana* (Dirlewanger et al. 1996), peach × *P. ferganensis* (Quarta et al. 1998), and myrobalan plum × (almond × peach hybrid) (Dirlewanger et al. 2004a), have been constructed (Table 14.4).

The 'Texas' (almond) × 'Earlygold' (peach) linkage map (T × E) is the first saturated linkage map constructed completely from transferable markers and is considered the reference map for *Prunus* L. (Joobeur et al. 1998; Dirlewanger et al. 2004a) (http://www.bioinfo.wsu.edu/gdr/). In addition to 826 markers currently placed on the T × E map (Dirlewanger et al. 2004b; Howad et al. 2005), Abbott et al. (2007) recently reported on mapping efforts that tentatively put an additional 600 EST sequences on this map. The existence of the T × E map has been very useful for the *Prunus* research community, providing a highly polymorphic population for linkage studies, establishing a common terminology for linkage groups, and providing a set of transferable markers ("anchor" markers) of known map position that facilitated the development of framework maps in other crosses. It also allowed the location of different major genes and QTLs in a unique map, the search for markers to saturate specific genomic regions, and the establishment of map comparisons with other *Prunus* species (Dirlewanger et al. 2004b).

7.3 Traits Tagged with Molecular Markers

Peach has a relatively small genome, estimated at 300 Mb in the haploid genome (Arumuganathan and Earle 1991; Baird et al. 1994), and is considered genetically the best characterized species in *Prunus* and among fruit trees (Mowrey et al. 1990). There are 43 morphological characters with simple Mendelian inheritance in peach (Dirlewanger et al. 2004b; Dirlewanger and Arús 2005) and for 23 of them linkage relationships with molecular markers have been determined (Table 14.5). So far, molecular markers are proposed for only 20 peach monogenic traits, and only for 12 of those the linkages are tight enough (less than 5 cM) to be sufficient for MAS (Table 14.5).

Molecular markers linked to six Mendelian characters have recently been reported (Dirlewanger et al. 2006): pollen sterility, peach or nectarine fruit, saucer or round fruit, clingstone or freestone fruit, low acidity in fruit, and fruit abortion. The character of trees bearing aborting fruit (*Af*) is recessive and linked to the saucer gene, and is bounded by two SSR markers, MA040a and MA014a. For the other five traits, linkage relationships were previously reported and placed on the *Prunus* reference map (Dirlewanger et al. 2004b), but tightly linked PCR based molecular markers were lacking. Although peach genomic and EST sequence databases are constantly expanding and a highly saturated *Prunus* reference map is available, there is still a need for markers, preferably PCR based ones such as SSRs, which are tightly linked to loci of agronomic importance. Wang et al. (2002b) identified SSR loci tightly linked to two important peach traits, root-knot nematode resistance and

Table 14.4 Peach intra- and interspecific linkage maps

Population	Type	Marker #	LG #	Map size (cM)	References
P. persica × *P. persica*					
Weeping clone (1161:12×2678:47) 1:55×'Early Summergrand'	F_2	52	8	350	Dirlewanger and Bodo (1994)
NC174RL×'Pillar'	F_2	88	15	396	Chaparro et al. (1994)
'New Jersey Pillar'× 'KV77119'	F_2	58	13	540	Rajapakse et al. (1995); Abbott et al. (1998); Sosinski et al. (2000)
'Suncrest'×'Bailey'	F_2	147	23	926	Abbott et al. (1998); Sosinski et al. (2000)
'Lovell'×'Nemared'	F_2	153	15	1,300	Abbott et al. (1998); Lu et al. (1998); Sosinski et al. (2000)
'Harrow Blood'× 'Okinawa'	F_2	76	10		Gillen and Bliss 2005
'Akame'×'Juseitou'	F_2	178	8	571	Shimada et al. (2000); Yamamoto et al. (2001, 2005)
'Ferjalou Jalousia'× 'Fantasia'	F_2	181	7	621	Dirlewanger et al. (1998, 2006); Etienne et al. (2002)
'Contender×Fla.92-2C'	F_2	127	8	535	Fan et al. 2010
'Guardian®'× 'Nemaguard' (*P persica*× *P. davidiana*)	F_2	158	11	737	Blenda et al. (2007)
P. dulcis×*P. persica*					
'Texas'×'Earlygold'	F_2	826	8	524	Joobeur et al. (1998); Aranzana et al. (2003b); Dirlewanger et al. (2004b); Howad et al. (2005)
'Padre'×'54P455'	F_2	161	8	1,144	Foolad et al. (1995); Bliss et al. (2002)
'Garfi'×'Nemared'	F_2	51	7[a]	438	Jáuregui et al. (2001)
P. persica×*P. ferganensis*					
IF7310828 ('J.H. Hale'×'Bonanza')× *P. ferganensis*	BC_1	216	8	665	Quarta et al. (1998, 2000); Verde et al. (2005)
P. persica×*P. davidiana*					
'Summergrand'×Clone P1908	F_1	23/97[b]	3/9	159/471	Dirlewanger et al. (1996); Viruel et al. (1998); Foulongne et al. (2002)
'Rubira×Clone P1908'	F_1	4/88[b]	0/8	454.2	Rubio et al. (2010)
(*P. cerasifera*)×(*P. dulcis*×*P. persica*)					
P.2175×GN22 ('Garfi'×'Nemared')	F_1	93/166[b]	8/7	525/716	Dirlewanger et al. (2004a)

[a]Linkage groups 6 and 8 of this map were mapped as a single group due to a reciprocal translocation

[b]Separate maps were created for each parent

Table 14.5 Molecular markers linked to monogenic traits in peach

Trait	Gene	Marker name	Distance[a] (cM)	Reference
Flower				
Double flower	*Dl*	*pchgms1*	7.8	Sosinski et al. (2000)
Flower color	*Fc*	*EACA/MCTG-220*	7.1	Yamamoto et al. (2001)
Male sterility	*Ps*	*FG40*	4.8	Dirlewanger et al. (2006)
Leaf				
Leaf color	*Gr*	*UDP96-015*	3.7	Yamamoto et al. (2001)
Leaf glands	*E*	*AG104*	2	Dettori et al. (2001)
Leaf shape	*Nl*	*EAC/MCAC-180*	12.0	Yamamoto et al. (2001)
Tree				
Dwarf plant	*Dw*	*EAC/MCAC-180*	12.0	Yamamoto et al. (2001)
Pillar growth habit	*Br*	*pchgms1*	12.5	Sosinski et al. (2000)
Evergrowing	*evg*	*EAT/MCAC*	1.0	Wang et al. (2002b)
		pchgms10	1.0	
		pchgms11	1.0	
		pchgms12	1.0	
		pchgms13	1.0	
		pchgms14	1.0	
Fruit				
Blood flesh	*bf*	*C41H*	10.3	Gillen and Bliss 2005
Saucer fruit	*S*	*MA040a*	0	Dirlewanger et al. (2006)
Aborting fruit	*Af*	*MA040a*	0	Dirlewanger et al. (2006)
Flesh adhesion	*F*	*UDAp-431/b*	1.2	Dirlewanger et al. (2006)
		BPPCT009/b	2.2	Dettori et al. (2001)
		AG12 & AG16b	2.0	
Flesh color	*Y*	*UDP98-407*	2.2	Mingliang et al. 2007
Flesh color around stone	*Cs*	*OPO2/0.6*	12.4	Yamamoto et al. (2001)
Nonacid fruit	*D*	*pTC-CTG/a*	0	Dirlewanger et al. (2006)
		pGT-TTG/a	0	
Skin color	*Sc*	*UDP96-015*	3.7	Yamamoto et al. (2001)
Skin pubescence	*G*	*eAC-CAA/a*	0	Dirlewanger et al. (2006)
		UDP96-018	4.5	Mingliang et al. (2007)
Pest resistance				
Nematode resistance	*Mij*	*EAA/MCAT10*	3.4	Lu et al. (1998)
		pchgms26	3.4	Wang et al. (2002a)
		ISSR834-1/0.4	4.8	Wang et al. (2002a)
	Mja	*EAA/MCAC-135*	4.8	Yamamoto et al. (2001)

[a]cM distance <5 is considered close enough for MAS

evergrowing, by using the high-throughput technique of AFLP mapping with subsequent direct targeting of SSRs identified in AFLP-marked regions of interest. However, this approach relies on the availability of a peach bacterial artificial chromosome (BAC) library resource. Examples of using bulk segregant analysis (BSA) in discovering markers tightly linked to disease resistance traits are also available in peach (Claverie et al. 2004a, b; Lecouls et al. 2004; Gillen and Bliss 2005).

Most agronomically important traits in which breeders are interested exhibit continuous phenotypic variation indicating more complex, polygenic control. There are 25 peach traits associated with QTLs, and most of them are related to fruit quality (Abbott et al. 1998; Dirlewanger et al. 1999; Quarta, et al. 2000; Etienne et al. 2002; Peace et al. 2006; Cantín et al. 2010b), adaptation (Abbott et al. 1998; Dirlewanger et al. 1999; Quarta et al. 2000; Etienne et al. 2002; Blenda et al. 2007) or disease resistance (Dirlewanger et al. 1996; Viruel et al. 1998). The detection of QTLs related with tolerance to abiotic stresses as iron chlorosis, and the search of candidate genes differentially expressed under iron deficiency is under development. Preliminary results have showed QTLs located in chromosomes 6 and 8 (Gonzalo et al. 2009), near other QTLs involved in fruit quality and nematodes resistance (Yamamoto et al. 2001; Dirlewanger et al. 2004a). The candidate genes approach has been implemented based on *in silico* screening of genes shown to be expressed in response to iron deficiency in roots (Gonzalo et al. 2011).

High level of synteny and colinearity between different *Prunus* maps and the existence of a reference map for the genus allowed integration of 28 major *Prunus* genes, mapped in populations of apricot, peach, almond, and myrobalan plum, into a single map (Dirlewanger et al. 2004b). The approximate position of these genes on the T×E map and information available from the interconnecting *Prunus* maps allows the discovery of additional markers in regions of interest and their usage in MAS.

7.4 Marker-Assisted Selection

Peach breeding has been very active in the last decade with hundreds of new cultivars released (Sansavini et al. 2006). The ability of breeders to generate large populations is almost unlimited, but the management, phenotyping and selection of these seedlings are the main limiting factors for the generation of new cultivars. Molecular markers linked to traits of interest are essential for both MAS and improvement of selection efficiency in standard breeding procedures, especially for economic traits that are difficult to select by phenotype early in the plant life cycle. MAS is particularly useful when the expression of the gene is recessive and the evaluation of the character is expensive or time-consuming or, in tree crops such as peach, with long juvenile periods (Scorza 2001; Luby and Shaw 2001). The low level of variation found in peach (Byrne 1990) and the narrow genetic base of modern peach cultivars (Scorza et al. 1985) impede implementation of MAS in peach breeding programs. One of the major impediments to using MAS in applied breeding of fruit crops identified by breeders in a survey was the lack of markers and simplified technology to screen progenies (Byrne 2007). These impediments are being addressed in several large collaborative research programs in both Europe (Audergon et al. 2009; Dirlewanger et al. 2009) and the USA (Iezzoni et al. 2009).

MAS is currently used in rootstock breeding programs for early selection for resistance genes to root-knot nematode (*Meloidogyne* spp.) from two sources: peach (Lu et al. 1998; Yamamoto and Hayashi 2002; Arús et al. 2004) and myrobalan plum (Lecouls et al. 1999, 2004; Claverie et al. 2004a, b). In addition, Claverie et al. (2004a, b) showed that plum and peach genes are nonallelic and thus can be pyramided into interspecific hybrid rootstocks based on the plum and peach species. Recently, another successful implementation of MAS has been reported for the recessive character *Af* that encodes the fruit abortion trait in peach trees (Dirlewanger et al. 2006). Despite the growing availability of genomic resources in peach, the existence of a highly saturated reference T×E map [826 markers with average density of 0.63 cM per marker, Dirlewanger et al. (2004a, b, c) and Howad et al. (2005)] and most of the simple characters being sufficiently marked for selection (Table 14.5), the use of markers for commercial breeding in peach is still in its infancy. In addition, most of the agronomic important traits are quantitatively inherited and although 28 QTL have been identified in *Prunus*, further work is necessary before QTL-associated markers can routinely be integrated into selection programs.

7.5 Genomics

The availability of whole genome sequences and expressed sequence tag (EST) databases for important crops is accelerating the process of gene discovery. The recently released first draft of the assembled peach genome sequence, *peach v1.0* (http://www.rosaceae.org/peach/genome), along with previously available *Prunus* Genome (http://www.rosaceae.org/) and ESTree databases (http://www.itb.cnr.it/estree/index.php), provides access to genomic data for peach and constitutes very useful sources of information for genome comparative studies and identification of important genes. Abbott et al. (2002) first reported on the initiative to build genomic resources using peach as a model for the identification, characterization, and cloning of important genes of Rosaceae species. Since then, several BAC libraries have been constructed for peach (http://www.bioinfo.wsu.edu/gdr/) (Wang et al. 2001; Georgi et al. 2002; Boudehri et al. 2009). The BAC library constructed from fruit mesocarp of the peach rootstock 'Nemared,' consisting of 44,160 clones with an average size of 70 kb and 8.8-fold genome coverage (Georgi et al. 2002), and the BAC library from a haploid of the peach rootstock "Lovell" with 34,560 clones having an average insert of 80 kb providing 9.2-fold genome coverage (L. Georgi, unpublished data), have been used for the development of a framework physical map for peach (Zhebentyayeva et al. 2008). Two hundred and fifty-two clones, out of 2,138 contigs that form the initial physical map of peach, were anchored to eight linkage groups of the *Prunus* reference map.

Approximately 100,000 EST sequences from different *Prunus* species have been sequenced and deposited to NCBI and ESTree databases (http://www.ncbi.nlm.nih.gov; http://www.itb.cnr.it/estree/index.php). Most of the *Prunus* ESTs originated from 19 peach libraries, representing nine cultivars and four tissues from four developmental stages (Lazzari et al. 2005). Peach EST-SSRs, originated from fruit transcriptome, have already been isolated and show transportability across other *Prunus* species (Vendramin et al. 2007) and/or were tentatively mapped to the *Prunus* reference map (Abbott et al. 2007). Additionally, development of a peach transcript map (Horn et al. 2005; Abbott et al. 2007) and its integration with the physical map (Zhebentyayeva et al. 2008) was reported. This provides the necessary foundation for the identification of candidate genes that control many important fruit tree characters.

Breeding for disease resistance is a major goal in most cultivar development programs. The identification of loci for pathogen resistance in peach would provide information about resistance loci, the organization of resistance genes throughout the genome, and permit comparison of resistance regions among other genomes in the Rosaceae. Lalli et al. (2005) generated a resistance map for *Prunus* using a candidate gene approach. Resistance gene analogs (RGAs) and resistance-associated genes (RAGs) were hybridized to a peach BAC library and mapped using the peach physical map database and the Genome Database for Rosaceae (GDR). More than 40 RGAs and RAGs are mapped in regions known to contain resistance to powdery mildew, plum pox virus, and parasitic nematodes (Lalli et al. 2005; Abbott et al. 2007).

7.6 Transgenics

Improvement of fruit tree species through traditional breeding methods is a long-term effort due to their lengthy juvenile periods. Genetic transformation presents a promising tool for genetic improvement of peach and other woody fruit species. The main obstacle to genetic engineering of fruit tree species is the regeneration of transformed tissues/plantlets. Therefore, to use genetic engineering techniques for germplasm improvement, reliable protocols for transformation, selection, and regeneration of transgenic plantlets are needed. There are several reports of using different peach explants for regeneration: immature zygotic embryos (Hammerschlag et al. 1985; Hammerschlag 1988), immature cotyledons (Mante et al. 1989), embryo cells (Smigocki and Hammerschlag 1991), mature cotyledons (Pooler and Scorza 1995), and recently, mature embryos (Pérez-Clemente et al. 2005) and in vitro leaf tissue (Gentile et al. 2002).

Most of the work on peach transformation was done using *Agrobacterium*-mediated transformation. However, gene delivery via particle bombardment to embryonic callus derived from immature embryos has also been reported (Ye et al. 1994). Although high levels of transformation were demonstrated, no regeneration was obtained from the transformed embryogenic callus. In spite of much work and dedication, peach regeneration remains difficult; for example, only two reports of stable peach plant transformants have been published, and a total of four transgenic plants have been produced (Smigocki and Hammerschlag 1991; Pérez-Clemente et al. 2005). In addition, successful transformation and in vitro regeneration of peach plants was reported mostly from zygotic tissue, which is not favored for fruit tree transformation because the ability to improve established cultivars is lost (Abbott et al. 2007). However, while *Prunus* breeders and geneticists wait for efficient and repeatable transformation methodology for peach somatic tissues, breeding programs are now using and benefiting from current technologies such as MAS to incorporate targeted genes into elite germplasm.

References

Abbott, A., Georgi G.L., Inigo, M., Sosinski B., Yvergniaux D., Wang Y., Blenda A. and Reighard G. (2002) Peach, the model genome for Rosaceae. Acta Hort. 575, 145–155.

Abbott, A.G., Rajapakse B., Sosinski, B., Lu Z.X., Sossey-Alaoui K., Gannavarapu M., Reighard G.L., Ballard R.E., Baird, W.V., Scorza, R. and Callahan, A. (1998) Construction of saturated linkage maps of peach crosses segregating for characters controlling fruit quality, tree architecture and pest resistance. Acta Hort, 465, 41–49.

Abbott A., Arús P. and Scorza R. (2007) Peach. In: Chittaranjan K. (ed.), *Genome Mapping and Molecular Breeding in Plants, Fruits and Nuts*. Springer Heidelberg, Berlin, pp. 137–156.

Albás, E.S., Jiménez, S., Aparicio, J., Betrán, J.A. and Moreno, M.A. (2004) Effect of several peach × almond hybrid rootstocks on fruit quality of peaches. Acta Hort. 658, 321–326.

Amador, M.L. (2010) Estudio de las bases bioquímicas y moleculares de la tolerancia a la asfixia radicular. Ph.D. Thesis. Facultad de Ciencias. Universidad de Zaragoza. Spain. Pp. 177 (in English).

Amador, M.L., Sancho, S., Rubio-Cabetas, M.J. (2009) Biochemical and molecular aspects involved in waterlogging tolerance in *Prunus* rootstocks. Acta Hort 814, 715–720.

Amador, M.L., Bielsa, B., Gómez-Aparisi, J., Sancho, S., Jaime S., and Rubio-Cabetas, M.J. (2010) Avances en el estudio de la tolerancia a la asfixia radicular en patrones de melocotonero. Revista de Fruticultura n 9. Especial Melocotonero, pp. 48–55.

Anderson, N. (2009). Diversity of low chill peaches (*Prunus persica*) from Asia, Brazil, Europe, and the USA. M.S. Thesis. Texas A&M University, College Station, TX.

Anderson, N., Byrne D.H., Sinclair, J. and Burrell, A.M. (2002) Cooler temperatures during germination improves survival of embryo cultured peach seed. HortScience 37, 402.

Aranzana, M.J., Carbo, J. and Arús, P. (2003a) Microsatellite variability in peach [*Prunus persica* (L.) Batsch], cultivar identification, marker mutation, pedigree inferences and population structure. Theor. Appl. Genet. 106, 1341–1352.

Aranzana, M.J., Pineda, A., Cosson, P., Dirlewanger, E., Ascasibar, J., Cipriani, G., Ryder, C.D., Testolin, R., Abbott, A., King, G.J., Iezzoni, A.F. and Arus, P. (2003b) A set of simple-sequence repeat (SSR) markers covering the *Prunus* genome. Theor. Appl. Genet. 106, 819–825.

Aranzana, M.J., Garcia-Mas, J., Carbo, J. and Arús, P. (2002) Development and variability analysis of microsatellite markers in peach. Plant Breeding 121, 87–92.

Arumuganathan, K. and Earle, E. (1991) Nuclear DNA content of some important plant species. Plant Mol. Biol. Rep. 9, 208–218.

Arús, P., Mnejja, M., Esmenjaud, D., Bosselut, N. and Dirlewanger, E. (2004). High marker density around the peach nematode resistance genes. Acta Hort. 658, 567–571.

Audergon. J.M., D. Ruiz, A. Bachellez, A. Blanc, M.N. Corre, C. Croset, A.M. Ferreol, P. Lambert, T. Pascal, J.L. Poëssel, V. Signoret, B. Quilot, K. Boudehri, C. Renaud, E. Dirlewanger, L. Dondini, B. Gouble, M. Grotte, M. Bogé, P. Reiling, M. Reich, S. Bureau, C. Deborde, M. Maucourt, A. Moing, S. Monllor, and P. Arús (2009) ISAFRUIT – Study of the genetic basis of Prunus fruit quality in two peach and apricot populations. Acta Hort. 814, 523–527.

Bacon, T.A. and Byrne, D.H. (2005) Percent dry weight of the ovule predicts peach seed viability. HortScience, 40(7), 2211–2212.

Badenes, M.L., Martínez-Calvo, J., and Llácer, G. (1998) Analysis of peach germplasm from Spain. Acta Hort. 465, 243–250.

Bailey, C.H. and Hough, L.F. (1959) A hypothesis for the inheritance of season of ripening in progenies from certain early ripening peach varieties and selections. Proc. Amer. Soc. Hort. Sci. 73, 125–133.

Bailey, J.S. and French, A.P. (1949) The inheritance of certain fruit and foliage characters in the peach. Mass. Ag. Expt. Sta. Bul. 452. Univ. Mass. (1949). 31 p.

Bailey, J.S. and French, A.P. (1933) The inheritance of certain characters in the peach. Proc. Amer. Soc. Hort. Sci. 29, 127–130.

Baird, W.V., Estager, A.S. and Wells, J.K. (1994) Estimating nuclear DNA content in peach and related diploid species using laser flow cytometry and DNA hybridization. J. Amer. Soc. Hort. Sci. 119, 1312–1316.

Bassi, D. (2006) Breeding for resistance to Plum pox virus in Italy. EPPO/OEPP Bulletin 36, 327–329.

Bassi, D. (Ed.) (2003) Growth habits in stone fruit trees. Il Divulgatore, Bologna, Italy.

Bassi, D. and Rizzo, M. (2000) Peach breeding for growth habit. Acta Hort. 538, 411–414.

Bassi, D., Rizzo, M. and Canton I.L. 1998 Assaying brown rot [(*Monilinia laxa* Aderh. Et Ruhl (Honey)] susceptibility in peach cultivars and progeny. Acta Hort. 465, 715–718.

Battistini, A. and De Paoli, G. (2002) Large scale micropropagation of several peach rootstocks. Acta Hort. 592, 29–33.

Beckman, T.G. (1998) Developing Armillaria resistant rootstocks in peach. Acta Hort. 465, 219–224.

Beckman, T.G. and Pusey, P.L. (2001) Field testing peach rootstocks for resistance to Armillaria root rot. HortScience 36, 101–103.

Beckman, T.G., Okie, W.R., Nyczepir, A.P., Pusey, P.L. and Reilly, C. C. (1998) Relative susceptibility of peach and plum germplasm to Armillaria root rot. HortScience 33, 1062–1065.

Beckman, T.G., Reighard, G.L., Okie, W.R., Nyczepir, A.P., Zehr, E.I. and Newall, W.C. (1997) History, current status and future potential of Guardian™ peach rootstock. Acta Hort. 451, 251–258.

Beckman, T.G., Rodríguez, J., Sherman, W.B. and Werner, D.J. (2005) Evidence for qualitative suppression of red skin color in peach. HortScience 40, 523–524.

Beckman, T.G. and Sherman, W.B. (1996) Non-melting semi-freestone genotype in peach. Fruit Var. J. 50,189–193.

Beckman, T.G. and Sherman, W.B. (2003) Probable quantitative inheritance of full red skin color in peach. Hortscience 38, 1184–1185.

Begheldo M., Manganaris G.A., Bonghi C. and Tonutti, P. (2008) Different postharvest conditions modulate ripening and ethylene biosynthetic and signal transduction pathways in Stony Hard peaches. Postharvest Biology and Technology, 48, 84–91.

Bellini, E., Surico, G. Mugnai, L. Natarelli, L. and Nencetti, V. (1993) Osservazioni su una progenie di pesco resistente a Taphrina deformans (Berck.) Tul. Italus Hortus 1, 11–13.

Bernhard, R. and Grasselly, C. (1981) Les pêchers x amandiers. L'Arboriculture Fruitière, 328, 37–42.

Bernhard, R. and Renaud, R. (1990) Le point sur les porte-greffes du prunier. L'Arboriculture Fruitière 432, 28–36.

Blake, M.A. (1937) Progress in peach breeding. Proc. Amer. Soc Hort. Sci, 35, 49–53.

Blake, M.A. (1932) The J.H. Hale as a parent in peach crosses. Proc. Amer. Soc. Hort. Sci. 29, 131–136.

Blenda, A.V., Verde, I., Georgi, L.L., Reighard, G.L., Forrest, S.D., Muñoz-Torres, M., Baird, W.V. and Abbott, A. (2007) Construction of a genetic linkage map and identification of molecular markers in peach rootstocks for response to peach tree short life syndrome. Tree Genetics and Genomes 3, 341–350.

Bliss, F.A., Arulsekar, S., Foolad, M.R., Becerra, V., Gillen, A.M., Warburton, M.L., Dandekar, A.M., Kocsisne, G.M. and Mydin, K.K. (2002) An expanded genetic linkage map of *Prunus* based on an interspecific cross between almond and peach. Genome 45, 520–529.

Boonprakob, U., Byrne, D.H. and Mueller, D.M.J. (1996) Anatomical differences of axillary bud development in blind nodes and normal nodes in peach. HortScience 31(5), 798–801.

Boonprakob, U., Byrne, D.H. and Rouse, R.E. (1994) A method for blind node evaluation. Fruit Var. J. 48, 213–215.

Boudehri, K., Bendahmane, A., Cardinet, G., Troadec, M.C., Moing, A. and Dirlewanger, E. (2009) Phenotypic and fine genetic characterization of the D locus controlling fruit acidity in peach. BMC Plant Biology 9, 59.

Bouhadida, M., Moreno, M.A., Gonzalo, M.J., Alonso, J.M., Gogorcena, Y. (2011) Genetic variability of introduced and local Spanish peach cultivars determined by SSRs markers. Tree Genetics & Genomes 7(2), 257–270.

Bouhadida M., Casas, A. M., Moreno, M. A. and Gogorcena, Y. (2007a) Molecular characterization of Miraflores peach variety and relatives using SSRs. Scientia Hort. 111, 140–145.
Bouhadida, M., Martín, J.P., Eremin, G., Pinochet, J., Moreno, M.A., and Gogorcena, Y. (2007b) Chloroplast DNA diversity in *Prunus* and its implication on phylogenetic relationships. J. Amer. Soc. Hort. Sci. 132, 670–679.
Bouhadida, M., Casas, A.M., Gonzalo M.J., Arús P., Moreno, M.A., Gogorcena, Y. (2009) Molecular characterization and genetic diversity of *Prunus* rootstocks. Sci. Hortic. 120, 237–245.
Brooks, S.J., Moore, J.N. and Murphy, J.B. (1993) Quantitative and qualitative changes in sugar content of peach genotypes [*Prunus persica* (L.) Batsch]. J. Amer. Soc. Hort Sci. 118, 97–100.
Brooks, R.M. and Olmo, H.P. (1997). Register of Fruit & Nut Varieties. 3rd ed. ASHS Press, Alexandria, Virginia, USA.
Brovelli, E.A., Brecht, J.K., Sherman, W.B. and Sims, C.A. (1998) Anatomical and physiological responses of melting-flesh and nonmelting-flesh peaches to postharvest chilling. J. Amer. Soc. Hort. Sci. 123, 668–674.
Brovelli, E.A., Brecht, J.K., Sherman, W.B. and Sims, C.A. (1995) Quality profile of fresh market melting and non-melting peach fruit. Proc. Fla. State Hort. Soc. 108, 309–311.
Burgos, L., Egea, J. and Dicenta, F. (1991) Effective pollination period in apricot (*Prunus armeniaca* L.) cultivars. Annals Applied Biology, Cambridge- England, V.119, 533–539.
Byrne, D.H. (1986) Mechanism of spring freeze injury avoidance in peach. HortScience 21, 1235–1236.
Byrne, D. H. (1990) Isozyme variability in four diploid stone fruits compared with other woody perennial plants. J. Hered. 81, 68–71.
Byrne, D.H. (2003) Founding clones of low chilling fresh market peach germplasm developed in the USA and Brazil. Acta Hort. 606, 17–21.
Byrne, D.H. (2005) Trends in stone fruit cultivar development. Hort. Technol.15, 494–500.
Byrne, D.H. (2007) Molecular marker use in perennial plant breeding. Acta Hort 751, 163–167.
Byrne, D. H. (2010). Environmental challenges of breeding peaches for low chill regions. Acta Hort. 872, 129–138.
Byrne. D.H. and Bacon, T.A. (1999) Founding clones of low-chill fresh market peach germplasm. Fruit Var. J. 53, 162–171.
Byrne, D.H., Noratto, G., Cisneros Zevallos, L., Porter, W. and Vizzotto, M. (2009) Health benefits of peaches and plums. Acta Hort., 841, 267–274.
Byrne, D.H., Nikolic, A.N. and Burns, E.E. (1991) Variability in sugars, acids, firmness, and color characteristics of 12 peach genotypes. J. Amer. Soc. Hort Sci. 116, 1004–1006.
Byrne, D.H. and Raseira, M.C.B. (2006) Inbreeding of the major commercial fresh market peach cultivars grown in Southern Brazil. Acta Hort. 713, 99–101.
Byrne, D.H., Sherman, W.B. and Bacon, T.A. (2000) Stone fruit genetic pool and its exploitation for growing under warm winter conditions. In: Erez, A. (Ed.). *Temperate Fruit Crops in Warm Climates*. Boston, Kluwer Academic Publishers, pp. 157–230.
Callahan, A., Scorza, R., Morgens, P., Mante, S., Cordts, J. and Cohen, R. (1991) Breeding for cold hardiness, searching for genes to improve fruit quality in cold-hardy peach germplasm. HortScience 26, 522–526.
Cambra, R. (1990). 'Adafuel', an almond x peach hybrid rootstock. HortScience 25, 584.
Cantín, C.M., Gogorcena, Y. and Moreno, M.A. (2009a) Analysis of phenotypic variation of sugar profile in different peach and nectarine [*Prunus persica* (L.) Batsch] breeding progenies. J Sci Food Agric 89, 1909–1917.
Cantín, C.M., Moreno, M.A. and Gogorcena Y (2009b) Evaluation of the antioxidant capacity, phenolic compounds and vitamin C content of different peach and nectarine [*Prunus persica* (L.) Batsch] breeding progenies. J Agric Food Chem 57, 4586–4592.
Cantín, C.M., Gogorcena, Y. and Moreno M.A. (2010a) Phenotypic diversity and relationships of fruit quality traits in peach and nectarine [*Prunus persica* (L.) Batsch] breeding progenies. Euphytica 171, 211–226.

Cantín C.M., Crisosto, C.H., Ogundiwin, E.A., Gradziel, T., Torrents, J., Moreno, M.A., Gogorcena, Y. (2010b) Chilling injury susceptibility in an intra-specific peach [*Prunus persica* (L.) Bastch] progeny. *Postharvest Biol. Technol.* 58, 79–87.

Cevallos-Casals, B., Byrne, D.H., Okie, W.R. and Cisneros-Zevallos, L. (2005). Selecting new peach and plum genotypes rich in phenolic compounds and enhanced functional properties. Food Chem. 96, 273–280.

Chaparro, J.X., Werner, D.J., O'Malley, D. and Sederoff, R.R. (1994) Targeted mapping and linkage analysis of morphological isozyme, and RAPD markers in peach. Theor. Appl. Genet. 87, 805–815.

Chen, W., Wang, L., Zhang, C., Chen, C., and Cao, K. (2007) Genetic diversity analysis of peach (*Prunus persica*) cultivars introduced from different countries by SSR (in Chinese). J. Fruit Sci. 24(5), 580–584.

Chibiliti G. and Byrne, D. H. (1989) Relative aluminum tolerance of *Prunus* rootstocks. HortScience 24(4), 657–658.

Citadin, I., Raseira, M.C.B., Herter, F.G. and Silva, J.B. (2001) Heat requirement for blooming and leafing in peach. HortScience 36, 305–307.

Citadin, I., Raseira, M.C.B., Quezada, A.C. and Silva, J.B. (2003) Herdabilidade da necessidade de calor para a antese e brotação em pesseguiero. Rev. Bras. Frutic. 25(1), 119–123.

Cinelli, F. and Loreti, F. (2004) Evaluation of some plum rootstocks in relation to lime-induced chlorosis by hydroponic culture. Acta Hort. 658, 421–427.

Cipriani, G., Lot, G., Huang, W.G, Marrazzo, M.T., Peterlunger, E. and Testolin, R. (1999) AC/GT and AG/CT microsatellite repeats in peach [*Prunus persica* (L) Batsch], isolation, characterisation and cross-species amplification in Prunus. Theor. Appl. Genet. 99, 65–72.

Claverie, M., Bosselut, N., Lecouls, A.C., Voisin, R., Lafargue, B., Poizat, C., Kleinhentz, M., Laigret, F., Dirlewanger, E. and Esmenjaud, D. (2004a) Location of independent root-knot nematode resistance genes in plum and peach. Theor. Appl. Genet. 108, 765–773.

Claverie M., Dirlewanger E., Cosson P., Bosselut N., Lecouls A.C., Voisin R., Kleinhentz M., Lafargue B., Caboche M., Chalhoub B., Esmenjaud D. (2004b) High-resolution mapping and chromosome landing at the root-knot nematode resistance locus Ma from Myrobalan plum using a large-insert BAC DNA library. Theor. Appl. Genet. 109, 1318–1327.

Colaric M, Veberic, R., Stampar, F., Hudina, M. (2005) Evaluation of peach and nectarine fruit quality and correlations between sensory and chemical attributes. J. Sci. Food Agric. 85, 2611 – 2616.

Connors, C. H. (1922) Peach breeding. A summary of results. Proc. Amer. Soc. Hort. Sci. 19, 108–115.

Connors, C. H. (1920) Some notes on the inheritance of unit characters in peach. Proc. Amer. Soc. Hort. Sci. 16, 24–36.

Cos, J., Frutos, D., García, R., Rodríguez, J. and Carrillo, A. (2004) In vitro rooting study of the peach-almond hybrid 'Mayor'. Acta Hort. 658, 623–627.

Couto, M. (2006) Efeito da temperatura durante a diferenciação de gemas, floração, crescimento e desenvolvimento de frutos em pessegueiro na região de Pelotas, RS. Dr.Thesis (Doutorado em Agronomia – Fruticultura de Clima Temperado) – Faculdade de Agronomia Eliseu Maciel, Universidade Federal de Pelotas, Pelotas, pp.122.

Couto, M., Raseira, M.C.B., Herter, F.G. and Silva, J.B. (2007) Influence of High temperatures at blooming time on pollen production and fruit set of peach cvs. Maciel and Granada. VIII Symposium of Temperate Zone fruits in the Tropics and Subtropics. Program and Abstracts, Florianópolis, Brazil, Oct. 2007, p.36.

Crisosto, C.H. (2006) Peach quality and postharvest technology. Acta Hort. 713, 479–488.

Crisosto, C.H. (2002). How to increase peach consumption? Acta Hort. 592, 601–605.

Crisosto, C.H. and Crisosto, G.M. (2005) Relationship between ripe soluble solids concentration (RSSC) and consumer acceptance of high and low acid melting flesh peach and nectarine [*Prunus persica* (L.) Batsch] cultivars. Postharv Biol. Technol. 38, 239–246.

Crisosto, C.H., Crisosto, G.M., Echevarria, G. and Puy, J. (2006) Segregation of peach and nectarine [*Prunus persica* (L.) Batsch] cultivars according to their organoleptic characteristics. Postharv. Biol. Technol. 39, 10–18.

Crisosto, C.H., Day, K.R., Crisosto, G.M. and Gardiner, D. (2001) Quality attributes of white flesh peaches and nectarines grown under California conditions. J. Amer. Pomol. Soc. 55, 45–51.
Crisosto, C.H., Mitchell, F.G. and Ju, Z. (1999) Susceptibility to chilling injury of peach, nectarine, and plum cultivars grown in California. HortSci. 34, 1116–1118.
Ctifl (1994) "Peche" – Les variétiés et leur conduite. Centre technique interprofessionnel des fruits et legumes. Paris. pp. 303.
Cumming, G.A. (1989) Effect of soil pH and calcium amendments on peach yield, tree growth and longevity. Acta Hort. 254, 179–184.
Dabov, S. (1983) Inheritance of peach resistance to powdery mildew. III. Leaf resistance in F1 of J.H. Hale × nectarine Ferganensis 2. Genet. Plant Breed. 16, 146–150.
D'Bov, S. (1975) Inheritance of the powdery mildew resistance in peach. II Resistance of some vegetative organs in F1 from crosses between freestone and clingstone varieties with pubescent fruit skin. Genetica i Selektjiva 8 (4), 267–271 (in Bulgarian English abstract).
Decroocq, V., Foulongne, M., Lambert, P., Le. Gall, O., Mantin, C., Pascal, T., Schurdi-Levraud, V. and J. Kervella. (2005) Analogues of virus resistance genes map to QTLs for resistance to sharka disease in *Prunus davidiana*. Mol. Gen. Genomics 272, 680–689.
DeJong, T., Johnson, R.S., Doyle, J. F., Weibel, A., Solari, L., Basile, B., Marsal, J., Ramming, D. and Bryla, D. (2004) Growth, yield and physiological behaviour of size-controlling peach rootstocks developed in California. Acta Hort. 658, 449–455.
Della Strada, G. and Fideghelli, C. (2003) Le cultivar di drupacee intridotte del 1991 al 2001. *L'Informatore Agrario* 41, 65–70.
Della Strada, G., Fideghelli, C., and Grassi, F. (1996) Peach and nectarine cultivars introduced in the world from 1980 to 1992. Acta Hort. 374, 43–51.
De Salvador, F.R., Liverani, A. and Fideghelli, C. (1991) La scelta dei portinnesti delle piante arboree da frutto, Pesco. L'lnformatore Agrario, supplemento, 36, 43–50.
De Salvador, F.R., Ondradu, B. and Scalas, B. (2002) Horticultural behaviour of different species and hybrids as rootstocks for peach. Acta Hort. 592, 317–322.
Dettori, M.T., Quarta, R. and Verde, I. (2001) A peach linkage map integrating RFLPs, SSRs, RAPDs, and morphological markers. Genome 44, 783–790.
Dichio, B., Xiloyannis, C., Celano, G., Vicinanza, L., Gómez-Aparisi, J., Esmenjaud, D. and Salesses, G. (2004). Performance of new selections of *Prunus* rootstocks, resistant to root knot nematodes, in waterlogging conditions. Acta Hort. 658, 403–405.
Dirlewanger, E., Cardinet, G., Boudehri, K., Renaud, C., Momllor, S., Illa, E., Howard, W., Arus, P., Croset, C., Poessel, J.L., Maucourt, M., Deborde, C. and Moing A. (2009) Detection of QTLs controlling major fruit quality components in peach within the European Project ISAFRUIT. Acta Hort 814, 533–538.
Dirlewanger, E., Cosson, P., Boudehri, K., Renaud, C., Capdeville, G., Tauzin, Y., Laigret, F. and Moing, A. (2006) Development of a second-generation genetic linkage map for peach [*Prunus persica* (L.) Batsch] and characterization of morphological traits affecting flower and fruit. Tree Genet. Genomes 3, 1–13.
Dirlewanger, E. and Arus, P. (2005) Markers in Fruit Tree Breeding, Improvement of Peach. In, Lörz, H. and Wenzel, G. (Eds), *Molecular marker systems in plant breeding and crop improvement*. Springer Verlag, Berlin, pp. 279–304.
Dirlewanger, E., Cosson, P., Howad, W., Capdeville, G., Bosselut, N., Claverie, M., Voisin, C. Pozat, R., Lafargue, B., Baron, O., Laigret, F., Kleinhentz, M., Arús, P. and Esmenjaud, D. (2004a) Microsatellite genetic linkage maps of myrobalan plum and an almond-peach hybrid-location of root-knot nematode resistance genes. Theor. Appl. Genet. 109, 827–838.
Dirlewanger, E., Graziano, E., Joobeur, T., Garriga-Caldere, F., Cosson, P., Howad, W. and Arús, P. (2004b) Comparative mapping and marker-assisted selection in Rosaceae fruit crops. Proc. Nat. Acad. Sci. (USA) 101, 9891–9896.
Dirlewanger, E., Kleinhentz, M., Voisin, R., Claverie, M., Lecouls, A.C., Poëssel, J.L., Faurobert, M., Arús, P., Gómez-Aparisi, J., Di Vito, M., Xiloyannis, C. and Esmenjaud, D. (2004c) Breeding for a new generation of *Prunus* rootstocks, an example of MAS. Acta Hort. 658, 581–590.

Dirlewanger, E., Cosson P., M. Tavaud P., Aranzana M., Poizat C., Zanettyo A., Arús P. and Laigret F. (2002). Development of microsatellite markers in peach (*Prunus persica* L.) and their use in genetic diversity analysis in peach and sweet cherry (*P. avium* L.). Theor. Appl. Genet. 105(1), 127–138.

Dirlewanger, E., Moing, A., Rothan, C., Svanella, L., Pronier, V., Guye, A., Plomion, C. and Monet, R. (1999) Mapping QTLs controlling fruit quality in peach [*Prunus persica* (L.) Batsch]. Theor. Appl. Genet. 98, 18–31.

Dirlewanger, E., Pronier, V., Parvery, C., Rothan, C., Guye, A. and Monet, R. (1998) Genetic linkage map of peach [*Prunus persica* (L.) Batsch] using morphological and molecular markers. Theor. Appl. Genet. 97, 888–895.

Dirlewanger, E., Pascal, T., Zuger, C. and Kervella, J. (1996) Analysis of molecular markers associated with powdery mildew resistance genes in peach [*Prunus persica* (L) Batsch] x *Prunus davidiana* hybrids. Theor. Appl. Genet. 93, 909–919.

Dirlewanger, E. and Bodo, C. (1994) Molecular genetic mapping of peach. Euphytica 77, 101–103.

du Plessis, H.J. (1988) Differential virulence of *Xanthomonas campestris pv. pruni* to peach, plum, and apricot cultivars. Phytopathology 78, 1312–1315.

Edin, M. and Garcin, A. (1994) Un nouveau porte-greffe du pêcher Cadaman®-Avimag. L'Arboriculture Fruitière 475, 20–23.

Edwards, G. R. (1987). Temperatures in relation to peach culture in the tropics. Acta Hort. 199, 61–62.

Egea, J., Burgos, L., García, J.E. and Egea, L. (1991) Stigma receptivity and style performance in several apricot cultivars. Annals applied Biology, Cambridge- England, V.66, n.2, p.19–25.

Egilla, J.N., and Byrne, D.H. (1989) The search for peach root-stocks tolerant to alkalinity. Fruit Var. J. 43, 7–11.

Espada J.L., Romero, J., Socias i Company, R. and Alonso, J.M. (2009) Preview of the second clonal selection from the autochthonous peach population "Amarillos Tardíos de Calanda" (late yellow peaches of Calanda). Acta Hort. 814, 251–254.

Esmenjaud, D. (2009) Resistance to root knot nematodes in *Prunus*, Characterization of sources, marker-assisted selection and cloning strategy for the Ma gene from myrobalan plum. Acta Hort 814, 707–714.

Esmenjaud, D., Minot, J.C., Voisin, R. Pinochet, J. and Salesses, G. (1994) Inter- and intraspecific resistance variability in myrobalan plum, peach and peach-almond rootstocks using 22 root-knot nematode populations. J. Amer. Hort. Sci. 119, 94–100.

Esmenjaud, D., Minot, J.C. and Voisin R. (1996) Effect of durable inoculum pressure and high temperature on root galling, nematode numbers and survival of Myrobalan plum genotypes (*Prunus cerasifera* Ehr.) highly resistant to *Meloidogyne* spp. Fund. Appl. Nematol. 19:85–90.

Etienne, C., Rothan, C., Moing, A., Plomion, C., Bodénès, C., Svanella-Dumas, L., Cosson, P., Pronier, V., Monet, R. and Dirlewanger, E. (2002) Candidate genes and QTLs for sugar and organic acid content in peach [*Prunus persica* (L.) Batsch]. Theor. Appl. Genet. 105, 145–159.

Fan S., Bielenberg D.G., Zhebentyayeva T.N., Reighard G.L., Okie W.R., Holland D., Abbott A.G. (2010) Mapping quantitative trait loci associated with chilling requirement, heat requirement and bloom date in peach (*Prunus persica*). New Phytologist 185, 917–930.

Faust, M. and Timon, B. (1995) Origin and dissemination of the peach. Hort. Rev. 17, 331–379.

Feliciano, A., Feliciano, A.J. and Ogawa, J.M. (1987) *Monilinia fructicola* resistance in the peach cv Bolinha. Phytopathology 77,776–780.

Felipe, A. (2009). 'Felinem', 'Garnem', and Monegro' almond x peach hybrid rootstocks. *HortScience* 44, 196–197.

Felipe, A., Carrera, M. and Gómez-Aparisi, J. (1997a) 'Montizo' and 'Monpol', two new plum rootstocks for peaches. Acta Hort. 451, 273–276.

Felipe, A.J., Gómez Aparisi, J., Socias i Company, R. and Carrera, M. (1997b). The almond × peach hybrid rootstocks breeding program at Zaragoza (Spain). Acta Hort. 451, 259–262.

Fernández, C., Pinochet, J., Esmenjaud, D., Salesses, G. and Felipe, A. (1994) Resistance among new Prunus rootstocks and selections to root-knot nematodes in Spain and France. HortScience 29, 1064–1067.

Fideghelli, C., Della Stada, G., Grassi, F. and Morico, G. 1998. The peach industry in the world, present situation and trend. Acta Hort. 465, 29–40.

Fideghelli, C., Della Strada, G., Quarta, R. and Rosati, P. (1979) Genetic semi-dwarf peach selections. Proceedings of Eucarpia Fruit Section Symposium, Tree Fruit Breeding. INRA, Angers, France, pp. 3–7.

Fogle, H.W. (1974) Evaluating combining ability in peach and nectarine. HortScience 9, 334–335.

Foolad, M.R., Arulsekar, S., Becerra, V. and Bliss, F.A. (1995) A genetic map of *Prunus* based on an interspecific cross between peach and almond. Theor. Appl. Genet. 91, 262–269.

Foulongne, M., Pascal, T., Pfeiffer, F. and Kervella, J. (2003a) QTLs for powdery mildew resistance in peach × *Prunus davidiana* crosses, consistency across generations and environments. Mol. Breed. 12, 33–50.

Foulongne, M., Pascal, T., Arús, P., and Kervella, J. (2003b) The potential of *Prunus davidiana* for introgression into peach [*Prunus persica* (L.) Batsch] assessed by comparative mapping. Theor. Appl. Genet. 107, 227–238.

Foulogne, M., Pascal, T., Pfeiffer, F. and Kevella, J. (2002) Introgression of a polygenic resistance to powdery mildew from a wild species *Prunus davidiana* into peach (*Prunus persica* (L.) Batsch), a case study of marker assisted selection in fruit tree. Acta Hort. 592, 259–265.

French, A.P. (1951) The peach, inheritance of time of ripening and other economic characters. Mass. Agric.Exp. Sta. Bull., 462.

Gentile, A., Monticelli, S. and Damiano, C. (2002) Adventitious shoot regeneration in peach [*Prunus persica* (L.) Batsch]. Plant Cell Rep. 20, 1011–1016.

Georgi, L.L., Wang, Y., Yvergniaux, D., Ormsbee, T., Inigo, M., Reighard, G.L. and Abbott, A.G. (2002) Construction of a BAC library and its application to the identification of simple sequence repeats in peach [*Prunus persica* (L.) Batsch]. Theor. Appl. Genet. 105, 1151–1158.

George, A.P., Nissen, R.J. and Sherman, W.B. (1988). Overlapping double and early single cropping in low chill peach in Australia. Fruit Var. J. 42, 91–95.

Gillen, A.M. and Bliss, F.A. (2005) Identification and mapping of markers linked to the mi gene for root-knot nematode resistance in peach. J. Amer. Soc. Hort. Sci. 130, 24–33.

Gonzalo M.J., M.A. Moreno, Y. Gogorcena (2011) Physiological responses and differential gene expression in *Prunus* rootstocks under iron deficiency conditions. Journal of Plant Physiology 168, 887–893.

Gonzalo, M.J., Dirlewanger, E., Legrait, F., Moreno, M.A. and Gogorcena, Y. (2009) Genetic analysis of iron chlorosis tolerance in a Myrobalan plum x almond peach hybrids. Acta Hort. 814, 799–804.

Gradziel, T.M. (2003) Interspecific hybridizations and subsequent gene introgression within *Prunus* Subgenus Amygdalus Acta Hort. 622, 249–255.

Gradziel, T.M., Thorpe, R.M., Bostock, R.M. and Wilcox, S. (1997) Breeding for brown rot (*Monilinia fructicola*) resistance in clingstone peach with emphasis on the role of fruit phenolics. Acta Hort. 465, 161–170.

Grasselly, C. (1983) Nouvelles obtentions INRA de pêchers porte-greffes, multiplies par semences. L'Arboriculture Fruitière 357, 50–55.

Grasselly, C. (1988) Les porte-greffes du pêcher, des plus anciens aux plus récents. L'Arboriculture fruitière 409, 29–34.

Guillaumin, J.J., Pierson, J. and Grasselly, C. (1991). The susceptibility to *Armillaria mellea* of different *Prunus* species used as stone fruit rootstocks. Scientia Hort. 46, 43–54.

Haji, T., Yaegaki, H. and Yamaguchi, M. (2005) Inheritance and expression of fruit texture melting, non-melting and stony hard in peach. Scientia Hort. 105, 241–248.

Hammerschlag, F.A., Bauchan, G. and Scorza, R. (1985) Regeneration of peach plants from callus derived from immature embryos. Theor. Appl. Genet. 70, 248–251.

Hammerschlag, F.A. (1988) Selection of peach cells for insensitivity to culture filtrates of *Xanthomonas campestris* pv. *pruni* and regeneration of resistant plants. Theor. Appl. Genet. 76, 865–869.

Hansche, P.E. (1990) Heritability of spring bloom and fall leaf abscission dates in Prunus persica. HortScience, 25, 1639–1641.
Hansche, P.E. (1988) Two genes induce brachytic dwarfism in peach. HortScience. 23, 604–606.
Hansche, P.E. (1986) Heritability of fruit quality traits in peach and nectarine breeding stocks dwarfed by dw gene. HortScience 21, 1193–1195.
Hansche, P.E., Hesse, C.O. and Beres, V. (1972) Estimate of genetic and environmental effects on several traits in peach. J. Amer. Soc. r Hort. Sci. 97, 9–12.
Hayama, H., Shimada T., Fujii H., Ito A. and Kashimura Y. (2006) Ethylene-regulation of fruit softening-related genes in peach. J. Exp. Bot. 57(15), 4071–7.
Hedrick, H.P. (1917). The peaches of New York. NY Agr Exp Sta. NY, EUA.
Hesse, C.O. (1975). Peaches. In: J. Janick and J.N. Moore (Eds.), *Advances in fruit breeding*, Purdue University Press, W. Lafayette, Indiana, pp. 285–335.
Hesse, C.O. (1971) Monoploid peaches, *Prunus persica* L. Batsch, description and meiotic analysis. J. Amer. Soc. Hort. Sci. 96, 326–330.
Horn, R., Lecouls, A.C., Callahan, A., Dandekar, A.M., Garay, L., McCord, P., Howad, W., Chan, H., Verde, I., Main, D., Jung, S., Georgi, L.L., Forrest, S., Mook, J., Zhebentyayeva, T.N., Yu, Y., Kim, H.R, Jesudurai, C., Sosinski, B., Arús, P., Baird, W.V., Parfitt, D., Reighard, G., Scorza, R., Tomkins, J., Wing, R. and Abbott, A.G. (2005) Candidate gene database and transcript map for peach, a model species for fruit trees. Theor. Appl. Genet. 110, 1419–1428.
Howard, B.H. (1987) Propagation. In: R.C. Rom, and R.F. Carlson (Eds.), *Rootstocks for Fruit Crops*. John Wiley & Sons, New York, pp. 29–77.
Howad, W., Yamamoto, T., Dirlewanger, E., Testolin, R., Cosson, P., Cipriani, G., Monforte, A.J., Georgi, L., Abbott, A.G. and Arús, P. (2005) Mapping with a few plants, using selective mapping for microsatellite saturation of the Prunus reference map. Genetics 171, 1305–1309.
Hu, D. and R. Scorza (2009) Analysis of the 'A72' peach tree growth habit and its inheritance in progeny obtained from crosses of 'A72' with columnar peach trees. J. Amer. Soc. Hort. Sci. 134, 236–243.
Hu, D., Zhang, Z. Zhang, D. Zhang, Q. and Li, J. (2005) Genetic relationships of ornamental peach determined using AFLP markers. HortScience 40, 1782–1786.
Hu, D., Zhang, Z., Zhang, Q., Zhang, D. and Li, J. (2006) Ornamental peach and its genetic relationship revealed by inter-simple sequence repeat (ISSR) fingerprints. Acta Hort. 713, 113–120.
Iezzoni, A., Peace, C., Bassil, N., Fazio, G., Luby, J., Main, D., Weebadde, C., Yue, C., van de Weg, E., Bink, M., Brown, S., Byrne, D., Clark, J., Crisosto, C., Davis, T., Evans, K., Finn, C., Gallardo, K. Gasic, K., Gradziel, T., Hancock, J., Jussaume, R., McCracken, V., Oraguzie, N., Reighard, G., Stone, A., Taylor, M., Wang, D. and Xu, K. (2009) RosBREED, Enabling marker-assisted breeding in Rosaceae. Abstract. ASHS meeting. Palm Desert, CA. August, 2009.
Indreias, A., Dutu, I. and Stefan, I. (2004) Peach rootstocks created and used in Romania. Acta Hort. 658, 505–508.
Infante R., Martínez-Gómez P. and Predieri S. (2008) Quality oriented fruit breeding, Peach [*Prunus persica* (L.) Batsch]. J. Food Agric. Environ. (JFAE) 6, 342–356.
Jacob, H. (1992) *Prunus pumila* L., eine geeignete schwachwachsende Pfirsichuntererlage. Erwerbsobstbau 34, 144–146.
Jáuregui, B., de Vicente, M.C., Messeguer, R., Felipe, A., Bonnet, A., Salesses, G. and Arús, P. (2001) A reciprocal translocation between 'Garfi' almond and 'Nemared' peach. Theor. Appl. Genet. 102, 1169–1176.
Jiang, W., Qu, D., Mu, D. and Wang, L. (2004) Protected cultivation of horticulturasl crops of China. Hort. Rev. 30, 115–162.
Jiménez, S., Pinochet, J., Abadía, A., Moreno, M.A. and Gogorcena, Y. (2008) Tolerance response to iron chlorosis of *Prunus* selections as rootstocks. HortScience 43(2), 304–309.
Jiménez S., Pinochet, J., Romero, J., Gogorcena, Y., Moreno, M.A., and Espada J.L. (2011) Performance of peach and plum based rootstocks of different vigour on a late peach cultivar in replant and calcareous conditions. Sci. Hortic. 129, 58–63.
Joobeur, T., Viruel, M.A., de Vicente, M.C., Jáuregui, B., Ballester, J., Dettori, M.T., Verde, I., Truco, M.J., Messeguer, R., Batlle, I., Quarta, R., Dirlewanger, E. and Arús, P. (1998)

Construction of a saturated linkage map for *Prunus* using an almond x peach F2 progeny. Theor. Appl. Genet. 97, 1034–1041.

Kader, A.A. (2002) Postharvest biology and technology, an overview. In: A.A. Kader (Ed.), *Postharvest technology of horticultural commodities.* University of California Press, Davis, pp. 39–47.

Kester, D.E. and Assay, R.N. (1986) 'Hansen 2168' and 'Hansen 536', two *Prunus* rootstock clones. HortScience 21, 331–332.

Khanizadeh, S., Tsao, R., Rekika, D., Yang, R., Charles, M. T. and Rupasinghe, V. (2007) Advances in fruit breeding in eastern Canada – Role of phytochemicals in designing specialty fruits. Acta Hort 814, 205–208.

Knight, R.L. (1969) Abstract bibliography of fruit breeding and genetics. Easter Press, London.

Kozai, N., Beppu, K., Mochioka, R., Boonprakob, U., Subhadrabandhu, S., Kataoka, I. (2004). Adverse effects of high temperature on the development of reproductive organs in 'Hakuho' peach trees. J. Hort. Sci. Biotech. 79(4), 533–537.

Kozai, N., Beppu, K., Kataoka, I. (2002) Adverse effects of temperature on the development of reproductive organs in ´Hakuto` peach trees. In: Reports of the First *International Workshop on Production Technologies for low chill temperate Fruits,* Chiang-Mai, Thailand, pp. 212–220.

Lalli, D.A., Decroocq, V., Blenda, A.V., Schurdi-Levraud, V., Garay, L., Le Gall, O., Damsteegt, V., Reighard, G.L. and Abbott, A.G. (2005) Identification and mapping of resistance gene analogs (RGAs) in *Prunus*, a resistance map for *Prunus*. Theor. Appl. Genet. 111, 1504–1513.

Lammerts, W. E. (1945) The breeding of ornamental edible peaches for mild climates. I. Inheritance of tree and flower characteristics. Am. J. Bot. 32, 53–61.

Layne, R.E.C. (1987) Peach rootstocks. In: R.C. Rom and R.F. Carlson (Eds.), *Rootstocks for Fruit Crops*. John Wiley & Sons, New York, pp. 185–216.

Layne, R.E.C. (1984) Breeding peaches in North America for cold hardiness and perennial canker (*Leucostoma* spp.). Review and outlook. Fruit Var. J. 38, 130–136.

Layne, R.E.C. (1982) Cold hardiness of peaches and nectarines following a test winter. Fruit Var. J. 36, 90–98.

Lazzari, B., Caprera, A., Vecchietti, A., Merelli, I., Barale, F., Milanesi, L., Stella, A. and Pozzi, C. (2008) Version VI of the ESTree db, an improved tool for peach transcriptome analysis. BMC Bioinformatics 9, S9.

Lazzari, B., Caprera, A., Vecchietti, A., Stella, A., Milanesi, L. and Pozzi, C. (2005) ESTree db, a tool for peach functional genomics. BMC Bioinformatics 6 Suppl 4, S16.

Lea, M., Ibeh, C., desBordes, C., Vizzotto, M., Cisneros-Zevallos, L., Byrne, D.H., Okie, W.R. and Moyer, M.P. (2008) Inhibition of growth and induction of differentiation of colon cancer cells by peach and plum phenolic compounds. Anticancer Res. 28, 2067–2076.

Lecouls, A.C., Bergougnoux, V., Rubio-Cabetas, M.J., Bosselut, N., Voisin, R., Poessel, J.L., Faurobert, M., Bonnet, A., Salesses, G., Dirlewanger, E. and Esmenjaud, D. (2004) Marker-assisted selection for the wide-spectrum resistance to root-knot nematodes conferred by the Ma gene from Myrobalan plum (*Prunus cerasifera*) in interspecific *Prunus* material. Mol. Breed. 13, 113–124.

Lecouls, A.C., Rubio-Cabetas, M.J., Minot, J.C., Voisin, R., Bonnet, A., Salesses, G., Dirlewanger, E. and Esmenjaud, D. (1999) RAPD and SCAR markers linked to the Ma1 root-knot nematode resistance gene in Myrobalan plum (*Prunus cerasifera* Ehr.). Theor. Appl. Genet. 99, 328–335.

Lecouls A.C., Salesses G., Minot J.C., Voisin R., Bonnet A. and Esmenjaud D. (1997) Spectrum of the Ma genes for resistance to *Meloidogyne* spp. in Myrobalan plum. Theor. Appl. Genet. 85:1325–1334.

Lee, M.H. and Bostock, R.M. (2007) Fruit exocarp phenols in relation to quiescence and development of *Monilinia fructicola* infections in *Prunus* spp., A role for cellular redox? Phytopath. 97, 269–277.

Lesley, J. (1957). A genetic study of inbreeding and crossing inbred lines in peaches. Proc. Amer. Soc. Hort. Sci. 45, 243–250.

Lesley, J. W. (1939) A genetic study of saucer fruit shape and other characteristics in the peach. Proc. Amer. Soc. Hort. Sci. 38, 218–222.

Liu, X., Reighard, G.L., Swire-Clark, G.A. and Baird, W.V. (2007) Peach rootstock identification by DNA-fingerprinting with microsatellite (SSR) markers. J. Amer. Pomol. Soc. 61, 162–166.

Liverani, A., Giovannini, D. and Brandi, F. (2004). Development of new peach cultivars with columnar and upright growth habit. Acta Hort. 663, 381–386.

Liverani, A., Giovannini, D. and Brandi, F. (2002) Increasing fruit quality of peaches and nectarines, the main goals of ASF-FO (Italy). Acta Hort. 592, 507–514.

Liverani, A. and Giovannini, D. (2000) The Peach Breeding Program at the Istituto sperimentale per la frutticoltura di Forli´ (Italy). Summaries. *Prunus* Breeders Meeting, Empresa Brasileira de Pesquisa Agropecuária, Clima Temperado. Pelotas (RS)Brazil. 29 Nov.–2 Dec. 2000, pp. 19–23.

Llácer G. (2009) Fruit breeding in Spain. Acta Hort 814, 43–56.

Llácer G., Alonso, J.M., Rubio-Cabetas, M.J., Batlle, I., Iglesias, I., Vargas, F.J., García-Brunton, J. and Badenes, M.L. (2009) Peach industry in Spain. J. Amer. Pomol. Soc. 63(3), 128–133.

López, G., Johnson, R. S. and Dejong, T. M. (2007) High spring temperatures decrease peach fruit size. California Agriculture Vol.61(1), 31–34.

Loreti, F. and Massai, R. (1994) Sirio, Nuovo portinnesto ibrido pesco x mandorlo. L'Informatore Agrario 28, 47–49.

Loreti, F. and Massai, R. (2006) 'Castore' and Polluce', Two new hybrid rootstocks for peach. Acta Hort. 713, 275–278.

Lu, M., Song, C., Huang, C. and Ou, S. (2008) Changes in flesh firmness and ethylene production of different peach types during fruit ripening. Acta Hort. 768, 153–159.

Lu, Z.X, Reighard, G.L., Nyczepir, A.P., Beckman, T.G. and Ramming, D.W. (2000) Inheritance of resistance to root-knot nematodes (*Meloidogyne* sp.) in *Prunus* rootstocks. HortScience 35, 1344–1346.

Lu, Z.X., Sosinski, B., Reighard, G.L., Baird, W.V. and Abbott, A.G. (1998) Construction of a genetic linkage map and identification of AFLP markers for resistance to root-knot nematodes in peach rootstocks. Genome 41, 199–207.

Luby, J.J. and Shaw, D.V. (2001) Does marker-assisted selection make dollars and sense in a fruit-breeding program? HortScience 35, 872–879.

Luchsinger, L., Ortin, P., Reginato, G. and Infante, R. (2002) Influence of canopy fruit position on the maturity and quality of Angelus peaches. Acta Hort. 592, 515–521.

Ma, R., Byrne, D.H., Yu, M., Du, P., and Shen, Z. (2006) Inbreeding and coancestry of the major commercial fresh market peach cultivars in China. Acta Hort. 713, 145–148.

Manaresi, A. and Draghetti, A. (1915) Influenza del germoglio ascellare sullo sviluppo e sulla composizione del frutto del pesco. Bollettino Associazione Orticola Professionale Italiana, 1, 1–4 (in Italian).

Mante, S., Scorza, R. and Cordts, J.M. (1989) Plant regeneration from cotyledons of *Prunus persica*, *Prunus domestica*, and *Prunus cerasus*. Plant Cell, Tiss. Organ Cult. 19, 1–11.

Marini, R.P. and Sowers, D.L. (1994) Peach fruit weight is influenced by crop density and fruiting shoot length but not position on the shoot. J. Amer. Soc. Hort. Sci. 119, 180–184.

Marchese, A., K.R. Tobutt, and T. Caruso, 2005. Molecular characterisation of Sicilian *Prunus persica* cultivars using microsatellites. J. Hort. Sci. Biotechnol. 80:121–129.

Martínez-Gómez, P. and Gradziel T.M. (2002). New approaches to almond breeding at the University of California – Davis program. Acta Hort. 591, 253–256.

Martínez-Gómez, P., Rubio, M., Dicenta, F. and Gradziel, T.M. (2004) Resistance to plum pox virus (Dideron isolate RB3.30) in a group of California almonds and transfer of resistance to peach. J. Amer. Soc. Hort. Sci 129, 544–548.

Martins, O.M. (1996) Evaluation of virulence of strains of *Xanthomonas campestris* pv. pruni on peach and plum cultivars. Fruit Var. J. 50, 221–225.

Marull, J., Pinochet, J., Felipe, A., and Cenis, J.L. (1994) Resistance verification in *Prunus* selections to a mixture of thirteen *Meloidogyne* isolates and resistance mechanisms of a peach-almond hybrid to M. *javanica*. Fundam. Appl. Nematol. 17, 85–92.

Massonie, G., Maison, P., Monet R. and Grasselly, C. (1982) Resistance to the green peach aphid *Myzus persicae* Sulzer (Homoptera, Aphididae) in *Prunus persica* (L.) Batsch and other *Prunus* species (in French). Agronomie 2, 63–69.

Mathais, C., Mayer, N. A., Mattiuz, B., and Pereira, F. M. (2008) Efeito de porta-enxertos e espaçamentos entre plantas na qualidade de pêssegos 'Aurora 1'. Rev. Bras. Frutic. Jaboticabal – SP 30, 165–170.

McFadden-Smith, W., Miles, N. and Potter, J. (1998) Greenhouse evolution of Prunus rootstocks for resistance or tolerance to the root lesion nematode (*Pratylenchus penetrans*). Acta Hort 465, 723–729.

Meader. E.M. and Blake, M.A. (1940) Some plant characteristics of second generation of *P. persica* x *P. kansuensis* crosses. Proc. Amer. Soc. Hort. Sci. 37, 223–231.

Mehlenbacher, S.A. and Scorza. R. (1986) Inheritance of growth habit in progenies of 'Compact Redhaven' trees. HortScience. 21, 124–126.

Mingliang, Y., Ruijuan, M., Zhijun, S. and Zhen, Z. (2007) Molecular markers linked to specific characteristics of *Prunus persica* (L.) Batsch. Acta Hort. 763, 147–154.

Monet, R. (1995) Il miglioramento genetico del pesco. In: Bellini, E. (Ed.) *State of the art and perspectives of world genetic improvement of fruit tree species*. ERSO, Faenza (Italy), pp. 13–27 (in Italian).

Monet, R. (1989) Peach genetics, past, present and future. Acta Hort. 254, 49–57.

Monet, R. (1985) Heredity of the resistance to leaf curl (*Taphrina deformans*) and green aphid (*Myzus persicae*) in the peach. Acta Hort. 173, 21–24.

Monet, R. (1979) Genetic transformation of the 'fruit sweetness' character-incidence on selection for quality (in French). Eucarpia Fruit Section Symposium, Tree Fruit Breeding, Angers, France. pp. 273–276.

Monet, R. (1967) A contribution to the genetics of peaches (in French). Ann. Amelior. Plant 17, 5–11.

Monet, R. and Bassi, D. (2008) Classical genetics and breeding. In: D. R. Layne and D. Bassi (Eds.), *The Peach. Botany, Production and Uses.* CAB International, Wallingford, UK, pp. 61–84.

Monet, R., Guye, A. and Massonie, G. (1998). Breeding for resistance to green aphid Myzus persica Sulzer in the peach. Acta Hort. 465, 171–175.

Monet, R., Guye, A., Roy, M. and Dachary, N. (1996) Peach Mendelian genetics, a short review and new results. Agronomie 16, 321–329.

Monet, R. and Massonié, G. (1994) Déterminisme génétique de la résistance au puceron vert (*Myzus persicae*) chez le pêcher. Résultats complémentaires. Agronomie 2, 177–182.

Moreno, M.A. (2004) Breeding and selection of *Prunus* rootstocks at the Estación Experimental de Aula Dei, Zaragoza, Spain. Acta Hort. 658, 519–528.

Moreno, M.A., Tabuenca, M.C. and Cambra, R. (1995) 'Adesoto 101', a plum rootstock for peaches and other stone fruits. HortScience 30, 1314–1315.

Moreno, M.A., Tabuenca, M.C. and Cambra, R. (1994) Performance of 'Adafuel' and 'Adarcias' as peach rootstocks. HortScience 29, 1271–1273.

Moreno, M.A., Moing, A., Lansac, M., Gaudillère, J.P. and Salesses, G. (1993) Peach/Myrobalan plum graft incompatibility in the nursery. J. Hort. Sci. 68, 705–714.

Mowry, J.B. (1964) Inheritance of cold hardiness of dormant peach flower buds. Proc. Amer. Soc. Hort. Sci. 85,128–133.

Mowrey, B.D., Werner, D.J. and Byrne, D.H. (1990) Inheritance of isocitrate dehydrogenase, malate dehydrogenase, and shikimate dehydrogenase in peach and peach x almond hybrids. J. Amer. Soc. Hort. Sci. 115, 312–319.

Mowrey, B.D. and Sherman, W.B. (1986) Flower bud set and relationship to vigor in 18 month-old peach seedlings. Proc. Fla. State Hort. Soc. 99, 209–210.

Myeki, J. and Sazabó, S. (1989) Effect of frost damage on peach varieties in Hungary. Acta Hort. 254, 255–256.

Nicotra, A. and Moser, L. (1997) Two new plum rootstocks for peach and nectarines, Tetra and Penta. Acta Hort. 451, 269–271.

Nicotra, A., Conte, L., Moser, L. and Fantechi, P. (2002) New types of high quality peaches (*Prunus persica* var, platicarpa) and Ghiaccio peach series with long on tree fruit life. Acta Hort. 592, 131–136.

Noratto, G, Porter, W., Byrne, D.H. and Cisneros-Zevallos, L. (2010) Identifying peach and plum polyphenols with chemopreventive potential against estrogen-independent breast cancer cells. J. Agric. Food Chem 57, 5219–5226.

Nyczepir, A.P., Beckman, T.G. and Reighard, G.L. (2006) Field evaluation of 'Guardian' ™ peach rootstock to different root-knot nematode species. Acta Hort. 713, 303–309.

Ogundiwin, E.A., Peace, C.P., Gradziel, T.M., Parfitt, D.E., Bliss, F.A. and Crisosto, C.H. (2009) A fruit quality gene map of *Prunus*. BMC Genomics 10,587 doi 10.1186/1471-2164-10-587.

Okie, W.R. (1998). Handbook of peach and nectarine varieties, performance in the Southeastern United States and Index of names. USDA/ARS, Agriculture handbook 714, pp. 808.

Okie, W.R. (1987) Plum rootstocks. In: R.C. Rom and R.F. Carlson (Eds.), Rootstocks for Fruit Crops. John Wiley & Sons, New York, pp. 321–360.

Okie, W.R. (1988) USDA peach and nectarine breeding at Byron, Georgia. In: N.F. Childers and W.B. Sherman (Eds.), The Peach. Horticultural Publications, Gainesville, Florida, pp. 51–56.

Okie, W.R. (1984) Rapid multiplication of peach seedlings by herbaceous stem cuttings. HortScience 19, 249–251.

Okie, W.R., Bacon, T. and Bassi, D. (2008). Fresh market cultivar development. In: D.R. Layne and D. Bassi (Eds.), The peach – Botany, Production and Uses. CAB International, 139–174.

Okie, W.R., Reighard, G.L., Beckman, T.G., Nyczepir, A.P., Reilly, C.C., Zehr, E.I. and Newall, W.C. Jr, Cain, D.W. (1994) Field-screening *Prunus* for longevity in the southeastern United States. HortScience 29, 673–677.

Pascal, T., Pfeiffer, F. and Kervella, J. (2010) Powdery mildew resistance in the peach cultivar Pamirskij 5 is genetically linked with the Gr gene for leaf color. HortScience 45, 150–152.

Pascal, T.F. Pfeiffer, J. Kervella, J.P. Lacroze and M.H. Sauge (2002) Inheritance of green peach aphid resistance in the peach cultivar Rubira. Plant Breeding 121, 459–461.

Pascal, T., Kervella, J., Pfeiffer, F.G., Sauge, M.H. and Esmenjaud, D. (1997) Evaluation of the interspecific progeny *Prunus persica* cv. Summergrand × *Prunus davidiana* for disease resistance and some agronomic features. Acta Hort. 465, 185–191.

Peace, C.P., Crisosto, C.H., Garner, D.T., Dandekar, A.M., Gradziel, T. and Bliss, F.A. (2006) Genetic control of internal breakdown in peach. Acta Hort. 713, 489–496.

Peace, C.P., Crisosto, C.H. and Gradziel, T.M. (2005) Endopolygalacturonase, a candidate gene for freestone and melting flesh in peach. Mol. Breed. 16, 21–31.

Peace, C.P., Callahan, A., Ogundiwin, E.A., Potter, D., Gradziel, T.M., Bliss, F.A. and Crisosto, C.H. (2007) Endopolygalacturaonase genotypic variation in Prunus. Acta Hort. 738, 639–646.

Pérez, S. (1989) Characterization of Mexican peach population from tropical and subtropical regions. Acta Hort. 254, 139–144.

Pérez, S. (1997) Breeding peaches for powdery mildew (*Sphaeroteca pannosa*) resistance in the subtropical regions of central Mexico. Acta Hort. 441, 87–92.

Pérez, S. and Moore J.N. (1985) Prezygotic endogenous barriers to interspecific hybridization in *Prunus*. J. Amer. Soc. Hort. Sci. 110, 267–73.

Pérez, S., Montez, S. and Mejía, C. (1993). Analysis of peach germplasm in Mexico. J. Am. Soc. Hort. Sci. 118, 145, 519–524.

Pérez-Clemente, R.M., Pérez-Sanjuán, A., García-Férriz, L., Beltrán, J.P. and Cañas, L.A. (2005) Transgenic peach plants (*Prunus persica* L.) produced by genetic transformation of embryo sections using the green fluorescent protein (GFP) as an in vivo marker. Mol. Breed. 14, 419–427.

Pérez, S. (1990) Relationship between parental blossom season and speed of seed germination in peach. HortScience 25, 958–960.

Pina A. and Errea P. (2005) A review of new advances in mechanism of graft compatibility–incompatibility. Scientia Hort. 106 (1): 1–11.

Pinochet, J., Calvet, C., Hernández-Dorrego, A., Bonet, A., Felipe, A. and Moreno, M.A. (1999) Resistance of peach and plum rootstocks from Spain, France, and Italy to rootknot nematode *Meloidogyne javanica*. HortScience 34, 1259–1262.

Pinochet, J. (2009) 'Greenpac' a new peach hybrid rootstock adapted to Mediterranean conditions. HortScience 44, 1456–1457.

Pinochet, J., Cunill, M., Felipe, A., Eremin, G., Eremin, V., Penyalver, R., López, M.M., Jiménez, S., Gogorcena, Y. and Moreno, M.A. (2005) Performance of new *Prunus* rootstocks for replant situations to biotic and abiotic stress in Spain. VI International Symposium on Peach, ISHS. Santiago de Chile. January 9–13. Abstracts book (O 5).

Pinochet, J., Fernández, C., Calvet, C., Hernández-Dorrego, A. and Felipe, A. (2000) Selection against *Pratylenchus vulnus* populations attacking *Prunus* rootstocks. HortScience 35, 1333–1337.

Pinochet, J., Fernandez, C., Cunill, M., Torrents, J., Felipe, A., López, M.M., Lastra, B. and Penyalver, R. (2002) Response of new interspecific hybrids for peach to root-knot and lesion nematodes, and crown gall. Acta Hort. 592, 707–716.

Pinto, A.C.Q., Rogers, S.M.D. and Byrne, D.H. (1994) Growth of immature peach embryos in response to media, method of ovule support and ovule manipulation. HortScience 29, 1081.

Pisani, P.L. and Roselli, G. (1983) Interspecific hybridazation of *Prunus persica* × *P. davidiana* to obtain new peach rootstocks. Genet. Agr. 37, 197–717.

Pooler, M.R. and Scorza, R. (1995) Regeneration of peach [*Prunus persica* (L.) Batsch] rootstock cultivars from cotyledons of mature stored seed. HortSci. 30, 355–356.

Quamme, H.A. and Sushnoff, C. (1983) Resistance to environment stress. In: J.N. Moore & J. Janick (Eds.), Methods in fruit breeding. Purdue University Press, pp.185–335.

Quarta, R., Dettori, M. T., Sartori, A. and Verde, I. (2000) Genetic linkage map and QTL analysis in peach. Acta Hort. 521, 233–241.

Quarta, R., Dettori, M.T., Verde, I., Gentile, A. and Broda, Z. (1998) Genetic analysis of agronomic traits and genetic linkage mapping in a BC1 peach population using RFLPs and RAPDs. Acta Hort. 365, 51–60.

Quilot, B., Wu, B.H., Kervella, J., Génard, M., Foulongne, M. and Moreau, K. (2004) QTL analysis of quality traits in an advanced backcross between *Prunus persica* cultivars and the wild relative species *P. davidiana*. Theor. Appl. Genet. 109, 884–897.

Rajapakse, S., Belthoff, L.E., He, G., Estager, A.E., Scorza, R., Verde, I., Ballard, R.E., Baird, W.V., Callahan, A., Monet, R. and Abbott, A.G. (1995) Genetic linkage mapping in peach using morphological, RFLP and RAPD markers. Theor. Appl. Genet. 90, 503–510.

Ramming, D.W. (1990) The use of embryo culture in fruit breeding. HortScience 25, 393–398.

Ramming, D.W. (1985) In ovule embryo culture of early maturing *Prunus*. HortScience 20, 419–420.

Ramming, D.W. and Tanner, O. (1983) 'Nemared' peach rootstock. HortScience 18 (3), 376.

Raseira, M.C.B. Baptista da Silva, J., Herter, F. and Peters, J. A. (1992a) Sensibilidade de gemas floríferas de pessegueiro, *Prunus persica* L. (Batsch) ao frio. Rev. Bras. Frutic., Cruz das Almas. 14, 167–172.

Raseira, M.C.B., Byrne, D.H., Franzon, R.C. (2008) Pessegueiro – Tradição e poesia-. In, Rosa Lía Barbieri, Elisabeth Regina Tempel Stumpf. (Org.). Origem e evolução de plantas cultivadas. 1 ed. Brasília/DF, Embrapa Informação Tecnológica, 1, 679–705.

Raseira, M.C.B. and Nakasu, B.H. (2006). Peach breeding program in Southern Brazil. Acta Hort. 713, 93–97.

Raseira, M.C.B. and Nakasu, B.H. (2003). Cultivares. In: EMBRAPA – Pêssego Produção, Embrapa Informação Tecnológica, Brasília, pp. 41–50.

Raseira, M.C.B., Silva, J.B., Herter, F. and Peters, J.A. (1992a) Sensibilidade de gemas floríferas de pessegueiro, *Prunus persica* L. (Batsch) ao frio. Rev. Bras. Frutic., Cruz das Almas, V.14, n. 1, 167–172.

Raseira, M.C.B., Nakasu, B.H., Santos, A.M., Fortes, J.F., Martins, O.M., Raseira, A. and Bernardi, J. (1992) The CNPFT/EMBRAPA fruit breeding program in Brazil. HortScience 27, 1154–1157.

Razeto, B. and Valdés, G. (2006) Effect of iron clorosis on yield, fruit size and fruit maturity in nectarine. 713, 227–230.

Reighard, G.L. and Loreti, F. (2008) Rootstock development. In: D. Layne, and D. Bassi (Eds.), The Peach, Botany, Production and Uses. CAB International, Wallingford, U.K, pp. 193–220.

Reighard, G.L. (2002) Current directions of peach rootstock programs worldwide. Acta Hort. 592, 421–427.

Reighard, G.L., Newall, W.C., Beckman, T.G., Okie, W.R., Zehr, E.I. and Nyczepir, A.P. (1997) Field performance of *Prunus* rootstock cultivars and selections on replant soils in South Carolina. Acta Hort. 451, 243–250.

Renaud, R., Bernhard, R., Grasselly, C. and Dosba, F. (1988) Diploid plum × peach hybrid rootstocks for stone fruit trees. HortScience 23, 115–117.

Riaz, A., Potter, D. and Southwick, S.M. (2004) Genotyping of peach and nectarine cultivars with SSR and SRAP molecular markers. J. Amer. Soc. Hort. Sci. 129, 204–210.

Richards, G.D., Porter, G.W. Rodríguez, J. and Sherman, W.B. (1994) Incidence of blind nodes in low-chill peach and nectarine germplasm. Fruit Var. J. 48(4), 199–202.

Ritchie, D.F. and Werner, D.J. (1981) Susceptibility and inheritance of susceptibility to peach leaf curl in peach and nectarine cultivars. Plant Dis. 65(9), 731–734.

Rivers, S. (1906) The cross-breeding of peaches and nectarines. Report on Third International Conference on Genetics, London, 463–467.

Robertson, J.A., Meredith, F.L., Forbus, W.R. and Lyon, B.G. (1992). Relationship of quality characteristics of peach (cv. Loring) to maturity. J. Food Sci. 57, 1401–1404.

Rodríguez, J., Sherman, W.B. and Jasso, J. (1992) Evaluation of peach and nectarine germplasm for powdery mildew resistance (*Sphaeroteca pannosa* (Wallr.) Lev.). Acta Hort 315, 163–169.

Rodríguez, A.J., Sherman, W.B., Scorza, R. and Wisniewski, M. (1994) 'Evergreen' peach, its inheritance and dormant behavior. J. Amer. Soc. Hort. Sci. 119, 789–792.

Rouse, R.E. and Sherman, W.B. (2002) Foliar rust resistance in low-chill peaches. Proc. Fla. State Hort. Soc. 115, 98–100.

Rouse, R.E. and Sherman, W.B. (2002) High night temperatures during bloom affect fruit set in peach. Proc. Fla. State Hort. Soc. 115, 96–97.

Rowe, R.N. and Catlin, P.B. (1971) Differential sensitivity to waterlogging and cyanogenesis by peach, apricot, and plum roots. J. Amer. Soc. Hort. Sci. 96, 305–308.

Rubio, M., García-Ibarra, A., Martínez-Gómez, P. and Dicenta, F. (2009) Analysis of the main factors involved in the evaluation of *Prunus* resistance to Plum pox virus (Sharka) in controlled greenhouse conditions. Sci. Hort. 123, 46–50.

Rubio, M., Pascal, T., Bachellez, A. and Lambert, P. (2010) Quantitative trait loci analysis of Plum pox virus resistance in *Prunus davidiana* P1908, new insights on the organization of genomic resistance regions. Tree Gen Genomes 6, 291–304.

Rubio-Cabetas, M.J. Lecouls, A.C. Salesses, G. Bonnet, A. Minot, J.C. Voisin, R. and Esmenjaud D. (1998) Evidence of a new gene for high resistance to *Meloidogyne* spp. in Myrobalan plum (*Prunus cerasifera*). Plant Breeding. 117 (6): 567–571.

Salesses, G., Dirlewanger, E., Bonnet, A., Lecouls, A.C. and Esmenjaud, D. (1998) Interspecific hybridization and rootstock breeding for peach. Acta Hort. 465, 209–217.

Salesses, G. and Bonnet A. (1992) Some physiological and genetic aspects of peach/plum graft incompatibility. Acta Hort. 315,177–186.

Salesses, G. and Alkai, N. (1985) Simply inherited grafting incompatibility in peach. Acta Hort. 173, 57–62.

Salesses, G. and Juste, C. (1970) Recherches sur l'asphyxie radiculaire des arbres fruitières à noyau. I- Rôle éventuel de certaines substances présentes dans les racines du pêcher *Prunus* persica. Ann. Amélior. Plantes 20, 87–103.

Salesses, G., Saunier, H. and Bonnet, A. (1970) L'asphyxie radiculaire chez les arbres fruitières. Bull. Tech. Infor. 251, 403–415.

Sansavini, S., Gamberini, A. and Bassi, D. (2006) Peach breeding, genetics and new cultivar trends. Acta Hort. 713, 23–48.

Sauge, M.H. (1998) Analysis of the mechanisms of resistance to the green peach aphidin several *Prunus* genotypes. Acta Hort. 465, 731–739.

Saunier, R. (1973) Contribution a l'étude des relations existant entre certains caractères a déterminisme génétique simple chez le pêcher et la sensibilité a l'oidium, *Sphaeroteca pannosa* (Wallr) Lev. des cultivars de cette espéce. Annales des Amélioration des Plantes, 23 (3), 235–243 (in French).

Scorza, R. (2001) Progress in tree fruit improvement through molecular genetics. HortSci. 36, 855–858.

Scorza, R., Bassi, D. and Liverani, A. (2002) Genetic interaction of pillar (columnar), compact and dwarf peach tree genotypes. J. Amer. Soc. Hort. Sci. 127, 254–261.

Scorza, R. and Sherman, W.B. (1996) Peaches. In: J. Janick and J.N. Moore (Eds.), Fruit breeding Vol. I. Tree and Tropical Fruits. John Wiley & Sons, Inc., New York, U.S.A., pp. 325–440.

Scorza, R. and Pooler, M. (1993) Development and testing of F1 hybrid peaches on alternative peach production strategy. HortScience 28, 95.

Scorza, R. and W. Okie. (1990). Peaches, In: J.N. Moore and J. R. Ballington Jr (Eds.), Genetic resources of temperate fruit and nut crops. ISHS-Wageningen, The Netherlands, pp. 175–232.

Scorza, R., Lightner, G.W. and Liverani, A. (1989) The Pillar peach tree and growth habit analysis of compact x Pillar progeny. J. Amer. Soc. Hort. Sci. 114, 991–995.

Scorza, R., Mehlenbacher, S.A. and Lightner, G.W. (1985) Inbreeding and coancestry of freestone peach cultivars of the eastern United States and implications for peach germplasm improvement. J. Amer. Soc. Hort. Sci. 110, 547–552.

Scorza, R., Miller, S., Glenn, D.M., Okie, W.R. and Tworkoski T. (2006) Developing peach cultivars with novel growth habits. Acta Hort. 713, 61–64.

Scorza, R., Sherman, W.B., and Lightner, G.W. (1988) Inbreeding and co-ancestry of low chill short fruit development period freestone peaches and nectarines produced by the University of Florida breeding program. Fruit Var. J. 42,79–85.

Scott, D.H. and Weinberger, J.H. (1944) Inheritance of pollen sterility in some peach varieties. J. Amer. Soc. Hort. Sci. 45, 229–232.

Scott, D. H. and Cullinan, F. P. (1942) The inheritance of wavy-leaf character in the peach. J. Hered. 33, 293–295.

Sharpe, R.H., Hesse, C.O., Lownsberry, B.F., Perry, V.G. and Hansen, C.J. (1970) Breeding peaches for root-knot nematode resistance. J. Amer. Soc. Hort. Sci. 94, 209–212.

Sherman, W.B. and Lyrene, P.M. (2003) Low chill breeding of deciduous fruit at the University of Florida. Acta Hort. 622, 599–605.

Sherman, W.B. and Lyrene, P. M. (1984). Biannual peaches in the tropics. Fruit Var. J. 38, 37–39.

Sherman, W.B. and Lyrene, P.M. (1981) Bacterial spot susceptibility in low chilling peaches. Fruit Var. J. 35, 74–77.

Sherman, W.B., Sharpe, R.H. and Janick, J. (1973) The fruiting nursery, ultrahigh density for evaluation of blueberry and peach seedlings. HortScience 8, 170–172.

Shi, Y. and Byrne, D.H. (1995) Tolerance of *Prunus* rootstocks to potassium carbonate-induced chlorosis. J. Amer. Soc. Hort. Sci. 120, 283–285.

Shimada, T., Yamamoto, T., Hayama, H., Yamaguchi, M. and Hayashi, T. (2000) A genetic linkage map constructed by using an interspecific cross between peach cultivars grown in Japan. J. Japan. Soc. Hort. Sci. 69, 536–542.

Sinclair, J.W. and Byrne, D.H. (2003). In vitro growth of immature peach embryos as related to carbohydrate source. HortScience 38(4), 582.

Smigocki, A.C. and Hammerschlag, F.A. (1991) Regeneration of plants from peach embryo cells infected with a shooty mutant strain of Agrobacterium. J. Amer. Soc. Hort. Sci. 116, 1092–1097.

Sosinski, B., Gannavarapu, M., Hager, L.D., Beck, L.E., King, G.J., Ryder, C.D., Rajapakse, S., Baird, W.V., Ballard, R.E. and Abbott, A.G. (2000) Characterization of microsatellite markers in peach [*Prunus persica* (L.) Batsch]. Theor. Appl. Genet. 101, 421–428.

Sosinski, B., Sossey-Alaoui, K., Rajapakse, S., Glassmoyer, K., Ballard, R.E., Abbott, A.G., Lu, Z.X., Baird, W.V., Reighard, G.L., Tabb, A. and Scorza, R. (1998) Use of AFLP and RFLP markers to create a combined linkage map in peach [*Prunus persica* (L.) Batsch] for use in marker assisted selection. Acta Hort. 465, 61–68.

Souza, V.A.B., Byrne, D.H., Taylor, J.F. (2000) Predicted breeding values for nine plants and fruits characteristics of 28 peach genotypes. J. Amer. Soc. Hort. Sci. 125(4), 460–465.
Souza, V.A. B., Byrne, D.H. and Taylor, J.F. (1998a) Heritability, genetic and phenotypic correlations, and predicted selection response of quantitative traits in peach. I. An analysis of several reproductive traits. J. Amer. Soc. Hort. Sci. 123(4), 598–603.
Souza, V.A. B., Byrne, D.H., Taylor, J.F. (1998b) Heritability, genetic and phenotypic correlations, and predicted selection response quantitative traits in peach. II. An analysis of several fruit traits. J. Amer. Soc. Hort. Sci. 123(4), 604–611.
Stushnoff, C. (1972) Breeding and selection methods for cold hardiness in deciduous fruit crops. HortScience 7, 10–13.
Szabó, Z. 1992. Evaluation of cold hardiness of peach cultivars based on freezing injury to twigs. Acta Hort. 315, 219–228.
Tagliavini, M. and Rombolà A.D. (2001) Iron deficiency and chlorosis in orchard and vineyard ecosystems. Eur. J. Agron. 15, 71–92.
Tatsuki M, Haji T, Yamaguchi M. (2006) The involvement of 1-aminocyclopropane-1-carboxylic acid synthase isogene, Pp-ACS1, in peach fruit softening. J Exp Bot. 57(6), 1281–9.
Trevisan, R., Gonçalves, E. D., Gonçalves, R., Antunes, L. E., and Herter, F. G. (2008) Influência do plastic branco, poda verde e amino quelant® na qualidade de pêssesgos 'Santa Áurea'. Bragantia, Campinas 67, 243–247.
Todorovic, R.R. and Misic, P.D. (1982) Susceptibility of peach cultivars and seedlings to *Taphrina deformans* (Berk.) Tul. J. Yugoslay Pomol. 16, 97–102.
Topp B. L., Sherman, W.B. and Raseira, M. C. B. (2008) Low-chill cultivar development. In, D. R. Layne and D. Bassi (Eds.), The Peach. Botany, Production and Uses. CAB International, Wallingford, UK, p. 106–138.
Topp, B.L. and Sherman W.B. (2000) Breeding strategies for developing temperate fruits for the subtropics, with particular reference to Prunus. Acta Hort. 522, 235–240.
Topp, B.L. and Sherman, W.B. (1989) Location influences on fruit traits of low-chill peaches in Australia. Proc. Florida State Horticultural Society 102,195–199.
Toyama, T.K. (1974) Haploidy in peach. HortScience 9, 187–188.
Tukey, H.B. (1934) Artificial culture methods for isolated embryos of decidous fruits. Proc. Am. Soc. Hort. Sci. 32, 313–322.
USDA/ARS (2008) Identification and correction of germplasm redundancy/deficiency in the NPGS peach and almond collection, Davis, CA.
Vendramin, E., Dettori, M. T., Giovinazzi, J., Micali, S., Quarta, R. and Verde, I. (2007) A set of EST-SSRs isolated from peach fruit transcriptome and their transportability across Prunus species. Mol. Ecol. Notes 7, 307–310.
Verde, I., Lauria, M., Dettori, M. T., Vendramin, E., Balconi, C., Micali, S., Wang, Y., Marrazzo, M. T., Cipriani, G., Hartings, H., Testolin, R., Abbott, A. G., Motto, M. and Quarta, R. (2005) Microsatellite and AFLP markers in the [*Prunus persica* (L.) Batsch] x *P. ferganensis* BC (1) linkage map, saturation and coverage improvement. Theor. Appl. Genet. 111, 1013–1021.
Viruel, M. A., Madur, D., Dirlewanger, E., Pascal, T. and Kervella, J. (1998) Mapping quantitative trait loci controlling peach leaf curl resistance. *Acta Hort.* 465,79–87.
Vizzotto, M., Cisneros-Zevallos, L., Byrne, D., Ramming, D. W, Okie, W R. (2007) Large Variation Found in the Phytochemical and Antioxidant Activity of Peach and Plum Germplasm. Journal of the American Society for Horticultural Science, v. 132, p. 1–7, 2007.
Wagner Jr., A, Raseira, M. C. B., Pierobom, C. R., Fortes, J. F., Silva, J. B. (2005a). Peach flower reaction to inoculation with *Monilinia fructicola* (Wint.) Honey) J. Amer. Pomol. Soc. 59(3), 141–147.
Wagner Jr., A, Raseira, M. C. B., Pierobom, C. R., Fortes, J. F., Silva, J B. (2005b). Non-Correlation of Flower and Fruit Resistance to Brown Rot (*Monilinia fructicola* (Wint.) Honey) Among 27 Peach Cultivars and Selections. J. Amer. Pomol. Soc. 59,148–152.
Wang, Y. (1985). Peach growing and germplasm in China. Acta Hort. 173, 51–55.
Wang, L., Zhu, G. and Fang, W. (2002) Peach germplasm and breeding programs at Zhengzhou in China. Acta Hort. 592,177–182.

Wang, Q., Zhang, K., Qu, X., Jia, J., Shi, J., Jin, D. and Wang, B. (2001) Construction and characterization of a bacterial artificial chromosome library of peach. Theor. Appl. Genet. 103, 1174–1179.
Wang Y, Georgi LL, Zhebentyayeva TN, Reighard GL, Scorza R, Abbott AG (2002a) High-throughput targeted SSR marker development in peach (*Prunus persica*). Genome 45,319–328.
Wang Y, Georgi LL, Reighard GL, Scorza R, Abbott AG (2002b) Genetic mapping of the ever growing gene in peach [*Prunus persica* (L.) Batsch]. J Hered 93,352–358.
Wang, Z. and Zhang, E. (2001) Chinese fruit documentation, Peach. Chinese Forestry Publisher. Beijing (In Chinese).
Warburton, M.L. and Bliss, F.A. (1996) Genetic diversity in peach (*Prunus persica* L Batch) revealed by randomly amplified polymorphic DNA (RAPD) markers and compared to inbreeding coefficients. J. Amer. Soc. Hort. Sci. 121, 1012–1019.
Wargovich, M. J. (2000). Anticancer properties of fruits and vegetables. HortScience 35, 573–575.
Watkins, R. Cherry, plum, peach, apricot and almond. (1995). In, Smartt, J. e Simmonds, N.W. (Ed.) Evolution of crop plants. London, Longman Scientific & Technical, 1995. p. 423–429.
Webster, A.D. (1995) Temperate fruit tree propagation. New Zealand J. Crop Hort. Sci. 23, 355–372.
Weinbaum, S.A. V.S. Polito, and D.E. Kester. (1986). Pollen retention following self pollination in peach, almond and peach almond × hybrids. Euphytica 35, 883–889.
Weinberger, J. H., Marth, P. C. and Scott, D. H. (1943) Inheritance study of root-knot nematode resistance in certain peach varieties. Proc. Amer. Soc. Hort. Sci. 42, 321–325.
Werner, D.J. and Chaparro, J.X. (2005) Genetic interaction of pillar and weeping peach genotypes. HortScience 40, 18–20.
Werner, D. J., Creller, M. A. and Chaparro, J. X. (1998) Inheritance of blood flesh in peach. HortScience 33, 1243–1246.
Werner, D. J. and Creller, M. A. (1997) Genetic studies in peach, Inheritance of sweet kernel and male sterility. J. Amer. Soc. Hort. Sci. 122, 215–217.
Werner, D.J, Mowrey, B.D. and Chaparro J.X. (1988) Variability in flower bud number among peach and nectarine cultivars. HortScience 23, 578–580.
Westwood, M.N. (1978). Temperate zone pomology, W.H. Freeman & Co. San Francisco.
Xiloyannis, C., Dichio, B., Tuzio, A.C., Gomez Aparisi, J., Rubio-Cabetas, M.J., Kleinhentz, M., Esmenjaud, D. (2007) Characterization and selection of *Prunus* rootstocks resistant to abiotic stresses: waterlogging, drought condition and iron chlorosis. Acta Hort. 732:247–250.
Xu, D.H., S. Wayhuni, y. Sato, M. Yamaguchi, H.T. Senematsu, and T. Ban. (2006). Genetic diversity and relationships of Japanese peach (*Prunus persica* L.) cultivars revealed by AFLP and pedigree tracing. Genet. Resources and Crop Evolution 53, 883–889.
Yamamoto, T. and Hayashi, T. (2002) New root-knot nematode resistance genes and their STS markers in peach. Scientia Hort. 96, 81–90.
Yamamoto, T., K. Mochida and T. Hayashi (2003) Shanhai Suimitsuto, one of the origins of Japanese peach cultivars. J. Japan. Soc. Hort. Sci 72,116–121.
Yamamoto, T., Shimada, T., Imai, T., Yaegaki, H., Haji, T., Matsuta, N., Yamaguchi, M. and Hayashi, T. (2001) Characterization of morphological traits based on a genetic linkage map in peach. Breed. Sci. 51, 271–278.
Yamamoto, T., Yamaguchi, M. and Hayashi, T. (2005) An integrated genetic linkage map of peach by SSR, STS, AFLP and RAPD. J. Japan. Soc. Hor. Sci. 74, 204–213.
Ye, X., Brown, S. K., Scorza, R., Cordts, J. and Sanford, J. C. (1994) Genetic transformation of peach tissues by particle bombardment. J. Amer. Soc. Hort. Sci. 119, 367–373.
Yoon, J., D. Liu, W. Song, W. Liu, A. Zhang, and S. Li. (2006). Genetic diversity and ecogeographical phylogenetic relationships among peach and nectarine cultivars based on simple sequence repeat (SSR) markers. J. Amer. Soc. Hort. Sci. 131,513–521.
Yoshida, M. (1976). Genetical studies on the fruit quality of peach varieties, texture and keeping quality. Bull. Fruit Tree Res. Sta. A3, 1–16.
Yu, M., Ma, R, and Tang, X. (1997) Inheritance of ripening season in F1 hybrids of peach. Jiangsu J. Agr. Sci. 13(3),176–181.

Yulin W. (2002). Peach, in Yulin W. (ed.) Genetic Resources of deciduous fruit and nut crop in China. China Agricultural Science and Technology Press, Beijing, p. 135–156.

Zarrouk, O., Gogorcena, Y., Moreno, M.A., and Pinochet, J. (2006) Graft compatibility between peach cultivars and *Prunus* rootstocks. HortScience 41, 1389–1394.

Zarrouk, O., Gogorcena Y., Gómez-Aparisi J., Betrán J.A. and Moreno M.A. (2005) Influence of peach x almond hybrids rootstocks on flower and leaf mineral concentration, yield and vigour of two peach cultivars. Sci. Hortic. 106(4), 502–514.

Zhebentyayeva, T., Swire-Clark, G., Georgi, L., Garay, L., Jung, S., Forrest, S., Blenda, A., Blackmon, B., Mook, J., Horn, R., Howad, W., Arús, P., Main, D., Tomkins, J., Sosinski., Baird, W. V., Reighard, G. L. and Abbott, A. G. (2008) A framework physical map for peach, a model Rosaceae species. Tree Genetics and Genomics 4, 745–756.

Chapter 15
Plum

Bruce L. Topp, Dougal M. Russell, Michael Neumüller, Marco A. Dalbó, and Weisheng Liu

Abstract There are 19–40 species of plum, depending on taxonomist, that have originated in Europe, Asia and America. From this great diversity only two species, the hexaploid European plum (*Prunus domestica*) and the diploid Japanese plum (*P. salicina* and hybrids), are of worldwide commercial significance. The European plums were cultivated in Roman times and stone remnants indicate human use 6,000 years ago. Their origin is uncertain but may have involved *P. cerasifera* and possibly *P. spinosa* as ancestors. The rich diversity and history of European plums is reflected in the many pomological groups including Prunes, Gages, Mirabelles, Damsons, Bullaces and St Juliens. Today, European plum breeding concentrates on selection for resistance to Sharka disease caused by the Plum Pox Virus which limits production in many countries. Resistant cultivars have been developed using both conventional

B.L. Topp (✉)
Queensland Alliance for Agriculture and Food Innovation, University of Queensland
Maroochy Research Station, SCMC, PO Box 5083, Nambour 4560, QLD, Australia
e-mail: b.topp@uq.edu.au

D.M. Russell
Horticulture & Forestry Science, Department of Employment
and Economic Development, Nambour, QLD, Australia
e-mail: dougal.russell@deedi.qld.gov.au

M. Neumüller
Technische Universität München, Fachgebiet Obstbau, Dürnast 2,
Freising 85354, Germany
e-mail: neumueller@wzw.tum.de

M.A. Dalbó
Epagri, Estação Experimental de Videira, C.P. 21, Videira 89560-000,
Santa Catarina, Brazil
e-mail: dalbo@epagri.sc.gov.br

W. Liu
Liaoning Institute of Pomology, Xiongyue, Yingkou 115009, Lianoning,
People's Republic of China
e-mail: weishengliu@yahoo.com.cn

M.L. Badenes and D.H. Byrne (eds.), *Fruit Breeding*, Handbook of Plant Breeding 8,
DOI 10.1007/978-1-4419-0763-9_15,

and genetic transformation techniques. Japanese plums originated in China but were introduced to the west, from Japan, only 150 years ago. Luther Burbank hybridised them with other plum species with the result that most modern cultivars are multi-species amalgams. This heterogeneity, plus the high heterozygosity from outcrossing, means that large seedling populations are required in cultivar development. Efficient cross-pollination and seedling management techniques are required for these large populations. The trend of interspecific hybridisation continues today with four of the top 20 Californian cultivars being interspecifics involving plum and apricot. Fruit quality, functional food value, productivity and adaptation through disease resistance, chilling requirement and phenology are selection criteria in both Japanese and European plum breeding. Molecular markers are used for selection of self-compatibility and nematode resistance and for diversity and taxonomic studies. Most new rootstock releases are clonally propagated and of interspecific origin. The priorities for plum and peach rootstock breeding are similar and rootstocks developed for peach are sometimes also used for plum. American plum species, ancient Oriental cultivars and autochthonous European cultivars represent important germplasm resources that require preservation for use in future breeding.

Keywords *Prunus salicina* and *P. domestica* • Japanese plum • European plum • Cultivars • Interspecific hybrids • Rootstocks • Heritability • Disease resistance, • Controlled pollination, • Molecular markers • Self-incompatibility • Genetic transformation • Germplasm • Production • Taxonomy • Progeny

1 Introduction

There are over 6,000 cultivars of plums from 19 to 40 species (Hedrick 1911; Rehder 1954; Blazek 2007) distributed across Asia, Europe and America. As such, it is little wonder that Hedrick (1911) considered that plums "give a greater range of flavour, aroma, texture colour, form and size, the qualities that gratify the senses and make fruits desirable, than any other of our orchard fruits". Watkins (1976) considered that plums hold the centre of the *Prunus* genetic stage because they have the largest diversity of any subgenus and are a link between the major subgenera.

From this great diversity, only two species predominate in modern commercial production: the hexaploid, European plum (*Prunus domestica* L.) and the diploid Japanese plum (*P. salicina* Lindl and hybrids). A challenge for breeders is in using the other plum and related *Prunus* species to develop new cultivars with broadened adaptation and innovative products.

European and Japanese plums although in the same taxonomic section are distinct crops in terms of their uses, adaptation, origin and domestication. They are usually grown in different locations, with European plums in cooler areas and Japanese plums in warmer areas. They have distinct cultural and historical backgrounds that have resulted in European plums being dominant in Europe and Japanese plums dominant elsewhere. They are sold as separate crops and marketed as distinct commodities.

Table 15.1 Plum production (1,000 t) in the top 13 countries (FAO 2006)

Country	2000	2001	2002	2003	2004	2005	2006	Mean
China[a]	3,942	4,061	4,397	4,435	4,835	5,229	5,326	4,604
The USA	819	591	668	728	295	432	645	597
Romania	550	557	221	910	476	622	599	562
Serbia and Montenegro	362	338	205	577	567	312	556	417
France	204	272	246	250	229	214	234	236
Chile[a]	172	211	215	255	250	245	260	230
Turkey	195	200	200	210	210	220	214	207
Spain	168	150	211	230	146	252	179	191
Italy	180	172	177	128	179	185	180	172
Iran[a]	143	143	145	147	147	147	147	146
Russia[a]	135	125	152	160	178	169	89	144
Ukraine	123	138	95	135	173	166	127	137
Argentina[a]	75	106	106	151	127	128	155	121
Total	*7,066*	*7,063*	*7,038*	*8,317*	*7,813*	*8,321*	*8,713*	*7,761*

[a]FAO estimate or unofficial figure

Plum genetics and breeding have been reviewed by Cullinan (1937), Weinberger (1975), Ramming and Cociu (1991), Okie and Weinberger (1996), Okie and Ramming (1999), Okie and Hancock (2008) and most recently by Hartmann and Neumüller (2009). During this time, there have been changes in the location and activity of breeding programmes, the spread of a major disease, the availability of breeding techniques and the wild germplasm resources available for breeding. In this chapter we focus on European and Japanese plum but will also provide details on other plum species that are used in breeding.

1.1 Economic Importance

Plums are grown worldwide, but mainly in temperate zones. Production was 9,431,322 tonne which ranked 12th of all fruits and fourth of deciduous fruits behind apple, pear and peach (FAO 2006). Most plums are produced in Asia (58%) followed by Europe (29%), America (10%), Africa (2%) and Oceania (1%).

China is the largest producer with over seven times the production of second ranked the USA (Table 15.1). European countries account for 8 of the top 13 producers, while southern hemisphere production is dominated by Chile and Argentina. These 13 countries accounted for 83% of production from 2000 to 2006. Production for five of these countries is based on unofficial or FAO estimates.

Almost all production in China is of Japanese plums. There has been a remarkable increase in production in China. During the 1980s, production rose gradually from about 500,000 to 1,000,000 tonne with a rapid increase during the 1990s up to about 4,000,000 tonne (FAO 2006). Liu (2007) attributes this rapid increase to national economic reform, involving the development of fruit processing industries in southern China and export of fresh fruit to neighbouring SE Asian countries.

Table 15.2 Commercial and traditional uses of plum and plum products

Application	Species	Remarks
Fresh fruit	*P. salicina, P. domestica*	Japanese plums are used mostly fresh
Dried fruit	Mostly *P. domestica*	70% of world prune production is from California and 80% of this production is from the 'D'Agen' cultivar
Spirits and wine	*P. domestica*	Brandy such as Slivovitz; also Cognac
Baking	*P. salicina, P. domestica*	Glace fruit used in baking; used as antimicrobials, fat replacers, flavourants and to retard lipid oxidation that causes rancidity (Nunez de Gonzalez et al. 2008)
Canning	*P. domestica*	Compotes using the Mirabelle cultivars are famous
Jams and jellies	*P. domestica, P. salicina, P. spinosa* and American species	Most plum species can be used for jams, but commercially there are *P. domestica* cultivars released specifically for this purpose such as 'Jam Session' and 'Blues Jam' (Andersen et al. 2006a, b)
Confectionery products including sugar plums	*P. domestica*	Sugar plum was a generic term from the 1600s for any large comfit or sweetmeat. They were usually sugar-covered seeds of almond or caraway. More recently, Janick and Paull (2008) described them as plums candied by boiling in a thick sugar syrup
Ornamental trees and flowers	*P. salicina, P. cerasifera, P. pissardii*	Ornamental value of flowers, bark and leaves. *P. pissardii* is a red leaf variant of *P. cerasifera*
Medicinal	*P. cerasifera*	Stimulate respiration, improve digestion, give sense of well-being
	P. spinosa	Flowers used as a blood cleaner and leaves are dried and used as a tea substitute. Dyes are obtained from the fruit, leaves and bark (Khoshbakht and Hammer 2005)
	P. domestica	Prunes and prune juice used as laxative

In the USA, California was the major producing state from 2000 to 2006 with 294,300 tonne comprising 57% Japanese plums and 43% European plums (USDA Yearbook 2007).

European plums are generally adapted to cool temperate climates and are the main commercial species in Europe. They have many uses, including fresh consumption and also as dried fruit known as prunes (Table 15.2). Prunes and prune juice have a long history of use for their digestive–laxative properties. European plums are also distilled, primarily in the Slavic regions of Central and Eastern Europe, to make brandy such as Slivovitz, the national drink of Serbia. The fruit are also used for production of jams, jellies, dumplings and paste. The Portuguese plums in syrup "Ameixa d'Elvas" (Nunes et al. 2008) are historically famous as are plum puddings and sugar plums. It should be noted that sugar plum was originally a generic term for small, round candies or sweetmeats (Fig. 15.1). They were called

Fig. 15.1 Eighteenth century depiction of a confectioner's shop and the popular candied sweetmeat "Sugar Plumbs" being eaten by the military officer on the left. Detail from James Gilray's 1797 "Heroes recruiting at Kelsey's"

sugar plums because they resembled plums in size and shape not because they contained plum (Ivan Day personal communication).

Japanese plums have a wide range of adaptation from temperate to subtropical regions. They are sufficiently cold hardy for commercial use in northeast Chinese provinces of Helongjiang, Liaoning and Jilin where the temperature falls to −40°C (Liu 2007). The fruit are mostly consumed fresh but are also preserved by canning. The trees are used for ornamental purposes and artworks celebrating the beauty of the flowers date back many centuries in China (Faust and Surányi 1999).

Outside of European and Japanese plums, there is limited use of the other species. *Prunus cerasifera* Ehrh. and *Prunus spinosa* L. are used in Europe for processing and spirit production (Table 15.2). North American species such as *P. angustifolia* Marshall, *P. hortulana* L.H. Bailey, *P. munsoniana* Wight & Hedrick (Reid and Gast 1993), *P. subcordata* Benth. (Roberts and Hammers 1951) and *P. maritima* Marshall (Bailey 1944) are collected locally for processing into jams and preserves.

1.2 Botany and Taxonomy

Plums are small to medium-sized, deciduous trees or shrubs growing to 2–10 m, with a range of tree habits from upright to spreading. Leaves are alternate, serrate,

Fig. 15.2 Variation in plum leaf type, from *left to right*, *P. hortulana*, *P. maritima*, *P.insititia* cv Pixy and (*P. salicina* × *P. angustifolia)* cv 'Robusto'

rarely entire and stipulate. They can be glaborous as for *P. salicina, P. cerasifera* and *P.angustifolia* or with pubescence beneath as for *P. domestica* and *P. insititia* L. and range in length from 2 to 10 cm (Fig. 15.2). The flowers are perfect, solitary and with five sepals and petals which are usually white. The flowers emerge prior to or with the new leaves. The fruit are a one-seeded drupe with a distinct furrow or suture line, no pubescence and a waxy bloom on the skin. There is a wide range of skin colours including black, blue, purple, red, orange, yellow and green. The flesh is generally yellow, green or red or combinations.

The phylogenetic classification of Rosaceae is still controversial (Katayama and Uematsu 2005; Rohrer et al. 2004; Shaw and Small 2004, 2005) and the taxonomy of plums has changed with time, an indication that the differences among species are not always great. Pre-Linnean botanists considered *Prunus* to include only plums but Linneaus (1707–1778) used the term *Prunus* to include plums along with peach, cherry, apricot and almond (Faust and Surányi 1999). The *Prunus* genus is within the Rosaceae family and depending on the taxonomist, contains from 19 to 40 species of plum.

Rehder (1954) divided *Prunus* into five subgenera with all the plums (and apricots) falling into the Prunophora subgenus (Fig. 15.3). Plums in Prunophora are characterised by having sulcate (grooved) fruit with an epicuticular bloom, solitary axillary buds and no terminal bud. He further divided the Prunophora subgenus into three sections. The Euprunus section contains six plum species native to Europe and Asia, the Prunocerasus section contains 13 plum species native to North America, and the Armeniaca section contains six apricot species. Okie and Weinberger (1996) considered that the Oriental species fitted better in Prunocerasus both taxonomically and horticulturally.

Euprunus species are distinguished from others because they bear convolute leaves in the bud stage, glabrous ovaries and fruit, and pedunculate flowers (Reales et al. 2010). Prunocerasus species by contrast bear conduplicate leaves in the bud

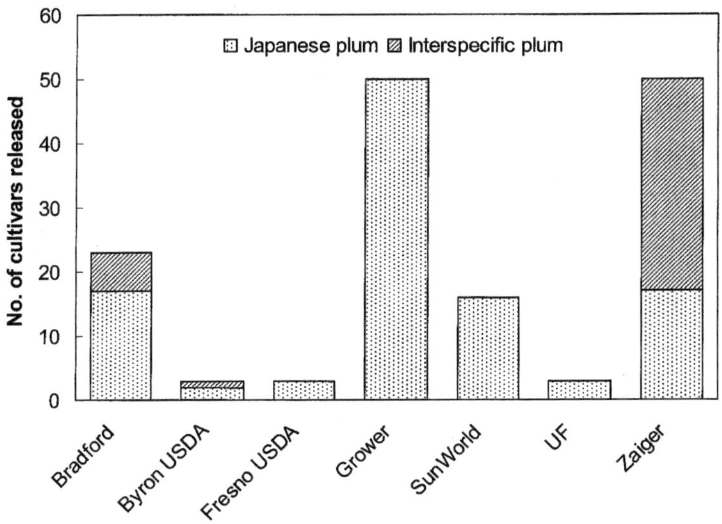

Fig. 15.3 Taxonomic classification of plum in relation to other *Prunus* according to Rehder (1954)

stage. Armeniaca species differ from the others by having pubescent ovaries and fruit, flowers that are sessile or shortly pedunculate and leaves rolled up in the bud stage. The use of nucleic acid based techniques will assist with future taxonomic studies.

2 Origin and Domestication

Different species of plums originated and have been domesticated independently on three continents. The centre of origin was Europe for *P. domestica*, western and central Asia for *P. cerasifera*, China for *P. salicina* and North America for the species of the Prunocerasus section such as *P. americana* Marshall (Table 15.3). Watkins (1976) considered that the Prunocerasus species have separated from the Euprunus species relatively recently. He argues that the similarities between them would not be as great if they had diverged at an early stage. There is a high degree of interspecific compatibility between most diploid species and they have been hybridised over the past two centuries to improve adaptation, enhance fruit quality and develop novel fruits.

2.1 European Species

P. domestica is called the European plum, garden plum or domesticated plum and is the most important species from Europe. It is a hexaploid ($2n=6x=48$) that is unknown in the wild and considered to be relatively young. It probably originated

Table 15.3 Plum species as classified by Rehder (1954) with their origin and traits of use to breeders

Species	Common name	Origin	Useful traits
American Species from the Section Prunocerasus			
P. alleghaniensis Porter.	Allegheny plum	North-east US	Crown gall resistance
P. americana Marsh.	Common wild plum	Central, east and south US	Wide climatic adaptation including cold hardy
P. angustifolia Marsh.	Chickasaw plum	South US	Bacterial spot resistance
P. gracilis Engelm. & Gr.	Oklahoma plum	West Arkansas, Oklahoma and north Texas	Similar to *P. maritima,* adapted to dry sites
P. hortulana Bailey	Hortulana plum	Central Kentucky, Tennessee to Iowa and Oklahoma	Bacterial spot resistance
P. lanata (Sudw.) Mack. & Bush.		Illinois to Texas	Similar to *P. americana* late bloom
P. maritima Marsh.	Beach plum	New Brunswick to Virginia	Late bloom; high heat to bloom requirement
P. mexicana S. Wats.	Big-tree plum	Mexico and south-central US	Large tree
P. munsoniana Wight & Hedr.	Wild Goose plum	Kentucky and Tennessee to Mississippi, Texas, Minnesota and Kansas	Fruit quality; late flowering; winter hardiness
P. nigra Ait.	Canada plum	Canada and north US	Cold hardy
P. reverchonii Sarg.	Hog plum	Oklahoma, Texas	Adapted to limestone soils; can withstand drought
P. subcordata Benth.	Pacific plum	California to Oregon	Drought tolerance
P. umbellata Ell.	Black sloe	South-east US	Crown gall resistance
Oriental species from the Section Euprunus			
P. salicina Lindl.	Japanese plum	China	Large fruit size, firmness; low-chilling requirement
P. simonii Carr.	Apricot plum, Simon plum	North China	High volatiles; firmness
European Species from the Section Euprunus			
P. cerasifera Ehrh.	Cherry plum, Myrobalan	West Asia, Caucasus	Nematode resistance; bright skin colours; earliness
P. domestica L.[a]	Garden plum, European plum	West Asia, Europe	Good fruit quality
P. institia L.[a]	Bullace, Damson	West Asia, Europe	
P. spinosa L.	Blackthorn, Sloe	Europe, north Africa, west Asia	Dwarfism and robustness in rootstock breeding

[a]All species are diploid except *P. domestica* (hexaploid) and *P. institia* (tetraploid)

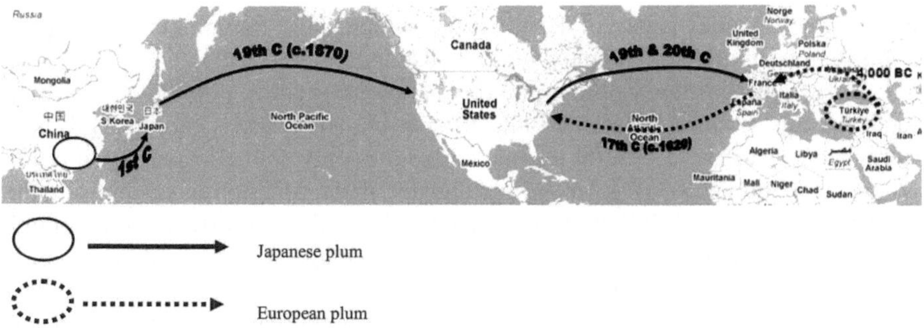

Fig. 15.4 The dissemination of Japanese and European plums

in western Asia in the region south of Caucasus Mountains through to the Caspian Sea, overlapping with the centre of origin of *P. cerasifera*, and from there moved into western Europe. Archaeological finds of European plum seeds indicate that it was used in Roman times about 2,000 years ago. The oldest remains are stones of damson plums from Ulm, Germany about 6,000 years ago (Faust and Surányi 1999). The pilgrims and early settlers transported *P. domestica* to America in the 1600s (Fig. 15.4).

The origin of *P. domestica* is uncertain. Crane and Lawrence (1934) proposed that it arose on many occasions in prehistoric and historic times as an allo-polypoloid. The diploid ($2n=2x=16$) *P. cerasifera* (syn. *P. divaricata* Led.) and tetraploid ($2n=4x=32$) *P. spinosa* naturally form hybrids where their centres of origin overlap in Turkey, Iran and Greece (Faust and Surányi 1999). They suggested that *P. domestica* was produced through doubling of the chromosomes of *P. spinosa* or from unreduced gametes of both *P. spinosa* and *P. cerasifera.*

Zohary (1992) considered this was unlikely on the grounds that there are wide morphological differences between *P. spinosa* and *P. domestica* and there is pronounced sterility in the hybrids. He proposed that *P. domestica* had arisen as an auto-polyploid from *P. cerasifera*. In a study of ribosomal DNA, Reynders-Aloisi and Grellet (1994) found there were two *P. cerasifera* ribosomal units present in *P. spinosa* but a third unit that was specific to *P. spinosa.*

A recent phylogenetic analysis of four sequenced chloroplast DNA regions indicated *P. domestica* and *P. insititia* clustered in a distinct monphyletic clade (Reales et al. 2010). They were closely related to *P. cerasifera*, indicating they may have been derived from a common *P. cerasifera* ancestor. However, a *P. cerasifera* and *P. spinosa* origin could not be rejected because the sequence data were from chloroplasts and so reflect only maternal lineage.

Wild populations of *P. domestica* were found in Xinjiang in northwest China along the Ili River (Lin and Shi 1989) and it was speculated that these may have been native (Zhang et al. 1997). Liu et al. (2006, 2007a) used RAPD and ISSR markers to study the variability of these populations and found a low level of diversity indicating that they were not native but were naturalised populations from other regions. The Ili valley is part of the silk route, which could explain their occurrence in this region through human movement (Faust and Surányi 1999).

P. cerasifera, the cherry plum or myrobalan plum, is native to middle Asia, Iran, Iraq, Caucasia, Crimea, Anatolia, Balkan Peninsula and sparsely through Slovakia, Moravia and Austria to Central Europe where it is only sporadic and possibly not indigenous (Buttner 2001). It has been cultivated in the Mediterranean area and the Balkan Peninsula since 200 BC. It has been widely used as a rootstock and, through the use of RAPD markers, has been found in the parentage of many modern Japanese plums (Boonprakob et al. 2001). Hedrick (1911) lists different horticultural variants of *P. cerasifera* including a weeping form, one with contorted twisted foliage, a narrow willow leaf form and a purple-leaved form (*P. pissardi* Carr. or *P. cerasifera* '*Atropurpurea*').

P. spinosa, the blackthorn or sloe, is native through Europe to the Urals, north of Africa, north of Anatolia, Caucasus, north of Iran and northwest of Turkmenistan (Khoshbakht and Hammer 2005). The fruit are small, often sour and always astringent (Ercisli 2004). It is used in folk medicine (Buttner 2001) and for distillation. The large-fruited selections of sloe, propagated and cultured in central Europe, often are hybrids of *P. domestica* and *P. spinosa*.

2.2 *Oriental Species*

There are two plums, both diploids ($2n=2x=16$) that are native to China. *P. salicina*, commonly called Japanese plum, is the commercially dominant Oriental species. It is almost certainly native to China but is unknown in the wild (Hedrick 1911). It is considered to have originated in the Yangtze River basin (Yoshida 1987) and is thought to occur wild in the Tsunglin range in Shensi and Kansu (Hedrick 1911).

Domestication involves a reduction in genetic variability because only a small number of genotypes are taken from the wild during initial cultivation before subsequent development of new cultivars. A triple bottleneck occurred with the domestication of Japanese plum as the crop was domesticated in China, moved to Japan and then transported to California (Fig. 15.4). It was transported from China to Japan over 2,000 years ago. Cultivation in Japan, based on stone remnants, dates back to the Yayoi period (tenth century BC to third century AD) (Yoshida 1987). Many Oriental crops moved into Europe, from east to west, via the overland silk route. Japanese plum is unusual in that it was commercialised in the west only within the past 150 years and did not follow this overland route but instead was transported by sea from Japan in an easterly direction to the USA and thence onto Europe.

The common name "Japanese plum" is used in the west because the plum was first introduced to the USA (and then on to Europe) from Japan (Fig. 15.4). It was imported to California in 1870 by Mr. Hough of Vacaville (Faust and Surányi 1999). John Kelsey of Berkeley, California obtained trees from Mr. Hough and produced the first crop in 1876. He propagated and recommended the fruit and in the following decade it was widely propagated and named in his honour (Hedrick 1911). 'Kelsey' is still grown commercially (Fig. 15.5).

Fig. 15.5 "Kelsey" was one of the first Japanese plums to fruit in the USA in 1876 and is still grown commercially

Luther Burbank was a pioneer in Japanese plum breeding. In 1884, he imported 210 seedlings of 12 cultivars from Japan to California (Okie and Weinberger 1996). From this population, he selected the cultivars 'Burbank' and 'Satsuma'. During his lifetime, Burbank produced hundreds of thousands of seedlings. He hybridised Japanese plums with native American species, imported germplasm and released many new cultivars (Howard 1945; Crow 2001). His 'Santa Rosa' released in 1906 was still ranked as the fourth most produced plum in California in 1994.

Prunus simonii Carr., the Simon plum or the apricot plum, is considered to be native to northern China. It has an upright habit and may be just an upright variant of *P. salicina* (Okie 2006; Liu et al. 2006, 2007a). It was used by Burbank to hybridise with *P. salicina* and contains the useful traits of cold hardiness and very firm flesh. The flavour is described as a combination of apricot, peach, pineapple and cantaloupe, but sometimes with an acrid skin (Starnes 1905). The unique flavour is explained by 23 aromatic compounds present in *P. simonii* that are absent in *P. salicina* (Gomez and Ledbetter 1994). Hexyl acetate which produces a characteristic apple aroma was 50-fold higher in *P. simonii* than *P. salicina* ('Blackamber').

2.3 American Species

There are 13 American species (Table 15.3) in the section Prunocerasus according to Rehder (1954), although Wight (1915a) describes 23 species and subspecies. Many species including *P. maritima*, *P. hortulana* and *P. americana* were collected and in some cases cultivated by early settlers and before them by native Americans

Fig. 15.6 "Sapa" is a hybrid of *P. besseyi* and *P. salicina* released in 1908 from South Dakota State University as cold hardy plum adapted to the northern USA

(Faust and Surányi 1999; Roberts and Hammers 1951; Hedrick 1911). The earliest European record of native American plums is by John de Verrazano who recorded in his diary for 8 July 1524 that they found "damson trees" growing wild around 41°N latitude (southern New York) which Wight (1915b) considered were *P. maritima*. Early settlers domesticated American species where they could not easily grow European plums. In the Mississippi Valley, the prairie and the southern states, *P. americana* was propagated and cultivated. In California, *P. subcordata* was a standard food product of native Americans eaten raw, cooked or sometimes dried and was also used by early European settlers such as trappers, gold seekers and ranchers (Hedrick 1911).

By the end of the nineteenth century, there were over 150 cultivars that had been named and disseminated, about half of which were direct selections from the wild (Bailey 1892). Public breeding programmes at Iowa, Minnesota and South Dakota using native American species and Japanese plums were active in the early to mid 1900s (Fig. 15.6). Trees were selected for adaptation to cold northern regions or resistance to specific pests and diseases, using *P. nigra* Aiton, *P. besseyi* L. H. Bailey and *P. americana* (Cullinan 1937). Many of these cultivars were hardy and well adapted but were not commercially successful and were replaced by European cultivars (Andersen and Weir 1967).

3 Genetic Resources

3.1 European Species

Within *P. domestica*, there are several pomological groups that have distinct histories, characteristics and uses. The names of the groups and the cultivars provide an insight into the rich diversity and history of plums. Principal component analysis confirms the phenotypic aggregation of cultivars into distinct groups (Kaufmane et al. 2002). RAPD marker analysis has confirmed a high degree of polymorphism among genotypes and clustering of cultivars with similar ancestry but not always into clearly defined groups (Ortiz et al. 1997). Despite this discrepancy, the pomological groups are still important today for historic and marketing reasons and include prunes, gages and yellow eggs.

The pomological group of prunes are *P. domestica* cultivars capable of being dried without fermentation due to their high sugar content. The fruit are generally small, oval to elongated and with a dark skin. Common cultivars are 'D'Agen Prune', 'Italian Prune' and 'German Prune'. Prunes are eaten fresh, used in baking (plum cakes), distillery, jam production or dried. The word "prune" can also be used to refer to the dried product.

In Europe, *P. domestica* cultivars that produce round to oval fruits with soft flesh are called plums. They are distinguished from prunes in that they lose their shape during cooking. Some large-fruited cultivars are important in middle Europe.

Gages or greengages (sometimes called *P. italic* Borkh. = *P. domestica* subsp. *italica* (Borkh.) Gams ex Hegi) or Reine-Claudes produce round, green, deeply sutured fruit with a firm green flesh that is sweet and scented. The Greengage originated in continental Europe, probably from Armenia and then to Greece and Italy where it was called 'Verdocchia', and then onto France in the early 1500s where it was named after Queen Rein Claude. The process of naming this plum after people of influence continued when the plum was sent to England by a Paris-based Roman Catholic priest named John Gage. He sent it to his brother at Bury St Edmunds in 1725 but the 'Reine Claude' label was lost and so the gardener called them "Green Gage" in honour of his employer (Roach 1985).

Yellow Eggs are a group of *P. domestica* that have yellow skin and are egg shaped. They are used primarily for canning and include 'Yellow Egg', 'Pershore Egg' and 'Golden Drop'.

P. insititia has been classified as a distinct species of the hexaploid European plum (Rehder 1954) but also as a subspecies of *P. domestica* (Buttner 2001). Faust and Surányi (1999) considered *P. insititia* as a separate species from *P. domestica* because the trees are more compact, leaves smaller and more ovate, fruit smaller and either purple or yellow without intermediate colours. *P. insititia* is divided into the pomological groups of Damsons, Bullaces, Mirabelles and St Juliens. The fruit are mostly used for preserving and distillation rather than for fresh consumption. The Damson and St Julien plums are also used as rootstocks for hexaploid scions.

Fig. 15.7 "Mirabelle de Nancy" is the most famous of the Mirabelle group of plums and is important in France and Germany since the eighteenth century for canning and distillation

Mirabelle plums grow naturally in southern Europe and the southern Causcasus (Salinero et al. 2003). They have been important in France and Germany since the eighteenth century for fresh fruit or used in canning and distillation (Jacob 2007a). The most famous cultivar is 'Mirabelle de Nancy' (Fig. 15.7). The fruit are small, about 32 mm diameter but sweet with sugar levels up to 30° brix and highly aromatic.

P. spinosa, the blackthorn or sloe, is a wild species native to southern Europe, Turkey and Armenia (Ramming and Cociu 1991; Erturk et al. 2009). The trees are small, seldom over 2 m high and with many thorns. The fruit are small with black skin and a blue waxy bloom and green, acidic flesh. They are not eaten fresh but used to make sloe wine and gin.

The 2,000 plus year history of plum domestication in Europe has resulted in a large number of cultivars. In the nineteenth century, pomologists Knight, Laxton and Rivers developed many new cultivars, although 'Victoria', the most important English cultivar developed during this period, was found as a seedling in Sussex (Roach 1985). Many landrace or autochthonous cultivars are still available in collections or are grown commercially. However they are being replaced by modern cultivars with improved quality and more regular production (Petrovic et al. 2002).

The hexaploid *P. domestica* can be easily hybridised with the tetraploid *P. spinosa* and the diploid *P. cerasifera* (Minev and Balev 2002; Neumüller et al. 2009) Hexaploid plums are generally incompatible with other diploids and while hybrids have been produced (Olden 1965; Hunter and Bragdo 1969) they have not yet resulted in commercial cultivars. Zhivondov and Djouvinov (2002) produced 42 interspecific hybrids involving *P. domestica* as the seed parent and pollen parents of *P. armeniaca* L., *P.* × *dascycarpa* Ehrh., *P. cerasifera*, *P. ussuriensis* Kovalev & Kostina and hybrids between *P. salicina* and *P. cerasifera*.

Information on genetic resources is available from the European Prunus Database, the Chinese Crop Germplasm Resources Information System (in Chinese) and the US National Plant Germplasm System (NPGS). The European Prunus database is maintained by the National Institute for Agronomical Research (INRA) in Bordeaux, France, under an initiative of the European Cooperative Programme for Crop Genetic Resources (ECPGR) Networks. This database includes data on the European collections of all *Prunus* species, both cultivated and wild. There are over 2,200 accessions with the majority being *P. domestica* followed by *P. cerasifera* (Blazek 2007) held at 67 institutes in 30 countries. Over 700 accessions of plum are held at the Nikita Botanical Gardens in Ukraine (Yezhov et al. 2005) and 566 accessions at two sites in Romania (Butac and Budan 2009).

3.2 Oriental Species

Most modern Japanese plums are predominately *P. salicina* but also include varying amounts of other species largely due to Luther Burbank's early breeding work and the ancestry predominance of his cultivars. In 1996, eight of the top ten Californian cultivars had Luther Burbank genetics in their ancestry (Okie and Ramming 1999).

Historical records of Burbank's breeding indicate he combined *P. salicina* with *P. cerasifera*, *P. simonii*, *P. americana*, *P. hortulana* and *P. munsoniana.* (Hedrick 1911; Howard 1945). More recent RAPD marker analysis confirmed that *P. salicina* (29–36%), *P. simonii* (21–26%) and *P. cerasifera* (21–28%) had the greatest contribution to the USA plum gene pool, but *P. americana* and *P. angustifolia* were also involved (Boonprakob and Byrne 2003).

This Burbank effect plus the natural outcrossing of Japanese plums has provided great variability. Byrne (1989) found that the mean inbreeding and coancestry coefficients for plum were one half or less than those for peach which is a self-pollinating, single species. Cultivated diploid plums have a similar level of diversity to almonds and greater diversity than peaches (Byrne 1990). Outcrossing resulting from self-incompatibility and multi-species ancestry are significant factors in creating this variability. Knowledge of how this variability is partitioned among and within species is useful for breeders.

In general, there is a high degree of interspecific cross-compatibility among the diploid plum and non-plum species within the subgenus Prunophora (Okie and Weinberger 1996). This includes *P. salicina*, *P. simonii* and *P. cerasifera* in Euprunus, the American plums in Prunocerasus and the apricots and mumes in Armeniaca (Jun and Chung 2007; Jun et al. 2009; Arbeloa et al. 2006).

The diploid plums can also be hybridised with species from the subgenera Amygdalus (peach and almond) and Cerasus (cherry) but with less fertility (Liu et al. 2007b; Wakana et al. 2006; Kataoka et al. 1988). These wide crosses are of particular importance in the breeding of new rootstocks (Lespinasse et al. 2003).

Qiao et al. (2007) used RAPD, ISSR and SSR markers to study 54 plums of which 40 were bred in China, 4 in Japan and 10 in the USA. Cultivars from the USA and Japan clustered in one group and were distinct from the 40 Chinese cultivars. The Chinese cultivars fell into three distinct groups. Two groups corresponded to the northern or southern regions from where they originated and provide useful insights into parent choices in breeding for adaptation to these regions. The third group consisted of six cultivars which were all ancient, including 'Furongli' and 'Hongxinli' which have been cultivated for over 400 years.

The Chinese National Germplasm Repository for Plums and Apricots is located in Liaoning province under the direction of its curator Dr Weisheng Liu and contains 717 accessions from nine plum species including *P. salicina* and *P. simonii*. Collection and preservation of ancient cultivars and wild species is important in providing a germplasm resource for future breeding. Since the importation of Japanese plums in Luther Burbank's era, introgression of new germplasm has been rare. Inclusion of ancient cultivars, such as those described Qiao et al. (2007), would be useful in increasing the diversity and range of traits available for selection in current breeding programmes.

3.3 North American Species

The North American species are a valuable resource for diploid plum and rootstock breeding. They are adapted to a wide range of environments and have many horticulturally important traits (Table 15.3). There is a high degree of cross-compatibility between *P. salicina* and American species which allows introgression of these traits. Useful traits include: late bloom from *P. umbellata* Elliot and *P. maritima*; limited root suckering and winter hardiness from *P americana* and *P. nigra*, late ripening from *P. hortulana*; drought tolerance from *P. subcordata* and bacterial leaf spot resistance from *P. angustifolia* (Dorsey and Bushnell 1925; Okie 2001; Okie and Hancock 2008).

A major disincentive to using these species is that fruit of the hybrids is almost uniformly poor quality in terms of size and flavour. It requires long-term commitment through multiple generations of breeding to introduce the desired traits. Use of marker techniques to accelerate breeding will require detailed study of the trait and its genetic control.

The US National Plant Germplasm System of the USDA-ARS is dedicated to the collection and preservation of plant germplasm from around the world. It includes the National Clonal Germplasm Repository for fruit and nut crops at Davis, California which was established in 1981. It contains 313 plum accessions, including 154 *P. domestica*, 45 *P. cerasifera*, 63 *P. salicina* and 39 American plum species (*Prunus* Crop Germplasm Committee 2010).

Several American species are considered threatened or endangered, including *P. geniculata* R. M. Harper, *P. minutiflora* Engelm. ex A. Gray, *P. alleghaniensis* Porter, *P. alleghaniensis* var. *davisii* W. Wight, *P. maritima* var. *gravesii* (Small) G. J. Anderson, *P. murrayana* E. J. Palmer and *P. texana* D. Dietr. (Ramming and Cociu 1991; Prunus Crop Germplasm Committee 2010).

3.4 Rootstocks

Rootstocks for plums are propagated clonally or by seed from many different species and from hybrids that combine the traits of species. A disadvantage of seedling rootstocks from outcrossing species is that they may be highly variable. Their advantage is that they are inexpensive to propagate and generally virus-free because most viruses are not seed transmitted.

Ashton (2008) describes 46 rootstocks for plum and Okie (1987) provides detailed summaries of the origin, propagation, productivity, compatibility and disease susceptibility of 33 important rootstocks for plum. Commonly used rootstocks for European plums are:

- Myrobalan (*P. cerasifera*), as seed or clonal selections such as 'H29C', 'GF31' and 'B'.
- Marianna (*P. cerasifera* × *P. munsoniana*), as seed or cloned as '2624', 'GF8-1' and 'Buck'.
- *P. insititia* cloned selections such as 'Pixy', 'St Julien A' and 'St Julien GF655-2'.
- *P. domestica* clones such as 'Black Damas', 'Brompton', 'Common Mussel', 'Prune GF43' and 'Wangenheims'. 'Wangenheims' is an old German cultivar that was originally released as 'Weiwa' and is also compatible with apricot.

Non-plum *Prunus* species from sections Armeniaca (apricot and mume) and Amygdalus (peach and almond) are also used for European plums but with varying compatibility. GF677 and GF557 are hybrids of almond and peach and are used to produce vigorous trees in alkaline soils. Peach seedlings are used as rootstocks for *P. domestica* cultivars such as 'Stanley' but are incompatible with the German Prune (Okie 1987).

Japanese plums are grown on Myrobalan and Marianna rootstocks particularly in orchards with poorly-drained soils. Peach seedlings such as 'Nemaguard', 'Elberta', 'Lovell' and 'Flordaguard' are also commonly used without the compatibility problems that occur under European plum and are reported to produce larger fruit (La Rue and Johnson 1989).

The use of such a wide germplasm pool for rootstocks means that compatibility is a key issue that requires further assessment. Incompatibility is rare when the same species is used as both rootstock and scion. Day (1953) presents data on compatibility for 178 plum cultivars on rootstocks of Myrobalan, peach, apricot and almond.

4 Major Breeding Achievements

Active breeding programmes exist in Asia, North America, Europe and the Southern Hemisphere and have been reviewed by Fogle (1978), Ramming and Cociu (1991), Okie and Weinberger (1996), Okie (2006) and Hartmann and Neumüller (2009). European plum breeding occurs mostly in Europe but also in California. Japanese plum breeding occurs mostly in Asia, North America and the Southern Hemisphere but also in Europe.

4.1 Japanese Plums

Seven active programmes in China are located in cold regions where plum is the main tree fruit crop. The National Germplasm Repository at Liaoning provides germplasm and resources for the programmes. The main objectives are for increased fruit size, flesh firmness and eating quality, with adaptation to each of the seven provinces. Four elite lines, '05-1-03', '06-14-03', '06-14-14' and '06-14-21', with firm flesh attractive appearance and strong aroma are in the final stages of evaluation.

A small breeding programme in Taiwan has produced five selections that are in the final stages of evaluation. The elite genotypes were selected from populations that combined large fruit size from Florida plums with non-bitter skin from locally adapted plums (Wen and Sherman 2003). The Florida plums contained a Taiwan plum ancestry as the source of their low-chilling but there was still significant genetic difference between them and the sweet skinned Taiwan parents 'Hei tao' and 'Shui li' (Wen and Liu 2004).

In Japan, about 80% of new cultivars are released by private breeders or growers. For example, Mr. Oishi Roshio in Fukushima prefecture has released many cultivars including 'Oishiwasesumomo', 'Oishinakate' and 'Gekkou'. (Masami Yamaguchi pers. comm.). The National Institute of Fruit Tree Science at Tsukuba began breeding Japanese plums in 1970 with the aim of extending the harvest season and improving quality and has released 'Honey Rosa' (Yamaguchi et al. 1998) and 'Honey Heart' (Yamaguchi et al. 1999). In Korea, the National Horticultural Research Institute is breeding for improved fruit quality and has released 'Purple Queen' and 'Honey Red' (Jun et al. 2008).

Both public and private breeders in California have influenced plum marketing with their cultivars widely planted in most countries that grow Japanese plums. Californian cultivars have replaced local cultivars in many locations because of their large fruit and firm flesh. In particular, the USDA-bred cultivars 'Friar', 'Blackamber', 'Queen Rosa' and 'Fortune' have been mainstays in many markets. Similarly, the privately bred 'Red Beaut' and 'Angeleno' have been extremely important because they extended the plum season with size and firmness.

The USDA breeding programme at Fresno, California released 'Black Splendor', 'Owen T' and 'John W' in 2002. Selection objectives are for larger fruit especially in the early season and for high eating quality. The programme is currently concentrating on evaluation of elite selections with no crosses in the past 5 years.

Red-skinned plums with yellow flesh dominated Californian production with 'Santa Rosa', 'Casselman', 'Late Santa Rosa' and 'Red Beaut' accounting for 49% of production in 1975. Black-skinned plums with yellow flesh such as 'Laroda', and 'Nubiana' were available in 1954 but the release of 'Friar' (Fig. 15.8) in 1968 and the earlier ripening 'Blackamber' in 1980 resulted in a major switch by the industry to the black-skinned types. These cultivars were high yielding and produced large fruit that did not readily show bruising due to the black skin. By 1994, these two cultivars plus the late ripening 'Angeleno' accounted for 42% of production in California.

Fig. 15.8 "Friar" was released by the USDA in 1968 and became a major cultivar around the world where Japanese plums could be grown without severe bacterial spot pressure. It has attractive black skin, large size and high productivity

The majority of new releases are from the private breeders in California, particularly from Sun World, Zaiger Genetics and Bradford Farms (Fig. 15.9). In 2008, 20 cultivars accounted for 65% of production and 14 of these cultivars were from private breeders (Table 15.4). The cultivar mix had been static for many years, but over the past decade has begun to change rapidly. In 2008, 16% of the crop was produced by cultivars that were not planted a decade earlier. The increased plantings of new cultivars (Table 15.4) indicate that this trend will continue.

Bradford Farms produce 10,000–15,000 seedlings each year with about 90% of them being controlled first generation crosses. Japanese plums are primarily used as seed parents, with apricots and interspecifics used as pollen parents. New interspecific germplasm will be used in the future for both seed and pollen parents. Bradford applied for 35 plant patents from 2000 to 2007. Zaiger Genetics has released about 50 cultivars described in HortScience from 1997 to 2008 (Fig. 15.9) and have seven cultivars listed in the top 20 for California in 2008 (Table 15.4). Sunworld has developed a series of black-skinned, red-fleshed plums that ripen through the season and are sold under the trademark of "Black Diamond". This is an important innovation that combines breeding, production and marketing to provide consumers with a continuous supply of plums with consistent eating traits.

Plumcots are hybrids of Japanese plum and apricot, and while cultivars were developed over 100 years ago (Hedrick 1911; Howard 1945), they were not commercial due to low yields and high acidity (Jun and Chung 2007; Ledbetter et al. 1994). In the first generation, the hybrid seedlings inherit pubescent skin from apricot

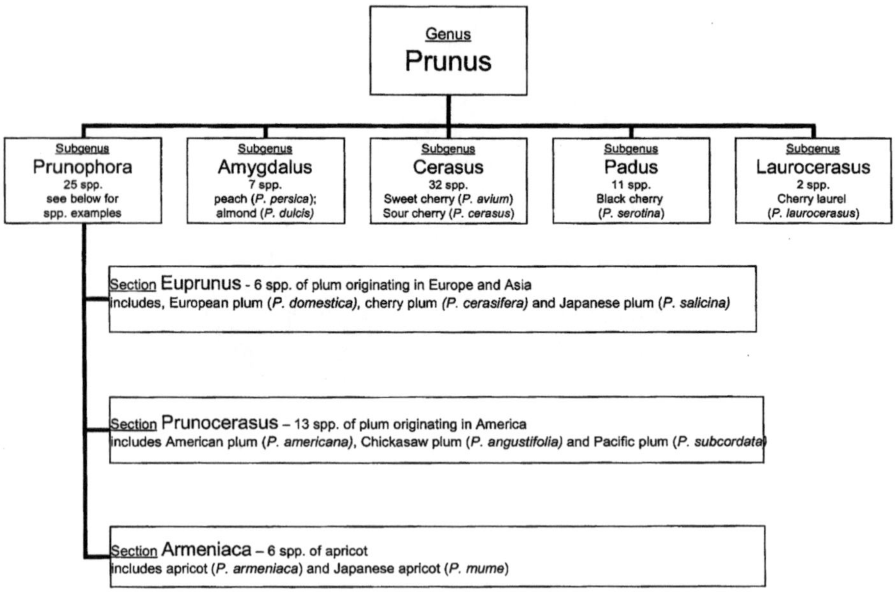

Fig. 15.9 Origin of plum cultivars released in California (Register of New Cultivar Lists, HortScience 1997–2008)

and clingstone seed from plum and there is variable fruit size, shape, colour, flavour and firmness (Jun et al. 2009; Ledbetter et al. 1994). One of the reasons for hybridising plum and apricot is to combine the firm flesh and wide range of bloom and ripe times of plum with the aroma and flavour of apricot. There are distinct differences in the aromatics of plum and apricot (Gomez-Plaza and Ledbetter 2010), with lactones and terpenic alcohols providing characteristic apricot aroma (Gomez and Ledbetter 1997). Plumcots produce high levels of volatile compounds typical of both plum and apricot (Gomez et al. 1993) and also unique volatiles that are not present in either parent (Gomez and Ledbetter 1993).

With continued selection and backcrossing to plum, plumcot defects have been ameliorated and there are now increasing numbers of plum interspecifics being planted (Fig. 15.9). Zaiger Genetics trademarked the "pluot" and "aprium" names which distinguish their interspecifics from regular plums. Pluots have predominately plum ancestry and morphology, and apriums are predominantly apricot. Other trademarked categories include 'Peacotum' for peach–plum–apricot hybrids and 'Nectaplum' for nectarine–plum hybrids. The interspecifics 'Flavorfall', 'Flavorich', 'Flavorosa' and 'Black Kat' are in the Californian top 20 production list for 2008 (Table 15.4). According to US Plant Patent pedigree records, the amount of apricot ancestry varies from 25% in cultivars such as 'Flavorfall' to 12.5% in 'Black Kat'. Morphologically, the fruit are similar to plum and have firm, juicy and sweet flesh; all traits favoured by consumers. Ahmad et al. (2004) studied the genetic similarity of plums, apricots and pluots using microsatellite markers.

Table 15.4 Important plum cultivars in California, arranged in order of production in 2008 (CTFA 2009)

Cultivar	Crop type[a]	Skin colour	Flesh colour	Production 2008 (%)	Production 1998 (%)	Tree sales 2000–2006 (%)	Year of release	Breeder
Angeleno	JP	Black	Yellow	12.1	19.4	4.2	1967	John Garabedian
Friar	JP	Black	Yellow	9.7	19.2	1.2	1968	USDA
Blackamber	JP	Black	Yellow	6.0	9.7	0.5	1980	USDA
Fortune	JP	Red	Yellow	4.8	7.0	0.7	1988	USDA
Flavor Fall	IS	Red	Yellow	3.7	0	na	2001	Zaiger Genetics
Howard Sun	JP	Black	Yellow	2.8	3.6	0.2	1986	Chamberlin
Yummy™beaut (36P205)	JP	Red	Yellow	2.8	0	19.7	2003	Bradford Farms
Owen T	JP	Black	Yellow	2.6	0	4.9	2002	USDA
Hiromi Red	JP	Red	Yellow	2.3	0	7.5	1997	Zaiger Genetics
Black Splendor	JP	Black	Red	2.2	0	15.3	2002	USDA
Suplumeleven (Black Diamond)	JP	Black	Red	2.1	na	na	1982	SunWorld
Catalina	JP	Black	Yellow	2.0	2.0	1.4	1982	Krause, Walter
Joanna Red	JP	Red	Yellow	2.0	0.1	2.2	1998	Zaiger Genetics
Flavorich	IS	Black	Yellow	1.7	na	na	1994	Zaiger Genetics
Earliqueen	JP	Red	Yellow	1.6	0.3	2.5	1994	Zaiger Genetics
Flavorosa	IS	Black	Red	1.4	na	na	1998	Zaiger Genetics
Tulare Giant	EP	Purple	Yellow	1.4	0	1.3	2000	UC Davis
Black Kat	IS	Black	Yellow	1.4	0	0.4	2002	Zaiger Genetics
Red Beaut	JP	Red	Yellow	1.3	2.9	3.00	1956	Fred Anderson
Black Beaut	JP	Black	Yellow	1.2	2.9	1.26	1975	Fred Anderson

[a] *EP* European plum, *IS* Interspecific hybrid plum, *JP* Japanese plum

They found that the apricots clustered distinctly from the plums; the pluots shared 12% of common plum markers but no markers in common with the apricots. The lack of apricot markers in the pluots may have been due to differences among the apricots used in the study compared with those in the pluot ancestry.

In the southern USA, there is a major breeding effort at the Agricultural Research Service of the USDA Southeastern Fruit and Tree Nut Research Laboratory that started in 1964. The aim of the breeding is to produce plums adapted to the humid southeastern region with late bloom for frost avoidance, high eating quality and resistance to bacterial canker, bacterial spot and plum leaf scald. Okie (2006) describes eight released cultivars including ‘Byrongold’, ‘Segundo’, ‘Ruby Queen’ and the plumcot ‘Spring Satin’. Low-chill plums adapted to warm climates are being bred by the University of Florida at Gainesville (Sherman et al. 1992) and by Texas A&M University at College Station. Releases from Florida include ‘Gulfruby’, ‘Gulfgold’, ‘Gulfrose’, ‘Gulfbeauty’ and ‘Gulfblaze’ (Sherman and Lyrene 1998).

The programme at the Agriculture Research Council, Infruitec-Nietvoorbij at Stellenbosch South Africa began in the 1950s. It has had great success as demonstrated with over 80% of national production being from local cultivars. The aim of the programme is to produce red, yellow and black skin cultivars that are suited for export to Europe and ripen over 25 weeks from November to March. Releases include ‘Harry Pickstone’, ‘Reubennel’ and ‘Laetitia’ and more recently the yellow skinned ‘Golden Kiss’ and ‘Sun Breeze’. The programme integrates information from exporters and growers in deciding on cultivar release.

In Australia, plum breeding programmes are run by the Department of Primary Industries in Queensland and the Agriculture Department in Western Australia. Bacterial spot resistant and early ripening cultivars (Topp and Russell 1989, 1990a, b) have been released in Queensland and the programme is now focussed on breeding low-chill cultivars. In Western Australia, the main focus is in developing new cultivars suited for export to south-east Asia. In South America, there is breeding in Brazil (Nakasu et al. 1981) and more recently in Uruguay at Las Brujas Experiment Station. The Santa Catarina State Agricultural Research Institute in Brazil is breeding for low-chill requirement and resistance to bacterial leaf scald and bacterial leaf spot. It is a large programme producing about 3,000 seedlings per year and has released ‘Camila’ and ‘Piuna’ (Ducroquet and Dalbó 2007). At Embrapa Clima Temperado at Pelotas a small programme producing about 400 seedlings per year was restarted in 2002.

As Japanese plum production has increased in Italy there have been increased efforts in breeding, a trend likely to continue in other Mediterranean regions. The breeding programme of the Fruit Tree Research Unit at Forli is breeding both European and Japanese plums and is selecting for high fruit quality and yield. At the University of Florence, about 300–1,000 seedlings are produced annually and they have released ‘DOFI-Guidy’ and ‘DOFI-Sandra’ (Bellini and Nencetti 2002a, b; Bellini et al. 2002)

4.2 European Plums

Atanasoff (1935) emphasised the importance of breeding in the battle against Plum Pox Virus (PPV) causing the Sharka disease and encouraged European breeders to establish new breeding programmes. One of the largest programmes is located at Cacak, Serbia. The two most successful cultivars released from that programme are 'Cacanska lepotica' and 'Cacanska rodna', the first being tolerant, the latter highly sensitive to PPV. 'Cacanksa lepotica' is still a leading cultivar in Germany, Hungary, Switzerland, Austria and Poland. It produces heavy crops of fruit that are very attractive but have poor taste. Breeding activities at the Cacak Station decreased during the last decades due to political instability and the retirement of the breeder but have been recently reactivated.

In Bulgaria, the Plovdiv Breeding Station has released ten cultivars but activities declined in the 1990s. In the late 1980s and early 1990s, *P. domestica* and *P. armeniaca* were hybridised to obtain new fruit characteristics and to transfer Sharka resistance (Zhivondov 2007). Some of the resulting hybrids were fertile and produced fruit with pubescent skin.

In Romania, plum breeding programmes are running at Pitest, Bistrita, Valcea and Voinesti with Pitest and Bistrita being the more active programmes. Several cultivars were released during the last decades including 'Pitestean', 'Valcean' and 'Ialomita'.

In Bulgaria and Romania where PPV is particularly devastating, breeding activities were low in the last two decades because no source of PPV resistance was available. Recently, the programmes have been revitalised due to European Union funded cooperative research and the availability of enduring and stable PPV resistant parents originating from German breeding.

A small breeding programme in Norway aims to produce cultivars with high fruit quality and large fruit for the North European market. In Italy, a breeding programme for high fruit quality was established at Bologna and smaller breeding programmes are still active at Firenze and Forli. Plum breeding in France and Poland (at Skierniewice) has been stopped recently.

In Germany, plum breeding started at the University of Hohenheim in the 1980s. Some of the most important plum cultivars in commercial use such as 'Katinka', 'Hanita' and 'Presenta' were bred at that station by Hartmann. His most important success, which influences plum breeding worldwide, is the breeding of 'Jojo', the first European plum which is completely resistant to PPV in the open field (Fig. 15.10). Further crossings with this and other completely PPV resistant genotypes were made and are under evaluation. The breeding programme at Hohenheim will soon close but the work is being continued at Technische Universität München at Freising-Weihenstephan by Neumüller. The most important goals in this programme are PPV resistance, fruit size, fruit firmness, stability to heat and inner quality. Interspecific hybrids between *P. domestica* and either *P. salicina*, *P. armeniaca*, *P. cerasifera* or *P. spinosa* have been generated to transfer PPV resistance to related

Fig. 15.10 "JoJo" the first completely Plum Pox Virus resistant cultivar obtained by conventional breeding from Germany

stone fruit species, to obtain new fruit characteristics and to breed PPV resistant rootstocks. The breeding programme at Geisenheim Research Station in Germany has produced large-fruited cultivars such as 'Tophit' and regular bearing cultivars such as 'Topper' (Jacob 2007b) but was closed in 2005.

The University of Davis in California is using traditional breeding methods to develop cultivars for drying (prunes). They are selecting for earlier and later ripening than 'Improved French' which is currently the dominant drying prune cultivar and have released 'Sutter Prune' and 'Tulare Giant' (DeJong et al. 2002). Since 1989, the USDA at Kearneysville, West Virginia has been working on developing PPV resistance through transformation of *P. domestica.* and has released the PPV resistant 'HoneySweet' (Scorza et al. 2007).

4.3 Rootstocks

Rootstock breeding receives less attention than scion breeding and plum rootstocks less than peach rootsocks. It requires large populations often in the order of 100,000 seedlings (Cummins and Aldwinckle 1983) and commitment of resources for up to 30 years (Table 15.7). Despite these obstacles, new rootstocks are being developed to provide improved economic performance and wider adaptation. The priorities for rootstocks for plum and peach are similar (Ramming and Cociu 1991) and rootstocks developed for peach are sometimes also used for plum.

Over the last decades, most new rootstock releases have been clonally propagated and many are of interspecific origin (Renaud and Salesses 1994). From 1970 to 2001, 11 new rootstocks for plum were introduced in the USA and all were clonally propagated (Beckman and Lang 2003). Testing new rootstocks is time consuming and expensive. Large testing programmes such as the NC-140 in the USA provide valuable information on performance in multiple regions and with a range of scions (NC-140 2002).

Major rootstock breeding programmes in Europe include INRA (Institut National de la Reserche Agronomique) in France which has released the GF series and 'Ishtara', 'Jaspi', 'Myran' and 'Mirabi' and the Vavilov Research Institute which has released 'Krymsk 1', 'Krymsk 86' and 'Krymsk 2' (Ashton 2008). There is breeding or selection in the UK, Germany and Russia (Okie 1987). In Spain, the Estación Experimental de Aula Dei at Zaragoza is breeding new rootstocks adapted to the limiting conditions of the Mediterranean including chlorosis, root asphyxia, replant disorders and nematodes (Moreno 2004). Releases have included 'Ademir', 'Adara' and 'Adesoto 101' (Moreno et al. 1995). In Italy, there has been a long running rootstock breeding programme at the Instituto Sperimentale per la Frutticoltura at Rome. They released 'Penta' and 'Tetra' which are both *P. domestica* and are compatible with plum as well as peach, apricot and almond (Nicotra and Moser 1997).

In 2005, a plum rootstock breeding programme was established at Technische Universität München at Freising-Weihenstephan, Germany. The aim is to develop semi-dwarfing and dwarfing rootstocks which are hypersensitive to the Plum pox virus. Inter- and intra-specific hybridisations are carried out. If budsticks latently infected with PPV are grafted upon rootstocks showing a strong hypersensitive response against PPV the budstick will either not grow or die after a short period of growth. In this way, it is guaranteed that only trees free from PPV will leave the nursery. Therewith the main way of distribution of PPV over long distances could be interrupted. Hypersensitive rootstocks could also be used for scions hypersensitive against PPV. The advantage of hypersensitivity resistance is that the cycle of virus spreading both over long and short distances can be interrupted. The hypersensitivity resistance trait has been introduced into rootstocks series 'Docera' (*P. domestica* × *P. cerasifera*) and 'Dospina' (*P. domestica* × *P. spinosa*) which are being currently under evaluation in Germany and will be released in the future.

In the USA, the USDA and University of California have jointly released 'Controller 5' and 'Controller 9' which are hybrids of Japanese plum and peach. These are tree size-reducing stocks for peach but are compatible also with Japanese plum (Ashton 2008). Zaiger Genetics released 'Citation' which also reduces tree size and is compatible with a range of *Prunus* including Japanese plum. At Byron, Georgia the USDA has a major rootstock breeding project that is breeding primarily for peach scions but also evaluates with Japanese and European plum. Recently, they released, jointly with the University of Florida, a clonal plum rootstock 'Sharpe' which is a *P. angustifolia* hybrid for use in Armillaria-infested sites. It is compatible with Japanese and European plum in short-term trials and long-term testing is continuing (Beckman et al. 2008).

5 Current Goals and Challenges of Breeding

Breeders aim to combine high levels of desirable characteristics in new cultivars. They wish to produce cultivars with the largest, sweetest, firmest fruit with high productivity and adaptation. Table 15.5 provides examples of cultivars that have high levels of important traits for the plum industry. In practice, it is difficult to combine high levels of all desirable traits in one cultivar but the cultivars in this table provide examples of the current commercial limits.

Table 15.5 Sources of useful traits in plum cultivars

Trait	Cultivar	Remarks/Reference
Fruit traits		
Aroma: high	Royal Zee, Joanna Red, Fortune, Flavorosa, Flavor King, Harmona	Crisosto et al. (2007), Lozano et al. (2009)
Firmness: high	Larry Ann	Lozano et al. (2009)
Flesh acidity: low	Wu mei	0.7% malic acid compared with 1.5% for Gulfbeauty (Wen and Sherman 2003)
Flesh colour: deep red	Queen Garnet, Rubysweet, Rubyqueen	USPP19630
Flesh colour: orange	Veeblue	Brooks and Olmo (1997)
Flesh colour: white	Black Torch (Suplumthirteen)	Brooks and Olmo (1997)
Flesh sugar: high	Avalon, October Sun, September Yummy	Avalon at 19% (Vangdal et al. 2007); October Sun at 19.8% (Crisosto et al. 2007)
High phenolics	Black Beaut, Angeleno	Important for functional food value (Tomas-Barberan et al. 2001)
Climacteric: suppressed	Shiro, Rubyred	Abdi et al. (1997). See also Yamaguchi and Kyotani (1985) for slow softening cultivars
Shape: oblate	Eldorado	Hedrick (1911)
Size: large	August Yummy, September Yummy, Dapple Dandy, Fortune	August Yummy 75 mm diameter (Peter Buchanan personal)
Skin: non-bitter	Chin hsien, Hei tao, Hei tzyy, Shui li	Wen and Sherman (2003)
Skin colour: black	Friar, Common Prune	Brooks and Olmo (1997)
Skin colour: green	Kelsey	Hedrick (1911)
Skin colour: red	Fortune	Brooks and Olmo (1997)
Skin colour: yellow	Shiro, Tipala	Hedrick (1911)
Free stone	Laetitia, Katinka	Brooks and Olmo (1997)
Stone: less	Miracle	Hedrick (1911)
Tree traits		
Chilling: low	Koushih, Hsing tsai and I Ian	All about 100 chill units (Wen and Sherman 2003)

(continued)

Table 15.5 (continued)

Trait	Cultivar	Remarks/Reference
Flowering: late	Italian Prune, Common Prune, Burbank, Alderman, Ozark Premier, Shirley	
Fruit development period: long	October Sun, October Gem, Ruby Red, Autumn Beaut, Flavor Fall, September Yummy, Presenta	First 4 are harvested in September with an FDP of about 220 days (CTFA 2009)
Fruit development period: short	Spring Flavor, Flavorosa, Red Beaut, Ruth Gerstetter, Juna	First 3 are harvested in first week of May in California with an FDP of about 75–80 days (CTFA 2009)
Self compatibility: high	Santa Rosa, Honey Rosa Reubennel, Harry Pickstone, Methley	Fruit set after selfing over 10% (Beppu et al. 2003, 2010)
Yield: high	Angeleno, Fortune	
Stature: reduced	Compact Friar	Brooks and Olmo (1997)

5.1 *Fruit Traits*

The fruit are usually harvested from a single cultivar at one location over 1–2 weeks. So to produce plums over 6 months, 12–24 cultivars are required. Ripening time, or fruit development period (FDP) when bloom time is fixed, is quantitatively inherited with high heritability (Table 15.6) with the progeny mean close to the parental mean (Weinberger and Thompson 1962). The occurrence of bud-sports such as 'Roysum', 'Casselman' and 'Late Santa Rosa' which ripen later than the non-mutated parent tree indicates there are genes with major effects on ripening time. In practice, it seems difficult to reduce FDP for Japanese plum below about 75 days. 'Red Beaut' released in 1958 has an FDP of about 80 days and despite a half century of selection, the modern Californian cultivars 'Ebony Jewel', 'Earliqueen' and 'Spring Red' ripen only 5–7 days earlier (CTFA 2009).

Size, colour and shape are visual traits that will influence a consumer's decision to purchase fruit (Fig. 15.11). Internal organoleptic traits of flavour, texture, firmness and juiciness will then influence repeat sales. Other fruit quality traits such as resistance to storage disorders and functional food attributes will also be important.

Japanese plums are generally larger than European plums, but compared with peach they are small. There is always selection for increased fruit size to reduce harvesting and packaging costs. Fruit weight is moderately heritable (Table 15.6), but only a small percentage of progeny can be expected to have fruit larger than the largest parent (Paunovic and Misic 1975), and so, parents with large fruit should be chosen in most instances.

Red and black are the most common skin colour in Japanese plum and are inherited quantitatively with neither dominant. Yellow skin colour is controlled by a single gene recessive to black and red (Weinberger and Thompson 1962).

Table 15.6 Heritability estimates for plum

Trait	Crop	Heritability	Reference
Phenological			
Bloom date	*P. domestica*	0.86	Hansche et al. (1975)
Ripe date	*P. domestica*	0.84	Hansche et al. (1975)
Ripe date	*P. domestica*	0.38	Botu (1998)
Ripe date	*P. salicina*	0.44	Bellini et al. (1998)
Fruit			
Weight (g)	*P. domestica*	0.97	Hansche et al. (1975)
Weight (g)	*P. domestica*	0.52	Botu (1998)
Soluble solids (%)	*P. domestica*	0.49	Hansche et al. (1975)
Shape (subjective scale)	*P. domestica*	0.19	Botu (1998)
Colour (subjective scale)	*P. domestica*	0.34	Botu (1998)
Stone adherence (subjective scale)	*P. domestica*	0.44	Botu (1998)
Tree			
Yield (subjective scale)	*P. domestica*	0	Hansche et al. (1975)
Bacterial leaf spot (subjective scale)	*P. salicina*	0.42	Topp et al. (1991)
Bacterial twig canker (mm)	*P. salicina*	0.58	Topp et al. (1991)
Plant vigour (cm)	*P. domestica*	0.12	Botu (1998)
Plant growth habit (subjective scale)	*P. domestica*	0.29	Botu (1998)
2-Month radius (mm)	*P. domestica*	0.44	DeBuse et al. (2007)
10-Month radius (mm)	*P. domestica*	0.2	DeBuse et al. (2007)
2-Month height (mm)	*P. domestica*	0.23	DeBuse et al. (2007)
10-Month height (mm)	*P. domestica*	0.08	DeBuse et al. (2007)

Fig. 15.11 Variability of fruit size, shape and colour available for selection for consumer visual appeal

Fruit should be symmetrical to avoid damage during harvest and packing. Japanese plums may be round, ovate, heart-shaped or oblate as for 'Durado' and 'Nubiana'. European plums are round or elliptical. Renaud (1975) noted that "round" was dominant to "oblong", but Weinberger and Thompson (1962) considered no shape to be dominant.

Good flavour or taste is important in all new cultivars, but it is difficult to evaluate because of the variability induced by the environment and the variation in individual preferences. Objectively measured traits related to consumer acceptance (Crisosto et al. 2004), such as soluble solid content (SSC) and acidity, allow comparison among genotypes and avoid selector bias. SSC was reported to have a moderate heritability (0.49) in European plum (Hansche et al. 1975).

In plums and many other stone fruits SSC varies greatly among individual fruit within a genotype (Walsh et al. 2007). One approach to this problem is to breed for higher SSC so that even with the variability the less-sweet fruit are still acceptable. An alternative approach will be to study the cause of the variation and select for genotypes with less variability. This may involve modification of tree architecture so that fruit are more evenly spaced and with improved light distribution throughout the canopy. It may also involve regulation of crop load to avoid overcropping that results in small fruit with low sugar and aroma.

5.2 *Functional Food Traits*

Plums and plum products are functional foods in that they provide health benefits beyond their value as a source of nutrients and may help maintain health and prevent chronic diseases.

In the past two decades, there have been many studies of germplasm variability for functional chemicals (Tomas-Barberan et al. 2001; Gil et al. 2002; Kim et al. 2003; Dikeman et al. 2004; Cevallos-Casals et al. 2006; Rupasinghe et al. 2006; Vizzotto et al. 2007; Byrne et al. 2009). Demonstrated effects on health include inhibition of cancer cells in human cancer cell lines (Lea et al. 2008) and in mice (Kim et al. 2008) and improved bone health measures in post-menopausal women (Arjmandi et al. 2002). Stacewicz-Sapuntzakis et al. (2001) provide a comprehensive review of the chemical composition and the potential health effects of prunes. Prunes contain high levels of fibre, sorbitol, potassium, copper, boron and phenolics which have potential biological functions in regard to glucose metabolism, cardiovascular health, bone metabolism, laxative effects and anti-tumour activity. The laxative effects of prunes and prune juice are such common knowledge that the Californian Prune marketing board directed a change from the name "dried prune" to "dried plum" to increase the appeal to young consumers (Janick and Paull 2008).

Phenolics are higher in red-fleshed plums than in yellow-fleshed plums and are more important in determining antioxidant activity than were the anthocyanins or carotenoids (Vizzotto et al. 2007). The skin contained 3–9 times higher concentration of phenolics than the flesh, but the flesh provided about 70% of the total because of its greater fraction (Cevallos-Casals et al. 2006). Selection for high phenolics is possible, but extreme levels may be associated with astringency (Gil et al. 2002). Breeding for the functional food benefit of plum will require further studies on the inheritance of these key traits.

5.3 *Flower Traits*

In some regions, late blooming is required to avoid spring frost damage. Full bloom date is highly heritable (0.86) in European plum (Hansche et al. 1975) and so can be altered with selection based on phenotype. Variability for late bloom exists in *P. maritima* and *P. besseyi*. Bloom date can be manipulated by altering the chilling requirement or the heat accumulation requirement of a cultivar, or both. In warm regions, delaying flowering by increasing the cultivar chilling requirement would result in poor productivity due to the lack of chilling. In such a situation, it is preferable to develop late flowering cultivars by increasing the requirement for heat unit accumulation prior to flowering. Werner et al. (1988) demonstrated there was variability for this trait in *P. besseyi*.

Most Japanese and some European plums are self-incompatible. Incorrect orchard design with regard to polleniser layout or the lack of pollen transfer due to low insect pollinator activity can result in low yields. The gametophytic self-incompatibility system in plum is controlled by a single polymorphic 'S' locus, which contains genes for pollen and pistil specifics such that pollen tube growth is arrested in the style with the same haplotype (Beppu et al. 2002; Sutherland et al. 2007). The S^e-RNase allele confers self-compatibility and can be screened for at the seedling stage using PCR markers (Beppu et al. 2005, 2010) thus allowing early selection for this trait.

5.4 *Tree Traits*

High and regular yields are important. An important component of yield is precocity and there is great variability for this trait. In Japanese plum, the most precocious cultivars are from the low-chill University of Florida selections which bear fruit on 1-year-old wood (Topp and Sherman 1990a) and have a high proportion of 2 year-old seedlings setting crops. European plums are less precocious than Japanese plums but newer cultivars have improved precocity compared to 'Italian Prune' and 'German Prune' (Hartmann and Neumüller 2009). *P. texana* may be a useful source of precocity as it is possible to go one generation from seed to seed in 12 months.

Many genetic and environmental factors influence yield. The heritability of yield when measured using a subjective scale at one location on a single seedling tree is very low or zero (Hansche et al. 1975). Compounding these difficulties, it is often

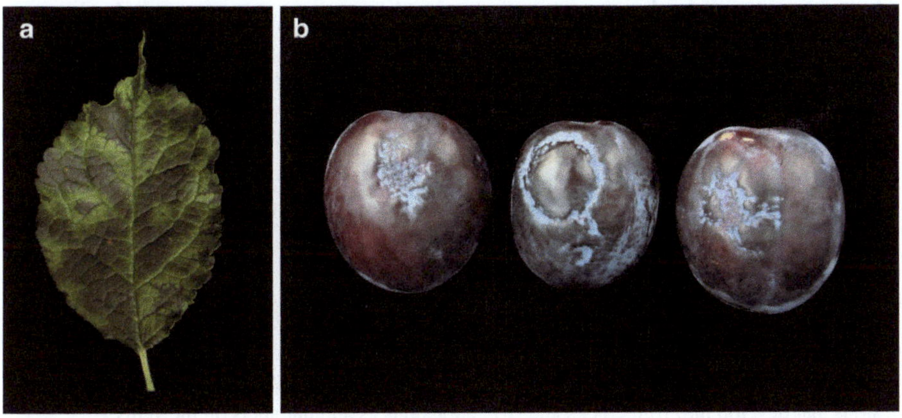

Fig. 15.12 Plum Pox Virus symptoms on (**a**) leaf of P. domestica and (**b**) fruit of "Zwintschers Früher" showing external depressions on surface. Internally fruit show browning, sugars are reduced and acid is increased

standard management practice to regulate the crop load by thinning the flowers and fruit. In these situations, the breeder should consider selecting for components of yield, such as fruit size and fruit set, that have higher heritabilities.

5.5 *Pest and Disease Resistance*

There are over 60 major pests and diseases that attack plums, including 4 bacteria, 19 fungi, 6 viruses, 4 nematodes and 36 insects (Janick and Paull 2008). Fortunately only a few are problems in any one region and in any one season. Dry locations suffer less pressure than wet locations. Breeders select for the economically most important problems in their location, but do so at a loss of selection pressure for the all important fruit quality traits and so must consider the benefits and costs of genetic versus management pest control. Genetic forms of control will become more important as pesticide resistance, declining access to registered chemicals and consumer demand for pesticide-free fruit will combine to remove current chemical solutions. Summaries of resistant cultivars (Ramming and Cociu 1991) and genetics of resistance (Okie and Weinberger 1996) are available for the economically important diseases and pests.

Sharka disease caused by the plum pox virus has devastated European plum industries (Fig. 15.12). Resistance, or at least tolerance, is essential for economic production in many locations. This disease is spreading around the world where host plants including European plum, Japanese plum, peach and apricot are cultivated. Two of the most promising directions for obtaining resistance are through use of a hypersensitive response against the virus from the cultivar 'JoJo' in conventional breeding (Neumüller et al. 2009) and transgenic manipulation using a rapid early

Fig. 15.13 Symptoms of bacterial spot on susceptible cultivar showing (**a**) greasy, water-soaked lesions that develop into fruit cracks and (**b**) leaf lesions as angular spots on the veins

flowering locus from poplar which speeds backcrossing (Srinivasan et al. 2011b). Progress in breeding for sharka resistance is reviewed in detail by Hartmann and Neumüller (2009). No sources of resistance in Japanese plum are known.

Brown rot, caused by the fungus *Monilinia laxa*, is a problem in all regions that experience rainfall during flowering or fruit development and is currently controlled by routine fungicide applications and the removal of infected twigs and fruit. There are no commercial cultivars with complete resistance that do not require fungicides under wet conditions. Crawford (1997) lists 23 European plum cultivars which have some level of resistance and describes 'Stanley', 'Ruth Gerstetter' and 'Ontario' as very resistant. In inoculation tests of injured and uninjured skin, Pascal et al. (1994) identified the *P. salicina* × *P. cerasifera* hybrids 'H11' and 'Pobieda' as having epidermal resistance and the *P. cerasifera* cultivar 'J74' as having flesh resistance.

Resistance to bacterial spot caused by *Xanthomonas campestris* pv. *pruni* (Fig. 15.13) is a selection objective in Japanese plum breeding programmes in summer rainfall areas, including South Africa, south-eastern USA, Brazil and Australia (Okie and Weinberger 1996). Californian cultivars are selected without bacterial spot pressure and are often susceptible (Topp et al. 1989). Resistance is available within commercial germplasm and high levels of resistance are available in the native American species *P. hortunlana*, *P. angustifolia* and *P. texana* (Werner et al. 1986). Resistant cultivars usually contain either *P. angustifolia* or *P. cerasifera* in their ancestry (Topp and Sherman 1990b) and fruit quality is not as high as predominantly *P. salicina* cultivars from California.

Plum leaf scald, caused by the bacteria *Xylella fastidiosa* is the most important disease in Brazil and southeast USA. Resistance is polygenic and predominantly recessive, and major QTLs for susceptibility have been identified to aid in screening progeny (Dalbó et al. 2010). Cultivars with high levels of resistance have not been produced but 'Sanguinea', 'Piamontesa' and 'Chatard' from Argentina are partially resistant (Okie and Weinberger 1996; Dalbó et al. 2010). Field observations suggest that higher levels of resistance are required in warm regions where disease pressure is high.

Bacterial canker caused by the bacteria *Pseudomonas syringae* is widespread in regions with high rainfall and cool weather. Japanese plums are generally more susceptible than European plums.

6 Breeding Methods and Techniques

Pollination methods used by breeders vary depending on the self-compatibility of the germplasm and the intended use of the seedlings. For marker, genetic or cytological studies, where correct identification of the parent is essential, the process involves pollen collection and storage, emasculation, hand pollination and protection of the flowers. For creation of hybrid seedling populations used for cultivar development several less expensive methods are available and commonly used by breeders.

6.1 Controlled Pollination

Pollen is collected from unopened flowers that are at the balloon or popcorn stage of development. In the laboratory, the petals are removed and the flower is gently rubbed over a sieve which separates the anthers from the remaining flower parts. The anthers are then dried overnight at room temperature which allows dehiscence and pollen release. Some breeders collect and dry the anthers in glass petri dishes after which the pollen is transferred by use of a small brush to a container. Other breeders collect the anthers into a zip lock plastic bag which is left open overnight, and then sealed prior to storage in a freezer. The sieves, pollen containers and brushes are cleaned with 70% alcohol between sampling.

If the pollen is to be used immediately, it can be stored at room temperature for 1–2 weeks without problems. We prefer to store all pollen in a freezer at −18°C where it will remain viable for a year and is available for use the following season. This is essential when the desired pollen parent flowers after the proposed seed parent. Some breeders adhere strictly to this regime and will only take to the field the amount of pollen to be used that day. Containers that hold 12 or so pollen applicators are used to carry pollen to the field and are kept out of direct sunlight or in a small insulated cooler.

The calyx, corolla and stamens all arise from the edge of the cup-shaped hypanthium (also called the calyx or floral cup). Cutting the calyx cup below this point of attachment will remove the stamens and leave the pistil exposed for pollination. The cut can be accomplished using a finger nail or one of a variety of emasculation tools (Fig. 15.14). All accomplish the same objective which is to make a complete cut around the calyx cup or to make the initial cut and then tear away the remainder.

Pollen is applied directly onto the stigma of the emasculated flower using a small brush, glass rod, pencil eraser stub or finger. Pistils remain functional for several days after emasculation but studies in peach have indicated that multiple applications

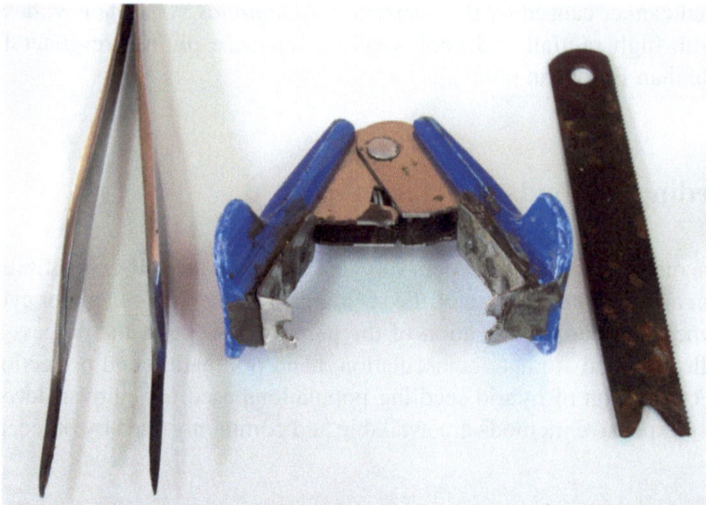

Fig. 15.14 Tools used for emasculation may include (from *left*) forceps, a converted staple remover and a hack-saw blade with a sharpened V-notch

of pollen on successive days are not an efficient way of improving fruit set (El-Agamy and Sherman 1987).

Individual flowers open over 7–14 days in temperate zones and over 2–6 weeks in the subtropics. Open flowers on the branch or tree selected for pollination are removed and then the "balloon stage" flowers are emasculated and pollinated. The branch or tree is then immediately covered with an insect-proof screen to prevent contamination. Emasculation reduces the incidence of insect visitation but covering is required to ensure parental fidelity, particularly when fruit set is low as with interspecific crosses (Arbeloa et al. 2006).

6.2 *Alternative Pollination Methods*

Plum flowers are small compared to peach or apricot and emasculation is time consuming and the pistil may be damaged resulting in reduced fruit set. When genetic purity is not essential it is more efficient to avoid emasculation by using one of the following methods:

- Self-infertile cultivars are used as the seed parent and the whole tree or sections of limbs are enclosed in insect-proof cages. As the flowers inside the enclosure open they are hand pollinated as described above, but without emasculation. Flowers which have recently opened require pollination but are difficult to distinguish from flowers that have opened and already been hand pollinated. In practice, this means that some flowers are pollinated more than once, but the time saved in emasculation outweighs this duplication.

Fig. 15.15 The production of hybrid seed using a small hive of bees inside an insect-proof tent, with two self-infertile cultivars as parents

- Bees are used as the pollinators on self-infertile seed parents that are enclosed in an insect-proof cage with bouquets of the desired polleniser (Fig. 15.15). The bees transfer pollen from dawn to dusk in this ideal system but must be cared for with adequate provision of water. Sugar syrup can be added to the water to encourage the bees to forage for pollen rather than for nectar.
- A polycross system as used by forage breeders may be implemented as an efficient method of combining alleles from several selected genotypes. This involves planting about 5–15 self-infertile seed parents in an area that is isolated from other plums. The trees will cross provided there is no cross-incompatibility and flowering overlaps. This system allows recording of the seed parent but not the specific pollen parent.
- Open pollinated seed are sometimes collected from commercial orchards which have plantings of the desired parents. Alternatively, a new selection can be planted or grafted into an established orchard. This is particularly useful for interspecific hybridisation where low fruit set is expected and a highly efficient pollination technique is required.

Fig. 15.16 A block of Japanese plum seedling progeny planted at high density in double row beds at 0.3 m between trees and 2.4 m between beds

6.3 Seed and Seedling Management

Seed from ripe fruit are collected and labelled with hybrid identification. The stony endocarp is removed and the seed are stratified (provided with moist chilling) in moist perlite at 4°C for from 1 to 3 months, depending on the chilling requirement of the parents. To avoid stratification, the testa and endosperm residues can be removed from the seeds just after harvest and immediately prior to sowing (Theiler 1971). Short fruit development period parents produce seed that is immature and will germinate poorly or not at all using conventional stratification techniques. These seed require embryo culture which involves removing the integuments and growing the embryo on a nutrient media prior to germination (Ramming 1983; Gercheva and Zhivondov 2002). Embryo rescue techniques are particularly important in breeding for early ripening and for developing interspecific hybrids whose embryos may abort due to post-pollination incompatibility (Daorden et al. 2004).

Germinated seed are planted into pots and grown in glasshouses over winter. Field planting occurs in spring about 1 year after pollination. Trees are generally planted at high density of 2,000–12,000 trees per ha (Fig. 15.16), trained to a central leader and encouraged to grow rapidly by careful attention to irrigation, nutrition and weed control. Seedlings generally produce their first crop 3–5 years after planting for Japanese plums and 4–6 years after planting for European plums. Most breeders record notes only on elite selections and remove the inferior ones as soon as possible. The most common reasons for culling include small fruit, low flesh

Table 15.7 Mean number of years (range in parentheses) for the release of European or Japanese plum cultivars and rootstocks by private and public breeders

Breeders	European plums	Japanese plums	Rootstocks
Private	13	10	
Public	19	16	28
Average	*16 (12–22)*	*11 (7–17)*	*28*

Data are based on 26 cultivars released from 7 breeding programmes noted in the HortScience "Register of New Fruit and Nut Variety" lists from 1991 to 2008

SSC, high flesh or skin acidity or bitterness, lack of flavour, lack of juice, soft flesh, unattractive skin colour, irregular shape, skin cracking, light crop, incorrect bloom or ripening time, disease susceptibility or short storage life.

Heterozygosity is high due to the outcrossing and heterogeneity is high due to the multi-species ancestry. This means that there is a great deal of variability in seedling populations. Selecting for 12 or more fruit and tree traits often controlled by many genes, requires large populations to provide reasonable chances of successfully selecting individuals with the desired combination of traits. Some breeders produce up to 15,000 seedlings annually and others only a few hundred. Weinberger (1975) estimated that as many as 5,000 seedlings may be required to produce a good cultivar. Critical aspects in the seedling stage are to reduce juvenility and to minimise environmental variability and cost as in an efficient fruiting nursery (Sherman and Lyrene 1983).

6.4 Advanced Testing and Release

The time from the hybridisation and seed sowing to final release varies greatly across crops and breeding programmes. Generally, breeding for scions is quicker than breeding for rootstocks and Japanese plum breeding is quicker than European plum breeding and private breeders are quicker than public breeders (Table 15.7).

There is a trade-off between the extent of testing and the time to release. Longer testing will sample more seasonal variability and allow more accurate estimation of mean selection performance. However this will extend the breeding cycle and the time to release. Extensive regional testing can be used to provide environmental replication via locations rather than years and thus speed up the process. Advanced testing must involve evaluation by industry and consumers.

Advanced testing can be as simple as propagation of two trees of the advanced selection planted at the same site as the seedlings and evaluated for three crops. In other instances it can involve randomised and replicated testing over multiple sites, with inclusion of industry standards. Plant Breeders Rights regulations require six replications, with inclusion of the closest variety of common knowledge as a control in a statistically-analysed experiment that proves distinctness, uniformity and stability. Naming and release should occur when the candidate is proven to fill the selected gap in the market or outperform its competitors.

7 Integration of New Biotechnologies in Breeding Programmes

7.1 Molecular Markers and Genetic Mapping Studies on Plums

Molecular markers have a wide range of possible applications in plum breeding but they have not been widely used compared to other crops. Practical use of markers depends on basic studies to find markers linked to specific alleles, but few of them have been reported for plums.

Early molecular studies on *Prunus* species were made with dominant markers, such as RAPD and AFLPs. They were used for studies of genetic diversity and genetic relationships among cultivars (Bellini et al. 2002; Boonprakob et al. 2001).

The majority of recent reports are based on microsatellite (SSR) markers developed from other *Prunus* species. The cross-transportability of microsatellite loci among *Prunus* species is usually high. More than 80% of microsatellite primers derived from cherry amplified products in plum (Ahmad et al. 2004); with 98% amplification for peach (Dirlewanger et al. 2002) and 84% for almond (Mneija et al. 2005). The reverse is also true, thus microsatellite primers derived from plum also amplified in peach (85%) and almond (78%) (Mneija et al. 2004). A relatively large number of microsatellite loci are now available from related *Prunus* species, such as peach (Cipriani et al. 1999; Sosinski et al. 2000; Yamamoto et al. 2002), almond (Testolin et al. 2004; Mneija et al. 2005), apricot (Lopes et al. 2002; Hagen et al. 2004; Messina et al. 2004) and cherry (Clarke and Tobutt 2003; Struss et al. 2003; Vaughan and Russell 2004). Most can be used in plum without further cost of marker development.

Aranzana et al. (2003) presented a genetic map of *Prunus* with genome-wide coverage based on a series of microsatellite markers evenly spaced as anchors for each of the eight linkage groups. Other genetic maps generated from related species (Joobeur et al. 2000; Dirlewanger et al. 2004, 2006) can also be used as references for plum mapping projects. Dondini et al. (2006) made a comparison of SSR-based maps created for different *Prunus* species. A substantial co-linearity and synteny was observed, indicating that large genomic rearrangements in *Prunus* are not likely.

Genetic maps of Japanese plums were constructed for a population created from the cross 'Chatard' × 'Santa Rosa' using a pseudo-test cross strategy and based mostly on AFLP markers (Vieira et al. 2005). The map was enhanced with the inclusion of additional microsatellite markers (Moraes 2005). The main objective of the study was to find QTLs for resistance to leaf scald (*X. fastidiosa*), the most important disease in Brazil. 'Chatard' is resistant and is widely used in the breeding programme at Epagri (Santa Catarina State, Brazil), while 'Santa Rosa' is highly susceptible. For an accurate evaluation of this trait it is necessary to grow adult plants in the field for up to 10 years. The possibility of developing markers for early selection of seedlings is therefore very attractive. However, no consistent QTLs were found in this study. The analysis of two other crosses with 'Chatard' were made using microsatellite markers chosen to be evenly spaced in the *Prunus* genome. Some strong QTLs were found but differences among progeny seedlings were associated with allele segregation only from the more susceptible parents, 'Harry Pickstone' and 'Angeleno' (Dalbó et al. 2010). These results indicated that leaf scald resistance is predominantly recessive, at least for this source of resistance. The absence of dominant alleles and the polygenic nature of leaf scald resistance prevented the practical use of marker assisted selection in breeding. Other sources of resistance are now under study.

Another genetic mapping study was carried out with microsatellite markers in an F_1 progeny of a cross between Myrobalan plum (*P. cerasifera*) and an almond–peach hybrid. Two root-knot nematode resistance genes (*Ma* from Myrobalan and R_{MiaNem} from peach) were mapped and specific markers (SCARs) linked to these genes were developed (Dirlewanger et al. 2004).

Microsatellite markers have also proven to be a powerful tool for cultivar identification (Ahmad et al. 2004) and can be used to solve cases of synonyms, misidentifications and patent issues. They have also been used for diversity analysis and genetic relationship studies among *Prunus* species (Rohrer et al. 2004). Most microsatellite loci are useful for these objectives because they are usually polymorphic. They can be chosen from the literature selecting those that have higher polymorphism (number of alleles, observed heterozygosity or discriminating power). Usually 10–15 microsatellite markers, with three or more alleles are enough to distinguish among genotypes.

7.2 Markers for Self-Incompatibility Alleles

PCR-based markers can be used for identification of self-incompatibility alleles and so reduce the labour and time of in-vivo pollination testing. Plums exhibit gametophytic self-incompatibility that is controlled by the highly polymorphic S-locus. Gametophytic incompatibility occurs when the allele in the pollen matches that of the style.

S-RNase alleles were first cloned in Japanese plums by Yamane et al. (1999). Subsequently, 14 S-alleles were identified by PCR in different cultivars and named S^a–S^n (Beppu et al. 2002, 2003, 2004). The primers were designed using sequences of conserved regions flanking the second intron of S-RNase stylar genes. Cross-compatibility among cultivars can be predicted by markers that identify the S-alleles present in each cultivar. Sapir et al. (2004) cloned five additional alleles (S^1, S^3–S^6), with S^1 corresponding to S^a allele and so on. Allele-specific primers were designed, including for the S^e allele, which was associated with self-compatibility in a range of Japanese plums tested by Beppu et al. (2005). Self-compatible cultivars have a horticultural advantage because no cross-pollination is required. Breeders can use a specific marker for the S^e allele to select seedlings for self-compatibility prior to field planting (Beppu et al. 2010). The genetic control of self-compatibility appears to be more complex and involve more than only the S^e allele. Not all self-compatible cultivars contain the S^e allele (Dalbó personal communication) and some, but not all, cultivars with the S^b allele are self-compatible (Beppu et al. 2010).

European plums have a similar self-incompatibility system, but the analysis of S-alleles is more complex due to the hexaploid nature of this species. Theoretically, up to six S-alleles can be present in one genotype which can make the PCR test less reliable. Two to six S-alleles per genotype were observed in a study with 19 cultivars (Sutherland et al. 2004a, b).

7.3 Genetic Transformation

Genetic transformation is a powerful technology for the genetic improvement of plants. In the case of fruit trees, it can overcome some limitations of conventional breeding, such as long generation cycle, long juvenile period, high heterozygosity and shortage of genetic variability in the gene pool.

Fruit trees, and particularly the *Prunus* species, are among the most recalcitrant plants for regeneration of adventitious shoots. This feature seriously limits the development of gene transfer technology. Plum has been among the most successful of the *Prunus* species to regenerate and transform (Petri et al. 2009b). A successful and reproducible system for plant regeneration in the *Prunus* genus is the use of hypocotyl segments of *P. domestica* described by Mante et al. (1991) and improved by Gonzalez-Padilla et al. (2003). TDZ and IBA are used to regenerate shoots following *Agrobacterium*-mediated transformation of hypocotyl segments. The protocol has recently been improved with the addition of 2,4D during co-cultivation allowing transformation efficiencies up to 42% and enabling the production of self-rooted transgenic plants after 6 months (Petri et al. 2008).

In the case of *P. salicina*, Tian et al. (2007) reported regeneration from hypocotyl segments as previously described in *P. domestica,* using IBA combined with various concentrations of TDZ, benzylaminopurine (BAP), or kinetin. Shoots were induced from hypocotyl segments of the *P. salicina* cultivars 'Shiro', 'Early Golden', and 'Redheart' and regenerated plants were established in a greenhouse. Urtubia et al. (2008) also reported an *Agrobacterium*-mediated transformation from hypocotyl slices of the *P. salicina* cultivars, 'Angeleno' and 'Larry Anne'. The protocol also included the use of TDZ and IBA and the regeneration rates reached 15% for 'Angeleno' and 6% for 'Larry Anne'.

The possibility of incorporating genes outside the plum gene pool is the most exciting application of gene transfer. Virus resistance is a classic example. The devastating impact of Sharka virus in Europe and the possibility of it spreading to other countries motivated efforts to develop resistant material by heterologous gene expression. This technology was successfully used in *P. domestica* to develop genetically engineered PPV (plum pox virus) resistant materials (Malinowski et al. 2006; Scorza et al. 2007; Câmara Machado et al. 2007; Damiano et al. 2007; Wang et al. 2009). Extensive work in this area was conducted at USDA, Kearneysville, WV. The PPV-CP (coat protein) gene was isolated, sequenced and cloned (Ravelonandro et al. 1992) and used for *Agrobacterium*-mediated transformation of plum. One clone ('C5') was highly resistant after 2 years of greenhouse tests. This clone was patented as 'Honeysweet' and submitted for field testing, under high disease pressure in Poland, Romania and Spain where it exhibited a high level of resistance (Malinowski et al. 2006). 'Honeysweet' was also tested in the USA to evaluate its horticultural performance as well as to initiate risk assessment studies. 'Honeysweet' was deregulated in the USA by the Animal and Plant Health Inspection Service in 2006 (Scorza et al. 2007) and approved by the US Food and Drug Administration (Petri et al. 2009a).

'Honeysweet' originated from a transformation with a sense PPV-CP construct but it was the only resistant clone obtained from this construct. Later, it was found that it contained a duplicated and rearranged transgene insert forming a hairpin-RNA structure that probably increased its effectiveness in RNA silencing of the PPV gene (Scorza et al. 2001). The PPV-CP hairpin structure from 'Honeysweet' was cloned and used to produce transgenic lines with high levels of resistance (Scorza et al. 2001). Another approach to PPV resistance by genetic transformation is the insertion of constructs with self-complementary sequences separated by an intron producing an intron hairpin-RNA (ihpRNA). Transgenic plum lines were obtained with this kind of construct (Hily et al. 2005) and some clones are under evaluation (Petri et al. 2009a).

Genetic engineering has also been conducted to improve fruit quality by manipulating genes controlling ethylene synthesis to delay fruit softening. Plum hypocotyls were transformed with an antisense construct of a peach ACC oxidase (ACCO) gene under control of CaMV35S promoter (Gonzalez-Padilla et al. 2003). Some clones derived from this project had delayed ethylene production and softening (Callahan and Scorza 2007).

Another important application of gene transfer technology is the development of disease resistant cultivars. An important initiative in this field is the transformation of *P. domestica* with the Gastrodia anti-fungal protein (GAFP) (Nagel et al. 2008). GAFP is a monocot mannose-binding lectin with anti-fungal action that was isolated from the Asiatic orchid, *Gastrodia elata*. *Agrobacterium*-mediated transformation with the GAFP-1 gene resulted in plum clones with increased tolerance to *Phytophtora* root rot, caused by *P. cinnamomi*, and to the root-knot nematode, *Meloidogyne incognita*. Long-term field experiments will be necessary to confirm these findings in an applied context (Petri et al. 2009a).

Research continues to improve the transformation protocols and increase the frequency of adventitious shoot formation. Srinivasan et al. (2011a) reported plum transformation with KNOX genes, a family of genes involved in the control of meristem formation. A high frequency of adventitious shoot regeneration (96%) was observed in cultures of leaf explants excised from corn KNOX1-expressing transgenic plum shoots. In contrast to tobacco, leaf and internode explants of corn KNOX1-expressing plum required synthetic cytokinin (thidiazuron) in the culture medium to regenerate adventitious shoots.

Transformation of *P. domestica* with the poplar flowering locus T1 (PtFT1) gene induced early flowering from transgenic plantlets within 2 months of transformation (Srinivasan et al. 2011b). The plants changed from upright and deciduous to bushy and capable of continual fruit bearing without the need for chilling. This regulation of flowering enables a breeding system where the limitations of juvenile period are overcome and makes possible a generation time of 1 year instead of the conventional 3–6 years. One possible application, documented in the Srinivasan et al. (2011b) patent application, is the introgression of desirable traits from wild or low quality genotypes into elite materials. Using the transgenic plums, 4–5 backcross generations may be completed in 4–5 years. In the final backcross, the desired phenotype is selected, moving back to the non-transgenic form. Srinivasan et al. (2011b) also suggest that the transgenic plants may be used in greenhouse conditions to produce a continuous supply of fruit, since a dormancy period is not required and the size of the modified plants are dramatically reduced.

Most transformation studies are conducted using hypocotyls instead of somatic tissues. This means that the resulting genotype is not the same of the original. This limits this technique, since commercial production is mainly based on well tested and sometimes traditional cultivars. There are a few reports of regeneration from clonal explants of *P. domestica* (Yancheva et al. 2002; Mikhailov and Dolgov 2007) but with low transformation rates. Petri et al. (2009a) report more promising results with regeneration rates up to 50% from 'Improved French' leaves.

The progress already made in the transformation of plums will also contribute to the development of functional genomics in the *Prunus* genus. The peach genome is available, but the absence of an efficient transformation system in this species is a limitation for gene function studies (Petri et al. 2009a). Since plum has the most efficient transformation system among the *Prunus*, it offers a solution to the bottleneck in functional genomic research in this genus. Progress in the development of transformation protocols for the diploid *P. salicina*, would be particularly helpful for improvement of this species through the development of genetically transformed cultivars and additionally to transform plums as a platform for functional genomics studies for all *Prunus* and Rosaceous species. Classical breeding techniques will remain the methods of choice in most breeding programmes during the next decades, but the opportunity to combine transgenic and naturally occurring resistance should be explored.

Acknowledgements The authors gratefully acknowledge the contributions of Tom Beckman, Kenji Beppu, Unaroj Boonprakob, Glen Bradford, Peter Buchanan, David Byrne, Jose Chaparro, Ivan Day, Kesi Kesavan, Richard Haas, Nik Hulse, Alessandro Liverani, Jean Clement Marcaillou, Debby Maxfield, Chris Menzel, Valter Nencetti, Antonino Nicotra, Dick Okie, David Ramming, Maria do Carmo Bassols Raseira, Wayne Sherman, Chris Smith, Jorge Soria, Ien-chie Wen and Masami Yamaguchi. DEEDI employed BL Topp during part of this chapter preparation.

References

Abdi, N., Holford, P., McGlasson, W.B., and Mizrahi, Y. 1997. Ripening behaviour and responses to propylene in four cultivars of Japanese type plums. Postharvest Biol. Tech. 12:21–34.

Ahmad, R., Potter, D., and Southwick, S.M. 2004. Identification and characterization of plum and pluot cultivars by microsatellite markers. J. Hort. Sci. Biotech. 79:164–169.

Andersen, E.T., and Weir, T.S. 1967. *Prunus* hybrids, selections and cultivars, at the University of Minnesota fruit breeding farm. Univ. Minn. Agric. Exp. Sta. Tech. Bull. 252.

Andersen, R.L., Freer, J., and Watson, J. 2006a. New plum Jam Session™. New York Fruit Quart. 14:24.

Andersen, R.L., Freer, J., and Watson, J. 2006b. New plum Blues Jam™. New York Fruit Quart. 14:26.

Aranzana, M.J. Pineda, A., Cosson, P., Dirlewanger, E., Ascasiber, J., Cipriano, G., Ryder, C.D., Testolin, R., Abbott, A., King, G.J., Iezzoni, A.F., Arus, P. 2003. A set of simple-sequence repeat (SSR) markers covering the *Prunus* Genome. Theor.Appl. Gen., 108:819–825.

Arbeloa, A., Daorden, M.E., Garcia, E., Wunsch, A., Hormaza, J.I., and Marin, J.A. 2006. Significant effect of accidental pollinations on the progeny of low setting *Prunus* interspecific crosses. Euphytica 147:389–394.

Arjmandi, B.H., Khalil, D.A., Lucas, E.A., Georgis, A., Stoecker, B.J., Hardin, C., Payton, M.E., and Wild, R.A. 2002. Dried plums improve indices of bone formation in postmenopausal women. J. Wom. Health Gender-Based Med. 11:61–68.

Ashton, R.W. 2008. Plums of North America. Third Millennium Publishing, Temple, Arizona.

Atanasoff, D. 1935. Mosaic of stone fruits. Phytopathologische Zeitschrift 8, 259–284.

Bailey, J.S. 1944. The beach plum in Massachusetts. Mass. Agric. Exp. Sta. Bull. 422.

Bailey, L.H. 1892. The cultivated native plums and cherries. Corn. Univ. Agric. Exp. Sta. Bull. 38.

Beckman, T.G. and Lang, G.A. 2003. Rootstock breeding for stone fruits. Acta Hort. 622:531–550.

Beckman, T.G., Chaparro, J.X. and Sherman, W.B. 2008. 'Sharpe', a clonal plum rootstock for peach. HortScience 43:2236–2237.

Bellini, E., and Nencetti, V. 2002a. "Dofi-Giudy": A new early red Japanese plum. Acta Hort. 577:221–222.

Bellini, E., and Nencetti, V. 2002b. "Dofi-Sandra": A new early black Japanese plum. Acta Hort. 577:223–224.

Bellini, E., Nencetti, V., and Nin, S. 2002. Genetic improvement of plum in Florence. Acta Hort. 577:19–24.

Bellini, E., Nencetti, V., Nin, S., and Paraluppi, S. 1998. Ripening time within a cross-derived population of Japanese plum. Acta Hort. 478:61–66.

Beppu, K., Yamane, H., Yaegaki, H., Yamaguchi, M., Kataoka, I., Tao, R. 2002. Diversity of S-RNase genes and S-haplotypes in Japanese plum (*Prunus salicina* Lindl). J. Hort. Sci. Biotech., 77:658–664.

Beppu, K., Takemoto, Y., Yamane, H., Yaegaki, H., Yamaguchi, M., Kataoka, I., Tao, R. 2003. Determination of S-haplotypes of Japanese plum (*Prunus salicina* Lindl.) cultivars by cross-pollination tests. J. Hort. Sci. Biotech., 78:315–318.

Beppu, K., Yamane, H., Yaegaki, H., Yamaguchi, M., Tao, R. and Kataoka, I. 2004. Analysis of S-RNase genes in self-compatible cultivars of Japanese plum, 'Methley','Karari' and 'Kosyu'. J. Japan. Soc. Hort. Sci. 73(Suppl. 2):253.(In Japanese).

Beppu, K., Komatsu, N., Yamane, H., Yaegaki, H., Yamaguchi, M., Tao, R., and Kataoka, I. 2005. S-e-haplotype confers self-compatibility in Japanese plum (*Prunus salicina* Lindl.). J. Hort. Sci. Biotech. 80:760–764.

Beppu, K., Syogase, K., Yamane, H., Tao, R., and Kataoka, I. 2010. Inheritance of self-compatibility conferred by the Se-haplotype of Japanese plum and development of Se-RNase gene-specific PCR primers. J. Hort. Sci. Biotech. 85:215–218.

Blazek, J. 2007. A survey of the genetic resources used in plum breeding. Acta Hort. 734:31–45.

Boonprakob, U., Byrne, D.H., Graham, C.J., Okie, W.R., Beckman, T., and Smith, B.R. 2001. Genetic relationships among cultivated diploid plums and their progenitors as determined by RAPD markers. J. Amer. Soc. Hort. Sci. 126:451–461.

Boonprakob, U., and Byrne, D.H. 2003. Species composition of Japanese plum founding clones as revealed by RAPD markers. Acta Hort. 622:473–476.

Botu, M. 1998. Inheritance of some characteristics to the offspring and evaluation of the genitors' value for plum. Acta Hort. 478:155–162.

Brooks, R.M., and Olmo, H.P. 1997. Register of Fruit & Nut Varieties. 3rd ed. ASHS Press, Alexandria, Virginia.

Butac, M., and Budan, S. 2009. Evaluation of local plum varieties (*Prunus domestica* L.) from the Romanian national collection. Acta Hort. 814:91–94.

Buttner, R. 2001. *Prunus*, p. 513–525, In P. Hanelt, ed. Mansfields Encyclopedia of Agricultural and Horticultural Crops. Inst. Plant Genet. Crop Plant Res.

Byrne, D.H. 1989. Inbreeding, coancestry, and founding clones of Japanese-type plums of California and the southeastern United States. J. Amer. Soc. Hort. Sci. 114:699–705.

Byrne, D.H. 1990. Isozyme variability in four diploid stone fruits compared with other woody perennial plants. J. Hered. 81:68–71.

Byrne, D.H., Noratto, G., Cisneros-Zevallos, L., Porter, W., and Vizzotto, M. 2009. Health benefits of peach, nectarine and plums. Acta Hort. 841:267–274.

Callahan, A. and Scorza, R. 2007. Effects of a peach antisense ACC oxidase gene on plum fruit quality. Acta Hort. 738:567–573.

Câmara Machado, A., Katinger, H. Câmara Machado, M.L. 2007. Coat protein-mediated protection against plum pox virus in herbaceous model plants and transformation of apricot and plum. Euphytica 77:129–134.

Cevallos-Casals, B.A., Byrne, D., Okie, W.R., and Cisneros-Zevallos, L. 2006. Selecting new peach and plum genotypes rich in phenolic compounds and enhanced functional properties. Food Chem. 96:273–280.

Cipriani, G., Lot, G., Huang, W-G., Marrazzo, M.T., Peterlunger, E., Testolin, R. 1999. AC/GT and AG/CT microsatellite repeats in peach (*Prunus persica* (L.) Batsch): isolation, characterization and cross-species amplification in *Prunus*. Theor. Appl. Gen. 99:65–72.

Clarke, J.B., Tobutt, K.R. 2003. Development and characterization of polymorphic microsatellites from *Prunus avium* 'Napoleon'. Mol. Ecol. Notes 3:578–580.

Crane, M.B., and Lawrence, W.J.C. 1934. The genetics of garden plants. MacMillan and Co., Ltd., London, UK.

Crawford, M. 1997. Fruit Varieties Resistant to Pests and Diseases. Agroforestry Res. Trust, Devon, UK.

Crisosto, C.H., Garner, D., Crisosto, G.M., and Bowerman, E. 2004. Increasing 'Blackamber' plum (*Prunus salicina* Lindell) consumer acceptance. Postharvest Biol. Technol. 34:237–244.

Crisosto, C.H., Crisosto, G.M., Echeverria, G., and Puy, J. 2007. Segregation of plum and pluot cultivars according to their organoleptic characteristics. Postharvest Biol. Technol. 44:271–276.

Crow, J.F. 2001. Plant breeding giants: Burbank, the artist; Vavilov, the scientist. Genetics 158:1391–1395.

CTFA. 2009. California Tree Fruit Agreement, Annual Report [Online] http://www.eatcalifornia-fruit.com/.

Cullinan, F.P. 1937. Improvement of stone fruits., p. 665–748 U.S. Dept. Agric. Yearbook of Agriculture. U.S. Govt. Printing Office, Washington.

Cummins, J.N. and Aldwinckle, H.S. 1983. Rootstock breeding. p. 294–327. In J.N. Moore and J. Janick, eds. Methods In Fruit Breeding. Purdue Univ. Press, West Lafayette, Indiana.

Dalbó., M.A., Klabunde, G.H.F., Nodari, R.O., Fernandes, D., Basso, M.F., 2010. Evolution of the response of segregating populations of plums and the association with microsatellite markers of leaf scald. Crop Breed. Appl. Biotech. 10:337–344.

Damiano, C., Gentile, A., Monticelli, S., Scorza, R., Kondakova, V., Todorovska, E., Kamenova, I., Badjiakov, I., Atanassov, A. 2007. Improving regeneration and transformation for resistance to Sharka in *Prunus*. Acta Hort. 738:583–587.

Daorden, M.E., Marin, J.A., and Arbeloa, A. 2004. Stratification temperature affects the in vitro germination of immature *Prunus* embryos. Acta Hort.658:135–140.

Day, L.H. 1953. Rootstocks for Stone Fruits. Calif. Agric. Exp. Sta. Ext. Bull. 736.

DeBuse, C., Shaw, D.V., and DeJong, T. 2007. Heritabilities of seedling traits in a *Prunus domestica* (L.) breeding population. Acta Hort. 734:63–67.

DeJong, T.M., Doyle, J.F. and DeBuse, C.J. 2002. Development of a prune breeding program in California. Acta Hort. 577:151–153.

Dikeman, C.L., Bauer, L.L., and Fahey, G.C. 2004. Carbohydrate composition of selected plum/ prune preparations. J. Agric. Food Chem. 52:853–859.

Dirlewanger, E., Cosson, P., Tavaud, M., Aranzana, M.J., Poizat, C., Zanetto, A., Arus, P., Laigret, F. 2002. Development of microsatellite markers in peach [*Prunus persica* (L.) Batsch] and their use in genetic diversity analysis in peach and sweet cherry (*Prunus avium* L.). Theor.Appl. Gen., 105:127–138.

Dirlewanger, E., Cosson, P., Howad, W., Capdeville, G., Bosselut, N., Claverie, M., Voisin, R., Poizat, C., Lafargue, B., Baron, O., Laigret, F., Kleinhentz, M., Arús, P., Esmejaud, D. 2004. Microssatelite genetic linkage maps of mirobalan plum and almond-peach hybrid - Location of root-knot nematode resistance genes. Theor. Appl. Genet. 109:827–838.

Dirlewanger, E., Cosson, P., Boudehri, K., Renaud, C., Capdeville, G., Tauzin, Y., Laigret, F., Moing A., 2006. Development of a second-generation genetic linkage map for peach [*Prunus persica* (L.) Batsch] and characterization of morphological traits affecting flower and fruit. Tree Genet. Genom. 3:1–3.

Dondini, L., Lain, O., Geuna, F., Banfi, R., Gaiotti, F., Tartarini, S., Bassi, D., Testolin, R. 2006. Development of a new SSR-based linkage map in apricot and analysis of synteny with existing *Prunus* maps. Tree Genet. Genom. 3:239–249.

Dorsey, M.J., and Bushnell, J. 1925. Plum investigations II. The inheritance of hardiness. Univ. Minn. Agric. Exp. Sta. Tech. Bull. 32.

Ducroquet, J.P., Dalbó, M.A. 2007. SCS 409 Camila e SCS 410 Piuna - Novas cultivares de ameixeira com resistência à escaldadura das folhas. Agropecuária Catarinense 20:67–70.

El-Agamy, S.Z., and Sherman, W.B. 1987. Sequence of pollination in relation to fruit set and progeny produced in peach (*Prunus persica* L. Batsch). J. Hort. Sci. 62:469–473.

Ercisli, S. 2004. A short review of the fruit germplasm resources of Turkey. Genetic Resources and Crop Evolution 51:419–435.

Erturk, Y., Ercisli, S., Maghradze, D., Orhan, E., and Agar, G. 2009. An assessment of genetic variability and relationships among wild-grown blackthorn (*Prunus spinosa* L.) plants based on RAPD markers. Genet. Mol. Res. 8:1238–1244.

FAO. 2006. Food and Agriculture Organisation of the United Nations - Crop Production [Online] http://faostat.fao.org.

Faust, M., and Surányi, D. 1999. Origin and dissemination of plums. Hort. Rev. 23:179–231.

Fogle, H.W. 1978. Plum improvement in the United States. Acta Hort. 74:35–40.

Gercheva, P., and Zhivondov, A. 2002. Embryo rescue of early ripening plum cultivars. Acta Hort. 577:165–168.

Gil, M.I., Tomas-Barberan, F.A., Hess-Pierce, B., and Kader, A.A. 2002. Antioxidant capacities, phenolic compounds, carotenoids, and Vitamin C contents of nectarine, peach and plum cultivars from California. J. Agric. Food Chem. 50:4976–4982.

Gomez, E., and Ledbetter, C. 1993. Transmission of Biochemical Flavor Constituents from Apricot and Plum to Their Interspecific Hybrid. Plant Breeding 111:236–241.

Gomez, E., and Ledbetter, C.A. 1994. Comparative-Study of the Aromatic Profiles of 2 Different Plum Species – *Prunus salicina* Lindl and *Prunus simonii* L. J. Sci. Food Agric. 65: 111–115.

Gomez, E., and Ledbetter, C.A. 1997. Development of volatile compounds during fruit maturation: Characterization of apricot and plum x apricot hybrids. J. Sci. Food Agric. 74:541–546.

Gomez-Plaza, E., and Ledbetter, C. 2010. The flavor of plums. p.415–430. In: Y.H. Hui (ed.) Handbook of fruit and vegetable flavors. John Wiley & Sons Inc.

Gomez, E., Ledbetter, C.A., and Hartsell, P.L. 1993. Volatile Compounds in Apricot, Plum, and Their Interspecific Hybrids. Journal of Agricultural and Food Chemistry 41:1669–1676.

Gonzalez-Padilla,I.M., Webb,K., Scorza, R. 2003. Early antibiotic selection and efficient rooting and acclimatization improve the production of transgenic plum plants (*Prunus domestica* L.).Plant Cell Reports 22:38–45.

Hagen, L.S., Chaib, J., Fady, B., Decroocq, V., Bouchet, J.P., Lambert, P., Audergon, J.M. 2004. Genomic and cDNA microsatellite from apricot (*Prunus armeniaca* L.). Mol. Ecol. Notes 4:742–745.

Hansche, P.E., Hesse, C.O., and Beres, V. 1975. Inheritance of Fruit Size, Soluble Solids, and Ripening Date in *Prunus domestica* Cv Agen. J. Amer. Soc. Hort. Sci. 100:522–524.

Hartmann, W., and Neumüller, M. 2009. Plum breeding, p. 1–71 Breeding Plantation Tree Crops: Temperate Species. Springer, New York.

Hedrick, U.P. 1911. The plums of New York N.Y. Agric. Exp, Sta. 18th Ann. Rep. Vol 3, Part II, Geneva.

Hily, J.-M., Scorza, R., Webb, K., Ravelonandro, M. 2005. Accumulation of the long class of siRNA is associated with resistance to Plum pox virus in a transgenic woody perennial plum tree. Mol Plant-Microbe Interact. 18:794–799.

Howard, W.L. 1945. Luther Burbank's plant contributions. Cal. Agric. Exp. Sta. Bull. 691.

Hunter, A.W.S., and Bragdo, M. 1958. The cytology of three hybrids between diploid and hexaploid plums (*Prunus* spp.). Proc. 10th Int. Congr. Genet., Montreal 2:128. Abstract from Knight, R.L. 1969. Abstract Bibliography of Fruit Breeding and Genetics to 1965 *Prunus*. CAB, London.

Jacob, H.B. 2007a. Experience with new mirabelle cultivars from Geisenheim: 'Bellamira'(R) and 'Miragrande'(R) as fruit for the fresh market and for distillation. Acta Hort. 734:347–351.

Jacob, H.B. 2007b. Twenty-five years plum breeding in Geisenheim, Germany: breeding targets and previous realisations. Acta Hort. 734:341–346.

Janick, J., and Paull, R.E. 2008. The Encyclopedia of Fruit & Nuts CABI, Wallingford UK.

Joobeur, T., Periam, N., de Vicente, M.V., King, G.J., Arus, P. 2000. Development of a second generation linkage map for almond using RAPD and SSR markers. Genome, 43:649–655.

Jun, J.H., and Chung, K.H. 2007. Interspecific cross compatibility among plum, apricot and plumcot. Korean J. Hort. Sci. Tech. 25:217–222.

Jun, J.H., Chung, K.H., Kang, S.J., Kwack, Y.B., Park, K.S., Yun, H.K., and Jeong, S.B. 2008. 'Honey Red', an early maturing Japanese plum. J. Amer. Pom. Soc. 62:27–29.

Jun, J.H., Kwon, J.H., and Chung, K.H. 2009. Morphological Characteristics of Interspecific Hybrids between Japanese Plum (*Prunus salicina* Lindl.) cv. Soldam and Apricot (*P. armeniaca* L.) cv. Harcot. Korean J. Hort. Sci. Tech. 27:269–274.

Kataoka, I., Sugiura, A., and Tomana, T. 1988. Interspecific Hybridization between *Microcerasus* and Other *Prunus* Spp. J. Jap. Soc. Hort. Sci. 56:398–407 (English summary).

Katayama, H., and Uematsu, C. 2005. Structural analysis of chloroplast DNA in *Prunus* (Rosaceae): evolution, genetic diversity and unequal mutations. Theor. App. Genet. 111:1430–1439.

Kaufmane, E., Ikase, L., Trajkovski, V., and Lacis, G. 2002. Evaluation and characterization of plum genetic resources in Sweden and Latvia. Acta Hort. 577:207–213.

Khoshbakht, K., and Hammer, K. 2005. Savadkouh (Iran) - an evolutionary centre for fruit trees and shrubs. Genet. Resources Crop Evol. 00:1–11.

Kim, D.O., Jeong, S.W., and Lee, C.Y. 2003. Antioxidant capacity of phenolic phytochemicals from various cultivars of plums. Food Chem. 81:321–326.

Kim, H.J., Yu, M.H., and Lee, I.S. 2008. Inhibitory effects of methanol extract of plum (*Prunus salicina* L., cv. 'Soldam') fruits against benzo(alpha)pyrene-induced toxicity in mice. Food Chem. Tox. 46:3407–3413.

La Rue, J.H., and Johnson, R.S. 1989. Peaches, Plums and Nectarines - Growing and Handling for Fresh Market. Univ. of Cal., Oakland, California.

Lea, M.A., Ibeh, C., desBordes, C., Vizzotto, M., Cisneros-Zevallos, L., Byrne, D.H., Okie, W.R. and Moyer, M.P. 2008. Inhibition of growth and induction of differentiation of colon cancer cells by peach and plum phenolic compounds. Anticancer Res. 28:2067–2076.

Ledbetter, C.A., Peterson, S.J., and Burgos, L. 1994. Variability of horticultural characteristics among plumcot progenies. J. Genet. Breed. 48:117–123.

Lespinasse, Y., Durel, C.E., Eskes, A., Esmenjaud, D., and Poessel, J.L. 2003. Resistance to biotic stress in fruit trees. Acta Hort. 622:303–315.

Lin, P., and Shi, L. 1989. The discovery and distribution of Ili wild *Prunus domestika* (*P. communis* Fritsch) in Xinjiang, p. 282–286 International Symposium on Horticultural Germplasm, Cultivated and Wild. Part 1. Fruit Trees. Internat. Acad. Pub. Beijing, China.

Liu, W. 2007. Plum production in China. Acta Hort. 734:89–92.

Liu, W.S., Liu, D.C., Feng, C.J., Zhang, A.M., and Li, S.H. 2006. Genetic diversity and phylogenetic relationships in plum germplasm resources revealed by RAPD markers. J. Hort. Sci. Biotech. 81:242–250.

Liu, W.S., Liu, D.C., Zhang, A.M., Feng, C.J., Yang, J.M., Yoon, J.H., and Li, S.H. 2007a. Genetic diversity and phylogenetic relationships among plum germplasm resources in China assessed with inter-simple sequence repeat markers. J. Amer. Soc. Hort. Sci. 132:619–628.

Liu, W., Chen, X.S., Liu, G.J., Liang, Q., He, T.M., and Feng, J.R. 2007b. Interspecific hybridization of *Prunus persica* with *P. armeniaca* and *P. salicina* using embryo rescue. Plant Cell Tissue and Organ Culture 88:289–299.

Lopes, M.S., Sefc, K.M., Laimer, M., Camara Machado, A. 2002. Identification of microsatellite loci in apricot. Molecular Ecology Notes, 2:24–26.

Lozano, M., Vidal-Aragon, M.C., Hernandez, M.T., Ayuso, M.C., Bernalte, M.J., Garcia, J., and Velardo, B. 2009. Physicochemical and nutritional properties and volatile constituents of six Japanese plum (*Prunus salicina* Lindl.) cultivars. European Food Res. Tech. 228:403-410.

Malinowski, T., Cambra, M., Capote, N., Zawadzka, B., Gorris, M.T., Scorza, R., Ravelonandro, M. 2006. Field trials of plum clones transformed with the Plum pox virus coat protein (PPV-CP) gene. Plant Dis., 90:1012–1018.

Mante, S., Morgens, P. H., Scorza, R., Cordts, J. M., Callahan, A. M. 1991. *Agrobacterium*-mediated transformation of plum (*Prunus domestica* L.) hypocotyl slices and regeneration of transgenic plants. BioTech. 9:853–857.

Messina, R., Lain, O., Marrazzo, M.T., Cipriani, G., Testolin, R. 2004. New set of microsatellite loci isolated in apricot. Mol. Ecol. Notes 4:432–434.

Mikhailov, R.V. and Dolgov, S.V., 2007. Transgenic plum (*Prunus domestica* L.) plants obtained by *Agrobacterium*-mediated transformation of leaf explants with use of various selective agents. Acta Hort. 738:613–623.

Minev, I., and Balev, M. 2002. Interspecific hybrids of the *Prunus* genus bred at RIMSA, Troyan, Acta Hort. 577:195–198.

Mneija, M., Garcia-Mas, J., Howad, W., Badenes, M.L., Arús, P. 2004. Simple-sequence repeat (SSR) markers of Japanese plum (*Prunus salicina*) are highly polymorphic and transferable to peach and almond. Mol. Ecol. Notes 4:163–166.

Mneija, M., Garcia-Mas, J., Howad, W., Arús, P. 2005. Development and transportability across *Prunus* species of 42 polymorphic almond microsatellites. Mol. Ecol. Notes 5:531–535.

Moraes, L.K.A., 2005. Avanços no mapeamento genético da resistência à escaldadura das folhas em ameixeira. MSc. Thesis, UFSC, 96 p.

Moreno, M.A., Tabuenca, M.C. and Cambra, R. 1995. Adesto 101, a plum rootstock for peaches and other stone fruit. HortScience 30:1314–1315.

Moreno, M.A. 2004. Breeding and selection of *Prunus* rootstocks at the Aula Dei Experimental Station, Zaragoza, Spain. Acta Hort. 658:519–528.

Nagel, A.K., Schnabel, G., Petri, C., Scorza, R. 2008. Generation and characterization of transgenic plum lines expressing the Gastrodia antifungal protein. HortScience, 43:1514–1521.

Nakasu, B.H., Bassols, M., and Feliciano, A.J. 1981. Temperate fruit breeding in Brazil. Fruit Var. J. 35:114–122. NC-140. 2002.

NC-140, 2002. NC-140 Regional Rootstock Research Project. [Online] http://www.nc140.org/.

Neumüller, M., Lanzl, S., Hartmann, W., Feucht, W., and Treutter, D. 2009. Towards an understanding of the inheritance of hypersensitivity resistance against the sharka virus in European plum (*Prunus domestica* L.): generation of interspecific hybrids with lower ploidy levels. Acta Hort. 814:721–726.

Nicotra, A. and Moser, L. 1997. Two new plum rootstocks for peach and nectarines: Penta and Tetra. Acta Hort. 451:269–271.

Nunes, C., Coimbra, M.A., Saraiva, J., and Rocha, S.M. 2008. Study of the volatile components of a candied plum and estimation of their contribution to the aroma. Food Chem. 111:897–905.

Nunez de Gonzalez, M.T., Hafley, B.S., Boleman, R.M., Miller, R.K., Rhee, K.S., and Keeton, J.T. 2008. Antioxidant properties of plum concentrates and powder in precooked roast beef to reduce lipid oxidation. Meat Sci. 80:997–1004.

Okie, W.R. 1987. Plum rootstocks, p. 321–360, In R. C. Rom and R. F. Carlson, eds. Rootstocks for fruit crops. Wiley, New York.
Okie, W.R. 2001. Plum crazy: Rediscovering our lost *Prunus* resources. HortScience 36:209–213.
Okie, W.R. 2006. Introgression of *Prunus* species in plum. New York Fruit Quart. 14:29–37.
Okie, W.R., and Hancock, J.F. 2008. Plums, pp. 337–358 Temperate Fruit Crop Breeding. Springer, New York.
Okie, W.R., and Ramming, D.W. 1999. Plum breeding worldwide. HortTechnology 9:162–176.
Okie, W.R., and Weinberger, J.H. 1996. Plums, p. 559–607, In J. Janick and J. N. Moore, eds. Fruit Breeding, Volume I: Tree and Tropical Fruits. John Wiley & Sons, Inc., New York.
Olden, E.J. 1965. Interspecific plum crosses. Res. Rep. Balsgard Fruit Breed. Inst. No. 1. 58pp. Abstract from Knight, R.L. 1969. Abstract Bibliography of Fruit Breeding and Genetics to 1965 *Prunus*. CAB, London
Ortiz, A., Renaud, R., Calzada, I., and Ritter, E. 1997. Analysis of plum cultivars with RAPD markers. J. Hort. Sci. 72:1–9.
Pascal, T., Levigneron, A., Kervella, J., and Nguyen-The, C. 1994. Evaluation of two screening methods for resistance of apricot, plum and peach to *Monilinia laxa*. Euphytica 77:19–23.
Paunovic, S.A. and Misic, P.D. 1975. The study of inheritance in the plum progenies. Acta Hort. 48:91–109.
Petri, C., Webb, K., Hily, J.-M., Dardick, C., Scorza, R. 2008. High transformation efficiency in plum (*Prunus domestica* L.): a new tool for functional genomics studies in *Prunus* spp.. Mol. Breed. 22:581–591.
Petri, C., Scorza, R., Dardick, C. 2009a. Genetic engineering of plum (*Prunus domestica* L.) for plant improvement and genomics research in Rosaceae. p.277–290, In: K.M. Folta and S.E. Gardiner (eds.) Genetics and genomics of Rosaceae, Plant genetics and genomics : crops and models Vol. 6. Springer, New York.
Petri, C., Webb, K., Dardick, C., Scorza, R. 2009b. A high-throughput transformation system in plum (*Prunus domestica* L.) useful for functional genomics in Rosaceae. Acta Hort. 839:375–379.
Petrovic, R., Miletic, R., and Mitrovic, M. 2002. The results of the study on autochthonous plum cultivars in Eastern Serbia. Acta Hort. 577:239–243.
Prunus Crop Germplasm Committee. 2010. *Prunus* vulnerability statement USDA - ARS National Plant Germplasm System [Online]. Available by http://www.ars-grin.gov/npgs/cgc_reports/prun2010.doc.
Qiao, Y.S., Fang, J.G., Cong, Y., Zhou, J., and Zhang, Z. 2007. Analysis of genetic diversity of Japanese plum cultivars based on RAPD, ISSR and SSR markers. Acta Hort. 763:177–183.
Ramming, D.W. 1983. Embryo culture, p. 136–144, In J. N. Moore and J. Janick, eds. Methods in Fruit Breeding. Purdue Univ. Press, West Lafayette, Indiana.
Ramming, D.W., and Cociu, V. 1991. Plums (*Prunus*), p. 235–287, In J. N. Moore and J. R. J. Ballington, eds. Genetic Resources of Temperate Fruit and Nut Crops. ISHS, Wageningen.
Ravelonandro, M., Monsion, M., Teycheney, P.Y., Delbos, R., Dunez, J. 1992. Construction of a chimeric viral gene expressing plum pox virus coat protein. Gene 120:167–173.
Reales, A., Sargent, D.J., Tobutt, K.R., and Rivera, D. 2010. Phylogenetics of Eurasian plums, *Prunus* L. section *Prunus* (Rosaceae), according to coding and non-coding chloroplast DNA sequences. Tree Genet. Genom. 6:37–45.
Rehder, A. 1954. Manual of cultivated trees and shrubs. 2nd ed. Dioscorides Press, Portland.
Reid, W., and Gast, K.L.B. 1993. The potential for domestication and utilization of native plums in Kansas, p. 520–523, In J. Janick and J. E. Simon, eds. New Crops. Wiley, New York.
Renaud, R. 1975. A study of inheritance in plum intraspecific cross-breeding. Acta Hort. 48:79–82.
Renaud, R., and Salesses, G. 1994. Interspecific hybridization and rootstocks breeding for European plums. Acta Hort. 359:97–100.
Reynders-Aloisi, S., and Grellet, F. 1994. Characterization of the ribosomal DNA units in two related *Prunus* species (*P. cerasifera* and *P. spinosa*). Plant Cell Reports 13:641–646.
Roach, F.A. 1985. Plums, p. 142–460 Cultivated Fruits of Britain, their Origin and History. Basil Blackwell, Oxford, UK.
Roberts, A.N., and Hammers, L.A. 1951. The native Pacific plum in Oregon. Oregon State Coll. Agric. Exp. Sta. Bull. 502.

Rohrer, J.R., Ahmad, R., Southwick, S.M., and Potter, D. 2004. Microsatellite analysis of relationships among North American plums (*Prunus* sect. *Prunocerasus*, Rosaceae). Plant Syst. Evol. 244:69–75.

Rupasinghe, H.P.V., Jayasankar, S., and Lay, W. 2006. Variation in total phenolics and antioxidant capacity among European plum genotypes. Scientia Hort. 108:243–246.

Salinero, C., Aguin, O., and Sainz, M.J. 2003. Fruit yield and characteristics of three cultivars of mirabelle plum (*Prunus insititia* var. syriaca) in northwest Spain. J. Amer. Pom. Soc. 57:70–75.

Sapir, G., Stern, R.A., Eisikowitch, D., Goldway, M. 2004. Cloning of four new Japanese plum S-alleles and determination of the compatibility between cultivars by PCR analysis. J. Hort. Sci. Biotech., 79:223–227.

Scorza, R., Callahan, A., Levy, L., Damsteegt, V., Webb, K., Ravelonandro, M. 2001. Post-transcriptional gene silencing in plum pox virus resistant transgenic European plum containing the plum pox potyvirus coat protein gene. Transgenic Res., 10:201–209.

Scorza, R., Hily, J.-M., Callahan, A., Malinowski, T., Cambra, M., Capote, N., Zagrai, I., Damsteegt, V., Briard, P., Ravelonandro, M. 2007. Deregulation of Plum Pox transgenic plum 'Honeysweet'. Acta Hort., 738:669–673.

Shaw, J., and Small, R.L. 2004. Addressing the "hardest puzzle in American pomology:" Phylogeny of *Prunus* sect. *Prunocerasus* (Rosaceae) based on seven noncoding chloroplast DNA regions. Am. J. Bot. 91:985–996.

Shaw, J., and Small, R.L. 2005. Chloroplast DNA phylogeny and phylogeography of the North American plums (*Prunus* subgenus *Prunus* section *Prunocerasus*, Rosaceae). Am. J. Bot. 92:2011–2030.

Sherman, W.B. and Lyrene, P.M. 1983. Handling seedling populations, p. 66–73, In J. N. Moore and J. Janick, eds. Methods In Fruit Breeding. Purdue Univ. Press, West Lafayette, Indiana.

Sherman, W.B. and Lyrene, P.M. 1998. 'Gulfbeauty' and 'Gulfblaze' Japanese-type plums. Fruit Var. J.52:19–19.

Sherman, W.B., Topp, B.L., and Lyrene, P.M. 1992. Breeding low-chill Japanese-type plums for subtropical climates. Acta Hort. 317:149–153.

Sosinski, B., Gannavarapu, M., Hage, L.D., Beck, E., King, G.J., Ryder, C.D., Rajapakse, S., Baird, W.V., Ballard, R.E., Abbott, A.G. 2000. Characterization of microsatellite markers in peach [*Prunus* persica (L.) Batsch]. Theor. Appl. Gen., 101:421–428.

Srinivasan, C., Liu, Z., Scorza, R. 2011a. Ectopic expression of class 1 KNOX genes induce adventitious shoot regeneration and alter growth and development of tobacco (*Nicotiana tabacum* L) and European plum (*Prunus domestica* L). Plant Cell Reports 30:655–664.

Srinivasan, C., Scorza, R., Callahan, A., Dardick, C. 2011b. Development of very early flowering and normal fruiting plum with fertile seeds. United States Patent Application Pub. US 2011/0067147.

Stacewicz-Sapuntzakis, M., Bowen, P.E., Hussain, E.A., Damayanti-Wood, B.I., and Farnsworth, N.R. 2001. Chemical composition and potential health effects of prunes: A functional food? Critical Rev. Food Sci. Nutrit. 41:251–286.

Starnes, H.N. 1905. Japan and hybrid plums. State Coll. Agric. Mech. Arts, Georgia Exp. Sta. Bull. 68.

Struss, D., Ahmad, R., Southwick, S.M. 2003. Analysis of sweet cherry (*Prunus avium* L.) cultivars using SSR and AFLP markers. J. Amer. Soc. Hort. Sci., 128:421–428.

Sutherland, B.G., Robbins, T.P., Tobutt, K.R. 2004a. Primers amplifying a range of *Prunus* S-alleles. Plant Breeding, 123:582–584.

Sutherland, B.G., Robbins, T.P., Tobutt, K.R., 2004b. Molecular genetics of self-incompatibility in plums. Acta Hort. 663:557–562.

Sutherland, B.G., Tobutt, K.R., and Robbins, T.P. 2007. Molecular genotyping of self-incompatible plum cultivars. Acta Hort. 734:47–51.

Testolin, R., Messina, R., Lain, O., Marrazzo, M.T., Huang, W-G., Cipriani, G. 2004. Microsatellites isolated in almond from an AC-repeat enriched library. Mol. Ecol. Notes 4:459–461.

Theiler, R., 1971: Embryonenkultur für die Anzucht neuer Kirschenhybriden (*Prunus avium* L.). Schweiz. Landwirt. Forsch. 10:65–93.

Tian, L.-N., Wen, Y., Jayasankar, S., Sibbald, S. 2007. Regeneration of *Prunus salicina* Lindl (Japanese plum) from hypocotyls of mature seeds. In Vitro Cellular Develop. Biol. 43:343–347.

Tomas-Barberan, F.A., Gil, M.I., Cremin, P., Waterhouse, A.L., Hess-Pierce, B., and Kader, A.A. 2001. HPLC-DAD-ESIMS analysis of phenolic compounds in nectarines, peaches, and plums. J. Agric. Food Chem. 49:4748–4760.

Topp, B.L. and Russell, D.M. 1989. Breeding early ripening Japanese plums. Acta Hort. 240:27–30.

Topp, B.L., and Russell, D.M. 1990a. 'Queensland Bellerosa' plum. HortScience 25:814.

Topp, B.L., and Russell, D.M. 1990b. 'Queensland Earlisweet' cherry plum. HortScience 25:713.

Topp, B.L., and Sherman, W.B. 1990a. Potential for low-chill Japanese plums in Florida. Proc. Fla. State Hort. Soc. 103:294–298.

Topp, B.L., and Sherman, W.B. 1990b. Sources of bacterial spot resistance in Japanese-type plum cultivars. Fruit Var. J. 44:32–35.

Topp, B.L., Heaton, J.B., Russell, D.M., and Mayer, R. 1989. Field susceptibility of Japanese-type plums to *Xanthomonas campestris* pv. *pruni*. Aust. J. Exp. Agric. 29:905–909.

Topp, B.L., Sherman, W.B., and Stall, R.E. 1991. Comparison of rating methods for bacterial spot resistance in Japanese-type plum. Fruit Var. J. 45:70–74.

Urtubia, C., Devia, J., Castro, A., Zamora, P., Aguirre, C., Tapia, E., Barba, P., Dell'Orto, P., Moynihan, M., Petri, C., Scorza, R., Prieto, H. 2008. *Agrobacterium*-mediated genetic transformation of *Prunus salicina*. Plant Cell Reports. DOI 10.1007/s00299-008-0559-0.

USDA. 2007. Fruit and Tree Nuts Situation and Outlook Yearbook. Market and Trade, Economics Division, Economic Research Service, U.S. Department of Agriculture, October 2007, FTS-2007.

Vangdal, E., Sekse, L., and Slimestad, R. 2007. Phenolics and other compounds with antioxidative effect in stone fruit - Preliminary results. Acta Hort. 734:357–361.

Vaughan S.P., Russell K. 2004. Characterization of novel microsatellites and development of multiplex PCR for large-scale population studies in wild cherry, *Prunus avium*. Mol. Ecol. Notes 4: 429–431.

Vieira, E.A., Nodari, R.O., Dantas, A.C.M., Ducroquet, J.P., Dalbó, M., Borges, C.V. 2005. Genetic mapping of the Japanese plum. Crop Breed. Appl. Biotech. 5:29–37.

Vizzotto, M., Cisneros-Zevallos, L., Byrne, D.H., Ramming, D.W., and Okie, W.R. 2007. Large variation found in the phytochemical and antioxidant activity of peach and plum germplasm. J. Amer. Soc. Hort. Sci. 132:334–340.

Wakana, A., Hanada, N., Torikai, Y., Fukudome, I., and Yasukochi, K. 2006. The extent of intersubgeneric cross compatibility between Japanese plum (*Prunus salicina* Lindl.) and peach (*P. percica* Batsch.). J. Fac. Agric. Kyushu Univ. 51:87–92 (English summary).

Walsh, K., Long, R.L. and Middleton, S.G. 2007. Use of near infra-red spectroscopy in evaluation of source-sink manipulation to increase the soluble sugar content of stonefruit. J. Hort. Sci. Biotech. 82:316–322.

Wang, A.M., Tian, L.N., Huang, T.S., Brown, D.C.W., Svircev, A.M., Stobbs, L.W., Kiki, B.L.A. and Sanfacon, H. 2009., 2009. The development of genetic resistance to Plum pox virus in transgenic *Nicotiana Benthamiana* and *Prunus Domestica*. Acta Hort. 839:665–672.

Watkins, R. 1976. Cherry, plum, peach, apricot and almond. *Prunus* spp., pp. 242–247, In N. W. Simmonds, ed. Evolution of crop plants. Longman., London UK.

Weinberger, J.H. 1975. Plums, p. 336–347, In J. Janick and J. N. Moore, eds. Advances in Fruit Breeding. Purdue University Press, West Lafayette, Indiana.

Weinberger, J.H., and Thompson, L.A. 1962. Inheritance of certain fruit and leaf characters in Japanese plums. Proc. Amer. Soc. Hort. Sci. 81:172–179.

Wen, I.C., and Liu, Y. 2004. Evaluation and genetic relationship analysis by RAPD on Oriental plum germplasm. Journal of Agricultural Research of China 53:97–110 (in Chinese).

Wen, I.C., and Sherman, W.B. 2003. Developing low chill, high quality Japanese plums in subtropical Taiwan. Acta Hort. 622:437–441.

Werner, D.J., Ritchie, D.F., Cain, D.W., and Zehr, E.I. 1986. Susceptibility of peaches and nectarines, plant introductions, and other *Prunus* species to bacterial spot. HortScience 21: 127–130.

Werner, D.J., Mowrey, B.D., and Young, E. 1988. Chilling requirement and post-rest heat accumulation as related to difference in time of bloom between peach and western sand cherry. J. Amer. Soc. Hort. Sci. 113:775–778.

Wight, W.F. 1915a. Native American species of *Prunus*. U.S. Dept. Agric. Bull. 179.
Wight, W.F. 1915b. The varieties of plums derived from native American species. U.S. Dept. Agric. Bull. 172.
Yamaguchi, M. and Kyotani, H. 1985. Differences in fruit ripening patterns of Japanese plum cultivars under high (30°C) and medium (20°C) temperature storage. Bull. Fruit Tree Res. Stn. A 13:1–19. (English summary).
Yamaguchi, M., Yoshida, M., Kyotani, H., Nakamura, Y., Nishimura, K., Haji, T., and Miyake, M. 1998. New Japanese plum cultivar 'Honey Rosa'. Bull. Nat. Inst. Fruit Tree Sci. 30–31:1–14.
Yamaguchi, M., Yoshida, M., Kyotani, H., Nakamura, Y., Nishimura, K., Haji, T., and Miyake, M. 1999. New Japanese plum cultivar 'Honey Heart'. Bull. Nat. Inst. Fruit Tree Sci. 32:15–29.
Yamane, H., Tao, R., Sugiura, A. 1999. Identification and cDNA cloning of S-RNases in self-incompatible Japanese plum (*Prunus salicina* Lindl. Cv. Sordum). Plant Biotech. 16:389–396.
Yamamoto, T., Mochida, K., Imai, T., Shi, Y.Z., Ogiwara, I., Hayashi, T. 2002. Microsatellite markers in peach [*Prunus persica* (L.) Batsch] derived from an enriched genomic and cDNA libraries. Mol. Ecol. Notes 2:298–301.
Yancheva, S.D., Druart, P., Watilon, B. 2002. *Agrobacterium*-mediated transformation of plum (*Prunus domestica* L). Acta Hort. 577:215–217.
Yezhov, V.N., Smykov, A.V., Smykov, V.K., Khokhlov, S.Y., Zaurov, D.E., Mehlenbacher, S.A., Molnar, T.J., Goffreda, J.C., and Funk, C.R. 2005. Genetic resources of temperate and subtropical fruit and nut species at the Nikita Botanical Gardens. HortScience 40:5–9.
Yoshida, M. 1987. The origin of fruits, 2: Plums. Fruit Japan 42:49–53.
Zhang, J., Lu, Z., and Guan, S. 1997. The plum germplasm resources in the cold area of Northeast China. China Fruits 4:44–45.
Zhivondov, A. 2007. Biometric studies of plum-apricot hybrids (*Prunus domestica* x *Armeniaca vulgaris*). Vocarstvo 41:9–12.
Zhivondov, A., and Djouvinov, V. 2002. Some results of the plum breeding programme at the Fruit-Growing Research Institute in Plovdiv. Acta Hort. 577:45–49.
Zohary, D. 1992. Is the European plum, *Prunus domestica* L., a *P. cerasifera* EHRH. × *P. spinosa* L. allo-polyploid? Euphytica 60:75–77.

Chapter 16
Citrus

Patrick Ollitrault and Luis Navarro

Abstract Citrus is the most extensively produced tree fruit crop in the world. There are two clearly differentiated markets: the fresh fruit and the processed juice market. Citrus species are essentially diploids ($2n=2x=18$) and were domesticated in Southeast Asia several thousand years ago and then spread throughout the world. Most of the cultivated citrus species are part of the *Citrus* genus containing, depending on the taxonomist, between 16 and 156 species. The relative complexity of these classifications results from the conjunction of a broad morphological diversity, total sexual interspecific compatibility within the genus and apomixis. There are four basic taxa on the basis of morphological descriptors and molecular data (*C. maxima* (Burm.) Merr., the pummelos; *C. medica* L., the citrons, *C. reticulata* Blanco, the andarins, and *C. micrantha Wester*). The other cultivated species (*C. aurantium* L., the sour orange; *C. sinensis* (L.) Osbeck, the sweet orange; *C. paradisi* Macf, the grapefruit; *C. limon* (L.) Burm. f., the lemon, and *C. aurantifolia* (Christm.) Swingle, the lime) appeared by recombination among the basic taxa. Most of the citrus scion cultivars result from the selection of spontaneous bud mutations identified in production orchards. Today several projects of ploidy manipulation are developed in different countries to select seedless triploid mandarins. Fruit quality (size, color, easy-peeling), seedlessness, and the extension of the harvest season are the main selection objectives for fresh market cultivars. The majority of the rootstocks used for propagation are original species or ancient natural hybrids. However, intergeneric hybrids (*Citrus* × *Poncirus*) such as citranges, citrumelos, and citrandarins have an increasing importance. The first rootstock breeding objective is adaptation to soil

P. Ollitrault (✉)
Centre de Coopération Internationale en Recherche Agronomique pour le Développement (CIRAD), UPR 75, Avenue Agropolis, TA A-75/02, Montpellier, Cedex 5 34398, France
e-mail: ollitrault@cirad.fr

L. Navarro (✉)
Centro de Protección Vegetal y Biotecnología, Instituto Valenciano de Investigaciones Agrarias (IVIA), Apartado Oficial, Moncada, Valencia 46113, Spain

M.L. Badenes and D.H. Byrne (eds.), *Fruit Breeding*, Handbook of Plant Breeding 8, DOI 10.1007/978-1-4419-0763-9_16,

conditions and soil pathogens. The most widespread needs would be tolerance to CTV. Biotechnology is strongly integrated into breeding and propagation schemes. Shoot tip grafting is widely used for sanitation of the mother plants in certification propagation schemes. Somatic hybridization is an effective tool used for scion and rootstock breeding. Functional genomics studies ESTs, microarray platforms, and a full genome sequence are available. Efficient and reliable transformation systems for several economically important citrus species exist.

Keywords Orange • Grapefruit • Mandarin • Lemon • Lime • Pummelo • Citron • Poncirus • Rootstock • Scion • Shoot tip grafting • Somatic hybridization • Ploidy manipulation, genomics

1 Introduction

Citrus is the most extensively produced tree fruit crop in the world. There are two clearly differentiated markets: the fresh fruit and the processed juice market. Sweet orange is the predominant species for both of these markets.

Production. The increase of citrus world production was relatively constant during the last decades of the twentieth century, and annual production has reached more than 105 million tons (2000–2004). Oranges constitute the major part of the citrus production (61%), followed by mandarins (20%), lemons and limes (14%), and grapefruits (5%) (FAO 2006). The production of citrus fruits is very widespread around the world, located approximately between 40°N and 40°S latitudes, with more than 140 producing countries. However, the major part of the production is concentrated in Brazil (20%), the Mediterranean countries (20%), China (16%), and the USA (11%) (FAO 2006). These areas account for about two thirds of the total citrus production. In the Mediterranean basin, citrus fruits are primarily produced for the fresh fruit market. Spain is the principal producer of the area with a surface of 305,000 ha and a production about 6 million tons.

Consumption. Citrus are mainly consumed within developed countries, although the consumption per capita is increasing in developing countries. According to FAO, fresh orange consumption is decreasing in the industrialized countries and increasing in the emergent developing countries such as Mexico, India, Argentina, Brazil, and China. One of the major market shifts during last two decades of the twentieth century was the increase of mandarin production, which includes the tangerines, clementines, and the satsumas, at the expense of fresh oranges. The citrus juice consumption also increased due to the improvements in quality and the price competitiveness associated with technological progress.

International market. Fresh citrus fruit exports account for approximately 8% of the world production. Most of the citrus (62% in 2003) is imported into the Northern hemisphere. The Mediterranean area is the major exporter of fresh citrus with 60% of the volume. Spain is the largest exporter with 25% of the total world exports. Countries of the Southern hemisphere, such as Argentina, Australia, and South Africa, are increasing their exports by providing out-of-season citrus fruits to the Northern market. The principal destinations of fresh citrus exports of the Mediterranean basin are European countries. In the case of the USA, the first destinations are Japan, Canada, and the countries of Southeast Asia. In the Asian countries, the production is primarily consumed in the domestic markets and their contribution to world trade is limited.

The citrus juice market uses mainly sweet oranges (80%) and accounts for approximately one third of the production. Two production areas, namely, the state of Florida in the USA and the state of Sao Paulo in Brazil supply approximately 85% of the world market. Brazil exports 99% of its production whereas 90% of the production of Florida is consumed in the US domestic market. The European Union is the largest orange juice importer with more than 80% of the world imports (source: UNCTAD http://www.unctad.org/infocomm/anglais/orange/market.htm).

Biotic and abiotic constraints. Citrus fruit production is confronted worldwide with increasing biotic and abiotic constraints. They are affected by nematodes, fungi, bacteria, phytoplasmas, spiroplasmas, viruses, and viroids. Some diseases just cause a reduction in production and quality, while others have the potential to destroy a citrus industry. Some diseases are present in the majority of the production regions, such as those caused by nematodes, the oomycete *Phytophtora* sp., or the *Citrus tristeza virus* (CTV), which precludes the use of some rootstocks with excellent horticultural behavior and reduces fruit production and quality of specific varieties. Others are restricted to specific growing areas, although some of them are spreading quickly to new areas. Citrus canker caused by the bacteria *Xanthomonas axonopodis* pv. *citri* (Hasse) Vauterin et al. was producing important damage in several South American and Asian citrus areas, but recently it has widely spread without control in Florida (The USA). Citrus Variegated Chlorosis, caused by the bacteria *Xylella fastidiosa* Wells et al. and Sudden death, probably caused by *Citrus sudden death associated virus,* are producing important damage in Brazil. Huanglongbing (*ex* greening), caused by the bacteria *Candidatus Liberibacter asiaticus*, is the main limitation for Citrus production in Asia, and recent outbreaks in Sao Paulo (Brazil) and Florida (The USA) are affecting millions of trees and seriously threatening the industry. In Africa, cercosporiosis (a fungal disease caused by *Phaemularia angolensis* (De Carvalho & O. Mendes) P.M. Kirk) is a major constraint and is spreading quickly. In the Mediterranean basin, Mal Secco (a fungal disease caused by *Phoma tracheiphila* (Petri) L.A. Kantsch. & Gikaschvili) causes important damage on lemon trees and constitutes a constraint for some rootstocks. The susceptibility of citrus production to emergent and invasive diseases is extremely important as a major portion of the production (sweet oranges in particular) is based on a very low genetic diversity.

Citrus is grown in a wide diversity of climatic and soil conditions and consequently affected by several abiotic stresses. The increasing scarcity of water and the degradation of its quality (mainly salinity) are part of the major abiotic constraints for many countries, but in others flooding is an important problem. Some rootstocks are very sensitive to the ferric chlorosis associated with calcareous and basic soils, whereas in some areas acid soils are a problem. Freezing and very high temperatures also cause important losses in many areas of production.

2 Origin and Domestication of Scion Cultivars

Although the area of origin for citrus is generally agreed to be wide, including south China, northeast India, the Indo-Chinese peninsula and Burma, the localization of the primary center of origin is still controversial. This center of origin varies from the mountainous region of southern China and northeast India (Tolkowsky 1938) to northeast India and Burma (Tanaka 1954) or to the Yunnan province of China (Gmitter and Hu 1990) according to different authors.

Citrus species were domesticated in Southeast Asia several thousand years ago and then spread throughout the world (Fig. 16.1). The first written documents about citrus came from China in the chapter "Tribute to Yu" (2205–2197 BC) that made references to various citrus types, probably mandarins and pummelos, and from India in the collection of religious text "Vajaseneyi sambita" written about 800 BC, that made references to citron and lemon. Citron was the first species known in the Mediterranean Basin, probably introduced to Persia and Greece from India by Alexander the Great around 300 BC. Old mosaics show that the Romans may have known lemons around 100 AD, although there is no proof that they were cultivated. The Arabs spread citrus throughout Europe and North Africa with the expansion of their empire. Citron, sour orange, lemons, limes, and pummelos are described in tenth and eleventh century books from Spain. Early types of sweet orange were grown in Europe at the beginning of the fifteenth century, probably introduced by the Genoese, but they were not spread until the Portuguese introduced more selected types in the early sixteenth century. Surprisingly, mandarins that were cultivated in Southeast Asia from ancient times were not introduced to Europe until the early nineteenth century.

There are written documents related to citrus in Japan since the first century, when some types were introduced by a Japanese expedition to China and other neighboring countries. Citrus were introduced to America by Christopher Columbus in his second trip in 1483. He took seeds of oranges, lemons, and citron to the island Hispaniola and from there they spread quickly to other islands and Central America. Citrus were introduced into Florida by Spanish explorers between 1515 and 1565 and approximately during the same period the Portuguese took citrus to Brazil. The Portuguese also introduced citrus to West Africa and the Dutch colony introduced sweet orange to South Africa in 1654. Colonists of the First Fleet brought oranges, limes, and lemons from Brazil to Australia in 1769. Franciscan missions introduced oranges and lemons to California around 1769.

Most of the cultivated citrus species are part of the *Citrus* genus containing, according to the taxonomists, between 16 (Swingle and Reece 1967) and 156 species (Tanaka 1961). The classification of Swingle and Reece (1967) distinguishes the Eucitrus subgenus where all the cultivated taxa are found, and the subgenus Papeda. The relative complexity of these classifications results from the conjunction of a broad morphological diversity, total sexual interspecific compatibility within the genus, and apomixis. This form of apomixis, fixing complex genetic structure through seedling propagation, has lead some taxonomists to consider clonal families of interspecific

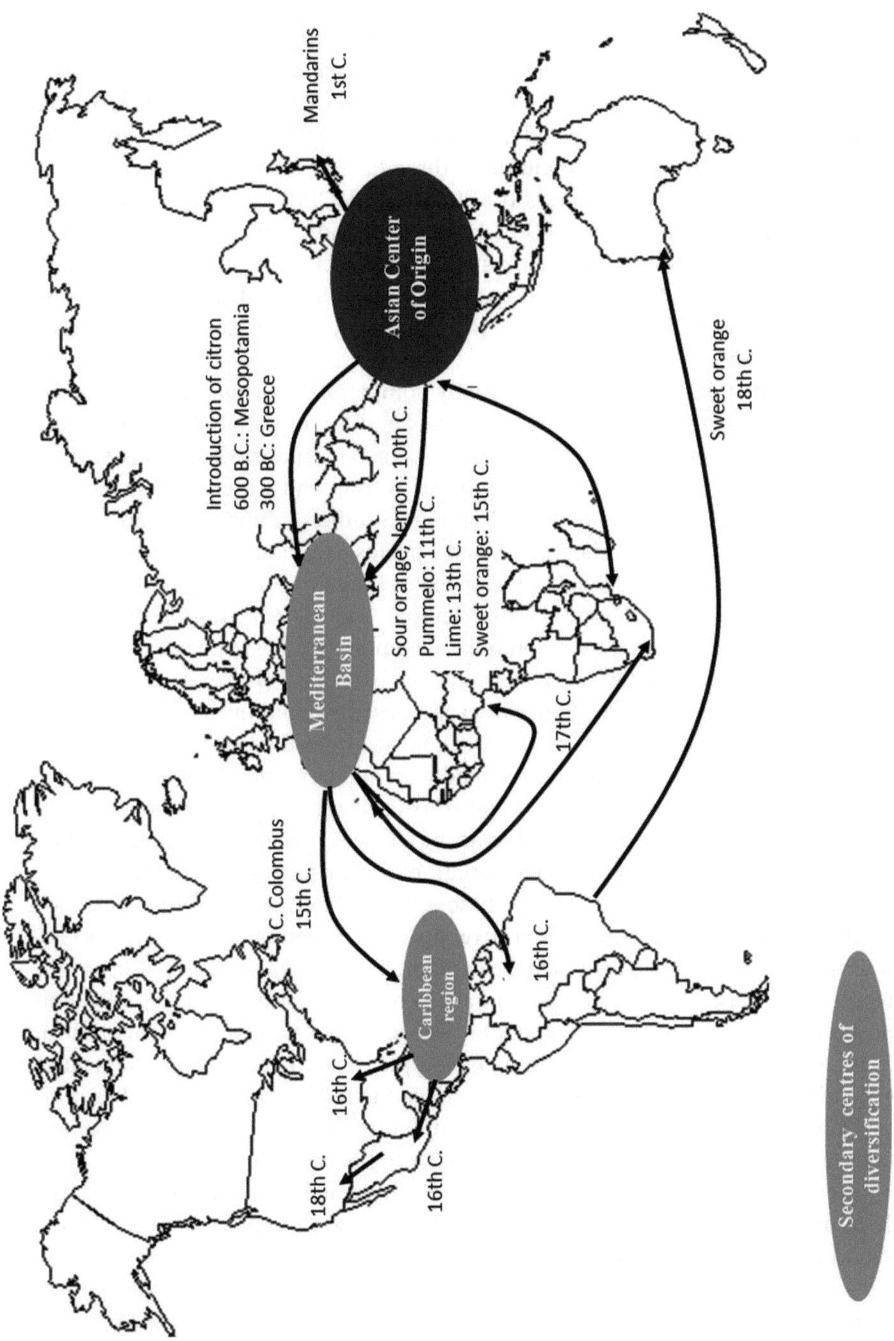

Fig. 16.1 Origin and domestication of citrus

origin as new species. Numerical taxonomy brought decisive information to better understand the domestication and the relationships between the various cultivated species of *Citrus*. Barret and Rhodes (1976) were the first to propose the existence of three basic taxa on the basis of morphological descriptors (*C. maxima* (Burm.) Merr.—the pummelos, *C. medica* L.—the citrons, and *C. reticulata* Blanco—the mandarins) from which all cultivated *Citrus* have originated. The strong organization of phenotypic variability around these three taxa was confirmed further with vegetative morphological characters (Ollitrault et al. 2003) and carotenoid flesh composition (Fanciullino et al. 2006). It was also validated with studies using total proteins (Handa et al. 1986), isozymes (Herrero et al. 1996, 1997; Ollitrault et al. 2003), RFLPs, RAPDs (Federici et al. 1998; Nicolosi et al. 2000; Fanciullino et al. 2007), SSRs (Luro et al. 2001; Barkley et al. 2006), and plastome data (Green et al. 1986).

The differentiation between these sexually compatible taxa can be explained by the foundation effect in three geographic zones and by an initial allopatric evolution. Pummelos originated in the Malay Archipelago and Indonesia, citrons evolved in northeastern India and the nearby region of Burma and China, and mandarins were diversified over a region including Vietnam, Southern China, and Japan (Scora 1975; Webber 1967). The other cultivated species (*C. aurantium* L.—the sour orange, *C. sinensis* (L.) Osbeck—the sweet orange, *C. paradisi* Macf.—the grapefruit, *C. limon* (L.) Burm. f.—the lemon, and *C. aurantifolia* (Christm.) Swingle—the lime) appeared by recombination among the basic taxa.

With codominant genetic markers, such as isozymes (Herrero et al. 1996; Ollitrault et al. 2003), RFLPs (Federici et al. 1998; Fanciullino et al. 2007) and SSRs (Luro et al. 2001; Barkley et al. 2006), most of these secondary species display a high level of heterozygosity associated with no intervarietal polymorphism at the intraspecific level. They are typical "false species" associated with clonal propagation (either by apomictic seeds or by grafting), where phenotypic varietal diversification occurred by an accumulation of variation from an ancestral hybrid prototype, without sexual recombination. Within these species there is a total disconnection between the phenotypic and molecular diversity. The important intraspecific pomological and phenological diversity can be explained by human selection of favorable budsport mutations. Bretó et al. (2001) have proposed that transposable elements could be an important source of variation in such vegetatively propagated citrus species. A good example of such rapid phenotypic diversification under vegetative propagation is given by Clementine, a spontaneous hybrid selected in Algeria by father "Clement" one century ago, for which numerous varieties have been selected, extending the harvesting period from September to February and displaying significant differentiation for quality (fruit size, color, sugar and acid content). In a larger scale of time, the diversification of sweet orange is also exemplary of such diversification based on the selection of budsport mutations. Although they were introduced to the Mediterranean basin relatively late, it constitutes the principal center of diversification of the modern sweet orange (Aubert 2001). Three subareas of diversification exist. The main area, the Iberian Peninsula, is associated with blond sweet oranges. The Cadenera orange was the most important in the development of the

Spanish citrus industry during the second half of the nineteenth century. Concurrently, a late Portuguese cultivar, Don Joao, was introduced in the Azores and then to the USA where it was renamed Valencia Late. Today, this cultivar is cultivated for both the fresh and the juice markets. Similarly, Navel oranges were initially cultivated in Portugal and Spain. After the Portuguese introduced the sweet orange to Bahia (Brazil), the navel cultivar Bahia (renamed in the USA as Washington navel) was described for its excellent quality and became the leading cultivar in the world for fresh fruit consumption. The origin of this cultivar is uncertain. Some authors consider that it was a budsport mutation in Bahia of a common orange, although a mutation of the original navel oranges or even a direct introduction from Portugal cannot be discarded.

The second area of diversification of sweet orange covers Tunisia, Malta and Sicily from where the blood and half-blood oranges originate. Three groups can be distinguished among these oranges. The Moro group originated in Liguria and then diversified in Sicily. The Tarocco cultivar is certainly the most cultivated of this group at present. The second group is the Maltese, largely cultivated in Tunisia. Early cultivars such as Bokobza or late cultivars such as Barlerin were selected within this group. The last group is the "doubles fines," which also diversified in Spain. The Near East constitutes the third center of sweet orange diversification from where blond oranges such as Shamouti and Beladi have been selected.

The conclusions on the origin of these secondary species drawn from the isoenzymatic (Herrero et al. 1997; Ollitrault et al. 2003), RFLP (Federici et al. 1998), RAPD, SCAR (Nicolosi et al. 2000), AFLP (Liang et al. 2006), and STMS (Luro et al. 2001; Barkley et al. 2006) data generally support the following conclusions (Nicolosi 2007):

C. sinensis and *C. aurantium* are close to *C. reticulata* but clearly contain introgressed fragments of the *C. maxima* nuclear genome. The higher proportion of *C. reticulata* indicate that they are not direct hybrids but probably backcross of first or second generation with the *C. reticulata* gene pool. Analysis of chloroplastic genome (Green et al. 1986; Nicolosi et al. 2000) shows that *C. maxima* was the female parent during the hybridization with *C. reticulata*.

C. paradisi is close to *C. maxima,* but alleles from the *C. reticulata* gene pool are also shared with *C. sinensis*. It could result from a hybridization between *C. maxima* and *C. sinensis* introduced into the Caribbean after the New World discovery. Thus, *C. paradisi* is the only species not native to Southeast Asia.

Genetic relationships between *C. medica*, *C. aurantifolia*, and *C. lemon* are clearly established. Chloroplastic and nuclear data indicate that the genetic pools of *C. reticulata* and *C. maxima* contributed to the genesis of the lemon tree. Nicolosi et al. (2000) proposed that this species resulted from a direct hybridization between *C. aurantium* and *C. medica*. This assumption is supported by Gulsen and Roose (2001) and Fanciullino et al. (2007). The origin of *C. aurantifolia* has been controversial, but molecular data (Federici et al. 1998; Nicolosi et al. 2000; Fanciullino et al. 2007) support the hypothesis of Torres et al. (1978) that it is a hybrid between

C. medica and a Papeda species. Nicolosi et al. (2000) proposed that *C. micrantha* might be the parental Papeda but it has yet to be confirmed.

When studying the diversity of cultivated forms, the strong organization still observed today, at the molecular and morphological levels, indicates that genetic exchange between the three original taxa was limited. The partial apomixis, linked with polyembryony of most of the secondary species, has certainly been an essential element in the limitation of gene flow. It could be supposed that other factors, such as structural differentiation of the genomes have also favored the maintenance of generalized linkage disequilibrium over the whole genome by limiting the recombination of large genomic portions. Indeed, flow cytometry analyses of nuclear genome size of Citrus species display a differentiation reaching 10% between *C. reticulata* (0.74 pg/2C) and *C. medica* (0.81 pg/2C) (Ollitrault et al. 2003). These two species have the smallest and largest genomes in the *Citrus* genus. Secondary species displayed intermediate genome size reflecting their interspecific origin. The differentiation of the nuclear genome size agrees with cytogenetic observations (Guerra 1993; Nair and Randhawa 1969; Raghuvanshi 1969). It probably indicates the advanced stage that the three basic taxa have reached toward speciation.

2.1 Varietal Groups

Four main varietal groups are distinguished in the international market:

Sweet orange is the main one. It is used both for fresh fruit and processing. It probably originated in China but its major center of diversification is the Mediterranean Basin (Aubert 2001). Major cultivars in this group are classified as navel oranges (Washington Navel, Navelina, Navelate, Powell, Rhode Navel, Cara Cara), blonde oranges (Shamouti, Valencia Late, Hamlin, Pineapple, Trovita, Salustiana, Delta Valencia, Pera), and blood oranges (Tarocco, Moro, Sanguinelli, Maltese).

Lemon and lime is the second group. Two main types of limes are distinguished: the small diploid (and seedy) Mexican lime and the big seedless triploid lime (Tahiti, Bears). Several lemon cultivars contribute to world production (Lisbon, Verna, Eureka, Feminello, Fino, Primofiori).

The easy peeling mandarins are becoming more important in the fresh fruit market. Clementines are the most important mandarins in the Mediterranean Basin, while Satsumas predominate in Japan. Other commercial mandarins include intraspecific or interspecific hybrids such as Fortune, Kinnow, and Minneola and several chance seedlings such as Ponkan, Ellendale, Ortanique, Murcott, and Nadorcott.

The last group is grapefruit which are divided into the yellow flesh cultivars (Marsh, Duncan) and the red flesh cultivars ('Hudson', 'Star Ruby', 'Ray Ruby' 'Rio Red').

In Southeast Asia and the Pacific, pummelo (*C. maxima*) and many traditional local mandarin cultivars are still important in the domestic market.

3 Genetic Resources

Citrus refers to all species of three sexually compatible genera within the tribe Citreae of the subfamily of Aurantioideae. The *Citrus* genus contains the majority of the consumed species. The *Fortunella* genus includes some commercial cultivars (the kumquats), and the monospecific *Poncirus* genus consists of *P. trifoliata* (L.) Raf., which plays a central role in rootstock breeding because of its resistance to many biotic stresses. Citrus constitutes the primary genetic pool of the cultivated forms. The number of basic chromosomes in the subfamily Aurantioideae is 9 (Krug 1943; Stace et al. 1993) and citrus and the related genera are mainly diploids, although there are a few polyploids of which the best known is the triploid Tahiti or Bears lime (*C latifolia* (Yu. Tanaka) Tanaka) producing the commercial large green lime.

The agro-morphologic variability of citrus is considerable. It relates to the pomological and organoleptic characters as well as to susceptibility, tolerance or resistance to biotic and abiotic stresses. This variability opens very broad prospects for the exploitation of citrus genetic resources for breeding. Many sources of tolerances for abiotic stresses have been identified (Krueger and Navarro 2007): tolerance to iron chlorosis of Rough lemon (*C. jambhiri* Lush) *C. macrophylla* Wester, Volkamer lemon (*Citrus limonia* Osbeck), and *C. amblycarpa* (Hassk.) Ochse; tolerance to salinity of the Rangpur lime (*C. limonia* Osb.) and Cleopatra mandarin; tolerance to cold of the Satsuma mandarins, the Kumquats and *P. trifoliata*; and tolerance to drought of the Rangpur lime. Resistance to important pest and diseases are also present: resistance to *Phytophthora* sp. of certain pummelos and mandarins, sour orange, Volkamer lemon, and *C. amblycarpa*; tolerance to the African cercosporiosis of pummelos, lemons, and Satsuma and Beauty mandarins; immunity to *Citrus tristeza virus* (CTV) of *P. trifoliata*, partial resistance to CTV of some pummelos and kumquats and tolerance of mandarins, lemons, and several rootstocks such as Cleopatra mandarin, *C. amblycarpa*, Rangpur lime, Rough lemon, and Volkamer Lemon; tolerance to the citrus canker of *C. junos* Siebold ex Tanaka and some mandarins (Satsuma, Dancy); and resistance to the nematode *Tylenchulus semipenetrans* Cobb of *P. trifoliata*.

The diversity present in pomological and organoleptic traits is also extensive. The diameter of the fruits varies from a few centimeters in mandarins, *Poncirus*, and kumquats to more than 30 cm for some pummelos. Albedo is not present in kumquats and is poorly developed in mandarins, while it constitutes the major part of the citron fruit and can be very thick in pummelos. The pulp of the fruits is green, orange, yellow, or red according to carotenoids (Fanciullino et al. 2006) or anthocyanins (blood oranges) present. Acidity can be almost absent in fruits of some sweet oranges grown in tropical areas, but it is very high in lemons and limes. The flavors and essential oils are diverse both qualitatively and quantitatively. The maturation period is very wide. In the Mediterranean basin, the harvest period ranges from early September for some Satsuma and Clementine varieties to July for sweet orange cv 'Valencia Late,' while flowering occurs at the same time for most cultivars. The number of seed in the fruit is a very important quality trait for the fresh market and ranges from some seedless genotypes to others that have more than 20 seeds per fruit.

Citrus species are sexually compatible with other genera (*Eremocitrus*, *Microcitrus*, and *Clymenia*), with which they constitute the "true citrus fruit trees" according to the terminology of Swingle and Reece (1967). Some of these, such as Eremocitrus, contain interesting characters of tolerance to drought and salinity. However, the intergeneric hybrids are generally strongly sterile (Iwamasa et al. 1988). This prohibits an effective use via sexual hybridization of this germplasm because the first generation hybrids present too many unfavorable traits of the wild relative. Other genera of the Aurantioideae family (tribe of Citreae and tribe of Clausenae) contain many biotic and abiotic stress resistance traits of which some are not found in the primary and secondary gene pools. For example, *Severinia buxifolia* (Poir.) Ten. is highly tolerant to salt and boron stress. However, this diversity is not exploitable by conventional breeding because of the sexual incompatibility between these genera and citrus. With some genera of the Citrinae subtribe, such as Citropsis, sexual hybridization produced seeds unable to germinate while the combinations with more distant genera do not produce seeds. In the case of the genera of Clauseniae subtribe (*Murraya, Clausena, and Glycosmis*), the sexual incompatibility is caused by the inability of the pollen tube to grow beyond the higher part of style (Iwamasa et al. 1988). Krueger and Navarro (2007) have reviewed the attributes of Aurantioideae genera.

Molecular marker studies have given a clear understanding of the origin and diversification process of the cultivated species and serve as a good base for the establishment of Citrus germplasm management strategies. Among the three basic taxa, the pummelos and mandarins show significant intraspecific polymorphism at the molecular and phenotypic levels. Intraspecific varietal improvement can be carried out traditionally by sexual recombination, taking advantage of the existing nonapomictic monoembryonic cultivars in the two species. Thus, intraspecific germplasm can be managed as a core collection that preserves the maximum allelic diversity. Analysis of genetic intraspecific organization with molecular markers should help in the definition of this core collection.

In contrast, the set of characters defining the secondary species (sweet and sour oranges, grapefruit, lemon, and lime) relies on complex genomic structures of interspecific origin with high heterozygosity that are stabilized by vegetative propagation (apomixis and more recently cuttings or grafting). The conservation of genetic resources of these species must be based on the collection of genotypes based on phenotypes because there is a strong disconnect between molecular marker diversity and morphological variation within these groups (Ollitrault et al. 2003). Exploitation of the intercultivar diversity by sexual crosses for intraspecific improvement is mostly precluded since it will recombine the traits defining the "species." These collections aim at preserving cultivars with greatest adaptive, morphological and phenological diversity within each species. They have direct application to preserve the best cultivars for the citrus industry according to their adaptation to environmental conditions and the changing demands and opportunities of the markets. They are also very useful for genetic improvement using biotechnological tools based on the introduction of specific traits into existing genotypes without altering their genetic background. They also can be used as "mutant collection" for functional genomic studies.

For the global preservation of Citrus diversity, germplasm can be managed to conserve the maximum amount of gene diversity. The three basic taxa (*C. reticulata, C. maxima*, and *C. medica*), identified as being the origin of most cultivated forms, constitute an essential reservoir as they contain the majority of the allelic diversity of cultivated citrus. Mandarins and pummelos from the original diversification areas (Southeast Asia), where the traditional methods of propagation by seeds should favor adaptive selection, is clearly the highest priority. The lime group displays an important level of phenotypic and genetic diversity and should be a priority for acid citrus preservation. Greater attention should also be given in the future to the Papeda species, whose potential was little explored until recently. Indeed, molecular data suggest they are involved in *C. aurantifolia* and *C. macrophylla* genesis and show potential as an important source of adaptive or tolerance traits for rootstock breeding. Finally, the development of biotechnologies such as somatic hybridization and genetic transformation considerably enlarge the gene pool that could be used for scion and rootstock breeding (Grosser et al. 1996a, 2000; Ollitrault et al. 2001; Peña et al. 2007, 2008). It is thus advisable today to extend the concept of citrus germplasm preservation at least to the Citreae tribe.

Systematic molecular characterization of collections has been done by several organizations. SSR markers will probably be the reference markers for accessions of the three basic taxa and sexual hybrids (Luro et al. 2001; Barkley et al. 2006) while alternative markers have to be developed for varietal characterization within species such as sweet orange, grapefruit, lemon, and some mandarin groups such as clementines and satsumas. Markers related to epigenetic variation such as MSAP (methylation-sensitive amplification polymorphism) (Hong and Deng 2005) or markers associated with transposition events (Asíns et al. 1999; Bretó et al. 2001; Bernet and Asins 2003) could be useful for this purpose. The systematic development of Single Nucleotide Polymorphism (SNPs) may also be an efficient way to generate useful markers.

4 Major Breeding Achievements

4.1 Scions

For sweet oranges, lemons, clementines, satsumas and grapefruits, new cultivars have only been selected from spontaneous or induced mutations, while both mutant selection and sexual breeding have allowed varietal progress in mandarins.

Most of the citrus cultivars result from the selection of spontaneous bud mutations identified in production orchards. Fruit quality (size, color, seedlessness) and the extension of the harvest season are the main selection objectives. New cultivars of this origin are constantly released, particularly for clementine (about 17 new cultivars protected in Spain during the last 10 years), satsumas and sweet oranges, particularly from the navel group. Induced seedless mutants of mandarins have been released recently in Israel (Mor) and the USA (Tango).

Interesting cultivars have been selected among mandarin × grapefruit hybrids, the tangelos (Nova, Orlando, Minneola, Page) and mandarin hybrids (Fairchild, Fremont, Fortune, Kara, Kinnow Wilking, Honey) from the old sexual mandarin breeding project conducted in the USA (Frost 1935). Kinnow is a very important cultivar for the subcontinent of India and Pakistan while the partial success of the late maturing cultivar Fortune in some growing areas is now tempered by its high susceptibility to *Alternaria alternata* pv. *citri*.

Natural hybrids have been an important source of new cultivars. Clementine first appeared at the beginning of the last century in the garden of an orphanage in Algeria as a natural cross between mandarin and sweet orange. The cultivar was imported to Spain in 1925 and since then many excellent bud mutants have been found and propagated. Clementines are now the main mandarins in the Mediterranean area and also are being grown in several countries of the Americas and South Africa. Other natural mandarin hybrids are Ponkan, widely cultivated in Asian countries and Brazil, Murcott, cultivated in Florida and Brazil, Imperial and Ellendale discovered in the nineteenth century in Australia, and Ortanique, discovered in Jamaica in 1920. A more recent natural hybrid that is reaching significant commercial development is the late maturing Afourer or Nadorcott mandarin, probably a hybrid between Murcott and a mandarin, discovered in Morocco.

More recently triploid seedless cultivars have been released in Italy, the USA, Japan, and Spain, and some of these cultivars are now being produced for the market. 'Tacle' (Starrantino 1999) is a tangor obtained in Sicily by crossing the diploid 'Clementine' with the autotetraploid 'Tarocco' sweet orange. Other triploid tangor and mandarin hybrids as well as some triploid lemon types have been recently released by the same group (Russo et al. 2004). In California, three mid and late season triploid mandarin hybrids resulting from the same cross [tetraploid ('Temple' tangor × 'Dancy' mandarin) × diploid 'Encore' mandarin] have been released (Williams and Roose 2004). In Japan, a triploid *C. sudachi* hybrid obtained from a cross of a diploid and a tetraploid genotype of this species has been submitted for registration (Tokunaga et al. 2005). In Spain 13 triploid hybrids obtained from $2x$X $2x$ crosses with 'Fortune' mandarin as female parent have been registered, and two of them have already been released for commercial propagation (Navarro et al. 2006; Aleza et al. 2010b; Cuenca et al. 2010).

4.2 Rootstocks

The majority of the rootstocks used for propagation are original species or ancient natural hybrids. This is the case of *P. trifoliata, C. macrophylla,* Rangpur lime, Cleopatra mandarin, Sunki mandarin, sour orange, Volkamer lemon, and Rough lemon.

The oldest traditional varietal improvement program in *Citrus* was undertaken by USDA in Florida in 1893. The objective was to breed for cold and disease resistance using *P. trifoliata* as a parent. Owing to the very severe decrease in fruit quality associated with *P. trifoliata*, this program did not result in edible cultivars. Nevertheless, some of these *Citrus* × *Poncirus* hybrids are now very important rootstocks used in many countries. These include some citrumelos (Grapefruit × *P. trifoliata* cvs. Swingle, Sacaton, 4475) and particularly the citranges (Sweet orange × *P. trifoliata* cvs. Troyer, Carrizo, C-35). However, most of them present some susceptibility to adverse abiotic conditions (alkalinity, salt) and a new range of intergeneric (*Citrus* × *P. trifoliata*) hybrids have been developed. Crosses between mandarins and *Poncirus* appear very promising to combine tolerances to abiotic and biotic constraints both by sexual breeding (Forner et al. 2003) or somatic hybridization (Grosser et al. 2000; Ollitrault et al. 2000a).

5 Currents Goals of Breeding

5.1 Scions

In relation with specific market demands and environmental conditions (biotic and abiotic constraints), the main goal of breeding may vary dramatically between the production areas. However, some general trends can be outlined.

Juice processing implies highly productive cultivars with a high percentage of juice and sugar content. The pigment composition is also important for high quality fresh juices. Expanding the harvesting period with high quality fruits is currently the main objective of scion breeding for the fresh fruit market. This market promotes organoleptic qualities (aroma, taste, acidity, sugar) and pomological qualities (easy to peel, seedlessness, external appearance). Since the definition of organoleptic quality varies with the consumer, citrus breeders must therefore endeavor to develop a wide range of varieties likely to meet these diverse needs. Nutritional quality based on vitamin C, carotenoid and polyphenol contents are now considered as breeding criteria in some projects.

Seedlessness is a major criterion for the fresh fruit market. Sterility can be classified into three types (Yamamoto et al. 1995): female sterility, male sterility, and self-incompatibility. Gametophytic self-incompatibility is common in pummelo and is found in several mandarin and mandarin hybrid cultivars (Soost 1968). Male and female sterility may be due to different genetic factors such as triploidy, sterility genes, and chromosomal abnormalities such as reciprocal translocations and inversions. Some male-sterile and self-incompatible accessions cannot produce seedless fruits because of the lack of parthenocarpy. Thus, parthenocarpy is an indispensable trait for seedless fruit production, and this character seems to be widely present in citrus germplasm. Autonomic parthenocarpy, where seedless fruit is produced without any external stimulation (pollination), is the main type of parthenocarpy in citrus ('Navel' orange or 'Satsuma' mandarin), but stimulative parthenocarpy has also been reported (Vardi et al. 1988). Male sterile or self-incompatible accessions have the ability to produce seedless fruits when cross pollination is prevented. However, those accessions may produce seedy fruits in mixed plantings particularly if pollinating insects are present. For areas such as the Mediterranean Basin (where the main seedless mandarins are the self-incompatible clementines) selection is required for both male and female sterility.

Some diseases cause considerable damage in orchards. These include Huanglongbing (*ex* citrus greening) in Asia, South Africa and recently in Brazil and Florida, citrus canker in most tropical and subtropical areas, cercosporiosis in Africa, Mal secco in the Mediterranean Basin for lemon and some rootstocks, citrus variegated chlorosis and Sudden death in Brazil. *Alternaria* is also a problem for some mandarin cultivars such as Fortune in Spain. Ranges of varietal susceptibility have been established for most of these diseases and tolerant parents are selected in some breeding projects. However, in the case of Huanglongbing, no exploitable tolerance source has been identified.

5.2 Rootstocks

The first rootstock breeding objective is adaptation to soil conditions and soil pathogens. The most widespread needs would be tolerance to CTV, *Phytophthora sp.* and nematodes (mainly *Tylenchulus semipenetrans* Cobb). Other important objectives include tolerance to mal secco, which decreases nursery survival of susceptible rootstocks, tolerance to salinity, tolerance to iron chlorosis in calcareous soils, and recently in Florida, blight tolerance has become a priority objective. In the Mediterranean Basin, breeding for rootstocks which allow better management of the scarce water resources is a major challenge.

Rootstocks are asexually propagated by apomictic seeds, and thus it is important that new rootstocks have a high degree of apomixis, to avoid the germination of sexual embryos, that are difficult and costly to eliminate at the nurseries.

The rootstock genotype can modify the behavior of the rootstock/scion interaction with respect to many characteristics. Breeding of rootstock conferring tolerance to CTV is needed for all citrus production areas. Resistance to exocortis (disease caused by the *Citrus exocortis viroid*) is less important than for CTV because exocortis can be controlled with certification schemes and rigorous nursery and orchard practices which eliminate exocortis transmission by grafting and pruning tools. In Southeast Asia, resistance to *Tatter leaf virus* is an important criteria. Cold tolerance can also be enhanced by the rootstock choice, and this is of particular interest to Japan and Georgia. Vigor of the tree and productivity are also dependant on the rootstock. In some areas, breeders seek dwarf rootstock to limit pruning and harvest costs in high density orchards.

Beyond productivity, the rootstock choice has a strong impact on fruit quality. It can affect the sugar and acid content, the fruit size, and the percentage of juice content. For example *Poncirus* and its hybrids are known to favor high quality of mandarin and orange production, while *C. macrophylla* is very productive but reduces fruit quality. Thus productivity and fruit quality of the associated scions are an essential criteria for rootstock selection.

6 Breeding Methods and Techniques

Citrus breeding strategies take advantage of the vegetative propagation and integrate the advances of biotechnology.

Conventional breeding in citrus has important limitations due to the complex reproductive biology of these species. Most genotypes are apomictic, and adventitious embryos develop directly from nucellar cells limiting or precluding the development of zygotic embryos. This limits the recovery of large sexual populations, and in practice, apomictic genotypes are avoided as female parents in many programs. Several high-quality genotypes have pollen and/or ovule sterility and thus cannot be used as parents in breeding programs. Self- and cross-incompatibility are relatively common among many genotypes, also limiting the possibilities to select parents for specific crosses. Citrus has a long juvenile phase and in most species at least 5 years are required to start flowering and many more to completely lose the undesirable characters, such as thorns, associated with juvenility. These biological factors and the difficulty to manage field evaluation for large progenies and multi-generational breeding schemes are the main reason for the relatively low success of conventional breeding programs.

Citrus breeders, taking advantage of vegetative propagation, put their main effort in the induction of polymorphism in one cycle from which they make clonal selections. Selection thus relates either to spontaneous mutations identified in the orchards, or to genotypes obtained by hybridizations, induced mutagenesis or after recourse to various biotechnological approaches (somatic hybridization, transformation) that have been developed in citrus in part to solve the problems found in conventional breeding. To improve the effectiveness of such strategies, the transfer of the important commercial traits of the parents to the progeny needs to be optimized. This depends on a systematic management of heterozygosity and a thorough knowledge of the inheritance of the selected characters. The implementation of marker-assisted selection is particularly important for all the traits that cannot be phenotyped during the juvenile phase.

Among the three basic taxa, pummelos and mandarins have an important inter-cultivar polymorphism both at molecular and morphological levels. The intraspecific breeding can be carried out by sexual recombination. However, when the objective is to diversify the varietal range around a precise ideotype (for example the Clementines), the recourse to sexual hybridization is precluded due to the highly heterozygous parental structures. Methods allowing punctual modification of the genome such as mutagenesis or somaclonal variation, asymmetrical somatic hybridization or genetic transformation must then be applied. Taking into account their genetic structures, the same approaches will be the only usable ones for the improvement of species such as sweet orange or grapefruit, if their specific characteristics need to be maintained.

6.1 Natural and Induced Mutants or Variants

Selection of spontaneous mutations is the oldest citrus breeding method and most of the varieties cultivated worldwide arose from this process. Work in Spain, Morocco, and Corsica has developed clementines with extended production periods and enhanced fruit size and color. On a much longer time scale, spontaneous mutations are responsible for the diversification of Satsuma mandarins in Japan as well as for sweet oranges, lemons, or grapefruits. Mutations in bud meristems can lead to chimera formation as has been reported for some red or pink grapefruits, the Shamouti sweet orange, and some acidless sweet orange and lemons. Some variegated chimeras are of interest for ornamental purposes.

Induced mutagenesis for cultivar improvement has been used repeatedly since 1935. Gamma irradiation has been the most common method of mutagenesis. It can induce a wide range of mutations (point mutation, chromosomes breaks, and rearrangements). The latter type is particularly interesting for seedless selection (Hearn 1984) because translocations and inversions may cause sterility. Roose and Williams (2007) recommend exposure of budwood to 30–50 Gy to induce seedless mutations. Classical examples of cultivars obtained by irradiation programs are the Star Ruby red flesh grapefruit obtained by irradiation of seeds of Hudson grapefruit (Hensz 1971), and more recently the low-seeded selection of Murcott, called 'Mor' mandarin produced by budwood mutation (Vardi et al. 1993), and the seedless mandarin Tango produced by irradiation of Afourer (Nadorcott) budwood (Roose and Williams 2007).

Somaclonal variation has been used with success in Florida for sweet orange breeding (Grosser et al. 2007) to extend the harvesting period, to improve some fruit quality traits, and to reduce seed content.

6.2 Sexual Breeding at Diploid Level

Sexual breeding is mainly used for diversification in mandarins and for rootstock improvement. As already mentioned, interesting rootstocks have been recently obtained by intergeneric hybridization between citrus species and *Poncirus* (Forner et al. 2003). Regarding mandarins, the main limitation of this strategy is that most of the diploid hybrids are fertile and thus seedy. The selection of seedless cultivars displaying high quality and good yield requires the evaluation of very large progenies. Moreover, if seedlessness of these new hybrids is based on self-incompatibility or male sterility, important production problems will be encountered in areas where self-incompatible but pollen producing varieties such as the clementine predominate.

6.3 Seedlessness and Ploidy Manipulation for Triploid Creation

The selection of triploid lines is the classic route to develop seedless cultivars, as triploids are generally both male and female sterile. Thus, most of the trees of a triploid progeny under field evaluation should be seedless, and consequently an efficient selection for other traits is possible. Moreover, larger fruit size associated with triploidy should correct the reduction of fruit size generally observed in seedless mutants of seedy cultivars. This strategy is being developed by several groups worldwide and new avenues have been opened by biotechnology (Ollitrault et al. 2008).

Several methods have been developed for triploid citrus creation (Aleza et al. 2010a; Ollitrault et al. 2008; Navarro et al. 2003). One of them exploits natural events of polyploidization such as $2n$ gametes, using embryo rescue and flow cytometry to select triploids in $2x$X $2x$ crosses. Second meiotic division restitution (SDR) has been proposed for diploid megagametophyte development in clementine (Luro et al. 2004), while Chen et al. (2008) reported first meiotic division restitution (FDR) in sweet oranges. The classic strategy is to cross diploid monoembryonic females with tetraploid males. Such tetraploid plants can be found in apomictic seedlings (natural doubling of the chromosome stock of nucellar cells) or are created by somatic hybridization (Grosser et al. 2000). Recently, tetraploid monoembryonic lines have been obtained by colchicine treatment of shoot tips grafted in vitro (Juárez et al. 2004; Navarro and Juárez 2007). These tetraploids open the avenue to $4x$X $2x$ crosses with the maternal tetraploid Clementine. As an example, in the Spanish mandarin triploid breeding program, more that 10,000 triploids have been recovered using the three crossing strategies and using in vitro embryo rescue to germinate the seeds followed by flow cytometry to determine the ploidy level of regenerated plants.

In the last 10 years, several new triploid cultivars resulting mostly from diploid × autotetraploid sexual crosses have been released in Italy, the USA, and Japan, More recently, two new triploid mandarin cultivars from $2x$X $2x$ crosses have been released in Spain. Some of these cultivars are now in commercial production.

6.4 Adding Dominant Traits by Somatic Hybridization for Rootstock Breeding

Interesting traits of tolerance for biotic and abiotic stresses are present in *Citrus* and the related gene pool. Complementary progenitors can be identified particularly between *Poncirus* for resistance to biotic constraints and some *Citrus* species for abiotic tolerances. Even if interesting sexual hybrid rootstocks such as citranges or citrumelo have been obtained, progress of citrus rootstock by conventional sexual breeding is a difficult task, mainly because of the reproductive biology (apomixis) and the heterozygosity of the progenitors. This heterozygosity leads to important segregations of characters in sexual progenies and the low probability of obtaining recombinant hybrids combining all the desirable genes and traits of the two parents (Grosser et al. 2000; Ollitrault et al. 2000a). Conversely, somatic hybridization is very promising in rootstock breeding because it allows combining the dominant genes for tolerance to biotic and abiotic factors of the two parents, irrespective to their heterozygosity level. The potential biodiversity available is also very large because somatic hybrids can be obtained between sexually incompatible species and genera (Grosser et al. 1996a).

6.5 Propagation

In Southeast Asia and the Pacific, there are traditional areas of citrus cultivation where the citrus trees are grown on their own roots and are multiplied by cutting, layering, or by seeds in apomictic cultivars. The two first methods present a high risk of disease dissemination, while the third one is hampered by a long juvenile phase and frequent susceptibility of cultivars to *Phytophthora* sp. In commercial plantings, citrus trees, like the majority of the fruit trees, are grafted plants. All commercial rootstocks are apomictic and are propagated by seed. Scions are multiplied at the nurseries by bud grafting. The change of varieties of adult trees planted in the field is done by top working. This last technique, frequently used in Spain, allows a rapid varietal change and thus a quicker response to the changing market demands. Thus, clonal propagation of highly heterozygous plants is the rule both scion and rootstock.

Many citrus genotypes are apomictic and contain seeds with several embryos (supernumerary embryos) derived from nucellar cells (maternal somatic tissue). Fecundation seems necessary for the development of the nucellar embryos (Soost 1987; Kepiro and Roose 2007). Only two species (*C. medica* and *C. maxima*) have only nonapomictic monoembryonic genotypes. For the others, the degree of polyembryony varies widely. The rate of partial apomixis, which results from the competition between the sexual embryo and the nucellar ones, is influenced by the genotype and the environmental conditions (Khan and Roose 1988). Although nucellar embryony is an obstacle to obtain sexual hybrids and self-sexual progenies and therefore to breeding programs, it does provide a simple and inexpensive method for clonal rootstock propagation by seed.

Citrus graft and insect-transmissible diseases produced by viruses, viroids, some bacteria, spiroplasmas, and phytoplasmas produce very important economic losses in most citrus growing areas. In general, they cause decline, loss of vigor, short commercial tree life, low yields, poor fruit quality and restrict the use of some rootstocks. Thus, they may become primary limiting production factors, and some severe diseases have the potential to destroy an industry. Only preventive measures are useful for the control of graft-transmissible pathogens, such as use of tolerant or resistant germplasm, exclusion of potential diseases from the citrus area, and the establishment of orchards using pathogen-free high-quality nursery trees. This last measure is accomplished with the implementation of certification programs for nursery trees (Navarro 1993).

The purpose of certification programs is to guarantee the sanitary status and trueness-to-type of the propagating material during the process of commercial propagation in the nurseries. In addition, they also control the horticultural quality of nursery plants (Lee et al. 1999; Navarro 1993). These programs have regulations governing nursery operations and require periodic indexing and inspection of trees of the different blocks used for nursery propagation. Usually, they are operated by a state or provincial agency having legal authority to impose restrictions and to inspect all steps of the propagation process in private nurseries. Pathogen-free plants are necessary to start propagation, and in citrus these programs are usually organized in four blocks of propagating material.

1. Protected foundation blocks of plants growing in containers under insect-proof screenhouses to maintain a supply of the original healthy plants used for propagation. One or two plants per accession are usually maintained and they are usually under the responsibility of public agencies.
2. Foundation blocks usually belong to private nurseries and are field or screenhouse plantings propagated with budwood from the protected foundation blocks. Two to six trees are maintained per accession, and they are annually evaluated for trueness to type and indexed for diseases as regulated.
3. Budwood multiplication blocks are nursery plants propagated directly from foundation trees to increase the number of buds for the propagation of certified trees. The establishment of these multiplication blocks allows an exponential increase in the amount of high quality budwood that can be produced from a limited number of foundation trees. This method enables better inspection and indexing of the limited number of foundation trees. Multiplication blocks may be established under normal field conditions or in greenhouses where they are protected from vector-borne pathogens and a faster and better growth can be obtained. Buds should only be collected from these blocks for a maximum period of 3–5 years to avoid the possible propagation of undetected bud sport mutations and to reduce the risk of pathogen infection over time. Consequently, new budwood increase blocks have to be periodically established with buds from the foundation trees.
4. Blocks of certified nursery trees are propagated with budwood from the multiplication blocks. They may be produced under regular field conditions or in different types of greenhouses according to the specific needs and technology available at each nursery. These trees are inspected to guarantee that they meet the horticultural quality required in the certification regulations. Nurseries must also keep records to show they have complied with the regulations.

Countries that have well-established certification programs have a higher production and better quality of fruit, as well as a system to avoid the dissemination of diseases with the planting material. The application of a certification program has allowed the total renewal of the Spanish citrus industry (Navarro et al. 2002) by eliminating graft-transmissible pathogens that were the main production limitation.

7 Integration of New Biotechnologies in Breeding Programs

7.1 Shoot Tip Grafting

Frequently, pathogen-free plants of many cultivars are not available and it is necessary to recover healthy plants from infected ones. The procedure used in the past to recover pathogen-free citrus plants was the selection of nucellar seedlings of apomictic cultivars (Weathers and Calavan 1959). Nucellar embryony is effective because most citrus pathogens are not transmitted through the process of embryogenesis, and nucellar plants are produced by asexual embryogenesis in vivo and thus have the same genotype as the mother plants that produced the seeds. The limitation of the procedure is that nucellar plants are juvenile, and consequently, they are excessively vigorous, thorny, and late in bearing, and they have to be grown for many years until these characters disappear and became acceptable for commercial propagation. Thermotherapy has also been used to recover pathogen-free citrus plants without juvenile characters. However, this technique is not effective for the elimination of pathogens that replicate well under warm conditions, such as viroids, citrus stubborn, or *Citrus leaf blotch virus* (Roistacher 1977).

The technique of shoot tip grafting in vitro (STG) consists on grafting very small shoot tips from diseased plants on young rootstock seedlings (Navarro et al. 1975; Navarro 1992; Navarro and Juarez 2007). The small shoot tips are usually free of pathogens and thus the technique allows the recovery of nonjuvenile plants free of all citrus pathogens. Owing to these advantages, STG is being used in most citrus growing countries to recover healthy plants for commercial propagation.

Movement of citrus species and varieties between different citrus areas for commercial and scientific purposes including breeding and germplasm maintenance and evaluation is often desirable. However, uncontrolled importation of budwood has the risk of introducing new pest and pathogens that, in some instances, may be devastating or may cause very important economic damage. This risk may be minimized by controlled introduction through quarantine stations that have the objective of importing foreign varieties avoiding the introduction of new pest and diseases that the original material may carry. In citrus, there are two different quarantine procedures that can be safely used for the importation of plant material. The classical method consists of propagating the imported budwood in quarantine greenhouses located outside the citrus growing areas. Then, the newly propagated plants can be indexed or subjected directly to STG followed by indexing. This procedure requires the availability of facilities and personnel trained on citrus pests, diseases, and cultural practices. It is used in some countries with a well-established quarantine system that have central facilities and personnel for the importation of plant material of several crops, but it is very expensive and difficult to establish only for citrus in most countries.

An alternative citrus tissue culture procedure developed for the safe introduction of citrus (Navarro et al. 1984, 1991) has been proved very efficient for the exclusion of citrus pests and diseases. Budsticks that are received from another country are thoroughly cleaned and surface sterilized and then cultured in vitro to induce the sprouting of lateral buds and formation of flushes from which shoot tips are isolated and micrografted in vitro. The only material really imported is a small shoot tip that usually is free of pests and pathogens. This process includes many controls to minimize the possibility of the escape of harmful pathogens and allows the rapid processing of new entries.

This tissue culture method has several advantages over the traditional quarantine methods. Pest and diseases that might be in the original material are eliminated at the early stages of introduction, thus shortening the quarantine period. With tissue culture, the test tubes serve as the substitute for the greenhouses located in isolated areas, and thus the quarantine station may be located at citrus research stations. At many of these stations, STG is being used for the sanitation of local cultivars and the needed facilities and personnel are usually available. Consequently, the tissue culture procedure can be easily established in many countries for the safe importation of citrus vegetative material. This method is recommended for the exchange of citrus germplasm (Frison and Taher 1991) and has been legally accepted in several countries, as those of the European Union. In Spain, it has been successfully used to import over 280 genotypes from various citrus areas. The introductions are included in the Citrus Germplasm Bank of IVIA in Valencia and some are being extensively used in breeding programs.

STG is increasingly being used as a tool in genetic transformation, somatic hybridization, and the recovery of haploid and tetraploid plants to regenerate elite genotypes or produce plants that cannot be recovered by other means (Navarro and Juarez 2007).

One of the major limitations of citrus transformation is the difficulty of rooting transgenic shoots. The use of STG to regenerate plants from transformed shoots or buds has become a routine procedure in most laboratories, allowing an important increase of the efficiency of the genetic transformation protocols.

In protoplast fusion experiments abnormal embryos are frequently produced. These include multiple fasciated cotyledons, embryos that only produce shoots, germinating embryos without a good vascular connection among shoot and root, and abnormal shoot proliferation, among others. These embryos do not produce plants that can be established in the greenhouse, thus reducing the efficiency of somatic hybrid recovery and losing potentially valuable genotypes. At IVIA, in vitro shoots produced by these embryos are routinely grafted in vitro to recover plants that are efficiently established in the greenhouse.

Haploids developed by in situ parthenogenesis induced by pollination with irradiated pollen can be recovered by embryo rescue from the resultant underdeveloped seeds. These haploid plants grow poorly in vitro and are very difficult to establish in the greenhouse. However, grafting in vitro shoots from these plants allowed to regenerate several haploid Nules clementine plants (Aleza et al. 2009a). Tetraploid plants of monoembryonic genotypes are very important as female parents in triploid mandarin breeding programs. IVIA has generated several tetraploid plants that are now widely used female parents in its triploid breeding program, by adding a drop of colchicine solution on a shoot tip 2 weeks after grafting in vitro (Aleza et al. 2009b).

The technique of STG is having a very important impact in the citrus industry worldwide. It has led to the elimination of some graft-transmissible diseases in several countries, has decreased the risks of introducing exotic pest and diseases with the international exchange of genotypes, and has reduced limitations of rootstock use related to scion infection. In addition, there is an increase of production and fruit quality as a consequence of controlling graft-transmissible diseases. In Spain alone, more than 130 million certified trees derived from healthy plants recovered by STG have been planted since 1982, representing about 95% of the Spanish citrus industry.

7.2 Embryo Rescue

Triploid embryos produced in interploid or diploid × diploid hybridization form abnormal or undeveloped seeds due to an unbalanced ploidy ratio between the embryo and the endosperm. The recovery of plants from these embryos is very difficult by conventional methods, but the problem can be overcome by in vitro embryo rescue. In vitro culture of excised fully developed embryos to obtain zygotic plantlets from apomictic cultivars was described 50 years ago (Maheshwari and Ranga Swamy 1958). Globular to early cotyledonary stage embryos can be rescued by extracting seeds of developing (3–4 months after anthesis) to mature fruits. Embryo rescue, most commonly using underdeveloped seeds from mature fruit, is now routine in the triploid breeding projects throughout the world (Navarro et al. 2003, 2004; Ollitrault et al. 2007a; Starrantino and Recupero 1982).

7.3 Somatic Embryogenesis and Embryogenic Callus Cryopreservation

Embryogenic callus lines are generally obtained by ovule culture of apomictic genotypes (Rangan et al. 1969; Ollitrault et al. 1995; Pérez et al. 1998, 1999). Plants regenerated from such callus display little or no somaclonal variation (Starrantino and Russo 1983) even after protoplast isolation (Kobayashi 1987; Olivares-Fuster et al. 2000). Conversely, induction of genetically uniform embryogenic callus lines from the ovules of nonapomictic cultivars is a difficult task, and regenerated plants do not have the same phenotype as the original plants (Navarro et al. 1985). Style and stigma culture (Carimi et al. 1995, 1999) or anther culture (Germanà 2003) has also been tested with some success. However, even with these last methods, examples of efficient embryonic callus system for nonapomictic cultivars remain rare.

Embryogenic callus lines have been used in induced mutagenesis projects to limit chimera formation (Kochba and Spiegel-Roy 1977). Application of in vitro selection at the regeneration step of irradiated callus allowed obtaining plants with higher tolerance to salt stress (Kochba et al. 1982; Spiegel-Roy and Ben Hayim 1985) or mal secco toxin (Gentile et al. 1992). Today, the main application of somatic embryogenesis is somatic hybridization.

Cryopreservation of nucellar callus lines allows the long-term conservation of germplasm resources in a form directly usable for biotechnology programs (Duran-Vila 1997) and is routinely employed by several groups (Sakai et al. 1991; Engelmann et al. 1994; Pérez et al. 1997, 1999; Hao et al. 2002).

7.4 Somatic Hybridization

Since the success by Ohgawara et al. (1985), the contribution of somatic hybridization to citrus cultivar improvement continues to expand (Grosser et al. 2000;

Ollitrault et al. 2007b). Although fusion can be efficiently accomplished between protoplasts derived from embryonic callus (Ollitrault et al. 1996a), citrus somatic hybrids are generally produced from the fusion of protoplasts isolated from embryogenic callus or suspension cultures of one parent with leaf-derived protoplasts of the second parent. The embryogenic parent provides the capacity for plant regeneration from the fusion products. The process of protoplast fusion, regeneration, and selection of hybrid plants has been visualized using a transgenic citrus plant expressing the Green Fluorescent Protein as one of the donors in somatic fusion experiments (Olivares-Fuster et al. 2002). Citrus protoplast fusion is induced either chemically using polyethylene glycol (Ohgawara et al. 1985; Grosser and Gmitter 1990) or electrically (Saito et al. 1991; Ling and Iwamasa 1994; Hidaka et al. 1995; Ollitrault et al. 1996a). Recently, Olivares-Fuster et al. (2005) have proposed an electrochemical protoplast fusion method combining chemical protoplast aggregation and a DC pulse to promote membrane fusion. After protoplast fusion, plant regeneration is accomplished without any selection pressure. However, in most of the published experiments, although some cybrid plants are generated, the majority of regenerated plants are allotetraploid somatic hybrids (Grosser et al. 2000). While chloroplasts have monoparental inheritance with random segregation (Moreira et al. 2000; Ollitrault et al. 2001; Liu et al. 2004), it has been demonstrated that in case of fusion between leaf and callus protoplasts, mitochondrial DNA was inherited from the embryonic callus parent (Moreira et al. 2000; Cabasson et al. 2001). These results suggest a possible relationship of mitochondria with regeneration ability (Grosser et al. 1996b). Some cases of potential mitochondrial recombination have also been reported (Moreira et al. 2000; Liu et al. 2004).

Somatic hybridization is a very valuable tool to manage parental heterozygosity and ploidy in citrus breeding schemes. It has three main applications:

Rootstock breeding. During the last decade, a primary application of somatic hybridization has been for citrus rootstock breeding. It allows the addition of all dominant traits, irrespective of the heterozygosity level of the breeding material. The most important project of somatic hybridization for rootstock breeding is in Florida (Grosser et al. 2000). Somatic hybrids between *Poncirus* and *Citrus* to combine tolerances to specific biotic and abiotic stresses of the Mediterranean Basin have been created in France (Ollitrault et al. 2000a). At the international level, evaluations are ongoing for several interspecific and intergeneric hybrids that combine promising horticultural traits and tolerance to pathogens such as *Citrus tristeza virus* and *Phytophthora* sp. (Grosser et al. 2000). Somatic hybrids have been obtained between sexually incompatible species, but their utility is limited by unfavorable traits and problems of nuclear genome instability for the intersubtribal and intertribal combinations.

Ploidy manipulation for triploid breeding. A second main application of somatic hybridization is the exploitation of apomictic and sterile cultivars, such as Satsuma or Navel sweet orange, for the synthesis of fertile tetraploid hybrids. The final objective of such ploidy manipulation is the synthesis of seedless triploid cultivars. Countries supporting programs based in this technology include the USA, Brazil, Japan, China, Spain, Italy, New Zealand, and France, and more than 100 combinations of allotetraploid parents have been produced (Grosser et al. 2000; Ollitrault et al. 2007b). Moreover, hundreds of triploids have been produced from interploid crosses using somatic hybrid parents. Somatic hybridization has also allowed the direct synthesis of triploid hybrids by protoplast fusion between diploid and haploid lines (Kobayashi et al. 1997; Ollitrault et al. 2000b).

Moreover allotetraploid citrus somatic hybrids constitute an original model to analyze the immediate effects of allopolyploidization on the regulation of gene expression and phenotype elaboration. Recent results of CIRAD and IVIA suggest that nonadditivity is frequent indeed at transcriptome (Bassene et al. 2009a), proteome (Gancel et al. 2006), and phenome level (Gancel et al. 2003; Bassene et al. 2009b, c) in interspecific combinations with *C. deliciosa.*

Cybrids and asymmetric hybrids. Several cybrid plants have been obtained after symmetric somatic hybridization. This material provides an opportunity for nucleus–cytoplasm interaction studies (Bassene et al. 2008) and should open new avenues for citrus breeding. One major application is the exploitation of the cytoplasmic male sterility of Satsuma mandarins by combining Satsuma cytoplasmic organelles with the diploid nucleus of good but seedy cultivars (Guo et al. 2004). Novel methodologies have been recently developed to enhance the efficiency of cybrid creation by electrochemical protoplast fusion (Olivares-Fuster et al. 2005) and by cytoplast fusion (Xu et al. 2006) and for the introgression of limited parts of the genome of one parent in asymmetric hybrids by microprotoplast fusion (Zhang et al. 2006; Louzada 2007) and UV treatments of protoplasts (Xu et al. 2007). These methodologies should find applications for varietal diversification of species such as sweet oranges, grapefruits, and lemons. The efficiency of the methods based on chromosome fragmentation and random introgression could be enhanced if combined with in vitro selection for specific tolerances during the regeneration phase.

7.5 Haplomethods

Production of citrus haploid plants has been recently reviewed (Germanà 2007). The first study of Citrus anther culture (*C. limon*) produced haploid callus but no plants (Drira and Benbadis 1975). Subsequent research has produced haploid embryos or plants for *Poncirus* (Hidaka et al. 1979), *C. microcarpa* (Chen et al. 1980), sweet orange 'Trovita,' Clementine (Germanà 1992), and *C. limon* (Germanà et al. 1992). Haploid embryos and plants have also been obtained from gynogenesis in monoembryonic cultivars. Haploid plants were obtained from fully developed seeds (2% were haploids) derived from the pollination of diploid Clementine and 'Lee' mandarin with triploid pollen (Oiyama and Kobayashi 1993) and some haploid plants have been obtained in diploid × diploid crosses in *C. maxima* (Toolapong et al. 1996; unpublished data) and in Clementine (unpublished data). Ollitrault et al. (1996b) demonstrated the efficiency of induced gynogenesis by pollination with irradiated pollen for monoembryonic cultivars (Clementine and *C. maxima*).

This method was used at IVIA to produce several haploid and dihaploid clementine plants and has been extended to several *C. reticulata* and tangor cultivars (Froelicher et al. 2007). Germanà and Chiancone (2001) induced gynogenesis in Clementine by in vitro pollination with triploid pollen. Haploid plants obtained by androgenesis or gynogenesis are generally weak and need to be grafted to survive. A haploid pummelo described by Yahata et al. (2005) had some fertile pollen grains and was able to produce diploid hybrid progenies when crossed with a diploid cultivar, suggesting that it produces unreduced pollen grains ($n=9$). At this time the only concrete application of haplomethods in citrus breeding has been the direct synthesis of triploid hybrids by somatic hybridization between haploid lines and diploid cultivars (Kobayashi et al. 1997; Ollitrault et al. 2000b). However, the development of structural genomics has renewed interest in haplomethods. Indeed, working with totally homozygous or monoploid lines presents a major advantage for genome sequencing projects, as well as for the analysis of copy number and allelic diversity of candidate genes. The international consortium of Citrus Genomics has decided to establish the reference whole sequence of Citrus from a haploid clementine obtained by induced gynogenesis at IVIA (Aleza et al. 2009a).

7.6 Citrus Genomics

Citrus, with a basic chromosome number of 9, has a relatively small genome size (372 Mb for the *C. sinensis* haploid genome). Therefore, Citrus is an interesting model for woody fruit tree genomics. Recent reviews in citrus genomics can be found in Roose and Close (2008) and Talón and Gmitter Jr (2008).

Structural genomics. Owing to the generally important heterozygosity of progenitors and the long juvenile period, the development of markers for focused introgression and early selection is a key for the improvement of the efficiency of citrus sexual breeding. Genetic mapping has been developed since the 1990s, and with the ongoing projects of physical mapping and whole genome sequencing, rapid progress should be expected in the near future.

Molecular markers and genetic mapping. Since isozymes markers (Torres et al. 1978), several kinds of nuclear markers have been used for citrus genetic studies such as Random Amplified Polymorphic DNA (RAPDs) and SCARs (Nicolosi et al. 2000), Restriction Fragment Length Polymorphism (RFLPs; Federici et al. 1998), Intersimple sequence repeat (ISSRs, Fang and Roose 1997), Amplified Fragment Lengh Polymorphism (AFLPs; Liang et al. 2006), and Cleaved Amplified Polymorphic Sequences (CAPs) from ESTs (Dr. Omura's group in Japan). Single-stranded conformational polymorphism (SSCP) analysis has been used for cytoplasm characterization (Olivares-Fuster et al. 2007). In the last 10 years a limited number of Simple Sequence Repeat (SSRs) have been derived from genomic libraries (Kijas et al. 1995; Barkley et al. 2006; Froelicher et al. 2008) 56 SSRs were obtained from the Genbank citrus EST data (Chen et al. 2006), and more than 200

SSR markers have been developed (Luro et al. 2008) from the 1,600 microsatellite sequences from 37,000 ESTs characterized by Terol et al. (2007). Recently, the same group has identified more than 7,600 SSRs from BAC end sequencing (Terol et al. 2008) that are used to develop SSR markers allowing direct anchoring of the genetic and physical maps. In addition to genetic mapping. SSRs have been used for the analysis of genetic diversity (Luro et al. 2001; Corazza-Nunes et al. 2002; Barkley et al. 2006), characterization of somatic hybrids, discrimination between zygotic and nucellar seedlings (Ruiz et al. 2000), control of the origin of plants obtained by induced gynogenesis (Froelicher et al. 2007), molecular characterization of triploid cultivars (Aleza et al. 2010b; Cuenca et al. 2010; our unpublished results), and the analysis of the origin of 2N gametes (Luro et al. 2004; Chen et al. 2008).

Because *P. trifoliata* has multiple stress-tolerance and disease-resistance traits, many of the genetic mapping projects have focused on *P. trifoliata* through intergeneric hybrids with *Citrus*. Owing to self-incompatibility of many nonapomictic or monoembryonic cultivars, most of the populations created for citrus mapping are crosses between two heterozygous parents. The first genetic maps were published in the early 1990s (Durham et al. 1992; Jarrell et al. 1992; Luro et al. 1995), which, as new marker systems have evolved, have been followed by maps with greater resolution and value. However, the population size and the number of codominant markers mapped remain generally low. Recently, collaborative efforts have been established to develop codominant markers (especially SSRs) and implement maps. The first such map for sweet orange and *P. trifoliata* was published in 2007 (Chen et al. 2007) and it is being expanded. New international, collaborative EST-SSR mapping efforts are currently using other families based on Clementine as part of a plan intended to lead to the full-length sequence of a haploid citrus genome, to be integrated with physical and genetic maps based on BAC end sequencing.

QTL and bulk segregant analysis have been conducted to identify markers linked with tolerance to abiotic stresses (cold and salt) (Cai et al. 1994; Tozlu et al. 1999; Weber et al. 2003), resistances to disease and pests (*Citrus tristeza virus*, nematodes, and leaf miner) (Asins et al. 2004; Bernet et al. 2005; Deng et al. 1997; Fang et al. 1998; Gmitter et al. 1996; Ling et al. 2000) as well as morphological or quality traits (fruit acidity, polyembryony, and apomixis) (Fang et al. 1997; García et al. 1999). High-throughput methods for marker saturation are needed for efficient QTL and association genetic studies, as well as for positional cloning of genes. For this purpose, arrays for SNP markers are being developed in the USA and Japan that should have an important impact of the progress of this work.

Physical mapping and whole genome sequencing. BAC libraries of *C. sinensis*, Clementines, and Satsumas have been established in the last few years in Spain, Japan, and the USA. The Spanish Citrus Genomic Consortium has constructed three BAC libraries from the Clementine mandarin (EcoR I, Hind III, and MboI) containing a total of 57,000 clones with an average insert size of 120 kb (19× coverage). 28,000 BAC clones were end-sequenced and these sequences analyzed (Terol et al. 2008). A physical map derived from the same 28,000-clone set of the Clementine BAC libraries is being constructed by restriction enzyme fragment fingerprinting.

The Citrus Genome Analysis Team from Japan is engaged in the construction of a physical map of citrus by HICF (High-Information-Content Fingerprinting) analysis of a BAC library from the Satsuma mandarin (*Citrus unshiu* Marc.) consisting of 37,000 clones, with 13.3× coverage of the citrus genome. A BAC library of Ridge Pineapple sweet orange was produced in the USA (USDA-ARS, Ft. Pierce, Fla, The USA) containing 18,432 clones (BamHI/Mbo I) with an average insert size of 145 kb, or an estimated 7× coverage. A total of 16,727 clones from this library have been fingerprinted and assembled into 472 contigs (http://phymap.ucdavis.edu:8080/citrus/).

A low-coverage (1.2X) shotgun sequence of the *C. sinensis* genome has revealed the difficulties associated with high heterozygosity, and lead the International Citrus Genomic Consortium to select a haploid Clementine as the model for whole citrus genome sequencing. International efforts are organized to establish a reference sequence of this haploid by 8–10X shotgun sequencing, a project that is almost finished (http://www.phytozome.net/clementine.php). This template will then be used to organize sequences of the highly heterozygous species such as *C. sinensis*, *C paradisi*, and *C. limon*.

It should be mentioned that the complete chloroplast genome sequence of *C. sinensis* was published by Bausher et al. (2006). It is 160,129 bp in length and contains 133 genes (89 protein-coding, 4 rRNAs, and 30 distinct tRNAs).

Functional genomics; ESTs and microarray platforms. Since the first work of Hisada et al. (1997), various groups have contributed to EST sequencing efforts using several species, mostly *C. sinensis* (sweet orange), *C. clementina* (Clementine mandarin), *C. paradisi* (grapefruit), *C. unshiu* (satsuma mandarin), *P. trifoliata,* and Carrizo citrange (Talón and Gmitter Jr 2008 for review). The total resource has reached 232,808 citrus sequences in the National Center for Biotechnology Information (NCBI) EST database. This EST collection includes a wide representation of sequences from many cDNA libraries derived from multiple reproductive and vegetative organs and tissues at different developmental stages and challenged with biotic and abiotic agents, and elicitor and hormonal treatments. Forment et al. (2005) generated 25 cDNA libraries covering different conditions and from 22,635 high-quality ESTs identified 11,836 putative unigenes. From an in-depth analysis of a collection of 54,000 single-pass ESTs, Terol et al. (2007) identified 13,000 putative unigenes. Dr. Machado's group in Brazil reported the contents of the CitEST Brazilian database including more than 260,000 valid reads contained unigene sets from several citrus species, but mainly sweet orange, mandarin, and *P. trifoliata.* BLAST searches against sequenced citrus ESTs are possible through several open database projects (i.e., http://harvest.ucr.edu, http://citest.centrodecitricultura.br/, http://cgf.ucdavis.edu, http://bioinfo.ibmcp.upv.es/genomics/cfgpDB/) or data deposited in GenBank.

Several microarray platforms have been developed in Japan (2,213 genes; Shimada et al. 2005 and 22K oligoarray containing 21,495 independent ESTs; Fujii et al. 2007, http://www.fruit.affrc.go.jp/index-e.html), Spain (12,672 probes corresponding to 6,875 putative unigenes; Forment et al. 2005, and a higher density citrus microarray composed of 24,000-element cDNA array containing 20,000 unigenes;

Martinez-Godoy et al. 2008), and the USA (designed for gene expression analysis using 30,264 probe sets; Close et al. 2006). Other research projects using cDNA citrus microarrays, or smaller custom arrays based on subtractive libraries, are in progress (Mozoruk et al. 2006).

Many candidate genes for tolerance to biotic and abiotic stresses, abscission, and quality elaboration have already been identified in the EST databases and with array experiments. Recent reviews are found in Gmitter et al. (2007), Tal6n and Gmitter Jr (2008), and Tadeo et al. (2008).

7.7 Genetic Transformation

Genetic transformation allows the introduction of specific traits into known genotypes without altering their elite genetic background. This applies to all citrus species, and particularly to improved commercial species such as *C. sinensis* (sweet orange), *C. limon* (lemon), and *C. paradisi* (grapefruit), and mandarin groups such as clementines and satsumas, where due to their highly heterozygous and complex genetic structure, genetic transformation should be considered as the most promising tool for improvement.

Genetic transformation methodologies in Citrus are based either on somatic embryogenesis (Vardi et al. 1990; Hidaka et al. 1990; Fleming et al. 2000; Duan et al. 2007) or more commonly organogenesis from in vitro growing seedling explants or internodes from greenhouse-grown plants (Moore et al. 1992; Kaneyoshi et al. 1994; Peña et al. 1995). Efficient and reliable transformation systems for many economically important citrus species exist. These are based on the selection of a proper *Agrobacterium* strain supertransforming citrus, the establishment of the optimal infection and cocultivation conditions and culture media, adequate selection conditions and culture media, the use of source plant material in a good ontological state, the determination of the competent cells for transformation in citrus explants, the use of appropriate marker genes, and the rapid production of whole transgenic plants through grafting of regenerating transgenic shoots onto vigorous rootstocks first in vitro and later in the greenhouse. (Peña et al. 2007, 2008). A very important step for utilization of genetic transformation for citrus breeding was the successful transformation of mature plant material to overcome the juvenile stage (Cervera et al. 1998, 2008). After 14–18 months in the greenhouse, the transgenic and control plants flowered, confirming their mature nature. This process greatly shortens the period of time until flowering and fruiting by years. The procedure for transformation of mature explants has been extended to several genotypes of interest, including several sweet orange varieties, sour orange, Mexican lime, Fino lemon, Cleopatra mandarin, *C. macrophylla*, and clementine Clemenules (Peña et al. 2008). Another important development related to public concern about transgenic crops is the development of procedures allowing the production of transgenic plants without using selective genes conferring herbicide or antibiotic resistance (Ghorbel et al. 1999; Domínguez et al. 2004; Ballester et al. 2007).

Genetic transformation has been used for incorporation of transgenes of potential interest into Citrus. For example Fagoaga et al. (2007) analyzed the effect of the expression of the sense and antisense forms of the *GA 20-oxidase* gene on gibberellin synthesis and plant phenotype. Gibberellins are considered key plant regulators for growth habit determination, and consequently the modulation of endogenous gibberellins affect the dwarfing ability of the rootstock and facilitate diverse cultural practices (e.g., pruning, pesticide applications, and harvesting). Wong et al. (2001) showed that citrus plants transformed with *ACC* synthase gene in antisense does not display the same ethylene synthesis as untransformed plants after chilling treatments. To develop a better understanding and reduce the juvenile phase, Carrizo seedlings constitutively overexpressing the *Arabidopsis* floral-regulatory genes *LEAFY* (*LFY*) or *APETALA1* (*AP1*) were generated (Peña et al. 2001). Both produced fertile flowers and fruits in their first year, considerably shortening the juvenile phase. Furthermore, sexual and nucellar seedlings derived from the transgenic plants had a very short juvenile period and flowered in their first spring. Transgenic *Poncirus* with ectopic expression of the *CiFT* gene (homolog to *FLOWERING LOCUS T*), another flowering time gene, also exhibited early flowering (Endo et al. 2005). These results open the way to very innovative approaches for citrus breeding over several recombining cycles by combining short juvenile phase and marker-assisted selection.

Tolerance or resistance to *P. citrophthora*, the most widely distributed fungal disease in citrus growing areas, was evaluated by introducing the *p23* gene that codes for a pathogenesis-related protein induced in tomato. The results provided evidence for the antifungal activity in vivo of the p23 pathogenesis-related protein against *P. citrophthora* (Fagoaga et al. 2001). Much research effort is in progress to understand the basis of the resistance to *Citrus tristeza virus* (CTV). CTV-resistant transformants have been obtained by genetically engineering the *p25* and *p23* genes from CTV (Domínguez et al. 2002; Fagoaga et al. 2006), although the stability of these pathogen-derived resistances over the large diversity of CTV strains still has to be demonstrated. An alternative strategy is to look for plant-derived resistance genes. The general resistance gene to CTV (*Ctv*) in *Poncirus trifoliata* has been characterized and localized to a region comprised of 10 predicted genes (Gmitter et al. 2007). Each of the 10 genes has been individually introduced in grapefruit (Rai 2006) and although results of the CTV challenge are still preliminary, transgenic lines expressing individually each one of these 10 genes were susceptible to CTV, suggesting that more than one gene in the locus are involved in resistance to CTV. Work on other pivotal diseases such as *Citrus mosaic virus* (Iwanami et al. 2004), citrus canker (Boscariol et al. 2006), and citrus blight, and for abiotic stress such as salt stress (Cervera et al. 2000) and osmotic stress (Molinari et al. 2004) are progressing. Regarding quality traits, a pectin methylesterase gene (*Cs-PME4*) isolated from sweet orange to prevent juice cloud separation was also introduced via protoplasts and subsequent regeneration through somatic embryogenesis (Guo et al. 2005).

The world citrus industry faces increasing biotic and abiotic constraints, while the market requires high quality products and more environmental friendly agricultural practices. In this context, citrus breeding appears as a main component to implement a sustainable citrus industry. Most of the scion cultivars cultivated in the world today have been selected from spontaneous mutants and natural hybridizations. However, intensive breeding of seedless mandarins in many of the countries producing citrus using both the triploid and induced mutation approaches should result in an increased importance of mandarins for the fresh fruit market. For rootstocks, diploid intergeneric *Citrus* × *Poncirus* hybrids will become more important in the control of the citrus Tristeza virus, and the first of the allotetraploid somatic hybrids will be released to the citrus industry within the next few years. In the long term, genomics and biotechnology will strongly modify the breeding efficiency both for conventional breeding with the integration of marker-assisted selection and for genetic transformation strategies.

References

Aleza, P., Juárez, J., Hernández, M., Ollitrault, P. and Navarro L. (2009a) Obtention and characterization of *Citrus clementina* Hort. ex Tan. `Clemenules´ haploid lines to establish the reference whole Citrus genome sequence. BMC Plant Biology. 9:110. doi:10.1186/1471-2229-9-110.

Aleza P., Juárez J., Ollitrault P., and Navarro L. (2009b) Production of tetraploid plants of non apomictic citrus genotypes. Plant Cell Rep 28:1837–1846. DOI 10.1007/s00299-009-0783-2.

Aleza, P., Juárez, J., Cuenca, J., Ollitrault, P. and Navarro, L. (2010a) Recovery of citrus triploid hybrids by embryo rescue and flow cytometry from 2x × 2x sexual hybridisation and its application to extensive breeding programs. Plant Cell Reports 29:1023–1034.

Aleza, P., Cuenca, C., Juarez, J., Pina, J.A. and Navarro, L. (2010b). 'Garbi' mandarin: a new late-maturing triploid hybrid. Hortscience 45(1): 139–141.

Asins, M.J., Bernet, G.P., Ruiz, C., Cambra, M., Guerri, J. and Carbonell E.A. (2004) QTL analysis of citrus tristeza virus-citradia interaction. Theor. Appl. Genet. 108: 603–611.

Asíns, M.J., Monforte, A.J., Mestre, P.F. and Carbonell, E.A.. (1999) Citrus and Prunuscopia-like retrotransposons. Theor. Appl. Genet. 99: 503–510.

Aubert, B. (2001) Genèse et développement de la culture des agrumes et patrimoine génétique méditerranéens de l'histoire naturelle des orangers. Dans la réédition de l'Histoire Naturelle des Orangers- Risseau et Poiteau, tome 2 Connaissance et mémoires éditeur, Paris.

Ballester, A., Cervera, M., and Peña, L. (2007) Efficient production of transgenic citrus plants using isopentenyl transferase positive selection and removal of the marker gene by site-specific recombination. Plant Cell Reports 26: 39–45.

Barkley, N.A., Roose, M.L., Krueger, R.R., and Federici, C.T. (2006) Assessing genetic diversity and population structure in a citrus germplasm collection utilizing simple sequence repeat markers (SSRs). Theor. Appl. Genet. 112: 1519–1531.

Barret, H.C. and Rhodes, A.M. (1976) A numerical taxonomic study of affinity relationships in cultivated *Citrus* and its close relatives. Syst. Bot. 1: 105–136.

Bassene, J.B., Berti, L., Carcouet, E., Dhuique-Mayer,C., Fanciullino, A.L., Bouffin, J., Ollitrault, P. and Froelicher, Y. (2008) Influence of mitochondria origin on fruit quality in a citrus cybrid. J. Agric. Food Chem. 56(18): 8635–8640.

Bassene, J., Froelicher, Y., Dubois, C., Ferrer, R., Navarro, L., Ollitrault, P., and Ancillo, G. (2009a) Nonadditive gene regulation in a citrus allotetraploid somatic hybrid between *C. reticulata* Blanco and *C. limon* (L.) Burm. Heredity. online publication 2 December 2009; doi: 10.1038/hdy.2009.162.

Bassene, J.B., Froelicher, Y., Dhuique-Mayer, C., Mouhaya, W., Ferrer, R., Ancillo, G., Morillon, R., Navarro, L. and Ollitrault, P. (2009b) Non additive phenotypic and transcriptomic inheritances in citrus allotetraploid somatic hybrid : the case of the pulp carotenoid biosynthesis pathway. Plant Cell Rep 28:1689–1697. DOI: 10.1007/s00299-009-0768-1.

Bassene, J.B., Berti, L., Costantino, G., Carcouet, E., Kamiri, M., Tomi, F., Dambier, D., Ollitrault, P. and Froelicher, Y. (2009c) Inheritance of characters involved in fruit quality in a citrus interspecific allotetraploid somatic hybrid. J. Agric. Food Chem., 57(11): 5065–5070.

Bausher, M.G., Singh, N.D., Lee, S.B., Jansen, R.K., and Daniell, H. (2006) The complete chloroplast genome sequence of *Citrus sinensis* (L.) Osbeck var 'Ridge Pineapple': organization and phylogenetic relationships to other angiosperms, BMC Plant Biology, 6: 21, doi:10.1186/1471-2229-6-21.

Boscariol, R.L., Monteiro, M., Takahashi, E. K., Chabregas, S. M., Vieira, M.L., Vieira, L.G., Pereira, L.F., Mourao Filho, F.A., Cardoso, S.C., Christiano, R.S., Bergamin Filho, A., Barbosa, J. M., Azevedo, F.A. and Mendes, B. M. (2006) Attacin A gene from Trichoplusia ni reduces susceptibility to *Xanthomonas axonopodis pv citri* in trangensic Citrus sinensis 'Hamlin'. Journal of the American Society for Horticultural Science, 131(4): 530–536.

Bernet G.P. and Asins, M.J. (2003) Identification and genomic distribution of gypsy like retrotransposons in *Citrus* and *Poncirus*. Theor. Appl. Genet.108: 121–130.

Bernet, G.P., Margaix, J., Jacas, J., Carbonell, E.A. and Asins, M.J. (2005) Genetic analysis of citrus leafminer susceptibility. Theor. Appl. Genet. 110: 1393–1400.

Bretó, M.P., Ruiz, C., Pina J.A., and Asins M.J. (2001) The diversification of *Citrus clementina* Hort. ex Tan., a vegetatively propagated crop species. Mol. Phylog. Evol. 21: 285–293.

Cabasson, C., Luro, F., Ollitrault, P. and Grosser, J.W. (2001) Non Random inheritance of mitochondrial genomes in *Citrus* hybrids produced by protoplast fusion. Plant Cell Rep., 20: 604–609.

Cai, Q., Guy, C.L. and Moore, G.A. (1994) Extension of the linkage map in *Citrus* using random amplified polymorphic DNA (RAPD) markers and RFLP mapping of cold-accimation-responsive loci. Theor. Appl. Genet. 89: 606–614.

Carimi, F., Pasquale, F. and De Crescimanno, F.G. (1995) Somatic embryogenesis in *Citrus* from styles culture. Plant-Science-Limerick. 105: 81–86.

Carimi, F., Pasquale, F. and De Crescimanno, F.G. (1999) Somatic embryogenesis and plant regeneration from pistil thin cell layers of Citrus. Plant-Cell-Reports.18: 935–940.

Cervera, M., Juárez, J., Navarro, A., Pina, J.A., Duran-Vila, N., Navarro, L., and Peña, L. (1998) Genetic transformation and regeneration of mature tissues of woody fruit plants bypassing the juvenile stage. Transgenic Research. 7: 51–59.

Cervera, M., Ortega, C., Navarro, L., Navarro, A., and Peña, L (2000) Generation of transgenic citrus plants with the tolerance-to-salinity gene HAL2 from yeast. J Hort. Sci. Biotech. 75: 26–30.

Cervera, M., Navarro, A., Navarro, L. and Peña, L. (2008) Production of transgenic adult plants from clementine mandarin by enhancing cell competence for transformation and regeneration. Tree Physiology 28: 55–66.

Chen, Z.Q., Wang, M.Q. and Huihua, L. (1980) The induction of Citrus pollen plants in artificial media. Acta Genet. Sin. 7: 189–191.

Chen, C., Zhou, P., Choi, Y.A., Huang, S. and Gmitter Jr., F.G. (2006) Mining and characterizing microsatellites from citrus ESTs. Theor. Appl. Genet. 112: 1248–1257.

Chen, C., Bowman, K.D., Choi, Y.A., Dang, P.M., Rao, M.N., Huang, S., Soneji, J.R., McCollum, T.G. and Gmitter, Jr F.G. (2007) EST-SSR genetic maps for *Citrus sinensis* and *Poncirus trifoliata*. Tree Genetics & Genomics, 4: 1–10.

Chen, C., Lyon. M.T., O'Malley, D., Federici, C.T., Gmitter, J., Grosser, J.W., Chaparro, J.X., Roose, M.L., Gmitter Jr., F.G. (2008) Origin and Frequency of 2n Gametes in *Citrus sinensis* × *Poncirus trifoliata* and Their Reciprocal Crosses, Plant Science174: 1–8.

Close, T.J., Wanamaker, S., Lyon, M., Mei, G.W., Davies, C., Roose, M.L. (2006) A GeneChip® for citrus. In: Plant & Animal Genome XIV Conference; January 2006; San Diego, Calif, USA. P. 82.

Corazza-Nunes, M.J., Machado, M.A., Nunes, W.M.C., Cristofani, M., Targon, M.L.P.N. (2002) Assessment of genetic variability in grapefruits (*Citrus paradisi* Macf.) and pummelos (*C. maxima* (Burm.) Merr.) using RAPD and SSR markers. Euphytica 126: 169–176.

Cuenca, J., Aleza, P., Juárez, J., Pina, J.A. and Navarro, L. (2010) 'Safor' mandarin: a new mid-late citrus triploid hybrid. HortScience 45(6): 977–980.

Deng, Z., Huang, S., Xiao, S. and Gmitter Jr., F.G. (1997) Development and characterization of SCAR markers linked to the citrus tristeza virus resistance gene from *Poncirus trifoliata*. Genome, 40: 697–704.

Domínguez, A., Hermoso de Mendoza, A., Guerri, J., Cambra, M., Navarro, L., Moreno, P. and Peña, L. (2002) Pathogen-derived resistance to citrus tristeza virus (CTV) in transgenic Mexican lime (*Citrus aurantifolia* (Christ.) Swing.) plants expressing its p25 coat protein gene. Molecular Breeding 10: 1–10.

Domínguez, A., Cervera, M., Pérez, R., Romero, J., Fagoaga, C., Cubero, J., López, M.M., Juárez, J., Navarro, L. and Peña, L. (2004) Characterisation of regenerants obtained under selective conditions after *Agrobacterium*-mediated transformation of citrus explants reveals production of silenced and chimeric plants at unexpected high frequencies. Molecular Breeding 14: 171–183.

Drira, N. and Benbadis, A. (1975) Analysis, by in vitro anther culture, of the androgenetic potential of two Citrus species (*Citrus medica* and *Citrus limon*). Comptes Rendus Hebdomadaires de Séances de l' Academie des Sciences 281:1321–1324.

Duan, Y.X., Guo, W.W., Meng, H.J., Tao, N.G., Li, D.D. and Deng, X.X. (2007) High efficient transgenic plant regeneration from embryogenic calluses of *Citrus sinensis* Biologia Plantarum. 51 212–216.

Duran-Vila, N. (1997) Cryoconservation of citrus genetic resources. In: Razdan, M.K., and Cocking, E.C. (eds.) Conservation of Plant Genetic Resources in vitro. Science Publishers, Inc., Enfield (N.H.), 1: 175–199.

Durham, R.E., Liou, P.C., Gmitter Jr., F.G. and Moore, G.A. (1992) Linkage of restriction fragment length polymorphisms and isozymes in Citrus. Theor. Appl. Genet., 84: 39–48.

Endo, T., Shimada, T., Fujii, H., Kobayashi, Y., Araki, T. and Omura, M. (2005) Ectopic expression of an FT homolog from citrus confers an early flowering phenotype on trifoliate orange (*Poncirus trifoliata* L. Raf.).Transgenic Research 14: 703–712.

Engelmann, F., Dambier, D. and Ollitrault, P. (1994) Cryopreservation of embryogenic cell suspensions and calluses of *Citrus* using a simplified freezing process. Cryo-Letters 15: 53–58.

Fagoaga, C., Rodrigo, I., Conejero, V., Hinarejos, C., Tuset, J.J., Arnau, J., Pina, J.A., Navarro, L. and Peña, L. (2001) Increased tolerance to *Phytophthora citrophthora* in transgenic orange plants constitutively expressing a tomato pathogenesis related protein PR-5 Molecular Breeding 7: 175–185.

Fagoaga, C., López, C., Hermoso de Mendoza, A., Moreno, P., Navarro, L., Flores, R. and Peña, L. (2006) Post-transcriptional gene silencing of the p23 silencing suppressor of Citrus tristeza virus confers resistance to the virus in transgenic Mexican lime. Plant Molec. Biol. 60: 155–167.

Fagoaga, C., Tadeo, F.R., Iglesias, D.J., Huerta, L., Lliso, I., Vidal, A.M., Talon, M., Navarro, L., García-Martinez, J.L. and Peña, L. (2007) Engineering of gibberellin levels in citrus by sense and antisense overexpression of a GA 20-oxidase gene modifies plant architecture. J. Exp. Botany, 58: 1407–1420.

Fanciullino, A.L., Dhuique-Mayer, C., Luro, F., Casanova, J., Morillon, R. and Ollitrault, P. (2006) Carotenoid diversity in cultivated *Citrus* is highly influenced by genetic factors. J. Agric. Food. Chem. 54: 4397–4406.

Fanciullino, A.L, Dhuique-Mayer, C., Luro, F., Morillon, R. and Ollitrault, P. (2007) Carotenoid Biosynthetic Pathway in the *Citrus* Genus: Number of Copies and Phylogenetic Diversity of Seven Genes. J. Agric. Food Chem., 55: 7405–7417.

Fang, D.Q. and Roose, M.L. (1997) Identification of closely related citrus cultivars with inter-simple sequence repeat markers. Theor. Appl. Genet. 95: 408–417.

Fang, D.Q., Federici, C.T. and Roose, M.L. (1997) Development of molecular markers linked to a gene controlling fruit acidity in citrus. Genome 40: 841–849.

Fang, D.Q., Federici, C.T. and Roose, M.L. (1998) A high resolution linkage map of the citrus tristeza virus resistance gene region in *Poncirus trifoliata* (L.) Raf. Genetics 150 : 883–890.

FAO (2006) Citrus Fruit fresh and processed. CCP.CI/ST/2006, FAO. 47 p.

Federici, C.T., Fang, D.Q., Scora, R.W. and Roose, M.L. (1998) Phylogenetic relationships within the genus *Citrus* (Rutaceae) and related genera as revealed by RFLP and RAPD analysis. Theor. Appl. Genet., 96: 812–822.

Fleming, G., Olivares-Fuster, O., Del-Bosco, S., Grosser, J. (2000) An alternative method for the genetic transformation of sweet orange. In Vitro Cellular Develop. Biol. - Plant, 36: 450–455.

Forment, J., Gadea, J., Huerta, L., Abizanda, L., Agustí, J., Alamar, S., Alos E., Andres, F., Berbel, A., Blázquez, M. A., Brumos, J., Cercós, M., Colmenero-Flores, J.M., Conesa, A., Estables, B., Gandia, M., García-Martinez, J.L., Gimeno, J. , Gisbert, A., Gómez, G., Gonzalez-Candelas, L., Granell, A., Guerra, J., Lafuente, T., Madueno, F., Marcos, J.F., Martinez, F., Martinez-Godoy, M.A., Miralles, S., Moreno, P., Navarro, L., Pallas, V., Perez-Amador, M.A., Perez-Valle, J., Pons, C., Rodrigo I., Rodríguez, P., Royo, C., Serrano, R., Soler, G., Tadeo, F., Talon, M., Terol, J., Trenor, M., Vaello, L., Vicente, O., Vidal, C.H., Zacarias, L. and Conejero, V. (2005) Development of a citrus genome-wide EST collection and cDNA microarray as resources for genomic studies. Plant Molecular Biology 57: 375–391.

Forner, J.B., Forner-Giner, M.A. and Alcaide, A. (2003) Forner-Alcaide 5 and Forner-Alcaide 13: Two new citrus rootstocks released in Spain. 38: 629–630.

Frison, E. and Taher, M.M. (1991) *FAO/IBPGR Technical guidelines for the safe movement of citrus germplasm.* Food and Agriculture Organization of the United Nations, Rome/ International Board for Plant Genetic Resources, Rome, 50 pp.

Froelicher, Y., Bassene, J.B., Jedidi-Neji, E., Morillon, R., Bernardini, G., Costantino, G. and Ollitrault, P. (2007) Induced parthenogenesis in mandarin: induction procedures and genetic analysis of plantlets. Plant Cell Rep. . 26: 937–944.

Froelicher, Y., Dambier, D., Costantino, G., Lotfy, S., Didout, C., Beaumont, V., Brottier, P., Risterucci, A.-M., Luro, F. and Ollitrault, P. (2008) Characterization of microsatellite markers in Citrus reticulata Blanco. Molecular Ecology Resources. 8: 119–122.

Frost, H.B. (1935) Four new citrus varieties—the Kara, Kinnow, and Wilking mandarins and the Trovita orange. Univ. Calif. Agr. Expt. Sta. Bul. 597: 14 pp.

Fujii, H., Shimada, T., Sugiyama, A., Nishikawa, F., Endo, T., Nakano, M., Ikoma, Y., Shimizu, T. and Omura, M. (2007) Profiling ethylene-responsive genes in mature mandarin fruit using a citrus 22K oligoarray. Plant Science 173: 340–348.

García, R., Asins, M.J., Forner, J. and Carbonell, E.A. (1999) Genetic analysis of apomixis in Citrus and Poncirus by molecular markers. Theor. Appl. Genet. 99: 511–518.

Gancel, A.L., Ollitrault, P., Froelicher, Y., Tomi, F., Jacquemond, C., Luro, F.and Brioullet, JM. (2003). Leaf volatile compounds of seven citrus somatic tetraploid hybrids sharing willow leaf mandarin (*Citrus deliciosa ten.*) as their common parent. J. Agric. Food Chem., 51: 6006–6013.

Gancel, A.L., Grimplet, J., Sauvage, F.X., Ollitrault, P. and Brillouet, J.M. (2006). Predominant expression of diploid mandarin leaf proteome in two citrus mandarin-derived somatic allotetraploid hybrids. J. Agric. Food Chem., 54(17): 6212–6218.

Gentile, A., Tribulato, E., Deng, Z.N., Galun, E., Fluhr, R. and Vardi A. (1992) In vitro selection of nucellar lemon callus and regeneration of plants tolerant to *Phoma tracheiphila* Toxin. Proc. Int. Soc. Citriculture 1: 150–153.

Germanà, M.A. (1992) Androgenesis in Citrus: A review. Proc. Int. Soc. Citriculture 1:183–189.

Germanà, M.A. (2003) Somatic embryogenesis and plant regeneration from anther culture of *Citrus aurantium* and *C. reticulata*. Biologia-Bratislava. 58: 843–850.

Germanà, M.A. (2007) Haploidy. In: I.A. Khan (ed.). *Citrus* genetics, breeding and biotechnology. CAB International, Wallingford, pp. 167–196.

Germanà, M.A. and Chiancone, B. (2001) Gynogenetic haploids of *Citrus* after in vitro pollination with triploid pollen grains. Plant Cell Tissue Organ Culture 66: 59–66.

Germanà, M.A., Crescimanno, F.G., De Pasquale, F. and Ying W.Y. (1992) Androgenesis in 5 cultivars of Citrus limon L. Burm. f. Acta Hort. 300: 315–324.

Ghorbel, R., Juárez, J., Navarro, L. and Peña, L. (1999) Green fluorescent protein as a screenable marker to increase the efficiency of generating transgenic woody fruit plants. Theor. Appl. Genet. 99: 350–358.

Gmitter Jr., F.G. and Hu, X. (1990) The possible role of Yunan, China, in the origin of contemporary *Citrus* species (Rutaceae). Economic Botany 44: 267–277.

Gmitter Jr., F.G., Xiao, S.Y., Huang, S., Hu, X.L., Garnsey, S.M. and Deng, Z. (1996) A localized linkage map of the citrus tristeza virus resistance gene region. Theor. Appl. Genet. 92: 688–695.

Gmitter Jr., F.G., Deng, Z. and Chen, C. (2007) Cloning and characterization of disease resistance genes. In: I.A. Khan (ed.). Citrus genetics, breeding and biotechnology. CAB International, Wallingford, pp. 287–305.

Green, R.M., Vardi, A. and Galun, E. (1986) The plastome of citrus. physical map, variation among citrus cultivars and species and comparison with related genera. Theor. Appl. Genet. 72: 170–177.

Grosser, J.W. and Gmitter Jr., F.G. (1990) Protoplast fusion and citrus improvement. Plant Breeding Rev. 8: 339–374.

Grosser, J.W., Mourao-Fo, F.A.A., Gmitter Jr., F.G., Louzada, E.S., Jiang, J., Baergen, K., Quiros, A., Cabasson, C., Schell, J.L. and Chandler, J.L. (1996a). Allotetraploid hybrids between citrus and seven related genera producted by somatic hybridization. Theor. Appl. Genet. 92: 577–582.

Grosser J.W., Gmitter Jr., F.G., Tusa, N., Recupero, G.R. and Cucinotta, P. (1996b) Further evidence of a cybridization requirement for plant regeneration from citrus leaf protoplasts following somatic fusion. Plant Cell Reports; 15: 672–676.

Grosser, J.W., Ollitrault, P. and Olivares-Fuster, O. (2000) Somatic hybridization in Citrus : an effective tool to facilitate cultivar improvement. In Vitro Cellular and Development Biology - Plant 36: 434–449.

Grosser, J.W., Deng, X.X. and Goodrich, R.M. (2007) Somaclonal variation in sweet orange: practical applications for variety improvement and possible causes. InI.A. Khan (ed.). Citrus genetics, breeding and biotechnology. CAB International, Wallingford, pp. 219–233.

Guerra, M.S. (1993) Cytogenetics of Rutaceae. V. High chromosomal variability in Citrus species revealed by CMA/DAPI staining. Heredity 71: 234–241.

Gulsen, O. and Roose, M.L. (2001) Lemons: diversity and relationships with selected Citrus genotypes as measured with nuclear genome markers. J. Am. Soc. Hort. Sci. 126: 309–317.

Guo, W.W., Prasad, D., Cheng, Y.J., Serrano, P., Deng, X.X. and Grosser, J.W. (2004) Targeted cybridization in citrus: transfer of satsuma cytoplasm to seedy cultivars for potential seedlessness. Plant Cell Reports 22: 752–758.

Guo, W.W., Duan, Y., Olivares-Fuster, O., Wu, Z., Arias, C.R.,. Burns, J.K, and Grosser, J.W. (2005) Protoplast transformation and regeneration of transgenic Valencia sweet orange plants containing a juice quality-related pectin methylesterase gene. Plant Cell Reports 24: 482–486.

Handa, T., Ishizawa, Y. and Oogaki, C. (1986) Phylogenetic study of fraction I protein of *Citrus* and its close related genera. J. Genet. 61: 15–24.

Hao, Y.J., You, C.X. and Deng, X.X. (2002) Effects of cryopreservation on developmental competency, cytological and molecular stability of citrus callus. CryoLetters, 23: 27–35.

Hearn, C.J. (1984) Development of seedless orange, *Citrus sinensis*, cultivar Pineapple and grapefruit, *Citrus paradisi*, cultivars through seed irradiation. J. Am. Soc. Hort. Sci. 109: 270–273.

Hensz, R.A. (1971) 'Star Ruby', a new deep-red-fleshed grapefruit variety with distinct tree characteristics. J. Rio Grande Valley Hort. Soc. 25: 54–58.

Herrero, R., Asíns, M.J., Carbonell, E.A., Navarro, L.(1996). Genetic diversity in the orange subfamilyAurantioideae. I. Intraspecies and intragenus genetic variability. Theor. Appl. Genet. 92: 599–609.

Herrero, R., Asins, M.J., Pina, J.A., Carbonell, E.A. and Navarro, L. (1997) Genetic diversity in the orange subfamily Aurantiodeae. II. Genetic relationships among genera and species. Theor. Appl. Genet. 93: 1327–1334.

Hidaka, T., Yamada, Y. and Shichijo, T. (1979) In vitro differentiation of haploid plants by anther culture in *Poncirus trifoliata* (L.) Raf. Japan. J. Breed. 29: 248–254.

Hidaka, T., Omura, M., Ugari, M., Tomiyama, M., Kato, A., Oshima, M. and Motoyoshi, F. (1990) Agrobacterium-mediated transformation and regeneration of *Citrus* spp. from suspension cells. Japan. J. Breed. 40: 199–207.

Hidaka, T., Moriguchi, T., Motomura, T., Katagi, S. and Omura, M. (1995) Development of a new electrode chamber and its efficiency in protoplast fusion in *Citrus*. Breeding Sci. 45: 237–239.

Hisada, S., Akihama, T., Endo, T., Moriguchi, T. and Omura, M. (1997) Expressed sequence tags of Citrus fruit during rapid cell development phase. J. Am. Soc. Hort. Sci. 122: 808–812.

Hong, L. and Deng, X. (2005) Analysis of DNA methylation in navel oranges based on MSAP marker. Scientia Agricultura Sinica. 38: 2301–2307.

Iwamasa, M., Nito, N. and Ling, J.T. (1988) Intra and intergeneric hybridization in the orange subfamily, Aurantioideae. Proc. Inten. Soc. Citriculture 1:123–130.

Iwanami T, Shimizu T, Ito T, Hirabayashi T (2004) Tolerance to citrus mosaic virus in transgenic trifoliate orange lines harboring capsid polyprotein gene. Plant Dis 88(8): 865–868.

Jarrell, D.C., Roose, M.L., Traugh, S.N. and Kupper, R.S. (1992) A genetic map of citrus based on the segregation of isozymes and RFLPs in an intergeneric cross. Theor. Appl. Genet. 84: 49–56.

Juárez, J., Aleza, P., Olivares-Fuster, O. and Navarro, L. (2004) Recovery of tetraploid clementine plants (*Citrus clementina* hort. ex Tan.) by in vitro colchicine treatment of shoot tips. Proc. Int. Soc. Citriculture 1:151–154.

Kaneyoshi, J., Kobayashi, S., Nakamura, Y., Shigemoto, N. and Doi, Y. (1994) A simple and efficient gene transfer system of trifoliate orange. Plant Cell Reports 13: 541–545.

Kepiro, J.L. and Roose, M.L. (2007) Nucellar embryony. In: I.A. Khan (ed.). Citrus genetics, breeding and biotechnology. CAB International, Wallingford, pp. 141–150.

Khan, I.A. and Roose M.L. (1988) Frequency and characteristics of nucellar and zygotic seedlings in three cultivars of trifoliate orange. J. Amer. Soc. Hort. Sci. 113: 105–110.

Kijas, J.M.H., Fowler, J.C.S. and Thomas, M.R. (1995) An evaluation of sequence tagged microsatellite site markers for genetic analysis within *Citrus* and related species. Genome 38:349–55.

Kobayashi, S. (1987) Uniformity of plants regenerated from orange (*Citrus sinensis* Osb.) protoplasts. Theor. Appl. Genet. 74:10–14.

Kobayashi, S., Ohgawara, T., Saito, W., Nakamura, Y. and Omura, M. (1997) Production of triploid somatic hybrids in *Citrus*. Journal of the Jpn. Soc. Hort. Sci. 66: 453–458.

Kochba, J. and Spiegel-Roy, P. (1977) Cell and tissue culture for breeding and developmental studies of Citrus. HortScience., 12 : 110–114.

Kochba, A.J., Ben Hayyim, G., Spiegel-Roy, P., Saad, S. and Neumann H., (1982) Selection of stable salt-tolerant callus cell lines and embryos in *Citrus sinensis* and *Citrus aurantium*. Z. Pflanzenphysiol. 106: 111–118.

Krueger, R.R. and Navarro, L. (2007) Citrus germplasm resources.. In: I.A. Khan (ed.). Citrus genetics, breeding and biotechnology. CAB International, Wallingford, pp. 45–140.

Krug, C.A. (1943) - Chromosome numbers in the subfamily Arantioideae, with special reference in the genus *Citrus*. Citrus Bot. Gaz., 104, 602–611.

Lee, R.F., Lehaman, P. S., Navarro, L. (1999) Nursery practices and certification programs for budwood and rootstocks, p.35-46. In: L.W. Timmer and L. W. Duncan (eds.), Citrus health management, APS Press, Minnesota, USA, 197 pp.

Liang, Xiong G., Guo, Q., He, Q. and Li X. (2006) AFLP analysis and the taxonomy of citrus. Acta Hort. 760: 137–142.

Ling, J.T. and Iwamasa, M. (1994) Somatic hybridization between *Citrus reticulata* and *Citropsis gabunensis* through electrofusion. Plant Cell Rep 13: 493–497.

Ling, P., Duncan, L.W., Deng, Z., Dunn, D., Xu, X., Huang, S. and Gmitter Jr., F.G. (2000) Inheritance of citrus nematode resistance and its linkage with molecular markers. Theor. Appl. Genet. 100:1010–1017.

Liu, J.H., Xu X.Y. and Deng, X.X. (2004) Advances on *Citrus* somatic hybrids and inheritance of their nuclear and cytoplasmic components. J. Agricultural Biotech. 12: 237–246.

Louzada, E.S. (2007) Microprotoplast-mediated chromosome transfer and its potential for citrus breeding. In: I.A. Khan (ed.). Citrus genetics, breeding and biotechnology. CAB International, Wallingford, pp. 261–273.

Luro, F., Lorieux, M., Laigret, F., Bové, J.M. and Ollitrault, P. (1995) Genetic Mapping of an Intergeneric Citrus Hybrid Using Molecular Markers. Fruits, 49: 404–408.

Luro, F., Rist, D. and Ollitrault, P. (2001) Evaluation of genetic relationships in Citrus genus by means of sequence tagged microsatellites. Acta Hort. 546: 237–242.

Luro, F., Maddy, F., Jacquemond, C., Froelicher, Y., Morillon, R., Rist, D. and Ollitrault, P. (2004) Identification and evaluation of diplogyny in clementine (*Citrus clementina*) for use in breeding. Acta Hort. 663: 841–847.

Luro, F., Costantino, G., Terol, J., Argout, X., Allario, T., Wincker, P., Talon, M., Ollitrault, P. and Morillon, R. (2008) Transferability of the EST-SSRs developed on Nules clementine (*Citrus clementina* Hort ex Tan) to other *Citrus* species and their effectiveness for genetic mapping. BMC genomics 9: 287.

Maheshwari, P. and Ranga Swamy, N.S. (1958) Polyembryonny and in vitro culture of embryos of *Citrus* and *Mangifera*. Indian J. Horticulture 15: 275–282.

Martinez-Godoy, M.A., Mauri, N., Juárez, J., Marques, M.C., Santiago, J., Forment, J. and Gadea, J. (2008) A genome-wide 20K citrus microarray for gene expression analysis. BMC Genomics 9: 318.

Molinari, H.B.C., Marur, C. J., Bespalhok, J.C.F., Kobayashi, A.K., Pileggi, M., Leite R.P., Junior, R.P., Pereira, L.F.P. and Vieira, L.G.E. (2004) Osmotic adjustment in transgenic citrus rootstock Carrizo citrange (*Citrus sinensis* Osb. x *Poncirus trifoliata* L. Raf.) overproducing proline, Plant Science 167: 1375–1381.

Moore, G.A., Jacono, C.C., Neidigh, J.L., Lawrence S.D. and Cline, K. (1992) Agrobacteruim mediated transformation of Citrus stem segments and regeneration of transgenic plants. Plant Cell Reports 11: 238–242.

Moreira, C.D., Chase, C.D., Gmitter Jr., F.G. and Grosser, J.W. (2000) Transmission of organelle genomes in citrus somatic hybrids. Plant-Cell,-Tissue-and-Organ-Culture. 61: 165–168.

Mozoruk, J., Hunnicutt, L.E., Cave, R.D., Hunter, W.B. and Bausher, M.G. (2006) Profiling transcriptional changes in *Citrus sinensis* (L.) Osbeck challenged by herbivory from the xylem-feeding leafhopper *Homalodisca coagulata* (Say) by cDNA macroarray analysis. Plant Science 170: 1068–1080.

Nair, P.K.R. and Randhawa, G.S. (1969) Chromosome morphology of the pachytene stage with respect to different *Citrus* types. In: H. Chapmn (ed.), Proc. First Int. Citrus Symp., (Univ. California, Riverside 1: 215–223.

Navarro, L. (1992) Citrus shoot tip grafting *in vitro*. In: J.P.S. Bajaj (ed.), Biotechnology in Agriculture and Forestry, Vol 18. High-Tech and Micropropagation. Springer-Verlag, Berlin, pp. 327–338.

Navarro, L. (1993) Citrus sanitation, quarantine and certification programs. In P. Moreno, J.V. da Graca and L.W. Timmer (eds.), Proc. 12th Conf. Intern. Organization Citrus Virol., IOCV, Riverside, pp. 383–391.

Navarro, L. and Juarez, J. (2007) Shoot-tip grafting in vitro: impact in the citrus industry and research applications. In: I.A. Khan (ed.). Citrus genetics, breeding and biotechnology. CAB International, Wallingford, pp. 353–364.

Navarro, L., Roistacher, C.N. and Murashige, T. (1975) Improvement of shoot tip grafting in vitro for virus-free citrus. J. Am. Soc. Hort. Sci. 100: 471–479.

Navarro, L., Juarez, J., Pina, J.A. and Ballester, J.F. (1984) The citrus quarantine station in Spain. In: Garnsey, S.M, Timmer, L.W. and Dodds, J.A. (eds), Proc. 9th Conf. Intern. Organization Citrus Virol., IOCV, Riverside, pp. 365–370.

Navarro, L., Ortiz, J.M. and Juarez, J. (1985) Aberrant citrus plants obtained by somatic embryogenesis of nucelli cultured in vitro. HortScience,20 (2): 214–215.

Navarro, L., Civerolo, E.L., Juárez, J., and Garnsey, S.M. (1991) Improving therapy methods for citrus germplasm exchange. In: Brlansky, R.H., Lee, R.F. and Timmer L.W. (eds.), Proc. 11st Conf. Intern. Organization Citrus Virol., IOCV, Riverside, pp. 400–408.

Navarro, L., Pina, J.A., Juárez, J., Ballester-Olmos, J.F., Arregui, J.M., Ortega, C., Navarro, A, Duran-Vila, N., Guerri, J., Moreno, P., Cambra, M. and Zaragoza, S. (2002) The Citrus Variety Improvement Program in Spain in the Period 1975–2001. In: N. Duran-Vila, R.G. Milne and J.V. da Graça (eds.), Proc. 15th Conf. Intern. Organization Citrus Virol, IOCV, Riverside. pp. 306–316.

Navarro, L., Juárez, J., Aleza, P. and Pina, J.A. (2003) Recovery of triploid seedless mandarin hybrids from 2n x 2n and 2n x 4n crosses by embryo rescue and flow cytometry, p.541-544. In: I.K. Vasil (ed.), Plant Biotechnology 2002 and Beyond, Kluwer Acad. Pu., Dordrecht.

Navarro, L., Olivares Fuster, O., Juarez, J., Aleza, P., Pina, J.A., Ballester Olmos, J.F., Cervera, M., Fagoaga, C., Duran Vila, N. and Peña, L. (2004) Applications of biotechnology to citrus improvement in Spain. Acta Hort. (632): 221–234.

Navarro, L., J. Juarez, P. Aleza, J., Cuenca, J.M., Julve, J.A., Piña and Olivares-Fuster, O. (2006) Selecci6n de nuevos mandarinos triploides. Agraria Comunitat Valenciana. 2ª época, año 2, 7: 23–26.

Nicolosi, E., Deng, Z.N., Gentile, A., La Malfa, S., Continella, G. and Tribulato, E. (2000) *Citrus* phylogeny and genetic origin of important species as investigated by molecular markers. Theor. Appl. Genet. 100: 1155–1166.

Nicolosi, E. (2007) Origin and taxonomy. In: I.A. Khan (ed.). *Citrus* genetics, breeding and biotechnology. CAB International, Wallingford, pp. 19–44.

Ohgawara, T., Kobayashi, S., Ohgawara, E., Uchimiya, H. and Ishii, S. (1985) Somatic hybrid plants obtained by protoplast fusion between *Citrus sinensis* and *Poncirus trifolata*. Theor Appl Genet: 71: 1–4.

Oiyama, I. and Kobayashi, S. (1993) Haploids obtained from diploid-triploid crosses of citrus. J. Japan. Soc. Hort. Sci. 62:89–93.

Olivares-Fuster, O., Asins, M.J., Duran-Vila, N. and Navarro, L. (2000) Cryopreserved callus, a source of protoplasts for citrus improvement. J. Hort. Sci. Biotech 75: 635–640.

Olivares-Fuster, O., Peña, L., Duran-Vila, N. and Navarro, L. (2002) Green fluorescent protein as visual marker in somatic hybridization. Ann. Bot. 89: 491–497.

Olivares-Fuster, O, Duran-Vila, N. and Navarro, L. (2005) Electrochemical protoplast fusion in citrus. Plant Cell Rep. 24: 112–119.

Olivares-Fuster, O., Hernández-Garrido, M., Guerri, J. and Navarro, L. (2007) plant somatic hybrid cytoplasm characterization by single strand conformation polymorphism. Tree Physiology 27: 785–792.

Ollitrault, P., Dambier, D., Cabasson, C., Allent, V. and Engelmann, F. (1995) optimized management of citrus embryogenic calli for breeding programmes. Fruits, 49: 394–397.

Ollitrault, P., Dambier, D., Sudahono, S. and Luro, F. (1996a) Somatic hybridisation in *Citrus*: some new hybrid and alloplasmic plants. Proc. Int. Soc. Citriculture 2: 907–912.

Ollitrault, P., Allent, V. and Luro, F. (1996b) Production of haploid plants and embryogenic calli of Clementine (*Citrus reticulata* Blanco) after in situ parthenogenesis induced by irradiated pollen. Proc. Int. Soc. Citriculture. 2: 913–917.

Ollitrault, P., Dambier, D., Seker, M. and Froelicher, Y. (2000a) Rootstock breeding by somatic hybridization for the Mediterranean citrus industry. Acta Hort. 535: 157–162.

Ollitrault, P., Dambier, D., Vanel, F. and Froelicher, Y. (2000b) Creation of triploid citrus hybrids by electrofusion of haploid and diploid protoplasts. Acta Hort. 535: 191–197.

Ollitrault, P., Dambier, D., Froelicher, Y., Luro, F. and Cottin, R. (2001) La diversité des agrumes ; structuration et exploitation par hybridation somatique. Comptes rendus de l'Académie d'Agriculture. 86: 197–221.

Ollitrault, P., Jacquemond, C., Dubois, C. and Luro, F. (2003) Citrus. In Genetic diversity of cultivated tropical plants. Hamon, Seguin, Perrier and Glaszmann (eds). Science Publishers Inc., pp. 193–217.

Ollitrault, P., Froelicher, Y., Dambier, D., Luro, F. and Yamamoto, M. (2007a) Seedlessness and ploidy manipulation. In: I.A. Khan (ed.). *Citrus* genetics, breeding and biotechnology. CAB International, Wallingford, pp. 197–218.

Ollitrault, P., Guo, W. and Grosser J.W. (2007b) Somatic hybridization. In: I.A. Khan (ed.). Citrus genetics, breeding and biotechnology. CAB International, Wallingford, pp. 235–260.

Ollitrault, P., Dambier, D., Luro, F. and Froelicher, Y. (2008) Ploidy manipulation for breeding seedless triploid citrus. Plant Breeding Review 30: 323–352.

Peña, L., Cervera, M., Juárez, J., Ortega, C., Piña, J.A., Durán-Vila, N. and Navarro, L. (1995) High efficiency Agrobacterium-mediated transformation and regeneration of citrus. Plant Science 104: 183–191.

Peña, L., Martin-Trillo, M., Juarez, J., Piña, J.A., Navarro, L. and Martinez-Zapater, J.M. (2001) Constitutive expression of Arabidopsis LEAFY or APETALA1 genes in citrus reduces their generation time. Nature Biotechnology. 19: 263–267.

Peña, L., Cervera, M., Ghorbel, R., Dominguez, A., Fagoaga, C., Juarez, J., Piña, J.A. and Navarro, L. (2007) In: I.A. Khan (ed.). *Citrus* genetics, breeding and biotechnology. CAB International, Wallingford, pp. 329–344.

Peña, L., Cervera, M., Fagoaga, C., Romero, J., Ballester, A., Soler, N., Pons, E., Rodríguez, A., Peris, J, Juárez, J. and Navarro, L. (2008) *Citrus*. In: C. Kole and T.C. Hall (eds.), Compendium of Transgenic Crop Plants, Vol. 5, Fruits & Nuts (Tropical & Sub-tropical). Willey-Blackwell, pp.

Pérez, R.M., Navarro, L. and Duran-Vila N. (1997) Cryopreservation and storage of embryogenic callus cultures of several Citrus species and cultivars. Plant Cell Reports 17: 44–49.

Pérez, R.M., Galiana, A.M., Navarro, L. and Duran-Vila, N. (1998) Embryogenesis *in vitro* of several *Citrus* species and cultivars. J. Hort. Sci. and Biotech. 73: 796–802.

Pérez, R.M., Mas, O., Navarro, L. and Duran-Vila, N. (1999) Production and cryoconservation of embryogenic cultures of mandarin and mandarin hybrids. Plant Cell Tissue Organ Culture 55:71–74.

Raghuvanshi, S.S. (1969) Cytological evidence bearing on evolution in *Citrus*. In: H. Chapmn (ed.), Proc. First Int. Citrus Symp., Univ. California, Riverside, 1: 207–214.

Rai , M. (2006) Refinement of the Citrus tristeza virus resistance gene (Ctv) positional map in *Poncirus trifoliata* and generation of transgenic grapefruit (Citrus paradise) plant lines with candidate resistance genes in this region. Plant Mol Biol 61: 399–414.

Rangan, T.S., Murashige, T. and Bitters, W.P. (1969) In vitro studies of zygotic and nucellar embryogenesis in citrus. In: H. Chapmn (ed.), Proc. First Int. Citrus Symp., Univ. California, Riverside 1: 225–229.

Roistacher, C.N. (1977) Elimination of citrus pathogens in propagative budwood. I. Budwood selection, indexing and thermotherapy. *Proc. Int. Soc. Citriculture* 3: 965–972.

Roose, M.L. and Williams, E. (2007) Mutation Breeding. In: I.A. Khan (ed.). Citrus genetics, breeding and biotechnology. CAB International, Wallingford, pp. 345–352.

Roose, M.L. and Close, T.J. (2008) Genomics of citrus, a major fruit crop of tropical and subtropical regions. In Book Series Plant Genetics and Genomics. Crops and Models. Paul H. Moore and Ray Ming Ed. Springer New York, 1: 187–202.

Russo, G., Recupero, S., Puglisi, A. and Recupero, G.R. (2004) New triploid citrus hybrids by Italian genetic improvement.. Rivista di Frutticoltura e di Ortofloricoltura 66: 14–18.

Ruiz, C., Bretó, P.M. and Asíns, M.J. (2000) A quick methodology to identify sexual seedlings in citrus breeding programs using SSR markers. Euphytica 112: 89–94.

Saito, W., Ohgawara, T., Shimizu, J. and Ishii, S. (1991) Acid citrus somatic hybrids between Sudachi (*Citrus sudachi* Hortex Shirai) and lime (*C. aurantifolia* Swing) produced by electrofusion. Plant Sci 77: 125–130.

Sakai, A., Kobayashi, S. and Oiyama, I. (1991) Survival by vitrification of nucellar cells of navel orange (*Citrus sinensis var. brasiliensis* Tanaka) cooled to -196^{0}C. J. Plant Physiol. 137: 465–470.

Scora, R.W. (1975) On the history and origin of citrus. In symposium on the biochemical systematics, genetic and origin of cultivated plants. Bul. Torrey Bot. Club, 102: 369–375.

Shimada, T., Fujii, H., Endo, T., Yakazi, J., Kishimoto, N., Shimbo, K., Kikuchi, S., and Omura, M.. (2005) Toward comprehensive expression profiling by microarray analysis in citrus: monitoring the expression profiles of 2213 genes during fruit development" Plant Science 168: 1383–1385.

Soost, R.K. (1968) The incompatibility gene system in citrus. In: H. Chapmn (ed.), Proc. First Int. Citrus Symp., Univ. California, Riverside 1: 189–190.

Soost, R.K. (1987) Breeding citrus-genetics and nucellar embryony. In Abbott A.J. and Atkin R.K. (eds.) Improving vegetatively propagated crops. Academic Press Limited, London, pp. 83–110.

Spiegel-Roy, P. and Ben Hayim, G. (1985) Selection and breeding for salt tolerance in vitro. Plant Soil 89: 243–252.

Stace, H.M, Armstrong, J.A. and James, S.H. (1993) Cytoevolutionary patterns in Rutaceae. Pl. Syst. Evol. 187: 1–28.

Starrantino, A. (1999) Tacle, a new triploid clementine x Tarocco hybrid. Rivista di Frutticoltura e di Ortofloricoltura 61(1): 45–47.

Starrantino, A. and Recupero, G. (1982) *Citrus* hybrids obtained in vitro from 2x females X 4x males. Proc. Int. Soc. Citriculture, 1: 31–32.

Starrantino, A. and Russo, F. (1983) Reproduction of seedless orange cultivars from undeveloped ovules raised 'in vitro'. Acta Hort. 131: 253–258.

Swingle, W.T. and Reece, P.C. (1967) The botany of *Citrus* and its wild relatives. In : The citrus industry. 1. History, world distribution, botany and varieties, W. Reuther et al. éd., Berkeley, Etats-Unis, University of California Press, pp. 190–430.

Tadeo, F., Cercos, M, Colmenero-Flores, J.M., Iglesias, D.J., Naranjo, M.A., Rios, G., Carrera, E., Ruiz-Rivero, O., Lliso I., Morillon, R., Ollitrault, P. and Talon, M. (2008) Molecular physiology of development and quality of citrus. Advances in Botanical Research, 47: 147–223.

Tal6n, M. and Gmitter Jr., F.G (2008) . *Citrus* Genomics. Intern. J. Plant Genomics. 2008, article ID528361, 17 pages, doi:10.1155/2008/528361.

Tanaka, T. (1954) Species problem in *Citrus* (Revisio aurantiacearum, IX). Japan Society Prom. Sci., Veno, Tokyo.

Tanaka, T. (1961) Citrologia : semi centennial commemoration papers on citrus studies. Citrologia supporting fondation, Osaka, Japan 114 p.

Terol, J., Conesa, A., Colmenero, J. M., Cercos, M., Tadeo, F., Agustí, J., Alos, E., Andres, F., Soler, G., Brumos, J., Iglesias, D. J., Gotz, S., Legaz, F., Argout, X., Courtois, B., Ollitrault, P., Dossat, C., Wincker, P., Morillon, R. and Talon, M. (2007) Analysis of 13000 unique *Citrus* clusters associated with fruit quality, production and salinity tolerance. BMC Genomics, vol. 8:31, 22 pp., doi:10.1186/1471-2164-8-31.

Terol, J.M., Naranjo, A., Ollitrault, P. and Talon, M. (2008) Development of genomic resources for *Citrus clementina*: characterization of three deep-coverage BAC libraries and analysis of 46,000 BAC end sequences. BMC Genomic. 9: 423.

Tokunaga, T., Nii, M., Tsumura, T. and Yamao, M. (2005) Production of triploids and breeding seedless cultivar 'Tokushima 3X No.1' from Tetraploid x diploid crosses in sudachi (*Citrus sudachi* Shirai). Horticultural Research Japan 4: 11–15.

Tolkowsky, S. (1938) Hesperides: a history of the Culture and Use of *Citrus* fruits. John Bale, Sons and Curnow Ltd, London.

Toolapong, P., Komatsu, H. and Iwamasa, M.. (1996) Triploids and haploid progenies derived from small seeds of Banpeiyu'. a pummelo, crossed with 'Ruby Red' grapefruit. J. Japan. Soc. Hort. Sci. 65: 255–260.

Torres, A.M., Soost, R.K.and Diedenhofen, U. (1978) Leaf isozymes as genetic markers in *Citrus*. Amer. J. Bot. 65: 869–881.

Tozlu, I., Guy, C.L. and Moore, G.A. (1999) QTL analysis of morphological traits in an intergeneric BC1 progeny of *Citrus* and *Poncirus* under saline and non-saline environments. Genome 42: 1020–1029.

Vardi, A., Frydman Shani, A. and Weinbaum, S.A. (1988) Assessment of parthenocarpic tendency in citrus using irradiated marker pollen. Proc. Int. Soc. Citriculture 1: 225–230.

Vardi, A., Spiegel-Roy, P. and Elchanati, A. (1993) Mandarin tree named Mor. US patent PP8, 378.

Vardi, A., Bleichman, S. and Aviv, D. (1990) Genetic transformation of citrus protoplasts and regeneration of transgenic plants. Plant Sci.: 69: 199–206.

Weathers, L.G. and Calavan, E.C. (1959) Nucellar embryony as a means of freeing citrus clones of viruses. In:Wallace JM (ed.) Citrus virus diseases. University of California, Division Agricultural Sciences, Berkeley, pp. 197–200.

Webber H.J. (1967) History and development of the citrus industry. In W. Reuther, L.D. Bachelor and H.J. Webber (eds),: The citrus industry. 1. History, world distribution, botany and varieties. University of California Press, Berkeley, pp. 1–39.

Weber, C.A., Moore, G.A., Zhanao, D. and Gmitter Jr., F.G. (2003) Mapping freeze tolerance quantitative trait loci in a *Citrus grandis x Poncirus trifoliata* F1 pseudo-testcross using molecular . J. Am. Soc. Hort. Sci. 128: 508–514.

Williams, T.E. and Roose, M.L. (2004) `TDE2´ Mandarin hybrid (Shasta Gold® Mandarin), `TDE3´ Mandarin hybrid (Tahoe Gold® Mandarin) and `TDE4´ Mandarin hybrid (Yosemite Gold® Mandarin): Three New Mid and Late-Season Triploid Seedless Mandarin Hybrids from California. Proc. Inten. Soc. Citriculture 1: 394–398.

Wong, W.S., Li, G.G., Ning, W., Xu, Z.F., Hsiao, W.L.W., Zhang, L.Y. and Li, N. (2001) Repression of chilling-induced ACC accumulation in transgenic citrus by over-production of antisense 1-aminocyclopropane-1-carboxylate synthase RNA. Plant Science 161: 969–977.

Xu, X.Y., Hu, Z.Y., Li, J.F., Liu, J.H. and Deng, X.X. (2007) Asymmetric somatic hybridization between UV-irradiated *Citrus unshiu* and *C. sinensis*: regeneration and characterization of hybrid shoots. Plant Cell Rep 26: 1263–1273.

Xu, X.Y., Liu, J.H. and Deng, X.X. (2006) Isolation of cytoplasts from Satsuma mandarin (*Citrus unshiu* Marc.) and production of alloplasmic hybrid calluses via cytoplast-protoplast fusion. Plant Cell Rep 25: 533–539.

Yahata, M., Harusaki, S., Komatsu, H., Takami, K., Kunitake, H., Yabuya, T., Yamashita, K. and Toolapong, P. (2005) Morphological characterization and molecular verification of a fertile haploid pummelo (*Citrus grandis* Osbeck). J. Am. Soc. Hort. Sci. 130: 34–40.

Yamamoto, M., Matsumoto, R. and Yamada, Y. (1995) Relationship between sterility and seedlessness in citrus. J. Japan. Soc. Hort. Sci. 64: 23–29.

Zhang, Q.H., Liu, J.H. and Deng, X.X. (2006) Isolation of microprotoplasts from a partially synchronized suspension culture of *Citrus unshiu*. J Plant Physiol 163: 1185–1192.

Chapter 17
Persimmon

Masahiko Yamada, Edgardo Giordani, and Keizo Yonemori

Abstract *Diospyros kaki* Thunb., the Oriental persimmon originated and was domesticated in Eastern Asia where many indigenous cultivars exist. Ninety percent of the worldwide production in 2006 is in China, Korea, and Japan. The largest producers outside Asia are Brazil, Israel, Spain, and Italy. Persimmon is classified into nonastringent and astringent cultivars depending on whether the fruit loses astringency on the tree at maturity. Astringency is caused by water-soluble tannins found in large "tannin" cells, which are scattered throughout the fruit flesh. Drying after peeling, treatment with carbon dioxide gas or ethanol vapor changes these soluble tannins into insoluble forms so that astringent fruit becomes nonastringent. The persimmon cross-breeding program at the national institute in Japan has been continuing since 1938, and has released 11 nonastringent and two astringent cultivars, including nonastringent ones with high eating quality, and with both early ripening and noncracking. There are breeding programs in Korea, Italy, and Spain. The current goals of breeding are: nonastringency, fruit quality and appearance, fruit ripening time, postharvest conservation, productivity, and disease and pest resistance. Due to the hexaploid nature of the persimmon, linkage maps are not available but molecular markers for nonastringency have been developed. The MAS for nonastringency is going on practically and effectively in Japan. There are a few reports of genetic transformation of persimmon using *Agrobacterium tumefaciens*.

M. Yamada (✉)
National Institute of Fruit Tree Science, 2-1 Fujimoto, Tsukuba, Ibaraki 305-8605, Japan
e-mail: kaki@affrc.go.jp

E. Giordani
Plant, Soil and Environmental Science, University of Florence, Florence, Italy
e-mail: edgardo.giordani@unifi.it

K. Yonemori
Graduate School of Agriculture, Kyoto University, Kyoto, Japan
e-mail: keizo@kais.kyoto-u.ac.jp

M.L. Badenes and D.H. Byrne (eds.), *Fruit Breeding*, Handbook of Plant Breeding 8,
DOI 10.1007/978-1-4419-0763-9_17, © Springer Science+Business Media, LLC 2012

Keywords Astringent • Cracking • *Diospyros* • Germplasm • Marker • Pollination • Quality • Ripening • Transformation

1 Introduction

The name persimmon was given to the American species, *Diospyros virginiana* L., by the Algonquin Indians of Virginia (Yonemori et al. 2000). In contrast to *D. virginiana*, *Diospyros kaki* Thunb. originated in Eastern Asia, and many indigenous cultivars were developed in China, Japan, and Korea (ARS 1912; Cho and Cho 1965; Wang et al. 1997; Yamada 2005). *D. kaki* is called the Japanese persimmon, the Oriental persimmon, the Chinese persimmon, kaki, or, simply, persimmon. Today, the term persimmon is commonly used for *D. kaki* because the global production of *D. kaki* is overwhelmingly higher than that of *D. virginiana*.

The worldwide production of persimmons based on FAO statistics in 2006 was estimated at approximately three million tons, of which ca. 90% was produced in China, Korea, and Japan. The largest producers outside Asia are Brazil, Israel, Spain, and Italy. The production of the European Mediterranean countries (Italy, Spain, Greece, and Portugal) in 2002 was estimated at 95,000 mt (Bellini and Giordani 2005).

Persimmon is classified into nonastringent and astringent cultivars depending on whether the fruit loses astringency on the tree at maturity. In addition, Hume (1914) developed a grouping based on whether the flesh color darkens or remains unchanged under the influence of pollination. The former is called pollination-variant (PV), and the latter, pollination-constant (PC). Strictly speaking, changes in flesh color are related to seed formation rather than to pollination. Therefore, cultivars are commonly classified into four groups, pollination-variant nonastringent (PVNA), pollination-constant nonastringent (PCNA), pollination-variant astringent (PVA), and pollination-constant astringent (PCA).

Astringency is caused by water-soluble tannins. They are found in large special “tannin” cells, which are scattered throughout the fruit flesh. Treatment with carbon dioxide gas, ethanol vapor, and drying after peeling change these soluble tannins into insoluble forms so that astringent fruit becomes nonastringent. In PV cultivars, seeds exude acetaldehyde and ethanol. Acetaldehyde causes the soluble tannins to condense or coagulate and to become insoluble and oxidized. As a result, many brown specks are formed in the flesh which darkens the flesh. Cultivars whose fruit have a large number of seeds that produce a considerable amount of acetaldehyde are nonastringent and are classified as PVNA, while those whose fruit have seeds that produce little acetaldehyde are classified as PVA and retain the astringency in the flesh. Even in PVNA cultivars, when seed formation is poor, a dark area develops only around the seeds and the remaining flesh is astringent. In PCA cultivars, seeds produce very little acetaldehyde and ethanol, and, as a result, the flesh color is not changed by seed formation.

In PCNA cultivars, the mechanism in which astringency is naturally lost while the fruit are on the tree is different from that in PV cultivars (Yonemori et al. 2000). Fruits in PCNA cultivars stop accumulating tannins at early stage of fruit growth, while other types (PVNA, PVA, and PCA) accumulate tannins until late fruit developmental stage (Yonemori and Matsushima 1985).

Mature fruit of nonastringent cultivars can be eaten at the firm stage, as can apples. In contrast, the fruit of astringent cultivars need to be treated after harvest to remove astringency with carbon dioxide gas or ethanol vapor or by drying after peeling before consumption. Without those treatments, the fruit cannot be eaten until they are over ripe with very soft flesh.

In the past, astringent cultivars were commonly eaten as over-ripened soft or dried fruit. Even now, in some countries over-ripened soft fruit (China, Italy, and Korea) or dried fruit (China, Korea, and Japan) are commonly consumed and commercially produced. In Japan and Israel, the fruit of astringent cultivars are commercially consumed as fresh table fruit after the astringency is removed.

Among the four types, the most desirable type is PCNA, in which fruit loses astringency naturally and stably when grown in warm areas. PCNA does not need any treatments for edibility. Breeding has been focused on the improvement of PCNA cultivars. However, PCNA genetic resources are very limited, which is a serious obstacle in the breeding.

Persimmon is widely distributed from the temperate to the subtropical regions in the world. Many regionally adapted cultivars are well adapted to warm, humid climates although these can be grown in a dry climate with irrigation. However, leading cultivars are generally adapted to a temperate climate where high-quality production is possible. The critical temperature for the cold injury of shoots and wood is between −15°C and −20°C.

PCNA cultivars require a relatively high temperature in summer and autumn for the natural loss of astringency. These cultivars are primarily grown in areas with a mean annual temperature of 14–16°C in Japan. PVNA cultivars can lose their astringency even under cooler temperatures and are consequently produced throughout the main three islands in Japan. PVA and PCA cultivars are also produced widely in Japan and these cultivars can be grown commercially in areas with a mean annual temperature as low as 11°C.

Oriental persimmon cultivars are grafted onto *D. kaki*, *D. lotus*, and *D. virginiana* rootstocks. *D. kaki* roots are well adapted to slightly acid to neutral soils (pH 5–7) that are well aerated especially when young but are sensitive to drought. Interestingly, older trees can develop deeper root systems in soil with less aerobic conditions as compared to young trees.

D. lotus is used as the rootstock extensively in the world, but some *D. kaki* cultivars, including 'Fuyu,' 'Shogatsu,' 'Yokono,' and 'Takura' are incompatible with *D. lotus* (Tanaka 1930). *D. virginiana* rootstock shows some problems with tree decline of *D. kaki* (Cohen et al. 1991).

2 Origin and Domestication of Cultivars

The origin of the oriental persimmon is in Eastern Asia where many indigenous cultivars have been developed (Figs. 17.1 and 17.2). More than 2,000, 1,000, and 500 cultivars have been reported of Chinese (Wang et al. 1997), Japanese (ARS 1912) and Korean origin (Cho and Cho 1965).

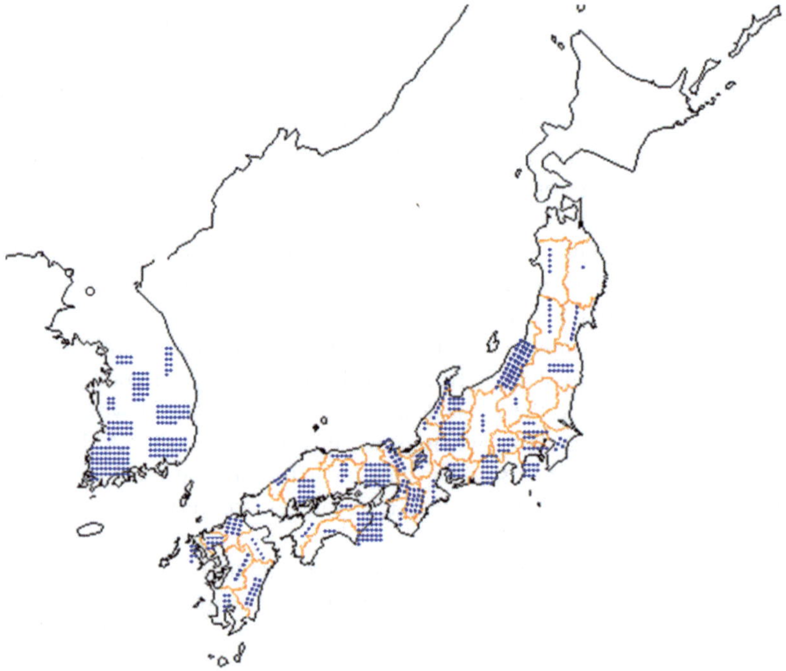

Fig. 17.1 Distribution of local persimmon cultivars of Japanese and Korean origin. Hiroshima Prefecture (1979) and Cho and Cho (1965)

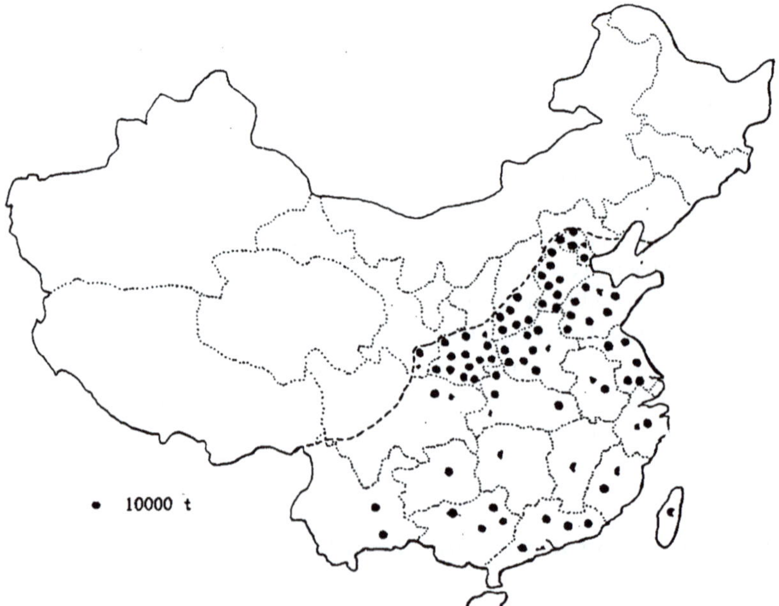

Fig. 17.2 Distribution of persimmon cultivars in China in 1992 (Wang et al. 1983)

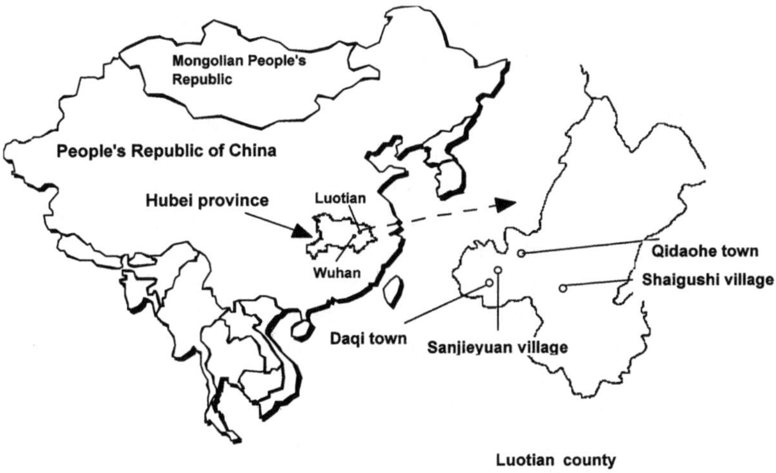

Fig. 17.3 Distribution sites of the fruits classified as nonastringent cultivars in Hubei province in China (Yonemori et al. 2009)

In Japan, two types of nonastringent cultivars (PCNA and PVNA) have been developed. Although in 1965, Cho and Cho (1965) did not report any nonastringent varieties in Korea, at present one PVNA cultivar, 'Johongsi' and a few PVA cultivars of Korean origin are known (Yamada unpublished data).

The number of PCNA cultivars of Chinese origin is very limited. The first PCNA cultivar 'Luo Tian Tian Shi' was reported in 1983 (Wang 1983; Yamada et al. 1993a). Since then, only five PCNA cultivars of Chinese origin have been reported. These are all found within a narrow band between Ba River and Ju River at the south of Dabie Mountains, around Luo Tian and Macheng counties in Hubei province (Fig. 17.3) (Wang et al. 2005; Yonemori et al. 2005). Studies with AFLPs (Kanzaki et al. 2000a, b) and RAPDs (Luo et al. 1999) have shown that the Chinese PCNA cultivars are not closely related to the Japanese PCNA cultivars.

The persimmon breeding program at the national institute in Japan began at Okitsu in 1938 and moved to Akitsu in 1968, where it now continues. The emphasis was placed on the improvement of PCNA cultivars. Since the leading PCNA cultivars ('Fuyu' and 'Jiro') were late ripening, one of the most important breeding objectives was to develop superior early ripening PCNA cultivars. In addition, the following characteristics have been important: high eating quality, large fruit, no fruit cracking, high productivity, long shelf life, good fruit appearance, and disease and pest resistance. Crosses have been mainly made among PCNA cultivars and selections to obtain PCNA offspring at National Institute of Fruit Tree Science (NIFTS) breeding because no PCNA offspring are generally yielded from crossing PCNA and non-PCNA types of local cultivars of Japanese origin. Nevertheless, due to problems with inbreeding depression caused by intercrossing the closely related

PCNA types, a backcross program to introgress the PCNA trait into the more diverse non-PCNA persimmon germplasm was begun in 1990.

In Korea, the Sweet Persimmon Experiment Station was founded by Gyeongnam province in 1994 to develop superior early ripening PCNA cultivars. A small program is ongoing in Taiwan (Wen 2003), but there are no reports of breeding in China.

In Italy, a wide number of progenies have been evaluated from different cross-combinations to obtain superior astringent and nonastringent types but no advanced astringent selections showed outstanding characteristics in comparison to 'Kaki Tipo' (University of Florence). Although most of the seedlings derived from PCNA × PCNA crosses showed weak growth and poor productivity, a few seedlings obtained from the backcross (PCNA × non-PCNA) × PCNA are under observation (Bellini and Giordani 1998).

In Spain, a strong persimmon industry has been developed with 'Rojo Brillante,' which is a productive astringent cultivar that is sold after treating it with carbon dioxide gas to eliminate the astringency. Recently, a breeding program has been initiated at Instituto Valenciano de Investigaciones Agrarias (IVIA). The main goals are to extend the harvest season with earlier and later cultivars with the same quality of 'Rojo Brillante' and to produce new PCNA cultivars with high quality and good adaptation. The program is currently producing hybrids from conventional breeding and variants derived from mutation breeding (X-ray or chemical mutagens) and somaclonal producing cultures.

3 Genetic Resources of Persimmon

3.1 Japan

There are many astringent and PVNA local cultivars throughout Japan. In contrast, only six PCNA cultivars are listed in the 1912 report after excluding those with synonyms (ARS 1912). The Japanese persimmon germplasm collection at the NIFTS (Akitsu, Hiroshima) includes about 600 genotypes, including many local cultivars and their strains, selections, and cultivars introduced from foreign countries. Even now, only 18 genotypes of PCNA cultivars of Japanese origin are conserved at Akitsu, excluding their bud-sports, synonyms, and newly released cultivars.

Persimmon is believed to be originally astringent because most cultivars of Chinese and Korean origin are PCA, and, even in Japan, local PCA cultivars have the widest distribution and variation. In a study of 188 cultivars of Japanese origin evaluated at NIFTS (Yamada et al. 1994a), the PCNA group matured later than the other groups and along with the PVA group had larger fruit size than the PCA and PVNA groups. The variance among cultivars within the PCNA for both fruit maturing time and soluble solids content (SSC) was smaller than that in the PCA (Table 17.1) further supporting the lack of genetic diversity reported in PCNA group using AFLPs (Kanzaki et al. 2000a).

The PVNA group had a slightly earlier fruit maturity and higher SSC than the PCA group. This appears to be due to the effect of seeds exuding volatile compounds

Table 17.1 Mean and variance of fruit maturing time, fruit weight, and soluble solids content in cultivars of oriental persimmon of Japanese origin (Yamada et al. 1994a)

Astringency type	No. of cultivars	Fruit maturing time		Fruit weight (g)		Soluble solids content (%)	
		Mean	Variance	Mean	Variance[1]	Mean	Variance
PCA	83	5.82 b[2]	2.17 a[3]	190 b	0.0258 a	16.8 b	2.39 a
PVA	25	4.88 c	1.36 ab	267 a	0.0113 a	15.6 c	1.81 a
PVNA	58	5.09 c	1.87 ab	188 b	0.0265 a	17.7 a	3.23 a
PCNA	22	6.50 a	1.12 b	247 a	0.0115 a	17.2 ab	0.89 b
Total	188	5.57	2.09	206	0.0265	17.0	2.61

[1]Variance in log-transformed value
[2]Mean separation between PCNA and others for fruit maturing time, and soluble solids content by Cochran's *t*-test ($P=0.05$). Mean separation for the others by Duncan's multiple range test ($P=0.05$)
[3]Variance separation by *F*-test ($P=0.05$)

removing astringency because highly seeded fruit has earlier fruit maturing time and higher SSC than the fruits with no or few seeds in a PVNA cultivar.

PVNA cultivars were first mentioned in a thirteenth century document (Kikuchi 1948). PVNA fruit needs seed formation for natural astringency loss on the tree. This group has the wide distribution and high genetic variation similar to that of the PCA group. The local PVNA cultivars throughout Japan are generally chance seedlings that have been selected from the open-pollinated seeds of PVNA genotypes throughout the country, suggesting a rapid development and dissemination (Yamada et al. 1994a).

In contrast, the origin of the PCNA cultivar group was limited to only one cultivar, 'Gosho' (Figs. 17.4 and 17.5), that was grown in a wide area 200 years ago. The existence of 'Gosho,' which naturally lost its astringency regardless of seed formation and had few brown specks in flesh was first mentioned in the seventeenth century (Kikuchi 1948). The nonastringent trait in the PCNA of Japanese origin is qualitative (Ikeda et al. 1985; Yamada and Sato 2002) and recessive to the other three non-PCNA types (PVNA, PVA, and PCA). *D. kaki* is primarily hexaploid with a few nonaploid cultivars. Thus, PCNA-type offspring are obtained by intercrossing PCNA genotypes but almost no PCNA offspring resulted from crossing PCNAs with non-PCNAs, or from crosses among non-PCNA types of native cultivars of Japanese origin. Consequently, native non-PCNA cultivars in Japan seem to carry no or a few alleles for the PCNA trait, probably due to the recent origin of the PCNA trait. However, in some backcrosses of PCNA × (non-PCNA × PCNA) around 15% of the offspring were PCNA (Ikeda et al. 1985).

It is probable that once 'Gosho' (Fig. 17.4) or a related PCNA genotype appeared, and natural crosses between it and non-PCNA cultivars and its descendants over some generations resulted in the present PCNA local cultivars. Currently, most local PCNA persimmon cultivars are found in a few prefectures in central Japan not far from the town of 'Gose' in Nara Prefecture where 'Gosho' originated (Fig. 17.5). Morphological characterization of 16 PCNA and 18 non-PCNA types representing leading and local varieties indicated that the PCNA types were distinct from non-PCNA and morphologically less diverse than the non-PCNA types (Yamada et al.

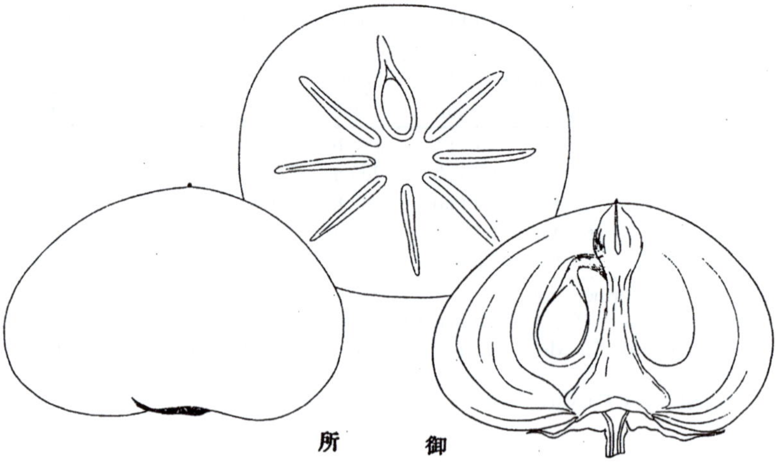

Fig. 17.4 Fruit shape of 'Gosho' persimmon (ARS 1912)

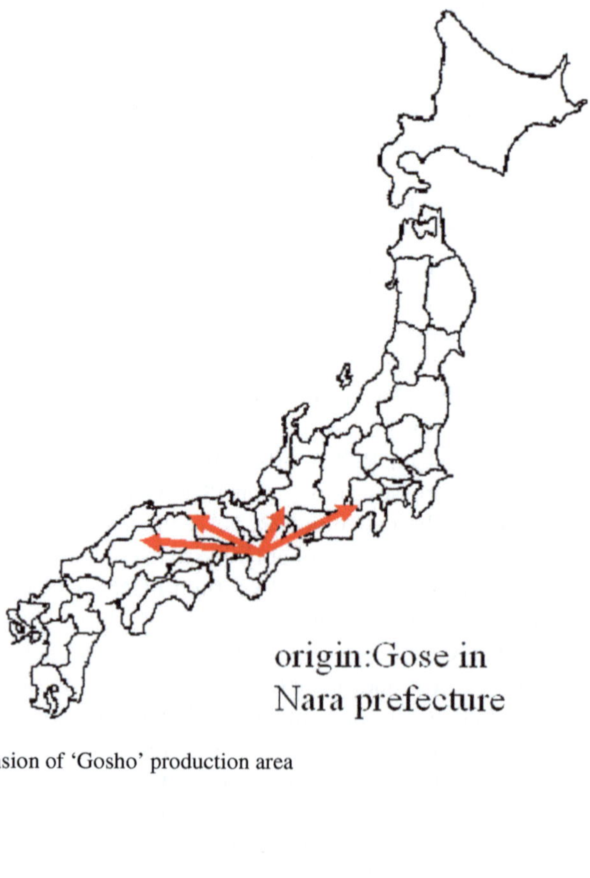

Fig. 17.5 Extension of 'Gosho' production area

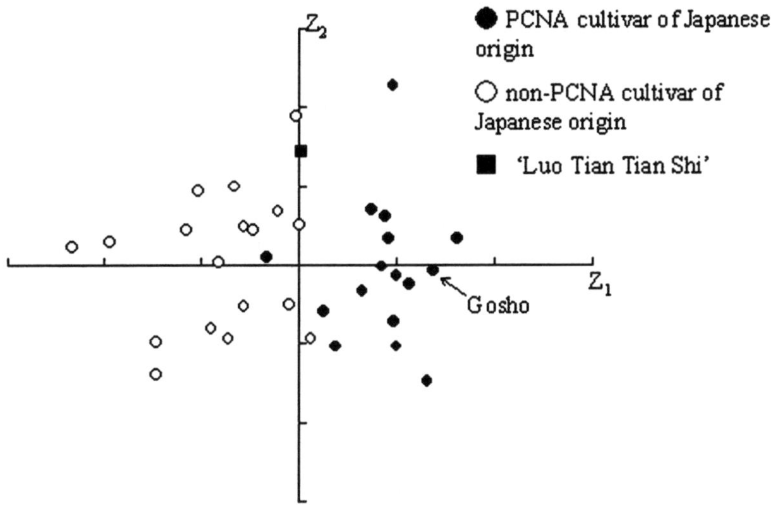

Fig. 17.6 Principal component scatter diagram on the first (Z_1) and second (Z_2) principal components for 18 fruit traits (Yamada et al. 1993a)

1993a) (Fig. 17.6). PCNA cultivars were distinguished by flat-shaped fruit, fruit cracking at the calyx and stylar ends, many wrinkles in the fruit skin at the calyx end, and dark-colored and thick seeds; all traits similar to those of 'Gosho.' These qualities are probably a result of the recent and localized appearance of the PCNA trait and its recessive nature which would require a degree of inbreeding to express the trait in descendants.

3.2 *China*

There are more than 900 local persimmon cultivars, most of which are astringent, in China (Wang et al. 1997). These are conserved at National Persimmon Germplasm Repository in China, North West Sci-Tech University, formerly the Experimental Farm of Pomology Institute of the Shaanxi Academy of Agricultural Sciences (SAAS), Meixian, Shaanxi Province. Their characteristics have been evaluated.

SAAS in China and NIFTS in Japan cooperated in comparing the variations in persimmon cultivars of Chinese and Japanese origin. Fifteen cultivars were evaluated for their maturity time, fruit weight and SSC at both Meixian (China) and Akitsu (Japan) (Yamada et al. 1995a). An analysis of variance revealed that the fruit maturity time and fruit weight, but not soluble solids content, were greatly affected by the location and cultivar but little by the cultivar × location interaction. On average, the cultivars evaluated at Akitsu matured 18 days later and weighed 70 g more than did those at Meixian. These results showed that cultivar performance in one location could be estimated for fruit maturity time and fruit weight from that in

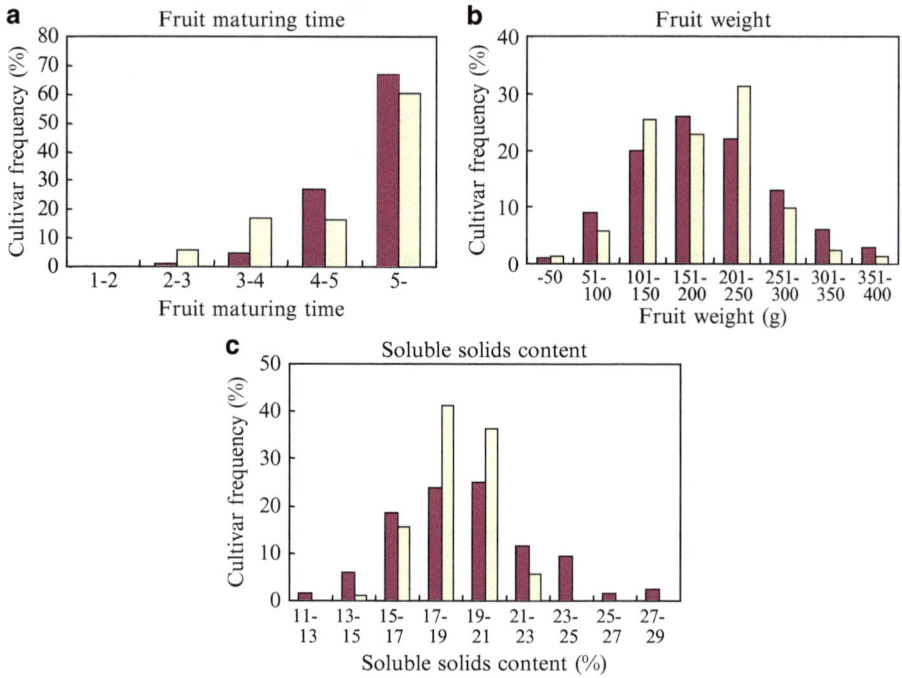

Fig. 17.7 Frequency distribution of PCA type persimmon cultivars of Chinese (*solid column*) and Japanese origin (*open column*) for fruit maturing time (**a**), fruit weight (**b**), and soluble solids content (**c**) (Yamada et al. 1995b). Data were adjusted for the location effect by adding correction constant. The fruit maturity time was scored: 1, late September; 2, early October; 3, mid-October; 4, late October; 5, early November

another location by incorporating the correction constant of the mean difference between the two locations.

Subsequent evaluations of PCA cultivars of Japanese origin (83 cultivars) at NIFTS, Akitsu, Japan, and of Chinese origin (132 cultivars) at SAAS, Meixian, China indicated that these cultivars, after adjusting for the location effect, were very similar for the means and variation among the cultivars within the Chinese and Japanese groups for fruit maturity time and fruit weight (Fig. 17.7) (Yamada et al. 1995b). There was a small difference in the mean value for SSC although the variance was larger in the Chinese cultivars than in the Japanese cultivars. When cultivars originated in more northern areas are compared to those from southern origin in a location, there is a higher frequency of early maturing cultivars, in cultivars from northern areas than those from southern areas (Yamada et al. 1994a). Most cultivars evaluated at Meixian and Akitsu originated at nearly the same latitudes, between 30° and 40°N latitude (Yamada et al. 1995b). The results suggest that selection pressures in China and Japan were similar for fruit maturity time and fruit weight during the development of the persimmon culture. Astringent persimmon production is currently common even in Japan because superior PCNA cultivars are limited and have a narrow genetic variability. Yamada et al. (2002b) tested carbon dioxide gas and ethanol vapor treatments to remove astringency in fruit, which is applied commonly for 'Hiratanenashi' in Japan, to 13 and 12 astringent persimmon cultivars of Japanese and Chinese origin, respectively. The reduction in soluble tannins and thus astringency was related to the specific cultivar and treatment combination and unrelated to the cultivar's Chinese or Japanese origin. As these techniques had not been used in the old times in China and Japan, there has been no selection for the cultivar's ability to respond to these techniques. It was difficult to completely remove astringency in a considerable number of cultivars with the current techniques.

3.3 Europe

The history of the appearance and development of persimmon (*D. kaki*) and related *Diospyros* species in Europe is not clear. Both persimmon (*D. kaki*) and *D. lotus* are often called in Italy and Greece also "loto" (from "lotus"). Given this, the oldest record probably associated to *D. lotus* can be attributed to AD for Pliny the Elder (24–79 AD), who, in the Historia Naturalis, quotes that "lotus trees in the Piazza del Tempio and by the Temple of Vulcano, well known for the development of shoots and for the nice shade they produced" were present in ancient Rome. Many centuries later, Ricci, a Jesuit Monk who traveled to China, mentioned persimmon in 1613 (Evreinoff 1948) and kaki was quoted by Trigault in 1615. One century later both *D. kaki* and *D. lotus* were not yet a well-known species for De Candolle (1778–1841) (Occhialini and Tirocco 1923).

In Greece, persimmon may have been present for a relatively long time (Morettini 1949). In the latter part of the nineteenth century, persimmon was reported in France (1860) (Occhialini and Tirocco 1923; Morettini 1949), Italy (1871–1876), Algeria (Evreinoff 1948; Morettini 1949), Spain (Climent and Llácer 2001), Russia in 1888 (Evreinoff 1948), and Turkey (Tuzcu and Seker 1997). It probably arrived to the Balkan countries from Russia during the twentieth century.

Little is known about the breadth and development of persimmon genetic resources in Europe. Bellini and Giordani (2005) presumed that European persimmon cultivars are probably derived from genotypes imported at the end of the nineteenth century mainly as seeds or seedlings and bud mutations of ancestral Oriental cultivars.

Bellini (1982) described 'Kaki Tipo' as a synonym of 'Trakankaki,' while 'Amankaki,' 'Akoumankaki,' 'Hyakume,' 'Kirakaki,' and 'Thiene' are considered very similar to it. Recent studies of Yamada and Sato (personal communication) indicate that 'Kaki Tipo' is morphologically very close to 'Amahyakume' of Japanese origin. Characterization of 'Kaki Tipo' selections from different sources with AFLPs clearly indicates that they are separate genotypes (Yonemori et al. 2008b).

Other accessions of presumed Italian origin, all of them belonging to the PVNA group like 'Kaki Tipo,' but bearing both male and female flowers, are 'Brazzale,' 'Moro,' and 'Rispoli.' All of them morphologically very close to 'Zenjimaru,' a Japanese cultivar (Sato and Yamada personal communication). Principal component analysis on morphological quantitative variables (leaves, flowers, 1-year shoots and fruits) showed a well-defined cluster with 'Brazzale,' 'Mercatelli,' 'Moro,' 'Mandarino,' 'Rispoli,' and 'Vainiglia' (Fig. 17.8), although the cultivars are distinct as indicated by RAPDs (Bellini et al. 2003).

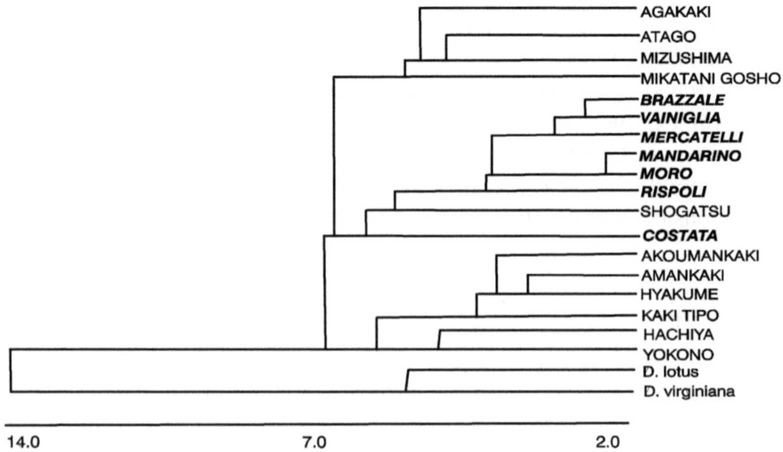

Fig. 17.8 Dendrogram obtained from the quantitative morphologic distance matrix (the presumed Italian accessions in *bold italics*) (Bellini et al. 2003)

The cultivars of Spanish origin are all astringent. Among them 'Rojo Brillante,' a pollination variant cultivar, is the most important cultivar in Spain, while 'Tomatero,' 'Cristalino,' and 'Xato del Bonrepós' are secondarily cultivated in the Province of Valencia. RAPD markers applied on a set of 40 cultivars, revealed no matching between Spanish cultivars and ancestral imported cultivars both from Oriental countries and Italy (Badenes et al. 2003). In a cooperative study with Japanese, Italian, and American scientists, persimmon genotypes of Italian, Spanish, Japanese, Korean, Chinese, and of unknown origin were evaluated for genetic differences by AFLP analysis (Yonemori et al. 2008b). The results suggested that Spanish and Italian cultivars could have evolved from a common gene pool, while Japanese, Chinese, and Korean cultivars formed distinct clusters. Furthermore, the groups with mixed cultivars of different origin (Japanese cultivars in the European set and some European cultivars nested between Chinese and Korean sets) suggest that similar, but different progenitors were used in the development of the present European cultivars (Fig. 17.9). 'Coroa de Rei,' an astringent type, is the most cultivated variety in Portugal (De Sousa and Gomes-Pereira 1995). There are no reports about local varieties in France. In Greece, most propagation material seems to come from local cultivars selected by farmers, but they have been neither inventoried nor described. In Turkey, many astringent local cultivars have been documented and collected (Tuzcu and Seker 1997). Since a current survey on persimmon germplasm has not been done in Europe and Mediterranean countries, the amount of *D. kaki* Thunb., *D. lotus* L., and *D. virginiana* accessions there is not known.

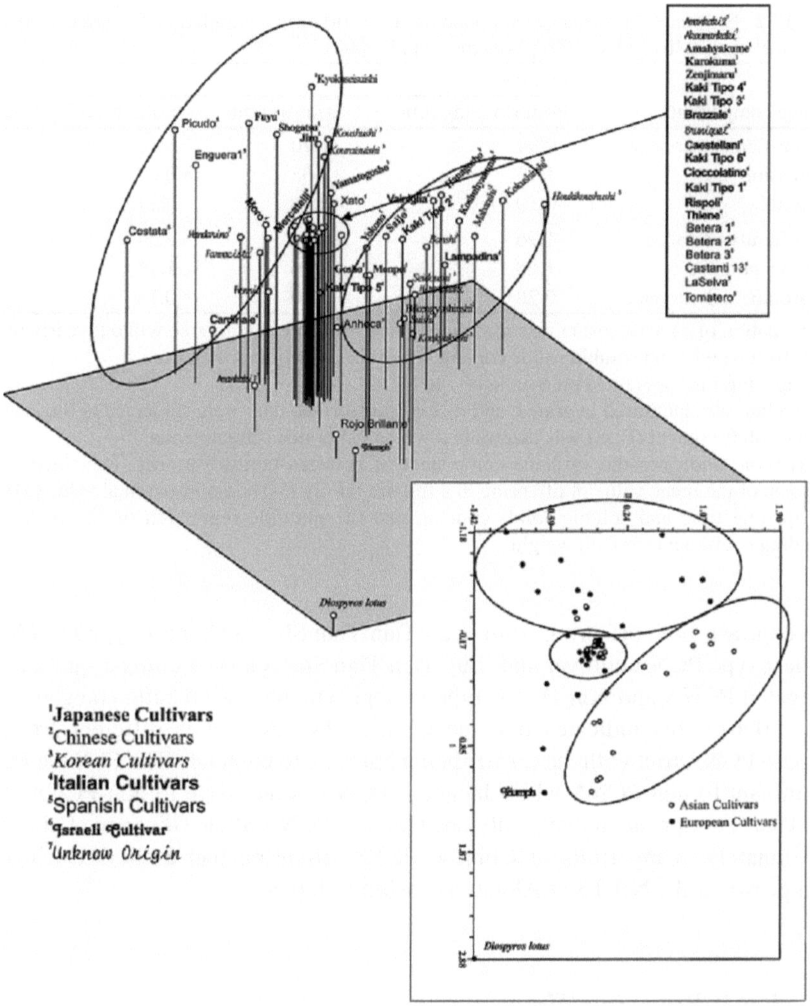

Fig. 17.9 Multidimensional scaling analysis from NTSYS2.0 for 61 persimmon cultivars and *D. lotus* (Yonemori et al. 2008b). A three dimensional figure is shown, representing the 1st three coordinates. The first 2 dimensions are also shown in the lower right insert figure, showing relative placement of the Asian cultivars to the European cultivars

4 Major Breeding Achievements

4.1 Astringency Type

As previously mentioned, the PCNA nonastringent trait of Japanese origin is qualitative and recessive to the other three non-PCNA (PVNA, PVA, and PCA) types (Ikeda et al. 1985). Thus, PCNA types are produced from intercrossing PCNA genotypes and in some backcrosses from PCNA × (non-PCNA × PCNA) (Ikeda et al. 1985).

Table 17.2 Estimates of variance components in an offspring population for three fruit traits (Yamada et al. 1994b, 1995c, 1997; Yonemori et al. 2000)

	Estimates		
Variance component	Fruit ripening time[a]	Fruit weight[b]	Soluble solids content[c]
Between-family variance	0.94	1.70	0.18
Regression[d]	1.13	1.65	0.07
Residual	−0.19 (0)	0.06	0.11
Within-family variance	1.96	10.96	2.54
Genetic variance	1.68	8.76	1.76
Environmental variance	0.28	2.20	0.78

The evaluation of each offspring was made using five fruits on a single tree without yearly repeating for fruit weight and soluble solids content, 10 fruits for fruit ripening time

[a]Fruit ripening time was rated on a scale of 1 to 8

[b]Fruit weight was measured in grams, and its log-transformed data were subjected to the analysis

[c]Soluble solids content (Brix) was determined with a calibrated refractometer

[d]"Regression" indicates the variance component of between-family variance, explained by the regression of the mean value of offspring in a full-sib family (Mf) on mid-parental value (MP) for fruit ripening time and soluble solids content, and the multiple regression of Mf on MP and inbreeding coefficient for fruit weight

In contrast, both offspring from 'Luo Tian Tian Shi' (a Chinese type PCNA) × a Japanese type PCNA cultivar and 'Luo Tian Tian Shi' × a PCA cultivar qualitatively segregated PCNA and non-PCNA type in approximately a 1:1 ratio (Ikegami et al. 2004, 2006). This indicates that the Chinese PCNA trait is different from the Japanese PCNA trait with the former being epistatic to the latter. The Chinese PCNA is dominant to non-PCNA while Japanese PCNA is recessive to non-PCNA. 'Luo Tian Tian Shi' appears to have only one Chinese PCNA allele (Ikegami et al. 2006). Unfortunately, many fruits of Chinese PCNA offspring had a slight astringency when grown at the NIFTS at Akitsu, Hiroshima, Japan.

4.2 *Fruit Ripening Time*

The inheritance of fruit ripening time (FRT) is under quantitative control (Yamada et al. 1995c) with a high broad-sense heritability (0.84) when the evaluation was made with five fruits on a single tree for 1 year for a population of cultivars/selections used as cross-parents in the 1970s and 1980s at NIFTS (Yamada et al. 1993b, 1994c). Therefore, it is easy to evaluate the genetic difference for FRT. Further analyses of full-sib families (Table 17.2) indicated that the genetic differences among families were explained solely by the mid-parental value (MP) under the assumption of homogeneous within-family variances (Yamada et al. 1995c). Thus, breeders can estimate the distribution of genotypic values of the offspring of specific crosses with accuracy (Yamada and Yamane 1997).

Most of PCNA cultivars of Japanese origin are late ripening as were the MPs in the progenies among PCNA cultivars. Since the within-family variance was small,

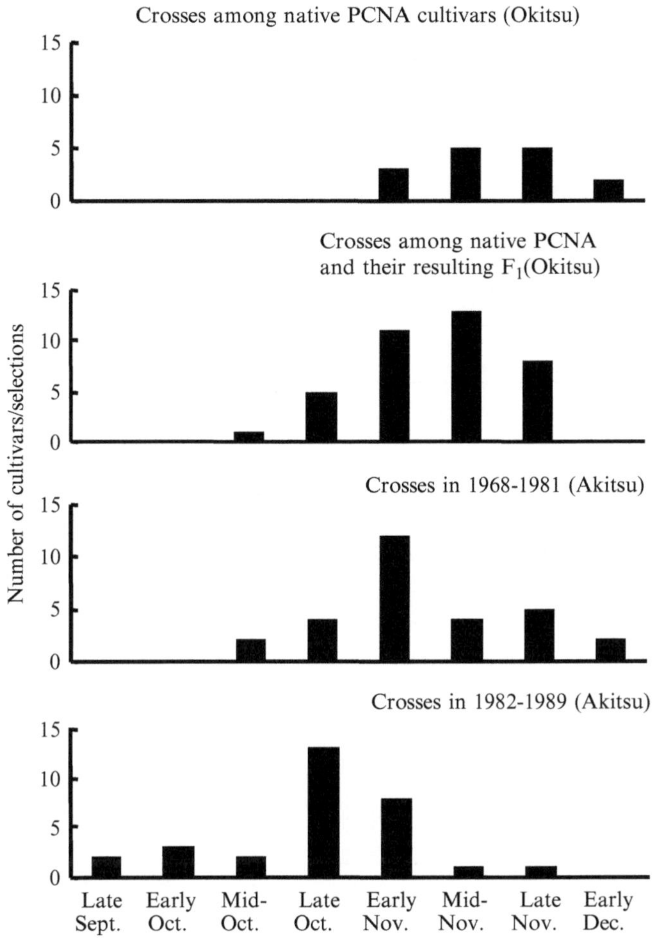

Fig. 17.10 Fruit ripening time of cultivars/selections used as cross-parents at Okitsu (1938–1967) and Akitsu (1968–1989) in Japan (Yamada 1993)

advances in the time of ripening was limited in progenies among the PCNA cultivars. After several cycles of selection over 50 years, the parental genotypes have gradually shifted toward early ripening (Fig. 17.10).

4.3 Fruit Cracking

The PCNA cultivars of Japanese origin are distinct from other types of cultivars as they have tendency to crack at the calyx and/or stylar ends (Yamada et al. 1988) (Table 17.3). Both cracking habits are independently and quantitatively inherited. Cultivars that do not crack are mostly homozygous, whereas cultivars that crack are

Table 17.3 Frequency distribution of cultivars of Japanese origin for calyx and stylar end fruit cracking (Yamada et al. 1988)

		Number of cultivars in each cracking score									
	Total number	Calyx-end cracking score[a]					Stylar-end cracking score[a]				
Astrin-gency type	of cultivars	0	1	2	3	4	0	1	2	3	4
PCNA	21	5	7	4	3	2	6	8	2	5	0
PVNA	35	28	7	0	0	0	30	4	1	0	0
PVA	15	12	1	1	0	1	15	0	0	0	0
PCA	37	34	2	0	0	1	33	4	0	0	0

[a]Cracking score: 0=none, 1=minute, 2=slight, 3=medium, 4=severe

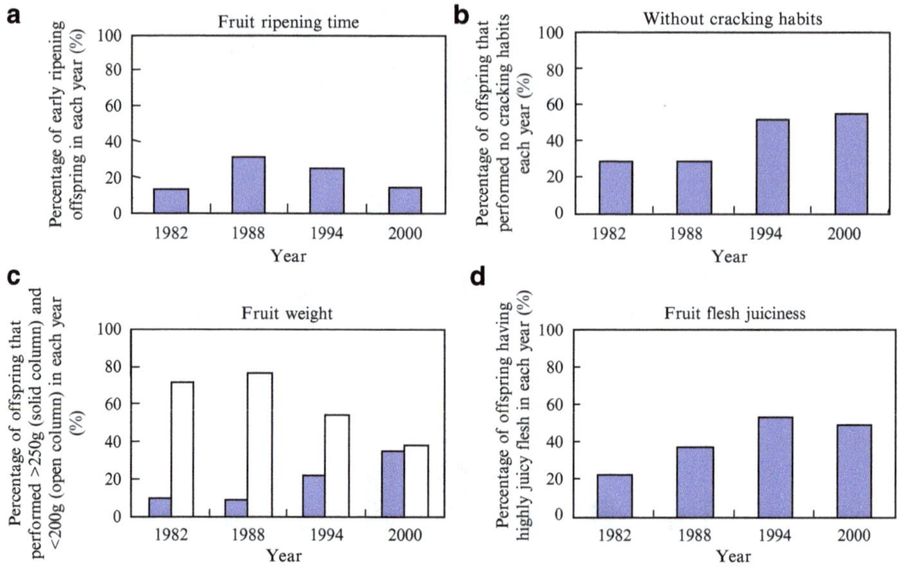

Fig. 17.11 Offspring frequency in breeding populations from 1982 to 2000 for fruit ripening time (**a**), crack resistance (**b**), fruit weight (**c**), and fruit flesh juiciness (**d**). Each column shows the percentage of offspring evaluated for fruit traits in each year. Number of offspring evaluated was 697 in 1982, 154 in 1988, 506 in 1994, and 839 in 2000 (Yamada and Sato 2003)

heterozygous (Yamada et al. 1988). Offspring derived from crosses among crack-resistant parents exhibit little or no cracking, whereas offspring from crack-susceptible parents exhibit a wide range of cracking. Consequently, crossing among crack-resistant parents is desirable but not easy to achieve because most PCNA cultivars are crack-susceptible. Many selections showed cracking at the calyx and/or stylar ends, especially in the early years of the breeding (Yamada et al. 1988). However, the effort to use PCNA cultivars/selections that have less propensity to crack as parents has resulted in the increase of offspring without cracking problems (Yamada and Sato 2003) (Fig. 17.11).

The magnitude of cracking is environmentally influenced and fluctuates greatly from year to year largely due to genotype and year interaction (Yamada et al. 1986b, 1987b, 2002a). The percentage of cracking fluctuates more widely for the more susceptible genotypes, in contrast, resistant genotypes show no cracking irrespective of the year. Therefore, breeders should discard genotypes exhibiting a severe cracking even in a single year (Yamada et al. 1987b).

Fruit cracking at the stylar end decreases remarkably in seedless fruit (Yamada et al. 1991) and thus can be controlled by preventing pollination for genotypes having a high parthenocarpic ability. Flower thinning is commonly used in Japanese persimmon production to greatly increase the fruit weight and parthenocarpy. 'Jiro' has a stylar-end cracking habit and is normally grown in Japan by thinning flowers without a pollinizer.

4.4 Fruit Weight

Fruit weight is a quantitative character with a high broad-sense heritability (Yamada et al. 1993b, 1994c). Yamada et al. (1994b) showed that genetic differences among families were explained mostly by the multiple regression of the mean value of a family on the inbreeding coefficient (F) and mid-parental value (MP) (Table 17.2), indicating that breeders can estimate the distribution of the genotypic values of offspring with accuracy (Yamada et al. 1994b, 1997).

Unfortunately, fruit weight is reduced greatly by inbreeding. Even at $F=0$, the family mean in the offspring was smaller than the MP. Since the F values are based on the assumption that parents with unknown ancestry are unrelated, it is likely that even crosses with $F=0$ do not exclude inbreeding. Although the repeated selection within a limited number of PCNA genotypes is effective for FRT(Fig. 17.10), which is not influenced by inbreeding, it resulted in a serious reduction in fruit weight for cross-parents (Fig. 17.12) and the resultant offspring. However, efforts to make crosses with an F-value of zero or a very small value and outcrosses using non-PCNA cultivars/selections has lessened the reduction in fruit weight (Fig. 17.11).

4.5 SSC and Juiciness

Soluble solids content, an indicator of sweetness, is a quantitatively inherited character which fluctuates markedly with environmental conditions (Yamada et al. 1986a, 1993b). It has a lower broad-sense heritability than either FRT or fruit weight (Yamada et al. 1993b) and a very large within-family genetic variance (Yamada et al. 1997). Thus, any cross has a wide range of SSC among its offspring. In addition, the mean SSC in PCNA cultivars is intermediate between those of the PCA and PVNA groups (Yamada et al. 1994a). Therefore, the restriction of parents in crosses to PCNA genotypes does not provide special obstacles in improving SSC, in contrast with those described with FRT, fruit weight, and the propensity of fruit to crack.

Juiciness is an important trait affecting fruit eating quality. An effort to use cultivars/selections with high eating quality (soft and juicy flesh with high sugar content) as cross-parents over two decades has resulted in an increased percentage of offspring having highly juicy flesh from 22% in 1982 to 53% in 1994 (Fig. 17.11).

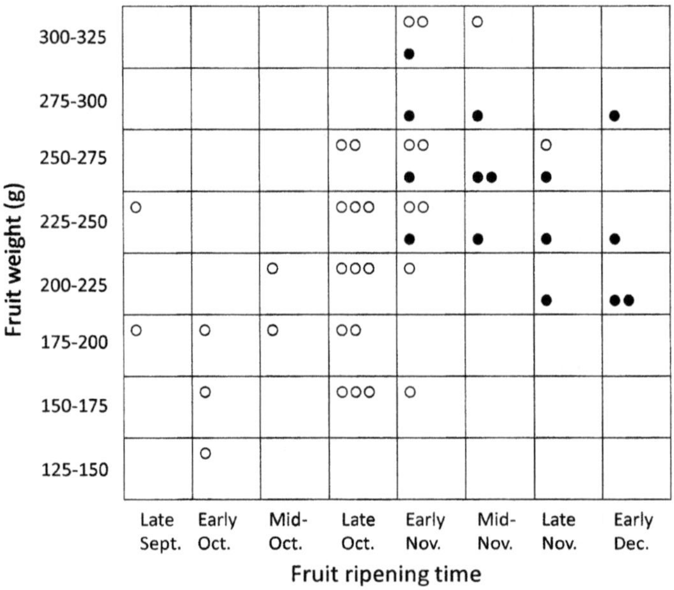

Fig. 17.12 Fruit ripening time and fruit weight in cross-parents at Okitsu and Akitsu in Japan (Yamada 1993). *Filled circles* indicate PCNA native cultivars used as parents at Okitsu. *Open circles* indicate PCNA cultivars and selections derived from intercrossing of native cultivars, used as parents at Akitsu in 1982–1989

4.6 Sex Expression

Persimmon exhibits three types of sex expression: pistillate-type (only pistillate flowers), monoecious-type (both pistillate and staminate flowers), and polygamomonoecious-type (hermaphroditic flowers in addition to pistillate and staminate flowers). The type of sex expression is determined genetically although the pistillate-type cultivar may very rarely produce staminate flowers (Yakushiji et al. 1995). The inheritance of the presence and quantity of staminate flowers appears to be inherited quantitatively (Oohata et al. 1964).

4.7 Parthenocarpy

The parthenocarpic ability in persimmon cultivars leads to stable fruit production, but it fluctuates greatly with the amount of flower thinning and climatic conditions in Japan, especially the light intensity in June and July (Yamada et al. 1987a). Thus the parthenocarpic ability of a genotype needs to be assessed by bagging flowers that have been thinned in a standard fashion and compared to control cultivars (Yamada et al. 1987a). Although the inheritance of parthenocarpy has not been elucidated, a high percentage of new PCNA cultivars released from NIFTS breeding, such as 'Suruga,' 'Yoho,' 'Yubeni,' 'Kanshu,' and 'Taigetsu' are highly parthenocarpic.

4.8 Cultivars Released from the NIFTS Breeding Program in Japan

Since NIFTS began its breeding program it has released 11 PCNA and 2 PVA persimmon cultivars (Tables 17.4 and 17.5). The first release in 1959, a cross between local Japanese cultivars was 'Suruga' (Iikubo et al. 1961). This was a late ripening PCNA cultivar with a dense and juiceless texture and susceptible to calyx cracking. The second PCNA cultivar, 'Izu,' released in 1970 (Hirose et al. 1971), was initially grown on over 600 ha in Japan because it ripened early and produced fruit with good eating quality. Since then, its production has gradually decreased mainly due to its short shelf life, susceptibility to calyx cracking and fruit skin blackening, and low productivity. Subsequent breeding within PCNA germplasm resulted in four new PCNA cultivars (early to mid season 'Shinshu,' mid season 'Yoho' and 'Taishu,' late season 'Yubeni') in the 1990s (Yamada et al. 2003; Yamane et al. 1991a, 1991b, 2001), and three early ripening and calyx-end crack resistant PCNA cultivars ('Soshu,' 'Kanshu,' and 'Kishu') in the early 2000s (Yamada et al. 2004, 2006, 2009). In addition, two PCNA cultivars 'Tanrei' and 'Kinshu' were released for the ornamental use of red-color leaves at defoliation (Yamane et al. 1998).

All of the releases for table use from the 1990s had good to excellent eating quality. The earliest ripening cultivar, 'Shinshu' was recommended only for greenhouse production due to its excessive softening when exposed to rain or wetness. Among the others, 'Taishu' produces large fruit with excellent eating quality with soft and juicy flesh (Table 17.4), 'Yubeni' has the reddish fruit skin, and 'Yoho' has the firmest fruit texture and longest shelf life. 'Yoho' also has high vigor, productivity, and a fruit parthenocarpic ability superior to 'Matsumotowase-Fuyu' (Table 17.5).

The earliest of the crack resistant releases of the 2000s and the one used most commercially is 'Soshu' which ripens in late September to early October. Although it is susceptible to anthracnose and only has medium yield with a higher early fruit drop in June and July than 'Fuyu' in Japan, it produces fruit with good size, moderate sweetness, and moderately fine and highly juicy flesh. 'Kanshu' is highly parthenocarpic and produces high sugar fruit but is prone to skin blackening which averaged 40% in the national trial. 'Kishu' produces larger fruit with juicy flesh than 'Fuyu.' Fruit drop unusually occurs late in fruit development.

Since 1990 the NIFTS has run a backcross (PCNA × non-PCNA) × PCNA program to increase the diversity of the breeding population and avoid problems with inbreeding depression resulting in low productivity and small fruit size. So far, two

Table 17.4 Fruit characteristics of persimmon cultivars grown commercially or released in Japan

Cultivar	Astringency type	Harvesting time	Fruit weight(g)	Fruit skin color	Quantity of brown specks in flesh	Flesh firmness	Flesh texture	Juiciness	Soluble solids content (%)	Fruit cracking at calyx-end	Fruit cracking at stylar-end	Stained fruit skin	Shelf life
Soshu	PCNA	Late Sept. to Early Oct	250	Red-orange	Few	Medium	Dense	Much	15	None	Very few	Little	Semi-short
Nishimurawase	PVNA	Late Sept. to Early Oct.	230	Red-orange	Many	Firm	Coarse	Medium	16	Few	None	Little	Long
Tonewase	PVA (seedless)	Early Oct.	230	Yellow-orange	None	Soft	Dense	Much	14	None	None	Very little	Medium
Izu	PCNA	Mid-Oct.	240	Red-orange	Few	Medium	Dense	Medium	15	Medium	None	Many	Short
Saijo (early ripening strain)	PCA	Late-Oct.	180	Yellow	None	Medium	Dense	Much	18	None	Very few	Many	Short
Maekawa-Jiro	PCNA	Early Nov.	270	Red-orange	Few	Medium	Dense	Medium	17	None	Medium	Little	Medium
Hiratanenashi	PVA (seedless)	Early Nov.	240	Yellow-orange	None	Soft	Dense	Much	14	None	None	Very little	Medium
Matsumotowase-Fuvu	PCNA	Early Nov.	270	Red-orange	Medium	Medium	Medium	Much	16	Few to Medium	None	Little	Semi-long
Yoho	PCNA	Early Nov.	280	Red-orange	Few	Firm	Dense	Medium	16	Few to Medium	Few	Little	Semi-long
Taigetsu	PVA	Early Nov.	400	Yellow-orange	None but many only around seeds	Soft	Medium	Much	14	None	None	Little	Short
Taishu	PCNA	Early Nov.	400	Orange	Few	Soft	Coarse	Much	17	Medium	Few	Medium	Medium
Taiten	PVA	Late Nov.	450	Yellow-orange	None but many only around seeds	Soft	Medium	Much	16	None	None	Little	Medium
Fuyu	PCNA	Late Nov.	280	Red-orange	Medium	Medium	Medium	Much	16	Few to Medium	None	Little	Long
Jiro	PCNA	Mid-Nov.	280	Red-orange	Few	Medium	Dense	Medium	17	None	Many	Little	Medium
Atago	PCA	Early Dec.	270	Orange	None	Medium	Medium	Medium	15	Few	None	Little	Long

Table 17.5 Tree characteristics of persimmon cultivars grown commercially or released in Japan

Cultivar	Tree vigor	Quantity of female flowers	Male flowers	Fruit drop in the early fruit developmental stage	Yield
Soshu	Medium	Many	None	Medium	Medium
Nishimurawase	Medium	Medium	Medium	Little	Medium
Tonewase	Vigorous	Many	None	Little	High
Izu	Not vigorous	Many	None	Medium	Low
Saijo (early ripening strain)	Very vigorous	Medium	None	Little	Medium
Maekawa-Jiro	Vigorous	Medium	None	Little	High
Hiratanenashi	Vigorous	Many	None	Little	High
Matsumotowase-Fuyu	Medium	Many	None	Little	High
Yoho	Vigorous	Many	None	Little	High
Taigetsu	Vigorous	Many	Very few	Little	Very high
Taishu	Medium	Medium	Medium	Little	Medium
Taiten	Vigorous	Many	Few	Little	Very high
Fuyu	Vigorous	Many	None	Little	High
Jiro	Vigorous	Medium	None	Little	High
Atago	Vigorous	Many	None	Little	Very high

large fruited and highly productive PVA cultivars ('Taiten' and 'Taigetsu') have been released from PCNA × non-PCNA crosses (Yamada et al. 2008a, b).

5 Current Goals of Breeding

The most important negative trait is flesh astringency and therefore an essential goal is the development of superior PCNA cultivars. Beyond this, cultivars need to combine a range of traits, such as high fruit quality, size, appearance, resistance to fruit cracking, good shelf life, high productivity, high ability to set parthenocarpic fruit, and resistance to various pests and diseases.

5.1 Fruit Quality and Appearance

The consumer preference and therefore the breeding goal in Japan is for persimmons with large sized, juicy, very sweet (high soluble solids), soft fruit. The flesh texture can range from dense to coarse but can never be mealy. The skin color preference is red although this is not essential as major commercial cultivars, such as 'Fuyu' and 'Hiratanenashi', which have red-orange and yellow-orange fruit skin, respectively. There is also selection for the low rate of fruit skin blackening. This defect, common in the 'Taishu' and 'Taigetsu' cultivars develops as shallow concentric cracks which although reduces fruit attractiveness actually increases the SSC by 2 degrees Brix under the shallow concentric cracks (Iwanami et al. 2002). Finally, although concentric circular cracks are acceptable, it is essential to develop cultivars that are resistant to calyx or stylar end cracking as a fruit with end cracking is not marketable.

5.2 FRT and Shelf Life

In the fresh fruit market, increases in consumption can be achieved by making the fruit available for a longer time. Thus, the extension of the harvest season is of high priority especially in the earlier seasons where fewer good varieties are available. Better postharvest life is also an important breeding target.

5.3 Productivity

In Japan, normal commercial yields of 'Fuyu,' 'Hiratanenashi,' and 'Atago' are 20–25 mt/ha, 30 mt/ha, 40 mt/ha, respectively. High productivity is an important target in breeding; however, inbreeding depression seems to be an obstacle in raising the productivity. A major determinant of yield in persimmon is photosynthesis ability, which should be further elucidated. In addition, other factors determining the productivity are the number of female flowers and the extent of physiological fruit drop. Physiological fruit drop in the early fruit development stage varies greatly among cultivars and is controlled by the two factors: parthenocarpic ability and seed formation ability (Kajiura 1941; Yamada et al. 1987a). High parthenocarpy leads to low fruit drop and stable production as does the ability to set many seeds in cultivars that are not parthenocarpic. Thus, the highly parthenocarpic 'Hiratanenashi,' which is a seedless cultivar, and the highly seeded 'Fuyu' show little early fruit drop and stable production.

5.3.1 Disease and Pest Resistance

Disease and pest resistance is an important breeding objective even though persimmon trees grow well without spraying in home gardens in Japan, Korea, and China.

Major diseases attacking persimmon are anthracnose (*Gloesporium kaki* Holi), angular leaf spot (*Cerocospora kaki* Ellis et Everhart), circular leaf spot (*Mycosphaerella nawae* Hiura et Ikata), and powdery mildew (*Phyllactinia kakicola* Sawada). 'Fuyu' and 'Soshu' is susceptible to anthracnose and 'Saijo' is tolerant.

Major pests are the persimmon fruit moth (*Stathmopoda masinissa* Meyrick), yellow tea thrips (*Scirtothrips dorsalis* Hood), fruit-piercing stink bugs, the Japanese mealybug (*Planococcus kraunhiae* Kuwana), and thrips (*Ponticulothrips diospyrosi* Haga et Okajima). 'Fuyu' fruit is resistant while 'Hiratanenashi' is susceptible to yellow tea thrips.

6 Breeding Methods and Techniques

The genome composition of persimmon (*D. kaki*) is poorly understood. Various lines of evidence suggest that *D. kaki* is an autoallohexaploid. Cytogenetic studies, including the physical mapping of 45S rDNA by fluorescent in situ hybridization (FISH) in persimmon and its wild relatives indicate that in *D. kaki* four chromosomes were homologous or at least some chromosomes are homoeologous among the different genome sets within the *D. kaki* genome (Choi et al. 2003). The possibility of autoallohexaploidy is also reported by Kanzaki et al. (2001, 2008, 2009) from the observation of polysomic inheritance of a molecular marker linked to astringency trait.

However, the possibility of autohexaploidy is not eliminated. A recent study using quantitative real-time PCR to determine the copy number for the astringency locus by a marker linked to natural astringency-loss in transgenic PCNA 'Jiro' suggested that 'Jiro' has six copies of the fruit astringency gene (Akagi et al. 2009). This result indicates a high possibility of autohexaploidy for persimmon.

6.1 Crossbreeding

The traditional crossbreeding approach is common. It is a lengthy process that takes 5–8 years after sowing the seeds for seedlings to begin to bear fruit. This process can be shortened to 3–5 years by top-grafting 1-year-old seedling scions in the spring to mature trees. The fruit obtained in top-grafting exhibit the same characteristics as those of adult trees even in early stages. In addition, this technique allows more seedlings to be evaluated in the selection field than when growing individual seedling trees.

Primary selection can be made with a small number of fruit from a small shoot evaluation done over 3–5 years after first fruiting. Selected clones are propagated and usually tested over 5–6 years in several locations. A clone selected in this trial is released as a new cultivar and registered for plant variety protection.

6.2 Nonaploid Breeding

Although persimmon, *D. kaki* Thunb., is generally hexaploid, a few cultivars are nonaploid, including 'Hiratenashi,' a commercially important astringent cultivar in Japan (Zhuang et al. 1990). Nonaploid cultivars bear seedless fruit via parthenocarpy. In spite of this, embryos in young fruit of nonaploid cultivars can be rescued by in vitro culture and grown into a plant (Ishida et al. 1980). So far, using this technique, one new astringent cultivar, 'Tokiotome,' developed from a 'Tonewase' (9×) × 'Nishimurawase' (6×) cross was released in 2001 by Niigata Prefecture in Japan.

Nonaploid genotypes are obtained by a union between a nonreduced gamete (6x) and a reduced gamete (3X). Although very few nonreduced gametes are normally produced in persimmon, this fluctuates yearly and with the cultivar. Low temperatures during flowering raise the percentage of nonreduced gametes (Yamada et al. 2005), and 'Fujiwaragosho' produces higher numbers of nonreduced gametes than other cultivars (Yamada and Tao 2006). To produce nonaploid seedlings, Sugiura et al. (2000) selectively used large sized pollens which are nonreduced gametes to pollinate hexaploid female flowers followed by in vitro embryo rescue.

7 Integration of New Biotechnologies in Breeding Programs

7.1 *Molecular Markers for Cultivar Identification and Genetic Diversity Studies*

Several SSR markers were developed for identifying cultivar and/or elucidating genetic relatedness in germplasm resources (Soriano et al. 2006; Guo and Luo 2006). Soriano et al. (2006) developed 37 pairs of SSR primers with annealing temperatures between 47 and 60°C by constructing a CT/AG-enriched genomic library from 'Rojo Brillante,' which is the most important commercial cultivar in Spain. According to their report, 22 pairs of SSR primers showed polymorphism when 37 pairs of SSR primers were tested on 12 persimmon cultivars. In addition, Guo and Luo (2006) developed 9 pairs of SSR primers from the sequences of inter-SSR-PCR amplification products using 8 ISSR primers from Chinese PCNA 'Luo tian tian shi' and Japanese PCNA 'Maekawa-Jiro' without making an SSR-enriched genomic library. They designed 12 primers from the sequences of inter-SSR-PCR products, but only 9 primers were showed high polymorphism when tested on 30 genotypes of *Diospyros* spp. Some primer pairs for amplifying inter-retrotransposon regions were also reported and these primers were used for analyzing cultivar relationships among persimmon cultivars and *Diospyros* spp. (Guo et al. 2006).

7.2 *Linkage Maps*

Linkage maps of persimmon are not available. Since the cultivated persimmon (*D. kaki*) is hexaploid ($2n=6x=90$ chromosomes) with a few nonaploid cultivars, it would be very difficult to make a linkage map. Nevertheless, some related species (*D. lotus* and *D. oleifera*) are diploid (Yonemori et al. 2008a) and may be useful in developing segregating populations suitable for linkage map development in persimmon.

7.3 *Marker-Assisted Breeding*

The main target of persimmon breeding is to obtain new superior PCNA type. Thus, markers to detect this recessive PCNA phenotype at a seedling stage without having to wait 3–5 years for the plant to fruit would be a tremendous advantage and save time, space, and money. Currently, the Japanese breeding program in NIFTS is using non-PCNA cultivars as parents to expand the genetic diversity of the breeding population. However, the frequency of PCNA type individuals in backcrossed progenies [PCNA × F1 selection (PCNA × non-PCNA)], is only about 15%. Thus, if you could detect the PCNA phenotype of a nonfruiting seedling with a molecular marker, it would be possible to handle 7 times the initial population size since only the 15%

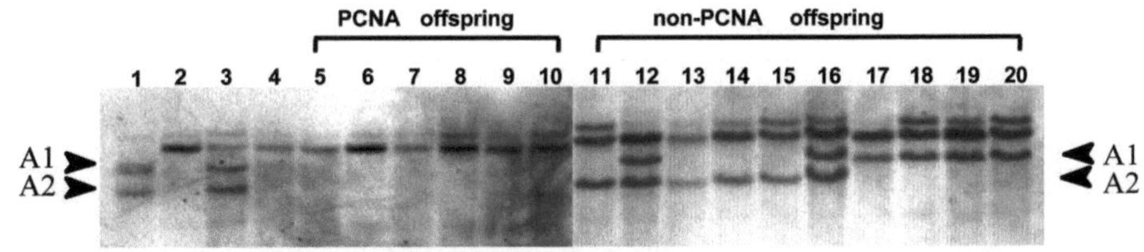

Fig. 17.13 RFLP analysis of genomic DNA digested with *Hin*dIII, using EACC/MCTA-400 as a probe in offspring from a backcross (PCNA) × (non-PCNA × PCNA). *Lanes*: 1 and 3, non-PCNA parent and ancestor; 2 and 4, PCNA parent and ancestor; 5–10, PCNA offspring used for BSA; 11–20, non-PCNA offspring used for BSA. *Arrows* indicate the 8 kb (*upper*) and 6.5 kb (*lower*) RFLP markers (Kanzaki et al. 2001)

of PCNA type are planted into the field for further evaluation. This would increase the efficiency and productivity of the breeding manifold.

The PCNA trait of Japanese type is controlled by a single locus but given that most persimmon cultivars are hexaploid, four or six copies of this locus are involved (Kanzaki and Yonemori 2007; Kanzaki et al. 2009). Thus, a PCNA genotype must have the recessive alleles in all copies of the gene whereas the non-PCNA (PVNA, PVA, and PCA) type only needs one dominant allele copy for expression. Consequently, the approach taken was to identify markers for the dominant non-PCNA alleles by using bulked segregant analysis (BSA) and amplified fragment length polymorphism (AFLP) (Kanzaki et al. 2001; Yonemori et al. 2003). One hundred and twenty eight AFLP primer combinations were screened and a reliable candidate of AFLP marker for selecting the PCNA type was identified, using a backcross population derived from 'Nishimura-wase,' a non-PCNA cultivar. This marker was absent in all 26 PCNA offspring and was present in 13 of the 25 non-PCNA offspring. And then, by converting this AFLP marker into RFLP markers they succeeded in distinguishing the PCNA type from non-PCNA types in the offspring of 42 genotypes in breeding population with 100% accuracy by the absence of 8.0 kb (A1) and/or 6.5 kb (A2) band(s) (Fig. 17.13). Furthermore, Kanzaki et al. (2008) designed a pair of primer sets, E4/E9r and E4/A2r, which could amplify the selected regions corresponding to 8.0 kb (A1) and 6.5 kb (A2) RFLP band, respectively (Fig. 17.14). This lead to the development of PCR based SCAR markers which simplifies the marker identification and makes the use of these markers with large breeding populations possible.

Currently, seedling selection by these PCR-based molecular markers is progressing at NIFTS, in cooperation with Kyoto University and Kinki University. However, some non-PCNA selections not derived from 'Nishimurawase' which was used in the original marker work, did not show the bands for either the PCR-based markers or the RFLP markers and therefore was indistinguishable from the PCNA type.

To solve this problem, fosmid libraries from PCNA 'Jiro,' non-PCNA 'Nishimura-wase' and *D. lotus* 'Mamegaki' were made, and each library was screened using the AFLP marker as a probe to locate the appropriate clone. An analysis of the clone sequences indicated that the marker problem was caused by a large deletion (ca. 16 kb)

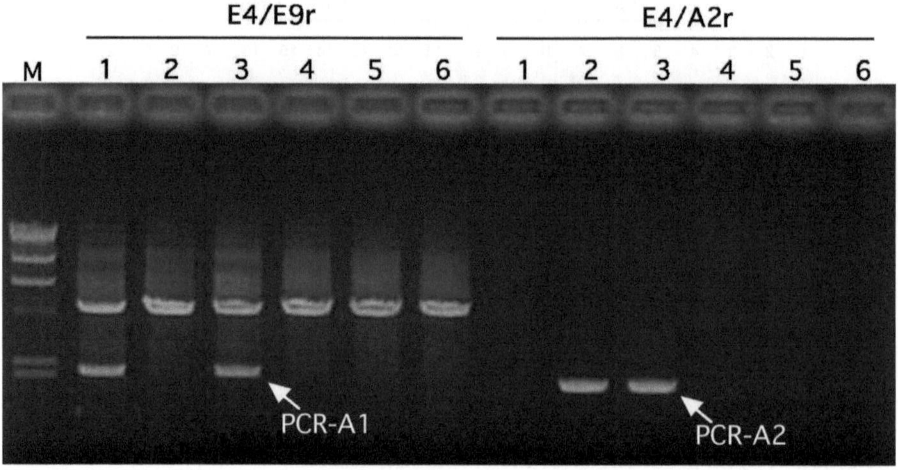

Fig. 17.14 The PCR-A1 and PCR-A2 SCAR markers generated by using the primer pairs E4/E9r and E4/A2r in offspring from a backcross (PCNA) × (non-PCNA × PCNA). *Lanes*: 1. A1-type offspring (non-PCNA), 2. A2-type offspring (non-PCNA), 3. A1A2-type offspring (non-PCNA), 4–6. a-type offspring (PCNA), and M: molecular marker (λ/*Hin*dIII digest) (Kanzaki et al. 2009)

around this marker region. In addition, a small insert found only in PCNA types was characterized and used to design a new primer set capable of detecting every non-PCNA type among the cultivars/selections tested (Yonemori et al. 2009).

Cooperative research (Kyoto and Kinki University and NIFTS) to identify molecular markers linked to the dominant Chinese PCNA alleles is also progressing. Thus far, using the BSA, ten F_1 individuals of PCNA or non-PCNA type derived from Chinese PCNA 'Luo tian tian shi' × Japanese PCNA 'Okugosho,' have been screened with 384 AFLP primer combinations. Three putative AFLP candidate bands have been identified and are currently being verified. One of these AFLP markers was successfully converted to PCR-based marker and is being used in marker-assisted selection (MAS) to distinguish Chinese-type PCNA from non-PCNA type at NIFTS in their breeding program to develop improved cultivars with Chinese-type PCNA trait.

7.4 Genomics

Nakagawa et al. (2008) generated 9,952 ESTs (expressed sequence tags) from randomly selected clones of two different cDNA libraries; one that was derived from the fruit of PCA 'Saijo' at an early stage of development and the other was from ripening fruit. These ESTs were clustered into 6,700 nonredundant sequences. Sixty-five percent (4,356) of these showed significant homology to known proteins when the deduced amino acid sequences were evaluated. Some putative genes were involved in proanthocyanidin and carotenoid synthesis. This research is important to provide basic information for genomic research.

7.5 Transgenics

There are two reports of genetic transformation of persimmon using *Agrobacterium tumefaciens*. The first trial introduced the *cry*IA(c) gene of *Bacillus thuringiensis* into persimmon with a leaf disc–*Agrobacterium* system. Plantlets regenerated from the calli derived from the leaf discs successfully produced an insecticidal protein and resisted damage from the Oriental moth (*Monema flavescens* Walker) in vitro (Tao et al. 1997).

The other research focused on enhancing the tolerance of persimmon to environmental stress by introducing the *cod*A gene of *Arthrobacter globiformis* encoding choline oxidase, an enzyme catalyzing complete oxidation of choline to glycine betaine (Gao et al. 2000) and the gene-encoding NADP-dependent sorbitol-6-phosphate dehydrogenase of apple cDNA (Gao et al. 2001). However, the efficacies of these genes against environmental stress have not yet been determined.

References

Akagi, T., Kanzaki, S., Gao, M., Tao, R., Parfitt, D.E., and Yonemori, K., 2009. Quantitative real-time PCR to determine allele number for the astringency locus by analysis of a linked marker in *Diospyros kaki* Thunb. Tree Genet. Genomics 5:483–492.

Agricultural Research Station (ARS). 1912. Investigation on persimmon cultivars. Bul. Agr. Res. Sta. (extra) 28:1–46. (in Japanese).

Badenes M, Garces A, Romero C, Romero M, Clave J, Rovia M, Llacer G. 2003. Genetic diversity of introduced and local Spanish persimmon cultivars revealed by RAPD markers. Genet. Resour. Crop Evol. 50:579–585.

Bellini E., 1982. Monografia delle principali cultivar di kaki introdotte in Italia. Florence, Italy. (in Italian).

Bellini E., Benelli C., Giordani E., Perria R. and Paffetti D. 2003. Genetic and morphological relationships between possible Italian and ancestral cultivars of persimmon. Acta Hort., 601:192–197.

Bellini E., Giordani E. 1998. Persimmon. In: Italian Contribution to Plant Genetics and Breeding. Ed. Scarascia Mugnozza G.T., Pagnotta M.A., University of Tuscia, Viterbo (Italy): 675–684.

Bellini E. and Giordani E. 2005. Germplasm and breeding of persimmon in Europe. Acta Hort. 685:65–75.

Climent C. and Llácer G. 2001. Caqui. In: Nuez, F. and Llácer, G. (eds.), La Horticultura Española, Sociedad Española de Ciencias Hortícolas, Madrid. (in Spanish).

De Sousa R.M. and Gomes-Pereira J. 1995. Notes sur quelques spèces fruitières sous-utilisées au Portugal. Cahiers Options Méditerranéennes, 13: 63–67. (in French with English summary).

Evreinoff V.A., 1948 – Le plaqueminier du Japon ou Kaki. Fruits d'Outre-Mer, 3(4): 124–132. (in French).

Cho, S.K., and Cho, T.H. 1965. Studies on the local varieties of persimmon in Korea. Res. Rep. RDA 8:147–190. (in Korean with English summary).

Cohen Y, Gur A, Barkai Z, Blumenfeld A, 1991. Decline of persimmon (*Diospyros kaki* L.) trees on *Diospyros virginiana* rootstocks. Scientia Hort. 48:61–70.

Choi, Y. A., R. Tao, K. Yonemori and Sugiura, A. 2003. Physical mapping of 45S rDNA by fluorescent in situ hybridization in persimmon (*Diospyros kaki*) and its wild relatives. J. Hort. Sci. Biotechnol. 78:265–271.

Gao, M., Sakamoto, A., Miura, K., Murata, N., Sugiura, A., and Tao, R. 2000. Transformation of Japanese persimmon (*Diospyros kaki* Thunb.) with a bacterial gene for choline oxidase. Mol. Breed. 6: 501–510.

Gao, M., Tao, R., Miura, K., Dandekar, A.M., and Sugiura, A. 2001. Transformation of Japanese persimmon (*Diospyros kaki* Thunb.) with apple cDNA encoding NADP-dependent sorbitol-6-phosphate dehydrogenase. Plant Sci. 160: 837–845.

Guo, D., Zhang, H., and Luo, Z., 2006. Genetic relationships of *Diospyros kaki* Thunb. and related species revealed by IRAP and REMP analysis. Plant Sci. 170: 528–533.

Guo, D.L., and Luo, Z.R., 2006. Development of SSR primers using ISSR-PCR in *Diospyros kaki* Thunb. Mol. Ecol. Notes 6: 886–887.

Hirose K, Yamamoto M, Sato T, Ohhata T, Nishida T, Ikeda I, Shimura I, Siba S, Yagi M, Tominaga N, 1971. New Japanese persimmon variety 'Izu'. Bul. Hort. Res. Sta. B11:1–17. (in Japanese with English summary).

Hiroshima Prefecture Fruit Tree Experimental Station, 1979. Survey on persimmon varieties in Japan for identification. pp.436. Nagao Printing Office, Akitsu, Hiroshima, Japan. (in Japanese).

Hume H H, 1914. A kaki classification. J. Hered. 5:400–406.

Iikubo S, Sato T, Nishida T, 1961. New Japanese persimmon variety 'Suruga'. Bull. Hort. Sta. Natl. Tokai-kinki Agr. Exp. Sta. 6:33–37. (in Japanese with English summary).

Ikeda, I., Yamada, M., Kurihara, A. and Nishida, T. 1985. Inheritance of astringency in Japanese persimmon (*Diospyros kaki* Thunb.). J. Japan. Soc. Hort. Sci. 54:39–45. (in Japanese with English summary).

Ikegami, A., Yonemori, K., Sugiura, A., Sato, A. and Yamada, M. 2004. Segregation of astringency in F_1 progenies derived from crosses between pollination-constant non-astringent persimmon cultivars. HortScience 39:371–374.

Ikegami A, Sai E, Yonemori K, Yamada M, Sato A, Mitani N, Kitajima A, 2006. Segregations of astringent progenies in the F_1 populations derived from crosses between a Chinese pollination-constant nonastringent (PCNA) 'Luo Tian Tian Shi', Japanese PCNA and pollination-constant astringent (PCA) cultivars of Japanese origin. HortScience 41:561–563.

Ishida M., S. Inaba, and Sobajima, Y. 1980. In vitro culture of young embryo in Hiratanenashi persimmon. Scientific Rep. Kyoto Pref. University, Agriculture. 32:2–24. (in Japanese with English summary).

Iwanami, H., M. Yamada, and Sato, A. 2002. A great increase of soluble solids concentration by shallow concentric cracks in Japanese persimmon. Scientia Hort. 94:251–256.

Kajiura M., 1941. Studies on physiological fruit drop in persimmon. 2. Relationships between fruit drop and pollination or parthenocarpic ability. J. Japan. Soc. Hort. Sci. 12:247–283. (in Japanese).

Kanzaki, S., Sato, A., Yamada, M., Utsunomiya, N., Kitajima, A. Ikegami, A. and Yonemori, K., 2008. RFLP markers for the selection of pollination-constant and non-astringent (PCNA)-type persimmon and examination of the inheritance mode of the markers. J. Japan. Soc. Hort. Sci. 77: 28–32.

Kanzaki, S., Yamada, M., Sato, A., Mitani, N., Ustunomiya, M., and Yonemori, K., 2009. Conversion of RFLP markers for the selection of pollination-constant and non-astringent type persimmons (*Diospyros kaki* Thunb.) into PCR-based markers. J. Japan. Soc. Hort. Sci. 78: 68–73.

Kanzaki, S., Yonemori, K., 2007. Persimmon. p.353-358. In: Kole C (ed) Genome mapping and molecular breeding. Vol. 4. Fruits and Nuts. Springer, Heidelberg, Berlin, New York, Tokyo.

Kanzaki, S., Yonemori, K., Sato, A., Yamada, M. and Sugiura, A. 2001. Identification of molecular markers linked to the trait of natural astringency loss of Japanese persimmon (*Diospyros kaki* Thunb.) fruit. J. Amer. Soc. Hort. Sci. 126:51–55.

Kanzaki, S., Yonemori, K., Sato, A., Yamada, M. and Sugiura, A. 2000a. Analysis of the genetic relationships among pollination-constant and non-astringent (PCNA) cultivars of persimmon (*Diospyros kaki* Thunb.) from Japan and China using amplified fragment length polymorphism (AFLP). J. Japan. Soc. Hort. Sci. 69:665–670.

Kanzaki, S., Yonemori, K., Sato, A., Yamada, M. and Sugiura, A. 2000b. Evaluation of RFLP analysis for discriminating PCNA genotype in some persimmon cultivars. J. Japan. Soc. Hort. Sci. 69:702–704.

Kikuchi, A. 1948. Pomology-part I., p. 347–400. Yokendo, Tokyo, Japan. (in Japanese).

Luo, Z., Li, F. and Cai, L. 1999. Molecular systematics of China native non-astringent persimmon based on random amplified polymorphic DNA. Acta Hort. Sinica 26:297–301. (in Chinese with English summary).

Morettini A. 1949. Il kaki. Ramo Editoriale degli Agricoltori, Roma. (In Italian).

Nakagawa, T., Nakatsuka, A., Yano, K., Yasugahira, S., Nakamura, R., Ningjing, S., Itai, A., Suzuki, T., Itamura, H. 2008. Expressed sequence tags (ESTs) from persimmon at different developmental stages. Plant Cell Reports 27: 931–938.

Occhialini O. and Tirocco G.B. 1923. Il *Diospyros kaki* (Loto del Giappone). Francesco Battiato, Catania. (in Italian).

Oohata T., I. Ikeda, and Nishida, T. 1964. The inheritance of some characters in Japanese persimmon (*Diospyros kaki* Thunb.) 2. The inheritance of bearing habit of staminate flowers Bul. Hort. Res. Sta. B3:51–66. (in Japanese with English summary).

Soriano, J.M., Pecchili, S., Romero, C., Vilanova, S., Llaalcer, G., Giordani, E., and Badenes, M. L., 2006. Development of microsatellite markers in polyploidy persimmon (*Diospyros kaki* L.f.) from an enriched genomic library. Mol. Ecol. Notes 6: 368–370.

Sugiura A., T. Ohkuma, Y. A. Choi, R. Tao and Tamura, M. 2000. Production of nonaploid (2n=9x) Japanese persimmon (*Diospyros kaki* Thunb.) by pollination with unreduced (2n=6x) pollen and embryo rescue culture. J, Amer. Soc. Hort. Sci. 125:609–614.

Tanaka C, 1930. Experiments on the rootstocks for the Kaki or Japanese persimmon (*Diospyros kaki* L.). J. Okitsu Hort. Soc. 25:1–30. (in Japanese).

Tao, R., Dandekar, A.M., Uratsu, S.L., Vail, P.V., and Tebbets. J.S., 1997. Engineering genetic resistance against insects in Japanese persimmon using *cry*1A(c) gene of *Bacillus thuringiensis*. J. Amer. Soc. Hort. Sci. 122: 764–771.

Tuzcu, O. and Seker, M. 1997. The situation of persimmon (*Diospyros kaki* L.) cultivation and germplasm resources in Turkey. Acta Hort. 441:107–114.

Wang R, 1983. The origin of 'Luo Tian Tian Shi'. Chinese Fruit Tree 2:16–19. (in Chinese).

Wang R, Yang Y, Li G, 1997. Chinese persimmon germplasm resources. Acta Hort. 436:43–50.

Wang R, Yang Y, Ruan X, Li G, 2005. Native non-astringent persimmons in China. Acta Hort. 685:99–102.

Wen IC, 2003. Evaluation and breeding of persimmon in Taiwan. Acta Hort. 601:233–237.

Yakushiji H., M. Yamada, K. Yonemori, A. Sato, and Kimura, N. 1995. Staminate flower production on shoots of 'Fuyu' and 'Jiro' persimmon. J. Japan. Soc. Hort. Sci. 64:41–46.

Yamada, A. and R. Tao. 2006. High frequency polyploidisation observed in hexaploid Japanese persimmon (*Diospyros kaki*) 'Fujiwaragosho'. J. Hort. Sci. Biotech. 81:402–408.

Yamada, A., R. Tao and Sugiura, A. 2005. Influence of low temperature before flowering on the occurrence of unreduced pollen in Japanese persimmon (*Diospyros kaki* Thunb.). HortScience 40: 24–28.

Yamada M, 2005. Persimmon genetic resources and breeding in Japan. Acta Hort. 685:51–64.

Yamada, M. 1993.Persimmon breeding in Japan. Japan. Agr. Res. Quart. 27:33–37.

Yamada, M., Ikeda, I., Yamane, H. and Hirabayashi, T. 1988. Inheritance of fruit cracking at the calyx and stylar end in Japanese persimmon (*Diospyros kaki* Thunb.). J. Japan. Soc. Hort. Sci. 57:8–16. (in Japanese with English summary).

Yamada, M., Kakutani, M., Yamane, H. and Yoshinaga, K. 1991. Influence of seed formation on the cracking at fruit apex in Japanese persimmon. Bul. Fruit Tree Res. Stn. 20:1–11. (in Japanese with English summary).

Yamada, M., Kurihara, A. and Sumi, T. 1987a. Varietal differences in fruit bearing in Japanese persimmon (*Diospyros kaki* Thunb.). J. Japan. Soc. Hort. Sci. 56:293–299. (in Japanese with English summary).

Yamada, M. and Sato, A. 2002. Segregation for fruit astringency type in progenies derived from crosses of 'Nishimurawase' x pollination constant non-astringent genotypes in oriental persimmon (*Diospyros kaki* Thunb.). Scientia Hort. 92:107–111.

Yamada, M. and Sato, A. 2003. Status of persimmon breeding at the National Institute of Fruit Tree Science, Japan. Options Mediterraneennes Series A 51:75–79.

Yamada, M., Sato, A. and Ukai, Y. 2002a. Genetic differences and environmental variations in calyx-end fruit cracking among Japanese persimmon cultivars and selections. HortScience 37:164–167.

Yamada, M., Sato, A., Yakushiji, H., Yoshinaga, K., Yamane, H. and Endo, M. 1993a. Characteristics of 'Luo Tian Tian Shi', a non-astringent cultivar of oriental persimmon (*Diospyros kaki* Thunb.) of Chinese origin in relation to non-astringent cultivars of Japanese origin. Bul. Fruit Tree Res. Stn. 25:19–32. (in Japanese with English summary).

Yamada, M., A. Sato, H. Yamane, N. Mitani, H. Iwanami, M. Shiraishi, N. Hirakawa, T. Ueno, A. Kono, M. Yoshioka and Nakajima, I. 2008a. New Japanese persimmon cultivar, 'Taigetsu'. Hort. Res. (Japan) 7 (suppl.1):309 (in Japanese).

Yamada, M., A. Sato, H. Yamane, N. Mitani, H. Iwanami, M. Shiraishi, N. Hirakawa, T. Ueno, A. Kono, M. Yoshioka and Nakajima, I. 2008b. New Japanese persimmon cultivar, 'Taiten'. Hort. Res. (Japan) 7 (suppl.1):310. (in Japanese).

Yamada, M., Taira, S., Ohtsuki, M., Sato, A., Iwanami, H., Yakushiji, H., Wang, R., Yang, Y. and Li, G. 2002b. Varietal differences in the ease of astringency removal by carbon dioxide gas and ethanol vapor treatments among Oriental astringent persimmons of Japanese and Chinese origin. Scientia Hort. 94:63–72.

Yamada, M., Wang, R., Yamane, H., Sato, A. and Hirakawa, N. 1995a. Variation in the performance of fruit maturing time, fruit weight, and soluble solid content in oriental persimmon grown at Akitsu, Japan and Meixian, China. J. Japan. Soc. Hort. Sci. 64:221–226.

Yamada, M., Wang, R., Yamane, H., Sato, A. and Hirakawa, N. 1995b. Comparison of the variations in fruit maturing time, fruit weight, and soluble solids content of Oriental persimmon cultivars of Chinese and Japanese origin. J. Japan. Soc. Hort. Sci. 64:227–233.

Yamada, M. and Yamane, H. 1997. Relationship between the observed and predicted distribution of offspring for fruit ripening time and fruit weight in Japanese persimmon. Scientia Hort. 69:157–167.

Yamada, M., Yamane, H. and Hirabayashi, T. 1986a. Studies on crossbreeding of Japanese persimmon (*Diospyros kaki* Thunb.). IV. A sampling method in determining refractometer index. Bul. Fruit Tree Res. Stn. E6:11–20. (in Japanese with English summary).

Yamada, M., Yamane, H. and Hirabayashi, T. 1986b. Studies on crossbreeding of Japanese persimmon (*Diospyros kaki* Thunb.). V. Variation of 'Hetasuki' (fruit cracking under the calyx). Bul. Fruit Tree Res. Stn. E6:21–30. (in Japanese with English summary).

Yamada, M., Yamane, H. and Hirabayashi, T. 1987b. Yearly fluctuations of two types of fruit cracking in seedling populations of Japanese persimmon (*Diospyros kaki* Thunb). J. Japan. Soc. Hort. Sci. 56:287–292.

Yamada, M., Yamane, H., Kurihara, A., Nagata, K., Sato, A., Kishi, T., Matsumoto, R., Yoshinaga, K., Hirakawa, N., Iwanami, H., Kakutani, M., Ozawa, T., Sumi, T., Hirabayashi, T., Kanato, K. and Nakajima, I. 2003a. New Japanese persimmon cultivar 'Yubeni'. Bul. Natl. Inst. Fruit Tree Sci. 2:65–75. (in Japanese with English summary).

Yamada, M., Yamane, H., Sato, A., Hirakawa, N. and Wang, R. 1994a. Variations in fruit ripening time, fruit weight and soluble solids content of oriental persimmon cultivars native to Japan. J. Japan. Soc. Hort. Sci. 63:485–492.

Yamada, M., Yamane, H., Sato, A., Iwanami, H., Hirakawa, N., Yoshinaga, K., Ozawa, T. and Nakajima, I. 2004a. New Japanese persimmon cultivar 'Soshu'. Bul. Natl. Inst. Fruit Tree Sci. 3:53–66. (in Japanese with English summary).

Yamada, M., Yamane, H., Sato, A., Hirakawa, N., Iwanami, H., Yoshinaga, K., Ozawa, T., Kakutani, M., Mitani, N., Yoshioka, M. and Nakajima, I. 2006. New Japanese persimmon cultivar 'Kanshu'. Bul. Natl. Inst. Fruit Tree Sci. 5:95–106. (in Japanese with English summary).

Yamada, M., Yamane, H., Sato, A., Yoshinaga, K., Hirakawa, N., Iwanami, H., Ozawa, T., Hirabayashi, T., Kakutani, M., Shiraishi, M., Mitani, N., Sumi, T., Yoshioka, M. and Nakajima, I. 2009. New Japanese persimmon cultivar 'Kishu'. Bul. Natl. Inst. Fruit Tree Sci. 8:25–38. (in Japanese with English summary).

Yamada, M., Yamane, H., Takano, Y. and Ukai, Y. 1997b. Estimation of the proportion of offspring having soluble solids content in fruit exceeding a given critical value in Japanese persimmon. Euphytica 93:119–126.

Yamada, M., Yamane, H. and Ukai, Y. 1994b.Genetic analysis of Japanese persimmon fruit weight. J. Amer. Soc. Hort. Sci. 119:1298–1302.

Yamada, M., Yamane, H. and Ukai, Y. 1994c. Efficiency of use of control genotypes for reducing yearly fluctuations of quantitative fruit characters in Japanese persimmon breeding. Bull. Fruit Tree Res. Stn. 26:29–37.

Yamada, M., Yamane, H. and Ukai, Y. 1995c. Genetic analysis of fruit ripening time in Japanese persimmon. J. Amer. Soc. Hort. Sci. 120:886–890.

Yamada, M., Yamane, H., Yoshinaga, K. and Ukai, Y. 1993b.Optimal spatial and temporal measurement repetition for selection in Japanese persimmon breeding. HortScience 28:838–841.

Yamane, H., Yamada, M., Kurihara, A., Sato, A., Yoshinaga, K., Nagata, K., Matsumoto, R., Hirakawa, N., Kakutani, M., Ozawa, T., Sumi, T., Hirabayashi, T. and Iwanami, H. 2001. New Japanese persimmon cultivar 'Taishuu'. Bul. Fruit Tree Res. Stn. 31:15–24. (in Japanese with English summary).

Yamane, H., Kurihara, A., Nagata, K., Yamada, M., Kishi, T., Yoshinaga, K., Matsumoto R., Ozawa, T., Sumi, T., Hirabayashi, T. and Kakutani, M. 1991a. New Japanese persimmon cultivar 'Shinshuu'. Bul. Fruit Tree Res. Stn. 19:13–27. (in Japanese with English summary).

Yamane, H., Kurihara, A., Nagata, K., Yamada, M., Kishi, T., Yoshinaga, K., Matsumoto, R., Kanato, K., Sumi, T., Hirabayashi, T., Ozawa, T., Hirose, K., Yamamoto, M. and Kakutani, M. 1991b. New Japanese persimmon cultivar 'Youhou'. Bul. Fruit Tree Res. Stn. 19:49–61. (in Japanese with English summary).

Yamane, H., Yamada, M., Kurihara, A., Yoshinaga, K., Nagata, K., Ozawa, T., Sumi, T., Hirabayashi, T., Hirakawa, N., Sato A., Matsumoto R. and Kakutani, M. 1998. Bul. Natl. Inst. Fruit Tree Sci. 30 and 31:15–24.

Yonemori, K., Akagi, T., and Kanzaki, S. 2009. Construction of a reliable PCR marker for selecting pollination constant and non-astringent (PCNA) type offspring among breeding population of persimmon (*Diospyros kaki* Thunb.). Acta Hort. 839:625–629.

Yonemori, K. Honsho, C., Kanzaki, S., Ino, H., Ikegami, A., Kitajima, A., Sugiura, A., and Parfitt, D.E. 2008a. Sequence analyses of the ITS regions and the *mat*K gene for determining phylogenetic relationships of *Diospyros kaki* (persimmon) with other wild *Diospyros* (Ebenaceae) species. Tree Genetics Genomes 4: 149–158.

Yonemori K, Honsho C, Kitajima A, Aradhya M, Giordani E, Bellini E, Parfitt D.E. 2008b. Relationship of European persimmon (*Diospyros kaki* Thunb.) cultivars to Asian cultivars, characterized using AFLPs. Genet. Resour. Crop Evol. 55:81–89.

Yonemori K, Ikegami A, Kitajima A, Luo Z, Kanzaki S, Sato A, Yamada M, Yang Y, Wang R, 2005. Existence of several pollination constant non-astringent type persimmons in China. Acta Hort. 685:77–83.

Yonemori K, Matsushima J, 1985. Property of development of the tannin cells in non-astringent type fruits of Japanese persimmon (*Diospyros kaki*) and its relationship to natural deastringency. J. Japan. Soc. Hort. Sci. 54:201–208. (in Japanese with English summary).

Yonemori, K., Sugiura, A., Sato, A., Yamada, M. and Kanzaki, S. 2003. Molecular marker for selecting pollination constant non-astringent type persimmon at juvenile stage. Acta Hort. 622:189–203.

Yonemori K, Sugiura A, Yamada M, 2000. Persimmon genetics and breeding. Plant Breeding Reviews 19:191–225, John Wiley & Sons Inc.

Zhuang, D., Kitajima, A., Ishida, M. and Sobajima, Y. 1990. Chromosome number of Diospyros kaki cultivars. J. Japan. Soc. Hort. Sci. 59:289–297. (in Japanese with English summary).

Part IV
Tree Nuts

Chapter 18
Almond

Rafel Socias i Company, José Manuel Alonso, Ossama Kodad, and Thomas M. Gradziel

Abstract The almond is economically the most important tree nut in the world. Its production is limited to areas characterized by a Mediterranean climate, including regions in the Mediterranean countries, the Central Valley of California, the Middle East, Central Asia, the Himalayan slopes, and the Southern Hemisphere, including Chile, Argentina, South Africa, and Australia. The main production region in the world is the Central Valley of California. The cultivation of almond in the eastern Mediterranean area occurred as early as the second millennium BC. Selection for domesticated almond types favored sweet kernels and larger nut size among these wild populations. Traditional seed propagation resulted in extensive genetic variability due, in part, to the obligate out-crossing nature of the self-incompatible almond. Local cultivars and landraces were selected over centuries of almond growing and in the twentieth century breeding activities began. Currently, there is active almond breeding programs in Spain, France, the USA and Israel. Self-compatibility has become the main objective along with late blooming, frost tolerance, resistance to diseases, and tree architecture. Despite the difficulties in defining a kernel quality ideotype due to the differences in consumer preferences, almond quality is currently an important breeding goal.

Keywords Almond • *P. amygdalus* • *P. dulcis* • Breeding • Cultivars • Roostocks • Germplasm • Self-compatibility • Quality • Nut • Stone fruit • Drupe

---

R. Socias i Company (✉) • J.M. Alonso • O. Kodad
Unidad de Fruticultura, Centro de Investigación y Tecnología Agroalimentaria de Aragón (CITA), Av. Montañana 930, 50059 Zaragoza, Spain
e-mail: rsocias@aragon.es; jmalonsos@aragon.es; okodad@aragon.es

T.M. Gradziel
Department of Plant Sciences, University of California, Davis, CA 95616, USA
e-mail: gradziel@plantsciences.ucdavis.edu

M.L. Badenes and D.H. Byrne (eds.), *Fruit Breeding*, Handbook of Plant Breeding 8, DOI 10.1007/978-1-4419-0763-9_18,

1 Introduction

An adaptation to harsh climates combined with an ability to develop a deep and extensive root system has allowed cultivated and wild almond [*Prunus dulcis* (Mill.) D.A. Webb syn. *P. amygdalus* Batsch] to exploit a wide range of ecological niches in its ancestral range in central Asia extending from the Takla Makan desert in western China to the Mediterranean (Kester et al. 1990; Ladizinsky 1999). Almond is also well adapted to mild winter and dry, hot summer conditions due to its low chilling requirement for early bloom, rapid early shoot growth, and high tolerance to summer heat and drought. It is the earliest temperate tree crop to bloom, which limits its production to areas relatively free from spring frosts. Late-winter and early-spring frosts can damage and even completely destroy almond crops. Because almond is mostly naturally self-incompatible, it requires cross-pollination, which further acts to promote genetic variability and adaptability to diverse environments.

The almond is the most important tree nut crop in terms of commercial production. This production is limited to areas characterized by a Mediterranean climate (Kester and Asay 1975), including regions in the Mediterranean countries, the Central Valley of California, the Middle East, Central Asia, the Himalayan slopes, and some equivalent areas in the Southern Hemisphere, including Chile, Argentina, South Africa, and Australia. The Mediterranean climate is also characterized by very low rainfall during late winter, summer, and early fall, which is required because of almond's high susceptibility to foliar diseases. World production is variable from year to year, depending primarily on climatic conditions which affect pollination success and as well as disease and insect damage. In the Mediterranean and Asiatic regions, most almond orchards are not irrigated, and rainfall, mainly in the winter and spring, is essential to ensure an acceptable crop. Rains during the fall also disrupt harvesting operations, while rains during bloom interfere with pollination by reducing the activity of pollinating insects and also increasing disease damage. As a consequence, a strong negative correlation has been found in California between total rainfall in February and final crop level (Alston et al. 1995). Therefore, consistent productivity has been an important goal of breeding in most production areas.

Production statistics for the last 8 years are shown in Table 18.1. In most new-world production regions (United States, Chile, Australia), many cultivars are soft-shell with a high shelling percentage, whereas in more traditional growing areas (Spain, Italy, Central Asia) most cultivars are hard-shell, with a lower shelling percentage but greater shell seal which provides greater resistance to insect pests of the kernel.

The main production region is the Central Valley of California, where high productivity is obtained from a combination of favorable soils, climate, and intensive management systems, including the proper combination of cross-compatible cultivars, the optimum utilization of honeybees for cross-pollination, extensive mechanization, and high water and fertilizer inputs (Kester and Gradziel 1996). More than 70% of the world's almond crop is produced here for distribution by a highly developed marketing system. Almond production also occurs in a number of additional countries, which, although important at the regional level, do not profoundly affect

Table 18.1 Almond kernel production (1,000 mt) in major producing countries (Almond Board of California 2009)

Country	2001	2002	2003	2004	2005	2006	2007	2008	Average
USA	373.8	491.6	468.1	452.7	413.5	506.5	627.3	732.4	508.2
Spain	57.0	66.0	44.0	26.2	63.5	82.6	56.9	52.6	56.1
Italy	18.0	9.0	5.0	12.0	12.0	6.0	12.0	12.0	10.8
Greece	13.0	17.1	10.0	17.1	14.0	15.0	10.0	12.0	13.5
Turkey	14.0	14.0	13.7	12.3	14.5	14.4	15.5	16.0	14.3
Australia	9.2	9.3	10.1	11.5	16.2	15.9	26.5	26.1	15.6
India	1.0	1.1	1.0	1.1	1.1	1.2	1.0	1.2	1.1
Total	486	608.1	551.9	532.9	534.9	641.6	749.2	852.3	619.6

Table 18.2 Manufactured almond products and applications

Common products	Benefits	Applications
Natural almonds whole; and whole and broken	Provides color contrast to lighter foods; adds visual appeal and texture; stronger flavor than blanched products	Roasting, flavoring, snack foods; complementary ingredients for confectionery, cake, bread, cookies, and cooking
Natural or blanched sliced almonds	Increases almond recognition; adds flavor, visual appeal; provides texture contrast	Cake, bread, and cooking garnish; cereal additive ingredients
Natural or blanched diced almonds	Adds almond flavor and characteristics, and visual appeal	Cake and confectionery fillings; additive ingredients for cooking
Blanched almonds whole and broken	Complementary flavor, high quality, garnishing, visual contrast	Ingredients for mixed dried fruits and nuts retail packing, and blanched manufactured products; cookie and cake garnish
Blanched slivered almonds	Adds crunchy, complementary flavor, and nutritional value	Ingredients for cake, cookie, bread, snack, and cereals; additive ingredients for cooking
Natural or blanched almond meal	Adds color, flavor, richness; fat replacement and binding agent	Cake and confectionery fillings; ingredients for fortified breads and cereals
Roasted almonds	Strengthen flavor and color	Fillings and garnish for dairy products, ice cream; ingredients for chocolate or energy bar

the world market and are characterized by less specialized production and marketing systems. In many of these countries, orchards are not irrigated, soils are poor, inputs are very low, and mechanization is reduced, resulting in productivity that often is one tenth or less of that obtained in California. Orchards in the Southern Hemisphere follow the California production model, as do many new orchards in Spain and Israel with consequent higher inputs and higher yields (Socias i Company 2001).

Almonds are consumed raw, roasted, blanched, unblanched, alone or mixed with other foods; in addition, kernel pieces (sliced, diced, etc.) are used in different confectioneries (Table 18.2). Almonds are occasionally consumed fresh, once the seed is filled but before ripening. Once fully mature, harvested kernels may be specially processed for many different kinds of "turrón" (nougat), marzipan, sweets, cakes, ice cream, chocolate bars, and almond milk. In addition, almond oil is widely utilized in the pharmaceutical and cosmetic industries because of its chemical stability and versatility (Felipe 2000; Schirra 1997). Each use requires kernels with a specific composition of fatty acids, proteins, sugars, and related phytochemicals (Berger 1969). A wide variety of specialized uses has also evolved in different regions and cultures. Almond's high nutritional value is becoming increasingly recognized by health-conscious consumers. Almond is not only an important source of macronutrients, such as lipids, proteins, fiber, and minerals, but also an important source of more unique phytonutrients, such as vitamin E (α-tocopherol), folate, and oleic acid.

Recent reviews have summarized earlier research on almond phytonutrient quality (Socias i Company et al. 2008a) and genetic improvement (Gradziel 2008). In addition, the rootstocks described for peach and related stone fruit species in the peach chapter, are often also appropriate for almond.

2 Origin and Domestication of Scion Cultivars

The almond is probably the oldest tree nut crop to be domesticated, possibly during the third millennium BC (Spiegel-Roy 1986). It has been suggested that this domestication took place in Central Asia (Kovalyov and Kostina 1935), where wild-type but sweet kernelled almond trees could still be found (Popov et al. 1929). Many wild species (Fig. 18.1) that are closely related to almond and which intercross freely with cultivated almond have also been described in this region (Browicz and Zohary 1996; Denisov 1988; Grasselly 1976). Among these species, *P. fenzliana* Fritsch., *P. bucharica* (Korsh.) Fedtsch., *P. kuramica* (Korsh.) Kitam. and *P. triloba* Lindl. may have been involved in natural hybridizations, giving rise to the current

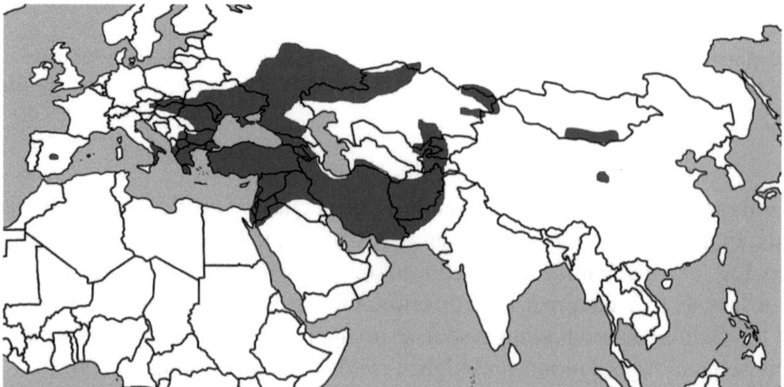

Fig. 18.1 Distribution area of the almond wild species in the subgenus *Euamygdalus* Schneid (modified from Browicz and Zohary 1996)

cultivated almond (Grasselly and Crossa-Raynaud 1980; Kester et al. 1990; Gradziel 2008). Furthermore, as almond cultivation spread into the Mediterranean region, new hybridizations occurred, especially with the wild Mediterranean species *P. webbii* (Spach) Vierh. (Godini 1979; Socias i Company 2004), resulting in unique self-compatible populations found along the northern shore of the Mediterranean Sea from Greece and the Balkans to Spain and Portugal. Ladizinsky (1999) suggested that these self-compatible almonds were derived from *P. fenzliana* rather than *P. webbii*, erroneously concluding that self-compatibility was not naturally found in *P. webbii*. However, Godini (2000) clearly demonstrated that *P. webbii* from the Italian region of Puglia was self-compatible and probably contributed self-compatibility to the Puglia pool of almond cultivars.

The genetic closeness of almond and peach led Watkins (1979) to suggest that both originated from the same primitive self-incompatible species but evolved separately following the mountain upheavals of the Central Asian massif. Peach evolved in the eastern regions of China, in a more humid climate and at lower elevations, where the more temperate climates encouraged inbreeding and self-compatibility (Tao et al. 2007). In contrast, almond evolved in the west, in arid steppes, deserts, and mountainous areas, under severe conditions that encouraged the original self-incompatibility of the species. Selection for domesticated almond types would have favored sweet kernel and larger nut size among these wild populations. Propagation was by seed, which is still common in many regions of the world, particularly the countries of Central Asia and the Middle East (Kester et al. 1990).

Based mainly on archaeological remains, Zohary and Hopf (1993) put forward the hypothesis that almond was originally taken into cultivation in the eastern part of the Mediterranean basin, more or less at the same time as the olive, grapevine, and date palm were domesticated. Later, Browicz and Zohary (1996) further promoted this hypothesis, suggesting that the sweet-kernel wild almond found in Central Asia were feral populations or resulted from natural hybridization between wild and cultivated almonds from previous domestication. Natural crossing among cultivated almond and related wild almond species has previously been reported (Kovalyov and Kostina 1935; Socias i Company 2004).

The cultivation of almond in the eastern Mediterranean area occurred as early as the second millennium BC as almond samples have been found in the tomb of Tutankhamen (Zohary and Hopf 1993) and in the Franchthi Cave in Greece (Hansen and Renfrew 1978). Almond cultivation must have existed in Greece long before the creation of the Greek myths to explain its origin (Graves 1955), and there is substantial evidence of almond trade in the western Mediterranean by the fourth century BC (Cerdá Juan 1973).

Cultivated almonds were presumably introduced into the Mediterranean and Central Asian areas through seeds carried by caravans on their travels between China and Europe. This manner of dispersal has also been suggested for other fruit trees (Juniper et al. 1999) and would work in both directions. Gustafson et al. (1989) reported that the primary sources of almond at Kashgar, Xinjiang (China) were old seedling trees which had originated from Central Asia across the Tian Shan Mountains. Kashgar is on the old Silk Road connecting China and the West.

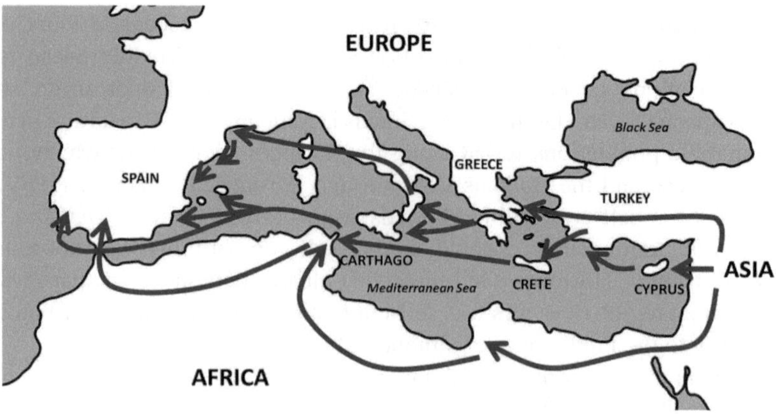

Fig. 18.2 Routes of almond dissemination in the Mediterranean Basin (modified from Felipe 2000)

Traditional almond culture utilized open-pollinated seedlings (Grasselly 1972; Rikhter 1972) which, together with self-incompatibility, created a very high heterozygosity in this species (Kester et al. 1990; Socias i Company and Felipe 1992). This large variability has provided a useful genetic pool for breeding, but has made it more difficult to determine cultivar origins and to introduce improved clonal selections in more traditional regions. Almond propagation by grafting was already known by the first century AD, however, allowing improved scion selections to be clonally propagated (Columella 1988).

Almond domestication and cultivar development appears to have taken place in three main stages (Kester et al. 1990). Stage 1 includes the Asian period, following almond introduction into cultivation. To a large extent, almond growing has not changed much in this region from the practices used for centuries. Many orchards are still from seedlings, often mixed with other crops and under minimum cultural management. Many of these local plantings have evolved through natural and human selection and represent distinct ecotypes and landraces.

Stage 2 or the Mediterranean stage (Fig. 18.2) involved the spread of cultivated almond further into European Mediterranean regions where, thanks to favorable climatic conditions, almond production thrived. Although this favorable climate is primarily found in the region extending 150–200 km from the seacoast, almond growing often expanded further inland in valleys, where the climate becomes more continental and so with greater extremes in temperatures. Phoenicians, Carthaginians, Greeks, and Romans played a significant role in this extension of almond production, which took place during their trade expansions. A second period of introductions occurred in the eleventh and twelfth centuries, accompanying the Arab conquests through North Africa and into southern Spain. Following these introductions, almond production became concentrated in specific areas, mostly with a traditional cultural system adapted to the drought-resistant and frost-sensitive characteristics of the available almond germplasm. Most of these orchards were seedlings from natural cross-pollinations. Subsequent centuries of selection resulted in the emergence of adapted land races associated with specific production areas (Grasselly and Crossa-Raynaud 1980). Selection of clonal cultivars occasionally occurred, but was not of great importance until the last 150 years when graft and bud-propagation became more widely practiced. Because of the ability of clonal propagation to capture locally adapted, elite genotypes, hundreds of local cultivars within specific production areas were identified and propagated (Estelrich 1907; Kester et al. 1990).

Stage 3 or the Californian stage initially began as an extension of Mediterranean culture, utilizing a hard-shelled germplasm brought by missionaries from Spain and, later, soft-shell types from France. However, new orchard practices soon differentiated Californian production from that of Europe and Asia. The expansion of this stage also reached the different growing areas of the Southern Hemisphere. Important cultural changes included the movement of almond production from more marginal coastal sites to the high-quality Central Valley soils, the development of new rootstocks and orchard management practices for these highly productive sites, the introduction of honeybees for efficient cross-pollination (Tufts 1919), the selection of consistently high-yielding cultivars, and the standardization of markets based upon cultivar type. The combination of highly adapted cultivars and rootstocks, favorable soil and climate, abundant water, and effective management has given California growers the highest productivity in the world. Production per hectare continues to show upward trends, with yields surpassing 5 mt/ha presently possible with some cultivar/site combinations.

Early almond production was concentrated to specific regions, where selection for good local adaptation resulted in the emergence of specific regional types. Seedling propagation resulted in the proliferation of a large number of highly variable local genotypes. However, because of their origin from a limited germplasm, they often lacked significant genetic diversity. This condition was particularly prevalent in more isolated areas, such as all the islands of the Mediterranean Sea, some remote mainland valleys, and the Canary Islands. The best performing seedlings were later selected for clonal propagation (Estelrich 1907), with some selections being several centuries old and others more recent.

Selections from Northern Africa, Sicily, and other Mediterranean Islands are characterized by a very early blooming season, mostly due to their very low chilling requirements related to the low frost risks in these areas. In addition, a very early bloom promotes earlier seed development, allowing the kernel to be completely formed before the hot, dry season. In Tunisia, the Sfax cultivars showed this general pattern, but also possessed thick but soft shells which also allowed fresh consumption of immature almond which is widely practiced in these regions. The short fruit development period has also been useful in avoiding climatic stresses in other regions (Grasselly and Crossa-Raynaud 1980).

The Italian region of Puglia is characterized by a group of cultivars showing late blooming, reduced branching, with flower buds mostly on spurs and often with two flowers, hard shells, a high proportion of double kernels, and self-compatibility in many genotypes. Some of these traits are clearly positive, but others are negative. In general, this germplasm has been considered valuable, and has been utilized as parents in many breeding programs as a source for the positive traits, such as late bloom ('Cristomorto'; Grasselly and Crossa-Raynaud 1980) and self-compatibility ('Tuono'; Socias i Company 2002).

Despite the large environmental variation within countries, common kernel types often predominate. For example, Provence cultivars are characterized by a late bloom and semi-soft shell, whereas most of the cultivars of Spain and many other Mediterranean countries are hard shelled (Kester et al. 1990). The Portuguese selections show an open growth habit while the Romanian selections generally have higher vigor. Even the more recent stage three or Californian selections show characteristic traits, such as their early-to mid-blooming period and soft to paper-shell.

3 Genetic Resources

3.1 Scions

Traditional seed propagation resulted in extensive genetic variability due, in part, to the obligate out-crossing nature of the naturally self-incompatible almond (Kester et al. 1990; Socias i Company and Felipe 1992). A large genetic variability may even be found within the more homogeneous localized ecotypes or landraces though the full genetic diversity may be limited (i.e., genetic variability within landraces may result from the efficient reshuffling of a very limited genetic diversity). The last century has seen a sharp decrease in overall diversity as large numbers of local cultivars and landraces have been replaced with commercially preferred breeding releases. In Europe, an effective replacement of cultivars began in many countries with the release of the late-blooming Puglia cultivars and later, with the high market quality French cultivar 'Ferragnès' (Grasselly and Crossa-Raynaud 1980). In Spain, 'Guara' has been a commercial success, favored by its self-compatibility (Felipe and Socias i Company 1987), and now represents more than 50% of new plantings (Socias i Company et al. 2004) (Table 18.3). The same trend is also occurring in

Table 18.3 Percentage of plants of each almond cultivar produced by the Spanish nurseries

Cultivar	Percentage
Guara	53.11
Ferragnès	13.34
Ferraduel	10.45
Desmayo Largueta	5.89
Marcona	4.54
Tuono	1.93
Ramillete	1.92
Others	8.82

Source: Web page of the Spanish Ministry of Agriculture, Fisheries and Food: http://www.mapya.es/agric/pags/semillas/vivero/almendro.pdf

other countries of the Mediterranean region as the increasingly global market places a premium on crop consistency and phenotypic uniformity. As has been well demonstrated in other "improved" agronomic crops, such market-driven cultivar changeovers can dramatically reduce crop genetic diversity and so increase vulnerability to later large-scale crop failures from pest, disease and/or climate changes (Socias i Company et al. 2003). An interesting example of such vulnerability is the recent findings by Alonso Segura and Socias i Company (2007) that the extensive use of 'Tuono' as a source of self-compatibility in most European breeding programs appears to result in inbreeding depression and reduced productivity in at least some of the new cultivar releases.

Local cultivars and landraces were selected over centuries of almond growing due to improved local performance in terms of production, resistance and/or quality traits. These highly selected traits represent very valuable germplasm for addressing future challenges and so need to be preserved, characterized, and incorporated into advanced breeding lines. One of the oldest and largest almond germplasm collections was established at the Nikistki Botanical Garden in Yalta (Crimea), at the end of the nineteenth century. The collection was based on the Russian scientist Vavilov's pioneering research on the value of diverse germplasm collections which, in turn, greatly contributed to the success of the similarly pioneering almond breeding program at Yalta (Rikhter 1972).

Based on these achievements, collections were established in many other countries to help preserve local and introduced germplasm, including new cultivars from other countries (Gagnard 1954).

In France, the collection made by Charles Grasselly since the early 1950s included over 700 accessions from Europe, Northern Africa, and the Middle East. In addition, this germplasm was thoroughly characterized, resulting in the rediscovery of almond self-compatibility and other useful agronomic traits and their subsequent incorporation into applied breeding programs (Grasselly and Crossa-Raynaud 1980).

In Spain, collections made by Antonio J. Felipe from the 1960s also led to the preservation and characterization of regional and more exotic germplasm (leading also to the independent rediscovery of self-compatibility), and incorporation of elite germplasm into his breeding program. The ongoing success of this program has led to its designation as the reference for excellence in breeding by the international crop improvement organization GREMPA (Espiau et al. 2002).

The situation in California differed in two major aspects. First, the rapid development of this industry in the early 1900s did not involve a long-term selection of locally adapted land races but rather was primarily based on only a small number of seedling selections chosen from an initially wide diversity of European and Asian germplasm. A second important difference is that California production quickly adopted intensive agricultural practices, where good soils and climates combined with high water and fertilizer inputs allowed very high crop returns. The traditional low input European and Asian cultivars and landraces were poorly adapted to such high input systems but did represent potential sources for disease and stress resistance. Recognizing that a richer genetic diversity was available in closely related almond and peach species germplasm, a diversity which could be readily incorporated into almond breeding lines because of the greater market plasticity in almond tree and nut characteristics, Kester and Gradziel (1996) and later, Gradziel (2008) placed a greater emphasis on the collection and introgression of interspecies traits into advanced breeding lines (see section below). This approach has allowed greater genetic options in solving breeding problems. For example, self-compatibility in California cultivars and advanced breeding lines has been independently derived from *P. persica* Batsch, *P. mira* Koehne, *P. davidiana* Carr. (Franch), and Yugoslavian accessions of *P. webbii* (Gradziel 2008).

The goal of germplasm conservation was thus, not only the preservation of specific agronomic traits, but also germplasm from related species. In Europe, a first initiative on documenting these collections was undertaken by the Nordic Genebank with the European Almond Catalogue (Niklasson 1989). Later, through the efforts of GREMPA and FAO, an extensive inventory of almond research, germplasm, and references was published by Monastra and Raparelli (1997). The European Cooperative Programme for Plant Genetic Resources, joining efforts of the European Union and Bioversity International (previously IBPGR and IPGRI), has developed a European *Prunus* Database and has promoted the collection and characterization of almond germplasm even in countries, where almond is not a major crop. For example, the current database covers not only the countries with significant almond acreage, such as France, Italy, and Spain, but also countries with more marginal production, including Albania, Bulgaria, Cyprus, the Czech Republic, Georgia, Greece, Hungary, Israel, Portugal, Romania, Serbia, and Slovakia (Maggioni and Lipman 2006).

An extensive list of cultivars was published by Kester et al. (1990), including those from major production regions. The large number of accessions, from many different geographical regions, makes almond among the most heterogeneous tree crop species (Socias i Company and Felipe 1992). To better deal with this variability and to facilitate its management, a European project is currently attempting to define the most appropriate core collection of almond germplasm (i.e., capturing the greatest genetic diversity within an inherently limited number of accessions) (Bacchetta et al. 2008).

While several reports have documented recovery of genes for self-compatibility from related almond species through either natural or controlled crosses (Denisov 1988; Felipe 2000; Gradziel and Kester 1998; Socias i Company and Felipe 1988, 1992), only Rikhter (1969), Grasselly (1972), Denisov (1988), Kester et al. (1990) and Socias i Company (1990) have previously reported on the use of wild species germplasm to create improved almond cultivars. The previous widespread use of these species and their hybrids as almond rootstocks would facilitate subsequent introgressions. The use of wild species directly as a rootstock for dry land almond production has also been reported, including *P. spartioides* (Spach) Schneid. in Iran; *P. bucharica* and *P. fenzliana* in Russia; *P. webbii* in Turkey; and *P. fenzliana*, *P. bucharica*, *P. kuramica*, *P. argentea* (Lam.) Rehd., *P. tangutica* Batal., and *P. kotschyi* (Boiss. et Hohen.) Nib. at lower incidence in these (Fig. 18.1) and nearby areas (Denisov 1988; Gradziel et al. 2001; Grasselly 1972; Rikhter 1969). More recently, crosses between almond and related species have been readily achieved under controlled conditions (Gradziel and Kester 1998; Gradziel et al. 2001; Gradziel 2003). While a wide variability in tree and branch architecture results, leaf and nut phenotypes of resultant hybrids are typically intermediate to the parents. Interspecific crosses between related species (mainly peach × almond, but also *P. webbii* × almond) have been used for almond rootstock breeding in France, the USA, Spain, and Yugoslavia (Denisov 1988; Felipe et al. 1997; Gradziel et al. 2001; Grasselly 1972; Rikhter 1969). In addition, Browicz and Zohary (1996) and Ladizinsky (1999) have reviewed evidence for a high level of spontaneous interspecific hybridization in the wild between species with overlapping ranges.

3.2 Rootstocks

In traditional dry land almond production, almond seedlings were used as rootstocks because of their deep growth and associated efficiency for mining nutrients and water. Originally, bitter almonds were used for producing seedling rootstocks, though sources of more uniform nursery plants, such as 'Mission' in California, 'Desmayo Largueta,' 'Atocha,' and 'Garrigues' in Spain (Felipe 1989), and the series 'Alnem' in Israel (also resistant to nematodes) were utilized later (Kochba and Spiegel-Roy 1976). More recently, almond × peach hybrids are showing promising performance under nonirrigation, due, in part, to the loss of the deeply mining almond seedling tap-root when first transplanted (Felipe 2000; Kester and Grasselly 1987). Under high input, irrigated conditions, however, the deeper almond-type tap roots are more susceptible to asphyxiation and disease in even occasionally saturated soils. In these cases, the shallower peach and plum rootstocks currently used for peach have been often shown more effective for almond. In irrigated soils, the proliferation of near-surface roots often suppress deeper taproot formation even in rootstock showing such growth under dry land production.

3.3 Molecular Characterization of Germplasm Diversity

Increasingly, different fruit species are being characterized with molecular markers, since morphological traits are often affected by environmental conditions. In addition, most horticultural traits need to be evaluated in fully mature trees, which is a very labor and time-consuming process. The usefulness of molecular markers has been increasing since its development, especially PCR-based markers, which have become an essential tool for variability analysis, pedigree assessment, and cultivar identification.

Historically, the main molecular markers used in almond studies have been isozymes, restriction fragment length polymorphisms (RFLPs), randomly amplified polymorphic DNAs (RAPDs), simple sequence repeats (SSRs), amplified fragment length polymorphism (AFLP), and markers based on unique DNA sequences. Isozymes were one of the first molecular markers utilized in almond studies and offered codominant expression and good reproducibility, but they were limited by the small number of loci which could be analyzed by conventional staining methods, and by a low genetic variation at most loci. Nonetheless, it was isozymes studies which first documented extensive genetic variability in almonds overall, as well as the limited genetic base of many almond breeding programs (Arulsekar et al. 1986; Cerezo et al. 1989; Foolad et al. 1995; Hauagge et al. 1987a, b; Sathe et al. 2001; Vezvaei et al. 1995). RFLPs are also codominant but can detect a virtually unlimited number of markers. RFLPs have been used in almond for discovering linkages between markers, for constructing genetic maps, for cultivar identification, and for the characterization of genetic variability (Viruel et al. 1995). RAPDs based on PCR amplification of arbitrary primers have been useful for characterizing germplasm variability (Bartolozzi et al. 1998; Martins et al. 2003, Shiran et al. 2007) but had limited application for cultivar identification and map construction since they are dominant markers with difficulties with repeatability. SSR or microsatellite markers have been obtained covering almost the whole genome of *Prunus*, and recently, the first set of almond SSRs has been published (Testolin et al. 2004). SSR markers have proven more useful for genetic relationships (Martínez-Gómez et al. 2003a), cultivar identification (Fernández i Martí et al. 2009a; Martínez-Gómez et al. 2003b; Martins et al. 2003, Testolin et al. 2004; Xie et al. 2006), and map construction (Dirlewanger et al. 2004a) due to their high polymorphism, codominant inheritance, abundance, and the frequent efficacy of SSR markers developed in related species (Martínez-Gómez et al. 2003c, Shiran et al. 2007). Finally, AFLPs have recently shown promise as a tool for molecular characterization of the genetic diversity among cultivated genotypes and related wild species of central Asian almonds (Sorkheh et al. 2007a, b). More recently, single nucleotide polymorphism (SNP) markers have also been specifically applied for almond identification (Wu et al. 2008, 2010).

3.4 Major Traits and Genetic Sources

Self-compatibility has become the main objective in many breeding programs. The main European source has been the pool of self-compatible cultivars from the Italian region of Puglia, mainly from 'Tuono' (Socias i Company 2002). In addition to the Puglia source, self-compatibility has been transferred from peach, *P. mira*, *P. davidiana*, and Yugoslavian *P. webbii* accessions (Gradziel 2008).

Late blooming is also an important objective in Europe but has been associated with lower productivity in high-input California plantings. The main sources have been late-blooming cultivars, such as those from Puglia, from the Nikitski breeding program or genotypes deriving from the 'Tardy Nonpareil' mutant (Socias i Company et al. 1999).

Resistance to various diseases seems to be quantitatively inherited from known resistant cultivars, such as 'Ardèchoise,' but the transfer of disease susceptibility from cultivars, such as 'Tuono' (Grasselly 1972) must also be guarded against. Resistance to frosts also seems be transmitted from 'Tuono' to progeny (Felipe 1988). Insect resistance is very important, particularly in the more vulnerable soft-shelled California-type cultivars. Improved tree-architectures, characterized by sufficient growth to allow fruit wood renewal while minimizing pruning (Socias i Company et al. 1998) is related to fruiting predominantly on spurs, as is characteristic of Puglia cultivars and their offspring (Grasselly 1972). This fruiting habit is also related to a high bud density (Kodad and Socias i Company 2006) and final tree productivity (Kodad and Socias i Company 2008). This approach is further advanced in the new California cultivar 'Winters' in which crop production is predominantly on longer, dard-type spurs which result in increased nuts per spur as well as greater spur longevity (Gradziel et al. 2007).

4 Major Breeding Achievements

4.1 Scions

General breeding progress to the mid-1990s has been summarized by Kester et al. (1990) and Kester and Gradziel (1996), and more recently Martínez-Gómez et al. (2006) and Gradziel (2008). While the last 10 years has seen a large proportion of genetic improvement efforts redirected from cultivar development to gene discovery and molecular characterization, new scion cultivars continue to be released from the remaining breeding programs, though the numbers differ by production region. Characteristics of the most important new releases from the different breeding programs are summarized in Table 18.4.

The Mediterranean region maintained its traditional cultivars until the late 1960s when the late-blooming Puglia cultivars became widely planted owing to their more consistent yields. Of these, 'Tuono' and 'Cristomorto' were the most heavily planted cultivars. However, because of lower quality, 'Cristomorto' was rapidly replaced by 'Ferragnès' and 'Ferraduel' released in 1967 by the French breeding program. 'Ferragnès' became the most successful new cultivar in European plantings with 'Ferraduel' often planted as a pollinizer. The 'Ferragnès' dominance was ultimately ended with the release of newly developed self-compatible cultivars (Table 18.3), during the last 20 years (Socias i Company 2001).

Self-compatibility has since become the main objective of most Mediterranean programs (Socias i Company 1990). The first self-compatible releases from a breeding program were 'Guara,' 'Aylés,' and 'Moncayo' (Felipe and Socias i Company 1987) from the Zaragoza (Spain) breeding program. 'Moncayo', while showing self-compatibility in laboratory conditions, was subsequently shown by Kodad et al. (2008) to be self-incompatible in the field. The success of 'Guara' has led to an important renewal of almond plantings in Spain, representing more than 50% of new plantings (Table 18.3). Consequently, all new releases from Zaragoza are self-compatible cultivars, including 'Blanquerna,' 'Cambra,' and 'Felisia' (Socias i Company and Felipe 1999). 'Felisia' also incorporated the late-blooming trait for avoiding spring frost damage. In 2006, 'Belona' and 'Soleta' were released as self-compatible cultivars with improved kernel quality (Socias i Company and Felipe 2007) and in 2008, 'Mardía' was released as an extra late-blooming cultivar (Socias i Company et al. 2008b).

Table 18.4 Characteristics of important new almond cultivars

Cultivar	Description
Spain	
CITA de Aragón (Zaragoza)	
Blanquerna	"Genco" OP, SC, mid-blooming, hard shell, large kernel of excellent quality, early ripening
Cambra	"Ferragnès" × "Tuono," SC, late blooming, hard shell, medium ripening
Felisia	"Titan" × "Tuono," SC, very late blooming, medium-hard shell, small kernel, very low alternance, early-medium ripening
Belona	"Blanquerna" × "Belle d'Aurons," SC, late blooming, hard shell, large kernel with an outstanding composition, medium ripening
Soleta	"Blanquerna" × "Belle d'Aurons," SC, late blooming, large kernel with an outstanding performance when roasted, medium-late ripening
Mardía	"Felisia" × "Bertina," SC, extremely late blooming, disease tolerant, early-medium ripening
CEBAS—CSIC (Murcia)	
Antoñeta	"Ferragnès" × "Tuono," SC, late blooming, hard shell, high vigor, spreading with dense branching, medium-late ripening
Marta	"Ferragnès" × "Tuono," SC, hard shell, high vigor, upright, late blooming, medium ripening
Penta	S5133 × "Lauranne," SC, extremely late blooming, hard shell, intermediate vigor and branching, early ripening
Tardona	S5133 × R1000, SC, extremely late blooming, hard shell, small kernel, intermediate vigor with dense branching, medium ripening
IRTA—Mas de Bover (Reus)	
Constantí	("Ferragnès" × "Ferraduel") OP, SC, late blooming, mid ripening, vigorous, mid branching
Marinada	"Lauranne" × "Glorieta," SC, very late blooming, mid ripening, mid vigor, mid branching
Tarraco	("Ferralise" × "Tuono") × Anxaneta, SI, very late blooming, mid ripening, mid vigor, large kernel, mid branching
Vairo	("Primorskij" × "Cristomorto") × "Lauranne," SC, late blooming, early ripening, high vigor, mid branching
France	
INRA (Avignon)	
Lauranne	"Ferragnès" × "Tuono," SC, medium-hard shell, medium vigor, late blooming, early-medium ripening, some double kernels
Steliette	"Ferragnès" × "Tuono," SC, semi-hard shell, medium vigor, late blooming, early ripening, some double kernels
Mandaline	"Ferralise" × "Tuono," SC, late blooming, medium ripening, hard shell, medium to upright growth
Israel	
Shefa	"Tuono" × local cross, SI, vigorous, early blooming, highly adapted to Israel conditions, soft shell, large kernel, early ripening
USA	
University of California (Davis)	
Avalon	Probably "Nonpareil" OP, SI, medium kernel, early blooming, paper-shell, harvest approx. 8 days after "Nonpareil"
Kahl	Chance seedling in a "Nonpareil," "Davey," and "Mission" planting, SI, mid-blooming, large kernel, semi-soft-shell, harvest 14 days after "Nonpareil"
Morley	"Mission" × late blooming almond seedling, SI, late blooming, medium kernel, semi-soft shell
Savanna	"Nonpareil" × late blooming almond seedling, SI, late blooming (2 weeks after "Nonpareil"), large kernel, semi-soft shell, harvest 14 days after "Nonpareil"
Sweetheart	SB3,54-39E [{"Lukens Honey" peach × "Mission"} × "Nonpareil"] × Sel 25–26. SC, mid-blooming, large "Marcona"-type kernel, harvest approx. 10 d after "Nonpareil", semi-soft shell, high kernel oil and roasting quality, resistant to postharvest worm damage
Winters	"3-1" ("Peerless" × "Harpereil") × "6-27" ("Nonpareil" × "Jordanollo"), SI, early blooming, large Carmel-type kernel, paper-shell, good bloom overlap with early "Nonpareil" bloom, harvest 3 weeks after "Nonpareil"

OP open pollinated, *SC* self-compatible, *SI* self-incompatible

The breeding program from IRTA (Reus, Spain) placed greater emphasis on fruit quality and late blooming traits, with self-compatibility as a secondary objective. The first cultivars released were 'Masbovera,' 'Glorieta,' and 'Francolí' (Vargas and Romero 1994). 'Masbovera' is the most successful among the three despite the self-compatibility of 'Francolí' (López et al. 2005). The 2006 releases further delay bloom time, with the self-compatible 'Constantí,' 'Marinada,' and 'Vairo,' and the self-incompatible 'Tarraco' cultivars (Vargas et al. 2008).

The breeding program from CEBAS-CSIC (Murcia, Spain) has as its main objectives self-compatibility and late-bloom time. The first releases in the late 1990s were 'Antoñeta' and 'Marta' (Egea et al. 2000), which resulted in increased plantings in Spanish orchards. Two more recent cultivars, 'Penta' and 'Tardona,' are characterized by their extremely late blooming time (Dicenta et al. 2009).

The French breeding program was the most successful program in Europe for many years. After the successful introduction of 'Ferragnès' and 'Ferraduel' in 1967, the later blooming self-incompatible cultivars 'Ferralise' and 'Ferrastar' were released in the late 1970s (Grasselly and Crossa-Raynaud 1980), though with less success. The 1989 releases 'Lauranne' and 'Steliette' were self-compatible (Grasselly et al. 1992), with 'Lauranne' being particularly successful in France (Duval and Grasselly 1994). A more recent release from this program has been 'Mandaline' (Duval 1999).

The only other recent release from the Mediterranean region has been the 'Shefa' cultivar from Israel, which is a self-incompatible seedling of 'Tuono' characterized by a very large nut (Holland et al. 2006).

Almond breeding in California is done by public programs at the University of California at Davis and the USDA program at Parlier, as well as by private breeders, such as Fredrick Anderson, Norman Bradford, and Floyd Zaiger of Modesto. Primary objectives included the development of improved pollinizers for 'Nonpareil,' particularly its early bloom, and resistance to the epigenetic disorder noninfectious bud failure. Novel clonal selection procedures first developed by Kester in the 1980s proved successful in developing a foundation source for the commercially important 'Carmel' cultivar which has allowed continued extensive plantings of this otherwise noninfectious-bud-failure prone pollinizer for Nonpareil late early bloom (Kester et al. 2004). Cross-compatible cultivars with good bloom overlap with the early 'Nonpareil' bloom include the cultivars 'Avalon,' released in 1999 and 'Winters' released in 2006 (Gradziel et al. 2007). 'Kahl' was released in 1995 as a pollinizer for the later 'Nonpareil' bloom. 'Morley' and 'Savanna' were previously released in 1993 as late and very late blooming cultivars, respectively. In 2007, the partially self-compatible 'Sweetheart' cultivar was released as a 'Marcona'-like, premium roasting quality almond with improved resistance to postharvest pests.

4.2 Rootstocks

Breeding efforts for rootstocks are mainly focused on peach, although most *Prunus* rootstocks may be utilized for several stone fruit species. In almond, there is an increasing utilization of almond × peach hybrids both in irrigated and nonirrigated conditions in the Mediterranean area, whereas in California peach seedlings remain

Table 18.5 Characteristics of the new rootstocks for almond

Rootstock	Description
Spain	
CITA de Aragón (Zaragoza)	
Felinem	"Garfi" almond × "Nemared" peach, red leaves, easy propagation, nematode resistant, good vigor, adapted to replanting and to poor and calcareous soils
Garnem	"Garfi" almond × "Nemared" peach, red leaves, easy propagation, nematode resistant, good vigor, adapted to replanting and to poor and calcareous soils
Monegro	"Garfi" almond × "Nemared" peach, red leaves, easy propagation, nematode resistant, good vigor, adapted to replanting and to poor and calcareous soils
EE Aula Dei—CSIC (Zaragoza)	
Adafuel	Natural hybrid selection (probably "Marcona" seedling), easy propagation, very vigorous, adapted to calcareous soils
Adarcias	Natural hybrid selection, easy propagation, low vigor, adapted to calcareous soils
Italy University of Pisa	
Sirio	"INRA GF 557" OP, low vigor, poor vegetative propagation, good root system
USA (California)	
Atlas	Interspecific cross to *Prunus blireiana,* vigorous, upright
Hansen 536	Almond × peach hybrid, vigorous, deep rooting, resistant to drought
Nickels	Almond × peach hybrid, vigorous, deep rooting, resistant to drought, soil fungi
Marianna M40	*P. cerasifera* × *P. munsoniana*, improved anchorage, fewer suckers
Viking	Interspecies cross to *P. blireiana*, vigorous, upright, tolerant wet soils

the principal rootstock. As a consequence, breeding efforts for almond rootstocks is limited by progress in the more general field of *Prunus* rootstock breeding.

In the Mediterranean area, some important almond × peach releases, such as 'Adafuel' and 'Adarcias' from the Aula Dei Experimental Station resulted from natural selection, not from controlled breeding (Moreno 2004). However, these hybrids are not well adapted to almond growing due to the high input requirements of 'Adafuel' and the low vigor of 'Adarcias.' The most important rootstock breeding lines have been the 'Garfi' × 'Nemared' crossing program at Zaragoza, giving rise to the nematode-resistant and red-leafed hybrid rootstocks 'Felinem,' 'Garnem,' and 'Monegro' (Felipe 2009). These rootstocks are characterized by a good adaptability to poor soils and easy propagation by hardwood cuttings (Gómez Aparisi et al. 2002). Another hybrid released by the University of Pisa is 'Sirio' (Loreti and Massai 1998) which, like 'Adafuel,' requires high-input growing conditions. Also from the University of Pisa is the ISG rootstock series primarily derived from myrobolan plum, and so of limited application to almond (Cinelli and Loreti 2004).

In California the 'Hansen' and 'Nickels' hybrid rootstocks, which exploited the hybrid vigor of the interspecies (almond × peach) cross, were bred to promote rapid tree growth in orchard replant situations (where the replanted trees would normally be out-competed by nearby, older, and very larger trees). The vigor of these rootstocks has also allowed successful commercial almond production on more marginal soils, resulting in their use as the primary rootstock in new plantings in previously unsustainable production regions (Kester et al. 2001). Characteristics of the most important new rootstocks for almond are summarized in Table 18.5.

5 Current Breeding Objectives

5.1 Scion

The ultimate success of a cultivar is dependent as much upon its freedom from deficiencies in any of the multitude of required productivity, resistance, and quality traits as it is upon the presence of desirable new traits, such as self-compatibility (Gradziel 2008; Socias i Company et al. 1998). The ultimate goal is the continued economic profitability of almond growing, either by increasing yields and prices and/or by reducing costs (Kester and Gradziel 1996). Traits of specific current relevance to almond breeding include self-compatibility and later flowering time.

High levels of self-fruitfulness, achieved through the incorporation of both pollen–pistil self-compatibility and the capacity for self-pollination (autogamy) within the flower, has become a primary goal in almost all modern almond breeding programs in order to minimize the problems associated with cross-pollination (Socias i Company 1990). Despite a detailed genetic characterization of the self-compatibility major gene, variable expression is often observed (Alonso and Socias i Company 2005) requiring a final field evaluations of productivity to determine cultivar value (Kodad and Socias i Company 2008; Socias i Company et al. 2004).

The goal of late-blooming cultivars is the avoidance of late-winter/early spring frosts which are recurring threats to almond production because of its very early flowering season. This is particularly important in those regions where new plantings are occurring in inland regions with a more continental climate, and thus with an increasing risk of frosts. Concurrent with selections for late-blooming is selection for genetic resistance to low temperatures damage, which is known to vary among cultivars at the same development stage (Felipe 1988).

Modified tree architectures which maximize fruit-wood renewal while significantly reducing pruning needs are also desirable in new cultivars (Socias i Company et al. 1998). This type of growth habit is characterized by the predominance of fruiting spurs, as found in the Puglia cultivars and their progeny (Grasselly 1972). The presence of many spurs is essential for a very high bud density (Kodad and Socias i Company 2006), resulting in a high potential fruit productivity, possibly also compensating for the damages from occasional frosts (Kodad and Socias i Company 2008).

Ripening time is becoming more important in almond as it is in other fruit species, in order to advance harvest to a period with more favorable weather conditions and for earlier marketing. A range of successively maturing cultivars is also desirable to extend the harvest period for more efficient farm operations. Resistance to pests and diseases is also an increasingly important goal as it allows reductions in costly chemical sprays as well as environmental contamination by pesticides.

Despite the difficulties in defining a kernel quality ideotype because of differences in consumer preferences (Janick 2005), almond quality has become an important new goal for breeding (Socias i Company et al. 2008a). Quality considerations include not only the chemical composition conferring a specific organoleptic quality, but also the physical traits related to industry processing. Thus, a different type of shell is preferred depending on the industry of each region; hard in most Mediterranean countries and soft in California.

The chemical composition of almond kernels represents a challenging goal for breeding because of the complex organoleptic aspect of quality and also their increasingly recognized contribution to human health. Documented phytonutrients include the antioxidant compounds of almond kernels, a high content of oleic acid among the fatty acids, and fiber content. Although these aspects have not yet been fully incorporated into the new releases, they are receiving increasing attention not only among almond breeders, but also among growers, processors and consumers (Socias i Company et al. 2008a). Recently released cultivars, such as ‘Sweetheart’ in California and ‘Belona’ and ‘Mardía’ in Spain, possess very high levels of the phytonutrient oleic acid which has been shown to have health benefits to consumers (Gradziel 2008; Socias i Company and Felipe 2007: Socias i Company et al. 2008b).

5.2 *Rootstock*

While most almond rootstocks are shared with other stone fruit species, primarily peach, almonds also have unique requirements from these other species. As European rootstock breeding efforts direct greater emphasis to peach, specific differences between these two crops need to be considered. For example, tree–size reducing rootstocks are not as desirable in almond as in peach. Similarly, most plum rootstocks, because of their size reducing tendency, are only appropriate for almond plantings in heavy soils with problems of asphyxia and greater disease presence. The use of plum for almond rootstocks also suffers from increased root suckering and possible graft-compatibility with many almond cultivars.

Almond seedling rootstocks, which were very popular in the past, are much less utilized at present due to the lack of homogeneity in most sources, and the better adaptability of almond × peach hybrids to most modern growing conditions. This is also true for peach seedling rootstocks which continue to predominate in California. In general, both peach and almond × peach hybrid rootstocks show a good graft-compatibility with almond. Almond × peach clonal hybrids also are becoming the most utilized rootstock in Europe. Desirable traits in new rootstocks include ease of propagation by hardwood cuttings and/or micro propagation, easy distinction of rootstock growth from scions (i.e., red leaves) to identify failed scion bud growth, tolerance to calcareous and/or otherwise poor soils, and high vigor. Increased tolerance to heavy soils and water saturated soils is also becoming an important goal for new hybrids (Xiloyannis et al. 2007).

6 Breeding Methods

Almond occupies a peculiar place among tree crops. Although it belongs to the genus *Prunus*, which comprises all the stone fruit species, it is generally placed among the nuts, which belong to several different botanical families. When attempting a genetic approach to almond improvement, it is useful to consider it within the general context of stone fruits, although almond has been much less studied than the other rosaceous fruits. However, when considered as a nut, almond would be rated as a well studied species, as the scientific approach to investigation of most of the other nut species has been more limited.

6.1 Major Traits

Major breeding traits include tree-productivity, kernel quality, self-compatibility, bloom time, and disease and pest resistance. With the exception of self-compatibility which is controlled primarily by a single major gene, most traits in almond are quantitatively inherited and still poorly characterized (Socias i Company 1998). Reported heritabilities for some important quantitative traits are summarized in Table 18.6. Quantitative inheritance has also been suggested for several other traits in almond, but difficulty in their measurement, the low number of observations and a typically small number of breeding offspring evaluated have frustrated accurate estimations of their heritability. These traits include the color and linear dimensions of the leaf (Grasselly 1972). This information may explain similarities for leaf traits among seedlings coming from the same cross (Bernad and Socias i Company 1994). Similar findings also relate to flower dimensions (Bernad and Socias i Company 1994; Grasselly 1972) and to stamen number (Grasselly 1972).

A sweet kernel is essential for commercial cultivars. Kernel sweetness is a qualitative dominant trait with all commercial cultivars having sweet to slightly bitter kernels. Interestingly, genetic studies confirm that most cultivars are heterozygous for this trait.

As previously discussed, self-compatibility has become a high priority objective. After assessing the transmission of self-compatibility, Socias i Company and Felipe (1977) and Socias i Company (1984) suggested that self-compatibility was dominant over self-incompatibility, and that the self-compatible cultivars used in most breeding programs were heterozygous for this trait. However, transmission may be affected by inbreeding (Alonso Segura and Socias i Company 2007) due to the reduced number of parents involved in many breeding programs (Socias i Company 2002). Many distinct self-incompatible alleles have also been identified (Barckley et al. 2006; Kodad et al. 2008; Ortega et al. 2006).

Blooming time also shows a qualitative component modified by quantitative elements with additive effects (Socias i Company et al. 1999). Since bloom time depends both on the bud chilling and subsequent heat requirements of the cultivar (Tabuenca et al. 1972), this breeding objective is more easily attained by selecting parents possessing appropriate chilling as well as heat requirements (Alonso et al. 2005).

A highly productive tree growth habit is strongly associated with the presence of many fruiting spurs, as found in the Puglia cultivars and their progeny (Grasselly 1972). This trait shows only moderate heritability (Sarvisé and Socias i Company 2005). The Puglia cultivars, however, have shown susceptibility to fungal diseases as well as frost resistance, which is also inherited in their progeny (Grasselly 1972; Felipe 1988).

Table 18.6 Heritability of some quantitative traits in almond

Trait	Heritability	Source
Physiological traits		
Blooming time	0.804	Kester et al. (1973)
	0.67	Dicenta et al. (1993)
Leafing time	0.829	Kester et al. (1973)
Blooming duration	0.20	Dicenta et al. (1993)
Blooming intensity	0.54	Dicenta et al. (1993)
Ripening season	0.69	Dicenta et al. (1993)
Production intensity	0.45	Dicenta et al. (1993)
Morphological traits		
Bud density	0.30	Sarvisé and Socias i Company (2005)
Branching habit	0.19	Sarvisé and Socias i Company (2005)
Nut and shell traits		
Weight	0.81	Kester et al. (1977)
Length	0.50	Arteaga and Socias i Company (2002)
Width	0.37	Arteaga and Socias i Company (2002)
Thickness	0.28	Arteaga and Socias i Company (2002)
Width/length ratio	0.46	Arteaga and Socias i Company (2002)
Thickness/length ratio	0.53	Arteaga and Socias i Company (2002)
Thickness/width ratio	0.30	Arteaga and Socias i Company (2002)
Shell hardness	0.55	Kester et al. (1977)
Shell thickness	0.51	Arteaga and Socias i Company (2002)
Kernel traits		
Weight	0.64	Kester et al. (1977)
Length	0.77	Kester et al. (1977)
Width	0.62	Kester et al. (1977)
Thickness	0.71	Kester et al. (1977)
Width/length ratio	0.46	Arteaga and Socias i Company (2002)
Thickness/length ratio	0.43	Arteaga and Socias i Company (2002)
Thickness/with ratio	0.21	Arteaga and Socias i Company (2002)
Double kernels	0.51	Kester et al. (1977)
Kernel composition		
Oil	0.57	Font i Forcada et al. (2011)
Palmitic acid	0.15	Font i Forcada et al. (2011)
Palmitoleic acid	0.12	Font i Forcada et al. (2011)
Stearic acid	0.11	Font i Forcada et al. (2011)
Oleic acid	0.11	Font i Forcada et al. (2011)
Linoleic acid	0.25	Font i Forcada et al. (2011)
α-Tocopherol	0.21	Font i Forcada et al. (2011)
γ-Tocopherol	0.60	Font i Forcada et al. (2011)
δ-Tocopherol	0.11	Font i Forcada et al. (2011)
Protein	0.12	Font i Forcada et al. (2011)

6.2 Breeding Methodology

Traditional breeding approaches have predominated in almond breeding. These are typically based on crosses between parents selected for desired traits. The breeding techniques applied are well known (Kester and Gradziel 1996), involving controlled crosses following bud emasculation, with previously collected donor pollen. Effective procedures for seed handling and field evaluation are also well established and previously described by Kester and Gradziel (1996).

The several breeding programs maintained in different European institutions follow the traditional approach utilizing a well established and highly selected genetic base. In California, most current breeding efforts are concentrated at the University of California at Davis, where the initial California germplasm was very limited (Bartolozzi et al. 1998; Martínez-Gómez et al. 2003a). Previous genetic and rootstock development studies by Kester, however, had made available a wide array of related species and associated interspecies hybrids with cultivated almond (Kester et al. 1990). Recognizing that the initial domestication of almond involved significant introgression of new traits from wild species (such as the self-compatibility gene from *P. webbii*), the California breeding effort is attempting to resynthesize the cultivated almond through appropriate hybridizations among selected accessions of different species followed by appropriate introgression to a desirable cultivated almond background. This approach avoids the genetic bottleneck of using cultivars developed for distinctly different environment, and has allowed the transfer a wide range of novel and desirable traits, including self-compatibility, autogamy, disease and pest resistance, and improved tree, nut, and kernel quality (Gradziel 2008).

Almond is characterized by a short juvenile period relative to other tree nut species. Most seedling plants, if well managed, should begin to initiate flower buds during 3rd year after the cross. Shortening of the breeding cycle is possible by germinating the immature embryo and top-budding the buds from the seedling onto established rootstocks (Kester and Gradziel 1996). Some methods of early screening have been applied to advance the selection process as well (Vargas et al. 2005). At present, detection of self-compatibility at the seedling stage by specific primers (Ma and Oliveira 2001) allows a discarding of self-incompatible seedlings during the first year of growth. Other new technologies are also being increasingly integrated in the breeding programs as summarized by Martínez-Gómez et al. (2003b, 2005).

The chromosome number of almond is $2n = 16$ (Darlington 1930), which is the same as many other *Prunus* species. Cultivars of almond with histories of reduced fertility have been associated with chromosome abnormalities (Almeida 1945). Recent studies have shown several types of changes in chromosomes, both genetic (deletions, point mutations, etc.) and chromosomal, including aneuploidy (Martínez-Gómez and Gradziel 2003), translocations (Jáuregui et al. 2001) and epigenetic (gene activation/silencing, etc.). These changes can be selected, though because the subsequent selections are vegetatively propagated, the specific nature of control is rarely scrutinized.

6.3 Propagation

Almond is a species which is difficult to propagate by hardwood cuttings. Thus, clonal propagation of selected cultivars has had to rely on grafting or budding on selected rootstocks. Seedling propagation is still practiced in some countries (Kester et al. 1990), as well as grafting in situ on field planted seedling rootstocks which maintains the vigorous taproot for adaptability to nonirrigated conditions (Felipe 2000).

Modern nursery production relies mostly on late spring or early fall budding onto vigorously growing rootstocks planted in the nursery the previous fall. The usual method is by T-budding, where the scion cultivar bud is first removed from the actively growing and so easily separable inner, woody tissue. Chip-budding can be done at other times of the year when growth is not active enough to ensure proper "slipping" of the bark from the inner wood (Gómez Aparisi and Felipe 1984). More recently, propagation by mini-chip budding on small in vitro propagated rootstocks is increasingly done, allowing the production of plants in pots, table-budding operations, rapid plant growth in greenhouses, in-container plant marketing and the possibility of field planting during an extended period of time (rather than just dormant season bare-root plantings). With T-budding, vigorous growth of the scion bud is encouraged through appropriate irrigation and fertilization. In this way, a marketable tree can be developed in a single growing season, which greatly reduces nursery tree production costs. With mini-chip budding, production can be reduced to 3–4 months and can be done at any time during the year.

In all budding approaches, once the inserted buds begin to grow, the rootstock shoot above the inserted bud is removed. For this operation, red-leafed rootstocks are most convenient as they allow the ready distinction of the scion from the rootstock.

Felipe (1984) has identified an almond cultivar with a very high rate of hardwood propagation and has further shown that this ability can be transmitted to its offspring (Felipe 1992). This material has greatly improved the hardwood-cutting propagation of almond × peach hybrid rootstocks (Gómez Aparisi et al. 2002). However, no further attempts of selecting almond genotypes for clonal propagation have been reported.

Micropropagation and callus regeneration have also been shown to be more difficult in almond than in most other species (Kester and Gradziel 1996), thus limiting the development of other propagation systems in almond. In vitro propagation, however, is widely utilized for rootstock production, especially the almond × peach hybrids. A large number of the plants produced in the Mediterranean countries and California are on this type of propagated rootstock.

7 Integration of New Biotechnologies

The recent development of powerful new biotechnologies has advanced plant breeding efforts through the direct incorporation of foreign genes using genetic engineering strategies, and through the ability to use a DNA molecule directly as a marker for desired traits. While almond cultivars are readily transformed using *Agrobacterium*-mediated approaches, the regeneration of plantlets from established cultivar cells has proven very difficult. This difficulty is believed to be due to the recalcitrance of cultivar cells to initiate the required organogenesis, presumably because they have lost their juvenility with their advanced clonal age. Molecular markers, however, because they offer the opportunity for fast, accurate, and environment-independent evaluation at the seedling stage, promise to dramatically increase breeding efficiency. In addition, specific markers offer the advantage of codominant expression, good reproducibility, and allow the ability to compare genetic variation among homologous regions of the same or different species (Martínez-Gómez et al. 2003a, 2003c). A detailed review of biotechnology research with almond has recently been provided by Martínez Gómez et al. (2005, 2006).

SSR analysis has confirmed previous isozymes studies which identified the almond as the most polymorphic species within the major *Prunus* tree crop species (Fernández i Martí et al. 2009a; Martínez-Gómez et al. 2006; Xie et al. 2006) making it an ideal candidate for map construction. Three main linkage maps have been developed for almond from the linkage analysis performed on three different progenies, the 'Ferragnès' × 'Tuono' F_1, corresponding to the FxT map (Viruel et al. 1995); the F_2 population of the interspecific cross between the almond cultivar 'Padre' × the peach selection '54P455', corresponding to the Px5 map (Foolad et al. 1995); and the reference for the *Prunus* species, the almond (cv. 'Texas', syn. 'Mission') × peach (cv. 'Earlygold') F_2 progeny, corresponding to TxE map (Arús et al. 1994a, 1994b). These maps have been progressively improved with the development of molecular markers, and more saturated versions have been produced, such as the FxT map (Joobeur et al. 2000) and the Px5 map (Bliss et al. 2002) and for the TxE map (Joobeur et al. 1998; Aranzana et al. 2003). The current version of the TxE map (Dirlewanger et al. 2004b) includes 562 markers (361 RFLPs, 185 SSRs, 11 isozymes and 5 STSs), which cover a total distance of 519 cM with high density (average density 0.92 cM/marker and largest gap of 7 cM). More recently, Sánchez-Pérez et al. (2007) studied an F_1 almond progeny 'R1000' × 'Desmayo Largueta' constructing a genetic linkage map with 56 SSRs.

The order of molecular markers observed in the almond map was similar to maps developed with other *Prunus* species, suggesting a high level of synteny within the genus (Dirlewanger et al. 2004a; Martínez-Gómez et al. 2006). This homology among *Prunus* genomes supports the opportunity for successful interspecific gene introgression as demonstrated by the successful transfer of traits from closely related species to almond (Gradziel et al. 2001; Martínez-Gómez et al. 2003c). The high level of synteny within the genus also supports the transferability of genetic information developed from linkage maps of other *Prunus* species (Arús et al. 2006).

The availability of high-density linkage maps has allowed recent successes in establishing the approximate map position of major genes in almond. Important examples include the use of bulk segregant analysis (BSA) to map, based on the F_1 progeny from the cross 'Felisia' × 'Bertina,' self-incompatibility (Ballester et al. 1998), shell hardness (Arús et al. 1999), and blooming time (Ballester et al. 2001), and based on the F_2 of 'Garfi' almond × 'Nemared' peach (Jáuregui et al. 2001), genes involved in rootknot nematode resistance (Dirlewanger et al. 2004b), and flower color (Jáuregui 1998). In addition, the physical mapping of rDNA genes by Corredor et al. (2004) has allowed the establishment of a more precise karyotype for almond.

Cloning of genes expressed during seed development has been reported by Garcia-Mas et al. (1996). Suelves and Puigdomènech (1998) have described the cloning of the mandelonitrile lyase gene responsible for the creation of both cyanide and the amaretto flavor of bitter almonds. A major effort has been directed toward cloning and characterizing the economically important self-incompatibility gene in almond (Bacarella et al. 1991). The cDNA-encoding almond *S*-RNase was first cloned by Ushijima et al. (1998). To better understand the nature of the self-incompatibility gene, Ushijima et al. (2001) later cloned and characterized the cDNA encoding mutated *S*-RNase from the almond cultivar 'Jeffries' which has a disfunctional *S*-allele haplotype in both pistil and pollen.

PCR-based markers of almond self incompatibility *S*-alleles have been successfully used to identify different self-incompatibility genotypes (Barckley et al. 2006; Bošković et al. 2007; Channuntapipat et al. 2003; López et al. 2004; Ortega et al. 2005; Tamura et al. 2000), and to identify more than 30 different *S*-RNases (Kodad et al. 2008, 2010; Barckley et al. 2006; Ortega et al. 2006). Similar results were obtained by Bošković et al. (2003, 2007) who identified major almond cultivar stylar *S*-RNases by electrophoresis in vertical polyacrylamide gels. PCR-based markers of almond self-incompatibility *S*-alleles have been employed to facilitate the integration of self-compatible *S*-alleles from related species (Gradziel et al. 2001). Screening efficiency and flexibility have been greatly increased with the development of successful multiplex PCR techniques by Sánchez-Pérez et al. (2004). Using advanced cloning strategies, Ushijima et al. (2003) have recently described the structural and transcriptional analysis of a pollen-expressed F-box gene with haplotype-specific polymorphism strongly associated with self-incompatibility. However, the identification of the S_f allele by specific primers and sequencing has created some confusion. Probably, some missequencings and misinterpretations have occurred during allele analysis, as shown by the fact that Bošković et al. (2007) incorrectly named a new allele, S_{30}, which is identical to S_f, although showing a different activity (Kodad et al. 2009). The two phenotypic expressions (pollen and pistil) of the S_f allele must be carefully considered when determining self-compatibility in almond because the two forms of the S_f allele can recognize each other (Fernández i Martí et al. 2009b).

Molecular markers are currently being employed to elucidate the genetic basis of plant processes controlled by multiple genes. For example, Campalans et al. (2001) have described a differential expression technique based on cDNA-AFLP (amplified restriction fragment polymorphism) derived technique for RNA fingerprinting to characterize genes involved in drought tolerance in almond. This work has identified increased drought tolerance in specific genes associated with leaf function.

Despite recent advances in the application of traditional and newer biotechnologies, almond, as well as other tree crops, lags behind the progress typically observed for annual crops. This is, in large part, the consequence of the inherent difficulties in doing genetic studies on such large-sized and long generation-time plants (Martínez-Gómez et al. 2003b; Socias i Company 1998). These inherent obstacles to traditional breeding make opportunities with the new technologies much more revolutionary when applied to tree crops. When fully integrated with the array of breeding methods developed to capitalize on the inherent advantages of tree crops, such as the capability to capture desirable genetic/epigenetic arrangements through vegetative propagation, breeding potential could be expected to surpass that for seed propagated annual crops. Almond is currently well positioned to be a leader in this effort.

References

Almeida, C.R.M. de. 1945. Âcerca da improdutividade na amandoeira. An. Inst. Agron. Lisboa 15:1–186.

Alonso, J.M. and Socias i Company, R. 2005. Self-incompatibility expression in self-compatible almond genotypes may be due to inbreeding. J. Amer. Soc. Hort. Sci. 130:865–869.

Alonso, J.M., Ansón, J.M., Espiau, M.T. and Socias i Company, R. 2005. Determination of endodormancy break in almond flower buds by a correlation model using the average temperature of different day intervals and its application to the estimation of chill and heat requirements and blooming date. J. Amer. Soc. Hort. Sci. 130:308–318.

Alonso Segura, J.M. and Socias i Company, R. 2007. Negative inbreeding effects in tree fruit breeding: self-compatibility transmission in almond. Theor. Appl. Genet. 115:151–158.

Alston, J., Carman, H., Christian, J.E., Doreman, J., Murua, J.R. and Sexton, R. 1995. Optimal reserve and export policies for the California almond industry: theory, econometrics and simulation. Univ. California, Giannini Foundation Monograph 42, 130 p.

Aranzana, M.J., Cosson, P., Dirlewanger, E., Ascasibar, J., Cipriani, G., Arús, P., Testolin, R., Abbott, A., King, G.J. and Iezzoni, A.F. 2003. A set of simple-sequence repeat (SSR) markers covering the *Prunus* genome. Theor. Appl. Genet. 106:819–825.

Arteaga, N. and Socias i Company, R. 2002. Heritability of fruit and kernel traits in almond. Acta Hort. 591:269–274.

Arulsekar, S., Parfitt, D.E. and Kester, D.E. 1986. Comparison of isozyme variability in peach and almond cultivars. J. Hered .77:272–274.

Arús, P., Olarte, C., Romero, M. and Vargas, F.J. 1994a. Linkage analysis of ten isozyme genes in F_1 segregating progenies of almond. J. Amer. Soc. Hort. Sci. 119:339–344.

Arús, P., Messeguer, R., Viruel, M.A., Tobutt, K., Dirlewanger, E., Santi, F., Quarta, R. and Ritter, E. 1994b. The European *Prunus* mapping project. Euphytica 77:97–100.

Arús, P., Ballester, J., Jáuregui, B., Joobeur, T., Truco, M.J. and Vicente, M.C. de. 1999. The European *Prunus* mapping project: update of marker development in almond. Acta Hort 484:331–336.

Arús, P., Yamamoto, T., Dirlewanger, E. and Abbott A.G. 2006. Synteny in the Rosaceae. Plant Breed. Rev. 27:175–211.

Bacarella A., Chironi, G. and Barbera, G. 1991. Aspetti tecnici economici e di mercato del mandorlo in Sicilia. Quarderni di Ricerca di Sperimentazione (Palermo, Sicily) 40:1–191.

Bacchetta, L., Avanzato, D., Drogoudi, P., Duval, H., Socias i Company, R. and Spera, D. 2008. The European SAFENUT project for improving almond genetic resources utilization. XIV GREMPA, 30 March – 4 April 2008, Athens, Greece.

Ballester, J., Bošković, R., Batlle, I., Arús, P., Vargas, F. and Vicente, M.C. de. 1998. Location of the self-incompatibility gene on the almond linkage map. Plant Breed. 117:69–72.

Ballester, J., Socias i Company, R., Arús, P. and Vicente, M.C. de. 2001. Genetic mapping of a major gene delaying blooming time in almond. Plant Breed. 120:268–270.

Barckley, K.K., Uratsu, S.L., Gradziel T.M. and Dandekar, A.M. 2006. Multidimensional analysis of *S*-alleles from cross-incompatible groups of California almond cultivars. J. Amer. Soc. Hort. Sci. 131:632–636.

Bartolozzi, F., Warburton, M.L., Arulsekar, S. and Gradziel, T.M. 1998. Genetic characterization and relatedness among California almond cultivars and breeding lines detected by randomly am-plified polymorphic DNA (RAPD) analysis. J. Amer. Soc. Hort. Sci. 123: 381–387.

Berger, P. 1969. Aptitude à la transformation industrielle de quelques variétés d'amandier. Bull. Techn. Inf. 241:577–580.

Bernad, D. and Socias i Company, R. 1994. Caracterización morfológica y bioquímica de algunas selecciones autocompatibles de almendro. Inf. Técn. Econ. Agrar. 90V:103–110.

Bliss, F.A., Arulsekar, S., Foolad, M.R., Becerra, A.M., Gillen, A., Warburton, M.L., Dandekar, A.M., Kocsisne, G.M. and Mydin, K.K. 2002. An expanded genetic linkage map of *Prunus* based on an interspecific cross between almond and peach. Genome 45:520–529.

Bošković, R., Tobutt, K.R., Batlle, I., Duval, H., Martínez-Gómez, P. and Gradziel, T.M. 2003. Stylar ribonucleases in almond: correlation with and prediction of self-incompatibility genotypes. Plant Breed. 122:70–76.

Bošković, R., Tobutt, K.R., Ortega, E., Sutherland, B.G. and Godini A. 2007. Self-(in)compatibility of the almonds *P. dulcis* and *P. webbii*: detection and cloning of 'wild-type *Sf*' and new self-compatibility alleles encoding inactive *S*-RNases. Mol. Genet. Genomics. 278:665–676.

Browicz K. and Zohary D. 1996. The genus *Amygdalus* L. (Rosaceae): species relationships, distribution and evolution under domestication. Genet. Resour. Crop Evol. 43: 229–247.

Campalans, A., Pagès, M. and Messeguer, R. 2001. Identification of differentially expressed genes by the cDNA-AFLP technique during dehydration of almond (*Prunus amygdalus*). Tree Physiol. 21:633–643.

Cerdá Juan, D. 1973. Economía antigua de Mallorca, p. 417–448. In: J. Mascaró Pasarius (ed.): Historia de Mallorca, vol I. Ed. J. Mascaró Pasarius, Palma de Mallorca, Spain.

Cerezo, M., Socias i Company, R. and Arús, P. 1989. Identification of almond cultivars by pollen isoenzymes. J. Amer. Soc. Hort. Sci. 114:164–169.

Channuntapipat, C., Wirthensohn, M., Ramesh, S.A., Batlle, I., Arús, P., Sedgley, M. and Collins, G. 2003. Identification of incompatibility genotypes in almond using specific primers based on the introns of the S-alleles. Plant Breed. 122:164–168.

Cinelli, F. and Loreti, F. 2004. Evaluation of some plum rootstocks in relation to lime-induced chlorosis by hydroponic culture. Acta Hort. 658:421–427.

Columella, L.J.M. 1988. De re rustica. Spanish edition by A. Holgado Redondo, MAPA, Madrid.

Corredor, E., Román, M., García, E., Perera, E., Arús, P. and Naranjo, T. 2004. Physical mapping of rDNA genes enables to establish the karyotype of almond. Ann. Appl. Biol. 144:219–222.

Darlington, C.G. 1930. Studies in Prunus. III. J. Genet. 22:65–93.

Denisov V.P. 1988. Almond genetic resources in the USSR and their uses in production and breeding. Acta Hort. 224:229–236.

Dicenta, F., García, J.E. and Carbonell, E.A. 1993. Heritability of flowering, productivity and maturity in almond. J. Hort. Sci. 68:113–120.

Dicenta, F., Ortega, E., Martínez-Gómez, P., Sánchez-Pérez, R., Gambín, M. and Egea, J. 2009. Penta and Tardona: two new extra-late flowering self-compatible almond cultivars. Acta Hort. 814:189–192.

Dirlewanger, E., Graziano, E., Joobeur, T., Garriga-Calderé, F., Cosson, P., Howad, W. and Arús, P. 2004a. Comparative mapping and marker-assisted selection in Rosaceae fruit crops. Proc. Natl. Acad. Sci. USA 101:9891–9896.

Dirlewanger. E., Cosson, P., Howad, W., Capdeville, G., Bosselut, N., Claverie, M., Voisin, R., Poizat, C., Lafargue, B., Baron, O., Laigret, F., Kleinhentz, M., Arús, P. and Esmenjaud, D. 2004b. Microsatellite genetic linkage maps of myrobalan plum and an almond-peach hybrid - location of root-knot nematode resistance genes. Theor. Appl. Genet. 109:827–832.

Duval, H. 1999. 'Mandaline', a new French almond vatiety. Nucis 8:36.
Duval, H. and Grasselly, C. 1994. Behaviour of some self-fertile almond selections in the southeast of France. Acta Hort. 373:69–74.
Egea, J., Dicenta, F., Berenguer, T. and García, J.E. 2000. Antoñeta and Marta almonds. HortScience 35:1358–1359.
Espiau, M.T., Ansón, J.M. and Socias i Company, R. 2002. The almond germplasm bank of Zaragoza. Acta Hort. 591:275–278.
Estelrich, P. 1907. El almendro y su cultivo en el mediodía de España e Islas Baleares. Hijos de J. Cuesta, Madrid - Antonio López, Barcelona.
Felipe, A.J. 1984. Enrancinement de l'amandier par bouturage ligneux. Options Méditerr. CIHEAM/IAMZ 84/II:97–100.
Felipe, A.J. 1988. Observaciones sobre comportamiento frente a heladas tardías en almendro. Rap. EUR 11557:123–130.
Felipe, A.J. 1989. Rootstock for almond. Present situation. Options Méditerr. Ser. A 5:13–17.
Felipe, A.J. 1992. Aptitude pour la propagation chez l'amandier 'Garrigues' et sa descendence. Rap. EUR 14081:73–79.
Felipe, A.J. 2000. El almendro: El material vegetal. Integrum, Lérida.
Felipe, A.J. 2009. 'Felinem', 'Garnem', and 'Monegro' almond×peach hybrid rootstocks. HortScience 44:196–197.
Felipe, A.J. and Socias i Company, R. 1987. 'Aylés', 'Guara', and 'Moncayo' almonds. HortScience 22:961–962.
Felipe, A.J., Gómez-Aparisi, J., Socias i Company, R. and Carrera, M. 1997. The almond×peach hybrid rootstock breeding program at Zaragoza (Spain). Acta Hort. 451: 259–262.
Fernández i Martí, À., Alonso, J.M., Espiau, M.T., Rubio-Cabetas, M.J. and Socias i Company, R. 2009a. Genetic diversity in Spanish and foreign almond germplasm assessed by molecular characterization with simple sequence repeats. J. Amer. Soc. Hort. Sci. 134:535–542.
Fernández i Martí, A., Hanada, T., Alonso, J.M., Yamane, H., Tao, R. and Socias i Company, R. 2009b. A modifier locus affecting the expression of the *S*–RNase gene could be the cause of breakdown of self-incompatibility in almond. Sex. Plant Reprod. 22:179–186.
Font i Forcada, C., Kodad, O., Juan, T., Estopañan, G. and Socias i Company, R. 2011. Genetic variability and pollen effect on the transmission of the chemical components of the almond kernel. Span. J. Agric. Res. 9:781–789.
Foolad, M.R., Arulsekar, S., Becerra, V. and Bliss, F.A. 1995 A genetic map of *Prunus* based on an interspecific cross between peach and almond. Theor. Appl. Genet. 91:262–269.
Gagnard, J.M. 1954. Recherches sur les caractères systematiques et sur les phénomènes de stérilité chez les variétés d'amandiers cultivées en Algérie. Ann. Inst. Agric. Serv. Rech. Exp. Agric. Algérie 8:1–163.
Garcia-Mas, J., Messeguer, R., Arús, P. and Puigdomènech, P. 1996. Accumulation of specific mRNAs during almond fruit development. Plant Sci. 113:185–192.
Godini A. 1979. Ipotesi sulla comparsa dell'autocompatibilità nel mandorlo. Riv. Sci. Tecn. Agrar. 19(2/3):3–10.
Godini A. 2000. About the possible relationship between *Amygdalus webbii* Spach and *Amygdalus communis* L. Nucis 9:17–19.
Gómez Aparisi, J. and Felipe, A.J. 1984. Surgreffage d'amandiers. Options Méditerr. CIHEAM/IAMZ 1984/II: 41–49.
Gómez Aparisi, J., Carrera, M., Felipe, A.J. and Socias i Company, R. 2002. 'Garnem', 'Monegro' y 'Felinem': nuevos patrones híbridos almendro × melocotonero resistentes a nematodos y de hoja roja para frutales de hueso. Inf. Técn. Econ. Agrar. 97V:282–288.
Gradziel, T.M. 2003. Interspecific hybridizations and subsequent gene introgression within *Prunus* subgenus *Amygdalus*. Acta Hort. 622: 249–255.
Gradziel, T.M. 2008. Almond (*Prunus dulcis*), p. 1–33. In: M. Priyadarshan and S.M. Jain (eds). Breeding of plantation crops. Springer Sci. Publ. Berlin.
Gradziel, T.M. and Kester, D.E. 1998. Breeding for self-fertility in California almond cultivars. Acta Hort. 470:109–117.

Gradziel, T.M., Martínez-Gómez, P., Dicenta, F. and Kester, D.E. 2001. The utilization of related almond species for almond variety improvement. J. Amer. Pomol. Soc. 55:100–108.
Gradziel, T.M., Lampinen, B., Connell, J.H. and Viveros, M. 2007. 'Winters' almond: an early-blooming, productive and high-quality pollenizer for 'Nonpareil'. HortScience 42: 1725–1727.
Grasselly, C. 1972. L'amandier: caractères morphologiques et phsyiologiques des variétés, modalité de leurs transmissions chez les hybrids de première génération. PhD Thesis, Univ. Bordeaux, France.
Grasselly C. 1976. Les espèces sauvages d'amandier. Options Méditerr. 32: 28–43.
Grasselly C. and Crossa-Raynaud P. 1980. L'amandier. G.P. Maisonneuve et Larose, Paris, XII + 446 pp.
Grasselly, C., Olivier, G. and Niboucha, A. 1992. Le caractère "autocompatibilité" de l'amandier dans le programme de l'I.N.R.A. Rap. EUR 14081:9–17.
Graves, R. 1955. The Greek myths. George Brazillier Inc. New York, USA.
Gustafson W.A., Morrisey T.M. and Bishop C. 1989. Plant exploration and germplasm collection of cold hardy woody plants for Nebraska from the People's Republic of China. Univ. Nebaraska, Lincoln, Neb.
Hansen, J. and Renfrew, J.M. 1978. Palaeolithic-Neolithic seed remains at Franchthi Cave, Greece. Nature 271:349–352.
Hauagge, R., Kester, D.E. and Asay, R.A. 1987a. Isozyme variation among California almond cultivars: II Cultivar characterization and origins. J. Amer. Soc. Hort. Sci. 112:693–698.
Hauagge, R., Kester, D.E. and Asay, R.A. 1987b. Isozyme variation among California almond cultivars: I. Inheritance. J. Amer. Soc. Hort. Sci. 112:687–693.
Holland, D., Bar-Ya'akov, I, Hatib, K., Albert, T., Mani, Y. and Spiegel-Roy, P. 2006. 'Shefa' almond. HortScience 41:1502–1503.
Janick, J. 2005. Breeding intractable traits in fruit crops: dream the impossible dream. Introduction. HortScience 40:1944.
Jáuregui, B. 1998. Localización de marcadores moleculares ligados a caracteres agronómicos en un cruzamiento interespecífico almendro × melocotonero. PhD Thesis, Univ. Barcelona, Spain.
Jáuregui. B., Vicente, M.C. de, Messeguer, R., Felipe, A., Bonnet, A., Salesses, G. and Arús, P. 2001. A reciprocal translocation between 'Garfi' almond and 'Nemared' peach. Theor. Appl. Genet. 102:1169–1176.
Joobeur, T., Viruel, M.A., Vicente, M.C. de, Jáuregui, B., Ballester, J., Dettori, M.T., Verde, I., Truco, M.J., Messeguer, R., Batlle, I., Quarta, R., Dirlewanger, E. and Arús, P. 1998. Construction of a saturated linkage map for *Prunus* using an almond × peach F_2 progeny. Theor. Appl. Genet. 97:1034–1041.
Joobeur, T., Periam, N., Vicente, M.C. de, King, G.J. and Arús, P. 2000. Development of a second generation linkage map for almond using RAPD and SSR markers. Genome 43:649–655.
Juniper, B.E., Watkins, R. and Harris, S.A. 1999. The origin of the apple. Acta Hort. 484: 27–33.
Kester, D.E. and Asay, R. 1975. Almonds, p. 387–419. In: J. Janick and J.N. Moore (eds.): Advances in fruit breeding. Purdue Univ. Press, West Lafayette, IN, USA.
Kester, D.E. and Gradziel, T.M. 1996. Almonds, p. 1–97. In: J. Janick and J.N. Moore (eds.): Fruit breeding, vol 3, John Wiley & Sons, New York, USA.
Kester, D.E. and Grasselly, C. 1987. Almond rootstocks, p. 265–293. In: R.C. Rom and R.F. Carlson (eds.). Rootstocks for fruit crops. Wiley, New York, USA.
Kester, D.E., Raddi, P. and Asay, R. 1973. Correlation among chilling requirements for germination, blooming and leafing in almond (*Prunus amygdalus* Batsch). Genetics 74:s135.
Kester, D.E., Hansche, P.E., Beres, V. and Asay R.N. 1977. Variance components and heritability of nut and kernel traits in almond. J. Amer. Soc. Hort. Soc. 102:264–266.
Kester D.E., Gradziel T.M. and Grasselly C. 1990. Almonds (*Prunus*). Acta Hort. 290:699–758.
Kester, D.E., Asay, R.A. and Gradziel, T.M. 2001. 'Nickels' almond × peach hybrid clonal rootstock. HortScience 37:415–417.

Kester, D.E., Shackel, K.A., Micke, W.C., Viveros, M. and Gradziel, T.M. 2004. Noninfectious bud failure in 'Carmel' almond: I. Pattern of development in vegetative progeny trees. J. Amer. Soc. Hort. Sci. 129:244–249.

Kochba, J. and Spiegel-Roy, P. 1976. Alnem 1, Alnem 88, Alnem 201: nematode resistant rootstock seed sources. HortScience 11:270.

Kodad, O. and Socias i Company, R. 2006. Influence of genotype, year and type of fruiting branches on the productive behaviour of almond. Scientia Hort. 109:297–302.

Kodad, O and Socias i Company, R. 2008 Significance of flower bud density for cultivar evaluation in almond. HortScience 43:1753–1758.

Kodad, O., Alonso, J.M., Sánchez, A., Oliveira, M.M. and Socias i Company, R. 2008. Evaluation of genetic diversity of *S*-alleles in an almond germplasm collection. J. Hort. Sci. Biotechnol. 83:603–608.

Kodad, O., Socias i Company, R., Sánchez, A. and Oliveira, M.M. 2009. The expression of self-compatibility in almond may not only be due to the presence of the Sf allele. J. Amer. Soc. Hort. Sci. 134:221–227.

Kodad, O., Alonso, J.M., Fernández i Martí, À., Oliveira, M.M. and Socias i Company, R. 2010. Molecular and physiological identification of new *S*-alleles associated with self-(in)compatibility in local Spanish almond cultivars. Scientia Hort. 123:308–311.

Kovalyov N.V. and Kostina K.F. 1935. A contribution to the study of the genus *Prunus* Focke. Questions of taxonomy and plant breeding (in Russian). Tr. Prikl. Bot. Genet. Selek. Ser. 8, 4:1–76.

Ladizinsky, G. 1999. On the origin of almond. Genet. Resour. Crop Evol. 46:143–147.

López, M., Mnejja, M., Rovira, M., Collins, G., Vargas F.J., Arús, P. and Batlle, I. 2004. Self-incompatibility genotypes in almond re-evaluated by PCR, stylar ribonucleases, sequencing analysis and controlled pollinations. Theor. Appl. Genet. 109:954–964.

López, M., Romero, M.A., Vargas, F.J. and Batlle, I. 2005. 'Francolí', a late flowering almond cultivar re-classified as self-compatible. Plant Breed. 124:502–506.

Loreti, F. and Massai, R. 1998. Sirio: new peach × almond almond × peach hybrid rootstock for peach. Acta Hort. 465:229–236.

Ma, R.C. and Oliveira, M.M. 2001. Molecular cloning of the incompatibility genes *S1* and *S3* from almond (*Prunus dulcis* cv. Ferragnès). Sex. Plant Reprod. 14:163–167.

Maggioni, L. and Lipman, E. 2006. Report of a working group on *Prunus*. Bioversity International, Rome, Italy, 128 p.

Martínez-Gómez, P. and Gradziel, T.M. 2003. Sexual polyembryony in almond. Sex. Plant Reprod. 16:135–139.

Martínez-Gómez, P., Arulsekar, S., Potter, D. and Gradziel, T.M 2003a. An extended interspecific gene pool available to peach and almond breeding as characterized using simple sequence repeat (SSR) markers. Euphytica 131:313–322.

Martínez-Gómez, P., Arulsekar, S., Potter, D. and Gradziel, T.M. 2003c. Relationships among peach and almond and related species as detected by SSR markers. J. Amer. Soc. Hort. Sci. 128:667–671.

Martínez-Gómez, P., Sozzi, G.O., Sánchez-Pérez, R., Rubio, M. and Gradziel, T.M. 2003b. New approaches to *Prunus* tree crop breeding. J. Food Agric. Environ. 1:52–63.

Martínez-Gómez, P., Sánchez-Pérez, R., Rubio, M., Gradziel, T. M., Sozzi, G.O. 2005. Application of recent biotechnologies to *Prunus* tree crop genetic improvement. Ciencia Investigacion Agraria 32:55–78.

Martínez-Gómez, P., Sánchez-Pérez, R., Dicenta, F., Howad, W. and Gradziel, T.M. 2006. Almond, p. 229–242. In: C. Kole (ed.). Genome mapping and molecular breeding, vol. 4, Fruits and nuts. Springer-Verlag, Heidelberg, Berlin.

Martins, M., Tenreiro, R. and Oliveira, M.M. 2003. Genetic relatedness of Portuguese almond cultivars assessed by RAPD and ISSR markers. Plant Cell Rep. 22:71–78.

Monastra, F. and Raparelli, E. 1997. Inventory of almond research, germplasm and references. FAO REUR Technical Series 51, Rome, 232 p.

Moreno, M.A. 2004. Breeding and selection of *Prunus* rootstocks at the Aula Dei Experimental Station, Zaragoza, Spain. Acta Hort. 658:519–528.

Niklasson, M. 1989. The European almond catalogue. Nordic Gene Bank, Alnarp, Sweden, 38 p.

Ortega, E., Sutherland, B.G., Dicenta, F., Bošković, R. and Tobutt, K.R. 2005. Determination of incompatibility genotypes in almond using first and second intron consensus primers: detection of new *S* alleles and correction of reported *S* genotypes. Plant Breed. 124:188–196.

Ortega, E., Bošković, R., Sargent, D.J. and Tobutt, K.R. 2006. Analysis of *S*-RNase alleles of almond (*Prunus dulcis*): characterization of new sequences, resolution of new synonyms and evidence of intragenic recombination. Mol. Genet. Genomics 276:413–426.

Popov, M.G., Kostina, K.F. and Poyarkova, A.I. 1929. Wild trees and shrubs in Central Asia (in Russian). Tr. Prikl. Bot. Genet. Selek. 2: 241–483.

Rikhter, A.A. 1969. Ways and methods of almond breeding (in Russian). Tr. Gos. Nikit. Sad 43:81–94.

Rikhter, A.A. 1972. Biological basis for the creation of almond cultivars and commercial orchards (in Russian). Ed. AN SSSR, Glavny Bot. Sad, Moscow, Russia.

Sánchez-Pérez, R., Dicenta, F. and Martínez-Gómez, P. 2004. Identification of *S*-alleles in almond using multiplex-PCR. Euphytica 138:263–269.

Sánchez-Pérez, R., Howad, W., Dicenta, F., Arús, P. and Martínez-Gómez, P. 2007. Mapping major genes and quantitative trait loci controlling agronomic traits in almond. Plant Breed. 126:310–318.

Sarvisé, R. and Socias i Company, R. 2005. Variability and heritability of bud density and branching habit in almond. Acta Hort. 663:401–404.

Sathe, S.K., Teuber, S.S., Gradziel, T.M. and Roux, K.H. 2001. Electrophoretic and immunological analyses of almond genotypes and hybrids. J. Agric. Food Chem. 49:2043–2052.

Schirra, M. 1997. Postharvest technology and utilization of almonds. Hort. Rev. 20:267–292.

Shiran, B., Amirbakhtiar, N., Kiani, S., Mohammadi, S., Sayed-Tabatabaei, B.E. and Moradi, H. 2007. Molecular characterization and genetic relationship among almond cultivars assessed by RAPD and SSR markers. Scientia Hort. 111:280–292.

Socias i Company, R. 1984. A genetic approach to the transmission of self-compatibility in almond (*Prunus amygdalus* Batsch). Options Méditerr. CIHEAM/IAMZ 84/II:123–127.

Socias i Company R. 1990. Breeding self-compatible almonds. Plant Breed. Rev. 8:313–338.

Socias i Company R. 1998. Fruit tree genetics at a turning point: the almond example. Theor. Appl. Genet. 96:588–601.

Socias i Company, R. 2001. Almendro, p. 271–274. In: F. Nuez and G Llácer (eds.): La horticultura española. SECH – Ed. Horticultura, Reus, Spain.

Socias i Company, R. 2002. Latest advances in almond self-compatibility. Acta Hort. 591:205–212.

Socias i Company, R. 2004. The contribution of *Prunus webbii* to almond evolution. Plant Genet. Resour. Newsl. 14:9–13.

Socias i Company, R. and Felipe, A.J. 1977. Heritability of self-compatibility in almond. III Coll. GREMPA, 3–7 October 1977, Bari, 181–183.

Socias i Company, R. and Felipe, A.J. 1988. Self-compatibility in almond: transmission and recent advances in breeding. Acta. Hort. 224:307–317.

Socias i Company, R. and Felipe, A.J. 1992 Almond: a diverse germplasm. HortScience 27:717–718, 803.

Socias i Company, R. and Felipe, A.J. 1999. 'Blanquerna', 'Cambra' y 'Felisia': tres nuevos cultivares autógamos de almendro. Inf. Técn. Econ. Agrar. 95V:111–117.

Socias i Company, R. and Felipe, A.J. 2007. 'Belona' and 'Soleta' almonds. HortScience 42:704–706.

Socias i Company, R., Felipe, A.J., Gómez Aparisi, J., García, J.E. and Dicenta, F. 1998. The ideotype concept in almond. Acta Hort. 470:51–56.

Socias i Company, R., Felipe, A.J. and Gómez Aparisi, J. 1999. A major gene for flowering time in almond. Plant Breed. 118:443–448.

Socias i Company, R., Felipe, A.J. and Gómez Aparisi, J. 2003. Almond bloom in a changing climate. J. Amer. Pomol. Soc. 57: 89–92.

Socias i Company, R., Alonso, J.M. and Gómez Aparisi, J. 2004. Fruit set and productivity in almond as related to self-compatibility, flower morphology and bud density. J. Hort Sci. Biotechnol. 79:754–758.

Socias i Company, R., Kodad, O., Alonso, J.M. and Gradziel, T.M. 2008a. Almond quality: a breeding perspective. Hort. Rev. 34:197–238.

Socias i Company, R., Kodad, O., Alonso, J.M. and Felipe, A.J. 2008b. 'Mardía' almond. HortScience 43: 2240–2242.

Sorkheh, K., Shiran, B., Aranzana, M.J., Mohammadi, S.A. and Martinez-Gomez, P. 2007a. Application of amplified fragment length polymorphism (AFLP) analysis to plant breeding and genetics: procedures, applications and prospects. J. Food Agric. Environ. 5:197–204.

Sorkheh, K., Shiran, B., Gradziel, T.M., Epperson, P., Martinez-Gomez, P., and Asadi, E. 2007b. Amplified fragment length polymorphism as a tool for molecular characterization of almond germplasm: Genetic diversity among genotypes and related wild species of almond, and its relationships with agronomic traits. Euphytica 1563:327–344.

Spiegel-Roy P. 1986. Domestication of fruit trees, p. 201–211. In: C. Barigozzi (ed.) The origin and domestication of cultivated plants. Elsevier, Amsterdam.

Suelves, M. and Puigdomènech, P. 1998. Molecular cloning of the cDNA coding for the (R)-(+)-mandelonitrile lyase of *Prunus amygdalus*: temporal and spatial expression patterns in flowers and mature seeds. Planta 206:388–393.

Tabuenca, M.C, Mut, M. and Herrero, J. 1972. Influencia de la temperatura en la época de floración del almendro. An. Estac. Exp. Aula Dei 11:378–395.

Tamura, M., Ushijima, K., Sassa, H., Hirano, H., Tao, R., Gradziel, T.M. and Dandekar, A.M. 2000. Identification of self-incompatibility genotypes of almond by allele-specific PCR analysis. Theor. Appl. Genet. 101:344–349.

Tao, R., Watari, A., Hanada, T., Habu, T., Yaegaki, H., Yamaguchi, M. and Yamane, Y. 2007. Self-compatible peach (*Prunus persica*) has mutant versions of the *S* haplotypes found in self-incompatible *Prunus* species. Plant Mol. Biol. 63:109–123.

Testolin, R., Messina, R., Lain, O., Marrazo, T., Huang, G. and Cipriani, G. 2004. Microsatellites isolated in almond from an AC-repeat enriched library. Mol. Ecol. Notes 4:459–461.

Tufts, W.P. 1919. Almond pollination. Calif. Agric. Sta. Bull. 306.

Ushijima, K., Sassa, H., Tao, R., Yamane, H., Dandekar, A.M., Gradziel, T.M. and Hirano, H. 1998. Cloning and characterization of cDNAs encoding *S*-RNases from almond (*Prunus dulcis*): primary structural features and sequence diversity of the *S*-RNases in Rosaceae. Mol. Gen. Genet. 260:261–268.

Ushijima, K., Sassa, H., Kusaba, M., Tao, R., Tamura, M., Gradziel, T.M., Dandekar, A.M. and Hirano, H. 2001. Characterization of the *S*-locus region of almond (*Prunus dulcis*): analysis of a somaclonal mutant and a cosmid conting for an *S* haplotype. Genetics 158:379–386.

Ushijima, K., Sassa, H., Dandekar, A.M., Gradziel, T.M., Tao, R. and Hirano, H. 2003. Structural and transcriptional analysis of the self-incompatibility locus of almond: identification of a pollen-expressed F-box gene with haplotype-specific polymorphism. Plant Cell 15:771–781.

Vargas, F.J. and Romero, M. 1994. 'Masbovera', 'Glorieta' and 'Francolí', three new almond varieties from IRTA. Acta Hort. 373:75–82.

Vargas, F.J., Romero, M.A., Clavé, J., and Batlle, I. 2005. Early selection in the almond breeding programme at IRTA Mas Bové. Options Méditerr. Ser. A 63: 17–21.

Vargas, F., Romero, M., Clavé, J., Vergés, J., Santos, J. and Batlle, I. 2008. 'Vayro', 'Marinada', 'Constantí', and 'Tarraco' almonds. HortScience 43:535–537.

Vezvaei, A., Hancock, T.W., Giles, L.C., Clarke, G.R. and Jackson, J.F. 1995. Inheritance and linkage of isozyme loci in almond. Theor. Appl. Genet. 91:432–438.

Viruel, M.A., Messeguer, R., Vicente, M.C. de, Garcia-Mas, J., Puigdomènech, P., Vargas, F. and Arús, P. 1995. A linkage map with RFLP and isozyme markers for almond. Theor. Appl. Genet. 91:964–971.

Watkins, R. 1979. Cherry, plum, peach, apricot and almond. *Prunus* spp., p. 242–247. In: N.W. Simmonds (ed.), Evolution of crop plants. Logman, London, UK.

Wu, S.B., Wirthensohn, M., Hunt, P., Gibson, J.P. and Sedgley, M. 2008. High resolution melting analysis of almond SNPs derived from ESTs. Theor. Appl. Genet. 118:1–14.

Wu, S.B., Franks, T.K., Hunt, P., Wirthensohn, M., Gibson, J.P. and Sedgley, M. 2010. Discrimination of SNP genotypes associated with complex haplotypes by high resolution melting analysis in almond: implications for improved marker efficiencies. Mol. Breed. 35:351–357.

Xie, H., Sui, Y., Chang, F.Q., Xu, Y. and Ma, R.C. 2006. SSR allelic variation in almond (*Prunus dulcis* Mill.). Theor. Appl. Genet. 112:366–372.

Xiloyannis, C., Dichio, B., Tuzio, A.C., Kleinhentz, M., Salesses, G., Gómez-Aparisi, J., Rubio-Cabetas, M.J. and Esmenjaud, D. 2007. Characterization and selection of *Prunus* rootstocks resistant to abiotic stresses: waterlogging, drought and iron chlorosis. Acta Hort. 732: 247–251.

Zohary, D. and and Hopf, M. 1993. Domestication of plants in the old world. Clarendon Press, Oxford, UK.

Chapter 19
Chestnut

Santiago Pereira-Lorenzo, Antonio Ballester, Elena Corredoira, Ana M. Vieitez, Sandra Agnanostakis, Rita Costa, Giancarlo Bounous, Roberto Botta, Gabriele L. Beccaro, Thomas L. Kubisiak, Marco Conedera, Patrik Krebs, Toshiya Yamamoto, Yutaka Sawamura, Norio Takada, José Gomes-Laranjo, and Ana M. Ramos-Cabrer

Abstract The genus *Castanea*, chestnuts and chinkapins, belongs to the family *Fagaceae*, which includes other important timber producing genera such as *Quercus* and *Fagus*. The genus *Castanea* is divided into three geographically delimited sections with at least seven consistently recognized interfertile species: 4 species in Asia (*C. mollissima*, *C. henryi*, *C. seguinii*, and *C. crenata*), two or more species in North America (*C. dentata*, *C. ozarkensis*, and *C. pumila*) and one in Europe and Turkey (*C. sativa*). The two most important diseases of chestnut are ink disease (*Phytophthora*) and chestnut blight (*Cryphonectria*). Resistance to these is the major objective for rootstock breeding in Europe and scion breeding in North America. In both cases, the source of resistance was Asian species. European breeding programs developed resistant hybrid rootstocks, which are propagated by stooling, cuttings, or in vitro culture. A major pest of chestnut is the gall wasp *Dryocosmus kuriphilus* whose control is based on the spread of parasitoids but also on the selection of resistant cultivars. For nut production, the most important breeding objectives include the following: good horticultural traits, product quality, suitability to storage and processing, and ease of peeling. For timber, important characters include wood quality, rapid growth, and nonchecking of wood (ring-shake). Molecular maps have been developed, which has expanded the genetic knowledge of the chestnut. An efficient genetic transformation protocol for *C. sativa* through the coculture of somatic embryos with different strains of *Agrobacterium tumefaciens* has been described.

S. Pereira-Lorenzo (✉) • A.M. Ramos-Cabrer
Departamento de Producción Vegetal, Universidad de Santiago de Compostela, Campus de Lugo, 27002, Lugo, Spain
e-mail: santiago.pereira.lorenzo@usc.es; ana.ramos@usc.es

A. Ballester • E. Corredoira • A.M. Vieitez
Instituto de Investigaciones Agrobiológicas de Galicia, CSIC. Apartado 122, 15080 Santiago de Compostela, Spain
e-mail: aballester@iiag.csic.es; elenac@iiag.csic.es; amvieitez@iiag.csic.es

M.L. Badenes and D.H. Byrne (eds.), *Fruit Breeding*, Handbook of Plant Breeding 8, DOI 10.1007/978-1-4419-0763-9_19,

Keywords *Castanea* • Taxonomy • Cultivars • Rootstocks • Genetic resources • Breeding • Genomics • Transgenic • Ecophysiology

1 Introduction

In the world there are about 349,000 ha of orchards which produce 1,140,332 mt of chestnuts (mean value for 2000–2007). Chestnut production in Asia is almost 8 times that of Europe, with China being the dominant producer with an average production for 2000–2007 of 803,213 mt. China is reported to have about 130,000 ha in chestnut orchards, but Liu and Zhou (1999) estimated that the figure should be 670,000 ha, five times the FAO figure. Japan, Turkey, and Korea produce about 25,000, 50,000, and 78,000 mt respectively (FAO 2009, faostat.fao.org).

A survey among European countries estimates that there are 2.22 million ha of chestnut dominated forest (Conedera et al. 2004a). The main chestnut production is in Italy (24,000 ha) and Portugal (30,000 ha), with about 51,000 and 29,000 mt, respectively (FAO 2009, faostat.fao.org). Although the FAO accounts only 7,000 ha in Spain, Spanish statistics estimate 45,000 ha and 60,000 mt in 2006 (http://www.mma.es).

Chestnuts are multipurpose trees valued for nuts, timber, tannins, and landscape. They were historically distributed only throughout the northern hemisphere, but have more recently been introduced into Chile, Argentina, Australia, and New Zealand.

S. Agnanostakis
The Connecticut Agricultural Experiment Station, Box 1106, New Haven, CT 06504, USA
e-mail: Sandra.Anagnostakis@ct.gov

R. Costa
Instituto Nacional de Recursos Biológicos I.P. Quinta do Marquês, Av. da República, 2780-159 Oeiras, Portugal
e-mail: rita.lcosta@inrb.pt

G. Bounous
Chairman ISHS Group on Chestnut, FAO/CIHEAM Liaison Officer Subnetwork on Chestnut, Department of Colture Arboree, University of Torino, Turin, Italy
e-mail: giancarlo.bounous@unito.it

R. Botta • G.L. Beccaro
Department of Colture Arboree, University of Torino, V. Leonardo da Vinci 44, Grugliasco (TO), Italy
e-mail: roberto.botta@unito.it; gabriele.beccaro@unito.it

T.L. Kubisiak
USDA Forest Service, Southern Research Station, Southern Institute of Forest Genetics, 23332 Highway 67, Saucier, MS 39574-9344, USA
e-mail: tkubisiak@fs.fed.us

M. Conedera • P. Krebs
WSL, Swiss Federal Institute for Forest, Snow and Landscape Research, Bellinzona, Switzerland
e-mail: marco.conedera@wsl.ch; patrik.krebs@wsl.ch

T. Yamamoto • Y. Sawamura • N. Takada
National Institute of Fruit Tree Science, Fujimoto 2-1, Tsukuba, Ibaraki 305-8605, Japan
e-mail: toshiya@affrc.go.jp; ysawa@affrc.go.jp; ntakada@affrc.go.jp

J. Gomes-Laranjo
CITAB, University of Trás-os-Montes and Alto Douro, Apt 1013, 5001-801 Vila Real, Portugal
e-mail: jlaranjo@utad.pt

1.1 Taxonomy and Distribution

The genus *Castanea* ($2n=24$), chestnuts and chinkapins, belongs to the family *Fagaceae*, which includes other important timber producing genera such as *Quercus* (oaks) and *Fagus* (beech) (Castroviejo et al. 1990). It is supported as a monophyletic clade most closely related to the genus *Castanopsis* (Manos et al. 2001).

The genus *Castanea* has been divided into three sections with at least seven consistently recognized species (Camus 1929; Johnson 1988). The section *Eucastanon* consists of five species characterized by three nuts per cupule; *C. crenata* Sieb. & Zucc. from Japan, *C. mollissima* Blume and *C. seguinii* Dode from China, *C. sativa* Miller from Europe, and *C. dentata* Borkhausen from North America. The section *Balanocastanon* is found exclusively in southeastern North America and characterized by a single nut per cupule. Although four to six imprecisely defined species are generally cited for this section (Graves 1961; Elias 1971; Jaynes 1972; Little 1979), more recently it has been proposed that it should be reduced to a single species, *C. pumila* Miller, with two varieties: var *pumila* and var. *ozarkensis* (Hardin and Johnson 1985; Johnson 1988). The section *Hypocastanon* consisting of *C. henryi* Rehder and Wilson from China is also characterized by a single nut per cupule.

UPGMA analysis of isozyme-based genetic distance estimates (Dane et al. 2003) and phylogenetic analysis based on cpDNA sequence data (Lang et al. 2006) suggest that *Castanea* species are geographically structured. This is inconsistent with the current phylogeny based on cupule characteristics. The section *Eucastanon* appears to be paraphyletic with the differentiation among species being best explained by their current geographical ranges. *C. crenata* appears to be the most basal taxa and sister to the remainder of the genus. The three Chinese species [*C. mollissima* and *C. seguinii* (*Eucastanon*) and *C. henryi* (*Hypocastanon*)] are supported as a single monophyletic clade and sister to a group containing the North American and European species. There appears to be weak but consistent support for a sister-group relationship between the North American and European species.

With cpDNA data, the chestnut appeared to expand westward from the extant *Castanea* species originating in eastern Asia, followed by intercontinental dispersion and divergence between the Chinese and European/North American species during the middle Eocene, followed by subsequent divergence between the European and North American species during the late Eocene (Lang et al. 2007). Morphological evolution of one nut per bur in the genus may have occurred independently on two continents.

1.2 Where Grown

There are three main areas where native chestnuts are found (Table 19.1, Fig. 19.1).

1. In Asia, mainly in China, where *C. mollissima* Blume, *C. henryi* (Skan) Rehd. & E.H. Wils., and *C. seguinii* Dode are found in wild and cultivated stands. In the

Table 19.1 Distribution and use of chestnut species (modified from Bounous and Torello Marinoni 2005)

Origin	Section	Species	Common name	Planted	Prevalent use
Europe	Eucastanon	*C. sativa* Mill.	European or sweet chestnut	Europe, Asia Minor, North Africa	Nut, timber
Asia	Eucastanon	*C. crenata* Seib & Zucc.	Japanese chestnut	Japan, Korea	Nut
		C. mollissima Blume	Chinese chestnut	China	Nut
		C. seguinii Dode		China	Firewood
		C. davidii Dode		China	Firewood
	Hypocastanon	*C. henryi* (Skan) Rehd. & E.H. Wils.	Willow leaf or pearl chestnut	China	Timber
USA	Eucastanon	*C. dentata* (Marsh.) Borkh.	American chestnut	North America	Timber
	Balanocastanon	*C. pumila* (L.) Mill. var. *pumila*	Allegheny chinkapin	Southeast USA	Nut
		C. pumila (L.) Mill. var. ozarkensis	Ozark chinkapin	USA (Arkansas, Missouri, Oklahoma)	Timber
		C. floridana Ashe (Sarg.)	Florida chinkapin	Southeast USA	Ornamental
		C. ashei (Sudw.) Ashe	Ashe chinkapin	Southeast USA	Ornamental
		C. alnifolia Nutt.	Creeping chinkapin	Southern USA (Alabama, Florida)	___
		C. paucispina Ashe		Southern USA (Texas, Louisiana)	___

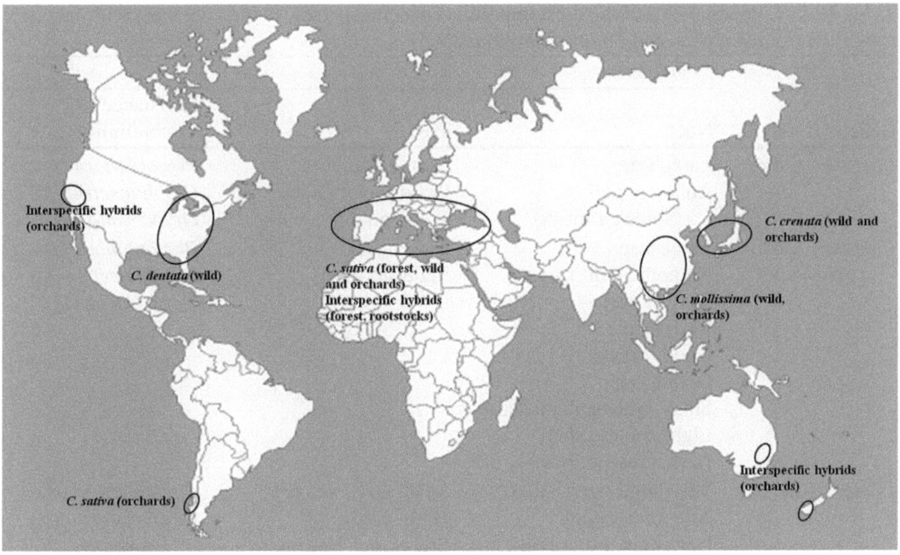

Fig. 19.1 Main areas where chestnut (*Castanea* spp.) is grown

Korean Peninsula, Japan, and the temperate region of east Asia, the Japanese chestnut (*Castanea crenata* Sieb. & Zucc.) is naturally distributed and cultivated.

2. In Europe and Turkey where *C. sativa* is predominant.
3. In North America where *C. dentata* (Marsh.) Borkh. and *C. pumila* (L.) Mill. var. *pumila* were once widespread throughout the Appalachian Mountain Range, and where *C. ozarkensis* Ashe and *C. alnifolia* Nutt occupied small niches on the Ozark Plateau and in northern Florida.

The main species cultivated for fruit are *C. mollissima*, *C. sativa*, and *C. crenata* due to their large nut size (Table 19.2). Marrone types (*C. sativa*) are considered the most valuable for nut production. *C. sativa* and *C. dentata* are the most vigorous species, and they are also used for timber production. Interspecific hybrids which have emerged from disease resistance work are used for nut, timber and as rootstock.

1.3 Limits to Adaptation

In Europe, *C. sativa* is commonly found between 400 and 1,000 m above sea level depending on the latitude. The lowest elevations are recommended for the highest latitudes and vice versa (Bounous 2002). The early leafing of hybrids (mid March) restricts their use due to spring frosts to altitudes lower than 500 m.

The minimum rainfall for chestnut is 800 mm. Plants from this species are moderately thermophilic and well adapted to ecosystems with a year mean temperature ranging between 8 and 15°C and monthly mean temperatures during 6 months over 10°C.

Characterized as a mesophilic species, chestnut tree growth actually shows some limitations to high temperatures. European chestnut trees do not thrive in soil rich

Table 19.2 Characteristics of the most important chestnut species (in bold the most relevant ones) (modified from Bounous and Torello Marinoni 2005)

	Characters		
Genetic resources	Nut	Tree	Resistance (R)/ susceptibility (s)
Castanea sativa	Large size Adherent pellicle (some cultivars)	Strong branches Good growth habit Wood quality	*Phytophthora* (s) *Cryphonectria* (s) *Dryocosmus* (s)
Castanea sativa (marrone)	Large size No pellicle intrusion Easy to peel Sweet flavor Good texture Ovoid shape Small, rectangular hylar scar Light colored shell Dark, close stripes	Lower yield Male sterility	*Phytophthora* (s) *Cryphonectria* (s) *Dryocosmus* (s)
Castanea crenata	Very large size (≥30 g) Adherent pellicle Not sweet, astringent	Small size (≤15 m) High yield Precocious bearing Early ripening	*Phytophthora* (R) *Cryphonectria* (R) (moderate) *Dryocosmus* (s) (high) Spring frost (s)
Castanea mollissima	Weight (10–30 g) Sweetness, flavor, protein content No pellicle intrusion Thin pellicle Easily removed pellicle High variable size	Medium size (≤20 m) Semiupright habit Early ripening (variable) Precocious (variable) Two crops/year (in subtropical areas) (variable) Good pollinizer	*Phytophthora* (R) *Cryphonectria* (R) (variable) *Dryocosmus* (s)
Castanea dentata	Very sweet Nonastringent Easy to peel Very small (300 nuts/kg)	Fast, straight growth with strong central leader Self-pruning Well coppiced	*Cryphonectria* (s) (high) Frost or cold (−35°C) (R)
Castanea seguinii	Small size Very prolonged blooming and ripening period Very precocious	Small, medium size Precocious flowering Ever bearing Two crops/year (some clones) Chain of 10–20 burs (some clones)	*Cryphonectria* (R) *Dryocosmus* (s)
Castanea pumila	Very small Single nut burs Sweet, flavorful Very precocious	Moderate size Stoloniferous clones Prolific suckering ability Soft spined burs Suitable for warm climate	*Cryphonectria* (R) (partial) Warmer temperate climates (R) Quickly replacing blighted stems
Castanea henryi	Single nut burs Very small	Fast growth Straight trunk Good wood Suitable for warm temperate or tropical climates	*Cryphonectria* (R)

in active calcium, basic pH or with poor drainage. These are commonly grown in poor sandy to loamy soil on slopes but also in volcanic islands (Sicily, Canary, Madeira, and Azores). Deep soil and a deep root system are important to help trees maintain their water potential during the dry hot summer months (June to September) (Martins et al. 2005).

Chestnut is a dim-light species with better adaptation to shade and cold north-facing slopes, rather than south-facing ones (Gomes-Laranjo et al. 2007). The latter have higher mean temperatures, earlier leafing and flowering, and consequently a greater frost risk relative to north-facing ones.

2 Origin and Domestication of Scion Cultivars

2.1 Origin of the Cultivars

In Europe, the most probable natural range of the native chestnut species *C. sativa*, is delimited by six macroregions (Fig. 19.2): the Transcaucasian region, north-western Anatolia, the hinterland of the Tyrrhenian coast from Liguria to Lazio along the Apennine range, the region around Lago di Monticchio (Monte Vulture) in southern Italy, the Cantabrian coast on the Iberian Peninsula, and probably also the Greek Peninsula (Peloponnese and Thessaly), and north-eastern Italy (Colli Euganei, Monti Berici, Emilia-Romagna) (Krebs et al. 2004).

The first evidence of active chestnut cultivation dates back to the third millennium before Christ in the eastern part of European range (Anatolian Peninsula, Northeastern Greece, and Southeastern Bulgaria). From there, Greeks first and then the Romans diffused the chestnut to the west (Pitte 1985, 1986; Conedera et al. 2004b) such that in the Middle Ages the cultivation of chestnut for timber production and as a staple food was a widespread component of the traditional farming system in most.

Mediterranean countries and southern parts of Central Europe (Conedera and Krebs 2008).

A large-scale molecular study based on simple sequence repeat (SSR) loci of the diversification process in chestnut cultivars from Portugal and Spain, from the northern Iberian Peninsula to the Canary Islands and the Azores, showed geographical and genetic structure in ten main cultivar groups (Pereira-Lorenzo et al. 2011). Cultivar origin and the diversification process was a combination of clonal propagation of selected seedlings, hybridization, and mutations. Mean value of clonality owing to grafting was 33%, mutations accounted for 6%, with hybridization being the main diversification process that can explain the great diversity found. Seedlings and graft sticks were transported in the colonization process, sometimes more than 3,000 km if we consider the Azores and the Canary Islands.

Although the phylogenetic map of the chestnut in Europe is not fully understood yet (Fineschi et al. 2000), the greater genetic similarity of chestnuts from the western Anatolia Peninsula to Italian and French populations than to the chestnut groves

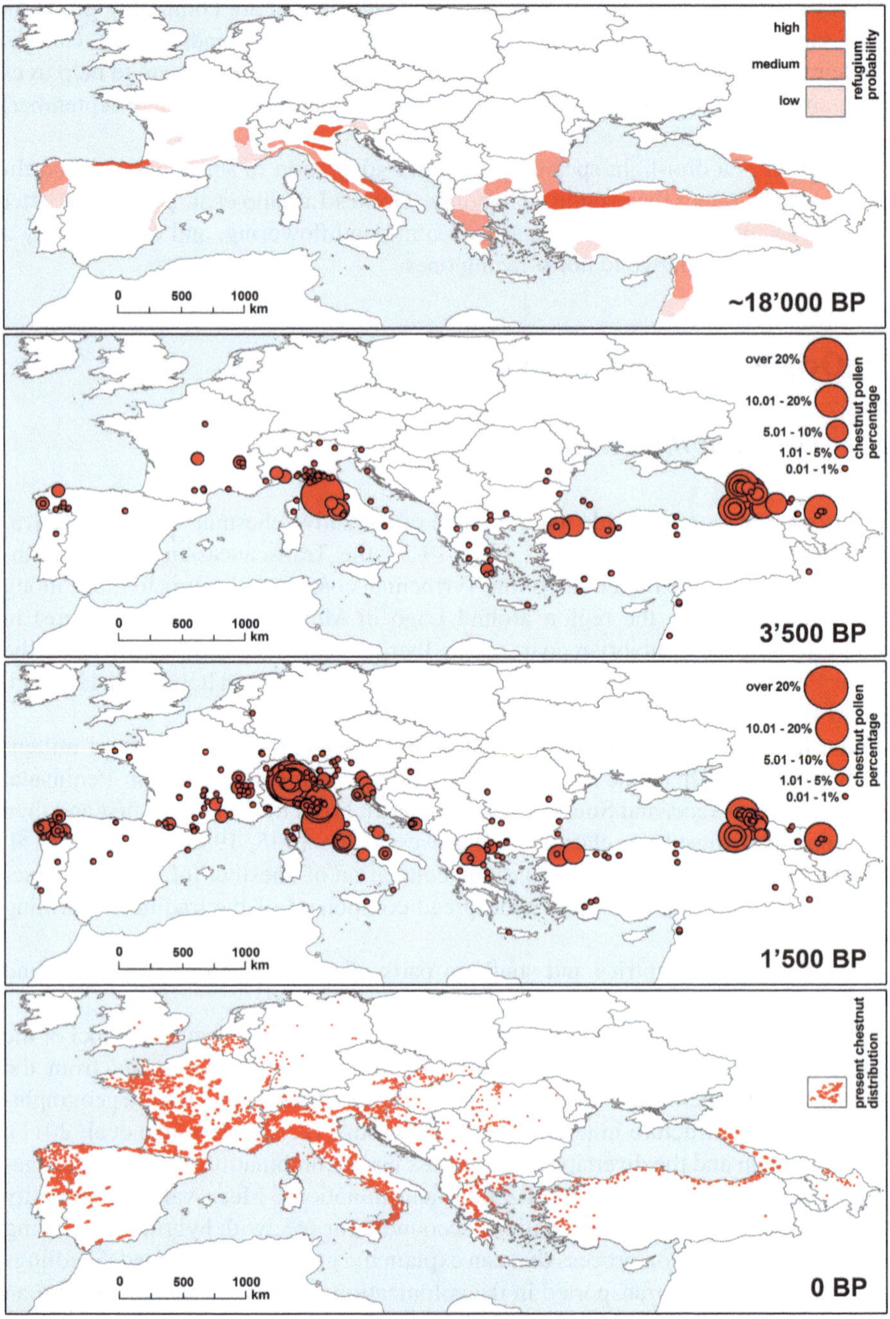

Fig. 19.2 Distribution of *Castanea sativa* in Europe

of eastern Anatolia facing the Black Sea (Villani et al. 1999) suggests that the flow of chestnut-related elements between Ancient Greece and the Italian world was not only cultural (Conedera et al. 2004b). It is probable that Greek colonists introduced chestnut cultivation in the Italian Peninsula from the main chestnut zones of Ancient Greece as they did for grapes (Dion 1977).

2.2 *Brief History of Breeding*

First hybridizations were made in 1884 in the USA (Van Fleet 1920), in 1926 in Spain (Gallastegui 1926) and in 1929 in Japan (Yamamoto et al. personal communication) (Table 19.3). Hybrid clones have been released in Japan, the USA, and Europe (France, Spain, and Portugal; Pereira-Lorenzo et al. 2010) and some of them are commercialized for nut, timber, and rootstocks.

2.3 *Varietal Groups*

The International Society for Horticultural Science maintains a registry of chestnut cultivars with names and their characteristics (Connecticut Agricultural Experiment Station, The USA, http://www.ct.gov/caes).

Most of the best cultivars recommended in France for new plantations are *C. sativa* cultivars, which produce marron nuts between 12 and 18 g (Table 19.4). French researchers have focused in the breeding of interspecific hybrids resistant to ink diseases suitable for fresh market use (Bergougnoux et al. 1978; Breisch 1995).

The most important cultivars in Spain are 'Parede' and 'Longal'. They have been propagated profusely during the last 300 years (Pereira-Lorenzo et al. 2001a, 2006a). In Northern Spain, the most popular cultivars are 'Amarelante', 'Negral', 'Famosa', 'Longal', 'Ventura', 'Garrida', 'Loura', and 'Luguesa' (Table 19.5). In Extremadura, Central Spain, 'Injerta' and 'Verata' are cultivated, and in Southern Spain, 'Planta Alajar', 'Temprana', and 'Pilonga' are the best. In the Canary Islands, the most widespread cultivars are 'Mulata' in Tenerife and 'Jabuda' in La Palma (Pereira-Lorenzo et al. 2001b, c). In Spain, hybrids are considered an alternative in Atlantic areas where they show very good adaptability (Pereira-Lorenzo and Fernandez-Lopez 2001) with sufficiently large and monoembryonic nuts which are harvested before the 20th September.

In Portugal, 'Longal', one of the most ancestral varieties, is widely spread over all the chestnut regions (Trás-os-Montes, located in the northeast) and has been promoted as the best cultivar for industry. 'Judía' and 'Martaínha', due to their larger nut size, are usually preferred for the fresh market. 'Judía frequently presents polyembryony (Table 19.6). Some cultivars such as 'Longal', 'Amarelal', and 'Verdeal' are found in North Spain and Portugal.

Table 19.3 Breeding summary on chestnut

Years	Country	Breeding activity	Achievements
USA			
1894	The USA	Crosses European × American chestnut	Cultivar Paragon
1895	The USA	First cross Japanese × American chestnut	Cultivar Daniel Boone
1900–1921	The USA	Crosses between *C. pumila* and European and Japanese cultivars	S-8 row plantation
1912	The USA	Wild seeds of *Castanea* species collected in Tientsin, China	Collection
1937	The USA	Van Fleets S-8 trees were crossed in Connecticut with a forest-type Japanese	Cultivar Essate Jap
1937	The USA	Making and testing hybrids for their resistance to chestnut blight	Testing fitness throughout the US
1946	The USA	Hybrids BC1 with timber form, good blight resistance, and acceptable nuts.	Cultivars Clapper and Graves
1962–1983	The USA	Making interspecific hybrids, looking for the ideal progeny that could be propagated clonally	Over 10,000 hybrid chestnut seedlings planted in the Lesesne State Forest in Virginia
1983 to present	The USA	Hybrids from resistant × susceptible trees, backcrossed again to the susceptible parent species	True to type offspring for reforestation with chestnut.
2009	The USA	First crosses between ozark chinquapins with Chinese chinquapins and Japanese chestnuts to produce blight resistant trees, and these will be back-crossed to C. ozarkensis	Trees that will have a better chance of surviving in their habitat
2009	The USA	Selection of some complex hybrids with the Chinese shrub *C. seguinii* in their background, which are compact dwarfs.	Dwarfing rootstocks (to get early flowers on short trees)
2006	The USA	After more than 20 years, BC3-F2 generation has been obtained	93% of seedlings showed morphological characteristics of American chestnut incorporating 100% Chinese resistance to blight
Europe			
End of the 19th century	France	Asian species resistant to *Phytophthora* were introduced in the Atlantic area, like in south-western France	First introductions to fight ink disease
1929	France	Comparative description between main species	Botanical classification and speciess description
1917–1940	Spain	Seedlings from Asian species *C. crenata* and *C. mollissima* were introduced between 1917 and 1940	First plantations with Asian species but refused by growers due to the lack of timber quality and nut characteristics

1926	Spain	Hybridization program between *C. crenata* and *C. sativa*	Spanish hybrids resistant to ink disease
1947	Portugal	Interspecific hybridizations were initiated in 1947 by Bernardino Barros Gomes	Portuguese hybrids resistant to ink disease
1952	France	began a breeding program to select the interspecific hybrids which offered characteristics more similar to the European species and species	French hybrids resistant to ink disease
1955	Italy	Established the base descriptors for chestnuts that are used today	Scheda Castanografica sul Castagno
1970	Italy	Breeding Euro-Japanese hybrids	New Euro-Japanese hybrids Primato and Lusenta
1990 to present	Spain	Selection and testing local cultivars hybrids for rootstocks, forest, and nut production	Selected cultivars available
2006 to present	Portugal	Hybridization program between *C. mollissima*, *C. crenata* and *C. sativa* to provide new hybrids resistant to ink and blight diseases	New hybrids
2004 to present	Italy	Testing European cultivars and hybrids to gall wasp. Breeding program for gall wasp resistance/tolerance.	Selection of resistant genotypes in *C. sativa*.
Asia			
1929–1947	Japan	Japanese chestnut was crossed with a Chinese chestnut in an attempt to introgress an easy-peeling kernel trait from Chinese chestnut into Japanese chestnut.	These efforts were unsuccessful
1959–1981	Japan	Screening of germplasms and selective breeding eventually led to the development of several cultivars resistant to the chestnut gall wasp.	Hybrids Norin No.1 to Norin No.5.
1981–1992	Japan	Previous resistance to wall gasp did not last	Shiho and Kunimi were developed as more highly resistant
2007	Japan	Cross combination of [(Moriwase × KairyouToyotamawase) × Kunimi] and Tanzawa	Easy to peel hybrid Porotan, screened removing chestnut pellicle after deep frying in cooking oil (HOP method, High-temperature Oil Peeling method)

Van Fleet (1920); Taylor and Gould (1914); Nienstaedt (1948); Diller and Clapper (1969); Burnham (1988); Diskin et al. (2006); Lafitte (1946); Schad et al. (1952); Camus (1929); Elorrieta (1949); Gallastegui (1926); Urquijo (1944, 1957); Gomes Guerreiro (1948, 1957); Vieira Natividade (1947); Breviglieri (1955); Bounous (2002)

Table 19.4 Recommended French cultivars for new orchards (modified from Breisch 1995)

Country/cultivar	Origin	Splitting of pericarp (%)	Weight (g)
Bouche de Betizac[F]	Bouche-Rouge (*C. sativa*) × CA04 (*C. crenata*)	8–25	15–>18
Maridonne[E]	Sardonne (*C. sativa*) × CA04 (*C. crenata*)	5	15–18
Marigoule[F]	*C. crenata* × *C. sativa*	5	15–>18
Precoce migoule[F]	*C. crenata* × *C. sativa*	20–40	15–18
Bournette[F]	*C. crenata* × *C. sativa*	5	12–18
Iphara[F]	*C. crenata*	5	15–>18
CA75[P]	*C. mollissima*	5	10–12
Merle,[F,R] Aguyane,[F] Dorée de Lyon,[F] Laguépie,[F] Précoce Ronde des Vans,[F] Sardonne,[F] Comballe,[F,I] Insidina,[F,I] Marron Comballe,[F,I] Imperiale[P]	*C. sativa*	≥12	12–18
Arizinca,[F,I] Bouche Rouge,[F,I] Belle Epine,[F,I,P] Marron de Goujounac,[F,P] Montagne,[F,P] Tricciuda,[I] Verdale (Delsol) [I,P], Marron de Chevanceaux,[M] Pellegrine[M]	*C. sativa*	<12	12–18

E experimental, *F* fresh, *R* rootstock, *I* industry, *P* pollinizer, *M* natural 'marron'

Table 19.5 Main quality characteristics of the most important Spanish chestnut cultivars of *C. sativa* (modified from Pereira-Lorenzo et al. 2006a, 2007)

Region	Cultivar	Nuts/kg	Splitting of pericarp (%)	Central nut weight	Lateral nut weight (g)
Andalucía	Comisaria,[M] Dieguina,[M] Helechal,[M] Pilonga,[M] Planta Alajar,[M] Temprana,[M] Tomasa,[M] Vazqueña[M]	70–87	4–14	10–22	14–25
Asturias	Chamberga/Valduna,[F] Grúa [F]	90	2–7	11	12–13
Castilla-León	Injerta,[I] Negral[F]	80–140	7	7–12	8–13
Extremadura	Injerta,[M] Verata[M]	78–90	0–6	11	12–14
Galicia	Amarelante,[M] Famosa,[M] Garrida,[M] Inxerta,[M] Loura,[M] Longal[I], Luguesa,[M] Negral,[I] Parede,[I] Presa,[F] Rapada,[I] Ventura[M]	74–130	0	8–14	9–14
Canary Islands	Arafero,[M] Castagrande,[F] Picudo,[M] Polegre[M]	72–100	3–6	13–15	13–15

F fresh, *R* rootstock, *I* industry, *P* pollinizer, *M* natural 'marron'

Table 19.6 Most important Portuguese cultivars of *C. sativa* (modified from Costa et al. 2008)

Region	Cultivar	Caliber (nuts/kg)	Hilum area (shape)	Fruit/length (shape)	Poly-embryony (%)	Flowering date[a] Male flower	Female flower
Beira Litoral	Martaínha	69–95	3.3	1.06	15	2	1
	Colarinha	84–96	2.51	1.27		2	2
	Verdeal	62–74	4.33	1.06	3.5	5	5
	Longal	67–87	2.7	1.15	0.1	5	2
	Negral	77		1.03	1.6	5	5
	Demanda	97		1.1	3.7		
	Passa	75.4		0.92	3.7		
Minho	Amarelal	68–76	4.5	0.98	2.5	5	5
	Misericórdia	153		1.28	0.0		
Tras-os-Montes	Lamela	71		1.01	0.4	5	5
	Zeive	73		0.98		5	2
	Redonda	80		0.97	0.0	7	5
	Judía	49–69	3.98	1.01	4.7	5	2
	Lada	78		1.09	0.0	5	5
	Longal	104		1.24	0.1	5	2
	Trigueira	111		1.04	2.5		
	Reborda	76–92	4.33	1.03	2.0	1	1
	Boaventura	82		1.09	0.8		
	Bebim	85		0.95	1.0		
	Benfeita	103		1.15	0.4		
	Aveleira	82–98	3	1.01	0.2	2	2
	Negral	100		1.06	4.6		
	Sousa	95		1.05	1.7		
	Côta	102		1.08	0.0	5	5
Marvao	Bária	111		1.16	2.0	5	5
	Enxerta	130		1.21	0.0		

[a]1 – very early, 2 – early, 5 – medium, 7 – late, 9 – very late

In California, the principal cultivar is 'Colossal' (*C. sativa* × *C. crenata*) grown with 'Silverleaf' (*C. sativa*), 'Nevada' (*C. sativa* × *C. crenata*), 'Eurobella' (*C. sativa* × *C. crenata*), or 'Colossal' seedlings as pollinizers (Vossen 2000).

In Italy, where the environmental conditions are favorable, the best cultivars are the *marron* type ('Chiusa Pesio', 'Luserna', 'Val Susa', 'Castel del Rio', 'Marradi', 'Fiorentino' from Italy) because of their large size which is preferred for the fresh and candy market (*marrons glacés*) (Table 19.7). Early maturing cultivars suitable for the premium market are 'Tempurive', 'Castagne della Madonna', and 'Precoce di Roccamonfina', while 'Garrone Rosso', 'Garrone Nero', 'Gioviasca', 'Bionda di Mercogliano', and 'Montemarano' produce large chestnuts for fresh market and candying. Many cultivars with small but very sweet and easy to peel nuts suitable for drying and flour production have been selected through the centuries in Italy: 'Frattona', 'Gabbiana', 'Siria', 'Pastinese', 'Carpinese' (Bounous 2002). In the 1980s, the University of Torino, Department of Arboriculture, released two Italian Euro-Japanese hybrids: 'Primato' and 'Lusenta'.

Table 19.7 Main Italian *C. sativa* cultivars

Origin	Cultivar name (Synonymies in brackets)	Nuts/kg	Pericarp splitting (%)
Campania	Castagna di Montella,[a] Palummina[a]	64–100	0–2
Piemonte	Marrone di Chiusa di Pesio,[b] Marrone di Val di Susa,[b] Marrone di Luserna,[b] Bracalla,[b] Castagna della Madonna[b]	50–80	0–2
Toscana	Marrone di Caprese Michelangelo,[b] Castagna Marzapanara,[b] Castagna Pastinese,[b] Castagna Pistolese[b]	60–80	0–2
Calabria	Curcia,[c] Inserta,[c] Nzerta, Ricciola[bc]	80–140	0
Piemonte	Frattona,[d] Gabbiana,[d] Garrone Nero,[b] Garrone Rosso,[b] Gentile,[b] Gioviasca,[b] Lusenta,[b] Marrubia,[b] Neirana,[b] Pelosa Grossa,[b] Pelosa Piccola,[b] Rossastra,[bd] Siria,[b] Solenga,[b] Spinalunga,[bd] Temporiva,[b] Verdesa[b]	50–154	0–6
Toscana	Marrone di Citta di Castello,[b] Marrone di Gavignano,[b] Marrone di Marradi,[b] Marrone di Montemarano,[b] Marrone Badia Coltibuono,[b] Marrone Borra Montesevero,[b] Marrone di Forlì,[b] Marrone di Monfenera,[b] Marrone Fiorentino (C-asentinese, Toscano)[ef], Marrone di Segusino,[b] Marrone di Stia,[b] Marrone. del Monte Amiata,[b] Marrone dellIsola dElba,[b] Marrone di Caprarola[b]	50–80	0–2
Emilia-Romagna	Marrone Castel del Rio,[a] Marrone di Castiglione dei Pepoli,[a] Marrone di Montepastore[a]	57–100	0–6
Veneto	Marrone di Combai[b]	50–80	0–2

[a]Bassi and Sbaragli (1984)
[b]Bounous (2002)
[c]Antonaroli et al. (1984)
[d]Bounous et al. (1989)
[e]Breviglieri (1955)
[f]14–22 g per nut, all uses

In southern Switzerland, only two cultivars are ubiquitous and widespread throughout the chestnut area: the 'Lüina', a tree producing small-sized very sweet fruits for drying, and the 'Verdesa', a late ripening cultivar that keeps the fruit inside the bur allowing the chestnuts to be conserved over months for fresh consumption. It is not surprising, given the former staple food function of the chestnut in this mountain region in southern Switzerland, that the most suitable cultivars have excellent fresh and dry storage traits (Conedera et al. 1993).

Liu and Zhou (1999) identified the best six Chinese cultivars (*C. mollissima*) out of 28 examined. These include 'Chu shu hong', 'Jiu jia zhong', 'Duan zha', 'Qin zha', 'Jiao zha', and 'Jian ding you li'. All of them produce nuts over 10 g (Jiao zha over 20 g), are easy to peel, and have excellent kernel quality for both cooking and roasting.

Table 19.8 Japanese cultivars

Origin	Cultivar name	Weight (g)	Polyembryonic nuts (%)	Pericarp splitting (%)
C.crenata	Tsukuba,[I] Tanzawa,[E] Ginyose,[I] Ishizuchi,[L] Kunmi,[E] Ganne[L], Toyotamwase,[VE] Moriwase,[VE] Ti-7,[E] Ibuki,[E] Ginrei, [E] Otomune,[I] Tajiriginyose,[L] Akatyuu,[I] Arima[I]	17.8–27.5	3.4–21.4	6.7–22.2
New Japanese cultivars, *C.crenata*	Shiho, Syuho, Porotan	23–30	3–9.7	4.7–8

Ripening time: *I*, intermediate; *E* early; *L* late; *VE* very early

In Japan, most commercial cultivars belong to Japanese chestnut (*Castanea crenata* Sieb. & Zucc.), with only a few cultivars, such as 'Riheiguri', being hybrids between Japanese and Chinese chestnuts. In 2004, the most widely cultivated chestnut cultivars in Japan were 'Tsukuba' (23.3%), 'Tanzawa' (12.0%), 'Ginyose' (11.4%), 'Ishizuchi' (4.1%), 'Riheiguri' (4.0%), 'Kunimi' (3.7%), and 'Ganne' (2.9%) (Table 19.8). 'Tsukuba' ('Ganne' × 'Hayadama'), the leading cultivar grown in Japan, was released as 'Norin No.3' in 1959 by NFIFTS (National Institute of Fruit Tree Science, formerly Fruit Tree Research Station). Although damage by gall wasps is serious, it is still a leading cultivar because of its high productivity. 'Tanzawa' ('Otomune' × 'Taisyowase') was named and released as 'Norin No.1' in 1959 by NFIFTS. 'Ginyose' is a rather old cultivar, which is thought to be derived from a chance seedling found at Toyono, Osaka around 1750, and is resistant to chestnut gall wasp. 'Kunimi' ('Tanzawa' × 'Taisyowase') was named and released as 'Norin No.5' in 1981 by NFIFTS. It is resistant to gall wasp, and suffers less damage from yellow peach moth than other cultivars. 'Ishizuchi' ('Ganne' × 'Kasaharawase') was named and released as 'Norin No.4' in 1968 by NFIFTS, and is resistant to chestnut gall wasp. 'Riheiguri' was selected and registered by K. Tsuchida in Gifu Prefecture in 1950. 'Ganne', a chance seedling, is resistant to chestnut gall wasp.

2.4 Rootstocks

Traditionally, growers have used as rootstocks, seedlings growing under grafted trees as well as seedlings from selected mother trees that gave seed with good emergence rate, growth and drought tolerance (Soylu and Serdar 2000) to establish the new orchards. These rootstocks have excellent graft compatibility and are suitable for poor soils; however, rootstocks from *C. sativa* are susceptible to ink disease.

At the beginning of the twentieth century, researchers from France, Spain, Italy and Portugal introduced seeds of *C. crenata* from Japan and *C. mollissima* from China into Europe. These species were resistant to ink disease, but their nuts were not appreciated by growers because of poor peeling, nor were they good for timber due

Table 19.9 Hybrid rootstock resistant to the ink disease recommended in Spain (modified from Pereira-Lorenzo and Ramos-Cabrer 2004)

Rootstock	Resist. to ink disease	Resist. to early frost	Compatibility	Vigor with the cultivar	Country of origin	References
Ferosacre CA90	3	0	2	5	France	Breisch (1995)
Maraval CA74	2	2	1	2	France	Breisch (1995)
Marigoule CA15	3	2	1	4	France	Breisch (1995)
Marlhac CA118	2	1	2	3	France	Breisch (1995)
Marsol CA07	1	2	2	4	France	Breisch (1995)
CHR-162 (7521)	3	2	3	5	Spain	Pereira-Lorenzo and Fernández-López (1997)
CHR-151 (HS)	2	2	3	4	Spain	Pereira-Lorenzo and Fernández-López (1997)
CHR-168 (110)	2	2	3	5	Spain	Pereira-Lorenzo and Fernández-López (1997)
CHR-161 (100)	2	2	3	–	Spain	Pereira-Lorenzo and Fernández-López (1997)

to poor vigor. Later, they were tried as rootstocks, but incompatibility was common. Interspecific hybrids were made in Portugal, Spain and France and some clones were selected as rootstocks, combining resistance to ink disease, easy propagation, and good compatibility with good growth and production in poor soils where chestnut is normally cultivated.

Currently, only France and Spain commercialize their resistant hybrid rootstocks propagated by stooling, cuttings, or in vitro culture. Five French (Breisch 1995) and four Spanish hybrid clones (Pereira-Lorenzo and Fernández-López 1997; Pereira-Lorenzo et al. 1999) are recommended. Resistance to ink disease varies from low to very high for French rootstocks (Breisch 1995), while the Spanish hybrids vary from medium to very resistant (Fernández-López et al. 2002) (Table 19.9) (Breisch 1995; Pereira-Lorenzo and Fernández-López 1997). Their compatibility is excellent. Among the Spanish rootstocks, 'CHR-151' ('HS') which is easily propagated via in vitro culture, has been broadly used (Miranda-Fontaíña and Fernández-López 1992). No data are available about the importance of hybrid rootstocks in new plantations.

3 Genetic Resources

Genetic resources of chestnut have been collected by different institutions throughout the world (Bounous 2002). Cultivars, seedlings and interspecific hybrids are preserved in different institutions in Austria, China, France, Hungary, Korea, Italy, Portugal, Slovak Republic, Slovenia, Spain, Switzerland, Turkey, and the UK (Bounous 2002; Tables 19.4–19.8).

Species-level genetic diversity within the genus. Based on isoenzymes, *C. dentata* appears to be the least variable of the North American species $H_e \sim 0.18$ (Huang et al. 1994b; Huang et al. 1998). Levels of genetic diversity are higher for *C. pumila* $H_e \sim 0.30$ (Fu and Dane 2003) as well as for *C. pumila* var. ozarkensis $H_e \sim 0.27$ (Dane et al. 1999). Although estimates of gene diversity appear to be lower for *C. dentata*, this level of diversity is similar to that found in other woody plant species (Hamrick and Godt 1989). *C. dentata* also appears to harbor less variation (11%) among populations than does *C. pumila* (30.4%) or *C. ozarkensis* (14.7%). Huang et al. (1998) showed evidence for possible geographic structure in *C. dentata*, with southern populations showing higher levels of genetic diversity possibly related to their glacial refugium; they studied RAPD and SSR variation in ~1,000 trees from 18 sample sites. Subsequent research using RAPDs and SSRs found that, although genetic differentiation among *C. dentata* populations has taken place, no disjunct regional pattern is apparent. *C. dentata* still exists as a highly variable species, even at the extremes of its natural range. Genetic variability in *C. dentata* follows a pattern consistent with the hypothesis of a single metapopulation where genetic drift will continue to play a major evolutionary role (Kubisiak and Roberds 2006). When compared to Asian and European *Castanea* (Huang et al. 1994a; Lang et al. 2007; Villani et al. 1991a, b), levels of genetic diversity based on isoenzymes in *C. dentata* appear to be similar to those reported for *C. seguinii* ($H_e \sim 0.20$), and levels reported for the two cultivars of *C. pumila* appear similar to those reported for all other *Castanea*: *C. mollissima* ($H_e \sim 0.31$), *C. henryi* ($H_e \sim 0.26$), and *C. sativa* ($H_e \sim 0.24$).

Levels of within and among natural diversity for specific species. In Europe, genetic variability in natural chestnut populations established that genes flow from East (Turkey) to the West (Italy) (Pigliucci et al. 1990a, b; Villani et al. 1991a, b, 1993; Aravanopoulos et al. 2002). Two main origins of variability in European cultivated chestnut were found in the Iberian Peninsula by SSRs, one in the North and a second in the Center (Pereira-Lorenzo et al. 2010).

The genetic diversity of wild chestnut (*C. crenata*) populations in northern Japan showed a high level of heterozygosity in wild populations (Tanaka et al. 2005). The H_O and H_E values in the chestnut (*C. crenata*) populations (H_O: 0.727 and H_E: 0.780) were similar to other *Fagaceae* such as *Fagus sylvatica* (0.727 and 0.753) (Pastorelli et al. 2003), *Fagus orientalis* (0.697 and 0.740) (Pastorelli et al. 2003), and *Quercus rubra* (0.679 and 0.737) (Aldrich et al. 2002).

Cultivar within species genetic diversity. Important efforts are being made in studying chestnut variability using morphological characteristics based on Breviglieri's (1955) 'Scheda Castanografica', after the UPOV chestnut guideline (1988) and, more recently, applied to the Spanish cultivars (Pereira-Lorenzo et al. 1996a, 2006a) and different chestnut species (Oraguzie et al. 1998).

The first molecular markers based studies in chestnut used isoenzymes. Sawano et al. (1984) studied 16 clones (10 Japanese, 3 Chinese, and 2 hybrids). Wen and Norton (1992) studied isoenzymes and identified 22 Chinese cultivars. Other genetic analyses with isoenzymes were performed by Bonnefoi (1984), Malvotti and Fineschi (1987), Fineschi et al. (1990a, b), Huang et al. (1994a), and Pereira et al. (1999).

RAPD markers were studied by Valdiviesso (1999), RAPD and ISSR by Goulao et al. (2001), and nuSSRs by Costa et al. (2008). The Portuguese cultivars showed great genetic variability, with multiple genotypes obtained per cultivar indicating their polyclonal origin (Costa et al. 2008). The different genotypes obtained for the main Portuguese cultivars have derived mainly from cross-pollination between them, but also from mutations. Four regions of Protected Denomination of Origin (DOP) were created to preserve the Portuguese cultivars: Castanha da Terra Fria, Castanha da Padrela, Castanha dos Soutos da Lapa and Castanha do Marvão. Molecular analysis showed the greatest variability (the largest number of genotypes) was located in the Northern regions (Castanha da Padrela) as compared to the southern region (Castanha do Marvão), which can be explained by the common practice of exchanging plant material for grafting in the Northern regions (Costa et al. 2008).

However, Fineschi et al. (1994) showed a relatively high degree of homogeneity both among individuals of the same cultivar and among cultivars of the same area in Italy, but a high genetic distance between geographic areas. Pereira-Lorenzo et al. (1996b, 2006a) studied the variability of the Spanish chestnut cultivars by isoenzymes and demonstrated that, in main cultivars. a main clone was predominant in orchards (over 60% of the samples), but intracultivar variability was important, surely due to the use of seedlings of those main cultivars by the growers. The H_O and H_E values obtained with isoenzymes in the Spanish chestnut cultivars were, on average, 0.398 and 0.333 respectively (Pereira-Lorenzo et al. 2006a). The excess of heterozygotes found in Galician chestnuts that were at least 300 years old was similar to that found in natural populations of *C. dentata* that were over 70 years old in Virginia, USA (Stilwell et al. 2003). The excess heterozygosity in these two populations may be due to the selection of heterozygous breeding material and by an absence of new recruits from other populations as suggested by Stilwell et al. (2003). Subsequent asexual propagation through grafting would maintain this situation. Subpopulations isolated in southern Spain had lower variation. Microsatellites confirmed the variability found with isoenzymes in Spanish cultivars (Pereira-Lorenzo et al. 2006a, 2010; Ramos-Cabrer et al. 2006) and heterozygosity was significantly higher as it occurs in other species.

In Italy, 33 microsatellite (SSR) loci were isolated in chestnut (Marinoni et al. 2003) and several oak loci (Steinkellner et al. 1997; Kampfer et al. 1998) were found to be polymorphic in *Castanea sativa* (Boccacci et al. 2004). Microsatellites are preferred for the DNA genotyping of cultivars aimed at identification, and were used in many studies of characterization of *Castanea sativa* Mill. cultivated germplasm, leading to the identification of over 70 cultivars.

Twenty cultivars from the North West Italian germplasm were characterized at 14 polymorphic loci (Marinoni et al. 2003). The total number of alleles was 90, and ranged from 4 to 10 per locus, with an average of 6.4. The mean expected heterozygosity was 0.72 (range: 0.65–0.83). The average observed heterozygosity (Ho) was 0.793 (range: 0.35–0.95). Further work was carried out within the EU project MANCHEST, and 121 North Italian accessions (Piemonte Region), including 39 Marrone individuals, were characterized (Botta et al. 2006) using a selected set of ten loci that included additional SSR markers isolated within the project by Buck et al. (2003). The loci (QrZAG96, QpZAG110, QpZAG119, CsCAT1, CsCAT3,

CsCAT4, CsCAT6, CsCAT16, CsCAT17, and EMC15) were chosen on the basis of their position and distribution in the genome. Fifty-two genotypes were identified by the markers and were described by chemical and morphological traits.

Thirty Japanese chestnut accessions (*Castanea crenata* Sieb. & Zucc.) were evaluated by SSR markers, including 12 cultivars and 6 wild landraces originated in Japan, and 6 cultivars and 6 wild landraces originated in the Korean Peninsula (Yamamoto et al. 2003). The 14 polymorphic SSR loci produced 2–16 alleles per locus. The average values of heterozygosity and polymorphic information content among the 14 loci were 0.50 (0.10–0.93) and 0.54 (0.10–0.89), respectively. No differences on allele composition were observed between cultivated and wild landtraces as well as between Japanese and Korean origins. The results could indicate that the Japanese chestnuts originating from Japan and the Korean Peninsula showed similar genetic background, and that cultivated chestnuts might have been selected from wild chestnuts.

SSRs were also used to identify cultivars from Italy (Martín et al. 2010), Portugal and Spain (Pereira-Lorenzo et al. 2011). In the Iberian Peninsula, ten main groups of cultivars have been found related with the two main origins of variability, the Northern and the Central Iberian Peninsula. This study demonstrated that cultivar origin and the diversification process was a combination of clonal propagation of selected seedlings, hybridization, and mutations, which allowed high levels of diversity to be maintained with respect to selected clones for fruit production.

4 Major Breeding Achievements

4.1 General Achievements

Diseases. Two main diseases, ink disease (*Phytophthora* spp.) and blight (*Cryphonectria parasitica*), threaten chestnut production. The European chestnut species *C. sativa* present less tolerance to main pest and diseases. Different genetic markers and different analytical approaches have shown a very significant amount of genetic variation for the whole range of species, pointing out the uniqueness of the Greek gene pool (Aravanopoulos et al. 2005).

Chestnut breeding in Europe began with the production of hybrids resistant to ink disease (*Phytophthora* spp.) to substitute the indigenous species. Initially, seedlings from Asian species *C. crenata* and *C. mollissima* were introduced between 1917 and 1940 (Elorrieta 1949) as a way to control ink disease, which was threatening the European chestnut orchards. Resistance on the Asian species was confirmed later, but these hybrids were in many traits inferior to the European species *C. sativa*; i.e. less vigor, lower quality of the nuts, bad affinity with the local cultivars, sensitivity to early spring frost and summer drought, and difficulty adapting to climatic characteristics of some areas in Europe (Elorrieta 1949; Pereira-Lorenzo and Fernandez-Lopez 2001). In 1989, a new program began to identify some hybrid clones that were interesting for timber, nut production or rootstocks (Pereira-Lorenzo and Fernández-López 1997, 2001).

In France, Asian species were introduced in 1925 and they showed high tolerance to ink disease but poor adaptation to soil and weather conditions. Schad et al. (1952) developed a breeding program to produce and select interspecific hybrids obtained by open or controlled crosses. Some of the French clones became very popular for nut production (Bergougnoux et al. 1978; Breisch 1995).

The first interspecific hybridizations in Portugal were initiated in 1947 by Bernardino Barros Gomes to introduce resistance to ink disease in *C. sativa* (Gomes Guerreiro 1948, 1957). More recently, interspecific crosses made between *Castanea sativa* 'Aveleira' (mother tree) with pollen of *C. crenata* (SC) or *C. mollissima* (SM) were done as a first step to identify molecular markers associated to ink and blight disease resistance in chestnut for developing marker assisted selection (MAS) and as a tool to identify genomic regions linked to resistance (QTLs) (Batista et al. 2008). Resistant *C. sativa* selections from COLUTAD in Portugal are being tested in a micropropagation program to rapidly make these available to the producers.

Chestnut blight is caused by *Cryphonectria parasitica* (Murr.) Barr (Syn. *Endothia parasitica* [Murr.] And.). It became the major disease of chestnut due to the sensitivity of *C. dentata* as well as, although to a lesser degree, of *C. sativa* to this fungus. Blight destroys the bark and the cambium causing the death of the branches or the tree above the wound when the disease girdles around them (Anagnostakis 1987; Heiniger and Rigling 1994). It was first observed in Europe in Genoa, Italy in 1938. The spread was quick through Italy and other European countries (Robin and Heiniger 2002), and less so in Southern UK, the Netherlands, Central and Southern Spain, and the Canary Islands. Blight almost eliminated the American chestnut (*C. dentata*) but European chestnut is recovering due to the natural occurrence of hypovirulence dsRNA hypovirus CHV1. Allemann et al. (1999) isolated five different CHV1 subtypes. Biological control is applied in Europe by hypovirulent strains of hypovirus growing cankers using Grente's method (Grente and Berthelay-Sauret 1978).

Only two loci conferring resistance to *Cryphonectria parasitica* have been identified within germplasm of *C. sativa*, *C. mollissima*, and *C. crenata* (Sisco et al. 2005). The main origin of resistance to blight is coming from Asian species, mainly *C. mollissima* (Hebard and Stiles 1996). The American Chestnut Foundation has developed a backcross-breeding program to restore the American chestnut *C. dentata*. By the third backcross, the progenies reach on average 96% American background, which eventually exhibit entirely American characteristics in later generations (Diskin et al. 2006).

Pests. Oriental chestnut gall wasp (*Dryocosmus kuriphilus* Yasumatsu) causes very serious loss of nuts production in Japan, China and Korea. This insect is a tiny gall-forming wasp endemic to China, and was accidentally introduced into Japan (1941), Korea (1963), and the USA (1974). The larvae in the gall can cause extensive bud loss, decreased shoot and leaf growth and reduced nut production. Severely affected trees can die. Therefore, the Japanese chestnut breeding programs have focused their work over the last several decades on the development of resistant cultivars to the chestnut gall wasp. These programs have developed several resistant cultivars through the screening of germplasm and selective breeding. They have bred 4 resistant cultivars: 'Tanzawa', 'Ibuki', and 'Tsukuba' in 1959 and 'Ishizuchi' in 1968. The cultivars were planted extensively throughout Japan allowing a recovery of the chestnut industry. However, these resistances broke down, and two cultivars with better resistance to gall wasp were released in 1981 ('Kunimi') and 1992 ('Shiho').

Another strategy to overcome damage by chestnut gall wasp was biological control by the introduction of the natural enemy of the chestnut gall wasp, *Torymus sinensis*. It was introduced into Japan in 1979 in Fukuoka City in Kyushu, the western island of Japan, and then in 1980 at NIFTS (Tsukuba City) in Ibaraki in eastern Japan. As a result, it is difficult to find chestnut gall wasps in chestnut orchards today. Breeding of resistant cultivars combined with the use of biological control by natural enemies have contributed to overcome chestnut gall wasp.

In 2002, gall wasp was reported for the first time in Europe in northwest Italy. The cynipid there causes the development of galls on leaves, buds and inflorescences, resulting in a decrease in both growth and yield of the European chestnut. Although the biological control with the parasitoid *Torymus sinensis* Kamijo, recently introduced in northwest Italy from Japan, may be a promising method for reducing the pressure of the pest in chestnut forests, this will most likely not be enough to guarantee high yield and good nut quality in orchards. For this reason, the University of Torino (Sartor et al. 2007) is assessing the level of susceptibility to *D. kuriphilus* in *Castanea sativa* Miller and hybrid cultivars for developing a breeding program. So far *C. sativa* cultivars tested are susceptible to gall wasp, although at different levels. Yet, resistance sources were recently discovered in the *C. sativa* germplasm. In this case plants do not show gall development. Among the Euro-Japanese hybrids, 'Bouche de Bétizac' and 'Marsol' showed opposite reactions to the insect: no gall development was observed in 'Bouche de Bétizac', while the highest level of infestation was observed in 'Marsol'. 'Bouche de Bétizac' has a hypersensitive response to infestation which results in larvae death at budburst. If the trait will remain stable, the selected materials will be used for orchard planting or for breeding resistant individuals. Preliminary work to understand the genetic mechanism of the susceptible response to the cynipid presence in chestnut tissues showed the expression of genes probably related to differentiation, nourishment, and the ability of the larvae to switch on parts of the seed development pathway.

Other two main insect pests are the moth larvae *Cydia* (=*Laspeyresia*) *splendana* Hb. and the weevil *Curculio* (=*Balaninus*) *elephas* Gyll. *Cydia* penetrate the nut through the bur as neonate larvae, and *Curculio* females oviposit through the bur. Debouzie et al. (1996) demonstrated that presence of chestnut moth larvae inhibited weevil egg-laying. It appears that *Curculio* ovoposits less in those cultivars with longest bur spines (Bergougnoux et al. 1978).

4.2 Scion

Nut quality. In a large study of Spanish cultivars (Pereira-Lorenzo et al. 2006a), only 7% of samples came from cultivars producing nuts over 15 g, and most cultivars produced smaller and less valued nuts (under 10 g). Of the four main nut shapes found, the most common are the elliptical-short (46%), the elliptical-triangular (30%), and elliptical-broad (18%). The most distinct and least common is the triangular shape (6%), which gives the name 'Longal' to the main cultivar in the Iberian Peninsula. Only 12% of the Spanish cultivars studied had more than 12% of multi embryo chestnuts. This indicates the strong selection made by growers to avoid peeling problems.

Asian species mature their nuts more quickly (early September) than does *C. sativa* (late October). This early harvest from the Asian species has been genetically transmitted to interspecific hybrids (Pereira-Lorenzo and Fernandez-Lopez 2001). However, important genetic variability is found in Spanish cultivars, a product of selection, with cultivars in South Spain production collected during the second half of September combined with cultivars harvested in October, and in the North with cultivars such as 'Negral' harvested at the beginning of October when the most frequent period is at the end of that month.

Chestnut pericarp split when the rainfall season is delayed till October, facilitating the development of fungus. This is not related to genetic variation (Ramos-Cabrer and Pereira-Lorenzo 2005) as embryo intrusions of the pericarp were found to be related to region and cultivar variation.

Another characterestic selected by growers is the shortness of the bur's spines, only present in 14% of cultivars such as 'Rapada' or 'Rapuga', which are easy to harvest by hand (Pereira-Lorenzo et al. 2006a).

Larvae and weevils develop inside mature nuts, *Cydia* (=*Laspeyresia*) *splendana* Hb. and *Curculio* (=*Balaninus*) *elephas* Gyll. *Cydia* penetrate in the nut through the bur as neonate larvae and *Curculio* females make their ovoposition using with her long snout also through the bur. In Spanish cultivars, we confirmed the relationship between the length of the burr's spines and lower attacks of *Curculio* as was previously reported by Bergougnoux et al. (1978). Burs with longest spines, longer than 1.5 cm, reduced the attacks up to 34%. Thus, the selection of long spur spines should be considered in breeding programs as an approach to develop *Curculio* resistant cultivars.

Chestnuts store starch in cotyledons, and the content is three- to fourfold higher than that found in other nuts (Ensminger et al. 1995). *C. dentata* and *C. mollissima* show a higher starch content (49%) than *C. sativa* (40%) (McCarthy and Meredith 1988). High starch content is important in cultivars for flour production or for animal feed. The highest average starch content in the Iberian Peninsula is produced by 'Longal' (Ferreria-Cardoso et al. 1993; Pereira-Lorenzo et al. 2006b) with 53–67% d.m. Another chemical characteristic that should be taken into account is high fiber content that reduces digestibility in some cultivars.

Concerning Spanish cultivars, Pereira-Lorenzo et al. (2006b) concluded that high variability in chemical composition between cultivars and regions corresponded to high genetic variability between cultivars. Correlations with environmental parameters were low, indicating that differences found between regions were probably due to the differences between cultivars. In Central and Southern Spain, some cultivars presented lowest moisture content due to the low summer rainfall in these regions. Lowest values of fiber content and ease of digestibility were found in cultivars from Galicia and Extremadura. No significant differences in

Fe, Zn, and Cu were found, although Zn content is twice the value reported for European chestnuts.

Japanese chestnuts produce the largest nuts among *Castanea* spp., sometimes exceeding 30 g. Nut size is one of the most important morphological characteristics and many newly bred cultivars have nuts weighing 25–30 g. By contrast, the thick pellicle is often extensively invaginated into the nuts, which makes peeling difficult. The easy peeling characteristic in kernel is a major objective in Japanese chestnut breeding programs.

4.3 Rootstocks

Compatibility. The most popular clonal rootstocks are the Euro-Japanese hybrids selected in France. They are easy to propagate by layering or softwood cuttings, are tolerant to *Phytophthora* spp. and *Cryphonectria parasitica* and have genetic compatibility with most of the best cultivars. Popular rootstocks include: 'CA 07' ('Marsol') (moderately resistant to *Phytophthora*); 'CA 74' ('Maraval') (*Phytophthora* resistant, low vigor); 'CA 118' ('Marlhac') (moderately resistant to *Phytophthora*, but able to grow at temperature <−10°C); 'CA 90' ('Ferosacre') (*Phytophthora* resistant, but sensitive to temperatures <−10°C). European chestnut cultivars are usually grafted onto seedlings of *C. sativa*.

The ease of vegetative propagation and stock–scion compatibility are of primary importance in rootstock breeding. Chapa et al. (1990) and Bounous et al. (1992) found that *C. crenata* hybrids (*C. crenata* × *C. sativa*) are easier to propagate by cuttings or layering than *C. sativa*. Ease of propagation by layering or cuttings and *Phytophthora*-resistance of the French hybrids ('Marsol', 'Marigoule', 'Maraval', 'Précoce Migoule') make them useful as rootstocks or as direct producers. Unfortunately, graft incompatibility problems with many European cultivars have limited their wider application (Chapa et al. 1990; Ferrini et al. 1992; Breisch 1992; Pereira-Lorenzo and Fernández-López 1997).

Although environmental and stress factors may have a role, the success of a particular graft, stock–scion compatibility is most certainly under genetic control (Anagnostakis 1991). Three peroxidase isozyme genes known for *Castanea* (six types) may be involved with graft compatibility (Santamour et al. 1986). Graft incompatibility is also affected by ChMV (Desvignes 1996).

5 Current Goals of Breeding

The chestnut ideotype is a function of the final use (nuts or timber), and production and processing technology (harvesting systems, fresh or processed uses) (Table 19.10). For nut production the most important breeding objectives include the following: good horticultural traits, product quality, suitability to storage and processing, and easy peeling.

Table 19.10 Main objectives of chestnut breeding (modified from Bounous and Torello Marinoni 2005)

Use	Characters required
Nut production	Tree: Medium–low vigor, strong branches, upright growth habit for mechanical harvesting, good pollinizer, self fertility, regular and high yields, precocious bearing, early ripening, ease of propagation, rootstock–scion compatibility, resistance to *Cryphonectria parasitica* and *Phytophthora* spp., resistance to *Dryocosmus kuriphilus*
	Nuts: Large size for fresh or confectionary uses, small or medium size for drying or flour, light color, shiny, shell with evident stripes, evenness of shape, no multiple embryos, ease of manual or machine pellicle removal, no hollow kernels, good flavor, sweetness, adequate texture, good adaptability to confectionary use, resistance to *Cydia* spp., *Curculio elephas*, *Cyboria batschiana*
	Bur: Dehiscent for manual harvesting, Nondehiscent for mechanical harvesting, long and dense spines for insect resistance
Wood production	Tree: Resistance to *Cryphonectria parasitica* and *Phytophthora* spp., resistance to wood-boring insects, resistance to frost and drought, suitable for poor soils, timber products, high vigor, straight trunk, fast growth, high wood production, high yields, self pruning ability, nonchecking wood, no ring shake

For timber, important characters include wood quality, rapid growth, and nonchecking of wood (ring-shake). Ease of propagation and resistance to major diseases and pests are common for nut and timber types. A few chestnut cultivars such as 'Garrida', 'Loura', and 'Parede' combine both timber and chestnut production under the appropriate climatic conditions.

Japanese chestnut breeding trials started in 1929 at the Agricultural Experiment Station in Niigata Prefecture. Originally, the Japanese chestnut was crossed with a Chinese chestnut in an attempt to introgress the easy peeling kernel trait from Chinese chestnut. However, these efforts were unsuccessful. In 1947, national chestnut breeding programs started at the Horticultural Research Station (now National Institute of Fruit Tree Science: NIFTS). The breeding target at that time was to produce a cultivar with the Japanese chestnut characters of high yield and big nut size combined with the character of easy peeling kernel of Chinese chestnut. Before this was achieved, the objective was changed to chestnut gall wasp resistance in 1952 because the chestnut gall wasp which appeared in Okayama Prefecture in western Japan around 1941 quickly spread to the whole country. Since the Chinese chestnut breeding material had little resistance to gall wasp, two gall wasp resistant native cultivars 'Ginyose' and 'Ganne' were used to breed gall wasp resistant cultivars.

Once the damage caused by the chestnut gall wasp decreased, the goal of Japanese chestnut breeding apparently shifted from insect resistance to the nut quality. The decrease of insect damage enabled the use of the Chinese chestnut for breeding. At the same time, a new rapid screening system for removing the chestnut pellicle after deep frying in cooking oil (HOP method, High-temperature Oil Peeling method) was developed and applied to chestnut breeding. By using the HOP method, easy peeling Japanese chestnut cultivars were identified and used in breeding. The newest promising cultivar with an easy peeling trait, 'Porotan', bred and registered in 2007. But it was developed from Japanese chestnuts without the easy peeling trait. The genetic control of the easy peeling trait of 'Porotan' is unknown.

5.1 Plant Characteristics

Semicompact, medium, or low vigor are the most suitable features for medium or high density plantations. Other valuable cultivar characteristics include the following: early maturity, precocious bearing, regular and high yields, strong branches, good pollinizer ability, and intercompatibility with the best cultivars. Harvesting is one of the most costly aspects of chestnut production. Harvest-related traits include upright habit for mechanical shaking and low detaching force to shake off burs from the tree. Mechanical harvesting of the nuts from the ground may be easier with nuts that fall closed in the burs (to prevent nuts from infection), than with nuts which fall free from dehiscent burs. For timber production, trees have to demonstrate high vigor, high wood production, straight trunk, self-pruning ability, and wood not subject to ring-shake or radial checking.

5.2 Nut Characteristics

Large nut size is desirable from the standpoint of harvesting, handling, fresh marketing and making candy (*marrons glacés*), while a small or medium size nut may be used for dried chestnuts or as a vegetable. The marketing of peeled or processed chestnuts puts less emphasis on size. Evenness of shape, shiny color, dark brown stripes, flavor, and firm texture are valuable traits for fresh marketing. Other desirable traits are easy pellicle removal, no pellicle intrusion, no hollow kernel, no multiembryo nuts, and resistance to pests (*Cydia*, *Curculio* and others) and to storage diseases (*Cyboria* and others).

Japanese chestnut breeding programs have focused on breeding excellent cultivars with high nut quality. The major selection criteria for the NIFTS programs are large nut size, low pericarp splitting, low polyembryony, white color of the steamed kernel, high sweetness, flavor, and kernel quality. In addition, an easy peeling pellicle is highly desired but a difficult characteristic to obtain for more than 50 years of breeding. The newly released cultivar 'Porotan' is the most prominent cultivar with good nut quality and an easy peeling trait for East Asian Markets.

6 Breeding Methods and Techniques

6.1 Ploidy, Karyotype, and Genome Size in Castanea

Castanea spp. are generally recognized as having $2n=24$ chromosomes (Jaynes 1962), the number characteristic of most of the Fagaceae studied to date (Mehra et al. 1972; Ohri and Ahuja 1991; D'Emerico et al. 1995). With only minor exceptions such as occasional triploidy and aneuploidy (Jaynes 1962), the normal somatic number of chromosomes in species and hybrids in the genus is 24. In general, a rather high degree of fertility has been observed among interspecific hybrids (Jaynes 1964, 1972). Although pairing appears to be normal in many interspecific hybrids, the presence of segregation distortion in some mapping populations (Kubisiak et al. 1997; Kubisiak, unpublished data) and abnormal pairing in F_1 hybrid pollen mother cells (Faridi et al. 2008), suggest that significant chromosomal differences such as translocations and/or inversions are likely to exist among *Castanea* species. A better understanding of chromosome-level genomic differences between species will be important for effective breeding using interspecific hybrids. Genome size appears to be fairly conserved among species ($2C=1.57$–1.67 pg), only five times that reported for *Arabidopsis*. A tractable genome size and abundant genetic and genomic resources make *Castanea* a good candidate for future targeted- or whole-genome sequencing (Kremer et al. 2007).

6.2 Mating System

Chestnut is a monoecious species presenting male flowers in catkins and female flowers that develop at the base of bisexual catkins. Some chestnut cultivars present morphological male sterility. Soylu (1990) proposed a genetic model based on two genes and five morphotypes: astaminate (xxzz), brachystaminate (xxZz), mesostaminate (Xxzz), and longistaminate long/short (XXZZ/XxZz). Astaminate flowers do not produce pollen, brachystaminate can produce very limited quantities of pollen, while longistaminate catkins are those that produce more pollen. Astaminate catkins are supposed to be more frequent in some of the best 'marron' type cultivars. Male sterility may be related to lower energy consumption during flowering. Male sterility has been found in up to 21% of Spanish cultivars. Up to 8% of the cultivars presented astaminate catkins, and 13% had brachystaminate catkins. Those cultivars require pollinizers, mainly with longistaminate catkins (Pereira-Lorenzo et al. 2006a).

Pistillate flowers have six to eight styles whose tips are hollow at full bloom. The ovary presents seven (rarely six or eight) carpels. Each flower has 10–16 anatropous ovules. The bearing monoembryonic seeds (marron type) have been related to a high occurrence of anomalies, such as delayed embryo sac differentiation and the presence of supernumerary nuclei in the embryo sac (Botta et al. 1995).

Very little is still known about the genetic system controlling mating and the self-incompatibility system in chestnut, although it is considered to be of gametophytic type (Breviglieri 1951; Brewbaker 1957; Jaynes 1975). Cross-pollination is compulsory.

6.3 *Hybridization*

Interspecific hybridizations between most of the chestnut species are possible. Main problems are related with the different flowering time between species. Asian species are more precocious than European chestnut. Hybridization was the main breeding method used to incorporate resistance to blight into American chestnut, ink disease into European chestnut and wall gasp into Japanese chestnut (Table 19.3). It was also used in Japan to incorporate the easy peeling trait into Japanese cultivars, which were normally difficult to peel.

Meiosis in pollen mother cells occurs 10 to 15 days before anthesis, in the first week of June in Italian cultivars (Botta et al. 1995). Pollen viability varied from $81.3 \pm 6.1\%$ based on fluorochromatic reaction, to $58.2 \pm 7.0\%$ on hanging drops and $50.1 \pm 4.5\%$ germination on agar media.

Pollen is easily collected from the longistaminate pollinizers and desiccated to be stored in a refrigerator for short-term storage or in a freezer for long-term storage. Emasculation and female flower isolation is needed to avoid unknown pollination. Male flowers from the bisexual catkins must be also removed. Bags are attached to the base of the catkins. Hand pollination with fresh catkins brushing over pistils has been detailed by Nienstaedt (1956). The bags are removed after setting. Nuts are collected when burs begin to crack and nuts begin to turn brown.

European breeding programs to incorporate resistance to ink disease (*Phytophthora* spp.) produced interspecific hybrids of first generation. In the USA, a backcrossing program using blight resistant Chinese species as the donor and American chestnuts (*C. dentata*) as the recurrent parent has produced blight resistant American chestnut (Diskin et al. 2006). Diskin et al. (2006) pointed out that, in the BC3-F2 generation, 93% of seedlings showed morphological characteristics of American chestnut incorporating 100% Chinese resistance to blight. The program to obtain such material required more than 20 years.

6.4 *Propagation*

Seeds. Seedlings are used in breeding programs based on hybridization (Table 19.3), as local rootstocks for main producing areas (Pereira-Lorenzo et al. 2007), but also to propagate selected forest progenies in Europe. Wild progenies from Europe have shown adaptive response to water stress such as a lower growth rate (Pliura and Eriksson 2002; Lauteri and Villani 2004; Eriksson et al. 2005) and different bud break (Blanco Silva et al. 2005).

Chilling releases embryos from endodormancy. To avoid dehydration which reduces germination, chestnuts are stratified in sand or in humid peat moss at 1–2°C for 4 or 5 months.

Pregerminated nuts are sown in raised beds. The apex can be cut off to permit the formation of a well expanded root system. The depth of seeding must not exceed 3–5 cm, spacing in the row is usually 30–40 cm, with rows 80–100 cm apart.

At the end of the first growing season (August–September), the seedlings are 100–150 cm tall with a diameter of 8–12 mm and are ready to be chip budded in fall or grafted or budded the following spring.

Grafting. Grafting is used not only for propagation of selected cultivars but also in breeding programs to establish core collections and germplasm banks, to propagate mature origin such as plus trees, and to rejuvenilize plant material for later propagation by cuttings, layering, and micropropagation. Chestnut cultivars have been traditionally propagated using bark graft in spring or flute graft, and continue to be used in new orchards when in situ graft is made. Nurseries use various budding and grafting techniques. Summer budding (patch and T-bud) are easier and more effective since the higher temperatures cause rapid healing (Pereira-Lorenzo and Fernández-López 1997). However, summer budded trees are not ready for the market until the following winter. By contrast, spring grafting (chip, cleft, and whip) can produce finished plants for the same winter, although they need a second year growth to get to comparable growth with those summer-budded plants.

Cuttings. Vegetative propagation of chestnut is limited by the age of the mother plant from which the cuttings are taken. Cuttings from material of mature origin are difficult, if not impossible, to root (Vieitez 1974). The most efficient system is through the use of greenwood or softwood cuttings collected from 3- to 4-year-old mother plants, which can be grown in the open or as potted plants in the greenhouse. Cuttings, bearing 3–4 buds (10–15 cm) and leaving half of the uppermost leaf, are collected from May to July and their bases treated with indole-3-butyric acid (IBA) or naphthalene-3-acetic acid (NAA) at concentrations ranging from 1,000 to 4,000 ppm (Rodríguez et al. 2005). The cuttings are subsequently placed in trays, containing suitable substrates, in rooting tunnels equipped with a fog system. Rodríguez et al. (2005) showed that the production efficiency of the system is related to the genotype.

Layering. Chestnut layering is another vegetative propagation method mainly used in European countries for the clonal propagation of rootstocks, resistant Euro-Asian hybrids or direct producer trees. Sprouts are developed annually from stumps and, regardless of the age of the mother plant, exhibit a juvenile physiological condition, allowing them to root after proper auxin treatments. When the sprouts reach a length of 30–35 cm (May), their basal parts (8–10 cm) are stripped of leaves and a paste containing IBA and NAA (or both) at a rate of 2 g/kg Vaseline® is applied to this part of the sprout, which is finally covered with soil. When the shoots reach the rest period (November), the rooted shoots are excised from the mother stump and planted in the nursery to strengthen their root system. The number of shoots produced yearly per stump (10–20) and the rooting rates (60–90%) are clearly genotype dependent.

In vitro. Efforts are being made to establish reliable in vitro regeneration systems that allow chestnut clonal propagation. The two principal micropropagation systems are based on somatic embryogenesis and on micropropagation through axillary shoot development. Although somatic embryogenesis is theoretically more efficient for clonal mass production than propagation via axillary shoot proliferation, several difficulties need to be overcome to render it commercially viable, particularly when cultures originate from adult tissues (Corredoira et al. 2006). By contrast, chestnut can currently be micropropagated from both juvenile and mature material using the axillary shoot multiplication method, although it is common for the protocol to require optimization for a specific cultivar, making the large-scale propagation in many cases challenging (Vieitez et al. 2007).

Juvenile plant material may be collected from seedlings conventionally obtained in greenhouse or climate chamber, as well as from seedlings obtained by in vitro culture of embryonic axes (Sánchez et al. 1997a). In the case of mature material, the use of stump sprouts (juvenile parts of mature trees) or basal shoots grown on the lower part of the trees as source of explants allows the micropropagation of chestnut. Cuttings taken from these materials are collected in winter, stored at 4°C and forced to flush in a climate chamber, after which the primary explants are taken from the flushed shoots. By contrast, the reactivity of crown-derived primary explants is poor, and reinvigoration methods must be applied using pretreatments such as etiolation (Ballester et al. 1989) or grafting onto seedling rootstocks (Sánchez et al. 1997b).

The primary explants from which chestnut shoot cultures are initiated are generally shoot tips and nodes bearing 1 or 2 axillary buds. After excision from the plant source (juvenile or mature trees), they must be sterilized and established in vitro on a number of different initiation media supplemented with cytokinins. After 6–8 weeks of culture in initiation medium, new shoots develop which can be subcultured at 4- to 5-week intervals for the shoot multiplication culture stage. From the different systems employed for rooting of micropropagated chestnut shoots, the culture for 24 h in a rooting medium containing 25–50 mg/l IBA followed by transference to either an auxin-free root expression medium or to a substrate mixture appears as the most appropriate for obtaining acceptable rooting frequencies.

In addition to clonal propagation, in vitro tissue culture is a useful technique for germplasm conservation. The cold storage of cultures represents a procedure for medium-term conservation, and Janeiro et al. (1995) reported the possibility of keeping chestnut cultures at 2–4°C for up to 1 year without subculture. In addition, a successful cryopreservation system of chestnut shoot tips has been reported (Vidal et al. 2005) allowing long-term storage of chestnut genotypes in liquid nitrogen. A detailed protocol for micropropagation of European chestnut, including storage and molecular marker analysis to determine the genetic stability of in vitro regenerated plants, has recently been published (Vieitez et al. 2007).

7 Integration of New Biotechnologies in Breeding Programs

7.1 Genomic Resources for Castanea

A number of molecular marker systems have been used in *Castanea* for applications such as cultivar identification, population genetics, linkage analysis, and marker-assisted selection. These marker systems consisted initially of isoenzymes, followed by random amplified polymorphic DNAs (RAPDs), inter-simple sequence repeats (ISSRs), and amplification fragment length polymorphisms (AFLPs), and more recently by simple sequence repeats (SSRs) and single nucleotide polymorphisms (SNPs). Given that SSRs and SNPs are rapidly becoming the markers of choice due to their informativeness, high transferability across related taxa, prevalence in the genome, and amenability to automated high-throughput analysis, we have chosen to focus primarily on the development and use of these newer marker systems.

There are currently four main sources from which SSR or SNP markers are being developed: genomic DNA libraries enriched for repeat-containing sequences; expressed sequence tags (ESTs); bacterial artificial chromosome (BAC) sequencing, or from whole genome sequencing efforts. Owing to a previous lack of DNA sequence data available for the Fagaceae, and in particular for *Castanea*, most SSRs currently available have been developed from repeat-enriched genomic DNA libraries. A small number of SSR markers developed from *Quercus* sequences are proving useful in *Castanea* (Aldrich et al. 2003; Barreneche et al. 2004). Of those developed from *Fagus* (Tanaka et al. 1999; Pastorelli et al. 2003), comparatively fewer are proving to be useful within the *Castanea* (Kremer et al. 2007; Kubisiak, unpublished data). The transferability of SSRs from other genera within the Fagaceae to *Castanea* will largely depend upon their evolutionary distance, with higher levels of transferability expected between more closely related genera such as *Castanopsis* (Manos et al. 2001).

Sequence data for at least 83 SSR primer pairs developed specifically from *Castanea* are currently publicly available. Forty-six SSR primer pairs developed from *C. sativa* have been characterized (Buck et al. 2003; Marinoni et al. 2003). Yamamoto et al. (2003) characterized 15 SSR primer pairs developed from *C. crenata*. More recently, 22 SSR primer pairs were developed from *C. mollissima* and characterized in both *C. mollissima* and *C. crenata* (Inoue et al. 2009). In general, these markers appear to be highly transferable across the *Castanea*. SSRs are already proving useful for cultivar identification and typing (Boccacci et al. 2004; Beccaro et al. 2004; Pereira-Lorenzo et al. 2010), linkage analysis (Sisco et al. 2005), and QTL analyses (Kubisiak, unpublished data; Casasoli et al. 2004, 2006).

Previously, the availability of DNA sequence data for *Castanea* was extremely limited (Connors et al. 2001). More recently, a genomic tool development project for various members of the Fagaceae was initiated (Sederoff et al. 2008), with *C. mollissima* being a key model species. A large component of this project is focused on the transcriptomes of *C. mollissima* and *C. dentata* (Carlson et al. 2007, 2008). Large EST databases are being created with significant numbers of sequence contigs showing similarity to predicted proteins in *Populus trichocarpa*. Additional SSRs and a significant number of SNPs are being identified in EST sequence data that should prove to be an invaluable resource for the community interested in all aspects of the genetics, breeding, and biotechnology of *Castanea*. An integrated Web-based resource for the *Castanea* genetics/genomics community [Fagaceae Genomic Database (FGD): http://www.fagaceae.org] has been developed (Ficklin et al. 2007), and relevent sequence information, homology results, genetic/physical map information, SSRs, SNPs, and other genomic data are being posted as it become available.

Another large component of the genomic tool development project is the production of genetic and physical mapping resources for *C. mollissima*. A BAC library has been constructed which consists of ~20× coverage (http://www.fagaceae.org/progress/NE1015/Tomkins_FingerprintingAndcDNAUpdate.ppt). The entire library is currently being fingerprinted by high information content fingerprinting. A subset of clones consisting of the minimum tiling path will be selected for BAC-end sequencing. Genetic markers developed from BAC-end sequence and hybridization of genetic marker probes to high-density BAC colony filters will be used to combine the genetic and physical maps. An integrated genetic/physical map will become the platform for future targeted genome sequencing of regions harboring resistance genes and will be an invaluable resource for gene cloning studies in *Castanea*.

Three partial gene encoding proteins described as pathogenesis-related were isolated and cloned from infected resistant chestnut plants: a cystatin, a beta 1,3 glucanase isoform, and a thaumatin-like protein gene using the RT-PCR technology (Serrazina 2004). The nucleotide sequences and amino acid deduced sequences have high homology with resistance genes' sequences of other plant species in the GenBank database. A partial gene encoding AOC (allene oxide cyclase) was also cloned, similar to the *Lycopersicum esculentum* gene. The gene expression analysis by Northern Blotting of *aoc, cist*, *gluc*, and *pttaum* points out to a straight relation of these genes with *C. sativa* resistance to *P. cinnamomi* (Serrazina 2004). Tobacco explants were transformed by particle bombardment, to study the overexpression effect of the isolated genes on plant resistance to *P. cinnamomi*. The observation of inoculated transformed and nontransformed plantlets suggests that the constitutive expression of *aoc*, *cist*, *gluc*, and *pttaum* genes (separately) attenuated the pathogenic effects of *P. cinnamomi* in transformed tobacco plants (Serrazina 2004).

To gain insight into genetic reorganization, which leads to the formation of ectomycorrhiza, a cDNA microarray was constructed and used to study the interaction of *Castanea sativa* roots and *Pisolithus tinctorius* during the first hours of contact. (Sebastiana 2006). Statistical analysis of microarray results identified a set of *32 C. sativa* genes and 8 *Pisolithus tinctorius* genes with altered expression in response to the interaction between the two organisms. Differentially expressed genes identified in *C. sativa* roots displayed significant sequence similarities to proteins involved in cellular processes such as defense response, protein maturation/degradation, cell wall modification, primary metabolism, signal transduction, and cytoskeletal organization. Fungal genes regulated by the interaction with *C. sativa* roots displayed significant sequence similarities to proteins involved in cell wall structure, protein maturation/degradation, and cellular organization (Sebastiana 2006).

Linkage relationships between isoenzymes and morphological traits in interspecific crosses were found (Huang et al. 1996). Molecular maps have been developed (Kubisiak et al. 1997; Casasoli et al. 2001) opening a new way to the genetic knowledge of chestnut.

7.2 *Transgenics*

A conventional chestnut breeding program requires around 15–20 years to incorporate a new disease resistance allele and to reach the BC3F2 generation. The great advantage of genetic transformation, as a complementary breeding system, is that a new set of genes can be transferred into chestnut somatic cell lines in a matter of 2–3 years. This is possible today because, after decades of research on in vitro tissue culture of both European and American chestnut (Vieitez and Merkle 2004), a repetitive and reproducible system useful for genetic transformation through somatic embryogenesis has been described for both species (Robichaud et al. 2004; Corredoira et al. 2006). Although in most cases the induction of somatic embryos was carried out from immature material (embryonic axes), Corredoira et al. (2003) have shown that embryogenic cultures can be initiated from leaf explants of seedlings, opening up the possibility of propagating and transforming mature material in the future.

The first attempts to transform European chestnut used hypocotyl segments from in vitro-germinated seedlings and stem segments of in vitro-grown shoots, which were cocultured with *Agrobacterium tumefaciens* (Seabra and Pais 1998, 1999). Unfortunately, no transgenic plants were obtained as there were a large number of escapes (97%) and gene integration was transient. On the other hand, some transformation experiments were carried out by coculturing cotyledonary node explants with *A. tumefaciens* harboring reporting genes, with 2.3% of explants developing kanamycin-resistant shoots, although no transgenic plants were recovered (Corredoira et al. 2005).

An efficient genetic transformation protocol for *C. sativa* has been described for the first time (Corredoira et al. 2004) through the coculture of somatic embryos with different strains of *Agrobacterium tumefaciens* carrying marker genes. The plasmids contain the *nptII* genes driven by the *nos* promoter for kanamycin selection and the β-glucuronidase reporter *uid*A gene (*gus*) driven by either ubiquitin (*Ubi-1*) or the *CaMV 35S* promoter. Following 4 days of coculture and after 12 weeks of culture in selection medium, cotyledonary stage-regenerated embryos were isolated from GUS-positive lines and subcultured on selection medium to establish and proliferate embryogenic transgenic lines. The presence of the *nptII* and the *uid*A genes in GUS-positive embryogenic lines was assessed by PCR and Southern blot analyses. Transformation efficiencies as high as 25%, were recorded. This transformation protocol was improved by studying the effect of both the genotype and the type of initial explant (Corredoira et al. 2007). In order to increase the tolerance of European chestnut to fungi diseases, attempts to transform embryogenic lines with a thaumatin-like protein (CsTL1), which has antifungal activity in vitro, are currently in progress (Maynard et al. 2008).

The first reports on transgenic American chestnut tissues were carried out by Carraway et al. (1994) who used microprojectile bombardment (biolistics) to transform proembryogenic masses derived from immature zygotic embryos, although no stable transformation events were recovered. Subsequently, a coculture transformation experiment was carried out with *Agrobacterium tumefaciens* containing the plasmid construct pΔVspB-OxO, which included a germin-like *oxalate oxidase* gene *(OxO)* to enhance blight resistance, a phosphinothricin acetyltransferase gene (*bar*) as a selectable marker, and a green fluorescent protein gene (*mgfp5-ER*) as a visual marker (Polin et al. 2006; Maynard et al. 2008). To increase the transformation rate, the plate flooding system approach was assayed (Rothrock et al. 2007).

One problem affecting both European and American chestnut is the relatively low conversion rate of somatic embryos into plants (Corredoira et al. 2003, 2004). However, germinating embryos with only shoot development are also produced in germination medium, and these shoots can be multiplied successfully by axillary shoot proliferation, giving rise to an unlimited number of transgenic shoots to be rooted. This constitutes a valuable alternative for plant regeneration from transformed germinating embryos in which plantlet conversion is not achieved. Both European and American transgenic chestnut plants were acclimatized in phytotron and grown in the greenhouse and in the open fields (Maynard et al. 2008).

In addition to this consolidated transformation procedure, Fernando et al. (2006) described a promising preliminary approach consisting in the transformation of America chestnut pollen. The method, in which in vitro tissue culture technology is not required, makes use of transformation via particle bombardment. Only transient green fluorescent protein (GFP) expression was recorded and the highest values were achieved using ungerminated pollen.

References

Aldrich, P.R., Michler, C.H., Sun, W., Romero-Severson, J. (2002) Microsatellite markers for northern red oak (Fagaceae: *Quercus rubra*). Mol. Ecol. Notes., 2:472–474.

Aldrich, P.R., Parker, G.R., Michler, C.H., Romero-Severson, J. (2003) Whole-tree silvic identifications and the microsatellite genetic structure of a red oak species complex in an Indiana old-growth forest. Can. J. Forest Res., 33:2228–2237.

Allemann, C., Hoegger, P., Heiniger, U., Rigling D. (1999) Genetic variation of Cryphonectria hypoviruses (CHV1) in Europe, assessed using restriction fragment length polymorphism (RFLP) markers. Mol. Ecol., 8:843–854.

Anagnostakis, S.L. (1987) Chestnut blight: The classical problem of an introduced pathogen. Mycologia, 29:23–37.

Anagnostakis S.L. (1991) Peroxidase allozyme phenotypes in *Castanea* and their segregation among progeny. HortSci., 26:1424.

Antonaroli, R., Bagnaresi, U., Bassi, D. (1984). Indagini sulla variazione di alcuni caratteri morfologici in popolazioni di castagno da frutto nella provincia di Bologna. Monti e Boschi, 2, 47–50.

Aravanopoulos, F.A., Drouzas, A.D., Alizoti, P.G. (2002) Electrophoretic and quantitative variation in chestnut (*Castanea sativa* Mill.) in Hellenic populations in old-growth natural and coppice stands. For. Snow Landsc. Res., 76(3):429–434.

Aravanopoulos, F.A., Bucci, G., Akkak, A., Blanco Silva, R., Botta, R., Buck, E.,Cherubini, M., Drouzas, A.D., Fernandez-Lopez, J., Mattioni, C., Marinoni, D., Papadima, A., Russel, K., Zas, R., Villani, F. (2005) Molecular population genetics and dynamics of chestnut (*Castanea sativa* Mill.) in Europe: inferences for gene conservation and tree improvement. Acta Hort., 693:403–411.

Ballester,A., Sánchez, M.C., Vieitez, A.M. (1989) Etiolation as a pretreatment for in vitro establishment and multiplication of mature chestnut. Physiol. Plant., 73:77(3):395–400.

Barreneche, T., Casasoli, M., Russell, K., Akkak, A., Meddour, H., Plomion, C., Villani, F., and Kremer, A. (2004) Comparative mapping between *Quercus* and *Castanea* using simple-sequence repeats (SSRs) Theor. Appl. Genet., 108:558–566.

Bassi, D., Sbaragli, E. (1984) Indagine pomologica su alcuni cloni di castagno da frutto (*C.sativa* Mill) Rivista di Frutticoltura N° 6. Italia.

Batista, D., Valdiviesso T., Santos, L. Paulo, O., Laranjo, J, Costa, R. (2008) Genotyping *Castanea sativa* × *C. crenata* and *C. sativa* x *C. mollissima* F1 Hybrids using Nuclear SSRs. Acta Hort., 784:107–112.

Beccaro, G. L., Botta, R., Torello Marinoni, D., Akkak, A., Bounous, G., (2004) Application and evaluation of morphological, phoenological and molecular techniques for the characterization of *Castanea sativa* Mill. cultivars. Acta Hort., 693:453–457.

Bergougnoux, F., Verlhac, A., Breisch, H., Chapa, J. (1978) Le châtaignier. INVUFLEC, Paris, 192 p.

Blanco Silva, R., Silva, R., Zas, R., Fernández López, J. (2005) Geographic differentiation in adaptive traits of wild chestnut Spanish populations (*Castanea sativa* Miller). Investigación agraria. Sistemas y recursos forestales, 14:13–26.

Boccacci, P., Akkak, A., Marinoni, D.T., Bounous, G., Botta, R. (2004) Typing European chestnut (*Castanea sativa* Mill.) cultivars using oak simple sequence repeat markers. HortSci., 39:1212–1216.

Bonnefoi, C. (1984) Etude du polymorphism enzymatique des populations forestières de chataignier, *Castanea sativa* Miller. Thèse de Doctorat, Université des Sciences et Techniques du Languedoc, Académie de Montpellier, 142 pp.

Botta, R., Vergano, G., Me, G., Vallania, R. (1995) Floral biology and embryo development in chestnut (*Castanea sativa* Mill.). HortSci., 30:1283–1286.

Botta, R., Guaraldo, P., Mellano, M. G., Bounous, G. (2006) DNA typing and quality evaluation of chestnut (*Castanea sativa* Mill.) cultivars. International Symposium "Optimization, Productivity and Sustainability of Chestnut Ecosystems in Mediterranean Europe" Catania (Italy), 23–26 February 2005. Advances Hort. Sci., 1:96–100.

Bounous, G. (2002) Propagazione. In: Il castagno: Coltura, ambiente ed utilizzazioni in Italia e nel mondo. Bounous, G (ed.) Edagricole, Bologna, Italy, pp 47–63.

Bounous, G., Barone, E., Gioffré, D., Inglese, P, Zappia, R., Peano, C. (1989) Primi risultati dell'indagine sulle cultivar di castagno da fruttodiffuse in Calabria. L'Informatore Agrario, 45:53–57.

Bounous G., Bouchet M., Gourdon L. (1992) Reconstruction of traditional chestnut orchard. Experiences in Piedmont and Southern France. L'Inf. Agr., 9:155–160.

Bounous, G., Torello Marinoni, D. (2005) Chestnut: Botany, Horticulture, and Utilization. Hort. Reviews, Edit. J. Janick; Wiley & Sons Inc., New Jersey, vol. 31:291–347.

Breisch H. (1992) Compatibility tests between the main French varieties of chestnut trees and ink-resistant hybrid rootstocks. Proc. World Chestnut Industry Conference, Morgantown, West Virginia: 41–53.

Breisch, H. (1995) Châtaignes et marrons. CTIFL, 239 p.

Breviglieri N. (1951) Research on the flower and fruit biology in *Castanea sativa* and *Castanea crenata* in Vallombrosa territory. Publication n.1, Centro di Studio sul Castagno (C.N.R.) La Ricerca Scientifica (Suppl.) 21:15–49.

Breviglieri, N. (1955) Indagini ed osservazioni sulle migliori varietà italiane di Castagno. Centro di Studio Sul Castagno, Pubblicazione N.2, Supplemento a la Ricerca Scientifica, pp. 27–164.

Brewbaker, J.L. (1957) Pollen cytology and incompatibility systems in plants. J. Hered., 48: 217–277.

Buck E. J, Hadonou M., James C.J., Blakesley D., Russell K. (2003) Isolation and Characterisation of polymorphic microsatellites in European chestnut (*Castanea sativa* Mill.) Mol. Ecol. Notes, 10:1046–1048.

Burnham, C. R. (1988) The restoration of the American chestnut. American Scientist 76:478–487.

Camus, A. (1929) Les châtaigniers. Encyclopédie économique de sylviculture. P. Lechevalier, 3, 600 p.

Carlson, J., Powell, W., Tomkins, J., Ficklin, S., and Smith, C. (2007) GS20 454 sequencing of the chestnut transcriptome: results of a pilot study. Plant & Animal Genomes XV Conference, January 13–17, 2007, San Diego, CA. W127.

Carlson, J., Barakat, A., DiLoreto, D.S., Smith, C. (2008) The chestnut transcriptome. Plant & Animal Genomes XVI Conference, January 12–16, 2008, San Diego, CA. P25.

Carraway, D.T., Wilde, H.D., Merkle, S.A. (1994) Somatic embryogenesis and gene transfer in American chestnut. J. American Chestnut Foundation, 8:29–33.

Casasoli, M., Mattioni, C., Cherubini, M., Villani, F. (2001) A genetic linkage map of European chestnut (*Castanea sativa* Mill.) based on RAPD, ISSR and isozyme markers. Theor. Appl. Genet., 102:1190–1199.

Casasoli, M., Pot, D., Plomion, C., Monteverdi, M.C., Barreneche, T., Lauteri, M., Villani, F. (2004) Identification of QTLs affecting adaptive traits in *Castanea sativa* Mill. Plant Cell Environ., 27:1088–1101.

Casasoli, M., Derory, J., Morera-Dutrey, C., Brendel, O., Porth, I., Guehl, J-M., Villani, F., Kremer, A. (2006) Comparison of QTLs for adaptive traits between oak and chestnut based on an EST consensus map. Genetics, 172:533–546.

Castroviejo, S., Lainz, M., López, G., Montserrat, P., Muñoz, F., Paiva, J., Villar, L. (1990) Flora Ibérica. Plantas Vasculares de la Península Ibérica e Islas Baleares. Real Jardín Botánico, C.S.I.C., 2, 10–15 pp.

Chapa, J., Chazerans, P., Coulie, J. (1990) Multiplication végétative du châtaignier, amelioration par greffage de printemps et bouturage semi-légneux. L'Arboriculture Fruitierè, 431:41–48.

Conedera, M., Krebs, P. (2008) History, Present Situation and Perspective of Chestnut Cultivation in Europe. In: Abreu, C.G.; Peixoto, F.P.; Gomes-Laranjo, J. (eds): Proceedings of the second Iberian Chestnut Congress, Vila Real (P), June 20-22th 2007. Acta Hort., 784:23–27.

Conedera, M., Marcozzi, M., Jud, B. (1993) Banque de données sur les incendies de forêt au Sud des Alpes suisses. Proceedings of the Symposium "Contribution of European Engineers to Reduction of Natural Disasters", Lausanne, 29–30 Sept. 1993. 165–171.

Conedera, M., Manetti, M-C., Giudici, F., Amorini, E. (2004a) Distribution and economic potential of the Sweet chestnut (*Castanea sativa* Mill.) in Europe. Ecologia mediterranea, 30: 2 p. 179–183.

Conedera, M., Krebs, P., Tinner, W., Pradella, M., Torriani, D. (2004b) The cultivation of *Castanea sativa* (Mill.) in Europe, from its origin to its diffusion on a continental scale. Veget Hist Archaeobot, 13:161–179.

Connors, B.J., Maynard, C.A., Powell, W.A. (2001) Expressed sequence tags from stem tissue of the American chestnut, *Castanea dentata*. Biotech. Letters, 23:1407–1411.

Corredoira, E., Ballester, A., Vieitez, A.M. (2003) Proliferation, maturation and germination of *Castanea sativa* Mill somatic embryos originated from leaf explants. Ann. Bot., 92:129–136.

Corredoira, E., Montenegro, D., San-José, M.C., Vieitez, A.M., Ballester, A. (2004) Agrobacterium-mediated transformation of European chestnut embryogenic cultures. Plant Cell Rep., 23: 311–318.

Corredoira, E., San-José, M.C., Ballester, A., Vieitez, A.M. (2005) Genetic transformation of *Castanea sativa* Mill. by Agrobacterium tumefaciens. Acta Hort., 693:387–393.

Corredoira, E., Ballester, A., Vieitez, F.J., Vieitez, A.M. (2006) Somatic embryogenesis in chestnut. In: Plant Cell Monographs, Vol. 2, Somatic Embryogenesis. Mujib, A. and Samaj, J. (eds.). Springer, Berlin, Heidelberg, 177–199.

Corredoira, E., San-José, M.C., Vieitez, A.M., Ballester, A. (2007) Improving genetic transformation of European chestnut and cryopreservation of transgenic lines. Plant Cell Tissue Organ Cult., 91:281–288.

Costa, R., Ribeiro, C., Valdiviesso, T., Afonso, S., Borges, O., Soeiro,,J., Costa, H., Fonseca, L., Augusta, C., Cruz, M.H., Salazar, M., Matos Soares, F., Sequeira, J., Assunção, A., Correia, P., Lima, M.J. (2008) Variedades de Castanha das Regiões Centro e Norte de Portugal. 80 pp. Ed. INRB.I.P.

Dane, F., Hawkins, L.K., Huang, H. (1999) Genetic variation and population structure of *Castanea pumila* var. *ozarkensis*. J. Amer. Soc. Hort. Sci., 124:666–670.

Dane, F., Lang, P., Huang, H., Fu, Y. (2003) Intercontinental genetic divergence of *Castanea* species in eastern Asia and eastern North America. Heredity, 91:314–321.

Debouzie, D., Heizmann, A., Desouhant, E., Manu, F. (1996) Interference of several temporal and spatial scales between two chestnut insects. Oecologia, 108:151–158.

D'Emerico, S., Bianco, P., Medagli, P., and Schirone, B. (1995) Karyotype analysis in *Quercus* spp. (Fagaceae). Silvae Genet., 44:66–70.

Desvignes J.C. (1996) L'incompatibilitè du châtaignier induite par le chestnut mosaic virus ChMV. Infos-Ctifl, n° 121.

Diller, J. D. and R. B. Clapper. (1969) Asiatic and hybrid chestnut trees in the eastern United States. J. For., 67:328–331.

Dion, R. (1977) Histoire de la vigne et du vin en France dês origines au XIX siècle. Flammarion, Paris.

Diskin, M., Steiner, K. C., Hebard, F. V. (2006) Recovery of American chestnut characteristics following hybridization and backcross breeding to restore blight-ravaged *Castanea dentata*. For. Ecol. Manage., 223:439–447.

Elias, T.S. (1971) The genera of Fagaceae in the southeastern United States. J. Arnold Arboretum 52:159–195.

Elorrieta, J. (1949) El castaño en España. *Ministerio de Agricultura Pesca y Alimentación* No. 48. Madrid, 303 pp.

Ensminger, A. H., Ensminger, M. E., Konlande, J. E., Robson, J. R. K. (1995) The Concise Encyclopedia of Foods and Nutrition. 2nd ed., CRC Press, Boca Raton, FL.

Eriksson, G., Jonson, A., Laureti, M., Pliura, A. (2005) Genetic Variation in Drought Response of *Castanea sativa* Mill. Seedlings. Acta Hort., 693:247–254.

FAO (2009) http://faostat.fao.org/.

Faridi, N., Nelson, C.D., Banda, H., Abdul Majid, M., Kubisiak, T.L., Hebard, F.V., Sisco, P.H., Phillips, R. (2008) Cytogenetic analysis of a reciprocal translocation in F_1 hybrid between American and Chinese chestnuts. Plant & Animal Genomes XVI Conference, January 12–16, 2008, San Diego, CA. W346.

Fernández-López, J., Vázquez-Ruíz-de-Ocenda, R.A., Díaz-Vázquez, R.D., Pereira-Lorenzo, S. (2002) Evaluation of resistance of *Castanea* sp. clones to *Phytophthora* sp. using excised chestnut shoots. For. Snow. Landsc. Res., 76:451–454.

Fernando, D.D., Richards, J.L., Kikkert, J.R. (2006) In vitro germination and transient GFP expression of American chestnut (*Castanea dentata*) pollen. Plant Cell Rep., 25:450–456.

Ferreria-Cardoso, J. V., Fontainhas-Fernandes A.A., Torres-Pereira, M.G. (1993) Nutritive value and technological characteristics of *Castanea sativa* Mill. fruits - comparative study of some Northeastern Portugal cultivars. In: International Congress on Chestnut, Spoleto. Proceedings. Spoleto, Italy.

Ferrini F., Mattii G.B., Nicese F.P., Pisani P.L. (1992) Investigation for realisation of clonal rootstock for chestnut trees. Proceedings of a scientific meeting SOI 1992, Ravello, Rome, Italy: 412–413.

Ficklin, S., Smith, C., Sisco, P., Wheeler, N., Carlson, J., Sederoff, R., Tomkins, J. (2007) Fagaceae genomic database (FGD): an integrated web-based resource for tree genomics. Plant & Animal Genomes XV Conference, January 13–17, 2007, San Diego, CA. P848.

Fineschi, S., Gillet, E., Malvolti, M.E. (1990a) Genetics of sweet chestnut (*Castanea sativa* Mill.). Silvae Genet., 39:188–193.

Fineschi, S., Malvolti, M.E., Morgante, M., Vendramin, G.G., Paciucci, M. (1990b) Genetic studies on cultivated chestnut. Abstract, Congreso ISHS, Florencia, Agosto 1990.

Fineschi, S., Malvolti, M.E., Morgante, M., Vendramin, G.G. (1994) Can. J. For. Res., 24:1160–1165.

Fineschi, S., Taurchini, D., Villani, F., Vendramin, G.G. (2000) Chloroplast DNA polymorphism reveals little geographical structure in *Castanea sativa* Mill.(Fagaceae) throughout southern European countries. Molecular Ecology, 9:1495–1503.

Fu, Y., Dane F. (2003) Allozyme variation in endangered *Castanea pumila* var. *pumila*. Annals Bot., 92:223–230.

Gallastegui, C. (1926) Técnica de la hibridación artificial del castaño. Boletín de la Real Sociedad Española de Historia Natural. Tomo XXVI, pp. 88–94.

Gomes-Laranjo, J., Coutinho, J.P., Peixoto, F., Araujo-Alves, J. (2007) Ecologia do castanheiro. In J Gomes-Laranjo, J Ferreira-Cardoso, E Portela, CG Abreu, eds, Castanheiros. UTAD, Vila Real, pp 109–149.

Gomes Guerreiro, M. (1948) Acerca do uso da análise discriminatória: comparação entre duas castas de castanhas. – Sep. Das Publicações da Direcção Geral dos Serviços Florestais e Aquícolas, Vol.XV, Tomo I e II, p.137–151.

Gomes Guerreiro, M. (1957) Castanheiros: alguns estudos sobre a sua ecologia e o seu melhoramento genético. Instituto Superior de Agronomía, Lisboa.

Goulao, L., Valdiviesso, T., Santana, C., Oliveira, C.M. (2001) Comparison between phenetic characterisation using RAPD and ISSR markers and phenotypic data of cultivated chestnut (*Castanea sativa* Mill.). Gen. Res. Crop Evol., 48:329–338.

Graves, A.H. (1961) Keys to chestnut species. Annu. Rep. North. Nut Grow. Assoc., 61:78–90.

Grente, J., Berthelay-Sauret, S. (1978) Biological control of chestnut blight in France. En: Proc. Am. Chestnut Symp. Morgantown: Virginia University Books pp 30–34.

Hamrick, J.L., Godt, M.J.W. (1989) Allozyme diversity in plants. *In:* Brown, A.H.D., Clegg, M.T., Kahler, A.L., and Weir, B.S., eds. Plant population genetics, breeding and genetic resources. Sunderland, MA: Sinauer Associates Inc.

Hardin, J.W., Johnson, G.P. (1985) Atlas of foliar surface features in woody plants, VIII. *Fagus* and *Castanea* (Fagaceae) of eastern North America. Bull. of the Torrey Bot. Club, 112:11–20.

Hebard, F.V., Stiles, S. (1996) Backcross breeding simplified. Journal of the American Chestnut Foundation, 10:35–39.

Heiniger,U., and Rigling, D. (1994) Biological Control of Chestnut Blight in Europe. Ann. Rev. Phytopathol., 32:581–589.

Huang, H., Dane, F., Norton, J.D. (1994a) Genetic analysis of 11 polymorphic isozyme loci in chestnut species and characterization of chestnut cultivars by multi-locus allozyme genotypes. J. Amer. Soc. Hort. Sci., 119:840–849.

Huang, H., Dane, F., Norton, J.D. (1994b) Allozyme diversity in Chinese, Seguin and American chestnut (*Castanea* spp.). Theor. Appl. Genet., 88:981–985.

Huang, H., Dane, F., Norton, J.D. (1996) Linkage relationships of isozymes and morphological traits in interspecific chestnut crosses. HortSci., 31:419–420.

Huang, H., Dane, F., Kubisiak, T.L. (1998) Allozyme and RAPD analysis of the genetic diversity and geographic variation in wild populations of the American chestnut (Fagaceae). Amer. J. Bot., 85:1013–1021.

Inoue, E., Ning, L., Hara, H., Ruan, S., Anzai, H. (2009) Development of Simple Sequence Repeat Markers in Chinese Chestnut and Their Characterization in Diverse Chestnut Cultivars. J. Am. Soc. Hort. Sci., 134:610–618.

Janeiro, L.V., Vieitez, A.M., Ballester, A. (1995) Cold storage of in vitro cultures of wild cherry, chestnut and oak. Ann. Sci. For., 52:287–293.

Jaynes, R.A. (1962) Chestnut chromosomes. For. Sci., 8:372–377.

Jaynes, R.A. (1964) Interspecific crosses in the genus *Castanea*. Silvae Genet., 13:146–154.

Jaynes, R.A. (1972) Genetics of chestnut. USDA Forest Serv. Res. Pap. WO-17, 13p.

Jaynes, R.A. (1975) Chestnuts. Pages 490–503 in: Janick, J and Moore, J.N. (eds.) Advances in Fruit Breeding. Purdue Univ. Press, West Lafayette, IN.

Johnson, G.P. (1988) Revision of *Castanea* sect. *Balanocastanon* (Fagaceae). J. Arnold Arboretum, 69:25–49.

Kampfer, S., Lexer, C., Glössi, J., Steinkellner, H. (1998) Characterization of (GA)n microsatellite loci from Quercus robur. Hereditas, 129:183–186.

Krebs, P., Conedera, M., Pradella, M., Torriani, D., Felber, M., Tinner, W. (2004) Quaternary refugia of the sweet chestnut (*Castanea sativa* Mill.): an extended palynological approach. Veget Hist Archaeobot, 13:145–160.

Kremer, A., Casasoli, M., Barreneche, T. [and others]. 2007. Chapter 5 Fagaceae Trees. In: Genome Mapping and Molecular Breeding in Plants Vol. 7; Forest Trees. (Kole, C. Ed.), Springer Publ, Leipzig, Germany.

Kubisiak, T.L., Roberds, J.H. (2006) Genetic structure of American chestnut populations based on neutral DNA markers. *In:* Steiner, K.C., and Carlson, J.E., eds. Restoration of American Chestnut to Forest Lands – Proceedings of a Conference and Workshop. May 4–6, 2004, The North Carolina Arboretum, Natural Resources Report NPS/NCR/CUE/NRR – 2006/001, National Park Service, Washington, DC.

Kubisiak, T.L., Hebard, F.V., Nelson, C.D., Zhang, J., Bernatzky, R., Huang, H., Anagnostakis, S.L., Doudrick, R.L. (1997) Molecular mapping of resistance to blight in an interspecific cross in the genus *Castanea*. Phytopathol., 87:751–759.

Lafitte G (1946) Le Châtaignier Japonais en Pays Basque. Mendionde 69.

Lang, P., Dane, F., Kubisiak, T.L. (2006) Phylogeny of *Castanea* (Fagaceae) based on chloroplast trnT-L-F sequence data. Tree Gen. and Genom., 2:132–139.

Lang P, Dane F, Kubisiak TL, Huang H. (2007) Molecular evidence for an asian origin and a unique westward migration of species in the genus *Castanea* via Europe and North America. Mol. Phylogen. Evol., 43:49–59.

Lauteri, M., Villani, F. (2004) Identification of QTLs affecting adaptive traits in *Castanea sativa* Mill. Plant Cell Environ., 27:1088–1101.

Little, E.L. Jr. (1979) Checklist of United States trees (native and naturalized). US Dept. of Agric. Handbook No. 541, Washington, DC. 375 p.

Liu, L.; Zhou, J.Y. (1999) Some considerations on chestnut development in the 21st century in China. Acta Hort., 494:85–88.

Malvotti, M.E., Fineschi, S. (1987) Analysis of enzyme systems in chestnut (*Castanea sativa* Mill.). Genet. Agr., 41:243–256.

Manos, P.S., Zhe-Kun Zhou, Z-K., Cannon, C.H. (2001) Systematics of *Fagaceae*: phylogenetic tests of reproductive trait evolution. Int. J. Plant Sci., 162:1361–1379.

Marinoni, D., Akkak, A., Bounous, G., Edwards, K.J., Botta, R. (2003) Development and characterization of microsatellite markers in *Castanea sativa* (Mill). Mol. Breed., 11:127–136.

Martín, M.A.., Mattioni, C., Cherubini, M., Taurchini, D., Villani, F. (2010) Genetic characterisation of traditional chestnut varieties in Italy using microsatellites (simple sequence repeats) markers. Annals Applied Biol., 157:37–44.

Martins, A., Linhares, I., Raimundo, F., Coutinho, J.P., Gomes-Laranjo, J., Borges, O., Sousa, V. (2005) The importance of deep soil layers to supply water to agro-forestry systems: A case study of a mature chestnut orchard in Northern Portugal. Acta Hort., 693:663–670.

Maynard, C.A., Powell, W.A., Polin-McGuigan, L.D., Viéitez, A.M., Ballester, A., Corredoira, E., Merkle, S.A. and Andrade, G.M. (2008) Chestnut. In: Kole, C. and Hall, T.C. (eds.) A Compendium of Transgenic Crop Plants: Transgenic Forest Trees Species. Volume 9. Blackwell Publishing, Oxford, UK, pp. 169–192.

McCarthy, M. A., Meredith, F.I. (1988) Nutrient data on chestnuts consumed in the United States. Econ. Bot., 42:29–36.

Mehra, P.N., Hans, A.S., Sareen, T.S. (1972) Cytomorphology of Himalayan Fagaceae. Silvae Genet., 21:102–109.

Miranda-Fontaíña, M.E., Fernández-López, J. (1992) Micropropagation as nursery technique in chestnut compared with stooling. Proceedings of the World Chestnut Industry Conference, July 8–10, 1992. Morgantown, West Virginia, USA, 54–60.

Nienstaedt, H. (1948) Notes on the Chestnut: Breeding, Culture, and Botanical Characters of Species and Hybrids. Master's Thesis, Yale School of Forestry, 104 p + XIII.

Nienstaedt, H. (1956) Receptivity of the pistillate flowers and pollen germination test in genus *Castanea*. Zeitschrift Forstgenetik Forstplanzenz, 5:40–45.

Ohri, D., Ahuja, M.R. (1991) Giesma C-banding in *Fagus sylvatica* L., *Betula pendula* Roth and *Populus tremula* L. Silvae Genet., 40:72–75.

Oraguzie, N.C., McNeil, D.L., Paterson, A.M., Chapman, H.M. (1998) Comparison of RAPD and MorphoNut markers for revealing genetic relationships between chestnut species (*Castanea* spp) and New Zealand chestnut selections. N.Z. J. Crop & Hort. Sci., 26:109–15.

Pastorelli, R., Smulders, M.J., Van't Westende, W.P.C, Vosman, B, Gianinini, R, Vettori, C, Vendramin G.G. (2003) Characterization of microsatellites markers in *Fagus sylvatica* L. and *Fagus orientalis* Lipsky Mol. Ecol. Notes, 3:76–78.

Pereira, M.J.P., Castro, L.F.T., Torres-Pereira, J.M.G., Pereira-Lorenzo, S. (1999) Isozyme polymorphisms in portuguese chestnut cultivars. Acta Hort., 494:283–286.

Pereira-Lorenzo, S., Fernández-López, J. (1997) Propagation of chestnut cultivars by grafting: methods, rootstocks and plant quality. J. Hort. Sci., 72:731–739.

Pereira-Lorenzo, S., Fernandez-Lopez, J. (2001) El castaño. In: La Horticultura Española (F. Nuez y G. Llácer, eds.), Madrid, 280–284.

Pereira-Lorenzo, S., Ramos-Cabrer, A.M. (2004) Chestnut, and ancient crop with future. Production Practices and Quality Assessment of Food Crops, Volume 1, Preharvest Practice (Ramdane Dris and S. Mohan Jain ed.), Kluwer Academic Publishers, 1:105–161. ISBN: 1-4020-1698-0.

Pereira-Lorenzo, S., Fernández-López, J., Moreno-González, J. (1996a) Variability and grouping of Northwestern Spanish Chestnut Cultivars (*Castanea sativa*). I. Morphological traits. J. Amer. Soc. Hort. Sci., 121:183–189.

Pereira-Lorenzo, S., Fernández-López, J., Moreno-González, J. (1996b) Variability and grouping of Northwestern Spanish Chestnut Cultivars (*Castanea sativa*). II. Isoenzyme traits. J. Amer. Soc. Hort. Sci., 121:190–197.

Pereira-Lorenzo, S.; Ascasibar, J., Ramos, A. (1999) Spanish scheme in chestnut improvement for nut and timber production. FAO-CIHEAM-Nucis- Newsletter, 8:32–34.

Pereira-Lorenzo, S., Díaz-Hernández, B., Ciordia-Ara, M., Ascasibar-Errasti, J., Ramos-Cabrer, A.M., Sau, F. (2001a) Spanish chestnut cultivars. HortSci., 36:344–347.

Pereira-Lorenzo, S., Ramos-Cabrer, A.M., Rios-Mesa, D., Perdomo, A., González-Pérez, J. (2001b) Update of the Spanish Chestnut Inventory of Cultivars. FAO-CIHEAM-Nucis-Newsletter, 10:34–37.

Pereira-Lorenzo, S., Rios, D., Cubas, F., Calzadilla, C., Ramos-Cabrer, A.M., Sánchez-Sánchez, A. (2001c) Chestnut cultivars on the Canary Islands. Forest Snow Lands. Res., 76:445–450.

Pereira-Lorenzo, S., Díaz-Hernández, M.B., Ramos-Cabrer, A.M. (2006a) Use of highly discriminating morphological characters and isoenzymes in the study of Spanish chestnut cultivars. J. Amer. Soc. Hort. Sci., 131:770–779.

Pereira-Lorenzo, S., Ramos-Cabrer, A.M., Díaz-Hernández, M.B., Ciordia-Ara, M., Rios-Mesa, D. (2006b) Chemical composition of chestnut cultivars from Spain. Sci. Hort., 107:306–314.

Pereira-Lorenzo, S., Ríos-Mesa, D., González-Díaz, A.J., Ramos-Cabrer, A.M. (2007) Los castañeros de Canarias. CCBAT - Cabildo de Tenerife; CAP - Cabildo de La Palma, 1–136.

Pereira-Lorenzo, S., Costa, R., Ramos-Cabrer, A.M., Ribeiro, C., Serra da Silva, C., Manzano, G., Barreneche, T. (2010). Variation in grafted European chestnut and hybrids by microsatellites reveals two main origins in the Iberian Peninsula. Tree Gen. Genom., 6:701–715.

Pereira-Lorenzo, S., Lourenço Costa, R.M., Ramos-Cabrer, A.M., Ciordia-Ara, M., Marqués Ribeiro, C.A., Borges, O., Barreneche, T. (2011). Chestnut cultivar diversification process in the Iberian Peninsula, Canary Islands and Azores. Genome, 54:301–315. Doi:10.1139/G10-122.

Pigliucci, M., Benedettelli, S., Villani, F. (1990a) Spatial patterns of genetic variability in Italian chestnut (*Castanea sativa*). J. Can. Bot., 68:1962–1967.

Pigliucci, M., Villani, F., Benedettelli, S. (1990b) Geographic and climatic factors associated with the spatial structure of gene frequencies in *Castanea sativa* Mill. from Turkey. J. Genet., 69: 141–149.

Pitte, J.R. (1985) Le châtaignier en Gaule et dans les provinces voisines. En: Le Bois dans la Gaule romaine et les provinces voisines, ERRANCE, 21:185–190.

Pitte, J.R. (1986) Terres de Castanide. Fayard, 479 pp.

Pliura, A., Eriksson, G. (2002) Genetic Variation in Juvenile Height and Biomasa of Open-pollinated Familias of six *Castanea sativa* Mill. Populations in a 2 x 2 Factorial Temperature x Watering Experiment. Silvae Gent., 51:152–160.

Polin, L.D., Liang, H., Rothrock, R.E., Nishii, M., Diehl, D.L., Newhouse, A.E., Nairn, C.J., Powell, W.A., Maynard, C.A. (2006) Agrobacterium-mediated transformation of American chestnut (*Castanea dentata* (Marsh.) Borkh.) somatic embryos. Plant Cell Tissue Organ Cult., 84:69–79.

Ramos-Cabrer, A.M., Pereira-Lorenzo, S. (2005) Genetic Relationship between *Castanea sativa* Mill. Trees from North-western to South Spain Based on Morphological Traits and Isoenzymes. Gen. Res. Crop Evol., 52:879–892.

Ramos-Cabrer, A.M., Díaz-Hernández, M.B., Ciordia-Ara, M., Rios-Mesa, D., Gonzalez-Díaz, J., Pereira-Lorenzo, S. (2006) Study of Spanish chestnut cultivars using SSR markers. Adv. Hort. Sci., 20:113–116.

Robichaud, R.L., Lessard, V.C., Merkle, S.A. (2004) Treatments affecting maturation and germination of American chestnut somatic embryos. J. Plant Physiol., 161:957–969.

Robin, C., Heiniger, U. (2002) Chestnut blight in Europe: diversity of Chryphonectria parasitica, hypovirulence and biocontrol. For. Snow Landsc. Res., 76:361–367.

Rodríguez, L., Cuenca, B., Pato, B., Cámara, M.J., Ocaña, L. (2005) Cost and efficiency of propagating *Castanea sativa* hybrids in vitro and by cutting in a commercial nursery. Acta Hort., 693:305–311.

Rothrock, R.E., Polin-Mcguigan, L.D., Newhouse, A.E., Powell, W.A., Maynard, C.A. (2007) Plate flooding as an alternative Agrobacterium-mediated transformation method for American chestnut somatic embryos. Plant Cell, Tissue Organ Cult., 88:93–99.

Sánchez, M.C., San-José, M.C., Ferro, E., Ballester, A., Vieitez, A.M. (1997a) Improving micropropagation conditions for adult-phase shoots of chestnut. J. Hortic. Sci., 72:433–443.

Sánchez, M.C., Ballester, A. and Vieitez, A.M. (1997b) Reinvigoration treatments for the micropropagation of mature chestnut trees. Ann. Sci. For., 54:359–370.

Santamour F.S. Jr., McArdle A.J., Jaynes R.A. (1986) Cambial isoperoxidase patterns in *Castanea*. J. Environ. Hort., 4:14–16.

Sartor, C., Mellano, M.G., Quacchia, A., Alma, A., Botta, R. (2007) Cinipide galligeno del castagno: prospettive di impiego di strategie da affiancare alla lotta biologica. Riassunto dei lavori, VIII giornate scientifiche SOI. Italus Hortus, 14:130–133.

Sawano, M., Ichii, T., Nakanishi, T., Kotera, Z. (1984) Studies on identification of chestnut species and varieties by isozyme analysis. Science Reports of Faculty of Agriculture, Kobe University, 16:67–71.

Schad, C., Solignat, G., Grente, J., Venot, P. (1952) Recherches sur le châtaignier à la Station de Brive. Ann. Amel. Plantes, 3:376–458.

Seabra, R.C., Pais, M.S. (1998) Genetic transformation of European chestnut. Plant Cell Rep., 17:177–182.

Seabra, R.C., Pais, M.S. (1999) Genetic transformation of European chestnut (*Castanea sativa* Mill.) with genes of interest. Acta Hort., 494:407–413.

Sebastiana, M. (2006) Identificação de genes envolvidos no reconhecimento hóspede/hospedeiro em ectomicorrizas. Ph.D. Dissertation.127 pp.

Sederoff, R., Nielson, D., Smith, C., Tomkins, J., Atkins, M., Blackmon, B., Staton, M., Ficklin, S., Hebard, F., Sisco, P., Carlson, J., Diloreto, S., Barakat, A., Powell, W., Baier, K., Anagnostakis, S., Kubisiak, T., Wheeler, N. (2008) Genomic tool development for the *Fagaceae*. Plant & Animal Genomes XVI Conference, January 12–16, 2008, San Diego, CA. p494.

Serrazina, S. (2004) Isolamento e Caracterização de Genes de Resistência à Doença da Tinta em *Castanea sativa* Mill. Ph.D. Dissertation. 301 pp.

Sisco, P.H., Kubisiak, T.L., Casasoli, M., Barreneche, T., Kremer, A., Clark, C., Sederoff, R.R., Hebard, F.V., Villani, F. (2005) An improved genetic map for *Castanea mollissima/Castanea dentata* and its relationship to the genetic map of *Castanea sativa*. Acta. Hort., 693:491–495.

Soylu A. (1990) Heredity of male sterility in some chestnut cultivars (*Castanea sativa* Mill.) XXIII Int. Congr. Florence.
Soylu, A., Serdar, U. (2000) Rootstock selection on chestnut (*Castanea sativa* Mill.) in the middle of black sea region in Turkey. Acta Hort., 538:483–487.
Steinkellner H., Fluch S., Turetschek E., Lexer C., Streiff R., Kremer A., Burg K., Glössl J. (1997) Identification and characterization of (GA/CT)n -microsatellite loci from *Quercus petraea*. Plant Mol. Biol., 33:1093–1096.
Stilwell, Kevin L., Wilbur, Henry M., Werth, Charles R., Taylor, Douglas R. (2003) Heterozygote advantage in the American chestnut, *Castanea dentata* (Fagaceae) Am. J. Bot., 90:207–213.
Tanaka, K., Tsumura, Y., Nakamura, T. (1999) Development and polymorphism of microsatellite markers for *Fagus crenata* and the closely related species, F. japonica. Theor. Appl. Genet., 99:11–15.
Tanaka, T, Yamamoto, T, Suzuki, M. (2005) Genetic Diversity of *Castanea crenata* in Northern Japan Assessed by SSR Markers. Breed. Sci., 55:271–277.
Taylor, W. A. and H. P. Gould. (1914) Promising new fruits. pp 122–124 IN: Yearbook of Agriculture for 1913. USDA, Washington, D.C.
UPOV (1988) Draft guidelines for the conduct of tests for distinctness, homogeneity and stability (CHESNUT). TG/124/1(proj.), 23 p.
Urquijo, P. (1944) Aspectos de la obtención de híbridos resistentes a la enfermedad del castaño. Bol. Veg. Ent. Agr. XIII: 447–462.
Urquijo, P. (1957) La regeneración del castaño. Bol. de Pat. Veg. y Entomología Agrícola, XXII:217–232.
Valdiviesso, M.T. (1999) Estudo sobre a reproduçao sexuada e caracterizaçao de cultivares de *Castanea sativa* Mill. Thesis Doctoral, Alcobaça.
Van Fleet, W. (1920) Chestnut work at Bell Experiment Plot. 11th annual report of the Northern Nut Growers Association, pp 16–21.
Vidal, N., Sánchez, C., Jorquera, L., Ballester, A., Vieitez, A.M. (2005) Cryopreservation of chestnut by vitrification of in vitro-grown shoot tips. In Vitro Cell. Dev. Biol.-Plant, 41:63–68.
Vieira Natividade, J. (1947) Quatro anos na defesa da campanha e Reconstituição dos Soutos. Edição da Junta Nacional das frutas. Lisboa.
Vieitez, E. (1974) Vegetative propagation of chestnut. N.Z. For. Sci., 4:242–252.
Vieitez, F.J., Merkle, S.A. (2004) Fagaceae. In: Litz RE (ed) Biotechnology of fruit and nut crops. CAB International, Wallinford, pp 263–296.
Vieitez, A.M., Sánchez, M.C., García-Nimo, M.L., Ballester, A. (2007) Protocol for micropropagation of *Castanea sativa*. In: Protocols for Micropropagation of Woody Trees and Fruits. AM Jain and H Häggman, eds. Springer, pp 299–312.
Villani, F.S., Pigliucci, M., Benedettelli, S., Cherubini, M. (1991a) Genetic differentiation among Turkish chestnut (*Castanea sativa* Mill.) populations. Heredity, 66:131–136.
Villani, F.S., Benedettelli, S., Paciucci, M., Cherubini, M., and Pigliucci, M. (1991b) Genetic variation and differentiation between natural populations of chestnut (*Castanea sativa* Mill.) from Italy. In: Fineschi, S., Malvolti, M.E., Cannata, F., Hattemer, H.H., eds. Biochemical markers in the population genetics of forest trees. The Netherlands: SPB Academic Publishing BV Press.
Villani, F.; Pigliucci, M.; Cherubini, M.; Sun, O.; Parducci, L. (1993) Genetic diversity of *Castanea sativa* Mill. in Europe: Theorical aspects and applied perspectives. Abstracts, International Congress on Chestnut, Spoleto, Italia.
Villani, F., Sansotta, A., Cherubini, M., Cesaroni, D., Sbordoni, V. (1999) Genetic structure of *Castanea sativa* in Turkey: evidence of a hybrid zone. J. Evol. Biol., 12:233–244.
Vossen, P. (2000) Chestnut culture in California. University of California, Division of Agriculture and Natural Resources, Publication 8010, 17 p.
Wen, H., Norton, J. (1992) Enzyme Variation in Chinese Chestnut Cultivars. Abstract, International Chestnut Conference, Morgantown, USA.
Yamamoto, T., Tanaka, T., Kotobuki, K., Matsuta, N., Suzuki, M., Hayashi, T. (2003) Characterization of simple sequence repeats in Japanese chestnut. J. Hort. Sci. Biotech., 78: 197–203.

Chapter 20
Pecan

Tommy E. Thompson and Patrick J. Conner

Abstract The pecan, *Carya illinoinensis* (Wangenh.) K. Koch, is the most economically important member of the *Carya* genus and is the most valuable native North American nut crop. The *Carya* genus is a member of the walnut family, Juglandaceae, and comprises 20 species. Over 98% of the world's annual pecan production is produced in the southern USA and northern Mexico. Pecan is a diploid ($n=16$), monoecious, long-lived tree species. Owing to its heterodichogamy, pecan is primarily cross-pollinated, resulting in high heterozygosity with severe inbreeding depression when selfed. Establishment of commercial pecan orchards during the nineteenth century was mainly by planting open-pollinated nuts from mother trees possessing desirable characteristics. These orchards consist of trees with widely varying production and quality attributes due to the heterozygosity of pecan. Vegetative propagation became popular ca. 1900, and most newly planted orchards consist of a chosen combination of clonally propagated superior varieties. Clonally derived orchards are more productive and produce nuts of much higher quality than remaining native or seedling orchards. Thirteen *Carya* species, including pecan, are native to the USA. The National Clonal Germplasm Repository for Pecans and Hickories which preserves over 300 pecan cultivars, landraces, and species accessions was established in 1984 to describe and preserve this underutilized resource. Objectives of pecan breeding are higher yields and nut quality, and resistance to diseases and insects. Pecans are attacked by a wide range of disease and insect pests causing substantial losses to the crop. Various levels of resistance to scab and aphids are available in improved pecan varieties, and breeding programs are focusing on developing new cultivars with high levels of resistance in combination with good

T.E. Thompson (✉)
USDA-ARS Pecan Genetics and Breeding Program, 10200 FM 50, Somerville, TX 77879, USA
e-mail: tommy.thompson@ars.usda.gov

P.J. Conner
University of GA, 4604 Research Way, Tifton, GA 31793, USA
e-mail: pconner@uga.edu

M.L. Badenes and D.H. Byrne (eds.), *Fruit Breeding*, Handbook of Plant Breeding 8,
DOI 10.1007/978-1-4419-0763-9_20,

horticultural attributes. Another major effort in pecan breeding is the development of earlier maturing cultivars with the potential to bear more consistently over years.

Keywords Pecan • Breeding • Genetics • Host plant resistance • Insect resistance • Disease resistance • Trees • Nuts • Hickory • Plant selection • *Carya illinoinensis*

1 Introduction

The pecan, *Carya illinoinensis* (Wangenh.) K. Koch, is the most economically important member of the *Carya* Genus, and is the most valuable native North American nut crop. Pecans are harvested from "native" trees throughout the natural range of the species (Fig. 20.1). The culture of "improved" trees has extended considerably beyond the native range; from Ontario, Canada, south to Oaxaca, Mexico, and from the Atlantic coast of Virginia and the Carolinas west to California (Fig. 20.2) In addition, the pecan is grown commercially to a minor extent in Israel, South Africa, Australia, Egypt, Peru, Argentina, and Brazil.

Over 98% of the world's annual pecan production is produced in 15 US southern states and northern Mexico (Pena 2007). This North American annual production averaged 176,443 metric tons (in shell basis) for 1998–2005. Mexico produced about 35% of this, followed by Georgia (19.2%), Texas (14.2%), and New Mexico

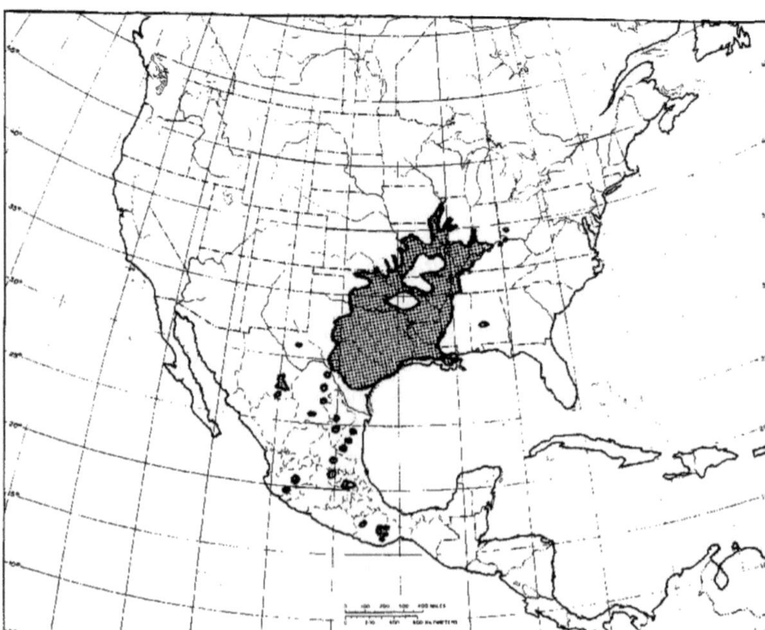

Fig. 20.1 Native pecan distribution (Grauke and Thompson 1996)

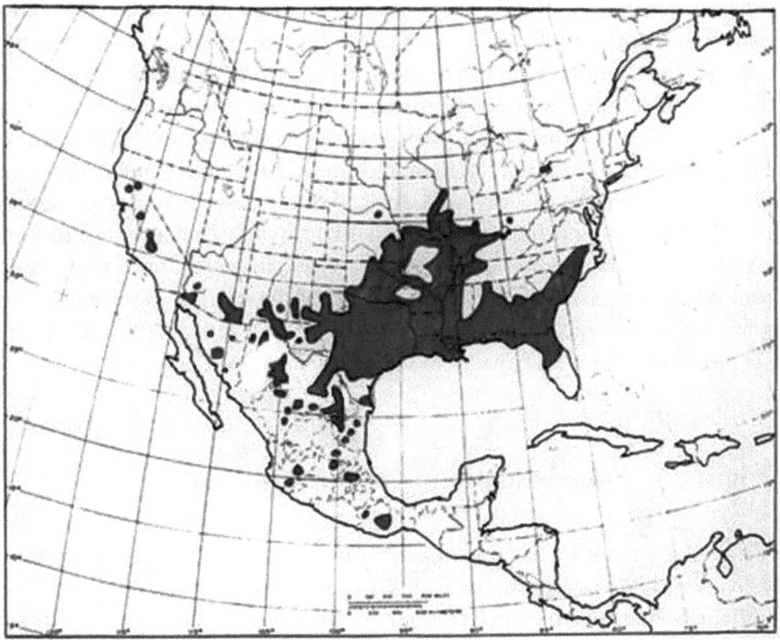

Fig. 20.2 Commercial pecan production in America (Grauke and Thompson 1996)

(12%). The total US production average for 1991–2001 was 121,545 metric tons. The production dropped to 104,682 metric tons for 2002–2005 (Pena 2007). Major recent production challenges such as disease problems in Texas and Georgia, hurricanes along the gulf coast, and droughts limited global production.

The *Carya* genus is a member of the walnut family, Juglandaceae, and comprises 20 species (Grauke and Thompson 1996). Thirteen *Carya* species, including pecan, are native to the USA. Of all *Carya* species, seven are reportedly cultivated for their nuts (Grauke and Thompson 1996), but pecan is the only economically important crop. Selection of superior genotypes and limited horticultural use has been made of two other species in North America: shagbark hickory [*C. ovata* (*Mill*) *K. Koch*] and shellbark hickory [*C. laciniosa* (*F. Michx.*) *Loudon*]. Culture of both shagbark and shellbark hickories is restricted by their long juvenile periods (>10 years) and low yields of hard-to-shell nuts. The Chinese reportedly cultivate some of their hickories for food to a small degree.

Many hickory species, including pecan, have a deserved reputation of producing tough useful wood for tool handles, flooring, veneer, among other products. Hickory wood is also much prized for use in smoking meats because of the distinctive flavor it imparts on the product. Because hickories are slow to grow to an economical size, naturally occurring trees are harvested for wood rather than plantation trees. As a result, the best specimen trees are often preferentially harvested, depleting the genetic potential of these populations over time.

Pecan is grown in a wide range of environments ranging across the arid Southwest, the humid Southeast, and the variable Midwest. Each of these geographic regions places unique environmental constraints on the cultivars that can succeed there. In addition, pecan culture has become more complex with the recent adoption of improved orchard techniques such as hedging and other forms of tree control and mechanical thinning of excess crop load. No single cultivar can meet all the requirements the industry now places on them. Instead, there is an increased demand for an array of regionally and horticulturally adapted cultivars. Orchards of inferior older cultivars or poorly adapted new cultivars are continually abandoned or updated with more profitable cultivars. A review and update of the current genetic status of this crop is needed since breeding objectives have become more refined, and available methods of genetic plant improvement have expanded.

2 Origin and Domestication of Scion Cultivars

Establishment of commercial pecan orchards during the nineteenth century was mainly by planting open-pollinated nuts from mother trees possessing desirable characteristics. Trees that produced large nuts with thin shells were especially prized by early growers for seedstock as this combination of traits greatly decreased the workload of obtaining the edible kernel, a process that was done by hand (Corbett et al. 1926). Other traits selected include resistance to scab disease, early maturity, and heavy yields (Taylor 1906, 1907). This system facilitated genetic improvement of cultivated germplasm since each tree in the orchard was genetically different, and superior trees were identified each cycle of growth. Seed from these superior trees could be used to establish the next orchard, and so on. Thus open-pollinated half-sib populations existed until clonal propagation of superior genotypes led to the widespread use of true cultivars. Currently, the few remaining seedling orchards in the Southeast, some of which have been abandoned, are being examined by researchers in the hopes of discovering genotypes with a high degree of insect and disease resistance (Goff et al. 1998).

The term cultivar was poorly defined early in the industry. Although experienced growers knew it not to be true, a large influx of new growers and a limited understanding of genetic science led to belief that pecan seed would come true to the female parent. This belief persisted in some locations even into the early twentieth century (Halbert 1909). This erroneous concept was disproved as seedling orchards began to bear and the variability of the nut characteristics of the seedlings became evident. Once improved methods of budding and grafting became widespread, the concept of a scion cultivar being a clone instead of an open pollinated collection of mainly half-sib trees was accepted. From that point on, vegetative propagation essentially established what a cultivar was in pecan production. This development allowed more accurate selection of superior pecan material since genetic variability of the scion was eliminated among tested trees, and environmental variability could be more adequately defined. Clonal propagation also vastly improved the uniformity and quality of the harvested crop, while simplifying management and nut processing.

Early clonal propagation of pecan essentially followed ideology common to pomology, but consistent success requires greater care and attention to details than in many other species. Many early pecan growers propagated favorite trees on a small scale with no record of their achievement. The first documented success was that by Abner Landrum of Edgefield, South Carolina in budding pecan scions onto hickory stocks in 1822 (True 1919). Later, in 1846, a slave gardener named Antoine propagated an orchard of 'Centennial' pecans at Oak Alley Plantation in Louisiana. The first record of a nursery selling grafted pecan trees was that of William Nelson of New Orleans, who began selling grafted trees in 1879 (Crane et al. 1937). E.E. Risien of San Saba, Texas developed a ring budding technique in the 1890s that increased the supply and decreased the price of grafted trees, precipitating an active period of pecan nursery sales and orchard establishment (McHatton 1957; Wood et al. 1990).

The period from the 1890s to 1930s was one of rapid proliferation of named clonally propagated pecan cultivars. The new-found ease of propagation allowed the owners of supposedly superior trees to attach a name, often the owner's, and propagate trees locally. This was an exciting era in pecan history because new orchards were being planted on a large scale and beginning to come into production. Also of note, the value of plant breeding and plant improvement in general was filtering down to the growers, and generating much enthusiasm for the use of new "improved" cultivars. Unfortunately, new cultivars were often developed after observing only a few years production of the parent tree, and were of dubious horticultural merit. Thompson and Young (1985) documented over a thousand pecan cultivars which have been listed over the years, and there are likely many more. Most of these were never widely popular and are now extinct, but a few exceptional cultivars from this period still comprise a major portion of current orchards. The latest national cultivar inventory (Thompson 1990) showed that 'Stuart,' which was first propagated in 1886, made up almost one quarter of all trees in USA grafted or budded orchards (Table 20.1). Approximately half (47.3%) of the improved trees in the USA consisted of three cultivars: 'Stuart' (22%), 'Western Schley' (14.6%), and 'Desirable' (10.9%), which were all developed in the late nineteenth or early twentieth century. Of the top 33 cultivars mentioned above, 5 are clones selected directly from native stands. Most others are only two or three generations from native parentage.

The original 'Stuart' tree was selected from seed from an Alabama seedling, while 'Desirable' was grown and selected by a nurseryman in an early breeding effort (Thompson and Young 1985).

These figures strongly reflect the permanence of pecan orchards and the understandable reluctance of growers to replace older trees with superior newer cultivars due to the nonproductive establishment years. An additional barrier to the adoption of new cultivars is the paucity of long-term yield data for new cultivars. The large size and long life-cycle of pecan place strong limits to the scope of cultivar trials that can be reasonably conducted. Planting new cultivars requires a leap of faith on the part of the grower that recently released cultivars that are successful in academic trials will do well as mature trees in his location. Mistakes in cultivar choice will require that the grower either replace the trees and once again endure the nonproductive establishment years, or adapt to the new cultivars' faults as best they can.

Table 20.1 Estimated hectares and percent of each cultivar in the USA (Thompson 1990)

Cultivar	Hectares	%	Cultivar	Hectares	%
Stuart	47,703	21.8	VanDeman	877	0.4
Western Schley	31,848	14.6	Maramec	830	0.4
Desirable	23,849	10.9	Cherokee	809	0.4
Wichita	22,168	10.1	Tejas	809	0.4
Schley	11,696	5.4	Delmas	767	0.4
Cheyenne	10,498	4.8	Sumner	735	0.3
Success	5,550	2.5	Barton	722	0.3
Cape Fear	4,786	2.2	Frotscher	707	0.3
Moneymaker	4,295	2.0	Elliott	682	0.3
Mohawk	3,099	1.4	Pabst	668	0.3
San Saba Imp.	2,873	1.3	Caddo	617	0.3
Mahan	2,856	1.3	Teche	615	0.3
Moore	2,825	1.3	Burkett	526	0.2
Choctaw	2,549	1.2	Shoshoni	454	0.2
Kiowa	1,788	0.8	Mobile	398	0.2
Sioux	1,649	0.8			
Ideal	1,097	0.5	Other	26,019	11.9
Chickasaw	1,084	0.5			
			Total	218,449	100.0

For this reason, many growers continue to replant with cultivars that they are familiar with even when new superior cultivars appear to be available.

Pecan trees are cultivated over a wide geographic area spanning from California to Virginia, and contributes to the economy of 24 states (Wood et al. 1990). Pecan production can be separated into four broad regions: the southeastern spanning from Virginia to Louisiana and Arkansas, the south central consisting of east and central Texas and southern Oklahoma, the northern containing northern Oklahoma and the Midwest, and the west which includes far west Texas and southern areas of New Mexico, Arizona, and California. Each of these production regions has environmental and economic constraints which must be met by the cultivar to be successful. Not surprisingly, orchards in each region consist of different sets of cultivars. In many cases, cultivars which are successful in one region cannot be grown profitably in other regions. Breeding programs must, therefore, target new cultivars to the regions and uses to which it is best adapted.

The southeastern region is typified by a long growing season with humid summers. Pecan scab, *Cladosporium caryigenum* (Ell. et Lang.) Gottwald (1982), is a fungal disease that infects pecan leaf and nut shuck tissue when they are wet. Commercial pecan plantings may require up to 11 fungicide applications annually to control the disease (Ellis et al. 2000). The frequent rainfall in this region during the growing season makes resistance to pecan scab a necessity in successful cultivars. Highly susceptible cultivars such as Wichita and Western Schley, which are extremely productive in the southwest, are not productive in normal years in the Southeast even with the use of fungicide sprays. The most profitable cultivars in this region mature their nuts early in the season (mid September to early October) allowing them to be processed in time for the holiday gift-pack trade (Sparks 1992). Historically, most successful cultivars in this region have moderate crop loads and a less pronounced alternate bearing intensity (Conner and Worley 2000). However, the adoption of mechanical fruit thinning may allow fruit loads to be adjusted so that cultivars which set heavier crops can be successful here in the future.

Two cultivars, Stuart and Desirable, make up over half of the mature trees in commercial orchards in Georgia (Florkowski et al. 1999), where the majority of the production lies in this region. 'Stuart' continues to be popular as a mature tree in Georgia, but new plantings have decreased due to its low precocity and inadequate kernel percentage. 'Desirable' is currently the most popular commercial cultivar in Georgia and comprised 49% of the trees planted in 1993–1997. 'Desirable' sets the standard for nut quality in the Southeast, but requires excellent cultural practices to perform well, and has also become increasingly more susceptible to pecan scab. A range of other cultivars are being planted in this region (Wells 2007), but no cultivar combines all the attributes of large nut size, early harvest date, high kernel quality, and scab resistance that is desired.

In the arid environments of the western region rainfall in the summer is sparse, and fungal diseases are a minor concern. This region has high light intensities and orchards managers often use mechanical pruning techniques to maximize light infiltration of the canopy. Because harvest in this region is later than that of the southeast, cultivars must be able to maximize production to make up for the lower prices received. This region has a shorter growing season, and early freezes can be a problem. Orchards in this region are often composed of 'Western Schley,' with 'Wichita' as a pollinizer. Both of these cultivars are capable of producing a high yields. 'Western Schley' was developed in the early twentieth century, and is popular because of its profuse branching which responds well to pruning, and it is less susceptible to zinc deficiency and water stress (Byford 2005). 'Wichita' is the most productive pecan cultivar ever developed, but requires optimum management to fulfill its potential (McEachern and Stein 1997).

The south central region is a transition zone between the southeastern and western regions. Scab resistance becomes a more important factor in cultivar choice as you move from western Texas to the south and east. 'Desirable,' 'Pawnee,' 'Wichita,' and 'Western Schley' are all grown in this region. Some very productive cultivars with high nut quality have been developed by the USDA for this region.

Older inferior cultivars lacking in productivity, nut quality, and disease and insect resistance are being replaced with superior newer cultivars. In central Texas, for example, 'Wichita' routinely out yields 'Western Schley,' producing at least twice as much kernel weight per acre (Thompson et al. 1981; Thompson and Hunter 1983). 'Pawnee,' released by USDA in 1984 (Thompson and Hunter 1985), is currently the most popular cultivar being propagated worldwide, probably followed by 'Western Schley,' 'Wichita,' and 'Desirable.'

The northern production region requires cultivars that have trees that are resistant to winter injury and can mature their fruits in a shorter growing season. Cultivars suited to this region generally have smaller sized nuts, which is a characteristic of most early maturing nuts (Sparks 1992). Most northern adapted cultivars also do not have the productivity of the southern cultivars. Cultivars can be chosen for either the in-shell market or the shelling market. The in-shell market is a direct market to the consumer, and requires a larger nut with an early harvest. When nuts are sold for the shelling market, size is less important than a good kernel percentage. Cultivars grown in the most northerly regions generally consist of selections from native stands which possess superior nut size and kernel development. Cultivars in the more southern end of this region are more likely from breeding programs. Recent USDA releases with northern adapted germplasm in their pedigree ('Pawnee,' 'Kanza,' 'Osage,' and 'Lakota') are currently gaining popularity in this region.

3 Genetic Resources

Louis D. Romberg, a former ARS pecan breeder, began a pecan and hickory collection in the 1930s at Brownwood, Texas to have parental material to use in the pecan breeding program. The collection of pecan cultivars and other clones were grafted to trees. This collection was designated the National Clonal Germplasm Repository for Pecans and Hickories in 1984, and a Crop Germplasm Committee was formed. Native pecan collections have since been added, as well as many clones of other *Carya* species. Presently, the Cultivar Collection maintains over 300 pecan cultivars as live trees, and nut specimens of many additional cultivars are also preserved. This collection represents all pecan growing regions of the USA and is the largest collection of pecan cultivars in the world. Supporting records of accession origin and characteristics are also available. Live accessions are maintained as grafted trees, targeting two trees of each cultivar at the Brownwood site, and duplicate collections at College Station, Texas. Accessions are provided upon request to researchers, and are provided to private growers when commercial nurserymen cannot provide propagation wood of a clone. Accessions are distributed as graftwood (typically five double graft sticks per accession) in January and February. In addition, seed is occasionally distributed from particular accessions for establishment of seedling rootstocks for subsequent grafting. Nut voucher specimens are maintained for each tree to verify identification. Additional nut samples from other orchards are maintained for many cultivars to provide a sample of the variation that exists across locations. This *ex situ* collection provides an abundance of readily available, verified, and well-documented plant materials for use in biochemical and molecular characterizations. Verified inventories of some pecan cultivars have been characterized with isozyme analysis (Marquard et al. 1995) to provide a method of biochemical verification. To aid cultivar identification, color photographs of many accessions of the cultivar collection are available on the internet at the site maintained by the USDA Pecan Breeding Program and the Georgia Breeding Program (http://extension-horticulture.tamu.edu/carya) and (http://www.caes.uga.edu/commodities/fruits/pecanbreeding/). Photos are color standardized (Thompson et al. 1996) and are linked to specific inventory trees for which additional evaluation information is available. In addition, the site provides passport information for the most commonly planted cultivars.

Collections of other *Carya* species are maintained either as grafted trees (in the case of selected hickory cultivars) or as own-rooted trees (in the case of native tree collections). Currently, all hickory cultivars maintained in the repository are available from commercial sources and have not been distributed. Seed collected from native trees has been sent to researchers, but seedlings in repository collections are still juvenile and are not disseminated. The collection provides an excellent foundation for the study of diversity in this genus. Some accessions are maintained of the sister genera Annamocarya, Juglans, Pterocarya, and Platycarya, providing resolution for the study of diversity in the Walnut Family, Juglandaceae.

Other collections of pecan and hickory exist in the USA and other countries (see Bettencourt and Konopka 1989). Notable US collections include (1) Southeastern Fruit and Tree Nut Lab, Byron, Ga., (2) Coastal Plain Experiment Station, Tifton, Ga., (3) Pecan Experimental Field, Chetopa, Kan., (4) Northern Pecan Research Planting, University of Nebraska, Lincoln, Neb., (5) Pecan Research-Extension Station, Louisiana State University Agricultural Center, Shreveport, La., (6) Alabama Pecan Collection, Fairhope, Ala, and (7) Pecan Provenance and Hybridity Test, Louisiana State University, Idlewild, La. Most collections of *Carya* in other countries are small collections of named US cultivars. Notable exceptions include (1) a collection of cultivars and seedlings of several US *Carya* species and interspecific hybrids, maintained at the Holden Arboretum, Kirtland, Ohio, (2) a collection of *C. laciniosa* from Canada, maintained at the University of Guelph Arboretum, Guelph, Ontario, Canada, and (3) a collection of commercial cultivars and landraces of pecan maintained at the Campo Agricola Experimental de La Laguna, Matamoros, Torreon, Mexico.

Major sources of superior genetic characteristics for nut quality and productivity are provided by superior new cultivars and selections produced in the USDA and the UGA (University of Georgia) breeding programs. These selections represent the forefront to pecan genetic improvement, but new selections are still only a few generations removed from wild trees.

Other potential sources of useful quality traits are provided by experienced growers who discover chance seedling trees with valuable characteristics. Traits which are commonly selected by growers include the following: high kernel percentage, early harvest date, large nut size, and resistance to scab. The UGA breeding program regularly trials grower selections and occasionally makes use of them as parents in the breeding program. Since most seedling trees developed from nuts from popular cultivars, these genotypes can have many favorable quality traits. However, long-term evaluation in replicated orchards often reveal flaws that prevent their use as new cultivars.

A plethora of diseases, insects, and mites attack pecan (Tables 20.2 and 20.3). Host plant resistance to diseases, especially scab, has been observed in many improved cultivars and native populations in the more humid pecan production areas (Table 20.4). Pecan clones exist in Louisiana on which scab has never been observed, even though they are grown in high scab environments (Goff, personal communication). However, the presence of a large number of scab races has been demonstrated, and most pecan cultivars, even those that are highly susceptible, have

Table 20.2 Pecan diseases of the USA and area of occurrence

Common name	Scientific name	Geographic area of occurrence
Fungi		
Scab	*Cladosporium caryigenum* (Eli. et Lang) Gottwald [=Fusicladium effusum (Wint.)]	E. of 98 Longitude
Vein spot	*Gnomonia nerviseda* Cole	Most production areas E. of C. Tex
Downy spot	*Mycosphaerella caryigena* Demaree and Cole	Most production areas E. of C. Tex.
Liver spot	*Gnomonia caryae* Wolfe var. pecanae Cole	Most production areas E. of C. Tex
Zonate leaf spot	*Cristulariella pyramidalis* Waterman and Marshall	Most production areas E. of C. Tex
Powdery mildew	*Microsphaera alni* de Candolle ex Winter	Most production areas
Pink mold	*Cephalothecium roseum* Corda	Most production areas E. of C. Tex
Leaf blotch	*Mycosphaerella dendroides* (Cooke) Demaree and Cole	Most production areas E. of C. Tex
Brown leaf spot	*Cercospora fusca* Rands	Most production areas E. of C. Tex
Clitocybe root rot	*Clitocybe tabescens* (Scop. ex Fr.) Bres.	Ga. and possibly other S.E. states
Phymatotrichum root rot	*Phymatotrichum omnivorum* (Shear) Duggar	C. Tex. and W
Bacteria		
Crown gall	*Agrobacterium tumefaciens* E.F. Smith and Townsend	All production areas
Bacterial Leaf Scorch	*Xylella fastidiosa*	All production areas
Unknown cause		
Shuck dieback		Most production areas
Stem-end blight		Red River and Mississippi River Valleys
Tumor disease		Humid Red River and Mississippi River Valleys
Bunch disease		Most production areas

Table 20.3 Pecan insects and mites in North America

Common name	Scientific name
Pecan nut casebearer	*Acrobasis nuxvorella* Neunzig
Hickory shuckworm	*Cydia caryana* Fitch
Pecan weevil	Curculio caryae Horn
Black pecan aphid	*Melanocallis caryaefoliae* Davis
Black margined aphid	*Monellia caryella* Fitch
Yellow hickory aphid	*Monelliopsis pecanis* Bissell
Pecan phylloxera	*Phylloxera devastatrix* Pergande
Pecan leaf phylloxera	*Phylloxera notabilis* Pergande
Southern pecan leaf phylloxera	*Phylloxera russellae* Stoetzel
Lesser pecan leaf phylloxera	*Phylloxera texana* Stoetzel

(continued)

Table 20.3 (continued)

Common name	Scientific name
Pecan budmoth	*Gretchena bolliana* Slingerland
Southern green stinkbug	*Nezara viridula* L.
Brown stinkbug	*Euschistus servus* Say
Fall webworm (2 races)	*Hyphantria cunea* Drury
Pecan leaf casebearer	*Acrobasis juglandis* LeBaron
Pecan cigar casebearer	*Coleophora laticornella* Clemens
Pecan nursery casebearer	*Acrobasis caryivorella* Ragonot
Walnut caterpillar	*Datana integerrima* Grote and Robinson
Serpentine leaf miner	*Stigmella juglandifoliella* Clemens
Upper southern leaf miner	*Cameraria caryaefoliella* Clemens
Lower southern leaf miner	*Phyllonorycter caryaealbella* Chambers
Pecan leaf scorch mite	*Eotetranychus hicoriae* McGregor
Top leaf southern. mite	*Oligonychus viridis* Banks
Vein mite	*Brevipalpus* sayedi Baker
Leaf roll mite	*Aceria caryae* Keifer
Pecan catocala (several spp.)	*Catocala maestosa* (Hulst) and C. spp.
May beetles (15 spp.)	*Phyllophaga* and *Anomala* spp.
Plant hoppers (4 spp.)	*Anormenis septentrionalis* Spinola and others
Myriads (3 spp.)	*Orthotylus ramus* (Knight) and others
Cicadas (2 spp.)	*Magicicada septendecim* L.
Hickory horned devil	*Citheronia regalis* F.
Sawfly	*Periclista marginicollis* Norton
	Megaxyela major Cresson
Obscure scale	*Melaspis obscura* Comstock
Hickory shoot curculio	*Conotrachelus aratus* Germar
Shoot curculio	*Conotrachelus pecanae*
Nut curculio	*Conotrachelus hicoriae* School
Cambium curculio	*Conotrachelus anaglypticus* Say
Red shoulder, shot hole borer	*Xylobiops basilaris* Say
Pinhole borer	*Xyleborus affinis* Eichhoff and others
American plum borer	*Euzophera semifuneralis* Walker
Flat headed appletree borer	*Chrysobothris femorata* Oliver
Banded hickory borer	*Knulliana cincta* Drury
Pecan borer	*Conopia scitula* Harr.
Pecan carpenter worm	*Cossula magnifica* Strecker
Oak pruner	*Hypermallus villosus* Fab.
Twig girdler	*Oncideres cingulata* Say
Giant bark aphid	*Longistigma caryae* Harris
Leaf-footed bug	*Leptoglossus phyllopus* L.
Northern leaf-footed bug	*Leptoglossus oppositus* Say
Pecan spittle bug	*Clastoptera achatina* Germar
Alder spittle bug	*Clastoptera obtusa* Say
Tile-horned Prionus	*Prionus imbricornis* L.
Broad-necked Prionus	*Prionus laticollis* Drury
Termites	*Reticulitermes* spp.

Table 20.4 Sources of genes for pest resistance in *Carya*

Pest	Resistant cultivars or clones	References
Diseases		
Fungi		
Cladosporium caryigenum	Deakle's Special, Dixie, Elliott, Gafford, Gloria Grande, Melrose, Sumner, Pioneer, USDA 61-6-67, USDA 56-6-148	Goff et al. (1993)
	Barton, Buchel I, Curtis, USDA 88-7-1	Goff et al. (2003)
	A-1, Bradley (or Bradley-2?)Cs-14, Cs-60, Elliot, Gloria Grande, Enloe, Pseudocarman, Russell	KenKnight (1968a, b)
	Barton, Candy, Curtis, Davis, Elliott, Farley, Gloria Grande, Jackson, Melrose, Peruque, Sumner	Hunter et al. (1986)
	Curtis, Dependable, Elliott, Gloria Grande	Payne et al. (1979)
Gnomonia nerviseda	Curtis, Choctaw, Mahan	KenKnight (1968a)
	Barton, Cape Fear, GraBohis, Jackson, Maramec, Mohawk, Sumner	Hunter et al. (1986)
Mycosphaerella caryigena	Jennings Elliott, Wichita	KenKnight (1968a), Hunter et al. (1986)
Gnomonia caryae var. *pecanae*	Carman, Curtis, Desirable, Gloria Grande, Jackson, Jennings, Moreland, Russell, Superdesirable	KenKnight (1968a)
Mycosphaerella dendroides	Most clones resistant, except Desirable	KenKnight (1968a)
Cercospora fusca	Carman, Candy, Curtis, Gloria Grande, Moreland, Natchez, Russell, A-93	KenKnight (1968a)
Cephalothecium roseum	Those clones resistant to scab	Payne et al. (1979)
Microsphaera alni	Most resistant, except Caspiana, Pabst, Superdesirable	KenKnight (1968a)
From unknown causes		
Shuck dieback	Success is susceptible	Payne et al. (1979)
Stem-end blight	Most cultivars seem resistant, except Success, Dunstan, Magenta, Barton, Desirable	Payne et al. (1979)
Bunch disease	Candy, Choctaw, Curtis, Farley, Gloria Grande, Jackson, Lewis, Mohawk, Stuart	KenKnight (1968a)
Tumor disease	Desirable, Stuart	Payne et al. (1979)
Leaf scorch	Barton, Choctaw, Curtis, Desirable, GraBohls, Kiowa, Maramec, Mohawk, Shawnee	Hunter et al. 1986
Insects/mites		
Cydia caryana	USDA Selections 44-15-51 and 44-4-135, Osage, GraBohls, Cape Fear, Chickasaw, Cherokee, Shoshoni, Brake	Calcote et al. (1976), Hansen et al. (1970)
Curculio caryae	Success, Mobile, Teche, Van Deman, Nugget, Mahan, Schley	Moznette (1948), Criswell et al. (1975), Boethel and Eikenbary (1979), Gill (1917)

(continued)

Table 20.4 (continued)

Pest	Resistant cultivars or clones	References
Hemipterans	Candy, Creek, Forkert, Grabohls, Gloria Grande, Kanza, Kiowa, Maramec, Owens, Pawnee, Sumner, Tejas, Western Schley	Dutcher et al. (2001)
Melanocallis caryaefoliae	Curtis, Moneymaker, Moore	Moznette et al. (1940)
	Cape Fear, Creek, Kiowa, Pawnee, Schley	Kaakeh and Dutcher (1994)
	Barton, Cape Fear, Cowley, Curtis, Farley, Grabohls, Mahan, Sioux	Wood and Reilly (1998)
Monellia caryella	Success, Schley	Carpenter et al. (1979)
	Gloria Grande, Pawnee	Kaakeh and Dutcher (1994)
Monelliopsis pecanis	Cape Fear, Pawnee	Kaakeh and Dutcher (1994)
Phylloxera notabilis	Delmas, Western Schley, 1983 Williamson, Success, Squirrel's Delight, Stuart Moneymaker, Burkett, plus many others	Boethel et al. (1976), Calcote (1983)
Phylloxera devastatrix	Many	Calcote and Hyder (1980)
Clastoptera achatina	Stuart, Lewis, Mahan	Neel et al. (1976)
Tetranychidae	Stuart	Gentry et al. (1976)
Boarmia selenaria	Moneymaker, Mahan, Schley	Wysoki and Yizhar (1976)

resistance to multiple scab races (Conner and Stevenson 2004). As a result, when newly selected clones displaying strong scab resistance at a single location are propagated and distributed on a wide scale, resistance often breaks down as they are exposed to a larger number of scab races (Goff et al. 1998; Thompson et al. 1995). Resistance to other diseases has been observed in many sources, but verification is lacking (Table 20.4).

The black pecan aphid *Melanocallis caryaefoliae* (Davis) and the yellow aphid complex [the black margined aphid. *Monellia caryella* (Fitch) and the yellow pecan aphid (*Monelliopsis pecanis* Bissell)] are major entomological pests of pecan. Several studies of host plant resistance to these aphid species have been undertaken (Table 20.4). Breeding for resistance to aphids is an integral part of the current pecan breeding programs, but is complicated by the fact that cultivars preferred by one aphid species are not necessarily preferred by another aphid species (Kaakeh and Dutcher 1994). Some cultivars do, however, seem to have resistance to more than one species. 'Pawnee' has been shown to have a high level or resistance to the yellow pecan aphid complex (Kaakeh and Dutcher 1994; Thompson and Grauke 1998; Thompson et al. 2000), and 'Cape Fear' appears resistant to black and yellow pecan aphids (Kaakeh and Dutcher 1994). A major source of the damage caused by the yellow pecan aphid complex is caused by the deposition of honeydew on leaf surfaces which leads to the growth of a fungal mat on the leaf surface which reduces photosynthesis (Tedders and Smith 1976). Adherence of this fungal mat appears to be controlled by leaf surface morphology which varies among cultivars (Sparks and Yates 1991). Sources of resistance to many other insects have been little studied, and most putative sources of resistance need to be validated (Table 20.4).

4 Major Breeding Achievements

There have been three foundation breeding locations for genetic improvement of pecan scion cultivars: Jackson County, Mississippi; San Saba County, Texas; and the USDA Pecan Breeding Station at Brownwood, Texas (Crane et al. 1937; Thompson and Grauke 1991).

Jackson County cultivars were the result of selections made by several area nurserymen and included 'Stuart,' 'Schley,' 'Desirable,' 'Success,' 'Pabst,' and 'Forkert' (KenKnight 1970). The first person to attempt controlled pollinations of pecan was C. Forkert of Jackson County, who planted seed from his first controlled crosses in 1903 and is responsible for 'Desirable' ('Success' × 'Jewett') and 'Forkert' ('Success' × 'Schley') (Forkert 1914). Jackson County cultivars have dominated orchards in the Southeast since the late 1800s.

E.E. Risien of San Saba County, Texas, was the first person to conduct a systematic survey of wild pecans for seedlings worthy of propagation (Crane et al. 1937). Around 1882, Risien discovered the tree that he later propagated as 'San Saba.' An orchard planted using nuts of 'San Saba' produced the trees 'San Saba Improved' and 'Squirrel's Delight' (Crane et al. 1937). Risien used controlled pollinations to produce the cultivars 'Banquet' ('Sovereign' × 'Attwater') and 'Commonwealth' ('Longfellow' × 'Sovereign'). He developed improved pecan propagation techniques during the 1890s and was a pioneer in top-working large pecan trees (Crane et al. 1937). A particularly significant contribution was his introduction of the technique of grafting juvenile buds from controlled crosses into large bearing trees to reduce the period of juvenility (Romberg and Smith 1950).

The third pecan cultivar "nursery" has been the USDA Pecan Breeding Program at Brownwood, and College Station, Texas. The program was initiated by L.D. Romberg, who worked from 1931 to 1968. The program was continued by G.D. Madden (1968–1977), and T.E. Thompson (1979–present). Early breeding objectives included increasing nut size, percent kernel, ease of shelling, scab resistance, and many minor genetic traits. Scab resistance screening was very limited due to lack of humidity and scab pressure at Brownwood, but many crosses of resistant parents produced progenies that were sent for evaluation in Louisiana and other higher scab pressure areas. This program released improved pecan cultivars for all pecan growing regions. Some cultivars were scab resistant, and could be grown in both southeastern US environments and western locations, while some cultivars were very susceptible to scab, and were released as "western cultivars." Few northern US cultivars were released until recently.

'Mahan' and 'Schley' have been the most productive parents used in the USDA program, in existence since ca. 1930. Each of these cultivars parented of six of the 26 USDA cultivars (Table 20.7). Both parents have a very thin shell, which leads to a high kernel percentage. Other commonly used parents include 'Success' which has a thin shell, 'Mohawk' which is large and early ripening, and 'Evers' which is very prolific and thin shelled. Cultivars released by the program are steadily gaining popularity, with many nurseries, especially in the south central region, selling mostly improved cultivars from this program. Highly popular recent releases from

Table 20.5 Rootstocks used in different US states (Thompson 1990)

State	Cultivar
Alabama	Elliott, Curtis, plus others
Arizona	Riverside and many others
Arkansas	Mainly natives
California	Riverside, Apache, VC1-68, plus others
Florida	Elliott, Curtis, Waukeenah, plus others
Georgia	Elliott, Curtis, plus others
Kansas	Giles, plus natives
Kentucky	Natives
Louisiana	Stuart, Moore, Elliott, Desirable, Candy, natives, plus others
Mississippi	Owens, Big Dan, Moore, water hickory
Missouri	Mainly natives
New Mexico	Riverside, Burkett
North Carolina	Cape Fear, plus others
Oklahoma	Riverside, Apache, Giles, plus others
South Carolina	Curtis, Stuart, Elliott
Tennessee	Gerardi, plus natives
Texas	Riverside, Apache, plus many others

this program include 'Pawnee,' 'Oconee,' 'Kanza,' and 'Creek.' 'Hopi,' 'Nacono,' 'Waco,' and 'Lakota' are more recent releases which are expected to gain popularity as growers become familiar to them.

Success in the improvement of pecan rootstocks has been mainly the identification of scion clones that produce superior half-sib and full-sib open pollinated populations of seedlings that are vigorous enough to be easily propagated to good scion cultivars, and at the same time are adapted to high-salt soils of the west or other specific industry requirements. Nurseries grow their pecan rootstocks from open-pollinated seed of favorite scion cultivars (Table 20.5). The seedlings from these families are genetically highly variable and produce many inferior seedlings that are nonvigorous and that must be removed prior to scion propagation. Techniques to produce clonal rootstocks have been attempted without commercially useful results (Gossard 1941; Romberg 1942, 1967; Pokorny and Sparks 1967; McEachern 1973; Gustafson 1978; Hansen and Lazarte 1984). Although rooted ramets have been produced by juvenile and adult phase cuttings, layerage, and in vitro techniques, ramets generally express low vigor and survival. The ramet trees generally lack the ability to establish a vigorous root system, and decline over time.

The objective of the nurserymen is to select a rootstock source (scion cultivar) that will produce a large proportion of rapidly growing seedlings. Seedling height, and especially lower trunk diameter (where most propagation occurs), are of prime importance. There is a recognized need for salt-resistant rootstocks for orchards west of central Texas. 'Riverside,' 'Burkett,' and 'Apache' are widely used in this area.

In the central and western USA, scions are propagated onto the seedling rootstocks mainly by patch budding, while in the eastern USA, many trees are whip grafted at or just below soil level. Traditionally all pecan orchards were established with bare root trees, but container grown trees are gaining popularity.

Container trees offer greater uniformity of establishment, and can be grown in nonsoil media if needed to circumvent soil import restrictions into western states.

The USDA rootstock breeding program is currently identifying parental material with low harmful ion uptake (sodium and chlorine), and high zinc uptake. The goal is to identify superior clones that can be released to serve as parents for open-pollinated seedling rootstocks. These superior clones would need to be grown in isolation to allow interpollination, and exclude other pollen sources. Controlling the male parentage in this way would add greatly to the genetic uniformity and value of rootstock seedlings.

There is a strong need in the pecan industry for a breeding program to produce synthetic populations of rootstock seedlings. This has never been attempted in pecan, except perhaps by E.E. Risien who had somewhat of a rootstock breeding program. 'Riverside' is a superior producer of rootstock seedlings, and is traceable to Risien's early work. This clone resulted from a scion tree that was transplanted, and when the scion died, it was replaced by rootstock growth. A rootstock breeding program should follow traditional synthetic crop breeding techniques with diligence given to shortening the sexual generation time using techniques outlined below. Inbreeding depression is very common when pecan is selfed, so simple recurrent selection should be used (Allard 1966).

5 Current Goals/Challenges of Breeding

Pecan is diploid ($n=16$), anemophilous, monoecious, and heterodichogamous. In pecan, male and female flowers are produced at different locations on the same tree. On each clone (cultivar), the male or the female flowers mature first (heterodichogamy). The complete heterodichogamy of pecan makes it almost completely cross-pollinated, resulting in high heterozygosity with severe inbreeding depression when selfed. Hybrid vigor has been selected naturally in the evolution of this species. Survival of pecan in its native environment depended greatly on growth potential. Therefore, it seems to be a naturally vigorous, wood-producing tree.

From a breeding standpoint, we know less about tree crops than agronomic crops, which are usually annuals. The reason for this greater knowledge of agronomic crops is that they lend themselves to breeding research, whereas tree crops have much longer generation times. It seems, however, that techniques for improvement through breeding may be equally effective in tree crops and annual agronomic crops, especially if compared on a generation basis. The genetic improvement of pecan is impressive considering that only one to five cycles of controlled crossing have been used. In other crops, breeding cycles usually mean more than one generation and usually involve selfing. In pecan a single improved clone takes years to test, but during this testing phase, plants are genetically stable since the genes of the clone are fixed and the trees are clonally propagated. As a result, genetic variability is zero in evaluation trials. This contributes greatly to the effectiveness of testing clonal fruit and nut crops like pecan.

As mentioned earlier, pecan is diploid. Genetically, this makes selection more direct for both qualitative and quantitative characters. Hopefully, we can determine segregation ratios for more simply inherited traits in the future. For example, a single gene determines the type of dichogamy in pecans (Thompson and Romberg 1985). This knowledge is used to produce either protandrous or protogynous clones in the breeding program as needed. There may also be specific genes conditioning resistance to different races of the scab organism. The inheritance of many other traits such as precocity, length and time of season of nut fill, and some insect-resistance mechanisms is probably quantitative.

Basic research related to the breeding program consists mainly of techniques to improve breeding efficiency and expand the genetic knowledge of pecan. One of the most direct needs is a technique to induce early flowering in juvenile clones at perhaps 2 or 3 years of age. Currently, most pecan seedlings flower at 6 or 7 years of age. Early pistillate flowering on 15-month-old clones (time of germination to pistillate flower production) has been accomplished (Thompson 1986). The frequency, however, was low, and to be useful as a breeding technique, the frequency must be greatly increased. Early juvenile flowering has been accomplished in some other tree species, but specific techniques to routinely induce female flowering in pecan has not been developed. The benefits of such techniques are obvious in selection programs to radically alter gene frequencies which control important traits, such as yield, nut maturity time, and disease and insect resistance.

Pecans are considered by some to be a relatively inefficient food production crop. We feel the main reason for this is its late nut-filling period. The pecan kernel begins to form about August 1 in early nut maturing cultivars like 'Pawnee' and 'Kanza.' This is a period of the year when days are shorter (less light for photosynthesis), the leaves have been damaged by insects and diseases all season, the roots are competing with the nuts for photosynthate to replenish root carbohydrate reserves for winter and spring growth and flowering, and perhaps soil moisture and nutrients have been exhausted by 6 months of active growth. This heavy masting effect late in the season also induces the absence of flower production the following spring which produces the alternate bearing syndrome in pecan. Perhaps this alternate cycle was needed in the wild to escape nut feeding insects, but it is definitely not needed in improved orchards.

The basic consideration here is that the pecan tree is designed wrong for maximum nut production. It is too much of a forest tree designed to effectively compete with other species for space in forest canopies. This is mainly related to fast vegetative growth which is needed for competitive survival in the wild, but exactly what is not needed in developed orchards where competition is artificially removed. The idea is to direct more photosynthate into the earlier production of nuts and less into the production of unneeded wood.

Late nut development in pecans may have resulted from selection induced by animals feeding on the earliest-maturing nuts. This effect is obvious in stands of clones, some of which mature early. These nuts are completely destroyed by feeding animals in the area. Clones with nuts maturing later partially escape this severe feeding pattern, and a portion of the nuts are stored underground by squirrels or otherwise allowed to germinate the following spring. It is interesting that pecan is one of the latest species, as far as developing nuts, in the *Carya* genus.

The nut-filling period may also be too short in pecan. Lengthening this period in some other crops has improved yield ability. We are accumulating data on this trait now and it may be related to yield.

The xenia effect or the immediate effect of the pollen on nut filling and development is also being determined. The presence of this pollen source effect on nut development in species related to pecans has been documented. In pecan pollen from some cultivars reduces premature nut sprouting or vivipary. We need to determine the value of the xenia effect so that specific cultivar recommendations can be made that maximize productivity and nut size when new orchards are established

A need to control or reduce tree size is generally recognized in pecan. There have been some past references in pecan literature to dwarf varieties that are currently available. For example, Cheyenne is sometimes considered "dwarf-like." This terminology is unfortunate because Cheyenne and some other clones are only slower-growing, and are not really dwarf-like at all. Whether tree size can be reduced most effectively by discovering and using dwarfing rootstocks or by developing dwarfed cultivar (scion) clones is debatable. There are advantages to each. In Persian walnut production in California, small tree size results from genetic characteristics of the scion growing on a very vigorous rootstock. This should also work in pecan production. In any event, hopefully future cultivars will be partially dwarfed by high nut production which will limit the photosynthate available for vegetative growth in the spring when most shoot extension growth occurs.

Heritability studies of genetic traits are also conducted as part of the breeding program. This knowledge allows the effectiveness of the breeding program to be improved by more accurate prediction of how many clones of each cross will be discarded due to inadequate yield potential, nut size, disease resistance, or other trait.

Pecans are attacked by a wide range of disease and insect pests causing substantial losses to the crop. In the humid growing conditions of the southeastern USA, the most economically damaging of these is pecan scab, caused by the fungus *Cladosporium caryigenum*. Foliar infections result in black circular lesions that under favorable conditions can result in severe leaf spotting, premature defoliation, and shoot death. Development of lesions on fruit shucks reduces yield and nut quality, and if not controlled it can result in total crop loss. Commercial pecan plantings in the southeastern USA may require up to 11 fungicide applications annually to control the disease (Ellis et al. 2000). Pecan scab has developed resistance to at least two separate classes of common fungicides (Stevenson 2005). The development of scab resistant cultivars with excellent commercial quality would greatly increase the profitability of pecan cultivation in the Southeast and is the focus of several cultivar development programs (Conner 1999; Goff et al. 1998; Thompson and Grauke 1994).

It is useful to study the history of pecan scab to better understand how to approach the development of scab resistant cultivars. In their 1929 paper, Demaree and Cole provide an interesting review of the history of pecan scab in the Albany, Ga., region.

Prior to 1910, scab was considered a relatively minor disease, of spotty incidence, primarily affecting seedlings or a few cultivars. Before 1920, the authors state that 'Georgia' was the only cultivar generally affected by scab. Beginning in 1920, however, 'Delmas' began to be affected, and in 3 years the fungus had spread to the entire region and became a serious problem on this cultivar. At the same time, 'Alley' also began to be affected. In 1923, 'Schley' began to be affected in Putney and Baconton Ga., located to the south of Albany. From there it spread so rapidly that by 1926 it had become extremely destructive throughout the region. In 'Van Deman' the amount of scab slowly increased during the 1920s and was causing some damage under favorable conditions. 'Pabst' was still free of the disease in Albany at the time the article was written. In contrast, in Ocean Springs, Miss., 'Pabst' was very susceptible but 'Schley' was relatively free of the disease. In a Louisiana orchard, 'Pabst' and 'Moneymaker' were scabbing, while trees of the very susceptible cultivars 'Delmas' and 'Georgia' were unaffected.

Two facts stand out from these early reports on scab incidence: (1) cultivars now considered quite susceptible, such as 'Schley' and 'Alley,' were at one time little affected by scab, and (2) cultivars can vary in susceptibility depending upon location. Both of these factors are explained by the existence of multiple races of the fungus. Indeed, the presence of multiple races of the scab fungus has been demonstrated experimentally by several authors including Demaree and Cole (1929) and Converse (1960).

Even with the pessimistic situation presented above, there are still many opportunities for a breeding program to assist in the control of this disease. Many new cultivars seem to have a grace period during which they are relatively free of the disease. For some cultivars, this period is relatively short, and for others it has lasted decades. By testing new selections in several locations breeders can hopefully select cultivars whose resistance will not be overcome quickly. An active breeding program can take advantage of this grace period by producing a continual supply of new cultivars. This will assist growers by giving them an opportunity to plant a new cultivar with new resistance genes when they turn over an orchard. Hopefully, by the time a current cultivar has become extremely susceptible to scab, there will be new cultivars with different resistance genes ready to replace it. Thus, the overall level of disease decreases and becomes more manageable. If resistant selections have nut quality equal or superior to the standard susceptible cultivars, then loss of resistance once it happens need not be catastrophic. Growers would begin controlling scab using the methods they use on susceptible varieties, and eventually rotate to newer resistant varieties when replanting.

Other projects include developing DNA markers for resistance genes and examining the physiological basis of scab resistance. DNA markers for scab resistance genes will be very useful in a breeding program. They will allow us to quickly identify resistance genes in our seedling progenies without laborious inoculation procedures. They may also allow us to pyramid multiple resistance genes into a single cultivar. Resistance based on several different resistance genes may be more difficult for the scab fungus to overcome and thus be more durable in the field. Currently we understand very little about how pecan protects itself from scab infection.

By studying the infection process microscopically, we hope to better understand this process and use this knowledge to select trees with higher levels of resistance.

Various levels of resistance to scab are available in pecan germplasm. However, few cultivars contain sufficient resistance so that fungicide applications are not necessary and these usually lack many of the nut quality traits desirable for commercial plantings. In addition, many important high quality cultivars such as ‘Stuart’ and ‘Desirable’ are becoming increasingly susceptible to the scab pathogen, due at least partly to the presence of multiple races of the fungus (Thompson and Grauke 1994). As a result, commercial pecan plantings require 8–11 applications of fungicides to remain profitable. Pecan scab has developed resistance to at least one common fungicide, Benlate. In addition, concern over negative environmental health effects of pesticides has resulted in pressure to increase regulation of other valuable chemical control agents. Development of varieties with combinations of disease and insect resistance would result in further savings. Resistant varieties could also reduce risks of epidemics when weather conditions are favorable for disease growth and unfavorable for pesticide application. The development of resistant cultivars will play a vital role in maintaining the profitability of pecan culture in the Southeast.

The basis of scab resistance in pecan is not well understood at the genetic level. In the only large-scale analysis of inheritance of scab resistance, Thompson and Grauke (1994) evaluated 948 seedlings derived from 15 controlled crosses for the presence of nut scab. Seedlings were grown in an unsprayed orchard at Brownwood, Texas, and evaluated for nut scab from naturally occurring infections in a year of high disease incidence. The heritability of resistance was determined by regressing progeny scab rating values on male, female, and midparent values. Midparent values gave the highest correlation (0.54) indicating a moderate level of additive gene action. This work also indicated that certain cultivars such as ‘Gloria Grande’ may transmit a higher level of scab resistance to their progeny, making them superior parents.

One of the most important factors to be considered by any breeding program aimed at producing resistant cultivars is the presence of multiple races of the scab pathogen. Many cultivars that were once highly resistant to scab are now widely considered susceptible. For example, the cultivars ‘Desirable’ and ‘Stuart’ are grown throughout the Southeast and were initially popular at least in part due to their high levels of scab resistance. Both cultivars are now commonly considered susceptible and the appearance and spread of a race of scab capable of infecting ‘Stuart’ was documented (Cole and Gossard 1956).

The presence of multiple races of the scab pathogen has been inferred from the wide range of scab susceptibility cultivars demonstrate when grown in different geographic locations (Sparks 1992; Demaree and Cole 1929). Demaree and Cole (1929) used orchard inoculations to demonstrate that at least four races of the pathogen exist which differ in their ability to infect cultivars. Converse (1960) further demonstrated the presence of four races on the basis of their pathogenicity in greenhouse and field tests on four pecan cultivars. In a recent study conducted in this laboratory, four scab isolates were inoculated onto each of the four cultivars from which they were isolated (Conner 2002). Detached leaves were then examined

Table 20.6 Summary of detached-leaf reactions of four pecan cultivars inoculated with *Cladosporium caryigenum* isolates from each of the same four host cultivars

	Scab isolate tested			
Cultivar tested	Wichita isolate	Desirable isolate	Cape Fear isolate	Elliot isolate
Wichita leaf	++	–	–	–
Desirable leaf	–	++	–	–
Cape Fear leaf	–	–	++	++
Elliot leaf	–	–	–	+

++=30–60% of conidia form subcuticular hyphae; +=10–15% of conidia form subcuticular hyphae; –=<5% of conidia form subcuticular hyphae

microscopically to determine the susceptibility of each cultivar to each isolate. Scab isolates differed in their ability to form subcuticular hyphae on the different cultivars, with the greatest amount of infection usually occurring when the isolate was placed back onto the cultivar from which it was isolated (Table 20.6). The cultivars in this test were generally highly resistant or immune to isolates from other cultivars. It is apparent from these studies that a range of genetic types of the pathogen exist and these differ markedly in their ability to cause disease on different pecan cultivars.

With this information in hand, the next question becomes how is resistance inherited in the progeny resulting from crosses between pecan cultivars with differential resistance to scab isolates? Testing with known isolates will allow us to further refine our knowledge of the inheritance of resistance by avoiding the two most common complications of previous studies (1) the possibility of escapes due to inadequate or variable inoculum and (2) variability in the genetic makeup of the inoculum challenging the seedlings. By evaluating resistance of the progeny of crosses between these cultivars to defined isolates of the pathogen the mode of action of resistance genes and their inheritance in the progeny can be determined. This information will be vital to designing future crosses aimed at achieving high levels of resistance in the progeny and for developing molecular marker tags for important resistance genes. This work will also provide information on those cultivars most likely to be useful as parents in breeding new resistant cultivars.

Effective breeding for resistance to *C. caryigenum* requires information on the pathogenic diversity of the fungus. There is a range of pathotypes of *C. caryigenum* exist that differ markedly in their ability to cause disease on different pecan cultivars. The work reported here was undertaken to further examine the extent of pathogenic variation among scab isolates using a larger number of cultivars and fungal isolates. These results may be useful in designing crosses to pyramid resistance genes into a single cultivar or in selecting combinations of cultivars to be included in an orchard.

The USDA-ARS pecan breeding program in concert with the UGA breeding program is conducted cooperatively across the entire US production area and consists of many varied and interrelated activities by breeders, geneticists, horticulturists, pathologists, and entomologists. To date (and in cooperation with state agricultural experiment stations), 26 improved cultivars (Table 20.7) have been

Table 20.7 Cultivars developed cooperatively by the US Department of Agriculture, Agricultural Research Service and cooperators

Cultivar	Parentage[a]	Selection number	Year released	Dichogamy[b]
Barton	Moore × Success	37-3-20	1953	I
Comanche	Burkett × Success	37 -8-22	1955	II
Choctaw	Success × Mahan	46-15-276	1959	II
Wichita	Halbert × Mahan	40-9-193	1959	II
Apache	Burkett × Schley	40-4-1 7	1962	II
Sioux	Schley × Carmichael	43-4-6	1962	II
Mohawk	Success × Mahan	46-15-195	1965	II
Caddo	Brooks × Alley	Philema 1175	1968	I
Shawnee	Schley × Barton	49-17-166	1968	II
Cheyenne	Clark × Odom	42-13-2	1970	I
Cherokee	Schley × Evers	48-22-27	1971	I
Chickasaw	Brooks × Evers	44-4-101	1972	II
Shoshoni	Odom × Evers	44-15-59	1972	II
Tejas	Mahan × Risien 1	44-10-293	1973	II
Kiowa	Mahan × Desirable	53-9-191	1976	II
Pawnee	Mohawk × Starking HG	63-1 6-125	1984	I
Houma	Desirable × Curtis	58-4-61	1989	I
Osage	Major × Evers	48-15-3	1989	I
Oconee	Schley × Barton	56-7-72	1989	I
Navaho	Apalachee × Wichita	74-1-11	1994	I
Kanza	Major × Shoshoni	55-11-11	1996	II
Creek	Mohawk × Western	61-6-67	1996	I
Hopi	Schley × McCulley	39-5-50	1999	II
Nacono	Cheyenne × Sioux	74-5-55	2000	II
Waco	Cheyenne × Sioux	75-5-6	2005	I
Lakota	Mahan × Major	64-6-502	2007	II
Mandan	BW-1 × Osage	85-1-2	2009	I
Apalachee	Moore × Schley	48-13-311	2009	I

[a]First parent is the female. Second parent is the male
[b]I = protandrous and II = protogynous

released. One of these, 'Pawnee,' is probably the most popular cultivar in the world, as far as the number of trees being propagated. The value of this one cultivar equals that of all USDA and UGA breeding program costs many times over. Public funding of pecan breeding research is therefore an excellent investment in the future well-being of our country and the world.

6 Breeding Methods and Techniques

There are two pecan scion breeding programs. The US Department of Agriculture, Agricultural Research Service (USDA-ARS), in cooperation with state agricultural experiment stations, state extension services, and private growers; conducts a

Table 20.8 Pecan selection technique in the USDA Breeding Program

Phase	Description	Years	Number of clones per year	Location or spacing (m)
I	BBP Seed production	1	1,000–2,000	Nuts harvested
II	BBP Scab screening	1	1,000–2,000	Potted seedlings, screenhouse/field
III	BBP Orchard	10	500–1,000	Seedling orchard, 4.6×9.1
IV	NPACTS	10–15	5–10	Grafted orchard, 10.7×10.7

national pecan breeding program headquartered in College Station and Brownwood, Texas. It is directed by the senior author. The University of Georgia also conducts a breeding program for that state that is directed by the junior author. Improved cultivars produced in these two programs are also widely grown in other countries.

A breeding system is used which combines desirable genetic characteristics from the two parents. The parents are controlled crossed, and the resultant seedlings are selected based upon desirable characteristics. Although thousands of seedlings are produced and selected, very few clones are produced that are considered worthy of release as new cultivars.

Considering the heritability estimates for major nut characteristics (Thompson and Baker 1993), and the reasonable probabilities for improvement of other traits, large populations of plants need to be produced. There are two selection cycles in the USDA program: the Basic Breeding Program (BBP) and the National Pecan Advanced Clone Testing System (NPACTS) (Table 20.8). Large numbers of seedlings are produced and eliminated in the BBP based upon highly heritable, easily selected characteristics. Only one or two clones per thousand are considered good enough to advance to NPACTS. For instance, elimination of inferior clones based upon yield, precocity, vigor, scab susceptibility, and nut quality, as well as resistance to insects, can be accomplished in the seedling cycle and continued in NPACTS.

In Phase I, the traditional crossing technique is used to produce up to 4,000 seed each year. Crosses are made at Brownwood and College Station, Texas. This large amount of seed is possible due to improved techniques of tree preparation and care so that each crossed cluster produces more seed. For example, some trees in our crossing program routinely produce two to four nuts per cluster, compared with the average of less than one per cluster a few years ago. All fruit on trees to serve as female parents should be removed early in the growing season of the year before crossing. This insures more and larger clusters at time of bagging. Other obvious cultural techniques such as adequate space for the tree, water, etc. are also needed. Bagged clusters should be pollinated twice, 1 day apart. The first pollination to all bags on each tree should be made when any nonbagged receptive flowers can be found on the tree. This insures that viable pollen is on all receptive bagged pistils throughout the pollination period.

All the seed produced by these hand crosses is stratified, then planted in the greenhouse in December and the seedlings are monitored for vigor and other characteristics. In the spring, the seedlings are placed under scab-susceptible trees and rated for resistance two or three times during the growing season. After each rating, the leaves are removed so that new scab-susceptible leaves are again produced. In the fall, one third to three quarters of the seedlings are discarded due to scab susceptibility (Phase II).

Planting seed directly into a disease garden or scab nursery should also be effective in eliminating most disease-susceptible clones. As above, this assumes that resistance in juvenile leaves is correlated with resistance in mature-phase leaves. Seedlings can be planted directly in the field under, or close to, disease-susceptible cultivars. Again, several susceptible cultivars need to be included to produce an array of diseases and sufficient races of different diseases. Seedlings can be rated for disease resistance for 2 or 3 years; then, resistant seedlings are replanted or grafted into the BBP, for Phase III evaluation.

Phase III is the initial field selection phase at College Station, Texas for yield, precocity, nut quality, desirable leaf and tree structure, and disease and insect resistance. Although most of these seedling trees are transplanted and grown on their own roots, some of these clones are grafted to pollarded large trees to hasten flowering. Trees grown on their own roots are grown at a relatively close spacing and the elimination of trees begins in the 6th or 7th year based upon precocity, nut size, scab resistance, and other traits. This early elimination allows more room for the more desirable clones to develop and be more adequately evaluated. Only about one or two of these clones are saved per thousand for Phase IV NPACTS testing.

In NPACTS, elite clones from Phase III are grafted into replicated trials across the entire pecan belt for environmental adaptation. These tests are conducted using standard extension recommendations for each test location. Testing is often done cooperatively with growers, state experiment stations, state agricultural extension services and universities. For instance, NPACTS tests are currently established at College Station and Amarillo, Texas, in cooperation with Texas Agrilife Research and Extension Service. Other Texas tests are conducted on private land in cooperation with pecan growers. Clones which perform well in these NPACTS tests are released as new USDA-State unpatented cultivars. A new cultivar could possibly be released every 2–5 years. This means that thousands of clones are screened to produce a single new cultivar. This is realistic from a genetic standpoint when projected heritabilities of different traits are considered. Table 20.7 shows the pedigree and other information for the USDA-ARS/state released cultivars.

In 1999, P.J. Conner initiated a new breeding program for Georgia based at the University of Georgia-Tifton Campus. The UGA pecan breeding program was initiated with the goal of releasing high quality cultivars adapted to the southeast region, and especially the state of Georgia (Conner 1999). Given the prevalence of rain during the growing season in this region, durable scab resistance is a primary objective of this program (Conner 2003). Other traits being targeted include early harvest date, large nut size, and high kernel percentage to capture the profitable gift-pack market. A previous breeding effort based at UGA-Athens Campus by D. Sparks has resulted in the 2008 release of 'Byrd' ('Wichita' × 'Pawnee'), an early maturing cultivar with high kernel quality. Several other selections are in the process of being released from this breeding effort.

The UGA pecan breeding program uses methods similar to those of the USDA. Seedlings are grown for 3–4 months in the greenhouse in root pruning flats. In April or May the seedlings are shifted up to 3-gallon root pruning containers and placed outside in a shade house underneath 50% shade cloth. Some sort of root pruning device is highly desirable since pecan has a dominant tap root that will circle a standard pot. The shade cloth is needed to keep seedlings actively growing in the heat of the summer. Starting in June, scabbed branches are cut from a wide variety of cultivars and selections and are rubbed over damp seedlings at dusk. Overhead irrigation is applied intermittently during night to keep the leaves wet. This process is repeated several times over the summer. Seedlings are then rated for leaf scab and, depending upon the progeny, anywhere from 20 to 80% may be eliminated. Seedlings have usually made sufficient growth at the end of the year that they are then planted into fields where they grow on their own roots at a spacing of 3 m between trees within the row and 4.6 m between rows. Seedling trees are monitored for approximately 10 years and superior selections are grafted into trial orchards at Tifton and in grower orchards in Georgia. Superior selections are released as patented cultivars to support the breeding program.

7 Integration of New Biotechnologies in Breeding Programs

The potential of molecular markers to increase our understanding of the pecan genetic diversity has been demonstrated in several studies. Pecan is a newly domesticated crop and many important historical and current cultivars are chance genotypes discovered by nurserymen and growers in seedling orchards or native groves. Understanding the genetic relationships between these cultivars can offer the pecan breeder insights into the best way of producing new favorable combinations of alleles. Protocols for the analysis of five isozyme systems: malate dehydrogenase, phosphoglucose isomerase, phosphoglucomutase, leucine aminopeptidase, and diaphorase have been developed (Marquard 1987, 1989, 1991; Marquard et al. 1995). Using these isozymes, 177 cultivars were sorted into 72 classes and the historical pedigree of some cultivars was called into question. These systems were then used by Grauke et al. (1995) in the evaluation of the pecan germplasm collection to designate a core subset. Conner and Wood (2001) demonstrated the value of randomly amplified polymorphic markers (RAPD) markers in determining genetic relationships among pecan cultivars. Genetic distances, based on the similarity coefficient of Nei and Li, varied from 0.91 to 0.46, with an average of 0.66 among all cultivars. Cerna-Cortes et al. (2003) used AFLP markers to study the genetic diversity of native pecan genotypes from Central Mexico. Genetic diversity in these genotypes was found to be relatively low, probably due to the relatively restricted geographical region sampled. Grauke et al. (2003) developed simple sequence repeats (SSRs) or microsatellite DNA markers and carried out an initial evaluation of SSR markers for use in genetic studies of pecan. The authors found 11 primers that produced polymorphisms among the 48 pecan and hickory accessions, but encountered difficulty in scoring many SSR profiles.

There is a great need in pecan genetics to develop an easy and robust marker system to reliably fingerprint pecan cultivars. Growers often find a few unknown cultivars mixed in with their purchase of grafted trees. These mistakes can come from mistakes in collecting or handling graftwood, mislabeling, or sorting errors of trees in the nursery. It is often difficult to identify these cultivars based on nut phenotype alone. In addition, molecular marker fingerprints could be produced as soon as tissue was available rather than waiting several years for the tree to produce fruit. Molecular fingerprints would also perhaps facilitate tracing the parentage of new seedling cultivars. However, currently developed marker systems in pecan suffer from irreproducibility between laboratories and require technology that is relatively cumbersome for breeding programs to apply on a routine basis.

Molecular marker based maps have the potential to facilitate pecan breeding in two main ways. First, maps can greatly facilitate genetic studies in pecan. Most horticulturally important traits in pecan appear to have a complex mode of inheritance, and genetic maps will allow us to tease apart the individual loci in control of these traits and describe their effects. Second, molecular markers linked to useful traits will facilitate marker-assisted selection of these traits. This is especially important in pecan because of the limitations that long juvenile periods and large plant size place on the number of seedlings that can be grown to fruition. Beedanagari et al. (2005) have produced the only linkage maps of pecan. Because of the outbred nature of pecan, separate maps were produced for both parents of the cross 'Pawnee' × 'Elliott' using a combination of amplified polymorphic DNA (AFLP) and random amplified polymorphic DNA (RAPD) markers. 'Pawnee' is a USDA release which has an exceptionally early harvest date and large, high-quality nut. 'Pawnee' is being used extensively in breeding programs to incorporate early harvest date into new cultivars. 'Elliott' is an older cultivar from Florida which is being used to incorporate scab resistance into new cultivars. The 'Pawnee' map is 2,227 cM in length and is estimated to cover 83% of the 'Pawnee' genome. The 'Elliott' map is 2,965 cM in length and is estimated to cover 57% of the 'Elliott' genome. Two phenotypic traits, dichogamy type and stigma color, were found to be tightly linked and were mapped to linkage group 16 of the 'Elliott' map. Mapping of other phenotypic traits was not attempted due to the young age of many of the progeny trees.

Molecular mapping appears to hold much potential for facilitating pecan breeding. However, the same limitations of large plant size, long juvenile periods, and complex inheritance of most important traits which make molecular mapping so attractive also make it difficult to proceed with the large scale mapping studies needed to produce results which will be useful to the breeding program. Added to these difficulties are the limited funding available to do molecular work in minor crops such as pecan and the severe inbreeding depression which prevents the formation of inbred lines which facilitate the genetic analysis of marker–trait associations. Near-term results are most likely to come from finding markers associated with simply inherited traits which are difficult to analyze phenotypically, such as resistance to pecan scab.

The development of transformation and regeneration protocols for pecan has been limited. Somatic embryogenesis has been accomplished from immature and mature zygotic embryos of several cultivars (Merkle et al. 1987; Obeidy and Smith 1993; Wetzstein et al. 1989; Yates and Reilly 1990). McGranahan et al. (1993) successfully used a gene transfer system for walnut (*Juglans regia* L.) on pecan. Embryogenic somatic embryos were cocultivated with an *Agrobacterium* strain which contains marker genes for beta-glucuronidase and resistance to kanamycin. Transgenic plants were obtained by grafting tissue cultured shoots onto seedling pecan rootstocks. Initial success in transformation has not been followed up in recent years for several reasons. Consumer acceptance of transgenic pecans is not assured, especially since there are no other transgenic nut crops on the market. Established regeneration protocols make use of zygotic starting material. This is undesirable since pecan cultivars are heterozygous and do not breed true from seed, thus preventing the addition of a transgene into an established cultivar. In addition, pecan is anemophilous, and wild trees exist in the forests surrounding many pecan orchards. This, in combination with nuts carried off by wildlife which can produce new trees, suggests that it would be very difficult to prevent the escape of transgenes into wild populations. The development of transgenic pecans will likely remain limited until methods are developed to overcome these limitations.

References

Allard, R. W. 1966. Principles of plant breeding. John Wiley and Sons. N.Y.

Beedanagari, S. R., S. K. Dove, B. W. Wood, and P. J. Conner. 2005. A first linkage map of pecan cultivars based on RAPD and AFLP markers. J. Applied Genetics 10(6):1127–1137.

Bettencourt, E. J. and J. Konopka. 1989. Directory of germplasm collections. 6. II. Temperate fruits and tree nuts. Int. Board Plant Genetic Resources, Rome.

Boethel, D. J. and R. D. Eikenbary. 1979. Pecan weevil: Seasonal emergence of larvae from three pecan cultivars in Oklahoma. J. Ga. Ent. Soc. 14:75–78.

Boethel, D. J., W. G. Grovenburg, R. D. Eikenbary, and R. W. McNew. 1976. Infestation levels of pecan leaf phylloxera, *Phylloxera notabilis* Pergande, on pecan cultivars. Env. Ent.5:637–639.

Byford, R. 2005. Pecan varieties for New Mexico. N.M. State Univ. Guide H-639.

Calcote, V. R. 1983. Southern pecan leaf phylloxera (Homopotera: phylloxeridae): clonal resistance and technique for evaluation. Env. Ent. 12(3):916–918.

Calcote, V. R. and D. E. Hyder. 1980. Pecan cultivars tested for resistance to pecan phylloxera. J. Ga. Ent. Soc. 15(4):428–431.

Calcote, V. R., G. D. Madden, and H. D. Petersen. 1976. Pecan cultivars resistant to hickory shuckworm. Proc. N. Nut Growers Assn. 67:19–21.

Carpenter, T. L., W. W. Neel, and P. A. Hedin. 1979. A preliminary study of host plant resistance of pecan to the blackmargined aphid, *Monellia caryella* (Fitch). Pecan Quart. 13(3):35–37.

Cerna-Cortes, J., J. Simpson, O. Martinez, and R. Martinez-Peniche. 2003. Measurement of genetic diversity of native pecan [*Carya illinoensis* (Wangenh.) K. Koch] populations established in Central Mexico and correlation with dichogamous flowering using AFLP. Food, Agr. & Env. 1:168–173.

Cole, J.R. and A. C. Gossard. 1956. Increased virulence of scab (*Cladosporium effusum* Wint. Demaree) on 'Stuart' pecan in Mississippi and its presence. Plant Disease Reporter 40:1120.

Conner, P.J. 1999. The Georgia pecan breeding program. Proc. Southeastern Pecan Growers Assn. 92:77–80.

Conner, P. and R. Worley. 2000. Alternate bearing intensity of pecan cultivars. HortScience 35:1067–1069.

Conner P, and B. Wood. 2001. Identification of pecan cultivars and their genetic relatedness as determined by Randomly Amplified Polymorphic DNA analysis. J. Amer. Soc. Hort. Sci. 126: 474–480.

Conner, P.J. 2002. A detached leaf technique for studying race-specific resistance to *Cladosporium caryigenum* in pecan. J. Amer. Soc. Hort. Sci. 127:781–785.

Conner, P.J. 2003. Breeding for scab resistance. Proc. Southeastern Pecan Growers Assn. 96:115–123.

Conner, P.J. and K.L. Stevenson. 2004. Pathogenic variation of *Cladosporium caryigenum* isolates and corresponding differential resistance in pecan. HortScience 39:553–557.

Converse, R. H. 1960. Physiologic specialization of *Fusicladium effusum* and its evaluation in vitro. Phytopathology 50:527–531.

Corbett, L., H. Gould, C. Hunn, T. Robinson, G. Darrow, G. Husmann, C. Reed, D. Shoemaker, J. Beattie, W. Beattie, J. Kincer, and L. Flohr. 1926. Fruit and Vegetable Production, The Pecans, p. 286–290. In: G.W. Hill (ed.). Yearbook of the USDA, 1925, Gov. Print. Off., Washington, D.C.

Crane, H. L., C. A. Reed, and M. N. Wood. 1937. Nut Breeding. USDA Yrbk. of Agr. 1937:827–889.

Criswell, J. T., D. J. Boethel, R. D. Morrison, and R. D. Eikenbary. 1975. Longevity, puncturing of nuts and ovipositional activities by the pecan weevil on three cultivars of pecan. J. Econ. Ent. 68:173–177.

Demaree, J. B. and J. R. Cole. 1929. Behavior of *Cladosporium effusum* (Wint.) Demaree on some varieties of Pecan. J. Agric. Res. 38:363–370.

Dutcher, J.D., R.E. Worley, P. Conner, and S.E. Dove. 2001. Pecan variety differences in the incidence of hemipteran kernel damage. J. Entomol. Sci. 36:445–452.

Ellis, H.C., P. Bertrand, and T.F. Crocker. 2000. 2000 Georgia pecan pest management guide. Univ. of Ga. Coop. Exten. Serv. Bull. 841.

Florkowski, W.J., T.F. Crocker, and G. Humphries. 1999. Commercial pecan tree inventory, Georgia, 1997. The University of Georgia, College of Agriculture and Environmental Sciences, Ga. Exp. Stations Research Report 678. Athens, GA.

Forkert, C. 1914. Twelve years' experience in hybridizing pecans. Proc. Natl. Nut Growers Assn. Bul. 13:28–30.

Gentry, C. R., J. S. Smith, J. A. Payne, E. J. Wehunt, J. M. Wells, and N.E. McGlohon. 1976. The first year of an integrated program for pest management on pecans. Proc. SE Pecan Growers Assn. 69:125–132.

Gill, J. B. 1917. Important pecan insects and their control. USDA Farmer's Bul. 834. 48pp.

Goff, W.D., L. Campbell, T.E. Thompson, J.S. Bannon, and A.J. Latham. 1993. Scab occurrence on pecan clones in Alabama in a year of high disease incidence. Fruit Var. J. 47:47–51.

Goff, W.D., M. Nesbitt, R. Mullenax, F. Rasberry, and B. Graves. 1998. Pest-resistant cultivars as a way to reduce input costs. Pecan South. 31(7):6–9.

Goff, W.D., M.L. Nesbitt, and C.L. Browne. 2003. Incidence of scab and foliage condition on pecan cultivars grown without fungicide or insecticide sprays in a humid region. HortTechnology 13:381–384.

Gossard, A. C. 1941. Rooting pecan stem tissue by layering. Proc. Amer. Soc. Hort. Sci. 38:213–2 14.

Gottwald, T. R. 1982. Taxonomy of the pecan scab fungus *Cladosporium caryigenum*. Mycologia 74:382–390.

Grauke L. J, M. J. Iqbal, A. S. Reddy, and T. E. Thompson. 2003. Developing microsatellite DNA markers in pecan. J. Amer. Soc. Hort. Sci. 128:374–380.

Grauke, L. J. and T. E.Thompson. 1996. Pecans and hickories. In: Janick, J. A. and Moore, J. N. (ed.). Fruit Breeding--III Nuts. pp. 185–239. John Wiley and Sons, Inc.

Grauke, L. J., T. E. Thompson, and Marquard, R. D. 1995. Evaluation of pecan germplasm collections, designation of a preliminary core subset, and suggestions for in situ conservation of native pecan populations. HortScience 30(5):950–954.

Gustafson, W. A. 1978. A propagation technique for rooting adult phase pecan, *Carya illi*noinensis (Wang.) K. Koch, stem cuttings. Ph. D. dissertation, Tex. A&M Univ., College Station, Tex.

Halbert, H. A. 1909. Should pecans be planted in west Texas? Proc. Tex. Nut Growers' Assoc. Twelfth Tex. Farmers' Congress. Tex. Dept. of Agric. Bul. Number 10. Austin.

Hansen, K. C. and J. E. Lazarte. 1984. In vitro propagation of pecan seedlings. HortScience 19:237–239.

Hansen, C. D., H. W. VanCleave, and J. J. Welch. 1970. Comparison of hickory shuckworm infestation rates of seven named varieties in central Texas. p. 42–43. In: Pecan Research 1965–1969, Tex. Agr. Expt. Sta. Prog. Rpt. 2718.

Hunter, R. E., T. E. Thompson, and R. S. Sanderlin. 1986. Control of pecan diseases through genetic resistance. Proc. SE Pecan Growers Assn. 79:51–54.

Kaakeh, W. and J.D. Dutcher. 1994. Probing behavior and density of *Monelliopsis pecanis, Monellia caryella*, and *Melanocallis caryaefoliae* (Homopter: Aphidae) on pecan cultivars. J. Econ. Ent. 87:951–956.

KenKnight, G. E. 1968a. Resistance of pecan to disease. Proc. SE Pecan Growers Assn. 61:168–173.

KenKnight, G. E. 1968b. Resistance of pecan to scab and bunch disease in Louisiana. Plant Dis. Rptr. 52 (4):307–309.

KenKnight, G. E. 1970. Pecan varieties "happen" in Jackson County, Mississippi. Pecan Quart. 4(3):6–7.

Marquard R. 1987. Isozyme inheritance, polymorphism, and stability of malate dehydrogenase and phosphoglucose isomerase in pecan. J. Amer. Soc. Hort. Sci. 112:717–721.

Marquard R. 1989. Rare allozymes of malate dehydrogenase in pecan. HortScience 24:156.

Marquard R. 1991. Inheritance of phosphoglucomutase isozymes in pecan. HortScience 26:1213–1214.

Marquard R, Grauke L, Thompson T, and R. Janos. 1995. Identifying pecan cultivars by isozymes and inheritance of leucine aminopeptidase. J. Amer. Soc. Hort. Sci. 120:661–666.

McEachern, G. R. 1973. The influence of propagation techniques, the rest phenomenon, and juvenility on the propagation of pecan, *Carya illinoinensis* (Wang.) K. Koch, stem cuttings. Ph.D. Diss., Texas A&M Univ., College Station.

McEachern, G. and L. Stein. 1997. Pecan Varietes for Texas, p. III-1 – III-13. In: R. Lyle (ed.). Texas Pecan Handbook. Texas Agr. Ext. Serv. , College Station, Tex.

McGranahan, G. H., C. A. Leslie, A. M. Dandekar, S. L. Uratsu, and I. E. Yates. 1993. Transformation of pecan and regeneration of transgenic plants. Plant Cell Reports 12:634–638.

McHatton, T. H. 1957. The history, distribution and naming of the pecan (*Hicoria pecan*). Proc. SE Pecan Growers Assn. 50:10–34. McLean, R. W. 1988. The effect of artificial drying temperature on the quality of early harvested pecan kernels. MS Thesis. Tex. A&M Univ., College Station.

Merkle, S.A., H.Y. Wetzstein, and H.E. Sommer. 1987. Somatic embryogenesis in tissue cultures of pecan. HortScience 22:128–130.

Moznette, G. F. 1948. The pecan weevil and latest developments in control. Proc. SE Pecan Growers Assn. 41:79–82.

Moznette, G. F., C. B. Nickels, W. C. Pierce, T. L. Bissell, J. B. Demaree, J. R. Cole, H. E. Parson, and J. R. Large. 1940. Insects and diseases of the pecan and their control. USDA Farmer's Bul. No. 1829.

Neel, W. W., C. H. Graves, Jr., and R. E. Coats. 1976. Pecan insect control test in 1975 with emphasis on yellow aphids, spittle-bugs, and weevils. Pecan S. 3:(4):430–435.

Obeidy, A.A. and M.A.L. Smith. 1993. Organogenesis from mature pecan cotyledons and embryonic axes. HortScience 28:213–215.

Payne, J. A., H. L. Malstrom, and G. E. KenKnight. 1979. Insect pests and diseases of the pecan. USDA Agr. Rev. and Manuals. ARM-5-5.

Pena, J. G. 2007. Trends in pecan production, consumption and trade. Texas Pecan Handbook. L. A. Stein and G. R. McEachern, ed. Texas A&M U. College Station.

Pokorny, F. A. and D. Sparks. 1967. Propagating Stuart pecans by air-layerage. Proc. SE Pecan Growers Assn. 60:120–123.

Romberg, L. D. 1942. Use of nurse seedlings in propagating the pecan from stem cuttings. Proc. Amer. Soc. Hort. Sci. 40:298–300.

Romberg, L. D. 1967. Clonal pecan rootstocks. Proc. Tex. Pecan Growers Assn. 46:72–75.

Romberg, L. D. and C. L. Smith. 1950. Progress report on the breeding of new pecan varieties. Proc. Texas Pecan Growers Assoc. 29:12–21.

Sparks, D. and I.E. Yates. 1991. Pecan cultivar susceptibility to sooty mold related to leaf surface morphology. J. Amer. Soc. Hort. Sci. 116:6–9.

Sparks, D. 1992. Pecan Cultivars: The Orchards Foundation. Pecan Production Innovations. Watkinsville, GA.

Stevenson, K.L. 2005. Fungicide sensitivity and performance on DMI fungicides for scab control. Proc. Southeastern Pecan Growers Assn. 98:93–97.

Taylor, W.A. 1906. Promising new fruits. Pecans, p. 504–508. In: G.W. Hill (ed.). Yearbook of the USDA, 1905, Gov. Print. Off., Washington, D.C.

Taylor, W.A. 1907. Promising new fruits. Pecans, p. 365–370. In: G.W. Hill (ed.). Yearbook of the USDA, 1905, Gov. Print. Off., Washington, D.C.

Tedders, W.L. and J.S. Smith. 1976. Shading effect on pecan by sooty mold growth. J. Econ. Entomol. 69:551–553.

Thompson, T. E. 1986. Induction of pistillate flowers on juvenile pecan clones. HortScience 21:528529.

Thompson, T. E. 1990 Update pecan cultivars: current use and recommendations. Pecan S. 24(1):1220.

Thompson, T. E. and Baker, J. F. 1993. Heritability and phenotypic correlations of six pecan nut characteristics. J. Amer. Soc. Hort. Sci. 118(3):415–418.

Thompson, T. E. and L. J. Grauke. 1991. Pecans and hickories (*Carya*). pp 839904. In Moore, J. N. and Ballington, J. R. (eds.). Genetic Resources of Temperate Fruit and Nut Crops, Intl. Soc. for Hort. Sci., Wageningen. 974 pp.

Thompson, T. E. and Grauke, L. J. 1994. Genetic resistance to scab disease in pecan. HortScience Vol. 29(9):1078–1080.

Thompson, T. E. and Grauke, L.J. 1998. Field resistance to yellow aphids in pecan. J. Amer. Soc. Hort. Sci. 123(1):85–90.

Thompson, T. E., Grauke, L.J., and Sibbett, G. S. 2000. Host plant resistance to blackmargined aphids on pecan. J. Amer. Pomological Soc. 54(4):193–198.

Thompson, T.E., Grauke, L.J. and Young, E.F., Jr. 1996. Pecan kernel color: standards using the Munsell system of color notation. J. Amer. Soc. Hort. Sci. 121(3):548–553.

Thompson, T. E. and R. E. Hunter. 1983. Further yield results from the Brownwood variety test. Pecan Quart. 17(4):13–19.

Thompson, T. E. and R. E. Hunter. 1985. 'Pawnee' pecan. HortScience 20(4):776.

Thompson, T. E., R. E. Hunter, G. D. Madden, and E. J. Brown. 1981. Performance of varieties and selections in a high density orchard establishment test. Pecan Quart. 15(2):14–19.

Thompson, T. E. and L. D. Romberg. 1985. Inheritance of heterodichogamy in pecan. J. Heredity 76:456–458.

Thompson, T. E. and E. F. Young, Jr. 1985. Pecan cultivars – past and present. Tex. Pecan Growers Assn., College Station.

Thompson, T.E., W.D. Goff, M. Nesbitt, L.J. Grauke, B. Wood, M. Smith, and M.T. Smith. 1995. Breeding scab resistant cultivars. Proc. SE Pecan Growers Assn. 88:162–165.

True, R. H. 1919. Notes on the early history of the pecan in America. Smithsonian Inst. Ann. Rept. 1917:435–448.

Wells, L. 2007. Cultivar trends in the Southeast. Pecan South 40(1):10,12.

Wetzstein, H.Y., S.A. Merkle, J.R. Ault, and H.E. Sommer. 1989. Further characterization of somatic embryogenesis and plantlet regeneration in pecan (*Carya illinoinensis*). Plant Sci. 64:193–201.

Wood, B.W., J.A. Payne, and L.J. Grauke. 1990. The rise of the U.S. pecan industry. HortScience 25:594,721–723.

Wood, B.W. and C.C. Reilly. 1998. Susceptibility of pecan to black pecan aphids. HortScience 33:798–801.

Wysoki, J. and J. Yizhar. 1976. Israeli experts study looper. Pecan Quart. 10:22–23.

Yates, I.E. and C.C. Reilly. 1990. Somatic embryogenesis and plant development in eight cultivars of pecan. HortScience 25:57.

Chapter 21
Pistachio

D.E. Parfitt, Salih Kafkas, Ignasi Batlle, Francisco J. Vargas, and Craig E. Kallsen

Abstract Within the Anacardiaceae family, the genus *Pistacia* L. consists of 11 or more species of which one, *P. vera* L. or pistachio, is commercially grown for its edible nut. Other *Pistacia* species are used as rootstocks or used in agroforestry. The cultivated pistachio is native to the Middle East and Central Asia. The center of diversity for wild *P. vera* is in Northern Iran and Southern Turkmenistan as well as parts of Afghanistan. Iran, the USA, Turkey, and Syria are the main pistachio producing countries, contributing over 90% of the world production. *Pistacia* species are dioecious with several isolated reports of monoecious individuals. Extensive collections of pistachio cultivars and germplasm resources were assembled at several experimental stations in the middle-southern former Soviet republics during the 1950s and 1960s. Selections of native cultivars in Iran, Italy, Greece, Syria, Turkey, and Tunisia were made and are now conserved. The number of described male and female pistachio cultivars is rather limited, and they are conserved in a few gene banks. The California pistachio industry was started with the introduction of the Kerman cultivar. California pistachios are grown primarily on three rootstocks, two species and

---

D.E. Parfitt (✉)
Department of Plant Sciences Mail Stop 2, University of California, One Shields Ave, Davis, CA 95616-8683, USA
e-mail: deparfitt@ucdavis.edu

S. Kafkas
Department of Horticulture, Faculty of Agriculture, University of Cukurova, Adana 01330, Turkey
e-mail: skafkas@mail.cu.edu.tr

I. Batlle • F.J. Vargas
Departament d'Arboricultura Mediterrània, IRTA-Centre de Mas Bové, Ap. 415, Reus-Tarragona 43280, Spain
e-mail: Ignasi.Batlle@irta.cat; Francisco.Vargas@irta.cat

C.E. Kallsen
University of California, Cooperative Extension, Kern County, 1031 S. Mt. Vernon Ave., Bakersfield, CA 93307, USA
e-mail: cekallsen@ucdavis.edu

M.L. Badenes and D.H. Byrne (eds.), *Fruit Breeding*, Handbook of Plant Breeding 8, DOI 10.1007/978-1-4419-0763-9_21,

one interspecific hybrid from the *Pistacia* genus. Beside the Californian breeding program, the only organized breeding programs at present are located in Spain, Turkey, and Israel. The California breeding program was formerly focused on precocity (early bearing), nut size, yield, split percentage, and early season harvest. Early season maturity is important to avoid navel orangeworm damage and to maximize the efficiency of harvest and processing facilities. Disease resistance, especially resistance to *Alternaria alternata*, has been a secondary objective in the program. Molecular markers have been used for genetic studies and determination of the origin of cultivars. While a number of molecular marker studies have been conducted, a molecular genetic marker map has not been constructed yet.

Keywords Pistachio • *Pistacia vera* L • *Pistacia* spp • Nut • Production • Cultivars • Varieties • Rootstocks • Genetics • Molecular markers

1 Introduction

The genus *Pistacia* L. is a member of the Anacardiaceae family that also includes cashew, mango, poison ivy, poison oak, pepper tree and sumac. The genus consists of eleven or more species (Zohary 1952; Whitehouse 1957; Kokwaro and Gillett 1980; Kafkas and Perl-Treves 2001; Parfitt and Badenes 1997). *Pistacia vera* L. has edible nuts and is the only commercially important species. The other species have been used for many years as rootstock sources for *P. vera*. *P. vera* is believed to be the most ancestral species and the other species are probably its derivatives (Zohary 1952). Wannan and Quinn (1991) compared the floral morphology of the genus with that of sister groups. All members of the genus are dioecious (note exceptions below), diploid dicots. Ila et al. (2003) reported a diploid chromosome number of 30 for *P. atlantica* Desf., *P. eurycarpa* Yalt., *P. terebinthus* L., and *P. vera*. These results were consistent with an analysis from Parfitt and Lin unpublished, of 11 species in which $2n=28$ or 30. Prior reports (Ghaffari and Harandi 2001) of $2n=24$ for *Pistacia khinjuk* Stocks appear to be incorrect. Molecular taxonomic descriptions of the genus have been reported by Parfitt and Badenes (1997), and Yi et al. (2008). Major differences from the 1952 Zohary classification include the definition of *P. integerrima* as a separate sister species to *P. chinensis* and classification of the species into only 2 sections.

The cultivated pistachio of commerce is the species *P. vera* L. It is native to the Middle East and Central Asia. There are two centers of diversity of cultivated pistachio: one comprises the Mediterranean region of Europe, Northern Africa, and the Middle East. The second comprises the Eastern part of Zagros Mountains from Crimea to the Caspian Sea (Maggs 1973; Hormaza et al. 1998). The Vavilov center of diversity for wild *P. vera* is located in Northern Iran and Southern Turkmenistan as well as parts of Afghanistan. The region straddling the border of Turkmenistan and Iran is referred to as the Baghtis region and is an area of rolling hills covered by grasslands

Fig. 21.1 Pistachio distribution in Europe and Asia. Textured area is the Badgtis region of wild pistachio savannah (Maggs 1973)

and scattered stands of *P. vera* trees. This is the area with the greatest present natural diversity, primarily because other areas of Iran with native pistachio stands have been topgrafted with clones of improved cultivars (Maggs 1972, 1973) (Fig. 21.1).

Pistacia species are dioecious with several isolated reports of monoecious individuals (Ozbek and Ayfer 1958; Crane 1974; Kafkas et al. 2000; İsfendiyaroğlu and Özeker 2009) and wind pollinated. Male and female apetalous flowers are borne in panicles on separate trees. Each panicle can have more than 100 flowers (typically 100–300 flowers per inflorescence, Fig. 21.2). *P. vera* produces a nut (classified as a semidry drupe) which is marketed as a dried in-shell product after removal of the husk. Pistachio nuts are drupes, the same classification as for almonds and stone fruits. All drupes consist of three parts: an exocarp, a fleshy mesocarp, and an endocarp that encloses a seed. The pistachio endocarp or shell encloses a single oil-rich seed and usually splits along its lateral suture, when the nut is ripe. The exocarp changes from green to white or white-purple color at maturity (Fig. 21.3). The kernel has a papery seed coat and two cotyledons.

The tree has a growth habit characterized by a strong apical dominance and lack of vegetative buds in old trees. Pistachios have an extensive root system. Under natural conditions, *P. vera* does not develop a central tap root, but produces a highly branched root system with many fine roots that allows the tree to efficiently extract water and nutrients. The tree has a pinnately compound leaf. Each leaf subtends a single axillary bud. Most of these lateral buds differentiate into inflorescences and produce female or male flower bearing rachi.

Fig. 21.2 Male and female *P. vera* inflorescences

Currently, Iran, the USA, Turkey, and Syria are the main pistachio producers in the world, contributing over 90% of the world production (FAOSTAT 2007) (Fig. 21.1). US production and planted areas have continued to expand, planted primarily with 'Kerman' and recently with a new University of California release, 'Golden Hills.' In 2006, 45,527 ha of pistachio were harvested in California and another 16,228 nonbearing ha were in the ground (CPC 2007; Pollack and Perez 2007). The average value of the California crop in 2005 and 2006 was approximately 518 million dollars (CPC 2007). Production is located primarily in southern San Joaquin valley of California (Fig. 21.4). The California industry is highly mechanized and is characterized by a few large, well-funded growers.

Fig. 21.3 'Golden Hills' pistachio, showing typical clusters

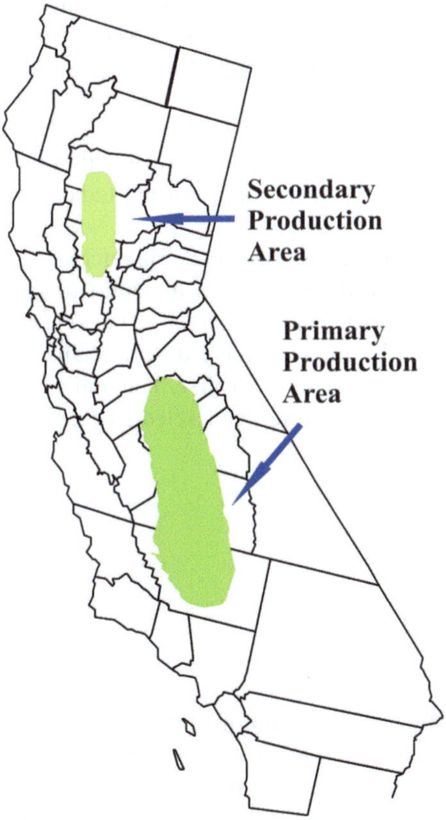

Fig. 21.4 Pistachio production areas in California

2 Origin and Domestication of Scion Cultivars

The natural distribution of wild *P. vera* is centered in Takzhikistan, Kirgizia, and north Afghanistan, and extends westward to the northern part of Khorasan district in Iran, and the Kopet mountain range of southern Turkmenistan (Zohary 1996). The Badgtis region of southeastern Turkmenistan and northeastern Afghanistan is an area where significant undomesticated *P. vera* forests remain (Popov 1994; Maggs 1973) and appears to be the center of diversity for this species (Whitehouse 1957). These wild pistachio nuts are typically much smaller than the cultivated pistachio, usually about 1 cm long and do not split. The trees are widely scattered among grass covered hills (Parfitt personal observation) in an area that is nominally protected as a nature preserve. Thus, geographically, wild *P. vera* represents the most northern wild pistachio taxon in Central Asia and it is spatially almost fully separated from the two other wild pistachio species that occur in this region (*P. atlantica* and *P. khinjuk*). The later two grow farther south, and overlap with wild *P. vera* only at the fringe of their distribution range in Khorasan and probably in the north of Afghanistan. In the middle part of Central Asia (Uzbekistan, Tadzhikistan, Kirgizia, and the southern most parts of Turkmenistan and Kazakhstan) *P. vera* forest extends over some 300,000 ha. Wild *P. vera* has been and is used as a source of nuts. The distribution area of the wild forms of *P. vera* and archaeological evidence indicate that this species was first brought into cultivation in Central Asia (Zohary and Hopf 2000; Zohary 2006). The species has been long propagated for nuts throughout the Mediterranean and Middle East. Several reports suggest that the Romans were responsible for the spread of *P. vera* within the Mediterranean basin. The total dependence of pistachio on grafting today suggests relatively late domestication.

About 100 cultivars from several regionally distinct groups have been described worldwide (Maggs 1973; Parfitt 1995). Pistachio cultivars from the middle East had good split shells and from Iran generally had large nut size and split shells, whereas those from Italy had small nuts with many unsplit shells and dark green kernels. Iranian pistachios are apparently more diverse than cultivar populations from other regions, probably because they have been selected from wild materials near the center of diversity.

3 Genetic Resources

3.1 Female Cultivars

The following descriptions are a partial list of the main cultivars grown in various regions of the world. In Europe and especially in Turkey and Iran, a substantial number of additional named cultivars exist, which may be synonymous with the cultivars described below or they may be distinct local varieties, analogous to land races of wheat. A single cultivar may have several local names; however, different discrete selections may be given the same name such as the name of a region or municipality. Therefore, reliance on cultivar names as the basis for accessing "trueness to type" may not be reliable. Patented or recently released cultivars are more likely to be correctly identified than old cultivars, especially those of Eurasian origin.

Old European cultivars are found distributed around the Mediterranian basin. 'Napoletana' is the predominant variety in Sicily and is considered to be synonymous with the 'Bianca' cultivar grown in Italy. Less often planted are 'Agostana,' 'Girasola,' 'Notaloro,' 'Cappuccia,' and 'Femminello.' 'Trabonella' and 'Bronte' are Sicilian cultivars with similar characteristics. Nut color is greenish and nut shape is longer and thinner than for 'Kerman' or Iranian cultivars. Nut size is considerably smaller than 'Kerman' and nut quality under California conditions is poor, with a high level of nonsplits in some seasons and significant disease and pest problems. The hulled nuts have a tendency to stain. 'Sfax,' 'Mateur,' and 'El Guettar' are grown in Tunisia. 'Mateur' may be the best of the Tunisian selections with the nut size and appearance similar 'Kerman.' 'Sfax' produces large tight nut clusters, but nut size, yield, and percent splits are inferior to 'Kerman' nuts. 'Aeginea' ('Aegenes') and the more recently released 'Pontikis' (Pontikis 1986) are grown in Greece. 'Aeginea' appears to be very susceptible to *Botryosphaeria dothidea*, perhaps due to its very early flowering in the spring. 'Pontikis' has a moderately large fruit, high kernel weight (55% of fruit weight) and an oblong-ovate shaped nut and kernel. It splits much better than 'Aegenes,' with 90–98% splits. Blank nut percentage is about 5–10%, the same as 'Aegenes' and has yields similar to 'Aegenes.' However, the adaptation of 'Pontikis' outside of the Athens area has not been determined. 'Lamarka' is the main cultivar in Cyprus.

A large number of named cultivars are grown in Turkey. 'Uzun,' 'Kirmizi,' 'Siirt,' 'Halebi' are favored. Less favored is the Turkish cultivar 'Red Aleppo' which is commonly grown in Syria. 'Achoury' is a major Syrian cultivar. 'Achoury,' 'Alemi,' 'El Bataury,' 'Obiad,' and 'Ayimi' are also grown in Turkey. 'Red Aleppo' was used as a cultivar during the early development of the California industry and produces a good quality nut with reasonable yield under California conditions. It has a high percentage of split nuts but nut size tends to be somewhat smaller than 'Kerman.'

Many of the best pistachio cultivars are found in Iran. Large nuts are preferred and many of the Iranian cultivars have relatively large nuts with good split percentages. Major pistachio cultivars grown in Iran are 'Ohadi,' 'Akbari' and 'Ahmad Aghaii.' Other cultivars are 'Momtaz,' 'Kalehghouchi,' 'Ghermeza,' 'Tbeahimi,' 'Ogah,' and 'Wahidi.' Another Iranian cultivar 'Rafsanjani,' considered to be a promising cultivar in Iran, is being tested for adaptation in Azerbaijan. In addition to these cultivars, many other named cultivars are grown in Iran, either locally or across the country. Some of these items may be the same as the aforementioned cultivars, but with different local names. 'Ohadi' produces attractive nuts that are slightly smaller than 'Kerman' nuts. 'Kalehghouchi' has very large nuts as well as a good yield and has attracted some interest in California because of its nut size and good split percentage, which may be better than 'Kerman' under California conditions. In a replicated trials budded in 1998, located on the west side of Kern County, 'Kalehghouchi' and 'Aria,' another Iranian variety were tested against 'Kerman.' Seventh to tenth leaf 'Kalehghouchi' trees yielded similarly to 'Kerman,' while 'Aria' yields were lower. Nut split percentages and weights were higher for both of these cultivars than for 'Kerman.' Both 'Kalehghouchi' and 'Aria' flower 5–10 days earlier than 'Kerman'. 'Kalehghouchi' matures at about the same time as 'Kerman' while 'Aria' matures about 2 weeks earlier than 'Kerman.' 'Kalehghouchi' produces excessive vegetative growth on mature trees under California management conditions. It has a tendency to produce many long 'whips' and 'hanger' branches on which a considerable fraction of the fruit is borne. This makes shaking difficult or requires considerable additional pruning to maintain tree structure. 'Aria,' when tested in the same trials, produced nuts with poor shell-hinge strength resulting in excessive loss of shells and kernels during hulling. The location of flower buds on new branches has made training this cultivar difficult and has resulted in sunburned nut clusters. These limited replicated trials have demonstrated that a particular cultivar's success in one region does not necessarily translate to successful performance in another.

Extensive collections of pistachio cultivars and germplasm resources were assembled at several experiment stations in the southern former Soviet republics during the 1950s and 1960s. With the dissolution of the Soviet Union, the independent republics have had difficulty supporting these stations and collections, so most of the materials have either disappeared or will disappear within the near term. The collections are in poor conditions and records for some of them are no longer available (Parfitt—personal observation-Kara Kala, Turkmenistan). Some of the selections maintained at these sites are of commercial quality, some are not. Several promising selections have been made at the Genetic Resources Institute of the National Academy of Sciences in Baku, Azerbaijan.

'Sirora' was developed in Australia (Maggs 1990) from a formal breeding program. The industry has remained small in Australia, so this cultivar has not been widely planted. 'Kastel' and 'Rashti,' grown in Israel, are similar in some aspects to the 'Kerman' variety. Neither 'Kastel' nor 'Rashti' have been tested directly against 'Kerman' in California yield trials. 'Rashti' has large nuts, high split percentage, and a good flavor. Tree structure is similar to 'Kerman' as is its alternate bearing characteristic. It is a late maturing cultivar, probably several weeks after 'Kerman.' This has been an issue during years when late summer and fall have been very cool. Under these conditions it may not ripen before the start of winter rains. 'Kastel' seems to be very similar to 'Kerman' in most characteristics. Nut size may be slightly larger.

Less than 20 named cultivars have been imported into the USA. Some of the female cultivars that were introduced into California by the USDA in the early 1900s were: 'Red Aleppo' from Syria, 'Bronte' and 'Trabonella' from Sicily, 'Sfax' from Tunisia, 'Kastel' and 'Rashti' from Israel and some from other countries. Additional unnamed *P. vera* germplasm was introduced through the USDA Plant Introduction Gardens at Chico which was closed in 1967. However, very little germplasm has been imported into the USA since the station was closed. A number of the introductions, both *P. vera* and other species, at the former Plant Introduction Gardens were collected and transferred to the National Clonal Germplasm Repository at Davis during the 1980s.

'Kerman,' a California cultivar, was collected in 1929 by W. E. Whitehouse in either Iran or Turkmenistan (Joley 1979), selected as a seedling in 1936, and released for trial by the USDA Plant Introduction Station, Chico CA in 1957 (Whitehouse 1957). The release of 'Kerman' occurred just prior to the major growth of the industry in California. 'Kerman' which is now the primary female cultivar commercially grown in California, produces large yields of attractive nuts with superior size (>1.2 g), especially in the primary growing areas of the southern San Joaquin valley. It is not a perfect cultivar. It has a strong alternate bearing tendency, a high percentage of blank nuts in some years, a relatively high level of nonsplit nuts, and a light greenish-yellow kernel with almost minimal flavor when dried at commercial temperatures. Kernel color and flavor has not been an issue with American consumers, who have not been exposed to European cultivars with different flavor characteristics. 'Kerman' is a relatively late maturing cultivar, which is not usually a problem in the San Joaquin valley with a large number of heat units. However, it's later maturity means that it can be exposed to a third flight of navel orangeworm (*Amyelois transitella* Walker). Beyond the direct losses from damaged nuts, navel orangeworm infestations have been implicated in the development of higher levels of *Aspergillus flavus var. flavus* (Klich and Pitt 1988) contamination and consequent aflatoxin contamination. In the Sacramento valley during years with low heat unit accumulation, 'Kerman' may not mature until after the first fall rains, with a concomitant increase in disease problems. 'Kerman' is relatively susceptible to *Alternaria alternata* (*tenuissima*, and *arborescens* species groups; Pryor and Michailides 2002), but under dry California conditions this is not usually a problem for growers, and can be controlled with fungicides where it is a problem.

In 1980, another open-pollinated seedling introduced as a seed from Damghan, Iran, was selected at the University of California, Davis, by Dr. J. Crane and named 'Joley,' in honor of the former director of the Chico USDA Plant Introduction Station (Gardens). 'Joley' has been planted in a few orchards in California and in the state of New Mexico, where there is a limited acreage of pistachios. The cultivar is considered by some to be one of the best tasting pistachios developed in California. The tree has moderate vigor and blooms and matures about 10 days earlier than 'Kerman.' It has almond-shaped nuts similar to 'Trabonella' or 'Bronte' (Sicilian varieties). The kernel color is greener than 'Kerman,' but the nut size is significantly smaller and the nonsplit percentage can be high in some years. There are few blanks; however, in some instances shell removal by consumers is not as easy as for 'Kerman.' On a few young trees grown in Kern County, the yield has never been high. It has a tendency to stain and does not appear to be a commercially viable cultivar.

'Lassen' was developed by Whitehouse from the same seed lot as 'Kerman' and released in 1962 from the USDA Plant Introduction Station at Chico CA. It is very similar to 'Kerman' with respect to nut characteristics. It has never been tested in replicated trials, but has good potential as a cultivar. Data from yield trials with 'Kerman' is not available. 'Damghan' was also developed from this seed collection. It has very large nuts but appears to be very low yielding under California conditions.

3.2 Male Cultivars

The California pistachio industry is based on one male cultivar, 'Peters.' 'Peters' is a good pollinizer and was found in the early 1900s by A. B. Peters at Fresno, California. However, the parentage is unknown. It produces abundant pollen, shed over a relatively long period (ca. 2+ weeks) and pollen viability over time (durability) is very good. In recent years, under low chill conditions it has shed pollen at the end of the receptive period for 'Kerman,' resulting in poor and irregular pollination. In addition to 'Peters,' the selections at Chico of '02-16' and '02-18' imported from Russia are early and late blooming compared to 'Peters.' Pollen from '2-18' is somewhat less durable than pollen from 'Peters.' 'Nazareth,' 'Ask,' and 'Chico' males are also grown sporadically in some locations in California. 'Chico' was introduced from the Chico, Calif., USDA Plant Introduction Station in 1962 as PI150646. It is reported to be a male seedling of *P. vera* selected from seed introduced under PI73396 from Aleppo, Syria; probably of species hybrid origin and tested as Chico 23. Leaf characters and bloom period observations suggest that it is probably an interspecific hybrid between *P. vera* L. and *Pistacia integerrima* L. (Parfitt personal observation). It is a prolific pollen producer and blooms early, courtesy of its *P. integerrima* parentage. 'Chico' sheds pollen prior to and during the earliest part of the 'Kerman' bloom period but often flowers too early to pollenize 'Kerman.' Some growers have used '02-16' to match the early part of the 'Kerman' bloom period. If there are any xenia effects on nut size (male contribution), then this cultivar should not be used. 'Ask' and 'Gazvin' were introduced from Israel a few years ago and may have some value as pollenizers. They flower earlier than 'Peters,' but have poor pollen durability.

3.3 Rootstocks

California pistachios are grown primarily on three rootstocks, two species and one interspecific hybrid, all members of the genus *Pistacia*. They include Atlantica (*P. atlantica*) Pioneer Gold I (*P. integerrima*), and UCBI, which is a hybrid between a *P. atlantica* female crossed with a *P. integerrima* male. Two other rootstocks have occasionally been grown in California; they include *P. terebinthus* and Pioneer Gold II (a *P. atlantica* female crossed with a *P. integerrima* male). All of these rootstocks are produced from seed. Pioneer Gold I (PG1) is distributed as seedling plants produced from a population of *P. integerrima* parents selected for resistance to

Verticillium wilt (*Verticillium dahliae*, http://broad.harvard.edu/annotation/genome/verticillium_dahliae/MultiHome.html). Consequently, these plants are expected to be genetically variable, although they appear to be quite uniform in the field. During the last 15 years, Pioneer nursery has continued to improve and select the best parent stock plants, so what is being released as Pioneer Gold I today may be genetically different than the material released in the past. They are all *P. integerrima*. Pioneer Gold II was released at about the same time as UCB1, but was not widely accepted by growers. UCB1 was released by L. Ashworth at the University of California Berkeley (Morgan et al. 1992). This rootstock is also distributed as seed or seedlings, but is the result of a closed cross between a *P. atlantica* female and a *P. integerrima* male. Originally two different *P. atlantica* females were used, but after incompatibility with 'Kerman' was detected in the seed lot derived from one of the females, UCB1 was refined to be the result of only one *P. altantica* female and one *P. integerrima* male.

4 Major Breeding Achievements

4.1 Cultivars

The only organized breeding programs at present are located in Spain, Turkey, and Israel. There is probably active work being conducted in Iran as well, but it is less well documented. California and Australia had breeding programs in the past, but these have been discontinued due to loss of funding.

The California program was a conventional breeding and genetics program using crosses among all possible combinations of 30 female and 45 male genotypes. Materials from the former Chico Plant Introduction Gardens were used as well as Iranian and Italian selections. Males from J. Cranes selection program of the 1970s were also used to make some of the crosses. Approximately 8,000 seedlings were produced and were evaluated initially at three locations on their own roots. Superior plants were selected from these seedlings and were tested in replicated trials on the two rootstocks used in California. Three cultivars, described below, were released from this program. A detailed description of the program is given in Chao et al. (1998). Some potential cultivar materials from the California program continue to be tested and several superior cultivars may be selected from among them.

Two new female cultivars 'Golden Hills' and 'Lost Hills', were released in 2005 from the pistachio breeding program of Parfitt et al. (Kallsen et al. 2009). Five years (6th through 11th leaf) of production and phenological data have been taken on these cultivars. 'Golden Hills' (Parfitt et al. 2007) is a new female cultivar that flowers 5–7 days before 'Kerman' and matures about 2 weeks earlier than 'Kerman'; permitting efficient use of harvesting and processing equipment in much the same way that cultivar maturity series for peaches facilitate limited resources for harvesting and marketing. Earlier harvest also allowed this cultivar to miss infestation by the (most damaging) September flight of navel orangeworm. This cultivar has produced 35% higher yield than 'Kerman' during the first 5 years of harvested yield trials in the southern San Joaquin valley. Nut size and weight were similar to 'Kerman,' but percentage of blank nuts was lower and split nut percentage about 25% greater than for 'Kerman.' 'Golden Hills' has more but smaller scaffold branches than 'Kerman,' producing a smaller more bushy tree after 3–4 years of training. 'Lost Hills' (Parfitt et al. 2008) is a new female cultivar from the breeding program of Parfitt et al. that flowers 4–7 days before 'Kerman' and matures about 2 weeks earlier than 'Kerman.' As with 'Golden Hills,' earlier maturity results in much less navel orangeworm damage at harvest (<0.2%) This cultivar produced 28% higher yield than 'Kerman' over the first five bearing years. Nut size and weight and percent split nuts were about 26% higher than for 'Kerman.' 'Lost Hills' produced more lost kernels and loose kernels than 'Kerman' during hulling as well (3.2%). Flowering was more uniform than 'Kerman' for both this cultivar and 'Golden Hills' during low chilling seasons. This translated into a more uniform maturity and less difficulty in determining the correct time to harvest for maximum splits and minimum staining. Evaluation of 'Lost Hills' and 'Golden Hills' is continuing.

A number of new pollen sources were evaluated as part of the pistachio breeding program conducted by Parfitt et al. They were evaluated for quantity of pollen produced, the period over which the pollen was shed, pollen viability percentage at pollen shed, and pollen durability (the length of time during which pollen viability remained high). Two selections were made to complement 'Peters,' flowering earlier and later than 'Peters.' The early flowering selection, 'Randy,' released in 2005, flowers 1–3 weeks earlier than 'Peters.' It is characterized by a long bloom period, in excess of 2 weeks which is twice that of most male pistachios, a characteristic that 'Peters' shares. Peak flowering is 1–2 weeks earlier than peak flowering for 'Kerman.' The pollen is more durable than 'Peters' pollen (75% viable declining to 35% viable after 29 days of storage vs. 'Peters' initial viability of 45%, declining to 15% to 5% after 27 days). Although it was selected as a pollenizer for early flowering cultivars from the breeding program of Parfitt et al., and blooms too early to serve as the primary pollenizer for 'Kerman,' 'Randy' may be useful as a 'Kerman' pollenizer during low chill seasons, when there is less overlap in flowering period between 'Kerman' and the later flowering 'Peters.' The bloom period for 'Randy' closely matches that of 'Kalehghouchi' and would be a better pollenizer choice for this cultivar than 'Peters.'

Small breeding and/or selection programs in Australia (Maggs 1990) and Greece (Pontikis 1986) have released the cultivars 'Sirora' and 'Pontikis.'

In Spain a pistachio scion breeding program has been conducted at IRTA Mas de Bover since 1989. A total of 31 controlled crosses among 10 female and 12 male parents were made between 1989 and 1990. Selections have been made from about 2000 seedlings to date (Vargas et al. 1996). Currently, 9 female selections and 7 male selections are under trial. Cross combinations were planned considering that they could not freely occur in nature due to their different geographical origins. The main female parents used were 'Aegina' (Greece), 'Batoury' and 'White Ouleimy'

(Syria), 'Kerman' (The USA), 'Larnaka' (Cyprus), 'Mateur' and 'Sfax' (Tunisia), and the main male parents chosen were: 'B' and 'C' (Greece), 'M-36' and 'M-38' (Syria), M-502 (Italy), 'Nazar' and 'Enk' (Israel) (Vargas et al. 1996). A number of traits have been studied in the progenies like vigor (Vargas et al. 1996; Vargas and Romero 1998a, b), blooming and leafing time (Vargas et al. 2001), flowering precocity (Vargas et al. 2002), and nut traits (Vargas and Romero 2005).

Even though *P. vera* was introduced into Italy from Syria by the Romans from selections developed by the Arabs, only 10+ female cultivars are grown together with even a more limited number of male selections (Barone and Caruso 1996). Currently, 'Bianca' (syn. 'Napolitana') is the main cultivar grown commercially in Bronte, Sicily which is the main Italian growing area. In 1984, the University of Palermo established a germplasm collection of 10 cultivars including 'Bianca' and 'Kerman' and 8 male pistachio selections (M1, M3, M4, M5, M7, M8, M9, and M10). M9, locally called 'Santagilisi,' is a putative hybrid between *P. vera* and *P. terebinthus*.

4.2 Rootstocks

Lloyd Joley identified several *P. atlantica* selections with nematode resistance at the former Plant Introduction Station in Chico CA (Joley 1979). Later J. Crane selected several *P. atlantica* seedlings that showed high levels of vigor. These selections were not used in California production, primarily because of the absence of a good clonal propagation system for *P. atlantica.* Seedling *P. atlantica* was used as the primary rootstock in the California production system until the development of UCB1, a *P. atlantica* × *P. integerrima* hybrid rootstock. This cross was identified by Ashworth (Morgan et al. 1992) as being resistant to Verticillium wilt, which was a major problem in California when using *P. atlantica* as a rootstock. Not only was this rootstock *Verticillium* sp. resistant, but it had exceptional vigor, and produced much improved yields of 'Kerman' at a relatively early point in the orchard development. Production could begin at 4–5 years, rather than 8–10 years, in the San Joaquin valley of California.

Pistachio rootstocks are produced from seeds except for a clonal selection of UCB1, being produced via micropropagation by Duarte nursery. Originally, the seed was produced from open pollinated tress, with control of the genetic composition of the seed being limited to the source tree. More recently closed crosses (control of both male pollen sources and female source trees) has become standard practice. PG1 was originally produced from open pollinated trees, but more recently has been reported to be produced from selected male and female trees. Both of these rootstocks were selected for resistance to Verticillium wilt. PG1 does not denote the specific genetic composition of the progeny, unlike UCB1, which is produced from an individual female clone and male clone. Thus, there is more potential for variability among the non-UCB1 hybrids. In the San Joaquin Valley, PG1 and UCB1 are the most commonly used rootstocks. In colder areas outside the San Joaquin Valley, *P. atlantica* and UCB1 are the most commonly planted rootstocks.

After a survey of native *P. terebinthus* located on the Spanish central plateau, a clonal selection from the best genotypes was made available to the nursery sector and used to produce seedling rootstocks (Guerrero et al. 2002). *P. atlantica* is thriving in the Canary Islands (Batlle et al. 2006).

5 Current Goals of Breeding

5.1 Scion Genotypes

The California breeding program was formerly focused on precocity (early bearing), nut size, yield, split percentage, and early season harvest. The program is currently limited to selection of additional clones from the original crossing programs and evaluation of progeny from several parents selected from that program. The primary objectives continue to be early season maturity, yield, nut size, and high split nut percentage (Kallsen et al. 2009). Early season maturity is important to avoid navel orangeworm damage and to maximize the efficiency of harvest and processing facilities. Disease resistance, especially resistance to *A. alternata* was a secondary objective in the program but is not being actively pursued at present. Several selections with high levels of resistance to *A. alternata* as well as nut size and potential yield have been retained for use in future breeding efforts.

The Spanish program has emphasized early bearing, nut quality, high productivity and strong vigor. Major traits of the female parents were early bearing, nut quality and productivity and, for the male parents the features were flowering precocity and trueness to type.

6 Breeding Methods and Techniques

Most of the genetic information that has been developed to date for pistachio has been derived from traditional qualitative and quantitative genetic analysis. The only simply inherited traits that have been described for pistachio are (a) a dwarfing gene that produces a genetic dwarf in a 1:3 ratio in the progeny of two specific parents from a *P. chinensis* × *P. integerrima* cross (Parfitt 2003) and (b) sex expression in the genus for which all progeny segregate 1:1. The dwarfing gene is a simple recessive, and probably acts through interruption of the gibberellic acid pathway, since application of gibberellin induces a normal phenotype in dwarfed progeny plants. Sex expression should be conditioned by a dominant gene in heterozygous condition in one of the sexes and a recessive condition in the other (Hormaza 1994). Interesting exceptions have been reported by Kafkas et al. (2004), who have described the occurrence of monoecy in an otherwise dioecious crop. Hormaza (1994) and Hormaza et al. (1994b) have described the characterization of a RAPD molecular

Table 21.1 Heritability estimates and estimates for nut weight, Alternaria resistance, and trunk cross sectional area

Traits	Location	Population	h_{ns}^2
Nut wt.	Bakersfield	Half-sib (OP)	0.76
% Split nuts	Bakersfield	Half-sib (OP)	0.44
Kernel wt.	Bakersfield	Half-sib (OP)	0.32
Suture	Bakersfield	Half-sib (OP)	0.30
Alt. resist.	Kearney 1995	Half-sib (OP)	0.48
Alt. resist.	Kearney 1997	Half-sib (OP)	0.11
Alt. resist.	Winters 1995	Half-sib (OP)	0.56
Alt. resist.	Winters 1997	Half-sib (OP)	0.56
Trunk XC area	Bakersfield 1995	Half-sib (OP)	0.20
Trunk XC area	Bakersfield 1996	Half-sib (OP)	0.21
Trunk XC area	Bakersfield 1997	Half-sib (OP)	0.25
Trunk XC area	Kearney 1995	Half-sib (OP)	0.29
Trunk XC area	Kearney 1996	Half-sib (OP)	0.28
Trunk XC area	Winters 1995	Half-sib (OP)	0.44
Trunk XC area	Winters 1996	Half-sib (OP)	0.56
Trunk XC area	Winters 1997	Half-sib (OP)	0.50
Precocity	Bakersfield	Half-sib (OP)	0.54
Precocity	Winters	Half-sib (OP)	0.93
Mid Flowering Date	Bakersfield and Winters	Half-sib (OP)	0.79
Mid Flowering Date	Bakersfield and Winters	Parent–Offspring	0.89
First Leafing Date	Bakersfield and Winters	Half-sib (OP)	0.75
First Leafing Date	Bakersfield and Winters	Parent–Offspring	0.60

marker for sex expression in *P. vera* from a bulked segregant analysis. RAPD primer OP008 was used to generate a band of 945 bp that was present in female but not male progeny from 'Kerman.' The marker was tested and confirmed against 14 other *P. vera* cultivars. This marker was subsequently tested by Kafkas et al. (2001) and was not found to be diagnostic for determination of sex in other *Pistacia* species. Kafkas et al. (2001) subsequently developed additional sex linked RAPD markers for several *Pistacia* species. A 1,300 bp sex linked band was identified in *P. eurycarpa* and cloned, but the cloned region was found to hybridize to both male and female plants. A 700 bp marker (amplified with primer BC346) and an 850 bp band (amplified with primer OPAK09) were also characterized in *P. atlantica*. Yakubov et al. (2005a) used the RAPD marker described by Hormaza (1994) as the basis for an improved RAPD marker which they converted to a SCAR marker. Yakubov et al. (2005b) have described the identification and cloning of a gene for a dehydrin-like protein from pistachio.

Several quantitatively inherited characters have been described for *P. vera*. Half sib family and parent–offspring regression analyses of flowering and leafing date (Table 21.1) showed that both are highly heritable (Chao et al. 2003). Leafing and flowering dates were strongly correlated (0.59–0.78), suggesting that early flowering could be selected for at the seedling stage. Resistance to *A. alternata* (Fries) Keissler was also shown to be heritable (Chao et al. 2001). Narrow sense heritability for precocity (early fruit production) was 0.54 and 0.93 at two locations, respectively (Parfitt and Chao unpublished). Other traits such as kernel weight, nut splitting, *Alternaria* resistance, and vigor (trunk cross-sectional area) have been shown to have low to medium (0.20–0.50) heritability (Table 21.1) (Chao et al. 1998; Parfitt et al. 1996).

Pistachio is a long generation tree crop. In the past, a generation time of 8–10 years was given for making crosses and obtaining progeny plant for grow out and evaluation. However, this can be shortened to a generation cycle of 4–6 years (Parfitt and Kallsen unpublished) by growing the plants on their own roots under ideal growing conditions with adequate water and selecting for early flowering plants.

Since males and females do not necessarily flower at the same time, so pollen may need to be stored. As is typical of many wind pollinated crops, pistachio pollen is not especially durable and is only useful for less than 4 days at room temperature, about 3 weeks at 4°C and stored under desiccation, and up to 8 months at −20°C (Polito and Luza 1988), although it should be noted that pollen germination and durability varies considerably among male genotypes (Parfitt unpublished).

Pollen is collected from male panicles brought from the field as soon as pollen shed is observed from the first open flowers. These are dried on a sheet of paper in the lab and the pollen is separated from the flowers and then stored at 4°C in a desiccator until used. Female inflorescences should be bagged in pollen and waterproof paper bags, tied at the base with cotton batting fitted around the branch to seal the base of the bags. Pollen may be applied by collecting pollen on a camel's hair brush and blowing the pollen off of the brush into the temporarily open mouth of the bag. Application of large amounts of pollen should be avoided. Once the receptive period of the last flowers in the bagged inflorescences has passed (usually about 4 weeks after pollination), these bags are removed and then replaced with breathable mesh bags to protect the developing fruit from birds and animals. Fruits should be collected in the fall after shell split when the husks are usually purplish-red or for some genotypes, white and easily separated (slipping) from the shells (Fig. 21.3).

An alternative mass selection strategy where only two or a few parents are to be used, is to set up a seed orchard with the female(s) and selected males or with only selected females and introduce the male pollen mechanically (broadcast across the orchard). Seed are harvested, planted, and evaluated. While this approach is more labor and cost efficient at the initial seed production stage, this efficiency is likely to be lost later in the grow out and evaluation process, which is the most expensive part of the breeding process, since this approach will result in a high level of redundancy and is ½ as efficient as a pedigree program for maximizing additive genetic variance.

Seed may be stored at 4°C under desiccated conditions for up to 1 year. After that time, viability will decrease but viable seed has been obtained after more than 3 years of storage. Frozen storage has not been used by the authors and may not be successful unless seed moisture can be carefully adjusted.

For germination the seed is hydrated (water soak) for 12–4 h at 4°C prior to planting into forestry pots (Jiffy 7 s were also used successfully). Once the plants are 15–20 cm tall, they are transferred to the field in the summer (if water can be applied immediately following planting), or the following spring when soil moisture is good.

If resources permit, seedling scion wood can be grafted to rootstocks in the field. This approach has the advantage of producing faster growing and fruiting plants (due to rootstock vigor), and permitting an evaluation of scion–rootstock compatibility at an early stage.

All breeding programs should include an advanced selection yield/performance trial using commercial production conditions with replicated clonal plants on rootstocks. Specific numbers of plants/clone and arrangement of replicates will be a function of the number of selections to be tested and the financial and/or human resources available to manage the trial. This is likely to be the part of the program requiring the most years to complete since multiple seasons with harvestable yields are needed for a robust evaluation of superior genotypes.

7 Integration of Biotechnologies into Breeding Programs

7.1 *Molecular Markers*

Isozyme markers (PGI, MDH, PGM, AAT, PER, and EST) in pistachio have been studied in California (Arulsekar and Parfitt 1986), Sicily (Dollo 1996; Barone et al. 1996) and Iran (Aalami and Nayeb 1996). Dollo (1996) studied isozyme polymorphism of *Pistacia* species and varieties growing in Sicily and analyzed the offspring obtained by controlled pollination of *P. vera* cv Bianca with pollen of *P. atlantica*, *P. vera*, *P. terebinthus*, and 'Santangilisi', a Sicilian pollinizer (previously published as a hybrid between *P. vera* and *P. terebinthus*). The Sicilian *Pistacia* species and *P. vera* cultivars showed high levels of polymorphism in the two systems studied. The results suggested that 'Santangilisi' is a hybrid between *P. vera* cultivars, and that the *P. vera* cv Insolia is a hybrid of *P. vera* × *P. terebinthus*. Isozyme analysis of eight male pistachio selections and 10 female pistachio cultivars (Barone et al. 1996). indicated that the male germplasm had a higher degree of polymorphism as compared to the female germplasm. Hence, using only three enzymes, it was possible to identify all of the male selections, but only 50% of the females.

Morphological descriptions and RAPD fingerprinting analysis were conducted on 24 cultivars of *P. vera* (8 male and 16 female) collected from Italy, Greece, Morocco, Spain and Turkey (Caruso et al. 1998). A high degree of polymorphism was detected both at the phenotypic and molecular levels. Among the female accessions, cluster analyses of both morphological and bio-molecular characters did not separate the Mediterranean from the Iranian–Caspian genotypes, in contrast to a study conducted by Hormaza et al. (1994a) which revealed two major clusters of *P. vera* germplasm: a Mediterranean cluster, which includes cultivars originating from the Mediterranean region of Europe, North Africa and the Middle East; and an Iranian–Caspian cluster, comprising germplasm originating from locations east of the Zagros Mountains. 'Kerman' is associated with Iranian cultivars, which is consistent with its selection from Iranian germplasm imported through the USDA Plant Introduction Garden at Chico California. The molecular data in combination with historical and geographical records, support the hypothesis that pistachio cultivation originated within, or near the present natural range of the species and was spread by cultivation to the Mediterranean region of the Middle East (Hormaza et al. 1994a, 1998).

Seedlings of *P. vera* developed from seeds of two separate populations in Turkmenistan, Kepele and Agachli, were characterized by Barazani et al. (2003) using RAPD markers. Genetic and morphological results showed some differences between 'Agachli' and 'Kepele' *P. vera* L. accessions. UPGMA cluster analysis divided 24 of the 27 accessions into two main groups according to their origins.

A genomic DNA library enriched for dinucleotide (CT)n and (CA)n, and trinucleotide (CTT)n microsatellite motifs was developed from 'Kerman' by Ahmad et al. (2003), who generated 14 polymorphic SSR primer pairs in pistachio with the objective of distinguishing US pistachios ('Kerman') from Iranian cultivars. The authors used them to characterize 25 commercially cultivated pistachio cultivars from Iran, Turkey, Syria, and the USA. Cluster analysis placed most of the Iranian samples in one group, while the Syrian samples were the most diverse and did not constitute a single distinct group.

Kafkas et al. (2006) characterized 69 pistachio cultivars and genotypes cultivated in Iran, Turkey, the USA, Syria, Greece, Italy, Israel, Cyprus and Tunisia by AFLP, ISSR, and RAPD analysis. Cluster analysis of the combined data formed two main groups correlated with the geographic origin of the pistachio genotypes. One group contained the cultivars originating from Iran, while the second group included cultivars originating from Turkey, Syria, Greece, Italy, Cyprus, and Tunisia. 'Siirt' (origin is the southeast part of Turkey) and its variants were placed between the two main groups. Turkish cultivars and the rest of the cultivars in the Mediterranean group were separated into two subgroups. One subgroup consisted of Turkish cultivars and the other subgroup contained Syrian, Italian and Tunisian cultivars. A recent paper by Afzadi et al. (2007) reports the use of AFLPs with UPGMA and PCA to characterize a number of Iranian pistachio cultivars.

Kafkas and Perl-Treves (2001) used RAPD markers and UPGMA cluster analysis of 41 accessions to show that the new species *P. eurycarpa*, formerly considered *P. khinjuk* in Turkey, and *P. altantica* accessions form separate clusters but were closely related when compared to *P. vera* and *P. terebinthus*, which was placed as the most distant of the four species studied. A considerable amount of morphological data was collected to support the analysis. A follow-up paper using RAPD markers and parsimony analysis supported the placement of *P. eurycarpa* with *P. atlanica* and separate from *P. khinjuk* (Kafkas and Perl-Treves 2002). The placement of *P. terebinthus*, distant from *P. atlantica*, did not agree with other phylogenetic analyses. An analysis of *Pistacia* species by Katsiotis et al. (2003) with RAPD and AFLP markers and grouped by UPGMA cluster and principle component analyses supported the conclusion of Kafkas and Perl-Treves (2002) associating *P. terebinthus* with *P. palaestina*. Male and female cultivars of *P. vera* were clearly separated into distinct clusters, suggesting that a number of the markers that were evaluated were closely associated with the gene (or genes) associated with sex expression. A limitation of all of the preceding studies is that a high level of polymorphism usually exists for these markers within species, such that samples to represent each species must be of sufficient number and composition (sampled from the full geographic species range) to correctly represent the genetic composition of that species.

Werner et al. (2001) used six RAPD primers and morphological analysis to show that *Pistacia* × *saporte* Burnat is a hybrid between *Pistacia lentiscus* and *Pistacia terebinthus*. This conclusion differs from Zohary's ((1972) evaluation of *P.* × *saporte* as being a hybrid with one of the parents consisting of *Pistacia palaestina* Boiss.

A combination of PCR amplification of chloroplast DNA, followed by RFLP analysis was used by Parfitt and Badenes (1997) to characterize species relationships among pistachio species when evaluated with distance and parsimony analyses. These results suggested that (a) *Pistacia* should be divided into two sections, *Lentiscus* and *Terebinthus*, rather than the four sections described by Zohary (1952), (b) *P. integerrima* and *P. chinensis* should probably be classified as separate species based both on molecular analysis, crossing behavior, and karyotype, and (c) the evolutionary rate for *Pistacia* is slow compared to annual crops. These conclusions were also supported by work using ITS and cpDNA sequence (Yi et al. 2008) and AFLP analysis Kafkas (2006). They also found evidence for reticulate evolution in the genus.

While a number of molecular marker studies have been conducted, a molecular genetic marker map has not been constructed. This is due to the need for well constructed crosses and the funding needed to place the markers on the map (e.g., by analysis of appropriate progeny populations). Funding for map development must be coupled with an effective breeding program from which relevant progeny populations can be obtained. Consequently, Marker Assisted Selection has not been practiced in pistachio improvement efforts. A potential MAS trait is the character of sex expression for which markers have been developed. Future directions for molecular genetic research should be focused on the identification of useful genes to permit the development of linked markers, development of a map with robust markers in *P. vera* for QTL or other applications, and more fundamental genetic analysis. Some areas for study are functional genomic analysis, identification and cloning of directly useful genes, and continued application of molecular markers to answer questions related to insect and disease control (characterization of specific genes or gene combinations for tolerance and/or resistance).

7.2 Micropropagation and Transformation Technology

Several research groups have developed micropropagation protocols for shoot tips of *P. vera* and rootstock species from shoot tips from seedlings (Barghchi 1982; Barghchi and Alderson 1983) and from clonal sources for several *Pistacia* species (Martinelli and Loretti 1988) , *Pistacia terebinthus* (Pontikis 1984), *P. vera* 'Mateur' using methyl jasmonate (Dolcet-Sanjuan and Claveria 1995), the male *P. vera* 'Atli' (Tilkat et al. 2008) and *P. vera*, *P. integerrima* and hybrids (Parfitt and Almehdi 1994). Variants of the medium developed by Parfitt and Almehdi (1994) are being used commercially to produce clonal pistachio rootstocks. Onay (2000a) reported a protocol for in vitro micropropagation of pistachio from mature trees. In vitro micrografting may be used in combination with in vitro propagation techniques to produce in vitro derived clonal trees from difficult to root pistachio genotypes (Onay et al. 2004b).

The development of protocols for regeneration of pistachio from somatic embryos and/or callus is a necessary first step for genetic engineering of pistachio, as the plants resulting from the protocol are generated from single cells, so that issues of generation of chimeric plants can be avoided. Genes for selection of single transformed cells in culture can be included in the transformation cassette, providing an efficient mechanism for selection of mutants. Onay (Onay et al. 1995; Onay 1996) developed a procedure for developing somatic embryos from pistachio kernels. The fact that preseedling materials were used as the source limited the value of these observations, however. Subsequently, the regeneration of plants ($2n=30$) from somatic embryos derived from callus of *P. vera* 'Siirt' pistachio flowers (Onay et al. 2004a) and from leaf explants (of the cv. Antep) (Onay 2000b) have been reported. Production of clonal plants via somatic embryogenesis can potentially be used to propagate other valuable cultivars, providing an opportunity to use genetic engineering approaches as well as conventional breeding strategies for cultivar improvement. Two major issues will limit genetic advance through this pathway.

1. Consumer acceptance: Lack of public acceptance and/or regulatory restrictions for genetically engineered food products may inhibit the introduction of new or modified genetically engineered cultivars. This is probably most likely to be an issue in Europe and since Europeans are major consumers of both Middle Eastern and US sourced pistachios, genetically engineered cultivars will be approached with caution by both US and Middle Eastern growers.
2. Control of Gene Expression: Introduction of a new trait via a genetic engineering approach requires stable expression of the introduced gene as well as expression during the appropriate developmental and/or seasonal stages of plant growth. In addition, expression of an introduced gene should not result in undesirable gene expression changes for nontransformed genes. Extensive testing of transformed products, probably to a greater extent and over a longer time period than for conventionally derived cultivars, will be needed prior to commercial utilization.

8 Conclusions: Status of Pistachio Improvement

Continued improvement of pistachio will require sources of stable, secure funding, since conventional programs require long term evaluations of selected materials. Costs for maintenance of collections and evaluation plots are significant due to the relatively large size of the plants (trees) and the need to maintain them for many years. Development of molecular markers and linkage of high resolution markers with important traits may help reduce the number of progeny that have to be evaluated in the field. Much more advanced and economical molecular marker technology is being developed in other fruit crops, and if resources become available, these technologies may change the way that researchers breed pistachio.

General information on pistachio production can be found in Joley (1979), Hormaza and Wünsch (2007), and Westwood (1993), as well as the "Pistachio Production Manual 4th edition (2005)," and the University of California Fruit and Nut Information Center (http://fruitsandnuts.ucdavis.edu).

References

Aalami, A. and M. Nayeb. 1996. Using isozyme for genetic diversity analysis of Iranian pistachio. M.Sc. Thesis, Faculty of Agriculture, Tarbiat Modares University, Iran.

Afzadi, M.A. B.E.S. Tabatabaei, S.A. Mohammadi, and A. Tajabadipur. 2007. Comparison of genetic diversity in species and cultivars of pistachio (*Pistacia* sp. L.) based on Amplified Fragment Length Polymorphism (AFLP) markers. Iranian J. Biotech. 5(3):147–152.

Ahmad, R., L. Ferguson, and S.M. Southwick. 2003. Identification of pistachio (*Pistacia vera* L.) nuts with microsatellite markers. J. Amer. Soc. Hort. Sci. 128:898–903.

Arulsekar, S. and D.E. Parfitt. 1986. Isozyme analysis procedures for stone fruits, almond, grape, walnut, pistachio, and fig. HortScience 21(4):928–933.

Barghchi, M. 1982. In vitro propagation of *Pistacia* species. Ph.D. Thesis. Univ. Nottingham, Nottingham, UK. 117 pp.

Barghchi, M. and P.G. Alderson. 1983. In vitro propagation of *Pistacia vera* L. from seedling tissues. J. Hort. Sci. 58(3):435–445.

Barazani, O, A. Atayev, B. Yakubov, V. Kostiukovsky, K. Popov, and A. Golan-Goldhirsh. 2003. Genetic variability in Turkmen populations of *Pistacia vera* L. Genet. Res. Crop Evol. 50:383–9.

Barone, E. L. Di Marco, F. P. Marra, and M. Sidari. 1996. Isozymes and canonical discriminant analysis to identify pistachio (*Pistacia vera*) germplasm. HortScience 31(1):134–138.

Barone, E. and T. Caruso. 1996. Genetic diversity within *Pistacia vera* in Italy. (In: Padulosi, S.; Caruso, T.; Barone, E. (eds.), Taxonomy, distribution, conservation and uses of *Pistacia* genetic resources. Report of a workshop 29–30 June:20–28.

Batlle, I., M.A. Romero, M. Rovira, and F.J. Vargas. 2006. *Pistacia* species genebank, conservation, characterization and use at IRTA: current situation and prospects in Spain. FAO-CIHEAM–Nucis Newsletter, (13):47–53.

California Pistachio Commission. California Pistachio Industry Annual Report, 2007. Crop Year 2006–2007. CPC. 1318 E. Shaw Avenue, Suite 420. Fresno, Calif. p.68.

Chao, C.T. , D. E. Parfitt, L. Ferguson, C. Kallsen, and J. Maranto. 1998. Breeding and genetics of pistachio: the California program.. Acta Hort. 470:152–161.

Chao, C.T. , D. E. Parfitt, L. Ferguson, C. Kallsen, and J. Maranto. 2003. Genetic analyses of phenological traits of pistachio (*Pistacia vera* L.). Euphytica. 129(3):345–349.

Chao, C.T., D.E. Parfitt, and T.J. Michailides. 2001. Alternaria late blight (*Alternaria alternata*) resistance in pistachio (*Pistacia vera*) and selection of resistant genotypes J. Amer. Soc. Hort. Sci. 126(4):481–485.

Caruso, T., C. Iannini, E. Barone, et al. 1998. Genetic and phenotypic diversity in pistachio (*P. vera* L.) germplasm collected in Mediterranean countries. Acta Hort. 470:168–78.

Crane, J.C. 1974. Hermaphroditism in *Pistacia*. Calif Agric. 28:3–4.

Dolcet-Sanjuan, R. and E. Claveria. 1995. Improved shoot-tip micropropagation of *Pistacia vera* L. and the beneficial effects of methyl jasmonate. J. Amer. Soc. Hort. Sci. 120(6):938–942. Dollo, L. 1996. An isozyme study of Sicilian *Pistacia* species varieties and offspring from artificial pollination. In: Caruso T, Barone F, Sottile F, (eds.), Proceedings of the IX GREMPA Meeting Pistachio 1993:May 20–21; Bronte, Italy. Renier Publisher Palermo; p. 112–8.

Dollo, L. 1996. An isozyme study of Sicilian *Pistacia* species varieties and offspring from artificial pollination. In: Caruso T, Barone F, Sottile F, editors. Proceedings of the IX GREMPA Meeting Pistachio 1993:May 20–21; Bronte, Italy. Renier Publisher Palermo; p. 112–8.

FAOstat 2007. data [databases on the Internet]. Available from http://www.apps.fao.org/faostat.

Ghaffari, S. M. and O.F. Harandi. 2001. Chromosome counts and assessment of two heterochromatic chromosomes in some species of *Pistacia* L. from Iran. III. Abstracts of International Symposium on Pistachios and Almond-XII GREMPA Colloquium, Zaragoza, Spain, 70.

Guerrero, J., J.F. Couceiro, and A. Moriana. 2002. Selection of terebinth (*Pistacia terebinthus* L.) trees as seed producers for pistachio (*Pistacia vera* L.) rootstocks in the Castilla-La Mancha region (Spain). NUCIS Newsletter FAO-CIHEAM, 11:25–29.

Hormaza, J.I. 1994. An analysis of sex expression, geographic distribution and genetic relatedness among clones and cultivars of pistachio (*Pistacia vera* L.). PhD Thesis Davis USA; University of California.

Hormaza, J.I. and A. Wünsch. 2007. Pistachio. In Genome Mapping and Molecular Breeding in Plants, Vol. 4. Fruits and Nuts, C. Kole (ed.), Springer, Berlin. p. 243–249.

Hormaza, J.I., L. Dollo and V.S. Polito. 1994a. Determination of relatedness and geographical movements of *Pistacia vera* (Pistachio; Anacardiaceae) germplasm by RAPD analysis. Economic Botany 48:349–358.

Hormaza J.I., L. Dollo, and V.S. Polito. 1994b. Identification of a RAPD marker linked to sex determination in *Pistacia vera* using bulked segregant analysis. Theor. Appl. Genet. 89:9–13.

Hormaza, J.I., L. Dollo, V.S. Polito. 1998. Genetic diversity of Pistachio (*Pistacia vera*, Anacardiaceae) germplasm based on randomly polymorphic DNA (RAPD) markers. Economic Botany 52:78–87.

İsfendiyaroğlu, M. and E. Özeker. 2009. Inflorescence features of a new exceptional monoecious *Pistacia atlantica* Desf. (*Anacardiaceae*) population in the barbaros plain of İzmir/Turkey. Intern. J. Plant Production 3:93–97.

Ila, H.B., S. Kafkas, and M. Topaktaz. 2003. Chromosome number of four *Pistacia* (Anacardiaceae) species. J. Hort. Sci. Biotech. 78:35–38.

Joley, L.E. 1979. Pistachios, p. 163–174. *In*: R.A. Jaynes (ed.), Nut Tree Culture in North America. Northern Nut Growers Assn., Hamden, CT.

Kafkas, S. 2006. Phylogenetic analysis of the genus *Pistacia* by AFLP markers. Pl. Syst. Evol. 262:113–124.

Kafkas S., I. Acar, H. Gozel, and S. Eti. 2004. Breeding monoecious pistachio cultivars. Nucis Newsletter FAO-CIHEAM; 2004 Sept; 12:21–3.

Kafkas, S., S. Cetiner, and R. Perl-Treves. 2001. Development of sex associated RAPD markers in wild *Pistacia* species. J. Hort. Sci. Biotechnology 76(2):242–246.

Kafkas, S., H. Ozkan, B.E. Ak, I. Acar, H.S. Atli, and S. Koyuncu. 2006. Detecting DNA polymorphism and genetic diversity in a wide pistachio germplasm: Comparison of AFLP, ISSR, and RAPD markers. J. Amer. Soc. Hort. Sci. 131(4):522–529.

Kafkas, S. and R. Perl-Treves. 2001. Morphological and molecular phylogeny of *Pistacia* species in Turkey. Theor. Appl. Genet. 102:908–915.

Kafkas, S. and R. Perl-Treves. 2002. Interspecific relationships in *Pistacia* based on RAPD fingerprinting. HortScience 37(1):168–171.

Kafkas, S, R. Perl-Treves, and N. Kaska. 2000. Unusual *Pistacia atlantica* Desf. (Anacardiaceae) monoecious sex type in the Yunt Mountains of the Manisa province of Turkey. Israel. J Plant Sci 48:277–280.

Kallsen, C.E., D.E. Parfitt, J. Maranto, and B. Holtz. 2009. New pistacho varieties show promise for California cultivation. Calif. Agric. 63(1):18–23.

Klich, M.A. and J.I Pitt. 1988. Differentiation of *Aspergillus flavus* from *A. parasiticus* and other closely related species. Trans . Br. Mycol. Soc. 1(1):9–108

Kokwaro, J.O. and J.B. Gillett. 1980. Notes on the Anacardiaceae of eastern Africa. Kew Bull. 34, 745–760.

Katsiotis, A. M. Haagidimitriou, A. Drossou, C. Pontikis, and M. Loukas. 2003. Genetic relationships among species and cultivars of *Pistacia* using RAPDs and AFLPs. Euphytica 132:279–286.

Maggs, D.H. 1972. Pistachios in Iran and California. 1972. Pl. Intro. Rev. 9:12–16.
Maggs, D.H. 1973. Genetic resources in pistachio. Plant Genetic Resources Newsletter. 29:7–15.
Maggs, D.H. 1990. The Australian pistachio Sirora. Fruit Var. J. 44:178–179.
Martinelli, A. and F. Loretti. 1988. Use of in vitro techniques for selection and cloning of different *Pistacia* species. Acta Hort. 227:436–437.
Morgan, D.P., L. Epstein, and L. Ferguson. 1992. Verticillium wilt resistance in pistachio rootsotck cultivars: Assays and an assessment of two interspecific hybrids. Plant Disease 76(3): 310–313.
Onay, A. 1996. *In vitro* organogenesis and embryogenesis of pistachio, *Pistacia vera* L. Ph.D.Thesis Univ. Edinburgh, Edinburgh, UK. 198 pp.
Onay, A. 2000a. Micropropagation of pistachio from mature trees. Plant Cell Tissue and Organ Culture. 60:159–162.
Onay, A. 2000b. Histology of somatic embryogenesis in cultured leaf explants of pistachio (*Pistacia vera* L.) Turkish J. Botany 24:91–95.
Onay, A., C.E. Jeffree, and M.M. Yeoman. 1995. Somatic embryogenesis in cultured immature kernels of pistachio, *Pistacia vera* L. Plant Cell Reports 15:192–195.
Onay, A., V. Piriç, E. Tilkat, Z. Aktürk, and H. Yildirim. 2004a. Somatic embryogenesis of pistachio from female flowers. J. Hort. Sci. Biotech. 79(6):960–964.
Onay, A., V. Piriç, H. Yildirim, and D. Basaran. 2004b. *In vitro* micrografting of pistachio (*Pistacia vera* cv. Siirt). Plant Cell Tissue Organ Culture. 77:215–219.
Ozbek S. and M. Ayfer. 1958. An hermaphrodite *Pistacia* found in the vicinity of Antep, Turkey. Proc. Amer. Soc. Hort. Sci. 72: 240–241.
Parfitt, D.E. 1995. Pistachio cultivars. *In* Pistachio Production Manual, 4th ed. 1995. L. Ferguson, ed. Fruit and Nut Information Center, Dept. of Plant Sci., Univ. Calif. Davis. p. 62–66.
Parfitt, D.E. 2003. 'Bonsai' ornamental Pistachio. HortScience. 38(6):1260–1261.
Parfitt, D.E. and A.A. Almehdi. 1994. Use of high CO_2 atmosphere and medium modifications for the successful micropropagation of pistachio. Scientia Hort. 56:312–329.
Parfitt, D.E. and M.L. Badenes. 1997. Phylogeny of the genus *Pistacia* as determined from analysis of the chloroplast genome. Proc. Nat. Acad. Sci., USA 94:7987–7992.
Parfitt, D.E., C.T. Chao, J. Maranto, C. Kallsen, and L. Ferguson. 1996. Pistachio Cultivar improvement. [CPC] California Pistachio Industry Annual Report, Crop Year 1996–1997. California Pistachio Commission. 1318 E. Shaw Avenue, Suite 420. Fresno, Calif. p.123–129.
Parfitt, D.E., C. Kallsen, J. Maranto, and B. Holtz. 2007. 'Golden Hills' Pistachio. HortScience : 42(3):694–696.
Parfitt, D.E. , C.E. Kallsen, B. Holtz, and J. Maranto. 2008. 'Lost Hills' A New Pistachio Cultivar for California. HortScience. 43(1):247.
Pistachio Production Manual, 4th ed. 2005. L. Ferguson, ed. Fruit and Nut Information Center, Dept. of Plant Sci., Univ. Calif. Davis. (http://fruitsandnuts.ucdavis.edu).
Polito, V.S. and J.G. Luza. 1988. Longevity of pistachio pollen determined by in vitro germination. J. Amer. Soc. Hort. Sci. 113:214–217.
Pollack, S. and A. Perez. 2007. Fruit and tree nuts situation and outlook yearbook 2007. USDA, ERS, FTS-2007.
Pontikis, C.A. 1984. *In vitro* propagation of *Pistacia terebinthus* L. Plant Propagator. 31:14–15.
Pontikis, C.A. 1986. 'Pontikis' pistachio. HortSci. 21(4):1074.
Popov, K.P. 1994. Trees, shrub, and semishrubs in the mountains of Turkmenistan. p. 173–186. In: V. Fet and K.I. Atamuradov (eds.), Biogeography and Ecology of Turkmenistan. Kluwer, Netherlands.
Pryor, B., and Michailides, T. J. 2002. Morphological, pathogenic, and molecular characterization of *Alternaria* isolates associated with Alternaria late of pistachio. Phytopathology. 92:406–416.
Tilkat, E., A. Onay, H. Yildirim, and H. Çetin Ozen. 2008. Micropropagation of mature male pistachio *Pistacia vera* L. J. Hortic. Sci. Biotech. 83(3):328–333.
Vargas, F.J. and M.A. Romero. 1998a. Vigour and juvenile stage in pistachio progenies. X GREMPA Seminar, Meknes, Morocco, 1996. Options Méditerranéennes, 33:105–111.

Vargas, F.J. and M.A. Romero. 1998b. Vigour in pistachio progenies. II International Symposium on Pistachios and Almonds, Davis (California), USA, 1997, L. Ferguson and D. Kester (eds). ISHS. Acta Horticulturae, 470:162–167.

Vargas, F.J. and M.A. Romero. 2005. Fruit characters in pistachio. IV Inter. Options Méditerranéennes, Série A, 63:49–56.

Vargas, F.J., M. Romero, M. Rovira, and I. Batlle. 1996. Pistachio cultivar improvement at IRTA–Mas Bové. Proceedings of the IX GREMPA Meeting on Almonds and Pistachios. Bronte-Sciacca, Italy, May 20–21:15–19.

Vargas, F.J., M.A. Romero, and I. Vargas. 2001. Blooming and leafing time in pistachio progenies. XI GREMPA Seminar on pistachios and almonds, Sanliurfa, Turkey, 1999, AK, B.A. (ed). Options Méditerranéennes, 56:41–46.

Vargas, F.J., M.A. Romero, and I. Vargas. 2002. Flowering precocity in pistachio progenies. III. Acta Horticulturae, 591:297–303.

Wannan, B.S. and C.J. Quinn, 1991. Floral structure and evolution in the Anacardiaceae Bot. J. Linn. Soc. 107:349–385.

Werner, O., P. Sanchez-Gomez, J. Guerra, and J.F. Martinez. 2001. Identification of *Pistacia x saportae* Burnat (Anacardiaceae) by RAPD analysis and morphological characters. Scientia Hort. 91:179–186.

Westwood, M.N. 1993. Temperate-Zone Pomology: Physiology and Culture, 3d ed. Timber Press, Portland Oregon. 523 p.

Whitehouse, W.E. 1957. The pistachio nut - A new crop for the Western United States. Economic Botany 11:281–321.

Yakubov, B., O. Barazani, and A. Golan-Goldhirsh. 2005a. Combination of SCAR primers and Touchdown-PCR for sex identification in *Pistacia vera* L. Scientia Hort. 103:473–478.

Yakubov, B. O. Barazani, A.. Shachack, L.J. Rowland, O. Shoseyov, and A. Golan-Goldhirsh. 2005b. Cloning and expression of a dehydrin-like protein from *Pistacia vera* L. Trees 19:224–230.

Yi, T., J. Wen, A. Golan-Goldhirsh, D.E. Parfitt. 2008. Phylogenetics and reticulate evolution in *Pistacia* (Anacardiaceae). American J. of Botany. 95(2):241–251.

Zohary, M. 1952. A monographical study of the genus *Pistacia*. Palestine J Bot. 5:187–228.

Zohary, M. 1972. *Pistacia* L. *Flora Palestina* 2 :297–300. Israel Academy of Sciences and Humanities, Jerusalem, Israel.

Zohary, D. 1996. The genus *Pistacia* L. (in Padulosi, S.; Caruso, T.; Barone, E. editors). Taxonomy, distribution, conservation and uses of *Pistacia* genetic resources. Report of a workshop 29–30 June:1–11.

Zohary, D. and M. Hopf. 2000. Domestication of Plants in the Old World. Oxford University Press: 316 pages.

Zohary, D. 2006. Domestication of the pistachio nut (*Pistacia vera*). FAO-CIHEAM – Nucis Newsletter, 13:53–55.

Chapter 22
Walnut

Gale McGranahan and Charles Leslie

Abstract All species of walnuts (*Juglans*) produce nuts, but the Persian or English walnut (*J. regia* L.) is the only species widely cultivated for nut production. Persian walnuts are grown in North and South America, Europe, and Asia. The considerable variation within *J. regia*, particularly in nut size and shape, led taxonomists to describe six additional species that have not been widely accepted but that illustrate some of the diversity. Persian walnuts are native to the mountain valleys of Central Asia. They were introduced into Europe by the Greeks and introduced into North America by the colonists. Breeding of walnuts is relatively recent, although it is probable that in the past, walnuts from the best genotypes were selected both for food and planting. The first breeding programs started at the University of California (The USA) and the Fruit and Vine Research Station in Bordeaux (France). Because Persian walnuts are native to the mountains of Central Asia, considerable effort in the USA has been directed toward collecting material from that area. The major breeding objectives are to increase yield, quality, and range of harvest dates while decreasing the amount of chemical input required to control pests and diseases. Isozymes and molecular markers have been used for identification of cultivars and genetic diversity analysis. Molecular markers have been developed for Walnut blackline disease that causes necrosis at the graft union. Gene transfer in walnut using *Agrobacterium tumefaciens* gene insertion into cells regenerated into plants via somatic embryo cultures has been successful. Traits of interest that have been tested in walnut include expression of a Bt gene from *Bacillus thuringiensis* for insect resistance and use of RNAi gene silencing.

Keywords Walnut • *Juglans* • *Juglans regia* • Breeding • Germplasm • Genetics • Distribution • Rootstock • Transgenic • Marker assisted breeding

G. McGranahan (✉) • C. Leslie
Walnut Improvement Program, Department of Plant Sciences, University of California, Mail Stop 2, Davis, CA 95616, USA
e-mail: ghmcgranahan@ucdavis.edu; caleslie@ucdavis.edu

M.L. Badenes and D.H. Byrne (eds.), *Fruit Breeding*, Handbook of Plant Breeding 8,
DOI 10.1007/978-1-4419-0763-9_22,

1 Introduction

All species of walnuts (*Juglans*) produce nuts, but the Persian or English walnut (*J. regia* L.) is the only species widely cultivated for nut production and is the focus of this chapter. Other species are grown for timber (e.g., *J. nigra* L., eastern black walnut) or are used as rootstocks for Persian walnut (e.g., *J. hindsii* Jeps. ex R. E. Sm., northern California black walnut and Paradox, the hybrid between *J. hindsii* and *J. regia*).

1.1 Economic Importance

Persian walnuts are grown in North and South America, Europe, Asia and the former Soviet Republics, and to a limited extent in Oceania and North Africa. Over 1.6 million metric tons were produced in 2007 (FAOSTAT data 2007). China leads world production, followed by the USA, Turkey, Iran, Ukraine, Mexico, India, France, Egypt, and Romania (FAOSTAT data 2007). The major exporters are the USA, which exports 115,000 MT, followed by Mexico (37,000 MT), France (27,000 MT), Chile (19,000 MT), and China (15,000 MT). Shelled walnuts make up 62% of the exports. Several of the major producers consume the bulk of their walnut production domestically, e.g., China, Iran, and Turkey. Chile, on the other hand, exports almost its entire production. China has encouraged its growers to plant high value crops like walnuts and expects to have over one million hectares of walnuts by 2012. New areas of production are also developing in Chile and Argentina.

1.2 Uses

Although now dried walnuts are consumed either as a snack or in baked goods and cereals, in times past they had a variety of uses. They were thrown by the groom in Roman weddings to signify maturity. In the middle ages, they were thought to ward off lightening, fevers, witchcraft, and epileptic fits. According to the Doctrine of Signatures (sixteenth–seventeenth centuries), tinctures of the husk were used for ailments of the scalp and the kernel could be used to sooth the brain (Rosengarten 1984). Currently, recipes can be found for green walnut pickles and walnut liqueurs, and in parts of the world undried walnuts, the "fresh walnuts," are eaten after peeling off the bitter seed coat.

Oils are the most prominent nutrient in walnuts. Recently, the health benefits of the oils, especially the omega-3 fatty acid, in walnuts have been investigated and found highly beneficial. In one study that compared a low fat, modified low fat and modified low fat that included 8–10 walnuts per day improved the HDL to total cholesterol ratio in men and women diagnosed with type 2 diabetes. The LDL was also decreased by 10% (Tapsell et al. 2004). In another study (Ros et al. 2004), a Mediterranean diet was compared to a similar diet in which 8–13 walnuts replaced approximately 32% of the energy from monounsaturated fat. The walnut diet increased endothelium-dependent vasodilation by 64% and reduced vascular cell adhesion molecule-1 by 20%. The diet also decreased total cholesterol and LDL cholesterol. Reiter et al. (2005), found significant levels of melatonin in walnuts. According to the author, R.J. Reiter, “the ingredients in walnuts would be expected to reduce the incidence of cancer, delay or make less severe neurodegenerative diseases of aging…and reduce the severity of cardiovascular disease.”

1.3 Taxonomy

The family *Juglandaceae* consists of seven genera and about 60 species of deciduous, monoecious trees with alternate, pinnately compound leaves. It has been extensively studied by Manning (1978) and Manos and Stone (2001). In addition to the genus *Juglans* (walnuts), the family includes *Carya* (pecans and hickories), *Pterocarya* (wingnuts), *Platycarya*, *Engelhardia*, *Alfaroa*, and *Oreomunnea*.

Members of the genus *Juglans* are trees or large shrubs possessing twigs with chambered piths, large aromatic compound leaves, generally solitary staminate catkins on 1-year-old wood and female flowers on current season’s wood. The husked fruit is a false drupe containing a large, woody-shelled nut. All *Juglans* produce edible nuts, although size and extractability differ considerably. Most species are highly regarded for their timber.

The genus *Juglans* consists of approximately 21 species native to parts of North America, the Andean region of South America, and the mountain ranges traversing Central Asia (Figs. 22.1–22.3). These species have been grouped taxonomically into four sections: *Juglans*, *Trachycaryon*, *Cardiocaryon*, and *Rhysocaryon*.

Section Juglans. The *Juglans* section consists of the commercially valuable Persian or English walnut, *Juglans regia* and the iron walnut *J. sigillata* Dode, which is thought to be the same species (Wang et al. 2008). This section is characterized by a four-celled nut, a husk that separates from the nut at maturity and seedlings with two opposite rows of buds immediately above the cotyledons and below the spirally arranged compound leaves. The typically large tree grows to a height of about 30 m and produces large, relatively smooth, and generally thin-shelled nuts.

Juglans regia selections have been identified in which nuts vary from nearly round to the greatly elongated “Barthere” and from pea sized to more than 5 cm diameter. Trees with a weeping growth habit have been identified in Belgium and California, and variations in leaf morphology and color have been identified. Cutleaf types include ‘Heterophylla’ and ‘Laciniata.’ ‘Monophylla’ has leaves with only an enlarged terminal leaflet occasionally with two greatly reduced side leaflets, ‘Adspersa’ produces mottled white leaves, and ‘Purpurea’ exhibits leaves of a dull red color (Rehder 1940). Cultivars with bright red seed coats have also been bred (McGranahan and Leslie 2004).

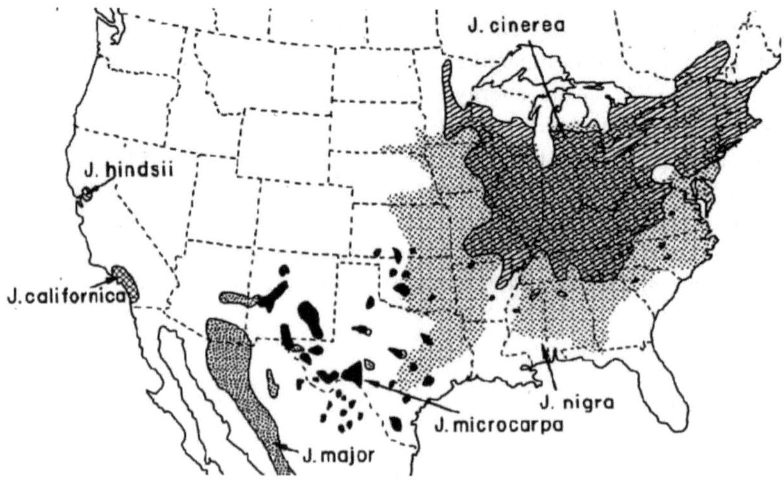

Fig. 22.1 Native range of *Juglans* species in North America

Fig. 22.2 Native range of *Juglans* species in Central and South America

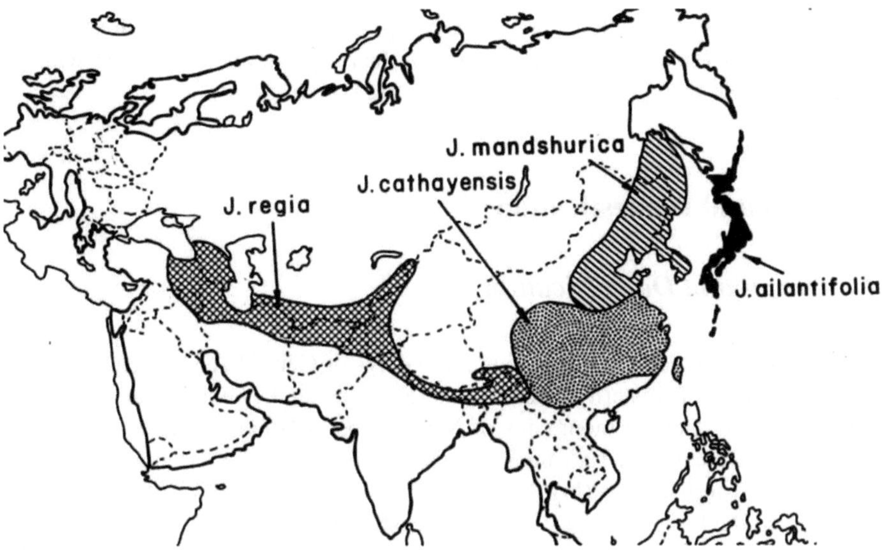

Fig. 22.3 Native range of *Juglans* species in Asia

The considerable variation within *J. regia*, particularly in nut size and shape, led taxonomists to describe six additional species that have not been widely accepted but that illustrate some of the diversity (Dode 1909). *Juglans sigillata*, a type from southern China and Tibet with a very thick rough-shelled nut, an adherent hull, and very dark colored kernels, is the most distinctive of the variations described and has been accepted as a separate species in the past. Known locally as the iron walnut, this type or species has been cultivated for a long time in Yunnan Province for its oil and edible nuts, and several cultivars have been developed.

1.4 Limits of Adaptation

Walnuts grow best on deep, fertile well drained soils. High or fluctuating water tables and flooding can injure the roots. Excessive cold temperatures restrict or prevent walnuts from growing in most parts of the USA, former USSR, and Europe. In many regions, walnuts are not limited by a steady winter cold, but by autumn and spring frosts. This can be partly overcome by using late leafing cultivars that avoid spring frosts when the new shoots and flowers are particularly vulnerable. Walnut culture can also be limited by insufficient winter chilling. Commercial walnuts require between 700 and 1,000 h of winter temperatures below 7°C for normal growth. Symptoms of lack of chilling include sporadic bud break, poor yield, and branch die-back. Excessively high summer temperatures, around 40°C, result in sunburned and darkened kernels, while cool summer temperatures reduce kernel size resulting in poorly filled nuts. This being said, the germplasm diversity of walnuts is poorly understood and expansion of the range for commercial growing is probably possible with germplasm evaluation and breeding.

2 Origin and Domestication

2.1 Origin and Domestication

Persian walnuts are native to the mountain valleys of Central Asia extending from Xinjiang province of western China, parts of Kazakhstan, Uzbekistan, and southern Kyrgyzstan, and from the mountains of Nepal, Tibet, northern India, and Pakistan west through Afghanistan, Turkmenia, and Iran to portions of Azerbaijan, Armenia, Georgia, and eastern Turkey (Fig. 22.3). Small remnant populations of *J. regia* may have survived the last glacial period in southeastern Europe but the bulk of the wild *J. regia* germplasm in the Balkan Peninsula and much of Turkey was most likely introduced from Iran and eastern Turkey by Greek commerce and settlement several thousand years ago (Zohary and Hopf 1993). From Greece, the cultivation spread to Rome, where walnuts were known as *Jovis Glans*, or Jupiter's acorn, from which comes the genus name *Juglans*. From Italy, *J. regia* spread to what are now France, Spain, Portugal, and southern Germany (Leslie and McGranahan 1998). The word walnut may be derived from "wealh nut," "wealh" meaning foreign in Anglo-Saxon or old German. Trees of the species were in England by 1562, and nuts were brought to America by the earliest settlers. The American colonists are said to have called the species "English" walnut to distinguish it from the native American eastern black walnut (*J. nigra*). *J. regia* germplasm in China is thought to have been introduced from central Asia about 2,000 years ago and in some areas became naturalized, although there appear to be natural stands in the Xinjiang Uygur Autonomous Region of China.

2.2 Breeding History

Breeding of walnuts is relatively recent, although it is probable that in the past walnuts from the best genotypes were selected both for food and planting. Many of the wild walnut trees have thick hard shells, a trait that must have been selected against. The first successful breeding program was developed by Gene Serr and Harold Forde (Tulecke and McGranahan 1994) at the University of California, Davis CA in 1948. Their goal was increased yield and quality. The target traits were lateral bud fruitfulness, originally found in one chance seedling in California, late leafing date, good shell quality, high percentage of kernel, and light colored nuts. Only selections with lateral bud fruitfulness and good performance relative to other traits were selected for release. A similar program designed to incorporate late leafing with lateral bud fruitfulness was undertaken later in France (Germain 1997).

3 Genetic Resources

3.1 Scion

A breeding program depends in part on a diverse collection of germplasm as a source of raw material from which traits of interest can be identified (McGranahan and Leslie 1990). For the past two decades, extensive evaluations of seedlings in orchards and naturalized trees have been undertaken in the Mediterranean countries of Europe and to a lesser extent North Africa [See Proceedings of Walnut Symposia: Acta Horticulturae numbers (1990) 284, (1993) 311, (1997) 442, (2001) 544, and (2006) 705]. From this work several new cultivars have been identified (Tomas 2000). China has also had an active nationwide search for new cultivars from seedling orchards. Because Persian walnuts are native to the mountains of Central Asia, considerable effort in the USA has been directed toward collecting material from that area (Leslie and McGranahan 1998). Funding and participation in this work has included a century-long plant introduction endeavor by US Department of Agriculture (USDA) plant collectors, and more recent trips by USDA and university researchers. Collecting has been funded in part by California growers, USAID exchanges, and USDA-ARS germplasm exploration funds. Material has also become available for use through international germplasm exchanges, private breeders, hobbyists, customs confiscations, and observant growers who have noticed interesting seedling trees.

A very useful book, "Inventory of Walnut Research, Germplasm and References," has been published by FAO and describes a great number of the germplasm collections in the world, especially in the European Union (Germain 2004). In the USA, both the University of California, Davis (UCD) and the USDA National Clonal Germplasm Repository, Davis, California (NCGR-Davis) maintain walnut germplasm collections. The content at the NCGR-Davis walnut collection is listed at http://www.ars-grin.gov/dav.

The intent of the USDA collection is to include as broad a diversity of all walnut species as possible and is maintained for public distribution of material. It will not accept proprietary material and is managed primarily for wood and nut distribution to researchers worldwide. The UC Davis collection includes a representation of California commercial cultivars, advanced selections, and some proprietary material and is focused primarily on material of interest for breeding purposes (Tulecke and McGranahan 1994). It is managed for a variety of activities, including crossing, breeding evaluations, and graft wood and seed distribution. While there is some overlap of material, duplication is generally avoided, and the two collections are used cooperatively.

3.2 Rootstock

The rootstock is the other half of the tree and provides anchorage, absorption of water and nutrients, hormone synthesis, and storage. Rootstocks are more difficult to study because they are mostly underground, and rootstock improvement is developing slowly because clonal propagation has only recently been commercialized. Given the susceptibility of common walnut rootstocks (Table 22.1), it is clear that genetic improvement is needed. To date, the Paradox rootstock (*J. hindsii* × *J. regia*), which exhibits hybrid vigor, is superior to pure species in most traits, but many other species combinations have not been tested (McGranahan and Catlin 1987).

4 Major Breeding Achievements

4.1 Scion

Prior to the Serr-Forde breeding program (1948–1978) in California, most cultivars grown in Northern California, where the industry now resides, were cultivars brought from France by Felix Gillet in the late 1800s or chance seedlings. Gene Serr and Harold Forde made remarkable progress in breeding new cultivars that revolutionized the industry. Their primary breeding objectives were to combine the late leafing and quality of the French types with the lateral fruitfulness and precocity of 'Payne.' They made 196 crosses, evaluated about 6,000 progeny and released 13 cultivars, ten in 1968, and three in 1978. The most important of these are 'Vina,' 'Serr,' 'Howard,' and 'Chandler.' In 1993, 'Tulare' was released from a cross made 27 years earlier by Serr and Forde (McGranahan et al. 1992). 'Tulare' is a vigorous, high yielding cultivar with some resistance to contamination by aflatoxin (Mahoney et al. 2003).

Recently, five new cultivars have been released. 'Robert Livermore' is a red-skinned walnut (McGranahan and Leslie 2004). 'Sexton,' 'Gillet,' and 'Forde' (US Plant Patents 16496, 17135, and 16495) are all precocious in bearing, laterally fruitful, high yielding, mid to late season harvesting, with low blight scores and high quality kernels. The latter two are protogynous, which is unusual in the cultivars available. 'Ivanhoe' was released just recently and is a very early harvesting cultivar.

The breeding program in France (Germain 1997) has resulted in two new cultivars ('Fernette' and 'Fernor'). Several promising selections are still being evaluated in the program, but funding has been cut, and breeding has essentially stopped.

4.2 Rootstock

The major achievement in development of rootstocks for walnuts was the determination that in many environments in California the hybrid, Paradox (*J. hindsii* × *J. regia*), was superior to either of the parents as a rootstock (McGranahan and Catlin 1987). California nurseries have actively sought *J. hindsii* that naturally produce

Table 22.1 Walnut rootstock response to pests and diseases

Rootstock	Crown gall	Phytophthora root and crown rot	Blackline disease	Armillaria root rot	Root lesion nematode	Root knot nematode	Salts
Persian walnut (*Juglans regia*)	Susceptible	Very susceptible	Symptomless	Susceptible	Very susceptible	Susceptible	Sensitive
Northern California Black walnut (*J. hindsii*)	Susceptible	Very susceptible	Hypersensitive	Variable	Susceptible	Resistant	Less sensitive
Paradox walnut (*J. hindsii*×*J. regia*)	Very susceptible	Susceptible	Hypersensitive	Variable	Very susceptible	Unknown	Sensitive
Wingnut (*Pterocarya stenoptera*)	Resistant	Resistant	Hypersensitive	Susceptible	Tolerant	Unknown	Unknown

For more information on pests and diseases see Ramos (1998), IPM (2003) or McGranahan and Catlin (1987)

the most hybrid seed and have successfully identified source trees that produce over 90% Paradox seedlings, although the species of the source trees has not been limited to *J. hindsii* (Potter et al. 2002b). Most other regions where walnuts are cultivated use *J. regia* as rootstock.

5 Current Goals of Breeding

5.1 *Scion*

The major breeding objectives are to increase yield, quality, and the range of harvest dates while decreasing the amount of chemical inputs required to control pests and diseases. The ideal walnut cultivar would be relatively late-leafing to escape frost and the rains that spread walnut blight (*Xanthomonas campestris* pv. *juglandis*), precocious (yielding more than 500 kg/ha in the fourth year), and vegetatively vigorous with bearing on both terminal and lateral shoots. It would have a low incidence of pistillate flower abscission and other drops and would not be alternate bearing. It would have high production capacity (>6 MT/ha) with low chemical inputs required. The harvest season would end in early October. The nutshell would be relatively smooth, well-sealed and make up no more than 50% of the nut weight. The nuts would fit the category of large or jumbo. The kernel would be plump and light colored, weighing about 8–9 g, and come out easily in halves. The tree would be at least moderately resistant to pests and diseases.

Marker-assisted backcross breeding is being used to develop a commercial quality, *J. regia*-like cultivar with resistance (hypersensitivity) to the cherry leafroll virus (CLRV), which causes blackline disease (Woeste et al. 1996). It has been shown that a single dominant gene from *J. hindsii* confers hypersensitivity and that progeny from backcrosses (*J. hindsii* × *J. regia*) × *J. regia* segregate 1:1 hypersensitive:tolerant (McGranahan et al. 1997). Currently, the BC4 generation is being evaluated. An anomaly in all the backcrosses is that they are male sterile; i.e., catkins, if formed, abscise when immature. We have selected three backcross genotypes, with close to commercial quality, for field trials. The field trials are designed to determine whether CLRV-infested pollen infects a hypersensitive flower, whether any damage to the flowers occurs at fertilization, and whether nut set is affected.

5.2 *Rootstock*

A study to evaluate the diversity of Paradox rootstocks was initiated in 1996. It was designed to examine variability among families of commercially available Paradox seedlings and controlled crosses between different black walnut species and *J. regia*. Eleven California walnut nurseries each donated about 500 seeds from each of three Paradox-producing black walnut source trees each year for 2 years. These were planted in replicate blocks in three nurseries, measured and divided into subsets. Four subsets were planted and grafted as orchard trees by UC Cooperative Extension Farm Advisors, two subsets were screened for nematode (*Pratylenchus vulnus*) resistance by M. McKenry (unpublished), and two were screened for crown gall (*Agrobacterium tumefaciens*) resistance (McKenna and Epstein 2003). Two subsets of ungerminated seed were provided for *Phytophthora* screening (Browne et al. 2006).

The work is ongoing in the four long-term field trials, but in the process of screening seedlings for various traits, it became apparent that certain individual seedlings were superior. Two genotypes from the crown gall screen were selected; one proved to be an escape rather than a resistant genotype, the other remains to be retested. Several genotypes have been identified that continue to have low susceptibility to *Phytophthora citricola* in repeated screens of micropropagated plants (Browne et al. 2006). No resistance to nematodes was found but one genotype that did not appear to be affected by infestation (tolerant response) was identified (M. McKenry, unpublished). Most of the selected genotypes in this study have been micropropagated for field trials. Patents have been applied for two rootstock clones: RX1 is a cross between *J. microcarpa* and *J. regia* that was selected for some resistance to *Phytophthora* and VX211, a cross between *J. hindsii* and *J. regia*, was selected for its vigor and response to nematodes. These have been repropagated and are undergoing further field trials.

Much more work is needed on rootstocks. Since the hybrids appear to have the most vigor, it is important to evaluate the performance of different species in hybrid combinations. One that is readily available in S. America and hybridizes easily with *J. regia* is *J. australis* Griseb. from Argentina. Other possibilities are *J. neotropica* Diels (northwestern S. America), *J. olanchana* Standl. & L. O. Williams (Mexico and Guatemala), and *J. cathayensis* Dode (East Asia, China).

In California, we have approached the blackline problem, caused by the cherry leafroll virus, through both cultivar hypersensitivity to the virus and rootstock tolerance. The latter, a short-term solution, is aimed at developing a rootstock combining the *J. regia* response to blackline disease with the vigor and other attributes of Paradox. This can be achieved, in theory, by selecting vigorous, tolerant individuals among seedlings of a backcross generation (*J. hindsii* × *J. regia*) × *J. regia*. In 1988, 13,000 Paradox offspring from 17 source trees were planted in a randomized complete block design with six blocks in *Phytophthora*-infested soil. Between 1992 and 1994, they were screened for vigor and tolerance to the virus. Five seedlings were selected in 1994, but it has taken until recently to establish grower trials to compare their performance in the field to Paradox and *J. regia* rootstocks because of the challenges of clonal propagation.

6 Breeding Methods and Techniques

Heritabilities are high for many traits of interest (Hanche et al. 1972; Forde and McGranahan 1996) (Table 22.2). However, it has been shown that many traits change with clone age; for example, leafing out, bloom, and harvest date all shift up

Table 22.2 Cultivar traits under evaluation and estimated heritabilities (Hanche et al. 1972)

Field	h^2	Crack out	h^2
Leafing date	0.96	Shell texture	
Female bloom: first, peak and last	0.93	Shell color	
Male bloom: first, peak and last	0.8	Shell seal	0.38
Dichogamy		Shell strength	
Percent overlap: male and female		Shell integrity	
Catkin abundance		Shell thickness	0.91
Female flower abundance		Packing tissue thickness	
Percent fruitful laterals	0.39	Nut weight	0.86
Yield	0.07	Kernel weight	0.87
Blight		Percent kernel	
Codling moth		Fill	
Sunburn		Plumpness	
Harvest date	0.85	Ease of kernel removal	
		Color (extra light, light, light amber, and amber)	0.52
		Shrivel	
		Veins	0.49

to 2 weeks earlier, stabilizing at age 15. Shells also thicken, and seals improve, but the in-shell weight, kernel weight, and percent kernel all decrease (McGranahan and Forde 1985).

The UC breeding program has used several distinct procedures for crossing parent material. In the first method, catkins are stripped and wind-blown pollen is excluded from flowering shoots by covering them with tightly secured bags that have small plastic windows (PBS International, The UK). Pollen is collected from the other parent of interest and stored frozen over saturated magnesium chloride until use.

When bagged female flowers open and are receptive, pollen is applied through the bags with a hypodermic needle and syringe. Care must be taken to avoid shooting too much pollen into the bag because excess pollen can result in pistillate flower abscission (McGranahan et al. 1994b). Bags are later removed after the flowers are no longer receptive (about 3 weeks) and shoots with control-pollinated nutlets are labeled for collection in the fall. Both parents are known with this method, but the costs are high and seedling production is low.

The second method is to locate geographically isolated young trees of the desired female parent. Using young trees is important, because as a cultivar matures, the female flowers are usually present 2–3 years before the male flowers. This often requires the cooperation of a grower with a recently planted orchard. Any male flowers on these trees are removed by hand before bloom to prevent selfing. Once the female flowers begin to bloom, pollen of the desired male parent or parents is applied by airbrush several times during the bloom period. At harvest the cooperating grower either donates or is compensated for the nuts. This method produces many more seed at lower cost but with a lower certainty of the male parent. Male parents of selections can be determined later by DNA analysis. Some selfing occurs which results in stunted, twisty trees with russetted hulls and small kernels.

A third method has been to plant advanced selections and superior cultivars in an isolated block and collect seed from them assuming that there has been natural crossing. Again, this method incurs lower costs and produces more seedlings, but the male parent can only be identified through DNA analysis.

Seed collected from controlled crosses or selected females is then stratified and grown for evaluation and to produce the next generation of seedlings. Recently, it has been found that *J. regia* seed can be germinated without stratification in the fall, grown to about 10–20 cm, and chilled at 7–10°C for 6–8 weeks, and planted dormant in a nursery row for the first year. This has been much more efficient than planting seed directly in nursery row due to poor survival of direct-planted seed. Commercial walnut nurseries have generously donated growing ground, time, resources, and expertise to assist this aspect of the program.

After one growing season in the nursery, trees are dug and replanted on wider spacing for evaluation. At this stage, trees are grown on their own roots, not grafted to rootstock. Most commonly, these trees are planted on UC Plant Sciences Department growing grounds and farmed by department staff supported by university and grower funding. In some cases, growers have assisted the program by donating orchard space for this purpose and have farmed these trees during the evaluation process. This has been done by planting between rows in an existing widely spaced orchard, or more effectively, by interplanting in available open space in a newly established orchard and then removing the breeding program trees as evaluations are completed and the orchard canopy fills in.

As seedling trees mature they are evaluated in the field for traits of interest, including leafing, flowering, and harvest dates, yield, disease presence and growth habit (McGranahan et al. 1994a) (Table 22.2). This usually begins at age three or four. When the trees are grown in university orchards they are left unsprayed so that variation in resistance to insects and disease can be observed. When grown within commercial orchards this is not normally possible. Nut samples ($n=10$) are hand collected from each tree at maturity before they fall from the tree. Samples are air-dried, cracked by hand, and evaluated for % kernel, kernel quality, kernel weight, shell characteristics, and yield of halves (Table 22.2). Data are entered into a database and summarized for multiple years. In addition, samples ($n=100$ nuts) of promising individuals are sent to commercial processors for their independent evaluation.

Collected data is presented to farm advisors, growers, and nurserymen in several ways. The first is at the annual Walnut Research Conference as part of the Walnut Improvement Program's annual report. Data on selections is presented orally to attendees and published in the annual proceedings of the conference. The annual reports are now available online (http://www.walnutresearch.ucdavis.edu).

The breeding program also holds an annual "Crackout Meeting" in the spring attended by farm advisors, handlers, nurserymen, and growers. The attendees generally have an expressed interest in development of new cultivars, are interested in assisting with evaluation of material, or are otherwise active in research activities and the marketing board. At this all-day meeting the data reports are distributed, and kernel samples and intact nuts of the material under evaluation are displayed.

Attendees are asked to review the material, examine the samples, and provide written comments. In an ensuing discussion period, they provide valuable input on priorities from their varying perspectives, help rank material, and suggest which seedlings/selections should continue in the program. The program also invites interested parties to view selections in the field, either through a formal field day or by scheduling informal visits at their convenience. Progress in the program and information about selections is also presented periodically to a wider range of growers at annual county grower meetings held around the state of California.

Once an individual seedling shows promise and is selected for further trials, graft wood is collected from the original seedling and grafted to rootstocks. Nurseries have often provided assistance at this stage by donating rootstock, supplying grafters, and, in many cases, growing the grafted trees for the program.

Grafted trees of each selection are then planted in test blocks on orchard spacing at diverse locations for further evaluation. Currently, these tests blocks are located at the Chico State University Farm in the northern part of the state, on the UC Davis campus in the central region, and at the UC Kearney Field Station in the south. These blocks are used to evaluate the performance of selections on rootstocks under a wide range of conditions, obtain a better look at yield, and allow farm advisors and growers to see selections in their local area.

In addition to the university plots, interested growers around the state have volunteered to establish trials ranging in size from several trees to several acres. Farm advisors assist in identifying suitable growers, establishing plots, and observing performance. Graft wood is distributed to these growers under test agreement, and they are asked to participate in its evaluation and to attend the crackout meeting. This gives the program valuable input on performance under a variety of conditions and in commercial settings from observers with extensive experience. Growers are assisting the process and get an early look at the material that is most interesting for their situation.

As new selections begin to show promise, commercial nurseries are encouraged to acquire graft wood from the program to test the cultivars for themselves and to begin increase-blocks of their own. This ensures nurseries have adequate input into final selection, firsthand knowledge of the material, particularly of its grafting performance, growth habit and training requirements and builds an adequate supply of production wood by the time the new cultivar is released. As with grower trials, nurseries receive wood under test agreement. This allows them to propagate for testing purposes, including grower trials, but selections cannot be produced for sale.

Selections that continue to show promise in test blocks and grower trials become candidates for patent and release as new cultivars. The patent disclosure process requires an extensive description of the selection, a summary of available data, and identification of attributes distinct from existing cultivars.

Once a selection is in the patent process as a new cultivar, nurseries may obtain a commercial license from the University of California that allows sale of trees. A per-tree royalty is assessed at the time of sale from the nursery and returned to the university. After patenting costs are recovered, part of this fee is assigned for overhead, and part is returned to the breeding program as well as to the breeders.

Patenting provides a return to the inventor and the university but also seeks to protect the growers from unlimited distribution. Patented material is not allowed to be sold or grown outside of California for 5 years after release.

7 Integration of New Biotechnologies in Breeding Programs

7.1 Markers

Isozymes were the first tools developed for walnuts (Arulsekar et al. 1985, 1986). Both glucose phosphate isomerase (GPI) and aspartate amino transferase (AAT) were simply inherited and were useful for differentiating scion cultivars (*J. regia*) from the Paradox rootstock (*J. hindsii* × *J. regia*). This was especially useful because the hybrid resembles *J. regia*. GPI and AAT could not differentiate among *J. regia* genotypes. Phosphoroglucomutase (PGM) and esterase (EST) were more useful in differentiating cultivars but only into groups. They were used then to differentiate between two fairly similar cultivars, 'Howard' and 'Chandler,' which had been mixed up in the nursery trade.

Later, restriction fragment length polymorphisms (RFLPs) were found to be superior to isozymes such as PGM in determining the origin of embryo cultures derived from "pollen-isolated" *J. regia* flowers. Cultures were in fact zygotic in origin rather than maternal as would have been expected if the flowers had truly been isolated and apomictic (Aly et al. 1992).

RFLPs were later successfully used to investigate the genetic diversity among 48 *J. regia* cultivars and germplasm introductions (Fjellstrom 1993; Fjellstrom et al. 1994). Cluster analysis of genetic differences among accessions along with principal component analysis of allelic genotypes revealed the presence of two major groups of walnut domesticates. The California germplasm was associated with germplasm from France, central Europe and Iran and had less genotypic similarity with germplasm from Nepal, China, Korea and Japan. This information was used for making breeding decisions and establishing germplasm collection priorities. A patent has been obtained for the first Chinese × California cultivar. RFLPs were also used to begin mapping the walnut genome (Fjellstrom 1993).

Randomly amplified polymorphic DNA (RAPD) loci from a walnut backcross population [(*J. hindsii* × *J. regia*) × *J. regia*], were used to improve the genetic map (Woeste et al. 1996). Segregation data from these polymorphisms were joined to the RFLP marker data set to expand the genetic map of walnut to 107 markers in 15 linkage groups. RAPD markers were also used for molecular characterization and confirmation of the genetic relatedness among walnut cultivars with known pedigrees (Nicese et al. 1998).

The utility of inter simple sequence repeat (ISSR) markers was examined in 2002 (Potter et al. 2002a). Like RAPD markers, ISSR markers are a quick, relatively inexpensive method for analyzing variability and developing fingerprints.

They have been considered more reliable than RAPD markers due to higher reproducibility. Eight ISSR primers were found in combination to provide a unique fingerprint for each of the 48 cultivars and germplasm accessions tested. In a dendrogram developed from these data some of the groupings corresponded to expected relationships from known pedigrees but others did not, suggesting that there is a limitation in using ISSRs for inferring genetic relationships.

A very useful study for rootstock breeding and selection involved DNA sequence markers (Potter et al. 2002b). The Paradox walnut rootstock is generally understood to be the hybrid of northern California black walnut (*J. hindsii*) and Persian walnut (*J. regia*). Almost every walnut nursery in California has several of their own confidential black walnut source trees. It was our intention to compare the Paradox from each nursery and evaluate them for vigor and resistance to pests and diseases. One of our questions to assist in breeding was whether the black parents were in fact *J. hindsii* or whether other black species (e.g., *J. microcarpa* Berland., *J. major* (Torr.) A. Heller, *J. nigra*, *J. californica* S. Watson) could be involved. Representatives of the five black walnut species were screened for variability in the internal transcribed spacer (ITS) regions of the nuclear ribosomal DNA and in three noncoding regions from the chloroplast genome. Unique sequence markers were identified for each species. Total DNA extracts from 27 nursery source trees were tested for those markers. Chloroplast DNA profiles were used to trace the maternal lineages of the source trees; the ITS data provided evidence as to whether the source trees were themselves hybrids. The results indicated that among industry Paradox sources, there is a considerable genetic contribution from black species other than *J. hindsii*. This information can now be used to reconstruct superior sources of walnut hybrid rootstock.

Recently, 14 microsatellite (SSR) markers have successfully been used to characterize the germplasm collection at the University of California (Dangl et al. 2005). Primer pairs originally designed to amplify microsatellites in eastern black walnut were used (Woeste et al. 2002). Among the 48 accessions, there were 44 unique multilocus profiles. The accessions with identical profiles were assumed to be either synonyms (as previously hypothesized) or bud sports. One French cultivar was also identified as a selection from the UC breeding program, and one grafting error was detected. This microsatellite method appears to be the method of choice for fingerprinting cultivars and germplasm.

7.2 Marker-Assisted Selection

Walnut blackline disease caused by the cherry leafroll virus causes a fatal necrosis at the graft union between *J. regia* which can be systemically infected without exhibiting symptoms and Paradox or black rootstock. This virus is pollen-borne and is transmitted through flowers of the Persian walnut scion in the spring. Over the years it moves down the stem to the graft union. The hypersensitive response of the rootstock to the virus kills the scion, and appears to be governed by a single dominant gene for hypersensitivity (McGranahan et al. 1997). We initiated a backcross breeding program in 1983 with two objectives in mind. The first is to introgress the hypersensitive resistance gene(s) from black walnut into Persian walnut over a sufficient number of generations to develop a tree with all the nut traits of the Persian walnut. The second objective is to select a tolerant rootstock with attributes superior to Persian walnut rootstock from the first backcross generation [(*J. hindsii*× *J. regia*)×*J. regia*]. In both cases, screening large numbers of progeny is necessary.

Our standard screening method was to graft the selected progeny onto both black and Persian rootstock, allowing a year for the grafts to take, and then patching infected wood into the scion if it was on the black rootstock and into the rootstock if the scion was on *J. regia* rootstock. The virus moves relatively slowly but a blackline at the graft union of the black rootstock and no blackline at the graft union of the *J. regia* rootstock indicates a tolerant response. A blackline at the graft union of the *J. regia* rootstock and failure of the patch to establish in the test material on black rootstock indicates a hypersensitive response in the scion. Obviously, this was long-term, labor-intensive, and subject to graft and patch failure.

Woeste et al. (1996) published the most useful molecular tool for our program. He identified a marker for hypersensitivity through bulked segregant analysis of backcross populations that were either tolerant or hypersensitive. It was later improved into a sequence characterized amplified region or SCAR marker (K. Woeste, unpublished). The marker has been found to be a reliable, efficient and cost- and time-effective means of screening large populations of progeny.

7.3 Genetic Engineering

Many of the new tools of biotechnology have been applied to walnuts as recently reviewed (Dandekar et al. 2005), but like many fruit and nut crops, walnuts lag behind the agronomic crops in this field. Genetic engineering was of particular interest due to the difficulties of conventional breeding. Gene transfer techniques have been in use for walnuts since 1988 (McGranahan et al. 1988, 1990), and field trials of mature transgenic trees have been completed (Leslie et al. 2001; Vahdati et al. 2002).

The protocol for gene transfer is based on inserting genes of interest into proliferating somatic embryo cultures via *Agrobacterium tumefaciens*. Somatic embryos are easily induced from immature cotyledons (Tulecke and McGranahan 1985), and with much more difficulty from immature catkins. The latter represent maternal tissue. We have been fortunate to induce embryogenic cultures from catkins of ‘Chandler,’ the most widely planted walnut cultivar in California, as well as Paradox ‘Burbank,’ the original Paradox clone developed by Luther Burbank (M. L. Mendum and G. McGranahan, unpublished). Healthy young white somatic embryos are easily infected by *A. tumefaciens*, kanamycin resistance from *nptII* is good for screening and GUS (β-glucuronidase) is used for confirmation. Green fluorescent protein (GFP) was clearly expressed in walnut somatic embryos but β-glucuronidase is easier to use (Escobar et al. 2000). The first generation of transgenic embryos may be chimeric, however, as each new embryo is derived from a single cell (Polito et al. 1989), the next generation is nonchimeric. Transgenic embryos are then dried to a popcorn stage and germinated. Because germination rates are low the shoot of a germinating embryo is usually excised and micropropagated for field trials. Micropropagation is the most time- and labor-consuming step in the process.

Traits of interest that have been tested and found effective in walnut include expression of a Bt gene from *Bacillus thuringiensis* for insect resistance (Dandekar et al. 1998; Leslie et al. 2001) and use of RNAi gene silencing to block the iaaM and ipt genes of *A. tumefaciens* which are responsible for the gall formation typical of crown gall disease (Escobar et al. 2002). Tree architecture has been modified by insertion of the *rolABC* genes from *A. rhizogenes* but the goal of increasing rootability was not achieved (Vahdati et al. 2002). When used as rootstock, the smaller stature and compressed internodes of the rol trees did not affect the phenotype of the scion.

The reticence of the public to accept genetically engineered organisms has prevented any commercialization of transgenic walnut trees, but it is expected that transgenic rootstocks will prove more acceptable.

References

Aly, M.A.M., Fjellstrom, R.G., McGranahan, G.H. and Parfitt, D.E. (1992) Origin of walnut somatic embryos determined by RFLP and isozyme analysis. HortScience 27, 61–63.

Arulsekar, S., Parfitt, D.E. and McGranahan, G.H. (1985) Isozyme gene markers in *Juglans* species. Inheritance of GPI and AAT in *J. regia* and *J. hindsii.* J. Hered. 76, 103–106.

Arulsekar, S., McGranahan, G.H. and Parfitt, D.E. (1986) Inheritance of phosphoglucomutase and esterase isozymes in Persian walnut. J. Hered. 77, 220–221.

Browne, G.T., McLaughlin, S.T., Hackett, W.P., McGranahan, G.H. and Leslie, C.A. (2006) Evaluation of resistance to *Phytophthora citricola* among diverse clones of Paradox hybrid rootstocks. Acta Hort. 705, 395–400.

Dandekar, A.M., McGranahan, G.H., Vail, P.V., Uratsu, S.L., Leslie, C.A. and Tebbets, J.S. (1998) High levels of expression of full-length cryIA(c) gene from *Bacillus thuringiensis* in transgenic somatic walnut embryos. Plant Science 131, 181–193.

Dandekar, A.M., Leslie, C.A. and McGranahan, G.H. (2005) *Juglans regia* walnut. In: R.E. Litz (ed.) *Biotechnology of Fruit and Nut Crops*. CABI Publishing. Oxfordshire, UK, pp. 307–323.

Dangl, G., Woeste, K.W., Aradhya, M., Koehmstedt, A., Simon, C., Potter, D., Leslie, C.A. and McGranahan, G.H. (2005) Characterization of 14 microsatellite markers for genetic analysis and cultivar identification of walnut. J. Amer. Soc. Hort. Sci. 130, 348–354.

Dode, L.A. (1909) Contribution to the study of the genus *Juglans*. Bull. Soc. Dendrologique de France 11, 22–90. Translated from the original by R.E. Cuendett.

Escobar, M.A., Park, J., Polito, V.S., Leslie, C.A., Uratsu, S.L., McGranahan, G.H. and Dandekar, A.M. (2000) Using GFP as a scorable marker in walnut somatic embryo transformation. Annals of Botany 85, 831–835.

Escobar, M.A., Leslie, C.A., McGranahan, G.H. and Dandekar, A.M. (2002) Silencing crown gall disease in walnut (*Juglans regia* L.). Plant Science 163, 591–597.

Fjellstrom, R.G. (1993) Genetic diversity of walnuts (*Juglans* L.) species determined by restriction fragment length polymorphisms. PhD Diss. Univ. California. Davis.

Fjellstrom, R.G., Parfitt, D.E. and McGranahan, G.H. (1994) Genetic relationships and characterization of Persian walnut (*Juglans regia* L.) cultivars using restriction fragment length polymorphisms (RFLPs). J. Amer. Soc. Hort. Sci. 119, 833–839.

Forde, H.I. and McGranahan, G.H. (1996) Walnuts In: Janick, J., and Moore, J.N. (Eds) *Fruit Breeding Volume III Nuts*. John Wiley and Sons, Inc. New York, USA, pp. 241–273.

Germain, E. (1997) Genetic improvement of the Persian walnut (*Juglans regia* L.). Acta Hort. 442, 21–31.

Germain, E. (2004) Inventory of walnut research, germplasm and references. *REU Technical Series 66*. Food and Agriculture Organization of the United Nations. Rome, Italy.

Hanche, P.E., Beres, V., and Brooks, R.M. (1972) Estimates of quantitative genetic properties of walnut and their implications for cultivar improvement. J. Am. Soc. Hort. Sci. 97, 279–285.

IPM Education and Publications Office (2003) *Integrated Pest Management for Walnuts*, 3rd Edition. University of California, Statewide Integrated Pest Management Program. Agriculture and Natural Resources. Oakland, CA, USA Publication 3270.

Leslie, C.A., and McGranahan, G.H. (1998) The origin of the walnut. In: Ramos, D.E. (ed), *Walnut Production Manual*. University of California. Division of Agriculture and Natural Resources. Publication 3373, pp. 3–7.

Leslie, C.A., McGranahan, G.H., Dandekar, A.M., Uratsu, S.L., Vail, P.V. and Tebbets, J.S. (2001) Development and field-testing of walnuts expressing the *cry*lA(c) gene for Lepidopteran insect resistance. Acta Hort. 544, 195–199.

Mahoney, N., Molyneux, R.J., McKenna, J., Leslie, C.A., and McGranahan, G. (2003) Resistance of 'Tulare' walnut (*Juglans regia* cv. Tulare) to aflatoxigenesis. Journal of Food Science: Food Microbiology and Safety 68, 619–622.

Manning, W.E. (1978) The classification within the *Juglandaceae*. Ann. Missouri Bot. Gard. 65, 1058–1087.

Manos, P.S. and Stone, D.E. (2001) Evolution, phylogyny and systematics of the Juglandaceae. Ann. Missouri Bot. Gard. 88, 231–269.

McGranahan, G.H. and Forde, H.I. (1985) Relationship between clone age and selection trait expression in mature walnuts. J. Amer. Soc. Hort. Sci. 110, 692–696.

McGranahan, G.H., and Catlin, P.B. (1987) *Juglans* rootstocks. In: Rom, R.C., and Carlson, R.F. (Eds.) *Rootstocks for Fruit Crops*. John Wiley and Sons. New York, pp. 411–450.

McGranahan, G.H., Leslie, C.A., Uratsu, S.L., Martin, L.A. and Dandekar, A.M. (1988) Agrobacterium-mediated transformation of walnut somatic embryos and regeneration of transgenic plants. Bio/Technology 6, 800–804.

McGranahan, G.H., Leslie, C.A., Uratsu, S.L. and Dandekar, A.M. (1990) Improved efficiency of the walnut somatic embryo gene transfer system. Plant Cell Reports 8, 512–516.

McGranahan, G.H., and Leslie, C.A. (1990) Walnuts (*Juglans*). In: Moore, J.N. and Ballington, J.R. (Eds.) *Genetic Resources of Fruit and Nut Crops*. Vol. 2. Acta Hort. 290: 907–951.

McGranahan, G.H., Forde, H.I., Snyder, R.G., Sibbett, G.S., Reil, W., Hasey. J.K. and Ramos, D.E. (1992) 'Tulare' Persian walnut. HortScience 27, 186–187.

McGranahan, G.H., Germain, E., Ramos, D.E., and Riggert, K. (1994a) *Descriptor list for walnut (Juglans* spp.*)*. International Plant Genetic Resources Institute, Rome.

McGranahan, G.H., Voyiatzis, D.G., Catlin, P.B. and Polito, V.S. (1994b) High pollen loads can cause pistillate flower abscission in walnut. J. Amer. Soc. Hort. Sci. 119, 505–509.

McGranahan, G.H., Leslie, C.A. and Woeste, K.W. (1997) Backcross breeding walnuts for resistance to the cherry leafroll virus. Acta Hort. 442, 121–127.

McGranahan, G.H., and Leslie, C.A. (2004) 'Robert Livermore', A Persian walnut cultivar with a red seedcoat. HortScience 39, 1772.

McKenna, J.R. and Epstein, L. (2003) Susceptibility of *Juglans* species and interspecific hybrids to *Agrobacterium tumefaciens*. HortScience 38, 435–439.

Nicese, F.P., Hormaza, J.I. and McGranahan, G.H. (1998) Molecular characterization and genetic relatedness among walnut (*Juglans regia* L.) genotypes based on RAPD markers. Euphytica 101, 199–206.

Polito, V.S., McGranahan, G.H., Pinney, K. and Leslie, C.A. (1989) Origin of somatic embryos from repetitively embryogenic cultures of walnut (*Juglans regia* L.): implications for *Agrobacterium*-mediated transformation. Plant Cell Reports 8, 219–221.

Potter, D., Gao, F.Y., Aiello, G., Leslie, C.A. and McGranahan, G.H. (2002a) Intersimple sequence repeat markers for fingerprinting and determining genetic relationships of walnut (*Juglans regia*) cultivars. J. Amer. Soc. Hort. Sci. 127, 75–81.

Potter, D., Gao, F.Y., Baggett, S., McKenna, J.R. and McGranahan, G.H. (2002b) Defining the sources of Paradox: DNA sequence markers for North American walnut (*Juglans* L.) species and hybrids. Scientia Hort. 94, 157–170.

Ramos, D.E. (Ed.) (1998) *Walnut Production Manual.* University of California, California. Division of Agriculture and Natural Resources. Publication 3373.

Rehder, A. (1940) *Manual of Cultivated Trees and Shrubs*. 2nd ed. Macmillan Co. New York.

Reiter, R.J., Manchester, L.C. and Tan, D. (2005) Melatonin in walnuts: Influence on levels of melatonin and total antioxidant capacity of blood. Nutrition: The Intl. J. Applied and Basic Nutr. Sci. 21, 920–924.

Ros, E., Nunez, I., Perez-Heras, A., Serra, M., Gilabert, R., Casals, E. and Deulofeu, R. (2004) A walnut diet improves endothelial function in hypercholesterolemic subjects: A randomized crossover trial. Circulation: J. Am. Heart. Assoc. 109, 1609–1614.

Rosengarten, F. Jr. (1984). *The Book of Edible Nuts*. Walker Publishing Company, Inc. New York.

Tapsell, L.C., Gillen, L.J., Patch, C.S., Batterham, M., Owen, A., Bare, M. and Kennedy, M. (2004) Including walnuts in a low fat/modified fat diet improves HDL cholesterol-to-total cholesterol ratios in patients with type 2 diabetes. Diabetes Care 27, 2777–2783.

Tomas, D.F. (2000) Walnuts (*Juglans regia* L.) in Mediterranean warm climates. In: Erez, A. (Ed.) *Temperate Fruit Crops in Warm Climates*. Kluwer Academic Publishers. Dordrecht, The Netherlands, pp. 405–427.

Tulecke, W. and McGranahan, G.H. (1985) Somatic embryogenesis and plant regeneration from cotyledons of walnut, *Juglans regia*. Plant Science 40, 57–63.

Tulecke, W. and McGranahan, G. (1994) The walnut germplasm collection of the University of California, Davis: A description of the collection and a history of the breeding program of Eugene L. Serr and Harold I. Forde. University of California. Genetic Resources Conservation Program, Division of Agriculture and Natural Resources, Report No. 13.

Vahdati, K., McKenna, J.R., Dandekar, A.M., Leslie, C.A., Uratsu, S.L., Hackett, W.P., Negri, P. and McGranahan, G.H. (2002) Rooting and other characteristics of a transgenic walnut hybrid (*Juglans hindsii* × *J. regia*) rootstock expressing *rolABC*. J. Amer. Soc. Hort. Sci. 127, 724–728.

Wang, H., Pei, D., Gu, R. and Wang, B. (2008) Genetic diversity and structure of walnut populations in central and southwestern China. J. Amer. Soc. Hort. Sci. 133, 197–203.

Woeste, K., McGranahan, G.H. and Bernatzky, R. (1996) The identification and characterization of a genetic marker linked to hypersensitivity to the cherry leafroll virus in walnut. Molecular Breeding 2, 261–266.

Woeste, K., Burns, R., Rhodes, O. and Michler, C. (2002) Thirty polymorphic nuclear microsatellite loci from black walnut. J. Hered. 93, 58–60.

Zohary, D. and Hopf, M. (1993) *Domestication of Plants in the Old World.* Clarendon Press. Oxford.

ERRATUM TO

Fruit Breeding

Marisa Luisa Badenes

Instituto Valenciano de Investigaciones, Agrarias (IVIA), Valencia, Spain
e-mail: mbadenes@ivia.es

David H. Byrne

Texas A&M University, College Station, TX, USA
e-mail: d-byrne@tamu.edu

M.L. Badenes and D.H. Byrne (eds.), *Fruit Breeding*, Handbook of Plant Breeding 8,
DOI 10.1007/978-1-4419-0763-9,

DOI 10.1007/978-1-4419-0763-9_23

In the front matter, there is a typo in the editor's name Marisa Luisa Badenes. The correct name is Maria Luisa Badenes.

The online version of the original front matter can be found at
http://dx.doi.org/10.1007/978-1-4419-0763-9

ERRATUM TO

Chapter 13
Cherry

Frank Kappel, Andrew Granger, Károly Hrotkó, and Mirko Schuster

M.L. Badenes and D.H. Byrne (eds.), *Fruit Breeding*, Handbook of Plant Breeding 8,
DOI 10.1007/978-1-4419-0763-9, © Springer Science+Business Media, LLC 2012

DOI 10.1007/978-1-4419-0763-9_23

In Chapter 13, page 459, Mirko Schuster's affiliation is incorrectly displayed as:

M. Schuster
Julius Kuehn Institute, Dossenheim, Germany
email: mirko.schuster@jki.bund.de

The correct affiliation is:

M. Schuster
Julius Kühn-Institut, Dresden, Germany
email: mirko.schuster@jki.bund.de

F. Kappel
Agriculture and Agri-Food Canada, 11305 Dale Meadows Rd., Summerland, BC, Canada V0H 1Z8
e-mail: Frank.Kappel@shaw.ca

The online version of the original chapter can be found at
http://dx.doi.org/10.1007/978-1-4419-0763-9_13

ERRATUM TO

Chapter 15
Plum

Bruce L. Topp, Dougal M. Russell, Michael Neuműller, Marco A. Dalbó, and Weisheng Liu

M.L. Badenes and D.H. Byrne (eds.), *Fruit Breeding*, Handbook of Plant Breeding 8,
DOI 10.1007/978-1-4419-0763-9, © Springer Science+Business Media, LLC 2012

DOI 10.1007/978-1-4419-0763-9_23

In Chapter 15, page 577, the image for Figure 15.3 is incorrect. The correct image is:

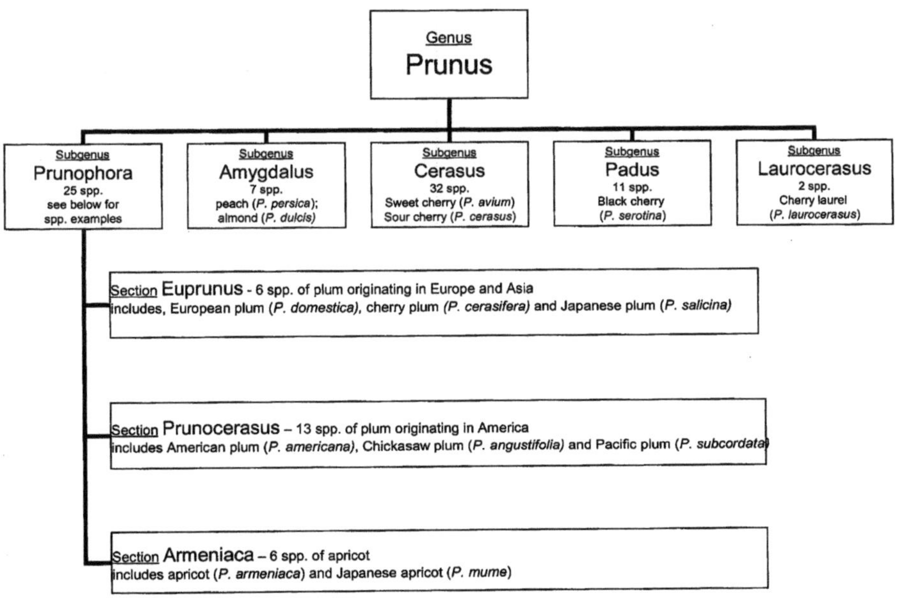

Fig. 15.3 Taxonomic classification of plum in relation to other Prunus according to Rehder (1954)

B. L. Topp
Queensland Alliance for Agriculture and Food Innovation , University of Queensland Maroochy Research Station, SCMC, PO Box 5083, Nambour 4560, QLD, Australia
e-mail: b.topp@uq.edu.au

The online version of the original chapter can be found at
http://dx.doi.org/10.1007/978-1-4419-0763-9_15

In Chapter 15, page 590, the image for Figure 15.9 is incorrect. The correct image is:

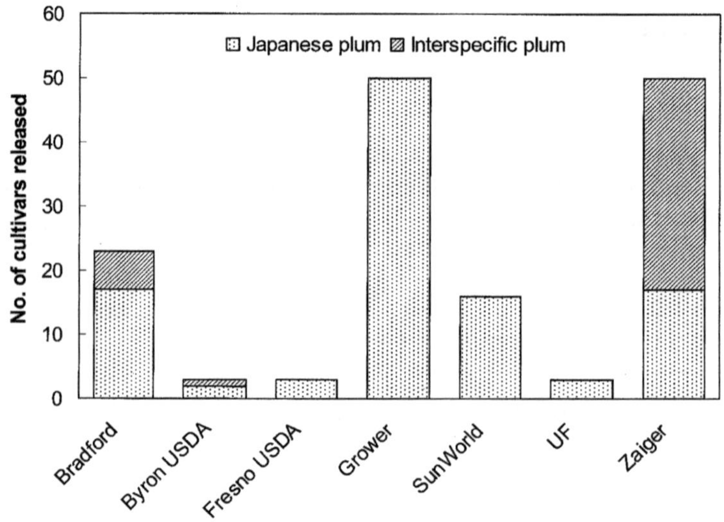

Fig. 15.9 Origin of plum cultivars released in California (Register of New Cultivar Lists, HortScience 1997–2008)

Index

A

M.L. Badenes and D.H. Byrne (eds.), *Fruit Breeding*, Handbook of Plant Breeding 8,
DOI 10.1007/978-1-4419-0763-9, © Springer Science+Business Media, LLC 2012

C

E

F

G

N

Q

R

S

MIX
Papier aus verantwortungsvollen Quellen
Paper from responsible sources
FSC® C105338

If you have any concerns about our products,
you can contact us on
ProductSafety@springernature.com

In case Publisher is established outside the EU,
the EU authorized representative is:
Springer Nature Customer Service Center GmbH
Europaplatz 3, 69115 Heidelberg, Germany

Printed by Libri Plureos GmbH
in Hamburg, Germany